The Genetic Code

Second Position

First Position (5' End)

	U	C	A	G	
U	UUU ⎤ UUC ⎦ Phe UUA ⎤ UUG ⎦ Leu	UCU UCC UCA UCG ⎤ Ser	UAU ⎤ Tyr UAC ⎦ UAA* Stop UAG* Stop	UGU ⎤ Cys UGC ⎦ UGA* Stop UGG Trp	U C A G
C	CUU CUC CUA ⎤ Leu CUG	CCU CCC CCA ⎤ Pro CCG	CAU ⎤ His CAC ⎦ CAA ⎤ Gln CAG ⎦	CGU CGC CGA ⎤ Arg CGG	U C A G
A	AUU ⎤ AUC ⎟ Ile AUA ⎦ AUG† Met	ACU ACC ACA ⎤ Thr ACG	AAU ⎤ Asn AAC ⎦ AAA ⎤ Lys AAG ⎦	AGU ⎤ Ser AGC ⎦ AGA ⎤ Arg AGG ⎦	U C A G
G	GUU GUC GUA ⎤ Val GUG†	GCU GCC GCA ⎤ Ala GCG	GAU ⎤ Asp GAC ⎦ GAA ⎤ Glu GAG ⎦	GGU GGC GGA ⎤ Gly GGG	U C A G

Third Position (3' End)

* Chain-terminating, or "nonsense," codons.

† Also used to specify the initiator formyl-Met-tRNA$_F^{Met}$. The Val triplet GUG is therefore "ambiguous" in that it codes both valine and methionine.

Ex Libris
Collegii Alberti Magni
in Novo Portu
Novae Angliae

COMPLETE

MOLECULAR BIOLOGY OF THE GENE FOURTH EDITION

James D. Watson   COLD SPRING HARBOR LABORATORY

Nancy H. Hopkins   MASSACHUSETTS INSTITUTE OF TECHNOLOGY

Jeffrey W. Roberts   CORNELL UNIVERSITY

Joan Argetsinger Steitz   YALE UNIVERSITY

Alan M. Weiner   YALE UNIVERSITY

The Benjamin/Cummings Publishing Company, Inc.

Menlo Park, California · Reading, Massachusetts · Don Mills, Ontario
Wokingham, U.K. · Amsterdam · Sydney · Singapore
Tokyo · Madrid · Bogota · Santiago · San Juan

Cover art is a computer-generated image of DNA interacting with the Cro repressor protein of bacteriophage λ. The image was prepared by the Graphic Systems Research Group at the IBM U.K. Scientific Centre, using coordinates courtesy of Brian W. Matthews, University of Oregon.

Editor: Jane Reece Gillen
Production Supervisor: Karen K. Gulliver
Editorial Production Supervisor: Betsy Dilernia
Cover and Interior Designer: Gary A. Head
Contributing Designers: Detta Penna, Michael Rogondino
Copy Editor: Janet Greenblatt
Art Coordinator: Pat Waldo
Art Director and Principal Artist: Georg Klatt
Contributing Artists: Joan Carol, Cyndie Clark-Huegel, Barbara Cousins, Cecile Duray-Bito, Jack Tandy, Carol Verbeek, John and Judy Waller

Library of Congress Cataloging-in-Publication Data
Molecular biology of the gene / James D. Watson ...... [et al.].—4th ed.
   p. cm.
  Includes bibliographies and indexes.
  ISBN 0-8053-9614-4
  1. Molecular genetics. 2. Molecular biology. I. Watson, James D., 1928–
QH447.M65 1988
574.87'328—dc19
          88-4115
          CIP

8 9 10 - CRW - 00 99 98 97

The Benjamin/Cummings Publishing Company, Inc.
2727 Sand Hill Road
Menlo Park, California 94025

About the Authors

James D. Watson is the Director of the Cold Spring Harbor Laboratory. He spent his undergraduate years at the University of Chicago and received his Ph.D. in 1950 from Indiana University. Between 1950 and 1953 he did postdoctoral research in Copenhagen and Cambridge, England. While at Cambridge, he began the collaboration that resulted in 1953 in the elucidation of the double-helical structure of DNA. (For this discovery, Watson, Francis Crick, and Maurice Wilkins were awarded the Nobel Prize in 1962.) Later in 1953 he went to the California Institute of Technology. He moved to Harvard in 1955, where he taught and did research on RNA synthesis and protein synthesis until 1976. While at Harvard he also wrote the first, second, and third editions of *Molecular Biology of the Gene,* which were published in 1965, 1970, and 1976, respectively. He has been at Cold Spring Harbor since 1968, where his major interest has been the induction of cancer by viruses.

Nancy H. Hopkins is a Professor of Biology at the Massachusetts Institute of Technology. She graduated from Radcliffe College in 1964 and did graduate work at Yale and Harvard, receiving her Ph.D. in Molecular Biology and Biochemistry from Harvard in 1971. After postdoctoral work at the Cold Spring Harbor Laboratory, she joined the faculty at M.I.T., where she teaches and does research on the molecular biology of retroviruses. She is the primary author of Chapters 23 through 27 in Volume II of this edition of *Molecular Biology of the Gene.*

Jeffrey W. Roberts is a Professor of Biochemistry at Cornell University. He received a B.A. in Physics and Liberal Arts from the University of Texas in 1964 and a Ph.D. in Biophysics from Harvard in 1970. He was a postdoctoral fellow at Harvard and also did research at the MRC Laboratory of Molecular Biology in Cambridge, England, before going to Cornell in 1974. His current research interests are genetic regulation in bacteria and phages, in particular the regulation of transcription and the control of DNA repair functions. He is the primary author of Chapters 11, 12, 13, 16, and 17 of this text.

Joan Argetsinger Steitz is a Professor of Molecular Biophysics and Biochemistry at Yale University. She graduated from Antioch College in 1963 and received a Ph.D. from Harvard in 1967. She did postdoctoral work at the MRC Laboratory of Molecular Biology before joining the Yale faculty in 1970. Her research interests have always focused on the structures and functions of RNA molecules; her current research is on gene expression in mammalian cells, with an emphasis on the roles of small RNA-protein complexes. A member of the National Academy of Sciences, she is a recipient of the National Medal of Science, among other awards. She is the primary author of Chapters 14, 15, 20, and 21 of this text.

Alan M. Weiner is a Professor of Molecular Biophysics and Biochemistry at Yale University. He graduated from Yale College in 1968 and received his Ph.D. from Harvard in 1973. After postdoctoral work at Stanford University and M.I.T., he returned to Yale as a faculty member in 1976. His current research concentrates on the structure, function, and evolution of mammalian genes for small nuclear RNA species. He is the primary author of Chapters 22 and 28 in Volume II of this text.

iii

Preface

Today no molecular biologist knows all the important facts about the gene. This was not the case in 1965 when the first edition of *Molecular Biology of the Gene* appeared. Then there were few practicing molecular biologists and not too many facts to learn. So what we knew about DNA and RNA could easily be explained to beginning college students. That year the final codons of the genetic code were being assigned, and everyone at the forefront of research could regularly assemble in the modest lecture hall at Cold Spring Harbor. Five years later, when the second edition appeared, our numbers were rising rapidly. Yet, despite the emerging popularity of molecular biology, it was still quite uncertain if the future would be as intellectually meaningful as the years just after the discovery of the double helix. The isolation of the first repressors and the demonstration that they bind specifically to control sequences in DNA seemed to some pioneers in DNA research to mark the end of the years of germinal discovery. With no means to isolate the genes of any higher organism, much less any way to know their nucleotide sequences, any pathway to understanding how genes guide the differentiation events that give rise to multicellular organisms seemed impossibly remote.

Happily, these worries did not last long. By the time the third edition of *Molecular Biology of the Gene* was published (1976), recombinant DNA procedures had given us the power to clone genes. Moreover, there was reason to believe that highly reliable methods to rapidly sequence long stretches of DNA would soon be available. As this new era of molecular biology began, however, there initially was widely voiced concern that recombinant DNA procedures might generate dangerous and pathogenic new organisms. It was not until after much deliberation that in 1977 the cloning of the genes of higher organisms began in earnest. The third edition could barely mention the potential of recombinant DNA, and of necessity its brief discussions of how genes function in eucaryotic organisms were tentative, and sometimes quite speculative.

It is only in this fourth edition that we see the extraordinary fruits of the recombinant DNA revolution. Hardly any contemporary experiment on gene structure or function is done today without recourse to ever more powerful methods for cloning and sequencing genes. As a result, we are barraged daily by arresting new facts of such importance that we seldom can relax long enough to take comfort in the accomplishments of the immediate past. The science described in this edition is by any measure an extraordinary example of human achievement.

Because of the immense breadth of today's research on the gene, none of us can speak with real authority except in those areas where our own research efforts are concentrated. Thus it was clear from the first discussions about the fourth edition that writing it would be beyond the capability of any one scientist who also had other major responsibilities. So the task of preparing this edition has required several authors. We also realized that it would be a formidable undertaking to keep the book within a manageable length; even by adopting a larger page format, we saw no way not to exceed a thousand pages. DNA can no longer be portrayed with the grandeur it deserves in a handy volume that would be pleasant to carry across a campus. Although this edition could have been shortened by eliminating the introductory material found in the first eight chapters, we never seriously considered this alternative. To do so would remove the background material that so many readers of previous editions have found valuable, and which has let many novices in molecular biology use this book as their first real introduction to gene structure and function.

Now that we are at last finished, we find that the book is even longer than we had planned. In part this happened because we are two years behind schedule, and 150 additional pages were needed to accommodate the immediate past. We also seriously underestimated how many words and illustrations would be required to describe the extraordinary variety of gene structures and functions that underlie the complexity of eucaryotic cells. We therefore made the decision to split the fourth edition into two volumes. In the first volume, first published in December 1986, we cover the general principles that govern the structure and function of both procaryotic and eucaryotic genes. The second volume, first published in August 1987, concentrates on those specialized aspects of the gene that underlie multicellular existence, and it concludes with a chapter on the evolution of DNA. Now, in response to popular demand, Benjamin/Cummings is bringing out this complete version, which includes the two volumes in one book.

In the fourth edition the second volume is appreciably smaller than the first. This will not be true of subsequent editions. Now that it is at last possible to study differentiation at the DNA level, we can easily foresee the time when, in fact, more than one volume will be required for even an introductory description of how genes are organized and expressed in the specialized cells of multicellular organisms.

We hope that the fourth edition, like its predecessors, will be found to be a highly suitable text for teaching at the undergraduate level, and that it also will provide all molecular biologists with an easy reference to the basic facts about genes. We have shown sections of the manuscripts to a variety of colleagues who are listed as reviewers. Their comments have been taken seriously, and we hope that the final manuscript faithfully reflects their expertise. Any mistakes that remain are, of course, our responsibility. Those who have made major contributions by writing or rewriting large sections of the text are Thomas Steitz (Chapter 6), Ira Herskowitz (Chapters 18 and 19), John Coffin (Chapter 24), and Brent Cochran (Chapter 25). Their generous contributions of specialized knowledge has vastly upgraded those portions of the book. In addition, John Coffin, Scott Powers, Haruo Saito, Lisa Steiner, and Parmjit Jat helped with the references for various chapters in Volume II. The excellent index was prepared by Maija Hinkle.

Equally important have been the efforts at Cold Spring Harbor of Andrea Stephenson, whose competent secretarial assistance helped coordinate our diverse labors, and Susan Scheib, whose intelligent attention to detail kept the manuscript and the galleys moving on a forward course. We also wish to acknowledge the pleasure of working with the staff of The Benjamin/Cummings Publishing Company, including Editor-in-Chief Jim Behnke and Production Supervisors Karen Gulliver and Betsy Dilernia. In particular we wish to thank Jane Gillen, who has functioned as the responsible editor during the entire writing and production of the book. An especially satisfying aspect of the process has been seeing rough drawings come alive through the efforts of the talented illustrator Georg Klatt, who has been responsible for the vast majority of the hundreds of new drawings prepared for this edition, and whose commitment and interest have greatly improved the book. And finally we gratefully acknowledge the strong support of our families throughout this endeavor, which was of course far more difficult and protracted than we ever foresaw.

James D. Watson

Nancy H. Hopkins

Jeffrey W. Roberts

Joan Argetsinger Steitz

Alan M. Weiner

Reviewers

John Abelson, California Institute of Technology
Bruce Alberts, University of California-San Francisco
Manny Ares, Yale University
Spyros Artanvanis-Tsakonas, Yale University
Piet Borst, The Netherlands Cancer Institute
Clifford Brunk, University of California-Los Angeles
Tom Cech, University of Colorado
Brent Cochran, Massachusetts Institute of Technology
John Coffin, Tufts University
Nick Cozzarelli, University of California-Berkeley
Don Crothers, Yale University
Steve Dellaporta, Cold Spring Harbor Laboratory
Ashley Dunn, Ludwig Institute for Cancer Research
Gary Felsenfeld, National Institutes of Health
John Fessler, University of California-Los Angeles
Wally Gilbert, Harvard University
Nigel Godson, New York University
Howard Green, Harvard Medical School
Nigel Grindley, Yale University
Carol Gross, University of Wisconsin
Ron Guggenheimer, Cold Spring Harbor Laboratory
Gary Gussin, University of Iowa
David Helfman, Cold Spring Harbor Laboratory
Roger Hendrix, University of Pittsburgh
Ira Herskowitz, University of California-San Francisco
Andy Hiatt, Cold Spring Harbor Laboratory
Bob Horvitz, Massachusetts Institute of Technology
David Housman, Massachusetts Institute of Technology
Martha Howe, University of Wisconsin
Richard Hynes, Massachusetts Institute of Technology
Amar Klar, Cold Spring Harbor Laboratory
Nancy Kleckner, Harvard University
Larry Klobutcher, University of Connecticut
Marilyn Kozak, University of Pittsburgh
Chuck Kurland, Uppsala University
Peter Laird, The Netherlands Cancer Institute
Richard Losick, Harvard University
Brenda Lowe, Cold Spring Harbor Laboratory
Tony Mahowald, Case Western Reserve University
Nancy Maizels, Yale University
Will McClure, Carnegie Mellon University
Bill McGinnis, Yale University
Jeffrey Miller, University of California-Los Angeles
Paul Modrich, Duke University

Peter Moore, Yale University
Steve Mount, University of California-Berkeley
Kim Mowry, Yale University
Tim Nelson, Yale University
Harry Noller, University of California-Santa Cruz
Bill Nunn, University of California-Irvine
Paul Nurse, Imperial Cancer Research Fund
Kim Nasmyth, Laboratory of Molecular Biology, Cambridge
Norm Pace, Indiana University
Carl Parker, California Institute of Technology
Sheldon Penman, Massachusetts Institute of Technology
Scott Powers, Cold Spring Harbor Laboratory
Mark Ptashne, Harvard University
Charles Radding, Yale University
David Raulet, Massachusetts Institute of Technology
Phil Robbins, Massachusetts Institute of Technology
Hugh Robertson, Rockefeller University
Lucia Rothman-Denes, University of Chicago
Earl Ruley, Massachusetts Institute of Technology
Haruo Saito, Harvard University
Robert Schleif, Brandeis University
Thomas Shenk, Princeton University
Gerry Smith, Hutchinson Cancer Center, University of Washington
Deborah Steege, Duke University
Lisa Steiner, Massachusetts Institute of Technology
Tom Steitz, Yale University
Bruce Stillman, Cold Spring Harbor Laboratory
Bill Studier, Brookhaven National Laboratory
Bob Symons, University of Adelaide
Susumu Tonegawa, Massachusetts Institute of Technology
Olke Uhlenbeck, University of Colorado
Axel Ullrich, Genentech
Chris Walsh, Massachusetts Institute of Technology
Jim Wang, Harvard University
Bob Webster, Duke University
Allan Wilson, University of California-Berkeley
Barbara Wold, California Institute of Technology
Sandra Wolin, University of California-San Francisco
Keith Yamamoto, University of California-San Francisco
Jorge Yunis, University of Minnesota

Brief Contents

Detailed Contents

VOLUME I   GENERAL PRINCIPLES

Part II
Chemical Facts and Principles ... 95

Part VI
Regulation of Gene Function in Bacterial Cells   463

CHAPTER 16
REGULATION OF PROTEIN SYNTHESIS AND FUNCTION IN BACTERIA   465

Part VII
Facing Up to Eucaryotic Cells   549

CHAPTER 19
RECOMBINANT DNA AT WORK   595

VOLUME II   SPECIALIZED ASPECTS

CHAPTER 23
THE GENERATION OF IMMUNOLOGICAL SPECIFICITY 832

Part X
Cancer at the Genetic Level   961

INDEX for both volumes in Vol. II p.I-1

Note to the Reader

The following features are intended to add to the usefulness of this book as both a text and a reference.

- **Pagination of Volumes I and II** of this text is consecutive, with the first chapter of Volume II (Chapter 22) beginning on page 745. (Cross-references refer to chapter or page numbers only.)

- **The index** at the end of Volume I covers that volume only; the index at the end of Volume II covers both volumes.

- **Key terms** within the text are highlighted by boldface type at the point in a chapter where the first full definition and major discussion of each term occur. Boldface type is also used in the index to identify the page where the full definition appears.

- **The concept headings,** which originated in the first edition of this text, have been retained. In addition, the longer chapters of Volume II have been subdivided into several major sections set off by briefer headings, to help organize the material for the reader. A complete list of all the headings in both volumes may be found in the Detailed Contents beginning on page x.

- **Summaries** follow the main text for each chapter.

- **Bibliographies** at the ends of the chapters provide a bridge to the scientific literature. Included are a relatively short list of recommended **General References,** which are mainly books and review articles, and a longer list of **Cited References.** The Cited References include the original papers in which important discoveries were first reported as well as a selection of more recent papers. The citations of these references within the text appear as **superscript numbers** that accompany text headings. Thus the Cited References provide a convenient way of finding more detailed information on specific topics.

- **Color plates** showing computer graphics follow page 64. They include

I

HISTORICAL BACKGROUND

CHAPTER 1

The Mendelian View of the World

It is easy to consider human beings unique among living organisms. We alone have developed complicated languages that allow meaningful and complex interplay of ideas and emotions. Moreover, great civilizations have developed and changed our world's environment in ways inconceivable for any other form of life. Hence, there has always been a tendency to think that something special differentiates us from every other species. This belief has found expression in the many religions through which we have sought out the reason for our existence and, in doing so, have tried to establish workable rules for conducting our lives. Until a little more than a century ago, it seemed natural to think that just as every human life begins and ends at a fixed time, the human species must also have been created at a fixed moment, with the same series of events leading to the creation of all other forms of life.

This belief was first seriously questioned 125 years ago, when Charles Darwin and Alfred R. Wallace proposed their theories of evolution, based on the selection of the most fit. They stated that the various forms of life are not constant, but are continually giving rise to slightly different animals and plants, some of which are adapted to survive and multiply more effectively. At the time of this theory, they did not know the origin of this continuous variation, but they did correctly realize that these new characteristics must persist in the progeny if such variations are to form the basis of evolution.

At first, there was a great furor against Darwin, most of it coming from people who did not like to believe that humans and the rather obscene-looking apes could have a common ancestor, even if this ancestor had lived some 10 to 20 million years ago. There was also initial opposition from many biologists who failed to find Darwin's evidence convincing. Among these was the famous Swiss-born naturalist Jean L. Agassiz, then at Harvard, who spent many years writing against Darwin and Darwin's champion, Thomas H. Huxley, the most successful of the popularizers of evolution. But by the end of the nineteenth century, the scientific argument was almost complete; both the current geographic distribution of plants and animals and their selective occurrence in the fossil records of the geologic past were explicable only by postulating that continuously evolving groups of organisms had descended from a common ancestor. Today, evolution is an accepted fact for everyone but a fundamentalist minority, whose objections are based not on reasoning but on doctrinaire adherence to religious principles.

Figure 1-1
Electron micrograph of a thin section from a cell of the African violet. The thin primary cellulose cell wall and the nucleus, containing a prominent nucleolus, are clearly visible. The cytoplasmic ground substance is heavily laden with ribosomes, visible as small black dots. The profiles of a network of hollow membranes, the endoplasmic reticulum, can be seen scattered throughout the cell. (Courtesy of K. R. Porter and M. C. Ledbetter.)

An immediate consequence of the acceptance of Darwinian theory is the realization that life first existed on our Earth some four billion years ago in a simple form, possibly resembling the bacteria—the simplest variety of life known today. Of course, the very existence of such small bacteria tells us that the essence of the living state is found in very small organisms. Evolutionary theory further suggests that the same basic principles of life apply to all living forms.

The Cell Theory[1]

The same conclusion was independently stated by the second great principle of nineteenth-century biology, the **cell theory.** This theory, first put forward convincingly in 1839 by the German microscopists Matthias Schleiden and Theodor Schwann, proposed that all plants and animals are constructed from small fundamental units called cells, and that all cells arise from other cells. It was later discovered that all cells are surrounded by a membrane and usually contain an inner body, the nucleus, which is also surrounded by a membrane, called the nuclear membrane (Figures 1-1 and 1-2). Most importantly, cells arise from other cells by the process of cell division. Most cells are capable of growing and of splitting roughly equally to give two daughter cells. At the same time, the nucleus divides so that each daughter cell can receive a nucleus.

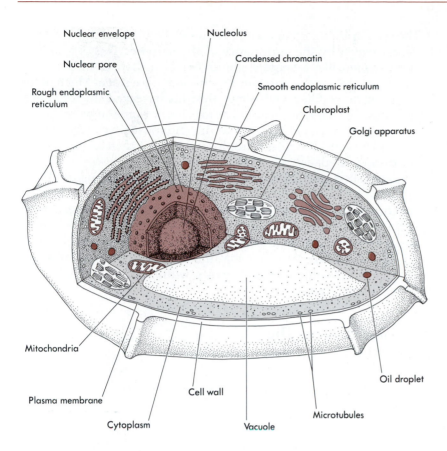

Nuclear envelope

Nuclear pore

Rough endoplasmic
reticulum

Nucleolus

Condensed chromatin

Smooth endoplasmic reticulum

Chloroplast

Golgi apparatus

Mitochondria

Plasma membrane

Cytoplasm

Cell wall

Vacuole

Microtubules

Oil droplet

Figure 1-2
A schematic view of the plant cell shown in Figure 1-1. The various components are not drawn to scale.

Mitosis Maintains the Parental Chromosome Number

Each nucleus encloses a fixed number of linear bodies called **chromosomes** (Figure 1-3). Before cell division, each chromosome duplicates to form two identical chromosomes. This process, first accurately observed by Walter Flemming in 1879, doubles the number of nuclear chromosomes. During nuclear division, one of each pair of daughter chromosomes moves into each daughter nucleus (Figure 1-4). As a result of these events (now collectively termed **mitosis**), the chromosomal complement of daughter cells is usually identical to that of the parental cell.

During most of a cell's life, its chromosomes exist in a highly extended linear form. Prior to cell division, however, they condense into much more compact bodies. The duplication of chromosomes occurs chiefly when they are in the extended state characteristic of interphase (the various stages of cell division are defined in Figure 1-4). One part of the chromosome, however, always duplicates during the contracted metaphase state; this is the **centromere,** a body that controls the movement of the chromosome during cell division. The centromere always has a fixed location specific to a given chromosome; in some chromosomes it is near one end, and in others it occupies an intermediate region.

When a chromosome is completely duplicated except for the centromere, it is said to consist of two **chromatids.** A chromatid becomes a chromosome as soon as its centromere has divided and is no longer shared with another chromatid. As soon as one centromere becomes two, the two daughter chromosomes begin to move away from each other.

Figure 1-3
Chromosomes of a human male. In this micrograph, the chromosomes have already duplicated in preparation for cell division. (Courtesy of M. Siniscalco, M.D., and B. Alhadeff, Memorial Sloan-Kettering Cancer Center, New York.)

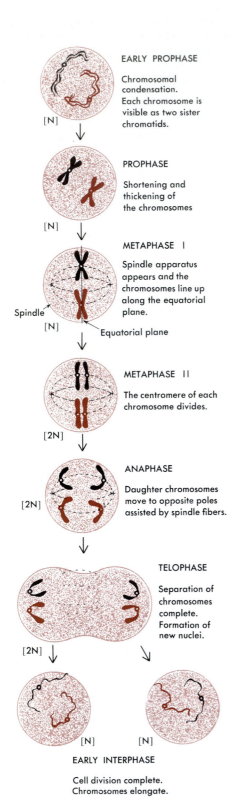

EARLY PROPHASE

Chromosomal condensation. Each chromosome is visible as two sister chromatids.

[N]

PROPHASE

Shortening and thickening of the chromosomes

[N]

METAPHASE I

Spindle apparatus appears and the chromosomes line up along the equatorial plane.

Spindle

[N]

Equatorial plane

METAPHASE II

The centromere of each chromosome divides.

[2N]

ANAPHASE

Daughter chromosomes move to opposite poles assisted by spindle fibers.

[2N]

TELOPHASE

Separation of chromosomes complete. Formation of new nuclei.

[2N]

[N]   [N]

EARLY INTERPHASE

Cell division complete. Chromosomes elongate.

Figure 1-4
Diagram of mitosis in the nucleus of a haploid cell containing two (nonhomologous) chromosomes. *N* refers to the number of chromosomes in the haploid cell.

The regular lining up of chromosomes during the metaphase stage is accompanied by the appearance of the **spindle.** This is a cellular region, shaped like a spindle, through which the chromosomes of higher organisms move apart during the anaphase stage. Much of the spindle region is filled with long, thin fibers, called microtubules. These fibers are largely responsible for chromosome movements on the spindle. Microtubules attach to the chromosomes at the centromeres, and as the daughter chromosomes move toward the spindle poles during anaphase, the centromeres are in the lead.

Objects called **nucleoli** are also present in the nucleus of almost every plant and animal cell. There is often one nucleolus per haploid set of chromosomes, and in some cells the nucleolus is connected to a specific chromosome. The functional role of the nucleolus was once thought by some biologists to be related to the formation of the spindle. Now, however, it is clear that the nucleolus is involved in the synthesis of ribosomes, small particles within the cell on which all proteins are synthesized.

Meiosis Reduces the Parental Chromosome Number

One important exception was found to the mitotic process. After conclusion of the two cell divisions that form the sex cells (the gametes: the sperm and the egg), the number of chromosomes is reduced to one-half of its previous number (Figure 1-5). This set of processes is called **meiosis.** In higher plants and animals, each specific type of chromosome is normally present as two copies of homologous chromosomes (the **diploid** state). In gamete formation, the resulting sperm or egg usually encloses only one of each chromosome type (the **haploid** state). Union of sperm and egg during fertilization results in a **zygote** containing one homologous chromosome from the male parent and another from the female parent. Thus, the normal diploid chromosome constitution is restored.

Although most cells are diploid in higher plants and animals, the haploid state is the most frequent condition in lower plants and bacteria, the diploid number existing only briefly following sex-cell fusion. In these organisms, meiosis usually occurs almost immediately after fertilization to produce haploid cells. Figure 1-6 shows a generalized life cycle that includes individual organisms in both haploid and diploid states.

The cell theory thus tells us that all cells come from preexisting cells. All the cells in adult plants and animals are derived from the division and growth of a zygote, itself formed by the union of two other cells, the sperm and the egg. All growing cells contain chromosomes, usually two of each type, and here again, new chromosomes always arise through division of previously existing bodies.

The Cell Theory Is Universally Applicable

Although the cell theory developed from observations about higher organisms, it holds with equal force for the more simple forms of life, such as protozoa and bacteria. Each bacterium or protozoan is a single cell, whose division ordinarily produces a new cell identical to its parent, from which it soon separates. In the higher organisms, on the

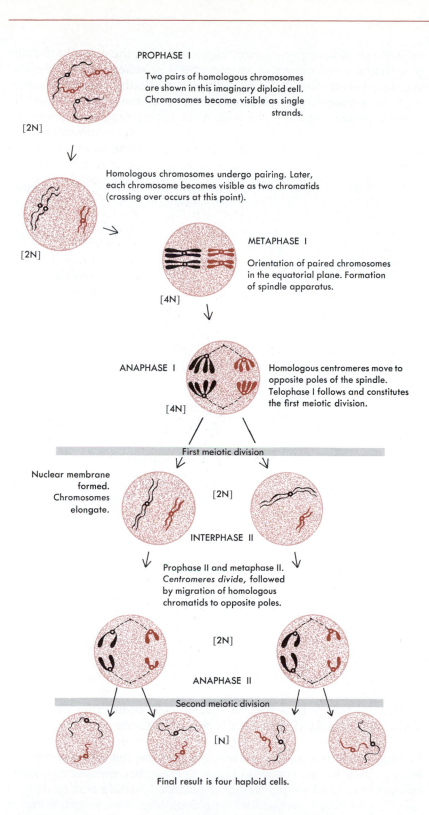

PROPHASE I

Two pairs of homologous chromosomes are shown in this imaginary diploid cell. Chromosomes become visible as single strands.

[2N]

Homologous chromosomes undergo pairing. Later, each chromosome becomes visible as two chromatids (crossing over occurs at this point).

[2N]

METAPHASE I

Orientation of paired chromosomes in the equatorial plane. Formation of spindle apparatus.

[4N]

ANAPHASE I

Homologous centromeres move to opposite poles of the spindle. Telophase I follows and constitutes the first meiotic division.

[4N]

First meiotic division

Nuclear membrane formed. Chromosomes elongate.

[2N]

INTERPHASE II

Prophase II and metaphase II. *Centromeres divide*, followed by migration of homologous chromatids to opposite poles.

[2N]

ANAPHASE II

Second meiotic division

[N]

Final result is four haploid cells.

Figure 1-5
Diagram of meiosis in an organism containing two pairs of homologous chromosomes in each diploid cell.

other hand, not only do the daughter cells often remain together, but they also frequently differentiate into radically different cell types (such as nerve or muscle cells) while maintaining the chromosome complement of the zygote. Here, new organisms arise from the highly differentiated sperm and egg, whose union initiates a new cycle of division and differentiation.

Thus, although a complicated organism like *Homo sapiens* contains a large number of cells (up to 5×10^{12}), all these cells arise initially from a single cell. The zygote contains all the information necessary for the growth and development of an adult plant or animal. Again, the living state per se does not demand the complicated interactions that occur in complex organisms; its essential properties can be found in single growing cells.

Mendelian Laws[2]

The most striking attribute of a living cell is its ability to transmit hereditary properties from one cell generation to another. The products of heredity must have been noticed by early humans, who witnessed the passing of characteristics, like eye or hair color, from parents to offspring. Its physical basis, however, was not understood until the first years of the twentieth century, when, during a remarkable period of creative activity, the chromosomal theory of heredity was established.

Hereditary transmission through the sperm and egg became known by 1860, and in 1868 Ernst Haeckel, noting that sperm consists largely of nuclear material, postulated that the nucleus is responsible for heredity. Almost 20 years passed before the chromosomes were singled out as the active factors, because the details of mitosis, meiosis, and fertilization had to be worked out first.

When this was accomplished, it could be seen that, unlike other cellular constituents, the chromosomes are equally divided between daughter cells. Moreover, the complicated chromosomal changes that reduce the sperm and egg chromosome number to the haploid number during meiosis became understandable as necessary for keeping the chromosome number constant. These facts, however, merely suggested that chromosomes carry heredity.

Proof came at the turn of the century with the discovery of the basic rules of heredity. These rules, named after their original discoverer, Gregor Mendel, had in fact been first proposed in 1865, but the climate of scientific opinion had not been ripe for their acceptance. They were completely ignored until 1900, despite some early efforts on Mendel's part to interest the prominent biologists of his time. Then Hugo De Vries, Karl Correns, and Erich Tschermak, all working independently, realized the great importance of Mendel's forgotten work. All three were plant breeders doing experiments related to Mendel's, and each reached similar conclusions before they knew of Mendel's work.

Principle of Independent Segregation

Mendel's experiments traced the results of breeding experiments (genetic crosses) between strains of peas differing in well-defined characteristics, like seed shape (round or wrinkled), seed color (yellow or green), pod shape (inflated or wrinkled), and stem length (long or short). His concentration on well-defined differences was of great importance; many breeders had previously tried to follow the inheritance of more gross qualities, like body weight, and were unable to discover any simple rules about their transmission from parents to offspring. After ascertaining that each type of parental strain bred true (that is, produced progeny with particular qualities identical to those of the parents), Mendel made a number of crosses between parents (P) differing in single characteristics (such as seed shape *or*

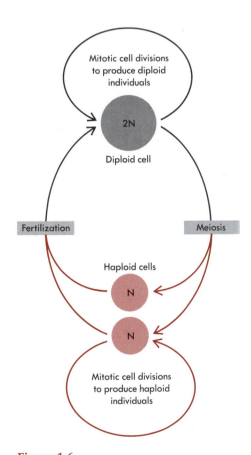

Figure 1-6
The alternation of haploid and diploid states that constitute the sexual life cycle. The haploid and diploid individuals may be unicellular or multicellular, depending on the organism.

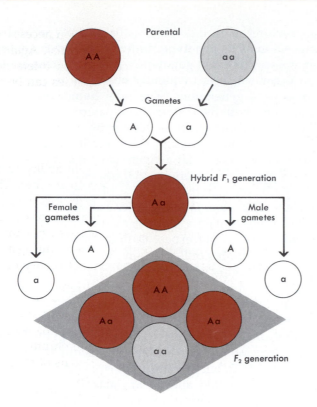

Parental

AA   aa

Gametes

A   a

Hybrid F₁ generation

Aa

Female gametes   Male gametes

A   A

a   a

AA

Aa   Aa

aa

F₂ generation

Figure 1-7
How Mendel's first law (independent segregation) explains the 3:1 ratio of dominant to recessive phenotypes among the F₂ progeny. *A* represents the dominant gene and *a* the recessive gene. The colored circles represent the dominant phenotype, the gray circles the recessive phenotype.

seed color). All the progeny (F₁ = first filial generation) had the appearance of *one* parent. For example, in a cross between peas having yellow seeds and peas having green seeds, all the progeny had yellow seeds. The trait that appears in the F₁ progeny is called **dominant,** whereas the trait that does not appear in F₁ is called **recessive.**

The meaning of these results became clear when Mendel made genetic crosses between F₁ offspring. These crosses gave the important result that the recessive trait reappeared in approximately 25 percent of the F₂ progeny, whereas the dominant trait appeared in 75 percent of these offspring. For each of the seven traits he followed, the ratio in F₂ of dominant to recessive traits was always approximately 3:1. When these experiments were carried to a third (F₃) progeny generation, all the F₂ peas with recessive traits bred true (produced progeny with the recessive traits). Those with dominant traits fell into two groups: one-third bred true (produced only progeny with the dominant trait); the remaining two-thirds again produced mixed progeny in a 3:1 ratio of dominant to recessive.

Mendel correctly interpreted his results as follows (Figure 1-7): The various traits are controlled by pairs of factors (which we now call **genes**), one factor derived from the male parent, the other from the female. For example, pure-breeding strains of round peas contain two genes for roundness (*RR*), whereas pure-breeding wrinkled strains have two genes for wrinkledness (*rr*).* The round-strain gametes each have one gene for roundness (*R*); the wrinkled-strain gametes each have one gene for wrinkledness (*r*). In a cross between *RR* and *rr*, fertilization produces an F₁ plant with both genes (*Rr*). The seeds look round because *R* is dominant over *r*. We refer to the appearance

---

*One or several letters may be used to represent a particular gene. The dominant variety of the gene (i.e., the dominant allele) may be indicated by a capital letter (e.g., *R*), a superscript + (e.g., *r*⁺), or a + standing alone.

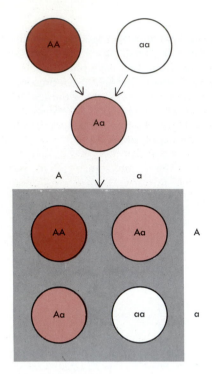

Figure 1-8
The inheritance of flower color in the snapdragon. One parent is homozygous for red flowers (*AA*) and the other homozygous for white flowers (*aa*). No dominance is present, and the heterozygous F₁ flowers are pink. The 1:2:1 ratio of red, pink, and white flowers in the F₂ progeny is shown by appropriate coloring.

(physical structure) of an individual as its **phenotype,** and to its genetic composition as its **genotype.** Individuals with identical phenotypes may possess different genotypes; thus, to determine the genotype of an organism, it is frequently necessary to perform genetic crosses for several generations. The term **homozygous** refers to a gene pair in which both the maternal and paternal genes are identical (e.g., *RR* or *rr*). In contrast, those gene pairs in which paternal and maternal genes are different (e.g., *Rr*) are called **heterozygous.**

It is important to notice that a given gamete contains only one of the two genes present in the organism it comes from (e.g., either *R* or *r*, but never both) and that the two types of gametes are produced in equal numbers. Thus, there is a fifty-fifty chance that a given gamete from an F₁ pea will contain a particular gene (*R* or *r*). This choice is purely random. We do not expect to find *exact* 3:1 ratios when we examine a limited number of F₂ progeny. The ratio will sometimes be slightly higher and other times slightly lower. But as we look at increasingly larger samples, we expect that the ratio of peas with the dominant trait to peas with the recessive trait will approximate the 3:1 ratio more and more closely.

The reappearance of the recessive characteristic in the F₂ generation indicates that recessive genes are neither modified nor lost in the **hybrid** (*Rr*) generation, but that the dominant and recessive genes are independently transmitted and so are able to segregate independently during the formation of sex cells. This **principle of independent segregation** is frequently referred to as Mendel's first law.

Some Genes Are Neither Dominant Nor Recessive

In the crosses reported by Mendel, one of each gene pair was clearly dominant, and the other recessive. Such behavior, however, is not universal. Sometimes the heterozygous phenotype is intermediate between the two homozygous phenotypes. For example, the cross between a pure-breeding red snapdragon (*Antirrhinum*) and a pure-breeding white variety gives F₁ progeny of the intermediate pink color. If these F₁ progeny are crossed among themselves, the resulting F₂ progeny contain red, pink, and white flowers in the proportion of 1:2:1 (Figure 1-8). Thus, it is possible here to distinguish heterozygotes from homozygotes by their phenotype. We also see that Mendel's laws do not depend on whether one *gene* of a gene pair is dominant over the other.

Principle of Independent Assortment

Mendel extended his breeding experiments to peas differing by more than one characteristic. As before, he started with two strains of peas, each of which bred pure when mated with itself. One of the strains had round yellow seeds; the other, wrinkled green seeds. Since round and yellow are dominant over wrinkled and green, the entire F₁ generation produced round yellow seeds. The F₁ generation was then crossed within itself to produce a number of F₂ progeny, which were examined for seed appearance (phenotype). In addition to the two original phenotypes (round yellow; wrinkled green), two new types (**recombinants**) emerged: wrinkled yellow and round green.

Again Mendel found he could interpret the results by the postulate of genes, if he assumed that each gene pair was independently trans-

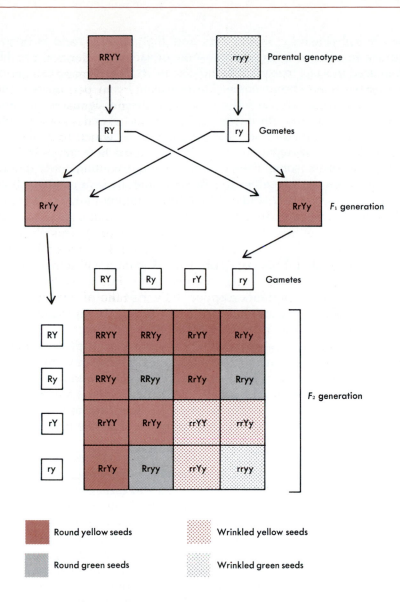

Figure 1-9

How Mendel's second law (independent assortment) operates. In this example, the inheritance of yellow (Y) and green (y) seed color is followed together with the inheritance of round (R) and wrinkled (r) seed shapes. The R and Y alleles are dominant over r and y. The genotypes of the various parents and progeny are indicated by letter combinations, and four different phenotypes are distinguished by appropriate shading.

mitted to the gamete during sex-cell formation. This interpretation is shown in Figure 1-9. Any one gamete contains only one type of inherited factor from each gene pair. Thus, the gametes produced by an F₁ (*RrYy*) will have the composition *RY, Ry, rY,* or *ry,* but never *Rr, Yy, YY,* or *RR.* Furthermore, in this example, all four possible gametes are produced with equal frequency. There is no tendency of genes arising from one parent to stay together. As a result, the F₂ progeny phenotypes appear in the ratio 9 round yellow, 3 round green, 3 wrinkled yellow, and 1 wrinkled green. This **principle of independent assortment** is frequently called Mendel's second law.

Chromosomal Theory of Heredity[3]

A principal reason for the original failure to appreciate Mendel's discovery was the absence of firm facts about the behavior of chromosomes during meiosis and mitosis. This knowledge was available, however, when Mendel's laws were reannounced in 1900, and was seized upon in 1903 by the American Walter S. Sutton. In his classic paper, "The Chromosomes in Heredity," Sutton emphasized the importance of the fact that the diploid chromosome group consists of two morphologically similar sets and that, during meiosis, every gamete receives only one chromosome of each homologous pair. He then used this fact to explain Mendel's results by assuming that genes are parts of the chromosome. He postulated that the yellow- and green-seed genes are carried on a certain pair of chromosomes and that the round- and wrinkled-seed genes are carried on a different pair. This hypothesis immediately explains the experimentally observed 9:3:3:1 segregation ratios. Although Sutton's paper did not prove the chromosomal theory of heredity, it was immensely important, for it brought together for the first time the independent disciplines of genetics (the study of breeding experiments) and cytology (the study of cell structure).

Chromosomal Determination of Sex

There exists one important exception to the rule that all chromosomes of diploid organisms are present in two copies. It was observed as early as 1890 that one chromosome (then called an accessory chromosome and now the X chromosome) does not always possess a morphologically identical mate. The biological significance of this observation was clarified by the American cytologist Edmund B. Wilson and his student, N. M. Stevens, in 1905. They showed that although the female carries two X chromosomes, the male carries only one. In addition, in some species (including humans), the male cells contain a unique chromosome, not found in females, called the Y chromosome. Wilson and Stevens pointed out how this situation provides a simple method of sex determination. Whereas every ovum (egg) will contain one X chromosome, only half the sperms will carry one. Thus, fertilization of an ovum by an X-bearing sperm leads to an XX zygote, which becomes a female; and fertilization by a sperm lacking an X chromosome gives rise to male offspring (Figure 1-10). These observations provided the first clear linking of a definite chromosome to a hereditary property. In addition, they elegantly explained how male and female zygotes are created in equal numbers.

The Importance of *Drosophila*[4, 5]

Initially, all breeding experiments used genetic differences already existing in nature. For example, Mendel used seeds obtained from seed dealers, who must have obtained them from farmers. The existence of alternative forms of the same gene (**alleles**) raises the question of how they arose. One obvious hypothesis states that genes can change (mutate) to give rise to new genes (mutant genes). This hypothesis was first seriously tested, beginning in 1908, by the great American biologist Thomas H. Morgan and his young collaborators, the geneticists Calvin B. Bridges, Hermann J. Muller, and Alfred H. Sturtevant. They worked with the tiny fly *Drosophila melanogaster*. This fly, which normally lives on fruit, was found to be easily maintained under laboratory conditions, where a new generation can be produced every 14 days. Thus, by using *Drosophila* instead of more slowly multiplying organisms like peas, it was possible to work at least 25 times faster and much more economically. The first mutant found was a male with white eyes instead of the normal red eyes. It spontaneously appeared in a culture bottle of red-eyed flies. Because essentially all *Drosophila* found in nature have red eyes, the gene lead-

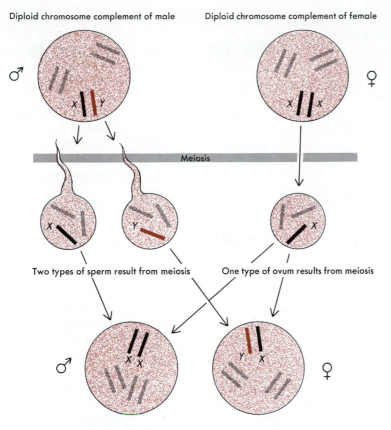

Diploid chromosome complement of male   Diploid chromosome complement of female

Meiosis

Two types of sperm result from meiosis   One type of ovum results from meiosis

Sex determined by the type of sperm entering the ovum

Figure 1-10
How sex chromosomes operate. Here is a case in which males carry one *X* and one *Y* chromosome, and females carry two *X* chromosomes. This is the situation in both humans and *Drosophila*. In certain species there is no *Y* chromosome, so that diploid male cells contain one less chromosome than diploid female cells.

ing to red eyes was referred to as the **wild-type gene;** the gene leading to white eyes was called a **mutant gene** (allele).

The white-eye mutant gene was immediately used in breeding experiments (Figure 1-11), with the striking result that the behavior of the allele completely paralleled the distribution of an *X* chromosome (i.e., was sex-linked). This immediately suggested that this gene might be located on the *X* chromosome, together with those genes controlling sex. This hypothesis was quickly confirmed by additional genetic crosses using newly isolated mutant genes. Many of these additional mutant genes also were sex-linked.

Gene Linkage and Crossing Over[6, 7]

Mendel's principle of independent assortment is based on the fact that genes located on different chromosomes behave independently during meiosis. Often, however, two genes do not assort independently, because they are located on the same chromosome (**linked genes**). Numerous examples of nonrandom assortment were found as soon as a large number of mutant genes became available for breeding analysis. In every well-studied case, the number of linked groups was identical with the haploid chromosome number. For example, there are four groups of linked genes in *Drosophila* and four morphologically distinct chromosomes in a haploid cell.

Linkage, however, is in effect never complete. The probability that two genes on the same chromosome will remain together during meiosis ranges from just less than 100 percent to about 50 percent.

This means that there must be a mechanism for exchanging genes on homologous chromosomes. This mechanism is called **crossing**

Figure 1-11
The inheritance of a sex-linked gene in *Drosophila*. Genes located on sex chromosomes can express themselves differentially in male and female progeny, because if there is only one X chromosome present, recessive genes on this chromosome are always expressed. Here are two crosses, both involving a recessive gene (*w*, for white eye) located on the X chromosome. In (a), the male parent is a white-eyed (*wY*) fly, and the female is homozygous for red eye (*WW*). In (b), the male has red eyes (*WY*) and the female white eyes (*ww*). The letter *Y* stands here not for an allele, but for the *Y* chromosome, present in male *Drosophila* in place of a homologous X chromosome. There is no gene on the *Y* chromosome corresponding to the *w* or *W* gene on the X chromosome.

(a)

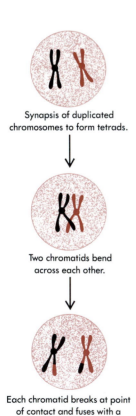

Synapsis of duplicated
chromosomes to form tetrads.

Two chromatids bend
across each other.

Each chromatid breaks at point
of contact and fuses with a
portion of the other.

Figure 1-12
Janssens' hypothesis of crossing over.

over. Its cytological basis was first described by the Belgian cytologist F. A. Janssens. At the start of meiosis, through the process of **synapsis,** the homologous chromosomes form pairs with their long axes parallel. At this stage, each chromosome has duplicated to form two chromatids. Thus, synapsis brings together four chromatids (a tetrad), which coil about one another. Janssens postulated that, possibly because of tension resulting from this coiling, two of the chromatids might sometimes break at a corresponding place on each. This could create four broken ends, which might rejoin crossways, so that a section of each of the two chromatids would be joined to a section of the other (Figure 1-12). Thus, recombinant chromatids might be produced that contain a segment derived from each of the original homologous chromosomes. Formal proof of Janssens' hypothesis that chromosomes physically interchange material during synapsis, however, had to await the passage of more than 20 years. Then, in 1931, Barbara McClintock and Harriet Creighton, working at Cornell University with the corn plant *Zea mays,* devised an elegant cytological demonstration of chromosome breakage and rejoining (Figure 1-13).

Chromosome Mapping[8]

Morgan and his students, however, did not await formal cytological proof of crossing over before exploiting the implication of Janssens' hypothesis. They reasoned that genes located close together on a

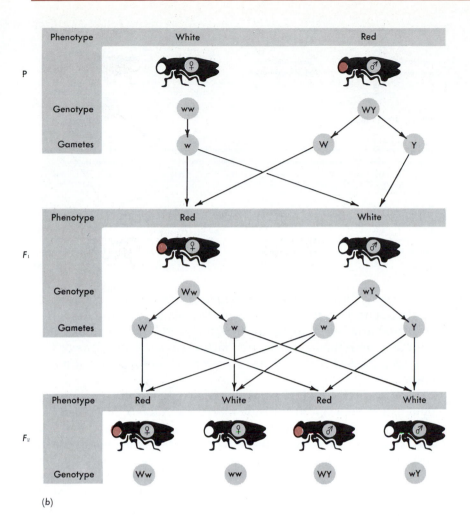

Figure 1-11
(*Continued*)

(b)

chromosome would assort with one another much more regularly (close linkage) than genes located far apart on a chromosome. They immediately saw this as a way to locate (map) the relative positions of genes on chromosomes and thus to produce a **genetic map.** The way they used the frequencies of the various recombinant classes is very straightforward. Consider the segregation of three genes all located on the same chromosome. The arrangement of the genes can be determined by means of three crosses, in each of which two genes are followed (two-factor crosses). A cross between *AB* and *ab* yields four progeny types: the two parental genotypes (*AB* and *ab*) and two recombinant genotypes (*Ab* and *aB*). A cross between *AC* and *ac* similarly gives two parental combinations as well as the *Ac* and *aC* recombinants, whereas a cross between *BC* and *bc* produces the parental types and the recombinants *Bc* and *cB*. Each cross will produce a specific ratio of parental to recombinant progeny. Consider, for example, the fact that the first cross gives 30 percent recombinants, the second cross 10 percent, and the third cross 25 percent. This tells us that genes *a* and *c* are closer together than *a* and *b* or *b* and *c* and that the genetic distances between *a* and *b* and *b* and *c* are more similar. The gene arrangement that best fits these data is *a-c-b* (Figure 1-14).

The correctness of gene order suggested by crosses of two gene factors can usually be unambiguously confirmed by three-factor crosses. When the three genes used in the preceding example are followed in the cross *ABC* × *abc*, six recombinant genotypes are found (Figure 1-15). They fall into three groups of reciprocal pairs. The rar-

Figure 1-13
Demonstration of physical exchanges between homologous chromosomes. In most organisms, pairs of homologous chromosomes have identical shapes. Occasionally, however, the two members of a pair are not identical, one being marked by the presence of extra-chromosomal material or compacted regions that reproducibly form knoblike structures. McClintock and Creighton, having found one such pair, used them to show that crossing over involves actual physical exchanges between the paired chromosomes. In the experiment shown here, the homozygous *c, wx* progeny had to arise by crossing over between the *C* and *wx* loci. When such *c, wx* offspring were cytologically examined, knob chromosomes were seen, showing that a knobless *Wx* region had been physically replaced by a knobbed *wx* region. The colored box in the figure identifies the chromosomes of the homozygous *c, wx* offspring.

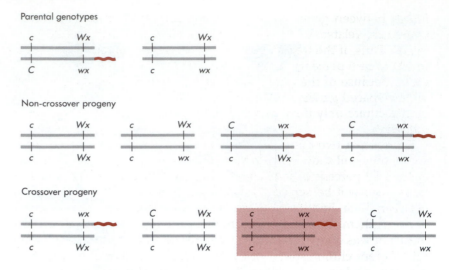

Parental genotypes

Non-crossover progeny

Crossover progeny

est of these groups arises from a double crossover. By looking for the least frequent class, it is often possible to instantly confirm (or deny) a postulated arrangement. The results in Figure 1-15 immediately confirm the order hinted at by the two-factor crosses. Only if the order is *a-c-b* does the fact that the rare recombinants are *AcB* and *aCb* make sense.

The existence of multiple crossovers means that the amount of recombination between the outside markers *a* and *b* (*ab*) is usually less than the sum of the recombination frequencies between *a* and *c* (*ac*) and *c* and *b* (*cb*). To obtain a more accurate approximation of the distance between the outside markers, we calculate the probability (*ac* × *cb*) that when a crossover occurs between *c* and *b*, a crossover also occurs between *a* and *c*, and vice versa (*cb* × *ac*). This probability subtracted from the sum of the frequencies expresses more accurately the amount of recombination. The simple formula

$$ab = ac + cb - 2(ac)(cb)$$

is applicable in all cases where the occurrence of one crossover does not affect the probability of another crossover. Unfortunately, accurate mapping is often disturbed by *interference* phenomena, which can either increase or decrease the probability of correlated crossovers.

Using such reasoning, the Columbia University group headed by Morgan had by 1915 assigned locations to more than 85 mutant genes in *Drosophila* (Table 1-1), placing each of them at distinct spots on one of the four linkage groups, or chromosomes. Most importantly, all the genes on a given chromosome were located on a line. The gene arrangement was strictly linear and never branched. Figure 1-16 shows the genetic map of one of the chromosomes of *Drosophila*. Dis-

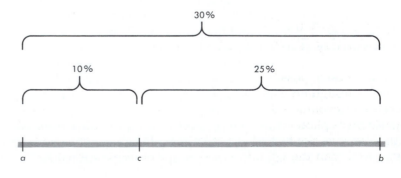

Figure 1-14
Assignment of the tentative order of three genes on the basis of 3 two-factor crosses.

tances between genes on such a map are measured in **map units,** which are related to the frequency of recombination between the genes. Thus, if the frequency of recombination between two genes is found to be 5 percent, the genes are said to be separated by five map units. Because of the high probability of double crossovers between widely spaced genes, such assignments of map units can be considered accurate only if recombination between closely spaced genes is followed.

Even when two genes are at the far ends of a very long chromosome, they will show not less than 50 percent linkage (assort together at least 50 percent of the time) because of multiple crossovers. The two genes will be separated if an odd number of crossovers occur between them, but they will end up together if an even number occur between them. Thus, in the beginning of the genetic analysis of *Drosophila*, it was often impossible to determine whether two genes were on different chromosomes or at the opposite ends of one long chromosome. Only after large numbers of genes had been mapped was it possible to convincingly demonstrate that the number of linkage groups equalled the number of cytologically visible chromosomes. It was then possible for Morgan, Sturdevant, Muller, and Bridges to publish their definitive book *The Mechanism of Mendelian Heredity*, which first announced the general validity of the chromosomal basis of heredity. We now rank this concept, along with the theories of evolution and the cell, as a major achievement in our quest to understand the nature of the living world.

Many Genes Control the Red Eye

Inspection of the list of mutant genes in Table 1-1 reveals an important fact. Many different genes act to influence a single characteristic. For example, 13 of the genes discovered by 1915 affect eye color. When a fly is homozygous for a mutant form of any one of these genes, the eye color is not red, but a different color, distinct for the mutant gene (e.g., carnation, vermilion). Thus, there is no one-to-one correspondence between genes and complex characteristics like eye color or wing shape. Instead, the development of each characteristic is controlled by a series of events, each of which is controlled by a gene. We might make a useful analogy with the functioning of a complex machine like the automobile. There are a number of separate parts—such as the motor, the brakes, the radiator, and the fuel tank—all of which are essential for proper operation of the vehicle. Although a fault in any one part may cause the car to stop functioning properly, there is no reason to believe that the presence of that component alone is sufficient for proper functioning.

The Origin of Genetic Variability Through Mutations[9–15]

It now became possible to understand the hereditary variation that is found throughout the biological world and that forms the basis of the theory of evolution. Genes are normally copied exactly during chromosome duplication. Rarely, however, changes (**mutations**) occur in genes to give rise to altered forms, most, *but not all*, of which function less well than the wild-type alleles. This process is necessarily rare;

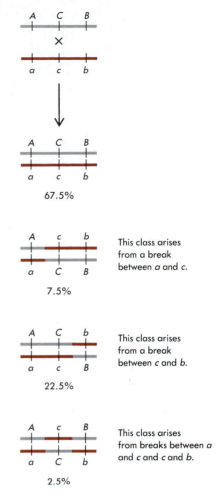

67.5%

7.5% This class arises from a break between *a* and *c*.

22.5% This class arises from a break between *c* and *b*.

2.5% This class arises from breaks between *a* and *c* and *c* and *b*.

Figure 1-15
The use of three-factor crosses to assign gene order. The least frequent pair of reciprocal recombinants must arise from a double crossover. The percentages listed for the various classes are the theoretical values expected for an infinitely large sample. When finite numbers of progeny are recorded, the exact values will be subject to random statistical fluctuations.

Table 1-1   The 85 Mutant Genes Reported in *Drosophila melanogaster* in 1915*

Name	Region Affected	Name	Region Affected
Group 1			
Abnormal	Abdomen	Lethal, 13	Body, death
Bar	Eye	Miniature	Wing
Bifid	Venation	Notch	Venation
Bow	Wing	Reduplicated	Eye color
Cherry	Eye color	Ruby	Leg
Chrome	Body color	Rudimentary	Wing
Cleft	Venation	Sable	Body color
Club	Wing	Shifted	Venation
Depressed	Wing	Short	Wing
Dotted	Thorax	Skee	Wing
Eosin	Eye color	Spoon	Wing
Facet	Ommatidia	Spot	Body color
Forked	Spine	Tan	Antenna
Furrowed	Eye	Truncate	Wing
Fused	Venation	Vermilion	Eye color
Green	Body color	White	Eye color
Jaunty	Wing	Yellow	Body color
Lemon	Body color		
Group 2			
Antlered	Wing	Jaunty	Wing
Apterous	Wing	Limited	Abdominal band
Arc	Wing	Little crossover	Chromosome 2
Balloon	Venation	Morula	Ommatidia
Black	Body color	Olive	Body color
Blistered	Wing	Plexus	Venation
Comma	Thorax mark	Purple	Eye color
Confluent	Venation	Speck	Thorax mark
Cream II	Eye color	Strap	Wing
Curved	Wing	Streak	Pattern
Dachs	Leg	Trefoil	Pattern
Extra vein	Venation	Truncate	Wing
Fringed	Wing	Vestigial	Wing
Group 3			
Band	Pattern	Pink	Eye color
Beaded	Wing	Rough	Eye
Cream III	Eye color	Safranin	Eye color
Deformed	Eye	Sepia	Eye color
Dwarf	Size of body	Sooty	Body color
Ebony	Body color	Spineless	Spine
Giant	Size of body	Spread	Wing
Kidney	Eye	Trident	Pattern
Low crossing over	Chromosome 3	Truncate intensf.	Wing
Maroon	Eye color	Whitehead	Pattern
Peach	Eye color	White ocelli	Simple eye
Group 4			
Bent	Wing	Eyeless	Eye

*The mutations fall into four linkage groups. Since four chromosomes were cytologically observed, this indicated that the genes are situated on the chromosomes. Notice that mutations in various genes can act to alter a single character, such as body color, in different ways.

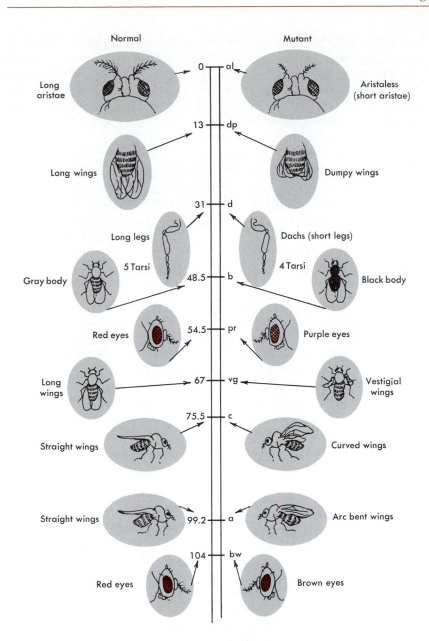

Normal   Mutant

Long aristae — 0 — al — Aristaless (short aristae)

Long wings — 13 — dp — Dumpy wings

Long legs — 31 — d — Dachs (short legs)

Gray body   5 Tarsi — 48.5 — b — 4 Tarsi   Black body

Red eyes — 54.5 — pr — Purple eyes

Long wings — 67 — vg — Vestigial wings

Straight wings — 75.5 — c — Curved wings

Straight wings — 99.2 — a — Arc bent wings

Red eyes — 104 — bw — Brown eyes

Figure 1-16
The genetic map of chromosome 2 of
Drosophila melanogaster.

otherwise, many genes would be changed during every cell cycle, and offspring would not ordinarily resemble their parents. There is, instead, a strong advantage in there being a small but finite mutation rate; it provides a constant source of new variability, necessary to allow plants and animals to adapt to a constantly changing physical and biological environment.

Surprisingly, however, the results of the Mendelian geneticists were not avidly seized upon by the classical biologists, then the authorities on the evolutionary relations between the various forms of life. Doubts were raised about whether genetic changes of the type studied by Morgan and his students were sufficient to permit the evolution of radically new structures, like wings or eyes. Instead, these biologists believed that there must also occur more powerful "macromutations," and that it was these that allowed great evolutionary advances.

Gradually, however, doubts vanished, largely as a result of the efforts of the mathematical geneticists Sewall Wright, Ronald A. Fisher, and J. B. S. Haldane. They showed that considering the great age of the earth, the relatively low mutation rates found for *Drosophila* genes, together with only mild selective advantages, would be sufficient to allow the gradual accumulation of new favorable attributes. By the 1930s, biologists began to reevaluate their knowledge on the origin of species and to understand the work of the mathematical geneticists. Among these new Darwinians were the biologist Julian Huxley (a grandson of Darwin's original publicist, Thomas Huxley), the geneticist Theodosius Dobzhansky, the paleontologist George Gaylord Simpson, and the ornithologist Ernst Mayr. In the 1940s, all four wrote major works, each showing from his special viewpoint how Mendelianism and Darwinism were indeed compatible.

Early Speculations About What Genes Are and How They Act[16–19]

Almost immediately after the rediscovery of Mendel's laws, geneticists began to speculate about both the chemical structure of the gene and the way it acts. No real progress could be made, however, because the chemical identity of the genetic material remained unknown. Even the realization that both nucleic acids and proteins are present in chromosomes did not really help, since the structure of neither was at all understood. The most fruitful speculations focused attention on the fact that genes must be, in some sense, self-duplicating. Their structure must be exactly copied every time one chromosome becomes two. This fact immediately raised the profound chemical question of how a complicated molecule could be precisely copied to yield exact replicas.

Some physicists also became intrigued with the gene, and when quantum mechanics burst on the scene in the late 1920s, the possibility arose that in order to understand the gene, it would first be necessary to master the subtleties of the most advanced theoretical physics. Such thoughts, however, never really took root, since it was obvious that even the best physicists or theoretical chemists would not worry about a substance whose structure still awaited elucidation. There was only one fact that they might ponder: Hermann Muller and L. J. Stadler's independent 1927 discoveries that X-rays induce mutations. Since there is a greater possibility that an X-ray will hit a larger gene than a smaller gene, the frequency of mutations induced in a given gene by a given X-ray dose yields an estimate of the size of this gene. But even here, so many special assumptions had to be made that virtually no one, not even the estimators themselves, took the estimates very seriously.

Bands (Genes?) Along the Giant Salivary Chromosomes of *Drosophila*[20, 21]

Unfortunately, no information on what genes are could be obtained by directly looking at ordinary chromosomes with the light microscope. Most chromosomes are far too small for such analysis, usually being just large enough to be clearly delineated from one another. Moreover, when they are visible during mitosis, they are highly supercoiled into arrangements about which there was much disagreement among the leading cytologists of that period. Much fresh hope was thus generated when in 1934 the chromosomes from the salivary glands of the fly *Chironomus tentans*, and soon thereafter of *Drosophila*, were found to be some thousand times larger than their equivalents in all other cells. Moreover, the individual chromosomes were not only highly extended, but marked by hundreds of discrete, dark-staining bands (Figure 1-17). Most importantly, the locations and sizes of the bands were highly reproducible, allowing very precise cytological maps to be made. These could be used to delineate the exact points where chromosomal segments of *Drosophila* switched partners during the rare chromosomal exchanges (translocations) that occur following accidental chromosomal breaks. Careful examination of the salivary chromosomes also showed that some mutations in fact represented discrete deletions or insertions of one or more bands.

By 1941, some 5059 different bands had been mapped, a number only severalfold larger than the number of genes that had been revealed by genetic analysis. This approximate similarity prompted the speculation that the bands in fact represented the genes, with the less darkly staining interbands representing the nonspecific chromosomal material. There was no way at that time, however, to follow up on this hypothesis, much less rule out the alternative possibility that the darkly stained masses were genetically inert, with the true genes being the less compacted interband material.

Preliminary Attempts to Find a Gene-Protein Relationship[22–25]

The most fruitful early endeavors to find a relationship between genes and proteins examined the ways in which gene changes affect which proteins are present in the cell. At first this study was difficult, since no one knew anything about the proteins that were present in structures such as the eye or the wing. It soon became clear that genes with simple metabolic functions would be easier to study than genes affecting gross structures. One of the first useful examples came from a study of a hereditary disease affecting amino acid metabolism. Spontaneous mutations occur in humans affecting the ability to metabolize the amino acid phenylalanine. When individuals homozygous for the mutant trait eat food containing phenylalanine, their inability to convert the amino acid to tyrosine causes a toxic level of phenylpyruvic acid to build up in the bloodstream. Such diseases, examples of "inborn errors of metabolism," suggested as early as 1909 to the English physician A. E. Garrod that the wild-type gene is responsible for the presence of a particular enzyme, and that in a homozygous mutant, the enzyme is congenitally absent.

Garrod's general hypothesis of a gene-enzyme relationship was extended in the 1930s by work on flower pigments by Haldane and Rose Scott-Moncrieff in England, studies on the hair pigment of the guinea pig by Wright in the U.S., and research on the pigments of insect eyes by A. Kuhn in Germany and by Boris Ephrussi and George W. Beadle, working first in France and then in California. In all cases, evidence revealed that a particular gene affected a particular step in the formation of the respective pigment whose absence changed, say, the color of a fly's eyes from red to ruby. However, the lack of fundamental knowledge about the structures of the relevant enzymes ruled out deeper examination of the gene-enzyme relationship, and no assurance could be given either that most genes control the synthesis of proteins (by then it was suspected that all enzymes were proteins) or that all proteins are under gene control.

(a)

(b)　　|←—5μm—→|

Figure 1-17
(a) Electron micrograph of chromosome 4 from a salivary gland of *Chironomus tentans*. [Reproduced with permission from B. Daneholt, *Cell* 4 (1975):1.] (b) Diagram of a portion of a salivary gland chromosome (the right arm of chromosome 3) of *Drosophila melanogaster*. [After P N. Bridges, *J. Heredity* 32 (1941).]

As early as 1936, it became apparent to the Mendelian geneticists that future experiments of the sort successful in elucidating the basic features of Mendelian genetics were unlikely to yield productive evidence about how genes act. Instead, it would be necessary to find biological objects more suitable for chemical analysis. They were aware, moreover, that contemporary knowledge of nucleic acid and protein chemistry was completely inadequate for a fundamental chemical attack on even the most suitable biological systems. Fortunately, however, the limitations in chemistry did not deter them from learning how to do genetic experiments with chemically simple molds, bacteria, and viruses. As we shall see, the necessary chemical facts became available almost as soon as the geneticists were ready to use them.

Uneasiness About Cytoplasmic Inheritance and Movable Genetic Elements[26–33]

The neat picture of the Mendelian world with its discrete nuclear chromosomes along which genes were sited at fixed locations was not without its exceptions. Some genetic markers, already well studied by the mid-1930s, did not map at all to chromosomes but rather showed "maternal inheritance," being transmitted only through the cytoplasm-rich egg—never through sperm. These misbehaving genetic traits must somehow reside in the cytoplasm, raising the question of whether such examples negate the general validity of the chromosomal theory of heredity. Since many of these maternally transmitted traits involve losses of photosynthetic pigments, speculations arose that the chloroplasts, the organelle carriers of these pigments, might in fact be tiny bacterium-like cells endowed with their own genetic material (chromosomes?). But there existed neither the means to purify away chloroplasts from other cellular constituents nor any way to show that one of their molecular constituents was a carrier of genetic information.

Equally puzzling were reports first emanating from research on *Drosophila* of extremely unstable genes that showed very high mutation rates. How such hypervariable genetic elements originated first began to be understood from Barbara McClintock's genetic analysis of the corn plant. This seminal research began soon after her arrival at Cold Spring Harbor in 1942 and culminated with her 1951 announcement that several key genetic elements, which she named "controlling elements," did not have fixed locations. Instead, they could apparently move from one chromosomal site to another, without in any way disturbing the general architecture of the chromosomes involved. When these controlling elements jumped, they frequently affected the functioning of the genes into which they became inserted; hence, their designation as control elements. Why only certain genes so moved, and in fact whether such movements played any role in the normal development of the corn plant, could not then be ascertained. So most geneticists saw no alternative but to sweep control elements under the rug (together with cytoplasmic inheritance) until the nature of more conventional chromosomal genes could be better explained.

Summary

The study of living organisms at the biological level has led to three great generalizations: (1) Darwin's and Wallace's theory of evolution by natural selection, which tells us that today's complex plants and animals are derived by a continuous evolutionary progression from the first primitive organisms; (2) the cell theory, the realization that all organisms are composed of cells; and (3) the chromosomal theory of heredity, the understanding that the function of chromosomes is what controls heredity.

All cells contain chromosomes, normally duplicated prior to a cell division process (mitosis) that produces two daughter cells, each with a chromosomal complement identical to that of the parental cell. In haploid cells, there is just one copy of each type of chromosome; in diploid cells, there are usually two copies (pairs of homologous chromosomes). A diploid cell arises by fusion of a male and a female haploid cell (fertilization), whereas haploid cells are formed from a diploid cell by a distinctive form of cell division (meiosis) that reduces the chromosome number to one-half its previous number.

Chromosomes control heredity because they are the cellular locations of genes. Hereditary factors were first discovered by Mendel in 1865, but their importance was not realized until the start of the twentieth century. Each gene can exist in a variety of different forms called alleles. Mendel proposed that a hereditary factor (now known to be a gene) for each hereditary trait is given by each parent to each of its offspring. The physical basis for this behavior is the distribution of homologous chromosomes during meiosis: One (randomly chosen) of each pair of homologous chromosomes is distributed to each haploid cell. When two genes are on the same chromosome, they tend to be inherited together (linked). Genes affecting different characteristics are sometimes inherited independently of each other; this is because they are located on different chromosomes. In any case, linkage is seldom complete, because homologous chromosomes attach to each other during meiosis and often break at identical spots and rejoin crossways (crossing over). Crossing over attaches genes initially found on a paternally derived chromosome to gene groups originating from the maternal parent.

Different alleles from the same gene arise by inheritable changes (mutations) in the gene itself. Normally, genes are extremely stable and are exactly copied during chromosome duplication; mutation occurs only rarely and usually has harmful consequences. It does, however, play a positive role, since the accumulation of the rare favorable mutations provides the basis for the genetic variability that the theory of evolution presupposes.

The most suitable chromosomes for assigning genes to specific chromosomal locations are the giant chromosomes of the salivary glands of insects. Such chromosomes are highly extended and are marked by hundreds (or thousands, depending on the species) of discrete, dark-staining bands. Through studies of the segregation patterns following chromosomal rearrangement, genes can be assigned locations in or adjacent to specific bands. Such maps, however, do not reveal whether the dense bands or the interband regions are the genetically active factors.

For many years, the structure of genes and the chemical way in which they control cellular characteristics were a mystery. As soon as large numbers of spontaneous mutations had been described, it became obvious that a one gene–one characteristic relationship does not exist and that all complex characteristics are under the control of many genes. The most sensible idea, postulated by Garrod in 1909, was that genes affect the synthesis of enzymes. However, the tools of Mendelian geneticists—organisms such as the corn plant, the mouse, and even the fruit fly *Drosophila*—were not suitable for detailed chemical investigations of gene-protein relations. For this type of analysis, work with much simpler organisms became indispensable.

The neat picture of discrete genes located linearly at fixed positions along a chromosome does not always hold true, since some genetic markers were found to reside in the cytoplasm; some organelles (e.g., chloroplasts) might contain their own genetic material. Moreover, genes with extraordinarily high mutation rates were found in rare instances. In maize (corn), these highly mutable elements were found to jump from one chromosome site to another, affecting the functioning of the genes into which they became inserted. However, most geneticists saw no alternative but to ignore these exceptions until the nature of conventional chromosomal genes was clarified.

Bibliography

General References

Alberts, B., D. Bray, J. Lewis, M. Raff, K. Roberts, and J. D. Watson. 1983. *Molecular Biology of the Cell.* New York: Garland.

Ayala, F. J., and J. A. Kiger, Jr. 1984. *Modern Genetics.* 2nd ed. Menlo Park, Calif.: Benjamin/Cummings.

Carlson, E. J. 1966. *The Gene Theory: A Critical History.* Philadelphia: Saunders. A survey of the development of genetic ideas beginning with Mendel.

Mayr, E. 1982. *The Growth of Biological Thought: Diversity, Evolution, and Inheritance.* Cambridge, Mass., and London: Harvard University Press. Traces the development of the earliest problems in biology, with historical and modern references to evolutionary and genetic ideas.

Moore, J. 1972. *Heredity and Development.* 2nd ed. Oxford, Eng., and New York: Oxford University Press. An elegant introduction to genetics and embryology, with emphasis on the historical approach.

Moore, J. 1972. *Readings in Heredity and Development.* Oxford, Eng.: Oxford University Press. Reproduces many of the key papers in the development of the chromosomal theory of heredity, including those of Sutton.

Peters, J. A. 1959. *Classic Papers in Genetics.* Englewood Cliffs, N.J.: Prentice-Hall. A collection of reprints of the most significant papers in the history of genetics, up to Benzer's fine-structure analysis of the gene.

Cited References

1. Wilson, E. B. 1925. *The Cell in Development and Heredity*. 3d ed. New York: Macmillan.
2. Olby, R. C. 1966. *Origins of Mendelism*. London: Constable and Company Ltd. The growth of ideas about inheritance and variation that eventually led to the Mendelian solution.
3. Sutton, W. S. 1903. "The Chromosome in Heredity." *Biol. Bull.* 4:231–251.
4. Sturtevant, A. H., and G. W. Beadle. 1962. *An Introduction to Genetics*. New York: Dover. Now available in paperback form; this book, originally published in 1939, remains a classic statement of the results of *Drosophila* genetics.
5. Morgan, T. H. 1910. "Sex-Linked Inheritance in *Drosophila*." *Science* 32:120–122.
6. McClintock, B., and H. B. Creighton. 1931. "A Correlation of Cytological and Genetical Crossing Over in *Zea Mays*." *Proc. Nat. Acad. Sci.* 17:492–497.
7. Sturtevant, A. H. 1913. "The Linear Arrangement of Six Sex-Linked Factors in *Drosophila* as Shown by Mode of Association." *J. Exp. Zool.* 14:39–45.
8. Morgan, T. H., A. H. Sturtevant, H. J. Muller, and C. B. Bridges. 1915. *The Mechanism of Mendelian Heredity*. New York: Holt, Rinehart & Winston.
9. Dobzhansky, T. 1941. *Genetics and the Origin of Species*. 2nd ed. New York: Columbia University Press.
10. Huxley, J. 1943. *Evolution: The Modern Synthesis*. New York: Harper & Row. The grandson of T. H. Huxley explores Mendelism, evolutionary trends, and genetic systems.
11. Mayr, E. 1942. *Systematics and the Origin of Species*. New York: Columbia University Press. A correlation of the evidence and points of view of systematics with those of other biological disciplines, particularly genetics and ecology.
12. Simpson, G. G. 1944. *Tempo and Mode in Evolution*. New York: Columbia University Press. A synthesis of paleontology and genetics.
13. Haldane, J. B. S. 1932. *The Courses of Evolution*. New York: Harper & Row.
14. Fisher, R. A. 1930. *The Genetical Theory of Natural Selection*. Oxford, Eng.: Clarendon Press.
15. Wright, S. 1931. "Evolution in Mendelian Populations." *Genetics* 16:97–159.
16. Muller, H. J. 1927. "Artificial Transmutation of the Gene." *Science* 46:84–87.
17. Carlson, E. 1981. *Genes, Radiation, and Society: The Life and Work of J. H. Muller*. Ithaca, N.Y.: Cornell University Press.
18. Stadler, L. J. 1928. "Mutations in Barley Induced by X-Rays and Radium." *Science* 110:543–548.
19. Lea, D. E. 1947. *Actions of Radiations on Living Cells*. New York: Macmillan.
20. Bridges, P. N. 1941. "A Revision of the Salivary Gland 3R-Chromosome Map of *Drosophila melanogaster*." *J. Heredity* 32:299–300.
21. Painter, T. S. 1934. "The Morphology of the X Chromosomes in the Salivary Glands of *D. melanogaster* and a New Type of Chromosome Map for this Element." *Genetics* 19:98–99.
22. Garrod, A. E. 1908. "Inborn Errors of Metabolism." *Lancet* 2:1–7, 73–79, 142–148, 214–220.
23. Beadle, G. W., and B. Ephrussi. 1937. "Development of Eye Color in *Drosophila*: Diffusible Substances and Their Interrelations." *Genetics* 22:76–86.
24. Scott-Moncrieff, R. 1936. "A Biochemical Survey of Some Mendelian Factors for Flower Color." *J. Genetics* 32:117–170.
25. Wright, S. 1941. "The Physiology of the Gene." *Physiol. Rev.* 21:487–527.
26. McClintock, B. 1951. "Chromosome Organization and Gene Expression." *Cold Spring Harbor Symp. Quant. Biol.* 16:13–57. First extensive report on movable elements in maize.
27. McClintock, B. 1984. "The Significance of Responses of Genome to Challenge." *Science* 226:792–800. This Nobel address elegantly traces and discusses how the concept of movable genes arose.
28. Caspari, E. 1948. "Cytoplasmic Inheritance." *Adv. Genetics* 2:1–66.
29. Correns, C. 1937. *Nicht Mendelnde Vererbung*. Edited by F. von Wettstein. Berlin: Borntraeger.
30. Rhoades, M. M. 1946. "Plastid Mutations." *Cold Spring Harbor Symp. Quant. Biol.* 11:202–207.
31. Sager, R. 1972. *Cytoplasmic Genes and Organelles*. New York: Academic Press.
32. Sonneborn, T. M. 1950. "The Cytoplasm in Heredity." *Heredity* 4:11–36.
33. Winge, O., and O. Laustsen. 1940. "On a Cytoplasmic Effect of Inbreeding in Homozygous Yeast." *Comptes Rend. Laboratoire de Carlsberg* 23:17–40.

CHAPTER 2

Cells Obey the Laws of Chemistry

In Darwin's time, chemists were already asking whether living cells work by the same chemical rules as nonliving systems. By then, cells had been found to contain no chemical elements unique to living material. Also recognized early was the predominant role of carbon, a major constituent of almost all types of biological molecules. The initial tendency to distinguish between carbon compounds, like those in living matter, and all other molecules is reflected in the division of modern chemistry into organic chemistry (the study of most compounds containing carbon atoms) and inorganic chemistry. We now know that this distinction is artificial and has no biological basis. There is no purely chemical way to decide whether a compound has been synthesized in a cell or in a chemist's laboratory.

Nonetheless, through the first quarter of this century, many scientists thought that some vital force outside the laws of chemistry differentiate the animate from the inanimate. Part of the reason for the persistence of this "vitalism" was that the success of the biologically oriented chemists (now usually called biochemists) was limited. Although the techniques of the organic chemists were sufficient to work out the structures of relatively small molecules like glucose (Table 2-1), there was increasing awareness that many of the most important molecules in the cell—the so-called macromolecules—were far too large to be pursued by even the best of organic chemists.

For many years, the most important group of macromolecules was believed to be the proteins because of the growing evidence that all enzymes are proteins. Initially, there was controversy as to whether enzymes are small molecules or macromolecules. It was not until 1926 that the enzymatic nature of a crystalline protein was demonstrated by the American biochemist James Sumner.[1] The controversy was then practically settled. But even this important discovery did not dispel the general aura of mystery about proteins. In those days, the complex structures of proteins were undecipherable by available chemical tools, so as late as 1940, many scientists still believed that these molecules would eventually be shown to have features unique to living systems.

There was also a general belief that genes, like enzymes, might be proteins. There was no direct evidence, but the high degree of specificity of genes suggested to most people who speculated on their nature that they could only be proteins, by then known to occur in the chromosomes. Another class of molecules, the nucleic acids, were also found to be a common chromosomal component, but these were

Table 2-1   Some Important Classes of Small Biological Molecules

Class	Characteristics	Example
Aliphatic hydrocarbon	Linear or branched molecules containing only carbon and hydrogen	Ethane
Aromatic hydrocarbon	Ring-shaped hydrocarbons containing alternating single and double bonds	Benzene
Pyrimidine	An aromatic compound of the formula $C_4H_4N_2$ (or a derivative thereof)	Uracil
Purine	An aromatic compound of the formula $C_5H_4N_4$ (or a derivative)	Guanine
Alcohol	A hydrocarbon skeleton substituted with one to several OH groups	Ethanol
Phosphate ester	Molecule formed from alcohols and phosphoric acid with the elimination of H_2O	Glucose-1-(P)*
Nucleoside	Contains a pentose sugar linked to either a purine or a pyrimidine base, through a C–N bond	Adenosine

*In this book, (P) is frequently used as an abbreviation for phosphoric acid (H_3PO_4), phosphate ion (PO_4^{3-}), or a phosphate group ($—PO_3^{2-}$). See Table 2-2.

Table 2-1 *(Continued)*

Class	Characteristics	Example
Nucleotide	Phosphate ester of a nucleotide	Adenosine-5-$\textcircled{P}$ or adenylic acid
Carboxylic acid	Hydrocarbon skeleton with one to several COO^- groups	Acetic acid
Hydroxy acid	Substituted hydrocarbon containing both an OH and a COOH group	Lactic acid
Keto acid	Substituted hydrocarbon containing both a keto group and a COOH group	Pyruvic acid
Amino acid	Substituted hydrocarbon containing both NH_2 and COOH groups; both groups are usually charged. The general formula is where R represents a group that varies from amino acid to amino acid. There are 20 amino acids, each with a distinctive R group.	Glycine
Sugar (monosaccharide)	Polyhydroxyl molecuole containing a C=O group (either an aldehyde or a ketone). The most common sugars have 3 (triose), 4 (tetrose), 5 (pentose), or 6 (hexose) carbon atoms.	Ribose

initially thought to be relatively small molecules and incapable of carrying sufficient information to function as genes.

Besides general ignorance of the structures of the large molecules in the cell, the feeling was often expressed that there is something unique about the three-dimensional organization of the cell that gives it its living feature. This argument was sometimes phrased in terms of the impossibility of ever understanding all the exact chemical interactions of the cell. More frequently, however, it took the form of the prediction that some new natural laws, as important as the cell theory or the theory of evolution, would have to be discovered before the essence of life could be understood. But these almost mystical ideas never led to meaningful experiments and, in their vague form, could never be tested. Progress was made instead only by biologically oriented chemists and physicists patiently attempting to devise new ways of solving the structures of more and more complex biological molecules. But for many years, there were no triumphs to shout. The chemists and biologists usually moved in different and sometimes hostile worlds, the biologist often denying that the chemist would ever provide the real answers to the important riddles of biology. Always not too far back in some biologists' minds was the feeling, if not the hope, that something more basic than mere complexity and size separated biology from the bleak, inanimate world of a chemical laboratory.

The Concept of Intermediary Metabolism

As soon as the organic chemists began to identify some of the various cellular molecules, it became clear that food molecules are extensively transformed after they enter an organism. In no case does a food source contain all the different molecules present in a cell. On the contrary, in some cases practically all the organic molecules within an organism are synthesized inside it. This point is easily seen by observing cellular growth on well-defined food sources, for example, the growth of yeast cells using the simple sugar glucose as the sole source of carbon. Soon after its entry into the cells, glucose is chemically transformed into a large variety of molecules necessary for the building of new cellular components. These chemical transformations rarely occur in one step; instead, intermediate compounds are produced. These intermediate compounds often have no cellular function besides forming part of a pathway leading to the synthesis of a necessary structural component, such as an amino acid.

The sum total of all the chemical reactions occurring in a cell is referred to as the **metabolism** of the cell. Correspondingly, the molecules involved in these transformations are often called **metabolites. Intermediary metabolism** is the term used to describe the chemical reactions involved in the transformation of food molecules into essential cellular building blocks.

Cellular Energy Is Generated by Oxidation-Reduction Reactions

By the middle of the nineteenth century, it was known that the food (initially of plant origin) eaten by animals and bacteria is only partly transformed into new cellular building blocks, some of it being burned by combustion with oxygen (O_2) to yield carbon dioxide (CO_2), water (H_2O), and energy. At the same time, it was becoming clear that in green plants, the reverse process also operates. These two processes, called **respiration** and **photosynthesis,** respectively, can be summarized by the following equations:

Respiration:

$C_6H_{12}O_6$ (glucose) + $6O_2$ $\longrightarrow$

$\qquad 6CO_2 + 6H_2O$ + energy (in the form of heat)   *(2-1)*

Occurs in both plants and animals.

Photosynthesis:

$6CO_2 + 6H_2O$ + energy (from the sun) $\longrightarrow$

$\qquad C_6H_{12}O_6$ (glucose) + $6O_2$ *(2-2)*

Occurs only in plants, algae, and certain bacteria.

Both these equations can be thought of as the sum total of a lengthy series of oxidation-reduction reactions.

In respiration, organic molecules such as glucose are oxidized by molecular oxygen to form C=O bonds (Table 2-2), which contain *less* usable energy (energy that can do work) than the starting C–H, C–OH, and C–C bonds. Energy is given off in respiration, just as it is when any organic compound burns at high temperatures outside the cell, to produce carbon dioxide, water, and energy in the form of heat. In contrast, during photosynthesis, the energy from the light quanta of the sun is used to reduce carbon dioxide to molecules that contain *more* usable energy.

When these relationships were first worked out, no one knew how the energy obtained during respiration is put to advantage. It was clear that somehow a useful form of energy has to be available to enable living organisms to carry out a variety of forms of work, such as muscular contraction and selective transport of molecules across cell membranes. Even then, it seemed unlikely that the energy obtained from food is first released as heat, because at the temperature at which life exists, heat energy cannot be effectively used to synthesize new chemical bonds. Thus, since the awakening of an interest in the chemistry of life, a prime challenge has been to understand the generation of energy in a useful form.

Most Biological Oxidations Occur Without Direct Participation of Oxygen[2,3]

Because oxygen is so completely necessary for the functioning of animals, it was natural to guess that oxygen would participate directly in all oxidations of carbon compounds. Actually, most biological oxidations occur in the absence of oxygen. This is possible because, as first proposed around 1912 by the German biochemist H. O. Wieland, most biological oxidations are actually dehydrogenations. A compound is *oxidized* when we remove a pair of hydrogen atoms from it (Figure 2-1). However, hydrogen atoms cannot merely be removed; they must be transferred to another molecule, which is then said to be *reduced* (Figure 2-2). In these reactions, as in all other **oxidation-reduction reactions,** every time one molecule is oxidized, another must be reduced. There are several different molecules whose role is to receive hydrogen atoms. All are medium-size (MW about 500) organic

Figure 2-1

Oxidation of an organic molecule by removal of a pair of hydrogen atoms. This figure shows the oxidation of lactate to pyruvate.

Figure 2-2

The participation of nicotinamide adenine dinucleotide (NAD^+) in the oxidation of lactate to pyruvate. The oxidizing agent here is NAD^+, and the hydrogen donor is lactate; lactate is oxidized and NAD^+ is reduced.

Table 2-2 Important Functional Groups in Biological Molecules

Group	Molecular Example	Biological Significance
$-\overset{\underset{\mid}{\text{H}}}{\underset{\mid}{\text{C}}}-\text{H}$ Methyl	Methyl group of alanine	Highly insoluble in water; does not form hydrogen bonds
—OH Hydroxyl	Ethanol (ethyl alcohol)	Water soluble; forms hydrogen bonds
$-\text{C}\overset{\text{O}}{\underset{\text{OH}}{}}$ Carboxyl	Acetate (acetic acid)	Usually charged: $-\text{C}\overset{\text{O}}{\underset{\text{OH}}{}} \rightleftharpoons -\text{C}\overset{\text{O}}{\underset{\text{O}^-}{}} + \text{H}^+$ Good acceptor of hydrogen bonds
$-\text{N}\overset{\text{H}}{\underset{\text{H}}{}}$ Amino	Glycine	Usually charged: $-\text{NH}_2 + \text{H}^+ \rightleftharpoons -\text{NH}_3^+$ Forms hydrogen bonds
$-\overset{\overset{\text{O}}{\parallel}}{\underset{\underset{\text{O}^-}{\mid}}{\text{P}}}-\text{O}^-$ Phosphate	Glyceraldehyde-3-phosphate	Always negatively charged; forms hydrogen bonds with water; thus very water soluble. Often abbreviated Ⓟ in this book.

molecules that associate with specific proteins to form active enzymes. The protein components alone (**apoenzymes**) have no enzymatic activity. Only when the smaller partner is present will activity be present. Hence, these smaller molecules are named **coenzymes.** (We should note that not all coenzymes participate in oxidation-reduction reactions; some coenzymes function in other types of metabolic reactions.)

Although the involvement of coenzymes in oxidative reactions was hinted at by 1910, it was not until the early 1930s that their cardinal significance was appreciated. Then the work of the great German biochemist Otto Warburg and the Swedish chemists Ulf S. von Euler and Theodor Theorell established the structure and action of several of the most important coenzymes: nicotinamide adenine dinucleotide (NAD$^+$, earlier called diphosphopyridine nucleotide or DPN, Figure 2-3), flavin mononucleotide (FMN), and flavin adenine dinucleotide (FAD).

Coenzymes, like enzymes, function over and over and are not used up in the course of a reaction. This is because the hydrogen atoms transferred to them do not remain permanently attached, but are transferred by a second oxidation-reduction reaction, usually to another enzyme, and sometimes to oxygen itself (Figure 2-4). Coenzymes are thus continually oxidized and reduced. Furthermore, although oxygen may not be directly necessary for a given reaction, we

Table 2-2 (*Continued*)

Group	Molecular Example	Biological Significance
$\diagdown \!\! C\!\!=\!\!O$ Carbonyl		
$\begin{smallmatrix}R\\\diagdown\\H\end{smallmatrix}C\!\!=\!\!O$ Aldehyde	CH_3 — $C\!\!=\!\!O$ Pyruvate (a ketone)	Forms hydrogen bonds; usually exists in keto form: $R\!-\!\underset{\underset{O}{\|\|}}{C}\!-\!CH_3$ as opposed to the enol form: $R\!-\!\underset{\underset{OH}{\|}}{C}\!\!=\!\!CH_2$
$\begin{smallmatrix}R\\\diagdown\\R\end{smallmatrix}C\!\!=\!\!O$ Ketone		
$\begin{smallmatrix}H\\\|\\-C\!-N\!\!-\!\!H\\\underset{O}{\|\|}\end{smallmatrix}$ Amide	Asparagine	Forms hydrogen bonds.
$-S\!\!-\!\!H$ Sulfhydryl	Cysteine	Two SH groups are easily oxidized to form S–S (disulfide bonds); SH groups form very weak hydrogen bonds.

see that it is often necessary indirectly, since it must be available to oxidize the coenzyme molecules to make them available for accepting additional pairs of hydrogen atoms (or electrons).

The Breakdown of Glucose

Much of the early work in intermediary metabolism dealt with the transformation of glucose into other molecules. Glucose was emphasized not only because it plays a central role in the metabolic economy of cells, but also for a practical reason; the alcohol (ethanol) produced when wine is made from grapes is derived from the breakdown of glucose. In 1810, the chemist J. Gay-Lussac demonstrated the production of ethanol by this process, and by 1837, the essential role of yeast was established. The production of the alcohol in wine is not a spontaneous process but normally requires the presence of living yeast cells.

Also important in the initial work with glucose was the French microbiologist Louis Pasteur, who discovered that the breakdown of glucose to alcohol does not require air; to distinguish it from reactions requiring oxygen, he used the term **fermentation.** He also showed that ethanol is not the only product of glucose fermentation but that there are other products, such as lactic acid and glycerol.

Figure 2-3
Nicotinamide adenine dinucleotide, a very important coenzyme. Shown here is the conversion between the oxidized form (abbreviated NAD$^+$), an acceptor of hydrogen atoms, and the reduced form (abbreviated NADH), a donor of hydrogen atoms. The release of hydrogen atoms decreases the free energy of a molecule; the acceptance of them increases the free energy.

The next great advance came with Eduard Buchner's discovery in 1897 that the living cell per se is not necessary for fermentation and that a cell-free extract from yeast can, by itself, transform glucose into ethanol. Not only was this step conceptually important, but it also provided a much more practical system for studying the chemical steps of fermentation. When working with cell-free systems, it is relatively easy to add or subtract components thought to be involved in the reaction; when living cells are being used, it is often very difficult, and sometimes impossible, to transfer specific compounds in an unmodified form across the cell membrane.

Over the next 40 years, cell-free extracts were used by a large number of distinguished biochemists, including the Englishmen Arthur Harden and William Young and the Germans Gustav Embden and Otto Meyerhof, to work out the exact chemical pathways of glucose degradation (Figure 2-5). The important generalization emerged that the reactions involved in the breakdown of glucose to pyruvate,* originally known as the *Embden-Meyerhof pathway* and now called **glycolysis,** are not peculiar to alcoholic fermentation in yeast but occur in many other cases of glucose utilization as well. Perhaps the most significant discovery was made by Meyerhof. He showed that when muscles contract in the absence of oxygen, the carbohydrate food reserve of glycogen is broken down via glucose to lactic acid (anaerobic glycolysis). Thus, it became clear that not only can microorgan-

---

*The terms *pyruvic acid* and *pyruvate* are used interchangeably. Technically, pyruvate refers to the negatively charged ion. Likewise, lactic acid is often called lactate; glutamic acid, glutamate; citric acid, citrate; and so on.

Figure 2-4
The transfer of a pair of hydrogen atoms from one coenzyme to another. In this series of reactions, the final hydrogen acceptor is molecular oxygen (O_2). FAD usually exists in combination with a specific protein, forming a flavoprotein.

isms obtain their energy and carbon via the Embden-Meyerhof pathway, but energy involved in muscle contraction is also generated by the same pathway.

The ubiquity of the Embden-Meyerhof pathway supported a growing feeling about "the unity of biochemistry." That is, the basic biochemical reactions upon which cell growth and division depend are the same, or very similar, in all cells—those of microorganisms as well as those of higher plants and animals. This unity was not surprising to the more astute biologists, many of whom were largely preoccupied with the consequences of evolutionary theory. Given that humans and fish are descended from a common ancestor, it should not be surprising that many of their cell constituents are similar.

Metabolic Importance of Phosphorus and the Generation of ATP[4]

As early as 1905, the phosphorus atom was implicated in a vital role in metabolism. Harden and Young found that alcoholic fermentation occurs only when inorganic phosphate (PO_4^{3-}) is present. This discovery was followed over the next two decades by the isolation of a large number of intermediary metabolites containing phosphate (Ⓟ) groups attached to carbon atoms by phosphate ester linkages,

$$
\begin{array}{ccc}
& \overset{\textstyle O}{\underset{\textstyle |}{\|}} & \\
-\overset{|}{\underset{|}{C}}-O-\overset{\textstyle}{\underset{\textstyle O^-}{P}}-O^- & \text{or} & -\overset{|}{\underset{|}{C}}-O-\text{Ⓟ}
\end{array}
$$

The significance of such phosphorylated intermediates was unclear for 25 years. Then, in the 1930s, Meyerhof and Fritz Lipmann had the crucial insight that phosphate esters enable cells to trap much of the energy of the chemical bonds present in their food molecules. During fermentation, several intermediates are created (e.g., D-1,3-diphosphoglyceric acid, Figure 2-6) that contain what are popularly known as "high-energy" phosphate ester bonds (see Chapter 6 for more details). These high-energy phosphate groups are usually transferred to acceptor molecules, where they can serve as sources of chemical energy for vital cellular processes, such as motion, generation of light, and (as we shall see in Chapter 6) the efficient biosynthesis of necessary cellular molecules. The most important of the acceptor molecules is *adenosine diphosphate* (ADP, Figure 2-7). Addition of a high-energy phosphate group to ADP forms **adenosine triphosphate (ATP)**.

$$\text{ADP} + \text{Ⓟ} \rightleftharpoons \text{ATP}$$

The discovery of the role of ADP as an acceptor molecule and that of ATP as a donor of high-energy phosphate groups was one of the most important landmarks of modern biology. Until the roles of these molecules were known, there was complete mystery about how cells obtained energy. There was constant speculation about how cellular existence was incompatible with the second law of thermodynamics, which states that in a closed system, the amount of disorder (entropy) invariably increases. The apparent paradox disappeared, however, as soon as it was seen how animal cells could trap and utilize the energy in food molecules. At that time, the mechanism by which the sun's

Figure 2-5
The stepwise degradation of glucose to pyruvate. This collection of consecutive reactions is called glycolysis. The name of the enzyme that catalyzes each step is shown in a colored rectangle. Additional enzymatic steps can convert pyruvate to other products, such as lactate or ethanol.

Figure 2-6
The formation of an energy-rich phosphate ester bond coupled with the oxidation of 3-phosphoglyceraldehyde by NAD to form 1,3-diphosphoglyceric acid. The symbol ~ signifies that the bond is of the high-energy variety.

energy is trapped in photosynthesis was not known. Here again, the primary action of the sun's energy is now known to be the generation of ATP.

Most Specific Cellular Reactions Require a Specific Enzyme

As the glycolytic pathway was being worked out, it became clear that each step requires a separate enzyme (see Figure 2-5). Each of the enzymes acts by combining with the molecules involved in the particular reaction (the **substrates** of the enzymes). For example, glucose and ATP are substrates for the enzyme hexokinase. When glucose and ATP molecules interact on the surface of hexokinase, in what is called the **active site** of the enzyme, the terminal phosphate of ATP is transferred to a glucose molecule to form glucose-6-phosphate (Figure 2-8).

The essence of an enzyme is its ability to speed up, or **catalyze,** a reaction involving the making or breaking of a specific covalent bond. In the absence of enzymes, most of the covalent bonds of biological

Figure 2-5
(Continued)

molecules are very stable and decompose only under high, nonphysiological temperatures; for example, only at several hundred degrees Celsius is glucose appreciably oxidized by oxygen in the absence of enzymes. Enzymes must therefore act by somehow lowering the temperature at which a given bond becomes unstable. A physical chemist would say that an enzyme lowers the "activation energy" of the reaction. How this is done is finally being understood at the molecular level. The three-dimensional structure of many enzymes is now known, and there are plausible chemical theories about how enzymes work (see Figure 2-25). Thus, there is no reason to suspect that still-undiscovered laws of chemistry underlie enzyme action. Numerous examples already exist in inorganic chemistry where well-defined molecules catalyze reactions between other molecules.

A very important characteristic of enzymes is that they are never consumed in the course of a reaction; once a reaction is complete, they are free to adsorb new molecules and function again. On a biological time scale (seconds to years), enzymes can work very fast, some being able to catalyze as many as 10^6 reactions per minute; often, no successful collision of substrates will occur for years when enzymes are absent.

The degree of specificity for particular substrates varies from enzyme to enzyme. Not all enzymatic reactions are highly specific. For example, a number of enzymes can break down a variety of different

Figure 2-7
The formation of ATP (adenosine-5'-triphosphate) from ADP and an energy-rich phosphate group. Here the donor of the high-energy phosphate is phosphoenolpyruvate.

proteins to their component amino acids. They are specific only in the sense that they catalyze the breakdown of a specific type of covalent bond, the peptide bond, and will not, for example, degrade the phosphodiester linkages of the nucleic acids.

The Key Role of Pyruvate: Its Utilization Via the Krebs Cycle[5, 6]

Attempts to understand the generation of ATP in the presence of oxygen occurred parallel with the study of fermentation and glycolysis. It was immediately obvious from the amounts of ATP generated by fermentation and glycolysis that these processes could account for only a small fraction of total ATP production in the presence of oxygen. This means that ATP production in the presence of oxygen does not cease once glucose has been degraded as far as pyruvate, but that pyruvate itself must be further transformed via energy-yielding reactions requiring the presence of oxygen.

The first real breakthrough in understanding how this happens came with discoveries made by the biochemists Albert Szent-Györgyi, Carl Martius and Franz Knoop, and Hans A. Krebs. Their work revealed the existence of a cyclic series of reactions (now usually called the **Krebs cycle,** citric acid cycle, or tricarboxylic acid cycle) by which pyruvate is oxidatively broken down to yield carbon dioxide and pairs of hydrogen atoms that attach to oxidized coenzyme molecules. Before pyruvate enters the Krebs cycle, it is transformed into a key molecule called acetyl coenzyme A or **acetyl-CoA,** known before its chemical identification as "active acetate" (Figure 2-9). This important intermediate, discovered in 1949 by Lipmann working in Boston, then combines with oxaloacetate to yield citrate. A series of at least nine additional steps then occur to yield four pairs of hydrogen atoms and two molecules of carbon dioxide (Figure 2-10). The pairs of hydrogen atoms never exist free, but are transferred to specific coenzyme molecules.

The Krebs cycle should be viewed as a mechanism for breaking down acetyl-CoA to two types of products: the completely oxidized carbon dioxide molecules, which cannot be used as an energy source, and the reduced coenzymes, whose further oxidation yields most of the energy used by organisms growing in the presence of oxygen.

Oxidation of Reduced Coenzymes by Respiratory Enzymes[7]

During the Krebs cycle, there is no direct involvement of molecular oxygen. Oxygen is involved only after the hydrogen atoms (or their electrons) have been transferred through an additional series of oxidation-reduction reactions that involve a series of closely linked enzymes, most of which contain iron atoms. These enzymes are often collectively called the respiratory enzymes. Their existence was hinted at late in the nineteenth century, but it was not until the period from 1925 to 1940 that their significance was appreciated, largely as a result of the work of Warburg and the Polish-born David Keilin, who spent most of his scientific life in Cambridge, England. The exact number of enzymes involved differs from organism to organism. Nevertheless, the general picture is similar in all cases. Figure 2-11 shows the basic features of the eucaryotic respiratory chain.

The chain operates by a series of coupled oxidation-reduction reactions, each of which releases energy. Cytochromes (iron-containing proteins) function as electron carriers in the respiratory chain. Thus, the energy present in the reduced coenzymes is released not all at once but in a series of small packets. If NADH were instead directly oxidized by molecular oxygen, a great amount of energy would be released, which would be impossible to couple efficiently with the formation of the high-energy bonds of ATP.

Oxidative Phosphorylation Generates Much More ATP Than Fermentation[8, 9, 10]

From 1925 to 1940, most biochemists concentrated on following the path of hydrogen atoms (or electrons) through the linked, energy-yielding oxidation-reduction reactions. Until the end of this period, only slight attention was given to how the energy is released in a useful form. Then H. M. Kalckar of Denmark and V. A. Belitzer of the Soviet Union observed ATP formation coupled with oxidation-reduction reactions in cell-free systems (1938–1940).

Further understanding did not come quickly, since most of the enzymes involved could not be obtained in pure soluble form. These troubles were not resolved until it was realized that the normal sites of **oxidative phosphorylation** (the synthesis of ATP in the presence of oxygen) in plant and animal cells are large, highly organized subcellular organelles, the **mitochondria** (singular, **mitochondrion**). Using

ATP

Glucose   Enzyme

Enzyme-substrate complex

ADP

Glucose-6-(P)   Enzyme

Figure 2-8
Simple diagram of enzyme action. The substrates glucose and ATP bind to the enzyme at its active site to form an enzyme-substrate complex. The substrates react to form products, which are released from the enzyme. In this example, the enzyme is hexokinase.

Figure 2-9
The transformation of pyruvate to acetyl-CoA. The transformation is written in a greatly simplified form; actually several steps are required, in which the coenzymes thiamine pyrophosphate and lipoic acid are involved. Acetyl-CoA is an extremely important intermediate, for it is formed not only from glucose via pyruvate, but also by the degradation of fatty acids.

intact mitochondria, it is easy to observe the oxidative generation of ATP; this was first demonstrated in 1949 by the Americans Albert L. Lehninger and Eugene Kennedy. Now we know that three ATP molecules are generated for each pair of hydrogen atoms that pass through the respiratory chain. It is believed, however, that at least six separate oxidation-reduction steps occur in the chain.

Roughly 20 times more energy is released by the respiratory chain than by the initial breakdown of glucose to pyruvate. This explains why the growth of cells under aerobic conditions is so much more efficient than growth without air: If oxygen is not present, pyruvate cannot accumulate, since its formation (see Figure 2-5) demands a supply of unreduced NAD^+. The amount of NAD^+ within cells, like that of other coenzymes, is, however, very small. In the absence of oxygen, it is rapidly converted to NADH as glucose is oxidized by the Embden-Meyerhof pathway. Thus, for the continued generation of ATP without oxygen (fermentation), a device other than the respiratory chain must be used to oxidize the reduced NADH. Often, pyruvate itself serves as the hydrogen acceptor. A typical reduced end product is lactic acid, which is formed during the anaerobic contraction of muscles as well as during the anaerobic growth of many bacteria (Figure 2-12).

ATP Is Generated During Photosynthesis (Photophosphorylation)[11, 12]

The ultimate source of the various food molecules used by most microorganisms and animals is photosynthesis, during which energy from sunlight is converted into energy present in covalent bonds. How this happens chemically did not become clear until the basic energy relations within animals and bacteria were understood. Largely as a result of the work of the American biochemist Daniel Arnon, it was discovered in 1959 that the primary action of the sun's light quanta is to phosphorylate ADP to ATP.

This photophosphorylation takes place in the **chloroplasts,** which are complicated, chlorophyll-containing organelles found in most cells capable of trapping the energy of the sun. Thus, the controlled release of energy in plants and photosynthetic bacteria depends on the same energy carrier important in nonphotosynthetic bacterial and animal cells—ATP. The first step is the capture of the light quanta by the green pigment molecule chlorophyll to excite one of its electrons to a high-energy state.

Then, this energy is released gradually by a series of coupled oxidation-reduction reactions mediated by specific electron carrier molecules (Figure 2-13). Oxidative phosphorylation and photophosphorylation thus turn out to be surprisingly similar processes, essentially differing only in the way energy is initially introduced.

Photosynthesis, however, is unique in an important way. It is the only significant cellular event that utilizes any energy source other than the covalent bond. All other important cellular reactions are accompanied by a decrease in the energy contained in covalent bonds. We might be tempted to guess that the ability to photosynthesize must have been a primary feature of the first forms of life. It is, however, difficult to imagine that very early forms possessed the complicated chloroplast structures necessary for photosynthesis. Instead, there is good reason to believe that early in the history of the Earth, a

Figure 2-10
The Krebs cycle, often called the citric acid cycle or tricarboxylic acid (TCA) cycle. Colored rectangles highlight the reduced coenzymes that carry pairs of hydrogen atoms stripped off the cycle intermediates.

unique chemical environment allowed the spontaneous creation of a large number of carbon-containing compounds. Energy derived from these organic molecules then served as the food energy supply for the first forms of life. As these original organic molecules were depleted and living matter increased, a strong selective advantage developed for those cells that evolved a photosynthetic structure to provide a means of increasing the amount of organic molecules. Today, essentially all glucose molecules are formed using chemical energy originating in photosynthesis.

The Chemiosmotic Generation
of ATP from ADP and Phosphate[13]

At first it seemed obvious that the energy released by the cytochrome-mediated oxidation-reduction reactions must momentarily be stored in high-energy compounds capable of phosphorylating ADP to ATP. But some 30 years of searching failed to yield convincing evidence for

Figure 2-11
The respiratory chain. The heavy black arrows represent the flow of electrons through the three major respiratory enzyme complexes. These oxidation-reduction reactions release, in small packets, the energy present in NADH molecules. Two electrons are transferred from NADH to oxygen. Ubiquinone and cytochrome *c* serve as carriers between the complexes.

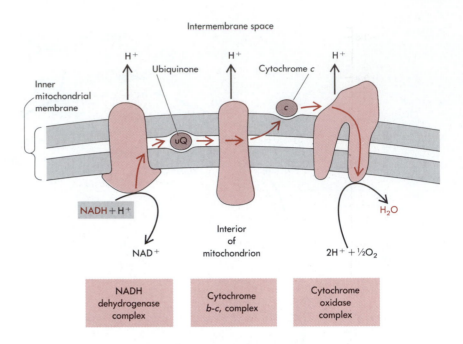

even one such intermediate. Equally troublesome was the observation that only intact mitochondria or chloroplasts make ATP. Unbroken surrounding membranes are required for the coupling of the oxidation-reduction reactions and ATP generation. Resolution of these unexpected complications came in the early 1960s from the work of the English biochemist Peter Mitchell. He proposed that the location of the cytochromes and other electron carriers within membranes allows the generation of gradients in which more hydrogen ions are present outside the mitochondrion or chloroplast than within. Such gradients form because the respiratory enzyme complexes are so placed that three hydrogen ion pairs are displaced outwardly as each pair of electrons goes down the respiratory chain from NAD to oxygen. These hydrogen ion gradients provide the energy for the union of ADP and phosphate to form ATP, a reaction catalyzed by the membrane protein ATP synthetase (Figure 2-14). This explanation, known as the **chemiosmotic theory,** was originally greeted with much skepticism by a biochemical world molded by a succession of new high-energy intermediates. Now, however, Mitchell's idea is recognized as one of the major conceptual advances in the development of modern biochemistry.

Figure 2-12
The fermentation of glucose to yield lactate. In the absence of oxygen, the NADH produced during pyruvate formation is oxidized to reduce pyruvate to lactate. When oxygen is present, the NADH is oxidized through the respiratory chain, and no lactate is produced.

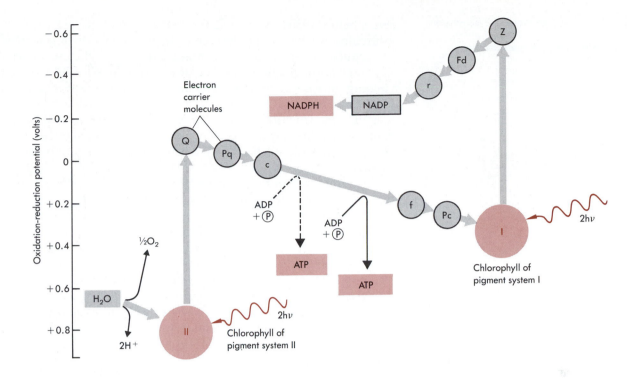

Vitamins and Other Dietary Requirements

Although some microorganisms, such as the bacteria *Escherichia coli*, can use glucose as their sole carbon and energy source, not all bacteria and none of the higher animals can use glucose to synthesize all the metabolites necessary for life. For example, rats are unable to synthesize 11 of the 20 amino acids present in their proteins; thus, their food supply must contain substantial amounts of these molecules.

In addition to dietary requirements for compounds with important structural roles, there is often a need for very small amounts of specific organic molecules. These molecules, needed in trace amounts only, are called **vitamins.** For many years, these molecules seemed quite mysterious. Now we realize that the vitamins are closely related to the coenzymes. Some are precursors of coenzymes and some are coenzymes themselves. For example, the vitamin niacin is used in the synthesis of NAD^+. The fact that coenzymes, like enzymes, are able to function over and over explains why vitamins are needed only in trace amounts.

Thus, there is nothing unusual about the fact that a particular molecule is sometimes required for growth; such requirements are fully explicable in chemical terms. It seems likely that the genes necessary for the synthesis of certain molecules were lost during evolution, resulting in the need for specific growth factors. There would be no selective advantage to an organism's retaining a specific gene if the corresponding metabolite were always available in its food supply.

The Polymeric Nature of Cellular Macromolecules

In striking contrast with the success of biochemists in understanding the behavior of small molecules like the amino acids and nucleotides, scientists interested in large molecules had arrived at only partial an-

Figure 2-13
A simplified scheme to show the passage of a pair of electrons (gray arrows) from water to NADP during photosynthesis, coupled with the formation of one or possibly two molecules of ATP. (NADP is a coenzyme similar to NAD; it has an additional phosphate group.) The NADPH and ATP formed are then used to incorporate carbon dioxide into carbohydrates. An electron, excited by the absorption of a photon by the pigment system II, is passed via a series of carrier molecules, by oxidation-reduction reactions, to the chlorophyll of pigment system I. This replaces an electron, which is emitted following the absorption of another photon, and which is passed by further carriers to reduce NADP. ATP is generated somewhere between the two carriers *c* and *f*. The electron emitted from system II is replaced by one taken from water, coupled with the release of oxygen. Q and Z are the primary electron acceptors of systems II and I, respectively. (Pq = plastoquinone, f = cytochrome *f*, Pc = plastocyanin, Fd = ferrodoxin, r = reductase, c = cytochrome b_{563}.)

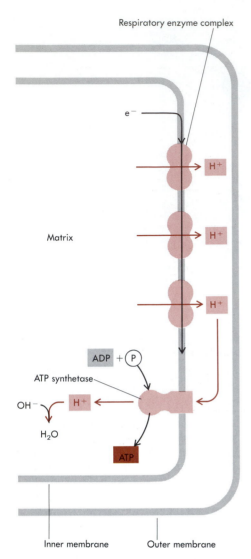

Figure 2-14
Simplified scheme showing the chemi-osmotic theory of the generation of ATP from ADP and phosphate. Electrons pass down a chain of membrane-bound, cytochrome-containing enzyme complexes and generate a hydrogen ion gradient across the membrane. This gradient then drives the synthesis of ATP, coupled with the formation of water (see Figure 2-11).

Figure 2-15
(a) General formula for an amino acid, where R is a side group that is different for each amino acid. Under physiological conditions, amino acids are usually in the ionized form (zwitterion). (b) Portion of a protein molecule. Three amino acids are shown linked together by peptide bonds (color). The chain of Cs and Ns forms the "backbone" of the protein.

swers before 1950 (see Table 2-3). Their only real success with macromolecules involved the polysaccharide glycogen, which is formed by the regular polymerization of large numbers of glucose molecules. Glycogen is enzymatically broken down when needed into its energy-yielding glucose monomers. The understanding of glycogen's structure, however, had little biological impact; as a highly regular polymer, it was structurally monotonous. The molecules that biologists really wanted to unravel were the proteins (because many of them are enzymes) and the nucleic acids (because they are chromosomally located).

Both proteins and nucleic acids are inherently much more complex than polysaccharides. Early in the twentieth century, the great German chemist Emil Fischer had established that proteins are polymers composed of large numbers of nitrogen-containing organic molecules called **amino acids**, linearly linked together by **peptide bonds** (Figure 2-15). (Hence chains of amino acids are called **polypeptides**.) Twenty different amino acids are found in proteins, each with a specific side group (Figure 2-16). The two nucleic acids, DNA and RNA, are also polymeric molecules, each consisting of a chain made up of four different nucleotides. Each nucleotide building block of DNA or RNA has a common sugar-phosphate group attached to a specific side group that is either a purine or a pyrimidine (see Table 2-3).

A feature common to all these biological polymers is that the individual subunits in the polymer chain contain two hydrogen atoms and one oxygen atom less than the simple monomers from which they are synthesized. Synthesis thus involves the release of water. When the polymers are degraded to yield smaller molecules, one water molecule is incorporated for each bond broken (Figure 2-17). Degradative reactions in which water uptake is required are known as hydrolytic reactions; the process is called **hydrolysis.** There are many types of hydrolytic reactions, since many small molecules can be broken down to still smaller products by the addition of water. Under normal physiological cell conditions, spontaneous hydrolysis of any of the important polymers or small molecules is *very rare*. Hydrolysis, however, is vastly speeded up by the presence of specific enzymes. For example, the enzymes pepsin and trypsin specifically catalyze the hydrolytic breakdown of proteins.

Distinction Between Regular and Irregular Polymers

Table 2-3 reveals an important difference between the polysaccharides, like glycogen, and the proteins and nucleic acids. Polysaccharides are invariably constructed by regular (or semiregular) joining of

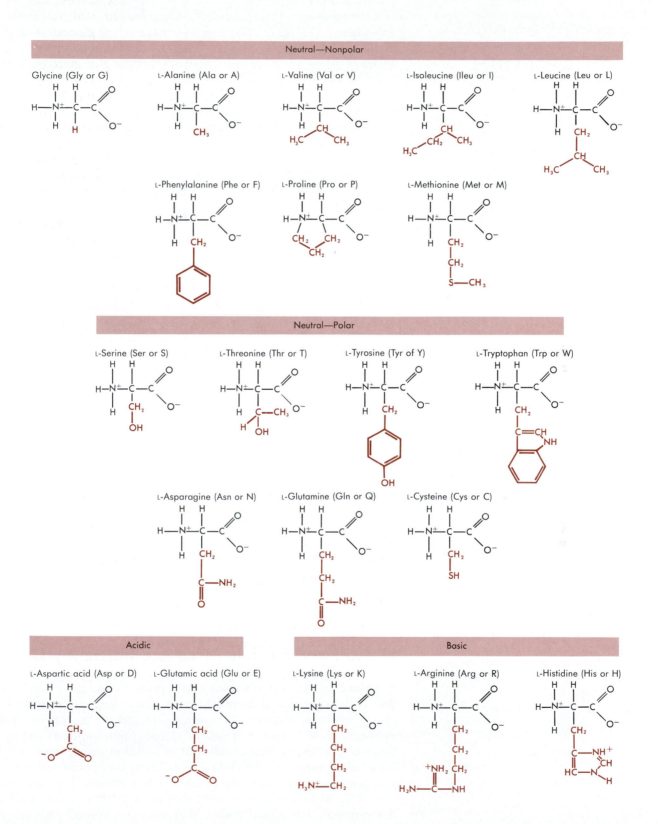

Figure 2-16

The 20 common amino acids found in proteins. The amino acids are grouped by the chemical characteristics of their side groups.

Figure 2-17

Hydrolysis of a tripeptide to form three amino acids. Breaking the two peptide bonds (shown in color) requires the addition of two water molecules.

Alanine   Valine   Aspartic acid

one or two different kinds of monosaccharide building blocks. In contrast, proteins may contain up to 20 different amino acids, and nucleic acids contain 4 different nucleotides. Moreover, in the nucleic acids and proteins, the order of subunits is highly irregular and varies greatly from one specific molecule to another. While polysaccharide synthesis from monosaccharides need only involve the making of one or two types of backbone bonds, the synthesis of nucleic acids and proteins demands, in addition, a highly efficient mechanism for choosing and ordering the correct subunits.

Proteins and Nucleic Acids Are Inherently Labile[14]

A major complication to working with proteins and nucleic acids was that they appeared to be much more labile (less stable) than most small molecules. Extremes of temperature and pH (acidity or alkalinity) cause them to lose their natural shapes (**denaturation**) and sometimes to precipitate irreversibly out of solution in an inactive form. Thus, great care had to be taken in isolating them, and sometimes it was necessary to perform the entire isolation process at temperatures near 0°C. At first it was thought that only proteins were subject to denaturation, but now it is clear that nucleic acid molecules can also denature during isolation if proper precautions are not taken.

Until 1946 (just after World War II ended), the techniques of organic chemistry were the main tools for studying most small molecules. Thus, the success of a biochemist working on intermediary metabo-

Table 2-3 Structural Organization of Several Important Biological Polymeric Macromolecules

Macromolecule	Monomeric Units	Number of Different Monomers	General Monomer Formula	Fixed or Irregular Chain Length	Linkage Between Monomers
Glycogen (a polysaccharide)	Glucose	1		Indefinite: may be > 1000	1–4-Glycosidic linkage
DNA (deoxyribonucleic acid)	Deoxyribonucleotides	4: deoxyadenylate deoxyguanylate deoxythymidylate deoxycytidylate	Purine-dexoyribose-$\circled{P}$ or pyrimidine-deoxyribose-$\circled{P}$	Genetically fixed: may be > 10^7	3′–5′-Phosphodiester linkage
RNA (ribonucleic acid)	Ribonucleotides	4: adenylate guanylate uridylate cytidylate	Purine-ribose-$\circled{P}$ or pyrimidine-ribose-$\circled{P}$	Genetically fixed: often > 3000	3′–5′-Phosphodiester linkage
Protein	L-Amino acids	20: glycine, alanine, serine, etc.		Genetically fixed: usually varies between 100 and 1000	Peptide linkage

lism often depended on his or her ability as an organic chemist. In work with proteins and nucleic acids, however, most of the initial stages of research did not rely on the analytical techniques of the organic chemist. Instead, the protein chemist, before even starting to worry about the detailed structure of a protein, needed to work very hard to be sure that the protein was both chemically pure and biologically active. Gentle isolation techniques had to be devised that avoided the usual strong acids and alkalies of organic analysis. Then additional techniques were needed to reveal whether the product was homogeneous and to provide data on molecular size. For these sorts of questions, the help of physical chemists was indispensable, and there developed a new line of research investigating the physical-chemical properties of macromolecules in solution. It concerned itself with topics like the osmotic properties of macromolecular solutions and the movement of macromolecules under electrical and centrifugal forces.

Perhaps the most striking contribution of physical chemistry to the study of biological macromolecules was the development, in the 1920s, of centrifuges that could rotate at very high speeds (**ultracentrifuges**) and cause the rapid sedimentation of proteins and nucleic acids. The initial development of the ultracentrifuge was the work of The Svedberg, of Sweden, after whom the unit of sedimentation (S for Svedberg) was named. Ultracentrifuges equipped with optical devices to observe exactly how fast the molecules sedimented

Nonpolar solvent in trough

A + B
B
A

Solvent movement

Water   Filter paper strip

Figure 2-18
Partition chromatography. Molecules in a mixture can be separated on the basis of their different solubilities in polar versus nonpolar solvents. Here we show one of many methods developed for paper chromatography of amino acids or peptides. A drop of the mixture is placed on a strip of filter paper, which is then hung in a closed vessel containing some water (the polar solvent). The water vapor moistens the paper. One end of the paper dips into a trough filled with the nonpolar solvent. As the nonpolar solvent moves along the moist surface of the paper, it carries solutes with it. Each substance in the original mixture moves a distance proportional to its relative solubilities in the two solvents. The finished chromatogram is dried and then stained with ninhydrin, which forms colored reaction products with the colorless amino acids.

were extremely valuable in obtaining data on the molecular weights of proteins and in establishing the concept that proteins, like smaller biological molecules, are of discrete molecular weights and shapes. This work revealed that the sizes of proteins vary greatly, with a continuous range in weights between the extremes of approximately 10,000 and 1,000,000.

Implications of Chromatography[15]

The means to determine the exact order in which amino acids are linked together to form a particular protein remained a great puzzle until 1951. This was partly because there are 20 different amino acids (see Figure 2-16) and their proportions vary from one type of protein to another. Until about 1942, the methodological problems involved in amino acid separation and identification were formidable, and most organic chemists chose to work with simpler molecules.

This state of affairs changed completely in 1942, when the Englishmen A. J. P. Martin and R. L. M. Synge developed separation methods that depended on the relative solubilities of the several amino acids in two different solvents (**partition chromatography**). Particularly useful were methods by which amino acids were separated on strips of paper (Figure 2-18). With these new techniques, it became a routine matter to separate quantitatively the 20 amino acids found in proteins.

Amino Acids Within Polypeptide Chains Are Linearly Linked Together in Unique Sequences[16, 17]

By breaking down polypeptide chains randomly into many smaller fragments and using the newly developed chromatographic methods to reveal their amino acid compositions, Frederick Sanger, working at Cambridge University in the early 1950s, established the first definitive order of amino acids within a protein. The molecule he examined, the hormone insulin, contains two separate polypeptide chains, A and B, held together by pairs of disulfide bonds (S–S) between sulfur atoms on the side groups of opposing cysteine residues. Sanger's work was a milestone in the study of proteins, for it was the first demonstration of the now accepted principle that a given type of protein contains a unique sequence of amino acids.

Synthesis of the S–S bonds holding the A and B insulin chains together occurs after the respective cysteine residues have been incorporated into their polypeptide chains. This was first suggested by experiments in which the separate chains were mixed together and then observed often to form the correct S–S bonds between appropriate partner cysteines. Many years later, D. F. Steiner at the University of Chicago showed that the A and B insulin chains are originally part of a single nonfunctional polypeptide chain (proinsulin) (Figure 2-19). As the chain is synthesized, it spontaneously adopts a three-dimensional configuration that brings close together the cysteine residues destined to form S–S bonds to each other. After these bonds form, the intact polypeptide chain is cleaved twice by a highly specific polypeptide-clipping enzyme, thereby releasing an inactive polypeptide from the now two-chained, functional insulin molecule.

The 25-Year Loneliness of the Protein Crystallographers[18, 19, 20]

An equally significant step in understanding macromolecules was the effective extension of X-ray crystallographic techniques to their study. **X-ray crystallography** utilizes the diffraction of X-rays by crystals to give precise data about the three-dimensional arrangement of the atoms in the crystals. The first successful use of X-ray diffraction was in 1912, when the Englishman Lawrence Bragg solved the structure of sodium chloride (NaCl). This success immediately initiated research on the structures of molecules of increasing complexity.

The technique used in the first X-ray diffraction studies of small molecules consisted of guessing the structure, calculating the theoretical diffraction pattern predicted by this structure, and then comparing the calculated pattern with the observed pattern. This method was practical for studying relatively simple structures, but was not often useful in the study of larger structures. It took much insight on the part of Bragg and the great American chemist Linus Pauling, in the 1920s, to solve the structures of some complicated inorganic silicate molecules. Clearly, however, proteins were too complicated for even the best chemist to guess their three-dimensional structures. Thus, the early protein crystallographers knew that until new methods for structural determination were found, they would have no results to present to the impatient biochemists, who were increasingly eager to know what proteins actually looked like.

The first serious X-ray diffraction studies on proteins began in the mid-1930s in J. D. Bernal's laboratory in Cambridge, England. Here it was found that although dry protein crystals gave very poor X-ray patterns, wet crystals often gave beautiful pictures. Unfortunately, however, there was no logical method available for their interpretation. Nonetheless, Bernal's student Max Perutz, an Austrian then in England, slowly increased the pace of his work (begun in 1937) with the oxygen-carrying blood protein **hemoglobin.** He had chosen hemoglobin for several reasons: Not only is it one of the most important of all animal proteins, but it is also easy to obtain, and it forms crystals that lend themselves well to crystallographic analysis. For many years, however, no very significant results emerged either from

Figure 2-19
Structure of bovine proinsulin. The amino acids removed during its conversion to insulin are shown in color.

Figure 2-20
The three-dimensional structure of
myoglobin as determined by X-ray crystallography. Each numbered circle represents the central carbon of one amino
acid; other atoms are not shown. Note
the eight segments of α helices, labeled
A through H. Also, NA is the amino
terminal, and HC is the carboxyl terminal. (Copyright Irving Geis.)

Perutz's work on hemoglobin structure or from the work begun in the
same laboratory in 1947 by John Kendrew on the structure of the
muscle protein myoglobin. This protein, which, like hemoglobin,
combines with oxygen, had the added advantage of being only a
fourth the size of hemoglobin (MW = 17,000).

During that lonely period of no real results, there was only one
triumph. Pauling correctly guessed from stereochemical considerations that amino acids linked together by peptide bonds would sometimes tend to assume helical configurations, and he proposed in 1951
that a helical configuration, which he called the alpha (α) helix, would
be an important element in protein structure. (The α helix will
be discussed later in this chapter and in Chapter 5.) Support for
Pauling's α-helix theory came soon after its announcement when
Perutz demonstrated that several synthetic polypeptide chains containing only one type of amino acid exist as α helices.

It was not until 1959 that Perutz and Kendrew got their answers.
An essential breakthrough, which occurred in 1953, showed how the
attachment of heavy atoms to protein molecules could logically lead
from the diffraction data to the correct structures. For the next several
years, these heavy-atom methods were exploited at a pace un-
dreamed of 20 years before, largely as a result of the availability of
high-speed electronic computers. Then, to everyone's delight, the
X-ray diffraction measurements could at last be translated into the
enormously complicated arrangement of atoms in myoglobin (Figure
2-20) and hemoglobin (see Chapter 3). Most importantly, their molecular configurations were found to obey in every respect the chemical
laws that govern the shape of smaller molecules. Absolutely no new
laws of nature are involved in the construction of proteins. This was
no surprise to the biochemists.

The Primary Structure of Proteins[21]

With the establishment of those first three-dimensional structures, it became possible to analyze the arrangements of atoms within proteins at three levels of order. The first level, the **primary structure,** is the description of the covalent bonds within a protein. It tells us the number of polypeptide chains and the sequence of amino acids within them. For example, the primary structure of myoglobin is a single polypeptide chain of 153 amino acids. Many other proteins also contain only a single type of chain. One such protein is chymotrypsinogen, with 246 amino acids. After the conversion of this protein to its active form, chymotrypsin, it catalyzes the hydrolysis of several highly specific peptide bonds. The determination of the sequence of chymotrypsinogen in the early 1960s required some 15 years of collective work by several talented chemists (Figure 2-21). Today, new experimental techniques allow such sequences to be established much more quickly. With moderate luck, less than a year may be sufficient for an expert protein chemist to work out a sequence 200 to 300 amino acids long.

Many other proteins are more complicated in that they possess several types of polypeptide chains. For example, hemoglobin contains both α and β chains, each present in two copies ($\alpha_2\beta_2$). The primary structure also specifies which cysteine residues are held together by disulfide bonds (S–S). Disulfide bonds are important in helping a protein maintain a particular shape by linking together either two parts of a single polypeptide chain or two different chains. As was shown in Figure 2-19, they are what hold together the two chains of the insulin molecule. In chymotrypsinogen, there are five disulfide bridges, each linking specific cysteine residues in the single chain. In chymotrypsin, which is derived from chymotrypsinogen by polypeptide chain cleavage, two of these disulfide bonds connect sep-

Figure 2-21
The amino acid sequence of the protein chymotrypsinogen. Key to the establishment of such protein sequences is the isolation of smaller, well-defined polypeptide fragments. The enzyme trypsin, for example, cleaves on the carboxyl side of arginine and lysine residues (light color). The so-called tryptic fragments have proved invaluable for working out almost all protein sequences. Larger fragments are generally made by treatment with the reagent cyanogen bromide, which cleaves the peptide bond on the carboxyl side of the much less common amino acid methionine (dark color).

arate chains. In other cases (e.g., hemoglobin), the several chains are held together by weaker, noncovalent bonds (see next section).

In addition, a number of proteins have attached to them nonprotein components called **prosthetic groups,** which play a vital role in their function. In contrast to coenzymes, prosthetic groups are bound very tightly to their protein, sometimes by covalent bonds. They are often metal-organic compounds. Both myoglobin and hemoglobin contain the prosthetic group *heme*, a metal-organic compound closely related to the porphyrin component of chlorophyll. Heme combines with oxygen and enables hemoglobin and myoglobin to bind oxygen.

A characteristic feature of prosthetic groups is that they possess very little functional activity unless they are attached to a polypeptide partner. Heme by itself, for example, combines with oxygen in an effectively irreversible fashion. Only when heme is attached to either myoglobin or hemoglobin does it possess the quality of reversibly binding oxygen. Then it can release bound oxygen when it is needed under conditions of oxygen scarcity.

Secondary Structures of Proteins May Be Sheets or Helices

Individual extended polypeptide chains are conformationally unstable and frequently either assume contracted helical configurations or aggregate side by side to form sheetlike structures. The driving force that leads to either of these two alternatives is the strong tendency of the carbonyl (CO) and imino (NH) groups that flank the peptide bonds to form hydrogen bonds, the most specific of the "weak" (noncovalent) secondary bonds (see Chapter 5 for an extensive discussion of their actions). The carbonyl group next to one peptide bond can hydrogen-bond either to an imino group flanking a peptide bond several residues removed from it on the same chain, or to an imino group flanking a peptide bond on a different polypeptide chain. The resulting hydrogen-bonding patterns are highly repetitive and lead to regular polypeptide conformations that are known as the **secondary structures** of proteins.

The stereochemically most satisfactory hydrogen-bonded helical conformation of the polypeptide backbone is the **α helix,** in which 3.6 amino acids are found for every turn of the helix (Figure 2-22). Significant sections of the polypeptide chains of myoglobin and hemoglobin were found to be so arranged, and now the α helix is recognized as a major secondary structural feature of many, if not most, proteins. Equally important are regular, hydrogen-bonded, lateral associations of extended polypeptide chains into sheets, called **β sheets** (Figure 2-23). β sheets are especially favored by the presence of large numbers of glycine residues. Their existence was first suggested by X-ray crystallographic analysis of the fibrous protein of silk, carried out in the 1930s by W. T. Astbury and his collaborators in Leeds, England. More recently, β sheets have also been found within a very large number of globular (nonfibrous) proteins.

Tertiary Structures of Proteins Are Exceedingly Irregular

The **tertiary structure** of a protein is its three-dimensional form. It is, in many cases, very irregular. Few proteins are built up from α helices or β sheets alone, or even from a simple combination of the two.

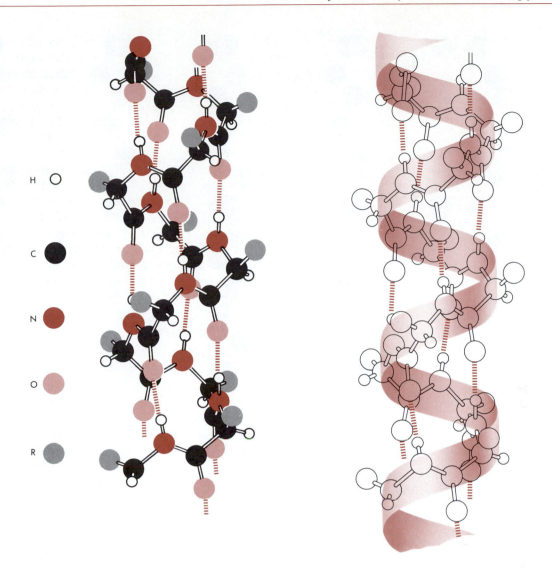

H ○

C ●

N ●

O ●

R ●

Instead, most proteins contain both regular and irregular regions. Some proteins, in fact, seem to have almost no regular structure. There are a number of stereochemical reasons why α helices, β sheets, and other regular secondary arrangements are not found more extensively. One reason is that the amino acid proline does not contain an amino group, and so wherever it occurs, the regular hydrogen bonding must be interrupted.

The most important reason for irregularity in protein structures, however, arises from the diverse chemical nature of the amino acid side groups (see Figure 2-16). Each of these side groups will tend to make the energetically most favorable secondary interactions with other chemical groups. As an example, the free hydroxyl group on tyrosine will tend to assume a position where it can form a hydrogen bond. The considerable energy of the bond would be lost if, for example, the hydroxyl group were next to an isoleucine side group, which is incapable of forming a hydrogen bond. Furthermore, the side groups of several amino acids, like valine and leucine, are insoluble in water, whereas others, like those of glutamic acid and lysine, are highly soluble. It thus makes chemical sense that the water-insoluble side groups are found stacked next to one another in the interior of a protein, whereas the external surface contains groups that mix easily with water.

Figure 2-22

A polypeptide chain folded into a helical configuration called the α helix. All the backbone atoms have identical orientations within the molecule. It may be looked at as a spiral staircase in which the steps are formed by amino acids. There is an amino acid every 1.5 Å along the helical axis. The distance along the axis required for one turn is 5.3 Å, giving 3.6 amino acids per turn. The helix is held together by hydrogen bonds between the carbonyl (CO) group of one residue and the imino (NH) group of the fourth residue down along the chain. [Molecular structure after L. Pauling, *The Nature of the Chemical Bond*, 3rd ed. (Ithaca, N.Y.: Cornell University Press, 1960), p. 500, with permission.]

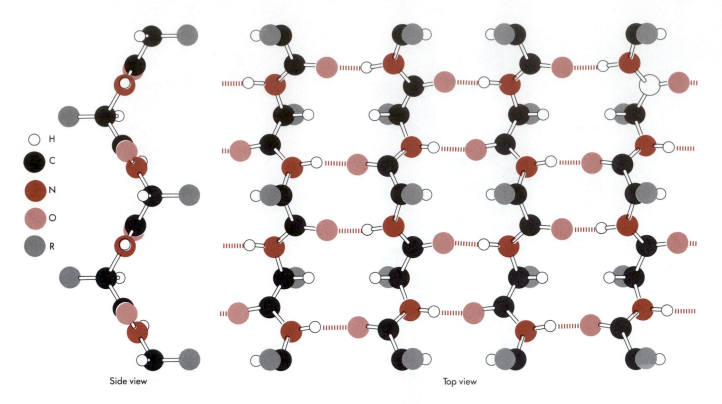

H   C   N   O   R

Side view   Top view

Figure 2-23
Several extended polypeptide chains held together by hydrogen bonds in a configuration called a β sheet. [Molecular structure after L. Pauling, *The Nature of the Chemical Bond*, 3rd ed. (Ithaca, N.Y.: Cornell University Press, 1960), p. 501, with permission.]

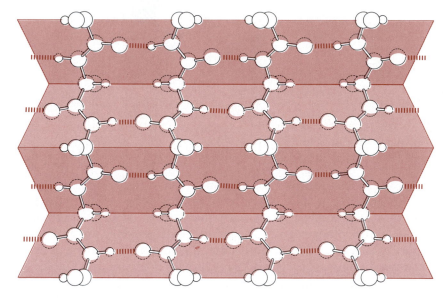

The tertiary configuration thus represents the energetically most favorable arrangement of the polypeptide chain. Each specific sequence of amino acids takes up the particular "native" arrangement that makes possible a maximum number of favorable atomic contacts between it and its normal environment. This concept is strongly supported by very striking experiments in which high temperature or some other unnatural condition denatures the protein to give randomly oriented, biologically inactive polypeptide chains. When the denatured chains are gently returned to their normal environment, some of them are found to **renature**—to resume their native conformation with full biological activity.

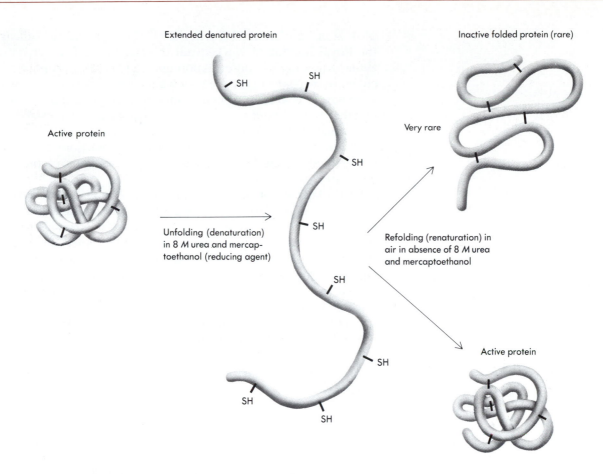

S–S Bonds Form Spontaneously Between Correct Amino Acid Residues[22]

In many cases, renaturation of a disordered protein to an active form involves not only the formation of thermodynamically favorable weak bonds, but also the making of specific disulfide (S–S) bridges. This was first shown by experiments with the enzyme ribonuclease, a protein consisting of one polypeptide chain of 124 amino acids, cross-linked by four specific S–S bonds. The native, active configuration of the enzyme can be destroyed by reducing the S–S groups to sulfhydryl (SH) groups in the presence of the denaturing agent 8 *M* urea and the reducing agent mercaptoethanol (Figure 2-24). When the urea is removed, oxygen in the air reoxidizes the SH groups to yield S–S bonds identical to those found in the original molecule. A given SH group reassociates, not randomly with any of the other seven SH groups in the molecule, but rather with a specific SH group brought into close contact with it by the folding of the polypeptide chain. Thus, S–S bridges are not a primary reason for the peculiar folding of the chain. They might be better viewed as a device for increasing the stability of an already stable configuration.

Visualization of the Active Sites of Enzymes[23, 24]

Although for decades organic chemists had speculated about how specific amino acid side groups might speed up certain chemical reactions, convincing explanations had to await detailed three-dimensional structural analysis. Such an analysis was first accom-

Figure 2-24
The fate of S–S bonds during protein denaturation and renaturation. When the denaturing agents are removed, most of the polypeptide chains resume the native configuration with the original S–S bonds. Only a few polypeptide chains fold up in an inactive form characterized by a different set of S–S bonds than those found in the native molecules.

plished in London in 1967 when D. C. Phillips and his colleagues at the Royal Institution worked out the structure of the enzyme lysozyme. Knowing its conformation enabled them to propose a precise chemical mechanism for the way lysozyme breaks down specific polysaccharide chains into their component sugar groups. This work, together with that of other scientists who have since established the conformations of over 35 additional enzymes, indicates that the chemical mechanisms of enzyme-catalyzed reactions can be as well understood as the best-known chemical reactions of pure organic chemistry.

Particularly pleasing has been the clear establishment of the way a number of proteolytic enzymes catalyze specific breaks in polypeptide chains. The working out of the three-dimensional structures of trypsin and chymotrypsin reveals very similar molecular shapes, with their active sites sharing many common amino acids. Three of these amino acids—serine, histidine, and aspartic acid—are directly involved in the catalytic process, with each residue playing identical roles in both enzymes (Figure 2-25). We now see that amino acid sequence data by themselves seldom can reveal profound information on how proteins function.

Lipid Bilayers Give Cellular Membranes Their Sheetlike Properties[25]

The cytoplasm of a cell is surrounded by a lipid-containing semipermeable membrane that prevents escape of cellular constituents to the outside environment as well as facilitates the selective entry into the cell of food molecules and inorganic ions needed for growth and cellular maintenance. Because these membranes are very thin and highly insoluble in aqueous environments, their underlying structural organization was long a mystery. Key to the working out of the so-called unit membrane structure was the early study of the red blood cells (erythrocytes) of mammals. These cells lack a chromosome-containing nucleus as well as any other internal membranous organelles (such as mitochondria); the only membranes they contain are their external plasma membranes. Most of their lipids, as well as those of all other membranes, are **phospholipids**; they contain phosphate "head" groups attached to long hydrocarbon tails. The four major phospholipids found in all higher cells are phosphatidylserine,

Figure 2-25 (Opposite)

Involvement of specific amino acid residues in the catalytic breakdown of peptide bonds by trypsin. The catalytic mechanism of trypsin accelerates the hydrolysis of peptide bonds by providing intermediate states for the reaction and by smoothing the transition from one intermediate to the next. Enzyme and substrate must first come together (1) to form a precisely oriented complex (2). The oxygen atom of serine 195 then bonds covalently to the substrate carbon, forming a tetrahedral intermediate (3); the proton, or hydrogen ion, from serine 195 is transferred to the substrate nitrogen. The proton transfer breaks the peptide bond, and the first product is liberated (4). The remaining complex is called an acyl enzyme; it breaks down to regenerate free enzyme in steps that are symmetrical with those of the first half of the process. A water molecule enters the reaction (5), its hydroxyl group forming with the substrate another tetrahedral intermediate (6). The hydrogen ion from the water molecule is transferred to serine 195, breaking the covalent bond between enzyme and substrate (7). The second product is now freed (8); its departure is hastened by repulsion between negatively charged carboxyl groups of product and aspartic acid 102. Histidine 57 and aspartic acid 102 participate in the proton transfers. [Redrawn from "A Family of Protein-Cutting Proteins," by R. Stroud. *Sci. Amer.* 231 (1974):76. Copyright © 1974 by Scientific American, Inc. All rights reserved.]

1

ENZYME

Asp 102   His 57   Ser 195

SUBSTRATE

2

ENZYME-SUBSTRATE
COMPLEX

3

CLEAVAGE

TETRAHEDRAL
INTERMEDIATE

4

PRODUCT 1

ACYL ENZYME

5

WATER
MOLECULE

ACYL ENZYME

6

TETRAHEDRAL
INTERMEDIATE

7

ENZYME-PRODUCT
COMPLEX

8

PRODUCT 2

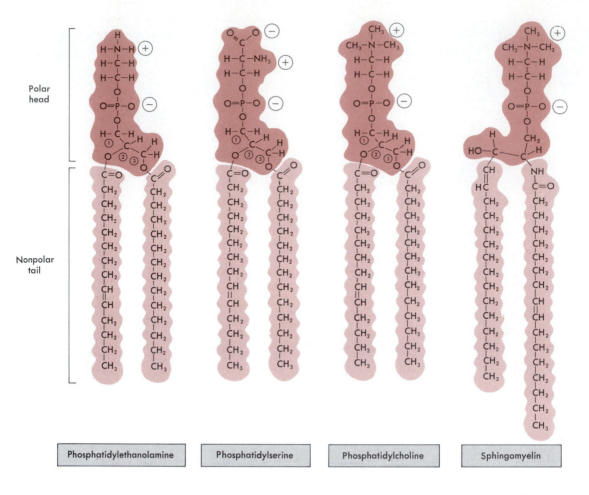

| Phosphatidylethanolamine | Phosphatidylserine | Phosphatidylcholine | Sphingomyelin |

Figure 2-26
The structure of the four major phospholipids found in eucaryotic cell membranes. The exact lengths of the fatty acid nonpolar tails are variable. ①, ②, and ③ indicate the three carbon atoms of glycerol, on which three of the phospholipids are based.

phosphatidylethanolamine, phosphatidylcholine, and sphingomyelin (Figure 2-26).

In the late 1930s, J. F. Danielli and H. Davson first postulated that lipid building blocks are organized into a 50 Å thick bilayer in which the charged phosphate groups are all on the outside. Such a bilayer arrangement neatly places all the water-insoluble hydrocarbon tails in contact with one another, leaving the highly charged phosphate head regions free to interact with the polar water environment. Probably from the very start of life on Earth, membranes possessed the same bilayer arrangement as that present today.

Most cell membranes also contain several types of **glycolipids,** in which one to several water-soluble sugar groups are attached to a hydrophobic tail made up of a fatty acid and the amino alcohol sphingosine (Figure 2-27). **Steroids** are also found within most bilayers, with cholesterol present in large amounts in virtually all vertebrate membranes. Like phospholipids, glycolipids and steroids are arranged so that their hydrophobic groups are in the center of the bilayer with their charged groups facing outward (Figure 2-28).

Insertion of Membrane Proteins into Lipid Bilayers[26]

Most membranes contain a variety of different protein molecules directly inserted into the lipid bilayer. Some pass right through the bilayer, having groups that protrude from both its sides. Those amino

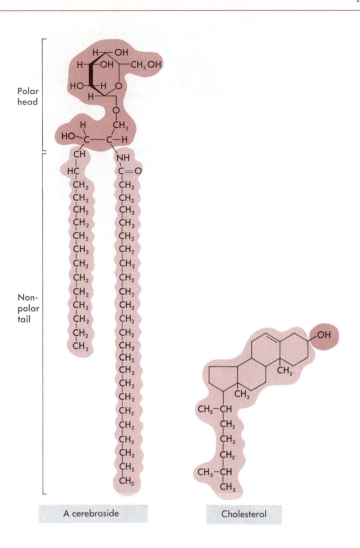

Figure 2-27
The structure of a glycolipid
(a cerebroside) and a steroid
(cholesterol).

A cerebroside

Cholesterol

acid side chains that lie within the bilayer are largely hydrophobic. In contrast, hydrophilic amino acids predominate in the regions that face out onto the surfaces (Figure 2-29). Depending on their function, membranes have very specific protein components. Some of the most highly specialized membranes contain only one major protein component (e.g., the protein of the retinal membrane is almost exclusively composed of the visual pigment rhodopsin).

Although most electron micrographs suggest that membranes are symmetrical, the two sides of a given membrane have fundamentally different chemical compositions. Phosphatidylserine and phosphatidylethanolamine face toward the cytoplasmic (inner) side, while phosphatidylcholine and sphingomyelin are present largely on the exterior surface. Those proteins that are inserted into membranes have specific regions that normally face exclusively inward or outward. The regions that face outward may have sugar groups covalently attached to them, making these proteins **glycoproteins.** The location of sugars only on the outer-facing side of membrane proteins suggests that the glycosylating enzymes that transfer sugars to membrane proteins are found exclusively on the outer membrane surface.

Until very recently, it has not been possible to crystallize membrane proteins, and conventional X-ray crystallographic techniques could not be used to work out their three-dimensional structures. Crystalline two-dimensional sheets of the purple membrane protein of the

Figure 2-28
How various lipids are arranged in the cell membrane. Hydrocarbon chains are in the center of the bilayer, and the polar heads face outward.

Figure 2-29
How glycophorin, a large glycoprotein, may be inserted in the red blood cell membrane.

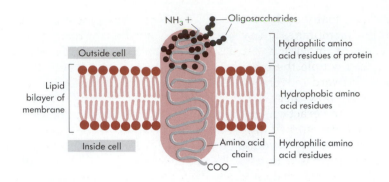

photosynthetic bacterium *Halobacterium halobium*, however, have been analyzed using electron micrograph image reconstruction techniques. In 1975, R. Henderson and P. N. T. Unwin in Cambridge, England, showed that the purple membrane protein contains seven closely packed α helices lying nearly perpendicular to the plane of the membrane (Figure 2-30).

Semifluidity of Cell Membranes[27, 28]

There is now very good evidence that the individual phospholipid and protein molecules do not maintain fixed orientations but are in constant movement within the two-dimensional fabric of a given membrane. Molecules that at one moment may be in touch, can move to opposite sides of a 10 μm diameter cell within an hour or so. Membranes thus can be thought of as two-dimensional liquids, virtually all of whose components are in constant diffusional motion. They have the essential quality of a self-sealing rubber tire, and holes quickly disappear owing to diffusion of nearby hydrocarbon molecules. An equally important property is the potential ability of adjacent membranous sacs to fuse on contact. Membranous sacs also have the capacity to divide when seemingly random motion creates unstable pinched-off regions. Probably most of these fusion and division events are not random occurrences but responses to well-defined stimuli. Conceivably, they are mediated by forces exerted on the lipid bilayers that momentarily create minute tears in the two-dimensional surface. If so, fusion and division events may be the result of rearrangements of such torn surfaces, whereby inner surfaces fuse with other inner surfaces while outer surfaces match up with outer surfaces. Figure 2-31 shows the probable mechanism by which membrane fusion and division events lead to **pinocytosis** (the uptake of macromolecular particles from the solution surrounding a cell) and the opposite process, **exocytosis.**

Figure 2-30
Purple membrane protein of *Halobacterium halobium*. This protein has seven α helices perpendicular to the membrane plane. Here the helices are represented by colored cylinders. [After R. Henderson, *Ann. Rev. Biophys. Bioeng.* 6 (1977):87.]

Membranes Contain Pumps and Gates[29, 30, 31]

The phospholipid component by itself provides a firm barrier to the transport across the membrane of most cellular molecules, the vast majority of which contain one to several polar groups that effectively cannot enter the hydrophobic hydrocarbon environment that characterizes the interior of a lipid bilayer. The functioning of membranes as agents that selectively transport food molecules, inorganic ions, and intracellular metabolites from one side of a membrane to the other depends on many of the hundreds of different proteins that characterize the membranes of most cells. Many of these proteins act as active pumps, moving molecules against concentration gradients. To do so, they couple the breakdown of intracellular ATP with the inward or outward movement of desired food molecules or inorganic ions. Exactly how these directed movements occur cannot be determined until the exact conformations of the respective membrane proteins have been worked out. Among the best-known cellular pumps are the several polypeptide chains that aggregate together to form the **Na$^+$-K$^+$pump** that maintains the high potassium and low sodium intracellular levels relative to the extracellular medium. The movement of these two ions in opposite directions is linked to ATP breakdown, most likely through cyclical conformational changes in the respective polypeptide chains induced by phosphorylation of key amino acid residues. Such pumps play essential roles in the cellular economy of many resting cells; about one-third of the total cellular ATP is used to maintain intracellular sodium and potassium levels.

In addition to such active-transport processes, other membrane proteins function as passive gates through which metabolites and ions can flow from regions of higher concentration to regions of lower concentration. In vertebrates, the best-known gates are the **gap junctions,** composed of six identical polypeptides surrounding a 15–20 Å diameter channel through which most small molecules (e.g., sugars, amino acids, and nucleotides) can flow virtually unimpeded from one cell to another (Figure 2-32). Gap junctions do not necessarily remain open. When the normally very low (10^{-8} *M*) intracellular calcium levels rise, the gap-junction proteins undergo conformation changes that close the gap.

Hormones and Their Receptors[32, 33]

By the late nineteenth century, it was realized that the cells of multicellular organisms are in part controlled by molecules emanating from far distant cells. These chemical agents, given the name **hormones** by the English physiologists W. M. Bayliss and E. H. Starling in 1909, fall into three main chemical groups: amino acid derivatives, polypeptides, and steroids (derived from cholesterol). Only over the past twenty years have their modes of action begun to be established at the molecular level. In all cases, they act by regulating preexisting cellular reactions, not by themselves acting as enzymes or coenzymes. To do so, they must first bind to specific receptors on their effector cells, unions which then lead to chains of events that dramatically alter preexisting patterns of enzyme synthesis or function.

The cellular receptors for steroid hormones (e.g., the female estrogenic hormones) are specific proteins located largely in the nucleus (Figure 2-33). As many as 30,000 such receptors exist in appropriate target cells. No specific receptors need exist on the cell's surface, since the highly lipid-soluble steroids passively penetrate lipid bilayers, pass through the cytoplasm, and enter the nucleus. After steroid hormones bind to their receptors, the hormone-receptor complexes affect gene function through their binding to specific chromosomal segments. In contrast, the receptors for hormones derived from polypeptides and amino acids are specific membrane proteins. The insulin receptor, for example, is a membrane-sited polypeptide with an approximate molecular weight of 135,000 daltons. In target fat cells, it is

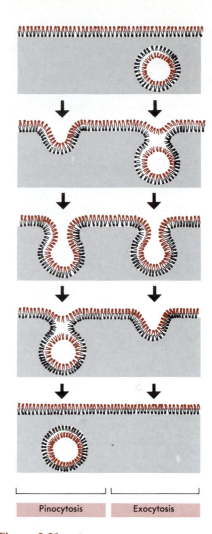

| Pinocytosis | Exocytosis |

Figure 2-31
Pinocytosis (left) and exocytosis (right) probably occur by symmetrical mechanisms of membrane fusion and division. (The outer half of the membrane lipid bilayer appears in color; the inner, cytoplasmic half, appears in black.)

Figure 2-32
A gap junction. Each gap-junction pore consists of an aggregation of identical polypeptides that form a channel through which small molecules can flow between two cells. [After L. Makowski et al., *J. Cell Biol.* 74 (1977):629.]

Six identical polypeptides

Gap between cells

Plasma membrane

Open channel

present in some 10^4 copies per cell and represents slightly less than 1 percent of the membrane protein.

cAMP as the Key Intermediate in the Action of Many Polypeptide Hormones[34, 35, 36]

Many surface-binding proteins act in part through raising or lowering the amount of the key cellular mediator **cyclic AMP** (**cAMP**). Its action as a "second messenger" was elucidated in the early 1960s by the classic investigations of the American biochemist E. W. Sutherland at Vanderbilt University in Nashville, Tennessee. Cyclic AMP is formed from ATP by the enzyme adenylate cyclase, which is bound to the cytoplasmic side of plasma membranes. So located, the activity of adenylate cyclase is influenced by the binding of hormones to their specific surface receptors (Figure 2-34). Such influences, however, are not direct, with the hormone-receptor complexes inducing changes in the conformations of a still incompletely understood class of proteins known as the G proteins (because of their ability to bind the ATP-like nucleotide GTP). In turn, the conformationally altered G proteins directly regulate adenylate cyclase activity.

After its synthesis, cAMP works by stimulating the activity of protein kinases, enzymes that transfer terminal phosphate groups from ATP to specific amino acids on target proteins. Phosphorylation alters the enzymatic activities of these proteins, either raising or lowering them. The skeletal muscle enzyme glycogen synthetase, for example, is inactivated after phosphorylation by a cAMP-activated protein kinase. In contrast, phosphorylation of glycogen phosphorylase makes it more active. Through these dual actions, cAMP simultaneously stimulates glycogen breakdown and inhibits its synthesis.

The use of second messengers like cAMP greatly amplifies the action of single-hormone molecules, since for every polypeptide- or amino acid-derived hormone bound by its receptor, hundreds of cAMP molecules are made. In turn, each cAMP molecule, by binding to an appropriate protein kinase, activates or inactivates many hundreds of specific target proteins. Of equal biological importance is the amplification of the signals carried by individual steroid hormone molecules. By making a specific gene work much faster or slower, the synthesis of its protein product is drastically increased or decreased.

Steroid hormone molecules enter the cell.

Receptor protein

Cytoplasm   Nucleus

Hormone-receptor complexes

Figure 2-33
Steroid hormones diffuse across the plasma membrane and bind to specific receptors within the cell. The hormone-receptor complex affects gene function by binding to chromatin in the nucleus.

The Classical Period of Biochemistry Is Drawing to a Close

Until recently, the study of the chemical reactions underlying the living processes has been dominated by the need to define the cast of key characters (molecules) and the enzymatic pathways by which they become transformed into one another. These pursuits have been successful beyond any reasonable anticipation. Now all major features of living cells—the generation of useful energy from food molecules and the sun, the ways energy-rich molecules promote the making of desired chemical bonds, and, as we shall relate in the next chapter, the functioning of genes—can be understood in terms of well-defined chemical principles.

Biochemistry has become an enormous discipline whose basic metabolic and biosynthetic principles will probably never be seriously modified. So over the next several decades, we shall be slowing down our search for still undiscovered key enzymes. Instead, more and more future effort will go to the working out of the mechanisms of enzymatic reactions at the molecular and atomic levels. For example, how light quanta are collected by the photosynthetic apparatus, or how a membrane protein pumps ions, increasingly have become the important biochemical questions that only chemically oriented scientists (as opposed to biologically oriented investigators) will be able to answer.

These changes should not, however, in any way distress those of us who have become interested in biochemistry through our curiosity about genes. As we shall see in the later chapters of this book, it is only because the basic facts of biochemistry are now so firm that we shall soon be able to move on from the way genes store and express hereditary information to attack the still unsolved major biological problem of how fertilized eggs develop into the marvelously complex forms of higher plants and animals. Only by feeling that the essence of the living single cell is now well within our grasp, can we so optimistically proclaim that the essence of multicellular existence may now be an achievable objective.

Figure 2-34
Cyclic AMP synthesis and breakdown. Cyclic AMP (cAMP) is synthesized by the membrane-bound enzyme adenylate cyclase. The synthesis of cAMP may be stimulated, via membrane-bound receptors and G proteins, by polypeptide- or amino acid-derived hormones outside the cell. Cyclic AMP is broken down inside the cell by a phosphodiesterase.

Summary

The growth and division of cells are based on the same laws of chemistry that control the behavior of molecules outside of cells. Cells contain no atoms unique to the living state; they can synthesize no molecules that the chemist, with inspired hard work, cannot someday make. Thus, there is no special chemistry of living cells. A biochemist is not someone who studies unique types of chemical laws, but a chemist interested in learning about the behavior of molecules found within cells (biological molecules).

The growth and division of cells depend on the availability of a usable form of chemical energy. The energy initially comes from the energy of the sun's light quanta, which is converted by photosynthetic plants into cellular molecules, some of which are used as food sources by various microorganisms and animals. The most striking initial triumphs of the biochemists told us how food molecules are transformed into other cellular molecules and into useful forms of chemical energy. The energy within food molecules largely resides within the covalent bonds of reduced carbon compounds; it is released when these molecules are transformed by oxidation-reduction reactions to carbon compounds of a higher degree of oxidation. For most forms of life, the ultimate oxidizing agent is molecular oxygen. The products of the complete oxidation of organic molecules like glucose are carbon dioxide and water.

Most organic molecules, however, are not oxidized directly by oxygen, but by diverse organic molecules, often coenzymes, such as the coenzyme NAD^+. The reduced coenzyme (e.g., NADH) is itself oxidized by another molecule (such as FAD) to yield a new reduced coenzyme ($FADH_2$) and the original coenzyme in the oxidized form (NAD^+). After several such cycles, molecular oxygen directly participates to end the oxidation-reduction chain, giving off water.

The energy released during the oxidation-reduction cycles is not released entirely as heat; in fact, more than half the energy is converted into new chemical bonds. Phosphorus atoms play a key role in the transformation. Phosphate esters are formed that have a higher usable energy content than most covalent bonds. These phosphate groups are transferred in a high-energy form to acceptor molecules such as ADP. A phosphate group is added to

ADP to form ATP—a primary step in photosynthesis called photophosphorylation. The ADP → ATP transformation is at the heart of energy relations in all cells.

Until a few decades ago, chemists' understanding of the cell's very large molecules, proteins and nucleic acids, was much less firm than it is today. Proteins and nucleic acids are complex structures, and only recently have the physical and chemical techniques needed for their study been developed. Among the most important techniques have been partition chromatography, analytical ultracentrifugation, and X-ray crystallography as extended to the study of large molecules. Now we know practically all the important features of many large proteins like myoglobin, hemoglobin, and trypsin.

The primary structure of proteins consists of amino acids linked together in a unique sequence by covalent bonds. These chains then acquire secondary structure by assuming helical configurations (α helix) or aggregating side by side to form β sheets. Both are held together by hydrogen bonds between adjacent carbonyl and imino groups. The tertiary structure is very irregular owing largely to the diverse side groups of the amino acid striving to form the most energetically stable configuration. Intra- and interchain cysteine (S–S) bonds also contribute to the stability of many proteins. The three-dimensional folding (tertiary structure) of a protein brings these cysteine residues into close contact, thereby allowing S–S bonds to form between correct partners.

All cellular membranes contain phospholipid bilayers arranged so that the hydrophilic phosphate groups are on the outside of the membrane and the hydrophobic hydrocarbon tails are on the inside. Proteins, many of which function as pumps to control the movement of molecules into and out of the cells, are dispersed throughout the membrane, usually extending through the membrane in a semifluid form in which they do not necessarily remain in fixed positions. Sodium and potassium concentrations are maintained against the concentration gradient via a Na^+-K^+ pump. In resting cells, such pumps collectively utilize approximately one-third the ATP generated. Small molecules can flow virtually unimpeded across the membranes of closely opposed cells via channels called gap junctions, which open and close, depending on physiological need. Hormones are molecules that function in cells at locations distant from their site of synthesis. They may be amino acid derivatives, polypeptides, or steroids. In acting, they bind to specific receptors on or in their target or effector cells and then function to regulate preexisting cellular reactions such as enzyme synthesis or function. Steroid hormone-receptor complexes affect cellular functions by binding intracellular receptors, which, in turn, bind to chromosomal segments and affect gene function. In contrast, polypeptide hormones, after binding to their membrane-located receptors, induce changes in the still little understood membrane-bound G proteins, which, in turn, regulate the activity of adenylate cyclase, the enzyme that forms cAMP from ATP. Each cAMP molecule can subsequently bind to an appropriate protein kinase, which, in turn, through its phosphate-adding capacity, activates or inactivates many specific target proteins.

Bibliography

General References

Fruton, J. S. 1972. *Molecules and Life*. New York: Wiley. A detailed guide to the history of biochemistry.
Lehninger, A. L. 1982. *Principles of Biochemistry*. New York: Worth.
Stryer, L. 1981. *Biochemistry*. 2nd ed. San Francisco: Freeman.
Zubay, G. 1983. *Biochemistry*. Reading, Mass.: Addison-Wesley.

Cited References

1. Sumner, J. 1926. "The Isolation and Crystallization of the Enzyme Urease." *J. Biol. Chem.* 69:435–439.
2. Krebs, H. 1981. *Otto Warburg: Biochemist and Eccentric*. Oxford, Eng., and New York: Oxford University Press.
3. Baldwin, E. 1959. *Dynamic Aspects of Biochemistry*. 3rd ed. New York: Cambridge University Press. This is one of the few texts in biochemistry deserving to be called a classic. Now it is best read for the way in which coenzymes participate in enzyme transfers.
4. Lipmann, F. 1971. *Wanderings of a Biochemist*. New York: Wiley. A look back at the author's key role in the discovery of high-energy bonds.
5. Krebs, H. A. 1970. "The History of the TCA Cycle." *Persp. Biol. Med.* 14:154–170.
6. Lipmann, F., and D. Novelli. 1950. "Coenzyme A in Citric Acid Synthesis." *J. Biol. Chem.* 182:213–228.
7. Keilin, D. 1966. *The History of Cell Respiration and Cytochromes*. New York: Cambridge University Press.
8. Kalckar, H. M. 1969. *Biological Phosphorylations*. Englewood Cliffs, N.J.: Prentice-Hall. Contains a collection of many of the classic papers that led to the understanding of the manyfold importance of high-energy bonds.
9. Kennedy, E. P., and A. L. Lehninger. 1949. "Oxidation of Fatty Acids and Tricarboxylic Acid Cycle Intermediates by Isolated Rat Liver Mitochondria." *J. Biol. Chem.* 179:957–972.
10. Lipmann, F. 1941. "Metabolic Generation and Utilization of Phosphate Bond Energy." *Adv. Enzymol.* 1:99–162.
11. Arnon, D. I. 1959. "Conversion of Light into Chemical Energy." *Nature* 184:10–21. Reviews the effect of sunlight in the conversion of ADP to ATP and traces the evolution of photochemical events.
12. Arnon, D. I. 1984. "The Discovery of Photosynthetic Phosphorylation." *Trends in Biochem. Sci.* 9:258–262.
13. Mitchell, P. 1979. "Keilin's Respiratory Chain Concept and Its Chemiosmotic Consequences." *Science* 206:1148–1159. The evolution of the chemiosmotic principle.
14. Svedberg, T., and K. O. Peterson. 1940. *The Ultracentrifuge*. Oxford, Eng., and New York: Oxford University Press.
15. Martin. A. J. P., and R. L. M. Syne. 1941. "A New Form of Chromotography Employing Two Liquid Phases." *Biochem. J.* 35:1358–1368. Original presentation of the theory of partition chromatography.
16. Sanger, F., and H. Tuppy. 1951. "The Amino Acid Sequence in the Phenylalanyl Chain of Insulin." *Biochem. J.* 49:463–490.
17. Steiner, D. F., O. Hallund, A. Rubenstein, S. Cho, and C. Bayliss. 1968. "Isolation and Properties of Proinsulin, Intermediate Forms, and Other Minor Components from Crystalline Bovine Insulin." *Diabetes* 17:725–736.
18. Pauling, L., R. B. Corey, and H. R. Branson. 1951. "The Structure of Proteins: Two Hydrogen Bonded Helical Configurations of the Polypeptide Chain." *Proc. Nat. Acad. Sci.* 37:205–211.
19. Perutz, M. F. 1962. *Proteins and Nucleic Acids*. Amsterdam, London, and New York: Elsevier.

20. Kendrew, J. 1966. *The Thread of Life: An Introduction to Molecular Biology*. Cambridge, Mass.: Harvard University Press.

21. Dickerson, R. E., and I. Geis. 1969. *The Structure and Action of Proteins*. New York: Harper & Row. A well-illustrated introduction to the ways polypeptide chains fold up.

22. Anfinsen, C. B., E. Haber, M. Sela, and F. H. White, Jr. 1961. "The Kinetics of Formation of Native Ribonucleases During Oxidation of the Reduced Polypeptide Chain." *Proc. Nat. Acad. Sci.* 47:1309–1314.

23. Phillips, D. 1968. "The Three Dimensional Structure of an Enzyme Molecule." In *The Molecular Basis of Life*, ed. R. H. Haynes and P. C. Hanawalt. San Francisco: Freeman. pp. 52–64. A description of the work that solved the structure of lysozyme, the first enzyme for which it was possible to correlate structure and function.

24. Stroud, R. M. 1974. "A Family of Protein-Cutting Proteins." *Sci. Amer.* 231:74–88. Describes the structure and functioning of trypsin and chymotrypsin-like enzymes.

25. Davson, H., and J. F. Danielli. 1952. *Permeability of Natural Membranes*. New York: Cambridge University Press.

26. Henderson, R., and P. N. T. Unwin. 1975. "Three Dimensional Model of Purple Membrane Obtained by Electron Microscopy." *Nature* 257:28–32.

27. Alberts, B., D. Bray, J. Lewis, M. Raff, K. Roberts, and J. D. Watson. 1983. *Molecular Biology of the Cell*. New York: Garland. See Chapter 6, "The Plasma Membrane," pp. 255–318.

28. Singer, S. and G. Nicholson. 1972. "The Fluid Mosaic Model of the Structure of Cell Membranes." *Science* 175:720–731.

29. Skou, J. C., and J. G. Norby, eds. 1979. *Na$^+$-K$^+$ ATPase: Structure and Kinetics*. New York: Academic Press.

30. Unwin, P. N. T., and P. D. Ennis. 1984. "Two Configurations of a Channel-Forming Membrane Protein." *Nature* 307:609–613. The protein oligimer forming the gap-junction channel is analyzed in two Ca^{++}-sensitive states.

31. Bennett, M. V. L., and D. C. Spray, eds. 1985. *Gap Junctions*. Cold Spring Harbor, N.Y.: Cold Spring Harbor Laboratory.

32. Jensen, E. V., G. L. Greene, L. E. Closs, E. R. DeSombre, and M. Nadji. 1982. "Receptors Reconsidered—A 20 Year Perspective." *Rec. Prog. Horm. Res.* 38:1–40.

33. Yamamoto, K. R., and B. M. Alberts. 1976. "Steroid Receptors: Elements for Modulation of Eukaryotic Transcription." *Ann. Rev. Biochem.* 45:721–746.

34. Sutherland, E. W., T. W. Rall, and T. Menon. 1962. "Adenyl Cyclase. I. Distribution, Preparation, and Properties." *J. Biol. Chem.* 237:1220–1227.

35. Sutherland, E. W. 1972. "Studies on the Mechanism of Hormone Action." *Science* 177:401–407. Nobel laureate speech on the discovery of cAMP as second messenger.

36. Gilman, A. G. 1984. "G Proteins and Dual Control of Adenylate Cyclase." *Cell* 36:577–579.

Plate 1

Computer Images of DNA

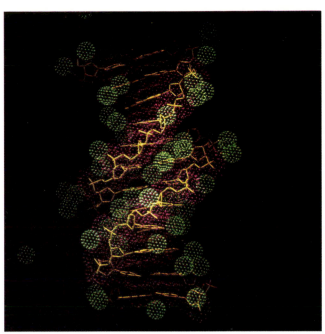

A.   Side view of B-DNA with the individual atoms represented as dot-covered spheres. Carbon is green, oxygen is red, nitrogen is blue, and phosphorus is yellow. The sugar-phosphate backbone on the far side of the structure is not shown. Image courtesy of the Computer Graphics Laboratory of the University of California at San Francisco.

B.   Model of B-DNA (skeleton in yellow and surface indicated with red dots) showing locations of water molecules (green spheres) bound to a crystallographically determined DNA dodecamer. Image courtesy of the Institut de Biologie Moleculaire, Strasbourg.

C.   Skeletal structure of a DNA octamer (pink) cocrystallized with the intercalator antibiotic triostin A (blue). Binding of the triostin causes the right-handed DNA double helix to partially unwind. In the right-hand image, aqua dots indicate the surfaces of the two triostin molecules. Images courtesy of A. Rich, Department of Biology, M.I.T.

Plate 2

Protein Structure at Increasing Levels of Complexity

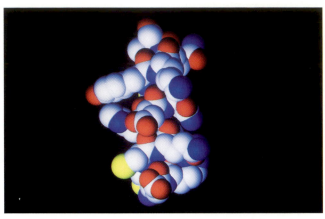

A.   Polypeptide backbone atoms of a portion of α helix, with the atoms shown as space-filling. Carbon is white, oxygen is red, and nitrogen is blue. Note how the oxygen and nitrogen atoms are placed for hydrogen bonding.

B.   As in A, but showing the amino acid side chains pointing outwards. Hydrogen atoms are not shown. The yellow atoms are sulfur.

C.   Skeletal model of the α helix backbone.

D.   Skeletal model of the α helix with side chains.

E.   Ball-and-stick model of hen egg-white lysozyme. The trisaccharide substrate NAM-NAG-NAM is shown in purple.

F.   α-carbon skeleton of lysozyme, tracing the course of the polypeptide backbone.

All images on plates 2 and 3 courtesy of the Graphic Systems Research Group, IBM U.K. Scientific Centre.

Plate 3

G. Lysozyme backbone with hydrophobic side chains (Val, Ile, Leu, Phe, Pro, Met) shown as space-filling.

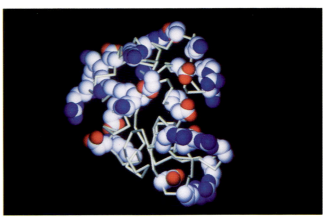

H. Lysozyme backbone with hydrophilic side chains (Asp, Glu, Lys, Arg) shown as space-filling.

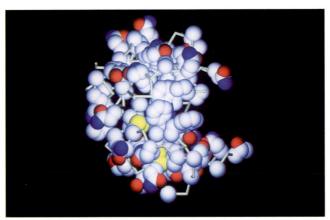

I. Lysozyme backbone with "hydro-indifferent" side chains (Ser, Thr, Gln, Asn, Cys, Gly, Ala, Trp, Tyr, His) shown as space-filling.

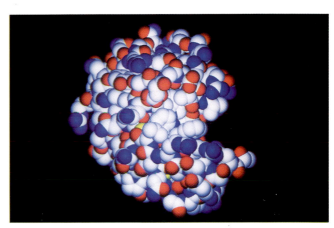

J. Lysozyme with all atoms (except hydrogens) shown as space-filling.

K. Secondary structure representation of lysozyme. The α-carbon—α-carbon "bonds" are white; α helices are shown as yellow spirals; two strands making up one β sheet are shown as blue arrows; three strands forming another β sheet are shown as red arrows. (The arrows point toward the carboxyl end of the polypeptide chain.) The substrate trisaccharide is shown as space-filling.

L. Lysozyme combined with its substrate trisaccharide, with all atoms shown as space-filling. The substrate atoms are in purple.

Plate 4

Interaction of Repressors with DNA

A. Model of the bacteriophage λ repressor protein (the *cI* gene product) viewed facing the side that binds to DNA. The white lines trace the α-carbon backbones of the two identical polypeptides that make up this dimeric molecule.

B. As in A, but highlighting the helix 2-turn-helix 3 motif (see page 473). The helices are represented by solid cylinders.

C. Helix 2-turn-helix 3 portion of the λ repressor, shown in a secondary-structure representation (white), bound to a space-filling model of DNA (sugar-phosphate backbone red, bases blue).

D. As in C, but showing the complete secondary structure of the λ repressor. The α helices are shown as cylinders and the rest of each polypeptide chain as an α-carbon skeleton. Domains of the repressor are differentiated with green and yellow color; helix 2-turn-helix 3 portions are highlighted in white. Images A–D courtesy of the Graphic Systems Research Group, IBM U.K. Scientific Centre.

E. The repressor protein of phage 434 interacting with DNA, as determined by direct crystallographic analysis. The orange region of the DNA is the 434 operator. Only the α-carbon skeleton of the repressor is shown.

F. As in E, except that the side chains of the repressor are shown. Images E and F courtesy of J.E. Anderson and S.C. Harrison, Department of Biochemistry, Harvard University.

Plate 5

G. Space-filling models of the CI repressor (left) and the Cro repressor (right) of λ. The two identical subunits of each dimeric molecule are shown in different shades of blue. Helix 3 is shown in red.

H. Helix 3 region (red) of CI and Cro repressors shown bound to their respective operator sequences. The phosphates which when ethylated reduce binding are colored yellow.

I. CI repressor bound to its operator with and without its helix 3 region (red). Images G–I courtesy of W.F. Anderson, Department of Biochemistry, University of Alberta.

Plate 6

The Complete Structure of a Multiprotein Complex

This spectacular computer image shows the membrane-spanning photosynthetic reaction center from the purple photosynthetic bacterium *Rhodopseudomonas viridis*. The four proteins that make up the complex are shown in a ribbon representation. The cytochrome subunit (Cyt) is shown in green, with its four covalently attached heme groups in brown. The light subunit (L) is red; the medium subunit (M) is blue; and the heavy subunit (H) is violet. The L and M subunits surround the pigments (4 bacteriochlorophyll *b* molecules, 2 bacteriopheophytin *b* molecules, 1 menaquinone molecule—all in yellow). The reaction center has a total of eleven membrane-spanning, hydrophobic α helices: 5 each from the L and M subunits and 1 from the H subunit. Image courtesy of H. Michel and J. Deisenhofer, Max-Planck-Institut für Biochemie, Munich.

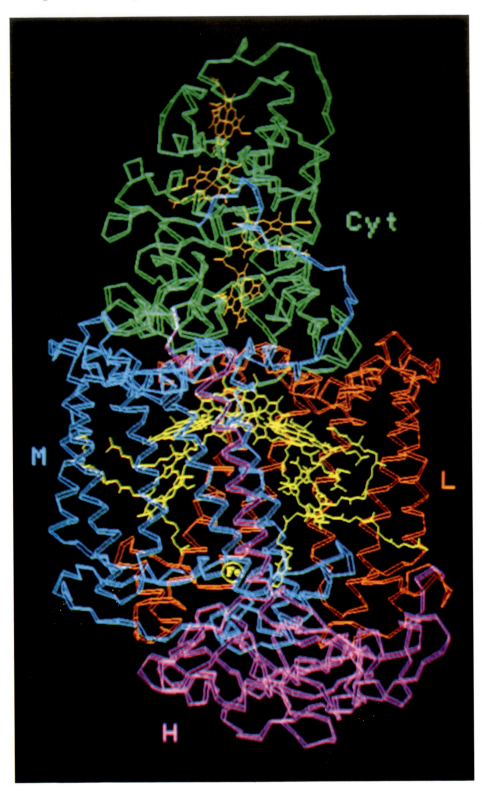

Plate 7

Immunoglobulin and Viral Hemagglutinin

A. Representation of an antibody molecule, human immunoglobulin G, showing its solvent accessible surface (light chain, blue; heavy chain, purple) and α-carbon backbone (light chain, yellow; heavy chain, red).

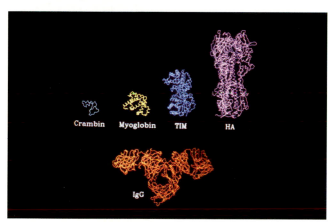

B. Size comparison of immunoglobulin G (IgG) with various other proteins. TIM is triose phosphate isomerase; HA is influenza hemagglutinin.

C. Model of antibody molecules interacting with a virus particle. The heavy chains of each immunoglobulin are light and dark blue; the light chains are purple. The virus is tomato bushy stunt virus, whose capsid (yellow) is made up of 180 identical protein subunits.

D. Hemagglutinin monomer of the influenza virus. Polypeptide HA_1 is blue, and HA_2 is red. (See pages 923–924.)

E. Trimer of the influenza hemagglutinin. Images on plate 7 courtesy of A.J. Olson, Scripps Clinic and Research Foundation.

Plate 8

Capsid Structure of Icosahedral Viruses

A.   Architecture of the tomato bushy stunt virus. Large spheres represent protein subunits on the surface of the virus. The colors blue, aqua, and red indicate the three different symmetry-packing environments of the subunits. Half of the spheres are removed to reveal the amino terminal arms (yellow) of the red subunits linked together to form an internal scaffold.

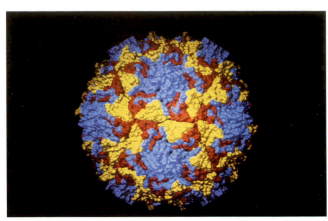

B.   Surface of the poliovirus. The solvent accessible surface of poliovirus consists of four capsid proteins, 60 copies each. Proteins VP1 (blue), VP2 (yellow), and VP3 (red) can be seen from the outside of the particle. The maximum radius of the virus particle is 160 Å. (See page 915.)

C.   Poliovirus capsid cut away to reveal the VP4 capsid protein (green) in the interior. The other capsid proteins are colored as in B. The minimum inside radius is 100 Å.

D.   Protein subunits of poliovirus. The α-carbon backbones of the three surface protein subunits (capsid proteins) are shown. VP1 is blue; VP2 is yellow; VP3 is violet.

E.   Pentameric assembly intermediate of the poliovirus capsid. Five assembled protomers are shown, with 4 subunits in each (color coded as in D). This image shows a view from the outside of the virus.

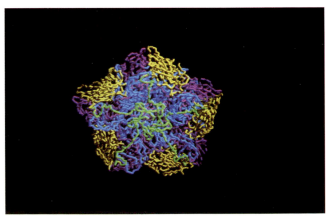

F.   As in E, but viewed from the inside of the virus. Images on plate 8 courtesy of A.J. Olson, Scripps Clinic and Research Foundation.

CHAPTER 3

Nucleic Acids Convey Genetic Information

That special molecules might carry genetic information was appreciated by geneticists long before the problem claimed the attention of chemists. By the 1930s, geneticists began speculating as to what sort of molecules could have the kind of stability that the gene demanded yet be capable of permanent, sudden change to the mutant states that must provide the basis of evolution. But until the mid-1940s, there appeared to be no direct way to attack the chemical essence of the gene. Though it was known that chromosomes possessed a unique molecular constituent, deoxyribonucleic acid (DNA), there was no way to show that this constituent carried genetic information, as opposed to serving merely as a molecular scaffold for a still undiscovered class of proteins especially tailored to carry genetic information.

So it made sense to approach the nature of the gene by asking how genes function within cells. Beginning in 1941, research on the mold *Neurospora* had begun to generate increasingly strong evidence supporting the then 30-year-old hypothesis of Garrod that genes work by controlling the synthesis of specific enzymes (the one gene–one enzyme hypothesis). Thus, given that all known enzymes had, by this time, been shown to be protein, the key problem was the way genes participate in the synthesis of proteins. From the very start of serious speculation, the simplest hypothesis was that genetic information within genes determines the order of the 20 different amino acids within the polypeptide chains of proteins.

In attempting to test this proposal, intuition was of little help to even the best biochemists, since there is no logical way to use enzymes as ordering tools. Such schemes would require, for the synthesis of a single type of protein, as many ordering enzymes as there are amino acids in the respective protein. But since all known enzymes are themselves proteins, still additional ordering enzymes would be necessary to synthesize the ordering enzymes. And so on. This is clearly a paradox, unless we assume a fantastically interrelated series of syntheses in which a given protein has many different enzymatic specificities. With such an assumption, it might be possible (and then only with great difficulty) to visualize a workable cell. It did not seem likely, however, that most proteins would be found to carry out multiple tasks. In fact, all our knowledge pointed to the opposite conclusion of one protein, one function.

Figure 3-1

Formation of a specific polymeric molecule on a template surface.

The Concept of Template Surfaces[1-4]

It was therefore necessary to throw out the idea of ordering proteins with enzymes and to predict instead the existence of a specific surface, the **template**, that attracts the amino acids (or their activated derivatives) and lines them up in the correct order (Figure 3-1). Then a specific enzyme common to all protein synthesis could make the peptide bonds. Furthermore it was necessary to assume that the templates also have the capacity of serving either directly or indirectly as templates for themselves (self-duplication). That is, in some way their specific surfaces must be exactly copied to give new templates. Again, scientists could not invoke the help of specific enzymes, for this immediately led back to the "enzyme cannot make enzyme" paradox.

Templates Must Be in the Size Range of Their Products

The finding that proteins possess unique amino acid sequences demanded the existence of specific templates on which their amino acid building blocks are laid down. These templates themselves had to be macromolecules, at least as large as their polypeptide products. This becomes clear when we examine the nature of the chemical forces that attract small molecules to their templates. It was expected that these forces would be the same as those that attract substrates to their enzymes. These are ionic bonds; hydrogen bonds in which an electropositive hydrogen atom is attracted to an electronegative atom, such as oxygen or nitrogen; and van der Waals forces (see Chapter 5 for a complete discussion of these forces).

Since all these forces operate only over very small distances (<5 Å), templates can order small molecules only when they are in close contact on the atomic level. Thus, it was correctly assumed that the specific attracting regions of the templates would be in the same size range as the amino acid side groups in the protein products.

Attraction of Opposites Versus Self-Attraction

Can a polypeptide chain serve as a template for its own synthesis? If it could, it would greatly reduce the chemical prerequisites for life. The problems of protein synthesis and template replication would then be the same, and the additional biochemical complexity required to maintain a special class of template molecules would be unnecessary. This conceptual possibility found no support, however, from close inspection of the amino acid side groups. There was no chemical reason why, for example, the occurrence of valine on a template should preferentially attract the specific side group of another valine molecule. In fact, none of the amino acid side groups have specific affinities for themselves. Instead, it was much easier to imagine molecules with opposite or complementary features attracting each other. Negative charges obviously attract positive groups, and hydrogen atoms can form hydrogen bonds only to electronegative atoms like oxygen or nitrogen. Similarly, molecules can be specifically attracted by van der Waals forces only when they possess complementary shapes, so that a cavity in one molecule can be filled with a protruding group of another molecule.

There is no way, however, to group the 20 amino acids into 10 pairs whose side groups have mutually self-complementary shapes. A formal scheme remained, however, that could conceivably allow the existence of protein templates. We could imagine the existence of 20 different specific molecules, which we call connectors. Each connector might possess two identical surfaces complementary in charge and/or shape to a given amino acid. The intervention of these connector molecules could then make possible the lining up of amino acids in a sequence identical to that of the template polypeptide chain. However, there was no evidence for such molecules. Instead, as we shall soon see, a specific template class (the nucleic acids) does in fact exist.

A Chemical Argument Against the Existence of Protein Templates[5]

The failure of proteins to evolve a template role may have originated in the composition of the amino acid side groups. The argument can be made that no template whose specificity depends on the side groups of closely related amino acids, like valine or alanine, can ever be copied with the accuracy demanded for efficient cellular existence. This follows from the fact that some amino acids are chemically very similar. For example, valine and isoleucine differ only by the presence of an additional methyl group in isoleucine. Likewise, glycine and alanine also differ by only one methyl group (see Figure 2-16). This close chemical similarity immediately posed the question of whether any copying process could be sufficiently accurate to distinguish between such closely related molecules. A good speculative guess was that each amino acid in a hereditary molecule would have to be copied with an accuracy of not more than one error in 10^8. On the other hand, a semirigorous chemical argument could be made that no chemical reaction could distinguish between molecules differing by only one methyl group with an accuracy of less than one error in 10^6. When the proper experiments were later carried out, amino acids like leucine were inserted into polypeptide chains with no greater than a 99.9 percent accuracy (one error in 10^3). Thus, proteins did not have the "smell" of a hereditary molecule.

Establishing the Chemical Constituents of DNA and RNA[6,7]

If proteins were not the genetic molecules, then the prime candidates had to be the nucleic acids. Though prominent features of most cells, they had yet to be ascribed any biological function. Dealing with them chemically, however, was initially difficult, since, like proteins, they are macromolecules, constructed from large numbers of their unique building blocks, the nucleotides. Moreover, in addition to DNA, which by the late nineteenth century had already been recognized as a major constituent of chromosomes, there also existed a second form of nucleic acid, called ribonucleic acid (RNA). Early on, it had been found to be a major constituent of yeasts, and for many years there was the erroneous belief that RNA was the plant nucleic acid while DNA was the nucleic acid of animals. However, with the development of better ways to distinguish DNA from RNA, all cells, including those of bacteria, were found to contain both DNA and RNA.

Both DNA and RNA are constructed from four main nucleotide building blocks. Each of these nucleotides contains a phosphate group linked to a five-carbon-atom sugar group, which, in turn, is joined to a flat aromatic molecule that can be either a double-ringed purine or a single-ringed pyrimidine (Figure 3-2). Since they contain the sugar deoxyribose, the nucleotides of DNA are called **deoxyribonucleotides,** while those of RNA, which contain the sugar ribose, are known as **ribonucleotides.** DNA and RNA are both built up from two purine-containing nucleotides and two pyrimidine-containing molecules. The purines of both DNA and RNA are the same, adenine and guanine. The pyrimidine cytosine is found in both nucleic acids, while the pyrimidine thymine is limited to DNA, being replaced in RNA by the very similar pyrimidine uracil.

In both DNA and RNA, the nucleotides are joined together by covalent bonds linking the phosphate group of one nucleotide to a hydroxyl group on the sugar of the adjacent nucleotide. Such bonds, called **phosphodiester linkages,** were known to exist, but the details, such as the exact atoms that linked together the nucleic acids, were long to remain unknown. Also initially confusing were the true sizes of DNA and RNA within cells. The original separation techniques used to isolate them seriously degraded them, and only beginning in the mid-1930s was it shown that both could be present in very large forms. Especially crucial were the experiments of chemists R. Signer, E. Hammarsten, and T. Caspersson, who found that gently prepared DNA had molecular weights even larger than most proteins (greater than 500,000). They also deduced that DNA was a very elongated molecule, a conjecture fully confirmed by 1950 with the advent of easy-to-use electron microscopes, which revealed DNA to be an extraordinarily thin molecule with a diameter of only 20 Å. Individual DNA molecules had lengths of many thousands of angstroms, telling us that they were built up from thousands of nucleotide building blocks (Figure 3-3).

Deoxyadenosine-5'-phosphate

Deoxythymidine-5'-phosphate

Deoxyguanosine-5'-phosphate

Uridine-5'-phosphate

Deoxycytidine-5'-phosphate

Figure 3-2
The nucleotides of DNA and RNA. The four nucleotides of DNA are shown, along with one of the RNA nucleotides. All RNA nucleotides have the sugar ribose (instead of deoxyribose), which has a hydroxyl group on carbon 2. In addition, RNA has the pyrimidine uracil instead of thymine.

Avery's Bombshell:
DNA Can Carry Genetic Specificity[8, 9]

That DNA might be the key genetic molecule emerged most unexpectedly from studies on pneumonia-causing bacteria. In 1928, the English microbiologist Frederick Griffith had made the startling observation that nonvirulent strains of the bacteria became virulent when mixed with their heat-killed pathogenic counterparts. That such transformations from nonvirulence to virulence represented hereditary changes was shown by using descendants of the newly pathogenic strains to transform still other nonpathogenic bacteria. This raised the possibility that when pathogenic cells are killed by heat, their genetic components remain undamaged and, liberated somehow from the heat-killed cells, can pass through the cell wall of the living recipient cells and subsequently undergo genetic recombi-

5' end

Adenine

Cytosine

Guanine

Thymine

(b)

(a)

3' end

Figure 3-3

(a) A portion of a DNA polynucleotide chain, showing the 3'-5' phosphodiester linkages that connect the nucleotides. Phosphate groups connect the 3' carbon of one nucleotide with the 5' carbon of the next. (b) Electron micrograph of a portion of a DNA molecule.

nation with the recipient's genetic apparatus (Figure 3-4). Subsequent research has confirmed this genetic interpretation. Pathogenicity reflects the action of the *S* (smooth) gene, which codes for a key enzyme involved in the synthesis of the carbohydrate-containing capsule that surrounds most pneumonia-causing bacteria. When the *R* (rough) allele of this gene is present, no capsule is formed and the respective cells are not pathogenic.

Within several years after Griffith's original observation, extracts of the killed bacteria were found capable of inducing hereditary transformations, and a search could begin for the chemical identity of the transforming agent. At that time, the vast majority of biochemists still believed that genes were proteins. It therefore came as a great surprise when in 1944, after some ten years of research, American microbiologist Oswald T. Avery and his colleagues Colin M. MacLeod and Maclyn McCarty, at the Rockefeller Institute in New York, made the momentous announcement that the active genetic principle was DNA (Figure 3-5). Supporting their conclusion were key experiments showing that the transforming activity of their highly purified active fractions was destroyed by pancreatic deoxyribonuclease, a recently purified enzyme that specifically degrades DNA molecules to their nucleotide building blocks and has no effect on the integrity of protein molecules or RNA. The addition of either pancreatic ribonuclease (which degrades RNA) or various proteolytic (protein-destroying) enzymes had no influence on transforming activity.

The Amount of Chromosomal DNA Is Constant[10, 11]

Even though the transformation results were clear-cut, there was initially great skepticism about their general applicability; many, if not the majority of, interested scientists doubted that they would be found relevant to anything but certain strains of bacteria. Thus, the momentous nature of Avery's discovery was only gradually appreciated.

One important confirmation came from studies on the chemical nature of chromosomes. DNA was found to be located almost exclusively in the nucleus and almost never where detectable chromosomes were absent. Moreover, the amount of DNA per diploid set of chromosomes was shown, through the cytochemical analysis of A. E. Mirsky and H. Ris of the Rockefeller Institute and by the Vendrelys in France, to be constant for a given organism and equal to twice the amount present in the haploid sperm cells (Table 3-1). Another type of evidence that favored DNA as the genetic molecule was the observation that it is metabolically stable. It is not rapidly made and broken down like many other cellular molecules; once atoms are incorporated into DNA, they do not leave it so long as healthy cell growth is maintained.

Viral Genes Are Also Nucleic Acids[12]

Even more important confirmatory evidence came from chemical studies with viruses and virus-infected cells. It was possible by 1950 to obtain a number of essentially pure viruses and to determine which types of molecules were present in them. This work led to the very important generalization that all viruses contain nucleic acid. Since there was at that time a growing realization that viruses contain genetic material, the question immediately arose as to whether the nucleic acid component was the viral chromosome. The first crucial test of the question came from isotopic study of the multiplication of T2, a bacterial virus (**bacteriophage,** or **phage**) containing a DNA core and a protective shell built up by the aggregation of a number of different protein molecules. In these experiments, performed in 1952 by A. D. Hershey and M. Chase working at Cold Spring Harbor on Long Island, the protein coat was labeled with the radioactive isotope ^{35}S, and the DNA with the radioactive isotope ^{32}P. The labeled virus was then used to follow the fates of the phage protein and nucleic acid as phage multiplication proceeded, particularly to see which labeled atoms from the parental virus entered the host cell and later appeared in the progeny phage.

Clear-cut results emerged from these experiments; much of the parental nucleic acid and none of the parental protein was detected in

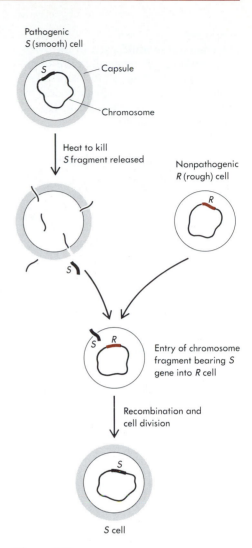

Figure 3-4
Transformation of a genetic characteristic of a bacterial cell (*Streptococcus pneumoniae*) by addition of heat-killed cells of a genetically different strain. Here we show an *R* cell receiving a chromosomal fragment containing the *S* gene. Since most *R* cells receive other chromosomal fragments, the efficiency of transformation for a given gene is usually less than 1 percent.

Figure 3-5
The chemical method used in the original isolation of a chemically pure transforming agent. [After F. W. Stahl, *The Mechanics of Inheritance* (Englewood Cliffs, N.J.: Prentice-Hall, 1964), Fig. 2.3, with permission.]

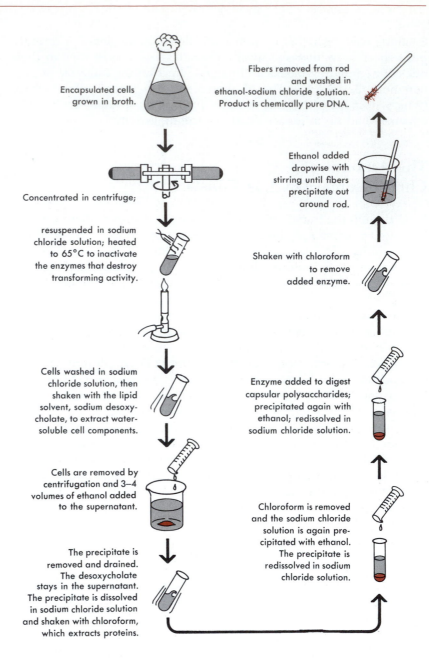

Encapsulated cells grown in broth.

Concentrated in centrifuge;

resuspended in sodium chloride solution; heated to 65°C to inactivate the enzymes that destroy transforming activity.

Cells washed in sodium chloride solution, then shaken with the lipid solvent, sodium desoxycholate, to extract water-soluble cell components.

Cells are removed by centrifugation and 3–4 volumes of ethanol added to the supernatant.

The precipitate is removed and drained. The desoxycholate stays in the supernatant. The precipitate is dissolved in sodium chloride solution and shaken with chloroform, which extracts proteins.

Chloroform is removed and the sodium chloride solution is again precipitated with ethanol. The precipitate is redissolved in sodium chloride solution.

Enzyme added to digest capsular polysaccharides; precipitated again with ethanol; redissolved in sodium chloride solution.

Shaken with chloroform to remove added enzyme.

Ethanol added dropwise with stirring until fibers precipitate out around rod.

Fibers removed from rod and washed in ethanol-sodium chloride solution. Product is chemically pure DNA.

Table 3-1   Amount of DNA per Nucleus in Beef Tissues

Organ	DNA (pg)*	Probable Chromosome Number
Thymus	6.6	Diploid
Liver	6.4	Diploid
Pancreas	6.9	Diploid
Kidney	5.9	Diploid
Sperm	3.3	Haploid

*1 picogram (pg) = 10^{-12} gram.
SOURCE: Data from R. Vendrely, "The DNA Content of the Nucleus," in E. Chargaff and J. N. Davidson, eds., *The Nucleic Acids*, vol. 2. New York: Academic Press, 1955.

the progeny phage (Figure 3-6). Moreover, it was possible to show that little of the parental protein ever enters the bacteria; instead, it stays attached to the outside of the bacterial cell, performing no function after the DNA component has passed in. This point was neatly shown by violently agitating infected bacteria after the entrance of the DNA; the protein coats were shaken off without affecting the ability of the bacteria to form new phage particles.

With some viruses it is now possible to do an even more convincing experiment. For example, purified DNA from the mouse virus polyoma can enter mouse cells and initiate a cycle of viral multiplication producing many thousands of new polyoma particles. The primary function of viral protein is thus to protect its genetic nucleic acid component in its movement from one cell to another. Thus no reason exists for the assignment of any genetic role to protein molecules.

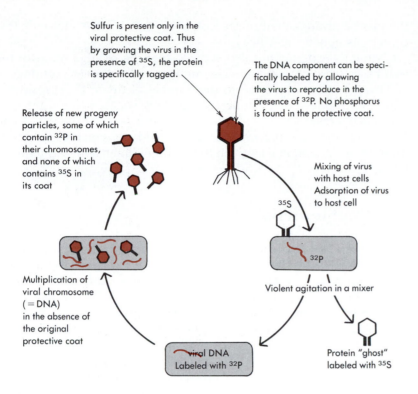

Sulfur is present only in the viral protective coat. Thus by growing the virus in the presence of ³⁵S, the protein is specifically tagged.

The DNA component can be specifically labeled by allowing the virus to reproduce in the presence of ³²P. No phosphorus is found in the protective coat.

Release of new progeny particles, some of which contain ³²P in their chromosomes, and none of which contains ³⁵S in its coat

Mixing of virus with host cells Adsorption of virus to host cell

³⁵S

³²p

Violent agitation in a mixer

Multiplication of viral chromosome (=DNA) in the absence of the original protective coat

viral DNA Labeled with ³²P

Protein "ghost" labeled with ³⁵S

Figure 3-6
Demonstration that only the DNA component of T2 carries the genetic information and that the protein coat serves only as a protective shell.

Chargaff's Rules[13, 14]

The paper chromatographic techniques that had first proved so useful in revealing the chemical makeup of proteins allowed the biochemist Erwin Chargaff, then working in New York at the College of Physicians and Surgeons of Columbia University, to analyze the nucleotide composition of DNA. By 1949 his data showed not only that the four different nucleotides are not present in equal amounts but also that the exact ratios of the four nucleotides vary from one species to another (Table 3-2). These findings opened up the possibility that it is

Table 3-2 Data Leading to the Formulation of Chargaff's Rules

Source	Adenine to Guanine	Thymine to Cytosine	Adenine to Thymine	Guanine to Cytosine	Purines to Pyrimidines
Ox	1.29	1.43	1.04	1.00	1.1
Human	1.56	1.75	1.00	1.00	1.0
Hen	1.45	1.29	1.06	0.91	0.99
Salmon	1.43	1.43	1.02	1.02	1.02
Wheat	1.22	1.18	1.00	0.97	0.99
Yeast	1.67	1.92	1.03	1.20	1.0
Hemophilus influenzae	1.74	1.54	1.07	0.91	1.0
E-coli K2	1.05	0.95	1.09	0.99	1.0
Avian tubercle bacillus	0.4	0.4	1.09	1.08	1.1
Serratia marcescens	0.7	0.7	0.95	0.86	0.9
Bacillus schatz	0.7	0.6	1.12	0.89	1.0

SOURCE: After E. Chargaff et al., *J. Biol. Chem.* 177 (1949).

Figure 3-7
The key X-ray photograph involved in the elucidation of the DNA structure. This photograph, taken by Rosalind Franklin at Kings College, London, in the winter of 1952–1953, confirmed the guess that DNA was helical. The helical form is indicated by the crossways pattern of X-ray reflections (photographically measured by darkening of the X-ray film) in the center of the photograph. The very heavy black regions at the top and bottom tell that the 3.4 Å thick purine and pyrimidine bases are regularly stacked next to each other, perpendicular to the helical axis. [Reproduced from R. E. Franklin and R. Gosling, *Nature* 171 (1953):740, with permission.]

the precise arrangement of nucleotides within a DNA molecule that confers its genetic specificity.

Chargaff's experiments also showed that the relative ratios of the four bases were not random. The number of adenine (A) residues in all DNA samples was equal to the number of thymine (T) residues, while the number of guanine (G) residues equaled the number of cytosine (C) residues. The fundamental significance of the A = T and G = C relationships (Chargaff's rules) could not emerge, however, until serious attention was given to the three-dimensional structure of DNA.

The Double Helix[15–23]

While work was proceeding on the X-ray analysis of protein structure, a smaller number of scientists were trying to solve the X-ray diffraction pattern of DNA. The first diffraction patterns were taken in 1938 by the Englishman Astbury and used DNA supplied from Sweden by Hammarsten and Caspersson. It was not until after the war (1950–1952), however, that high-quality photographs were taken by M. H. F. Wilkins and Rosalind Franklin, working in London at King's College (Figure 3-7). These photographs suggested not only that the underlying DNA structure was helical but that it was composed of more than one polynucleotide chain—either two or three. Only then, however, were all the covalent bonds of DNA being unambiguously established. In 1952, a group of organic chemists working in the Cambridge, England, laboratory of Alexander Todd, showed that 3'-5' phosphodiester bonds regularly link together the nucleotides of DNA (see Figure 3-3).

Because of interest in Linus Pauling's α helix, in 1951 an elegant theory of diffraction of helical molecules was developed in England by W. Cochran, Francis Crick, and V. Vand. This theory made it easy to test possible DNA structures on a trial-and-error basis. The correct solution, a complementary double helix (see Chapter 9 for details), was found in 1953 by Crick and James D. Watson, then working in England in the laboratory of Perutz and Kendrew. Their arrival at the correct answer was in large part dependent on finding the stereochemically most favorable configuration compatible with the X-ray diffraction data of the King's College group.

In the double helix, the two DNA chains are held together by hydrogen bonds (a weak noncovalent chemical bond; see Chapter 5 for details) between pairs of bases on the opposing strands (Figure 3-8). This base pairing is very specific: The purine adenine only base-pairs to the pyrimidine thymine, while the purine guanine only base-pairs to the pyrimidine cytosine. In double-helical DNA, the number of A residues must thus be equal to the number of T residues, while the number of G and C residues must likewise be equal. Hence, Chargaff's rules. As a result, the sequence of the bases of the two chains of a given double helix have a complementary relationship with the sequence on one strand exactly defining that of its partner strand.

The establishment of the double helix immediately began a profound revolution in the way many geneticists analyzed their data. The gene was no longer a mysterious entity whose behavior could be investigated only by breeding experiments. Instead, it quickly became a real molecular object about which chemists could think objectively, as they did about smaller molecules such as pyruvate and

NAD. Most of the excitement, however, came not merely from the fact that the structure was solved, but also from the nature of the structure. Before the answer was known, there had always been the worry that it would turn out to be dull, revealing nothing about how genes replicate and function. Fortunately, however, the answer was immensely exciting. The two intertwined strands of complementary structures suggested that one strand serves as the specific surface (template) upon which the other strand is made (Figure 3-8). If this hypothesis were true, then the fundamental problem of gene replication, about which geneticists had puzzled for so many years, was, in fact, conceptually solved.

Figure 3-8
The replication of DNA. The newly synthesized strands are shown in color.

Finding the Polymerases That Make DNA[24]

Rigorous proof that a single DNA chain is the template upon which a complementary DNA chain is laid down had to await the development of test-tube (in vitro) systems for DNA synthesis. These came much faster than anticipated by the molecular geneticists, whose world until then had been far removed from that of the biochemist well versed in the procedures needed for enzyme isolation. Leading this biochemical assault on DNA replication was (and still is) the American biochemist Arthur Kornberg, who by 1956 had demonstrated DNA synthesis in cell-free extracts of bacteria. Over the next several years, Kornberg went on to show that a specific polymerizing enzyme was needed to catalyze the linking together of the building block precursors of DNA. The nucleotide building blocks utilized (dATP, dGTP, dCTP, and dTTP) had earlier been revealed by him to be energy rich and ultimately dependent on the generation of ATP for their synthesis. The enzyme involved, DNA polymerase I, was the first enzyme found to be involved in the synthesis of polydeoxynucleotides. It links the nucleotide precursors by 3'-5' phosphodiester bonds (Figure 3-9). Furthermore, it works only in the presence of DNA, which is needed to order the four nucleotides in the polynucleotide product.

DNA polymerase I recognizes only the regular sugar-phosphate portion of the nucleotide precursors and so cannot determine sequence specificity. This is neatly demonstrated by allowing the enzyme to work in the presence of DNA molecules that contain varying amounts of A-T and G-C base pairs. In every case, the enzymatically synthesized product has the base ratios of the template DNA (Table 3-3). During this cell-free synthesis, no synthesis of proteins or any other molecular class occurs, unambiguously eliminating any non-DNA compounds as intermediate carriers of genetic specificity. There is thus no doubt that DNA is the direct template for its own formation. As we might expect, the enzymatic product, like the primer, has a double-helical structure.

The ability of DNA polymerase I to make complementary copies of DNA chains immediately suggested that this might be the major enzyme in *E. coli* involved in linking up nucleotides during DNA replication. But as we shall see later, DNA replication is a complex process involving several different polymerizing enzymes. Discovery of the first DNA polymerase, however, was of immense importance in showing that the assembly of complementary nucleotide sequences was amenable to relatively straightforward test-tube analysis.

Figure 3-9
Enzymatic synthesis of a DNA chain catalyzed by DNA polmerase I.

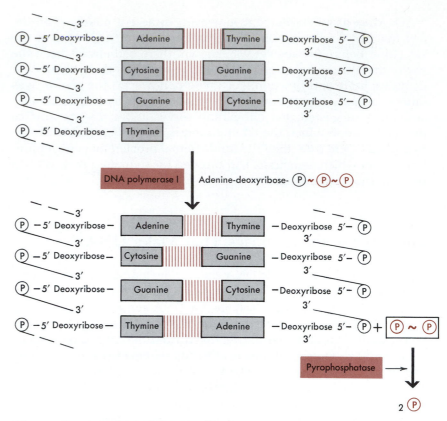

Experimental Evidence Favors Strand Separation During DNA Replication[25]

Simultaneously with Kornberg's research, in 1958 M. Meselson and F. W. Stahl, then at the California Institute of Technology, carried out an elegant experiment in which they separated parental from daughter DNA molecules and in so doing showed that the two strands of the double helix permanently separate from each other during DNA replication (Figure 3-10). Their success was due in part to the use of the heavy isotopes ^{15}N and ^{13}C as tags to differentially label the parental and daughter DNA strands. Bacteria grown in a medium containing the heavy isotope ^{15}N have denser DNA than bacteria grown

Table 3-3 A Comparison of the Base Composition of Enzymatically Synthesized DNA and Their DNA Templates

Source of DNA Template	Base Composition of the Enzymatic Product				$\dfrac{A+T}{G+C}$ in Product	$\dfrac{A+T}{G+C}$ in Template
	Ade-nine	Thy-mine	Gua-nine	Cyto-sine		
Micrococcus lysodeikticus (a bacterium)	0.15	0.15	0.35	0.35	0.41	0.39
Aerobacter aerogenes (a bacterium)	0.22	0.22	0.28	0.28	0.80	0.82
Escherichia coli	0.25	0.25	0.25	0.25	1.00	0.97
Calf thymus	0.29	0.28	0.21	0.22	1.32	1.35
Phage T2	0.32	0.32	0.18	0.18	1.78	1.84

Figure 3-10
Use of a cesium chloride (CsCl) density gradient to demonstrate the separation of complementary strands during DNA replication.

under normal conditions with ^{14}N. Also contributing to the success of the experiment was the development of procedures for separating heavy from light DNA in density gradients of heavy salts like cesium chloride. When high centrifugal forces are applied, the solution becomes more dense at the outside of the centrifuge cell. If the correct initial solution density is chosen, the individual DNA molecules will move to the central region of the centrifuge cell, where their density equals that of the salt solution. There the heavy molecules will band at a higher density than the light molecules. If bacteria containing heavy DNA are transferred to a light medium (containing ^{14}N) and allowed to grow, the precursor nucleotides available for use in DNA synthesis will be light; hence, DNA synthesized after transfer will be distinguishable from DNA made before transfer.

If DNA replication involves strand separation, definite predictions can be made about the density of the DNA molecules found after various growth intervals in a light medium. After one generation of growth, all the DNA molecules should contain one heavy strand and one light strand and thus be of intermediate hybrid density. This result is exactly what Meselson and Stahl observed. Likewise, after two generations of growth, half the DNA molecules were light and half hybrid, just as strand separation predicts.

DNA was thus proved to be a "semiconservative" process in which the single strands of the double helix remain intact (are conserved) during a replication process that distributes one parental strand into each of the two daughter molecules. Ruled out by these experiments were any "conservative" replication schemes in which the two parental strands only temporarily separate when they function as templates for daughter DNA strands. While conservative DNA replication always seemed stereochemically implausible, only through the Meselson-Stahl experiments was the semiconservative replication of DNA transformed from a very attractive hypothesis into a solid fact on which to base further experimentation.

The Genetic Information Within DNA Is Conveyed by the Sequence of Its Four Nucleotide Building Blocks

The finding of the double helix had effectively ended any controversy about whether DNA was the primary genetic substance. Even before strand separation during DNA replication was experimentally verified, the main concern of molecular geneticists had turned to how the genetic information of DNA functions to order amino acids during protein synthesis. With all DNA chains capable of forming double helices, the essence of their genetic specificity had to reside in the linear sequences of their four nucleotide building blocks. Thus, as information-containing entities, DNA molecules were by then properly regarded as very long words (as we shall see later, they are now best considered very long sentences) built up from a four-letter alphabet (A, G, C, and T). Even with only four letters, the number of potential DNA sequences (4^N, where N is the number of letters in the sequence) is very very large for even the smallest of DNA molecules; a virtually infinite number of different genetic messages can exist. Later it will be demonstrated that many bacterial genes contain some 1500 base pairs. The number of potential genes of this size is 4^{1500}, a value much larger than the number of different genes that could have existed in all the chromosomes present since the origin of life.

Proof That Genes Control Amino Acid Sequences in Proteins[26]

The first experimental demonstration that genes (DNA) control amino acid sequences arose from the study of the hemoglobin present in humans suffering from the genetic disease sickle-cell anemia. If the sickle (*s*) gene is present in both homologous chromosomes, a severe anemia results characterized by the red blood cells having a sickle shape. If only one *s* gene is present and the allele in the homologous chromosome is normal (+), then the anemia is less severe and the red blood cells appear almost normal in shape. The type of hemoglobin in red blood cells is likewise correlated with the genetic pattern. In the *ss* case, the hemoglobin is abnormal, characterized by a solubility different from that of normal hemoglobin, whereas in the *+s* condition, half the hemoglobin is normal and half sickle.

Wild-type hemoglobin molecules are constructed from two kinds of polypeptide chains: α chains and β chains (see Figure 5-32). Each chain has a molecular weight of about 16,100 daltons. Two α chains and two β chains are present in each molecule, giving hemoglobin a molecular weight of about 64,650 daltons. The α and β chains are controlled by distinct genes so that a single mutation will affect either the α or the β chain but not both. In 1957, V. M. Ingram at Cambridge University showed that sickle hemoglobin differs from normal hemoglobin by the change of one amino acid in the β chain; at position 6, the glutamic acid found in wild-type hemoglobin is replaced by valine. Except for this one change, the entire amino acid sequence is identical in normal and mutant hemoglobin. This result reveals that the sickle gene mutation involves a very specific change in the template for hemoglobin and strongly hints that all the information required to order hemoglobin (if not all other protein) sequences is present in DNA. Subsequent studies of amino acid sequences in hemoglobin isolated from other forms of anemia completely supported this proposal; here sequence analysis showed that each specific anemia is characterized by a single amino acid replacement at a unique site along the polypeptide chain (Figure 3-11).

DNA Cannot Be the Template That Directly Orders Amino Acids During Protein Synthesis[27, 28, 29]

Although DNA must carry the information for ordering amino acids, it was quite clear that the double helix itself could not be the template. Ruling out a direct role for DNA were experiments showing that protein synthesis occurs at sites where DNA is absent. The location of protein synthesis in all eucaryotic cells appeared to be the cytoplasm, which is separated by the nuclear membrane from the chromosomally located DNA.

A second information-containing molecule therefore had to exist that obtains its genetic specificity from DNA, after which it moves to the cytoplasm to function as the template for protein synthesis. Attention from the start focused on the still functionally obscure second class of nucleic acids, RNA. Sweden's Caspersson and Belgium's J. Brachet had found RNA to reside largely in the cytoplasm; and it was easy to imagine single DNA strands, when not serving as templates for complementary DNA strands, acting as templates for complementary RNA chains.

RNA Is Chemically Very Similar to DNA

Mere inspection of RNA structure shows how it can be exactly synthesized on a DNA template. Chemically, it is very similar to DNA. It, too, is a long, unbranched molecule containing four types of nucleotides linked together by 3'-5' phosphodiester bonds (Figure 3-12). Two differences in its chemical groups distinguish RNA from DNA. The first is a minor modification of the sugar component (see Figure

Figure 3-11

A summary of some established amino acid substitutions in human hemoglobin variants.

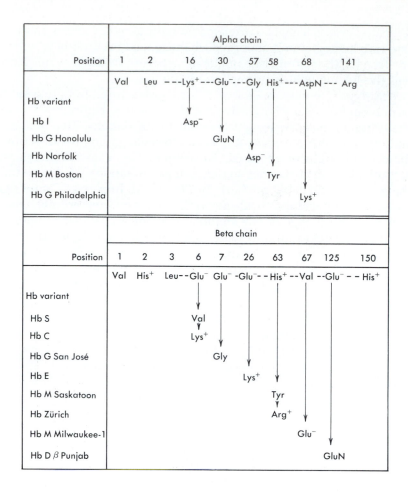

3-2). The sugar of DNA is deoxyribose, whereas RNA contains ribose, identical to deoxyribose except for the presence of an additional OH (hydroxyl) group. The second difference is that RNA contains no thymine, but instead contains the closely related pyrimidine uracil. Despite these differences, however, polyribonucleotides have the potential for forming complementary helices of the DNA type. Neither the additional hydroxyl group nor the absence of the methyl group found in thymine affects RNA's ability to form double-helical structures held together by hydrogen-bonded base pairs.

RNA Is Usually Single-Stranded

RNA molecules do not usually have complementary base ratios (Table 3-4). The amount of adenine does not often equal the amount of uracil, and the amounts of guanine and cytosine also usually differ from each other. This tells us that most RNA does not possess a regular hydrogen-bonded structure but, unlike double-helical DNA, exists as single polyribonucleotide strands. Because of the absence of regular hydrogen bonding, these single-stranded molecules do not have a simple, regular structure like DNA. This structural uncertainty initially caused much pessimism, for it was generally believed that we would have to know the template's structure before we could attack the problem of how the template selects amino acids during protein synthesis. Fortunately, as we will see later, this hunch was wrong.

Figure 3-12
A portion of a polyribonucleotide
(RNA) chain.

5′ end

Adenine

Cytosine

Guanine

Uracil

The Central Dogma[30]

By the fall of 1953, the working hypothesis was adopted that chromosomal DNA functions as the template for RNA molecules, which subsequently move to the cytoplasm, where they determine the arrangement of amino acids within proteins. In 1956, Francis Crick referred to
this pathway for the flow of genetic information as the **central dogma.**

Duplication $\quad$ DNA $\xrightarrow{\text{Transcription}}$ RNA $\xrightarrow{\text{Translation}}$ Protein

Here the arrows indicate the directions proposed for the transfer of
genetic information. The arrow encircling DNA signifies that DNA is
the template for its self-replication. The arrow between DNA and
RNA indicates that all cellular RNA molecules are made on ("transcribed off") DNA templates. Correspondingly, all proteins are determined by ("translated on") RNA templates. Most importantly, the
last two arrows were presented as unidirectional; that is, RNA sequences are never determined by protein templates, nor was DNA
then imagined ever to be made on RNA templates. That proteins

Table 3-4 The Base Composition of RNA from Various Sources

RNA Source	Proportion of the Four Main Bases			
	Adenine	Uracil	Guanine	Cytosine
E. coli	0.24	0.22	0.32	0.22
Proteus vulgaris (a bacterium)	0.26	0.19	0.31	0.24
Euglena (an alga)	0.22	0.21	0.30	0.27
Turnip yellow mosaic virus	0.23	0.22	0.17	0.38
Poliomyelitis virus	0.30	0.25	0.25	0.20
Rat kidney	0.19	0.20	0.30	0.31

never serve as templates for RNA has stood the test of time. However, as related later in this chapter, RNA chains sometimes do act as templates for DNA chains of complementary sequence. But such reversals of the normal flow of information are very rare events compared with the enormous number of RNA molecules made on DNA templates. Thus, the central dogma as originally proclaimed some 30 years ago still remains essentially valid.

The Adaptor Hypothesis of Crick[31]

At first it seemed simplest to believe that the RNA templates for protein synthesis were folded up to create cavities on their outer surfaces specific for the 20 different amino acids. The cavities would be so shaped that only one given amino acid would fit, and in this way RNA would provide the information to order amino acids during protein synthesis. By 1955, however, Crick became disenchanted with this conventional wisdom, arguing that it would never work. In the first place, the specific groups on the four bases of RNA (A, U, G, and C) should mostly interact with water-soluble groups. Yet, the specific side groups of many amino acids (e.g., leucine, valine, and phenylalanine) strongly prefer interceptions with water-insoluble (hydrophobic) groups. In the second place, even if somehow RNA could be folded so as to display some hydrophobic surfaces, there was no way for an RNA template to discriminate accurately between chemically very similar amino acids like glycine and alanine or valine and isoleucine, both of which differ only by the presence of single methyl (CH_3) groups. Crick thus proposed that prior to incorporation into proteins, amino acids are first attached to specific adaptor molecules, which in turn possess unique surfaces that can bind specifically to bases on the RNA templates.

The Test-Tube Synthesis of Proteins[32, 33, 34]

The discovery of how proteins are synthesized required the development of cell-free extracts capable of carrying on the essential synthetic steps. These were first effectively developed beginning in 1953 at the Massachusetts General Hospital in Boston by P. C. Zamecnik and his collaborators. Key to their success were the recently available radioactively tagged amino acids, which they used to mark the trace amounts of newly made proteins, as well as high-quality, easy-to-use, preparative ultracentrifuges for fractionation of their cellular extracts. Early on, the cellular site of protein synthesis was pinpointed to be the

Figure 3-13
Electron micrograph (105,000×) of a portion of a cell in the pancreas of a bat, showing a portion of a mitochondrion (upper right) and large numbers of ribosomes. Some ribosomes exist free; others are attached to a membranous component, the endoplasmic reticulum. (Courtesy of K. R. Porter.)

ribosomes, small RNA-containing particles in the cytoplasm of all cells actively engaged in protein synthesis (Figure 3-13).

Also to emerge from these first experiments was the realization that ATP is required as the energy source for peptide bond formation. Several years later, Zamecnik, by then collaborating with M. B. Hoagland, went on to make the seminal discovery that prior to their incorporation into proteins, amino acids are first attached to what we now call **transfer RNA** (**tRNA**) molecules by a class of enzymes called aminoacyl synthetases. Transfer RNA accounts for some 10 percent of all cellular RNA (Figure 3-14).

This was a totally unexpected discovery except to Crick, who had previously speculated that his proposed "adaptors" might be short RNA chains, since their bases would be able to base-pair with appropriate groups on the RNA molecules that served as the templates for protein synthesis. As we shall relate later in greater detail (Chapter 14), the transfer RNA molecules of Zamecnik and Hoagland are in fact Crick's adaptors. They each contain several adjacent bases (the anticodon) that bind specifically during protein synthesis to successive groups of several bases (codons) along the RNA templates.

The Paradox of the Unspecific-Appearing Ribosomes[35, 36]

About 85 percent of cellular RNA is found in ribosomes, and since its absolute amount is greatly increased in cells engaged in large-scale protein synthesis (e.g., pancreas and liver cells and rapidly growing bacteria), **ribosomal RNA** (**rRNA**) was initially thought to be the template for ordering amino acids. But as soon as the ribosomes of *E. coli* began to be carefully analyzed, several disquieting features emerged. First, all *E. coli* ribosomes, as well as those from all other organisms, are composed of two unequally sized subunits, each containing RNA, that reversibly either stick together or fall apart, depending on the surrounding ion concentration. Second, all the rRNA chains within the small subunits are of similar chain lengths (about 1500 bases in *E. coli*), as are the rRNA chains of the large subunits (about 3000 bases). Third, the base composition of both the small and large rRNA chains is approximately the same (high in G and C) in all known bacteria, plants, and animals despite wide variations in the AT/GC ratios of the respective DNAs. This was not to be expected if the rRNA chains were in fact a large collection of different RNA templates made of a large number of different genes. If that were the case, the basic composition of the rRNA chains should be strongly correlated with that of their respective genomes. Moreover, we might expect the chain lengths of the various RNA templates to be dissimilar and correlated with the sizes of their polypeptide products. Thus, neither the small nor large class of rRNA had the feel of template RNA.

Discovery of Messenger RNA (mRNA)[37–40]

Cells infected with phage T4 provided the ideal system to find the true template. Following infection by this virus, cell synthesis of host *E. coli* RNA stops, and the only RNA synthesized is transcribed off the infecting T4 DNA. Most strikingly, not only does T4 RNA have a base composition very similar to T4 DNA, but it does not become assembled into a ribosomal component by binding to the ribosomal proteins that normally associate with rRNA to form ribosomes.

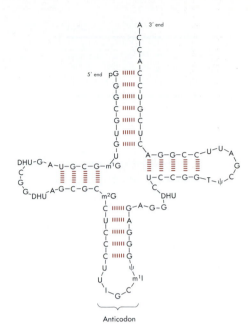

Figure 3-14
Yeast alanine tRNA structure, as determined by R. W. Holley and his associates. The anticodon in this tRNA recognizes the codon for alanine in the mRNA. Several modified nucleosides exist in the structure: ψ = pseudouridine, T = ribothymidine, DHU = 5,6-dihydrouridine, I = inosine, m^1G = 1-methylguanosine, m^1I = 1-methylinosine, and m^2G = N,N-dimethylguanosine.

Figure 3-15

Transcription and translation. The nucleotides of mRNA are assembled to form a complementary copy of one strand of DNA. Each group of three is a codon that is complementary to a group of three nucleotides in the anticodon region of a specific tRNA molecule. When base pairing occurs, an amino acid carried at the other end of the tRNA molecule is added to the growing protein chain.

Instead, T4 RNA, after first attaching to previously existing ribosomes, moves across their surface to bring its bases into positions where they can bind to the appropriate tRNA–amino acid precursors for protein synthesis (Figure 3-15). In so acting, T4 RNA orders the amino acids and is thus the long-sought-for RNA template for protein synthesis. Because it carries the information from DNA to the ribosomal sites of protein synthesis, it is called **messenger RNA (mRNA)**. The observation of T4 RNA binding to *E. coli* ribosomes, first made in the spring of 1960, was soon followed with evidence for a separate messenger class of RNA within uninfected *E. coli* cells, thereby definitively ruling out a template role for any rRNA. Instead, in ways that we shall discuss more extensively in Chapter 14, the rRNA components of ribosomes, together with some 50 different ribosomal proteins that bind to them, serve as the factories for protein synthesis, functioning to bring together the tRNA–amino acid precursors into positions where they can read off the information provided by the messenger RNA templates.

Only some 4 percent of total cellular RNA is mRNA, whose sizes show the expected large variations in length, depending on the proteins for which they code. Hence, it is easy to understand why mRNA was first overlooked. Because only a small segment of mRNA is attached at a given moment to a ribosome, which is moving across it, a single mRNA molecule can simultaneously be read by several ribosomes. Most ribosomes are found as parts of **polyribosomes,** ranging in size from 6 to more than 50 members (Figure 3-16).

Messenger RNA molecules are always translated in the same direction, starting near the 5′ end and finishing near the 3′ end. It is thus the 5′ end of an mRNA molecule that first attaches to a ribosome.

Enzymatic Synthesis of RNA Upon DNA Templates[41, 42, 43]

As messenger RNA was being discovered, the first of the enzymes that transcribe RNA off DNA templates was being independently isolated in the labs of American biochemists J. Hurwitz, A. Stevens, and

Figure 3-16

Diagram of a polyribosome. Each ribosome attaches at a start signal at the 5′ end of an mRNA chain and synthesizes a polypeptide as it proceeds along the molecule. Several ribosomes may be attached to one mRNA molecule at one time; the entire assembly is called a polyribosome.

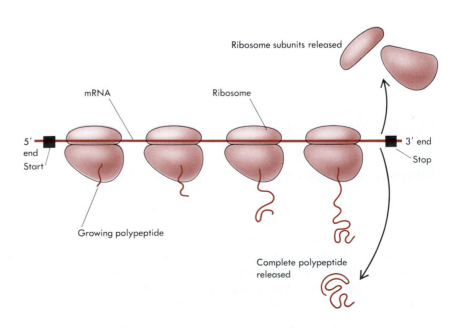

Figure 3-17

Enzymatic synthesis of RNA upon a DNA template, catalyzed by RNA polymerase. As the RNA is synthesized, pyrophosphate is split off each incoming nucleotide; the enzyme pyrophosphatase catalyzes the breakdown of pyrophosphate to phosphate ions.

Ribonucleotide triphosphates

RNA

S. B. Weiss. Called **RNA polymerases,** these enzymes function only in the presence of DNA, which serves as the template upon which single-stranded RNA chains are made, and use the nucleotides ATP, GTP, CTP, and UTP as precursors (Figure 3-17). In bacteria, the same enzyme makes each of the major RNA classes (ribosomal, transfer, and messenger), using appropriate segments of chromosomal DNA as their templates. Direct evidence that DNA lines up the correct ribonucleotide precursors came from seeing how the RNA base composition varied with the addition of DNA molecules of different AT/GC ratios. In every enzymatic synthesis, the RNA AU/GC ratio was roughly similar to the DNA AT/GC ratio (Table 3-5).

Table 3-5 Comparison of the Base Composition of Enzymatically Synthesized RNAs with the Base Composition of Their Double-Helical DNA Templates

| Source of DNA Template | Composition of the RNA Bases | | | | $\dfrac{A + U}{G + C}$ Observed | $\dfrac{A + T}{G + C}$ in DNA |
	Adenine	Uracil	Guanine	Cytosine		
T2	0.31	0.34	0.18	0.17	1.86	1.84
Calf thymus	0.31	0.29	0.19	0.21	1.50	1.35
E. coli	0.24	0.24	0.26	0.26	0.92	0.97
Micrococcus lysodeikticus (a bacterium)	0.17	0.16	0.33	0.34	0.49	0.39

During transcription, only one of the two strands of DNA becomes translated into RNA. This makes sense, because the messages carried by the two strands, being complementary but not identical, might be expected to code for completely different polypeptides. The synthesis of RNA always proceeds in a fixed direction, beginning at the 5' end and concluding with the 3'-ended nucleotide (see Figure 3-16). Transcription and translation thus occur in the same direction.

By this time, there was firm evidence for the postulated movement of RNA from the DNA-containing nucleus to the ribosome-containing cytoplasm. By briefly exposing cells to radioactively labeled RNA precursors, the site of RNA synthesis was shown to be the nucleus. Within an hour, most of this RNA had left the nucleus to be observed in the cytoplasm (Figure 3-18).

Figure 3-18
Demonstration that RNA is synthesized
in the nucleus and moves to the cytoplasm. (a) Autoradiograph of a cell (*Tetrahymena*) exposed to radioactive cytidine for 15 minutes. Superimposed on
a photograph of a thin section of the
cell is a photograph of an exposed silver emulsion. Each dark spot represents the path of an electron emitted
from a ^{3}H (tritium) atom that has been
incorporated into RNA. Almost all the
newly made RNA is found within the
nucleus. (b) Autoradiograph of a similar
cell exposed to radioactive cytidine for
12 minutes and then allowed to grow
for 88 minutes in the presence of nonradioactive cytidine. Practically all the
label incorporated into RNA in the first
12 minutes has left the nucleus and
moved into the cytoplasm. [Courtesy of
D. M. Prescott, University of Colorado
Medical School; reproduced from *Progr.
Nucleic Acid Res.* III, 35 (1964), with permission.]

Repressors and Activators of Gene Expression[44-49]

How cells control the exact amounts of their proteins became open for
detailed analysis as the general pathway for the flow of genetic information became clear. Particularly intriguing were those proteins
whose amounts changed greatly, depending on their need. Long
serving as the key model system was the bacterial enzyme
β-galactosidase, which breaks the sugar lactose down into the smaller
sugars glucose and galactose. In the absence of lactose, the bacterium
E. coli makes an exceedingly small number of molecules of this enzyme, while in its presence, each *E. coli* cell contains thousands of
such molecules. Leading this research were the French microbiolo-
gists J. Monod and F. Jacob, working at the Institut Pasteur in Paris.
In the late 1950s, they observed that the functioning of the gene that
codes for the order of amino acids in β-galactosidase is itself controlled by a separate regulatory gene. Monod and Jacob postulated its
gene product to be a **repressor** that in the absence of lactose specifically turns off the functioning of the structural gene. Only very small
numbers of this repressor are normally present in an *E. coli* cell, and it
was not until 1966 that it was first isolated at Harvard by B. Mueller-
Hill and W. Gilbert. Soon afterward, they demonstrated that it bound
just before the beginning of the β-galactosidase structural gene and
prevented its transcription into RNA.

Meanwhile, M. Ptashne and N. Hopkins, also at Harvard, showed
that a key regulatory gene of the bacterial virus λ also coded for a
protein repressor that acted by binding to specific DNA segments and
preventing their functioning. Not all DNA-binding regulatory proteins act negatively. Some turn up, or **activate,** DNA expression by
binding to other DNA sequences. After binding to cyclic AMP, the
catabolic activator protein (CAP) can stimulate the functioning of several sugar-utilizing genes (e.g., arabinose, lactose, and galactose) by
binding to appropriate segments of DNA.

Establishing the Genetic Code[50-55]

Given the existence of 20 amino acids but only four bases, groups of
several nucleotides must somehow specify a given amino acid.
Groups of two, however, would specify only 16 (4 × 4) amino acids.
So from 1954, the start of serious thinking about what the genetic
code might be like, most attention was given to how triplets (groups
of three) might work, even though they obviously would provide
more permutations (4 × 4 × 4) than needed if each amino acid was
specified by only a single triplet. The assumption of colinearity was
then very important. It held that successive groups of nucleotides
along a DNA chain code for successive amino acids along a given
polypeptide chain. That colinearity does in fact exist was shown by
elegant mutational analysis on bacterial proteins, carried out in the
early 1960s at Stanford University by C. Yanofsky and at Cambridge,
England, by A. Brenner. Equally important were the genetic analyses
by Brenner and Crick, which in 1961 first established that groups of
three nucleotides are used to specify individual amino acids.

But which specific groups of three bases (codons) determine which
specific amino acids could only be learned by biochemical analysis.
Here the major breakthrough came when M. W. Nirenberg and J. H.
Matthaei, then both working together at the National Institute of

Table 3-6 The Genetic Code

First Position →

		Second Position				
		U	C	A	G	
U	UUU ⎤ Phe UUC ⎦ UUA ⎤ Leu UUG ⎦	UCU ⎤ UCC UCA Ser UCG ⎦	UAU ⎤ Tyr UAC ⎦ UAA Stop UAG Stop	UGU ⎤ Cys UGC ⎦ UGA Stop UGG Trp	U C A G	
C	CUU ⎤ CUC CUA Leu CUG ⎦	CCU ⎤ CCC CCA Pro CCG ⎦	CAU ⎤ His CAC ⎦ CAA ⎤ Gln CAG ⎦	CGU ⎤ CGC CGA Arg CGG ⎦	U C A G	
A	AUU ⎤ AUC lle AUA ⎦ AUG Met	ACU ⎤ ACC ACA Thr ACG ⎦	AAU ⎤ Asn AAC ⎦ AAA ⎤ Lys AAG ⎦	AGU ⎤ Ser AGC ⎦ AGA ⎤ Arg AGG ⎦	U C A G	
G	GUU ⎤ GUC GUA Val GUG ⎦	GCU ⎤ GCC GCA Ala GCG ⎦	GAU ⎤ Asp GAC ⎦ GAA ⎤ Glu GAG ⎦	GGU ⎤ GGC GGA Gly GGG ⎦	U C A G	

Third Position →

Health in Bethesda, Maryland, observed, also in 1961, that the addition of the synthetic polynucleotide poly U (UUUUU . . .) to a cell-free system capable of making proteins leads to the synthesis of polypeptide chains containing only the amino acid phenylalanine. The nucleotide groups UUU thus must specify phenylalanine. Use of increasingly more complex, defined polynucleotides as synthetic messenger RNAs rapidly led to the identification of more and more codons. Particularly important in completing the code was the use of polynucleotides like AGUAGU, put together by the Indian organic chemist H. G. Khorana, then working in Madison, Wisconsin. Completion of the code in 1966 revealed that 61 out of the 64 possible permuted groups corresponded to amino acids, with most amino acids being coded by more than one nucleotide triplet (Table 3-6).

Start and Stop Signals Are Also Encoded Within DNA[56–59]

Initially, it was guessed that translation of an mRNA molecule would commence at one end and finish when the entire mRNA message had been read into amino acid sequences. But, in fact, translation both starts and stops at internal positions. Thus, signals must be present within DNA (and its mRNA products) to initiate and terminate translation. First to be worked out were the stop signals. Three separate codons (UAA, UAG, and UGA), first known as **nonsense codons,** do not correspond to any amino acids but instead serve as chain-termi-

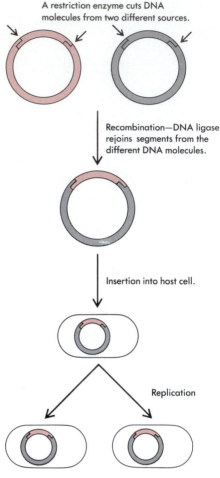

A restriction enzyme cuts DNA molecules from two different sources.

Recombination—DNA ligase rejoins segments from the different DNA molecules.

Insertion into host cell.

Replication

Figure 3-19
The joining of DNA from two different sources via recombinant DNA techniques. DNA ligase joins DNA segments previously cut with restriction enzymes.

nating (stop) signals. More complicated is the way translational start signals are encoded. The amino acid methionine starts all polypeptide chains, but the triplet (AUG) that codes for these initiating methionines also codes for methionine residues that have internal locations. The AUG codons at which polypeptide chains start are preceded by specific purine-rich blocks of nucleotides that serve to attach mRNA to ribosomes (see Chapters 14 and 15 for details).

The Extreme Utility of Enzymes That Cut DNA into Well-Defined Fragments[60, 61, 62]

Signals to start and stop the synthesis of RNA chains also are encoded within DNA. These are more complicated than the signals involved in translational starts and stops and could only be revealed when techniques became available for DNA sequencing. For all practical purposes, we could not observe it in action, as even the smallest of natural DNA molecules were much too big to be individually looked at by the methods of chemistry. This seemingly hopeless picture fortunately did not last long. By 1970, Hamilton Smith at Johns Hopkins Medical School had isolated the first enzyme that cuts DNA at a very specific nucleotide sequence. Over the next few years, a number of different such enzymes were isolated, each from a distinct type of bacterium, and there are now over a hundred different specificities.

In general, each such cutting enzyme (**restriction enzyme**) recognizes a unique sequence of four or six nucleotides (Table 3-7) and so cuts a given DNA molecule about once every several hundred or thousand bases. The availability of these enzymes quickly changed the course of DNA research, enabling the biochemist for the first time to experiment on well-defined DNA fragments of the size that, for example, has permitted the development of high-powered methodologies for sequencing DNA molecules. These cutting tools were first most effectively used in the isolation of distinct DNA fragments from viral DNAs ranging from several thousand to 50,000 nucleotides in length. Such relatively small DNAs generally yield fragments of all different lengths, which can be easily separated from each other by newly developed gel separation techniques.

Recombinant DNA[63–66]

The most important early consequence of restriction enzyme availability was the impetus it gave to the development of what are now called **recombinant DNA** procedures. Prior research on the enzymology of DNA replication had resulted in the discovery of the enzyme **DNA ligase,** which joins together (ligates) DNA chains. Using this ligase, the "restriction" DNA fragments generated by restriction enzyme cutting of a preexisting DNA molecule can be randomly linked to each other.

The possibility was thus conceived of joining together pieces arising from different parental molecules, and the first such attempts were performed in the laboratory of P. Berg at Stanford University in 1972. The first effective insertion of recombinant DNA into cells occurred the following year, when S. Cohen of Stanford and H. Boyer at San Francisco used a tiny bacterial chromosome (plasmid) as one of the DNA partners to be joined together (Figure 3-19). Following entry into a suitable bacterium, the tiny recombinant chromosome multi-

Table 3-7 Recognition Sequences of Several Restriction Enzymes*

Microbial Origin	Enzyme	Recognition Site
Escherichia coli KY13	*Eco* RI	Cut bond ↓ $^{5'}$GAA | TTC CTT | AAG ↑ (Axis of symmetry)
Hemophilus influenzae Rd	*Hin* dII	↓ $^{5'}$GT Py | Pu AC CA Pu | Py TG ↑
	Hin dIII	↓ $^{5'}$AAG | CTT TTC | GAA ↑
Hemophilus parainfluenzae	*Hpa* I	↓ $^{5'}$GTT | AAC CAA | TTG ↑
	Hpa II	↓ $^{5'}$CC | GG GG | CC ↑
Hemophilus aegyptius	*Hae* III	↓ $^{5'}$GG | CC CC | GG ↑

*Note that each sequence has twofold symmetry.

plied, and a clone of the recombinant DNA piece was generated as its host bacterium itself multiplied.

It is impossible to overstate the effect of recombinant DNA procedures, not only in their promotion of ongoing DNA research itself but also through their generation of new fields of research (such as cellular immunology at the molecular level).

Extraordinary New Techniques Aid in Sequencing DNA[67, 68, 69]

Only a decade ago, there was no simple way to determine the sequences of short DNA chains containing only a few nucleotides. Today, the sequences of restriction-enzyme-generated DNA fragments several hundred nucleotides in length can be routinely accomplished in two to three days. In fact, the complete sequences of the bacterial viruses T7 (39,936 base pairs) and λ (48,513 base pairs) are already worked out. And even longer viral genomes (e.g., herpesvirus, about 1.5×10^5 base pairs) can be established now within a one- to two-year interval. Two powerful new techniques for DNA sequencing, one developed at Harvard by A. M. Maxam and Gilbert, the other worked out in Cambridge, England, by Sanger, underlie these spectacular advances (see Chapter 9 for details). Both these procedures depended on the availability of restriction enzymes. In obtaining the necessary amounts of needed DNA fragments, recombinant DNA procedures are now routinely employed.

Many DNA sequences are already known that code from proteins whose amino acid sequences had previously been worked out. They not only completely confirm the codon assignments worked out using synthetic polynucleotides but also afford us the opportunity to determine the actual codons used when a given amino acid is specified by more than one codon. By now, DNA sequencing is a much more routine task than protein sequencing, and the sequences of many proteins are in fact first being known through the sequences of their corresponding genes.

RNA Can Also Function as a Genetic Molecule[70–73]

In addition to its role as the template for protein synthesis, RNA has the capacity to act, like DNA, as a genetic molecule. That RNA can store genetic information was first hinted in the late 1930s, when several plant viruses were found to contain RNA but no DNA. Proof that this RNA component was indeed the viral genetic material did not emerge, however, until 1956, when Alfred Gierer, working in Tübingen, Germany, showed that highly purified RNA from tobacco mosaic virus was infectious in the absence of any viral protein (Figure 3-20). Since then, a variety of other viruses (including influenza and polio) have been determined also to have RNA as their genetic molecule.

Because the structure of RNA is very similar to that of DNA, it seemed logical to assume that RNA replication followed the same principles established for DNA. Proof that RNA can be so replicated came first from experiments with a class of small viruses that multiply in bacteria and which contain single-stranded RNA molecules as their genetic component. A key breakthrough came from the 1965 isolation at the University of Illinois by I. Haruna and S. Spiegelman of a virally coded *replicase*, an enzyme that catalyzes the formation of new infectious viral RNA molecules in the presence of the appropriate precursors. How these new molecules are made was at first debatable, but by 1968, C. Weissmann and J. August, independently working in different New York laboratories, conclusively demonstrated the involvement of a double-stranded RNA intermediate. The self-replication of RNA as well as DNA thus involves base pairing between chains of complementary sequences.

Reverse Transcription of RNA Yields DNA[74, 75]

All RNA viruses do not replicate using double-stranded RNA intermediates. The infecting RNA of **retroviruses** (earlier known as RNA tumor viruses; Chapter 24) serves as a template for the synthesis of a single-stranded DNA complement that in turn functions as a template for a complementary DNA chain. The resulting double-stranded DNA, generally after insertion in a host cell chromosome, then acts as a template for the synthesis of an RNA chain identical to that which

Figure 3-20
Proof that RNA is the genetic component of TMV. The rod-shaped TMV particles can easily be separated into their protein and RNA components, which can then be separately tested for their ability to initiate virus infection. Only the RNA molecules have this ability. The virus particles produced by infecting with pure RNA are identical to those resulting from infection with intact virus.

initiated the viral life cycle (Figure 3-21). The flow of genetic information back from RNA to DNA does not present any conceptual problems, since RNA-DNA hybrid double helices also exist, though only fleetingly, when RNA is transcribed off single-stranded DNA templates. The direction of information flow may be determined by the particular enzyme available. Catalyzing the making of DNA on RNA during retroviral infection is the viral-specific enzyme **reverse transcriptase.** It was its discovery in 1970, by S. Mizutani and H. M. Temin at Madison and by D. Baltimore at MIT, that first led to general acceptance of the idea that information can flow back from RNA to DNA. In the absence of retrovirus infection, very little, if any, reverse transcriptase activity is present in normal cells; thus, under normal conditions, effectively very little DNA is made on RNA templates.

Interrupted Genes Within Eucaryotic Chromosomes[76, 77]

Whereas sequencing of the first bacterial DNA fragment confirmed previous structural speculations about bacterial genes, a completely different picture emerged when the first eucaryotic genes were cloned and sequenced. Suprisingly, it was found that the nucleotide sequences that specify amino acids for hemoglobin and ovalbumin are frequently interrupted by blocks of nucleotides that have no coding information. Those nucleotide segments that code for amino acids are called **exons,** while those segments that are not translated into amino acids are called **introns.** Most genes coding for mRNA products in higher eucaryotic cells are split. In the cell, they are transcribed into very large precursor RNA molecules, which subsequently have their introns removed by specific enzymes within the nucleus. In the process, the remaining exons are spliced together to form smaller RNAs, which then move into the cytoplasm where they function as mRNA. Later it was found that many tRNAs and some tRNA genes are also split, with their introns also being removed by RNA splicing events. Why so many genes of higher cells are split while their bacterial equivalents are not is not yet understood. In any case, an inevitable consequence is that eucaryotic genes are frequently very large, with many such genes having much more of their DNA existing as introns than as exons.

Unlimited Vistas for Genetics at the Molecular Level

Until 1953, genes were perceived to be at the heart of the existence of life and yet very mysterious. Once the double helix was revealed, however, biologists had the reference point for further research. In particular, biochemists could now effectively focus their highly developed skills for working out enzymatically catalyzed chemical reactions that underlie genetic phenomena. The discovery of the genetic code by 1966 was an enormous scientific achievement, and the rapidity with which molecular genetics has advanced over the past three decades has surprised and pleased virtually everyone. Now, with powerful new recombinant DNA techniques and DNA-sequencing procedures in hand, we stand in marvel at the genetic knowledge revealed in just the last decade, and we predict with awe where future DNA research will take us.

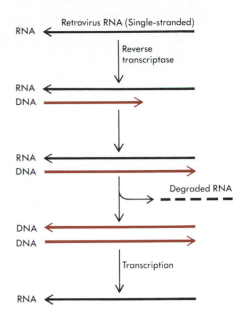

Figure 3-21
Reverse transcription of RNA to yield DNA. The DNA-RNA hybrid strand is highly unstable and exists only fleetingly.

The long-mysterious genetic signals that govern eucaryotic cells and that are now being discovered will open up new, direct avenues into the nature of diseases like cancer, as well as begin to reveal the essence of cell differentiation. Moreover, we can soon expect to uncover those molecular tricks that underlie the organization of nerve cells into the intricate cellular networks that in turn underlie perception, memory, and thinking. The gene will thus remain the key to future efforts to probe the nature of life.

Summary

Specific classes of molecules within cells carry the genetic information that determines the order of the 20 different amino acids in proteins. Such genetic molecules function by presenting specific surfaces, or templates, to which the correct amino acids (activated derivatives) are sequentially attracted and ordered prior to their polymerization. A specific template exists for each specific protein. When a cell grows and divides, the number of protein templates must also double. Each template must thus itself be formed on the surface of another template.

The templates that carry the genetic information of cellular chromosomes are always DNA molecules. All DNA molecules are very large in comparison to all other cellular molecules, being very long linear polymers built from large numbers of 4 different nucleotide building blocks connected by phosphodiester linkages (polynucleotide molecules). Nucleotide sequence along a given chain is very irregular, with a specific DNA sequence conveying each genetic message. Two of the nucleotides contain purine bases (adenine and guanine), while the other two nucleotides contain pyrimidine bases (thymine and cytosine). In virtually all DNA molecules, two polynucleotide chains are twisted around each other to form a regular double helix. The two chains within the double helix are held together by hydrogen bonds between pairs of bases. Adenine is always joined to thymine, and guanine is always bonded to cytosine. The existence of the base pairs means that the sequence of nucleotides along the two chains are not identical, but complementary. If the sequence of one chain is known, then the sequence of its partner is automatically known. The finding of this relationship suggested a mechanism for the replication of DNA in which each strand serves as a template for its complement. Proof for this hypothesis came from (a) the observation that the two strands of each double helix do separate during each round of DNA replication in cells and (b) the discovery of a specific enzyme (DNA polymerase I) that uses single-stranded segments of DNA as templates for the polymerization of the immediate precursors of DNA (the deoxynucleotide triphosphates dATP, dGTP, dTTP, and dCTP) into chains of complementary sequence.

The sequence of base pairs does not serve as the direct template for protein synthesis. Instead, it is used as a template to make RNA the other form of nucleic acid, which is primarily found in the cytoplasm. In turn, many of these RNA products function as templates for protein synthesis. RNA is chemically very similar to DNA, also being composed of four different nucleotide building blocks, two of which contain the purines adenine and guanine and two of which contain the pyrimidines uracil and cytosine. The synthesis of an RNA chain upon a DNA chain utilizes the same base-pairing rules involved in DNA self-duplication (except that uracil base-pairs like thymine). And as with the synthesis of DNA, a specific enzyme, in this case an RNA polymerase, is needed to catalyze the formation of covalent bonds between the immediate precursors, the ribonucleotide triphosphates ATP, GTP, UTP, and CTP.

During the synthesis of RNA upon DNA, a process called transcription, only one of the two strands of a given double helix functions as a template. The RNA products of transcription are thus single-stranded. Prior to amino acid polymerization into proteins, the 20 different amino acids are first transformed into energy-rich precursors in which a given amino acid is attached to a small transfer RNA (tRNA) molecule. Amino acids do not have specific regions that bind to RNA templates. Instead, they are recognized by the specific enzymes (the aminoacyl synthetases) that attach them to specific tRNA molecules. Each tRNA molecule contains a group of three nucleotides (the anticodon) that uses base pairing to seek out a group of three adjacent nucleotides (a codon) on a messenger RNA molecule, the RNA class that carries the ordering information. Successful codon-anticodon interactions only occur on the surfaces of ribosomes, the factories for protein synthesis. Ribosomes themselves consist of large amounts of ribosomal RNA (rRNA) as well as many different protein components. During protein synthesis, mRNA molecules move across the active surfaces of ribosomes, bringing successive codons into position to determine successive amino acids along polypeptide chains (the process of translation). Out of the 64 ($4 \times 4 \times 4$) potential codons, 61 are used to specify amino acids, while 3 are used to provide chain-terminating signals. Many amino acids are thus determined by more than one codon. The amino acids specified by the various codons (the genetic code) are essentially the same for all forms of life.

Many factors determine the rate at which specific segments of DNA function. Much control is effected at the level of transcription by specific DNA-binding proteins, some of which repress RNA synthesis (repressors, or negative control proteins), while others are necessary for its initiation (activators, or positive control proteins).

Over one hundred different enzymes are now known that recognize and cut DNA at very specific nucleotide sequences. Almost all such "restriction enzymes" recognize unique sequences of four to six nucleotides. Their availability allowed the isolation of DNA fragments of well-defined length and hastened the development of high-powered methods for precisely sequencing DNA. The availability of restriction enzymes also quickly led to the generation of "recombinant DNA" molecules and the means to clone specific DNA fragments (i.e., isolate specific genes). Sequencing of genes led to the realization that many eucaryotic genes are split. The nucleotide segments (exons) that code for amino acids are often interrupted by long DNA segments (introns) that do not code for amino acids. Both exons and introns are transcribed into long RNA precursors from which the introns are spliced out to yield functional RNA products.

RNA molecules are the genetic components of many small viruses as well as the templates for their viral-specific proteins. Replication of many RNA chromosomes involves the formation of complementary RNA chains. Other RNA viruses (retroviruses) replicate via complementary DNA intermediates using the enzyme reverse transcriptase. The initially single-stranded DNA products are then converted into double-stranded DNA molecules; then this DNA, generally after insertion into a host chromosome, acts as a template for the synthesis of RNA chains identical to those that started the viral life cycle.

Bibliography

General References

Cairns, J., G. S. Stent, and J. D. Watson, eds. 1966. *Phage and the Origins of Molecular Biology.* Cold Spring Harbor, N.Y.: Cold Spring Harbor Laboratory.

Judson, H. F. 1979. *The Eighth Day of Creation.* New York: Simon & Schuster.

Kornberg, A. 1980. *DNA Replication.* San Francisco: Freeman.

McCarty, M. 1985. *The Transforming Principle: Discovering that Genes Are Made of DNA.* New York: Norton.

Olby, R. 1975. *The Path to the Double Helix.* Seattle: University of Washington Press.

Portugal, F. H., and J. S. Cohen. 1980. *A Century of DNA: A History of the Discovery of the Structure and Function of the Genetic Substance.* Cambridge, Mass.: MIT Press.

Stent, G. S., and R. Calendar. 1978. *Molecular Genetics: An Introductory Narrative.* 2nd ed. San Francisco: Freeman.

Watson, J. D. 1968. *The Double Helix.* New York: Atheneum (text and paperback editions); New York: New American Library, 1969 (paperback).

Watson, J. D. 1980. *The Double Helix: A Norton Critical Edition.* Ed. G. S. Stent. New York: Norton.

Cited References

1. Friedrich-Freksa, H. 1940. "Bei der Chromosomen Konjugation Wirksame Krafte und ihre Bedeutung für die Identische Verdopplung von Nucleoproteinen." *Naturwissenshaffen* 28:376–379.
2. Pauling, L., and M. Delbruck. 1940. "The Nature of the Intermolecular Forces Operative in Biological Processes." *Science* 92:77–79.
3. Muller, H. J. 1947. "The Gene." *Proc. Roy. Soc. London* (B) 134:1–37. A lecture given in 1945 in which a distinguished geneticist traces the history of the gene concept and speculates about how it might function as a template.
4. Dounce, A. L. 1952. "Duplicating Mechanisms for Peptide Chain and Nucleic Acid Synthesis." *Enzymologia* 15:251–258.
5. Loftfield, R. B. 1963. "The Frequency of Errors in Protein Biosynthesis." *Biochem. J.* 89:82–91.
6. Signer, R., T. Caspersson, and E. Hammarsten. 1938. "Molecular Shape and Size of Thymonucleic Acid." *Nature* 141:122.
7. Chargaff, E., and J. N. Davidson, eds. 1955. *The Nucleic Acids.* Vols 1 and 2. New York: Academic Press.
8. Griffith, F. 1928. "The Significance of Pneumonococcal Types." *J. Hyg.* 27:113–159.
9. Avery, O. T., C. M. MacLeod, and M. McCarty. 1944. "Studies on the Chemical Nature of the Substance Inducing Transformation of Pneumonococcus Types." *J. Exp. Med.* 79:137–158.
10. Mirsky, A. E., and H. Ris. 1949. "Variable and Constant Components of Chromosomes." *Nature* 163:666–667.
11. Boivin, A., R. Vendrely, and C. Vendrely. 1948. *Compt. Rend. Acad. Sci.* 226:1061.
12. Hershey, A. D., and M. Chase. 1952. "Independent Function of Viral Protein and Nucleic Acid on Growth of Bacteriophage." *J. Gen. Physiol.* 36:39–56.
13. Chargaff, E., E. Vischer, R. Doniger, C. Green, and F. Misani. 1949. "The Composition of the Desoxypentose Nucleic Acids of Thymus and Spleen." *J. Biol. Chem.* 177:405–416.
14. Chargaff, E. 1951. "Structure and Function of Nucleic Acids as Cell Constituents." *Fed. Proc.* 10:654–659.
15. Astbury, W. T. 1951. "X-Ray Studies of Nucleic Acids." *Symp. Soc. Exp. Biol. I. Nucleic Acid* 1:66–76.
16. Brown, D. M., and A. R. Todd. 1952. "Nucleotides Part X: Some Observations on Structure and Chemical Behavior of the Nucleic Acids." *J. Chem. Soc.* Pt. 1: 52–58.
17. Dekker, C. A., A. M. Michaelson, and A. R. Todd. 1953. "Nucleotides Part XIX. Pyrimidine Deoxyribonucleoside Diphosphates." *J. Chem. Soc.* Pt. 1: 947–951.
18. Cochran, W., F. H. C. Crick, and V. Vand. 1952. "The Structure of Synthetic Polypeptides I. The Transform of Atoms on a Helix." *Acta Cryst.* 5:581–586.
19. Franklin, R. E., and R. G. Gosling. 1953. "Molecular Configuration in Sodium Thymonuclease." *Nature* 171:740–741.
20. Watson, J. D., and F. H. C. Crick. 1953. "Molecular Structure of Nucleic Acids: A Structure for Deoxyribose Nucleic Acid." *Nature* 171:737–738.
21. Watson, J. D., and F. H. C. Crick. 1953. "Genetical Implications of the Structure of Deoxyribonucleic Acid." *Nature* 171:964–967.
22. Wilkins, M. H. F., A. R. Stokes, and H. R. Wilson. 1953. "Molecular Structure of Deoxypentose Nucleic Acid." *Nature* 171:738–740.
23. Crick, F. H. C., and J. D. Watson. 1954. "The Complementary Structure of Deoxyribonucleic Acid." *Proc. Roy. Soc.* (A) 223:80–96.
24. Kornberg, A. 1960. "Biological Synthesis of Deoxyribonucleic Acid." *Science* 131:1503–1508.
25. Meselson, M., and F. W. Stahl. 1958. "The Replication of DNA in *Escherichia coli.*" *Proc. Nat. Acad. Sci.* 44:671–682.
26. Ingram, V. M. 1957. "Gene Mutations in Human Hemoglobin: The Chemical Difference Between Normal and Sickle Cell Hemoglobin." *Nature* 180:326–328.
27. Brachet, J. 1947. *Embryologia Chimique.* Paris, France: Masson.
28. Brachet, J. 1957. *Biochemical Embryology.* New York: Academic Press.
29. Caspersson, T. 1950. *Cell Growth and Cell Function.* New York: Norton.
30. Crick, F. H. C. 1958. "On Protein Synthesis." *Symp. Soc. Exp. Biol.* 12:548–555.
31. Crick, F. H. C. 1955. "On Degenerate Template and the Adaptor Hypothesis." A note for the RNA Tie Club, unpublished. Mentioned in Crick's 1957 discussion, pp. 25–26, in "The Structure of Nucleic Acids and Their Role in Protein Synthesis." *Biochem. Soc. Symp.* no. 14, Cambridge University Press.
32. Hoagland, M. B., E. B. Keller, and P. C. Zamecnik. 1956. "Enzymatic Carboxyl Activation of Amino Acids." *J. Biol. Chem.* 218:345–358.
33. Hoagland, M. B., M. L. Stephenson, J. F. Scott, L. I. Hecht, and P. C. Zamecnik. 1958. "A Soluble Ribonucleic Acid Intermediate in Protein Synthesis." *J. Biol. Chem.* 231:241–257.
34. Holley, R. W., J. Apgar, G. A. Everett, J. T. Madison, M. Mar-

quisse, S. H. Merrill, J. R. Penswick, and A. Zamir. 1965. "Structure of a Ribonucleic Acid." *Science* 147:1462–1465.

35. Tisséires, A., J. D. Watson, D. Schlesinger, and B. R. Hollingworth. 1959. "Ribonucleoprotein Particles from *Escherichia coli.*" *J. Mol. Biol.* 1:221.

36. Watson, J. D. 1963. "Involvement of RNA in Synthesis of Proteins." *Science* 140:17–26.

37. Volkin, E., and L. Astrachan. 1956. "Phosphorus Incorporation in *E. coli* Ribonucleic Acid After Infection with Bacteriophage T2." *Virology* 2:146–161.

38. Brenner, S., F. Jacob, and M. Meselson. 1961. "An Unstable Intermediate Carrying Information from Genes to Ribosomes for Protein Synthesis." *Nature* 190:576–581.

39. Gros, F., H. Hiatt, W. Gilbert, C. G. Kurland, R. W. Risebrough, and J. D. Watson. 1961. "Unstable Ribonucleic Acid Revealed by Pulse Labelling of *Escherichia coli.*" *Nature* 190:581–585.

40. Hall, B. D., and S. Spiegelman. 1961. "Sequence Complementarity of T2 DNA and T2-Specific RNA." *Proc. Nat. Acad. Sci.* 47:137–146.

41. Hurwitz, J., A. Bresler, and R. Diringer. 1960. "The Enzymatic Incorporation of Ribonucleotides into Polyribonucleotides and the Effect of DNA." *Biochem. Biophys. Res. Comm.* 3:15–19.

42. Weiss, S. B. 1960. "Enzymatic incorporation of ribonucleotide triphosphates into the interpolynucleotide linkages of ribonucleic acid." *Proc. Nat. Acad. Sci.* 46:1020–1030.

43. Stevens, A. 1960. "Incorporation of the Adenine Ribonucleotide into RNA by Cell Fractions from *E. coli* B." *Biochem. Biophys. Res. Comm.* 3:92–96.

44. Jacob, F., and J. Monod. 1961. "Genetic Regulatory Mechanisms in the Synthesis of Proteins." *J. Mol. Biol.* 3:318–356.

45. Gilbert, W., and B. Muller-Hill. 1966. "Isolation of the Lac Repressor." *Proc. Nat. Acad. Sci.* 56:1891–1898.

46. Ptashne, M. 1967. "Isolation of the λ Phage Repressor." *Proc. Nat. Acad. Sci.* 57:306–313.

47. Ptashne, M., and N. Hopkins. 1968. "The Operators Controlled by the Phage Repressor." *Proc. Nat. Acad. Sci.* 60:1282–1287.

48. Zubay, G., D. Schwartz, and J. Beckwith. 1970. "Mechanism of Activation of Catabolite-Sensitive Genes: A Positive Control System." *Proc. Nat. Acad. Sci.* 66:104–110.

49. Beckwith, J. R., and D. Zipser, eds. 1970. *The Lactose Operon.* Cold Spring Harbor, N.Y.: Cold Spring Harbor Laboratory.

50. Nirenberg, M. W., and J. H. Matthaei. 1961. "The Dependence of Cell-Free Protein Synthesis in *E. coli* upon Naturally Occurring or Synthetic Polyribonucleotides." *Proc. Nat. Acad. Sci.* 47:1588–1602.

51. Crick, F. H. C. 1963. "The Recent Excitement in the Coding Problem." *Prog. Nucleic Acid Res.* 1:164–217. A superb analysis of the state of the coding problem as of 1962.

52. Sarabhai, A. S., A. O. W. Stretton, S. Brenner, and A. Bolte. 1964. "Co-Linearity of the Gene with the Polypeptide Chain." *Nature* 201:13–17.

53. Yanofsky, C., B. C. Carlton, J. R. Guest, D. R. Helinski, and U. Henning. 1964. "On the Colinearity of Gene Structure and Protein Structure." *Proc. Nat. Acad. Sci.* 51:266–272.

54. Nishimura, S., D. S. Jones, and H. G. Khorana. 1965. "The In-Vitro Synthesis of a Copolypeptide Containing Two Amino Acids in Alternating Sequence Dependent upon a DNA-Like Polymer Containing Two Nucleotides in Alternating Sequence." *J. Mol. Biol.* 13:302–324. Demonstration of the fact that each codon contains three nucleotides.

55. 1966. "The Genetic Code." *Cold Spring Harbor Symposia on Quantitative Biology*, Vol. 31 (1966), Cold Spring Harbor publ., Cold Spring Harbor, N.Y.

56. Adams, J. M., and M. R. Capecchi. 1966. "N-Formyl-methionyl-sRNA as the Initiator of Protein Synthesis." *Proc. Nat. Acad. Sci.* 55(1):147–155.

57. Brenner, A., A. O. W. Stretton, and S. Kaplan. 1965. "Genetic Code: The Nonsense Triplets for Chain Termination and Their Suppression." *Nature* 206:994–998.

58. Zipser, D. 1967. "UGA: A Third Class of Suppressible Polar Mutants." *J. Mol. Biol.* 29:441–445.

59. Sambrook, J. F., D. P. Fan, and S. Brenner. 1967. "A Strong Suppressor Specific for UGA." *Nature* 214:452–453.

60. Linn, S., and W. Arber. 1968. "Host Specificity of DNA Produced by *E. coli.* X. In Vitro Restriction of Phage fd Replicative Form." *Proc. Nat. Acad. Sci.* 59:1300–1306.

61. Smith, H. O., and K. W. Wilcox. 1970. "A Restriction Enzyme from *Hemophilus Influenzae*. I. Purification and General Properties." *J. Mol. Biol.* 51:379–391.

62. Roberts, R. J. 1983. "Restriction and Modification Enzymes and Their Recognition Sequences." *Nucleic Acid Res.* 11:r135–r167.

63. Mertz, J. E., and R. W. Davis. 1972. "Cleavage of DNA by RI Restriction Endonuclease Generates Cohesive Ends." *Proc. Nat. Acad. Sci.* 69:3370–3374.

64. Jackson, D., R. Symons, and P. Berg. 1972. "Biochemical Method for Inserting New Genetic Information into DNA of Simian Virus 40: Circular SV40 DNA Molecules Containing Lambda Phage Genes and the Galactose Operon of *E. coli.*" *Proc. Nat. Acad. Sci.* 69:2904–2909.

65. Lobban, P., and A. D. Kaiser. 1973. "Enzymatic End-to-End Joining of DNA Molecules." *J. Mol. Biol.* 79:453–471.

66. Cohen, S., A. Chang, H. Boyer, and R. Helling. 1973. "Construction of Biologically Functional Bacterial Plasmids in Vitro." *Proc. Nat. Acad. Sci.* 70:3240–3244.

67. Sanger, F., and A. R. Coulson. 1975. "A Rapid Method for Determining Sequences in DNA by Primed Synthesis with DNA Polymerase." *J. Mol. Biol.* 94:444–448.

68. Maxam, A. M., and W. Gilbert. 1977. "A New Method of Sequencing DNA." *Proc. Nat. Acad. Sci.* 74:560–564.

69. Sanger, F., S. Nicklen, and A. R. Coulson. 1977. "DNA Sequencing with Chain-Terminating Inhibitors." *Proc. Nat. Acad. Sci.* 74:5463–5467.

70. Spiegelman, S., I. Haruna, I. B. Holland, G. Beaudreau, and D. Mills. 1965. "The Synthesis of a Self-Propagating and Infectious Nucleic Acid with a Purified Enzyme." *Proc. Nat. Acad. Sci.* 54:919–927. The first report of the in vitro replication of infectious viral (Qβ) RNA.

71. August, J. T., A. K. Banerjee, L. Eoyang, M. T. F. de Fernandez, K. Hori, C. H. Kuo, U. Rensing, and L. Shapiro. 1968. "Synthesis of Bacteriophage Qβ RNA." *Cold Spring Harbor Symp. Quant. Biol.* 33:73–83.

72. Weissmann, C., G. Felix, and H. Slor. 1968. "In Vitro Synthesis of Phage RNA: The Nature of the Intermediates." *Cold Spring Harbor Symp. Quant. Biol.* 33:83–101.

73. Zinder, N., ed. 1975. *RNA Phages.* Cold Spring Harbor, N.Y.: Cold Spring Harbor Laboratory. A collection of articles bringing together new facts about these fascinating viruses.

74. Temin, H. M., and S. Mizutani. 1970. "Viral RNA-Dependent DNA Polymerase." *Nature* 226:1211–1213.

75. Baltimore, D. 1970. "Viral RNA-Dependent DNA Polymerase." *Nature* 226:1209–1211.

76. Breathnach, R., J. L. Mandel, and P. Chambon. 1977. "Ovalbumin Gene Is Split in Chicken DNA." *Nature* 270:314–319.

77. Jeffreys, A. J., and R. A. Flavell. 1977. "The Rabbit Beta-Globin Gene Contains a Large Insert in the Coding Sequence." *Cell* 12:1097–1108.

II

CHEMICAL FACTS
AND PRINCIPLES

CHAPTER 4

A Chemist's Look at the Bacterial Cell

The smallest self-contained living entities governed by the genetic information of DNA are the bacteria. Because they are visible only when viewed under the microscope and do not contain morphologically distinct chromosomes that separate on mitotic spindles, their true nature was at first obscure. The question long persisted as to whether bacteria, like all other cells, have a conventional hereditary apparatus based on genes or whether they represent some radically different form of life governed by interrelated metabolic pathways that utilize the laws of chemical kinetics to generate the cellular molecules needed for cell growth and division.

Today, the latter speculation appears at best a silly aberration by biologically naive chemical kineticists. However, only in 1943 did convincing evidence begin to appear supporting the existence in bacteria of discrete genes capable of spontaneous mutation.[1] Soon afterward, DNA was pinpointed as the carrier of genetic information and then localized in bacteria within distinct "nucleoid" bodies by the methods of cytochemistry. When the 1946 finding of genetic recombination in the bacterium *E. coli* opened up the possibility of systematic genetic crosses, research on bacteria acquired a momentum that only recently has begun to be challenged by efforts on other forms of life.[2]

Bacterial Cells Do Not Have Nuclei[3, 4]

Although the cells of most bacteria are smaller than other types of cells, there is considerable size variation between the smallest and largest bacteria, whose cells approach the size of simple unicellular fungi (e.g., yeast). It is thus not size per se that makes a cell a bacterium. Instead, it is the absence of a discrete nucleus that defines bacteria and their very close relatives, the blue-green algae (now often called cyanobacteria), as evolutionarily distinct forms of life. No nuclear membrane separates the DNA (chromosomes) of bacteria and blue-green algae from the cytoplasm in which protein synthesis occurs.

Organisms lacking nuclei are called **procaryotes,** while those organisms whose cells contain nuclei are called **eucaryotes** (*caryon* means "nucleus" in Greek). Even the smallest of eucaryotic cells differ quite fundamentally from procaryotic cells in the way their genetic information is organized as well as in their patterns of RNA and protein synthesis (see Table 4-1). Bacteria accomplish these ends in more straightforward ways than eucaryotes, and it is unlikely that this sim-

Table 4-1 Differences Between Procaryotic and Eucaryotic Cells

Feature	Procaryotes	Eucaryotes
Genetic Organization		
Nuclear membrane	Absent	Present
Number of different chromosomes	1	>1
Chromosomes with histones	Absent	Present
Nucleolus	Absent	Present
Genetic exchange	Plasmid-mediated, unidirectional	By gamete fusion
Cell Structures		
Endoplasmic reticulum	Absent	Present
Golgi apparatus	Absent	Present
Lysosomes	Absent	Present
Mitochondria	Absent	Present
Chloroplasts	Absent	Present in plants
Ribosome size	70 S	80 S
Microtubules	Absent	Present
Cell wall with peptidoglycan	Present, except mycoplasma and archaebacteria	Absent
Some Functional Attributes		
Phagocytosis	Absent	Sometimes present
Pinocytosis	Absent	Sometimes present
Site of electron transport	Cell membrane	Organelle membranes
Cytoplasmic streaming	Absent	Present

SOURCE: After Stanier et al., *Introduction to the Microbiol World* (Englewood Cliffs, N.J.: Prentice-Hall, 1976), pp. 86–87.

plicity is a matter of chance. While evolving to multiply more rapidly than any other cells, bacteria have become as simple as their nutritional requirements allow. It is this economy in their molecular components that made them the obvious organisms with which to initially establish the detailed chemical pathways through which genes control the life of the cell.

Bacteria Grow Under
Simple, Well-Defined Conditions[5, 6, 7]

There exists an enormous number of different types of bacteria, and they vary not only in size and shape but also in the nutritional conditions best suited for their growth and survival. Some bacteria, for example, are aerobes and grow only in the presence of oxygen; others are anaerobes and multiply only in the absence of oxygen; and still others, facultative anaerobes, can change their exact mixture of enzymes to allow growth in both environments. While most bacteria derive their energy from breaking down externally derived food molecules, others have evolved photosynthetic pigments to let them use sunlight to make ATP. Independent of their nutritional specialization, most bacteria grow free as single cells, separating from each other as soon as cell division occurs. In general, it has proved easy to grow them in the laboratory once their nutritional requirements have been worked out. Most importantly, in contrast to mammalian cells, which require a large variety of growth factors, many bacteria will grow well on a simple, well-defined diet, or medium. For example, the bacteria *Escherichia coli (E. coli)* will grow in an aqueous solution containing just glucose and several inorganic ions (Table 4-2).

Table 4-2 A Simple Synthetic Growth Medium for *E. coli**

NH_4Cl	1.0 g
$MgSO_4$	0.13 g
KH_2PO_4	3.0 g
Na_2HPO_4	6.0 g
Glucose	4.0 g
Water	1000 ml

*Traces of other ions (e.g., Fe^{2+}) are also required for growth. Usually, these are not added separately, since they are normally present as contaminants in either the added inorganic salts or the water itself.

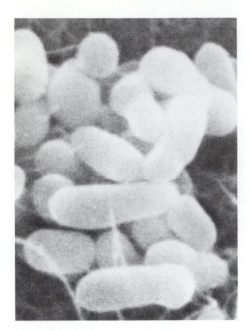

Figure 4-1
Electron micrograph of a group of intact
E. coli cells.

The growth of a specific bacterium is usually not dependent on the availability of a specific carbon source. Most bacteria are highly adaptable as to which organic molecules they can use as their carbon and energy sources. Glucose can be replaced by a number of other organic molecules, and the greater the variety of food molecules supplied, the faster a bacterium generally grows. For example, if *E. coli* grows only on glucose, about 60 minutes are required at 37°C to double the cell mass. But if glucose is supplemented by the various amino acids and purine and pyrimidine bases, then only 20 minutes are necessary for the doubling of cell mass. This shortening of **generation time** is due to the direct incorporation of the dietary components into proteins and nucleic acids, sparing the cell the task of carrying out the synthesis of their building blocks. There is a lower limit, however, to the time necessary for a cell generation; no matter how favorable the growth conditions, an *E. coli* cell is unable to divide more than once every 20 minutes.

E. coli Is the Best-Understood Organism at the Molecular Level[8, 9]

The most intensively investigated and correspondingly best-known bacterium at both the biochemical and genetic levels is *E. coli*. Its prominence as the premier organism used to probe the essential features of life started in the early 1940s through its adoption by a group of young physicists, chemists, and geneticists. These scientists believed that only by working on the simplest of biological systems would they ever be able to come to grips with the gene at its most fundamental level. So they initiated research on *E. coli*, a common inhabitant of the human intestine. This organism possesses the very desirable properties of small size, lack of pathogenicity to most other organisms, and extreme ease of cultivation under laboratory conditions. First extensively used to study the multiplication of its viruses, the bacteriophages (phages), *E. coli* became most interesting in its own right when it became the first bacterium shown to have discrete sexes that genetically recombine.

E. coli quickly became the obvious bacterium in which to search for mutations that blocked the synthesis of essential metabolites. Such mutations were relatively easy to find and map, and they soon proved indispensable in the working out of the enzymatic steps through which bacterial molecules are either made or broken down. Today, virtually a thousand different *E. coli* genes have already been assigned chromosomal locations, and it would make no sense to start serious work on another bacterium if *E. coli* can be used to solve a particular problem. As we shall see in subsequent chapters, the combined methods of genetics and biochemistry are so powerful that it is undesirable to perform biochemical studies on an organism with which genetic analysis is not possible.

Figure 4-2
The multiplication of single bacterial cells to form colonies. *E. coli* cells are usually not motile. Thus, when a cell divides on a solid surface, the two daughter cells and all their descendants tend to remain next to one another. After 24 hours at 37°C, each initial living cell gives rise to a solid mass of cells.

The Growth Parameters of *E. coli*[10–14]

The average *E. coli* cell is rod shaped and about 2 μm in length and 1 μm in diameter (Figure 4-1). It grows by increasing in length, followed by a fission process that generates two cells of equal length.

Growth occurs best at temperatures around 37°C, suiting it for existence in the intestines of higher mammals, where it is frequently found as a harmless parasite. It will, however, regularly grow and divide at temperatures as low as 20°C. Cell growth proceeds much more slowly at low temperatures; the generation time under otherwise optimal conditions is about 120 minutes at 20°C.

Cell number and size are often measured by observing the bacteria under the light microscope (and occasionally the electron microscope). Viewing a cell in this way, however, does not tell us whether the cell is alive or dead. We can determine whether a bacterium is alive or dead only by seeing whether or not it forms daughter cells. This observation is usually made by spreading a small number of cells on top of a solid agar surface, which has been supplemented with the nutrients necessary for cell growth (Figure 4-2). If a cell is alive, it will grow to form two daughter cells, which in turn give rise to subsequent generations of daughter cells. The net result after 12 to 24 hours of incubation at 37°C is discrete masses, or **colonies,** of bacterial cells. Provided that the colonies do not overlap, each must have arisen from a single bacterial cell.

The growth of bacteria may also be followed in liquid nutrient solutions. If a nutrient medium is inoculated with a small number of rapidly dividing bacteria from a similar medium, the bacteria will continue dividing with a constant division time, doubling the number of bacteria each generation time. Thus, the number of bacteria increases in an exponential (logarithmic) fashion (Figure 4-3). *Exponential growth* continues until there are so many cells that the initial optimal nutritional conditions no longer exist. One of the first factors to limit growth is usually the supply of oxygen. When the number of cells is low, the oxygen available by diffusion from the liquid surface is sufficient; but as the number of cells increases, additional oxygen is needed. It is often supplied by bubbling oxygen through the solution or shaking the solution rapidly. But even with violent aeration, growth rates begin to slow down after the cell density reaches about 10^9 cells per milliliter, and a tendency develops for the cells being produced to be shorter. Finally, at cell densities of about 5×10^9 cells per milliliter, cell growth is discontinued for nutritionally related reasons that are not yet clear. The term **growth curve** is frequently used to describe the increase of cell numbers as a function of time.

In most growing bacterial cultures, the exact division time of the cell varies, so that even if a culture has started from a single cell, after a few generations, cells can be found at various stages of the division cycle at any given moment. Such growth is frequently called *unsynchronized growth.* Over the past ten years, tricks have been developed to isolate bacterial cells at the same stage of the cell cycle. These tricks can be used to obtain several generations of **synchronized cell growth** (Figure 4-4). Then, because of slightly unequal division times, the growth curve gradually acquires an unsynchronized appearance.

Even Small Cells Are Very Complex

Even cells as small as those of *E. coli* present great difficulties when we study them at the molecular level. At first sight, the problem of understanding the essential features of *E. coli* seems insurmountable, for on a chemical scale, even the smallest cells are fantastically large.

Figure 4-3
Growth curve of *E. coli* cells at 37°C. The black line shows the increase in cell number following the inoculation of a sterile, nutrient-rich solution (glucose, salts, amino acids, purines, pyrimidines) with 10^5 cells from an *E. coli* culture in an exponential phase of growth. If this growth curve had been started instead from cells in a slow-multiplying, nearly saturated culture, the growth would not have begun immediately; rather, a lag period of approximately 1 hour would have preceded exponential growth (colored line).

Figure 4-4
The growth curve of a synchronized *E. coli* culture growing on glucose as the sole carbon source is given by the colored line. The black line shows the increase in cell number of an unsynchronized culture. In this example, the degree of synchronization lessens noticeably in the second and third cycles of growth.

Table 4-3 The Main Classes of Biological Molecules in Bacteria

	General Description	Functions	Building Blocks
Proteins	Molecules containing C, H, O, N, and sometimes S, which are built up from amino acids.	Most proteins are enzymes. Some, usually present in very large numbers per cell, are used to build up essential structures, such as the cell wall, the cell membrane, the ribosomes, muscle fiber, and nerves.	Amino acid
Lipids	Molecules insoluble in water. They include triglycerides, which are built up by the combination of glycerol and three long-chain fatty acids, and lecithins, in which one fatty acid is replaced by choline. Sometimes, glycerol is replaced by sphingosine. Phosphorus is present in many of the lipids (phospholipids).	Triglycerides are a main storehouse of energy-rich food. They degrade to give acetyl-CoA. Phospholipids are an essential component of all membranes. Their insolubility in water is related to their control of permeability.	Glycerol; Fatty acids (n usually 16 or 18); Sphingosine; Choline
Carbohydrates	Molecules containing C, H, and O, usually in ratios near 1:2:1. Polysaccharides are built up from simple sugars (monosaccharides) such as glucose and galactose. In some cases, the sugars contain amino groups (e.g., glucosamine).	Some, such as cellulose and pectin, are used to construct strong, protective cell walls; others, such as glycogen, provide a form of storing glucose.	β-D-Glucose; β-D-Galactose; β-D-Glucosamine
Nucleic Acids	Long, linear molecules containing P, as well as C, H, O, and N, which are built up from nucleotides	There are two main classes of nucleic acids in cells: DNA is the primary genetic component of all cells; RNA usually functions in the synthesis of proteins. In some viruses, RNA is the genetic material.	Bases: Purine, Pyrimidine; Pentose sugar: Ribose, Deoxyribose; + Phosphate → Nucleotide

Although an *E. coli* cell is about five hundred times smaller than an average cell in a higher plant or animal (which has a diameter of approximately 10 μm), it nonetheless has a wet weight of approximately 10^{-12} gram (5×10^{11} daltons, where a dalton is the weight of one hydrogen atom). This number, which initially may seem very small, is immense on a chemist's scale, since it is 3×10^{10} times greater than the weight of a water molecule (MW = 18 daltons). Furthermore, this mass reflects the highly complex arrangement of a large number of different carbon-containing molecules.

There is also seemingly infinite variety in the chemical nature of the molecules contained in an *E. coli* cell. Fortunately, it is possible to distribute most of the larger molecules into several well-defined classes possessing common arrangements of atoms. These classes are the proteins, lipids, carbohydrates, and nucleic acids (Table 4-3). Many molecules possess chemical groups common to several of these categories, so the classification of such molecules is necessarily arbitrary. Also in the cell are many smaller organic molecules (such as amino acids, purine and pyrimidine nucleotides, and various coenzymes), very small inorganic molecules (e.g., O_2 and CO_2), and numerous electrically charged inorganic ions (Table 4-4). Among the very small molecules is water (H_2O), the most common molecule in all cells and a solvent for most biological molecules, through which diffusion from one cellular location to another can occur quickly.

At present, we can make only an approximate guess of the number of chemically different molecules within a single *E. coli* cell. Each year, many new molecules are discovered, so the known molecules clearly represent only a fraction of those we shall eventually know. Already over 800 small molecules are implicated in its metabolism (Table 4-5), and with a slightly larger number of genes (and their proteins) already identified, we can say with confidence that *E. coli* must easily possess over 2000 different molecules. This is surely an underestimate, and later we shall give reasons for suspecting that as many as 4500 different molecules might in fact be found within the average growing cell. Thus, we must immediately admit that the structure of a cell will never be understood in the same way that we understand water or glucose molecules. Not only will the three-dimensional structures of most cellular proteins remain unsolved, but their location within cells often remain imprecisely defined.

It is thus not surprising that many chemists, after periods of enthusiasm for studying "life," silently return to the world of pure chemistry. Increasingly, however, others are becoming lifelong converts to the fascination of biology once they realize (1) that all macromolecules are polymeric molecules built up from smaller monomers, (2) that there exist well-defined chains of successive chemical reactions in cells (metabolic pathways), and (3) that there is a limit to the number of proteins that can exist in a cell owing to the fact that each cell contains a finite amount of DNA. This number, in turn, leads to an estimate of the number of enzymes and hence small molecules that a cell can possess. Most likely, at least half the proteins in a cell are enzymes, each of which catalyzes a specific metabolic reaction. The approximate number of different types of small molecules could be estimated if we knew, on the average, how many specific enzymes are needed for the metabolism of the average small molecule. At present, it seems a good guess that the number lies between one and two.

Table 4-4   Inorganic Ions Found in a Bacterial Cell After Growth in a Glucose-Minimal Medium

Ion	Function
K^+	Principal cation, cofactor for certain enzymes
NH_4^+	Principal form of inorganic nitrogen for assimilation
Mg^{2+}	Cofactor for a large number of enzymes
Ca^{2+}	Cofactor for certain enzymes
Fe^{2+}	Present in cytochromes and other enzymes
Mn^{2+}	Cofactor for several enzymes
Mo^{2+}	Present in several enzymes
Co^{2+}	Present in vitamin B_{12} and its coenzyme derivatives
Cu^{2+}	Present in several enzymes
Zn^{2+}	Present in several enzymes
Cl^-	Not required for many bacteria
SO_4^{2-}	Main source of sulfur in most media
PO_4^{3-}	Participant in many metabolic reactions

SOURCE: Reproduced from J. L. Ingraham et al., *The Growth of the Bacterial Cell* (Sunderland, Mass.: Sinauer Associates, 1983), p. 20, with permission.

Table 4-5   The Various Molecules in a Bacterium Growing in Glucose-Minimal Medium

Molecule	Approximate Number of Kinds
Amino acids and their precursors and derivatives	120
Nucleotides and their precursors and derivatives	100
Fatty acids and their precursors	50
Sugars, carbohydrates, and their precursors	250
Quinones, poly-isoprenoids, porphyrins, vitamins, other coenzymes and prosthetic groups, and their precursors	300

SOURCE: Reproduced from J. L. Ingraham et al., *The Growth of the Bacterial Cell* (Sunderland, Mass.: Sinauer Associates, 1983), p. 19, with permission.

Cellular Proteins Can Be
Displayed on Two-Dimensional Gels[15, 16]

The amount of DNA in the *E. coli* chromosome is about 2.5×10^9 daltons, representing some 4×10^6 base pairs. If all this DNA codes for amino acids (see Chapter 15), then a total of $(4 \times 10^6) \div 3 = 1.33 \times 10^6$ amino acids are coded for by the nucleotide sequences of *E. coli* DNA. As the average *E. coli* protein (MW 4×10^4) contains some 360 amino acids, $(1.33 \times 10^6) \div (3.6 \times 10^2) = {\sim}3600$ represents the maximum number of proteins that are ever likely to be described within *E. coli*.

This number can now be compared with the actual number of different *E. coli* proteins revealed by newly developed procedures for two-dimensional gel electrophoresis. (The basic methods of gel electrophoresis are described in Figure 4-5.) The new procedures make possible the display of the total cell protein as discrete spots on a thin rectangular layer of polyacrylamide gel. First, the proteins are separated on the basis of their net charge by isoelectric focusing in one dimension; then they are separated on the basis of their apparent size as revealed by their rates of movement in the presence of SDS (sodium dodecyl sulfate, an anionic detergent) when the current is applied in the second dimension. Figure 4-6 shows a gel in which the acidic proteins of a growing *E. coli* cell are separated into some 1400 spots. With few exceptions, each spot represents a distinct polypeptide chain coded by a separate gene. The basic proteins were not separated in that experiment, but these can be visualized by using different gel conditions to reveal the existence of still another 300 different proteins. In these gel systems, any *E. coli* protein present in at least 30 molecules per cell can be seen. By measuring and calibrating the intensity of individual spots, the number of copies of individual proteins can be seen to vary over a 10^5-fold range, with some proteins present in over 100,000 copies per cell.

Some 200 different spots have already been identified as known enzymes and structural (e.g., ribosomal) proteins, and no serious obstacle (except for tedium) prevents the identification of still many others. The exact pattern shown by a given two-dimensional gel varies with the exact nutritional conditions. *E. coli* cells growing under nutritionally rich conditions, for example, contain many more copies of the various proteins used to construct ribosomes than when they subsist on a simple, minimal medium.

The number of spots (proteins) observed to date (about 1700) is much less than the number predicted (about 3500) if all the *E. coli* DNA codes for protein products. Only time will tell whether there exist large numbers of proteins still undetected because they exist in only a few copies or are only present under nutritional conditions not normally encountered.

The Anatomy of *E. coli*
as Revealed by the Electron Microscope[17, 18, 19]

A direct way to appreciate the morphological simplicity of *E. coli* is through electron microscope examination of very thin sections cut from a rapidly growing cell. These sections reveal *E. coli* to have a simple, saclike structure in which an external envelope surrounds a membrane-free, dense granular cytoplasm (called the cytosol), which

(a) Isoelectric focusing

(b) SDS polyacrylamide gel electrophoresis

Figure 4-5

Gel electrophoresis separates macromolecules on the basis of their migration through a gel under the influence of an electric field. This figure illustrates the principles underlying two electrophoretic methods used to separate polypeptides. In both cases, proteins are first treated to denature them and to separate their subunits, and the electrophoresis is carried out in a gel of polyacrylamide. (a) Isoelectric focusing depends on the fact that the net charge on a polypeptide changes with pH; the electrophoresis is performed in a gel in which a pH gradient has been established. At a pH characteristic for each polypeptide (its isoelectric point), the net charge on the molecule is zero, and it ceases to migrate. (b) Electrophoresis in the presence of the negatively charged detergent sodium dodecyl sulfate (SDS) separates polypeptides by size. The SDS coats each polypeptide, giving it a roughly uniform negative charge. The smaller the polypeptide, the more readily it moves through the pores in the gel toward the positive pole. (Nucleic acid molecules, which intrinsically have uniform negative charge, can also be separated by size with gel electrophoresis; an agarose gel is usually used.) The locations and amounts of the macromolecules on the completed gel may be determined by using special dyes or autoradiography.

Figure 4-6

Two-dimensional gel of *E. coli* K-12, strain W3110, cell extract showing the acidic proteins separated into approximately 1400 spots. The grid overlay provides coordinates for individual spots. Letters A through H denote zones of increasing isoelectric points. [Courtesy of Dr. Frederick C. Neidhardt, *Microbiol. Revs.* 47 (1983):232.]

Figure 4-7

(a) Electron micrograph of an *E. coli* B cell prepared by freeze substitution with no pretreatment. [Courtesy of J. A. Hobot et al., *J. Bacteriol.* 162 (1985):964.] (b) Schematic diagram of such a cell, highlighting the nucleoid regions (color) and showing various other cellular components (not to scale). Most of the DNA resides in the relatively ribosome-free nucleoid region, which is in close association with the ribosomes, theoretically to provide for optimal efficiency in translation.

(a)

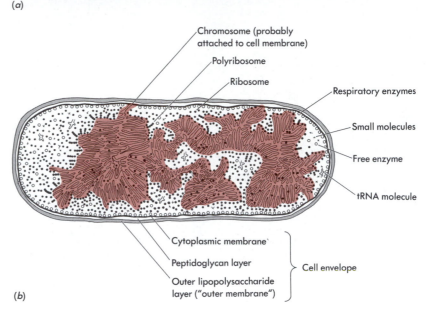

Chromosome (probably attached to cell membrane)

Polyribosome

Ribosome

Respiratory enzymes

Small molecules

Free enzyme

tRNA molecule

Cytoplasmic membrane

Peptidoglycan layer

Outer lipopolysaccharide layer ("outer membrane")

Cell envelope

(b)

contains an apparently less dense fibrous DNA component (Figure 4-7). The granular appearance of the cytosol reflects the very large number of ribosomes found in all growing bacterial cells. More ribosomes per unit dry mass are found in bacteria than in any other kind of cell. This makes possible the high rate of protein synthesis needed to allow bacteria to double their number as rapidly as once every 20 minutes at 37°C. Very high concentrations of free enzymes also exist in bacteria, and the dry/wet ratio of bacteria can be as high as 30 percent, much greater than the 10 percent ratio characteristic of typical higher eucaryotic cells. The metabolic rate of bacteria is correspondingly much higher, with the amount of oxygen and substrates consumed per unit of time frequently 10 to 100 times greater than that consumed by eucaryotic cells.

Bacteria Display Their Individuality Through Their Envelopes[20–25]

The envelopes of bacteria are extremely complex compared to their tightly packed, relatively undifferentiated cytosol (Figure 4-8). The innermost layer of the envelope is always a phospholipid bilayer into which many functionally distinct proteins are inserted. Surrounding this "cytoplasmic facing membrane" is a rigid shell of covalently

(a)

(b)

Figure 4-8
Comparison of Gram-negative and Gram-positive cell envelopes. (a) Electron micrograph of a cross-section of a Gram-negative *E. coli* cell envelope with outer membrane (lipopolysaccharide) and cytoplasmic membrane. [Courtesy of J. A. Hobot et al., *J. Bacteriol.* 160 (1985):143–152.] (b) Electron micrograph of the peptidoglycan layer and cytoplasmic membrane of Gram-positive *Bacillus cereus*. (Courtesy of M. T. Silva.) Bars = 0.1 μm.

linked carbohydrates and amino acids that has been given the name **peptidoglycan** (Figure 4-9). It is through their peptidoglycan component that bacteria obtain their structural integrity. If it is enzymatically removed (e.g., by lysozyme), the effectively naked bacteria assume spherical shapes. Depending on the specific bacterium, the peptidoglycan shell can be relatively thin (as in *E. coli*) or composed of many effective layers (as in bacteria belonging to the *Bacillus* groups).

Those bacteria that have thick peptidoglycan shells are very readily stained by the dyes crystal violet and iodine, first used by the Danish microbiologist Christian Gram. Such peptidoglycan-rich bacteria are known as Gram-positive bacteria, while those bacteria possessing only thin layers of rigid peptidoglycan are collectively known as Gram-negative bacteria. To compensate for their thin peptidoglycan shells, Gram-negative cells possess a second phospholipid-containing membrane that surrounds the peptidoglycan shell. This **outer membrane** is characterized by unique lipopolysaccharides that have their highly specific (immunogenic) polysaccharide side groups projecting outward, as well as by the presence of a very abundant lipoprotein that anchors the outer membrane to the peptidoglycan shell. Inserted into the outer membrane also are the matrix proteins, or **porins,** which form specific pores through which extracellular molecules must pass to reach the inner cytoplasmic membrane (Figure 4-10). Usually, such passages are passive processes governed by the laws of diffusion. In other cases, movement through these pores is facilitated by at least 30 different **transport proteins,** or receptors, that make possible the entry of molecules too large for simple passage through the more common porin channels. How such transport proteins work remains to be established, with the possibility still open that they form specialized pores.

Figure 4-9

(a) The chemical structure of a peptidoglycan unit from *E. coli*. Within the peptidoglycan layer, the two sugars alternate with each other in chains. A tetrapeptide is attached to each N-acetylmuramic acid residue. Adjacent tetrapeptides may be directly bonded to each other or linked by an oligopeptide bridge (not shown) to form a crosslink between two chains. These crosslinks make the peptidoglycan layer in effect one giant molecule. Other amino acids are occasionally found in some bacteria. (b) The likely arrangement of peptidoglycan within the *E. coli* cell envelope. The peptidoglycan is thought to form a periplasmic gel that is more crosslinked and thus denser toward the outer membrane. (After a drawing by J. A. Hobot.)

(b)

Until recently, little was known about the structure of the "periplasmic space" between the inner and outer lipid-containing membranes. This region was known to contain three important classes of vital proteins: (1) hydrolytic enzymes, which initiate the degradation of food molecules; (2) specific binding proteins that help initiate the transport of certain food molecules across the inner cytoplasmic membrane; and (3) specific chemoreceptors used to measure the concentration of nutrients (or poisons) in the environment so that their respective cells can move toward (or away) from them. Also, the periplasmic region was known to contain peptidoglycan, which was thought to exist in a thin, discrete layer near the outer membrane. Now, however, we believe that peptidoglycan actually fills the periplasmic space, forming a periplasmic gel that is highly fluid near the inner membrane and more compact near the outer membrane (see Figure 4-9b). The gel is heavily hydrated, and periplasmic proteins can freely diffuse within it.

Although it is generally thought that Gram-positive cells do not have an outer lipopolysaccharide layer, regular-appearing patterns suggestive of proteins have recently been seen on their outer surfaces. Whether these patterns are, in fact, proteins in a thin lipopoly-

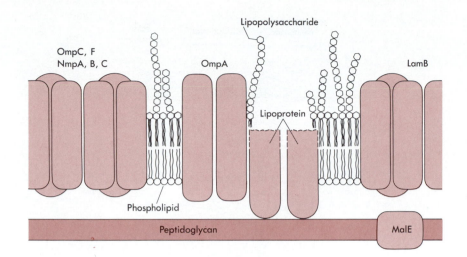

Figure 4-10
The major outer-membrane proteins of Gram-negative bacteria. Notice that OmpA and the porins OmpC, OmpF, NmpA, NmpB, NmpC, and LamB extend through the membrane. LamB is an outer-membrane protein that functions as a porin for transportation of maltose and maltodextrins (it is also a receptor for bacteriophage λ). MalE is a periplasmic maltose-binding protein. [After M. J. Osborn and H. C. P. Wu, *Ann. Rev. Microbiol.* 34 (1980).]

saccharide bilayer remains to be worked out. It is speculated that such a layer might not have been discovered earlier because of its relative sparcity compared to the thick peptidoglycan layer found in Gram-positive cells. If this speculation proves correct, the difference between Gram-negative and Gram-positive cells will simply be a quantitative difference in the amount of peptidoglycan in each.

The Inner Cytoplasmic Membrane Carries Out the Chemiosmotic Generation of ATP[26, 27]

Unlike eucaryotic cells, where the aerobic generation of ATP is made possible by the passage of protons through the membranes of mitochondria, the site of oxidative phosphorylation in bacteria is the inner, phospholipid-containing cytoplasmic membrane. Inserted within it are all the major proteins involved in the translocation of hydrogen ions, including the flavoproteins, quinones, iron-sulfur ("FeS") proteins, and cytochromes. Their combined action leads to lower hydrogen ion concentrations in the cytosol compared to the exterior medium as well as differences in electrical charge across the inner membrane. Together they create the proton motive force, which provides the energy for a number of vital cellular processes that occur on the cytoplasmic membrane, including the active transport of food molecules and certain inorganic ions into the cytosol, the turning of flagella, and the generation of ATP from ADP by cytoplasmic membrane-bound ATP synthetase (Figure 4-11). This latter enzyme is a complex molecular aggregate composed of at least ten different polypeptide chains, six of which combine to form knobs that protrude into the interior of the cell. The action of this membrane-based complex is reversible. The complex can either generate electrical charge differences across the cytoplasmic membrane (membrane potential) by hydrolysis of ATP to ADP (hence its alternative name, ATPase) or generate ATP from ADP and inorganic phosphate by the proton motive force.

We know very little about the relative locations of the various proteins of the cytoplasmic membrane. Studies using two-dimensional gel electrophoresis indicate that at least 200 different proteins are so located in the membrane; but this number is probably an underestimation of the true complexity. Clearly, we are just at the beginning of

Figure 4-11
Vital cellular processes that occur on the cytoplasmic membrane. Proton transfer and the generation of ATP provide the energy for activities such as active transport of food molecules and the turning of flagella. [After "How Cells Make ATP," by P. C. Hinkle and R. E. McCarthy. *Sci. Amer.* 238 (1978). Copyright © 1978 by Scientific American, Inc. All rights reserved.]

a real understanding of how membranes function at the molecular level. Fortunately, for the first time, true crystals suitable for X-ray crystallographic examination have recently been obtained for several integral membrane proteins. Through the elucidation of their three-dimensional conformations, we may know within the next decade how molecules selectively move through membranes.

Flagella and Pili Extend Outward from *E. coli* Surfaces[28-32]

E. coli, like many other bacteria, possesses two forms of extended surface appendages, flagella and pili. Though they are both long, linear bodies that arise from the cytoplasmic membrane, they are neither structurally nor functionally related. The bacterial **flagellum** (plural, **flagella**) is a very long, thin (200 Å diameter) organelle of locomotion, which is formed mainly by the helical aggregation of

(a)

(b)

large numbers of a protein building block called flagellin. In reality, it is structurally quite complex. It is composed of three parts: (1) the long filament made up of flagellin, which can extend 15 to 20 μm into the medium, (2) the basal structure that anchors the flagellum to the cell envelope, and (3) the hook, a short, curved structure that connects the filament to the basal structure (Figure 4-12). Flagella serve as locomotory organelles by rotating the basal structures in either a clockwise or counterclockwise direction. In *E. coli*, counterclockwise rotation leads to the formation of a stable bundle of the several (about five) rotating flagella that propels the cell smoothly forward. In contrast, clockwise rotation disperses the bundle and leads to a chaotic tumbling of the cell in many directions (see Figure 4-23).

While the rotating filaments clearly function as propellers, the exact function of the hook is less clear. Like the filament, it is built up by the aggregation of a number of identical protein subunits. Much more complicated is the basal structure, which consists of at least ten different types of protein subunits that come together to form rings surrounding a thinner rod that is inserted into the cytoplasmic membrane. Exactly how the basal structure rotates remains to be worked out, though it is clear that the movement of hydrogen atoms across the cell membrane (the proton motive force) is involved. How specific attractants or repellents set into motion the reversals between clockwise and counterclockwise rotations has yet to be discovered.

Much simpler both in structure and in function are the **pili** (singular, **pilus**) (Figure 4-13). Composed of only single protein building blocks called pilin, they are adhesive organs that can extend 10 μm or more into the medium. They function by attaching bacteria to other surfaces, often the glycoprotein components on the exterior of higher cells (such as the epithelial cells that line our intestines). In addition to the more common type of pili present in large numbers on every bacterial cell, most male bacteria contain one to several morphologically similar pili called **sex pili,** which function to initiate contacts with female cells.

Condensation of Bacterial DNA into Nucleoid Bodies[33]

All the essential genes of *E. coli* are present on a single chromosome, a single DNA molecule, which genetic experiments first revealed to be circular. Single cells, growing exponentially under favorable nutritional conditions, generally contain two to four such molecules in the

Figure 4-12
(a) The basal end of a flagellum from *E. coli*. The top arrow marks the junction between the hook and the filament. The L, P, S, and M rings (see part b) are also indicated (500,000×). (Courtesy of M. L. DePamphlis and J. Adler.) (b) Diagram of the hook–basal body complex.

Figure 4-13
Escherichia coli possessing type 1 pili, as seen with the electron microscope. (Courtesy of Steven Clegg.)

act of duplicating themselves, beginning at origins of replication present at one unique site on each chromosome. In such multiplying cells, the cell division process lags behind the DNA replication process, and only when cell growth has virtually ceased do many *E. coli* cells contain only one chromosome. The reason why rapidly growing cells contain multiple copies of the same chromosome only recently became clear. It relates to the fact that there is a fundamental upper limit to the rate of DNA replication. For reasons not yet understood, only 750 base pairs and no more can be polymerized per second at a given replicating fork (see Chapter 10). Even given the bidirectional pattern of the DNA replication, some 40 minutes are needed to complete the replication of the *E. coli* chromosome (4×10^6 base pairs). We might thus predict that an *E. coli* cell can never divide faster than once every 40 minutes. But in fact, as we mentioned earlier, optimally fed cells multiply once every 20 minutes. Resolution of this paradox came with the realization that a new round of DNA replication can be initiated before the previous round is completed. The replication of a given *E. coli* chromosome can thus occur during two successive cell cycles.

The length of a fully extended *E. coli* chromosome would be $4 \times 10^6 \times 3.4$ Å, or approximately 1 mm. However, it never even exists partially extended; electron microscopic examination reveals its compaction into irregularly shaped bodies, the **nucleoids** (see Figure 4-7). Despite their similarity in name, nucleoids bear no relation to conventional nuclei, since they are not surrounded by any form of phospholipid-containing membrane. Also distinguishing all procaryotic cells from their eucaryotic equivalents is the absence of conventional *his-*

tones, the highly conserved class of basic proteins around which eucaryotic DNA coils to form the nucleosome particles that give to chromatin (the underlying DNA-protein complex of chromosomes) its granular (beaded) appearance. Despite their lack of typical histones, however, gently prepared *E. coli* chromosomes initially display sections of beaded appearance that quickly transform into thinner fibers resembling pure DNA. So *E. coli* DNA must also, in part, be complexed with proteins that lead to its compaction. In fact, two small histonelike basic proteins have recently been found in *E. coli*, semitightly bound to the DNA. Their amounts, however, are insufficient to compact all the *E. coli* DNA into nucleosome-like particles. Instead, the most important factor that leads to the formation of nucleoid bodies may be enzymatically induced supercoiling (see Chapter 9), which of necessity leads to the compaction of the circular double helical DNA.

Also still a mystery is how progeny bacterial chromosomes correctly partition themselves into their respective daughter cells. Evidence suggests that regions of each bacterial chromosome are associated with the inner, or cytoplasmic, membrane of the bacterial envelope, with the occasional claim that the attached regions include the unique points where replication initiates (origins of replication). But even today, we have no good techniques to cleanly separate the cytoplasmic membrane from other cellular components.

Transcription and Translation Occur in the Same Cellular Compartment[34]

At any given moment, the *E. coli* chromosome is being transcribed at the rate of 60 nucleotides per second (at 37°C) into some 400 to 800 unique mRNA chains, some 100 different tRNA molecules, and some 700 precursor molecules for rRNA. Each of these transcription events requires the participation of a single RNA polymerase molecule. So at a given moment, some 1600 RNA polymerase molecules, representing about 1 percent of the total bacterial protein, are at work on a single *E. coli* chromosome. Even though only some 4 percent of the total bacterial RNA is mRNA, its average half-life of perhaps no more than 1½ minutes requires the participation of approximately one-half of a bacterial cell's RNA polymerase for its synthesis.

Very soon after an mRNA chain begins to be made, its leading (5') end becomes attached to a free ribosome, which in turn commences to translate its message and in so doing to move toward the 3' end still in the process of extending itself along its DNA template. By the time a given mRNA chain is completely transcribed, it usually has attached to it a string of ribosomes successively carrying polypeptide chains of ever-increasing length (Figure 4-14). With transcription occurring at approximately 60 nucleotides per second, the time required to make an average-size mRNA chain containing 2000 to 3000 nucleotides approaches 1 minute, an interval not much shorter than the average time (1.5 minutes) a given nucleotide remains incorporated within a given mRNA molecule. Thus, by the time the synthesis of many large mRNA molecules is completed, they may already have begun to be broken down. A large fraction, if not a majority, of the ribosomes collected together on a single mRNA molecule (the assembly is called a *polyribosome* or *polysome*) may thus be physically attached to the bacterial chromosomes. Such linkages, however, do not

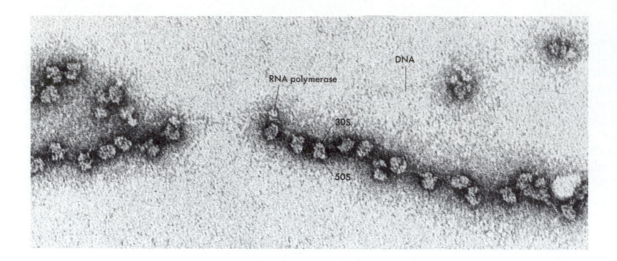

Figure 4-14
Groups of ribosomes (polysomes) moving across mRNA chains being formed on *E. coli* DNA. (Courtesy of Barbara Hamkalo.)

trap significant numbers of ribosomes within the DNA-containing nucleoid regions, since the extended lengths of average-size polysomes (10 to 15 ribosomes) approach the diameter of the *E. coli* cell itself.

The dilemma remains, however, as to how the nascent mRNA molecules with their attached strings of ribosomes separate from their DNA templates. In principle, this could occur either through rotation of the DNA or by the untwisting of RNA around DNA. Arguing against massive rotatory movements by the mRNA is the fact that many polyribosomes become securely fastened to the surrounding inner membrane so that their protein products can be built into the cell envelope. But it is equally unlikely for RNA to separate away from DNA by a coupling of transcription to the rapid rotation events required for DNA replication. Not only does DNA replication occur at a rate some ten times faster than RNA synthesis, but also various genes along the *E. coli* chromosome are transcribed in different directions. Perhaps those DNA segments being transcribed contain temporary single-stranded cuts along their backbone, which give them the capacity for the unimpeded rotation needed to separate away from their RNA products.

Membrane-Bound Polyribosomes Make the Protein Components of the Bacterial Envelope[35]

Slightly over half the dry mass of *E. coli* is protein. Of this, some 20 percent is used to form the ribosomes, about 50 percent consists of the thousand or so enzymes found floating in the cytosol, and the remaining 30 percent is used to construct the cell envelope. Both the cytosolic proteins and the ribosomal proteins are made on free polyribosomes. In contrast, the envelope proteins are made on polyribosomes attached to the inner (cytoplasmic) membrane of the envelope. Their peptide products are thus directly extruded into the surrounding envelope (Figure 4-15). Some of the completed protein ends up in the inner membrane and some in the periplasmic space. In addition, some of the protein first transferred to the inner membrane moves to the outer membrane, perhaps through tubular connections between the two membrane systems. Not only proteins, but all newly made phospholipids, appear first in the inner membrane before diffusing to the outer membrane.

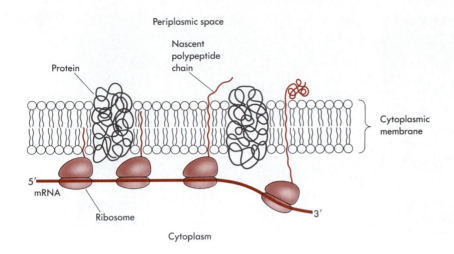

Periplasmic space

Figure 4-15
Synthesis of envelope proteins on polyribosomes attached to the cytoplasmic
membrane. As translation proceeds, the
nascent chain is transferred vectorially
across the membrane bilayer. Its final
destination may be the cytoplasmic
membrane, the periplasmic space, or
the outer membrane.

Metabolic Pathways Interconnect a Thousand Different Small Molecules[36, 37]

We can see that all the molecules in a bacterial cell arise from cellular transformations of food molecules when we grow *E. coli* on a simple, well-defined medium containing glucose. Under these conditions, glucose is the only organic source of carbon, and all the cellular carbon compounds (except for a few derived from carbon dioxide) result from enzyme-mediated chemical transformations that commence as the glucose molecules are broken down to smaller fragments. These smaller molecules, in turn, are used to form the many diverse building blocks whose polymerization creates the macromolecules from which major cellular organelles are assembled. The exact way in which all these transformations occur (collectively known as **metabolism**) is enormously complex, and most biochemists concern themselves with studying only a small fraction of the total interactions.

Fortunately, most of the major metabolic features of *E. coli* are common not only to all bacteria but also to all of life; therefore, what can be learned by focusing on *E. coli* is widely applicable. Figure 4-16 shows some of the more important types of chemical reactions that interconnect the various small molecules of *E. coli*, starting with its intake of glucose. Much information about this "intermediary metabolism" comes from experiments utilizing radioactively labeled food molecules. For example, if we expose *E. coli* for several seconds (a "pulse") to ^{14}C-labeled glucose, the radioactive atoms can be detected almost immediately in molecules chemically similar to glucose, such as glucose-6-phosphate. Only later do labeled atoms find their way into amino acids and nucleotides. The amount of time before radioactivity appears in the various compounds corresponds roughly to the number of enzymatic reactions separating glucose from the various intermediate metabolites.

Such experiments are relatively easy to interpret because most of the small molecules within cells are present in relatively small amounts. Only about 4 percent of the mass of *E. coli* is occupied by its collection of a thousand or so intermediary metabolites. Individual molecules thus have only fleeting existence before transformation into other small molecules or polymerization into macromolecules.

Figure 4-16 shows that various key intermediates in glucose degradation have several possible fates. They may be completely degraded

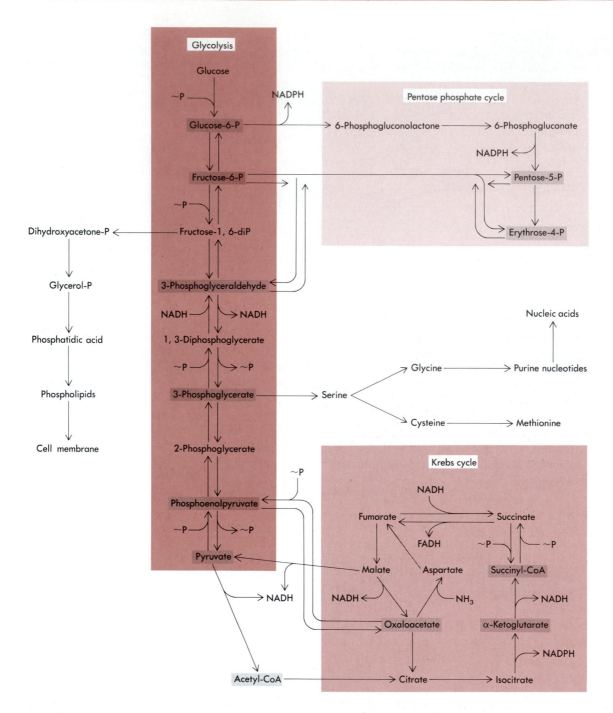

Figure 4-16
Some of the main metabolic pathways of *E. coli*. The 12 key precursor metabolites are highlighted.

via the Embden-Meyerhof pathway (glycolysis), the Krebs cycle, and the respiratory chain to yield carbon dioxide and water. During this process, ADP is converted to ATP. Alternatively, intermediates may be used to initiate a series of chemical reactions that end with the biosynthesis of vital building-block molecules such as amino acids or nucleotides. For example, dihydroxyacetone phosphate is used as a precursor for the lipid constituent glycerol, whereas 3-phosphoglyceric acid is the beginning metabolite in a series of reactions that lead to the formation of the amino acids serine, glycine, and cysteine.

Connected groups of biosynthetic or degradative reactions are referred to as **metabolic pathways.** Once a molecule has started on a pathway, it often has no choice but to undergo a series of successive transformations. Not all pathways, however, are necessarily linear. Some are branched, with the intermediates at branch points being transformed into one of two possible compounds. Sometimes, the intermediates that have several alternative fates are very important in cellular metabolism. Perhaps the most important such compound is acetyl-CoA; not only is it the main precursor of lipids, but it also supplies the acetate residues consumed by the Krebs cycle.

The metabolic fates of the majority of molecules, however, are much more limited. An average molecule can be either broken down to compound x or used as an intermediate in the biosynthesis of compound y; thus, each such metabolite is able to combine specifically with only two different enzymes.

Biosynthetic Reactions Commence with Twelve Key Precursor Metabolites

All the biosynthetic pathways begin with one or another of a small group of molecules called *key precursor metabolites* (Table 4-6). There exist just 12 of these precursor metabolites, from which some 75 building blocks (e.g., amino acids, purines and pyrimidines, and fatty acids) and coenzyme products are derived by specific sets of enzymatic reactions. These pathways, however, are not universally present in all bacteria. Many bacteria lack one or more specific biosynthetic routes and correspondingly require the respective end product(s) to be externally supplied for growth. Not surprisingly, it is those bacteria normally found in environments rich in organic material that have lost one or more specific biosynthetic pathways. When such a pathway is present, however, it is almost always composed of the same set of enzymatic reactions, which either begin directly with one of the 12 key precursor metabolites or branch off from an intermediate or end product of another pathway.

Degradative Pathways Are Distinct from Biosynthetic Pathways

When *E. coli* is growing with glucose as its *sole* carbon source, all its amino acids must be synthesized from metabolites derived from glucose. There is a distinct biosynthetic pathway for each of the 20 amino acids (one example is shown in Figure 4-17). But *E. coli* can also grow in the absence of sugar, using any of the 20 amino acids as a sole carbon source. This means that there must also exist 20 pathways of amino acid degradation by which the carbon and nitrogen atoms of the amino acids are usefully freed to form key metabolite compounds such as α-ketoglutarate (Figure 4-18). These compounds can then be used in the synthesis of other amino acids. Degradative pathways also exist for the various lipids, the purine and pyrimidine nucleotides, many pentose and hexose sugars, and so on.

That distinct degradative pathways are almost never simply the reverse of biosynthetic pathways is to be expected. As we shall see in Chapter 6, most biosynthetic reactions require energy and often involve the breakdown of ATP, whereas degradative reactions by their very function must eventually generate ATP in addition to supplying carbon and nitrogen skeletons.

The Extraordinary Diversity of Fueling Reactions

The degradative reactions that produce the 12 key precursor metabolites, ATP, and the needed reducing power (NADH) are called *fueling reactions*. While many fueling reactions are common to all bacteria,

Table 4-6 The Twelve Key Precursor Metabolites

Glucose-6-phosphate
Fructose-6-phosphate
3-Phosphoglyceraldehyde
3-Phosphoglycerate
Phosphoenolpyruvate
Pyruvate
Acetyl-CoA
α-Ketoglutarate
Succinyl-CoA
Oxaloacetate
Pentose-5-phosphate
Erythrose-4-phosphate

Figure 4-17

The biosynthesis of histidine from 5-phosphoribosyl-1-pyrophosphate (PRPP). The resulting histidine contains the five carbons of PRPP, the amide-nitrogen of glutamine, and the one nitrogen and one carbon of the pyrimidine ring of ATP.

others are unique to particular species. Those fueling reactions common to all cells, called the *central pathways,* are involved in the derivation of the key metabolic precursors from each other. They can function linearly, as in glycolysis, to replace key metabolites drained off for biosynthesis, or they can function cyclically, as in the Krebs cycle, to produce carbon dioxide and water as well as to generate ATP and NADH. Thus, there must be very great flexibility in the way the central pathways operate.

The real diversity of the fueling reactions is expressed by the peripheral pathways used when bacteria grow on compounds that are not intermediates in the central pathways. The enormous variety of potentially valuable sources of carbon, nitrogen, energy, and reducing power has led to the evolution of the approximately 4000 known forms of bacteria that can utilize almost every carbon compound naturally existing on Earth with the exception of diamonds and coal. Even within a single type of bacterium, the number of alternative fueling reactions that can be called into action, depending on the nutritional environment, is enormous. *E. coli* probably possesses at least 75 different peripheral pathways made up of at least 200 to 300 different enzymatic reactions. These numbers must be even larger in many other bacteria, particularly the pseudomonads, which can grow on a much larger, more varied collection of inorganic molecules than can *E. coli.*

Figure 4-18
Degradative pathway of histidine.

Peripheral Pathways Are Turned On or Off as Units[38]

At any given time under constant nutritional conditions, a bacterial cell may possess the enzymes for only a single peripheral pathway. Even when several different sugars are available as food sources, an *E. coli* cell tends to have only the enzymes for the pathway that generates the most energy. For example, in the presence of both glucose and galactose, *E. coli* only possesses the enzymes needed to break down glucose. Only when the glucose supply is exhausted does it make the enzymes needed to metabolize galactose.

The set of enzymes that forms a peripheral pathway is thus present or absent as a unit. This makes obvious biological sense, for it would be wasteful to possess only half a pathway and thereby produce intermediate compounds that cannot be used. Likewise, it would make no sense to possess the enzymes needed to metabolize food molecules that are not present in the environment. The genes coding for enzymes of peripheral pathways constitute a significant fraction of the *E. coli* genome, and if they functioned in the absence of need, a significant fraction of the cell's protein-synthesizing capacity would be tied up to no avail.

Rapid Turnover of Bacterial mRNA Molecules Allows Speedy Changes in Gene Expression

Given both their potentially short generation times as well as the rapid speed at which their nutritional environment can change, bacteria need to quickly call into action new peripheral pathways as well as to stop the synthesis of enzymes no longer needed. They accomplish

Threonine

α-Ketobutyrate

Acetohydroxybutyrate

Dihydroxyisoleucine

α-Ketoisoleucine

Isoleucine

Figure 4-19
The pathway of isoleucine biosynthesis starting from threonine. The dotted colored line shows that isoleucine inhibits the enzyme (threonine deaminase) that transforms threonine into α-ketobutyrate.

this by having the synthesis of many of their mRNA molecules dependent on the receipt of specific signals indicating that given nutrients are available to be broken down (this is control at the level of transcription). Moreover, bacterial mRNA molecules, unlike those of higher eucaryotic cells, are unstable, with average half-lives of only 1 to 2 minutes. When a given type of mRNA is no longer needed and its synthesis stops, its inherent instability leads quickly to the disappearance of the preexisting molecules (see Chapter 16 for more details). Even finer adaptation to need is achieved by a second set of control signals that operates at the level of translation. A number of proteins (e.g., certain ribosomal proteins), if present in excess, block the functioning of their mRNA templates by binding to them and thereby preventing further ribosomal attachment and commencement of translation.

Regulation of Protein Function by Feedback Inhibition[39, 40, 41]

The catalytic activity of many proteins is affected by their ability to bind to specific small molecules. Because of this property, the activity of enzymes may be blocked when they are not needed. Consider, for example, what happens when an *E. coli* cell growing on minimal glucose medium is suddenly supplied with the amino acid isoleucine. Immediately, the synthesis of the mRNA molecules that code for the specific enzymes utilized in isoleucine biosynthesis ceases. Without a further control mechanism, preexisting enzymes will cause continued isoleucine production, now unnecessary because of the extracellular supply. Wasteful synthesis, however, almost never occurs, because high levels of isoleucine block the activity of the enzyme involved in the first step of its biosynthesis from threonine (Figure 4-19). This inhibition is due to the binding of isoleucine to the enzyme threonine deaminase. Thus bound, the enzyme is unable to convert threonine to α-ketobutyrate. Because the association between the enzyme and isoleucine is weak and reversible, relatively high isoleucine concentrations must exist before most of the enzyme molecules are inactivated. This very specific inhibition is called **feedback inhibition** or end-product inhibition because accumulation of a product prevents its further formation. Usually, only the first step in a metabolic chain is blocked. With the first reaction blocked, there is no accumulation of unwanted intermediates; so inhibition of the remaining enzymes would serve no purpose.

The final enzymatic step in the synthesis of a feedback inhibitor is often separated by several intermediate metabolic steps from the substrate (or product) of the enzyme involved in the first step of its biosynthesis. Figure 4-20 gives another example of this situation. The structure of the inhibitor may thus only loosely resemble that of the substrate of the inhibited enzyme, so that we would not expect an end-product inhibitor to combine with the active site (region that binds the substrate) of the enzyme it inactivates. Instead, the end product reversibly binds to a second site on the enzyme, yet nevertheless causes the enzyme activity to be blocked. The inhibitor acts by causing a change in the precise enzyme shape (an **allosteric transformation**), which prevents the enzyme from combining with its substrate (Figure 4-21). Such proteins whose shapes are changed by the

Figure 4-20
Diagram showing how feedback inhibition controls the biosynthesis of pyrimidines in *E. coli*. Note that the end product, cytidine triphosphate (CTP), has no structural similarity to either the substrate or the product of the first reaction in the pathway.

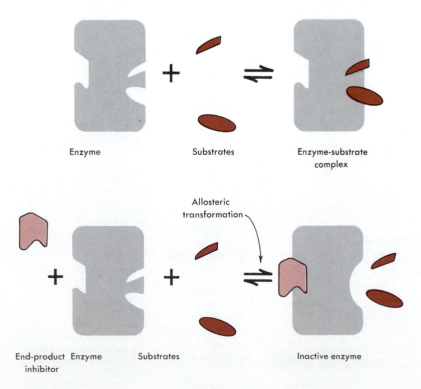

Figure 4-21
Schematic view of how the binding of an end-product inhibitor inhibits an enzyme by causing an allosteric transformation.

binding of specific small molecules at sites other than the active site are called *allosteric proteins;* correspondingly, the small molecules that bring about allosteric transformations are called *allosteric effectors*. The chemical forces binding specific feedback inhibitors to proteins are weak secondary forces such as hydrogen bonds, salt linkages, and van der Waals forces and do not involve covalent bonds. Hence, feedback inhibition can be quickly reversed once the end-product concentration is again reduced to a low level.

Protein Modifications That Lead to Modulations in Enzymatic Activity[42]

A second way enzyme activity can be regulated is by reversible covalent modification. Here the enzyme is not directly sensitive to the concentration of a small molecule, but instead to the action of another enzyme that adds or removes a modifying group. Phosphate, adenyl, uridyl, and methyl groups all are known to modify different proteins and thus to affect their function. In *E. coli*, the activity of the major enzyme for ammonia (and thus nitrogen) uptake, glutamine synthetase, is blocked if the enzyme is adenylated. Its activity decreases progressively as more of its 12 identical subunits acquire an adenyl group on a particular tyrosine side chain. As is true for feedback-inhibited enzymes, the activity of glutamine synthetase ultimately depends on the cellular concentration of its product, glutamine, although very indirectly. The enzyme that adds and removes adenyl groups from glutamine synthetase (adenyltransferase) is controlled by a protein (P_{II}) that in turn is controlled (through addition and removal of uridyl groups) by another enzyme, uridyltransferase, which finally is directly responsive to glutamine concentration (Figure 4-22). Why glutamine synthetase is controlled by this succession of steps rather than directly by glutamine we do not know, but it is a reasonable guess that the regulatory pathway branches in the cell; thus, adenyltransferase, P_{II}, and uridyltransferase also might affect other enzymes responsive to glutamine concentration. Also, other conditions besides the glutamine concentration might affect the

Figure 4-22
Modulation of glutamine synthetase activity. (a) The synthetic pathway from α-ketoglutarate to glutamine. (b) When ammonia is scarce, the concentration of α-ketoglutarate is high and that of glutamine low. The concentrations of glutamine and α-ketoglutarate indirectly affect glutamine synthetase activity by influencing the activity of the enzyme that controls the uridylation of a protein called P_{II}. (c) In turn, the state of the P_{II} protein influences the activity of the enzyme that controls the conversion between the active and inactive (adenylated) states of glutamine synthetase.

adenyltransferase and thus change the activity of glutamine synthetase.

Methylation of Transducer Proteins in Chemotaxis[43-46]

It is through the addition and removal of methyl groups that chemical attractants or repellents control the functioning of bacterial flagella. This movement toward or away from certain chemicals is called **chemotaxis.** *E. coli* cells are propelled in one direction when a set of flagella rotating together counterclockwise collects in a bundle that acts as a propeller. If the concentration of an attractant increases, this rotation is prolonged, so that a cell tends to "run" ("swim") toward the attractant. If a repellent is encountered, on the other hand, the direction of rotation is reversed to clockwise, destroying the efficiency of the propeller and causing the cell to "tumble" for a while (Figure 4-23). When tumbling stops, the cell is likely to move off in a different direction and so escape the repellent. When no chemical is present, the cell is in a neutral state in which it alternately runs and tumbles.

For many attractants and repellents, the signal to swim or to tumble is transmitted from the outside into the cell by **transducer proteins** that span the cell's cytoplasmic membrane. The protein's NH_2-terminal portion extending outside the cell either binds the attractant (or repellent) directly or is bound by other periplasmic proteins that detect the chemical. In either case, when the transducer protein is contacted from the outside, it undergoes some change in its intracellular (COOH-terminal) portion that ultimately signals the flagellar motor to go either counterclockwise (in response to an attractant) or clockwise (in response to a repellent). What this change might be is completely unknown; one guess, however, is that an attractant or repellent stimulates an enzymatic activity in the intracellular domain of the transducer to synthesize a signal molecule that diffuses to the flagellar motor and switches it from one direction to the other. How the cell actually behaves in a complex environment depends on the combined

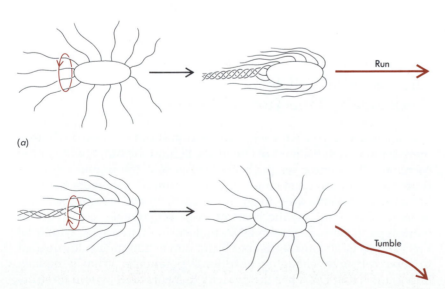

(a)

(b)

Run

Tumble

Figure 4-23
Mechanism of movement during chemotaxis in *E. coli.* (a) An attractant causes a counterclockwise rotation of the flagella, which form a bundle to propel the cell forward. (b) A repellent causes the bundle to rotate clockwise, the bundle falls apart, and the bacterium tumbles.

Figure 4-24
The mechanism of chemotaxis in *E. coli*. Small molecules such as glucose, ribose, galactose, and oxygen may serve as chemoeffectors that pass through outer-membrane pores and interact with receptor proteins in the periplasmic space (P_1, P_2, etc.) and in the cytoplasmic membrane (C_1, C_2, etc.). Messages generated by these interactions are focused through the Tar, Trg, and Tsr proteins (methyl-accepting transducer proteins), which become methylated to a degree determined by the amount of chemoeffector present. These signals are then processed by cytoplasmic proteins (S_1, S_2, etc.) to alter the level of certain response regulators, which in turn interact with proteins R_1 and R_2. Changes in these proteins, which surround the bases of the flagella, somehow regulate the flagellar rotation. [After D. E. Koshland, Jr., *Bacterial Chemotaxis as a Model Behavioral System* (New York: Raven Press, 1980.)]

effect of all signals sent to its motor by all transducing proteins (and certain other signaling systems) that may be sensing a variety of chemicals (Figure 4-24).

A further property of the transducer protein is **adaptation.** If the new concentration of attractant or repellent persists, the transducer soon adapts to this new condition and no longer transmits a signal. This means that the transducer responds to *change* and not to a given concentration of an attractant or repellent. Adaptation is caused by modification of a transducer through methylation and demethylation of about four glutamyl groups on its cytoplasmic side. Methylation favors the state in which a transducer signals clockwise rotation (tumbling), whereas demethylation promotes counterclockwise rotation (running). The modification that occurs opposes the last chemotactic signal, but takes about a minute to occur. Thus, exposure to an attractant eventually leads to greater methylation, allowing more tumbling again. And exposure to a repellent leads to demethylation, so that tumbling once again alternates with periods of swimming. Probably the same conformational change of a transducer protein that produces a signal to the flagellar motor also allows it to bind either the methylation or demethylation enzyme, so that the initial effect of the attractant or repellent is inevitably neutralized.

The Bacterial Cell Is a Precisionally Fine-Tuned Machine

The day has long passed when the question should be asked whether there is more the laws of chemistry behind the functioning of the bacterial cell. We now see the bacterium as an extraordinarily sophisticated set of interrelated molecules that harmoniously work together in highly predictable ways to ensure the growth and selective survival of more of its kind. At the heart of these remarkable, almost clockwork-type, machines are the DNA molecules that encode, with total precision, sets of commands that bring into action molecules needed to cope with ever-changing nutritional potentials. What is equally important is that DNA has the capacity to incorporate within its structure new changes that will permit further evolution of the cell into forms needed to prosper successfully upon the Earth's continuously changing face.

Mycoplasmas Are the Smallest Living Organisms[47, 48]

Existing with genomes some four times smaller than those of *E. coli* are the smallest of procaryotic cells, the mycoplasmas (Figure 4-25). They were at one time referred to as pleuropneumonia-like organisms owing to their initial discovery as disease-causing agents responsible for many of the pneumonia-like respiratory illnesses that affect higher vertebrates, including humans. These tiny, irregularly shaped, flexible organisms essentially have no cell walls and usually live in close contact with cells of higher eucaryotic organisms, on whose surfaces they survive as parasites, deriving many of their essential metabolites from body fluids like those found in our nasal passages. Because they live only in such nutritionally rich environments, these organisms have lost many of the key biosynthetic pathways common to most cells and, in so doing, have lost much of the DNA needed by the more freely living microbes. The smallest of mycoplasmas have chromosomes containing no more than 8 to 10×10^5 base pairs, limiting them to at most 800 to 1000 different proteins. We still do not know how these proteins function (enzymatically versus structurally). Because of their complex nutritional requirements, mycoplasmas are not easy to grow in culture. So despite the obvious attractiveness of their small genomes, they are still studied largely for their disease-causing properties and not because they are probably the smallest organisms on Earth today.

The present restriction of mycoplasmas to very specific ecological niches no doubt reflects the fact that they have too little DNA to compete on equal footing with the true bacteria, most of which possess roughly the same amount of DNA as *E. coli*. Bacterial genomes, as they now exist, represent trade-offs between the advantages of a low DNA content (with the corresponding relatively short time needed for chromosome replication) and the possession of sufficient metabolic pathways to ensure survival in nutritionally unpredictable environments.

Figure 4-25
Electron micrograph of *Mycoplasma mycoides*. (Courtesy of Alan Rodwell.)

Summary

The simplest cells to use in studying the essential chemical features of cell growth and division are the bacteria. They are generally 500 times smaller than the average cell of a higher plant or animal. By far the best-understood bacterium is *Escherichia coli* (*E. coli*), which weighs about 2×10^{-12} gram (10^{12} daltons), of which some 70 percent is water. Even with this small size, an *E. coli* cell is quite complicated chemically, containing at least 3000, and more likely 6000, different molecules. At least 1000 of these components are "small" molecules, and the remainder are macromolecules. The vast majority are either nucleic acids or proteins.

All the essential genes of *E. coli*, if not of all bacteria, are present on a single circular DNA molecule whose replication starts at a fixed point and proceeds bidirectionally. Each chromosome is greatly compacted by supercoiling

into irregularly arranged nucleoid bodies that occupy about 10 percent of the cytoplasmic volume. No membranes separate the chromosomal DNA from the remainder of the cytoplasm, with the result that in all bacteria, the processes of DNA replication, DNA transcription into RNA, and protein synthesis all occur in the same cellular compartment. Ribosomes attach to newly made mRNA molecules soon after their synthesis, and most mRNA molecules are covered by ribosomes carrying nascent polypeptide chains of ever-increasing length. How given RNA molecules become released from their constantly rotating DNA templates is still very much a mystery, as is the way progeny DNA molecules become partitioned into daughter cells.

The cytoplasm of *E. coli*, like that of all cells, is surrounded by a phospholipid-containing cytoplasmic membrane into which are inserted hundreds of different proteins. These proteins become part of the membrane through a complex set of events, in which ribosomes position themselves on the cytoplasm-facing membrane surface so that their polypeptide products grow into and then through it. Among the cytoplasmic membrane proteins are the vital collections of proteins involved in generating ATP through the proton motive force created by movements of electrons through the respiratory chain. The immediate space outside the cytoplasmic ("inner") membrane is occupied by a peptidoglycan gel, which, in turn, is attached to a second "outer" phospholipid membrane. Major components of this outer membrane include highly immunogenic lipopolysaccharide-containing proteins as well as matrix proteins (porins) that form specific holes through which extracellular molecules must pass to reach the inner cytoplasmic membrane.

The holes within the peptidoglycan gel constitute the periplasmic space. In it are found hydrolytic enzymes that initiate the degradation of food molecules, binding proteins involved in initiating transport across the cytoplasmic membrane, and chemoreceptors that sense attractants and repellents.

Many bacteria, including *E. coli*, contain two forms of linear surface appendages that arise from the cytoplasmic membrane, penetrate the periplasmic space, and extend out through the external surface. The most conspicuous are the flagella, organelles of locomotion. They generate movement by using the energy produced by the proton motive force to rotate. Rotation in a counterclockwise direction leads to formation of bundles of flagella that propel cells smoothly forward; clockwise rotation, on the other hand, disperses the bundle and causes the bacteria to move erratically (tumble). A flagellum is an inherently complex structure containing a basal structure and hook as well as an extended filament made up of very large numbers of the basic protein building block flagellin. Much less complex are the even thinner pili, filaments made from single protein building blocks called pilin. Pili often function to attach bacteria to other surfaces, often glycoproteins on the surfaces of higher cells.

All bacterial molecules arise from metabolic transformation, usually starting from simple food molecules like glu-

cose. Individual small molecules usually exist only fleetingly before transformation into other small molecules or polymerization into macromolecules. Most compounds cannot be directly transformed into a large number of other compounds. Instead, each compound is a step in a series of reactions leading either to the degradation of a food molecule or to the biosynthesis of a necessary cellular molecule such as an amino acid or fatty acid. Cellular metabolism is the sum total of a large number of such pathways connected in such a way that products of degradative pathways can be used to initiate specific biosynthetic pathways.

Connecting the biosynthetic and degradation pathways are the so-called central pathways that lead to the production of the 12 key precursor metabolites that initiate individual biosynthetic pathways. While central pathways operate virtually all the time, peripheral biosynthetic or degradative pathways are turned on or off as units, depending on need. Such decisions are often made for biosynthetic reactions by feedback (end-product) inhibition of enzyme function. An end-product metabolite can reversibly combine with the first enzyme involved in its specific biosynthetic pathway. This combination transforms the enzyme into an inactive form. The end-product inhibitor does not combine with the enzymatically active site but binds instead to a second site on the enzyme, causing a change in the enzyme shape. Proteins whose shapes and activities are changed by combination with other molecules are called allosteric proteins. Modulation in enzyme activity can also be effected by reversible covalent modifications, which are affected by still other enzymes. Additions of adenyl, phosphate, and uridyl groups can drastically influence the functioning of given enzymes.

The complexity of the metabolic map of an organism is related to the amount of genetic information (DNA) in the organism. The amount of DNA in a cell places an upper limit on the different number of enzymes the cell can produce. *E. coli* possesses sufficient DNA to code for the amino acid sequences of 3000 to 4500 different proteins. About 1700 different small molecules have now been detected in *E. coli*. The metabolic reactions that we already know of in *E. coli* must account for at least one-fifth of its total metabolism.

Bibliography

General References

Brock, T. D. 1984. *Biology of Microorganisms*. 4th ed. Englewood Cliffs, N.J.: Prentice-Hall. A good basic introduction to general microbiology.

Davis, B. D., R. Dulbecco, H. N. Eisen, and H. S. Ginsberg. 1980. *Microbiology*. 3rd ed. New York: Harper & Row. An excellent text that is medically oriented but also contains good general information.

Ingraham, J. L., O. Maaløe, and F. C. Neidhardt. 1983. *The Growth of the Bacterial Cell*. Sunderland, Mass.: Sinauer Associates. An excellent, more advanced text.

Joklick, W. K., H. P. Willet, and D. B. Amos, eds. 1984. *Zinsser Microbiology*. New York: Appleton-Century-Crofts. Text emphasizes medical microbiology but also contains good general information.

Losick, R., and L. Shapiro. 1984. *Microbial Development*. Cold Spring Harbor, N.Y.: Cold Spring Harbor Laboratory.

Stanier, R. Y., J. L. Ingraham, M. L. Wheelis, and P. R. Painter. 1986. *The Microbial World*. 5th ed. Englewood Cliffs, N.J.: Prentice-Hall. A superb treatment of microbiology that is somewhat more detailed than the text by Brock.

Cited References

1. Luria, S. E., and M. Delbrück. 1943. "Mutations in Bacteria from Virus Sensitivity to Virus Resistance." *Genetics* 28:491–511.
2. Lederberg, J., and E. L. Tatum. 1946. "Novel Genotypes in Mixed Cultures of Biochemical Mutants of Bacteria." *Cold Spring Harbor Symp. Quant. Biol.* 11:113–114.
3. Deley, J., and K. Kersters. 1975. "Biochemical Evolution in Bacteria." *Comp. Biochem.* 29B:1.
4. Margulis, L. 1970. *Origin of Eukaryotic Cells*. New Haven, Conn.: Yale University Press.
5. Bauchop, T., and S. R. Elsden. 1960. "The Growth of Microorganisms in Relation to Their Energy Supply." *J. Gen. Microbiol.* 23:457–469.
6. Marr, A. G., E. H. Nilson, and D. J. Clark. 1963. "The Mainte-

nance Requirements of *Escherichia coli.*" *Ann. N.Y. Acad. Sci.* 102:536–548.

7. Van Niel, C. B. 1971. "Techniques for the Enrichment, Isolation and Maintenance of the Photosynthetic Bacteria." *Methods in Enzymol.* 23:3–28. Includes a broad overview of the ecology of photosynthetic bacteria.

8. Delbrück, M. 1946. "Experiments with Bacterial Viruses (Bacteriophages)." *Harvey Lecture Series* 41:161–187.

9. Delbrück, M. 1966. "A Physicist Looks at Biology." In *Phage and the Origins of Molecular Biology*, ed. J. Cairns, G. S. Stent, and J. D. Watson. Cold Spring Harbor, N.Y.: Cold Spring Harbor Laboratory, pp. 9–22.

10. Helmstetter, C. E., and J. D. Cummings. 1963. "Bacterial Synchronization by Selection of Cells at Division." *Proc. Nat. Acad. Sci.* 50:767–774.

11. Herendeen, S. L., R. A. van Bogelen, and F. C. Neidhart. 1979. "Levels of Major Proteins of *E. coli* During Growth at Different Temperatures." *J. Bacteriol.* 139:185–194.

12. Anderson, K. B., and K. von Meyenburg. 1980. "Are Growth Rates of *E. coli* in Batch Culture Limited by Respiration?" *J. Bacteriol.* 144:114–123.

13. Mendelson, N. H. 1982. "Bacterial Growth and Division: Genes, Structures, Forces and Clocks." *Microbiol. Revs.* 46:341–375.

14. Ratkowsky, D. A., J. Olley, T. A. Meekin, and A. Ball. 1982. "Relationship Between Temperature and Growth Rate of Bacterial Cultures." *J. Bacteriol.* 149:1–5.

15. O'Farrell, P. H. 1975. "High Resolution Two-Dimensional Electrophoresis of Proteins." *J. Biol. Chem.* 250:4007–4021.

16. Phillips, T. A., P. L. Bloch, and F. C. Neidhart. 1980. "Protein Identifications on O'Farrell's 2-Dimensional Gels: Locations of 55 Additional *E. coli* Proteins." *J. Bacteriol.* 144:1024–1033.

17. Kellenberger, E. 1960. "The Physical State of the Bacterial Nucleus." In *Microbial Genetics*, ed. W. Hayes and R. C. Clowes. New York: Cambridge University Press, pp. 39–60.

18. Costerton, J. W. 1979. "The Role of Electron Microscopy in the Elucidation of Bacterial Structure and Function." *Ann. Rev. Microbiol.* 33:459–479.

19. Dubochet, J., A. W. McDowall, B. Menge, E. N. Schmid, and K. G. Lickfeld. 1983. "Electron Microscopy of Frozen-Hydrated Bacteria." *J. Bacteriol.* 155:381–390.

20. Cronan, J. 1978. "Molecular Biology of Bacterial Membrane Lipids." *Ann. Rev. Biochem.* 47:163–189.

21. Osborn, M. J., and H. C. P. Wu. 1980. "Proteins of the Outer Membrane of Gram-Negative Bacteria." *Ann. Rev. Microbiol.* 34:369–422.

22. Shockman, G. D., and J. F. Barrett. 1983. "Structure, Function and Assembly of Cell Walls of Gram-Positive Bacteria." *Ann. Rev. Microbiol.* 37:501–527.

23. Hobot, J. A., E. Carlemalm, W. Villiger, and E. Kellenberger. 1985. "Periplasmic Gel: New Concept Resulting from the Reinvestigation of Bacterial Cell Envelope Ultrastructure by New Methods." *J. Bacteriol.* 160:143–152.

24. Nogami, T., and S. Mizushima. 1983. "Outer Membrane Porins Are Important in the Maintenance of the Surface Structure of *E. coli* Cells." *J. Bacteriol.* 156:402–408.

25. Sutcliffe, J., R. Blumenthal, A. Walter, and J. Foulds. 1983. "*E. coli* Outer Membrane Protein K Is a Porin." *J. Bacteriol.* 156:867–872.

26. Hinkle, P. C., and R. E. McCarthy. 1978. "How Cells Make ATP." *Sci. Amer.* 238(3):104–123.

27. Davis, B. D., and P. C. Tai. 1980. "The Mechanism of Protein Secretion Across Membranes." *Nature* 283:433–438.

28. Berg, H. C., and R. A. Anderson. 1973. "Bacteria Swim by Rotating Their Flagellar Filaments." *Nature* 245:380–382.

29. Silverman, M., and M. I. Simon. 1977. "Bacterial Flagella." *Ann. Rev. Microbiol.* 31:397–419.

30. Achtman, M., G. Morelli, and S. Schwuchow. 1978. "Cell-Cell Interactions in Conjugating *E. coli*: Role of F Pili and Fate of Mating Aggregates." *J. Bacteriol.* 135:1053–1061.

31. McMichael, M. C., and J. T. Ou. 1979. "Structure of Common Pili from *E. coli.*" *J. Bacteriol.* 138:969–975.

32. Mett, H., L. Kloetzlen, and K. Vosbeck. 1983. "Properties of Pili from *E. coli* SS142 that Mediate Mannose-Resistant Adhesion to Mammalian Cells." *J. Bacteriol.* 153:1038–1044.

33. Hobot, J. A., W. Villiger, J. Escaig, M. Maeder, A. Ryder, and E. Kellenberger. 1985. "Shape and Fine Structure of Nucleoids Observed on Sections of Ultrarapidly Frozen and Cryosubstituted Bacteria." *J. Bacteriol.* 162:960–971.

34. Millar, O. L., B. R. Beatty, B. A. Hamkalo, and C. A. Thomas, Jr. 1970. "Electron Microscopic Visualization of Transcription." *Cold Spring Harbor Symp. Quant. Biol.* 35:505–512.

35. Silhavy, T. J., S. A. Benson, and S. D. Emr. 1983. "Mechanisms of Protein Localization." *Microbiol. Revs.* 47(3):313–344.

36. Atkinson, D. E. 1977. *Cellular Energy Metabolism and Its Regulation.* New York: Academic Press.

37. Gottschalk, G. 1979. *Bacterial Metabolism.* New York: Springer-Verlag.

38. Beckwith, J. R., and D. Zipser, eds. 1970. *The Lactose Operon.* Cold Spring Harbor, N.Y.: Cold Spring Harbor Laboratory.

39. Umbarger, H. E. 1956. "Evidence for a Negative Feedback Mechanism in the Biosynthesis of Isoleucine." *Science* 123:848.

40. Yates, R. A., and A. B. Pardee. 1956. "Control of Pyrimidine Biosynthesis in *E. coli* by a Feedback Mechanism." *J. Biol. Chem.* 221:757–770.

41. Umbarger, H. E. 1978. "Amino Acid Synthesis and Its Regulation." *Ann. Rev. Biochem.* 47:533–606.

42. Ginsberg, A., and E. R. Stadtman. 1973. "Regulation of Glutamine Synthetase in *E. coli.*" In *The Enzymes of Glutamine Metabolism*, ed. S. Prusiner and E. R. Stadtman. New York: Academic Press, pp. 9–43.

43. Alder, J. 1969. "Chemoreceptors in Bacteria." *Science* 166:1588–1597.

44. DeFranco, A. L., and D. E. Koshland, Jr. 1980. "Multiple Methylation in Processing of Sensory Signals During Bacterial Chemotaxis." *Proc. Nat. Acad. Sci.* 77(5):2429–2433.

45. Koshland, E. E., Jr. 1981. "Biochemistry of Sensing and Adaptation in a Simple Bacterial System." *Ann. Rev. Biochem.* 50:765–770.

46. Boyd, A., and M. I. Simon. 1982. "Bacterial Chemotaxis." *Ann. Rev. Physiol.* 44:501–517.

47. Smith, P. F. 1971. *The Biology of Mycroplasmas.* New York: Academic Press.

48. Kawauchis, Y., A. Muto, and S. Osawa. 1982. "The Protein Composition of *Mycoplasma capriola.*" *Mol. Gen. Genetics* 188:7–11.

CHAPTER 5

The Importance of Weak Chemical Interactions

Until now, we have focused our attention on discrete organic molecules and, following classical organic chemistry, have emphasized the covalent bonds that hold them together. It takes little insight, however, to realize that this type of analysis is inadequate for describing a cell and that we must also concern ourselves with the exact shapes of molecules and with the forces that bind them together in an organized fashion. The distribution of molecules within cells is not random, and we must ask ourselves what chemical laws determine this distribution. Clearly, covalent bonding cannot be involved; by definition, atoms united by covalent bonds belong to the same molecule.

The arrangement of distinct molecules within cells is controlled by chemical bonds much weaker than covalent bonds. Atoms united by covalent bonds are capable of weak interactions with nearby atoms. These interactions, sometimes called "secondary bonds," occur not only between atoms of different molecules, but also between atoms within the same molecule. Weak bonds are important in determining which molecules lie next to each other and in giving shape to flexible molecules such as the polypeptides and polynucleotides. It is therefore useful to understand the nature of weak chemical interactions and to understand how their "weak" character makes them indispensable to cellular existence. The most important such bonds include van der Waals bonds, hydrophobic bonds, hydrogen bonds, and ionic bonds.

Definition and Some Characteristics of Chemical Bonds[1]

A **chemical bond** is an attractive force that holds atoms together. Aggregates of finite size are called molecules. Originally, it was thought that only covalent bonds hold atoms together in molecules; but now, weaker attractive forces are known to be important in holding together many macromolecules. For example, the four polypeptide chains of hemoglobin are held together by the combined action of several weak bonds. It is thus now customary also to call weak positive interactions chemical bonds, even though they are not strong enough, when present singly, to effectively bind two atoms together.

Chemical bonds are characterized in several ways. An obvious characteristic of a bond is its strength. Strong bonds almost never fall apart at physiological temperatures. This is why atoms united by co-valent bonds always belong to the same molecule. Weak bonds are easily broken, and when they exist singly, they exist fleetingly. Only when present in ordered groups do weak bonds last a long time. The strength of a bond is correlated with its length, so that two atoms connected by a strong bond are always closer together than the same two atoms held together by a weak bond. For example, two hydrogen atoms bound covalently to form a hydrogen molecule (H:H) are 0.74 Å apart, whereas the same two atoms held together by van der Waals forces are 1.2 Å apart.

Another important characteristic is the maximum number of bonds that a given atom can make. The number of covalent bonds that an atom can form is called its **valence.** Oxygen, for example, has a va-lence of two: It can never form more than two covalent bonds. There is more variability in the case of van der Waals bonds, where the limiting factor is purely steric. The number of possible bonds is lim-ited only by the number of atoms that can simultaneously touch each other. The formation of hydrogen bonds is subject to more restric-tions. A covalently bonded hydrogen atom usually participates in only one hydrogen bond, whereas an oxygen atom seldom partici-pates in more than two hydrogen bonds.

The angle between two bonds originating from a single atom is called the **bond angle.** The angle between two specific covalent bonds is always approximately the same. For example, when a carbon atom has four single covalent bonds, they are directed tetrahedrally (bond angle = 109°). In contrast, the angles between weak bonds are much more variable.

Bonds differ also in the **freedom of rotation** they allow. Single cova-lent bonds permit free rotation of bound atoms (Figure 5-1), whereas double and triple bonds are quite rigid. Bonds with partial double-bond character, such as the peptide bond, are also quite rigid. For that reason, the carbonyl (C−O) and imino (N−H) groups bound to-gether by the peptide bond must lie in the same plane (Figure 5-2). Much weaker, ionic bonds, on the other hand, impose no restrictions on the relative orientations of bonded atoms.

Chemical Bonds Are Explainable in Quantum-Mechanical Terms

The nature of the forces, both strong and weak, that give rise to chemical bonds remained a mystery to chemists until the quantum theory of the atom (quantum mechanics) was developed in the 1920s. Then, for the first time, the various empirical laws about how chemi-cal bonds are formed were put on a firm theoretical basis. It was realized that all chemical bonds, weak as well as strong, are based on electrostatic forces. Quantum mechanics provided explanations for covalent bonding by the sharing of electrons and also for the forma-tion of weaker bonds.

Chemical-Bond Formation Involves a Change in the Form of Energy

The spontaneous formation of a bond between two atoms always involves the release of some of the internal energy of the unbonded atoms and its conversion to another energy form. The stronger the

Figure 5-1
Rotation about the C_5–C_6 bond in glu-cose. This carbon-carbon bond is a sin-gle bond, and so any of the three con-figurations, (a), (b), or (c), may occur.

Figure 5-2
The planar shape of the peptide bond. Shown here is a portion of an extended polypeptide chain. Almost no rotation is possible about the peptide bond be-cause of its partial double-bond charac-ter:

All the atoms in the gray area must lie in the same plane. Rotation is possible, however, around the remaining two bonds, which make up the polypeptide configurations. [After L. Pauling, *The Nature of the Chemical Bond*, 3rd ed. (Ithaca, N.Y.: Cornell University Press, 1960), p. 498, with permission.]

bond, the greater the amount of energy released upon its formation. The bonding reaction between two atoms A and B is thus described by

$$A + B \longrightarrow AB + energy \qquad (5\text{-}1)$$

where AB represents the bonded aggregate. The rate of the reaction is directly proportional to the frequency of collision between A and B. The unit most often used to measure energy is the calorie, the amount of energy required to raise the temperature of 1 gram of water from 14.5°C to 15.5°C. Since thousands of calories are usually involved in the breaking of a mole of chemical bonds, most energy changes within chemical reactions are expressed in kilocalories per mole.

However, atoms joined by chemical bonds do not remain together forever, since there also exist forces that break chemical bonds. By far the most important of these forces arises from heat energy. Collisions with fast-moving molecules or atoms can break chemical bonds. During a collision, some of the kinetic energy of a moving molecule is given up as it pushes apart two bonded atoms. The faster a molecule is moving (the higher the temperature), the greater the probability that, upon collision, it will break a bond. Hence, as the temperature of a collection of molecules is increased, the stability of their bonds decreases. The breaking of a bond is thus always indicated by the formula

$$AB + energy \longrightarrow A + B \qquad (5\text{-}2)$$

The amount of energy that must be added to break a bond is exactly equal to the amount that was released upon formation of the bond. This equivalence follows from the first law of thermodynamics, which states that energy (except as it is interconvertible with mass) can be neither made nor destroyed.

Equilibrium Between Bond Making and Breaking

Every bond is thus a result of the combined actions of bond-making and bond-breaking forces. When an equilibrium is reached in a closed system, the number of bonds forming per unit time will equal the number of bonds breaking. Then the proportion of bonded atoms is described by the following mass action formula:

$$K_{eq} = \frac{conc^{AB}}{conc^{A} \times conc^{B}} \qquad (5\text{-}3)$$

where K_{eq} is the **equilibrium constant**, and $conc^{A}$, $conc^{B}$, and $conc^{AB}$ are the concentrations of A, B, and AB, respectively, in moles per liter. Whether we start with only free A and B, with only the molecule AB, or with a combination of AB and free A and B, at equilibrium the proportions of A, B, and AB will reach the concentrations given by K_{eq}.

The Concept of Free Energy[2, 3]

There is always a change in the form of energy as the proportion of bonded atoms moves toward the equilibrium concentration. Biologically, the most useful way to express this energy change is through the physical chemist's concept of **free energy,** denoted by the symbol *G*.* We shall not give a rigorous description of free energy in this text nor show how it differs from the other forms of energy. For this, the reader must refer to a chemistry text that discusses the second law of thermodynamics. It must suffice to say here that *free energy is energy that has the ability to do work.*

The second law of thermodynamics tells us that a decrease in free energy (ΔG is negative) always occurs in spontaneous reactions. When equilibrium is reached, however, there is no further change in the amount of free energy ($\Delta G = 0$). The equilibrium state for a closed collection of atoms is thus the state that contains the least amount of free energy.

The free energy lost as equilibrium is approached is either transformed into heat or used to increase the amount of entropy. We shall not attempt to define entropy here except to say that the amount of entropy is a measure of the amount of disorder. The greater the disorder, the greater the amount of entropy. The existence of entropy means that many spontaneous chemical reactions (those with a net decrease in free energy) need not proceed with an evolution of heat. For example, when sodium chloride (NaCl) is dissolved in water, heat is absorbed rather than released. There is, nonetheless, a net decrease in free energy because of the increase in disorder of the sodium and chlorine ions as they move from a solid to a liquid phase.

K_{eq} Is Exponentially Related to ΔG

Clearly, the stronger the bond, and hence the greater the change in free energy (ΔG) that accompanies its formation, the greater the proportion of atoms that must exist in the bonded form. This common-sense idea is quantitatively expressed by the physical-chemical formula

$$\Delta G = -RT \ln K_{eq} \quad \text{or} \quad K_{eq} = e^{-\Delta G/RT} \qquad (5\text{-}4)$$

where *R* is the universal gas constant, *T* is the absolute temperature, ln is the logarithm (of K_{eq}) to the base *e*, K_{eq} is the equilibrium constant, and *e* = 2.718.

Insertion of the appropriate values of *R* (1.987 cal/deg-mol) and *T* (298 at 25°C) tells us that ΔG values as low as 2 kcal/mol can drive a bond-forming reaction to virtual completion if all reactants are present at molar concentrations (Table 5-1).

Covalent Bonds Are Very Strong

The ΔG values accompanying the formation of covalent bonds from free atoms such as hydrogen or oxygen are very large and negative in sign, usually -50 to -110 kcal/mol. Equation 5-4 tells us that K_{eq} of the bonding reaction will be correspondingly large, and so the concentration of hydrogen or oxygen atoms existing unbound will be very small. For example, with a ΔG value of -100 kcal/mol, if we start with 1 mol/L of the reacting atoms, only one in 10^{40} atoms will remain unbound when equilibrium is reached.

Table 5-1 The Numerical Relationship Between the Equilibrium Constant and ΔG at 25°C

K_{eq}	ΔG, kcal/mol
0.001	4.089
0.01	2.726
0.1	1.363
1.0	0
10.0	-1.363
100.0	-2.726
1000.0	-4.089

---

*It was formerly the custom in the United States to refer to free energy by the symbol *F*. Now, however, most texts use the international symbol *G*, which honors the great nineteenth-century physicist Josiah Gibbs.

Weak Bonds Have Energies Between 1 and 7 kcal/mol

The main types of weak bonds important in biological systems are the van der Waals bonds, hydrophobic bonds, hydrogen bonds, and ionic bonds. Sometimes, as we shall soon see, the distinction between a hydrogen bond and an ionic bond is arbitrary. The weakest bonds are the van der Waals bonds. These have energies (1 to 2 kcal/mol) only slightly greater than the kinetic energy of heat motion. The energies of hydrogen and ionic bonds range between 3 and 7 kcal/mol.

In liquid solutions, almost all molecules form a number of weak bonds to nearby atoms. All molecules are able to form van der Waals bonds, whereas hydrogen and ionic bonds can form only between molecules that have a net charge (ions) or in which the charge is unequally distributed. Some molecules thus have the capacity to form several types of weak bonds. Energy considerations, however, tell us that molecules always have a greater tendency to form the stronger bond.

Weak Bonds Are Constantly Made and Broken at Physiological Temperatures

The energy of the strongest weak bond is only about ten times larger than the average energy of kinetic motion (heat) at 25°C (0.6 kcal/mol). Since there is a significant spread in the energies of kinetic motion, many molecules with sufficient kinetic energy to break the strongest weak bond always exist at physiological temperatures.

Enzymes Are Not Involved in Making or Breaking Weak Bonds

The average lifetime of a single weak bond is only a fraction of a second. Thus, cells do not need a special mechanism to speed up the rate at which weak bonds are made and broken. Not surprisingly, therefore, enzymes never participate in reactions involving only weak bonds.

The Distinction Between Polar and Nonpolar Molecules

All forms of weak interactions are based on attractions between electric charges. The separation of electric charges can be permanent or temporary, depending on the atoms involved. For example, the oxygen molecule (O:O) has a symmetric distribution of electrons between its two oxygen atoms, so each of its two atoms is uncharged. In contrast, there is a nonuniform distribution of charge in water (H:O:H), in which the bond electrons are unevenly shared (Figure 5-3). They are held more strongly by the oxygen atom, which thus carries a considerable negative charge, whereas the two hydrogen atoms together have an equal amount of positive charge. The center of the positive charge is on one side of the center of the negative charge. A combination of separated positive and negative charges is

van der Waals radius of hydrogen = 1.2 Å

Covalent bond length = 0.95 Å

van der Waals radius of oxygen = 1.4 Å

105°

Direction of dipole moment

Figure 5-3
The structure of a water molecule.

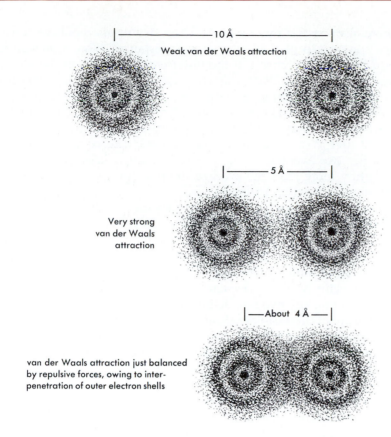

|————— 10 Å —————|
Weak van der Waals attraction

|———— 5 Å ————|
**Very strong
van der Waals
attraction**

|——About 4 Å ——|
**van der Waals attraction just balanced
by repulsive forces, owing to inter-
penetration of outer electron shells**

Figure 5-4
Variation of van der Waals forces with distance. The atoms shown in this diagram are atoms of the inert rare gas argon. (Redrawn from *General Chemistry*, 2nd ed., by Linus Pauling. W. H. Freeman and Company, copyright © 1953, p. 322, with permission.)

called an electric **dipole moment.** Unequal electron sharing reflects dissimilar affinities of the bonding atoms for electrons. Atoms that have a tendency to gain electrons are called **electronegative** atoms. **Electropositive** atoms have a tendency to give up electrons.

Molecules (such as H_2O) that have a dipole moment are called **polar molecules. Nonpolar molecules** are those with no effective dipole moments. In methane (CH_4), for example, the carbon and hydrogen atoms have similar affinities for their shared electron pairs, so neither the carbon nor the hydrogen atom is noticeably charged.

The distribution of charge in a molecule can also be affected by the presence of nearby molecules, particularly if the affected molecule is polar. The effect may cause a nonpolar molecule to acquire a slightly polar character. If the second molecule is not polar, its presence will still alter the nonpolar molecule, establishing a fluctuating charge distribution. Such induced effects, however, give rise to a much smaller separation of charge than is found in polar molecules, resulting in smaller interaction energies and correspondingly weaker chemical bonds.

Van der Waals Forces[4]

Van der Waals bonding arises from a nonspecific attractive force originating when two atoms come close to each other. It is based not on the existence of permanent charge separations, but rather on the induced fluctuating charges caused by the nearness of molecules. It therefore operates between all types of molecules, nonpolar as well as polar. It depends heavily on the distance between the interacting groups, since the bond energy is inversely proportional to the sixth power of distance (Figure 5-4).

Table 5-2 Van der Waals Radii of the Atoms in Biological Molecules

Atom	van der Waals radius (Å)
H	1.2
N	1.5
O	1.4
P	1.9
S	1.85
CH_3 group	2.0
Half thickness of aromatic molecule	1.7

There also exists a more powerful van der Waals repulsive force, which comes into play at even shorter distances. This repulsion is caused by the overlapping of the outer electron shells of the atoms involved. The van der Waals attractive and repulsive forces balance at a certain distance specific for each type of atom. This distance is the so-called **van der Waals radius** (Table 5-2 and Figure 5-5). The van der Waals bonding energy between two atoms separated by the sum of their van der Waals radii increases with the size of the respective atoms. For two average atoms, it is only about 1 kcal/mol, which is just slightly more than the average thermal energy of molecules at room temperature (0.6 kcal/mol).

This means that van der Waals forces are an effective binding force at physiological temperatures only when several atoms in a given molecule are bound to several atoms in another molecule. Then the energy of interaction is much greater than the dissociating tendency resulting from random thermal movements. For several atoms to interact effectively, the molecular fit must be precise, since the distance separating any two interacting atoms must not be much greater than the sum of their van der Waals radii (Figure 5-6). The strength of interaction rapidly approaches zero when this distance is only slightly exceeded. Thus, the strongest type of van der Waals contact arises when a molecule contains a cavity exactly complementary in shape to a protruding group of another molecule. This is the type of situation thought to exist between an antigen and its specific antibody (Figure 5-7). In this instance, the binding energies sometimes can be as large as 20 to 30 kcal/mol, so that antigen-antibody complexes seldom fall apart. The bonding pattern of polar molecules is rarely dominated by van der Waals interactions, since such molecules can acquire a lower energy state (lose more free energy) by forming other types of bonds.

Hydrogen Bonds[5, 6]

A hydrogen bond is formed between a covalently bound donor hydrogen atom with some positive charge and a negatively charged, covalently bound acceptor atom (Figure 5-8). For example, the hydrogen atoms of the imino (NH) group are attracted by the negatively charged keto (CO) oxygen atoms. Sometimes, the hydrogen-bonded atoms belong to groups with a unit of charge (e.g., NH_3^+ or COO^-). In other cases, both the donor hydrogen atoms and the negative acceptor atoms have less than a unit of charge.

The biologically most important hydrogen bonds involve hydrogen atoms covalently bound to oxygen atoms (O–H) or nitrogen atoms (N–H). Likewise, the negative acceptor atoms are usually nitrogen or oxygen. Table 5-3 lists some of the most important hydrogen bonds. In the absence of surrounding water molecules, bond energies range between 3 and 7 kcal/mol, the stronger bonds involving the greater charge differences between donor and acceptor atoms. Hydrogen bonds are thus weaker than covalent bonds, yet considerably stronger than van der Waals bonds. A hydrogen bond, therefore, will hold two atoms closer together than the sum of their van der Waals radii, but not so close together as a covalent bond would hold them.

Hydrogen bonds, unlike van der Waals bonds, are highly directional. In the strongest hydrogen bonds, the hydrogen atom points directly at the acceptor atom (Figure 5-9). If it points more than 30° away, the bond energy is much less. Hydrogen bonds are also much

Acetate

Glycine

Guanine

Figure 5-5
Drawings of several molecules with the van der Waals radii of the atoms shown in color.

Figure 5-6
The arrangement of molecules in a layer of a crystal formed by the amino acid glycine. The packing of the molecules is determined by the van der Waals radii of the groups, except for the N–H⸺O contacts, which are shortened by the formation of hydrogen bonds. [After L. Pauling, *The Nature of the Chemical Bond*, 3rd ed. (Ithaca, N.Y.: Cornell University Press, 1960), p. 262, with permission.]

more specific than van der Waals bonds, since they demand the existence of molecules with complementary donor hydrogen and acceptor groups.

Some Ionic Bonds Are Hydrogen Bonds

Many organic molecules possess ionic groups that contain one or more units of net positive or negative charge. The negatively charged mononucleotides, for example, contain phosphate groups (PO_3^{3-}) with three units of negative charge, whereas each amino acid (except proline) has a negative carboxyl group (COO^-) and a positive amino group (NH_3^+), both of which carry a unit of charge. These charged groups are usually neutralized by nearby, oppositely charged groups. The electrostatic forces acting between the oppositely charged groups are called ionic bonds. Their average bond energy in an aqueous solution is about 5 kcal/mol.

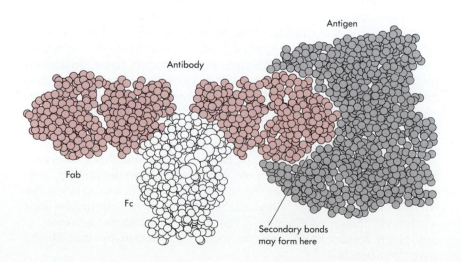

Figure 5-7
Antibody-antigen interaction. Complementary structure allows van der Waals contact between antibody and antigen. Note that the antigen may totally enclose the Fab portion (colored) of the antibody. The other, unoccupied Fab arm may interact with another antigen of similar type.

Hydrogen bond between peptide groups

Hydrogen bond between two hydroxyl groups

Hydrogen bond between a charged carboxyl group and the hydroxyl group of tyrosine

Hydrogen bond between a charged amino group and a charged carboxyl group

Figure 5-8
Example of hydrogen bonds in biological molecules.

Table 5-3   Approximate Bond Lengths of Biologically Important Hydrogen Bonds

Bond	Approximate H bond length. (Å)
$O-H \cdots O$	2.70 ± 0.10
$O-H \cdots O^-$	2.63 ± 0.10
$O-H \cdots N$	2.88 ± 0.13
$N-H \cdots O$	3.04 ± 0.13
$N^{\pm}H \cdots O$	2.93 ± 0.10
$N-H \cdots N$	3.10 ± 0.13

In many cases, either an inorganic cation like Na^+, K^+, or Mg^{2+} or an inorganic anion like Cl^- or SO_4^{2-} neutralizes the charge of ionized organic molecules. When this happens in aqueous solution, the neutralizing cations and anions do not carry fixed positions because inorganic ions are usually surrounded by shells of water molecules and so do not directly bind to oppositely charged groups. Thus, it is now believed that in water solutions, electrostatic bonds to surrounding inorganic cations or anions are usually not of primary importance in determining the molecular shapes of organic molecules.

On the other hand, highly directional bonds result if the oppositely charged groups can form hydrogen bonds to each other. For example, COO^- and NH_3^+ groups are often held together by hydrogen bonds. Since these bonds are stronger than those that involve groups with less than a unit of charge, they are correspondingly shorter. A strong hydrogen bond can also form between a group with a unit charge and a group having less than a unit charge. For example, a hydrogen atom belonging to an amino group (NH_2) bonds strongly to an oxygen atom of a carboxyl group (COO^-).

Weak Interactions Demand Complementary Molecular Surfaces[7]

Weak binding forces are effective only when the interacting surfaces are close. This proximity is possible only when the molecular surfaces have **complementary structures,** so that a protruding group (or positive charge) on one surface is matched by a cavity (or negative charge) on another. That is, the interacting molecules must have a lock-and-key relationship. In cells, this requirement often means that some molecules hardly ever bond to other molecules of the same kind, because such molecules do not have the properties of symmetry necessary for self-interaction. For example, some polar molecules contain donor hydrogen atoms and no suitable acceptor atoms, whereas other molecules can accept hydrogen bonds but have no hydrogen atoms to donate. On the other hand, there are many molecules with the necessary symmetry to permit strong self-interaction in cells, water being the most important example.

Water Molecules Form Hydrogen Bonds

Under physiological conditions, water molecules rarely ionize to form H^+ and OH^- ions. Instead, they exist as polar $H-O-H$ molecules with both the hydrogen and oxygen atoms forming strong hydrogen bonds. In each water molecule, the oxygen atom can bind to two external hydrogen atoms, whereas each hydrogen atom can bind to one adjacent oxygen atom. These bonds are directed tetrahedrally (Figure 5-10), so in its solid and liquid forms, each water molecule tends to have four nearest neighbors, one in each of the four directions of a tetrahedron. In ice, the bonds to these neighbors are very rigid and the arrangement of molecules fixed. Above the melting temperature (0°C), the energy of thermal motion is sufficient to break the hydrogen bonds and to allow the water molecules to change their nearest neighbors continually. Even in the liquid form, however, at any given instant most water molecules are bound by four strong hydrogen bonds.

(a)   (b)

Figure 5-9
Directional properties of hydrogen bonds. (a) The vector along the covalent O–H bond points directly at the acceptor oxygen, thereby forming a strong bond. (b) The vector points away from the oxygen atom, resulting in a much weaker bond.

Weak Bonds Between Molecules in Aqueous Solutions

The average energy of a secondary bond, though small compared to that of a covalent bond, is nonetheless strong enough compared to heat energy to ensure that most molecules in aqueous solution will form secondary bonds to other molecules. The proportion of bonded to nonbonded arrangements is given by Equation 5-4, corrected to take into account the high concentration of molecules in a liquid. It tells us that interaction energies as low as 2 to 3 kcal/mol are sufficient at physiological temperatures to force most molecules to form the maximum number of strong secondary bonds.

The specific structure of a solution at a given instant is markedly influenced by which solute molecules are present, not only because molecules have specific shapes, but also because molecules differ in which types of secondary bonds they can form. Thus, a molecule will tend to move until it is next to a molecule with which it can form the strongest possible bond.

Solutions, of course, are not static. Because of the disruptive influence of heat, the specific configuration of a solution is constantly changing from one arrangement to another of approximately the same energy content. Equally important in biological systems is the fact that metabolism is continually transforming one molecule into another and so automatically changing the nature of the secondary bonds that can be formed. The solution structure of cells is thus constantly disrupted not only by heat motion, but also by the metabolic transformations of the cell's solute molecules.

Organic Molecules That Tend to Form Hydrogen Bonds Are Water Soluble

The energy of hydrogen bonds per atomic group is much greater than that of van der Waals contacts; thus, molecules will form hydrogen bonds in preference to van der Waals contacts. For example, if we try to mix water with a compound that cannot form hydrogen bonds, such as benzene, the water and benzene molecules rapidly separate from each other, the water molecules forming hydrogen bonds among themselves while the benzene molecules attach to one another by van der Waals bonds. It is therefore impossible to insert a non-hydrogen-bonding organic molecule into water.

On the other hand, polar molecules such as glucose and pyruvate, which contain a large number of groups that form excellent hydrogen bonds (e.g., =O or OH), are soluble in water (i.e., they are hydrophilic as opposed to hydrophobic). While the insertion of such groups into a water lattice breaks water-water hydrogen bonds, it results simultaneously in the formation of hydrogen bonds between the polar organic molecule and water. These alternative arrangements, however, are not usually as energetically satisfactory as the water-water arrangements, so that even the most polar molecules ordinarily have only limited solubility.

Thus, almost all the molecules that cells acquire, either through food intake or through biosynthesis, are somewhat insoluble in water. These molecules, by their thermal movements, randomly collide with other molecules until they find complementary molecular surfaces on which to attach and thereby release water molecules for water-water interactions.

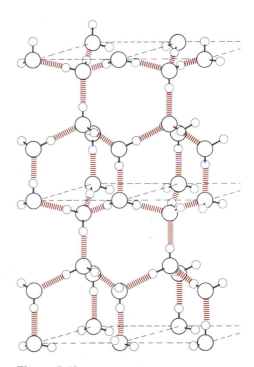

Figure 5-10
Diagram of a lattice formed by water molecules. The energy gained by forming specific hydrogen bonds between water molecules favors the arrangement of the molecules in adjacent tetrahedrons. Oxygen atoms are indicated by large circles, hydrogen atoms by small circles. Although the rigidity of the arrangement depends on the temperature of the molecules, the pictured structure is nevertheless predominant in water as well as in ice. [After L. Pauling, *The Nature of the Chemical Bond*, 3rd ed. (Ithaca, N.Y.: Cornell University Press, 1960), p. 465, with permission.]

L-Alanine

D-Alanine

Figure 5-11
The two stereoisomers of the amino acid alanine. (Redrawn from *General Chemistry*, 2nd ed., by Linus Pauling. W. H. Freeman and Company, copyright © 1953, p. 598, with permission.)

The Uniqueness of Molecular Shapes and the Concept of Selective Stickiness

Even though most cellular molecules are built up from only a small number of chemical groups, such as OH, NH_2, and CH_3, there is great specificity as to which molecules tend to lie next to each other. This is because each molecule has unique bonding properties. One very clear demonstration comes from the specificity of stereoisomers. For example, proteins are always constructed from L-amino acids, never from their mirror images, the D-amino acids (Figure 5-11). Although the D- and L-amino acids have identical covalent bonds, their binding properties to asymmetric molecules are often very different. Thus, most enzymes are specific for L-amino acids. If an L-amino acid is able to attach to a specific enzyme, the D-amino acid is unable to bind.

Most molecules in cells can make good "weak" bonds with only a small number of other molecules, partly because all molecules in biological systems exist in an aqueous environment. The formation of a bond in a cell depends not only on whether two molecules bind well to each other, but also on whether the bond will permit their water solvent to form the maximum number of hydrogen bonds.

Hydrophobic "Bonds" Stabilize Macromolecules[8, 9]

The strong tendency of water to exclude nonpolar groups is frequently referred to as **hydrophobic bonding.** Some chemists like to call all the bonds between nonpolar groups *in a water solution* hydrophobic bonds (Figure 5-12). In a sense this term is a misnomer, for the phenomenon that it seeks to emphasize is the absence, not the presence, of bonds. (The bonds that tend to form between the nonpolar groups are due to van der Waals attractive forces.) On the other hand, the term *hydrophobic bond* is often useful, since it emphasizes the fact that nonpolar groups will try to arrange themselves so that they are not in contact with water molecules. Hydrophobic bonds are important both in the stabilization of proteins and complexes of proteins with other molecules and in the partitioning of proteins into membranes. They may account for as much as one-half the total free energy of protein folding.

Consider, for example, the different amounts of energy generated when the amino acids alanine and glycine are bound, in water, to a third molecule that has a surface complementary to alanine. A methyl group is present in alanine but not in glycine. When alanine is bound to the third molecule, the van der Waals contacts around the methyl group yield 1 kcal/mol of energy, which is not released when glycine is bound instead. From Equation 5-4, this small energy difference alone would give only a factor of 6 between the binding of alanine and glycine. However, this calculation does not take into consideration the fact that water is trying to exclude alanine much more than glycine. The presence of alanine's CH_3 group upsets the water lattice much more seriously than does the hydrogen atom side group of glycine. At present, it is still difficult to predict how large a correction factor must be introduced for this disruption of the water lattice by the hydrophobic side groups. A current guess is that the water tends to exclude alanine, thrusting it toward a third molecule, with a hydrophobic force 2 to 3 kcal/mol larger than the forces excluding glycine.

Figure 5-12
Examples of van der Waals (hydrophobic) bonds between the nonpolar side groups of amino acids. The hydrogens are not indicated individually. For the sake of clarity, the van der Waals radii are reduced by 20 percent. The structural formulas adjacent to each space-filling drawing indicate the arrangement of the atoms. (a) Phenylalanine-leucine bond. (b) Phenylalanine-phenylalanine bond. [After H. A. Scheraga, in *The Proteins*, 2nd ed., vol. 1, ed. H. Neurath (New York: Academic Press, 1963), p. 527, with permission.]

We thus arrive at the important conclusion that the energy difference between the binding of even the most similar molecules to a third molecule (when the difference between the similar molecules involves a nonpolar group) is at least 2 to 3 kcal/mol greater in the aqueous interior of cells than under nonaqueous conditions. Frequently, the energy difference is 3 to 4 kcal/mol, since the molecules involved often contain polar groups that can form hydrogen bonds.

The Advantage of ΔG Between 2 and 5 kcal/mol

We have seen that the energy of just one secondary bond (2 to 5 kcal/mol) is often sufficient to ensure that a molecule preferentially binds to a selected group of molecules. Moreover, these energy differences are not so large that rigid lattice arrangements develop within a cell; that is, the interior of a cell never crystallizes, as it would if the energy of secondary bonds were several times greater. Larger energy differences would mean that the secondary bonds seldom break, resulting in low diffusion rates incompatible with cellular existence.

Weak Bonds Attach Enzymes to Substrates

Secondary forces are necessarily the basis by which enzymes and their substrates initially combine with each other. Enzymes do not indiscriminately bind all molecules, having noticeable affinity only for their own substrates.

Since enzymes catalyze both directions of a chemical reaction, they must have specific affinities for both sets of reacting molecules. In some cases, it is possible to measure an equilibrium constant for the binding of an enzyme to one of its substrates (Equation 5-4), which consequently enables us to calculate the ΔG upon binding. This calculation in turn hints at which types of bonds may be involved. For ΔG values between 5 and 10 kcal/mol, several strong secondary bonds are the basis of specific enzyme-substrate interactions. Also worth noting is that the ΔG of binding is never exceptionally high; thus, enzyme-substrate complexes can be both made and broken apart rapidly as a result of random thermal movement. This explains why enzymes can function quickly, sometimes as often as 10^6 times per second. If enzymes were bound to their substrates, or more importantly to their products, by more powerful bonds, they would act much more slowly.

Most Molecular Shapes Are Determined by Weak Bonds

The shapes of numerous molecules are automatically given by the distribution of covalent bonds. This inflexibility occurs when groups are attached by covalent bonds about which free rotation is impossible. Rotation is possible only when atoms are attached by single bonds. (For example, the methyl groups of ethane, H_3C-CH_3, rotate about the carbon-carbon bond.) When more than one electron pair is involved in a bond, rotation does not occur; in fact, the atoms involved must lie in the same plane. Thus, the aromatic purine and pyrimidine rings are planar molecules 3.4 Å thick. There is no uncertainty about the shape of any aromatic molecules. They are almost always flat, independent of their surrounding environment.

On the other hand, for molecules containing single bonds, the possibility of rotation around the bond suggests that a covalently bonded molecule exists in a large variety of shapes. This theoretical possibility, however, is seldom realized, because the various possible three-dimensional configurations differ in the number of weak bonds that can be formed. Generally, there is one configuration that has significantly less free energy than any of the other geometric arrangements.

It is no simple matter to guess the correct shape of a large molecule from its covalent-bond structure; although one configuration for a protein or a nucleic acid may be energetically much more suitable than any other, we cannot yet derive it from our knowledge of bond energies. The number of possible configurations of a nucleic acid or even a small protein molecule is immense. Given present techniques for building molecular models, a single person (or even a small group of people) armed with the best computers cannot rapidly calculate the sum of the energies of the weak bonds for each possible configuration. The reason for our present inability to derive protein and nucleic acid structures is that our knowledge about the nature of weak bonds is still very incomplete. In many cases, we are not sure about either the exact bond energies or the possible angles the bonds can form to each other.

Thus, protein and nucleic acid shapes still can be revealed only by X-ray diffraction analysis. Fortunately, these experimental structure determinations are beginning to yield some general rules that tell us which weak chemical interactions are most important in governing the molecular shapes of large molecules. In particular, these rules emphasize the vital importance of interactions of the macromolecules with water, by far the most common molecule in all cells.

Helical Arrangements of Polypeptide Backbones[10–17]

In Chapter 2 we saw that polypeptide chains have regular linear backbones in which specific groups (e.g., –CO–NH–) repeat over and over along the molecule. Often, these regular groups are arranged in helical configurations held together by secondary bonds. The helix is a natural conformation for a regular linear polymer like the polypeptide backbone, since it places each monomer group in an identical orientation within the molecule. Each monomer thereby forms the same group of secondary bonds as every other monomer. It is important to emphasize that helical symmetry does not evolve from the particular shape of the monomer, but is instead the natural consequence of a unique monomer arrangement that is significantly more stable than all other arrangements.

The most stable arrangement of a polypeptide backbone is the α helix. It is a right-handed helix, repeating every 5.4 Å along the helical axis, containing a regular pattern of hydrogen bonding between carbonyl and imino groups on the same chain (see Figure 2-23). Not only are its hydrogen bonds so oriented for maximum stability, but its atoms pack together to form a stereochemically perfect central core lacking energetically unfavorable holes (Figure 5-13). The nonintegral number of amino acids per turn (3.6) is a consequence of the precise geometry of the polypeptide chain. If, for example, four amino acids were used per turn, the hydrogen bonds would not be so neatly formed, nor would the individual backbone atoms fit together so well.

There are two energetically favorable forms of maximally extended (β) polypeptide backbones. In both forms, adjacent amino acids are related by 180° rotations, thereby projecting their respective side groups in opposite directions (see Figure 2-24). With such extended chains, there is no possibility of hydrogen bonds forming between carbonyl and imino groups relatively close to each other on the same chain. Instead, hydrogen bonding is restricted to groups belonging to different chains or to very distant parts of the same chain, which because of their twofold symmetry can only come together as sheets.

The two different forms of β sheets differ in the relative orientations of their chains (Figure 5-14). In one, the adjacent chains run in the same amino to carboxyl direction to produce a parallel β sheet. In the other, the adjacent chains run in opposite directions to yield an antiparallel β sheet. In both the parallel and antiparallel β sheets, all

Figure 5-13
Space-filling model of an α helix constituting part of the structure of myoglobin. The model on the top shows the α helix without its amino acid side chains. (Courtesy of Richard J. Feldmann.)

Figure 5-14
Two different forms of β sheets. (a) An antiparallel β sheet, in which the adjacent chains run in opposite directions. (b) A parallel β sheet, in which the adjacent chains run in the same direction. The peptide groups (several are shaded with color) lie in the plane of the sheet, while the R groups (not shown here for clarity) extend perpendicularly to the plane of the sheet. Actual β sheets in globular proteins are usually less regular than that shown here, as many sheets are slightly twisted (see Figure 5-20).

(a)

(b)

Figure 5-15

Structural representation of hen egg-white lysozyme by a ribbonlike model that emphasizes secondary structure—α helices and β sheets. The arrows on the ribbons representing the β chains point toward the carboxyl end of the polypeptide, and they make it immediately apparent that the β chains form an antiparallel β sheet. In this molecule, we see the tendency of amino acid side groups to twist the backbone α helices and β sheets into much less regular configurations. These changes in the structure of the regular backbone serve to maximize the strength of the secondary bonds formed by the side groups. [After Zubay, Biochemistry, p. 95.]

α Helix

β Sheet

Egg lysozyme

the peptide groups lie approximately in the plane of the sheet. Therefore, the interchain bonds that link them together can be energetically very favorable and of equal strength to those that hold together α helices.

In no case, however, are polypeptide chains of proteins regular polymers containing identical monomers. Instead, they invariably have irregular side groups attached to the regular backbone. Thus, we need not necessarily expect either a helical structure or a β sheet. A three-dimensional arrangement that is energetically very satisfactory for the backbone groups often produces very unsatisfactory bonding of the side groups. The three-dimensional structures of the polypeptide chains of proteins are thus compromises between the tendency of regular backbones to form either α helices or β sheets and the tendency of the side groups to twist the backbone into much less regular configurations that maximize the strength of the secondary bonds formed by the side groups (Figure 5-15).

This argument leads to the interesting prediction that in aqueous solutions, proteins containing very large numbers of nonpolar side groups will tend to be more stable than proteins containing mostly polar groups. If we disrupt a polar molecule held together by a large number of internal hydrogen bonds, the decrease in free energy is often small, since the polar groups can then hydrogen-bond to water. On the other hand, when we disrupt molecules having many nonpolar groups, there is usually a much greater loss in free energy because the disruption necessarily inserts nonpolar groups into water.

Interactions Between Nonpolar Side Groups Stabilize Proteins[18]

Almost one-half of the amino acid residues in proteins have nonpolar side groups that interact with each other rather than lying next to either polar side groups or the polar water molecules. Polypeptide chains within the aqueous portions of cells (as opposed to within the lipid environments of membranes) will thus be energetically driven to assume configurations that effectively isolate their nonpolar side groups within internal regions into which water molecules cannot penetrate. The free energy that would be lost by failure to so internalize the many hydrophobic groups of a given protein is very large and provides by far the major force that drives disordered, newly synthesized polypeptide chains into their "native," usually highly compacted, states. In contrast, very little real energy is gained by water-soluble proteins when they make their correct hydrogen bonds, since their polar backbone atoms and polar side groups can make nearly as favorably hydrogen bonds to water in the disordered (denatured) state as they can to themselves in the native state.

The Specific Conformation of a Protein Results from Its Pattern of Hydrogen Bonds

While the energy stabilizing a protein is provided by hydrophobic interactions, the specific conformation of a protein structure is provided by hydrogen bonds. The energy of hydrophobic "bonds" has no directional component, whereas hydrogen bonds require proper distances and angles. In general, all hydrogen-bond donors and acceptors within a protein's interior have suitable mates. Presumably, failure to make a hydrogen bond in a protein interior is energetically costly, at the rate of a few kilocalories per hydrogen bond. The vitally important role of hydrogen bonds in proteins is thus to destabilize the incorrect structures rather than to stabilize the correct one.

The necessity of satisfying all the hydrogen-bond donors and acceptors on the polypeptide backbone (two per residue) provides the energetic driving force that leads to the creation of the large sections of α helices and β sheets found in most proteins. The only way that a polypeptide can traverse the nonaqueous interior of a protein, as it must, and satisfy the hydrogen-bonding necessity is through formation of regular secondary structures. Side chains do not have enough donors and acceptors to do the job. Thus, all large proteins contain significant regions of β sheets, α helices, or both. Of course, some polypeptide sections must be less regular to allow their chains to turn at the ends of α helices and β chains.

Proteins embedded in the nonaqueous interior of membranes are stabilized in the opposite way. Hydrophobic interactions partition these proteins into membranes, but do not stabilize their structures within the lipid bilayer. Hydrogen bonds, on the other hand, must assume a very important role in stabilizing the correct structures of membrane proteins (Figure 5-16).

α Helices Come Together at Specific Angles[19, 20, 21]

A number of long, thin fibrous proteins (e.g., keratins and myosin tails) are constructed by the supercoiling of α helices about each other with their amino acid sequences arranged so that the nonpolar side groups are internally located. The twisting of the chains around each other reflects the nonintegral (3.6 residues per turn) nature of the α helix, which allows the side groups to neatly pack together only when the α helices interact at an angle of 18° from parallel. If the α helices remained perfectly rigid, they could stay in contact for only a few residues. But by supercoiling in a left-handed direction, neatly

Figure 5-16
An acetylcholine receptor subunit. Each polypeptide chain of this membrane protein traverses the membrane bilayer five times. The charges shown represent the average charges on the side chains. Outside the membrane the protein is predominantly hydrophilic; inside the membrane, it is mostly hydrophobic.

Helix axis

Fiber axis

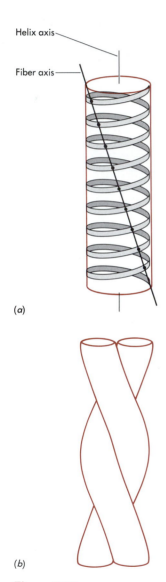

(a)

(b)

Figure 5-17
The coiled coil of α helices in α keratin. (a) One α helix and the positions within the helix of the fiber and helix axes when the two α helices are coiled about each other. (b) In the coiled coil, the side groups of each helix pack neatly together when the helices interact at an angle of 18°.

packed, highly stable, coiled coils are created (Figure 5-17). Fibrous molecules containing such coiled coils are frequently elastic, with their α helices becoming slightly uncoiled when subjected to tension.

Much shorter sections of interlocked α helices exist in a number of globular proteins. The signal that frequently leads the α helices to bend back on themselves is the presence of a proline residue, whose ringed structure prevents it from forming an intramolecular hydrogen bond along the α helix. Its occasional occurrence in the middle of a helix causes the helix to bend or kink. As a result of packing considerations, interacting α helices in globular proteins are related by one of three angles between their helix axes. Ridges formed by the side chains of one helix will pack into the grooves between side chains of another helix if the helix axes are inclined at 18°, 60°, or 90° to each other (Figure 5-18).

β Sheets Twist in a Right-Handed Fashion[22, 23, 24]

At one time, β sheets were thought to be flat. However, when they became visualized within known proteins by X-ray crystallography, they invariably appeared to have right-handed twists. The origin of this twisting lies in the slightly twisted nature of fully extended (β) polypeptide chains. They do not have the perfect twofold rotational symmetry initially postulated, but have slight intrinsic right-handed twists (Figure 5-19). To form an antiparallel β sheet, the direction of an extended β chain must suddenly reverse, and as with the α helix, proline residues are often used to effect such directional switches. The formation of parallel β sheets is more complicated, involving crossover sequences that usually assume α-helical conformations (Figure 5-20).

Twisted parallel β sheets are the dominant structural motif of many proteins. In some proteins, they are organized into saddlelike arrangements, while in other proteins they come together to form barrel-shaped structures (Figure 5-21). No protein is stable as a single-layered twisted β sheet or α helix; at least two layers ($\beta\beta$ or $\alpha\alpha$) are

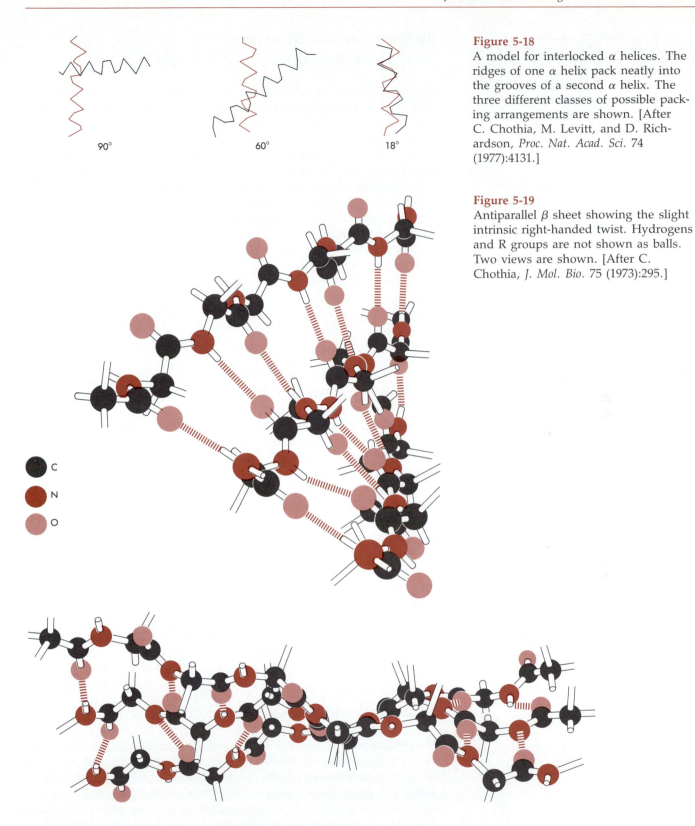

Figure 5-18
A model for interlocked α helices. The ridges of one α helix pack neatly into the grooves of a second α helix. The three different classes of possible packing arrangements are shown. [After C. Chothia, M. Levitt, and D. Richardson, *Proc. Nat. Acad. Sci.* 74 (1977):4131.]

Figure 5-19
Antiparallel β sheet showing the slight intrinsic right-handed twist. Hydrogens and R groups are not shown as balls. Two views are shown. [After C. Chothia, *J. Mol. Bio.* 75 (1973):295.]

required to isolate the hydrophobic residues that must come together in a central core. Although pairs of twisted antiparallel β sheets form the essential structure of many proteins (Figure 5-22), a twisted parallel sheet needs to be surrounded by additional sequences on both its sides. These sequences principally form α helices, which pack neatly with each other and with the two surfaces of the β sheet.

Figure 5-20

The formation of parallel β sheets. (a) The tendency of β sheets to twist right-handedly (b) produces sheets with right-handed crossover connections, although (c) left-handed connections also exist. (d) Frequently, crossover sequences assume helical conformation. [After G. Zubay, *Biochemistry* (Reading, Mass.: Addison-Wesley, 1983), p. 95].

Supersecondary Structures That Have Functional Implications

Although an oversimplification, a protein's structure can often be described as repeating patterns of supersecondary structural motifs. For example, myohemerythrin, cytochrome *c'*, and tobacco mosaic virus (TMV) coat protein each contain *four antiparallel α helices* (Figure 5-23), and these can be constructed from two αα units. Some proteins that are virtually all antiparallel β sheets, such as rubredoxin and immunoglobulin, consist of three (or more) ββ motifs. And all parallel β sheets are built up from βαβ units.

Some supersecondary structures have very specific functions and are therefore found in only certain kinds of proteins. For example, a supersecondary structure (helix-turn-helix) with two α helices at nearly right angles to each other was initially found in the DNA binding region of both the catabolite gene activator protein (CAP) of *E. coli* (see Figure 5-26) and the Cro repressor of phage λ (see Figure 5-30) and may well be a structural motif for most, if not all, gene regulatory proteins.

Most Proteins Are Modular, Containing Two or Three Domains[25, 26]

The subunits of soluble proteins vary in size from less than 100 amino acids for some peptide hormones to as large as 2000 amino acid residues for myosin. The smallest enzyme polypeptides have molecular weights of about 13,000 daltons (120 residues), but most are between 20,000 and 70,000 daltons for a single subunit. Protein subunits larger than about 20,000 daltons are usually formed from two or more domains (Figure 5-24). The term **domain** is used to describe a compact, folded part of the structure that appears separate from the rest, as if it would be stable in solution on its own (which is often demonstrated to be the case). There is a colinear correspondence between the structural domains and the amino acid sequence of a protein. That is, a single domain is formed from a continuous amino acid sequence and not portions of sequence that are scattered throughout the polypeptide. This is an important point when considering possible genetic mechanisms by which multidomain proteins have evolved.

A Surprisingly Small Number of Structural Motifs Are Found in Domains[27, 28]

Determination of the first half dozen protein structures showed a bewildering variety of protein folding motifs, implying the existence of an infinite number of protein structures. However, now that we know the three-dimensional structures of more than 200 proteins, it appears that a relatively small number of domain structural motifs account for most of the large variety of protein activities. Although an accurate estimate is not possible, the number of truly unique domain motifs might lie somewhere between 50 and 100 rather than in the thousands.

Specific kinds of domain motifs are often associated with particular kinds of activities. One frequently observed motif has been termed the *dinucleotide fold* because it is always found in enzymes, such as lactate dehydrogenase, that bind the coenzyme NAD^+ (Figure 5-25).

(a)

Twist — Flat rectangular sheet → Twisted sheet, saddle shape → Top view

(b)

Flat staggered sheet → β barrel → Top view

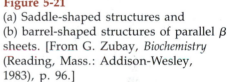

Figure 5-21
(a) Saddle-shaped structures and (b) barrel-shaped structures of parallel β sheets. [From G. Zubay, *Biochemistry* (Reading, Mass.: Addison-Wesley, 1983), p. 96.]

An analogous domain is found to bind the ATP substrate in those kinases whose structure is known, and indeed, it seems to be observed in all enzymes that use ATP as a substrate. What is common to the domains that bind ATP or NAD$^+$ is a central, parallel β sheet with α helices on both sides and the nucleotide binding site on the carboxyl end of the β strands. What varies is number and detailed arrangement of the α helices and, to a far lesser extent, the order of the β strands. These observed similarities and differences raise interesting questions about the evolution of domains like the dinucleotide fold. Did all ATP and NAD$^+$ binding domains evolve from a common ancestral precursor, or is this dinucleotide fold the only kind of protein fold that will function correctly?

Different Protein Functions Arise from Various Domain Combinations

The various functional properties of proteins appear to arise from their modular construction in much the same way as computers with different specifications can be assembled from the appropriate modular components. Numerous examples can be given. Dehydrogenases consist of two domains. One is a common dinucleotide binding do-

Figure 5-22
Concanavalin A, a protein formed from twisted antiparallel β sheets. (Courtesy of Jane Richardson.)

Cytochrome c'

Tobacco mosaic virus protein

Myohemerythrin

Figure 5-23
Proteins with the similar structural motif of four antiparallel α helices. [After G. Zubay, *Biochemistry* (Reading, Mass.: Addison-Wesley, 1983), p. 101.]

main that binds the coenzyme NAD^+. The other domain binds substrate and has the catalytic site; its structure varies among dehydrogenases.

The gene regulatory repressor and activator proteins provide another example of modular construction. The *lac* repressor and the catabolite gene activator protein (CAP) of *E. coli* both have subunits containing multiple domains. The crystal structure of CAP shows two domains: A larger domain is seen to bind a molecule of cyclic AMP in its interior, while the smaller domain appears to be involved in the recognition and binding of CAP to specific DNA sequences (Figure 5-26). There are very significant amino acid sequence similarities between the cAMP-binding domain of CAP and the regulatory subunit of cAMP-dependent protein kinase, suggesting that the cAMP-binding domain of both proteins evolved from the same precursor. In CAP, this cAMP-binding domain is attached to the DNA-binding domain, so that changes in cAMP levels control transcription levels. In the kinase, the cAMP-binding domain regulates the activity of the first enzyme in a cascade of enzymes that result in the breakdown of stored glycogen. The DNA-binding domain of CAP, on the other hand, shows structural and amino acid sequence similarities to many other gene regulatory proteins, suggesting a common ancestral precursor to the DNA binding domains of regulatory proteins.

The *lac* repressor has a similar modular construction. It has a small domain whose sequence is related to that portion of CAP that binds to specific DNA sequences, and it has a large domain that displays amino acid sequence similarity to the galactose binding protein. Once again, domains with different functions (galactose binding and specific-DNA binding) are combined into a protein with another function: control of gene expression.

We Can Now Determine Most Protein Structures by Crystallography[29, 30]

It took several research workers 7 to 15 years to solve the three-dimensional structures of the first proteins. Thanks to advances in crystallization techniques, computers and programming techniques, and methods of X-ray diffraction data measurement, it is now possible for one or two people to solve an average protein structure in one or two years. A great deal less time is needed to determine the structure of a modification of a known protein, such as a mutant protein. Therefore, ascertaining precisely how small molecule substrates or inhibitors bind to proteins of known structure can usually be accomplished in a few weeks if the small molecule diffuses into the crystals and binds without disturbing the crystal.

As a result of gene-cloning technology and improved methods of crystallization, almost all globular proteins, however rare normally, can now be studied by crystallographic techniques. For example, the crystal structure of the antiviral agent interferon is currently being determined, when only a short time ago only the minutest quantities were available for clinical and biochemical tests. The same is true for an entire host of proteins and nucleic acids of particular interest to molecular biologists—proteins involved in replication, recombination, transcription, and gene regulation, to name a few.

Even membrane proteins are now amenable to three-dimensional structural analysis. Using appropriate nonionic detergents like octal-glucoside to solubilize membrane proteins, researchers have already

Figure 5-24
The three domains of hexokinase. Domains I and II are related by a 180° rotation, which aligns points 1, 2, and 3. Domains II and III are connected by a continuous piece of α-helical backbone. (Courtesy of Jane Richardson.)

crystallized the pore-forming protein (porin); the proton pump bacteriorhodopsin; and a major component in photosynthesis, the photosynthetic reaction center. The last, a complex of four different proteins, has been solved at high resolution, giving us our first detailed view of the molecular architecture of an integral transmembrane protein complex (see color section).

In theory, there is no limit to the size of the macromolecular assembly whose detailed molecular structure can be determined by diffraction methods. Already, the structures of several plant viruses with molecular weights of millions of daltons have been solved, and even the much more complex ribosomes have been crystallized.

Protein Structure Is the Same in Crystals and Solution

Is the structure of a protein the same in the crystal as it is in solution under similar conditions, or does the process of crystallization significantly alter the structure? Macromolecule crystals, unlike crystals of

Figure 5-25
Schematic drawing of an idealized dinucleotide binding domain. Each β chain and α helix is assigned a letter.

small molecules, are very soft and actually consist of 40 to 60 percent solvent. The aqueous channels in crystals are so large that in many cases substrate molecules can diffuse into crystals and bind to enzyme active sites. The enzymatic activities of ribonuclease and carboxypeptidase A crystals have been measured and are only a few times smaller than their activities in solution. Thus, these enzymes must have very similar active-site structures in both the solution and crystalline states. Further, many protein structures have been solved in several, very different crystal forms, sometimes using crystals ob-

Figure 5-26
CAP complex with cAMP interacting with a bent DNA. The DNA binding domain is colored and the cAMP binding domain is gray. Each vertex on the polypeptide chain represents the α carbon of an amino acid residue. (Courtesy of T. Steitz.)

Figure 5-27
The hydrogen-bonded base pairs of DNA. Adenine is always attached to thymine by two hydrogen bonds, whereas guanine always bonds to cytosine by three hydrogen bonds. The obligatory pairing of the smaller pyrimidine with the larger purine allows the two sugar-phosphate backbones to have identical helical configurations. All the hydrogen bonds in both base pairs are strong, since each hydrogen atom points directly at its acceptor atom (nitrogen or oxygen).

tained from solvents as different as ammonium sulfate and alcohol. The molecular structures are found to be identical within experimental error.

DNA Can Form a Regular Helix

At first glance, DNA looks even more unlikely to form a regular helix than does an irregular polypeptide chain. Not only does DNA have an irregular sequence of side groups, but in addition, all its side groups are hydrophobic. Both the purines (adenine and guanine) and the pyrimidines (thymine and cytosine), even though they contain polar C=O and NH_2 groups, are quite insoluble in water because their flat sides are completely hydrophobic.

Nonetheless, DNA molecules usually have regular helical configurations. This is because most DNA molecules contain two polynucleotide strands that have complementary structures (see Chapter 9 for more details). Both internal and external secondary bonds stabilize the structure. The two strands are held together by hydrogen bonds between pairs of complementary purines and pyrimidines (Figure 5-27). Adenine is always hydrogen-bonded to thymine, whereas guanine is hydrogen-bonded to cytosine. In addition, virtually all the surface atoms in the sugar and phosphate groups form bonds to water molecules.

The purine-pyrimidine base pairs are found in the center of the DNA molecule. This arrangement allows their flat surfaces to stack on top of each other, thereby limiting their contact with water. This stacking arrangement would be much less satisfactory if only one polynucleotide chain were present. A single chain could not have a regular backbone because the pyrimidines are smaller than the purines, and so the angle of helical rotation would have to vary with the sequence of bases. The presence of complementary base pairs in double-helical DNA makes a regular structure possible, since each base pair is of the same size.

DNA Molecules Are Stable at Physiological Temperatures

The double-helical DNA molecule is very stable at physiological temperatures for two reasons. First, disruption of the double helix would bring the hydrophobic purines and pyrimidines into greater contact with water, which is very unfavorable. Second, individual DNA molecules have a *very large number of weak bonds,* arranged so that most of them cannot break without the simultaneous breaking of many others. Even though thermal motion is constantly breaking apart the terminal purine-pyrimidine pairs at the ends of each molecule, the two chains do not usually fall apart because the hydrogen bonds in the middle are still intact (Figure 5-28). Once a break occurs, the most likely next event is the reforming of the same hydrogen bonds to restore the original molecular configuration. Sometimes, of course, the first breakage is followed by a second one, and so forth. Such multiple breaks, however, are quite rare, so that double helices held together by more than ten nucleotide pairs are very stable at room temperature. The same principle also governs the stability of most protein molecules. Stable protein shapes are never due to the presence of just one or two weak bonds, but always represent the cooperative result of a number of weak bonds.

Ordered collections of hydrogen bonds become less and less stable as their temperature is raised above physiological temperatures. At physiologically abnormally high temperatures, the simultaneous breakage of several weak bonds is more frequent. After a significant number have broken, a molecule usually loses its original form (the process of denaturation) and assumes an inactive (denatured) configuration.

Weak Bonds Correctly Position Proteins Along DNA Molecules[31]

DNA never exists completely naked within cells, but at all times has a significant fraction of its surface covered with a large variety of protein molecules. The bonds that hold these proteins onto DNA are the same collection of weak bonds that give to DNA and the various proteins their own specific three-dimensional configurations. The most common DNA-binding proteins play a structural role by packaging and compacting the huge amount of DNA that must be fitted into the cell. For example, the nucleus of a human cell is only 10 μm (10^{-5} meter) across but contains roughly 3 meters of double-stranded DNA.

Best known of the chromosomal proteins are the **histones,** a group of small, positively charged proteins found in association with the DNA of all eucaryotic cells. The histones bind to DNA largely through ionic bonds between the negatively charged phosphate

Figure 5-28
The breaking of terminal hydrogen bonds in DNA by random thermal motion. Because the internal hydrogen bonds continue to hold the two chains together, the immediate reforming of the broken bonds is highly probable. Also shown is the very rare alternative: the breaking of additional hydrogen bonds and the consequent disentangling of the chains.

groups of DNA and the positively charged side groups of arginine and lysine, the most common amino acids found in the histones. More importantly, because the four major histone species (H2A, H2B, H3, and H4) form a disk-shaped octameric complex $(H2A_2H2B_2H3_2H4_2)$ that acts like a threaded spool for duplex DNA, the DNA is able to wrap itself around the histones rather than the histones wrapping around the DNA (see Figure 9-31). In this way, the DNA can be compacted while much of the DNA surface, and in particular many of the hydrogen-bond-forming groups of its purine-pyrimidine base pairs, remain accessible for interactions with other DNA binding proteins.

Prominent among these nonhistone partners of DNA are the DNA polymerases needed for its replication, the RNA polymerases used in its transcription, and a large variety of regulatory proteins that help ensure that individual genes function (are transcribed) when needed. All these proteins recognize the specific DNA segments on which they are to work through bonding to specific groups of base pairs. In contrast to the histones, most regulatory proteins that bind to DNA have more negatively than positively charged amino acids on their surfaces. Such electrostatic charge interaction is exemplified by the interaction between DNA and its enzyme CAP (Figure 5-29). Like CAP, the vast majority of cytoplasmic proteins have net negative charges; they may have evolved to prevent unwanted nonspecific aggregation between positively and negatively charged molecules.

Sequence Recognition Results
from Protein-DNA Complementarity[32, 33]

Sequence-specific protein-DNA interactions arise in large part from a complementarity in shape that permits weak interactions between protein side chains and the amino and keto groups of the bases exposed in the major groove (see Chapter 9). For a regulatory protein to make a specific interaction, the protein must move closer to the DNA than in its unspecific complex, pushing a polypeptide α helix into the

Figure 5-29
CAP binding to DNA, showing the electrostatic charge interaction between enzyme and DNA. The negatively charged DNA interacts with the positively charged amino acid residues (colored area) of the protein. (Courtesy of T. Steitz.)

major groove of the DNA. Side chains emanating from this helix can make hydrogen bonds with the edges of the bases. If the two macromolecules are complementary in hydrogen-bond donors and acceptors, a stable, specific complex is made; if not, the protein diffuses away.

There are at least two ways in which a protein might interact with duplex DNA. The structures of several such proteins (λ phage repressor, Cro repressor, and CAP) complexed with DNA suggest that protruding α helices fit into the DNA major groove. (In Figure 5-30, we see protrusions on the Cro protein fitting into grooves in the DNA.) In contrast, the *E. coli* DNA polymerase I has a handlike structure with a deep cleft into which duplex DNA can bind (Figure 5-31). In this case, it appears that the protein may partially or totally surround the DNA. Likewise, some repressors may have flexible "arms" that may be able to embrace the DNA.

Figure 5-30
Interaction of DNA with Cro repressor, a dimer, as proposed by D. H. Ohlendorf, W. F. Anderson, R. G. Fisher, Y. Takeda, and B. W. Matthews, *Nature* 298 (1982):718–723. (Figure courtesy of Jane M. Burridge.)

Figure 5-31
The α-carbon backbone of a fragment of DNA polymerase I (black) complexed with DNA (color). The protein can nearly surround the DNA, which is shown looking down its axis. (Courtesy of T. Steitz.)

Sliding Diffusion of Proteins Along DNA Molecules[34]

Many DNA binding proteins can bind nonspecifically to the DNA backbone as well as specifically to the base pairs. Mediating their nonspecific attachments are patches of positively charged amino acids located at sites very close to those that bind to the base pairs. In these nonspecific bindings, the proteins usually lie some 10 Å to 15 Å farther from the DNA than in specific binding, ruling out even the occasional hydrogen-bond interaction while allowing significant electrostatic attraction. Though such nonspecific binding may be 10^6 to 10^7 times less tight, there is always a much larger number of nonspecific than specific attachment sites, particularly for those proteins that recognize only a single DNA site. At a given moment, therefore, many more copies of a given binding protein are likely to be attached nonspecifically than are found at their biologically important destination. Now it is strongly suspected that these nonspecific unions are not unavoidable nuisances but a device to speed up the rate at which a given regulatory protein reaches its appropriate target. This would be the case if such nonspecifically bound proteins are constrained by their charge interaction to diffuse linearly along DNA and do not randomly pop on and off DNA in their "search" for their specific binding sites. By being so restricted to linear movements, they will reach their targets much faster than if they had the freedom for three-dimensional diffusion.

Not all the movements of proteins bound to DNA occur by purely thermal events. Some DNA binding proteins involved in DNA replication (see Chapter 10) utilize the breakdown of ATP to produce cycles of reversible conformational changes that propel them in fixed directions until they meet their targets.

Modifying Proteins with Lipid Side Groups Promotes Their Association to Membranes[35, 36]

Transmembrane proteins possessing separate domains protruding onto the inner and outer sides of lipid bilayers are known as **integral membrane proteins.** They become so sited by being synthesized on membrane-bound polyribosomes that can directly place them into appropriate membrane sections. Other "membrane proteins" are much less tightly associated and may have none of their amino acids directly inserted into the hydrophobic lipid interior. Instead, they bind to the cytoplasmic surface of the membrane. Such "nonintegral" membrane proteins frequently contain single lipid molecules covalently attached to an amino acid on their surface. Such lipid groups, being totally insoluble in water, automatically seek out other lipid groups and do so most often by directly inserting themselves into the inner side of the lipid bilayer. In some proteins (e.g., some protein kinases), lipids are attached via amide bonds to amino terminal glycine groups. In other proteins, lipids are attached by ester linkages to the hydroxyl groups of amino acids like serine and threonine.

Most Medium-Size and Almost All Large Protein Molecules Are Aggregates of Smaller Polypeptide Chains

Earlier we pointed out how the realization that macromolecules are all polymers constructed from small, regular monomers, such as the amino acids, greatly simplified the problem of solving macromolecular structure. It has recently become clear that most of the very large proteins are regular aggregates of polypeptide chains, containing up to 400 amino acids apiece. (The relationship among the polypeptide chains making up such a protein is termed its **quaternary structure.**) For example, the protein ferritin, which functions in mammals to store iron atoms, has a molecular weight of about 480,000. It contains, however, not just one long polypeptide chain of 4000 amino acids, but 20 identical smaller polypeptide chains of about 200 amino acids each. Similarly, the protein component of tobacco mosaic virus (TMV) was originally thought to have the horrendous molecular weight of 36 million. Most fortunately, it was subsequently discovered that each TMV particle contains 2150 identical smaller protein molecules, each containing 158 amino acids. Even much smaller protein molecules are frequently constructed from a number of polypeptide chains. Hemoglobin, which has a molecular weight of only 64,500, contains four polypeptide chains, two α chains and two β chains, each of which has a molecular weight of about 16,000 (Figure 5-32).

In all three examples, as with most other protein aggregates, the polypeptide subunits are held together by secondary bonds. This fact is known because the polypeptides can be dispersed by the addition of reagents (e.g., urea) that tend to break secondary bonds but not covalent bonds. Moreover, weak bonds are not the only force holding macromolecular units together. In some proteins, such as insulin, disulfide bonds (S–S) between cysteine residues are also a binding force.

β_2

β_1

α_2

α_1

Figure 5-32
The four polypeptide chains of hemoglobin. (Copyright Irving Geis.)

Subunits Are Economical

Both the construction of polymers from monomers and the use of polymeric molecules themselves as subunits to build still larger molecules reflect a general building principle applicable to all complex structures, nonliving as well as living. This principle states that it is much easier to reduce the impact of construction mistakes if we can discard the faulty subunits before they are incorporated into the final product. For example, let us consider two alternative ways of constructing a molecule with a million atoms. In scheme 1, we build the structure atom by atom; in scheme 2, we first build a thousand smaller units, each with a thousand atoms, but subsequently put the subunits together into the million-atom product. Now consider that our building process randomly makes mistakes, inserting the wrong atom with a frequency of 10^{-5}. Let us assume that each mistake results in a nonfunctional product.

Under scheme 1, each molecule will contain, on the average, ten wrong atoms, and so almost no good products will be synthesized. Under scheme 2, however, mistakes will occur in only 1 percent of the subunits. If there is a device to reject the bad subunits, then good products can be easily made, and the cell will hardly be bothered by the presence of the 1 percent of nonfunctional subunits. This concept is the basis of the assembly line, in which complicated industrial products, such as radios and automobiles, are constructed. At each stage of assembly, there are devices to throw away bad subunits. In industrial assembly lines, mistakes were initially removed by human hands; now, automation often replaces manual control. In cells, mistakes are sometimes controlled by the specificity of enzymes. If a monomeric subunit is wrongly put together, it usually will not be

Figure 5-33
Ligand-induced conformational changes. The ligand is represented by a color circle.

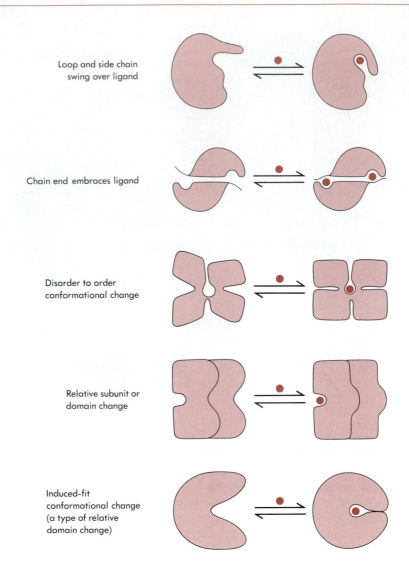

Loop and side chain swing over ligand

Chain end embraces ligand

Disorder to order conformational change

Relative subunit or domain change

Induced-fit conformational change (a type of relative domain change)

recognized by the polymer-making enzyme and hence will not be incorporated into a macromolecule. In other cases, faulty substances are rejected because they are unable to spontaneously become part of stable molecular aggregates.

Allostery: Alternative Shapes for Single Proteins[37, 38, 39]

The binding of either small or large molecules (*ligands*) to proteins often results in a substantial change in the conformation of the protein. These ligand-induced conformational changes can alter a protein's activity, enabling some proteins to function like an on-off light switch. Control of metabolism and gene expression is often modulated by such molecular protein switches (see Figure 4-20).

At least four different kinds of ligand-induced protein conformational changes have been observed (Figure 5-33). The simplest involve alterations around the binding site of the ligand itself, either a loop and side chain swinging over the ligand, as occurs when substrates bind to carboxypeptidase, or the polypeptide chain end moving to interact with the ligand. The latter change occurs upon the

(a)

▨ Catalytic polypeptide	☐ Substrates
▮ Regulatory polypeptide	▨ CTP (inhibitor)

Allosteric inhibitor site

Active sites

10Å

CTP site

(b)

binding of several repressors to DNA. These changes at the ligand binding site allow the protein to surround or embrace the ligand in a way that would not be possible if the protein were rigid. A third effect of ligand binding is often to stabilize a fixed protein conformation in a small part of the protein that is very flexible or disordered in its absence. When RNA binds to TMV coat protein to form the intact virus particle, such a disorder-to-order transition is induced at the RNA binding site. Once again, flexibility in the protein structure allows the RNA to bind before a binding site that surrounds the RNA is completed.

Perhaps the largest category of protein conformational change and the one that is most significant for regulation is the relative subunit or domain change. In this case, the binding of a ligand to one subunit results in a change in the relative orientation of subunits—a change in the quaternary structure of an oligomeric protein. Such changes in quaternary structure can have profound effects on the properties of binding sites that are very distant from the site to which the first ligand binds. The allosteric modification of the enzyme aspartate transcarbamoylase by its ligand, CTP, provides a good example of such an interaction (Figure 5-34). Also, ligand binding can alter the

Figure 5-34
The allosteric modification of aspartate transcarbamoylase (ATCase) by CTP. ATCase is a large molecule consisting of six catalytic subunits that bind its substrates and six regulatory subunits that are devoid of catalytic activity but bind CTP. (a) When CTP is bound to a regulatory subunit, small changes occur that compact the polypeptides, thereby closing the active sites to substrates. (b) Shown here are the structure of one of the regulatory chains and the site where CTP is thought to bind. The cylinders represent α helices. [Part (b) after Monaco, H. L., J. L. Crawford, and W. N. Lipscomb. *Proc. Nat. Acad. Sci.* 75 (1978): 5278.]

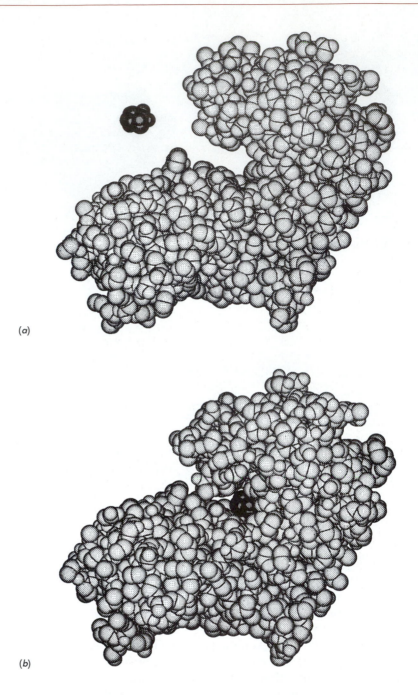

(a)

(b)

relative orientation of two domains in some multidomain proteins. For example, binding of the substrate glucose to hexokinase A results in a movement of two domains relative to each other, closing off the cleft in which the glucose binds and burying the substrate (Figure 5-35). Such a conformational change brought about by a substrate is called an **induced fit.** In this case, the correct binding site for ATP does not exist until glucose, the acceptor of the high-energy phosphate of ATP, is already bound. In this way, the shape change prevents the kinase from hydrolyzing the high-energy ATP molecule.

Ligand-induced quaternary structure changes often allow the binding of a ligand of one kind to alter the affinity or activity of a protein for a ligand of a different kind. The "interaction" between binding sites for ligands of different kinds is called **allosteric interaction.** When the binding of one ligand (e.g., cAMP to CAP) increases the protein's affinity for another ligand (e.g., DNA), the allosteric interaction is termed **positive cooperativity.** Alternatively, **negative cooperativity** describes the allosteric effect of one ligand reducing the protein's affinity for another ligand. An example of negative cooperativity is the binding of inducer to the *lac* repressor protein, which reduces the protein's affinity for operator DNA. Because of these allosteric proteins, the binding of a protein to DNA and thus its ability to alter gene expression can be switched on or off, depending on the presence or absence of a small-molecule metabolite.

The Principle of Self-Assembly[40, 41, 42]

With ΔG values of 1 to 5 kcal/mol, not only are single weak bonds spontaneously made, but also structures held together by several weak bonds are spontaneously formed. For example, an unstable, folded polypeptide chain tends to assume a large number of random configurations as a result of thermal movements. Most of these conformations are thermodynamically unstable. Inevitably, however, thermal movements bring together groups that can form weak bonds. These groups tend to stay together because more free energy is lost when they form than can be regained by their breakage. Thus, by a random series of movements, the polypeptide chain gradually assumes a configuration in which most, if not all, the atoms have fixed positions within the molecule.

Aggregation of separate molecules also occurs spontaneously. The protein hemoglobin furnishes a clear example (Figure 5-36). It can be broken apart by the addition of reagents such as urea, which break secondary bonds to yield half-molecules with molecular weights of 32,000. If, however, the urea is removed, the half-molecules quickly aggregate to form functional hemoglobin molecules. The surface structure of the half-molecules is very specific; they bind only to each other and not with any other cellular molecules.

This same principle of self-assembly operates to build even larger and more complicated structures, such as the cell membrane and the cell wall. Both are mosaic surfaces containing large numbers of various molecules, some large, like proteins, and others much smaller, like lipids. At present, practically nothing is known about the precise arrangement of the molecules in these very large, complex structures. Nonetheless, there is every reason to believe that the constituent molecules form stable contacts only with other molecules in the cell membrane (or wall). This situation is easy to visualize in the case of lipids, which are extremely insoluble in water because of their long, nonpolar hydrocarbon chains. Newly synthesized lipids have a stronger tendency to attach to other lipids in the cell membrane or cell wall by van der Waals forces than to enter some other, polar area, such as the aqueous (polar) interior of the cell.

The nucleic acid and protein components of viruses also come together spontaneously in the correct shape during the formation of new (progeny) viral particles. Enzymes often play no role in the aggregation of these components, since with many viruses, the formation of new covalent bonds is not needed either to build a stable protein coat or to inject its nucleic acid core. Only weak secondary bonds are needed. This last point was first shown for tobacco mosaic virus (TMV). Here the rod-shaped particles can be gently broken down and their free RNA and coat protein components separated. When they are again mixed together, new infectious particles, identical to the original rods, quickly form (Figure 5-37).

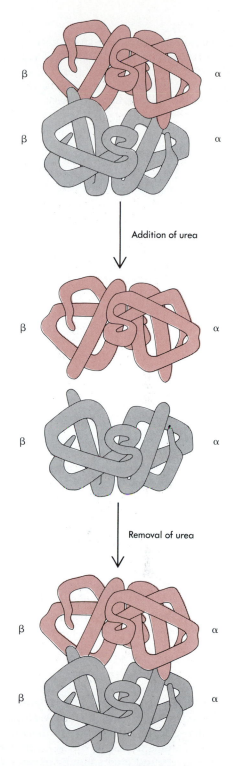

Figure 5-36

Formation of an active hemoglobin molecule from two half-molecules. Each hemoglobin molecule contains two α chains and two β chains. When placed in urea (a reagent that destabilizes weak bonds), the native molecule falls apart to two halves, each containing one α chain and one β chain. Upon removal of the urea, the halves reassociate to form the complete molecule.

Figure 5-37
Formation of a TMV particle from its protein subunits and its RNA molecule. [After H. Fraenkel-Conrat, *Design and Function at the Threshold of Life: The Viruses* (New York: Academic Press, 1962), Figure 18.]

Total length = 130 turns

Summary

Many important chemical events in cells do not involve the making or breaking of covalent bonds. The cellular location of most molecules depends on weak, or secondary, attractive or repulsive forces. In addition, weak bonds are important in determining the shape of many molecules, especially very large ones. The most important of these weak forces are hydrogen bonds, van der Waals interactions, hydrophobic bonds, and ionic bonds. Even though these forces are relatively weak, they are still large enough to ensure that the right molecules (or atomic groups) interact with each other. For example, the surface of an enzyme is uniquely shaped to allow specific attraction of its substrates.

The formation of all chemical bonds, weak interactions as well as strong covalent bonds, proceeds according to the laws of thermodynamics. A bond tends to form when the result would be a release of free energy (negative ΔG). For the bond to be broken, this same amount of free energy must be supplied. Because the formation of covalent bonds between atoms usually involves a very large negative ΔG, covalently bound atoms almost never separate spontaneously. In contrast, the ΔG values accompanying the formation of weak bonds are only several times larger than the average thermal energy of molecules at physiological temperatures. Single weak bonds are thus frequently being made and broken in living cells.

Molecules having polar (charged) groups interact quite differently from nonpolar molecules (in which the charge is symmetrically distributed). Polar molecules can form good hydrogen bonds, whereas nonpolar molecules can form only van der Waals bonds. The most important polar molecule is water. Each water molecule can form four hydrogen bonds to other water molecules. Although polar molecules tend to be soluble in water (to various degrees), nonpolar molecules are insoluble because they cannot form hydrogen bonds with water molecules.

Every distinct molecule has a unique molecular shape that restricts the number of molecules with which it can form strong secondary bonds. Strong secondary interactions demand both a complementary (lock-and-key) relationship between the two bonding surfaces and the involvement of many atoms. Although molecules bound together by only one or two secondary bonds frequently fall apart, a collection of these weak bonds can result in a stable aggregate. The fact that double-helical DNA never falls apart spontaneously demonstrates the extreme stability possible in such an aggregate. The formation of such aggregates can proceed spontaneously, with the correct bonds forming in a step-by-step fashion (the principle of self-assembly).

The shape of polymeric molecules is determined by secondary bonds. All biological polymers contain single bonds about which free rotation is possible. They do not, however, exist in a variety of shapes, as might be expected, because the formation of one of the possible configurations generally involves a maximum decrease in free energy. This energetically preferred configuration is thus exclusively formed. Some polymeric molecules have regular helical backbones held in shape by sets of regular internal secondary bonds between backbone groups. Regular helical structures cannot be formed, however, if the specific side groups occur in positions in which they cannot form favorable weak bonds. This situation occurs in many proteins where an irregular distribution of nonpolar side groups forces the backbone into a highly irregular conformation, permitting the nonpolar groups to form van der Waals bonds with each other. Irregularly distributed side groups do not always lead, however, to nonhelical molecules. In the DNA molecule, for example, the specific pairing of purines with pyrimidines in a double-stranded helix allows the nonpolar aromatic groups to stack on top of each other in the center of the molecule.

Polypeptide chains in the aqueous environments of cells are energetically driven to assume unique configurations that isolate their nonpolar side groups into internal regions of their proteins into which water cannot penetrate. Such hydrophobic bonding provides the major force that converts newly made polypeptide chains into "native," usually much more compacted, shapes. The exact configuration adopted results from hydrogen-bond formation between suitable donor and acceptor groups within a protein's interior. The need for these internal groups to be properly hydrogen-bonded creates the large sections of α helices and β sheets found in most proteins.

Many native proteins contain several discrete folded sections (domains) that are stable by themselves and which arise from a continuous amino acid sequence. Combinations of such domains account for a large variety of all known proteins. The number of truly unique domains probably falls between 50 and 100. Each domain is often associated with a specific functional activity (e.g., DNA binding).

Binding of specific proteins to specific sequences along DNA molecules also involves the formation of weak bonds, usually hydrogen bonds between groups on DNA bases and appropriate acceptor (or donor) groups on proteins. Many important regulatory proteins have a protruding two α-helix domain that fits into the major groove of DNA. Other DNA binding proteins contain deep clefts that can effectively surround sections of a double helix. Individual DNA binding proteins also contain regions that allow nonspecific bonding to the DNA backbone. This permits their linear diffusion along DNA, allowing them to reach their specific target sequences more quickly.

Bibliography

General References

Blum, H. F. 1962. *Time's Arrow and Evolution*. New York: Harper & Row. Chapter 3 provides a clear introduction to the thermodynamics applicable to biological systems.

Creighten, T. E. 1983. *Proteins*. San Francisco, Calif.: Freeman.

Dickerson, R. E., and I. Geis. 1969. *The Structure and Action of Proteins*. Menlo Park, Calif.: Benjamin/Cummings.

———. 1983. *Hemoglobin: Structure, Function, Evolution and Pathology*. Menlo Park, Calif.: Benjamin/Cummings.

Fersht, A. R. 1983. *Enzyme Structure and Mechanism*, 3rd ed. San Francisco, Calif.: Freeman.

Fletterick, R. J., T. Schroer, and R. J. Matela. 1985. *Molecular Structure; Macromolecules in Three Dimensions*. Palo Alto, Calif.: Blackwell Scientific Publishers. Includes basic principles and a guide to model building.

Lehninger, A. L. 1971. *Bioenergetics*. 3rd ed. Menlo Park, Calif.: Benjamin/Cummings. The laws of thermodynamics are neatly outlined for the beginning biology student in the first several chapters.

Pauling, L. 1960. *The Nature of the Chemical Bond*. 3rd ed. Ithaca, N.Y.: Cornell University Press. A great classic of chemical literature emphasizing the hydrogen bond.

Richardson, J. S. 1981. "The Anatomy and Taxonomy of Protein Structure." *Adv. Prot. Chem.* 34:168–364.

Schultz, G., and R. H. Schirmer. 1979. *Principles of Protein Structure*. New York: Springer-Verlag.

Zubay, G. 1983. *Biochemistry*. Reading, Mass.: Addison-Wesley. Includes an excellent treatment of the principles of protein structure.

Cited References

1. Gray, H. B. 1964. *Electrons and Chemical Bonding*. Menlo Park, Calif.: Benjamin/Cummings.
2. Klotz, I. M. 1967. *Energy Changes in Biochemical Reactions*. New York: Academic Press. Easy-to-read work on energetics, including the concepts of entropy, free energy, and the dependence of chemical potential on concentration of reactants and products.
3. Morowitz, H. J. 1970. *Entropy for Biologists*. New York: Academic Press.
4. Kabat, E. A. 1978. "The Structural Basis of Antibody Complimentarity." *Adv. Prot. Chem.* 32:1–76.
5. Donohue, J. 1968. "Selected Topics in Hydrogen Bonding." In *Structural Chemistry and Molecular Biology*, ed. A. Rich and N. Davidson. San Francisco, Calif.: Freeman.
6. Marsh, R. E. 1968. "Some Comments on Hydrogen Bonding in Purine and Pyrimidine Bases." In *Structural Chemistry and Molecular Biology*, ed. A. Rich and N. Davidson. San Francisco, Calif.: Freeman.
7. Pauling, L. 1944. Appendix. In *The Specificity of Serological Reactions* by K. Landsteiner. Cambridge, Mass.: Harvard University Press.
8. Hildebrand, J. H. 1979. "Is There a Hydrophobic Effect?" *Proc. Nat. Acad. Sci.* 76:194.
9. Hagler, A. T., and J. Moult. 1978. "Computer Simulation of Solvent Structure Around Biological Macromolecules." *Nature* 272:222–226.
10. Chothia, C. 1985. "Principles of Proteins That Determine the Three-Dimensional Structure." *Ann. Rev. Biochem.* In press.
11. Ramachandran, G. N., and V. Sasisekharan. 1968. "Conformations of Polypeptides and Proteins." *Adv. Prot. Chem.* 28:283–437. Describes the conformational space of polypeptide backbones.
12. Anfinsen, C. B., and H. A. Scheraga. 1975. "Experimental and Theoretical Aspects of Protein Folding." *Adv. Prot. Chem.* 29:205–299.
13. Schultz, G., and R. H. Shirmer. 1979. *Principles of Protein Structure*. New York: Springer-Verlag, pp. 67–78.
14. Lesk, A. M. O., and C. Chothia. 1980. "How Different Amino Acid Sequences Determine Similar Protein Structures. I. The Structure and Evolutionary Dynamics of the Globins." *J. Mol. Biol.* 136:225–270.
15. Hol, W. G., L. M. Halie, and C. Sander. 1982. "Dipoles of the Alpha Helix and Beta Sheet: Their Role in Protein Folding." *Nature* 294:532–536.
16. Lesk, A. M., and K. D. Hardman. 1982. "Computer-Generated Schematic Diagrams of Protein Structures." *Science* 216:539–540.
17. Novotny, J., R. Bruccoleri, and M. Karplus. 1984. "An Analysis of Incorrectly Folded Protein Models: Implications for Structure Predictions." *J. Mol. Biol.* 177:787–818.

18. Privilov, P. L. 1979. "Stability of Proteins." *Adv. Prot. Chem.* 33:167–236.

19. Crick, F. H. C. 1952. "Is α-Keratin a Coiled Coil?" *Nature* 170:882–883.

20. Pauling, L. 1953. "Compound Helical Configurations of Polypeptide Chains: Structure of Proteins of the α-Keratin Type." *Nature* 171:59–61.

21. Levitt, M., and C. Chothia. 1976. "Structural Patterns in Globular Proteins." *Nature* 261:552–557.

22. Chothia, C., M. Levitt, and D. Richardson. 1977. "Structure of Proteins: Packing of α-Helices and β Pleated Sheets." *Proc. Nat. Acad. Sci.* 74:4130–4134.

23. Sternberg, M. J. E., and J. M. Thornton. 1977. "On the Conformation of Proteins: The Handedness of the Connections Between Parallel β Strands." *J. Mol. Biol.* 110:269–283.

24. Cohen, F., and M. Sternberg. 1980. "Packing of α-Helices onto β Pleated Sheets and the Anatomy of αβ Proteins." *J. Mol. Biol.* 143:95–128.

25. Rose, G. E. 1979. "Hierarchic Organization of Domains in Globular Proteins." *J. Mol. Biol.* 134:447–470.

26. Novokhatny, V. V., S. A. Kudinov, and P. L. Privalov. 1984. "Domains in Human Plasminogen." *J. Mol. Biol.* 179:215–232.

27. Adams, M. J., G. C. Ford, R. Koekoek, P. J. Lentz, Jr., A. McPherson, Jr., M. R. Rossmann, I. E. Smiley, R. W. Schevitz, and A. J. Wonacott. 1970. "Structure of Lactate Dehydrogenase at 2.8 Å Resolution." *Nature* 227:1098–1103.

28. Steitz, T. A., I. T. Weber, and J. B. Matthew. 1982. "Catabolite Gene Activator Protein: Structure, Homology with Other Proteins, and Cyclic AMP and DNA Binding." *Cold Spring Harbor Symp. Quant. Biol.* 47:419–426.

29. Sherwood, D. 1976. *Crystals, X-rays, and Proteins.* New York: Wiley.

30. Blundell, T. L., and L. N. Johnson. 1976. *Protein Crystallography.* New York: Academic Press. An advanced text.

31. Sperling, R., and E. J. Wachtel. 1981. "The Histones." *Adv. Prot. Chem.* 34:1–52.

32. Ohlendorf, D. H., W. F. Anderson, R. G. Fisher, Y. Takeda, and B. W. Matthews. 1982. "The Molecular Basis of DNA-Protein Recognition Inferred from the Structure of the Cro Repressor." *Nature* 298:718–723.

33. Frederick, C. A., J. Grable, M. Melia, C. Samudzi, L. Jen-Jacobson, B.-C. Wang, P. Greene, H. W. Boyer, and J. M. Rosenberg. 1984. "Kinked DNA in Crystalline Complex with EcoRl Endonuclease." *Nature* 309:327–331.

34. von Hippel, P. H., A. Revzin, C. A. Gross, and A. C. Wang. 1974. "Non-Specific DNA Binding of Genome Regulating Protein as a Biological Control Mechanism. I. The *lac* Operon. Equilibrium Aspects." *Proc. Nat. Acad. Sci.* 71:4808–4812.

35. Carr, S. A., K. Biemann, S. Shoji, D. C. Parmelei, and K. Titani. 1982. "n-Tetradecanoyl Is the NH2-Terminal Blocking Group of the Catalytic Subunit of Cyclic AMP-Dependent Protein Kinase from Bovine Cardiac Muscle." *Proc. Nat. Acad. Sci.* 79:6128–6131.

36. Marchildon, G. A., J. E. Casnellie, K. A. Walsh, and E. G. Krebs. 1984. "Covalently Bound Myristate in a Lymphoma Tyrosine Protein Kinase." *Proc. Nat. Acad. Sci.* 81:7679–7682.

37. Gerhart, J. C., and A. B. Pardee. 1963. "The Effect of the Feedback Inhibitor CTP on Subunit Interactions in Aspartate Transcarbamylase." *Cold Spring Harbor Symp. Quant. Biol.* 28:491–496. Study of an enzyme system that was important in developing the concept of allostery.

38. Monod, J., J.-P. Changeux, and F. Jacob. 1963. "Allosteric Proteins and Cellular Control Systems." *J. Mol. Biol.* 6:306–329.

39. Monaco, H. L., J. L. Crawford, and W. N. Lipscomb. 1978. "Three-Dimensional Structures of Aspartate Carbamoyltransferase from *E. coli* and of Its Complex with Cytidine Triphosphate." *Proc. Nat. Acad. Sci.* 75:5276–5280.

40. Fraenkel-Conrat, H., and R. C. Williams. 1955. "Reconstitution of Active Tobacco Mosaic Virus from Its Inactive Protein and Nucleic Acid Components." *Proc. Nat. Acad. Sci.* 41:690–698.

41. Crick, F. H. C., and J. D. Watson. 1957. "Virus Structure: General Principles." In Ciba Foundation Symposium on *The Nature of Viruses*, ed. G. E. W. Wolstenholme and E. C. P. Millar. London: Churchill, pp. 5–13.

42. Caspar, D. L. D., and A. Klug. 1962. "Physical Principles in the Construction of Regular Viruses." *Cold Spring Harbor Symp. Quant. Biol.* 27:1–24.

CHAPTER 6

Coupled Reactions and Group Transfers

In the previous chapter we looked at the formation of weak bonds from the thermodynamic viewpoint. Each time a potential weak bond was considered, the question was posed, Does its formation involve a gain or a loss of free energy? because only when ΔG is negative does the thermodynamic equilibrium favor a reaction. This same approach is equally valid for covalent bonds. The fact that enzymes are usually involved in the making or breaking of a covalent bond does not in any sense alter the requirement of a negative ΔG.

On superficial examination, however, many of the important covalent bonds in cells appear to be formed in violation of the laws of thermodynamics, particularly those bonds joining small molecules together to form large polymeric molecules. The formation of such bonds involves an increase in free energy. Originally, this fact suggested to some people that cells had the unique ability to work in violation of thermodynamics and that this property was, in fact, the real "secret of life."

Now, however, it is clear that these biosynthetic processes do not violate thermodynamics but rather are based on different reactions from those originally postulated. Nucleic acids, for example, do not form by the condensation of nucleoside phosphates; glycogen is not formed directly from glucose residues; proteins are not formed by the union of amino acids. Instead, the monomeric precursors, using energy present in ATP, are first converted to high-energy "activated" precursors, which then spontaneously (with the help of specific enzymes) unite to form larger molecules. In this chapter, we shall illustrate these ideas by concentrating on the thermodynamics of peptide (protein) and phosphodiester (nucleic acid) bonds. First, however, we must briefly look at some general thermodynamic properties of covalent bonds.

Food Molecules Are Thermodynamically Unstable

There is great variation in the amount of free energy possessed by specific molecules. This is because covalent bonds do not all have the same bond energy. As an example, the covalent bond between oxygen and hydrogen is considerably stronger than the bond between hydrogen and hydrogen or oxygen and oxygen. The formation of an O—H bond at the expense of O—O or H—H will thus release energy. Energy considerations therefore tell us that a sufficiently concentrated mixture of oxygen and hydrogen will be transformed into water.

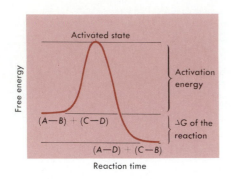

Figure 6-1
The energy of activation of a chemical reaction:
(A–B) + (C–D) → (A–D) + (C–B).

A molecule thus possesses a larger amount of free energy if linked together by weak covalent bonds than if it is linked together by strong bonds. This idea seems almost paradoxical at first glance, since it means that the stronger the bond, the less energy it can give off. But the notion automatically makes sense when we realize that an atom that has formed a very strong bond has already lost a large amount of free energy in this process. Therefore, the best food molecules (molecules that donate energy) are those molecules that contain weak covalent bonds and are therefore thermodynamically unstable.

For example, glucose is an excellent food molecule, since there is a great decrease in free energy when it is oxidized by oxygen to yield carbon dioxide and water. On the other hand, carbon dioxide is not a food molecule in animals, since in the absence of the energy donor ATP, it cannot spontaneously be transformed to more complex organic molecules, even with the help of specific enzymes. Carbon dioxide can be used as a primary source of carbon in plants only because the energy supplied by light quanta during photosynthesis results in the formation of ATP.

The Distinction Between Direction and Rate of a Reaction

The chemical reactions by which molecules are transformed into other molecules that contain less free energy do not occur at significant rates at physiological temperatures in the absence of a catalyst. This is because even a weak covalent bond is, in reality, very strong and is only rarely broken by thermal motion within a cell. For a covalent bond to be broken in the absence of a catalyst, energy must be supplied to push apart the bonded atoms. When the atoms are partially apart, they can recombine with new partners to form stronger bonds. In the process of recombination, the energy released is the sum of the free energy supplied to break the old bond plus the difference in free energy between the old and the new bond (Figure 6-1).

The energy that must be supplied to break the old covalent bond in a molecular transformation is called the **activation energy.** The activation energy is usually less than the energy of the original bond because molecular rearrangements generally do not involve the production of completely free atoms. Instead, a collision between the two reacting molecules is required, followed by the temporary formation of a molecular complex (the **activated state**). In the activated state, the close proximity of the two molecules makes each other's bonds more labile, so that less energy is needed to break a bond than when the bond is present in a free molecule.

Most reactions of covalent bonds in cells are therefore described by

$$(A\text{–}B) + (C\text{–}D) \longrightarrow (A\text{–}D) + (C\text{–}B) - \Delta G \qquad (6\text{-}1)$$

The mass action expression for such reaction is

$$K_{eq} = \frac{conc^{A\text{-}D} \times conc^{C\text{-}B}}{conc^{A\text{-}B} \times conc^{C\text{-}D}} \qquad (6\text{-}2)$$

where $conc^{A\text{-}B}$, $conc^{C\text{-}D}$, and so on, are the concentrations of the several reactants in moles per liter. Here, also, the value of K_{eq} is related to ΔG by Equation 5-4 (Table 6-1).

Since energies of activation are generally between 20 and 30 kcal/mol, activated states practically never occur at physiological temperatures. High activation energies should thus be considered barriers preventing spontaneous rearrangements of cellular covalent bonds.

These barriers are enormously important. Life would be impossible if they did not exist, for all atoms would be in the state of least possible energy. There would be no way to temporarily store energy for future work. On the other hand, life would also be impossible if means were not found to selectively lower the activation energies of certain reactions. This also must happen if cell growth is to occur at a rate sufficiently fast so as not to be seriously impeded by random destructive forces, such as ionizing or ultraviolet radiation.

Enzymes Lower Activation Energies

Enzymes are absolutely necessary for life. The function of enzymes is to speed up the rate of the chemical reactions requisite to cellular existence by lowering the activation energies of molecular rearrangements to values that can be supplied by the heat of motion (Figure 6-2). When a specific enzyme is present, there is no longer an effective barrier preventing the rapid formation of the reactants possessing the lowest amounts of free energy. Enzymes never affect the nature of an equilibrium: They merely speed up the rate at which it is reached. Thus, if the thermodynamic equilibrium is unfavorable for the formation of a molecule, the presence of an enzyme can in no way bring about the molecule's accumulation.

Since enzymes must catalyze essentially every cellular molecular rearrangement, knowing the free energy of various molecules cannot by itself tell us whether an energetically feasible rearrangement will, in fact, occur. The rate of the reactions must always be considered. Only if a cell possesses a suitable enzyme will the reaction be important.

A Metabolic Pathway Is Characterized by a Decrease in Free Energy

Thermodynamics tells us that all biochemical pathways must be characterized by a decrease in free energy. This is clearly the case for degradative pathways, in which thermodynamically unstable food molecules are converted to more stable compounds, such as carbon dioxide and water, with the evolution of heat. All degradative pathways have two primary purposes: (1) to produce the small organic fragments necessary as building blocks for larger organic molecules and (2) to conserve a significant fraction of the free energy of the original food molecule in a form that can do work, by coupling some of the steps in degradative pathways with the simultaneous formation of high-energy molecules such as ATP, which can store free energy.

Not all the free energy of a food molecule is converted into the free energy of high-energy molecules. If this were the case, a degradative pathway would not be characterized by a decrease in free energy, and there would be no driving force to favor the breakdown of food molecules. Instead, we find that all degradative pathways are characterized by a conversion of at least one-half the free energy of the food

Table 6-1 The Relationship Between K_{eq} and ΔG ($\Delta G = -RT \ln K_{eq}$)

K_{eq}	ΔG (kcal/mol)
10^{-6}	8.2
10^{-5}	6.8
10^{-4}	5.1
10^{-3}	4.1
10^{-2}	2.7
10^{-1}	1.4
10^{0}	0.0
10^{1}	−1.4
10^{2}	−2.7
10^{3}	−4.1

Figure 6-2
Enzymes (color curve) lower activation energies and thus speed up the rate of the reaction. Note that ΔG remains the same, since the equilibrium position remains unaltered.

molecule into heat or entropy. For example, it is now believed that in cells, approximately 40 percent of the free energy of glucose is used to make new high-energy compounds, the remainder being dissipated into heat energy and entropy.

High-Energy Bonds Hydrolyze with Large Negative ΔG

A high-energy molecule contains one or more bonds whose breakdown by water (*hydrolysis*) is accompanied by a large decrease in free energy (5 kcal/mol or more). The specific bonds whose hydrolysis yields these large negative ΔG values are called **high-energy bonds,** a somewhat misleading term, since it is not the bond energy but the free energy of hydrolysis that is high. Nonetheless, the term *high-energy bond* is generally employed, and for convenience, we shall continue this usage by marking high-energy bonds with the symbol $\sim$.

The energy of hydrolysis of the average high-energy bond (7 kcal/mol) is very much smaller than the amount of energy that would be released if a glucose molecule were to be completely degraded in one step (688 kcal/mol). A one-step breakdown of glucose would be inefficient in making high-energy bonds. This is undoubtedly the reason why biological glucose degradation requires so many steps. In this way, the amount of energy released per degradative step is of the same order of magnitude as the free energy of hydrolysis of a high-energy bond.

The most important high-energy compound is ATP. It is formed from inorganic phosphate ⓟ and ADP, using energy obtained either from degradative reactions or from the sun (photosynthesis). There are, however, many other important high-energy compounds. Some are directly formed during degradative reactions; others are formed using some of the free energy of ATP. Table 6-2 lists the most important types of high-energy bonds. All involve either phosphate or sulfur atoms. The high-energy pyrophosphate bonds of ATP arise from the union of phosphate groups. The pyrophosphate linkage (ⓟ$\sim$ⓟ) is not, however, the only kind of high-energy phosphate bond: The attachment of a phosphate group to the oxygen atom of a carboxyl group creates a high-energy acyl bond. It is now clear that high-energy bonds involving sulfur atoms play almost as important a role in energy metabolism as those involving phosphorus. The most important molecule containing a high-energy sulfur bond is acetyl-CoA. This bond is the main source of energy for fatty acid biosynthesis.

The wide range of ΔG values of high-energy bonds (see Table 6-2) means that calling a bond "high-energy" is sometimes arbitrary. The usual criterion is whether its hydrolysis can be coupled with another reaction to effect an important biosynthesis. For example, the negative ΔG accompanying the hydrolysis of glucose-6-phosphate is 3 to 4 kcal/mol. But this ΔG is not sufficient for efficient synthesis of peptide bonds, so this phosphate ester bond is not included among high-energy bonds.

High-Energy Bonds Necessary for Biosynthetic Reactions

The construction of a large molecule from smaller building blocks often requires the input of free energy. Yet, a biosynthetic pathway, like a degradative pathway, would not exist if it were not character-

Table 6-2 Important Classes of High-Energy Bonds

Class	Molecular Example	ΔG of Reaction, kcal/mole
Pyrophosphate	(P) ~ (P) pyrophosphate	(P) ~ (P) $\rightleftharpoons$ (P) + (P) $\Delta G = -6$
Nucleoside diphosphates	Adenosine—(P) ~ (P) (ADP)	ADP $\rightleftharpoons$ AMP + (P) $\Delta G = -6$
Nucleoside triphosphates	Adenosine—(P) ~ (P) ~ (P)(ATP)	ATP $\rightleftharpoons$ ADP + (P) $\Delta G = -7$ ATP $\rightleftharpoons$ AMP + (P) ~ (P) $\Delta G = -8$
Enol phosphates	Phosphoenolpyruvate (PEP)	PEP $\rightleftharpoons$ pyruvate + (P) $\Delta G = -12$
Aminoacyl adenylates		AMP ~ AA $\rightleftharpoons$ AMP + AA $\Delta G = -7$
Guanidinium phosphates	Creatine phosphate	Creatine ~ P $\rightleftharpoons$ Creatine + P $\Delta G = -8$
Thioesters	Acetyl-CoA	Acetyl CoA $\rightleftharpoons$ CoA-SH + acetate $\Delta G = -8$

ized by a net decrease in free energy. This means that many biosynthetic pathways demand an external source of free energy. These free-energy sources are the high-energy compounds. The making of many biosynthetic bonds is coupled with the breakdown of a high-energy bond, so that the net change of free energy is always negative. Thus, high-energy bonds in cells generally have a very short life. Almost as soon as they are formed during a degradative reaction, they are enzymatically broken down to yield the energy needed to drive another reaction to completion.

Not all the steps in a biosynthetic pathway require the breakdown of a high-energy bond. Often, only one or two steps involve such a bond. Sometimes this is because the ΔG, even in the absence of an externally added high-energy bond, favors the biosynthetic direction. In other cases, ΔG is effectively zero or may even be slightly positive. These small positive ΔG values, however, are not significant so long as they are followed by a reaction characterized by the hydrolysis of a

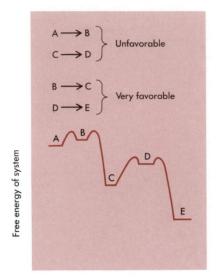

Figure 6-3
Free-energy changes in a multistep metabolic pathway, $A \rightarrow B \rightarrow C \rightarrow D \rightarrow E$. Two steps ($A \rightarrow B$ and $C \rightarrow D$) do not favor the $A \rightarrow E$ direction of the reaction, since they have small positive ΔG values. However, they are insignificant owing to the very large negative ΔG values provided in steps $B \rightarrow C$ and $D \rightarrow E$. Therefore, the overall reaction favors the $A \rightarrow E$ conversion.

high-energy bond. Rather, it is the *sum* of all the free-energy changes in a pathway that is significant (Figure 6-3). It does not really matter that the K_{eq} of a specific biosynthetic step is slightly (80:20) in favor of degradation if the K_{eq} of the succeeding step is 100:1 in favor of the forward biosynthetic direction.

Likewise, not all the steps in a degradative pathway generate high-energy bonds. For example, only two steps in the lengthy glycolytic (Embden-Meyerhof) breakdown of glucose generate ATP. Moreover, there are many degradative pathways that have one or more steps requiring the breakdown of a high-energy bond. The glycolytic breakdown of glucose is again an example. It uses up two molecules of ATP for every four that it generates. Here, of course, as in every energy-yielding degradative process, more high-energy bonds must be made than consumed.

Peptide Bonds Hydrolyze Spontaneously

The formation of a dipeptide and a water molecule from two amino acids requires a ΔG of 1 to 4 kcal/mol, depending on which amino acids are being bound. This ΔG value decreases progressively if amino acids are added to longer polypeptide chains; for an infinitely long chain, the ΔG is reduced to about 0.5 kcal/mol. This decrease reflects the fact that the free, charged NH_3^+ and COO^- groups at the chain ends favor the hydrolysis (breakdown accompanied by the uptake of a water molecule) of nearby peptide bonds.

These positive ΔG values by themselves tell us that polypeptide chains cannot form from free amino acids. In addition, we must take into account the fact that water molecules have a much, much higher concentration than any other cellular molecules (generally over 100 times higher). All equilibrium reactions in which water participates are thus strongly pushed in the direction that consumes water molecules. This is easily seen in the definition of equilibrium constants. For example, the reaction forming a dipeptide,

$$\text{amino acid(A)} + \text{amino acid(B)} \longrightarrow \text{dipeptide(A–B)} + H_2O \quad (6\text{-}3)$$

has the following equilibrium constant:

$$K_{eq} = \frac{\text{conc}^A \times \text{conc}^B}{\text{conc}^{A-B} \times \text{conc}^{H_2O}} \quad (6\text{-}4)$$

where concentrations are given in moles per liter. Thus, for a given K_{eq} value (related to ΔG by the formula $\Delta G = -RT \ln K_{eq}$), a much greater concentration of water means a correspondingly smaller concentration of the dipeptide. The relative concentrations are therefore very important. In fact, a simple calculation shows that hydrolysis may often proceed spontaneously even when the ΔG for the nonhydrolytic reaction is -3 kcal/mol.

Thus, in theory, proteins are unstable and, given sufficient time, will spontaneously degrade to free amino acids. On the other hand, in the absence of specific enzymes, these spontaneous rates are too slow to have a significant effect on cellular metabolism. That is, once a protein is made, it remains stable unless its degradation is catalyzed by a specific enzyme.

Coupling of Negative with Positive ΔG

Free energy must be added to amino acids before they can be united to form proteins. How this happens became clear with the discovery of the fundamental role of ATP as an energy donor. ATP contains three phosphate groups attached to an adenosine molecule (adenosine $-O-\textcircled{P}\sim\textcircled{P}\sim\textcircled{P}$). When one or two of the terminal $\sim\textcircled{P}$ groups are broken off by hydrolysis, there is a significant decrease of free energy.

Adenosine$-O-\textcircled{P}\sim\textcircled{P}\sim\textcircled{P} + H_2O \longrightarrow$

$\qquad$ adenosine$-O-\textcircled{P}\sim\textcircled{P} + \textcircled{P} \qquad (\Delta G = -7$ kcal/mol$) \quad$ (6-5)

Adenosine$-O-\textcircled{P}\sim\textcircled{P}\sim\textcircled{P} + H_2O \longrightarrow$

$\qquad$ adenosine$-O-\textcircled{P} + \textcircled{P}\sim\textcircled{P} \qquad (\Delta G = -8$ kcal/mol$) \quad$ (6-6)

Adenosine$-O-\textcircled{P}\sim\textcircled{P} + H_2O \longrightarrow$

$\qquad$ adenosine$-O-\textcircled{P} + \textcircled{P} \qquad (\Delta G = -6$ kcal/mol$) \quad$ (6-7)

All these breakdown reactions have negative ΔG values considerably greater in absolute value (numerical value without regard to sign) than the positive ΔG values accompanying the formation of polymeric molecules from their monomeric building blocks. The essential trick underlying these biosynthetic reactions, which by themselves have a positive ΔG, is that they are coupled with breakdown reactions characterized by negative ΔG of greater absolute value. Thus, during protein synthesis, the formation of each peptide bond ($\Delta G = +0.5$ kcal/mol) is coupled with the breakdown of ATP to AMP and pyrophosphate, which has a ΔG of -8 kcal/mol. This results in a net ΔG of -7.5 kcal/mol, more than sufficient to ensure that the equilibrium favors protein synthesis rather than breakdown.

Activation Through Group Transfer

When ATP is hydrolyzed to ADP and phosphate, most of the free energy is liberated as heat. Since heat energy cannot be used to make covalent bonds, a coupled reaction cannot be the result of two completely separate reactions, one with a positive ΔG, the other with a negative ΔG. Instead, a coupled reaction is achieved by two or more successive reactions. These are always *group-transfer* reactions: reactions, not involving oxidations or reductions, in which molecules exchange functional groups. The enzymes that catalyze these reactions are called transferases. Consider the reaction

$$(A-X) + (B-Y) \longrightarrow (A-B) + (X-Y) \qquad (6-8)$$

In this example, group X is exchanged with component B. Group-transfer reactions are arbitrarily defined to exclude water as a participant. When water is involved,

$$(A-B) + (H-OH) \longrightarrow (A-OH) + (BH) \qquad (6-9)$$

This reaction is called a hydrolysis, and the enzymes involved are called hydrolases.

Figure 6-4
Important group transfers involving
ATP.

The group-transfer reactions that interest us here are those involving groups attached by high-energy bonds. When such a high-energy group is transferred to an appropriate acceptor molecule, it becomes attached to the acceptor by a high-energy bond. Group transfer thus allows the transfer of high-energy bonds from one molecule to another. For example, Equations 6-10 and 6-11 show how energy present in ATP is transferred to form GTP, one of the precursors used in RNA synthesis:

Adenosine—(P)~(P)~(P) + guanosine—(P) ⟶

adenosine—(P)~(P) + guanosine—(P)~(P) *(6-10)*

Adenosine—(P)~(P)~(P) + guanosine—(P)~(P) ⟶

adenosine—(P)~(P) + guanosine—(P)~(P)~(P) *(6-11)*

The high-energy (P)~(P) group on GTP allows it to unite spontaneously with another molecule. GTP is thus an example of what is called an **activated molecule;** correspondingly, the process of transferring a high-energy group is called **group activation.**

ATP Versatility in Group Transfer

In Chapter 2, we emphasized the key role of ATP synthesis in the controlled trapping of the energy of food molecules. In both oxidative and photosynthetic phosphorylations, energy is used to synthesize ATP from ADP and phosphate:

Adenosine—(P)~(P) + (P) + energy ⟶

adenosine—(P)~(P)~(P) *(6-12)*

Since ATP is thus the original biological recipient of high-energy groups, it must be the starting point of a variety of reactions in which high-energy groups are transferred to low-energy molecules to give them the potential to react spontaneously. ATP's central role utilizes the fact that it contains two high-energy bonds whose splitting releases specific groups. This is seen in Figure 6-4, which shows three important groups arising from ATP: $P{\sim}P$, a pyrophosphate group; ${\sim}AMP$, an adenosyl monophosphate group; and ${\sim}P$, a phosphate group. It is important to notice that these high-energy groups retain their high-energy quality only when transferred to an appropriate acceptor molecule. For example, although the transfer of a ${\sim}P$ group to a COO^- group yields a high-energy $COO{\sim}P$ acyl-phosphate group, the transfer of the same group to a sugar hydroxyl group ($-C-OH$), as in the formation of glucose-6-phosphate, gives rise to a low-energy bond (less than 5 kcal/mol decrease in ΔG upon hydrolysis).

Activation of Amino Acids by Attachment of AMP

The activation of an amino acid is achieved by transfer of an AMP group from ATP to the COO^- group of the amino acid:

$$\text{H}-\underset{\underset{\text{H}}{|}}{\overset{\overset{\text{H}}{|}}{\text{N}^{\pm}}}-\underset{\underset{\text{H}}{|}}{\overset{\overset{\text{R}}{|}}{\text{C}}}-\text{C}\!\!\begin{array}{c}{\nearrow}{\text{O}}\\[-2pt]{\searrow}{\text{O}^-}\end{array} + \text{adenosine}-\!P{\sim}P{\sim}P\ \longrightarrow$$

$$\text{H}-\underset{\underset{\text{H}}{|}}{\overset{\overset{\text{H}}{|}}{\text{N}^{\pm}}}-\underset{\underset{\text{H}}{|}}{\overset{\overset{\text{R}}{|}}{\text{C}}}-\text{C}\!\!\begin{array}{c}{\nearrow}{\text{O}}\\[-2pt]{\searrow}{\text{O}{\sim}P-\text{adenosine}}\end{array} + P{\sim}P \qquad (6\text{-}13)$$

(R represents the specific side group of the amino acid.) The enzymes that catalyze this type of reaction are called aminoacyl synthetases. Upon activation, an amino acid (AA) is thermodynamically capable of being efficiently used for protein synthesis. Nonetheless, the AA~AMP complexes are not the direct precursors of proteins. Instead, for a reason we shall explain in Chapter 14, a second group transfer must occur to transfer the amino acid, still activated at its carboxyl group, to the end of a tRNA molecule:

$$\text{AA}{\sim}\text{AMP} + \text{tRNA} \longrightarrow \text{AA}{\sim}\text{tRNA} + \text{AMP} \qquad (6\text{-}14)$$

A peptide bond then forms by the condensation of the AA~tRNA molecule onto the end of a growing polypeptide chain:

AA~tRNA + growing polypeptide chain (of n amino acids) $\longrightarrow$

tRNA + growing polypeptide chain (of $n + 1$ amino acids) (6-15)

Thus, the final step of this "coupled reaction," like that of all other coupled reactions, necessarily involves the removal of the activating group and the conversion of a high-energy bond into one with a lower free energy of hydrolysis. This is the source of the negative ΔG that drives the reaction in the direction of protein synthesis.

Nucleic Acid Precursors Activated by the Presence of Ⓟ~Ⓟ

Both types of nucleic acid, DNA and RNA, are built up from mononucleotide monomers (nucleoside phosphate). Mononucleotides, however, are thermodynamically even less likely to combine than amino acids. This is because the phosphodiester bonds that link the former together release considerable free energy upon hydrolysis (−6 kcal/mol). This means that nucleic acids will spontaneously hydrolyze, at a slow rate, to mononucleotides. Thus, it is even more important that activated precursors be used in the synthesis of nucleic acids than in the synthesis of proteins.

It has been found that the immediate precursors for both DNA and RNA are the nucleoside-5′-triphosphates. For DNA, these precursors are dATP, dGTP, dCTP, and dTTP (d stands for deoxy); for RNA, the precursors are ATP, GTP, CTP, and UTP. ATP thus not only serves as the main source of high-energy groups in group-transfer reactions, but is itself a direct precursor for RNA. The other three RNA precursors all arise by group-transfer reactions like those described in Equations 6-10 and 6-11. The deoxytriphosphates are formed in basically the same way: After the deoxymononucleotides have been synthesized, they are transformed to the triphosphate form by group transfer from ATP:

Deoxynucleoside—Ⓟ + ATP $\longrightarrow$

deoxynucleoside—Ⓟ~Ⓟ + ADP   *(6-16)*

Deoxynucleoside—Ⓟ~Ⓟ + ATP $\longrightarrow$

deoxynucleoside—Ⓟ~Ⓟ~Ⓟ + ADP   *(6-17)*

These triphosphates can then unite to form polynucleotides held together by phosphodiester bonds. In this group-transfer reaction, a pyrophosphate bond is broken and a pyrophosphate group released:

Deoxynucleoside—Ⓟ~Ⓟ~Ⓟ

+ growing polynucleotide chain (of *n* nucleotides) $\longrightarrow$

Ⓟ~Ⓟ + growing polynucleotide chain (*n* + 1 nucleotides)   *(6-18)*

This reaction, unlike that which forms peptide bonds, does not have a negative ΔG. In fact, the ΔG is slightly positive (about 0.5 kcal/mol). This immediately poses the question, since polynucleotides obviously form: What is the source of the necessary free energy?

The Value of Ⓟ~Ⓟ Release in Nucleic Acid Synthesis

The needed free energy comes from the splitting of the high-energy pyrophosphate group that is formed simultaneously with the high-energy phosphodiester bond. All cells contain a powerful enzyme, pyrophosphatase, which breaks down pyrophosphate molecules almost as soon as they are formed:

$$ \text{P\~P} \longrightarrow 2\ \text{P} \qquad (\Delta G = -7\ \text{kcal/mol}) \qquad (6\text{-}19) $$

The large negative ΔG means that the reaction is effectively irreversible. This means that once P~P is broken down, it never reforms.

The union of the nucleoside monophosphate group (Equation 6-16), coupled with the splitting of the pyrophosphate groups (Equation 6-19), has an equilibrium constant determined by the combined ΔG values of the two reactions: (0.5 kcal/mol) + (−7 kcal/mol). The resulting value ($\Delta G = -6.5$ kcal/mol) tells us that nucleic acids almost never break down to reform their nucleoside triphosphate precursors.

Here we see a powerful example of the fact that often it is the free-energy change accompanying a *group of reactions* that determines whether a reaction in the group will take place. Reactions with small, positive ΔG values, which by themselves would never take place, are often part of important metabolic pathways in which they are followed by reactions with large negative ΔG values. At all times we must remember that a single reaction (or even a single pathway) never occurs in isolation; rather, the nature of the equilibrium is constantly being changed through the addition and removal of metabolites.

P~P Splits Characterize Most Biosynthetic Reactions

The synthesis of nucleic acids is not the only reaction where direction is determined by the release and splitting of P~P. In fact, the generalization is emerging that essentially all biosynthetic reactions are characterized by one or more steps that release pyrophosphate groups. Consider, for example, the activation of an amino acid by the attachment of AMP. By itself, the transfer of a high-energy bond from ATP to the AA~AMP complex has a slightly positive ΔG. Therefore, it is the release and splitting of ATP's terminal pyrophosphate group that provides the negative ΔG that is necessary to drive the reaction.

The great utility of the pyrophosphate split is neatly demonstrated when we consider the problems that would arise if a cell attempted to synthesize nucleic acid from nucleoside diphosphates rather than triphosphates. Phosphate, rather than pyrophosphate, would be liberated as the backbone phosphodiester linkages were made. The phosphodiester linkages, however, are not stable in the presence of significant quantities of phosphate, since they are formed without a significant release of free energy. Thus, the biosynthetic reaction would be easily reversible; as soon as phosphate began to accumulate, the reaction would begin to move in the direction of nucleic acid breakdown (mass-action law). Moreover, it is not possible for a cell to remove the phosphate groups as soon as they are generated (thus preventing this reverse reaction), since all cells need a significant internal level of phosphate to grow. Thus, the use of nucleoside triphosphates as precursors of nucleic acids is not a matter of chance.

This same type of argument tells us why ATP, and not ADP, is the key donor of high-energy groups in all cells. At first this preference seemed arbitrary to biochemists. Now, however, we see that many reactions using ADP as an energy donor would occur equally well in both directions.

Summary

The biosynthesis of many molecules appears, at a superficial glance, to violate the thermodynamic law that spontaneous reactions always involve a decrease in free energy (ΔG is negative). For example, the formation of proteins from amino acids has a positive ΔG. This paradox is removed when we realize that the biosynthetic reactions do not proceed as initially postulated. Proteins, for example, are not formed from free amino acids. Instead, the precursors are first enzymatically converted to high-energy activated molecules, which, in the presence of a specific enzyme, spontaneously unite to form the desired biosynthetic product.

Many biosynthetic processes are thus the result of "coupled" reactions, the first of which supplies the energy that allows the spontaneous occurrence of the second reaction. The primary energy source in cells is ATP. It is formed from ADP and inorganic phosphate, either during degradative reactions (e.g., fermentation or respiration) or during photosynthesis. ATP contains several high-energy bonds whose hydrolysis has a large negative ΔG. Groups linked by high-energy bonds are called high-energy groups. High-energy groups can be transferred to other molecules by group-transfer reactions, thereby creating new high-energy compounds. These derivative high-energy molecules are then the immediate precursors for many biosynthetic steps.

Amino acids are activated by the addition of an AMP group, originating from ATP, to form an AA~AMP molecule. The energy of the high-energy bond in the AA~AMP molecule is similar to that of a high-energy bond of ATP. Nonetheless, the group-transfer reaction proceeds to completion because the high-energy Ⓟ~Ⓟ molecule, created when the AA~AMP molecule is formed, is broken down by the enzyme pyrophosphatase to low-energy groups. Thus, the reverse reaction, Ⓟ~Ⓟ + AA~AMP → ATP + AA, cannot occur.

Almost all biosynthetic reactions result in the release of Ⓟ~Ⓟ. Almost as soon as it is made, it is enzymatically broken down to two phosphate molecules, thereby making impossible a reversal of the biosynthetic reaction. The great utility of the Ⓟ~Ⓟ split provides an explanation for why ATP, not ADP, is the primary energy donor. ADP cannot transfer a high-energy group and at the same time produce Ⓟ~Ⓟ groups as a by-product.

Bibliography

General References

Dugas, H., and C. Penney. 1981. *Biorganic Chemistry: A Chemical Approach to Enzyme Action*. New York: Springer-Verlag.

Jencks, W. P. 1969. *Catalysis in Chemistry and Enzymology*. New York: McGraw-Hill.

Kornberg, A. 1962. "On the Metabolic Significance of Phosphorolytic and Pyrophosphorolytic Reactions." In *Horizons in Biochemistry*, ed. M. Kasha and B. Pullman. New York: Academic Press, pp. 251–264.

Krebs, H. A., and H. L. Kornberg. 1957. "A Survey of the Energy Transformation in Living Material." *Ergeb. Physiol. Biol. Chem.* *Exp. Pharmakol.* 49:212. A classical comprehensive review which, although many of its facts have been modified by subsequent research, can still be used with great profit.

Lehninger, A. L. 1982. *Principles of Biochemistry*. New York: Worth.

Smith, E. L., and R. L. Hill. 1983. *Principles of Biochemistry*. 7th ed. New York: McGraw-Hill.

Stryer, L. 1981. *Biochemistry*. 2nd ed. San Francisco: Freeman.

Walsh, C. 1979. *Enzymatic Reaction Mechanisms*. San Francisco: Freeman.

Wood, W. B., J. H. Wilson, R. M. Benbow, and L. Hood. 1981. *Biochemistry: A Problems Approach*. 2nd ed. Menlo Park, Calif.: Benjamin/Cummings.

III

BACTERIAL GENETICS

CHAPTER 7

The Genetic Systems Provided by *E. coli* and Its Viruses

The genes of *E. coli* and its viruses, the bacteriophages (phages), are now understood better than those of any other form of life. They provide the simplest examples of what genes are and how the genetic code operates. Moreover, skillful exploitation of the *E. coli* genetic properties revealed that the information needed to properly regulate the rate at which specific genes work is also encoded within the nucleotide sequences of DNA. The more we learn about *E. coli*, the more we understand the extraordinary plasticity of its genetic systems and the more we appreciate how superbly programmed this organism is to face the highly changeable environments that all bacteria encounter daily.

Key to much, if not most, of our knowledge about the genetic systems of *E. coli* has been the use of mutants. By noting a specific aberration in a particular mutant, we can often reveal the function of the normal cell. For example, if a mutant *E. coli* unable to synthesize lactose is found to lack a specific gene group, we can surmise that that gene group is responsible for lactose synthesis. Without mutants, we would still be in the dark as to how gene expression is regulated, and we would still be unaware of many essential features of DNA replication.

The Intrinsic Advantages of Using Microorganisms for Genetic Research[1, 2, 3]

That microorganisms were the obvious route to discovering the nature of the gene became clear as soon as geneticists began to focus on the gene-protein relationship. This happened in the 1930s, when biosynthetic processes were discovered to be series ("pathways") of enzymatically catalyzed steps resembling the multistep procedures that skilled chemists invariably needed to make sophisticated organic molecules. Directly pursuing gene-enzyme connections, however, was virtually impossible. No way was known to select for mutations in genes that controlled already known biochemical pathways. Even when apparently suitable mutations were found in the fruit fly *Dro-*

sophila, it was not clear how to proceed. Direct comparison of the chemical compositions of mutant and wild-type flies was impractical because of the small size of the flies and the difficulties of breeding enough of them for chemical analysis (even though their generation time is relatively short—about 14 days). Furthermore, since fruit flies are diploid organisms, most mutations are not detected; this is because most mutations are recessive, and recessive mutations are not usually detected when the dominant wild-type allele is also present.

In contrast, single-celled eucaryotic microorganisms, such as the green alga *Chlamydomonas*, the yeast *Saccharomyces cerevisiae*, and the mold *Neurospora crassa*, which were among the first subjects of genetic analysis, were more attractive experimental organisms. First, they are usually haploid, so that the ease of genetic analysis does not depend on whether a mutation is dominant or recessive. In work with diploids, several generations of genetic crosses must often be carried out to detect the presence of a particular mutant gene; in a haploid organism, the mutant gene can express itself almost immediately, since there is no masking dominant allele present. Second, microorganisms can multiply very rapidly. There are enormous advantages to working with organisms that can produce a new cell generation every 2 hours rather than working with a plant like corn, which under the most favorable conditions can only produce two generations per year for study. When necessary, grams (and even kilograms) of desired mutant microorganisms can be grown under simple laboratory conditions.

At first, the fact that many microorganisms do not have easily recognizable features except for the morphology of their colonies on solid media made genetic analysis difficult. It was hard to know when the microorganisms contained mutations. But as soon as they could be characterized by their metabolic defects, the study of their genetics could blossom.

The first microorganism successfully used in studying gene-enzyme relationships was the mold *Neurospora crassa*, which, like *E. coli*, grows on a simple minimal glucose-containing medium. The essential trick was to seek out genes that controlled the biosynthetic pathways for vital metabolites that could either be ingested directly from the surrounding environment or made internally starting from one of the key precursor metabolites. Such genes were found by subjecting a population of *Neurospora* cells to ultraviolet (UV) light to increase the number of mutant cells. Searches were then made for descendants of treated cells that could only multiply in minimal medium when it was supplemented with one of a variety of known metabolites (e.g., amino acids, purine and pyrimidine bases, and vitamins). Such nutritionally restricted cells might, if all went well, lack the ability to synthesize one of the supplemented compounds. They were therefore tested to see if their growth on minimal medium could be restarted by the addition of a *single* known metabolite. The first gene to be so identified controlled the synthesis of one of the enzymes involved in the biosynthesis of the vitamin niacin. Over the next several years, many other *Neurospora* **growth factor** mutants were found that blocked the synthesis of many additional metabolites. In some cases, the specific enzyme activity lacking in the mutant cell could be identified, thereby directly demonstrating that a given gene controls the synthesis of a specific enzyme (the **one gene–one enzyme hypothesis**).

Five separate *E. coli* B cultures are inoculated onto plates covered with phage T1

Incubation 24 hrs

Colonies resistant to T1

28   35   13   38   107

Figure 7-1

Fluctuation analysis of bacterial resistance to phage. A total of 0.05 cc of each original culture was plated on each of five nutrient agar plates previously plated with an excess of bacteriophage T1. Microscopic analysis revealed that lysis of *E. coli* B cells followed rapidly. However, upon further incubation, colonies resulting from division of surviving resistant bacteria were formed. If the resistant bacteria represented adaptive responses to actual physical contact with T1, the same number of surviving colonies should be seen on each plate. The large fluctuation of the numbers of resulting colonies, however, proves that the mutants arose spontaneously prior to exposure to the phage.

Bacteria Have Genes That Mutate Spontaneously[4]

The first "biochemical" mutant in *Neurospora* was found in 1941, several years before convincing evidence appeared of a conventional hereditary mechanism within bacteria. Before that time, many variants of bacterial colony morphology were already known, and there was preliminary evidence for the existence of rare stable strains resistant to such antibacterial agents as antibiotics, viruses, and poisons. Nevertheless, there was much speculation that these variants were not the result of spontaneous genetic alterations but were direct adaptive (Lamarkian) responses to the environment. Today, such Lamarkian explanations appear to be totally without merit. However, in the absence of any evidence for chromosomes within bacteria, a novel form of proof had to be devised showing that bacterial variants occur spontaneously in the absence of discrete signals from the environment.

Such proof first arose out of experiments carried out in 1943 that measured the frequencies of newly occurring *E. coli* variants resistant to attack by the bacteriophage T1. Such variants arise with a frequency of approximately 1 in 10 million (10^{-7}). If they represented adaptive responses to actual physical contact with T1, then roughly the same number of phage-resistant cells should be detected in duplicate cultures of bacteria containing some 10^8 cells, each such culture arising from the multiplication of a single cell introduced into the pure medium. If, on the other hand, they represented spontaneous mutations, then an occasional culture would contain large jackpots of mutants, because they would result from mutations that occurred early in the growth of the culture. Moreover, there would also occur other cultures totally lacking any phage-resistant mutants. The re-

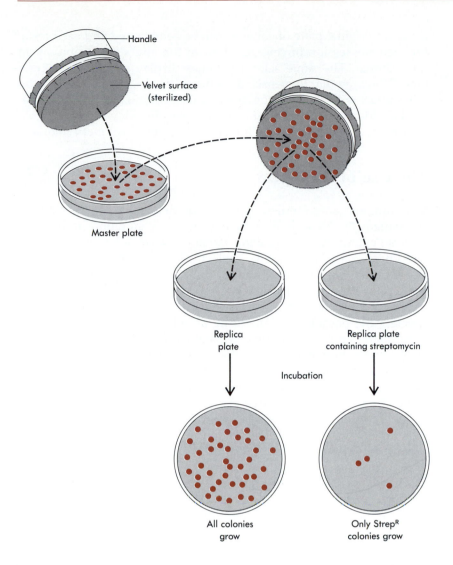

Master plate

Handle

Velvet surface
(sterilized)

Replica
plate

Replica plate
containing streptomycin

Incubation

All colonies
grow

Only Strep^R
colonies grow

Figure 7-2
Replica plating. A piece of sterile velvet cloth is pressed down on an agar plate containing many colonies, picking up cells in the process. The velvet is then pressed against another agar plate to transfer the cells, which then grow into colonies oriented exactly like those of the original (master) plate. Antibiotic-resistant colonies, such as streptomycin-resistant (Strep^R) colonies, can be identified by adding streptomycin to a second replica plate. Only Strep^R cells will grow on the replica plate with streptomycin.

sults of such "fluctuation analysis" unequivocally showed that resistance to phages as well as to antibiotics originated before the bacteria had contact with phages or antibacterial agents (Figure 7-1). Through these elegant experiments, adaptive (Lamarkian) explanations for bacterial variability lost all credibility. It became more logical to believe that all stable hereditary variations in bacteria represented changes in genetic material similar to that possessed by eucaryotic organisms.

The Great Power of Replica Plating[5]

An even simpler demonstration of the spontaneous nature of bacterial variants came a decade later through the development of **replica plating** (Figure 7-2). In this procedure, a flat, sterile piece of velvet cloth is placed face down on top of an agar plate containing a hundred to a thousand tiny bacterial colonies. Through this process, some living bacteria from each colony are transferred to the velvet. Subsequent application of the velvet to a sterile fresh agar surface allows each parent colony to give rise to a new colony positioned identically to those found on the original plate. Many identical replicas of an original plate of colonies can be so obtained and each tested for their possession of cells resistant to the antibiotic streptomycin, for example. The same sets of colonies display streptomycin resistance, showing that they existed on the original agar plate long before any contact with the streptomycin. Since its original use, replica plating has advanced bacterial (*E. coli*) genetics by allowing single colonies to be rapidly screened for large numbers of genetic markers.

The Use of Mutagens

Most mutant genes studied by the early Mendelian geneticists arose spontaneously. Now there is increasing use of mutations induced by external agents such as ionizing radiation, ultraviolet light, and certain chemicals. These agents, collectively called **mutagens,** greatly increase the rate at which mutant genes arise. Mutagens act quite indiscriminately; no presently known physical or chemical mutagen increases the possibility of mutating a given gene without also increasing the possibility of mutating all other genes.

For many years, the various forms of radiation were the most powerful known mutagens. Now chemical mutagens are usually used because they require no special radiation sources and because some of them produce a much higher percentage of mutated genes. Mutagenic chemicals fall into three classes (see Chapter 11 for more details): (1) nucleotide-base analogues, which because of their close structural similarity to the normal four bases of DNA can be incorporated into DNA (afterward their different base-pairing possibilities lead to mistakes in DNA replication); (2) chemicals that react with DNA to change the bases into forms that base-pair abnormally; and (3) DNA intercalating agents, which by inserting themselves between the base pairs also lead to frequent mistakes in DNA replication. The most powerful known mutagens are certain alkylating agents like ethyl methane sulfonate, which add methyl or ethyl groups to the rings of the DNA bases. One of these agents, nitrosoguanidine, acts preferentially at the DNA replicating forks (see Chapter 10), frequently producing clusters of closely linked changes in DNA base sequences.

Mutations That Lead to *E. coli* Growth Factor Requirements[6]

A further key step in the development of *E. coli* as a system for the study of genetics came in 1944 with the realization that mutations similar to those earlier obtained in *Neurospora* could be obtained that affected its ability to synthesize essential metabolites. For example, *E. coli* ordinarily grows well with only glucose as a carbon source; but as a result of specific mutations, there are mutant *E. coli* strains that will grow only when their normal medium is supplemented with specific metabolites.

Such **growth factor** mutations are very easy to work with: To test for their presence, we only need to grow a suspected mutant both in the presence and in the absence of a metabolite, for example, the amino acid arginine (Figure 7-3). If a mutation inhibiting arginine biosynthesis has occurred, the bacteria will grow only in the presence of arginine as a supplement. The use of this approach quickly led

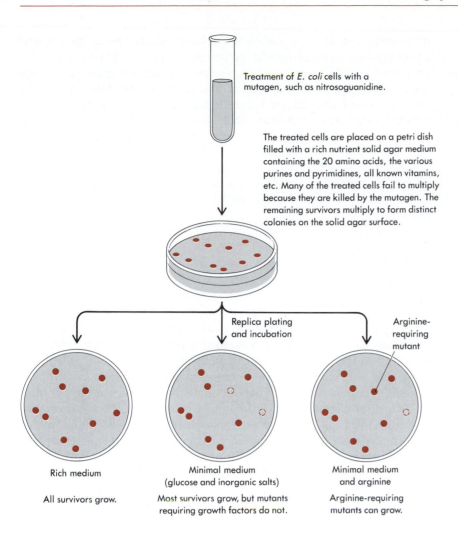

Treatment of *E. coli* cells with a mutagen, such as nitrosoguanidine.

The treated cells are placed on a petri dish filled with a rich nutrient solid agar medium containing the 20 amino acids, the various purines and pyrimidines, all known vitamins, etc. Many of the treated cells fail to multiply because they are killed by the mutagen. The remaining survivors multiply to form distinct colonies on the solid agar surface.

Replica plating and incubation

Arginine-requiring mutant

Rich medium

All survivors grow.

Minimal medium (glucose and inorganic salts)

Most survivors grow, but mutants requiring growth factors do not.

Minimal medium and arginine

Arginine-requiring mutants can grow.

Figure 7-3
The isolation of mutant *E. coli* cells with a specific growth factor requirement.

(with the help of mutagens) to the isolation of a large number of different growth factor mutations. Their existence has vastly speeded up elucidation of the enzymatic steps involved in the biosynthesis of many important intermediary metabolites.

The Highly Diverse Phenotypes of Bacterial Mutants[7]

Mutant cells that have lost their ability to synthesize essential metabolites are called **auxotrophs** ("increased growth requirements"), while wild-type cells lacking such requirements are called **prototrophs**. For example, there exists a class of *E. coli* auxotrophs unable to metabolize a diverse collection of sugars (for example, lactose, glucose, maltose) as well as other auxotrophs unable to metabolize one of these food molecules.

Some cells contain mutations that are only expressed under certain conditions (**conditional mutants**), as in the case of cells that can grow at 30°C but are unable to grow at 37°C. Most temperature-sensitive mutants synthesize enzyme products that are unstable at the restrictive temperature (37°C in our example), yet sufficiently active at the permissive temperature (30°C) to yield a wild-type phenotype. Conditional mutations can occur in virtually all genes, including those that control the steps in macromolecular synthesis, modification, and assembly into supermolecular structures.

Many valuable mutations involve the resistance to poisonous compounds like antibiotics. For example, most *E. coli* cells are rapidly killed by small amounts of streptomycin. Very rarely (one in 10^9 cell generations) there occur mutations (Strep^R) that make the cells resistant to certain amounts of this compound. Also easily observed are mutations that cause bacteria to be resistant to their viruses, the phages. One such mutation in *E. coli* strain B confers resistance to phage T1; these mutant cells are designated B/1. Correspondingly, *E. coli* strain B cells resistant to T2 and T4 are called B/2 and B/4, respectively.

Mutant Phenotypes Need Not Be Immediately Expressed[8]

Although they are genetically haploid, bacterial cells tend to contain more than one chromosome (two to four identical chromosomes); therefore, the time taken for a mutant gene to be expressed depends on whether the mutant allele is dominant or recessive. If it is dominant, the cellular phenotype becomes mutant as soon as sufficient protein products become synthesized. This interval may be no more than several minutes. However, most mutant alleles are recessive, and here the time taken for expression can be considerable. To start with, at least two cell divisions must occur before cells containing four chromosomes become genetically homogeneous. Even then, the resulting cells are likely to contain significant amounts of the wild-type protein product in the cytosol. Consider, for example, what happens when an *E. coli* cell becomes resistant to phage T4. Attachment of T4 to the surface of *E. coli* occurs by its binding to a specific outer membrane protein, a porin called OmpC, which normally is present in some 10^5 copies per cell. Resistance develops when the gene coding for OmpC is inactivated. As only one such porin receptor is necessary for T4 attachment, many cell divisions (up to 17) may have to occur before the mutant cell becomes totally resistant to T4.

Enriching Mutants Through Direct Selection, Counterselection, or Physical Selection[9, 10, 11]

Advancement of the genetic analysis of *E. coli* has depended on three different enrichment techniques for increasing the percentage of mutant cells in comparison to their wild-type equivalents: (1) **direct selection,** which utilizes growth conditions that greatly favor the growth of the desired mutant; (2) **counterselection,** which makes use of conditions that kill the parent cells; and (3) **physical selection,** which involves unique properties of mutant cells that enable them to be physically separated from their parent cells.

Direct selection is usually employed to isolate mutant cells relatively resistant to well-defined chemical (e.g., antibiotic), physical (e.g., ultraviolet radiation), or biological (e.g., phage) agents. Since most such agents kill the parent cells, at least several cycles of cell division must occur following exposure to a mutagen in order to increase the numbers of mutants before the direct selective agent is applied (Figure 7-4). Direct selection can also be used to isolate rever-

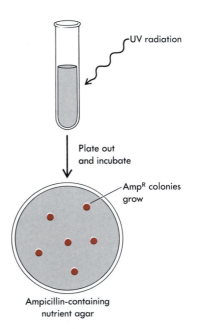

UV radiation

Plate out and incubate

Amp^R colonies grow

Ampicillin-containing nutrient agar

Figure 7-4
An example of direct selection. Bacterial cells are selected for ampicillin resistance after exposure to a mutagen.

tants, cells that have regained a function lacked by either auxotrophic parents or by carbon or nitrogen source mutants.

Counterselection usually involves application of agents that selectively kill *growing* cells (Figure 7-5). The first such method ever used utilized the fact that penicillin kills only growing cells. In the presence of penicillin, the crosslinks that tie together newly made peptidoglycan precursors cannot be made. This leads to weakened cell walls that burst from exposure to the cell's internal osmotic pressure. Thus, if mutagenized cells are placed in a minimal medium and then exposed to penicillin, the only cells not killed by the penicillin should be new auxotrophic mutants that are unable to synthesize essential metabolites. Another method for killing growing cells involves exposure to the thymine analogue 5-bromouracil. When incorporated into bacterial DNA, it makes the cells particularly sensitive to killing by ultraviolet light. Still another procedure utilizes parent cells that are unable to synthesize thymine. In the absence of thymine, these cells die once they begin to grow (thymineless death). If, however, they have another defect that prevents them from initiating growth, they remain alive. Selection for specific amino acid auxotrophs can thus be achieved by suspending mutagenized thymine-requiring cells in an enriched medium lacking thymine and the respective amino acid.

Physical selection often distinguishes mutant cells from normal cells on the basis of size. For example, cells unable to divide at high temperatures because of a temperature conditional mutation tend to be smaller at high temperatures. These cells can then be detected by cell-sorting systems that discriminate size differences on the basis of the amount of light they intercept. Another physical separation technique selectively divides a group of cells into different locations within a medium. For example, nonmotile mutants unable to respond to chemical attractants cannot swim away from nutrient-depleted regions in which their relative concentration becomes much enriched over their normal equivalents (Figure 7-6).

Colored Dyes Are Used to Detect Desired Metabolic Mutants[12, 13]

Carbon source (e.g., lactose) mutants of *E. coli* can be easily detected using identification media that dye the medium surrounding bacteria differentially, depending on whether the sugar in question can be metabolized. These methods utilize the fact that when the sugar is fermented sufficient acid is produced to lower the pH of the surrounding medium. The lower pH then triggers a change in the color of the dye (Figure 7-7). Such media can support equally well the growth of cells that can or cannot ferment the sugar. Colonies unable to utilize lactose, for example, can be selected for by looking for colonies growing in the presence of lactose that do not acidify the medium and thus do not change the color of the medium.

Brute Force Isolation[14, 15]

In many other cases, detection of desired mutants is much more laborious and depends on direct measurements of the amount of desired enzymes or cellular components. Such brute force approaches work only following application of highly efficient mutagens. Their use

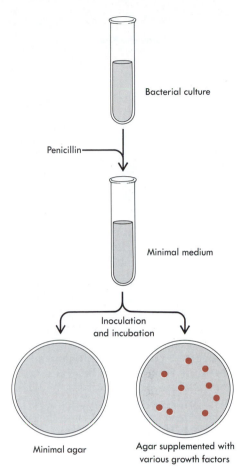

Figure 7-5

An example of counterselection. Selection is for a bacterial mutant unable to grow in the absence of a specific growth factor (e.g., arginine). Cells are added to minimal medium (lacking growth factors), and penicillin is added to kill all actively growing cells. Mutants that were not growing owing to a lack of growth factors are not killed by the penicillin and thus are selected for. They can then be seen by plating them on a supplemented agar plate.

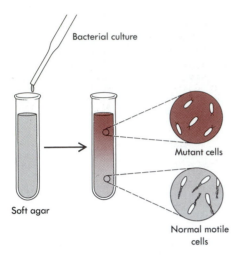

Figure 7-6
An example of physical selection. Cells with mutant flagella rendering them nonmotile can be selected for by inoculating a tube of soft agar with a bacterial culture previously exposed to a mutagen. Normal cells will move toward the bottom of the tube as nutrients in the upper portions of the tube are used up. The mutant nonmotile cells remain in the upper portion. An indicator dye can be used to detect the areas of nutrient depletion (colored area).

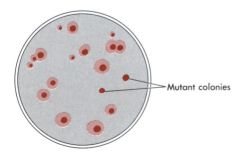

Figure 7-7
The use of pH-sensitive dyes to detect metabolic mutants. Mutants unable to ferment lactose in the medium do not release acid and thus do not change the color of the medium surrounding them.

now allows the isolation of virtually any "loss-of-function" mutation by screening some 10,000 treated colonies. An *E. coli* strain lacking the enzyme ribonuclease I was the first mutant isolated by massive brute force screening, and soon after a much sought after strain lacking virtually all of its DNA polymerase I was so obtained.

Proof of the Existence of Genetic Recombination Within *E. coli*[16]

The isolation of mutations involving growth factors, antibiotic resistance, and phage resistance was quickly followed by experiments demonstrating the occurrence of genetic recombination (and hence a sexual process) in *E. coli*. Twofold use was made of the mutant genes. First, they were used as conventional genetic markers since their segregation patterns revealed the chromosomal arrangement of genes. Second, they provided a method of detecting a genetic recombination process occurring in only a small fraction of the population during a given time. In *E. coli*, for example, simple morphological examination of bacterial cells gave no clues that cell fusion and genetic recombination existed. To detect recombination, it was necessary to devise an experiment in which only the recombinant cells would be able to multiply. This was done by using parent strains with specific growth requirements such that they could not multiply in minimal media.

In these experiments, conducted in 1946, two strains of bacteria, each possessing different growth requirements, were mixed together (Figure 7-8). Neither strain alone was able to grow in the absence of its specific growth factors; the amino acids threonine and leucine were required by one strain, the vitamin biotin and the amino acid methionine by the other. After the two strains were mixed together, a small number of cells were able to grow without any growth factors added to the medium. This result showed that the growing cells had somehow acquired wild-type copies of each of their mutant genes, strongly suggesting that *E. coli* has a sexual phase that can bring together the chromosomes of two different cells. Crossing over could then take place between two chromosomes carrying mutations in different genes and lead, in some cases, to a chromosome with wild-type alleles for all necessary genes. Genetic analysis confirmed this hypothesis, and through the genetic analysis done over the last three decades, *E. coli* has become one of the best understood organisms.

Bacterial Viruses (Phages) Provide Simple, Easy-to-Study Chromosomes[17, 18]

Our knowledge of the genetics of several bacterial viruses (usually called bacteriophages or phages) has made enormous strides along with the expansion of our knowledge of bacteria. Their discovery in 1914 produced great excitement, for it was hoped that they might afford an effective and simple way to combat bacterial diseases in the days before the discovery of antibiotics. Phages, however, were never medically useful because their bacterial hosts readily mutate into forms resistant to viral infection. Thus, almost everybody lost interest in the phages, and few studied their replication until the late 1930s. Then, a small group of biologists and physicists (who came to be known as "the phage group"), intrigued by the problem of gene rep-

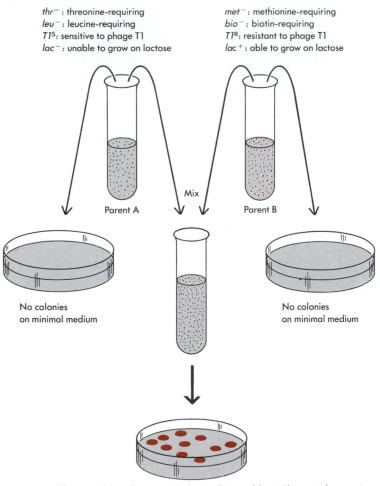

thr⁻ : threonine-requiring
leu⁻ : leucine-requiring
T1ˢ: sensitive to phage T1
lac⁻ : unable to grow on lactose

met⁻ : methionine-requiring
bio⁻ : biotin-requiring
T1ᴿ: resistant to phage T1
lac⁺ : able to grow on lactose

Parent A   Mix   Parent B

No colonies
on minimal medium

No colonies
on minimal medium

A very small fraction of the cells are *met⁺* , *bio⁺* , *thr⁺* , and *leu⁺* . They arise by genetic recombination, as shown by examination of the *lac* and *T1* markers. In addition to the parent *lac⁻ T1ˢ* and *lac⁺ T1ᴿ* genotypes, there are found *lac⁻ T1ᴿ* and *lac⁺ T1ˢ* cells.

Figure 7-8
The use of growth factor requirements to demonstrate sexuality in *E. coli*. Parent A, parent B, and a mixture of the two are plated out on minimal medium. Only the mixture gives rise to colonies.

lication, began to investigate the reproduction of several phages that multiplied in *E. coli*. They chose to work with phages rather than with the equally tiny plant or animal viruses because under laboratory conditions, it is far easier to grow bacterial cells than plant or animal cells.

Before the phage group came into being, the biological significance of phages, as well as that of animal and plant viruses, was obscure, and the question most often asked was, Are phages (and other viruses) living? Now we realize that all viruses are small pieces of genetic material (DNA or RNA), each enclosed within a protective protein-rich coat that facilitates its transportation from one cell to another. Progeny viruses closely resemble their parents because they contain identical chromosomes. Viruses are no more alive than isolated chromosomes; both the chromosomes of cells and those of viruses can duplicate only in the complex environment of a living cell. Nonetheless, viruses have been of immense value in our understanding of how cells live. For more than 30 years (1940–1973), they afforded the only convenient systems for studying the consequences of the sudden introduction of new genetic material into a cell.

Most work with phages has been—and still is—concentrated on a small collection that multiplies in *E. coli* or closely related bacteria. Arbitrarily, they have been given names like T1, T2, P1, F1, or λ. Now

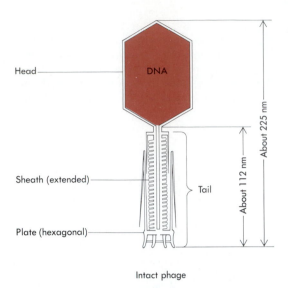

Intact phage

Phage after DNA expulsion

Figure 7-9
Electron micrograph and diagrammatic representation of bacteriophage T4.

among the best known of the DNA-containing phages are T4 (Figure 7-9), T7, and ϕX174. Also well characterized are the group of small spherical *E. coli* phages that use RNA as their genetic material. Among these are the phages f2, R17, and MS2.

Phages Do Not Grow by Gradually Increasing in Size[19]

The life cycle of an average cell involves a gradual increase in size to approximately twice its initial size, followed by a division process (mitosis) producing two identical daughter cells. Phages, however, do not multiply in this fashion. They are not produced by the fission of a large, preexisting particle; on the contrary, all the virus particles of a given variety have approximately the same (and in some cases identical) masses.

Upon infection of a bacterial cell, there is a temporary disappearance of the original parent virus particle. During this phase, which is called the **eclipse period,** the parent virus breaks down and releases its chromosome into the bacterial cytosol, leaving the empty protective protein coat, or **capsid,** outside the bacterial cell. Then the free viral chromosome serves as a template to direct the synthesis of new viral components. This breakdown process is an obligatory feature of viral multiplication. As long as the phage chromosome is tightly enclosed within the capsid, it cannot provide the information to direct the synthesis of either new chromosomes or new coat protein molecules.

The growth cycles of *E. coli* phage start when a phage particle collides with a sensitive bacterium and attaches to a specific receptor on the surface of its outer membrane (Figure 7-10). Different phages attach to different receptors. Phage λ, for example, attaches to a protein involved in the transport of maltose into *E. coli*, while T4 attaches to one of the porin proteins (OmpC) used to construct narrow channels spanning the outer membrane. How the viral chromosome passes through either the outer or inner cell membrane remains to be worked out. Many phage particles possess enzymes capable of breaking down components of bacterial envelopes. It is probable that they have specifically evolved to make the necessary holes.

Following its entry into the cytosol, the phage chromosome and the subsequent daughter chromosomes continue to duplicate to eventually form 100 to 10,000 new chromosomes. These, in turn, become encapsulated into newly synthesized capsids, forming a large number of new bacteriophage particles. The protein coats are encoded by phage genes that instruct the protein-synthesizing machinery of the host cell to construct the new capsids. The growth cycle is complete when the bacterial cell wall breaks open (lyses) and releases the progeny particles into the surrounding medium.

Phages (Viruses) Are Parasites at the Genetic Level

Viruses are of necessity parasites. It is impossible for them to reproduce without host cells, on which they are totally dependent for metabolic precursors and for ribosomes, the cellular organelles necessary for making proteins. It is not their small size per se that makes them viruses but the fact that they do not possess the capacity to synthesize their components independently. To replicate, they must always insert their nucleic acid into a functional cell. This feature enables us to distinguish the larger viruses, such as that causing smallpox, from very small cellular organisms, including the mycoplasmas and such intracellular disease-causing parasites as the rickettsias. The latter two organisms are distinct from viruses, since they contain both DNA and RNA and grow by increasing in size and then splitting into two smaller cells. At no time do their cell membranes break down, and all their proteins are synthesized upon their own protein-synthesizing machinery.

Phages Form Plaques

The presence of viable phage particles in a solution can be quickly demonstrated by adding some of the solution to the surface of a nutrient agar plate on which bacteria susceptible to the phage are rapidly multiplying. If no phage particles are present, the rapidly dividing bacteria will form a uniform surface layer of bacteria. But if even one phage particle is present, it will attach to a bacterium and multiply to form several hundred new progeny phage particles, which are then released by dissolution (lysis) of the cell wall some 15 to 60 minutes after the start of phage infection. Each of these several hundred progeny particles can then attach to a new bacterium and multiply. After several such cycles of attachment, multiplication, and release, all the bacteria in the immediate region of the original phage particles are killed. These regions of killed bacteria appear as circular holes, or **plaques,** in the lawn of healthy bacteria (Figure 7-11).

Phages Also Mutate[20, 21, 22]

The plaques formed by a given type of phage are quite characteristic and can often be distinguished from those of genetically distinct phages. For example, the plaques of phage T2 can easily be distin-

Figure 7-10
Injection of bacteriophage T4 DNA into an *E. coli* cell. Phage T4 attaches to the outer membrane porin OmpC.

Figure 7-11
Photograph of phage T2 plaques on a lawn of *E. coli* bacteria growing in a Petri plate. (From *Molecular Biology of Bacterial Viruses* by Gunther S. Stent. W. H. Freeman and Company, copyright © 1963, p. 41, with permission.)

Figure 7-12
Photograph of mutant phage plaques, showing a mixture of T2 r^+ (wild-type) plaques and T2 r (rapid-lysis mutant) plaques. The mottled plaques result from the simultaneous growth of both r and r^+ phages in the same plaque. (From *Molecular Biology of Bacterial Viruses* by Gunther S. Stent. W. H. Freeman and Company, copyright © 1963, p. 177, with permission.)

guished from those made by phage λ or by phage f2. More significantly, mutations occur that change the morphology of phage plaques. We do not usually know the biochemical basis of these plaque differences, but this does not matter. The important fact is that these differences are usually reproducible and simple to score. It is easy to examine the morphology of thousands of plaques to see if any look different from the plaques made by the wild-type phage. In this way, a large number of different plaque-type mutations have been found (Figure 7-12).

Another class of mutations changes the ability of the phage to adsorb to bacteria. For example, wild-type T2 cannot multiply on *E. coli* strain B/2, because the B/2 mutation changes the cell surface, thereby preventing the attachment of T2. Mutant T2 particles can, however, multiply on B/2: They are called T2h (h for host), and they are able to adsorb because they possess altered tail fibers.

Another very large and important class of phage mutants consists of phages that can multiply at 25°C but not at 42°C (i.e., they are temperature-sensitive conditional lethals). Although we still cannot always pinpoint the exact reason why these mutants do not multiply at the higher temperature, in many cases the high temperature destroys the three-dimensional structure of a protein necessary for their reproduction. This type of mutation has proven very useful because a large number of viral genes can mutate to a temperature-sensitive form.

The existence of conditional lethal mutations has allowed phage geneticists to find mutations in essentially all the genes of any phage they wish to investigate. Although genetic recombination between mutant phages was not discovered until 1945, phages are among the most completely characterized genetic objects yet studied.

Phage Crosses[23, 24]

More than one phage particle at a time can grow in a single bacterium. If several particles adsorb at once, the chromosome from each enters the cell and duplicates to form a large number of new copies. So long as the chromosomes exist free (unenclosed by a protective coat), they can cross over (recombine) with similar chromosomes. This is shown by infecting cells with two or more genetically distinct phage particles and finding recombinant genetic types among the progeny particles. For example, it is easy to obtain mutant T2 particles differing from the wild type by two mutations, one in an *h* gene, which allows the particles to grow on *E. coli* strain B/2 (a strain resistant to wild-type T2 particles), and the other in an *r* gene, which causes the particles to form larger and clearer plaques than wild-type T2 particles. These double mutant phages are designated T2 *hr* and the wild type is called T2 h^+r^+. When an *E. coli* cell is infected simultaneously with a T2 h^+r^+ and a T2 *hr* phage, four types of progeny particles are found: the parent genotypes h^+r^+ and *hr* and the recombinant genotypes hr^+ and h^+r (Figure 7-13).

The frequency with which recombinants appear depends on the particular mutants used in the cross. Crosses between some pairs of markers give almost 50 percent recombinant phages; crosses between others give somewhat lower recombinant values, and sometimes almost no recombinants are found. This immediately suggests that viruses also have unbranched genetic maps, a suggestion confirmed by

Figure 7-13
Phage recombination in cells infected with two different strains of phage T2. Within the cell, crossing over occurs between the free phage chromosomes, both identical and different. Recombinant and parent phages are identified by plating and observing the plaques. The technique used to see all four progeny types (*hr*, *hr⁺*, *h⁺r*, *h⁺r⁺*) is to look for plaques on a mixture of strain B and B/2. Only phages possessing the *h* gene can kill both B and B/2 cells. Phages with the *h⁺* gene kill only B cells, and their plaques look turbid owing to the presence of live B/2 cells. The *r* gene causes the formation of larger, clearer plaques than those formed by the wild type. (From *Molecular Biology of Bacterial Viruses* by Gunther S. Stent. W. H. Freeman and Company, copyright © 1963, p. 185, with permission.)

intense analysis of a large number of independently isolated mutations. The best-known phage genetic maps are those of T4 (Figure 7-14), T7, φX174, and λ. As we shall see in Chapter 17, the locations of viral genes are far from random, with genes grouped together according to when they function during their respective phage life cycles.

Phage Crosses Involve Multiple Pairings

The genetic structure of phage chromosomes is quite similar to that of cellular chromosomes. Nonetheless, it is important to note a distinct feature of phage crosses. In conventional meiosis, each chromosome pairs just once, whereas in a viral cross, pairing and crossing over may occur repeatedly throughout the period when free chromosomes are present in the infected cell. Thus, a given phage chromosome may participate in several pairing and crossing-over events. We can demonstrate this fact by infecting a cell with three phage particles, each with a distinct genetic marker, and finding single progeny particles that have derived chromosomal regions from all three original phage particles. Thus, the products of a phage cross are essentially the products of a large number of distinct pairings and crossings over. Moreover, there is great variation among viruses in the amount of crossing over; for example, a chromosome of phage T4 crosses over an average of 5 to 10 times during each growth cycle, while a phage λ chromosome crosses over only 0.5 time. The occurrence of many distinct crossovers does not, however, restrict our ability to map the genetic markers, since the general rule still holds that genes located close together seldom recombine.

Another way in which a phage cross differs from recombination in conventional meiosis is that phages are not separated into male and female particles. Differentiation into sexes can be considered a device to bring about cell fusion between genetically distinct organisms. Phage particles, however, do not have to fuse, since genetic recombination occurs when two phage chromosomes of different genotypes enter a single host bacterial cell. Moreover, in cells infected with several genetically distinct phages, crossing over can occur between both genetically identical and genetically different chromosomes; thus, only a fraction of the crossovers in a cell infected with different phages result in recombinant chromosomes of a new genotype. There

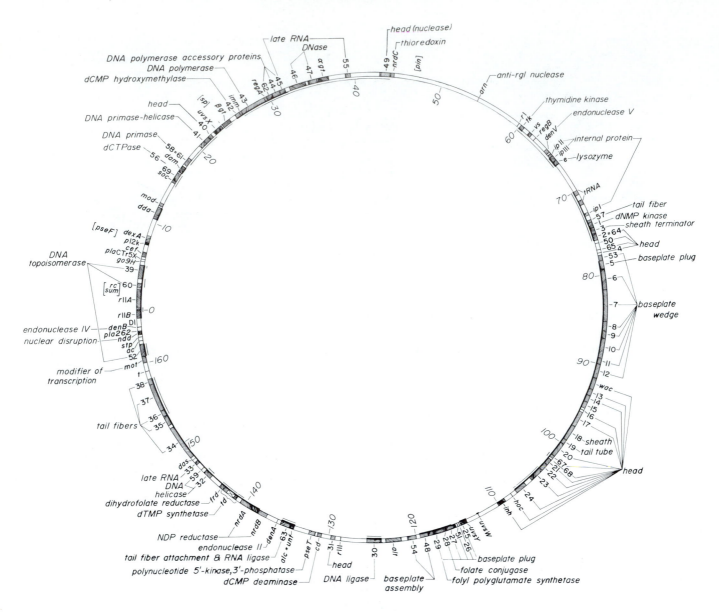

Figure 7-14
The genetic map of bacteriophage T4. The units (inner circle) are kilodaltons. (Courtesy of Dr. Burton S. Guttman and Dr. Elizabeth M. Kutter.)

is also evidence that crossing over occurs after infection by a single phage particle. However, this phenomenon cannot be revealed by genetic analysis, since all the progeny chromosomes are genetically identical; thus, experiments with isotopically labeled phage chromosomes are needed for its demonstration.

Two Distinct Sexes Are Found in *E. coli*[25-33]

When sexuality in *E. coli* was first discovered, it seemed simplest to believe that a conventional cycle of cell fusion occurred followed by meiotic segregation. Now we know, however, that the *E. coli* sexual cycle has distinctive features complicating conventional analysis. As in higher organisms, there exist male and female cells, but these do not fuse completely, allowing their two sets of chromosomes to intermix and form two complete diploid genomes. Instead, the transfer is always unidirectional, with male chromosomal material moving into female cells; the converse movement of female genes into male cells never occurs.

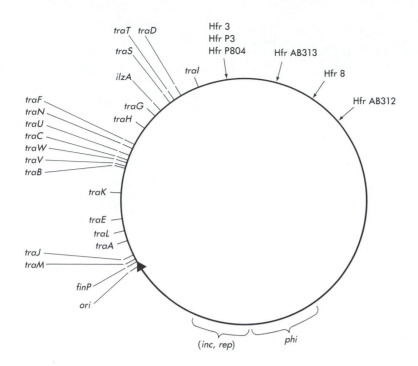

Figure 7-15
The genetic map of the F factor of *E. coli*. Note that about a third of the F$^+$ DNA consists of genes involved in the transfer of the F factor to female cells (*tra* genes). Also present are genes associated with fertility inhibition (*fin*), phage inhibition (*phi*), incompatibility (*inc*), replication (*rep*), and immunity to lethal zygosis (*ilz*). The origin of transfer replication (*ori*) and the positions where insertion elements on F recombine with the bacterial chromosome to create Hfrs are also indicated.

Male and female *E. coli* cells are distinguished by the presence of a distinct supernumerary sex chromosome called the **F (fertility) factor.** When it is present as a discrete body, the cells are male (F$^+$) and capable of transferring genes into female cells. In its absence, *E. coli* cells are female (F$^-$) and act as recipients for gene transfer from male cells. Like the main *E. coli* chromosome, the F factor is a circular double-helical DNA molecule present in one copy per cell. It only carries approximately 94×10^3 base pairs—$\frac{1}{40}$ the amount of genetic information contained in the main chromosome. About one-third of the F$^+$ DNA consists of 19 transfer (*tra*) genes specifically involved in the transfer of male genetic material into female cells (Figure 7-15). Among these genes are those responsible for synthesis of the sex-specific **F pili.**

When F$^+$ cells are mixed with F$^-$ cells, conjugal pairs form by the attachment of a male sex (F) pilus to the surface of a female cell (Figure 7-16). Attachment then triggers a set of still not understood events that initiates DNA replication of the F$^+$ chromosome and leads to the transfer of a progeny F$^+$ chromosome to the F$^-$ cell (see Chapter 10 for details). It is important to note that the only chromosomal material transferred by F$^+$ cells is the F$^+$ factor itself (Figure 7-17). None of the genes present on the main chromosome are transferred.

Pair formation and F$^+$ transfer occur very efficiently between F$^+$ and F$^-$ cells. As a result, all the cells in mixed cultures rapidly become male (F$^+$) donor cells.

Integration of F (Fertility) Factors into Main Bacterial Chromosomes Creates Hfr Strains[34]

Transfer of most bacterial genes from male to female cells only occurs when, through a crossing-over process, the F$^+$ factor becomes integrated into the cell's single main chromosome (Figure 7-18). Such **Hfr**

Figure 7-16
The attachment of a male F pilus to the surface of a female cell. This is followed by the replication of the F chromosome and its transfer to the female cell, which then becomes F$^+$. (Courtesy of C. C. Brinton, Jr., and Judith Carnahan.)

Figure 7-17
The transfer of F⁺ DNA to an F⁻ cell. The size of the F DNA is exaggerated for clarity. One strand of the F factor is cleaved, and replication commences by a rolling circle mechanism (see Chapter 11). The newly replicated F factor passes into the F⁻ cell at a point of direct contact between the two cells after they are linked by the F pilus. Note that only the F DNA is transferred.

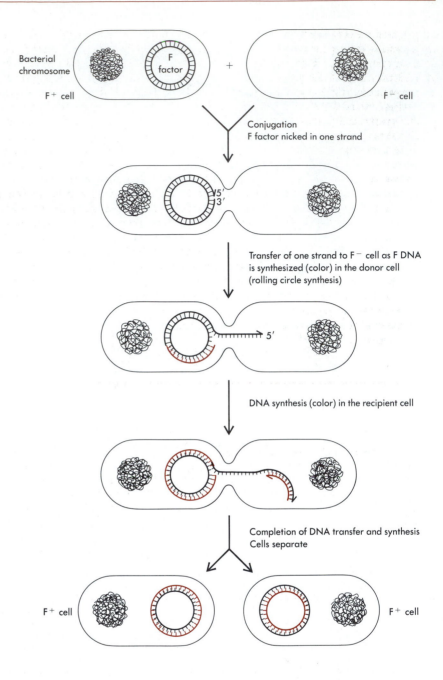

Bacterial chromosome

F⁺ cell

F factor

+

F⁻ cell

Conjugation
F factor nicked in one strand

Transfer of one strand to F⁻ cell as F DNA is synthesized (color) in the donor cell (rolling circle synthesis)

DNA synthesis (color) in the recipient cell

Completion of DNA transfer and synthesis
Cells separate

F⁺ cell

F⁺ cell

(**high frequency of recombination**) cells remain male, possess F sex pili, and fuse with F⁻ cells as frequently as F⁺ cells do. When an Hfr cell joins to an F⁻ cell, conjugation-induced replication of F DNA begins, and because the leading edge of the F⁺ factor is now attached to the main chromosome, transfer of the main chromosome follows (Figure 7-19). Because DNA replication is involved, transfer is inherently a lengthy process, requiring some 100 minutes at 37°C to bring a complete male genome into a female cell. Complete transfer, however, rarely occurs. The conjugal unions between male and female cells are very fragile, usually breaking and terminating transfer before it has finished. Since the main chromosomal genes are inserted into the F⁺ factor, only a portion of the F⁺ genome is usually transferred into recipient F⁻ cells (Figure 7-20). Consequently, the recipient cells usually remain female, in contrast to F⁺ × F⁻ matings, where the recipient cell invariably becomes converted to a male (F⁺) cell.

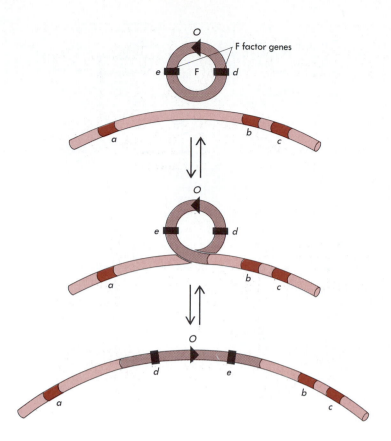

Figure 7-18
Integration of the F factor into the main chromosome, leading to the creation of an Hfr state. The F factor gene *O* is the origin of oriented chromosome transfer from the male cell to the recipient cell; *d* and *e* stand for two other F factor genes, and *a*, *b*, and *c* stand for genes on the main, circular chromosome. A single crossing-over event, as shown, leads to the integration of F between genes *a* and *b*. The resulting Hfr chromosome transfers *a* as one of its first genes and *b* as one of its last when conjugating with an F⁻ cell. In this diagram and the next two, double-stranded DNA is represented by a single line.

At first, the fact that only part of the male main chromosome enters the female cell made genetic analysis more difficult. Then it was realized that a fixed end of the male chromosome, specific for a given Hfr strain, always enters the female cell first and that the relative frequency with which male genes are incorporated into the recombinant chromosome is a measure of how close they are to the entering end. Moreover, it is possible to break apart the male and female cells artificially by violent agitation (this was first done in a mixing machine, the Waring Blender); matings can be made and the couples violently agitated at fixed times during the process. (These experiments were known as "the blender experiments"). If the pairs are disrupted soon after mating, only the genes very close to the forward end will have entered the cell (Figure 7-21). It is thus possible to roughly map *E. coli* gene positions merely by observing the time intervals required for various alleles carried by the male cells to enter female cells.

Bacterial Chromosomes Are Circular[35]

The genetic map obtained through analysis of interrupted matings is the same as that arrived at by analysis of frequencies of various recombinant classes. As in the chromosomal maps of higher organisms, the bacterial genes are arranged on an unbranched line. However, there is one important distinction: The genetic map of *E. coli* is a circle.

Within a given Hfr strain, the male chromosome always initiates DNA duplication prior to DNA transfer at the point of F factor insertion. The place where replication initiates, however, is not the same

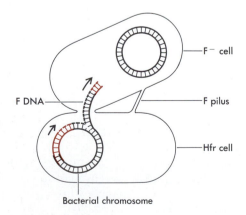

Figure 7-19
Movement of the bacterial chromosome from an Hfr cell into an F⁻ cell. The chromosome is led by a segment of the F factor DNA. The F DNA is actually only a tiny fraction of the total cell DNA.

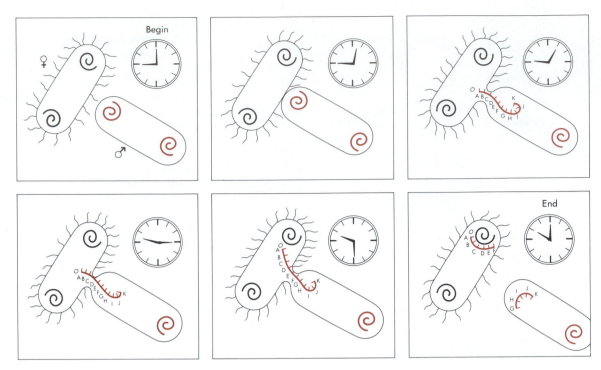

Figure 7-20

Conjugation between F⁻ and Hfr bacteria, as shown in a classic diagram, drawn before the bacterial chromosome was known to be circular. Since the transfer is a lengthy process, the fragile conjugal unions are usually broken before transfer is complete. Genes on the donor chromosome are denoted by capital letters. [After F. Jacob and E. L. Wollman, *Sexuality and the Genetics of Bacteria* (New York: Academic Press, 1961).]

in all strains, since the F factor can become integrated at any one of several preferred sites. There is one Hfr strain, for example, in which some of the genes involved in the synthesis of threonine and leucine are transferred soon after mating; in another strain, however, a gene involved in methionine synthesis is among the first to enter the female cell (Table 7-1). The existence of many different Hfr strains, each with its own specific sites of integration, has been very important in assigning *E. coli* gene locations. If there was only one entering point, it would be nearly impossible to assign even a rough order to those genes that are transferred last, since they would enter the female cell only very rarely.

The biological significance of circular chromosomes is still somewhat of a mystery. However, circular chromosomes are also found in many viruses, so the *E. coli* form should not be viewed as a strange

Figure 7-21

The frequency of donor Hfr marker genes (*azi, T1, lac, gal*) appearing among the resulting recombinants after an interrupted mating experiment. This procedure makes it possible to roughly map *E. coli* gene positions. Here the relative gene order would be *azi-T1-lac-gal*. [After F. Jacob and E. L. Wollman, *Sexuality and the Genetics of Bacteria* (New York: Academic Press, 1961).]

Table 7-1 Order of Genes in Conjugal Transfer in Different Hfr Strains

Hfr Strain	Order of Gene Transfer
Hayes	O-thr-leu-azi-ton-pro-lac-pur-gal-trp-his-gly-str-mal-xyl-mtl-ile-met-thi
Hfr 1	O-leu-thr-thi-met-ile-mtl-xyl-mal-str-gly-his-trp-gal-pur-lac-pro-ton-azi
Hfr 2	O-pro-ton-azi-leu-thr-thi-met-ile-mtl-xyl-mal-str-gly-his-trp-gal-pur-lac
Hfr 3	O-pur-lac-pro-ton-azi-leu-thr-thi-met-ile-mtl-xyl-mal-str-gly-his-trp-gal
Hfr 4	O-thi-met-ile-mtl-xyl-mal-str-gly-his-trp-gal-pur-lac-pro-ton-azi-leu-thr
Hfr 5	O-met-thi-thr-leu-azi-ton-pro-lac-pur-gal-trp-his-gly-str-mal-xyl-mtl-ile
Hfr 6	O-ile-met-thi-thr-leu-azi-ton-pro-lac-pur-gal-trp-his-gly-str-mal-xyl-mtl
Hfr 7	O-ton-azi-leu-thr-thi-met-ile-mtl-xyl-mal-str-gly-his-trp-gal-pur-lac-pro
AB311	O-his-trp-gal-pur-lac-pro-ton-azi-leu-thr-thi-met-ile-mtl-xyl-mal-str-gly
AB312	O-str-mal-xyl-mtl-ile-met-thi-thr-leu-azi-ton-pro-lac-pur-gal-trp-his-gly
AB313	O-mtl-xyl-mal-str-gly-his-trp-gal-pur-lac-pro-ton-azi-leu-thr-thi-met-ile

SOURCE: From F. Jacob and E. L. Wollman, *Sexuality and the Genetics of Bacteria* (New York: Academic Press, 1961).

exception. We shall return to the topic of circles when we look at the precise chemistry of the chromosome.

Over 900 Genes Have Already Been Mapped on the *E. coli* Chromosome[36, 37, 38]

By now, some 900 different genes have been located on the *E. coli* chromosome (Figure 7-22). Early on in gene mapping, it became clear that many genes with related functions (e.g., components of a biosynthetic or degradative pathway) were closely grouped on the chromosome. Such related gene clusters, called **operons,** are transcribed into single messenger RNA molecules, thereby ensuring that if one enzyme member of a metabolic pathway is present, its partner enzyme will also be present (see Chapter 16). Already, some 260 of the *E. coli* genes have been assigned to some 75 different operons, and the figure will probably go much higher, since less than half the *E. coli* genes have been identified so far. Moreover, we still remain ignorant of much of *E. coli*'s metabolism and so cannot yet spot the related functions of many adjacent genes.

Even without such definitive knowledge, most, if not all, the genes coding for certain key metabolic functions have been found (Table 7-2). Well characterized are the genes coding for the enzymes needed to synthesize amino acids, purines and pyrimidines, fatty acids, and vitamins, as well as the genes controlling the catabolism of most known carbon and nitrogen sources. The genes that code for more than 50 different protein components of ribosomes are also fully characterized.

Except for their grouping into operons, there appears no obvious reason for the relative position of most genes on the *E. coli* map. Genes controlling the biosynthesis of small molecules, fueling reactions, and macromolecular synthesis and assembly are found in all sections of the map. Moreover, while the genetic maps of *E. coli* and close enteric (intestinal) relatives (e.g., *Shigella, Salmonella, Enterobacter*) are quite similar, inversions of gene order occur. The map of *Salmonella typhimurium* is virtually identical to that of *E. coli* except that a segment containing some 10 percent of the genome is inverted with

Table 7-2 The Genes for Proteins Involved in Glycolysis

Gene Protein	Map Position
aceE pyruvate dehydrogenase (decarboxylase component)	3
aceF pyruvate dehydrogenase (dihydrolipoyltransferase component)	3
cxm methylglyoxal synthesis	6
eno enolase	59
fda fructose diphosphate aldolase	63
fpk fructose-1-phosphate kinase	46
fdp fructose diphosphatase	95
gap glyceraldehyde-3-phosphate dehydrogenase	39
glk glucokinase	51
kba ketose-*bis*-phosphate aldolase	69
pgi glucose phosphate isomerase	91
pfkA 6-phosphofructokinase I	88
pfkB suppressor of *pfkA*	38
pfkC modifier of 6-phosphofructokinase activity	58
pfl pyruvate-formate lyase	20
pgk phosphoglycerate kinase	63
pgm phosphoglucomutase	15
pps phosphoenolpyruvate synthase	37
tpiA triosephosphate isomerase	88

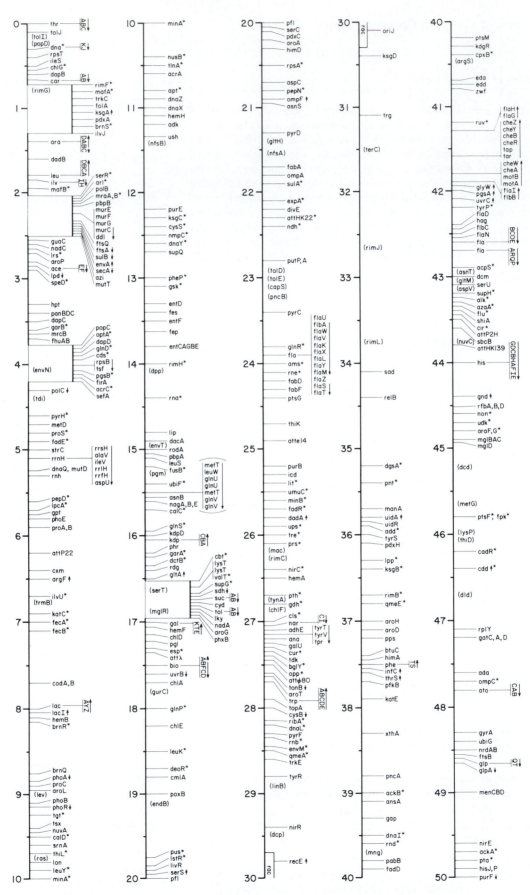

Figure 7-22

The complete genetic map of *E. coli*. [Courtesy of Barbara J. Bachman, *Microbiol. Rev. 47* (1983):180]

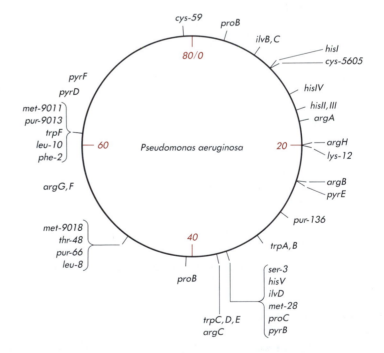

Figure 7-23
The genetic maps of *E. coli*, *B. subtilis* and *P. aeruginosa*, showing some of the comparable genes for these three bacteria. Note the different gene orders. (Data from J. L. Ingraham et al., *The Growth of the Bacterial Cell* pp. 178, 179, with permission.)

no obvious physiological consequence. Bacteria much more distantly related to *E. coli*, such as *Bacillus subtilus* and *Pseudomonas aeruginosa*, have completely different gene orders (Figure 7-23).

The orientation of a given operon along the chromosome also appears to be inconsequential. Because only one strand of a given operon's DNA is transcribed, it is either transcribed in the clockwise or counterclockwise direction. Of the 50 operons or genes whose transcription direction is known, 27 are transcribed in the clockwise direction and 23 in the counterclockwise direction.

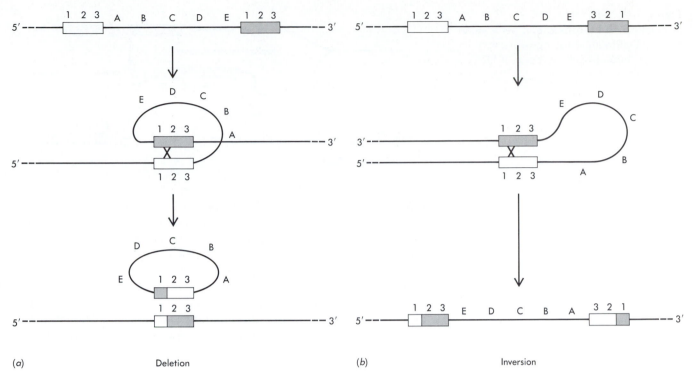

(a)   Deletion   (b)   Inversion

Figure 7-24
Crossing over between two identical genes on the same chromosome. (a) When two genes oriented in the same direction recombine, one gene and the DNA segment between them are deleted. (b) However, when the genes are oriented in opposite directions, there is no deletion, and the segment between them becomes inverted.

Bacterial Genes Are Usually Present as Single Copies[39]

Virtually all *E. coli* genes are present in only one copy per genome. Multiple copies on the same chromosome are inherently unstable, since crossing over can occur between them, leading to either deletions or inversions of the chromosomal segments lying between. Deletion results when crossing over occurs between two genes oriented in the same direction, while inversions result when two genes oriented in opposite directions recombine (Figure 7-24). Multiple copies not only of the same gene but also of two different genes with highly similar structures will tend to recombine. A gene's inherent stability thus demands that it not be too similar to any other genes. Duplications of minuscule chromosomal regions, however, are not strongly selected against, since the chance of crossing over occurring between them is almost negligible.

Despite their inherent long-term instability, duplications of certain genes arise with frequencies as high as one in 10^4 cells. Thus, a significant fraction of the cells in any culture can be expected to carry duplicated genetic material somewhere. Such duplications are usually generated by illegitimate crossing-over events that involve short, partially homologous sections of DNA on dividing daughter chromosomes (Figure 7-25). Such duplication events, by creating new stretches of homologous DNA on the same chromosome, automatically tend to be followed by further gene amplification or elimination events.

The fact that so little duplicated genetic material exists, despite its continual new generation, tells us that it seldom conveys any selective advantage to those bacteria possessing it. Imagine, for example, a duplication that adds 10 percent more DNA to the *E. coli* genome. Because of this added DNA, bacteria carrying it will have their mini-

Figure 7-25

A deletion and a duplication generated by an illegitimate crossing-over event between homologous sections of DNA on daughter chromosomes.

mum division time raised by 10 percent and so will be at a great disadvantage unless this additional DNA codes for particularly needed gene products. Duplications of the lactose operon, for example, have a positive selective advantage in cells growing solely on very limited amounts of lactose or on lactose derivatives that are very inefficiently metabolized. However, the moment such selective pressures are removed, the disadvantages of an unnecessary, if not actually harmful, genetic duplication manifest themselves. Then new variants are selected that eliminate the duplicated DNA through new crossing-over events.

There Are Multiple Copies of Genes Coding for Ribosomal RNA[40]

Most *E. coli* strains contain seven sets of the genes coding for the ribosomal RNA precursor that becomes enzymatically cleaved to 16S, 23S, and 5S mature ribosomal RNAs (Figure 7-26). Only through their possession of multiple rRNA genes can *E. coli* cells make enough rRNA chains to assemble the very large number of ribosomes needed to double their protein content every 20 minutes. It is important to note a very basic difference between those relatively few genes that code for rRNA and tRNA and the many more numerous genes that code for the many kinds of mRNA. The number of rRNA and tRNA molecules in a cell is inherently limited by the rate at which these genes are transcribed. In contrast, the number of products of those genes that code for proteins is potentially much higher. This is because each mRNA molecule is subsequently translated many times, yielding at least 50 to 100 identical protein products. There is thus no need for multiple copies of genes for proteins, even in the case of genes that code for ribosomal proteins, each of which accounts for about 0.5 percent of the total cellular protein.

Because the rRNA genes (rDNA) of necessity have virtually identical sequences coding for their functional rRNA products, unequal crossing over between two daughter *E. coli* chromosomes will occasionally occur. The result will either be a contraction or expansion of the number of rRNA genes. The deletion events that lead to contraction, however, will simultaneously delete essential genes lying between and so lead to the death of a small fraction of daughter *E. coli* cells each cell generation. Correspondingly, there will be produced a small number of cells with a duplicated DNA section containing one or more rRNA genes. At high growth rates, these tend to survive, but soon lose the duplicated DNA section when their growth rate is slowed down.

Most rRNA Genes Are Located Near the Origin of Replication

In *E. coli* cells duplicating their DNA, there is a greater concentration of genes near the point at which the bidirectional replication of DNA starts (map position 83) than near the terminus of replication (map position 33). Within cells dividing every 40 minutes, where the time required for DNA replication equals that required for a cell division, the copy number of genes near the origin of replication is almost twice that of those near the terminus. In very rapidly growing cells, where DNA replication reinitiates at the origin before the previous round of replication has been completed, genes near the origin will have copy numbers nearly four times those at the terminus. It is thus not surprising that most of the rRNA genes cluster around the origin of replication. If, on the other hand, they were located near the termination point, they would have to be present in even higher numbers to maintain an equivalent rate of protein synthesis.

The various control circuits that determine the level at which a given gene functions are thus assumed to have evolved with some regard for the gene's relative position on the chromosome. This may be the reason why gene location is so conserved among related enteric bacteria, despite the obvious ease with which new aberrantly rearranged chromosomes appear.

Transposable Genetic Elements Appear in *E. coli*[41, 42, 43]

Long sacred to genetic thinking was the conviction that genes have fixed chromosomal locations that change only as a consequence of chromosomal rearrangements resulting from illegitimate crossing over between incompletely homologous short sections of DNA. Then, in the late 1940s, genetic experiments with maize (corn) showed that certain genetic elements regularly "jump" to new locations. In so jumping, or **transposing,** they often inactivate the genes into which they become inserted. Thus, the existence of movable elements is often first revealed by their capacity to disturb the functioning of other previously characterized genes.

For almost 20 years, the significance of the jumping genes of maize was unclear. Then a number of apparently random mutations that blocked gene function in *E. coli* were found to arise by the insertion of large blocks of genetic material (insertion sequences). Soon after, several of these sequences were found to have the jumping property of the maize elements. Then, through the use of the newly developed methods of recombinant DNA and DNA sequencing (discussed in Chapter 19), the nature of the *E. coli* transposing elements quickly became established at the molecular level. They were found to be DNA sequences that code for enzymes that bring about the insertion of an identical copy of themselves into a new DNA site. Transposition events involve both recombination and replication processes that frequently generate two daughter copies of the original transposable element; one remains at the parental site, while the other appears at the target site.

The insertion of transposable elements invariably disrupts the integrity of their target genes. Since transposable elements carry signals for the initiation of RNA synthesis, they sometimes activate previously dormant genes. Insertion of these elements also may cause deletions, inversions, and chromosomal fusions. By possessing these elements, the *E. coli* genome has thus acquired the potential for much genetic plasticity.

Two main classes of transposable elements have been identified in bacteria, insertion sequences and transposons (Figure 7-27). **Simple transposons** (also called **insertion sequences,** or **IS,**) are transposable elements that carry no genetic information except that necessary for transposition. Key components include specific terminal DNA sequences as well as the coding region for an enzyme (transposase) that recognizes the ends. **Complex transposons** are transposable elements in which the terminal sequences enclose additional genetic material unconnected with transposition. In some complex transposons, two complete IS elements enclose one to several functional bacterial genes. By being contained within a transposon, a specific gene, such as one that confers resistance to an antibiotic, acquires a greatly heightened capacity to move with the transposon from one DNA molecule to another, if not from one bacterium to another. Antibiotic resistance genes that have been observed to spread quickly among a population of previously susceptible bacteria are often found to be inserted within transposons. For further discussion of transposons, see Chapter 11.

Plasmids

DNA molecules that possess their own origins of replication and so are capable of self-duplication are called **replicons.** The circular main bacterial chromosome is a replicon, as are the much smaller autonomously replicating circular DNA molecules that we call **plasmids** (Figure 7-28). Plasmids are not merely tiny chromosomes that in every other way resemble the much larger DNA molecules we designate as chromosomes. Plasmid DNAs usually encode no essential functions, and bacteria lacking them usually multiply normally. Rather, plasmids act as dispensable accessory sources of DNA that provide unique functions to bacteria. The F factors, for example, are plasmids that encode the proteins that allow bacteria to conjugate and donate

(a) Insertion sequence

IS1

IR Transposase gene IR

Tn1681

(b) Transposon

IR   IR   IR   IR

IS1   *E. coli* heat-stable   IS1
           Toxin 1 gene

Figure 7-27
(a) A simple transposon (also called an insertion sequence) carries only the genetic information necessary for transposition (known as the transposase gene) and is often enclosed by short regions of inverted repeat (IR) sequences. (b) The complex transposon Tn1681 is composed of two insertion sequences and contains genetic information in addition to that needed for transposition. In this case, the transposon carries the gene for a heat-stable toxin.

genetic material. The resistance transfer factors (RTF) carry the several genes that confer simultaneous multiple resistances to many different antibiotics (e.g., ampicillin, streptomycin, and tetracycline). While some plasmids, such as the F factor, are present in 1:1 correspondence with the bacterial chromosomes, other plasmids, particularly those very small ones that carry only one to several genes, can be present in many more copies (up to 50) per bacterium and replicate independently of the main chromosome.

Small plasmids are present in many different bacteria, since they exist in multiple copies per cell and thus allow gene amplification. Any gene that they carry is present in correspondingly large numbers, thereby opening up the possibility that very large numbers of their protein products are made. Such plasmids thus provide the ideal vehicles to carry the many multiple genes needed for high-level antibiotic resistance, where detoxifying enzymes are required in massive amounts. Likewise, they are ideal vehicles for producing many copies of enzymes needed to break down inefficiently metabolized food sources.

It is important to note that a given gene is present only once on a given plasmid. Unlike the multiple copies of ribosomal RNA genes present on the main *E. coli* chromosome, the copy number of a gene present on a plasmid cannot be expanded or contracted by crossing over, since it is present in only one copy (i.e., it is haploid). It is possible, however, for a *whole plasmid*, even when present in multiple copies, to be lost by chance segregation events during cell division. In the absence of positive selective pressure, small numbers of plasmid-free cells are constantly generated. Accessory genes thus tend to survive best when they are included within a transposon that gives them the potential to move onto other plasmids, if not into other cells by first jumping onto F factors.

Genetic Elements That Invert[44]

Long puzzling to scientists were certain bacterial traits that alternate between one phenotype and another at frequencies far higher than would be expected if they arose through mutational mechanisms (Table 7-3). The best-studied case involved the bacterium *Salmonella* (a close relative of *E. coli*), whose flagella are constructed either from the protein designated H1 or the protein designated H2, but never both. Changes from H1 to H2 (or vice versa) occur as frequently as once in every 1000 cell divisions, much too frequently to be caused by sponta-

Figure 7-28
Electron micrograph of plasmid pSC101. (Courtesy of Stanley N. Cohen, Stanford University.)

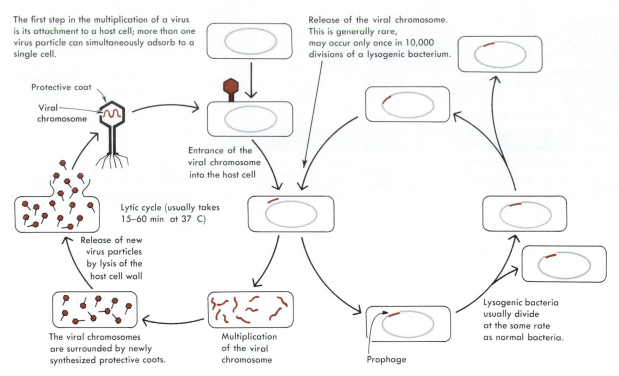

The first step in the multiplication of a virus is its attachment to a host cell; more than one virus particle can simultaneously adsorb to a single cell.

Protective coat

Viral chromosome

Entrance of the viral chromosome into the host cell

Release of the viral chromosome. This is generally rare, may occur only once in 10,000 divisions of a lysogenic bacterium.

Lytic cycle (usually takes 15–60 min at 37 C)

Release of new virus particles by lysis of the host cell wall

The viral chromosomes are surrounded by newly synthesized protective coats.

Multiplication of the viral chromosome

Prophage

Lysogenic bacteria usually divide at the same rate as normal bacteria.

Figure 7-29
The life cycle of a lysogenic bacterial virus. After its chromosome enters a host cell, it sometimes immediately multiplies like a lytic virus and at other times becomes transformed into a prophage. The lytic phase of its life cycle is identical to the complete life cycle of a lytic (nonlysogenic) virus. Lytic bacterial viruses are so called because their multiplication results in the rupture (lysis) of the bacteria.

Table 7-3 Examples of Bacterial Traits That Frequently Alternate Between Two Different Phenotypes

Phenotype	Organism
Colony morphology	*Escherichia coli*
Pigmentation	*Streptomyces reticuli*
Pigmentation	*Myxococcus xanthus*
Pilin	*Escherichia coli*
Surface antigen	*Borrelia hermsii*
Vi antigen	*Citrobacter freundii*
H and O antigens	*Salmonella*
Bioluminescence	*Vibrio harveyi*
Host range	*Bacteriophage Mu*

neous mutations. Only within the last five years has it become clear that chromosomal flip-flops (inversions) lead to the alternative expression of the H1 or H2 gene. When the controlling DNA segment is oriented in one direction, synthesis of the H1 protein occurs and H2 synthesis is repressed. After the controlling DNA segment inverts, H2 synthesis starts up while H1 synthesis slows down (see Chapter 11 for details).

Insertion of Phage Chromosomes into the Chromosomes of *E. coli*[45, 46, 47]

Some phages (e.g., phage λ) do not always multiply upon entering a host cell. Instead, their DNA becomes inserted into a host bacterial chromosome. Then the phage chromosome is, for all practical purposes, an integral part of its host chromosome and is duplicated, like the bacterial chromosome, just once every cell generation (Figure 7-29). A phage chromosome that is integrated into a host chromosome is called a **prophage.** Those bacteria containing prophages are called lysogenic bacteria, and those phages whose chromosomes can become prophages are known as **lysogenic phages.**

In contrast, some phages (e.g., T2, T7) always multiply when they enter a host cell and thus often lyse the cell because of the large numbers of phage particles created. These are called **lytic phages.** It is often difficult to know when a bacterium is lysogenic. We can only be sure when the phage chromosome is released from the host chromosome and the multiplication process that forms new progeny particles commences. Why only certain phages have evolved to form lysogenic associations and what advantages result from such associations are still open questions.

How phage chromosomes are inserted into bacterial chromosomes was at first very mysterious. Now we realize that such insertions do not always follow the same pathways and that two very different routes are used. The best-understood route is that used by phage λ. Integration of λ is achieved by crossing over between the *E. coli* chromosome and a circular form of the λ chromosome. Prior to integration, the originally linear λ chromosome circularizes and attaches to a very homologous short section of the *E. coli* chromosome. Both the host chromosome and the phage λ chromosome then break and rejoin in such a way that the broken ends of the λ chromosome join to the broken ends of the bacterial chromosome, instead of to each other, thereby inserting the λ chromosome into the *E. coli* chromosome (Figure 7-30). The prophage form of λ detaches from the host chromosome by the reverse process. The prophage-host chromosome junctions pair prior to a crossover event that ejects the λ chromosome. The now free λ genome can then begin to multiply as if it were the chromosome of a lytic virus.

The sections of genetic similarity between the λ chromosome and the *E. coli* chromosome are minuscule compared to the length of the *E. coli* chromosome. So the probability that ordinary crossing-over events occur between them is very small. Frequent crossing over between them is made possible by specific phage proteins that not only bring together the homologous sections, but also make the necessary cuts that allow the respective DNAs to exchange partners (see Chapter 11).

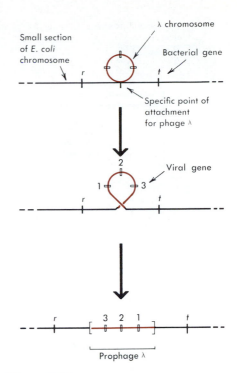

Figure 7-30
Insertion of the chromosome of phage λ into the *E. coli* chromosome by crossing over.

Phage Mu Is a Transposable Genetic Element[48]

At first, phage Mu (mutator) was thought to behave in a completely unique fashion. It never multiplies in a simple lytic way, but always commences its life cycle by inserting itself into the *E. coli* chromosome at totally random locations. The resulting prophagelike insertions often inactivate (mutate) the genes into which they become inserted (Figure 7-31). Thus, Mu was originally known for its mutagenic properties. Later in the life cycle, upon receipt of an appropriate signal, the Mu prophage somehow generates many new copies of itself, which then reinsert themselves at random into the *E. coli* chromosome. Subsequent DNA-cutting events release the new copies, which are then encapsulated by protein, thereby forming large numbers of progeny Mu particles. The nature of phage Mu became clear only when transposable elements were discovered and compared in their behavior to phage Mu. Then phage Mu was ascertained to be an especially large transposon that encodes not only the enzyme (transposase) that inserts it into the *E. coli* chromosome but also the large number of proteins necessary to construct the outer protective shell that surrounds the Mu chromosome. For further discussion of phage Mu, see Chapter 17.

Figure 7-31
Insertion of Mu DNA into a *lac*⁺ gene renders the gene inactive (*lac*⁻).

Phages Occasionally Carry Bacterial Genes[49, 50, 51]

Not only can bacterial genes be transferred through mating or transposition, but they can also be passively carried from one bacterium to

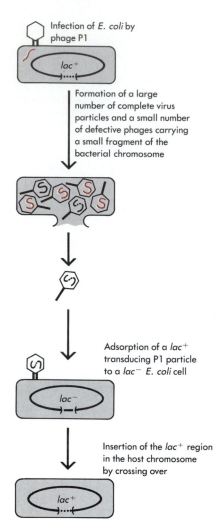

Figure 7-32
Transduction, the passive transfer of genetic material from one bacterium to another by means of carrier phage particles.

another by phage particles, a process called **transduction.** This happens when a virus particle (a transducing phage) accidentally encapsulates sections of a host chromosome. When this virus particle (usually biologically inactive because its own viral chromosome is incomplete or totally missing) attaches to a host cell, the fragment of bacterial chromosome is injected into the cell. It then can engage in crossing over with the host chromosome. If the transducing phage has been grown on a bacterial strain genetically different from the strain subsequently infected with the phage, a genetically altered bacterium may be produced (Figure 7-32). For example, a suspension of P1 phage particles grown on a strain of *E. coli* that is able to metabolize lactose will contain a small number of particles carrying the gene (*lac*$^+$) required to utilize lactose. Addition of such phages to an *E. coli* strain unable to use lactose for growth (*lac*$^-$) will transform a small number of the *lac*$^-$ bacteria to the *lac*$^+$ form by means of genetic recombination.

Because transduction is very rare, one might think that it would not be a useful tool for probing chromosomal structure. On the contrary, it has been most helpful in telling us whether two genes are located close to each other and what their exact order is. This is because the segment of the bacterial chromosome carried by a single transducing particle is relatively short, so that only genes located very close to each other will be enclosed in the same transducing particle. Thus, by determining the frequencies with which groups of genes can be transduced by the same phage particle, we can very accurately establish their relative locations.

In addition to transducing phages like P1, which randomly carry small sections of the *E. coli* chromosome, there are also specialized transducing phages that carry specific bacterial genes. Certain defective λ phage particles possess hybrid chromosomes in which a section of the λ chromosome is linked to a fragment of *E. coli* chromosome. Such hybrid chromosomes result when the λ prophage improperly recombines out of the *E. coli* chromosome to generate a λ DNA fragment still attached to bacterial genes that lie immediately adjacent to the site into which the λ DNA inserts on the *E. coli* chromosome (Figure 7-33). Purification of these specialized transducing phages therefore affords a way to obtain large numbers of certain *E. coli* genes, and it was through such a procedure that the genetically well-studied gene involved in the utilization of the sugar lactose was first molecularly investigated. Unfortunately, most prophages preferentially insert themselves at only a few chromosomal sites, and most genes cannot easily be isolated within specialized transducing particles.

Transfer of Genetic Traits Via Incorporation of Purified DNA[52, 53, 54]

Bacterial genes can also move from one cell to another in the form of pure DNA. The process by which a bacterium absorbs pure DNA from its surroundings and functionally integrates the exogenous DNA into its own genetic material is called **transformation.** As mentioned in Chapter 3, it was through the discovery of this phenomenon that DNA was pinpointed as the key genetic molecule. First observed in the Gram-positive bacterium *Streptococcus pneumoniae*, transforma-

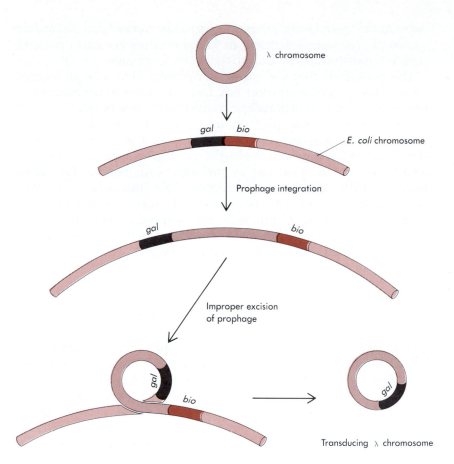

Figure 7-33
Specialized transducing phage λ can integrate into an *E. coli* cell and later recombine out (excise) carrying specific bacterial genes.

tion was found to occur naturally in many Gram-positive and Gram-negative bacteria. While initially the argument could be made that transformation is accidental and plays no significant role in nature, this is no longer a tenable belief. Indeed, some bacteria (e.g., *Hemophilus* and *Neisseria*) possess special proteins that mediate the entry of DNA into cells and that only recognize and take up DNA from related species (Figure 7-34). *Hemophilus* cells do this by binding specific short DNA sequences (e.g., 5′-AAGTGGGTCA-3′) that are characteristic of the *Hemophilus* genome and occur some 600 times on its chromosome (or once per 4000 base pairs). Only double-stranded DNA is so absorbed, after which it somehow enters into the host bacterial chromosome as single-stranded segments that base-pair to appropriate single-stranded regions of the host DNA, perhaps at the replicating forks.

Many other bacteria, including *E. coli*, do not seem to have evolved or retained ways of taking up functionally active DNA. A powerful trick, however, is now available for "artificially" introducing functional DNA. In 1970, it was discovered that DNA added in the presence of high amounts of Ca^{2+} can transform. This Ca^{2+}-induced artificial transformation works in *E. coli* only when the added DNA is a complete replicon capable of self-duplication. Plasmid DNAs and intact viral chromosomes transform with high efficiencies, while linear fragments of bacterial DNA transform at much lower levels. The failure of linear DNAs to transform well is thought to result from their

Figure 7-34

Transformation, in which a bacterial cell takes up free DNA. (a) Development of competence. Competence factor (▲) interacts with a cell-surface receptor (M) and causes certain competence-specific proteins to be expressed. One of these is an autolysin (■), which exposes a membrane-associated DNA-binding protein (●) and a nuclease (○). (b) Transformation. A long fragment of dsDNA is bound at several sites to the cell surface. It is progressively nicked and cut. One strand of the cut fragment is degraded by the nuclease, a process that releases nucleotides (−) into the medium. The other strand becomes associated with a competence-specific protein (◆). In this form it enters the cell and recombines by a process that involves replacement of one strand of the endogenote by the donor. (After Smith, 1981.)

digestion by periplasmic DNases before they can be brought into the cytosol. *E. coli* strains deficient in such nucleases are much more effectively transformed by exogenous DNA fragments.

The realization that Ca^{2+} treatment renders virtually all bacteria susceptible to transformation has had enormous consequences. It provides a way to functionally introduce any piece of DNA, after its insertion into a suitable plasmid, into any desired bacterial genome. Thus, bacterial transformation, which first told the world of science that DNA was the genetic molecule, now gives us a most versatile alternative to viral infection as the means to monitor the consequences of the sudden entry of specific DNA into cells. As we shall see in the next section, because it allows the introduction of test-tube-made recombinant DNA into cells, transformation provides a means of genetically engineering cells as well as making available highly powerful techniques for the isolation of desired genes.

Recombinant DNA Techniques Allow the Isolation of Single Genes[55, 56]

Until a few years ago, genetic crosses in their many manifestations were the key tools for revealing what genes are and how they function. Through genetic manipulations, not only could the location of genes on their respective chromosomes be found, but in a few cases, desired genes could be recombined on the phage λ chromosome. There they could be used both for molecular analysis and to search for postulated regulatory molecules (e.g., repressors) whose binding to specific genetic elements was believed to control the functioning of adjacent genes. Whether a desired gene could be so isolated on a transducing phage, however, was not only very chancy, but could easily consume many years of hard work.

The scope of genetic research, even with the simplest bacterial and viral systems, has thus dramatically improved as a result of the development of recombinant DNA procedures. These procedures make possible the isolation and **cloning** (duplication) of any gene essential for cellular growth and for which mutations exist that block its proper functioning. With genetically well-characterized microorganisms, the isolation of a given desired gene is no longer a major challenge.

Particularly straightforward are the techniques for the isolation of many *E. coli* genes (Figure 7-35). They start with the fragmentation of the *E. coli* chromosome (4×10^6 base pairs) into some thousand different discrete pieces by using an appropriate DNA **restriction** (cutting) enzyme that recognizes a specific nucleotide sequence that occurs on the average every several thousand base pairs along the *E. coli* chromosome. These chromosomal fragments are then mixed with an excess of DNA from an *E. coli* plasmid that has been converted into a linear form by the same sequence-specific restriction enzyme. Addition of the DNA-joining enzyme, DNA ligase, randomly links the two sets of fragments together, leading to the generation of many recombinant plasmid DNA circles into which have been inserted discrete fragments of the *E. coli* chromosome. Found within this recombinant DNA "library" are some thousand different recombinant plasmids, collectively containing all the DNA of the *E. coli* chromosome.

To use such an *E. coli* library to isolate a desired gene, we must first mix the library in the presence of Ca^{2+} with the respective mutant

Figure 7-35
Formation of a recombinant DNA "library."

bacteria (e.g., a strain that requires a specific amino acid for growth) under conditions that favor the uptake of the plasmid DNA. The bacteria are then placed under the appropriate selective conditions for growth (absence of the needed amino acid). The only cells that can multiply to form bacterial colonies will be those that have taken up specific recombinant plasmids that carry the normal gene needed for the biosynthesis of the amino acid. These colonies can in turn be used to grow a much larger number of genetically identical bacteria, from which the small plasmid DNA circles can be easily separated from the much larger main bacterial chromosomes. By cleaving these recombinant plasmids with the right restriction enzyme, pure copies of the desired *E. coli* gene can be separated away from the plasmid DNA.

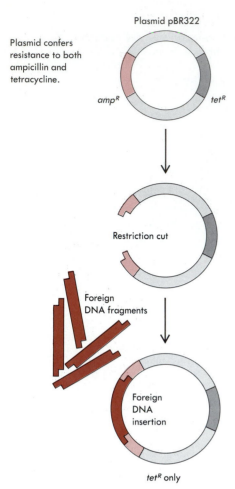

Plasmid confers resistance to both ampicillin and tetracycline.

Plasmid pBR322

amp^R

tet^R

Restriction cut

Foreign DNA fragments

Foreign DNA insertion

tet^R only

Figure 7-36

Plasmid pBR322 contains restriction sites into which foreign DNA can be inserted. A number of these sites are in genes conferring antibiotic resistance on the bacterium containing them. When foreign DNA is inserted into one of these genes, the gene is usually inactivated. Here, foreign DNA is shown inserting in amp^R, the gene for ampicillin resistance. The resulting loss of ampicillin resistance is a preliminary indicator that foreign DNA has been inserted.

The Development of Increasingly Sophisticated Cloning Vectors[57-61]

Vectors are autonomously replicating DNA units into which DNA fragments are inserted for gene cloning. Genes taken up by these vectors are multiplied (cloned) as the vector replicates, to yield numbers suitable for molecular analysis. Now the best-known plasmid vector is **pBR322,** which has genetically tailored cutting sites into which DNA fragments can be inserted without affecting plasmid self-replication (Figure 7-36). DNA fragments of up to 5 to 10 kilobase pairs (kbp) are very efficiently cloned in this vector. Larger fragments, however, often tend to be unstable and lose most of their DNA. The multiplication rates of plasmids slow down as they get larger, with those that have lost large chunks of their foreign DNA eventually predominating. When larger DNA pieces must be cloned, they often are placed in specially constituted λ vectors into which foreign fragments as large as 15 kbp are easily inserted without disturbing the normal packaging of the λ chromosome into functional virus particles (Figure 7-37). Still larger DNA pieces of up to 50 kbp can now be cloned in the so-called **cosmid vectors,** constructed from the two ends of the λ chromosome. Thus, depending on the size of the DNA needed to be cloned, DNA fragments are inserted into plasmid, λ phage, or cosmid vectors.

The initial plasmid vectors multiplied only in their natural host cells. DNA sequences around the origins of replication control this specificity. More recently, newly designed vectors have been made using recombinant DNA procedures that contain two distinct origins of replication derived from, for example, *E. coli* and yeast or *E. coli* and human. These vectors offer extended host specificity and are called **shuttle vectors.** They are now extensively used to transfer genes between very different cell types, and in so doing are helping to more clearly define the rules governing gene expression.

In Vitro Mutagenesis[62, 63]

Until recently, genetic analysis started with the finding of mutant organisms followed by a series of genetic crosses to locate the genes in which the mutations had occurred. Now an alternative approach is possible. First the desired gene (e.g., that for a repressor molecule) is isolated and then sequenced, after which mutations are created within it. Then the phenotypic consequences of these in vitro-made mutations are observed by reinserting the mutant gene back into suitable cells. Underlying this "reverse genetics" is a set of recently developed procedures for in vitro mutagenesis. Both small deletions and nucleotide substitutions can now be easily made around the sites of restriction enzyme cuts. Becoming even more powerful is a technique that uses well-defined oligonucleotides as primers onto which DNA chains can be enzymatically extended. By synthesizing primers containing specific base changes from the normal sequence, mutant genes can be created bearing these same changes. Through this technique, a much deeper analysis of how specific proteins are constructed and function will soon be possible. Equally important is the possibility of further pinpointing the DNA nucleotide sequences that control transcription and translation.

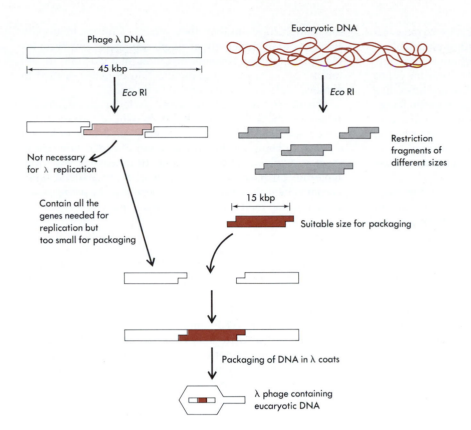

Figure 7-37
Phage λ DNA can act as a cloning vector for eucaryotic genes or other foreign DNA fragments of approximately 15 kbp in length.

Summary

The genes whose structure and functioning we now understand the best are those of bacteria, in particular *E. coli* and its viruses (phages). Like other haploid organisms, *E. coli* has the advantages of a very short life cycle and ease of manipulation under controllable laboratory conditions. Vital for the accumulation of this precise genetic knowledge has been the use of mutants. Through them we have been able to reveal sexual phenomena in bacterial cells, once thought too simple to make use of genetic recombination, and also to establish genetic maps revealing the nature of their chromosomes. Bacterial mutations involving growth factors and resistance to specific antibiotics and viruses are particularly useful, since they allow easy separation of wild-type and mutant organisms. There are numerous ways to increase the number of newly arising mutants through the application of increasingly sophisticated mutagenic agents. Many powerful techniques also exist to select desired rare mutants from very large populations of wild-type organisms. Most useful are agents like penicillin, which selectively kills growing wild-type cells, thereby enriching the population of mutant cells unable to multiply under certain environmental conditions (e.g., in the absence of a growth factor).

Bacterial conjugation involves the attachment of male and female cells followed by the movement of male chromosomal material into recipient female cells. The essence of a male cell is the presence of a specific DNA segment, which can exist free as a tiny supernumerary chromosome (the F factor) or as a segment integrated into the main and much larger bacterial chromosome. When the F factor exists free (as in F⁺ cells), only F factor DNA is transferred into female (F⁻) cells following mating between a male (F⁺) and a female (F⁻) cell. If, however, the F factor has become part of the *E. coli* chromosome (as in Hfr—high frequency of recombination—cells), then the entire male chromosome has the potential to move into the female cell, with such movement commencing within a section of F DNA. The movement of a male chromosome into a female cell often aborts in mid-passage, leading to the insertion of incomplete male chromosomal fragments into female cells. Thus, conjugation between Hfr cells and F⁻ cells rarely leads to conversion of the female cell into a male cell.

Bacterial viruses (phages), like all other viruses, contain single tiny chromosomes surrounded by protein-containing protective coats that allow the chromosomes to move from one cell to another. Cycles of phage replication always start with the injection of phage chromosomes into the host cells. Viruses can multiply only in cells and are best characterized as parasites at the genetic level. The growth of many phages leads to the obligatory death of their host cells (lytic phages). Other phages may have lysogenic life cycles in which their chromosomes become inserted into their host bacterial chromosomes in apparently passive (prophage) form. For the replication of such lysogenic phages, the prophage forms must detach from the respective bacterial chromosomes and commence cycles of phage chromosome multiplication. Occasionally, phage particles become assembled containing bacterial DNA instead of phage DNA. Such transducing phages, following infection of a suitable bacterial host, provide an alternate way to transfer genes from one bacterium to another. Phage chromosomes, like bacterial chromosomes, are capable of genetic exchange (crossing over) when present in the same bacterial cell. There is no differentiation into male and female phages, with genetic recombination being generally assured by simultaneous infection of a given cell by more than one genetically distinct phage.

Genetic recombination can also be initiated by the introduction of free DNA. How such free DNA enters recipient cells is still a mystery. Ca^{2+} somehow increases efficiency of such transformation events. In some bacteria, special proteins and DNA sequences are necessary for such DNA transfers, suggesting that movement of free DNA is frequently an important means of achieving genetic recombination.

Small, autonomously replicating DNA molecules like the F factor are present in many bacteria. Now known as plasmids, these relatively tiny DNA molecules do not segregate regularly at cell division and are invariably lost from a small fraction of all daughter cells. Thus, they usually are not the sole bearers of essential genes required for growth and multiplication. Because, however, they frequently are present in high copy number, they provide a means to amplify the number of any essential single-copy gene that becomes inserted into their genome. Insertion on a plasmid, for example, can lead to rapid amplification of genes needed to achieve high-level resistance to antibiotics.

The location of genes on a given bacterial chromosome is relatively fixed, with those genes whose products are needed in abnormally large amounts (e.g., ATP synthetase) frequently being located at the origin of replication. However, specialized genetic segments (transposons) exist that have the capacity to insert copies of themselves at other chromosomal sites. Other genetic segments have the capacity to flip-flop their relative orientation along a chromosome and in so doing to alter their functioning. Such flip-flop events demand the occurrence of inverted duplicated sequences at the termini of such segments. Their high frequencies reflect the existence of site-specific recombination, which promotes enzymes that recognize the inverted terminal sequences.

Traditional genetic procedures are increasingly being supplemented by recombinant DNA techniques. In these analyses, new in vitro procedures for site-specific mutagenesis are allowing mutant DNAs to be made for a "reverse" genetic approach in which mutations become produced before their phenotypic consequences are known.

Bibliography

General References

Adelberg, E. A., ed. 1966. *Papers on Bacterial Genetics*. 2nd ed. Boston: Little, Brown.

Davis, R. W., D. Botstein, J. R. Roth. 1980. *Advanced Bacterial Genetics*. Cold Spring Harbor, N.Y.: Cold Spring Harbor Laboratory. A manual for genetic engineering.

Hayes, W. 1968. *The Genetics of Bacteria and Their Viruses*. 2nd ed. New York: Wiley.

Ingraham, J. L., O. Maaloe, and F. C. Neidhardt. 1983. *Growth of the Bacterial Cell*. Sunderland, Mass.: Sinauer.

Miller, J. H. 1972. *Experiments in Molecular Genetics*. Cold Spring Harbor, N.Y.: Cold Spring Harbor Laboratory. Concentrates on the molecular genetics of *E. coli*.

O'Brien, S. J., ed. 1984. *Genetic Maps 1984*. Vol. 3. Cold Spring Harbor, N.Y.: Cold Spring Harbor Laboratory. A compilation of linkage and restriction maps of genetically studied organisms.

Silhavy, T. J., M. L. Berman, and L. W. Enquist. 1984. *Experiments with Gene Fusions*. Cold Spring Harbor, N.Y.: Cold Spring Harbor Laboratory. Demonstrates the use of gene fusions, transposable elements, and methods of recombinant DNA for genetic analysis in *E. coli*.

Stent, G. S., ed. 1965. *Papers on Bacterial Viruses*. 2nd ed. Boston: Little, Brown.

Stent, G. S., and R. Calendar. 1978. *Molecular Genetics: An Introductory Narrative*. 2nd ed. San Francisco: Freeman.

Timmis, K. N., and A. Putiler, eds. 1979. *Plasmids of Medical, Environmental, and Commercial Importance*. Amsterdam, London, and New York: Elsevier.

Cited References

1. "Heredity and Variation in Microorganisms." 1946. *Cold Spring Harbor Symp. Quant. Biol*. Vol. 11. Cold Spring Harbor, N.Y.: Cold Spring Harbor Laboratory.

2. Beadle, G. W., and E. L. Tatum. 1941. "Genetic Control of Biochemical Reactions in *Neurospora*." *Proc. Nat. Acad. Sci*. 27:499–506.

3. Beadle, G. W., and E. L. Tatum. 1945. "*Neurospora*. II. Methods of Producing and Detecting Mutations Concerned with Nutritional Requirements." *Amer. J. Bot*. 32:678–686.

4. Luria, S. E., and M. Delbruck. 1943. "Mutations of Bacteria from Virus Sensitivity to Virus Resistance." *Genetics* 28:491–511.

5. Lederberg, J., and E. M. Lederberg. 1952. "Replica Plating and Indirect Selection of Bacterial Mutants." *J. Bacteriol*. 63:399–406.

6. Gray, C. H., and E. L. Tatum. 1944. "X-Ray Induced Growth Factor Requirements in Bacteria." *Proc. Nat. Acad. Sci*. 30:404–410.

7. Newcombe, H. B. 1948. "Delayed Phenotypic Expression of Spontaneous Mutations in *E. coli*." *Genetics* 33:447–476.

8. Hayes, W. 1957. "The Phenotypic Expression of Genes Determining Various Types of Drug Resistance Following Their Inheritance by Sensitive Bacteria." In *Drug Resistance in Microorganisms*, ed. G. E. W. Wolstenhlome and C. M. O'Connor. Ciba Foundation Symposium. London: Churchill, p. 197.

9. Davis, B. D. 1948. "Isolation of Biochemically Deficient Mutants of Bacteria by Penicillin." *Amer. Chem. Soc*. 70:4267.

10. Lederberg, J. 1950. "Isolation and Characterization of Biochemical Mutants of Bacteria." In *Methods in Medical Research*, vol. 3, ed. J. H. Comrie, Jr. Chicago: Year Book, p. 5.

11. Bonhoeffer, F., and H. Schaller. 1965. "A Method for Selective Enrichment of Mutants Based on the High Ultraviolet Sensitivity of DNA Containing 5-bromouracil." *Biochem. Biophys. Res. Comm*. 20:93–97.

12. Lederberg, J. 1948. "Detection of Fermentation Mutants with Tetrazolium." *J. Bacteriol*. 56:695.

13. Lin, E. C. C., S. A. Lerner, and S. E. Jorgensen. 1962. "A Method for Isolating Constitutive Mutants for Carbohydrate-Containing Enzymes." *Biochem. Biophys. Acta* 60:422.

14. Gesteland, R. F. 1966. "Isolation and Characterization of Ribonuclease I Mutants of *E. coli*." *J. Mol. Biol.* 16:67–84.

15. De Lucia, P., and J. Cairns. 1969. "Isolation of an *E. coli* Strain with a Mutation Affecting DNA Polymerase." *Nature* 224:1164–1166.

16. Lederberg, J., and E. L. Tatum. 1946. "Novel Genotypes in Mixed Cultures of Biochemical Mutants of Bacteria." *Cold Spring Harbor Symp. Quant. Biol.* 11:113–114.

17. Stent, G. S. 1963. *Molecular Biology of Bacterial Viruses.* San Francisco: Freeman.

18. Mathews, C. K., E. M. Kutter, G. Mosig, and P. B. Berget. 1983. *Bacteriophage T4.* Washington, D.C.: Amer. Soc. for Microbiol.

19. Doermann, A. 1953. "The Vegetative State in the Life Cycle of Bacteriophage: Evidence for Its Occurrence, and Its Genetic Characterization." *Cold Spring Harbor Symp. Quant. Biol.* 18:3–11.

20. Luria, S. E. 1945. "Mutations of Bacterial Viruses Affecting Their Host Range." *Genetics* 30:84–99.

21. Hershey, A. D. 1946. "Mutations of Bacteriophage with Respect to Types of Plaque." *Genetics* 31:620–640.

22. Epstein, R. H., A. Bolle, C. M. Steinberg, E. Kellenberg, E. Boy de la Tour, R. Chevalley, R. S. Edgar, M. Susman, G. H. Denhardt, and A. Lielausis. 1963. "Physiological Studies on Conditional Lethal Mutants of the Bacteriophage T4D." *Cold Spring Harbor Symp. Quant. Biol.* 28:375–394.

23. Hershey, A. D., and R. Rotman. 1949. "Genetic Recombination Between Host Range and Plaque-Type Mutants of Bacteriophage in Single Bacterial Cells." *Genetics* 34:44–71.

24. Hershey, A. D. 1958. "The Production of Recombinants in Phage Crosses." *Cold Spring Harbor Symp. Quant. Biol.* 23:19–46.

25. Hayes, W. 1952. "Recombination in *Bact. coli* K-12: Unidirectional Transfer of Genetic Material." *Nature* 169:118–119.

26. Jacob, F., and E. L. Wollman. 1961. *Sexuality and the Genetics of Bacteria.* N.Y.: Academic Press.

27. Brinton, C. C., P. Gemski and J. Carnahan. 1964. "A New Type of Bacterial Pilus Genetically Controlled by the Fertility Factor of *E. coli* K12." *Proc. Nat. Acad. Sci.* 52:776–783.

28. Crawford, E. M., and R. A. Gesteland. 1964. "The Adsorption of Bacteriophage R-17." *Virology* 22:165–167.

29. Broda, P., J. R. Beckwith, and J. Scaife. 1965. "The Characterization of a New Type of F-Prime Factor in *E. coli* K-12." *Genet. Res.* 5:489–493.

30. Ou, J. T., and T. F. Anderson. 1970. "Role of Pili in Bacterial Conjugation." *J. Bacteriol.* 102:648–654.

31. Manning, P., and M. Achtman. 1979. "Cell-to-Cell Interactions in Conjugating *Escherichia coli:* The Involvement of the Cell Envelope." In *Bacterial Outer Membranes,* ed. M. Inouye. New York: Wiley, pp. 409–447.

32. Minkley, E. G., Jr., S. Polen, C. C. Brinton, Jr., and K. Ippen-Ihler. 1976. "Identification of the Structural Gene for F-Pilin." *J. Mol. Biol.* 108:111–121.

33. Achtman, M., G. Morelli, and S. Schwuchow. 1978. "Cell-Cell Interactions in Conjugating *E. coli:* Role of F Pili and Fate of Mating Aggregates." *J. Bacteriol.* 135:1053–1061.

34. Wollman, E. L., F. Jacob, and W. Hayes. 1956. "Conjugation and Genetic Recombination in *E. coli*." *Cold Spring Harbor Symp. Quant. Biol.* 21:141–162.

35. Cairns, J. 1963. "The Bacterial Chromosome and Its Manner of Replication as Seen by Autoradiography." *J. Mol. Biol.* 6:208–213.

36. Henner, D. J., and J. A. Hoch. 1980. "The *Bacillus subtilis* Chromosome." *Microbiol. Revs.* 44:57–82.

37. Royle, P. L., H. Matsumoto, and B. W. Holloway. 1981. "Genetic Circularity of the *Pseudomonas aeruginosa* PAO Chromosome." *J. Bacteriol.* 145:145–155.

38. Bachman, B. J. 1983. "Linkage Map of *E. coli* K-12, ed. 7." *Microbiol. Revs.* 47:180–230.

39. Anderson, R. P., and J. R. Roth. 1977. "Tandem Genetic Duplications in Phage and Bacteria." *Ann. Rev. Microbiol.* 31:473–506.

40. Lindahl, L., and J. M. Zengel. 1982. "Expression of Ribosomal Genes in Bacteria." *Adv. Genetics* 21:53–121.

41. Bukhari, A. I., J. A. Shapiro, and S. L. Adhya. 1977. *DNA Insertion Elements, Plasmids and Episomes.* Cold Spring Harbor, N.Y.: Cold Spring Harbor Laboratory.

42. Shapiro, J. A. 1983. *Mobile Genetic Elements.* New York: Academic Press.

43. "Moveable Genetic Elements." 1980. *Cold Spring Harbor Symp. Quant. Biol.* Vol. 45. Cold Spring Harbor, N.Y.: Cold Spring Harbor Laboratory.

44. Simon, M. I., and M. Silverman. 1983. "Recombinational Gene Expression in Bacteria." In *Gene Function in Procaryotes,* ed. J. Beckwith, J. Davies, and J. A. Gallant. Cold Spring Harbor, N.Y.: Cold Spring Harbor Laboratory, pp. 211–228.

45. Lwoff, A. 1953. "Lysogeny." *Bacteriol. Revs.* 17:269–337.

46. Campbell, A. 1962. "Episomes." *Adv. Genetics* 11:101–145.

47. Herskowitz, I., and D. Hagen. 1980. "The Lysis-Lysogeny Decision of Phage λ: Explicit Programming and Responsiveness." *Ann. Rev. Genetics* 14:399–446.

48. Bukhari, A. I. 1976. "Bacteriophage Mu as a Transposable Element." *Ann. Rev. Genetics* 10:389–412.

49. Zinder, N. D., and J. Lederberg. 1952. "Genetic Exchange in *Salmonella*." *J. Bacteriol.* 64:679–699.

50. Campbell, A., and E. Balbinder. 1959. "Transduction of the Galactose Region of *E. coli* K-12 by the Phages of λ and λ-434 Hybrid." *Genetics* 44:309–319.

51. Shapiro, J., L. MacHattie, L. Eron, G. Ihler, K. Ippen, and J. Beckwith. 1969. "Isolation of Pure lac Operon DNA." *Nature* 224:768–774.

52. Smith, H. O., D. B. Danner, and R. A. Deich. 1981. "Genetic Transformation." *Ann. Rev. Biochem.* 50:41–68.

53. Raiband, O., M. Mock, and M. Schwartz. 1984. "A Technique for Integrating Any DNA Fragment into the Chromosome of *E. coli*." *Gene* 29:231–241.

54. Mandel, M., and A. Higa. 1970. "Calcium-Dependent Bacteriophage DNA Infection." *J. Mol. Biol.* 53:159–162.

55. Old, R. W., and S. B. Primrose. 1981. *Principles of Gene Manipulation.* 2nd ed. Berkeley and Los Angeles: University of California Press.

56. Watson, J. D., J. Tooze, and D. T. Kurtz. 1983. *Recombinant DNA: A Short Course.* San Francisco: Freeman.

57. Bolivar, F., R. L. Rodrigues, P. J. Greene, M. C. Betlach, H. L. Hayneker, H. W. Boyer, J. Crosa, and S. Falkow. 1977. "Construction and Characterization of New Cloning Vehicles. II. A Multipurpose Cloning System." *Gene* 2:95–113.

58. Collins, J., and B. Hohn. 1978. "Cosmids: A Type of Plasmid Gene Cloning Vector That Is Packageable in Vitro in Bacteriophage Heads." *Proc. Nat. Acad. Sci.* 75:4242–4246.

59. Hohn, B., and J. Collins. 1980. "A Small Cosmid for Efficient Cloning of Large DNA Fragments." *Gene* 11:241–298.

60. Hicks, J. B., J. N. Strathern, A. J. S. Klar, and S. L. Dellaporta. 1982. "Cloning by Complementation in Yeast: The Mating Type Genes." In *Genetic Engineering—Principles and Methods,* vol. 4, ed. J. K. Stelow and A. Hollaender. New York: Plenum, pp. 219–248.

61. Sullivan, M. A., R. E. Yasbin, and F. E. Young. 1984. "New Shuttle Vectors for *Bacillus subtilis* and *Escherichia coli* Which Allow Rapid Detection of Inserted Fragments." *Gene* 29:21–26.

62. Watson, J. D., J. Toose, and D. T. Kurtz. 1983. "In Vitro Mutagenesis." In *Recombinant DNA: A Short Course.* San Francisco: Freeman, Chap. 8.

63. Shortle, D., D. DiMaio, and D. Nathans. 1981. "Directed Mutagenesis." *Ann. Rev. Genetics* 15:265–294.

CHAPTER 8

The Fine Structure of Bacterial and Phage Genes

Until the advent of recombinant DNA procedures, genetic crosses constituted the only real tool we had to investigate genes. Not only did they provide the means of ordering genes along chromosomes, but they also gave us our first detailed knowledge of the topology (fine structure) of the gene itself. This elucidation of the structure of the gene was possible because crossing over occurs not only between genes, but also *within a single gene.* In fact, in bacteria and their phages, virtually all crossing over occurs within, rather than between, the genes. This fact was recognized in the early 1950s, just after the double helix was discovered. Until then, the chromosome was generally viewed as a linear collection of genes held together by some nongenetic material, somewhat like a string of pearls. We now know that all chromosomes are likely to contain only single, usually immensely large, DNA molecules, each carrying the information needed to order the amino acids in many different protein molecules (see Chapter 9). All crossing over must thus occur by the breakage and reunion of DNA itself.

The probability of any crossing-over event occurring within a given gene is much lower than the probability of crossing over occurring somewhere within an entire chromosome. Its detection therefore demands that very large numbers of progeny (greater than 100,000) be looked at. Even with the relative ease of manipulating the fruit fly *Drosophila,* intragenic recombination long remained a controversial issue. It is very impractical and tedious to look at the more than 10,000 progeny arising from a single *Drosophila* cross. Only when microorganisms were adopted as the preferred tools for molecular genetics did intragenic crossing over become established as a powerful means for revealing important details about gene structure and function.

Recombination Within Genes Allows Construction of a Gene Map[1, 2]

The first incisive results on the genetic structure of the gene came from work with the *rIIA* and *rIIB* genes of the bacterial virus T4. These are two adjacent genes that influence the length of the T4 life cycle; mutations in either therefore affect the size of plaques produced in a

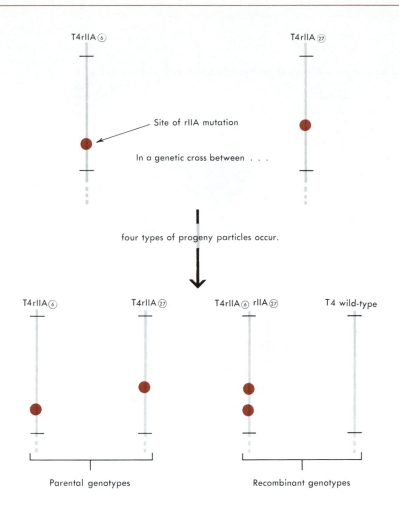

Figure 8-1
The use of T4 *rII* mutations in the demonstration of crossing over within the gene. Equal numbers of wild-type and double-*r* recombinants occur. The wild-type recombinants are easily found, because they are the only progeny genotype that will form plaques on K12(λ). It is much harder to identify the double *rIIA₆rIIA₇* recombinants, inasmuch as their plaques are indistinguishable from single-*r* plaques. To detect them, it is necessary to isolate a large number of progeny *r* viruses with the *r* phenotype and use them for new genetic crosses with both parent *r* strains. The double-*r* mutants will not produce wild-type recombinants with either of the single-*r* mutants.

bacterial layer growing on an agar plate. The presence of either an *rIIA* or *rIIB* mutation causes a shorter life cycle of T4 phage within an *E. coli* cell. T4-infected cells on an agar plate normally do not break open and release new progeny phage until several hours after they have been infected. Cells infected with *rII* mutants, however, always break open more rapidly. Thus, *rII* mutants produce larger plaques than wild-type phage. The *rII* mutants were chosen to work with because of the possibility of detecting a very small number of wild-type particles among a very large number of mutants. Although the wild type and the *rII* mutants grow equally well on *E. coli* strain B, there is another strain, *E. coli* K12(λ), on which only the wild type can multiply. Thus, when the progeny of a genetic cross between two different *rII* mutants are added to K12(λ), only the wild-type recombinants form plaques. Even as few as one wild-type recombinant per 10^6 progeny is easily detected.

Over 2000 independent mutations in the *rIIA* and *rIIB* genes were isolated and used in breeding experiments. In a typical cross, *E. coli* strain B bacteria were infected with two phage particles, each bearing an independently isolated *rIIA* or *rIIB* mutation. As the virus particles multiplied, genetic recombination occurred. The progeny were then grown on *E. coli* K12(λ) to detect wild-type recombinants. Normal particles were found in a very large fraction of the crosses, indicating recombination within the *rIIA* or *rIIB* gene (Figure 8-1). If recombination occurred only between genes, it would be impossible to produce wild-type recombinants by crossing two phage particles with muta-

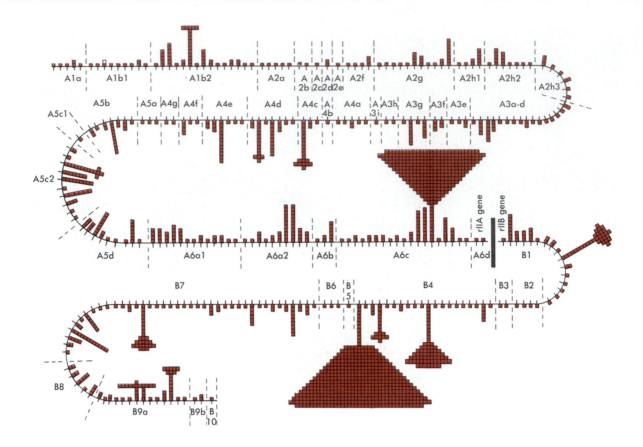

Figure 8-2

The genetic map of the *rIIA* and *rIIB* genes of phage T4. This map shows the assignment of a large number of different mutations in specific regions (A1a, A1b₁, etc.) of the chromosome. In most cases, the order of the mutations within a given region has not yet been determined. Each square corresponds to an independent occurrence of a mutation at a given site on the chromosome. Thus, some regions ("hot spots") mutate much more frequently than others. [From S. Benzer, *Proc. Nat. Acad. Sci.* 47, 410 (1961), with permission.]

tions in the same gene. A large spectrum of recombination values was found, just as in crosses between mutants in separate genes. This spread of values indicates that some mutant genes lie closer together than others and allows for the construction of genetic maps for the two *rII* genes. Examination of these genetic maps yields the following striking conclusions:

- A large number of different sites of mutation (mutable sites) occur within a gene. This number exceeds some 500 collectively for the *rIIA* and *rIIB* genes (Figure 8-2).

- The *rIIA* and *rIIB* genetic maps are unambiguously linear, strongly hinting that the gene itself has a linear construction.

- Most mutations are changes at only one mutable site. Many genes containing such mutations are able to be restored to the original wild-type gene structure by the process of undergoing a second (reverse) mutation at the same site as the first mutation.

- Other mutations can cause the deletion of one or more mutable sites. These reflect a physical loss of part of the *rII* gene (Figure 8-3). Mutations deleting more than one mutable site are highly unlikely to mutate back to the original gene form.

The genetic fine structure of a number of other viral and bacterial genes has since been extensively mapped. The lengths of these maps vary from gene to gene, suggesting that some gene products are larger than others. Although no other study has been as extensive as the *rII* work, each points to the same conclusion: that all genes have a very large number of sites at which mutation can occur and that these mutable sites are arranged in a strictly linear order. The geneticist's

Figure 8-3
Deletion mutations within the *rII* region of T4. About 10 percent of the spontaneous *rII* mutations do not map at a distinct point. They are due to deletions of a large number of adjacent mutable sites. Some deletions, for example 1272, involve both the *rIIA* and *rIIB* genes. The small rectangles indicate that the deletion most likely extends into the adjacent gene. The existence of deletion mutations has considerably facilitated genetic mapping. By crossing a newly isolated *rII* mutant with a number of deletion mutations covering increasingly larger regions, it is possible to assign an approximate map location to the new mutant. In this map, the size of the *rIIB* gene has been arbitrarily reduced.

view of a gene is thus a discrete chromosomal region that is responsible for a specific cellular product and that consists of a linear collection of potentially mutable units or sites, each of which can exist in several alternative forms and between which crossing over can occur.

The Complementation Test Determines If Two Mutations Are in the Same Gene

Since mutations in both the *rIIA* and *rIIB* genes result in a larger plaque, it is natural to ask why they are considered two genes. Would it not be simpler to consider them parts of the same gene? The answer is straightforward. If *E. coli* strain K12(λ) is infected simultaneously with T4 *rIIA* and T4 *rIIB* mutants, the chromosomes of both viruses multiply, and progeny virus is produced (Figure 8-4). In contrast, simultaneous infection of the K12(λ) with either two different T4 *rIIA* or two different T4 *rIIB* mutants results in no viral multiplication. This demonstrates that *rIIA* and *rIIB* genes carry out two different functions, each necessary for T4 multiplication on K12(λ).

In infection with a single T4 *rIIA* mutant, the *rIIB* gene functions normally, but no active form of the *rIIA* product is available, so that no viral multiplication is possible. Likewise, in infection with a single *rIIB* mutant, only an active *rIIA* product arises. Viral multiplication occurs only when we simultaneously infect a bacterium with an *rIIA* phage mutant and an *rIIB* phage mutant, because each mutant chromosome is able to produce the gene product that its infecting partner is unable to make. Two chromosomes can thus complement each other when the mutations are present in distinct genes, thereby giving a positive **complementation test** result. (This experiment is not affected by the possibility of intragenic crossing over between mutants to produce wild-type phage because the number of such recombinants is very small; in contrast, complementation between different mutant genes yields a normal number of progeny particles.)

It is easy to perform complementation tests with phage mutants. All that is necessary is to infect a bacterium with two different mutants at once. This automatically creates a cell containing one copy of each mutant chromosome. These tests were often useful in telling us

Figure 8-4
The demonstration that the *rII* region consists of two distinct genes that can complement each other during simultaneous infection.

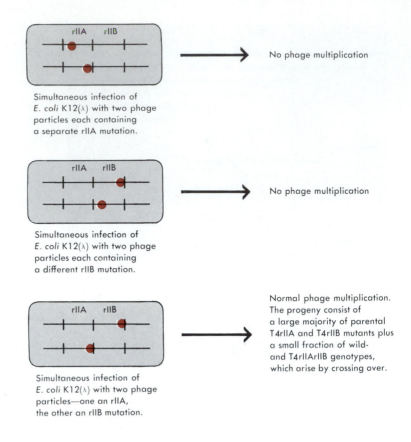

Simultaneous infection of *E. coli* K12(λ) with two phage particles each containing a separate rIIA mutation.

No phage multiplication

Simultaneous infection of *E. coli* K12(λ) with two phage particles each containing a different rIIB mutation.

No phage multiplication

Simultaneous infection of *E. coli* K12(λ) with two phage particles—one an rIIA, the other an rIIB mutation.

Normal phage multiplication. The progeny consist of a large majority of parental T4rIIA and T4rIIB mutants plus a small fraction of wild- and T4rIIArIIB genotypes, which arise by crossing over.

that a chromosomal region thought to contain only one gene actually produced several gene products and so must contain a corresponding number of genes. Until recently, it was more difficult to carry out complementation tests with normally haploid cells like *E. coli*. Laborious genetic tricks, too complicated to be described here, had to be employed to construct special strains with some chromosome sections present twice (partially diploid strains).

Now, recombinant DNA techniques greatly facilitate the use of complementation tests. The essential trick is to place one of the two mutant genes on a plasmid that can be introduced through the Ca^{2+} technique into a recipient strain carrying the second mutant gene (see Chapter 7). Since it is arranged that the plasmid also carry an antibiotic resistance gene, cells that have taken up DNA can be selected for by growing them in the presence of that antibiotic to ensure that the recipient cell carries both mutant genes. If the wild-type phenotype is restored, then the two mutations must have occurred in two different genes, each producing different polypeptide products (Figure 8-5).

Genetic Control of Protein Function[3, 4, 5]

It was not until the 1940s, when a variety of growth factor mutants became available, first in *Neurospora* and then in *E. coli*, that the idea that genes control the synthesis of single proteins (enzymes) became generally adopted. One of the first proofs involved the biosynthesis of the amino acid arginine. Its pathway starts with glutamic acid and proceeds via eight chemical reactions, each catalyzed by a distinct

Figure 8-5
A complementation test using recombinant DNA techniques. A plasmid carrying a mutant gene affecting lactose utilization (*lac⁻*) and an ampicillin resistance gene is introduced via the calcium technique into a bacterium also carrying a *lac⁻* mutation. The bacterium is then plated on an agar plate containing an indicator dye to distinguish between the two cell phenotypes Lac⁻ and Lac⁺ (see Figure 7-8). Since the medium also contains ampicillin, only cells that have taken up the plasmid DNA grow. If the wild-type phenotype (Lac⁺) is restored, then the plasmid and chromosomal mutations must be located in two different *lac* genes.

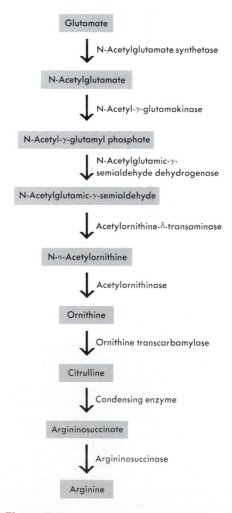

Figure 8-6
Pathway of arginine biosynthesis in *E. coli.*

enzyme (Figure 8-6). For each of these eight steps, mutants have been isolated that fail to carry out that specific enzymatic reaction. This suggests that a separate gene controls the presence of each enzyme, a hypothesis confirmed by the absence of the respective enzyme activity in cell extracts prepared from such mutant cells.

The biosynthesis of histidine provides another beautiful example of the direct relationship between genes and enzymes (Figure 8-7). Specific bacterial mutations result in the absence of each of the ten enzymes necessary for histidine biosynthesis. In this case, all the genes are tightly clustered together in one operon (set of functionally related genes; Figure 8-8), which is transcribed into one very long mRNA molecule. At first it was thought that the order of the genes in such operons correspond to the order in which their respective enzymes act in biosynthesis. However, the first two enzymes in histidine biosynthesis were found to be located at opposite ends of the cluster. When we discuss operons in detail in Chapter 16, we shall learn that those enzymes at the beginning of operons often tend to be

Figure 8-7
The pathway for histidine biosynthesis in *Salmonella typhimurium*. This bacterium, which is closely related to *E. coli*, appears to have a chromosome with a similar gene arrangement.

made in larger amounts than those present at the termini. Thus, the order of genes may have evolved to reflect the relative activities (turnover rates) of the enzymes.

One Gene–One Polypeptide Chain

Initially, the preceding ideas on gene-protein relationships were summarized by the slogan one gene–one protein (enzyme). Now we realize that this concept is more correctly stated as **one gene–one polypeptide.** A number of important proteins are composed of several polypeptide chains, each of which is the product of a separate gene. As later discussed, the *E. coli* enzyme tryptophan synthetase is composed of two distinct chains, each made from a separate gene. Likewise, *E. coli* RNA polymerase has separate genes for each of the four main polypeptides (α, β, β', and σ) used to construct it. However, the existence of several distinct polypeptide chains within functional proteins does not necessarily imply that separate genes code for each of their chains. Many such multichained proteins arise through proteolytic cleavage of single polypeptide chains. The A and B chains of insulin, for example, arise through two cuts in their single-chained proinsulin precursor (see Figure 2-19). Therefore, we now see that the complementation test tells us not whether two genes code for different proteins, but whether they control two different polypeptide chains.

Figure 8-8
Clustering of the genes involved in the biosynthesis of histidine by the bacterium *Salmonella typhimurium*. The gene order was determined by transduction experiments. Each gene is responsible for the synthesis of one of the ten enzymes needed to transform phosphoribosyl pyrophosphate into histidine. Here the genes are designated by numbers that correspond to the enzyme numbers of Figure 8-7. Enzyme 1 is responsible for catalyzing the first reaction in the biosynthesis, enzyme 2 for the second step, and so on. The names of these enzymes are given in Figure 8-7.

Identifying the Protein Products of Genes[6]

Even now, it is sometimes difficult to find the protein product of a given gene. Just a few years ago, this was often a particularly de-

(a)

(b)

(c)

Figure 8-9
Almost all the polypeptides of *E. coli* K-12 can be resolved as radioactive spots on two-dimensional O'Farrell focusing gels. (a) The gel from an extract of wild-type cells. The box encloses the *groEL* product; the circle encloses the *groES* gene (very faint). These gene products are essential for the morphogenesis of coliphages, and for the growth of the uninfected cell. (b) Gel from a mixture of wild-type extract and extract from cells with a missense mutation in *groEL*. The arrows point to a double spot; the left, lowermost being the product of the normal gene, and the one displaced upward and to the right being the product of the mutated gene. (c) Gel from wild-type cells harboring a multicopy plasmid (a derivative of pBR322) carrying the *groESL* operon. Both the *groEL* (box) and the *groES* (circle) protein products are overproduced as a result of the high gene dosage. By such means, the products of even essential genes (for which nonsense or deletion mutations are lethal) can be tentatively identified. (After Dr. Frederick C. Neidhardt.)

manding objective if the protein played only a minor structural role or possessed an enzymatic activity that had not yet been discovered. Now, with two-dimensional gel electrophoretic separation techniques, spotting the right protein is much easier. This method works particularly well for those genes whose mutated forms either fail to produce any product, yield a truncated protein product, or code for amino acid differences that change the electric charge of the protein product, because these mutants form spots that differ in location or are entirely absent from the gel, and thus the change is immediately apparent (Figure 8-9). Fortunately, a large fraction of the mutant genes that we isolate produce radically altered products that can easily be spotted on gels. This is not a matter of chance, since the best mutants to analyze are usually those with totally nonfunctional gene products. Those mutant proteins that differ from their wild-type equivalents only by a single amino acid replacement often retain par-

Figure 8-10
Last steps in the pathway of trypto-phan biosynthesis.

tial biological activity (they are said to be leaky) and thus are not as easy to work with as those that produce either no product or a drasti-cally rearranged product.

For a long time, the identity of the two proteins coded by the *rIIA* and *rIIB* genes of phage T4 remained unknown. Now they are known to be membrane proteins of molecular weights 86,000 (*rIIA*) and 30,000 (*rIIB*), both present in minor amounts. Unfortunately, still un-clear are their exact metabolic functions within T4-infected cells. Thus, for many years there has been a strong tendency to restrict intensive genetic analysis to those genes whose protein products are easy to isolate and whose metabolic or structural roles are well estab-lished.

Recessive Genes Frequently Do Not Produce Functional Products

Most mutant genes are recessive with respect to wild-type genes. This fact, puzzling to early geneticists, is now partially understood in terms of the gene-protein relationship. The recessive phenotype often results from the failure of mutant genes to produce any functional protein (enzyme). In heterozygotes, however, there is often present one "good" gene and, correspondingly, a number of "good" gene products. Because the wild-type gene is present in only one copy in heterozygotes, it is possible that there are always fewer good copies of the relevant protein in heterozygotes than in individuals with two wild-type genes. If this were the case, we might guess that the hetero-zygous phenotype would tend to be intermediate between the two homozygous phenotypes. Usually, however, this does not happen for one of two reasons. Either there are still enough good enzyme molecules to catalyze the metabolic reaction of concern, even though the total number of molecules is reduced, or the recessive gene is not noticeable because control mechanisms cause the single wild-type gene in a heterozygote to produce more gene product than does each wild-type gene in a homozygote. In Chapter 16, we shall discuss how the rates at which bacterial genes act are controlled.

Colinearity of the Gene and Its Polypeptide Product[7]

The best-understood example of the relationship between the order of the mutable sites in a gene and the order of their corresponding amino acid replacements involves the *E. coli* enzyme tryptophan syn-thetase, one of the several enzymes involved in tryptophan synthesis (Figure 8-10). This enzyme consists of two easily separated polypep-tide chains, A and B, neither of which is enzymatically active by itself. A large number of mutants unable to synthesize tryptophan lack a functional A chain in their tryptophan synthetase molecules. When these mutants were genetically analyzed, it was found that changes at a large number of different mutable sites could give rise to inactive A chains. Accurate mapping of these mutants revealed that they all could be unambiguously located on the linear genetic map shown in Figure 8-11. It was possible to isolate the inactive A chains from many of these mutants and to begin to compare their amino acid sequences with the sequence of the wild-type A chain, which contains 267

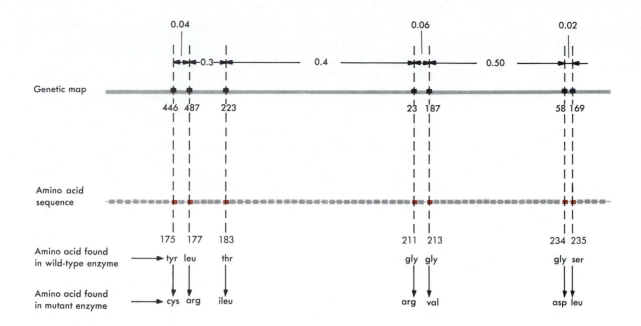

amino acids. This sequence allows us to see how the location of a mutation within a gene is correlated with the location of the replaced amino acid in its polypeptide chain product. Since both genes and polypeptide chains are linear, the simplest hypothesis is that amino acid replacements are in the same relative order as the mutationally altered sites in the corresponding mutant genes. This was most pleasingly demonstrated in 1964. The location of each specific amino acid replacement is exactly correlated with its location along the genetic map, a property called **colinearity.** Thus, successive amino acids in a polypeptide chain are controlled, or coded, by successive regions of a gene.

Mutable Sites Are the Base Pairs Along the Double Helix

In all bacterial genes extensively mapped, the large number of linearly arranged mutable sites that have been found in each gene, and between which genetic recombination (crossing over) is possible, leaves us no choice but to conclude that these sites are the specific base pairs along the DNA of the respective gene (Figure 8-12). A given mutable site can thus exist in any of four different states, AT, TA, GC, or CG. Many mutations are therefore likely to represent simple switches from one state to another. The genetic data that reveal deletions and insertions of genetic material must now be thought of in terms of the addition or deletion of discrete blocks of one to very many base pairs. The three classes of mutations resulting from changes in the sequence of nucleotide bases are illustrated in Figure 8-13.

By carefully studying the fine details of genetic maps, we should be able to obtain important information about the corresponding DNA. However, not every change in base sequence leads to easily observed changes in the corresponding protein. In the genetic code, many amino acids are specified by more than one **codon** (set of three adja-

Figure 8-11
Colinearity of the gene and its protein product. Here is the genetic map for one-fourth of the gene coding for the amino acid sequences in the *E. coli* protein tryptophan synthetase A. The designation 0.04, for example, refers to map distances (frequencies of recombination) between tryptophan synthetase mutations A446 and A487. The numbers in the amino acid sequence refer to their position in the 267 residues of the A protein. Following convention, the amino terminal end of the segment is on the left.

Figure 8-12

The relationship of mutations in the *rII* region of the phage T4 chromosome to the structure of DNA.

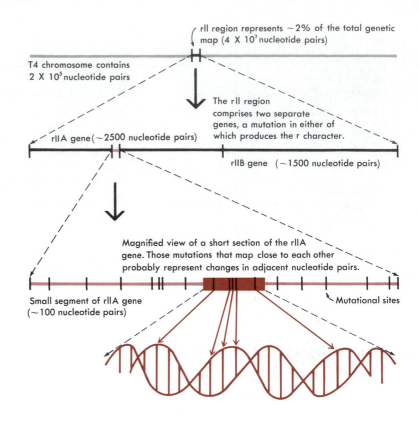

rII region represents ~2% of the total genetic map (4 X 10³ nucleotide pairs)

T4 chromosome contains 2 X 10⁵ nucleotide pairs

The rII region comprises two separate genes, a mutation in either of which produces the r character.

rIIA gene (~2500 nucleotide pairs)

rIIB gene (~1500 nucleotide pairs)

Magnified view of a short section of the rIIA gene. Those mutations that map close to each other probably represent changes in adjacent nucleotide pairs.

Small segment of rIIA gene (~100 nucleotide pairs)

Mutational sites

cent bases), which means that in many cases, base-pair substitutions will not lead to any amino acid replacements. Moreover, as we document later, many of the amino acids in proteins are not essential, and when they are replaced by somewhat similar amino acids, the proteins often retain full activity. The number of observed mutable sites therefore seriously underrepresents the number of base pairs within the corresponding gene.

There Are Four Alternative Structures for Each Mutable Site[8,9]

As anticipated, enzymatically inactive tryptophan synthetase molecules resulting from independent mutations at the same mutable site (as shown by failure to give wild-type recombinants) do not always contain the same amino acid replacement. For example, changes in a single mutable site that specifies the amino acid at position 213 results in the replacement of glycine by either glutamic acid or valine. Inspection of the genetic code (see Chapter 15) indicates that in the wild-type strain, this glycine must be specified by either GGA or GGG codons and that the mutable site under study specifies the G in the middle position of this codon. When this G is replaced by U, valine (GUA or GUG) becomes inserted into the glycine site while its replacement by A generates the glutamic acid (GAA or GAG) substitution. Further study of this particular mutable site might eventually turn up the anticipated third replacement in which a G to C switch leads to the appearance of alanine (GCA or GCG).

Figure 8-13
Three classes of mutations result from introducing defects in the sequence of bases (A, T, G, C) attached to the backbone of the DNA molecule. In one class, a base pair is simply changed from one into another (i.e., GC to AT). In the second class, a base pair is inserted (or deleted). In the third class, a block of base pairs is deleted (or inserted).

Single Amino Acids Are Specified by Several Adjacent Nucleotide Bases

We expected to find that given amino acids within a particular protein are specified by adjacent mutable sites. This point was first demonstrated in the tryptophan synthetase A gene, where the relevant evidence came from study of the tryptophan synthetase fragment illustrated in Figure 8-14. Treatment of the wild-type strain with a mutagen had given rise to mutant A23, in which arginine replaces glycine (this time at position 211), and mutant A46, in which glutamic acid replaces glycine at the same position. The difference between A23 and A46 does not represent changes to alternative forms of the same mutable site, since a genetic cross between A23 and A46 yields a number of wild-type recombinants (glycine in position 211). If these changes were at the same mutable site, no wild-type recombinants would be produced. Moreover, the very low observed frequency of the wild-type recombinants is compatible with the prediction from the genetic code that these mutable sites are adjacent to each other.

Additional genetic evidence that confirms the separate locations of the A23 and A46 mutable sites comes from observing how A23 and A46 themselves mutate upon treatment with mutagens. After exposure to a mutagen, both strains give rise to new strains, some of which contain active tryptophan synthetase A chains with glycine in position 211. These reverse mutations most likely involve changing the altered mutable sites back to the original wild-type configuration. However, strains containing active tryptophan synthetase also arise

Figure 8-14
Demonstration that a single amino acid is specified by more than one mutable site. We now know that the mutable sites are DNA bases and the codons are actually bases complementary to these in mRNA. (After Emanual J. Murgola.)

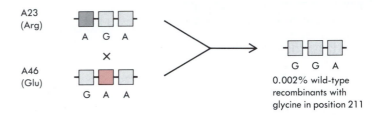

in which the amino acid in position 211 is replaced by another amino acid. Most significantly, the type of replacement differs for strains A23 and A46. Besides back-mutating to glycine, strain A23 mutates to threonine and serine, whereas A46 mutates to alanine and valine in addition to glycine. The failure of A23 ever to give rise to alanine or valine and the failure of A46 ever to mutate to threonine or serine is very difficult to explain if their differences from wild type are based on alternative configurations of the same mutable site. But these mutational patterns make perfect sense if glycine at the 211 position is coded by GGA with the A23 mutation to arginine representing a G to A change at the first position of the codon to give rise to AGA and the A46 mutation to glutamic acid occurring at the middle (second) position to give rise to GAA. Their divergent subsequent mutations to serine and threonine and to alanine and valine, respectively, can also be understood by inspecting the genetic code (Figure 8-15).

Single Amino Acid Substitutions Usually Do Not Alter Enzyme Activity

The ability of a polypeptide chain to be enzymatically active does not require an exactly specified amino acid sequence. This is shown by examination of the new mutant strains obtained by treating strains A23 and A46 with mutagens. The possession of either glycine or serine in position 211 yields a fully active enzyme, whereas threonine in

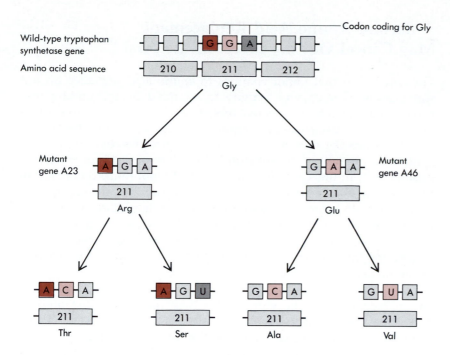

Figure 8-15
Formation of mutants A23 and A46 and their subsequent mutations. Notice that Thr and Ser cannot result from a single base change to the codon for Glu; likewise, Ala and Val cannot result from only one base change to the codon for Arg. Therefore, the A23 and A46 mutants must occur from mutations at two different mutable sites, as shown in Figure 8-14.

the same position yields an enzyme with reduced activity, demonstrating that the activity of an enzyme does not demand a perfectly unique amino acid sequence (Figure 8-16). In fact, evidence now indicates that amino acid replacements in many parts of a polypeptide chain can occur without seriously modifying catalytic activity. However, one sequence may often be best suited to a cell's particular needs, and it is this sequence that is encoded by the wild-type allele. Even though other sequences are almost as good, they will tend to be selected against in evolution.

Figure 8-16
Evidence that many amino acid replacements do not result in loss of enzymatic activity.

A Second Amino Acid Replacement
May Cancel Out the Effect of the First[10]

The conclusion that minor changes to amino acid sequence do not significantly alter enzyme activity is extended by the finding that some mutations that convert inactive mutant enzymes to active forms may work by causing a second amino acid replacement in the mutant enzyme. Consider mutant A46, which produces inactive tryptophan synthetase because of the substitution of glutamic acid for glycine at position 211. In this case, distant second-site mutations that result in the active enzyme occasionally emerge. For example, the second-site mutation A446 is located one-tenth of a gene length away from the first mutation. The double mutant A46A446 produces active enzyme molecules containing two amino acid replacements: the original glycine-to-glutamic acid shift and a tyrosine-to-cysteine shift located 36 amino acids away (Figure 8-17).

The second shift can be studied independently of the first by obtaining recombinant cells with only the A446 mutation. Most interestingly the A446 change, when present alone, also results in an inactive enzyme. We thus see that a combination of two wrong amino acids can produce an enzyme with an active three-dimensional configuration. However, only occasionally do two wrong amino acids cancel out each other's faults. For example, double mutants containing A446 and A23, or A446 and A187, do not produce active enzyme. At this time, it does not seem wise to speculate on how the various amino acid residues are folded together in the three-dimensional configuration and why only some combinations are enzymatically active. This kind of analysis must await the establishment of the three-dimensional structure of tryptophan synthetase.

The Very Drastic Consequences of the
Insertion or Deletion of Single Base Pairs[11, 12]

Early on in the analysis of mutant proteins, it became clear that the vast majority of mutants being isolated did not yield the minimally altered proteins, bearing single amino acid replacements, that would arise through the change of one type of base pair into one of its three alternatives. Instead, most mutants represented changes that led to drastically altered gene products, often containing many fewer amino acids and with many of their amino acid sequences bearing no relationship to the wild-type polypeptide products. The nature of these mutants first became apparent through the proposal that such mutations usually represented either insertions or deletions of single nucleotide pairs. The drastic effect of these insertion or deletion events is a consequence of the fact that mRNA molecules are read in successive blocks of three nucleotides, called codons. AUG codons, which code for the methionine residues found at the amino terminal ends of newly synthesized polypeptide chains, are the signal for ribosomes to begin reading the mRNA molecule about to be translated into a protein. Since reading always begins at the appropriate AUG condon, the mRNA molecules are aligned on the ribosomes so that their messages are read in the correct **reading frame.**

If, however, a single base pair is inserted or deleted in a coding sequence, the triplets that designate amino acids become completely changed beginning at the site of insertion or deletion (Figure 8-18).

Figure 8-17
Reversal (suppression) of mutant phenotype by a second mutation at a second site in the same gene.

For example, if normally the gene sequence ATTAGACAC . . . is read as (ATT)(AGA)(CAC) . . . , then the insertion of a new nucleotide C in the fourth position of that sequence creates ATTCAGACAC, which is read as (ATT)(CAG)(ACA)(C . . .). These new triplets may code for entirely different amino acids. A similar consequence follows from a deletion. Moreover, the crossing of two deletion or two insertion mutants yields double mutants in which the reading frame is still misplaced.

Reversion of Insertion or Deletion Mutants

Active (or partially active) genes are regenerated by crossing over between an insertion and a nearby deletion. Such events restore the correct reading frame except in the short region between the mutations (see Figure 8-18). If the affected gene region is nonessential (e.g., the early section of the T4 *rIIB* gene), then the resulting protein product is fully functional. In other cases, the short segments of inappropriate amino acids are only mildly disadvantageous, and partial activity results. No activity, however, will usually be found if the inappropriate codons include any of the three that signify chain termination (UAA, UAG, or UGA). Their presence inevitably results in incomplete fragments of the wild-type polypeptide.

It is also sometimes possible to obtain functional genes by producing recombinants containing three closely spaced insertions or deletions (Figure 8-19). In contrast, recombinants containing four nearby insertions or deletions produce only nonfunctional polypeptides. These later experiments were performed in 1961, before the basic outlines of the genetic code were known. They in fact provided the first good evidence that the genetic code was likely to be read in groups of three as opposed to groups of two or four.

Figure 8-18
Mutations that add or remove a base shift the reading frame of the genetic message.

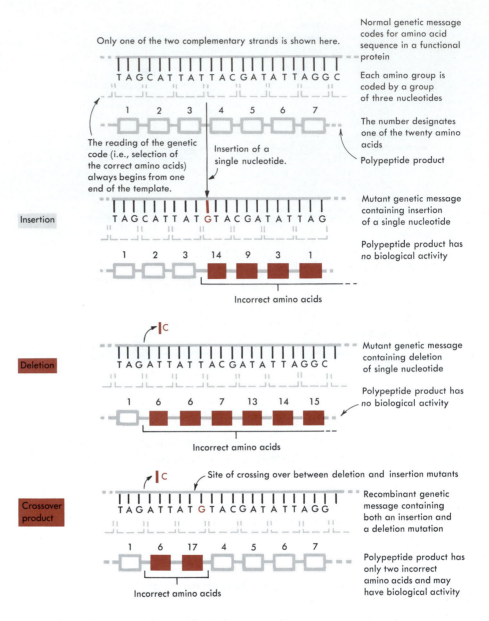

Only one of the two complementary strands is shown here.

Normal genetic message codes for amino acid sequence in a functional protein

Each amino group is coded by a group of three nucleotides

The number designates one of the twenty amino acids

The reading of the genetic code (i.e., selection of the correct amino acids) always begins from one end of the template.

Polypeptide product

Insertion of a single nucleotide.

Insertion

Mutant genetic message containing insertion of a single nucleotide

Polypeptide product has no biological activity

Incorrect amino acids

Deletion

Mutant genetic message containing deletion of single nucleotide

Polypeptide product has no biological activity

Incorrect amino acids

Site of crossing over between deletion and insertion mutants

Crossover product

Recombinant genetic message containing both an insertion and a deletion mutation

Polypeptide product has only two incorrect amino acids and may have biological activity

Incorrect amino acids

Cloned Genes Can Be Sequenced[13–17]

Virtually all the essential features of the genetic code were deduced by 1966 from the coding properties of either enzymatically or chemically synthesized mRNA molecules and from the accumulated knowledge of genetic fine structure that we have just detailed. No real genes were directly analyzed, however, since at that time there were no procedures either to sequence DNA or to isolate desired genes. But with the arrival of recombinant DNA and of powerful methods for DNA sequencing, the nature of genetic research has dramatically changed. No longer are genetic crosses the prime vehicle for probing genes. The quickest and most direct way to proceed is now the cloning and sequencing of relevant genetic material. As indicated in the previous chapter, it is now a relatively straightforward matter to isolate any *E. coli* gene that codes for a function that can be selected for by one of the many enrichment procedures.

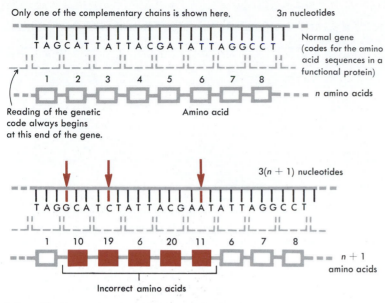

Only one of the complementary chains is shown here.

3n nucleotides

T A G C A T T A T T A C G A T A T T A G G C C T

Normal gene (codes for the amino acid sequences in a functional protein)

1 2 3 4 5 6 7 8

n amino acids

Reading of the genetic code always begins at this end of the gene.

Amino acid

3(n + 1) nucleotides

T A G G C A T C T A T T A C G A A T A T T A G G C C T

1 10 19 6 20 11 6 7 8

n + 1 amino acids

Incorrect amino acids

Polypeptide chain contains five incorrect amino acids; its chain length is increased by one amino acid. It may have some biological activity depending upon how the five wrong amino acids influence its 3-D structure.

Figure 8-19
When three nucleotides are added close together, the genetic message is scrambled only over a short region. The same type of result is achieved by the deletion of three nearby nucleotides.

Already, a large number of *E. coli* genes have been completely or partially sequenced. In all cases, the codons found to specify given amino acids are those predicted by the genetic code (Figure 8-20). This agreement between prediction and result, though inherently very satisfying, surprised no one, since the experimental evidence used to deduce the genetic code was effectively unassailable (see Chapter 15). Also as predicted, the coding segments of virtually all mRNAs start with the AUG codon and always conclude with a chain-terminating codon (UAA, UAG, or UGA).

Untranslated Sequences at the Beginnings and Ends of mRNA Molecules[18–23]

When mRNA was first discovered, it seemed simplest to assume that the translation events would begin at one end of the molecule and then move along in steps of three nucleotides until the other end was reached. This was a very naive view, adopted before the discoveries that methionine initiates all polypeptide chains and that specific codons specify chain termination. Now we realize that untranslated sequences exist at both the 5' end of the mRNA, near which translation begins, and at the 3' end, near which translation stops (Figure 8-21). Hence, there must be internal signals in mRNA that mark the starting and stopping sites for translation. With the exception of a small purine-rich block of nucleotides that functions to position ribosomes at the correct AUG start codon, the untranslated regions probably play no role in translation and are of variable lengths, ranging from 20 to more than 100 nucleotides, depending on the particular mRNA species.

These seemingly unnecessary extra sequences only make sense

Figure 8-20

The nucleotide sequence and corresponding amino acids of the *trpC* gene of the tryptophan operon of *E. coli*. Note that the codons found to specify given amino acids are those predicted by the genetic code. [Courtesy of Dr. Charles Yanofsky, *Nucleic Acid Res.* 9 (1981):6657.]

Figure 8-21

Untranslated sequences at the beginning (a) and the end (b) of the tryptophan operon. (a) The untranslated sequence (in color) at the beginning of the *trpE* gene of the operon showing the codon for methionine (boxed) signalling the beginning of translation. (b) The untranslated sequences (in color) at the end of the *trpA* gene at the end of the operon. The stop codon is boxed and the termination region is marked by an arrow. Other evidence indicates that termination may sometimes occur farther down the molecule. (After Terry Platt, *Nucl. Acids Res.* 9 (1981):6651–6653.

(a)

```
      1                    20                    40
AAGTTCACGTAAAAAGGGTATCGACA ATG AAAGCAATTTT
```

(b)

```
            6690               6710              ↓
CGTTGCGCT TAA TCCCACAGCCGCCAGTTCCGCTGGCGGCATTTTAA
```

when we consider how mRNA molecules are synthesized upon their DNA templates. Signals must be encoded into DNA that not only say to start RNA synthesis or to stop RNA synthesis but also that control the rates at which these processes occur. Those signal sequences that start RNA synthesis are called **promoters.** Their existence was first predicted through the discovery of mutants blocked in the initiation of transcription. Only a decade later, through DNA sequencing, was their fine structure revealed. In general, *E. coli* promoters are positioned just before the starting sites for transcription (at the 5' end) and so do not become transcribed into mRNA. Often situated between promoters and amino acid coding sequences are regulatory sequences called **operators,** to which specific DNA-binding proteins known as repressors can bind and thereby prevent RNA polymerase from attaching and initiating transcription (Figure 8-22). Operator sequences usually straddle RNA start sites and so when present have their 3'-terminal sections transcribed to form the beginning segments of the 5' untranslated "leader" segments. See Chapter 16 for details.

Equally complicated are the signals used to stop RNA synthesis, a process whose frequency also can be regulated. These stop signals

Repressed state of an operon

Figure 8-22
Repressors bind to operators to prevent transcription. When repressor molecules are not present, RNA polymerase can bind to DNA at the promoter site and initiate transcription. Shown here is an operon showing the relative locations of a promoter (P), operator (O), and structural gene (Z).

generally are located before the actual termination sites and when so sited are found as components of the 3' untranslated regions (see Chapter 13). The 3' untranslated regions may also be important in conferring stability on mRNAs. (The structure of mRNA is discussed in more detail in later chapters.)

Transcriptional Units Are the Fundamental Segments of Chromosomal Activity

Not all the DNA in bacteria (or in any other cells) codes for polypeptide chains. Some 1 percent of the *E. coli* chromosome codes for RNA chains that never become translated (e.g., tRNA and rRNA). The discrete DNA segments that code for these untranslatable RNAs are also regarded as genes, since they mutate and recombine like all other DNA regions. The designations rDNA and tDNA refer to those genes coding respectively for ribosomal and transfer RNA.

DNA molecules (chromosomes) should thus be functionally regarded as linear collections of discrete transcriptional units, each designed for the synthesis of a specific RNA molecule. Whether such "transcriptional units" should now be redefined as genes, or whether the term *gene* should be restricted to the smaller segments that directly code for individual mature rRNA or tRNA molecules or for individual polypeptide chains, is now an open question. This dilemma would not exist if all transcriptional units coded for either a single polypeptide chain or one mature RNA chain. But many, if not most, mRNA molecules in bacteria code for a number of functionally interrelated proteins (e.g., the several enzymes that make up a biosynthetic pathway). Likewise, most mature rRNA and tRNA molecules are synthesized as parts of larger RNA molecules that later become enzymatically cleaved into smaller functional units (see Chapter 14). Yet, the literature of genetics that has developed over the past three decades has defined genes in terms of their final polypeptide (structural RNA) products. Therefore, replacing the term *operon* with *gene* would probably cause more confusion than enlightenment at this time. Instead, it may actually make more sense to restrict the term *gene* to those DNA sequences that code for amino acids (and polypeptide chain termination) or that code for functional RNA chains, treating all transcriptional control regions as extragenic elements. DNA chromosomes, so defined, would be linear collections of genes interspersed with promoters, operators, and RNA chain-termination signals.

trpL	AAAGGGTATCGACA**ATG**AAA
trpE	ATTAGAGAATAACA**ATG**CAA
trpD	CAGGAGACTTTC**TGA**TGGCT
trpC	CACGAGGGTAAA**TGA**TGCAA
trpB	TAAGGAAAGGAACA**ATG**ACA
trpA	CGAGGGGAAATC**TGA**TGGAA

Figure 8-23
Short nucleotide segment between adjacent genes in the *trp* operon. Initiator and stop codons are boxed. [Courtesy of Dr. Terry Platt, *Nucl. Acids Res.* 9 (1981):6659]

This nontraditional definition is now the preferred way of viewing the bacterial chromosome, and we shall adopt this definition in our subsequent discussion of bacterial and phage genes. However, when the genes of eucaryotic organisms are later described, we shall see that they, too, present serious obstacles to being defined in terms of single gene products. Here again, logic may tell us that if we wish to continue to designate the gene as the unit of chromosomal function, then it should refer to those DNA segments that become transcribed into discrete RNA chains rather than to those DNA segments that provide the genetic code for given polypeptide chains (or are directly transcribed into the sequences of mature rRNA or tRNA molecules).

Gaps Between Genes Can Be Very Short[24, 25]

Usually, there is much less crossing over between mutations located at adjacent ends of two contiguous genes within a given bacterial operon (transcriptional unit) than between two mutations at extreme ends of even relatively short genes. This suggested that adjacent genes within operons are often separated by only a few base pairs (Figure 8-23). Confirmation of this point has come with the advent of DNA-sequencing procedures. They reveal, for example, that in the tryptophan operon, gaps of at most several nucleotides exist between adjacent genes. Also being accumulated are data on the distances in base pairs between different transcriptional units, which also are not large in bacteria. This makes biological sense, since such spacer DNA would be carrying out no function and would increase the time needed to replicate the bacterial chromosome containing it.

There Is Agreement Between the Genetic Map and the Corresponding Distance Along a DNA Molecule[26]

Ever since crossing-over analysis began, geneticists have wondered how closely their genetic maps correspond to actual physical lengths along chromosomes. The visualization, in the 1930s, of *Drosophila*'s highly extended salivary chromosomes first permitted the physical mapping of the sites of mutations. Good correlations between the physical and genetic distances along some chromosomal stretches were shown. But these views of chromosomes in the light microscope told us nothing about what was happening within the constituent DNA molecules. Firm answers at the molecular level came only when it became possible to locate the sites of specific mutations along genetically well-defined DNA molecules.

The first high-resolution data emerged through the mapping of well-defined deletions of large numbers of nucleotides from phage λ DNA. This initially was achieved using the technique of DNA-DNA hybridization. In this procedure, the two strands of the λ DNA double helix are separated by heating the DNA to near 100°C. At that temperature, the DNA will denature; virtually all the hydrogen bonds holding the double helix together break, and the resulting free single strands unwind from each other. Because the + and − strands have sufficiently different base compositions, they can be separated by density differences in a cesium chloride (CsCl) solution. Reformation

Figure 8-24
Mapping deletions by electron microscope visualization of renatured DNA molecules.

of the double helices from separated single strands (renaturation) occurs if the single strands are mixed together, heated to near 100°C, and then gradually cooled.

If, however, renaturation occurs between a normal + strand and a − strand containing a deletion, the section of the + strand complementary to the deleted − section cannot form hydrogen bonds. It exists as a single-stranded loop extended out from the predominantly double-stranded molecule (Figure 8-24). These loops are easily detected in the electron microscope (Figure 8-25), allowing different deletions to be localized precisely along the λ DNA molecule. Such data, taken together with the extensively studied λ genetic map, give the picture shown in Figure 8-26. It reveals a much closer correspondence between the genetic map and the actual physical structure of the DNA than had been anticipated. Thus, at medium resolution, the probability of crossing over is approximately equal throughout the DNA molecule. Exceptions to the correspondence between the genetic and physical maps are obvious in Figure 8-26, and caution must be taken not to overinterpret genetic linkage data.

Figure 8-25
(a) An electron micrograph (left) of a heteroduplex formed between one strand of wild-type λ and the complementary strand of λb2b5, and (b) an interpretive drawing. λb2b5 contains a deletion (b2) and a nonhomologous section (b5) unable to pair with its λ counterpart. [Reproduced with permission from B. C. Westmoreland, W. Szybalski, and H. Ris, *Science*, 163, 1343 (1969).]

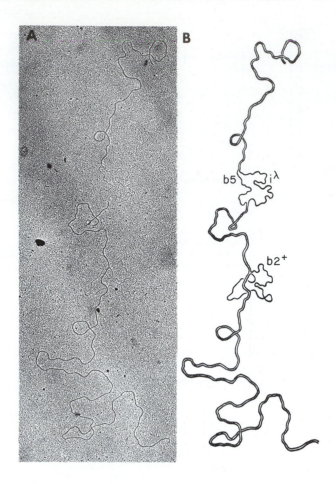

The Eventual Sequencing of the Entire *E. coli* Chromosome[27]

Major new insights into the genetic organization of *E. coli* and of bacteria in general may now have to await the sequencing of sufficiently long chromosomal segments that will reveal the actual density of coding sequences along the DNA. Preferably, this will be initiated on segments containing many genes whose polypeptide products are known and whose amino acid sequences can be compared with DNA sequences so obtained. It should not, however, be necessary to have prior knowledge of a gene to spot its presence along a DNA sequence. Its existence should be revealed by the occurrence of an extensive open reading frame uninterrupted by the presence of chain-terminating codons. The three chain-terminating codons (UAA, UAG, and UGA) occur by chance once in approximately every 20 codons, and inspection of known genes invariably reveals that the two unused reading frames are filled with chain-terminating codons (Figure 8-27). Computers are now routinely employed to search out long stretches of open reading frame that start with AUG codons preceded by ribosome binding site sequences. These long, open reading frames (ORFs) very likely are coding sequences for still undiscovered polypeptides.

At this time, we cannot predict when the complete *E. coli* sequence of some 4×10^6 base pairs will be determined. Since this task is

Genetic Map obtained from
crossing-over experiments

Physical Map showing absolute location of genes

clearly too big for any average-size laboratory, the *E. coli* sequence is emerging in small bits of ever-increasing size. Still unforeseen improvements in cloning and sequencing procedures may have to occur before one or more individuals decide that the rewards of obtaining and collecting such massive sequence data are commensurate with the tedium that would have to be endured with such an undertaking. Hopefully, this moment will not be too far in the future. The way we view *E. coli* is undoubtedly affected by our knowledge of the number of polypeptides encoded by its DNA. If we were to find that it has the potential to make 5,000 different polypeptides, it is bound to appear an infinitely harder objective to characterize completely than if it could synthesize only some 2,000 different polypeptides. We already know the approximate functions carried out by some 1000 different *E. coli* genes.

Figure 8-26

Comparison of the λ (recombination) map with the true physical map. [Recombination map after P. Amati and M. Meselson, *Genetics* 51 (1965):369–379; physical map after a map prepared by W. Szybalski, University of Wisconsin.]

Figure 8-27

Unused reading frames of the *trpC* gene in the *trp* operon of *E. coli* showing many chain terminating codons (boxed). (After Dr. Terry Platt.)

```
TRP C
       3330             3340             3350             3360             3370             3380             3390             3400             3410
ATG ATG CAA ACC GTT TTA GCG AAA ATC GTC GCA GAC AAG GCG ATT TGG GTA GAA GCC CGC AAA CAG CAG CAA CCG CTG GCC AGT TTT CAG
MET GLN THR VAL LEU ALA LYS ILE VAL ALA ASP LYS ALA ILE TRP VAL GLU ALA ARG LYS GLN GLN GLN PRO LEU ALA SER PHE GLN
 1                            10                                           20
       3420             3430             3440             3450             3460             3470             3480             3490             3500
GAG GTT CAG CCG AGC ACG CGA CAT TTT TAT GAT GCG CTA CAG GGT GCG CGC ACG GCG TTT ATT CTG GAG TGC AAG AAA GCG TCG CCG TCA
GLU VAL GLN PRO SER THR ARG HIS PHE TYR ASP ALA LEU GLN GLY ALA ARG THR ALA PHE ILE LEU GLU CYS LYS LYS ALA SER PRO SER
30                                           40                                           50
       3510             3520             3530             3540             3550             3560             3570             3580             3590
AAA GGC GTG ATC CGT GAT GAT TTC GAT CCA GCA CGC ATT GCC GCC ATT TAT AAA CAT TAC GCT TCG GCA ATT TCG GTG CTG ACT GAT GAG
LYS GLY VAL ILE ARG ASP ASP PHE ASP PRO ALA ARG ILE ALA ALA ILE TYR LYS HIS TYR ALA SER ALA ILE SER VAL LEU THR ASP GLU
60                                           70                                           80
```

Summary

Crossing over occurs within the gene, thereby making possible the mapping of mutations within it. Like the order of genes on a chromosome, the arrangement of mutable sites in a gene is strictly linear. Within average-size bacterial genes, there are many hundreds of sites in which changes lead to detectable mutations. Sometimes, it is initially difficult to determine whether a chromosomal region contains more than one gene. This uncertainty can usually be resolved by introducing into a given cell more than one chromosome (or chromosomal fragment) containing mutations within that region. If the two mutants can complement each other to restore function, their mutations must be in different genes (the complementation test).

Almost all bacterial genes (the exception being those DNA segments that code for rRNA and tRNA) are responsible for the synthesis of a specific polypeptide chain and contain the information that specifies the sequence of its amino acids. Genes and their polypeptide products are colinear, with successive regions along a gene controlling the successive amino acids along its polypeptide chain product. The discovery of the double helix led to the postulation that mutable sites are the base pairs along DNA molecules. There are four alternative conformations (AT, TA, GC, and CG) for each mutable site, with genetic analysis of well-defined polypeptides showing that each amino acid is coded by adjacent groups of base pairs. Genetic experiments exploring the consequence of mutations that insert or delete individual base pairs found that groups of three base pairs code for each amino acid. Insertions or deletions of base pairs often have more drastic consequences than simple base-pair replacement; they frequently lead to the utilization of the wrong group of codons (reading frame) on a given mRNA template.

Such genetic inferences have now been completely confirmed by the sequence analysis of genes whose polypeptide products already have been sequenced. In all cases, the amino acid coding segments inferred from knowledge of the genetic code were completely confirmed. DNA sequencing provided important new insights by revealing the existence of untranslated base pairs at the beginnings and ends of all mRNA molecules. Gaps between bacterial genes are often very short. The vast majority of the bacterial chromosomes are now thought to be composed of base pairs that code for amino acids. Exactly how many genes are present in the *E. coli* chromosome awaits complete sequencing (4×10^6 base pairs). Until recently, this task seemed beyond the capability of molecular biology. Now there is reason to believe that this task will be completed by the mid-1990s.

Bibliography

General References

Glass, R. E. 1982. *Gene Function: E. coli and Its Heritable Elements.* Berkeley: University of California Press.

Hayes, W. 1968. *The Genetics of Bacteria and Their Viruses.* 2nd ed. New York: Wiley.

Inouye, M. 1983. *Experimental Manipulation of Gene Expression.* New York: Academic Press.

Lewin, B. M. 1974. *Gene Expression I.* New York: Wiley.

Lewin, B. M. 1985. *Genes.* 2nd ed. New York: Wiley.

Pontecorvo, G. 1958. *Trends in Genetic Analysis.* New York: Columbia University Press. A classical essay on the fine-structure analysis of the gene.

Stent, G. S., and R. Calendar. 1978. *Molecular Genetics: An Introductory Narrative.* 2nd ed. San Francisco: Freeman.

Cited References

1. Benzer, S. 1955. "Fine Structure of a Genetic Region in Bacteriophage." *Proc. Nat. Acad. Sci.* 41:344–354.
2. Singer, B. S., S. T. Shinedling, and L. Gold. 1983. "The rII Genes: A History and a Prospectus." In *Bacteriophage T4*, ed. C. K. Mathews, E. M. Kutter, G. Mosig, and P. B. Berget. Washington, D.C.: Amer. Soc. for Microbiol., pp. 327–333.
3. Ames, B. N., B. Garry, and L. A. Herzenberg. 1960. "The Genetic Control of the Enzymes of Histidine Biosynthesis in *Salmonella typhimurium.*" *J. Gen. Microbiol.* 22:369–378.
4. Maas, W. K. 1961. "Studies on Repression of Arginine Biosynthesis in *E. coli.*" *Cold Spring Harbor Symp. Quant. Biol.* 31:151–162.
5. Gorini, L., W. Gundersen, and M. Burger. 1961. "Genetics of Regulation of Enzyme Synthesis in the Arginine Biosynthetic Pathway of *E. coli.*" *Cold Spring Harbor Symp. Quant. Biol.* 26:173–182.
6. Pribnow, D., D. C. Sigurdson, L. Gold, B. S. Singer, C. Napoli, J. Brosius, T. J. Dull, and H. F. Noller. 1981. "rII Cistrons of Bacteriophage T4. DNA Sequence Around the Intercistronic Divide and Positions of Genetic Landmarks." *J. Mol. Biol.* 149:337–376.
7. Yanofsky, C., B. C. Carlton, J. R. Guest, D. R. Helinski, and U. Henning. 1964. "On the Colinearity of Gene Structure and Protein Structure." *Proc. Nat. Acad. Sci.* 51:226–272.
8. Yanofsky, C., J. Ito, and V. Horn. 1966. "Amino Acid Replacements and the Genetic Code." *Cold Spring Harbor Symp. Quant. Biol.* 31:151–162.
9. Helinski, D. R., and C. Yanofsky. 1962. "The Correspondence Between Genetic Data and the Position of Amino Acid Alteration in A Protein." *Proc. Nat. Acad. Sci.* 48:173–183.
10. Helinski, D. R., and C. Yanofsky. 1963. "A Genetic and Biochemical Analysis of Second Site Reversion." *J. Biol. Chem.* 238:1043–1048.
11. Crick, F. H. C., L. Barnett, S. Brenner, and J. Watts-Tobin. 1961. "General Nature of the Genetic Code for Proteins." *Nature* 192:1227–1232.
12. Streisinger, G., Y. Okada, J. Emrich, J. Newton, A. Tsugita, E. Terzaghi, and M. Inouye. 1966. "Frameshift Mutations and the Genetic Code." *Cold Spring Harbor Symp. Quant. Biol.* 31:77–84.
13. Yanofsky, C., T. Platt, I. P. Crawford, B. P. Nichols, G. E. Christie, M. VanCleemput, and A. M. Wu. 1981. "The Complete Nucleotide Sequence of the Tryptophan Operon of *E. coli.*" *Nucleic Acid Res.* 9:6647–6668.
14. Farabaugh, P. J. 1978. "Sequence of the *lac I* Gene." *Nature* 274:765–771.
15. Buchel, D. E., B. Gronneborne, and B. Muller-Hill. 1980. "Sequence of the Lactose Permease Gene." *Nature* 283:541–545.
16. Wallace, R. G., N. Lee, and A. V. Fowler. 1980. "The *araC* gene of *Escherichia coli*: Transcriptional and Translational Start Points and Complete Nucleotide Sequence." *Gene* 12:179–190.
17. Wurtzel, E. T., M.-Y. Chou, and M. Inouye. 1982. "Osmoregulation of Gene Expression. I. DNA Sequences of the *ompR* Gene of the *ompB* Operon of *E. coli* and Characterization of Its Gene Product." *J. Biol. Chem.* 257:13685–13691.
18. Bronson, M. J., C. Squires, and C. Yanofsky. 1973. "Nucleotide Sequences from Tryptophan Messenger RNA of *Escherichia coli*. The Sequences Corresponding to the Amino-Terminal Region of the First Polypeptide Specified by the Operon." *Proc. Nat. Acad. Sci.* 70:2335–2339.
19. Platt, T. 1978. "Regulation of Gene Expression in the Tryptophan Operon of *Escherichia coli.*" In *The Operon*, ed. J. H. Miller and W. S. Reznikoff. Cold Spring Harbor, N.Y.: Cold Spring Harbor Laboratory, pp. 263–302.
20. Guarente, L. P., D. H. Mitchell, and J. Beckwith. 1977. "Transcription Termination at the End of the Tryptophan Operon of *E. coli.*" *J. Mol. Biol.* 112:423–436.
21. Mitchell, D. H., W. S. Reznikoff, and J. Beckwith. 1976. "Genetic Transfusions That Help to Define a Transcription Termination Region in *E. coli.*" *J. Mol. Biol.* 101:441–457.
22. Beckwith, J. R., and D. Zipser. 1970. *The Lactose Operon.* Cold Spring Harbor, N.Y.: Cold Spring Harbor Laboratory.
23. Miller, J. G., K. Ippen, J. G. Scaife, and J. R. Beckwith. 1968. "The Promoter and Operator Region of the *lac* Operon of *E. coli.*" *J. Mol. Biol.* 38:413–420.
24. Platt, T., and C. Yanofsky. 1975. "An Intercistronic Region and Ribosome Binding Site in Bacterial Messenger RNA." *Proc. Nat. Acad. Sci.* 72:2399–2403.
25. Steitz, J. A. 1978. "Genetic Signals and Nucleotide Sequences in Messenger RNA." In *Biological Regulation and Control*, ed. R. Goldberger. New York: Plenum.
26. Westmoreland, B. C., W. Szybalski, and H. Ris. 1969. "Mapping of Deletions and Substitutions in Heteroduplex DNA Molecules of Bacteriophage Lambda by Electron Microscopy." *Science* 163:1343–1348.
27. Staden, R., and A. D. McLaughlin. 1982. "Codon Preference and Its Use in Identifying Protein Coding Regions in Long DNA Sequences." *Nucleic Acid Res.* 10:141–156.

IV

DNA IN DETAIL

CHAPTER 9

The Structures of DNA

The realization that DNA is the prime genetic molecule, carrying all the hereditary information within chromosomes, immediately focused attention on its structure. Only by understanding how its atoms are linked together by covalent bonds and then establishing their three-dimensional arrangements in space would we be able to learn how DNA carries the genetic messages that are replicated when chromosomes divide to produce two identical copies of themselves. Not surprisingly, there initially were fears that DNA might have very complicated and perhaps bizarre structures that differ radically from one gene to another. Great relief, if not general elation, was thus generated when the fundamental DNA structure was found to be the double helix. It told us that all genes have roughly the same three-dimensional form and that the differences between two genes reside in the order and number of their four nucleotide building blocks along their complementary strands.

Now, some 30 years after the discovery of the double helix, we realize that the DNA structure is not as simple as first thought. For example, the chromosomes of some small viruses have single-stranded, not double-helical, molecules. Moreover, some DNA sequences permit the double helix to twist in the left-handed sense, as opposed to the right-handed sense originally formulated for DNA's general structure. And while some DNA molecules are linear, others are circular. Still additional complexity comes from the supercoiling (further twisting) of the double helix, often around cores of DNA-binding proteins. The supercoiling can so distort the basic double helix that sections of it may untwist to form sections of single-stranded DNA.

These complexities were often first perceived as annoying deviations from perfection. By now, however, they have been so firmly documented that we realize there is no one structure of DNA and will henceforth refer to the "structures" of DNA. Clearly, it will only be through understanding why these "variant" DNAs exist that we will finally comprehend the myriad devices used by DNA to govern the life of the cell.

DNA Is Usually a Double Helix[1–5]

The most important feature of DNA is that it usually consists of two complementary polymeric chains twisted about each other in the form of a regular right-handed double helix (see Figure 3-7). Each chain is a polynucleotide, a polymeric collection of nucleotides in which the sugar of each nucleotide is linked by a phosphate group to the sugar of the adjacent nucleotide. There are four main nucleotides, each containing a deoxyribose residue, a phosphate group, and a purine or pyrimidine base (see Figure 3-2). There are two single-ringed pyrimidines, thymine (T) and cytosine (C), and two double-ringed purines, adenine (A) and guanine (G). In the polynucleotide chain, the sugar and phosphate groups are always linked together by the same chemical bonds (3'–5' phosphodiester linkages). Hence, this part of the molecule, called the sugar-phosphate backbone, is very regular. In contrast, the order of the purine and pyrimidine residues along a chain is highly irregular, varying from one molecule to another. Both the purine and pyrimidine bases are flat, relatively water-insoluble molecules that tend to stack above each other roughly perpendicular to the direction of the helical axis.

The Two Chains of the Double Helix Have Complementary Sequences

The two chains are joined together by hydrogen bonds between pairs of bases. Adenine is always paired with thymine and guanine with cytosine (see Figure 5-28). Only these arrangements are possible, for two purines would occupy too much space to allow a regular helix, and two pyrimidines would occupy too little space. The strictness of these pairing rules results in a complementary relation between the sequences of bases on the two intertwined chains. For example, if we have a sequence 5'-ATGTC-3' on one chain, the opposite chain must have the sequence 3'-TACAG-5'. A stereochemical consequence of the way the AT and GC base pairs come together is that the phosphodiester linkages of the two polynucleotide chains are oriented in opposite directions. Thus, if the double helix is inverted by 180°, it superficially looks the same (Figure 9-1).

The two glycosidic bonds that attach a base pair to its sugar rings do not lie directly opposite each other. As a result, the two sugar-phosphate backbones of the double helix are not equally spaced along the helical axis, and the grooves that form between the backbones are not of equal size (Figure 9-2). The floor of the major groove is filled with nitrogen and oxygen atoms that belong to the "top" of the base pairs (as drawn in the figure) and that project inward from their respective sugar-phosphate backbones. In contrast, the floor of the minor groove is filled with base-pair nitrogens and oxygens that project outward toward their backbones. The hydrogen-bonding potential of the major groove shows much greater dependence on base sequence than that of the minor groove. This finding led to the speculation that proteins attach to sequence-specific DNA segments by forming hydrogen bonds predominantly to specific groups positioned along the major groove.

Each Base Has Its Preferred Tautomeric Form[6, 7]

An important chemical feature of DNA is the position of the hydrogen atoms in the purine and pyrimidine bases. Before 1953, many chemists thought that some of these hydrogen atoms randomly moved from one ring nitrogen or oxygen atom to another and so

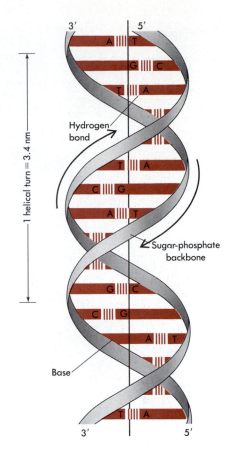

Figure 9-1
If inverted 180°, the double helix looks superficially the same, owing to the complementary nature of its component DNA strands. Turn the book upside down to demonstrate this.

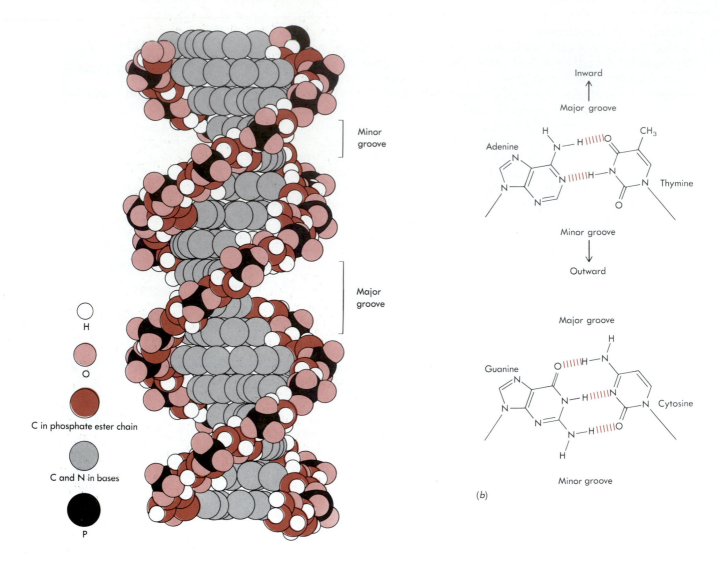

H

O

C in phosphate ester chain

C and N in bases

P

(a)

(b)

Figure 9-2

(a) The positions of the major and minor grooves of the double helix and (b) the relative directions in which the base-pair nitrogens and oxygens project in each groove. The nitrogens and oxygens of the floor of the major groove project toward the center of the DNA molecule, while those of the minor groove project toward the outer edge of the DNA molecule. (The structure in this figure is the form of DNA called B-DNA, discussed later.)

could not be assigned a fixed location. Now we realize that although such movements, called **tautomeric shifts,** do occur, these hydrogens have preferred atomic locations. The nitrogen atoms attached to the purine and pyrimidine rings are usually in the *amino* (NH_2) form and only rarely assume the *imino* (NH) configuration (Figure 9-3a). Likewise, the oxygen atoms attached to the C6 atoms of guanine and thymine normally have the *keto* (C–O) form and only rarely take up the *enol* (COH) configuration (Figure 9-3b).

It is essential to the biological functioning of DNA that the hydrogen atoms have relatively stable locations. If the hydrogen atoms had no fixed positions, adenine could often pair with cytosine and guanine with thymine, and the sequence of the bases on the two intertwined chains would not necessarily be complementary. If that were the case, then DNA could not function as a genetic molecule, for it is the complementary relationship between the opposing chain sequences that gives DNA its capacity for self-replication. As we shall discuss in great detail in the next chapter, replication of the double helix involves strand separation followed by formation of complementary DNA chains using the now free single strands as templates to attract the appropriate partner bases dictated by AT and GC base-pairing rules. The imino forms of adenine and cytosine and the enol forms of guanine and thymine must indeed occur only very rarely

(a) Amino   Imino (two rotational forms) Rare

←— Donor for H-bond
—→ Acceptor for H-bond

(b)   Keto   Enol (two rotational forms) Rare

compared to their amino and keto alternatives. Otherwise, the many errors (mutations) incurred during DNA replication would be incompatible with orderly cell growth and division.

DNA Renatures as Well as Denatures[8-11]

DNA molecules with a high GC content are more resistant to thermal melting than AT-rich molecules. When double-helical DNA molecules are heated above physiological temperatures (to near 100°C), their hydrogen bonds break and the complementary strands often

Figure 9-3
Amino ⇌ imino and keto ⇌ enol tautomerism. (a) The A and C nucleotide bases are usually in the amino form but rarely form the imino configuration. (b) The bases G and T are usually in the keto form but are rarely found in the enol configuration. Atoms shown in color may participate in hydrogen-bond formation.

Figure 9-4

Dependence of DNA denaturation on G + C content. The greater the G + C content, the higher the temperature must be to denature the DNA strand. DNA from different sources was dissolved in two different solutions at pH 7.0. The points represent the temperature at which the DNA denatured, graphed against the G + C content. [Data from J. Marmur and P. Doty, *J. Mol. Biol.* 5 (1962):120.]

separate from each other, a process called DNA denaturation. Because each GC base pair is held together by three hydrogen bonds rather than the two hydrogen bonds holding together each AT base pair, higher temperatures are necessary to separate GC-rich strands than to break apart AT-rich molecules (Figure 9-4). At intermediate temperatures or in the presence of intermediate amounts of destabilizing agents such as alkali or formamide, DNA molecules are partially denatured. In such molecules, those regions of the double helix that are relatively rich in AT base pairs melt apart, while regions rich in GC base pairs retain their double-helical character (see Figure 9-23).

Even complete denaturation is not necessarily an irreversible phenomenon. When heated solutions of totally denatured DNA are slowly cooled, single strands often meet their complementary strands and reform regular double helices. This ability to renature DNA molecules permits artificial hybrid DNA molecules to be formed by slowly cooling mixtures of denatured DNA from two different species. For example, hybrid molecules can be formed containing one strand from a man and one from a mouse. Only a fraction (25 percent) of the DNA from a man can form hybrids with mouse DNA. This is not surprising, since it merely means that some genes of man are very similar to those of a mouse, whereas others have quite different nucleotide sequences. Before the development of rapid DNA-sequencing methods, this molecular technique was widely employed in establishing the genetic similarity of the various taxonomic groups.

Many Very Small Viruses Have Single-Stranded DNA Chromosomes[12]

At first it was thought that all DNA molecules are double-stranded except during replication, when a small region around the replicating fork (see Figure 10-10) is temporarily in a non-hydrogen-bonded, single-stranded form. It therefore came as quite a surprise when experiments revealed that the DNA of several groups of small bacterial viruses exists as single-stranded molecules in which the amount of A is not equal to the amount of T and the amount of G does not equal the amount of C. Among these single-stranded phages are the spherically shaped ϕX174 and S13 viruses and the rod-shaped f1 and M13 viruses. The chromosomes of the parvoviruses, minute viruses that infect many vertebrate organisms, are also single-stranded DNA molecules.

When these single-stranded viral chromosomes enter their host cells, they serve as templates for the formation of complementary strands. The resulting double helices in turn serve as templates for new single strands that then become incorporated into new virus particles (Figure 9-5). Thus, the fundamental mechanism for ordering nucleotides during the synthesis of single-stranded DNA is basically the same as that used for double-helical DNA. Nucleotide selection always occurs by attraction of the complementary base. Single-stranded DNA replication usually differs from double-helical replication in that it uses only one of the two complementary strands (the $-$ strand) as a template for the progeny ($+$) strand.

Although single-stranded DNA can serve as the chromosome of a small virus, it would not be very effective in storing genetic information within the chromosomes of cells. In bacteria and in many eucaryotic cells (e.g., the haploid phase of yeast cells), there usually is only one copy of a given gene. If interphase chromosomes contained single-stranded DNA, with the double helix existing only briefly during mitosis, then the daughter cells would contain two completely different sets of genetic information. This follows from the fact that the two complementary chains do not have identical base sequences and hence would code for entirely different amino acid sequences. The double-strand ensures that each daughter cell will contain the same genetic information. The double-stranded form also permits effective DNA repair mechanisms to exist. For example, when one strand is damaged by exposure to X-rays, the remaining strand can provide the genetic information to rebuild the damaged section. It is probably no accident that only very small viral chromosomes are single-stranded. If they were to become, say, the size of T2, they would present too large a target for chain-damaging events.

Single-Stranded DNA Has a Compact Structure[13, 14]

Single-stranded DNA molecules have a strong tendency to fold back on themselves to form irregular double-helical hairpin loops whenever their sequences permit significant numbers of nucleotides to base-pair (Figure 9-6). Imperfect as these loops are, they nonetheless are energetically favored under physiological salt conditions, since they allow much more effective stacking of the flat hydrophobic surfaces of the bases than is possible in any fully extended single-stranded structure. Often, as many as half the bases of single-stranded DNA are so hydrogen-bonded, and their molecules are highly compacted. If, however, there are only minimal numbers of neutralizing cations (e.g., less than or equal to 0.01 M Na$^+$), then the electrostatic repulsion between the partially unneutralized phosphates will keep the single-stranded chains highly extended, and there will be fewer hairpin loops.

Figure 9-5
(a) The replication of a single-stranded DNA molecule. Each double helix generally serves as the template for the formation of a large number of new + strands. (b) Electron micrograph of a single-stranded (upper) and double stranded (lower) circular DNA molecule from phage φX174. (Courtesy of David Dressler and Kirston Koths, Harvard Medical School.)

Single-stranded DNA is always denser than double-helical DNA. Much more complete hydrogen bonding to water is possible for the atoms of the highly regular double helix than for the atoms of the inherently irregularly shaped single-stranded DNA. In the absence of base-pairing hydrogen bonds, van der Waal's interactions take over and bring the atoms of DNA closer together than would be the case if they formed regular hydrogen bonds with each other. An analogy can be made with the behavior of water, which becomes less dense as it forms the regular, hydrogen-bonded lattice of crystalline ice.

Rigorous Crystallographic Proof of the Double Helix[15, 16, 17]

X-ray diffraction patterns from parallel-oriented fibers of DNA provided the first clues that the polynucleotide chains of DNA have helical conformations. They also told us that DNA was a multichained molecule composed of two or three polynucleotide chains. However, the double-helical nature of DNA did not directly emerge from X-ray diffraction analysis alone. Equally important was the use of model building in which the question was asked, Given the chemical features of a single DNA chain, into what three-dimensional helical conformations could it fold? The simplest models were those in which the sugar-phosphate backbones were on the outside with the bases stacking in the center. The double helix emerged when it was realized that thymine and guanine had keto, not enol, configurations and that by base-pairing adenine to thymine and guanine to cytosine, a stereochemically pleasing, regular helical molecule resulted (i.e., the AT base pairs had the same "shape" as the GC base pairs). Immediately, a structural explanation was provided for the equivalence of the num-

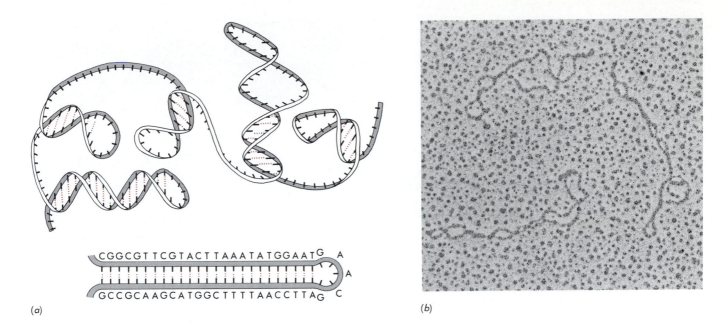

CGGCGTTCGTACTTAAATATGGAATG A
GCCGCAAGCATGGCTTTTAACCTTAG C

(a)

(b)

bers of adenine and thymine residues and of guanine and cytosine residues in most DNA molecules. Since the complementary relationship between the sequences of bases on the two opposing chains also suggested an obvious replication mechanism for DNA, there seemed every reason to assume that the double helix was the correct proposal.

Rigorous proof, however, could only come using X-ray diffraction procedures, and yet for many years it was only possible to say that the experimental data were compatible with the base-paired right-handed double helix derived by model building. The diffraction patterns obtainable from imprecisely oriented DNA fibers formed from DNA molecules of highly heterogeneous sequences could not provide sufficient information to locate individual atoms and thus establish definitive molecular structures like those obtained for protein crystals (e.g., hemoglobin). So while no molecular biologist seriously doubted that DNA was a base-paired double helix, for a long time there was no way to confirm that it was right-handed or that the base pairs were in fact held together by the originally proposed hydrogen bonds.

Solid proof came more than 25 years later, when organic chemists finally devised methods for the synthesis of short DNA chains called oligonucleotides. By mixing synthetic oligonucleotides with other oligonucleotides of complementary sequence, identical, well-defined double helices could be generated and crystallized. X-ray analysis of such complementary oligonucleotides containing all four bases

for example, $\dfrac{\text{GGTATACC}}{\text{CCATATGG}}$

revealed right-handed double helices held together by the long-postulated base pairs (see the A-DNA in Figure 9-9).

In the right-handed helices, the two members of each base pair frequently do not lie exactly in the same plane but display a ''propeller twist'' arrangement that stabilizes their respective helices by in-

Figure 9-6

(a) Schematic folding of a single-stranded DNA molecule to form irregular double-helical hairpin loops in regions of sequence homology. (b) Electron micrograph of single-stranded bacteriophage fd DNA showing regions of double-helical hairpin loops. [Courtesy of John E. Hearst, see C.-K. J. Shen, A. Ikoku, and J. E. Hearst. *J. Mol. Biol.* 127 (1979):167.]

Figure 9-7
The propeller twist between the purine and pyrimidine base pairs of a right-handed helix. The C1′ atoms of deoxyribose are shifted up and down by the twist and may serve to stabilize the helix by increasing the stacking overlaps between bases. The angle of the twist varies and may be as large as 45° in some cases.

creasing the stacking overlaps between bases (Figure 9-7). However, simple propeller twists generate too-close contacts between the individual purines, since, as the larger members, they extend out beyond the centers of their respective base pairs. Such unacceptable purine overlaps are avoided either by an alteration in the rotation angle between the closely opposing base pairs or by a sliding of one of the base pairs over its neighbor so that purines are pushed out of the main stack of bases. How these factors compensate for each other depends on the exact sequence of bases. As a result, individual DNA molecules are never perfectly regular double helices. Instead, their exact conformations depend on which base pair (AT, TA, GC, or CG) is present at each position along the double helix (Figure 9-8).

Alternative Forms of Right-Handed DNA[18–22]

There are two well-characterized forms of right-handed DNA, referred to as A and B. They differ in the distance required to make a complete helical turn (the **pitch**), in the way their deoxyribose groups are bent, or puckered, and in the angle of the tilt that the base pairs make with the helical axis (Table 9-1). The B conformation is the longer, thinner form with its 3.4 Å thick purine-pyrimidine base pairs oriented approximately perpendicular to the helical axis. Separating its sugar-phosphate backbones are two very noticeable grooves (the major groove and the minor groove) on its external surface. B-DNA converts to A-DNA by tilting the base pairs some 30° so that successive base pairs occur every 2.7 Å along the helical axis. A-DNA molecules are thus shorter and squatter than their B equivalents, with their base pairs pushed away from the helical axis and their sugar-phosphate backbones separated by a much deeper and thinner major groove than found in B-DNA (Figure 9-9).

In solution, DNA usually assumes the B form. Ordinarily the A conformation is only found when relatively little water is available to interact with the double helix. Relatively unhydrated DNA fibers tend to assume the A form, as do most crystals formed from complementary oligonucleotides, because the chemical reagents used in crystallization effectively dehydrate the DNA by binding water molecules. It is still unclear whether any significant sections of pure A-form DNA exists within cells. However, the DNA-RNA hybrid double helices, which exist only fleetingly when RNA is synthesized on single-stranded DNA chains, probably have A-like configurations; RNA chains cannot fold up into B-form double helices because they possess an extra hydroxyl group on each of their sugar residues. Instead, all double-helical RNA regions so far examined (such as the double-stranded RNA molecules from certain viruses or the double-helical sections of tRNA molecules) take up an A-like form in which the base pairs are strongly tilted away from the perpendicular position.

Polypurine-Polypyrimidine Double Helices Have Mixed A and B Properties[23]

Sections of double helices in which one strand is exclusively polypurine (e.g., poly dA) and its complementary strand is exclusively polypyrimidine (e.g., poly dT) show mixed A and B features. While the purine and pyrimidine bases have essentially the same tilt, their sugar-phosphate backbones have very different conformations. The sugar-phosphate backbone of the polypurine strand is A-like, while that of the polypyrimidine strand is B-like, marked by different puckering of the deoxyribose groups. Such "heteronomous" (obeying different rules) DNA sections appear to be unable to bind normally with histones to form nucleosomes (see Chapter 18). If so, the appearance of polypurine or polypyrimidine sequences along DNA molecules may be signals that mark nucleosome-free regions of DNA.

Alternating Anti and Syn Conformations Allow Transition into Left-Handed Helices[24-30]

DNA containing alternating purine and pyrimidine residues can fold up into left-handed as well as right-handed helices. In the left-handed helix, the fundamental repeating unit usually is a purine-pyrimidine dinucleotide, with the glycosyl bond connecting the base to the deoxyribose group oriented in the *anti* form at pyrimidine residues and in the *syn* configuration at purine residues. In right-handed DNA, the glycosyl bond is always oriented anti; so it is the different configuration at the purine nucleotides that usually generates left-handed helices (Figure 9-10). The alternating anti-syn configurations give the backbone of left-handed DNA a zigzag look (hence its designation of Z-DNA), which distinguishes it from the right-handed A and B forms (see Figure 9-9).

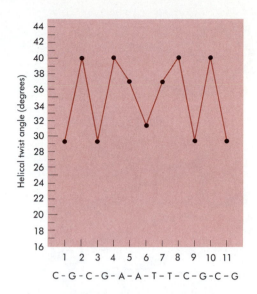

Figure 9-8
A given DNA molecule is never perfectly regular; the helical twist angle depends on the specific nucleotide bases present, reflecting the different stacking of each nucleotide base. [Data from Richard E. Dickerson, *Cold Spring Harbor Symp. Quant. Biol.* 47 (1982), p. 15.]

Table 9-1 A Comparison of the Structural Properties of A-, B-, and Z-DNAs as Derived from Single-Crystal X-Ray Analysis

	Helix Type		
	A	B	Z
Overall proportions	Short and broad	Longer and thinner	Elongated and slim
Rise per base pair	2.3 Å	3.32 Å (0.19 Å)*	3.8 Å
Helix-packing diameter	25.5 Å	23.7 Å	18.4 Å
Helix rotation sense	Right-handed	Right-handed	Left-handed
Base pairs per helix repeat	1	1	2
Base pairs per turn of helix	~11	~10	12
Mean rotation per base pair	33.6°	35.9° (4.2°)*	−60°/2
Pitch per turn of helix	24.6 Å	33.2 Å	45.6 Å
Tilt of base normals to helix axis	+19°	−1.2° (4.1°)*	−9°
Base-pair mean propeller twist	+18°	+16° (7°)*	~0°
Helix axis location	Major groove	Through base pairs	Minor groove
Major-groove proportions	Extremely narrow but very deep	Wide and of intermediate depth	Flattened out on helix surface
Minor-groove proportions	Very broad but shallow	Narrow and of intermediate depth	Extremely narrow but very deep
Glycosyl-bond conformation	anti	anti	anti at C, syn at G

SOURCE: R. E. Dickerson et al., *Cold Spring Harbor Symp. Quant. Biol.* 47 (1982):14. Reproduced by permission.

A-DNA

B-DNA

Z-DNA

Figure 9-9
Comparison of A, B, and Z DNA. The bases are colored and the back-bones are in grey. Note the zigzag course of the sugar-phosphate back-bone chain in the Z-DNA. [Drawings by Irving Geis, reproduced from R. E. Dickerson's "The DNA Helix and How It Is Read," *Sci. Am.* 249 (1983):100–104, by permission.]

Deoxyguanosine as in B-DNA

Anti position
of guanine

Deoxyguanosine as in Z-DNA

Syn position
of guanine

● C   ● N   ● O   ○ H   Ⓟ P

Overall, Z-DNA is more elongated and slimmer than B-DNA, with the base pairs of Z-DNA tilted in such a way that their successive distances along the helical axis are increased to 3.8 Å (see Figure 9-10). Twelve base pairs are found within each helical turn of 45.6 Å. Only one deep groove, corresponding to the minor groove of B-DNA, characterizes the external surface of Z-DNA. The relative narrowness of the groove places the base pairs as much on the outside of the helix as the sugar-phosphate backbone. The sugar-phosphate backbone of Z-DNA is not as fully extended as that of A- or B-DNA. In solution, the alternating purine-pyrimidine residues assume the left-handed configuration only in the presence of high concentrations of positively charged ions (e.g., Na$^+$) that neutralize the negatively charged phosphate groups. At lower salt concentrations, they form typical B configurations.

Because GC base pairs are much more stable than AT base pairs, it has proved much easier to obtain crystals of short, alternating dG-dC double helices than of corresponding short dA-dT repeats. Left-handed DNA was thus first crystallized using alternating dG-dC oligonucleotides. More complex sequences such as

CGCATGCG
GCGTACGC

were subsequently shown to form perfect crystals of Z-DNA, and now it is believed that any alternating purine-pyrimidine sequence of at least six base pairs under suitable ionic conditions can assume the Z form.

At first it was thought that only sections of DNA containing regularly repeating purine-pyrimidine sequences could form Z-DNA. Then, in 1985, the sequence

^{m5}C GAT^{m5}C G
G^{m5}CTA G^{m5}C   (m = methylated)

was found by X-ray methods to crystallize as Z-DNA. This means that alternating purine-pyrimidine sequences are not necessary to make Z-DNA if the cytosine residues are methylated at the 5' position (see following section for discussion of cytosine methylation). Thus, it is

Figure 9-10
In right-handed B-DNA, the glycosyl bond (color) connecting the base to the deoxyribose group is always in the anti position, while in left-handed Z-DNA it rotates in the direction of the arrow, forming the syn conformation at the purine (here guanine) residues but remains in the regular anti position (no rotation) in the pyrimidine residues. It is this change to the syn position in the purine residues that generates the zig-zag (anti to syn) conformation and thus allows the transition from B-DNA to Z-DNA. The rotation causing the change from anti to syn also causes the ribose group to undergo a change in its pucker. Note that C3' and C2' switch locations. [After A. H.-J. Wang, et al. *Cold Spring Harbor Symp. Quant. Biol.* (1982), p. 41.]

Z-DNA

B-DNA

Figure 9-11
Demonstration of antibody binding in Z-DNA and B-DNA in the X polytene chromosome of *Drosophila melanogaster.* In the Z-DNA (left), (b) addition of anti-Z antibodies results in no detectable binding, but when first treated with 45 percent acetic acid for (c) 5 seconds and (d) 30 seconds, some anti-Z binding may be seen to occur. The acid removes histones, exposing naked supercoiled DNA. In contrast, B-DNA (right) antibodies (a) bind readily even without acid treatment, and binding may be even slightly decreased in long (30-second) exposure to acid treatment. [Courtesy of R. J. Hill and B. D. Stollar, *Nature* 305 (1983):338–340.]

not yet possible to pinpoint potential Z-DNA segments simply by scanning nucleotide sequences.

The amount of Z-DNA within cells remains unknown, with the best guess now that it accounts for, at most, only tiny segments of a given cell's DNA. That some cellular DNA may in fact adopt this conformation is suggested by experiments that purport to identify proteins that bind specifically to Z-DNA and not to B-DNA. However, only one such protein has yet been even semirigorously characterized. Conceivably, by so binding, such proteins help stabilize in the Z form DNA that would otherwise assume the B form.

Perhaps because of its rarity, Z-DNA is much more immunogenic than B-DNA, against which antibodies are only infrequently produced. Addition of anti-Z antibodies to native *Drosophila* chromosomes results in no detectable binding (Figure 9-11). However, after weak-acid treatment, which removes the histone components from the chromosomes and exposes naked, supercoiled DNA, some anti-Z antibody binding occurs. (No binding occurs to DNA whose supercoiling has been removed, suggesting that DNA must be under torsional strain to go left-handed.) In contrast, anti-B antibodies readily attach to many regions of native chromosomes, confirming early evidence, obtained by X-ray diffraction techniques, that the vast majority of chromosomal DNA is in the B form.

Methylation of Specific Cytosine and Adenine Residues After Their Incorporation into DNA

A small fraction of the cytosine residues in the DNA of most chromosomes contain methyl groups attached to the C5 atoms (5-methyl cytosine, or m^5C). These methyl groups are added by specific enzymes, the **DNA methylases,** after cytosine residues are incorporated into DNA chains (Figure 9-12).

Methyl groups also can be added to adenine groups to form

6-methyladenine. Such groups are added after DNA synthesis by a DNA methylase distinct from that which modifies cytosine residues. In procaryotic cells, methylated adenine is more common than methylated cytosine, while in eucaryotes, almost all methyl groups are added only to cytosine residues.

In eucaryotic cells, DNA methylases modify predominantly those cytosines linked on their 3' sides to guanine ($^{5'}CG^{3'}$), a dinucleotide combination that is present much less often in eucaryotic genomes than would be expected by chance alone. Not all $^{5'}CG^{3'}$ groups become so methylated, and vast differences in the extent of methylation of given genes have been detected among the different cell types (tissues) of multicellular organisms. Now it is suspected that undermethylation of certain CG sequences in key control regions often promotes gene expression. There are, however, important exceptions to this rule: Heavily methylated genes may be expressed, unexpressed genes may be unmethylated, and the methylation level may gradually decrease over a period of hours after transcription of a gene has begun. Moreover, methylated DNA has not been found in yeast or in *Drosophila*. Thus, demethylation may be a consequence of gene activation rather than its cause or prerequisite (see Chapter 21).

DNA Methylation Favors the B to Z Transition[31-34]

Methylation of key cytosine residues may influence gene functioning through its potentiation of the conversion of B-DNA to Z-DNA. At the cation levels found in most cells, alternating CG segments are most likely to exist as B-DNA. However, after the cytosines have been methylated, the equilibrium switches to favor the Z form. The reason for this switch is that in B-DNA, the methyl groups protrude into the aqueous environment of the major groove, a destabilizing feature; but in the Z form, the same methyl groups form a stabilizing hydrophobic patch that otherwise must be filled with water (Figure 9-13).

Figure 9-12
Action of the eucaryotic DNA methylase which modifies cytosine.

Figure 9-13
Crystal structures of (dC–dG)$_3$ and (m^{5}dC–dG)$_3$ showing the structure of Z-DNA in both its unmethylated (left) and methylated (right) forms, respectively. The black line traces the sugar-phosphate backbone, and the carbons of the methyl groups are shown in black. The arrow marks a depression in the unmethylated form, which is filled with the hydrophobic methyl group in the methylated form, which thus stabilizes the structure. [Courtesy of Dr. A. H.-J. Wang, reproduced from *Cold Spring Harbor Symp. Quant. Biol.* 47 (1982):41.]

time ⟶

Figure 9-14
Spontaneous shape fluctuation and deformation of the double helix in solution. Shown here is a single folded T4 DNA molecule. Micrographs were taken at 0.1-second intervals. [Courtesy of Dr. M. Yanagida, reproduced from *Cold Spring Harbor Symp. Quant. Biol.* 47 (1982):179, Section 4.]

Spontaneous Deformations of the Double Helix in Solution[35–38]

In solution, B-DNA is quite a plastic structure. It is constantly subjected to localized thermally induced fluctuations in the arrangements of its atoms, which cause individual molecules to bend, twist, and stretch. Such distortions are not generally the consequence of momentary losses of hydrogen bonds between base pairs. Instead, they often represent changes in the rotational angles of the covalent bonds along the polynucleotide backbones that arise from sequence-dependent variations in the twist per base pair, reflecting the different ways the four base pairs optimally stack on each other. These can lead to gentle bends of the helical axis that partially close up the major groove and generate minor kinking of the stacked base pairs, which pushes them into the minor groove. As a result, the direction in which a given double helix points can change greatly over distances as short as several hundred angstroms (100 base pairs or 10 turns of the double helix). Thus, when individual DNA molecules are visualized in solution through the binding of a suitable fluorescent dye, they do not have long, linear shapes. Instead, they assume roughly spherical shapes, as expected for randomly coiled semirigid rods (Figure 9-14). Most dramatic are the occasional, drastic changes in the overall shape of DNA, which suggest abrupt kinking. In any case the fact that bends and kinks arise spontaneously indicates that relatively little energy is needed to deform the double helix. If the double helix can more effectively bind DNA binding proteins by bending, we can assume it will do so.

Sequence-Specific Bending and Kinking of DNA[39, 40]

The ability of all DNA to bend gently or to kink is accentuated by the presence of specific nucleotide sequences. Examination of a DNA molecule that moves anomalously fast through gels (and which thus has a more collapsed or kinked form than most DNA) reveals a set of four CAAAAAT (or CAAAAAAT) segments. These were found to be separated by single turns (ten nucleotides) of the double helix (Figure 9-15). Conceivably, the CAAAAAT sequence has adopted an unconventional (perhaps "heteronomous") conformation whose borders with conventional B-DNA are regions of predictable kinking.

Bending can also be the result of binding to specific proteins. The binding of the regulatory protein CAP to its normal binding site near the promoter of the *lac* operon leads to a more collapsed DNA conformation in that region. Even more direct evidence for kinking now comes from X-ray crystallographic analysis of DNA complexes with histones in nucleosomes and with the restriction enzyme *Eco* RI (see Figure 9-36).

Unwinding of the Double Helix by the Insertion of Flat, Ringed Molecules[41–44]

A number of flat, ring-shaped molecules bind to DNA by inserting themselves (**intercalating**) between the base pairs of the double helix. In this manner, they often act as inhibitors of DNA or RNA synthesis and thus act like drugs in living systems, killing cells or inhibiting

...GAATTCC<mark>CAAAAAT</mark>GT<mark>CAAAAAAT</mark>AGG<mark>CAAAAAAT</mark>GCC<mark>AAAAAT</mark>CCCAAAC...

their growth. They were originally discovered through their antimicrobial and anticancer attributes. Each time these intercalating agents become inserted into a DNA molecule, they necessarily *increase* the spacing of successive base pairs along the helical axes to roughly 7 Å, almost the distance between phosphate atoms in a fully extended polynucleotide chain (Figure 9-16). In this situation, very little rotation is possible around the helical axis, and so intercalating agents not only extend double helices, but extensively unwind them. For example, when a molecule of either ethidium bromide (an inhibitor of DNA synthesis) or actinomycin D (a powerful inhibitor of RNA synthesis) intercalates, the rotation angle between the two adjacent base pairs is reduced from 36° to 10°. The fact that intercalation occurs so readily indicates that it must be energetically favored, with the van der Waals bonds holding the inserted molecules to the base pairs being stronger than those found between conventionally stacked base pairs. Intercalation is additional evidence for the *metastability* of the double-helical structure—its ability to temporarily assume many inherently unstable configurations that normally quickly revert back to the standard B conformation. One such metastable variant must be short stretches of extended chains with gaps separating adjacent base pairs, into which an intercalating agent may bind.

The Chromosomes of Viruses, *E. coli,* and Yeast Are Single DNA Molecules[45–48]

The first estimates of the average molecular weights of DNA centered at about a million, the size needed to encompass an average-size bacterial gene coding for some 300 to 400 amino acids. Therefore, it

Figure 9-15
Segment of a kinetoplast DNA molecule found to move anomalously fast through gels. Note that the four $CA_{5-6}T$ sequences start about every ten bases—approximately a single turn of the double helix. These sequences allow the DNA to kink and thus become more compact. [Data from H.-M. Wu and D. M. Crothers, *Nature* 308 (1984):509.]

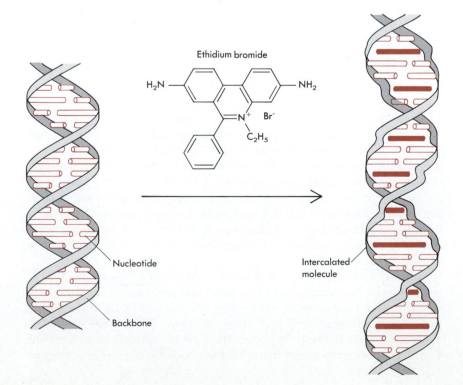

Ethidium bromide

Nucleotide

Backbone

Intercalated molecule

Figure 9-16
The intercalation of ethidium bromide into a DNA molecule. Note that the ethidium bromide increases the spacing of successive base pairs, distorts the regular sugar-phosphate backbone, and decreases the pitch of the helix.

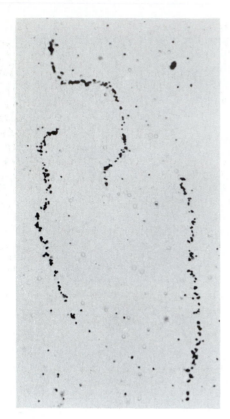

Figure 9-17
Autoradiograph of several T2 chromosomes. The total length of each is 52 μm. (Courtesy of J. Cairns.)

seemed natural to equate single DNA molecules with single genes. However, these early reports were inaccurate owing to DNA breakage during its isolation and study. Now it is clear that virtually all undegraded DNA molecules contain the information of at least several genes. The most certain molecular weight values come from DNA-containing viruses. Regardless of whether the DNA content is relatively small or large, each virus particle contains a single DNA molecule. For example, all the DNA of the small monkey (simian) virus SV40 is present within a single molecule of molecular weight $\sim 3 \times 10^6$ daltons (5×10^3 base pairs, or 5 kbp), while the DNA molecule of the large bacterial virus T2 has a molecular weight of 1.2×10^8 daltons (2×10^2 kbp) (Figure 9-17). In these cases, the entire viral chromosome, rather than separate genes, corresponds to a single DNA molecule. Likewise, the chromosome of an *E. coli* cell is a single DNA molecule whose molecular weight is about 2.5×10^9 daltons (about 4×10^3 kbp) and whose extended length is roughly 1 mm.

DNA molecules of the yeast *Saccharomyces cerevisiae* range in size between the T2 and *E. coli* DNAs, with their average size being that expected if each of the yeast's 17 chromosomes contain one DNA molecule. Thus, the centromeres of yeast chromosomes (if not those of all eucaryotic chromosomes) do not represent discontinuities between separate DNA molecules, but instead are specialized regions of DNA evolved to interact with the spindle bodies upon which chromosomal segregation occurs during cell division. As of yet, there is no direct proof that the very much larger chromosomes of higher plants and animals (often 50 times larger than the *E. coli* chromosomes) also contain only one DNA molecule. However, we now expect that the one chromosome—one DNA molecule rule will hold for the chromosomes of all organisms. In any case, some DNA molecules are larger than any other biological molecules by several powers of ten.

Circular Versus Linear DNA Molecules[49, 50, 51]

Our evidence, which came largely from autoradiography and electron microscopy, initially suggested that all DNA molecules are linear and have two free ends. But when it became possible to take a better look at undegraded DNA, many DNA molecules were found to be circular (see Figure 7-29). For instance, the small monkey DNA virus SV40 has a 5000-base-pair circular double-helical chromosome; the similarly short chromosomes of single-stranded phages are likewise circular, as are almost all autonomously replicating plasmid DNAs. Most, if not all, bacterial chromosomes are also circular. We suspect that the DNA found in the rare circular chromosomes of higher cells are likewise circular molecules. Circular shapes were initially puzzling, since they seemed to present obstacles to the untwisting of double helices during DNA replication. Only when specific enzymes that first snip and then join DNA chains were discovered did the untwisting dilemma disappear. (These enzymes, the topoisomerases, are discussed later in this chapter.)

Equally important is the realization that linear DNA molecules do not have the uncomplicated structures first envisioned for them. As we see in the next chapter, replicating the ends of DNA molecules is not a straightforward process. Thus, all linear DNA molecules have evolved special tricks to replicate their ends, as well as to prevent their free ends from either being nibbled back by cellular enzymes or being enzymatically ligated together to form even larger chromosomes. For example, the ends of the linear adenovirus chromosomes have specialized proteins covalently attached to the 5′ ends of strands; and poxvirus chromosomes have closed hairpin loops at their ends, with their basic structures being that of largely self-complementary circular single strands. Still other linear viral chromosomes (e.g., T2 and T7) have the same sequences at both ends, allowing the ends of the chromosome to recombine together during DNA replication to form very long, end-to-end aggregates (concatamers) (Figure 9-18).

Moreover, some DNA molecules that are linear when isolated from a virus particle (e.g., phage λ) are found as circles inside the host cell. This tells us that the linear and circular forms of such DNA molecules are interconvertible. A molecular mechanism that converts a DNA circle to the linear form (or vice versa) is shown in Figure 9-19. Starting with a closed circle, a specific enzyme introduces two breaks, one in the + strand and one in the − strand. As these breaks are very close to each other, the intervening hydrogen bonds occasionally are broken by thermal agitation, and then the circle unfolds. Most importantly, the resulting linear form contains single-stranded ends that have complementary nucleotide sequences ("*sticky ends*"). Thus, at a later time, the linear form can base-pair and resume a circular configuration. If the missing phosphodiester bonds are then reformed, the covalent circle is regenerated.

There is solid evidence that phage λ DNA interconverts between its circular and linear forms using precisely this mechanism. Later we shall see that the duplication of such "sticky" DNAs involves circular replicative intermediates. Thus, the linear form is most likely an adaptation for injection of the viral chromosome through the narrow phage tail.

Supercoiling of Circular DNA Molecules[52–56]

As DNA is a very flexible structure, its exact molecular parameters are a function of both the surrounding ionic environment and the nature of the DNA-binding proteins with which it is complexed. Because

Figure 9-18
Structures of some viral DNAs.
(a) Adenovirus DNA is linear and double-stranded and has specialized proteins covalently attached to its 5′ ends. (b) Poxvirus DNA has closed hairpin loops at its ends. (c) The linear DNA of certain phages has several hundred base pairs of identical sequences at the two ends, allowing the formation of long, end-to-end concatamers. All three molecules are actually very much longer than shown in this figure.

Figure 9-19
Interconversion of the linear and circular forms of λ DNA.

Figure 9-20
Generation of supercoiled DNA by cutting and ligating. In a circular DNA molecule, the number of times the DNA strands twist about each other (called the linkage number) is fixed. If both strands are cut (1) and rotated in the opposite direction from that of the twist of the helix (2), and if the cut ends are then rejoined (3), then the DNA molecule negatively supercoils (4). The upper electron micrograph is a relaxed (nonsupercoiled) DNA molecule of bacteriophage PM2. The lower electron micrograph shows the phage in its supertwisted form. [Electron micrographs courtesy of J. C. Wang, *Sci. Am.* 247 (1982):97.]

their ends are free, linear DNA molecules can freely rotate to accommodate changes in the number of times the two chains of the double helix twist about each other. But once the two ends become covalently linked to form circular DNA molecules, the absolute number of times the chains twist about each other (the **linkage number**) cannot change. For the most part, changes in the average number of base pairs per turn of the double helix will necessarily be accommodated by the formation of the appropriate number of supercoils in the opposite direction. **Supercoiling** refers to the further twisting of double-helical DNA molecules. Untwisting of the double helix usually leads to supercoiling in the negative (left-handed) direction (Figure 9-20), while overtwisting leads to positive (right-handed) supercoiling.

The supercoiled state is inherently less stable than uncoiled DNA, and the cutting, or **nicking,** of a single strand instantly converts a supercoiled molecule into its simple (**relaxed**) circular state (Figure 9-21). Because circular DNA molecules become increasingly compacted as they become more supercoiled, relaxed DNA is easily distinguished from supercoiled DNA by its slower sedimentation in a centrifugal field and by the longer time it takes to move through an agarose gel. At the present time, gel electrophoresis affords the most direct way to measure supercoiling, since it can separate molecules that differ by only single turns (Figure 9-22).

Localized Denaturation Within Supercoiled DNA[57, 58, 59]

The stress present within negatively supercoiled DNA molecules sometimes leads to localized denaturation, in which the complementary strands come apart. When this happens, the free energy lost by breaking the hydrogen bonds between base pairs is less than that gained by the unbending of the double helix. Those regions that denature tend to be rich in the relatively weak AT base pairs, while segments rich in the much stronger GC base pairs are usually the last to denature (Figure 9-23).

Localized denaturation should be especially favored for DNA segments containing self-complementary regions capable of forming hydrogen-bonded hairpin structures called **cruciform loops** (Figure 9-24). In such cases, relatively few net hydrogen bonds are lost and considerable energy is gained from the loss of one or more supercoils dominating the final equilibrium. There is considerable debate, however, about whether this phenomenon takes place in living cells. For example, in recent experiments, when a supercoiled plasmid DNA containing a perfect 68-base-pair inverted palindromic sequence was isolated very gently from bacteria, none of it was found as stable cruciforms. Rigorous proof for the absence of this cruciform could be obtained because a restriction enzyme cutting site is located at the exact center of the inverted palindrome: If a cruciform were present, this *Eco* RI recognition site (GAATTC) would be located within the terminal single-stranded hairpin loop and thus be resistant to digestion with *Eco* RI endonuclease, which can only cut double-stranded DNA (Figure 9-25).

The absence of any DNA uncut by *Eco* RI proved that none of the supercoiled plasmid DNA had rearranged into stable cruciforms. Apparently, within the bacterial cell, neither thermal energy alone nor the combined action of various enzymes (such as topoisomerases and proteins that bind to single-stranded DNA) is able to denature a sufficient number of base pairs in the supercoiled plasmid molecule so that the cruciform structure can begin to form. Thus, while cruciforms may be the most *energetically* stable configuration for certain negatively supercoiled DNAs, such structures appear to be *kinetically* forbidden in at least some situations.

Negative supercoils can also be lost through the conversion of any alternating purine-pyrimidine residues within the DNA molecules from the B form to the Z form of the DNA double helix. When supercoils are present, conversion into the Z form occurs at much lower salt concentrations and is much less dependent on methylation of key cytosine residues than when supercoils are absent.

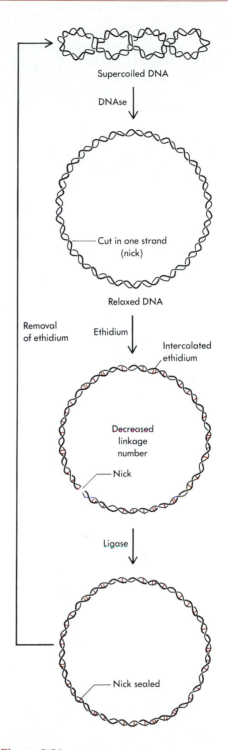

Figure 9-21
Use of nicking and intercalation to interconvert supercoiled and relaxed DNA. When one strand is nicked, a supercoiled circular DNA molecule relaxes. Ethidium intercalated into the DNA molecule increases the spacing of base pairs, unwind the double helix by 26°, and thus decrease the linkage number. DNA ligase can rejoin the free ends. When ethidium is removed, the DNA molecule supercoils negatively.

Figure 9-22
Gel electrophoresis separation of re-laxed and supercoiled DNA. The speed with which the DNA molecules migrate increases as the number of superhelical turns increases. (Courtesy of J. C. Wang.)

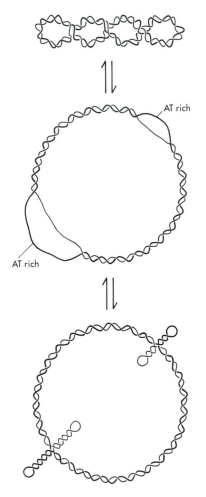

Figure 9-23
Upon partial denaturation, supercoiled DNA may convert to its unsupercoiled relaxed form. Areas of the DNA mole-cules relatively rich in the AT base pair tend to denature first, since it is held together by two hydrogen bonds rather than the three hydrogen bonds holding together the GC base pair. The single-stranded areas may then convert to more stable cruciform structures if there are inverted homologous sequences in the denatured regions.

Most Cellular DNA Exists as Protein-Containing Supercoils[60–63]

Just because isolated DNA molecules would energetically prefer to be in the relaxed state does not mean that the "natural state" of DNA is relaxed. In fact, virtually all DNA within both procaryotic and eucaryotic cells exists in the negative supercoiled state. As such it seldom, if ever, occurs as free DNA, but is complexed with specific DNA-bind-ing proteins to form compacted molecules called **chromatin.** How this compaction occurs is best understood for eucaryotic chromatin, whose most prominent DNA-binding components are the histones. *Histones* are relatively small, positively charged, arginine- and lysine-rich proteins that aggregate together to form discrete ellipsoid-shaped packets (histone cores) around which the DNA supercoils. The result-ing **nucleosomes** (see Chapter 20) give to chromatin its beaded ap-pearance (Figure 9-26).

Since the same collection of histones binds to nearly all sections of DNA, histones are thought to play essentially structural roles, as op-posed to enzymatic or regulatory roles. Only about half the mass of chromatin proteins is histone. The remaining proteins are not nearly as well characterized, with their exact functions yet to be determined.

Viral DNA chromosomes are also complexed with proteins, both when multiplying within cells and when packaged into virus parti-cles. The SV40 viral chromosome is at all times compacted into 24 histone-containing nucleosomes identical in structure to those exist-ing in the chromosomes of their host cells (Figure 9-27). When freed from its protein components, SV40 DNA is found as a negative super-coil. In contrast, no histones are complexed with the DNA found in adenovirus particles. These vertebrate viruses encode a unique DNA binding protein to help package their DNA within their protein cap-sules. When, however, an adenovirus DNA functions within a cell, its own DNA-binding protein is released and replaced with the same histone components that bind to all other cellular DNA.

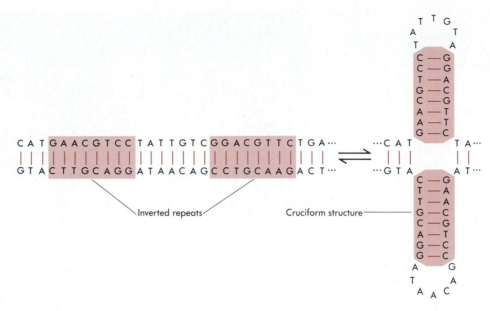

Figure 9-24
DNA segments containing self-complementary regions, such as the inverted repeat sequences shown here, can form hydrogen-bonded cruciform loops.

DNA Supercoils Twice Around Each Nucleosome[64, 65]

In forming the nucleosomes of eucaryotic cells, 200-base-pair-long segments of DNA negatively (left-handedly) supercoil twice around the histone cores (Figure 9-28). The energy lost by supercoiling may

Bacterium with a multicopy plasmid having a perfect 68-base-pair inverted palindromic sequence with an *Eco* RI recognition site at the center of symmetry

Gentle lysis to prevent thermal rearrangement of supercoiled plasmid DNA

"Native" supercoiled plasmid with duplex *Eco* RI site that is susceptible to *Eco* RI cutting

Native structure rearranges into cruciform after partial denaturation by heating or treatment with phenol.

Single-stranded *Eco* RI recognition site is resistant to *Eco* RI digestion in cruciform structure

Figure 9-25
Formation of stable cruciform structures from inverted palindromic sequences. It is now questionable whether these do indeed form within cells.

Figure 9-26
Nucleosomes give chromatin strands a beadlike appearance. Magnification 100,000. (Courtesy of Victoria Foe.)

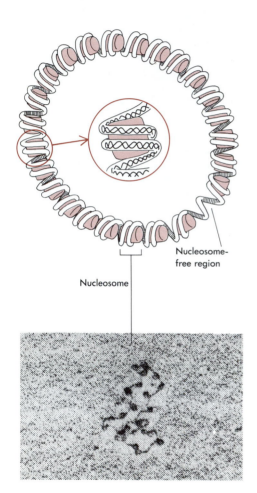

Nucleosome-free region

Nucleosome

Figure 9-27
SV40 chromosome showing its 24 histone-containing nucleosomes. Often such minichromosomes contain a nucleosome-free region located near the origin of replication (see Chapter 24).

Histone core

DNA

Figure 9-28
DNA coiling about a histone core.

be partially compensated by the energy gained from the ionic and hydrogen bonds formed between the histones and the DNA. The number of supercoils in solution need not be the same as found in chromatin. In chromatin, the double helix is slightly more twisted than in solution, with 10 base pairs per helical turn in chromatin versus 10.5 base pairs per helical turn in naked DNA in solution. As a result, there are fewer superhelices found in solution than found in chromatin.

Procaryotic Cells Contain Histonelike DNA-Binding Proteins[66, 67]

For many decades, it was believed that bacterial DNA, unlike eucaryotic DNA, occurs naked and does not have a compacted chromatin-like arrangement. But recently, using more gentle preparative procedures, condensed *E. coli* chromosomal sections have been seen that are organized into beadlike packets from which small, basic proteins analogous to the histones can be extracted. That bacterial DNA is complexed with proteins that have histonelike properties undoubtedly reflects its need to be regularly compacted in order to function. Crystallization of one of these proteins, DNA-binding protein II, reveals that its 9500-dalton chains associate in pairs as dimers containing extended arginine-containing arms able to interact with the phosphates of one turn of the DNA backbone. So it is likely that dimers sited adjacently along procaryotic DNA are oriented into a helical arrangement with the DNA bound on the outside (Figure 9-29). Interestingly, the average *degree of supercoiling* (i.e., the number of superhelical twists per ten base pairs) is about 0.05 for all naturally occurring DNA supercoils, regardless of their source. This may in part be because the underlying chromatin structure is so similar within both procaryotic and eucaryotic cells.

Topoisomerases Change the Linkage Numbers of Supercoiled DNAs[68–75]

That supercoiling not only occurs but is biologically very important is affirmed by the existence of the enzyme topoisomerase II, which specifically generates the negative supercoils that characterize so much of chromosomal DNA. **Topoisomerases** (Table 9-2) are a group of enzymes that convert (isomerize) one topological version of DNA into another. They do so by changing the linkage number, which is the number of times two DNA chains twist around each other. Procary-

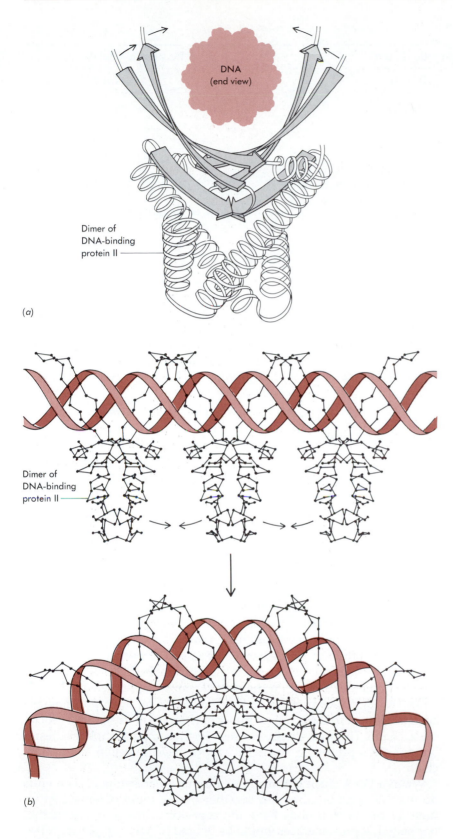

(a)

(b)

Figure 9-29
The histonelike DNA binding protein II of *E. coli*. (a) The structure of the dimer of DNA binding protein II bound to DNA viewed end-on. (b) How DNA binding protein II might induce super-coiling of DNA to form nucleosomes. The arms of the binding proteins are thought to interact with the DNA and each other, while electrostatic interactions at the other end of the proteins draw them together and in so doing bend the DNA. It is thought that 8 to 10 protein dimers and 80 to 100 nucleotide bases are needed to generate one turn of the supercoil. Here the DNA is not shown supercoiled, to prevent a confusing superimposition of the protein dimers.

otic topoisomerase II (sometimes called **gyrase**) uses the energy of ATP to generate negative supercoils by untwisting DNA in the left-handed direction. Its eucaryotic counterpart, however, may be capable of using ATP to generate negative supercoils only in the presence of other, yet to be described proteins. The supercoiling action of topo-

Table 9-2　DNA Topoisomerases

	Procaryotic (*E. coli*)		Eucaryotic (Hela)	
	Type I	Type II	Type I	Type II
Approximate MW	100,000	400,000	100,000	309,000
Subunits	Monomer	Tetramer (A₂B₂)	?	Dimer (subunit MW = 172,000)
Gene	*topA*	*gyrA* Polypeptide MW 105,000; activity inhibited by nalidixic acid and oxolinic acid *gyrB* Polypeptide MW 95,000; activity inhibited by coumermycin and novobiocin	?	?
Covalent intermediate with DNA	At 5′ Ⓟ	Subunit A Tyrosine covalent bond to 5′ Ⓟ	At 3′ Ⓟ	At 5′ Ⓟ
Relaxation*	Mg²⁺ required; ATP-independent	ATP-independent	No Mg²⁺ required; ATP-independent	ATP-dependent
Negative supercoiling	None	ATP-dependent	None	None†

*Eucaryotic type I topoisomerases usually catalyze relaxation of both negative and positive supercoils, but *E. coli* type I topoisomerase relaxes positive supercoils only when it is presented with a single-stranded DNA region.
†It may participate in supercoiling only in the presence of additional proteins.

isomerase II is counterbalanced by a second enzyme, topoisomerase I, which converts supercoiled DNA to the unstrained, energetically more favorable relaxed state (Figure 9-30). The relative amounts of topoisomerase I and topoisomerase II in cells are finely tuned, tending to create just the right amount of negative supercoiling. Thus, mutations that lower the number of topoisomerase I molecules are viable only if the number of topoisomerase II molecules also decreases.

Both topoisomerase I and II work by catalyzing the breakage and rejoining of DNA phosphodiester bonds. Unlike virtually all other enzymes, their action does not lead to new patterns of covalent bonds. Rather, their role is to create temporary gaps in polynucleotide chains. When so acting, topoisomerases do not create free ends but instead become themselves covalently attached to one of the two broken ends (depending on the specific enzymes, either the 3′ or 5′ end). Through such catalysis, a tyrosine group on the topoisomerase becomes linked to the terminal phosphate group of the cut polynucleotide chain. The free energy present in the original phosphodiester bond is thus preserved so that it can be used to rejoin the broken chain.

When a DNA chain is broken by a topoisomerase, its broken ends do not fall apart but are held together by the topoisomerase. At no time do the broken ends have the capacity to rotate freely. If that happened, topoisomerases would be limited to only completely relaxing double helices. Instead, individual topoisomerases function to make discrete changes of either one positive turn (topoisomerase I) or two negative turns (topoisomerase II) in the linkage number. Such discrete steps result from DNA chains passing through either transient single-stranded breaks (topoisomerase I) or transient double-stranded breaks (topoisomerase II) (Figures 9-31 and 9-32).

Figure 9-30
The counterbalancing actions of topoisomerase I and topoisomerase II (gyrase).

Linkage number = n

Linkage number = $n + 1$

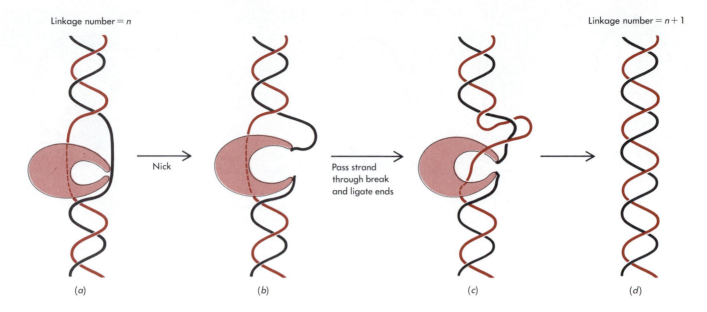

(a)   Nick   (b)   Pass strand through break and ligate ends   (c)   (d)

Topoisomerases are constructed in such a way that they can undergo reversible conformational changes that create cavities through which DNA chains can pass. The ATP requirement for bacterial topoisomerase II (gyrase)-induced supercoiling most likely reflects the use of ATP in mediating the necessary conformational changes that lead to chain passage (Figure 9-33). Topoisomerase I, however, relaxes supercoiled DNA without requiring a cyclical input of energy. Complete understanding of how topoisomerase I and II work can occur only when their precise structures are determined through X-ray crystallographic procedures. Even now, however, the prediction can be made that double helices wrap themselves around topoisomerase II in positive supercoils prior to the start of the cutting-rejoining cycle. Without this restraint on the orientation of the DNA, topoisomerase II would break as well as make negative supercoils, and the net result of its action would be the eventual relaxation of the double-helical substrate to the uncoiled state.

Figure 9-31
The mechanism of action of procaryotic topoisomerase I. (a) The enzyme binds to DNA, unwinding the double helix. (b) It then nicks one strand and prevents the free rotation of the helix by remaining bound to each broken end. (c) The enzyme passes the other strand through the break and ligates the cut ends, thereby increasing the linkage number of the DNA by 1. (d) The enzyme falls away and the strands renature, leaving a DNA with one less negative supercoil. [Redrawn from F. Dean, et al. *Cold Spring Harbor Symp. Quant. Biol.* 47 (1982):773.]

Long, Linear DNA Molecules May Be Divided into Looped, Supercoiled Domains[76–80]

The fact that linear DNA molecules have free ends would seem to preclude their possession of supercoiled segments. Yet, examination of gently isolated linear eucaryotic chromosomes shows that their chromatin is organized into a large number of successive looped domains organized along a scaffold containing two major DNA-binding proteins. Somehow, attachment to the scaffold prevents the rotation of one domain from being transmitted to adjacent sections, since the DNA within each of these loops independently supercoils and may be under different torsional strain. Now there is much evidence both from *Drosophila's* giant salivary chromosomes and from the looped lampbrush chromosomes of the salamander that the individual loops represent functional units of chromatin, with given loops usually being transcribed into one or more very long RNA molecules. How the protein scaffold might maintain independently supercoiled loops was a complete mystery until recently. Now it is less so, following the discovery that a major component of the scaffold is none other than topoisomerase II.

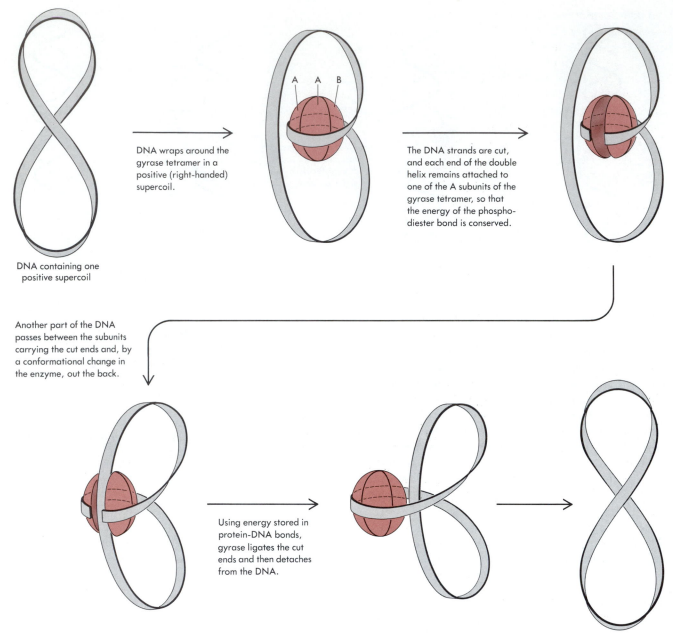

Within the figure:

DNA wraps around the gyrase tetramer in a positive (right-handed) supercoil.

DNA containing one positive supercoil

The DNA strands are cut, and each end of the double helix remains attached to one of the A subunits of the gyrase tetramer, so that the energy of the phospho-diester bond is conserved.

Another part of the DNA passes between the subunits carrying the cut ends and, by a conformational change in the enzyme, out the back.

Using energy stored in protein-DNA bonds, gyrase ligates the cut ends and then detaches from the DNA.

DNA containing one negative supercoil

Figure 9-32

A possible mechanism of action for procaryotic topoisomerase II (gyrase). For simplicity, the starting DNA is shown to have one positive supercoil, which then wraps around the enzyme. In fact, the starting DNA might originally have numerous positive or negative supercoils. The action of topoisomerase II is to decrease the linking number by two, a change that, in this case, is represented by the difference between one positive and one negative supercoil. (Redrawn from A. J. H. Wang et al. *Sci. Amer.* 247:97.)

Not only are the long, linear chromosomes of higher organisms divided into looped, functional domains, but the circular *E. coli* DNA chromosome also is somehow divided into some 50 looped domains (each about 20 μm long) whose supercoiling is independently controlled. No scaffold has yet been isolated from any bacterial chromosome. Indeed, the best way now to search for it may be to look for topoisomerase II activity in very gently ruptured cells.

Generation of Unique DNA Fragments by Restriction Enzymes[81, 82]

Enzymes that cut the phosphodiester bonds of polynucleotide chains are called **nucleases.** Those nucleases that preferentially break internal bonds are known as **endonucleases,** while those that clip off terminal nucleotides are referred to as **exonucleases** (Figure 9-34).

For a long time, it was thought that all endonucleases were relatively unspecific, cutting polynucleotide bonds regardless of the surrounding nucleotide sequence. Over the past decade, however, a new class of endonucleases that cleaves only specific nucleotide sequences has been discovered in procaryotic cells. These specific nucleases, which collectively are referred to as *restriction enzymes,* generally make breaks only within "recognition" sequences that exhibit twofold symmetry around a given point:

$$\text{for example, } \frac{\text{AAG CTT}}{\text{TTC GAA}}.$$

Because the same sequence (running in opposite directions) is found on both strands, such enzymes always create double-stranded breaks (Table 9-3). These breaks can occur almost simultaneously, since many, if not most, restriction enzymes bind as dimers to both component DNA strands at their recognition sites.

Restriction enzymes were initially discovered through their ability to break down ("restrict") foreign, as opposed to native, DNA. A given restriction enzyme can distinguish between DNA native to the cells in which it is synthesized and DNA from another species. Within cells, the main (and possibly only) function of restriction enzymes is to destroy foreign DNA segments before they can recombine into host chromosomal DNA. Their discovery was totally unanticipated, since there was no reason to believe that the introduction of foreign DNA was either a common event or so damaging that specific enzymes would have evolved to nullify their genetic potential.

By now, over 100 different endonuclease specificities are known from examination of restriction enzymes isolated from more than 300 different procaryotic organisms. While restriction enzymes are found in most procaryotic cells, they are absent from eucaryotic cells. What accounts for this difference remains a mystery. Most restriction enzymes recognize groups of four to six specific nucleotides, although a few recognize groups of eight. The inverted identical sequences on the two DNA strands are not always continuous; one or more unspecified base pairs often separate them. The sites of cleavage usually lie within the recognition sequences. Sometimes, blunt-ended fragments are produced, but more often, the cleavage sites are positioned off center, generating short, single-stranded tails (Figure 9-35). Quite frequently, two different enzymes that bind to the same recognition site cut between different pairs of nucleotides. So far, a preponderance of G and C bases have been found at recognition sites, and only recently has a restriction enzyme (*Aha* III) been found that recognizes a sequence composed entirely of A and T residues (TTTAAA). Whether this bias reflects the stronger base pairing between guanine and cytosine residues is not known.

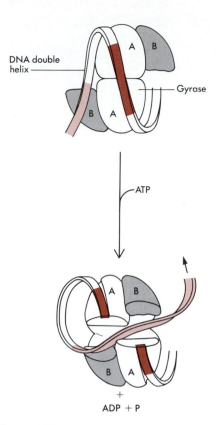

Figure 9-33

Conformational changes in procaryotic topoisomerase II (gyrase) allow the DNA duplex to pass through itself. A number of different models for gyrase action are consistent with current data. Figure 9-32 shows a model in which the DNA duplex passes through the front of the enzyme and out the back, thus requiring two "gates" for DNA. A simpler model (whose first two steps are shown here) requires only one "gate" since the DNA duplex passes through the front face of the enzyme into an internal cavity, and then out again through the same face. The A subunits first cut the DNA duplex and separate to allow another region of the same duplex to pass between the cut ends into the internal cavity. Later the A subunits reclose, the cut ends are ligated, and finally the A subunits separate again to let the intact DNA duplex back out of the internal cavity. The B subunits serve as a protein hinge for opening the DNA "gate." Both the A and B subunits are shown schematically, and are known to have a much more complex structure. [Adapted from A. Morrison and N. R. Cozzarelli, *Proc. Natl. Acad. Sci.* 78 (1981):1420.]

Figure 9-34

The action of endonucleases and exonucleases.

Table 9-3 The Recognition Sequences of Various Restriction Enzymes

Enzyme	Recognition Site		Enzyme	Recognition Site	
	Axis			Axis of symmetry	
	Cut bond ↓			Cut bond ↓	
Eco RI	5'-GAA	TTC	*Hin* dII	5'-GTPy↓	PuAC
	CTT	AAG ↑		CAPu	↑PyTG
Hin dIII	↓AAG	CTT	*Hpa* I	GTT↓	AAC
	TTC	GAA ↑		CAA	↑TTG
Hpa II	↓CC	GG	*Hae* III	GG↓	CC
	GG	CC ↑		CC	↑GG
Acy I	↓GPuC	GPyC	*Afl* III	↓ACPu	PyGT
	CPyG	CPuG ↑		TGPy	PuCA ↑
Aha II	↓GPuC	GPyC	*Ava* I	↓CPyC	GPuG
	CPyG	CPuG ↑		GPuG	CPyC ↑
Cfr I	↓PyGG	CCPu	*Gdi* II	↓PyGG	CCG
	PuCC	GGPy ↑		PuCC	GGC ↑
Hae II	PuGC	GCPy ↓	*Hgi* CI	↓GGPy	PuCC
	PyCG	CGPu ↑		CCPu	PyGG ↑
Hgi JII	GPuG	CPyC ↓	*Nsp* CI	PuCA	TGPy ↓
	CPyC ↑	GPuG		PyGT ↑	ACPu
Alu I	AG↓	CT	*Asu* II	↓TTC	GAA
	TC	↑GA		AAG	CTT ↑
Cla I	↓ATC	GAT	*Bse* PI	↓GCG	CGC
	TAG	CTA ↑		CGC	GCG ↑

Arrows denote the site of cleavage; Pu = purine; Py = pyrimidine. All base sequences are shown with the 5' end of the upper strand at the left.

A given combination of four bases will occur by chance only once every several hundred base pairs, while specific groups of six will randomly occur only once every several thousand base pairs. Thus, restriction enzyme cutting generally leads to fragments of significant length. For example, the six-base-pair recognition sequence of the enzyme *Eco* RI occurs only once in the SV40 DNA (5243 base pairs), while the enzyme *Hin* dIII breaks SV40 at five different sites. Moreover, sequences present in one DNA are frequently absent in other DNAs. For instance, phage T7 DNA (39,936 base pairs) does not contain even one example of the six-base-pair sequences recognized by *Eco* RI.

Kinking in *Eco* RI-DNA Recognition Site Complexes[83]

How a restriction enzyme binds to DNA was first revealed by X-ray analysis of the complex formed between *Eco* RI and the DNA segment

TCGCGAATTCGCG
GCGCTTAAGCGCT

containing within it the recognition sequence GAATTC. Such static complexes are made in the absence of Mg^{2+}, whose presence is necessary for *Eco* RI to hydrolyze its substrate. The two polypeptide subunits of *Eco* RI effectively wrap around the double helix, making tight complementary contacts with bases lining the major groove. No contacts are made with the minor groove. The DNA conformation differs from both the classical A and B forms; the distortions are concentrated in three abrupt kinks that separate four blocks of three base pairs each. Because of the symmetry of the DNA sequence, the first and third kinks are identical and differ from the central kink. The two identical terminal DNA segments have A-like conformations, while the central block resembles B-DNA. The terminal kinks reflect transition states separating A- and B-DNA. The central kink within the block of B-DNA contains a dramatic unwinding of the double helix by 25° and a 12° bend of the DNA that enlarges the major groove to permit access by *Eco* RI (Figure 9-36). This structure differs significantly from that found for the same sequence when not bound to a restriction enzyme. Especially unique to the complex is the central kink, which probably only exists very briefly in the absence of binding by *Eco* RI.

Figure 9-35
The action of restriction enzymes. (a) Restriction enzymes often cut the DNA molecule off center from the central axis of the recognition site, resulting in short, single-stranded tails. Here a sequence contains the recognition site for *Hin* dIII, which cleaves the sequence at the points marked by the arrows. (b) Other restriction enzymes, such as *Hpa* I shown here, cut at the central axis, resulting in blunt-ended DNA fragments. In both (a) and (b), the restriction enzyme recognition sequence is shaded.

Figure 9-36
Interaction of B-DNA with *Eco* RI. The arrow points to the central kink in the block of B-DNA interacting with the *Eco* RI protein. The central kink unwinds the double helix by 25° and bends the DNA by 12°, which consequently enlarges the major groove. (Courtesy of J. Rosenberg.)

B-DNA–*Eco* RI complex   B-DNA

Methylated Recognition Sites Protect Cells from Their Own Restriction Enzymes[84, 85]

Restriction enzymes are prevented from acting on the DNA native to their organism of origin by companion enzymes that add methyl groups to specific adenine or cytosine residues within their recognition sites (Figure 9-37). These highly specific methylases, unrelated to the more general methylases responsible for creating the m^5C and the m^6A found within procaryotic cells, place methyl groups only on appropriate bases within their recognition sites rather than elsewhere on the DNA molecule. Thus, a given bacterial chromosome is protected only from cutting by its own restriction enzymes, not from cutting by enzymes that bind to different recognition sites. Only one of the two complementary strands of DNA at a recognition site need be methylated for protection. Such **hemimethylation** is necessary, since if hemimethylated DNA were susceptible to cleavage, newly replicated chromosomes might often be cleaved before the cell's methylase had acted to form fully methylated DNA.

Although a given restriction enzyme and its companion methylase both bind to the same specific groups of bases, there is no obvious amino acid homology between their polypeptides, and they are encoded by completely different genes. The functional forms of methylases are monomers, perhaps because their substrates are singlechain segments.

Separating DNA Fragments on Agarose Gels[86, 87]

The DNA fragments that result from restriction enzyme cutting are easily separated and displayed by electrophoresis through agarose gels. The rate at which DNA pieces move in such gels depends on their lengths; the distances traveled decrease as the fragment lengths increase. After DNA has moved through a gel, it is easily visualized by the addition of dyes like ethidium bromide, which bind to DNA and generate visible bands corresponding to discrete fragments whose lengths can be established through calibration with DNA molecules of known weight (Figure 9-38). DNA moves unharmed

Figure 9-37
Specific adenine and cytosine residues in restriction enzyme recognition sites may become methylated to prevent the restriction enzymes from acting on DNA native to the cell. *S*-adenosylmethionine is the methyl donor.

through agarose gels and can be recovered as biologically functional double helices.

The ordering of DNA restriction fragments along a chromosome or chromosomal segment gives rise to a **restriction map** (Figure 9-39). The first such maps were obtained when viral chromosomes were digested with enzymes such as *Hin* dII, *Hin* dIII, and *Eco* RI. Ordering the fragments was accomplished by studying the temporary appearance of incompletely cut fragments as digestion went to completion. For example, SV40 is first converted into a linear molecule that is in turn progressively cut up into 11 smaller fragments. Maps produced by different enzymes are necessarily different; initially, the most useful are those made by enzymes whose recognition sequences are rare in DNA and which thereby generate small numbers of easily separable fragments.

Using a Methylase to Create Extended Restriction Enzyme Recognition Sequences[88, 89]

Through the clever employment of two highly specific methylases, extended recognition sequences for the restriction enzyme *Dpn* I can be constructed in vitro that contain eight and ten base pairs. Even the shortest of these sequences should occur by chance only once every 65,000 (4^8) base pairs in a given DNA molecule, thereby creating the opportunity to break even very large DNA molecules into a relatively small number of pieces. This ability to make abnormally long recognition sequences rests on a unique property of the restriction enzyme *Dpn* I to cut only DNA that has methylated adenine (^mA) within its restriction site:

$$G^mA \ TC$$
$$C \ T^mAG$$

Lambda DNA
Hind III DIGEST

Base Pairs

23,130
9,419
6,557
4,371

2,322
2,028

Figure 9-38
Separation of DNA restriction fragments by electrophoresis through an agarose gel. The smaller fragments move more rapidly than the larger ones; thus, the sizes of the restriction fragments can be estimated by comparing the distances they travel through the gel to the distances traveled by DNA molecules of known size. (Courtesy of D. Helfman.)

Figure 9-39

Cleavage map of the SV40 genome. The zero point of the map is the *Eco* R1 site. For clarity, the circular genome is shown opened at the R1 site, and the cleavage sites (and resulting fragments) for each restriction enzyme are indicated on a separate line.

(a)

(b)

Figure 9-40

Creation of long restriction enzyme recognition sequences. (a) *M. Taq* I is a methylase that adds a methyl group to the adenine of its recognition sequence of repeated TCGA. This creates an eight-base-pair recognition sequence for the restriction enzyme *Dpn* I. (b) An even longer, ten-base-pair *Dpn* I recognition sequence is generated by methylase *M. Cla* I, which recognizes and methylates two partially overlapping ATCGAT sequences. Such long restriction sequences occur very rarely in DNA and thus allow the cleavage of large DNA molecules into relatively few fragments. The recognition sequences of the methylases are shaded on one DNA strand. [After M. McClelland et al., *Proc. Nat. Acad. Sci.* 81 (1984):984]

Thus, *Dpn* I ordinarily lacks the ability to cut DNA from most organisms (e.g., vertebrate DNA is not broken by it). Such DNA, however, becomes susceptible to *Dpn* I following exposure to certain of those methylases whose recognition sequences overlap the *Dpn* I sequence and which add methyl groups to the appropriate adenine residues. For example, methylation by *M. Taq* I at a direct repeat of its TCGA methylation recognition sequence produces an eight-base-pair *Dpn* I restriction recognition sequence (Figure 9-40a). An even longer, ten-base-pair *Dpn* I recognition sequence is generated by methylation by *M. Cla* I at two partially overlapping *M. Cla* I methylation recognition sequences (Figure 9-40b).

Cleavage at these inherently rare recognition sequences should prove very useful in the generation of restriction maps of DNAs containing more than 10^5 base pairs. Until very recently, there was no way to physically separate the very large DNA molecules that can result from the construction of these sites. Now, however, through the use of pulse field electrophoresis (electrophoresis using short pulses of electricity in two different directions) through agarose gels, DNA molecules containing up to 10^6 base pairs can be routinely handled. Already, the *E. coli* chromosome (4×10^6 base pairs) has been cut into some 20 different fragments (Figure 9-41), and its restriction map should soon be available.

Ligating DNA Fragments to Create Recombinant DNA[90–91]

The complementary single-stranded tails produced by restriction enzymes that make staggered cuts tend to associate briefly with each other by base-pairing. Such unstable unions can be made permanent by adding the joining (ligating) enzyme DNA ligase (Figure 9-42). This enzyme plays a vital role in DNA replication and recombination within cells by catalyzing the formation of new phosphodiester bonds. Ligase acts only in the presence of ATP (or NAD$^+$), which provides the energy needed to make the backbone bonds. At first, ligase was thought capable of linking together only those fragments bearing complementary tails. If, however, all reactants are present in high concentrations, then blunt-ended DNA fragments can also be joined together. The frequency of such random bonding, however, is very low, and frequently complementary tails are enzymatically added to blunt-ended fragments to promote their joining (Figure 9-43).

Since any two fragments bearing complementary tails will tend to base-pair with each other, the pieces that ligase permanently joins together need not have been previously linked. Ligase, together with restriction enzymes that make staggered cuts (such as *Eco* RI), thus provides a simple way to produce recombinant DNA, recombining in the test tube DNA pieces originating from different DNA molecules. The generation of recombinant DNA molecules by itself would have had little immediate impact if it had not soon been followed by the development of procedures that allow the selective amplification of such newly created DNA (a process known as *cloning*). The heart of gene-cloning procedures involves the insertion of desired DNA sequences into autonomously replicating DNA molecules (**cloning vectors**) that multiply to form many identical clones after their reintroduction back into suitable host cells (see Figure 7-29). Most frequently used as cloning vectors today are several autonomously replicating plasmid and viral DNAs.

Libraries of Cloned DNA Fragments[92]

Although electrophoresis through agarose gels now provides a practical way to isolate restriction fragments from viral chromosomes, bacterial chromosomes, or plasmids, it cannot be used to isolate single fragments from the much bigger chromosomes of higher cells. Even those enzymes that cut within relatively rare recognition sequences produce so many fragments from any higher cell DNA that the fragments move in gels as smears rather than as discrete bands. For example, the 8×10^9 base-pair human genome is cut into approximately 2 million different fragments by *Eco* R1.

Study of individual pieces of cellular DNA only became possible with the advent of recombinant DNA procedures. These allow the separation of even very similar-size DNA pieces by individually inserting them into cloning vectors. All the restriction fragments from genomes even as large as *Drosophila* can be inserted in only some 40,000 different *E. coli* plasmids. The subsequent multiplication of these individual vectors (known as a "gene library") can generate all the pure DNA ever needed for exhaustive molecular characterization. Now the making of a gene library containing all the DNA sequences from a given cell is a routine task. Often far from trivial, however, is the identification of the library member that interests us. Fortunately, there is a large number of sophisticated, non-brute-force methods for selecting the desired genes. Some of these techniques are described in Chapter 19. However, it is worth emphasizing here that only through the development of gene libraries have recombinant DNA procedures so drastically raised the sights of biological research.

Figure 9-41
Pulse field electrophoresis gel of a *Not1* restriction enzyme digest of *E. coli* DNA. The 20 fragments range in size from 25 kb to 450 kb, not including the largest fragment, whose size is unknown. [Courtesy of C. L. Smith.]

Figure 9-42
In the presence of ATP or NAD$^+$, DNA ligase can covalently join DNA fragments that were previously in only a transitory, unstable union.

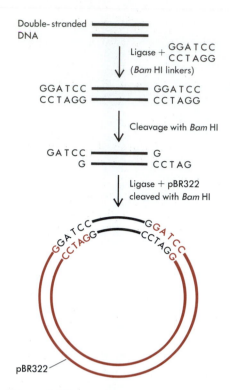

Figure 9-43
Double-stranded oligonucleotides called *Bam* HI linkers are used to add complementary ends to blunt-ended double-helical DNA. The linkers are first ligated to both ends of the given DNA. Then the product is treated with *Bam* HI, which generates "sticky ends" to promote joining to other DNA fragments, such as the plasmid pBR322 shown here.

Very Long DNA Segments Can Be Rapidly Sequenced[93-96]

Following close upon the discovery of restriction enzymes and development of recombinant DNA methodologies has been the formulation of two powerful methods for accurately establishing the order of the four bases (A, G, C, and T) along given DNA fragments. One method uses chemical reagents (e.g., dimethylsulfate) that react with specific bases to break DNA preferentially at given nucleotides. By correlating the appearance of fragments of specified length with the specific base destroyed by chemical attack, the exact order of bases along the original unbroken DNA strand can be determined (Figure 9-44). This method, as well as the enzymatic method described next, makes use of the fact that electrophoresis through polyacrylamide gels can resolve single chains of DNA differing in length by only a single nucleotide. On a given gel, over 300 bands can be resolved. These chemical procedures, first used to reveal the sequences of key regulatory regions in bacterial DNA (e.g., the *lac* operon), soon allowed the entire sequence of the plasmid cloning vector pBR322 (4362 base pairs) to be worked out by a single scientist in only one year.

The second widely used method for DNA sequencing uses an enzyme, DNA polymerase I, to produce different-size DNA fragments that when subsequently separated on gels allow, as in the first method, the direct reading of the nucleotide order. In this method, the single-stranded DNA being analyzed is used as a template for the synthesis of complementary DNA chains using as substrates chain-terminating 2'3'-dideoxynucleotides as well as the four normal nucleotide precursors of DNA. These nucleotide analogs become incorporated at the growing ends of DNA chains, but do not allow further elongation, since they cannot form phosphodiester bonds with incoming precursors. Inclusion of a small amount of a given dideoxynucleotide in a reaction mixture leads to its occasional random incorporation at all the sites usually occupied by its normal equivalent. A series of incompletely elongated DNA chains are thereby produced that are specifically terminated by the given dideoxy residue. To determine a sequence, four separate DNA synthesis reactions, each involving a different chain-terminating nucleotide, are run and their product chains electrophoresed side by side on a polyacrylamide gel to display bands of ever-increasing length (Figure 9-45). This method quickly revealed the complete sequence of the phage φx174 DNA (5577 base pairs) and more recently has established that of phage λ DNA (48,513 base pairs).

Now the dideoxy method is increasingly used in preference to the chemical fission approach. Once dideoxynucleotides became commercially available, the ease with which this technique could be mastered became apparent, and it is likely to dominate sequencing efforts for the next decade.

The Significance of Many Palindromic Sequences Remains Unclear[97, 98]

An unexpected consequence of the first serious DNA sequencing was the discovery of many inverted repetitious sequences in which the same (or almost the same) sequence runs in opposite directions if one

Figure 9-44

DNA-sequencing method developed by Maxam and Gilbert. This method uses chemical reagents to destroy specific nucleotide bases and thus break the DNA molecule at specific sites. First the strands of the DNA molecule are labeled radioactively at one end (usually the 5′ end), and the two strands are separated (only one will be sequenced). Then aliquots of the chosen strand are treated with four different chemical reagents that break the strand at one or two specific nucleotides; the treatment is limited so that at most a single residue of the susceptible base(s) in the molecule will react. Thus, in each reaction mixture, a nested set of radioactive fragments is generated, as shown here for only the reaction mixture that destroys C residues. Finally, gel electrophoresis is used to separate the products of each reaction by size. The pattern of radioactive bands seen on X-ray film immediately reveals the sequence.

switches from one strand to its complement at the central axis of symmetry, for example

GTATCC GGATAC
CATAGG CCTATG·

Figure 9-45
DNA-sequencing method developed by Sanger. A dideoxynucleotide is incorporated into a growing DNA strand, subsequently stopping chain growth, since it cannot form a phosphodiester bond with the next incoming nucleotide. Four different reactions are run, each with a different dideoxynucleotide. The products of each reaction are a series of incompletely elongated segments, which are separated by gel electrophoresis. As in the Maxam-Gilbert method, the sequence can be read from the bands produced in the gel.

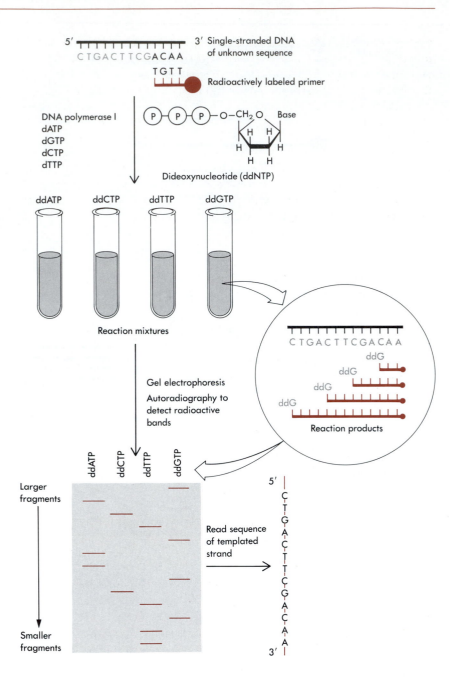

Some inverted palindromes (a palindrome is a word that reads the same backward and forward), like those recognized by certain restriction enzymes (see Table 9-3), are relatively short (three to ten bases in a given direction). Others are much longer, usually being somewhat imperfect as well as being interrupted by long stretches of unrelated sequences. Inverted sequences separated by hundreds to thousands of unrelated nucleotides mark the ends of certain movable genetic elements, while others permit site-specific crossing-over events that allow flip-flop genetic elements to function in alternative orientations.

Many of the medium-size, continuous, inverted palindromes probably serve as DNA recognition sites for **multimeric proteins** composed of identical subunits (e.g., repressors). The significance of the longer continuous palindromes is less obvious. As mentioned earlier in this chapter, when they are nearly perfect and present on super-coiled DNA, they have the potential to undergo semistable steric rearrangements in which the complementary sequences on each DNA strand base-pair to form cruciform (hairpin) loops extending out from the main helical axes. Such looping would remove some of the stress of supercoiling as well as generate structures that might be differentially recognized by specific DNA binding proteins.

In any case, those sections of RNA chains transcribed off palindromic regions will tend to form hydrogen-bonded hairpin loops, the stability of which will be directly related to how perfectly symmetrical the palindromic sequences are. Some palindromes may have evolved to endow structural compactness upon their single-stranded RNA products.

The Storing of DNA Sequence Information in Computers[99]

Computers are now routinely used to store DNA sequences. Such information is directly fed into the computer as the correct bases are read off gels, thereby avoiding errors that inevitably would occur if very long sequences were manually copied. Once in the computer, sequences can be scanned for restriction enzyme recognition sites, start and stop signals for RNA synthesis, the presence of inverted palindromes, preferred segments of potential Z-DNA (alternating purine and pyrimidine stretches), and homologies to other known DNA sequences. They can also be read to give amino acid sequences of proteins. High-powered software programs to carry out these searches are available from many sources, and the very large amounts of DNA sequence information already available requires the use of increasingly more powerful computers. The time has passed when a small, primitive desk-top computer can do the job, and much recombinant DNA research requires access to sophisticated minicomputers (e.g., VAX).

DNA sequencing itself can be speeded up through the use of a powerful computer. The working out of restriction maps for long sequences can consume much more time than DNA sequencing itself. It thus often makes sense to sequence large numbers of potentially overlapping DNA fragments and then use a computer to search for the sequence overlaps that directly order the fragments along their respective chromosomes. This approach only became practical when the M13 variation of the dideoxy sequencing method was developed, in which the DNA fragments to be analyzed are inserted at a precise spot in the single-stranded DNA chain of phage M13 (Figure 9-46). The subsequent DNA synthesis (sequencing) reactions all commence by elongating a single, short (15 to 17 bases) DNA **primer** that is complementary in sequence to the M13 region immediately adjacent to the foreign DNA insert. By using the same primer for all reactions, sequencing by the M13 method becomes almost child's play, allowing sequencing to be determined at rates never even imagined only a few years ago (about 2000 base pairs a week). Viral chromosomes containing more than 10^5 base pairs are now being sequenced, and the *E. coli* chromosome of about 4×10^6 base pairs is currently in the process of being worked out. Still seemingly too large are the chromosomes of higher vertebrates, but even here, selected key segments will eventually be sequenced.

Figure 9-46

The M13 DNA-sequencing method. First the foreign DNA strand to be sequenced is cloned into the single-stranded M13 phage genome using *Eco* RI. A short DNA sequence (primer) that is complementary to the M13 region immediately adjacent to the foreign DNA insert then elongates to form complementary DNA, which allows the sequence of the foreign DNA to be determined by the Sanger method.

Summary

Most DNA occurs as right-handed double helices held together by hydrogen bonds between the bases adenine and thymine and between the bases guanine and cytosine. The strict base-pairing rules (AT and CG) underlying all double helices reflect the highly fixed locations of the hydrogen atoms in the purine and pyrimidine bases. Adenine and cytosine almost always exist in the amino as opposed to the imino tautomeric forms, while guanine and thymine almost always exist in the keto as opposed to enol forms. The biological functioning of DNA depends on these highly preferred tautomeric forms for the accurate synthesis of complementary DNA or RNA chains upon single-stranded DNA templates.

The two strands of the double helix fall apart (denature) upon exposure to high temperature, extremes of pH, or any agent that causes the breakage of the individual hydrogen bonds. Upon slow return to normal cellular conditions, the denatured single strands can specifically reassociate to biologically active double helices (renature).

While all cellular DNA occurs as double helices, the genetic component of certain small viruses is single-stranded DNA. Such single-stranded (+) DNA chromosomes can become converted into +/− double helices after entry into cells. They then may use their − strands as templates to make new single-stranded + progeny chains identical to those found in the parent viral chromosomes. Single-stranded DNA molecules are more compact than double helices and often form imperfectly base-paired hairpin loops that allow more effective stacking of the hydrophobic flat surfaces of the purine and pyrimidine bases. True crystals of DNA can be obtained from short, chemically synthesized polydeoxynucleotides. The precise determination of their structures is then possible using X-ray crystallographic procedures. The structures determined by this method confirm the essential features of DNA postulated in 1953 from model building. Right-handed DNA has two major forms: a compacted A form containing tilted base pairs, and a longer and narrower B form, in which the almost flat base pairs lie approximately perpendicular to the helical axis. Both the A and B forms have major grooves that expose the insides and minor grooves that expose the outsides of the base pairs to the surrounding environment.

DNA segments containing alternating purine and pyrimidine bases can become converted to left-handed Z helices. In Z-DNA, the purines have syn conformations and the pyrimidines have the anti conformations characteristic of right-handed DNA. Transition into the Z form is favored by methylation of cytosine residues. Methylation also allows some nonalternating purine-pyrimidine sequences to become left-handed. Z-DNA is more elongated and slimmer than either A- or B-DNA and contains only one surface indentation (the minor groove). Almost all double helices in solution or in cells are in the B form and only very rarely are found to be in the Z form.

In solution, B-DNA constantly bends and twists. Abrupt kinks sometimes form, often favored by DNA sequences that affect the exact ways the four different base pairs can stack on each other. Some highly transient kinks allow the insertion (intercalation) of flat, ringed heterocyclic molecules into DNA. This insertion increases the translation of successive base pairs along the DNA axis and thereby extensively unwinds the DNA molecule.

Almost all cellular DNA molecules are extremely large, with only one DNA molecule now thought to reside within a given chromosome. The smallest DNA molecules are found within viruses; the tiniest contains only several thousand base pairs. Even the smallest cellular chromosomes (bacterial) contain at least a million base pairs. Some cellular DNA molecules within cells exist in linear forms, but these are usually of viral origin. The remainder are circular and often exist as supercoils; their two chains are slightly more (positive supercoils) or slightly less (negative supercoils) twisted around each other than in unconstrained linear double helices. The extent of DNA supercoiling within cells is in part controlled by enzymes called topoisomerases, which either add or remove supercoils. Within cells, most DNA is negatively supercoiled, partially owing to its attachment to specific DNA binding proteins. The best-understood of these proteins are the core histones that bind to DNA to form the nucleosomes that dominate chromatin structures. When bound to DNA, these core histones exist as ellipsoid-shaped octomers around which the DNA makes two superhelical turns, which slightly untwist the double helices. When separated from the histones, circular DNA becomes negatively supercoiled. It was originally assumed that linear DNA molecules would never appear supercoiled within cells. However, supercoiled regions of linear chromosomes exist, perhaps reflecting the binding of currently undetected specific protein and possibly RNA molecules.

Negative supercoiling of circular DNA molecules is sometimes partially relieved by the creation of localized denatured regions within DNA segments rich in the relatively weak AT base pairs. Localized denaturation is especially favored for segments containing self-complementary regions capable of forming hydrogen-bonded hairpin (cruciform) loops. Negative supercoils may also be lost through the conversion of sections of right-handed DNA into left-handed DNA.

Key tools for the analysis of DNA molecules are a group of bacterial enzymes that cut at specific nucleotide sequences. These enzymes, known as restriction enzymes, generally make breaks only within specific "recognition" sequences that exhibit twofold symmetry. As a result, they usually make cuts through both strands. Enzymes with over 100 different specificities have been isolated from various bacteria. Within a cell, a given restriction enzyme may function to degrade foreign DNA that may have entered the cell. The enzyme is prevented from breaking down the DNA native to the cell by very specific DNA methylases, which add methyl groups only to specific adenine or cytosine residues within the recognition sequences. All bacterial cells that contain a restriction enzyme thus also contain a corresponding methylase. Only one strand within a recognition sequence needs to be methylated to prevent its degradation. In this way, newly replicated hemimethylated DNA is protected from destruction. Restriction enzymes and their corresponding methylases have no common sequences and are coded by different genes.

The DNA fragments created by restriction enzyme cuts can be easily separated by electrophoresis on agarose gels, which separates fragments by their different rates of movement. The larger the fragment, the more slowly it moves. Movement through such gels does not harm DNA, which can be recovered in biologically active form. The original order of DNA restriction fragments along a chromosome (or chromosomal segment) is known as a restriction map. Maps produced by different enzymes are usually different; the most useful are those maps made by enzymes whose recognition sequences are rare in DNA.

Addition of the joining enzyme, DNA ligase, to mixtures of restriction fragments emanating from two genetically different DNA molecules creates recombinant DNA molecules. Such joining (ligation) occurs much more readily with those fragments containing single-stranded tails generated by cuts that occur outside of the center of the recognition sequence. However, blunt-ended fragments can also be joined together, although at a much slower rate. The most useful recombinant DNA molecules are those in which the desired DNA segments are inserted into autonomously replicating DNA molecules (cloning vectors). Collections of large numbers of different DNA segments from unique cellular or organismal sources that are each inserted into separate cloning vectors are referred to as "gene libraries."

Restriction fragments containing several hundred base pairs now can be rapidly sequenced by either of two very efficient DNA-sequencing methods. Continuous sequences of several thousand base pairs can now be determined within a week, and the sequences of viral chromosomes of up to several hundred thousand base pairs can be established in a year or two. Efforts are now being made to work out the sequence of the *E. coli* chromosome, a task that may be accomplished by 1990.

Bibliography

General References:

Kornberg, A. 1980. *DNA Replication*. San Francisco, Calif.: Freeman.

Kornberg, A. 1982. *1982 Supplement to DNA Replication*. San Francisco, Calif.: Freeman.

Linn, S. M., and R. J. Roberts, eds. 1982. *Nucleases*. Cold Spring Harbor, N.Y.: Cold Spring Harbor Laboratory.

Saenger, W. 1984. *Principles of Nucleic Acid Structure*. New York: Springer-Verlag.

Sarma, R. H., ed. 1981. *Bimolecular Stereodynamics*. Vols. 1 and 2. Guilderland, N.Y.: Adenine Press.

"Structures of DNA." 1982. *Cold Spring Harbor Symposium on Quantitative Biology*. Vol. 47. Cold Spring Harbor, N.Y.: Cold Spring Harbor Laboratory.

Cited References

1. Franklin, R. E., and R. G. Gosling. 1953. "Molecular Configuration in Sodium Thymonucleate." *Nature* 171:740–741.

2. Watson, J. D., and F. H. C. Crick. 1953. "Molecular Structure of Nucleic Acids: A Structure for Deoxynucleic Acids." *Nature* 171:737–738.

3. Wilkins, M. H. F., A. R. Stokes, and H. R. Wilson. 1953. "Molecular Structure of Deoxypentose Nucleic Acids." *Nature* 171:738–740.

4. Crick, F. H. C., and J. D. Watson. 1954. "The Complementary Structure of Deoxyribonucleic Acid." *Proc. Roy. Soc.* (A) 223:80–96.

5. Langridge, R., H. R. Wilson, C. W. Hooper, M. H. F. Wilkins, and L. D. Hamilton. 1960. "The Molecular Configuration of Deoxyribonucleic Acid. I. X-Ray Diffraction Study of a Crystalline Form of the Lithium Salt." *J. Mol. Biol.* 2:19–37.

6. Wolfenden, R. V. 1969. "Tautomeric Equilibria in Inosine and Adenosine." *J. Mol. Biol.* 40:307–310.

7. Pieber, M., P. A. Kroon, J. H. Prestegard, and S. I. Chan. 1973. "Erratum: Tautomerism of Nucleic Acid Bases." *J. Amer. Chem. Soc.* 95:3408.

8. Doty, P., J. Marmur, J. Eigner, and C. Schildkraut. 1960. "Strand Separation and Specific Recombination in Deoxyribonucleic Acids: Physical Chemical Studies." *Proc. Nat. Acad. Sci.* 46:461–476.

9. Marmur, J., and L. Lane. 1960. "Strand Separation and Specific Recombination in Deoxyribonucleic Acids: Biological Studies." *Proc. Nat. Acad. Sci.* 46:453–461.

10. Schildkraut, C. L., J. Marmur, and P. Doty. 1961. "The Formation of Hybrid DNA Molecules and Their Use in Studies of DNA Homologies." *J. Mol. Biol.* 3:595–617.

11. Marmur, J., R. Rownd, and C. L. Schildkraut. 1963. "Denaturation and Renaturation of Deoxyribonucleic Acid." *Prog. Nucleic Acid Res.* 1:231–300.

12. Denhardt, D. T., D. Dressler, and D. S. Ray. 1978. *The Single Stranded DNA Phages.* Cold Spring Harbor, N.Y.: Cold Spring Harbor Laboratory.

13. Schaller, H., H. Voss, and S. Gucker. 1969. "Structure of the DNA of Bacteriophage fd. II. Isolation and Characterization of a DNA Fraction with Double Strand-Like Properties." *J. Mol. Biol.* 44:445–458.

14. Shen, C.-K. J., A. Ikoku, and J. E. Hearst. 1979. "A Specific DNA Orientation in the Filamentous Bacteriophage fd as Probed by Psoralen Crosslinking and Electron Microscopy." *J. Mol. Biol.* 127:163–175.

15. Wing, R. M., H. R. Drew, T. Takano, C. Broka, S. Tanaka, K. Itakura, and R. E. Dickerson. 1980. "Crystal Structure Analysis of a Complete Turn of B-DNA." *Nature* 287:755–758.

16. Shakked, Z., D. Rabinovich, W. B. T. Cruse, E. Egert, O. Kennard, G. Sala, S. A. Salisbury, and M. A. Viswamitra. 1981. "Crystalline A-DNA: The X-Ray Analysis of the Fragment d(G-G-T-A-T-A-C-C)." *Proc. Roy. Soc. Lond.* (B) 213:479.

17. Dickerson, R. E. 1983. "The DNA Helix and How It Is Read." *Sci. Amer.* 249:94–111.

18. Arnott, S., F. Hutchinson, M. Spencer, M. H. F. Wilkins, W. Fuller, and R. Langridge. 1966. "X-Ray Diffraction Studies of Double Helical Ribonucleic Acid." *Nature* 211:227–232.

19. Wang, A. H.-J., S. Fujii, J. H. van Boom, and A. Rich. 1982. "Molecular Structure of the Octamer d(G-G-C-C-G-G-C-C): Modified A-DNA." *Proc. Nat. Acad. Sci.* 79:3968–3972.

20. Wang, A. H.-J., S. Fujii, J. H. van Boom, G. A. van der Marel, S. A. A. van Boeckel, and A. Rich. 1982. "Molecular Structure of r(GCG)d(TATACGC): A DNA-RNA Hybrid Helix Joined to Double Helical DNA." *Nature* 299:601–604.

21. Dickerson, R. E. 1983. "Base Sequence and Helix Structure Variation in B and A DNA." *J. Mol. Biol.* 166:419–441.

22. Conner, B. N., C. Yoon, J. L. Dickerson, and R. E. Dickerson. 1984. "Helix Geometry and Hydration in an A-DNA Tetramer: CCGG." *J. Mol. Biol.* 174:663–695.

23. Arnott, S., R. Chandrasekaran, I. H. Hall, and L. C. Puigjaner. 1983. "Heteronomous DNA." *Nucleic Acid Res.* 11:4141–4155.

24. Pohl, F. M., and T. M. Jovin. 1972. "Salt-Induced Co-Operative Conformational Change of a Synthetic DNA: Equilibrium and Kinetic Studies with Poly(dG-dC)." *J. Mol. Biol.* 67:375–396.

25. Wang, A. H.-J., G. J. Quigley, F. J. Kolpak, J. L. Crawford, J. H. van Boom, G. van der Marel, and A. Rich. 1979. "Molecu-

lar Structure of a Left-Handed DNA Fragment at Atomic Resolution." *Nature* 282:680–686.

26. Hill, R. J., and B. D. Stollar. 1983. "Dependence of Z DNA Antibody Binding to Polytene Chromosomes on Acid Fixation and DNA Torsional Strain." *Nature* 305:338–340.

27. Rich, A., A. Nordheim, and A. H.-J. Wang. 1984. "The Chemistry and Biology of Left-Handed Z-DNA." *Ann. Rev. Biochem.* 53:791–846.

28. Azorin, F. and A. Rich. 1985. "Isolation of Z-DNA Binding Proteins from SV40 Minichromosomes: Evidence for Binding to the Viral Control Region." *Cell* 41:365–374.

29. Kmiec, E. B., K. J. Angelides, and W. K. Holloman. 1985. "Left-handed DNA and the Synaptic Pairing Reaction Promoted by Ustilago Rec1 Protein." *Cell* 40:139–145.

30. Wang, A. H.-J., R. Gessner, G. A. van der Marel, J. H. van Boom, and A. Rich. 1986. "A Crystal Structure of Z-DNA Without an Alternating Purine-Pyrimidine Sequence." *Nature.* In press.

31. Razin, A., and A. D. Riggs. 1980. "DNA Methylation and Gene Function." *Science* 210:604–610.

32. Behe, M., and G. Felsenfeld. 1981. "Effects of Methylation on Synthetic Polynucleotides. The B-Z Transition in Poly(dG-m5dC) Poly(dG-m5dC)." *Proc. Nat. Acad. Sci.* 78:1619–1623.

33. Fujii, S., A. H.-J. Wang, G. van der Marel, J. H. van Boom, and A. Rich. 1982. "The Molecular Structure of (m5dC-dG)3: The Role of the Methyl Group on 5-Methyl Cytosine in Stabilizing Z-DNA." *Nucleic Acid Res.* 10:7879–7892.

34. Doerfler, W. 1983. "DNA Methylation and Gene Activity." *Ann. Rev. Biochem.* 52:93–124.

35. Sarma, R. H., ed. 1980. *Nucleic Acid Geometry and Dynamics.* New York: Pergamon Press.

36. Fratini, A. V., M. L. Kopka, H. R. Drew, and R. E. Dickerson. 1982. "Reversible Bending and Helix Geometry in a B-DNA Dodecamer: CGCGATT^{Br}CGCG." *J. Biol. Chem.* 257:14686–14707.

37. Levitt, M. 1982. "Computer Simulation of DNA Double Helix Dynamics." *Cold Spring Harbor Symp. Quant. Biol.* 47:251–262.

38. Yanagida, M., Y. Hiraoka, and I. Katsura. 1982. "Dynamic Behaviors of DNA Molecules in Solution Studied by Fluorescence Microscopy." *Cold Spring Harbor Symp. Quant. Biol.* 47:177–187.

39. Marini, J. C., S. D. Levene, D. M. Crothers, and P. T. Englund. 1982. "Bent Helical Structure in Kinetoplast DNA." *Proc. Nat. Acad. Sci.* 79:7664–7668.

40. Wu, H.-M., and D. M. Crothers. 1984. "The Locus of Sequence-Directed and Protein-Induced DNA Bending." *Nature* 308:509–513.

41. Bauer, W., and J. Vinograd. 1968. "The Interaction of Closed Circular DNA with Intercalative Dyes." *J. Mol. Biol.* 33:141–171.

42. Wang, J. C. 1974. "The Degree of Unwinding of the DNA Helix by Ethidium. I. Titration of Twisted PM2 DNA Molecules in Alkaline Cesium Chloride Density Gradients." *J. Mol. Biol.* 89:783–801.

43. Wang, A. H.-J., J. Nathans, G. van der Marel, J. H. van Boom, and A. Rich. 1978. "Molecular Structure of a Double Helical DNA Fragment Intercalator Complex Between DeoxyCpG and a Terpyridine Platinum Compound." *Nature* 276:471–474.

44. Record, M. T., Jr., S. J. Mazur, P. Melancon, J.-H. Roe, S. L. Shaner, and L. Unger. 1981. "Double Helical DNA: Conformations, Physical Properties and Interactions with Ligands." *Ann. Rev. Biochem.* 50:997–1024.

45. Cairns, J. 1961. "An Estimate of the Length of the DNA Molecule of T2 Bacteriophage by Autoradiography." *J. Mol. Biol.* 3:756.

46. Cairns, J. 1962. "A Minimum Estimate for the Length of the DNA of *Escherichia coli* Obtained by Autoradiography." *J. Mol. Biol.* 4:407.

47. Fangman, W. L. 1978. "Separation of Very Large DNA Molecules by Gel Electrophoresis." *Nucleic Acid Res.* 5:653–665.

48. Carle, G. F., and M. V. Olson. 1984. "Separation of Chromosomal DNA Molecules from Yeast by Orthogonal-Fixed-Alternative Gel Electrophoresis." *Nucleic Acid Res.* 12:5647–5664.

49. Fiers, W., and R. L. Sinsheimer. 1962. "The Structure of the DNA of the Bacteriophage ϕX174. I. The Action of Exopolynucleotidases." *J. Mol. Biol.* 5:408–419.

50. Dulbecco, R., and M. Vogt. 1963. "Evidence for a Ring Struc-

ture of Polyoma Virus DNA." *Proc. Nat. Acad. Sci.* 50:236–243.

51. Weil, R., and J. Vinograd. 1963. "The Cyclic Helix and Cyclic Coil Forms of Polyoma Viral DNA." *Proc. Nat. Acad. Sci.* 50:730–738.

52. Vinograd, J., J. Lebowitz, R. Radloff, R. Watson, and P. Laipis. 1965. "The Twisted Circular Form of Polyoma Viral DNA." *Proc. Nat. Acad. Sci.* 53:1104–1111.

53. Vinograd, J., J. Lebowitz, and R. Watson. 1968. "Early and Late Helix-Coil Transitions in Closed Circular DNA: The Number of Superhelical Turns in Polyoma DNA." *J. Mol. Biol.* 33:173–197.

54. Keller, W. 1975. "Determination of the Number of Superhelical Turns in SV40 DNA by Gel Electrophoresis." *Proc. Nat. Acad. Sci.* 72:4876–4880.

55. Crick, F. H. C. 1976. "Linking Numbers and Nucleosomes." *Proc. Nat. Acad. Sci.* 73:2639–2643.

56. Bauer, W. R., F. H. C. Crick, and J. H. White. 1980. "Supercoiled DNA." *Sci. Amer.* 243:118–133.

57. Lilley, D. M. J. 1980. "The Inverted Repeat as a Recognizable Structural Feature in Supercoiled DNA Molecules." *Proc. Nat. Acad. Sci.* 77:6468–6472.

58. Courey, A. J., and J. C. Wang. 1983. "Cruciform Formation in a Negatively Supercoiled DNA May Be Kinetically Forbidden Under Physiological Conditions." *Cell* 33:817–829.

59. Wells, R. D., R. Brennan, K. A. Chapman, T. C. Goodman, P. A. Hart, W. Hillen, D. R. Kellogg, M. W. Kilpatrick, R. D. Klein, J. Klysik, P. F. Lambert, J. E. Larson, J. J. Miglietta, S. K. Neuendorf, T. R. O'Connor, C. K. Singleton, S. M. Stirdivant, C. M. Veneziale, R. M. Wartell, and W. Zacharias. 1982. "Left Handed DNA Helices, Supercoiling, and the B-Z Junction." *Cold Spring Harbor Symp. Quant. Biol.* 47:77–84.

60. McGhee, J. D., and G. Felsenfeld. 1980. "Nucleosome Structure." *Ann. Rev. Biochem.* 49:1115–1156.

61. Kornberg, R. D., and A. Klug. 1981. "The Nucleosome." *Sci. Amer.* 244:52–79.

62. Burlingame, R. W., W. E. Love, B.-C. Wang, R. Hamlin, N.-H. Xuong, and E. N. Moudrianakis. 1985. "Crystallographic Structure of the Octameric Histone Core of the Nucleosome at a Resolution of 3.3 Å." *Science* 228:546–553.

63. Morse, R. H. and C. C. Cantor. 1985. "Nucleosome Core Particles Suppress the Thermal Untwisting of Core DNA and Adjacent Linker DNA." *Proc. Nat. Acad. Sci.* 82:4653–4657.

64. Klug, A., L. C. Lutter, and D. Rhodes. 1982. "Helical Periodicity of DNA On and Off the Nucleosome as Probed by Nucleases." *Cold Spring Harbor Symp. Quant. Biol.* 47:285–292.

65. Richmond, T. J., J. T. Finch, B. Rushton, D. Rhodes, and A. Klug. 1984. "Structure of the Nucleosome Core Particle at 7 Å Resolution." *Nature* 311:532–537.

66. Tanaka, I., K. Appelt, J. Dijk, S. W. White, and K. S. Wilson. 1984. "3-Å Resolution Structure of a Protein with Histone-Like Properties in Procaryotes." *Nature* 310:376–381.

67. Rouviere-Yaniv, J., and M. Yaniv. 1979. " *E. coli* DNA Binding Protein HU Forms Nucleosome-Like Structures with Circular Double Stranded DNA." *Cell* 17:265–274.

68. Brown, P. O., and N. R. Cozzarelli. 1979. "A Sign Inversion Mechanism for Enzymatic Supercoiling of DNA." *Science* 206:1081–1084.

69. Gellert, M. 1981. "DNA Topoisomerases." *Ann. Rev. Biochem.* 50:879–910.

70. Dean, F., M. A. Krasnow, R. Otter, M. M. Matzuk, S. J. Spengler, and N. R. Cozzarelli. 1982. "*Escherichia coli* Type-1 Topoisomerases: Identification, Mechanism and Role in Recombination." *Cold Spring Harbor Symp. Quant. Biol.* 47:769–777.

71. DiNardo, S., K. A. Voelkel, R. Sternglanz, A. E. Reynolds, and A. Wright. 1982. "*Escherichia coli* DNA Topoisomerase I Mutants Have Compensatory Mutations at or Near DNA Gyrase Genes." *Cold Spring Harbor Symp. Quant. Biol.* 47:779–784.

72. Wang, J. C. 1982. "DNA Topoisomerases." *Sci. Am.* 247:94–109.

73. Liv, L. 1983. "DNA Topoisomerases—Enzymes That Catalyze the Breaking and Rejoining of DNA." *CRC Crit. Rev. Biochem.* 15:1–9.

74. Drilica, K. 1984. "Biology of Bacterial Deoxyribonucleic Acid Topoisomerases." *Microbiol Revs.* 48:273–302.

75. Kirchhausen, T., J. C. Wang, and S. C. Harrison. 1985. "DNA Gyrase and its Complexes with DNA: Direct Observation by Electron Microscopy." *Cell* 41:933–943.

76. Worcel, A., and E. Burgi. 1972. "On the Structure of the Folded Chromosome of *Escherichia coli*." *J. Mol. Biol.* 71:127–147.

77. Pettijohn, D. E., and R. Hecht. 1973. "RNA Molecules Bound to the Folded Bacterial Genome Stabilize DNA Folds and Segregate Domains of Supercoiling." *Cold Spring Harbor Symp. Quant. Biol.* 38:31–41.

78. Luchnik, A. N., V. V. Bakayev, and V. M. Glaser. 1982. "DNA Supercoiling: Changes During Cellular Differentiation and Activation of Chromatin Transcription." *Cold Spring Harbor Symp. Quant. Biol.* 47:793–801.

79. Earnshaw, W. C., B. Halligan, C. A. Cooke, M. M. S. Heck, and L. F. Liu. 1985. "Topoisomerase II Is a Structural Component of Mitotic Chromosome Scaffolds." *J. Cell Biol.* 100:1706–1715.

80. Earnshaw, W. C., and M. M. S. Heck. 1985. "Localization of Topoisomerase II in Mitotic Chromosomes." *J. Cell Biol.* 100:1716–1725.

81. Danna, K., and D. Nathans. 1971. "Sequence Specific Cleavage of Simian Virus 40 DNA by Restriction Endonucleases of *Hemophilus influenzae*." *Proc. Nat. Acad. Sci.* 68:2913–2917.

82. Smith, H. 1979. "Nucleotide Sequence Specificity of Restriction Endonucleases." *Science* 205:455–462.

83. Frederick, C. A., J. Grable, M. Melia, C. Samudzi, L. Jen-Jacobsen, B.-C. Wang, P. Greene, H. W. Boyer, and J. M. Rosenberg. 1984. "Kinked DNA in Crystalline Complex with EcoR1 Endonuclease." *Nature* 309:327–330.

84. Arber, W., and S. Linn. 1969. "DNA Modification and Restriction." *Ann. Rev. Biochem.* 38:467–500.

85. Smith, J. D., W. Arber, and U. Kuhnlein. 1972. "Host Specificity of DNA Produced by E. coli. XIV. The Role of Nucleotide Methylation in *in Vivo* B-Specific Modification." *J. Mol. Biol.* 63:1–8.

86. Sharp, P. A., B. Sugden, and J. Sambrook. 1973. "Detection of Two Restriction Endonuclease Activities in *Hemophilus Parainfluenzae* Using Analytical Agarose-Ethidium Bromide Electrophoresis." *Biochemistry* 12:3055–3062.

87. Aaij, C., and P. Borst. 1972. "The Gel Electrophoresis of DNA." *Biochem. Biophys. Acta* 269:192–200.

88. McClelland, M., L. G. Kessler, and M. Bittner. 1984. "Site-Specific Cleavage of DNA at 8- and 10-Base-Pair Sequences." *Proc. Nat. Acad. Sci.* 81:983–987.

89. Schwartz, D. C., and C. R. Cantor. 1984. "Separation of Yeast Chromosome-Sized DNAs by Pulsed Field Gradient Gel Electrophoresis." *Cell* 37:67–75.

90. Mertz, J. E., and R. W. Davis. 1972. "Cleavage of DNA by R1 Restriction Endonuclease Generates Cohesive Ends." *Proc. Nat. Acad. Sci.* 69:3370–3374.

91. Cohen, S., A. Chang, H. Boyer, and R. Helling. 1973. "Construction of Biologically Functional Bacterial Plasmids *in Vitro*." *Proc. Nat. Acad. Sci.* 70:3240–3244.

92. Maniatis, T., R. C. Hardison, E. Lacy, J. Lauer, C. O'Connell, D. Quon, G. K. Sim, and A. Efstratiadis. 1978. "The Isolation of Structural Genes from Libraries of Eucaryotic DNA." *Cell* 15:687–701.

93. Maxam, A. M., and W. Gilbert. 1977. "A New Method of Sequencing DNA." *Proc. Nat. Acad. Sci.* 74:560–564.

94. Sanger, F., G. M. Air, B. G. Barrell, N. L. Brown, A. R. Coulson, J. C. Fiddes, C. A. Hutchison III, P. M. Slocombe, and M. Smith. 1977. "Nucleotide Sequence of Bacteriophage φX174." *Nature* 265:687–695.

95. Sanger, F., S. Nicklen, and A. R. Coulson. 1977. "DNA Sequencing with Chain Terminating Inhibitors." *Proc. Nat. Acad. Sci.* 74:5463–5467.

96. Sutcliffe, G. 1979. "Complete Nucleotide Sequence of the *E. coli* Plasmid pBR322." *Cold Spring Harbor Symp. Quant. Biol.* 43:77–90.

97. Thomas, C. A., Jr., R. E. Pyeritz, D. A. Wilson, B. M. Dancis, C. S. Lee, M. D. Bick, H. L. Haung, and B. H. Zimm. 1973. "Cyclodromes and Palindromes in Chromosomes." *Cold Spring Harbor Symp. Quant. Biol.* 38:353–370.

98. Collins, J. 1981. "Instability of Palindromic DNA in E. coli." *Cold Spring Harbor Symp. Quant. Biol.* 45:409–416.

99. Soll, D., and R. J. Roberts, eds. 1984. *The Applications of Computers to Research Nucleic Acids II. Parts 1 and 2.* Oxford, Eng.: IRL Press.

CHAPTER 10

The Replication of DNA

When the double helix was first discovered, the feature that most excited the world of genetics was the complementary relationship between the sequences of bases on its intertwined polynucleotide chains. It seemed unimaginable that such a complementary structure would not be utilized as the basis for DNA replication. In fact, it was the self-complementary nature of DNA that finally led most geneticists and biologists to accept Avery's conclusion that DNA, not some form of protein, was the carrier of genetic information.

Earlier in our discussion of how templates act, we emphasized that two identical surfaces will not attract each other and that it is much easier to visualize the attraction of groups with opposite shape or charge. Thus, without any detailed structural knowledge, we might guess that a molecule as complicated as the gene could not be directly copied. Instead, replication would involve the formation of a molecule complementary in shape, and this, in turn, would serve as a template to make a replica of the original molecule. So some geneticists, in the days before detailed knowledge of protein or nucleic acid structure, wondered whether DNA served as a template for a specific protein that in turn served as a template for a corresponding DNA molecule.

But as soon as the self-complementary nature of DNA became known, the idea that protein templates might play a role in DNA replication was discarded, for it was immensely simpler to postulate that each of the two strands of every parental DNA molecule served as a template for the formation of a daughter strand complement. Although from the start this hypothesis seemed too good not to be true, experimental support nonetheless had to be generated. Happily, within five years of the discovery of the double helix, there emerged both decisive evidence for the separation of the complementary strands during DNA replication and firm enzymological proof that DNA alone can function as the template for the synthesis of new DNA strands.

With these results, the problem of how genes replicate was in one sense solved. But in another and equally valid sense, the study of DNA replication had only begun. As this chapter will show, the replication of even the simplest DNA molecule is a complicated, multistep process, involving many more enzymes than was initially anticipated following the discovery of the first DNA polymerizing enzyme. Fortunately, it now seems likely that all the key enzymatic steps involved in the replication of both linear and circular DNA have been described and that the general picture of DNA replication presented here will not be seriously modified.

Strand Separation Requires Untwisting of the Double Helix[1,2]

In even the smallest DNA molecules, the two chains of the double helix are twisted around each other many times and must be untwisted an equal number of times for DNA replication to occur. This feature initially troubled many scientists who were otherwise enamored of the implications of DNA's double-helical structure. But untwisting linear DNA molecules need not lead to insuperable problems if the unreplicated sections of DNA can rotate around their axes, much like the rotating cables of speedometers (Figure 10-1). Since the DNA molecule is very thin, it should encounter virtually no frictional resistance, and thus the energy input needed for rotation should be negligible.

No conceptual difficulty arises from the need to break the hydrogen bonds joining the base pairs in the parental templates. Although these bonds are highly specific, they are also relatively weak, and no enzymes are required to make or break them. Hydrogen bonds thus provide highly suitable means of specifically holding together a template and its complementary replica.

Base Pairing Should Permit Very Accurate Replication[3]

In Chapter 3, we stated that the chemistry of the various amino acid side groups argued against their employment as templates. Just the opposite, however, is true of the purine and pyrimidine bases, because each of them can form several hydrogen bonds. These bonds are ideal template forces, for they are highly specific, in contrast to van der Waals interactions, whose attractive forces are both weaker and virtually independent of particular chemical groupings.

The average energy of a hydrogen bond is about 3 kcal/mol, or eight times the average energy of thermal motion of molecules at room temperature. This value gives us an estimate of how often a given group (e.g., an amino group) will be hydrogen-bonded. Under ordinary cellular conditions, the ratio of bonding to nonbonding is about $10^4 : 1$. This means that two molecules held together by several hydrogen bonds are almost never found in the nonbonded form. Because of the great specificity of the hydrogen bonds, the frequency with which adenine will attach to cytosine during DNA replication might be as low as 10^{-8} times the frequency of its bonding to thymine ($10^{-4} \times 10^{-4}$, since two hydrogen bonds are involved). The replication of the GC pair should be even more accurate, since it is held together by three hydrogen bonds rather than two. However, the actual selection process never reaches this high level of accuracy, because occasionally (perhaps about once in 10^4 times) the base selected will be in the "wrong" (imino or enol) tautomeric form, which must lead to mispairing. Now we realize that only through proofreading of the initially selected bases does the DNA replication process achieve its measured level of accuracy (error levels between 10^{-8} and 10^{-12}). (Mispairing and proofreading are discussed in detail later in the chapter.)

These arguments are not diminished by the fact that the bases can also form almost equally strong hydrogen bonds with the hydrogen and oxygen atoms of water. It does not matter if a potential thymine

Figure 10-1
Untwisting of linear DNA strands during replication. The strands untwist by rotating about the axis of the unreplicated DNA double helix.

Figure 10-2

Denaturation mapping. (a) Electron micrograph of a partially denatured T7 DNA molecule. (b) Mapping of the location of the denatured regions shown in (a). (c) A histogram summarizing the location of single-stranded regions in a large number of denatured molecules. The arrow in (a) points to a single-stranded loop-out that is present in the DNA molecule even before partial denaturation. The loop-out occurs because this was not just an ordinary T7 DNA molecule but a "heteroduplex" molecule (see Chapter 9) derived by re-annealing separated strands from wild-type T7 DNA with strands from T7 DNA molecules that contained a deletion known by genetic mapping to be located near the left end. It is this additional physical landmark that allows us to say that the AT-rich, partial denaturation sites are located in the left half of the bacteriophage chromosome. (Courtesy of Drs. John Wolfson and David Dressler.)

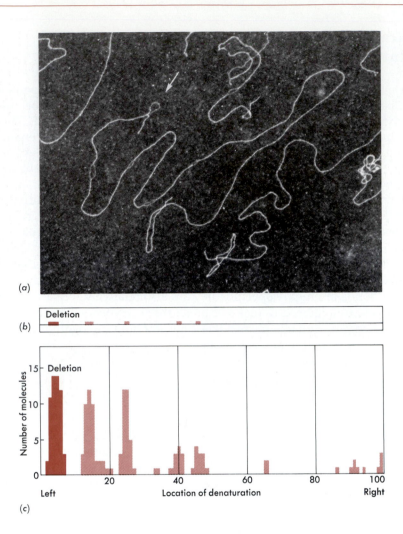

site is temporarily filled with several water molecules. These water molecules cannot be chemically joined to the growing polynucleotide chain, and they will soon diffuse away and be replaced by the suitable hydrogen-bonding nucleotide.

It is not yet possible to give a similar *quantitative* argument for why two purines or two pyrimidines never accidentally bond to each other. Both a purine-purine or a pyrimidine-pyrimidine pair will distort the sugar-phosphate backbone and thus may be energetically unfavorable. However, the large number of atoms involved makes the precise calculation of energy differences impossible. Part of the difficulty arises from the fact that the precise locations of the atoms in DNA are not known to an accuracy of 0.1 Å. Furthermore, it is unclear what effect such distortions of the double helix would have on the action of the enzyme that catalyzes the formation of the internucleotide covalent bonds.

Partial Denaturation Maps[4, 5]

Brief exposure of linear DNA molecules to denaturing conditions (e.g., high temperature, high or low pH, or hydrogen-bond breaking reagents) preferentially unwinds those sections rich in AT base pairs, leaving intact regions held together largely by the much stronger GC base pairs. When such partially denatured molecules are treated with

Figure 10-3

T7 DNA replication. (a) Electron micrograph of a T7 DNA molecule that has just begun replication and has a small replicating bubble centered about a point 17 percent from the left end of the genetic map. (b) A more advanced replicating molecule whose Y shape is the result of the left replicating fork having run off its end. (Courtesy of J. Wolfson and D. Dressler.)

formaldehyde, the single-stranded regions become stabilized, since formaldehyde combines with the free amino groups of adenine and prevents formation of the AT base pairs. Visualization of such treated molecules in the electron microscope reveals the melted regions as single-stranded "bubbles" that occur reproducibly at fixed positions along a given DNA molecule (Figure 10-2). Such "denaturation maps" now provide an easy way to distinguish one end of a linear DNA molecule from the other and so precisely define the locations where DNA replication is initiated.

Visualization of the Replication of a Linear DNA Molecule[6, 7]

It is easy to isolate DNA in the process of replication and visualize it by electron microscopy. When this was first done for the linear T7 viral DNA, it was expected that replication would be seen to commence at either one or both ends, since it is easier to imagine strand separation occurring at terminal, rather than internal, sections of a double helix. Surprisingly, however, replication was found to commence internally, always at a point (called the **origin of replication**) some 17 percent along the molecule from the left end (Figure 10-3). Once DNA synthesis starts, it moves in both directions, generating eye (-O-)-shaped intermediates that convert to Y-shaped molecules when the left-hand replicating fork reaches the end of the DNA. These structures clearly rule out a two-step replication process in which the parental strands first completely unwind and only later act as templates to form new double helices. In all such replication intermediates, only traces of unwound single-stranded regions can be seen. Thus, daughter polynucleotide strands are synthesized almost as soon as the parental strands separate.

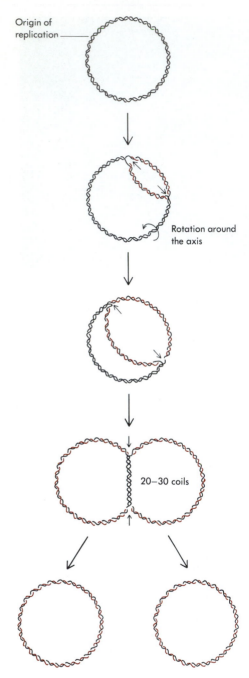

Origin of replication

Rotation around the axis

20–30 coils

Figure 10-4
Simplified version of the bidirectional replication of circular DNA molecules. Only a fraction of the replication intermediates observed with electron microscopy resemble these structures, since the molecules with intact parent strands take up supercoiled configurations.

Θ-Shaped Intermediates Form During the Replication of Circular DNA[8, 9]

Duplication of circular DNA does not involve the temporary creation of linear DNA. Visualization of the replicating intermediates of many ring chromosomes reveals that the parental strands maintain a circular form throughout replication. They always possess a θ-like shape that comes into existence by the initiation of a replicating bubble at the origin of replication (Figure 10-4). Circular λ DNA, for example, has its replication origin within gene O, and synthesis proceeds outwardly in both the clockwise and counterclockwise directions.

From the moment of its discovery, circular DNA posed the question of how two covalently closed intertwined strands could unravel during replication. As long as each chain remains intact, even minor untwisting of one section of a circular double helix forces the creation of positive supercoils in the opposing direction. The semiconservative replication of such DNA therefore demands temporary cuts in one or both of its polynucleotide backbones. The problem thus becomes how and when the breaks are made. About this there could only be speculation until the discovery of the topoisomerases, DNA nicking/closing enzymes that can either further supercoil or relax DNA (see Chapter 9). The transient single-stranded cuts made by topoisomerase I, for example, can create a molecular swivel around which supercoiled parental DNA strands can twist.

Now it seems clear that both parental strands remain intact throughout most of the replication cycle (Figure 10-5). The unreplicated portion of the θ-shaped intermediate becomes periodically supercoiled in either of two possible ways: Negative superhelices (untwists) may be made by topoisomerase II (gyrase) in anticipation of subsequent chain elongation, and positive superhelices may be made by elongation of replicating bubbles in the absence of chain cuts. The latter, positive supertwisting is clearly limited in scope and should halt replication bubble growth owing to the steric impossibility of creating further positive supercoils. Such halted bubbles could resume growth if a topoisomerase type I molecule made the single-stranded cuts that convert the highly supercoiled unreplicated section into the relaxed configuration. Alternatively, the essential cutting agent needed for further movement may be topoisomerase II (gyrase), acting to untwist the positively supertwisted DNA.

Overall Chain Growth Occurs in Both 5′→3′ and 3′→5′ Directions

The opposing chain directions (5′→3′ and 3′→5′) of the two strands of a double helix mean that the two daughter strands being synthesized at each replicating fork must also run in opposite directions. Therefore, the overall direction of chain growth must be 5′→3′ for one daughter strand and 3′→5′ for the other daughter strand (Figure 10-6). However, all known forms of DNA polymerase, the only type of enzyme that can add nucleotide precursors to DNA, extend chains only in the 5′→3′ direction, since the chemical reaction catalyzed by these enzymes allows a nucleotide triphosphate to react only with the free 3′-OH end of a growing polynucleotide strand (Figure 10-7). The question thus arose whether other replicating enzymes had been missed, specifically, those that would add nucleotides onto free 5′

(a)   (b)

Figure 10-5

Twisted replicating molecules of circular SV40 DNA: (a) Electron micrograph of an early phase of replication. (b) Diagrammatic representation showing the unreplicated, supercoiled parent strands. L1 and L2 indicate portions already replicated, while L3 indicates the unreplicated portion, which takes up a supercoiled configuration. Note that through the replication cycle, the DNA remains intact. [From N. P. Salzman et al., *Cold Spring Harbor Symp. Quant. Biol.* 48 (1973):257.]

ends. Despite a very extensive search, no enzyme with such specificity has ever been found.

Small Fragments Are Precursors of Many Long Chains: The Leading Versus the Lagging Strand[10, 11]

Resolution of the apparently paradoxical 3'→5' direction of synthesis came with the discovery that deoxyribonucleotide substrates are not always added directly to the very long daughter polynucleotide. Instead, they are sometimes found as parts of much shorter DNA chains (100 to 1000 bases long) that later link up to the main daughter strands through the action of the polydeoxyribonucleotide-joining enzyme, DNA ligase. This finding opened up the possibility that a chain whose final direction of growth is 3'→5' might, in fact, be formed by the joining of smaller chains (often called **Okazaki fragments,** after their discoverer), each of which grows in the conventional 5'→3' manner (Figure 10-8). In this way, the same set of enzymes could be responsible for making all progeny strands. Proof that this indeed is the case came from direct studies of the ends of growing Okazaki fragments to which precursor nucleotides had recently been added. All such experiments show growth exclusively at the 3' end.

Now it is very clear that the Okazaki fragments are precursors only to half of the daughter chains—those that grow overall in the 3'→5' direction. Synthesis of the strand that grows overall in the 5'→3' direction for the most part precedes that of the strand appearing to grow 3'→5'. This difference in rates of growth between the leading (5'→3') and lagging (3'→5') strands produces a small section of single-stranded parental DNA at the replication fork (Figure 10-9). It becomes converted to double-stranded DNA upon the initiation of an Okazaki fragment whose synthesis continues until it reaches the terminus of the previously synthesized Okazaki fragment, to which it then becomes joined. Okazaki fragments only commence synthesis at

Figure 10-6

Overall direction of progeny chain growth at a replicating fork.

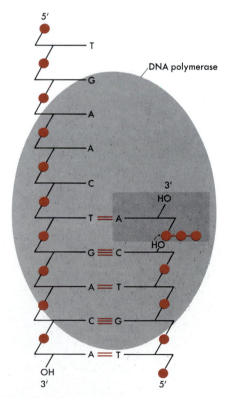

Figure 10-7

Reaction of a nucleoside triphosphate with the 3'-OH end of a growing DNA chain.

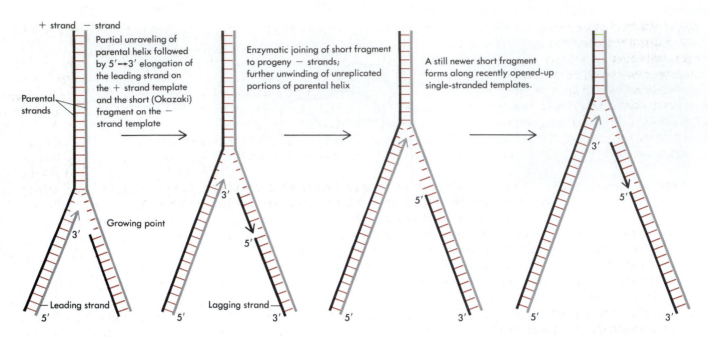

Figure 10-8
Hypothesis that replication of lagging progeny strands involves the joining together of short fragments, each synthesized in the 5'→3' direction.

precise points along single-stranded DNA templates, and it is the distance between these signals that determine the average length of Okazaki fragments formed during replication of any particular DNA.

Test-Tube Replication of Complete Double Helices[12–16]

While biochemists were initially very pleased to find any test-tube incorporation of nucleotides into DNA-like molecules, now their objectives are much more ambitious. They not only want systems where DNA synthesis occurs rapidly, but also aim for conditions that carefully reflect the process by which in vivo DNA synthesis occurs. One goal was the replication of complete DNA molecules rather than the mere addition of nucleotides to growing chains of an undefined structure. Now, after much intermittent progress, conditions have been found that clearly mimic those present in growing cells. DNA synthesis now occurs rapidly in cell-free extracts, and the products are often indistinguishable from intracellularly produced DNA molecules. Extracts made from *E. coli* cells, for example, not only can make linear T7-length molecules, but also can carry out the complete semiconservative replication of circular plasmid DNA. In the latter case, replicating bubbles start at the same sites in vitro that are used as origins in vivo, and the θ-shaped intermediates grow until replication is complete and the resulting circular double helices separate (see Figure 10-4). Most importantly, these daughter circles can serve as templates for still additional rounds of semiconservative replication. *E. coli* extracts are now being fractionated to reveal the nature and mode of action of each enzyme involved in the replication process. With these observations, our all-too-speculative ideas about certain key steps in DNA replication and its control may soon be replaced by a rigorous description of the proteins and reactions involved.

Precise conditions are also being worked out for the in vitro conversion of the single-stranded DNA of viruses like φX174 and M13 into their double-helical replicative intermediates. Surprisingly, the enzymology involved in converting single-stranded φX174 DNA into its duplex replicative form shows major differences from that involved in making an M13 DNA duplex. In particular, several bacterial DNA replication genes are required for φX174 replication but not for M13 replication. So we must face the complication that apparently similar DNA molecules may employ different modes of replication.

There Are Three Types of *E. coli* DNA Polymerase[17, 18, 19]

The exact ways in which enzymes are used to make complete DNA chains are much more complicated than initially suspected when the original DNA polymerase (now called DNA polymerase I) was first investigated. At least three different DNA polymerases have been identified within *E. coli*, and two appear to have distinct roles in DNA replication (Table 10-1). While DNA polymerase I was long thought to be the major enzyme that joins together deoxyribonucleotides, we now realize that a more recently discovered enzyme, DNA polymerase III, is the main polymerizing agent. In contrast, DNA polymerase I is largely used for filling in gaps between the small Okazaki fragment precursors that are formed during lagging-strand synthesis (see Figure 10-8). No specific task in DNA replication has been assigned to DNA polymerase II, and since mutant cells that have very few (if any) molecules of DNA polymerase II survive, this enzyme may be dispensable and its role tricky to define.

Correction of Mistakes by 3'→5' Exonuclease Action[20, 21]

Extensive purification of all three types of DNA polymerase reveals that each enzyme has a 3'→5' exonuclease activity as well as a polymerizing ability. Possession of this exonuclease specificity means that every DNA polymerase molecule has the ability to cut back nascent polynucleotides as well as to extend them. In the presence of even moderate levels of the deoxyribonucleotide triphosphate precursors, however, synthesis is overwhelmingly favored over degradation. The question thus arose why this specific degradative potential should be

Table 10-1 Properties of Polymerases I, II, and III of *E. coli*

	Pol I	Pol II	Pol III
Functions			
Polymerization: 5'→3'	+	+	+
Exonuclease: 3'→5'	+	+	+
Exonuclease: 5'→3'	+	−	+
Polymer synthesis de novo	−	−	−
General			
Size, kdal	109	120	140
Molecules per cell, estimated	400		10–20
Turnover number, estimated*	(1)	0.05	15
Structural genes known	*polA*	*polB*	*dnaE, N, Z*
Conditional lethal mutants	Yes	No	Yes

*Nucleotides polymerized at 37°/min/molecule of enzyme, relative to Pol I, which is near 600.
+ and − represent the presence and absence, respectively, of the property listed.
▨ = distinction from Pol I
▨ = distinction between Pol II and Pol III
SOURCE: After A. Kornberg, *DNA Replication* (San Francisco: Freeman, 1980), p. 169.

Single-stranded DNA

Figure 10-9
Different rates of movement of the leading and lagging strands lead to a small section of single-stranded parental DNA at the replication fork. (Courtesy of David Dressler and John Wolfson, Harvard Medical School.)

Figure 10-10
Binding of DNA to the Klenow fragment (a fragment of DNA polymerase I) as shown (left) from above and (right) from the side. [Courtesy of T. A. Steitz, D. L. Ollis, and P. Brick from *Nature* 313 (1985):765.]

coupled (within the same polypeptide chain) to polymerizing action. At first, possession of a specific degradative action made no sense. But the reason for its presence became obvious when it was shown that the exonuclease acted preferentially on incorrectly paired bases. If, by chance, the wrong base becomes added onto the 3′ end of a growing chain, it has a very high probability of being removed before the next base is added. The 3′→5′ exonuclease activity thus provides a proofreading ability that gives DNA replication much higher fidelity than it would have if selection were the result of only the initial base-pairing step.

A constant need for proofreading may be the reason that no enzyme has evolved that adds deoxyribonucleotides onto the 5′ end of a DNA chain. If growth were to occur in this direction, the growing end would always be terminated by a triphosphate group, whose high-energy bonds are necessary for the addition of the next nucleotide. Removal of the most recently added deoxyribonucleotide from such a triphosphate-terminated molecule during proofreading would thus create a 5′-monophosphate-ended chain that would be energetically incapable of being extended by DNA polymerase. Some form of ligaselike reaction could conceivably do the job, but such a step would probably slow down DNA replication markedly. Thus, a polymerase that adds nucleotides to the 5′ end of a growing DNA chain would never make sense.

DNA Polymerases Act Processively[22]

Once a DNA polymerase has begun adding nucleotides to growing DNA chains, it remains attached to the chain, continuing to add new nucleotides until a signal is reached that tells it to detach. How such processive action occurs remained a complete mystery until the elucidation of the first three-dimensional structure of a DNA polymerase. Such structural work requires large amounts of highly purified polymerases, a condition now possible through the advent of recombi-

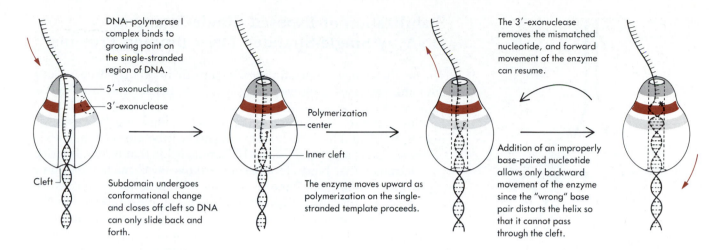

DNA–polymerase I complex binds to growing point on the single-stranded region of DNA.

Cleft

5′-exonuclease

3′-exonuclease

Subdomain undergoes conformational change and closes off cleft so DNA can only slide back and forth.

Polymerization center

Inner cleft

The enzyme moves upward as polymerization on the single-stranded template proceeds.

The 3′-exonuclease removes the mismatched nucleotide, and forward movement of the enzyme can resume.

Addition of an improperly base-paired nucleotide allows only backward movement of the enzyme since the "wrong" base pair distorts the helix so that it cannot pass through the cleft.

nant DNA procedures that attach desired genes to control regions that permit their high-level expression. The first polymerase so examined was a large carboxy-terminal fragment of DNA polymerase I called the **Klenow fragment,** which possesses polymerase activity as well as the 3′→5′ exonuclease activity that edits out mismatched nucleotides. This fragment contains two distinct domains: The larger domain contains a deep, positively charged, 20 Å diameter cleft into which DNA binds (Figure 10-10), while the smaller domain may contain a nucleotide-binding region. A flexibly attached subdomain most likely has the ability to close off the cleft after DNA binds. When bound in such a manner, the DNA may only be able to slide either backward or forward through the cleft. Conceivably, DNA containing a misincorporated base that cannot properly base-pair has a distorted shape that will not allow it to slip easily through the semirigid 20 Å diameter cleft. When so blocked, such abnormally terminated DNA may only diffuse backward, coming into extended contact with the 3′→5′ clipping (proofreading) exonuclease that will remove the misincorporated base (Figure 10-11).

Helicases Unwind the Double Helix in Advance of the Replication Fork[23–26]

The two strands of the double helix at the replicating fork do not spontaneously come apart frequently enough to permit the rapid rates of DNA polymerization observed in vivo. To facilitate strand separation, cells contain a class of specific proteins called **helicases,** which bind to ATP and to single-stranded sections of DNA. They use energy derived from the breakdown of their bound ATP to ADP to move progressively along and increase the rate of DNA strand separation. Two different helicases are often involved. One (helicase II or III) attaches to the template for the lagging strand and moves in the 5′→3′ direction. The other helicase (Rep protein) binds to the strand directing synthesis of the leading strand and moves in the 3′→5′ direction (Figure 10-12). Somehow, the conversion of the bound ATP to ADP must lead to conformational changes that propel the helicases unidirectionally along the DNA strands to which they have become attached.

Figure 10-11
Diagrammatic view of proofreading after incorporation of a mismatched nucleotide base into DNA.

Rep protein (a helicase)

Another helicase

Leading strand

Lagging strand

Figure 10-12
Helicases unwind the DNA double helix in advance of a replicating fork. The rep protein (a helicase) moves on one strand in the 3′→5′ direction, while another helicase (e.g., helicase II, or III) moves on the complementary strand in the 5′→3′ direction. (From *1982 Supplement to DNA Replication* by Arthur Kornberg, W. H. Freeman and Company, copyright © 1982, with permission.)

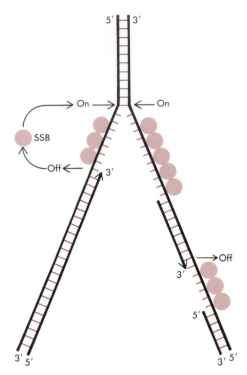

Figure 10-13
Hypothetical scheme for the action of single-stranded binding protein (SSB) at a replicating fork. The protein binds to single-stranded regions, facilitates replication, and is then recycled. (Courtesy of B. Alberts.)

Stabilization of Exposed Single-Stranded DNA by Single-Stranded DNA Binding Proteins[27]

The newly created single-stranded regions do not remain free for long after the helicases progress along a double helix. They quickly become covered by large amounts of specific single-stranded DNA binding proteins (designated **SSB**), which, unlike the helicases, do not bind ATP and have no known enzymatic function. The major SSB within *E. coli* is a polypeptide of 177 amino acids that normally exists as a tetramer that binds to adjacent groups of 32 nucleotides on a given DNA chain. The binding of one SSB tetramer promotes the binding of another to an adjacent section of a single-stranded DNA, a process called **cooperative binding** (Figure 10-13). Single-stranded DNA that is tightly covered by SSB has a rigid, semiextended form without bends or kinks. This rigid configuration is essential for the functioning of DNA as a template upon which complementary DNA chains can first be initiated and then elongated. Were the single-stranded DNA not held extended, its template capability would be repeatedly compromised by its tendency to fold back on itself to form imperfectly base-paired hairpin loops.

Initiation of Okazaki Fragments by RNA Primers: Primase and Primosomes[28–31]

As the reaction requirements for the several DNA polymerases became established, the unexpected fact emerged that each such enzyme could only add nucleotides onto preexisting polynucleotide chains. None of these polymerases could initiate new DNA chains, which raised the question of whether still undiscovered DNA polymerases might exist whose only role was to start new DNA chains. Now we suspect that such enzymes play only peripheral roles and that almost all DNA synthesis commences by the use of **RNA primers.** Starting points for DNA synthesis are thus recognized not by specific DNA polymerases but by enzymes that transcribe DNA to generate RNA chains that serve as primers. However, the classic RNA polymerases, the cellular enzymes normally used to make RNA off DNA templates, are involved in making only a small fraction of such primers. The primers for Okazaki fragments are made by a special class of enzymes called **primases,** which recognize specific DNA sequences along single-stranded DNA chains. The *E. coli* primase is a single polypeptide of 60 kilodaltons (kdal), present in 50 to 100 copies per cell. By itself, it is relatively inactive and only functions effectively when complexed with six or seven other polypeptides to form an aggregate called a **primosome.**

This multipolypeptide complex moves processively 5′→3′ on the lagging-strand template to prime the repeated initiation of Okazaki fragments needed to put together the lagging strands. Exactly what functions of the primosome are carried out by its various components largely remains unknown. These components somehow must work cooperatively to ensure the movement of the primosome along the DNA, the displacement of SSBs, the recognition of appropriate start sites, and the polymerization of ribonucleotides into RNA.

Thus, the primosome moves along the DNA template strand in the direction opposite to the direction in which it must move when cata-

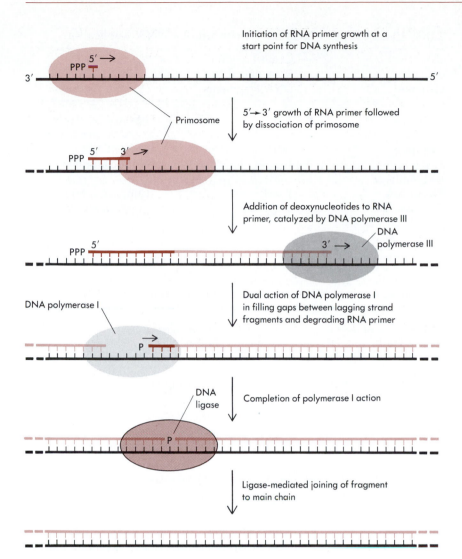

lyzing synthesis of its RNA chain products. This may explain why only very short RNA primers, usually containing only three to five bases, are normally made (Table 10-2). In a given organism (such as a virus), all the RNA primers have very similar sequences, reflecting the fact that primases synthesize RNA at specific sequences along their lagging-strand DNA templates.

Removal of the RNA primer occurs later through the action of other enzymes. One enzyme that has this capacity is DNA polymerase I, whose primary task may be to fill in the gaps between RNA-primed DNA fragments (Figure 10-14). In doing so, it adds nucleotides to the 3' end of the fragment at its back and simultaneously cuts away the RNA primer that lies ahead. The snipping off of ribonucleotides is catalyzed by a completely different active site from that involved in its 3'→5' proofreading capacity (see Figure 10-14). This 5'→3' exonuclease activity can remove deoxyribonucleotides as well as ribonucleotides, and so its use is not restricted to primer removal. As we shall see later in Chapter 12, DNA polymerase I is also employed to remove sections of DNA damaged by radiation attack.

Why RNA primers are used to start virtually all DNA chains probably has its origin in the need to prevent a lethal number of mistakes near the starting points of DNA chains. The laying down of the first few nucleotides along a DNA template may be inherently much less

Table 10-2 Primer RNAs for Discontinuous DNA Replication

Genome	Chain Length of RNA linked to DNA Fragments	Primary Products
T7 phage	1–5	pppApCpC/A(pN)$_{1-2}$* (N: A and C rich)
T4 phage	1–5	pppApC(pN)$_3$
E. coli	1–3	Not determined
ϕX174 phage	1–5†	‡
Sea urchin	1–~8	(p)ppA/G(pN)$_7$
Polyoma, SV40 (isolated nuclei)	~10	pppA/G(pN)$_{\sim 9}$
Animal cells	~9	

*The complementary template sequence that signals the primer is: 3'-pCpTpGpG/Tp–; G/T signifies that either G or T is found in this position.
†Of primers linked to the complementary strand in the SS→RF conversion in vitro, 39% were 1 residue long, 18% 2 residues, and 83% between 1 and 5 residues; comparable values for the reaction in vivo were: 1 residue 48%, 2 residues 23%, and 1 to 5 residues 92%.
‡The 5'-terminal residue is A in about 80% of the primers in vitro.
SOURCE: Data from T. Okazaki in A. Kornberg, *1982 Supplement to DNA Replication* (San Francisco: Freeman, 1982).

accurate than when nucleotides add onto long sections of an already formed double helix. So there may be great selective advantages for an initiation mechanism that allows easy discrimination and subsequent elimination of the starting nucleotide sequences.

DNA Polymerase III Is an Aggregate of Seven Different Polypeptides[32–36]

The active form of DNA polymerase III (pol III) is an assembly, or **holoenzyme,** of seven different polypeptides, all of which must be present for correct biological functioning (Table 10-3). One of these polypeptides, α, carries the 5'→3' exonuclease activities as well as the polymerizing function, while the ϵ polypeptide possesses the 3'→5' exonuclease activities. One or more of the remaining polypeptides bind an ATP molecule needed for DNA Pol III to commence synthesis at the end of an RNA primer. The functions of the remaining polypeptides are not known. The number of DNA Pol III holoenzyme molecules in an *E. coli* cell is small, and given the slow rate at which they initiate synthesis, it appears that once one such molecule has commenced polymerization, it generally continues processively until the end of its template is reached.

Two copies of the DNA pol III holoenzyme may be present at the replicating fork, giving rise to the possibility that one such dimeric complex can simultaneously carry out the synthesis of both the leading and lagging progeny strands. This could happen if the lagging-strand template is looped backward (perhaps around the polymerase) so that it can associate with a polymerizing component in the same orientation as the leading-strand template (Figure 10-15). According to this scheme, the primer made by the primosome would be extended by DNA pol III as the lagging-strand template is drawn through the holoenzyme complex. Then, when the elongating chain reaches the previously synthesized small Okazaki fragments, the loop would be relaxed. By this time, leading-strand synthesis would make available a new stretch of unpaired lagging-strand template that

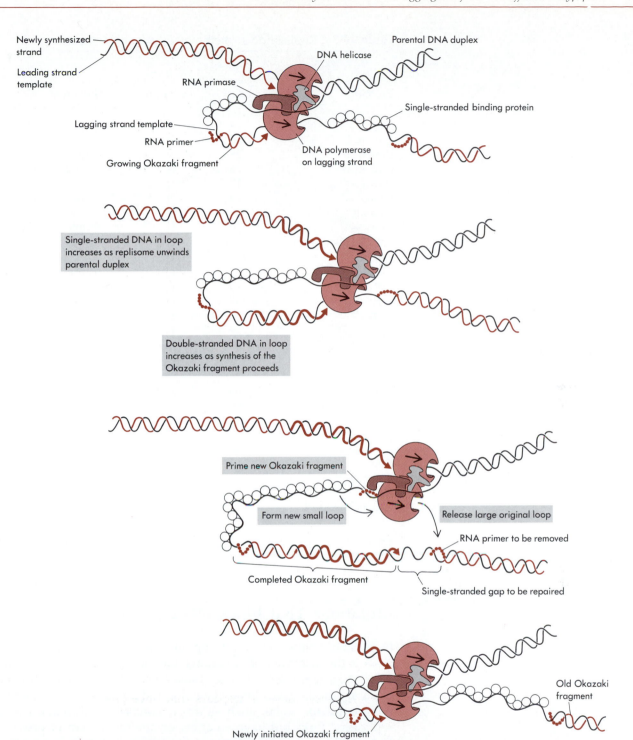

Newly synthesized strand

Leading strand template

RNA primase

Lagging strand template

RNA primer

Growing Okazaki fragment

DNA helicase

Parental DNA duplex

Single-stranded binding protein

DNA polymerase on lagging strand

Single-stranded DNA in loop increases as replisome unwinds parental duplex

Double-stranded DNA in loop increases as synthesis of the Okazaki fragment proceeds

Prime new Okazaki fragment

Form new small loop

Release large original loop

RNA primer to be removed

Completed Okazaki fragment

Single-stranded gap to be repaired

Old Okazaki fragment

Newly initiated Okazaki fragment

could loop backward and participate in the formation of the next small lagging-strand Okazaki fragment. By such a mechanism, leading-strand synthesis would never rush too far ahead of lagging-strand synthesis. Moreover, the primosome would be automatically translocated along the replicating DNA with the same speed as the DNA polymerizing components of DNA pol III.

A complex of DNA replication proteins (see Table 10-3)—the dimeric polymerase, a primosome, and a helicase—is of ribosome size. The name **replisome** has been proposed for this collective entity, which carries out the above-mentioned replication process (see Figure 10-15).

Figure 10-15
The simultaneous synthesis of both leading and lagging progeny strands by a dimeric complex of the DNA pol III holoenzyme in association with the primosome and helicases (the collection of which is known as a replisome.) [After Keith Roberts.]

Table 10-3   Replication Proteins of *E. coli*

Protein	Native Mass, kdal	Subunits	Function
SSB	74	4	Single-strand binding
Protein i	66	3 ⎫	
Protein n	28	2 ⎪	Primosome assembly
Protein n'	76	1 ⎪	and function
Protein n''	17	1 ⎬	
DnaC	29	1 ⎪	
DnaB	300	6 ⎪	
Primase	60	1 ⎭	Primer synthesis
Pol III Holoenzyme	(760)	(2)	
α	140	1 ⎫	
ϵ	25	1 ⎪	Processive chain
θ	10	1 ⎪	elongation
β	37	1 ⎬ × 2	
γ	52	1 ⎪	
δ	32	1 ⎪	
τ	83	1 ⎭	
Pol I	102	1	Gap filling, primer excision
Ligase	74	1	Ligation
Topoisomerase II (gyrase)	400	4	Supercoiling
GyrA	210	2	
GyrB	190	2	
Rep	65	1	Helicase
Helicase II	75	1	Helicase
DnaA	48		Origin of replication
Topoisomerase I	100	4	Relaxing negative supercoils

SOURCE: After A. Kornberg, *1982 Supplement to DNA Replication* (San Francisco: Freeman, 1982), Table S11-2.

Mutations That Block DNA Synthesis[37–40]

Increasingly important in finding out how enzymes like DNA pol III work is the discovery of mutations that block their functioning (Table 10-4). A large number of genes that can contain mutations that block DNA synthesis have already been located on the *E. coli* chromosome (Figure 10-16). Some, such as *dnaA,* control the formation of new replicating forks but have no effect on the movement of preexisting forks. Other genes participate in the movement of preexisting forks either by coding for components of the polymerization process (*dnaE* codes for the α polypeptide of DNA pol III) or by coding for components of the primosome (*dnaG* codes for primase). Mutations have also been found in the gene *polA,* which codes for DNA polymerase I, and in *lig,* the gene coding for DNA ligase. These mutations do not stop most DNA synthesis, but lead to the accumulation of unusually large numbers of unsealed Okazaki fragments. By now, a number of the genes involved in DNA synthesis have been cloned, allowing their polypeptide products to be expressed in abnormally large amounts.

Table 10-4 Replication Genes of *E. coli*

Gene	Map Location, minutes	In Vivo Phenotype of Mutant	Protein and in Vitro Function
dnaA	82	Slow stop; defective origin initiation	Binds to *oriC*; required for initiation of chromosome replication
dnaB	91	Quick stop	DnaB protein; prepriming
dnaC	99	Slow or quick stop (depending on the mutation)	DnaC protein complexes with DnaB protein
dnaE (polC)	4	Quick stop	Subunit α of Pol III holoenzyme
dnaG	66	Quick stop, defective initiation of fragments	Primase
dnaI	39	Slow stop	Unknown
dnaJ	0.5	Slow stop	Unknown
dnaK	0.5	Slow stop	Unknown
dnaL	28	Quick stop	Unknown
dnaP	84	Phenethyl alcohol resistance; slow stop	Unknown
dnaT	95–99	Regulates termination	Unknown
dnaZ	10	Quick stop	Subunit γ of Pol III holoenzyme; chain growth
cou	82	Coumermycin (and novobiocin) resistance	DNA topoisomerase II subunit β; nicking/closing
lig	51	Accumulation of replication fragments	DNA ligase; covalently seals DNA nicks
nalA	48	Nalidixate (and oxolinate) resistance	DNA topoisomerase II subunit α; ATPase
polA	85	Defective in DNA repair	DNA polymerase I; gap filling, RNA excision
rep	83	Slowed fork movement	ATP-using helicase
ssb	91	Quick stop	SS DNA binding protein

SOURCE: Adapted from A. Kornberg, *DNA Replication* (San Francisco: Freeman, 1980).

Figure 10-16
Genetic map of *E. coli* K12 showing loci involved in DNA replication.

Figure 10-18

Alternative schemes for how RNA polymerase, at the origin of DNA replication, acts to make RNA primers. (a) RNA polymerase makes RNA, which, after processing by RNase H (not shown), acts as primer. (b) RNA made by RNA polymerase opens the double helix to allow binding of a primosome, which will make the actual primer.

Figure 10-17

Consensus nucleotide sequence of the minimal origin of bacterial chromosomes. The minimal origin of *E. coli* is enclosed within the box. A large capital letter means that the same nucleotide base is found in all six origins; the small capital letter is used when the same nucleotide is present in five of the six sequences; the lowercase letter is used when the same nucleotide is found in three or four of the six origins but only two different nucleotides are found at the site; the letter "n" is used when three or four of the four possible nucleotides, or two different nucleotides plus a deletion, are found at the site. A dot (·) means that the nucleotide in the bacterial sequence is the same as that in the consensus sequence. Bold capital letters locate positions where single-base substitutions produce an *oriC⁻* phenotype in *E. coli*. Restriction enzyme recognition sites are indicated, and arrows denote the A-protein binding sites, R1, R2, R3, R4. [Courtesy of J. Zyskind and D. Smith, *Proc. Nat. Acad. Sci.* 80 (1983):1164–1168.]

Initiating Replicating Forks at Origins of Replication[41–50]

Origins of replication are by definition distinct from the many sites at which Okazaki fragments are initiated. Bacterial and viral DNA molecules usually contain just one such origin. The length of DNA comprising an origin can now be sharply delineated using cloned F plasmid DNA, since its size is known and it inserts at replication start sites. In this way, the unique *E. coli* origin has been narrowed to a region of 245 base pairs (Figure 10-17) lying between the gene for asparagine synthetase and the operon controlling ATP synthetase (see Figure 7-23). In vitro mutagenesis has shown that only a fraction of the bases within this origin of replication are essential for its functioning. Perhaps it is these bases that are recognized by the protein(s) specifically involved in the initiation of the replicating fork.

The key event in initiation of a replicating fork is the commencement of leading-strand synthesis. Once the polymerization of the leading strand has started, polymerization of the lagging strand soon follows, using the mechanism described earlier. An obligatory step in the starting of all leading strands is the transcription by RNA polymerase molecules of origin DNA sequences into short RNA chains. For some origins (e.g., that of the plasmid Col E) these RNA chains may function as the primers for leading strands after unneeded regions are removed by digestion with the enzyme **RNase H** (H stands for "hybrid"), which specifically digests the RNA strand of DNA-RNA hybrid molecules. However, most of these short origin RNA transcripts probably have no primer role. Instead, their only function seems to be to separate the two DNA strands of the origin so as to expose those specific sequences around which primosomes assemble prior to the synthesis of primers (Figure 10-18). This latter process is called **transcriptional activation.**

No matter how they are made, the presence of an RNA primer does not automatically mean that a DNA chain will begin growth upon it, since RNA chains usually dissociate immediately from their DNA templates. Normally favoring the release of most newly made RNA chains is the strong tendency of the separated DNA strand to reassociate to form double-helical DNA. Therefore, there must be molecules that specifically prevent such DNA-DNA reassociations, perhaps by binding to origins of replication.

Initiation of leading strands in many (but not all!) cases requires the presence of one or more additional proteins that function only during initiation. Among these are the dna A protein of *E. coli* and the O protein of phage λ, both of which bind to four highly conserved nine-base-pair-long recognition sequences. How either of these DNA-binding proteins functions is unknown. It is possible that they act by providing a site favoring the assembly of the seven different DNA pol III components into their functional holoenzyme form.

Even less clear are the tricks used to ensure bidirectional replication once DNA synthesis has commenced. We might naively predict that the elongation at the 3′ end of the first lagging-strand (Okazaki) fragment would routinely proceed past the origin of replication, becoming in effect the leading strand for a new replicating fork moving in the opposite direction. If this were true, replication would be bidirectional for virtually all replication starts. In fact, however, in a minority of replication starts, synthesis does not proceed bidirectionally. These unidirectional starts may reflect the proposal that the same DNA pol III complex (replisome) normally replicates both the leading and lagging strands simultaneously. For replication to proceed in the opposite direction, a second DNA pol III holoenzyme complex may need to be assembled.

Completing the Ends of Linear DNA Molecules[51, 52, 53]

The realization that RNA primers for DNA synthesis are ordinarily excised raises the question of how the extreme ends of linear DNA can be completed. Not only must a small RNA fragment initiate at the terminal 3′ nucleotide of the template strand, but then some way must be found to fill in the sequence after ribonucleotide removal. A resolution of this dilemma was first understood for T7 DNA, all of whose terminal deoxyribonucleotides conventionally base-pair, but which has the surprising property of having redundant ends; the sequence (about 160 base pairs) found at the left end is exactly repeated at the right end of the molecule. Moreover, when such linear T7 molecules replicate, they do not generate unit-length progeny molecules, but instead generate very long **concatamers** (units linked end to end) containing the complete genome sequence repeated over and over (Figure 10-19).

Why linear DNA might pass through a concatamer stage can be seen when we consider the incompleteness of the two daughter mole-

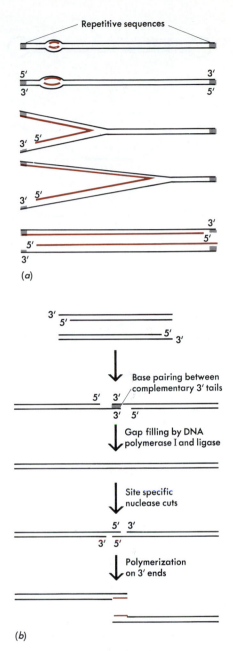

Figure 10-19

A model for replication of T7 DNA molecules. (a) The 3′ ends of the linear DNA are first incompletely replicated. (b) The formation of concatamers then makes possible the completion of replication by DNA polymerase working 5′→3′.

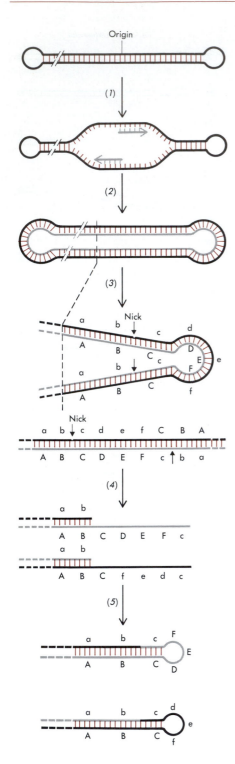

Figure 10-20
Replication of linear viral DNA molecules whose ends are joined by hairpin loops. Bidirectional replication beginning at the origin (1) converts the single-stranded structure to a double-helical equivalent (2). Two staggered cuts are made (3), and the molecules are rearranged (4) and ligated (5) to complete the replication cycle. In (3) through (5), half molecules are shown.

cules produced by the initial replication of a linear double helix. One end of each daughter molecule will be completely replicated, while the other end will have a 3'-ended single-stranded tail. If there are redundant terminal sequences, then the two tails will be complementary and can base-pair to form a two-unit length concatamer (see Figure 10-19). Following gap filling and ligase action, this two-unit DNA can be replicated to again produce unfinished progeny that will tend to stick together to produce a four-unit molecule; and so forth. Unit-length molecules can be produced from such concatamers by the action of specific endonucleases that create staggered nicks. These can easily be filled in by a DNA polymerase synthesizing in the 5'→3' direction to yield double helices identical to those present in the original parental phage particle.

A second suggested way the ends of linear viral chromosomes might be preserved is illustrated by the poxviruses, which have hairpin loops joining together their DNA strands at both their termini. Replication of such viral DNAs might begin at single internal initiation sites with subsequent bidirectional replication converting the single-stranded hairpins into double-helical equivalents. How such double-stranded hairpins become resolved into daughter single-stranded hairpins is unclear. One proposal starts with two staggered cuts displaced about the center of the hairpin, followed by the denaturation of the central DNA that leads to the separation of the two ends. The resulting single-stranded tails snap back into floppy loops, which DNA ligase can act upon to create new, totally closed hairpins (Figure 10-20).

Still much less clear is how the linear chromosomes that characterize all eucaryotic organisms replicate their ends. Each such chromosome is now thought to contain a single DNA molecule; yet there is no morphological evidence suggesting end-to-end linkage during DNA replication. At first it seemed logical to believe that their ends (also called telomeres) would be found to have closed hairpin loops of fixed length like the pox viruses. However, recent evidence favors the alternative hypothesis that the ends of eucaryotic chromosomes are maintained by enzymes that add new terminal DNA sequences to replace those that otherwise would be lost during each replication cycle. Such a terminal repair process was first suggested by experiments with the small, gene-sized DNA molecules present in the macronucleus of the ciliated protozoan *Tetrahymena* (Chapter 22). The ends of all these linear macronuclear DNA molecules have tandemly repeated blocks of the sequence

$$3'\text{-AACCCC-}5'$$
$$5'\text{-TTGGGG-}3'$$

that vary in number from 30 to 70. Variation in the number of these repetitive units prompted the search for—and recent discovery of—an enzyme that adds 5'-TTGGGG-3' to the 3' end of any preexisting single-stranded 5'-TTGGGG-3' sequence. The function of this completely new type of enzymatic activity is surely to prevent the ends of chromosomes from growing progressively shorter because of failure to completely replicate their 5'-terminal ends. Once the TTGGGG-adding enzyme has extended the 3' end of the telomere, synthesis of the complementary 3'-AACCCC-5' repeats could be primed by a conventional primase, by looping back of the 3' end of the 5'-TTGGGG-3' strand, or by a specialized priming protein (see Figure 10-26).

Complete Separation of Two Parental Circular Strands Requires the Participation of an Enzyme Like Topoisomerase II[54, 55]

There is no end-completing dilemma to contend with for the replication of circular DNA; there the gaps created by removal of the RNA primers each eventually become filled in by extension of the 3' termini growing around the circle. The occurrence of so many DNAs in circular form may therefore be explained, at least in part, by the ease with which circles ensure their own complete replication.

Bidirectional replication normally proceeds around circular DNA molecules until the two opposing replicating forks meet each other, usually at some site approximately 180° across from the origin of replication. No special termination signals are thought to exist, since deletions of DNA that remove the regions where the replicating forks usually meet do not prevent union of the daughter strands and their subsequent separation from each other. The final separation of the daughter double helices, however, is not an automatic affair. The completion of replication of the individual strands usually leaves the daughter double helices intertwined by 20 to 30 coils about each other (see Figure 10-4). The only way to separate such intertwined circles is the creation of double-stranded cuts that allow the double helices to pass through each other. In vitro, topoisomerase II catalyzes the separation of such intertwined double helices, and it would be most surprising if this capability were not used by cells to complete their replication.

Rolling Circles Are Alternative Means for Replicating Circular DNA[56, 57, 58]

An alternative way to replicate circular DNA is the **rolling circle mechanism.** This replication scheme underlies the multiplication of many viral DNAs, of bacterial F (fertility) factors during mating, and of the DNA in certain cases of gene amplification (see Chapter 22). In this mechanism, DNA synthesis starts with a specific cut at the origin of replication in one specific strand (+) of the parental duplex circle. The 5' end then becomes displaced from the duplex, thereby permitting DNA Pol III to add deoxyribonucleotides to the free 3' OH. As replication proceeds, the 5' end of the cut strand is rolled out as a free tail of increasing length (Figure 10-21), which soon becomes covered by the single-stranded DNA-binding proteins. Such a replicating structure is called a rolling circle because the unraveling of the SSB-bound single-stranded DNA is accompanied by rotation of the double-helical template about its axis. Such unraveling does not create any topological problem, since the circle is held together by only one intact DNA strand, which will freely rotate about many of the covalent bonds contained in its backbone.

When the rolling circle mechanism is used to replicate double-stranded DNA, the 5'-ended tails quickly become converted to double-stranded DNA through the formation of Okazaki fragments that grow from RNA primers started by primase molecules contained in the locomotory primosomes. The active force that then continues to unwind the 5' tail from the rolling circle is probably not the polymerization reactions occurring at the 3' end but the movement of the Pol III–primosome complex (replisome) propelled by its helicase com-

(a)

(b)

Figure 10-21
Electron micrographs of the rolling circle intermediates in (a) single-stranded φX174 DNA and (b) the replication of double-stranded phage P2 DNA. In (b), partial denaturing conditions were used to show sequence homologies between circle and tail sections. [Part (a) courtesy of D. Dressler; part (b) courtesy of R. B. Inman.]

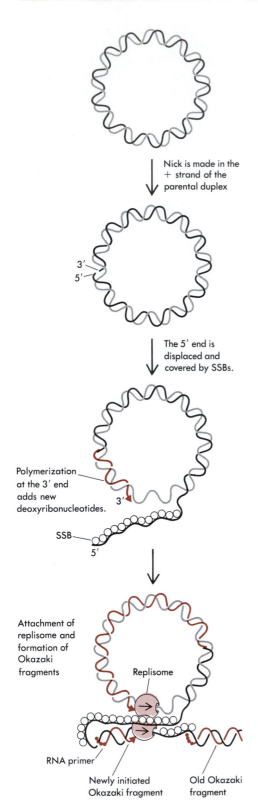

Nick is made in the + strand of the parental duplex

3'
5'

The 5' end is displaced and covered by SSBs.

Polymerization at the 3' end adds new deoxyribonucleotides.

3'

SSB

5'

Attachment of replisome and formation of Okazaki fragments

Replisome

RNA primer

Newly initiated Okazaki fragment

Old Okazaki fragment

Figure 10-22
Replication of double-stranded circular DNA molecules through the rolling circle mechanism. The active force that unwinds the 5' tail is the movement of the replisome propelled by its helicase components.

ponents (Figure 10-22). Elongation sometimes goes on to produce tails many times the complete length of the original circle. This happens, for example, in the later stages of phage λ DNA replication. How these long tails become converted into progeny-length DNA molecules is not yet established. The best guess is that the tails are cut by specific endonucleases like those that convert λ circles into λ linear forms (see Figure 9-21). The resulting unit-length progeny linear forms could remain linear and become encapsulated into mature virus particles. Alternatively, they could use their single-stranded complementary ends to form new circular double helices capable of becoming new circles.

Conversion of Circular Single-Stranded Viral DNA Chromosomes into Double-Helical Replicative Forms[59, 60]

RNA primers are needed to convert the circular single-stranded (+) DNA chromosomes of certain small viruses (the phages φX174, G4, and M13) into their double-helical (+/−) replicating forms (RFs). Replication of these viral DNAs commences after the infecting + strands enter suitable host cells and become coated with single-stranded DNA-binding proteins (SSBs). Subsequently, the small RNA chains that prime synthesis of the respective − strands are made. No one mechanism is used to produce these primers; the route followed depends on the specific + strand template (Figure 10-23). Full primosomes assemble on φX174 DNA, while primase alone appears necessary to synthesize phage G4 RNA primers. In phage M13 DNA, host *E. coli* RNA polymerase recognizes a small hairpin loop to which SSBs cannot adhere and uses this site to make a small primer. The RNA primers made in each of these processes are removed after the − strand synthesis catalyzed by DNA Pol III goes almost to completion around the circular template. The small gap remaining is filled in by DNA polymerase I, whose 5'→3' exonuclease also removes the RNA nucleotides lying immediately ahead.

Synthesis of Circular + Strands upon Circular Phage Replicating Forms[61]

All the RFs of single-stranded DNA phages use rolling circles both to produce more copies of themselves and to produce the + strand progeny that becomes encapsulated in new phage particles. How this happens is best understood for phage φX174 whose nicking enzyme, the **A protein,** is necessary to start the rolling circle and is encoded by the phage chromosome. This enzyme recognizes a single origin site on the covalently closed RF, and in making its cut, becomes covalently bound at the 5' end of the cut DNA. As the replicating fork proceeds around the + strand, the 5' tail of the displaced + strand does not hang loose, but instead remains complexed through its attached nicking enzyme to the DNA Pol III holoenzyme (perhaps to a replisome) acting at the replicating fork (Figure 10-24). Movement of the replicating fork one complete turn around the rolling circle brings the attached nicking enzyme into position to make the next specific cut at the origin of replication. In making such a cut and generating a new round of rolling circle synthesis, the enzyme becomes detached

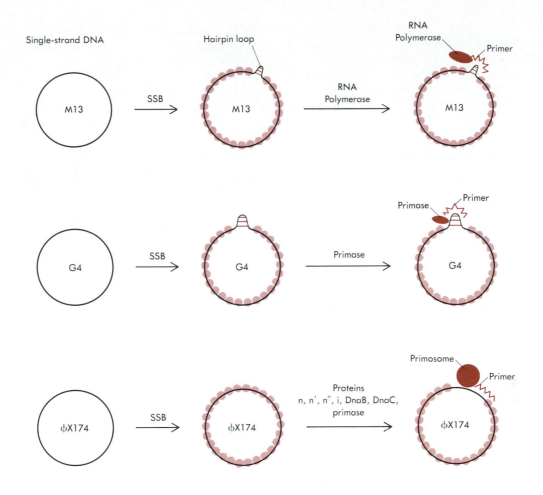

from its previously bound 5′ tail. The newly free 5′ end in turn links onto the newly made 3′ end, thereby yielding a circular unit-length + progeny strand. At no time in these nicking/closing cycles is any ATP employed, so the nicking (initiation) enzyme bears a resemblance to classical topoisomerase I, except that much longer intervals pass between the cutting and rejoining events in the case of the nicking enzyme.

Initially, all the + strand progeny produced during phage infection become converted into daughter RFs through the use of RNA primers. Later, as phage infection proceeds, the daughter circular + strands become covered with phage-specific proteins, which not only prevents the formation of RNA primers but serve as nucleation sites for assembly of new progeny phage particles.

Figure 10-23

Three mechanisms by which RNA primers are synthesized in single-stranded viral DNA molecules. (After Arthur Kornberg, *1982 Supplement to DNA Replication* p. S109, W. H. Free- and Company, copyright © 1982, with permission.)

Synthesis and Transfer of Single-Stranded DNA During Bacterial Mating[62]

When bacteria mate, the DNA that the male transfers into the female is single-stranded, not double-stranded. It is synthesized by the rolling circle mechanism started with the nicking of one specific strand of the male chromosome. This strand is then elongated at its 3′ end, leading to the displacement of growing 5′-ended single-stranded DNA. Most likely, the enzyme that makes the specific cut, like those that start off other rolling circles, remains attached to the 5′ end and

Figure 10-24

φX174 DNA replication involves two replicating forms (RFs). RF I, a double-stranded closed circle, results from the synthesis of a − strand complementary to the phage particle's + strand. RF II, a double-stranded open circle, is the rolling circle intermediate for the synthesis of new + strands. The phage's A protein plays a major role in the latter process.

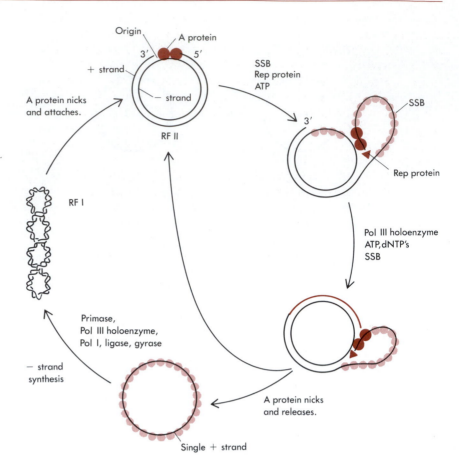

somehow serves to guide the single-stranded DNA into the female cell (Figure 10-25).

This transfer mechanism couples DNA replication and mating in such a way that genetically deficient male parents are never produced. What happens once the male DNA is inside the recipient cells remains unclear. Some of it may be converted into a normal double-helical molecule, or, as single-stranded DNA, it may be exchanged for a portion of the original female chromosome.

Linear Viral DNAs That Replicate Without Using RNA Primers[63–71]

A number of viruses, some of which multiply in procaryotic cells and others in eucaryotic cells, possess linear DNA molecules whose synthesis does not utilize RNA primers. Best known are the linear DNAs found in *Bacillus* phage φ29 as well as in the adenoviruses, a group of viruses that infect vertebrate cells. Unlike the superficially similar-looking T7 DNA, φ29 and adenovirus DNAs do not replicate using bidirectional replicating forks starting from an internal origin. Instead, replication always begins directly at the ends of the DNA, with equal probability of starting at either end. Although each adenoviral and φ29 DNA molecule possesses identical sequences at its two ends, these redundant sections are oriented in opposite directions, so their DNA replication cannot generate concatamers during replication. Nor do the two complementary DNA strands both serve as templates when replication proceeds from a given end. Instead, only one strand

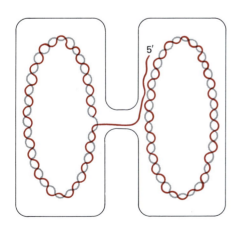

Figure 10-25

Transfer of single-stranded DNA from a male bacterium into a female via the rolling circle mechanism.

at each end acts as a template. This form of DNA synthesis involves a displacement process that leads to the generation of single-stranded tails representing the released 5'→3' oriented parental strands. Following their complete displacement and hydrogen bonding of their complementary 5'- and 3'-ended sequences, they in turn can serve as templates for new 5'→3' progeny strands starting at their 3' termini (Figure 10-26).

The first real clue to how such DNA synthesis can start de novo was the finding that the 5' ends of all growing adenoviral chains contain a specific protein attached to the terminal cytosine nucleotide. Before DNA replication starts, terminal protein becomes covalently attached to the 5' phosphate of deoxycytidine monophosphate, soon to be the 5'-terminal nucleotide in the newly synthesized DNA. The deoxycytidine base-pairs with the opposing 3'-end G residue on the template strand. This process requires a unique, virus-coded DNA polymerase that utilizes the terminal protein–dCTP complex as a primer and then synthesizes DNA in a processive fashion down the remaining 36,000 nucleotides needed to complete the chain. Complete replication requires the action of a cellular type I topoisomerase to remove torsional strain induced by the displacement synthesis. This is a case where even a linear DNA, presumably not attached at its ends, cannot replicate without the action of a topoisomerase. Still remaining to be worked out is the way the 5'→3'oriented parental strand is displaced. Conceivably, the adenovirus-coded 72 kdal single-stranded DNA binding protein that is abundantly made during adenovirus multiplication has some helicase-like function.

Regulation of DNA Synthesis[47, 72, 73]

How DNA synthesis is regulated in various cells is almost as mysterious today as it was when the first DNA polymerases were isolated. Until the general principles by which DNA synthesis is initiated became established, there was no way to probe this problem at the level of defined molecules. Now, however, it looks as if the way the leading DNA chains are started may soon be worked out at the molecular level. Then we may be able to search for the tricks that restrict DNA replication to limited periods within the cell cycle and somehow ensure that a given origin is used only once during a given cycle of chromosome replication. Perhaps key initiating proteins, like the dnaA protein of *E. coli*, somehow modify the DNA on which they act so that the origins cannot be used again as starting sites for DNA synthesis until they become modified back to their original condition (e.g., a methylation-demethylation cycle). But so far, there is no evidence of clear-cut chemical modifications occurring specifically at replication origins.

Incisive answers may come from analysis of plasmid DNAs. The replication of the *E. coli* plasmid ColE1, for example, is regulated at least in part by the synthesis of RNA molecules, transcribed in the opposite direction from the same DNA sequences that give rise to the RNA primers for leading strand synthesis. Such "anti-sense" RNA somehow blocks primer initiation of DNA synthesis, conceivably by annealing to the primers to form RNA-RNA double helices. The signals used to control when this anti-sense RNA is made are yet to be discovered.

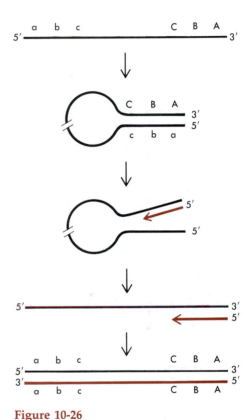

Figure 10-26
A model for replication of linear DNA (such as adenovirus and phage φ29) that does not use RNA primers. Replication begins at either end of the DNA, and only one strand serves as a template. The displaced single-stranded tail separates from the parental DNA molecule and forms a hairpin loop, a result of complementary sequences oriented in the opposite direction at either end. This tail then serves as a template for a new progeny strand starting at the 3' terminus.

Equally valuable information may come from study of eucaryotic plasmid DNAs that replicate only once per cell cycle and thus may be subject to the same general control mechanisms that prevent origins of replication from functioning more than once per cell cycle. The study of bovine papilloma virus (BPV) DNA should be particularly useful as it normally multiplies as a plasmid whose copy number is tightly regulated along with the copy number of the host cell's chromosomes.

Are Membranes Involved in DNA Replication and Segregation?[74]

The question of how DNA molecules (chromosomes) are correctly partitioned during cell division (mitosis) so that each daughter cell receives one of each parental chromosome still has no clear answer. It has always been natural to postulate a connection between the cell (or nuclear) membrane and specific regions of chromosomes, but despite many claims, no firm evidence yet exists for well-defined molecules that bind DNA to surrounding membranes. On the contrary, it is now clear that none of the stages in DNA replication requires the presence of any membrane component. This fact, however, does not mean that membranes and DNA do not need to specifically bind to each other. But as long as it remains so difficult to purify cellular membranes away from other cellular fractions, this area of investigation may continue to fail to yield convincing answers.

There Are Several Functionally Distinct DNA Polymerases Within Every Eucaryotic Cell[75]

Only recently has a major effort been made to understand DNA replication at the enzymatic level within eucaryotic cells. Three distinct forms of DNA polymerase (α, β, and γ) have been characterized in all well-studied cell systems (Table 10-5). The α DNA polymerase, the form responsible for most DNA synthesis, carries out the role played by DNA polymerase III in bacteria. Like pol III, it is a holoenzyme with a molecular weight of 220 kdal made up of several polypeptides, the largest of which (about 150 kdal) carries the polymerizing activity. So far, no form functionally equivalent to pol I has been identified, although a newly discovered δ form found in the calf thymus may fit the bill. Still functionally obscure is the much smaller β DNA polymerase, although there are hints that it may be involved in DNA repair. Already much better understood is the γ DNA polymerase, which is exclusively located within mitochondria (and perhaps chloroplasts) where it functions to replicate the mitochondrial DNA molecules. It catalyzes a displacement-type synthesis (Figure 10-27) in which only

Table 10-5 Eucaryotic DNA Polymerases α, β, and γ

	α	β	γ
Location	Nucleus	Nucleus	Mitochondria
Function	DNA replication	Repair	mtDNA replication
Mass, kdal	120–220	30–50	150–300

Origin for new H strand
Start of H strand
New H strand
Displacement loop
L
H
L
H
L
H
New L strand
Supercoil
Origin for new L strand

one strand (H) is initially copied. Not until the replicating fork moves halfway around the genome does synthesis commence on the L-strand template. As expected, topoisomerase molecules are found within mitochondria. They can catalyze the nicking and closing events that make possible the untwisting of the parental strands and subsequent separation of the daughter double helices.

Figure 10-27
Replication of mitochondrial DNA molecules. Replication of the H strand commences first. When the expanding displacement loop has traveled halfway around the circle, the replication of the L strand begins. (After Arthur Kornberg, *DNA Replication* p. 605, W. H. Freeman and Company, copyright © 1980, with permission.)

The Existence of a Cell-Free System for Replicating SV40 DNA[76, 77]

Working out the exact enzymology of eucaryotic DNA synthesis will require the development of well-defined in vitro systems similar to those now being used to reveal how the *E. coli* chromosome is replicated. In the case of *E. coli*, real progress was first made using the well-defined DNA's of *E. coli*'s viruses and plasmids. In the same way, real progress on the eucaryotic front has come through studying eucaryotic systems that replicate adenovirus DNA, even though adenovirus DNA replicates by a strand displacement mechanism rather than through the growth of replicating bubbles, as is typical for eucaryotic chromosomes. Much excitement has thus been generated by the recent working out of conditions for the in vitro replication of biologically active SV40 chromosomes, whose synthesis is initiated at a single site along its circular DNA molecule. Most importantly, this system works only in the presence of the SV40 T-antigen (see Chapter 24), a virally coded protein that binds to several sites surrounding the SV40 origin of replication and which from in vivo experiments is known to play an essential role in the initiation of SV40 DNA replication. Further fractionation of this system should soon reveal important facts regarding the exact composition of the α DNA polymerase holoenzyme, which is responsible for most of the synthesis. Also available in the near future will be data on how the RNA primers are made for both leading- and lagging-strand synthesis as well as the roles played by the several cellular topoisomerases.

From such analysis we should at last begin to understand what role, if any, the several major histones play during DNA replication. In the absence of hard facts, it had seemed most plausible that nucleosomes would dissociate from DNA in the course of its replication. This is not the case, however. As related later, nucleosomes are conserved intact during DNA replication, with the surprising result that they are not equally partitioned to the two daughter double helices, but remain attached to only one of the two strands along a given stretch of a double helix.

Finally, in understanding how T-antigen promotes SV40 synthesis, we should acquire clues that will let us look at how the multiple origins of replication along a given eucaryotic chromosome (see Chapter 21) are signaled to commence DNA replication. Although this may not be an easy goal to achieve, its accomplishment should provide many important dividends in understanding how the growth of cells is controlled.

Summary

Replication of DNA double helices to yield two identical daughter molecules, besides requiring the breakage of the hydrogen bonds holding base pairs together, necessarily involves untwisting of the intertwined parental chains. For this to happen, unreplicated sections must rotate around their axes, a process that need not require significant energy because the double helix is very thin. Strand separation and untwisting go hand in hand with synthesis of new strands to generate moving replicating forks that move away from the starting point (origins) for DNA replication. Origins of replication on a given DNA molecule are very precisely defined. On linear viral DNA molecules, they usually occur internally to generate bidirectionally growing replicating bubbles (⟷-shape molecules), which become Y-shaped molecules, until replication is completed. Circular DNA molecules usually have θ-shaped replication intermediates containing two replicating forks that move away from the origin until they meet on the opposite side of the DNA circle.

Within each cell, there are several types of DNA polymerase, which add on deoxyribonucleotides to preexisting DNA chains in a $5' \rightarrow 3'$ direction. In *E. coli*, three different DNA polymerases have been identified. Besides their polymerizing activity, DNA polymerases usually contain a $5' \rightarrow 3'$ exonuclease on one side of the molecule, and on the other side a $3' \rightarrow 5'$ exonuclease. The $3' \rightarrow 5'$ exonucleases are capable of removing misincorporated nucleotides and thus have a vital proofreading role, which reduces the error level of DNA replication. After proofreading, the average probability of an error in the insertion of a new nucleotide may be as low as 10^{-8} to 10^{-12}. DNA polymerase I (and probably all other polymerases) is constructed with a 20 Å diameter cleft through which a DNA double helix moves. Conceivably, the incorporation of a wrong base prevents this movement and thereby allows the $3' \rightarrow 5'$ exonuclease component the opportunity to remove the latest incorporated base.

The opposing chain directions of the two strands of the double helix ($5' \rightarrow 3'$ and $3' \rightarrow 5'$) requires the two daughter strands also to run in opposite directions. All known DNA polymerases, however, extend individual chains only in the $5' \rightarrow 3'$ direction. Resolution of this paradox came from the realization that while the $5' \rightarrow 3'$ strand of a replicating fork grows continuously, the $3' \rightarrow 5'$ strand grows discontinuously, being assembled from a group of DNA fragments each individually made in the $5' \rightarrow 3'$ direction. The strand growing continuously $5' \rightarrow 3'$ (the leading strand) moves slightly ahead of the strand that is assembled discontinuously $3' \rightarrow 5'$ (the lagging strand).

Separation of the two strands at the growing fork is speeded up by DNA-binding proteins called helicases, which use the energy of ATP to move along the just-separated DNA chains at growing forks. Soon after the two parental chains have separated through the helicase action, they become covered by specific proteins (single-stranded DNA-binding proteins) that keep the individual chains in semiextended, conceivably helical configurations. New deoxyribonucleotides are added onto leading daughter strands by the DNA polymerase III holoenzyme, a complex assembly of seven different polypeptides, all of which must be present for rapid processive growth. Although the DNA polymerase III holoenzyme also makes most of each lagging-strand fragment, it does not initiate the growth of these fragments. Each lagging-strand fragment is initiated by an RNA primer made by a second complex assembly of polypeptides called the primosome. Primosomes start RNA synthesis at very specific recognition sequences. The distance between the recognition sequences regulates the size of the lagging-strand fragments. DNA polymerase I fills in the remaining gaps between fragments and catalyzes, through its $5' \rightarrow 3'$ exonuclease component, removal of the RNA primer at the 5' end of the adjacent fragment.

Still unclear is whether any universal mechanism exists for the RNA priming of the leading strands. To start a replicating bubble, RNA polymerase–directed RNA synthesis must occur at the origin region. This raises the possibility that such RNA chains (after processing?) become the primers for leading strands. The alternative possibility, however, is that the function of this synthesis is to temporarily separate the two DNA chains at the origin so that a primosome can recognize an appropriate start signal (the transcriptional activation hypothesis). The starting of a leading strand also frequently demands the presence of specific proteins that bind sequences very close to the starting point for DNA synthesis.

The replication of the DNA of many single-stranded phages occurs in two main steps: (1) conversion of a circular single-stranded + template, using a primosome and the DNA polymerase III holoenzyme, into a double-helical +/− replicative form, and (2) the synthesis off an RF template of a + single-stranded circle using a rolling circle mechanism. Rolling circles are initiated by the cutting at the origin of replication of the + strand by a specific nicking enzyme. This cut is followed by a continuous displacement of the 5'-ended strand accompanied by DNA growth onto the 3'-ended fragment. DNA synthesis progresses completely around the circle to the origin. There a nicking enzyme makes a new cut that releases a complete + single strand, which immediately adopts a circular configuration. Circular double-helical DNA can also replicate by the rolling circle mechanism. Double-stranded tails are generated, which then can be enzymatically cut into unit-length linear fragments with sticky ends that make possible their subsequent circularization.

Initiation of replicating bubbles, however, does not ensure the successful synthesis of linear DNA molecules. Unless some special trick is used, each cycle of replication will produce shorter and shorter 5'-ended chains. Phage T7 DNA has identical sequences at its ends, which allow end-to-end union of progeny molecules to form a very long molecule, which subsequently can be cut up into single genomes with complete ends. A completely different device is used by the pox viruses, which have closed hairpin loops whose cutting provide the needed 3' primers. How the linear DNAs of cellular chromosomes maintain their ends remains unclear. One possibility is that their integrity is maintained by specific enzymes, which during every cycle of chromosome duplication add specific repetitive sequences (e.g., GGGGTT) to the 3' ends. If such sequences are recognized by appropriate primosomes, then these chromosomes will stay approximately constant in length and never become nonfunctional.

Certain linear viruses (e.g., adenoviruses and φ29 phage)

that have proteins covalently attached to their 5' ends do not start DNA synthesis with replicating bubbles. Instead, they employ a displacement process in which the 5' end of each strand is displaced by growth of a new 5'-end strand starting at the 3' end of the parental strand. RNA primers are not used; instead, 5'-ended growth is initiated by the nucleotide covalently bound to a terminal protein molecule.

Continued progress in understanding the essential mechanisms of DNA replication demands the development of cell-free systems for making biologically active DNA molecules. While systems that replicate procaryotic DNAs have been in use for over a decade, only during the past several years has it been possible to generate biologically active eucaryotic DNA in vitro. Particularly important is the recent development (1984) of cell-free systems in which circular SV40 DNA is replicated. This synthesis requires the presence of the SV40 T-antigen, a protein previously known to be essential for in vivo SV40 synthesis and which binds to the origin of replication. Hopefully, the study of DNA synthesis in these systems will begin to provide insights into how eucaryotic chromosomal DNA is initiated, elongated, and terminated.

Bibliography

General References

Boyer, P. D., ed. 1981. "Nucleic Acids." *The Enzymes.* Vol. 14(A). New York: Academic Press.

Denhardt, D. T., D. Dressler, and D. S. Ray. 1978. *The Single-Stranded DNA Phages.* Cold Spring Harbor, N.Y.: Cold Spring Harbor Laboratory.

DePamphilis, M. L. and P. M. Wassarman. 1982. "Organization and Replication of Papovavirus DNA." In *Organization and Replication of Viral DNA,* ed. A. S. Kaplan, Boca Raton, FL: CRC Press, pp. 38–113.

Kornberg, A. 1980. *DNA Replication.* San Francisco: Freeman.

Kornberg, A. 1982. *1982 Supplement to DNA Replication.* San Francisco: Freeman.

"DNA: Replication and Recombination." 1978. *Cold Spring Harbor Symposium on Quantitative Biology.* Vol. 43. Cold Spring Harbor, N.Y.: Cold Spring Harbor Laboratory.

"Structures of DNA." 1982. *Cold Spring Harbor Symposium on Quantitative Biology.* Vol. 47. Cold Spring Harbor, N.Y.: Cold Spring Harbor Laboratory.

Cited References

1. Delbrück, M. 1954. "On the Replication of Deoxyribonucleic Acid (DNA)." *Proc. Nat. Acad. Sci.* 40:783–788.
2. Delbrück, M., and G. Stent. 1957. "On the Mechanism of DNA Replication." In *The Chemical Basis of Heredity,* ed. W. D. McElroy and B. Glass. Baltimore: Johns Hopkins Press, pp. 699–736.
3. Brown, T., O. Kennard, G. Kneale, and D. Rabinovich. 1985. "High-Resolution Structure of a DNA Helix Containing Mismatched Base-Pairs." *Nature* 315:604–606. Gives an opposing view as to the extent of helix deformation due to mismatched base pairs.
4. Inman, R. B. 1966. "A Denaturation Map of the λ Phage DNA Molecule Determined by Electron Microscopy." *J. Mol. Biol.* 18:464–476.
5. Mulder, C., and H. Delius. 1972. "Specificity of the Break Produced by Restricting Endonuclease R1 in Simian Virus 40 DNA, as Revealed by Partial Denaturation Mapping." *Proc. Nat. Acad. Sci.* 69:3215–3219.

6. Dressler, D., J. Wolfson, and M. Magazin. 1972. "Initiation and Reinitiation of DNA Synthesis During Replication of Bacteriophage T7." *Proc. Nat. Acad. Sci.* 69:998–1002.
7. Wolfson, J., D. Dressler, and M. Magazin. 1972. "Bacteriophage T7 DNA Replication: A Linear Replicating Intermediate." *Proc. Nat. Acad. Sci.* 69:499–504.
8. Cairns, J. 1966. "The Bacterial Chromosome." *Sci. Amer.* 214:36–44.
9. Salzman, N. P., G. C. Fareed, E. D. Sebring, and M. M. Thoren. 1973. "The Mechanism of SV40 DNA Replication." *Cold Spring Harbor Symp. Quant. Biol.* 38:257–265.
10. Okazaki, T., and R. Okazaki. 1969. "Mechanism of DNA Chain Growth. IV. Direction of Synthesis of T4 Short DNA Chains as Revealed by Exonucleolytic Degradation." *Proc. Nat. Acad. Sci.* 64:1242–1248.
11. Wolson, J., and D. Dressler. 1972. "Regions of Single-Stranded DNA in the Growing Points of Replicating Bacteriophage T7 Chromosomes." *Proc. Nat. Acad. Sci.* 69:2682–2686.
12. Goulian, M., A. Kornberg, and R. L. Sinsheimer. 1967. "Enzymatic Synthesis of DNA, XXIV. Synthesis of Infectious Phage φX174 DNA." *Proc. Nat. Acad. Sci.* 58:2321–2328.
13. Wickner, R. B., M. Wright, S. Wickner, and J. Hurwitz. 1972. "Conversion of φX174 and fd Single-Stranded DNA to Replicative Forms in Extracts of *Escherichia coli.*" *Proc. Nat. Acad. Sci.* 69:3233–3237.
14. Sakakibara, Y., and J.-I. Tomizawa. 1974. "Replication of Colicin E1 Plasmid DNA in Cell Extracts." *Proc. Nat. Acad. Sci.* 71:802–806.
15. Masker, W. E., and C. C. Richardson. 1976. "Bacteriophage T7 Deoxyribonucleic Acid Replication *in Vitro.* VI. Synthesis of Biologically Active T7 DNA." *J. Mol. Biol.* 100:557–567.
16. Tsurimoto, T., and K. Matsubara. 1982. "Replication of Bacteriophage λ DNA." *Cold Spring Harbor Symp. Quant. Biol.* 47:681–691.
17. Knippers, R. 1970. "DNA Polymerase II." *Nature* 228:1050–1053.
18. Kornberg, T., and M. L. Gefter. 1971. "Purification and DNA Synthesis in Cell-Free Extracts: Properties of DNA Polymerase II." *Proc. Nat. Acad. Sci.* 68:761–764.

19. Kornberg, A. 1978. "Aspects of DNA Replication." *Cold Spring Harbor Symp. Quant. Biol.* 43:1–9.

20. Brutlag, D., and A. Kornberg. 1972. "Enzymatic Synthesis of Deoxyribonucleic Acid. XXXVI. A Proofreading Function for the 3'→5' Exonuclease Activity in Deoxyribonucleic Acid Polymerases." *J. Biol. Chem.* 247:241–248.

21. Livingston, D. M., and C. C. Richardson. 1975. "Deoxyribonucleic Acid Polymerase III of *Escherichia coli*: Characterization of Associated Exonuclease Activities." *J. Biol. Chem.* 250:470–478.

22. Ollis, D. L., P. Brick, R. Hamlin, N. G. Xuong, and T. A. Steitz. 1985. "Structure of Large Fragment of *Escherichia coli* DNA Polymerase I Complexed with dTMP." *Nature* 313:762–765.

23. Scott, J., and A. Kornberg. 1978. "Purification of the Rep Protein of *E. coli*. An Enzyme Which Separates Duplex Strands in Adenovirus Replication." *J. Biol. Chem.* 253:3292–3304.

24. Yarranton, G. T., R. H. Das, and M. L. Gefter. 1979. "Enzyme-Catalyzed DNA Unwinding." *J. Biol. Chem.* 254:11997–12001.

25. Geider, K., and H. Hoffmann-Berling. 1981. "Proteins Controlling the Helical Structure of DNA." *Ann. Rev. Biochem.* 50:233–260.

26. Arai, N., K. Arai, and A. Kornberg. 1981. "Complexes of Rep Protein with ATP and DNA as a Basis for Helicase Action." *J. Biol. Chem.* 256:5287–5293.

27. Sancar, A., K. R. Williams, J. W. Chase, and W. D. Rupp. 1981. "Sequences of the *ssb* Gene and Protein." *Proc. Nat. Acad. Sci.* 78:4274–4278.

28. Brutlag, D., R. Schekman, and A. Kornberg. 1971. "A Possible Role for RNA Polymerase in the Initiation of M13 DNA Synthesis." *Proc. Nat. Acad. Sci.* 68:2826–2829.

29. McMacken, R., K. Ueda, and A. Kornberg. 1977. "Migration of *Escherichia coli* dnaB Protein on the Template DNA Strand as a Mechanism in Initiating DNA Replication." *Proc. Nat. Acad. Sci.* 74:4190–4194.

30. Arai, K., and A. Kornberg. 1981. "Unique Primed Start of Phage φX174 DNA Replication and Mobility of the Primosome in a Direction Opposite Chain Synthesis." *Proc. Nat. Acad. Sci.* 78:69–73.

31. Ogawa, T., T. A. Baker, A. van der Ende, and A. Kornberg. 1985. "Initiation of Enzymatic Replication at the Origin of the *Escherichia coli* Chromosome: Contributions of RNA Polymerase and Primase." *Proc. Nat. Acad. Sci.* 82:3562–3566.

32. Wickner, S. H. 1978. "DNA Replication Proteins of *Escherichia coli*." *Ann. Rev. Biochem.* 47:1163–1191.

33. McHenry, C., and A. Kornberg. 1977. "DNA Polymerase III Holoenzyme of *Escherichia coli*: Purification and Resolution into Subunits." *J. Biol. Chem.* 252:6478–6484.

34. Alberts, B. M., B. P. Bedinger, T. Formosa, C. V. Jongeneel, and K. N. Kreuzer. 1982. "Studies on DNA Replication in the Bacteriophage T4 in Vitro Systems." *Cold Spring Harbor Symp. Quant. Biol.* 47:655–668.

35. Kornberg, A. 1982. *1982 Supplement to DNA Replication.* San Francisco: Freeman, pp. 124–127.

36. Scheuermann, R. H. and H. Echols. 1984. "A Separate Editing Exonuclease for DNA Replication: The ε Subunit of *Escherichia coli* DNA Polymerase III Holoenzyme." *Proc. Nat. Acad. Sci.* 81:7747–7751.

37. DeLucia, P., and J. Cairns. 1969. "Isolation of an *E. coli* Strain with a Mutation Affecting DNA Polymerase." *Nature* 224:1164–1166.

38. Wechsler, J. A., and J. D. Gross. 1971. *Escherichia coli* Mutants Temperature-Sensitive for DNA Synthesis." *Mol. Gen. Genetics* 113:273–284.

39. Dumas, L. B. 1978. "Requirements for Host Gene Products in Replication of Single-Stranded Phage DNA in Vivo." In *The Single-Stranded DNA Phages,* ed. D. T. Denhardt, D. Dressler, and D. S. Ray. Cold Spring Harbor, N.Y.: Cold Spring Harbor Laboratory, pp. 341–359.

40. Wechsler, J. A. 1978. "The Genetics of *E. coli* Replication." In *DNA Synthesis: Present and Future,* ed. M. Kohiyama and I. Molineux. New York: Plenum Press, p. 49.

41. Hirota, Y., M. Yamada, A. Nishimura, A. Oka, K. Sugimoto, K. Asada, and M. Takanami. 1981. "The DNA Replication Origin (*ori*) of *Escherichia coli*: Structure and Function of the *ori*-Containing DNA Fragment." *Prog. Nucleic Acid Res.* 26:33–48.

42. Von Meyenburg, K., and F. G. Hansen. 1980. "The Origin of Replication, *ori* C, of the *Escherichia coli* Chromosome: Genes Near *ori* C and Construction of *ori* C Deletion Mutations." *ICN-UCLA Symp.* 19:137–159.

43. Kaguni, M. and A. Kornberg. 1984. "Replication Initiated at the Origin (*oriC*) of the *E. coli* Chromosome Reconstituted with Purified Enzymes." *Cell* 38:183–190.

44. Marians, K. J. 1984. "Enzymology of DNA in Replication in Prokaryotes." *CRC Crit. Rev. Biochem.* 17:153–215.

45. Nossal, N. G. 1983. "Prokaryotic DNA Replication Systems." *Ann. Rev. Biochem.* 53:581–615.

46. Grosschedl, R., and G. Hobom. 1979. "DNA Sequences and Structural Homologies of the Replication Origins of Lambdoid Bacteriophages." *Nature* 277:621–627.

47. Tomizawa, J. 1984. "Control of ColE1 Plasmid Replication: The Process of Binding of RNAI to the Primer Transcript." *Cell* 38:861–870.

48. Tsurimoto, T., and K. Matsubara. 1984. "Multiple Initiation Sites of DNA Replication Flanking the Origin Region of λdv Genome." *Proc. Nat. Acad. Sci.* 81:7402–7406.

49. LeBowitz, J. H., M. Zylicz, C. Georgopoulos, and R. McMacken. 1985. "Initiation of DNA Replication on Single-Stranded DNA Templates Catalyzed by Purified Replication Proteins of Bacteriophage λ and *Escherichia coli*." *Proc. Nat. Acad. Sci.* 82:3988–3992.

50. Van der Ende, A., T. A. Baker, T. Ogawa, and A. Kornberg. 1985. "Initiation of Enzymmatic Replication at the Origin of the *Escherichia coli* Chromosome: Primase as the Sole Priming Enzyme." *Proc. Nat. Acad. Sci.* 82:3954–3958.

51. Watson, J. D. 1972. "Origin of Concatemeric T7 DNA." *Nature New Biol.* 239:197–201.

52. Baroudy, B. M., S. Venkatesan, and B. Moss. 1982. "Structure and Replication of Vaccinia Virus Telomeres." *Cold Spring Harbor Symp. Quant. Biol.* 47:723–729.

53. Shampay, J., J. W. Szostak, and E. H. Blackburn. 1984. "DNA Sequences of Telomeres Maintained in Yeast." *Nature* 310:154–157.

54. Sundin, O., and A. Varshavsky. 1981. "Arrest of Segregation Leads to Accumulation of Highly Intertwined Catenated Dimers: Dissection of the Final Stages of SV40 DNA Replication." *Cell* 25:659–669.

55. DiNardo, S., K. Voelkel, and R. Sternglanz. 1984. "DNA Topoisomerase II Mutant of *Saccharomyces cerevisiae*: Topoisomerase II Is Required for Segregation of Daughter Molecules at the Termination of DNA Replication." *Proc. Nat. Acad. Sci.* 81:2616–2620.

56. Gilbert, W., and D. Dressler. 1968. "DNA Replication: The Rolling Circle Model." *Cold Spring Harbor Symp. Quant. Biol.* 33:473–484.

57. Dressler, D., D. Hourcade, K. Koths, and J. Sims. 1978. "The DNA Replication Cycles of the Isometric Phages." In *The Single-Stranded DNA Phages,* ed. D. T. Denhardt, D. Dressler, and D. S. Ray. Cold Spring Harbor, N.Y.: Cold Spring Harbor Laboratory, pp. 187–214.

58. Bastia, D., N. Sueoka, and E. Cox. 1975. "Studies in the Late Replication of Phage Lambda: Rolling Circle Replication of the Wild Type and a Partially Suppressed Strain *O*am29*P*am80." *J. Mol. Biol.* 98:305–320.

59. McMacken, R., L. Rowen, K. Ueda, and A. Kornberg. 1978. "Priming of DNA Synthesis on Viral Single-Stranded DNA in Vitro." In *The Single-Stranded DNA Phages,* ed. D. T. Denhardt, D. Dressler, and D. S. Ray. Cold Spring Harbor, N.Y.: Cold Spring Harbor Laboratory, pp. 273–285.

60. Wickner, S. 1978. "Conversion of Phage Single-Stranded DNA to Duplex DNA in Vitro." In *The Single-Stranded DNA Phages,* ed. D. T. Denhardt, D. Dressler, and D. S. Ray. Cold Spring Harbor, N.Y.: Cold Spring Harbor Laboratory, pp. 255–271.

61. Eisenberg, S., J. F. Scott, and A. Kornberg. 1978. "Enzymatic Replication of φX174 Duplex Circles: Continuous Synthesis." *Cold Spring Harbor Symp. Quant. Biol.* 43:295–302.

62. Warren, G. J., A. J. Twigg, and D. J. Sherratt. 1978. "ColEl Plasmid Mobility and Relaxation Complex." *Nature* 274:259–261.

63. Sussenbach, J. S., P. C. van der Vliet, D. J. Ellens, and H. S.

Jansz. 1972. "Linear Intermediates in the Replication of Adenovirus DNA." *Nature New Biol.* 239:47–49.

64. Rekosh, D. M. K., W. C. Russell, A. J. D. Bellett, and A. J. Robinson. 1977. "Identification of a Protein Linked to the Ends of Adenovirus DNA." *Cell* 11:283–295.

65. Kelly, T. J., Jr., and R. L. Lechner. 1978. "The Structure of Replicating Adenovirus DNA Molecules: Characterization of DNA-Protein Complexes from Infected Cells." *Cold Spring Harbor Symp. Quant. Biol.* 43:721–728.

66. Challberg, M. D., and T. J. Kelly, Jr. 1979. "Adenovirus DNA Replication in Vitro." *Proc. Nat. Acad. Sci.* 76:655–659.

67. Lichy, J. H., K. Nagata, B. R. Friefeld, T. Enomoto, J. Field, R. A. Guggenheimer, J.-E. Ikeda, M. S. Horwitz, and J. Hurwitz. 1982. "Isolation of Proteins Involved in the Replication of Adenoviral DNA in Vitro." *Cold Spring Harbor Symp. Quant. Biol.* 47:731–740.

68. Sogo, J. M., J. A. García, M. A. Peñalva, and M. Salas. 1982. "Structure of Protein-Containing Replicative Intermediates of *Bacillus subtilis* Phage ϕ29 DNA." *Virology* 116:1–18.

69. Stillman, B. W., and F. Tamanoi. 1982. "Adenoviral DNA Replication: DNA Sequences and Enzymes Required for Initiation in Vitro." *Cold Spring Harbor Symp. Quant. Biol.* 47:741–750.

70. Blanco, L., and M. Salas. 1984. "Characterization and Purification of a Phage ϕ29-encoded DNA Polymerase Required for the Initiation of Replication." *Proc. Nat. Acad. Sci.* 81:5325–5329.

71. Watabe, K., M. Leusch, and J. Ito. 1984. "Replication of Bacteriophage ϕ29 DNA *in Vitro:* The Roles of Terminal Protein and DNA Polymerase." *Proc. Nat. Acad. Sci.* 81:5374–5378.

72. Grummt, F. 1978. "Diadenosine 5',5'''-P^i,P^4-Tetraphosphate Triggers Initiation of in Vitro DNA Replication in Baby Hamster Kidney Cells." *Proc. Nat. Acad. Sci.* 75:371–375.

73. Kornberg, A. 1980. "Regulation of Replication." In *DNA Replication.* San Francisco: Freeman, Chap. 13.

74. Kornberg, A. 1982. *1982 Supplement to DNA Replication.* San Francisco: Freeman, pp. S144–S146.

75. Kornberg, A. 1982. "Eukaryotic DNA Polymerase." In *1982 Supplement to DNA Replication.* San Francisco: Freeman, Chap. S6.

76. Li, J. J., and T. J. Kelly. 1984. "Simian Virus 40 DNA Replication in Vitro." *Proc. Nat. Acad. Sci.* 81:6973–6977.

77. Stillman, B. W., and Y. Gluzman. 1985. "Replication and Supercoiling of SV40 DNA in Cell-Free Extracts from Human Cells." *Mol. Cell. Biol.* 5:2051–2060.

CHAPTER 11

Recombination at the Molecular Level

All DNA is recombinant DNA. In engineering relatively simple changes in gene expression by combining and modifying DNA segments in the test tube, we are only simulating natural processes of recombination and mutation that have acted throughout evolution to construct complex chromosomes. Genetic exchange works constantly to blend and rearrange chromosomes, most obviously during meiosis, when homologous chromosomes cross over in every generation. Genetic maps derived from early measurements of the frequency of crossing over between different genes gave the first real information about chromosome structure: that genes are arranged in a fixed, linear order. A most important recent discovery is that gene order does change, although rarely: Movable DNA segments called transposons occasionally jump around chromosomes, thus fundamentally altering chromosomal structure. In addition to neatly moving genes, transposons also scramble DNA, making deletions, inversions, and other rearrangements. It is becoming clear that such changes are a critical feature of chromosome evolution, particularly in eucaryotic cells.

We now appreciate that recombination is not accidental, but is instead an essential cellular process catalyzed by enzymes that cells encode and regulate for the purpose. Besides providing genetic variation, recombination enzymes allow cells to retrieve sequences lost when DNA is damaged by radiation or chemical accidents, by replacing the damaged section with an undamaged DNA strand from a homologous chromosome. Furthermore, special types of recombination regulate gene expression. By switching specific segments within chromosomes, cells put dormant genes into sites where they can be expressed, even creating new protein-coding regions. Moreover, recent discoveries, in revealing the molecular mechanisms of recombination, have provided new tools for the deliberate manipulation of genes.

Crossing Over Is Due to Breakage and Rejoining of Intact DNA Molecules[1]

Until recently, there was not even a superficial understanding of the molecular basis of crossing over. The classical picture of crossing over, developed in the 1930s from cytological observations, hypothesized that during meiosis, the paired, coiled chromosomes were sometimes physically broken at the chromatid level as a result of ten-

313

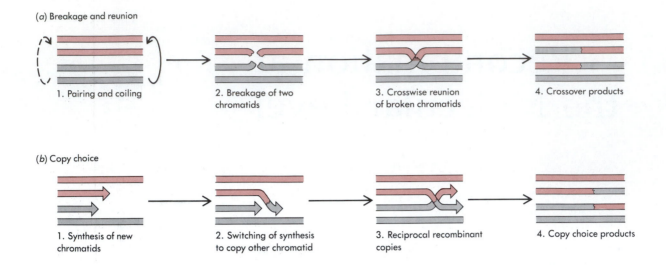

(a) Breakage and reunion

1. Pairing and coiling

2. Breakage of two chromatids

3. Crosswise reunion of broken chromatids

4. Crossover products

(b) Copy choice

1. Synthesis of new chromatids

2. Switching of synthesis to copy other chromatid

3. Reciprocal recombinant copies

4. Copy choice products

Figure 11-1

Representation of two possible mechanisms originally considered for crossing over between chromosomes. Each line represents a separate chromatid. According to the current idea that each chromatid contains a single double-helical DNA molecule, and our understanding of semiconservative replication of the double helix, a simple copy-choice mechanism with no strand breakage would not be possible.

sion created by their contraction. The broken ends could then relieve the tension by crossways reunion, creating two reciprocally recombinant chromatids (Figure 11-1a). According to this model, recombination occurs after chromosome duplication is complete—that is, when there are four chromatids in each pair of synapsed chromosomes. This hypothesis fell into disfavor about 1955, when geneticists found that crossing over occurs within the gene, by then realized to be part of a DNA molecule. A seemingly inconvenient consequence was that to produce exactly reciprocal recombinants, the breakage points must necessarily lie between the same nucleotides in the two homologous chromatids. Otherwise, recombination would generate new DNA molecules differing in length from the parental molecules.

As a result of this dilemma, enthusiasm developed for a hypothesis relating recombination to chromosome duplication. This alternative hypothesis proposes that during replication of the paired chromosomes, the new DNA strand being formed along the paternal chromosome (for example) switches to the maternal one that it thereafter copies. If the complementary replica of the maternal strand also switches templates when it reaches the same point, two reciprocally recombinant strands are formed. This hypothetical process is called **copy choice.** A fundamental distinction between the two hypotheses lies in their prediction of the physical origin of recombinant chromosomes: Either the recombinant chromosomes inherit physical material from the two parental chromosomes following breakage and reunion, or the recombinant chromosomes produced by copy choice are synthesized from new material.

These two hypotheses were tested by experiments using isotopically heavy (^{13}C, ^{15}N) parental phage λ particles (Figures 11-2 and 11-3). Here, again, the heavy isotopes were used so that a cesium chloride (CsCl) gradient could distinguish between parental and daughter DNA strands. Genetic crosses between heavy (or between heavy and light) phage particles were made in *E. coli* cells growing in a light (^{12}C, ^{14}N) medium. Under these conditions, all of the newly synthesized viral DNA molecules are derived from light precursors. Thus, if copy choice is the correct mechanism, all the recombinant particles should be light. On the other hand, if recombinants are derived by breakage and reunion, some of the recombinant phage particles will contain heavy atoms derived from the parental chromo-

Light bacteria growing in a light medium are infected with heavy phage of genotype ABC and light phage of genotype abc.

After injection of DNA into the bacteria, DNA duplication occurs, and most of the original parental strands become part of either hybrid or light molecules.

Most progeny are completely light and band here.

A few progeny phages move to denser regions, because they contain atoms derived from the heavy parent (ABC). The partially heavy recombinants of genotype ABc band in a denser region than those of genotype abC, because these classes arise by crossing over between B and C (b and c).

Many parental strands, however, do not remain intact, because they are fragmented by crossing over. This is seen by looking at the density of the recombinant phages.

Distribution of progeny phage after prolonged centrifugation in a dense CsCl solution

Key

— Light DNA strand
— Heavy DNA strand
ϙϙ Progeny phage containing some heavy atoms
ϙϙ Progeny phage containing only light atoms
·   CsCl molecules

somes. The progeny particles of these crosses were placed in dense CsCl solutions and rapidly centrifuged to separate particles of different densities. Phage particles of various densities were then collected and genetically tested to see which were recombinants. The results were clear-cut, and to the surprise of most molecular biologists, showed that some recombinant particles contained heavy atoms (see Figure 11-2). Moreover, further experiments revealed that recombination can occur between nonreplicating DNA molecules (see Figure

Figure 11-2
The employment of heavy isotopes to study the mechanism of crossing over.

Figure 11-3
The demonstration that crossing over and DNA duplication are independent phenomena.

Light bacteria growing in a light medium are infected with several particles, heavy phage, of genotype AB and several particles, heavy phage, of genotype ab.

After injection of the DNA molecules into the bacteria, most heavy strands become part of hybrid molecules. Rarely, however, an infecting molecule fails to duplicate and, when the new progeny particles are formed, these unreplicated molecules, in which the parental strands have never separated, become enclosed in new protein shells. This phenomenon enables us to ask whether the rare, completely heavy, DNA molecules are ever recombinants. Again use is made of a CsCl gradient to separate progeny particles of different density.

Most progeny are light and band here.

This is where the very rare progeny particles, containing completely heavy DNA, band. Some of these phage are recombinants of the classes

Distribution of progeny phage.

This shows that crossing over occurs between intact double helices and that extensive DNA synthesis is not involved in crossing over.

Key

— Light DNA strand
— Heavy DNA strand
·   CsCl molecule
⟋ϙ Progeny phage containing only light atoms
ϙϙ Progeny phage containing some heavy atoms

Figure 11-4
Reciprocal crossing over directly visualized in hamster cells in tissue culture. Chromosomes whose DNA contains bromodeoxyuridine in place of thymidine in both strands appear light after treatment with Giemsa stain, whereas those containing DNA substituted in only one strand appear dark. After two generations of growth in bromodeoxyuridine, one newly replicated chromatid has only one of its strands substituted, whereas its sister has both substituted. Thus, sister chromatids can be distinguished by staining. Then crossovers, called sister chromatid exchanges, are easily detected as alternating lengths of light and dark (top). Such exchanges increase dramatically when cells are treated by agents that damage DNA, like the drug adriamycin (bottom). (This is not *meiotic* crossing over between homologous chromosomes, which serves to create genetic diversity, but is an equivalent process occurring between *mitotic* sister chromosomes before they separate, which serves to repair DNA damage.) (Courtesy of Sheldon Wolff and Judy Bodycote.)

11-3). Breakage and reunion of intact double helices must therefore be the primary mechanism of crossing over in bacteriophage. In fact, the same is true in bacteria and higher organisms. Physical exchange of DNA between sister chromatids of mammalian cells can be visualized dramatically in tissue culture cells by means of a staining procedure that allows the chromatids to be distinguished (Figure 11-4).

Homologous Recombination Is Guided by Base Pairing

In retrospect it is obvious what mechanism most precisely aligns DNA molecules in crossing over, because we can hardly imagine any other: Complementary base-pairing between strands unwound from two different chromosomes puts the chromosomes in exact register. Crossing over thus generates **homologous recombination;** that is, it occurs between two regions of DNA containing identical or nearly identical sequences. Usually, these are sequences on two equivalent regions of homologous chromosomes, but crossing over also can be detected between homologous segments in nonequivalent regions, as long as the recombinant survives. Such unequal crossing over gives rise to duplications and deletions in chromosomes (Figure 7-25). In the laboratory, recombination between homologous segments on different DNAs is now used to construct new genetic variants. Thus, when a yeast DNA fragment cloned in a circular bacterial plasmid is introduced into a yeast cell, it finds and crosses over with its homologous sequence, thereby inserting the plasmid into a yeast chromosome (see Figure 19-3). At least in bacteria and yeast, any two homologous DNA segments will recombine in the cell, given one condition: One of them must have a break or a region of single-stranded DNA.

Recombination Is Initiated from Breaks or Gaps in DNA[2, 3]

A strong hint about the way crossing over begins is that breaks in DNA greatly stimulate it. Thus, ultraviolet irradiation, X-irradiation, and reactive chemicals that create double-strand breaks or gaps (single-stranded regions in otherwise double-stranded DNA) all increase crossing over. Mutations in bacterial genes for DNA ligase and DNA polymerase I that give rise to breaks and gaps by preventing proper sealing of DNA during replication or repair have the same effect. More direct proof is that a plasmid bearing a yeast DNA segment can be made to recombine with a yeast chromosome (as shown in Figure 19-3) up to a thousand times more efficiently by cutting the yeast segment with a restriction enzyme before the plasmid is introduced into the cell. Now we suspect that in a similar way, cellular enzymes actively induce recombination in meiosis by making occasional breaks along a pair of synapsed chromosomes. In fact, it is likely that all homologous recombination is initiated from discontinuities in DNA, whether their source is apparent or not. The usual semi-random nature of crossing over that allows a genetic map to be constructed probably results from these initiating sites occurring nearly at random.

RecA Protein Matches Single-Stranded DNA to a Complementary Sequence in a Target Chromosome[4, 5, 6]

From what we know about the physical properties of DNA, there is no reason to expect that two molecules will break and rejoin simply because one is already gapped or broken. Indeed, genetic analysis shows that phage, bacteria, and yeast have enzymes to catalyze the biochemical steps in recombination, a fact that is no doubt true of all organisms. Our most detailed understanding of the mechanism of recombination comes from studies of the *recA* gene product of *E. coli*, an enzyme that carries out the fundamental reaction in recombination—that of making the base-paired hybrid that joins two DNA molecules.

The enzymatic properties of **RecA protein** show why gapped or broken DNA is required to initiate recombination. No reaction occurs if two homologous but completely base-paired helices are mixed with RecA protein. However, RecA protein specifically recognizes single-stranded DNA and anneals it to a complementary sequence in a homologous duplex, simultaneously displacing the resident strand (Figures 11-5 and 11-6). RecA protein first binds single-stranded DNA in a fixed ratio of about one polypeptide per five nucleotides, thus forming a DNA-protein filament; if ATP is present, RecA protein in the filament may then unwind successively different parts of the duplex, attempting to anneal the bound single strand to unwound regions until a complementary sequence is found. Once a small portion is

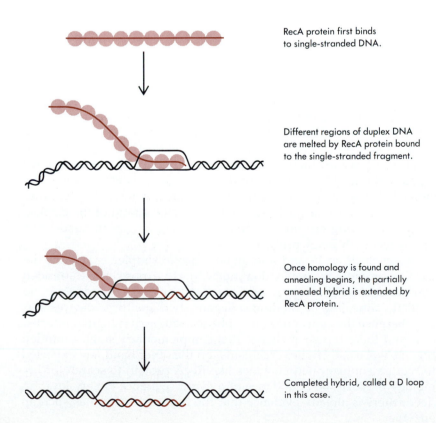

RecA protein first binds to single-stranded DNA.

Different regions of duplex DNA are melted by RecA protein bound to the single-stranded fragment.

Once homology is found and annealing begins, the partially annealed hybrid is extended by RecA protein.

Completed hybrid, called a D loop in this case.

Figure 11-5
Activity of RecA protein as demonstrated in vitro in a model reaction using DNA fragments.

Figure 11-6
Electron micrograph of RecA protein exchanging DNA strands in vitro. In this case, the reacting DNAs are a circular single strand and a linear double-stranded fragment that is homologous to a portion of the single-stranded circle. In the reaction, one strand of the double-stranded fragment is displaced by its equivalent segment in the circle. The final products are this displaced linear single strand, probably bound to the excess RecA protein that was present in the reaction mixture, and the mostly single-stranded circle that now contains a double-stranded region. Preparation of the sample for microscopy could have disrupted part of the structure, for example, the hypothetical complex of RecA protein and three DNA strands shown in Figure 11-7. The accompanying diagram identifies the structures seen in the electron micrograph. [Adapted from C. Radding, J. Flory, A. Wu, R. Kahn, C. Das Gupta, D. Gorda, M. Bianchi, and S. Tsang, *Cold Spring Harbor Symp.* 47 (1983):824, with permission.]

(a)

0.5 μm

Duplex DNA that has not yet reacted

Single-stranded circular DNA complexed with RecA protein

Newly formed duplex

RecA protein bound to the DNA strand displaced from the duplex (because there is excess RecA protein available)

Duplex DNA that has not yet reacted

Beginning of hybrid formation

(b)

properly annealed, the energy of ATP drives the pairing reaction to completion, moving in the 5'→3' direction relative to the single strand. RecA protein is displaced as new hybrid DNA is formed. We describe below how this same reaction can form a bridge between two separate duplex DNAs, if one of them contains a single-stranded region. Breaks as well as gaps in DNA initiate recombination because they provide sites for nucleases to degrade one strand of the duplex, or for unwinding enzymes to enter, exposing single-stranded DNA that reacts with RecA protein.

Phages such as T7 and λ (lambda) encode simpler enzymatic devices to promote recombination: nucleases to expose single-stranded DNA and DNA-binding proteins to extend single-stranded DNA and facilitate annealing by random collisions. Perhaps these devices succeed because the concentration of phage sequences is high in infected cells and there is a good chance that complementary single-stranded regions will exist. Higher organisms, on the other hand, are expected to require sophisticated enzymes like RecA protein to search for homology because the abundance of a given sequence is very low. Indeed, there is already evidence that the fungus *Ustilago* has such an enzyme.

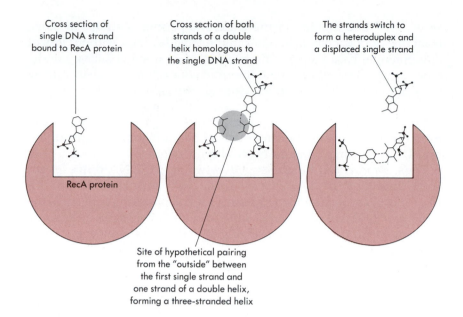

Cross section of single DNA strand bound to RecA protein

RecA protein

Cross section of both strands of a double helix homologous to the single DNA strand

The strands switch to form a heteroduplex and a displaced single strand

Site of hypothetical pairing from the "outside" between the first single strand and one strand of a double helix, forming a three-stranded helix

Figure 11-7
How a three-stranded helix might be involved in DNA strand exchange mediated by RecA protein. [Adapted from Howard-Flanders, P., S. West, and A. Stasiak, *Nature* 309 (1984):215–220, with permission.]

The Possibility of Three-Stranded and Four-Stranded DNA Helices in Recombination[7]

Initially it seemed likely that the specific pairing reaction that RecA protein promotes between single-stranded DNA and its complementary sequence in one strand of a double helix could occur only if the double helix first unwound so that new base pairing could occur. However, there is another, rather intriguing possibility: The single-stranded DNA might align in specific register with the duplex by wrapping around it. Model building shows that base-pair-specific contacts can be made between two separate duplexes, entwined in a four-stranded structure that is aligned by chemical groups exposed in the grooves of the two helices. Similarly, a single DNA strand bound to RecA protein might make these same specific contacts in a groove of the target duplex DNA and thus recognize its complementary sequence from the "outside" (Figure 11-7). Such a three-stranded helix would represent the first base-specific recognition and would be followed by unwinding of the duplex and formation of normal base pairing between the invading single strand and its complement. Clearly, initial pairing in a three-stranded helix should speed the search for homology, since each segment examined need not be first unwound.

Recombination Involves a Crossed-Strand Intermediate[8, 9]

The cross over intermediate that joins two DNA duplexes in recombination contains four DNA strands that switch pairing at the joining point to form a crossed-strand junction called the **Holliday structure,** after R. Holliday, who first proposed it (Figure 11-8). The existence of the Holliday structure was originally suggested by genetic experiments that detected heteroduplex regions in both recombining DNAs. **Heteroduplexes** are regions on recombinant DNA molecules where the two strands are not exactly complementary. Model building

(a)

Exchange site

(b)

Figure 11-8
The Holliday structure. (a) Photograph of a molecular model of sections of two double helices, showing that crossed-strand connections can be made without disruption of the individual DNA molecules. [Courtesy of N. Sigel and B. Alberts, *J. Mol. Biol.* 71 (1972):789.] (b) The mutual interchange of parallel strands between double helices (branch migration) as a consequence of right-handed axial rotation. [After T. Broker, *J. Mol. Biol.* 81 (1973):1.]

Crossed strands in a recombination joint

Diffusion of branch rightward ("branch migration"), creating extensive hybrid regions in both duplexes

Figure 11-9

Diffusion of a recombination joint.

shows that free rotation of the backbone permits a strand to move from one double helix to another in the Holliday joint without loss of any potential base pairs or the creation of strained chemical bonds. Once two helices are so joined, the cross connection can easily diffuse by a zipperlike action in which equivalent bases on the two original molecules exchange places, a process called **branch migration** (see Figure 11-9). In this way, extensive sections of a given chain can move from one helix to another, frequently generating long sections of hybrid DNA that may contain heteroduplex regions.

Thus, strands can be exchanged between two DNA duplexes for some distance from the initiating break. How fully recombinant duplexes form can be understood from a second physical property of the structure: its ability to undergo isomerization, in which a simple steric rearrangement converts the bridging strand pair into outside strands

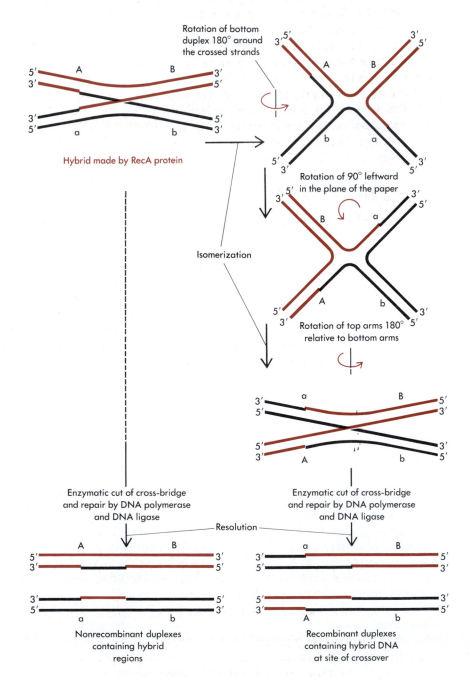

Figure 11-10

Isomerization of the recombination joint and resolution of two different isomers.

and vice versa (Figure 11-10). Such isomerizations can occur very frequently, so that a nuclease cut that completes crossing over (a process called resolution) can lead to a double exchange of two progeny strands or to single exchanges of all four strands, thus recombining the flanking arms of the chromosomes (Figure 11-10).

Direct proof that strand exchange occurs comes from electron microscopic visualization of phage DNA molecules isolated from infected cells, where recombination is frequent. The clearest results are obtained from phage λ, where denaturation mapping shows that cross-bridges always link together homologous regions of the pairing partners (Figure 11-11). Moreover, the individual DNA strands can be visualized in the bridging region, showing unambiguously their passage from one double helix to another.

Heteroduplexes

The frequent generation of heteroduplex segments during crossing over provides direct experimental support for the hypothesis that the fundamental recombination event involves pairing between regions of single-stranded DNA. Heteroduplexes arise when the primary pairing region, together with the adjacent region of branch migration (see above), encompasses sites of genetic differences between the two parental chromosomes. There is evidence of heteroduplex regions in all viruses whose genetics have been extensively analyzed. There is usually one such region, several thousand nucleotides long, on each T4 molecule. Heteroduplexes have also been well characterized in phage λ and again are much longer than might be expected if they reflected only the primary pairing event. So most of their length probably arises by branch migration transferring strands from one double helix to another.

Heteroduplex lifetime can be very short, because when the recombinant DNA molecules duplicate, the alternative alleles segregate out (Figure 11-12). This was first observed in phage systems, where mere

Figure 11-11
Electron micrograph visualization of crossover between two homologous regions of λ DNA. Here the Holliday joint has been partly unwound so that all four strands can be seen; this is the intermediate stage in isomerization (Figure 11-10). (Courtesy of R. Inman, University of Wisconsin.)

A heteroduplex recombinant containing
two mispaired bases a' and b'

Progeny

gregation following replication
th no repair event

Segregation following replication with a
single repair event b'→ b

Figure 11-12
Heteroduplex segregation with and without repair event.

Figure 11-13
Photograph of several mottled plaques arising from segregation of a heteroduplex DNA molecule containing both *rII*⁺ and *rII* markers on opposing strands. (Courtesy of A. D. Hershey.)

inspection of recombinant plaques reveals mixtures of two genetically distinct types. When the DNA of a parental T4 phage particle contains a heteroduplex *rII* region, the resulting plaque has a mottled appearance, with some sections characteristic of the *rII* phenotype and others of the wild-type *rII*⁺ phenotype (Figure 11-13).

RecA Protein Makes Holliday Structures in Vitro[5]

Although RecA protein initially transfers only one strand into a DNA duplex, a second exchange follows naturally from the first. Figure 11-14 shows how RecA protein makes a complete recombination joint in vitro from two identical duplexes, one of which has a gap at one end. When annealing has been driven to the edge of the gap by RecA protein (see Figure 11-14c), the joint diffuses into the wholly duplex region as the top helix to the left of the gap unwinds, and simultaneously the second hybrid helix to the right of the crossover forms (see Figure 11-14d). It has been suggested that RecA protein may act in a second way at this stage, guiding the formation of a four-stranded helix composed of the two DNA duplexes in the region to the left of the exchange site, thereby accelerating the exchange of strands that allows the joint to move through the wholly duplex region (Figure 11-15).

To explain how recombination could be initiated in the cell when RecA protein binds to gaps in duplex DNA that are distant from an end, a refinement of the model reaction shown in Figure 11-14 is needed. A break must be made in the target DNA duplex, giving a free end from which the strands interwind to form a duplex. We sup-

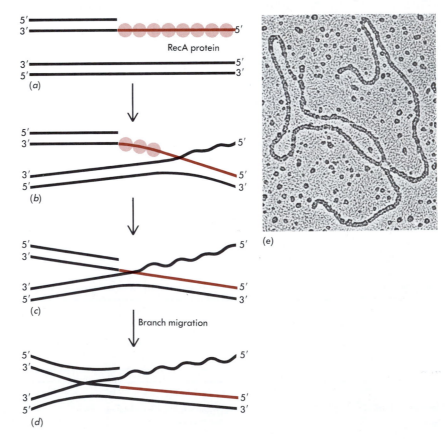

Figure 11-14
A model reaction by which RecA protein makes recombination joints in vitro. The two starting DNAs are identical, except that one is single-stranded at an end. The electron micrograph shows the product. (Courtesy of C. Radding.)

(a)

Paired duplexes
(four-stranded DNA helix)

Heteroduplex
DNA

Direction of branch migration

Cross section of a
double helix bound
to RecA protein

RecA protein

Site of hypothetical pairing
between two DNA duplexes
to form a four-stranded helix

The strands switch to
form two heteroduplexes

(b)

Figure 11-15

Suggested role of a (hypothetical) four-stranded DNA helix in recombination. (a) This structure is equivalent to the fully double-stranded portion of Figure 11-14d, but with two differences. A segment of the two homologous duplexes to the left of the exchange site is drawn as a four-stranded helix, and RecA protein is present in the four-stranded region. (b) How RecA protein might act to promote reciprocal strand exchange from a four-stranded intermediate. [Adapted from Howard-Flanders, P., S. West, and A. Stasiak, *Nature* 309 (1984):215–220, with permission.]

pose that a nuclease recognizes an incipient hybrid and breaks the displaced strand, allowing the crossed-strand joint to form as it does in the original model reaction (Figure 11-16).

Alternatively, RecBC enzyme (described in the next section) may provide single-stranded DNA with a free end, which can be annealed to a homologous duplex as in the in vitro reaction illustrated in Figure 11-5. A subsequent cut in the DNA strand displaced from the duplex, followed by branch migration, would allow the Holliday joint to form.

RecBC Enzyme Promotes Recombination by Unwinding DNA and Nicking It at Specific Sites Named Chi[10, 11]

Although RecA protein carries out the fundamental pairing step of recombination, other proteins are involved in other steps of the process. Paramount among these is the RecBC enzyme, specified by the *E. coli* genes *recB* and *recC*, whose absence reduces the yield of recombinants after bacterial mating about a hundredfold. This complex enzyme with a molecular weight of about 300,000 daltons has both DNA-unwinding and nuclease activities that combine to expose single-stranded DNA with a free end; possibly RecA binds this DNA to begin the pairing reaction. An important feature of the RecBC en-

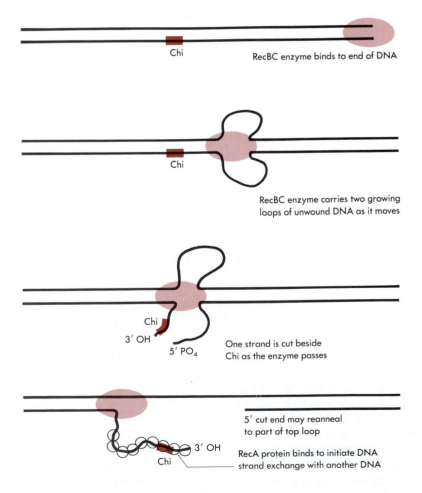

zyme is that it initiates unwinding only on DNA containing a free duplex end. First binding to the duplex end, it uses energy in ATP to travel along the duplex, unwinding and rewinding DNA as it goes. Since the RecBC enzyme unwinds DNA faster than it rewinds the strands, the unwound segment becomes progressively longer as the enzyme moves down the DNA. This unwound structure appears in electron micrographs as two single-stranded loops ("rabbit ears"), because the sites of unwinding and winding are both held by the enzyme as it moves (Figure 11-17).

A second activity of RecBC enzyme that acts while it is unwinding DNA is critical to its function in recombination. Genetic experiments show that the RecBC enzyme promotes recombination most frequently in DNA containing a site named Chi, which has the nucleotide sequence 5'-GCTGGTGG-3'. When Chi is present in DNA being unwound, a specific nuclease activity of RecBC enzyme cuts the exposed single strand near Chi, thus preventing its being rewound as the enzyme progresses (Figure 11-17). The consequence is to leave behind a free single-stranded tail and a gap, both of which are sites where RecA protein can bind to initiate DNA strand exchange with a homologous sequence.

E. coli DNA has about a thousand Chi sequences, or about one per five genes, and thus offers many opportunities for RecBC enzyme action. These sequences are not usually available to the enzyme in a normal cell, whose DNA has no free ends, but are particularly important in bacterial conjugation. Here a DNA end is introduced by the

Figure 11-16

A likely pathway for initiation of recombination from a gap in vivo. Here the gap is shown to be adjacent to a **DNA lesion,** which is a chemical change in a DNA base that can destroy its ability to act as a template in replication and thus can interrupt synthesis of the complementary strand; the resulting gap is a region where the bare template DNA strand is exposed. DNA lesions and their repair are discussed in Chapter 12.

Figure 11-17

The enzymatic activity of RecBC enzyme, and a model suggesting its role in recombination is to provide single-stranded DNA that RecA protein binds.

male bacterium, and it is thought that RecBC enzyme binds to the injected DNA at this end, moving along it and cutting at Chi sequences, thereby promoting recombination with the recipient cell's DNA. Ends also are introduced into a cell by phage P1 transduction and are present in phage λ–infected cells because free ends are made by the enzyme that cuts λ DNA for packaging into the phage particle; in both cases, the RecBC enzyme is required for efficient recombination. In fact, its function in phage λ recombination led to the discovery of Chi.

Other Proteins That Participate in Homologous Recombination[12]

Even in *E. coli,* we know only some of the genes and enzymes that contribute to recombination. Besides *recA, recB,* and *recC,* there are several other genes whose products are required for efficient recombination, although the specific functions of those products are unknown. One unidentified enzyme is the nuclease that cuts apart Holliday junctions and thus resolves recombinant DNA molecules. It is possible that the RecBC enzyme does this in addition to Chi-directed cutting, because it also has other, less specific nuclease activities.

Several proteins necessary for DNA synthesis also act in recombination. Single-stranded DNA-binding protein, which facilitates DNA synthesis by holding single-stranded DNA in an extended form accessible to replication enzymes (Chapter 10), also assists RecA protein in binding in an orderly way to single-stranded DNA. DNA polymerase fills gaps left when recombinant DNAs are cut apart, and finally, DNA ligase seals the nicks left by DNA polymerase.

Recombination Is Not Always Reciprocal at the Site of Crossing Over

Early investigation of crossing over between different genes revealed the seemingly obligatory occurrence of reciprocal recombinants (Figure 11-18). However, exceptions were discovered when the recombinants studied arose between nearby sites on the same gene. Then, nonreciprocal behavior was often observed. This phenomenon, called **gene conversion,** is best studied in organisms such as yeast or *Neurospora,* where all the products of a single meiotic event can be seen. Here, instead of always observing equal (2:2) segregation of the entering alleles, cases of 3:1 segregation are found. In these cases, one allele is lost, and an extra copy of the other appears.

The crossing-over hypothesis outlined earlier permits such exceptions. As shown in Figure 11-12, the final segregation pattern of genes localized around the crossover site may be affected by repair enzymes that recognize heteroduplex distortions in the double helix and randomly remove one of a mismatched base pair. Depending on the bases removed, either 2:2 or 3:1 ratios will be found. Seen from a broad perspective, however, the most striking thing about recombination, even at the molecular level, is still the prevalence of reciprocal recombinants. This is because heteroduplexes are short, relative to the distance between most recombining genes, so that neither of two genes under study is likely to be included in the heteroduplex that gives rise to a crossover between them.

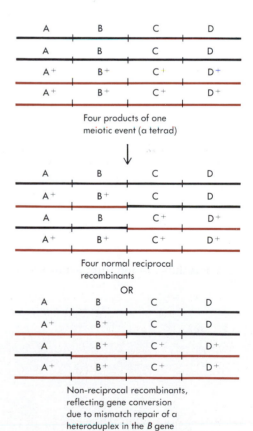

Four products of one meiotic event (a tetrad)

Four normal reciprocal recombinants

OR

Non-reciprocal recombinants, reflecting gene conversion due to mismatch repair of a heteroduplex in the *B* gene

Figure 11-18
Reciprocal and nonreciprocal recombination following crossover.

Insertions and Deletions Arise from Errors in Crossing Over

The extreme accuracy of most crossing-over events depends on the correct juxtaposition of the complementary single-stranded regions during reformation of the hydrogen bonds. This correct alignment is assured by the uniqueness of most polynucleotide sequences. When a random chain of polynucleotides contains more than 10–12 nucleotides, it will almost never have the same sequence as another fragment of similar length. Thus, as long as the single-stranded segments are relatively long, it is very, very unlikely that a fragment of one gene will be mistakenly linked up to the wrong part of its own gene or to a different gene. In those rare cases, however, where two different chromosomal regions have considerable homology, we should expect occasional misjoinings, leading to the insertion or deletion of large blocks of nucleotides (Figure 11-19). Such accidental rearrangements are usually catalyzed by RecA protein, although other, unknown enzymes may sometimes be responsible.

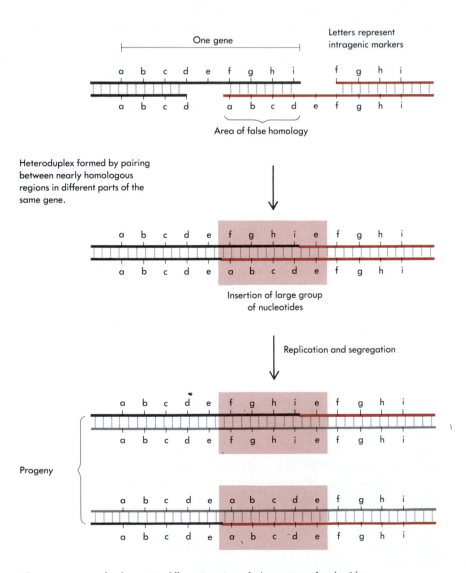

Figure 11-19

Origin of a multinucleotide insertion through pairing of nearly homologous sequences within the same gene.

The two progeny molecules contain different insertions of a large group of nucleotides.

Recombination Repairs Damaged DNA Molecules[3, 13]

We often think of crossing over as a mechanism that creates genetic diversity, but its most vital function is probably to repair DNA damage. This is revealed in studies of *recA⁻* bacteria and recombination-defective mutants of yeast, which are extremely sensitive to killing by radiation and by many chemicals that damage DNA. How recombinational repair works is implied by our description of how a gap in DNA leads to the recombination intermediate (see Figure 11-16). The gap is filled by DNA from a homologous duplex and is thus repaired. Repair is achieved whether or not the intermediate is cut so as to exchange flanking arms of the two helices; it only matters that the gap is filled with DNA of the correct sequence. Although a simple gap can be filled by DNA polymerase, a more serious problem is posed by gaps like the one shown in Figure 11-16 that contain DNA lesions such as thymine dimers that no longer can base-pair; these arise when replication necessarily stops at the damaged site and must reinitiate beyond it. The genetic information at the lesion is lost from both DNA strands and can be retrieved only by extracting it through recombination from a homologous duplex. The newly replicated sister DNA molecule is the most accessible source of the correct sequence, although a bacterial cell also usually has other copies of its chromosome.

A simple extension of this mechanism explains how RecA protein can repair the ultimate damage, a double-stranded break, which may be caused, for example, by X-irradiation. Degradation from each end of the break by an exonuclease leaves single-stranded ends, which then separately can invade a homologous sequence, forming two adjacent recombination joints (Figure 11-20). Alternatively, the RecBC enzyme might provide initiation sites by entering at each free end. When the two joints are cut, the parental helices are separated and

Double-strand break in DNA, with a missing segment

After unwinding, degradation by a nuclease exposes single-stranded DNA at each broken end; RecA protein binds and initiates strand exchange with homologous DNA

A double recombination joint is formed

The resulting gaps are filled by DNA polymerase

Both recombination joints are resolved, yielding two intact duplexes

Figure 11-20
Model for repair of a double-stranded DNA break by RecA protein. The fate of the single-stranded DNA that initiates the strand exchange is marked by colored shading.

the resulting gaps are filled by DNA polymerase and sealed by DNA ligase.

Site-Specific Recombination Gives Precise DNA Rearrangements[14–18]

By its very nature, crossing over generally preserves the order of DNA sequences in homologous chromosomes. However, in exceptional cases, cells also use a carefully regulated recombinational process that has the opposite consequence, rearranging sequences by recombination directed between special sites. DNA segments can be moved by **site-specific recombination,** often with the result that different genes or sets of genes are then expressed. The premier example, found in higher cells, is the construction of a great variety of specific antibody genes through site-specific DNA rearrangements of a single set of precursor sequences (Chapter 23).

Unlike crossing over, site-specific recombination is guided not by DNA sequence homology (although short homologies are involved), but by the location of binding sequences for specialized enzymes that catalyze breakage and rejoining. Site-specific recombination is initiated by regulatory processes that make available the correct enzymes, in contrast to the random breakage events that may expose any sequence to bind proteins like RecA and initiate crossing over.

Site-specific recombination was recognized first from genetic studies of bacteriophage λ, whose DNA integrates by recombination into a specific site on the *E. coli* chromosome and is carried thereafter in a passive state as a prophage (Chapter 17). There are two principal features of λ integration that are true of site-specific recombination in general: The exchange is reciprocal and preserves all preexisting DNA, and it occurs at a specific nucleotide within a short region of homology between the phage and bacterial DNA.

The Interacting Targets for Site-Specific Recombination by λ Integrase Share a Short Homologous Sequence But Differ Dramatically in Protein-Binding Sites[16, 17, 18]

Bacteriophage λ encodes an enzyme, **λ integrase,** that directs insertion of phage DNA into the *E. coli* chromosome by recombination across specific sites on both DNAs, generating one circular molecule out of two (Figure 11-21). Integrase is made abundantly in the early stage of phage infection, so that integration can occur in almost every infected cell. The reason λ integration is so well understood is that it occurs in vitro in a simple reaction mixture containing only four components: purified integrase protein, an *E. coli* accessory protein named **IHF** (for integration host factor), magnesium ions, and DNA containing the specific sites on phage and bacterial DNA (named *attP* and *attB*) across which recombination occurs. A convenient assay for integration uses a plasmid artificially constructed so that this single DNA contains both *attB* and *attP*, oriented such that the integration reaction recombines out the DNA between them, yielding two smaller DNA circles from the original circular plasmid (Figure 11-

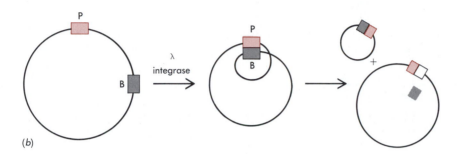

Figure 11-21
Integration of bacteriophage λ DNA into the *E. coli* chromosome by site-specific recombination. (a) The natural reaction, in which recombination joins two molecules. (b) An artificial substrate made by in vitro recombination to contain both phage and bacterial attachment sites. When integrase aligns and recombines P and B, the result is to produce two smaller DNA circles. Since these can be detected quickly by agarose gel electrophoresis, this reaction provides an easy assay for integrase activity.

21b). All steps of the integration reaction are coupled: Four strands are cut, exchanged, and religated without any stable intermediates appearing. This is much like a type II topoisomerase reaction (Chapter 9) in that both strands of each DNA duplex are cut and the phosphodiester bonds are remade without energy from ATP being required. In fact, integrase can act as a topoisomerase, relaxing supercoils that carry an *att* site or a similar sequence, in what is thought to be an aberrant variation of the normal integration reaction. As is true for topoisomerases, integrase makes a staggered cut, in this case with a seven-nucleotide overlap, in both the phage and bacterial sites (Figure 11-22).

Figure 11-22
Nucleotide sequence of both *attP* and *attB* core regions and of the hybrid recombinant sites. The core sequence is boxed, and vertical arrows mark the sites of exchange. The horizontal arrows between the strands are symmetrical integrase-binding sites that overlap the core sequence and serve to align integrase correctly at the sites of exchange.

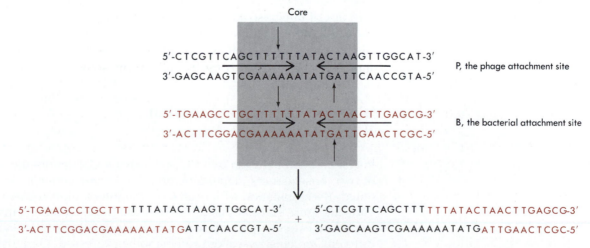

Core

5'-CTCGTTCAGCTTTTTTTATACTAAGTTGGCAT-3'
3'-GAGCAAGTCGAAAAAAATATGATTCAACCGTA-5'   P, the phage attachment site

5'-TGAAGCCTGCTTTTTTTATACTAACTTGAGCG-3'
3'-ACTTCGGACGAAAAAAATATGATTGAACTCGC-5'   B, the bacterial attachment site

5'-TGAAGCCTGCTTTTTTTATACTAAGTTGGCAT-3'   5'-CTCGTTCAGCTTTTTTTATACTAACTTGAGCG-3'
3'-ACTTCGGACGAAAAAAATATGATTCAACCGTA-5'   + 3'-GAGCAAGTCGAAAAAAATATGATTGAACTCGC-5'

Two hybrid sites resulting
from recombination
between P and B

(a)

(b)

Integrase-binding regions
IHF-binding regions

Figure 11-23

Binding of integrase and IHF to DNA. (a) Structure of phage and bacterial attachment sites, showing where the integrase enzyme and the integration host factor bind to DNA. The scale is length in nucleotides. (b) Electron micrograph of integrase protein bound to a DNA restriction fragment containing *attP*. The condensation of DNA into the complex is evident from the greater apparent length of an identical DNA fragment (at the lower left) that is not bound by integrase. (Courtesy of Marc Better and Harrison Echols).

How does integrase choose these specific sites? Experiments to delimit the phage and bacterial sites by asking how much DNA adjacent to the actual exchange sites is required for integrase to bind and act show that *attP* encompasses a region of 250 base pairs, whereas *attB* consists of only about 20 base pairs. Both integrase and the host factor IHF bind at distinct places throughout *attP*, so that the integrase activity is fixed to phage DNA near the exchange site (Figure 11-23). The entire 250-nucleotide *attP* segment appears to be wound around integrase in a condensed nucleosome-like structure that may contain as many as eight integrase monomers of 40,000 daltons each (Figure 11-23b). The shorter bacterial sequence consists mostly of a 15-base-pair core that also is present in *attP*. Integrase also binds both core sequences, which are the actual crossover sites. If the core sequence is changed even slightly in *attP* or *attB* alone, the rate of recombination is greatly reduced; but if the core is changed identically in *attP* and *attB*, exchange is still efficient. Thus, integrase requires a sequence homology in the core as well as a particular sequence to which it binds. It is not known if integrase detects this homology between separate duplexes before it cuts and unwinds, requiring, for example, a specifically paired four-stranded helix of the type previously discussed, or if it cuts first but then tends to reverse the reaction unless it detects homology in the emerging seven-base-pair overlap.

When a λ prophage is induced to grow, integration is reversed (a process called **excision**), and both phage and bacterial DNAs are reconstituted intact. Prophage λ invokes excision by expressing a second protein, **excisionase,** which allows integrase now to catalyze recombination between the hybrid attachment sites of the prophage. A complex of integrase and excisionase binds tightly to a hybrid bacterial-prophage attachment site, unlike integrase alone, and this changed specificity undoubtedly underlies its ability to carry out the reverse reaction.

Site-Specific Recombination Regulates Gene Expression[19]

Recombination between two sites on a single DNA molecule has one of two consequences, depending on how the interacting sites are oriented. Recombination either removes the intervening segment or inverts it (Figure 11-24). Cells sometimes use recombinational inversion to choose between two alternate DNA arrangements that allow two different proteins or sets of proteins to be expressed. Curiously, this mechanism often regulates proteins that appear on the outside of the organism, for example, the tail protein of phage Mu (mutator), regulated by the invertible *gin* segment (Chapter 17), and the flagellar

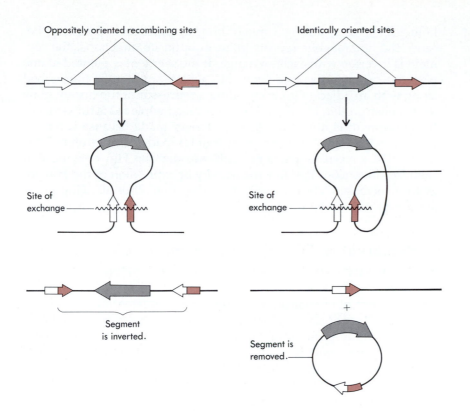

Figure 11-24
How recombination between two sites on one DNA molecule can either invert or delete the intervening segment.

antigen of the bacterium *Salmonella*. In a very complex case reminiscent of the construction of antibody genes, variation in the surface proteins of trypanosomes is achieved by a succession of DNA rearrangements. Perhaps recombination is used to control surface proteins because the switch can be absolute, allowing the organism to make one surface antigen exclusively and thus avoid even a slight reaction to antibodies directed by an infected host against other antigens.

Salmonella phase variation results from the alternate expression of its two flagellar proteins H1 and H2. At a given time, a cell expresses one of these, but never both together. The promoter for the *H2* gene is found on an adjacent invertible DNA segment 970 base pairs long, bounded by repeats (of 14 base pairs) oriented oppositely (Figure 11-25). This short homology appears to be used as the core sequence

Figure 11-25
Control of alternate expression of *Salmonella* flagellar genes *H1* and *H2* by an invertible DNA segment. IRL and IRR are inverted repeats that make the inversion possible in the presence of the enzyme coded by *hin*.

homology is used during λ integration, to guide recombination. Because the homologous repeats point in opposite directions, the segment is inverted with each exchange. If the segment is pointed in one way, the promoter is beside gene *H2*, which can then be transcribed along with an adjacent gene encoding a repressor of the distant gene for flagellar protein H1; thus, H1 is repressed while H2 is transcribed. If the segment is inverted, H2 is no longer made because it has no promoter, but neither is the repressor of H1 made; thus, protein H1 is made. The invertible segment itself encodes the Hin enzyme that catalyzes inversion. Possibly the activity or expression of the Hin enzyme is increased when the cell's growth suffers, thus allowing a new surface protein to be tried.

A Remarkable Discovery: Transposons Move Genes to New and Unrelated Sites[20–27]

The existence of specialized mechanisms for site-specific recombination did not seriously affect our conviction that recombinational processes generally preserve the order of genes on chromosomes; site-specific rearrangements are limited and conservative and in higher organisms often are not inherited because they occur only in somatic cells. However, a further discovery showed the chromosomes do rearrange in a completely unexpected way. The most dramatic example is provided by genes specifying bacterial resistance to antibiotics such as tetracycline and penicillin, which can be carried on plasmids that have no homology with the bacterial chromosome. In a cell carrying such a plasmid, the resistance gene occasionally appears in the bacterial chromosome or in the progeny of an infecting phage, apparently by recombination in the absence of homology. Once resident in a new genome, it then can move again (e.g., from the bacterial DNA to another plasmid). These are rare events, typically occurring less than once per million cell divisions, but they are easily detectable because a genome carrying the resistance gene can be selected. Electron microscopy of heteroduplexes and restriction enzyme analysis both show that a new DNA sequence is inserted when the resistance gene is acquired. Such DNA segments that move without benefit of homology are called transposons. Known transposons range in length from 750 to 40,000 base pairs, about the same as small- to medium-size phages.

Two types of transposons have been found in bacteria: complex and simple, the latter also called **insertion sequences.** Complex transposons contain one to several genes besides those essential for the process of transposition; they are most easily recognized when they carry identifiable genetic markers such as genes that specify antibiotic resistance or toxin production. Insertion sequences carry only the genes necessary for their own transposition and can be detected in two ways. First, they interrupt and inactivate genes into which they insert. Second, they may contain promoters that allow RNA polymerase to transcribe and thus turn on adjacent genes. Although no specific function of insertion sequences in bacteria is known, they may act as natural agents of genetic change through their ability to revise genome structure and alter the expression of genes.

Complex transposons often have insertion sequences or remnants of insertion sequences at their ends. This suggests that complex transposons can be created when insertion sequences jump to both sides of a cellular gene, so that the entire assembly afterwards moves

Table 11-1 Some Insertion Sequences and Complex Transposons of *E. coli*

	Size (bp)	Target DNA Repeat (bp)	Known Functions or Proteins Encoded (Besides Transposase)
Insertion Sequences			
IS1	768	9	—
IS2	1,327	5	—
IS10-R	1,329	9	—
Complex Transposons			
Tn3	4,957	5	Ampicillin resistance
Tn5	5,700	9 (IS50 at end)	Kanamycin resistance
Tn10	9,300	9 (IS10 at end)	Tetracycline resistance
Tn681	2,100	9 (IS1 at end)	Heat-stable enterotoxin
Tn2571	23,000	9 (IS1 at end)	Resistance to chloramphenicol, fusidic acid, streptomycin, sulfonamides, and mercury

Adapted from N. Kleckner, *Ann. Rev. Genetics.* 15 (1981):354.

from the outer insertion sequence boundaries. The bounding insertion sequences may no longer be able to move independently, because they have changed in evolution to favor movement of the complex as a whole.

There is good evidence that transposons can be found in all organisms. Some transposons of *E. coli* are given in Table 11-1. Transposons of higher organisms include the Ty elements of yeast (Chapter 18), the copia elements of *Drosophila* (Chapter 20), and the retroviruses that inhabit cells of many species (Chapter 24).

Resistance Transfer Factors Are Constructed Mostly of Transposons and Can Specify Resistance to Several Drugs at Once

A serious medical problem arises when infecting bacteria become resistant to several antibiotics simultaneously by acquiring a plasmid carrying separate genes that confer resistance to each antibiotic. The evolution of multiple-resistance plasmids is made easy by the ability of transposons to recombine without homology and thus to gather together in a single plasmid, whose survival is assured by frequent use of several antibiotics together. The naturally occurring complex plasmid R1 evolved from transposons encoding resistance to chloramphenicol (Cm) and kanamycin (Km), and it also contains the intact transposon Tn4 carrying resistance to streptomycin (Sm), sulfonamide (Su), and ampicillin (Ap) (Figure 11-26). Tn4 itself includes a separate and independent transposon, Tn3, which is responsible only for the resistance to ampicillin. The whole assembly of resistance genes (an **r-determinant**) is attached to other segments containing genes for plasmid replication and for **transfer functions,** which promote conjugal contact of the host bacterial cell with other bacteria, thus allowing the plasmid to be transferred readily between cells. The very plastic nature of this modular assembly gives rise to many plasmid variants, both in the laboratory and in nature.

Figure 11-26

R1, a naturally occurring 94-kilobase complex plasmid that confers resistance to five antibiotics: chloramphenicol (Cm), streptomycin (Sm), sulfonamide (Su), ampicillin (Ap), and kanamycin (Km). The plasmid has two parts, which also can exist separately in other plasmids: the resistance determinant (r-determinant) and a resistance transfer factor (RTF), which encodes proteins that promote movement of plasmids between cells through a mating process. R1 was formed when an IS1 insertion sequence on one of these elements transposed to the other, fusing them into a stable cointegrate bearing a copy of IS1 at each of the junctions. (Adapted from S. N. Cohen in *DNA Insertion Elements, Plasmids, and Episomes*, Cold Spring Harbor Laboratory, 1977, p. 672, with permission.)

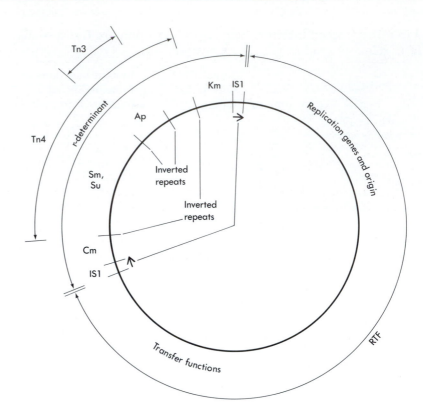

Transposons Encode Genes That Direct and Regulate Their Transposition

An antibiotic resistance gene is a passenger carried by transposon DNA that otherwise is devoted to directing and controlling the frequency of its own movement. Transposons have two features that are crucial to their behavior. First, 20 to 40 nucleotides of the sequence at one end is repeated nearly identically, but oriented oppositely, at the other end; these are called **inverted repeats.** Second, many (and probably all) transposons encode an enzyme called a **transposase,** which catalyzes their insertion into new sites.

The rarity of transposition results from the very small concentration of transposase in the cell. On average, less than one molecule of transposase is made by transposon Tn10 in each generation, an amount that allows Tn10 to transpose naturally only once in every 10^7 cell generations. However, if more transposase is provided by a plasmid engineered to contain the Tn10 transposase gene, then Tn10 transposes a thousand times more frequently. It seems likely that the natural transposition rate is also controlled by changes in transposase level.

One important determinant of transposase expression and function has been discovered recently. The transposase promoter of Tn10 and some other transposons contains the DNA sequence GATC, which is methylated in *E. coli* at the adenine nucleotide by the cellular *dam* methylase (Chapter 12). Since the sequence complementary to GATC is also GATC (when read in the 5'→3' direction), both DNA strands are methylated at this site. In this condition, the promoter is relatively inactive. However, after DNA replicates, the adenine in the newly made GATC sequence is not methylated by the *dam* methylase for a minute or so, and in this *hemimethylated* state, the promoter is much more active; thus, a small burst of transposase is made just after repli-

Figure 11-27
Structure of transposon Tn3. Tn3 encodes β-lactamase, an enzyme that destroys β-lactam antibiotics like penicillin and ampicillin, and the two enzymes transposase and resolvase; its length is 4957 nucleotides. The arrows above the genes show the directions of transcription. [Adapted from F. Heffron, B. J. McCarthy, H. Ohtsubo, and E. Ohtsubo, *Cell* 18 (1979):1153, with permission.]

cation. Similarly, the site in Tn10 DNA on which transposase acts contains GATC, and it is a much better substrate for transposase during the short time when only one strand is methylated. Therefore, Tn10 tends to transpose just after DNA replication. It is not yet known if other cellular conditions affect DNA methylation and in so doing change the transposition rate, or if transposition is also regulated in other ways.

Some transposons encode a second enzyme, a **resolvase,** that catalyzes a second stage of transposition (as we shall see in Figure 11-29). For Tn3, the resolvase protein also has a second and independent activity: It is a repressor of both its own gene and the transposase gene, serving to keep their normal levels of expression low (Figure 11-27). It represses both genes by binding a site between them in transposon DNA.

How Transposons Move[23, 27]

DNA sequence analysis of bacterial transposons like Tn3 and their junctions with target DNA provides strong clues to the mechanism by which they are transposed. First, we know that transposons move precisely, carrying all of the sequence inserted at the old site and none of the adjacent DNA. This could occur only if an enzyme, almost certainly transposase itself, recognizes both transposon ends; thus, the ends probably are identical because they are the common binding sites of transposase. Second, a precise duplication of 3 to 12 bases of the target DNA (depending on the transposon) is created at the site of insertion, one copy staying at each end of the transposon sequence (Figure 11-28). Some DNA synthesis must therefore occur during transposition to create the target site duplication, a clear difference from the site-specific recombination event of λ integration. Third, although most transposons move to almost any region of a new genome, they do not move completely at random, preferring instead to invade certain DNA sequences. Some attack specific four-

Figure 11-28
Nucleotide sequence of one target site of Tn3, with the crucial five-nucleotide segment highlighted. After Tn3 insertion, one such five-nucleotide sequence directly adjoins each of the identical (but oppositely directed) Tn3 ends.

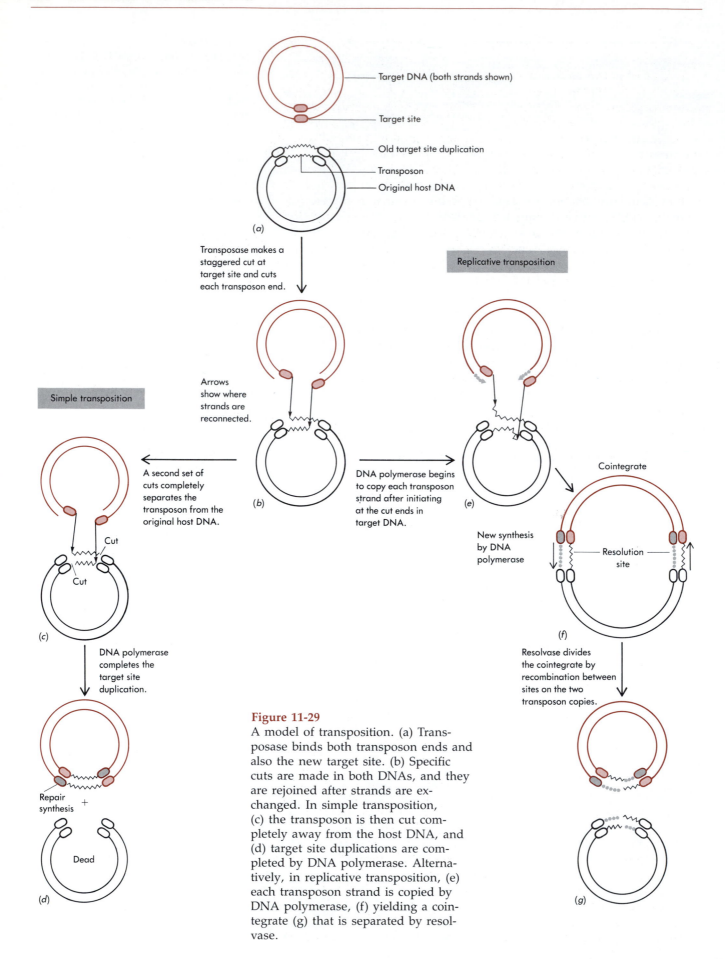

Target DNA (both strands shown)

Target site

Old target site duplication

Transposon

Original host DNA

(a)

Transposase makes a staggered cut at target site and cuts each transposon end.

Replicative transposition

Arrows show where strands are reconnected.

Simple transposition

A second set of cuts completely separates the transposon from the original host DNA.

(b)

DNA polymerase begins to copy each transposon strand after initiating at the cut ends in target DNA.

(e)

Cointegrate

New synthesis by DNA polymerase

Resolution site

Cut

Cut

(c)

DNA polymerase completes the target site duplication.

(f)

Resolvase divides the cointegrate by recombination between sites on the two transposon copies.

Repair synthesis +

Dead

(d)

Figure 11-29

A model of transposition. (a) Transposase binds both transposon ends and also the new target site. (b) Specific cuts are made in both DNAs, and they are rejoined after strands are exchanged. In simple transposition, (c) the transposon is then cut completely away from the host DNA, and (d) target site duplications are completed by DNA polymerase. Alternatively, in replicative transposition, (e) each transposon strand is copied by DNA polymerase, (f) yielding a cointegrate (g) that is separated by resolvase.

(g)

or six-base-pair symmetrical sequences, just as restriction enzymes do.

Studies of enzymes that break and rejoin DNA strands, like λ integrase and DNA topoisomerases, suggest a model to explain how transposase may work (Figure 11-29). We suppose that transposase binds both ends of the transposon and also the target DNA sequence in which the transposon will insert. It then makes staggered cuts in the target like a restriction enzyme and simultaneously one cut at each transposon end, but on different strands. The freed transposon ends are ligated to the target site ends, thus connecting the two DNA molecules, which are bridged at each end by one strand of the original transposon (Figure 11-29). Here, there are two choices, depending on the transposon or its particular conditions of growth. In **simple transposition,** a second set of cuts is made at the other transposon ends, transferring the transposon entirely to the new DNA and leaving a lethal gap in the old host DNA (Figure 11-29c). Insertion of a few nucleotides by DNA polymerase completes the target site duplication (Figure 11-29d). Thus, simple transposition is a conservative process that moves the transposon to a new site. The second possibility is **replicative transposition,** in which the transposon is duplicated: One copy moves, but also one copy stays so that the original host DNA is not disrupted. In this case, no second cuts are made, but instead the second transposon strand remains to be copied by DNA polymerase, which initiates at the first cut end (Figure 11-29e). The resulting structure is called a **cointegrate,** and its existence is strong evidence for this pathway of transposition. A resolvase enzyme can then catalyze site-specific recombination across the two transposon copies in a reaction like λ integration, producing separate DNA molecules, each with one transposon copy.

It is clear how variations of these specific steps account for other activities of transposons. For example, if resolvase is absent, a cointegrate (like plasmid R1, Figure 11-27) is the final product. Or a cointegrate might be divided incorrectly, yielding two independent and rearranged DNAs. A close look at the model in Figure 11-29 reveals that transposition into a second site of the same DNA causes an inversion or a deletion, again radically changing the structure of a chromosome.

A thorough understanding of the mechanism of transposition will require that the individual steps be studied in vitro. Fortunately, transposition by bacteriophage Mu, which is both a phage and a transposon (Chapter 17), does occur in cell-free extracts, and we expect that this discovery will quickly lead to an understanding of exactly how transposons move.

Recent experiments show that some transposons of higher organisms, particularly those related to retroviruses, move in a very different way from bacterial transposons like Tn3. Thus, Ty elements of yeast (Chapter 18) are transcribed first into RNA, which is copied to DNA by a reverse transcriptase enzyme; this DNA copy is then recombined into a new site of the host chromosome by a mechanism that is not yet understood.

Summary

Crossing over, or homologous recombination, is the exchange of nearly identical segments between DNAs that share regions of sequence homology, for example, two homologous chromosomes. Crossing over creates genetic diversity by producing different combinations of particular genetic alleles and also allows repair of DNA damaged by radiation or chemicals through reconstruction of intact DNAs from segments. Recombinant chromosomes are made by joining separate preexisting pieces, rather than by template switching during a copying process. Combining segments are joined accurately at heteroduplex regions, which contain one strand derived from each parental DNA and which thus ensure correct alignment. Homologous recombination is initiated from gaps in DNA or from breaks that lead to single-stranded regions. One initiation pathway is provided in *E. coli* by the RecBC enzyme, which unwinds DNA and leaves single-stranded ends by cutting strands at sites named Chi. The *E. coli* recombination enzyme RecA protein binds single-stranded DNA and pairs it with a complementary strand in a separate DNA molecule, thus establishing a bridge between DNAs. At the completed bridge, called a Holliday joint, two DNAs exchange a homologous strand. The Holliday joint can move by branch migration, thereby extending the region of heteroduplex DNA. Recombinant DNAs are made when the Holliday joint is broken by nucleases.

Site-specific recombination is an exchange of particular segments of DNA that is guided not (primarily) by homology at the boundaries, but instead by binding sites for proteins that cut and rejoin segments at precise sites. Generally, site-specific recombination serves regulatory functions. Phage λ integrase inserts phage into bacterial DNA by site-specific recombination, creating the passive prophage state in which phage DNA is replicated by the host. Phase variation in *Salmonella* is the result of a site-specific recombination event that inverts a DNA segment and thereby determines the orientation of a promoter that controls genes for two different flagellar antigens.

Transposons are small, mobile DNA sequences that move around chromosomes with near total disregard for homology, producing insertions, deletions, and more complicated rearrangements. Often they carry genes that allow them to be selected, for example, genes specifying antibiotic resistance. Transposons also encode transposases, enzymes that carry out the DNA-cutting and ligation reactions by which transposons move.

Bibliography

General References

"Recombination at the DNA Level." 1984. *Cold Spring Harbor Symp. Quant. Biol.* 49. Recent research reports on most topics covered in this chapter.

Shapiro, J. A., ed. 1983. *Mobile Genetic Elements*. New York: Academic Press.

Stahl, F. 1979. *Genetic Recombination: Thinking About It in Phage and Fungi*. San Francisco: Freeman. An advanced and detailed but very readable text on classical genetic analysis, published just before RecA function was understood.

Cited References

1. Meselson, M., and J. J. Weigle. 1961. "Chromosome Breakage Accompanying Genetic Reconstruction in Bacteriophage." *Proc. Nat. Acad. Sci.* 47:857–868.
2. Orr-Weaver, T., J. Szostak, and R. Rothstein. 1981. "Yeast Transformation: A Model System for the Study of Recombination." *Proc. Nat. Acad. Sci.* 78:6354–6358.
3. Szostak, J., T. Orr-Weaver, R. Rothstein, and F. Stahl. 1983. "The Double-Strand Break Repair Model for Recombination." *Cell* 33:25–35.
4. Cox, M., and I. Lehman. 1981. "*RecA* Protein of *Escherichia coli* Promotes Branch Migration, a Kinetically Distinct Phase of DNA Strand Exchange." *Proc. Nat. Acad. Sci.* 78:3433–3437.
5. Das Gupta, C., A. Wu, R. Kahn, R. Cunningham, and C. Radding. 1981. "Concerted Strand Exchange and Formation of Holliday Structures by *E. coli* RecA Protein." *Cell* 25:507–516.
6. Radding, C. 1982. "Homologous Pairing and Strand Exchange in Genetic Recombination." *Ann. Rev. Genetics* 16:405–437.
7. Howard-Flanders, P., S. West, and A. Stasiak. 1984. "Role of RecA Protein Spiral Filaments in Genetic Recombination." *Nature* 309:215–220.
8. Holliday, R. 1964. "A Mechanism for Gene Conversion in Fungi." *Genet. Res.* 5:282–304.
9. Sigal, N., and B. Alberts. 1972. "Genetic Recombination: The Nature of a Crossed Stranded Exchange Between Two Homologous DNA Molecules." *J. Mol. Biol.* 71:789–793.
10. Kobayashi, I., M. Stahl, and F. Stahl. 1984. "The Mechanism of the Chi-*cos* Interaction in RecA-RecBC-Mediated Recombination in Phage λ." *Cold Spring Harbor Symp. Quant. Biol.* 49:497–506.
11. Ponticelli, A. S., D. W. Schultz, A. F. Taylor, and G. R. Smith. 1985. "Chi-Dependent DNA Strand Cleavage by RecBC Enzyme" *Cell* 41:145–151.
12. Clark, A. J. 1973. "Recombination Deficient Mutants of *E. coli* and Other Bacteria." *Ann. Rev. Genetics* 7:67–86.
13. Howard-Flanders, P. 1981. "Inducible Repair of DNA." *Sci. Amer.* 245:72–80.
14. Nash, H. 1975. "Integrative Recombination of Bacteriophage λ DNA *in Vitro*." *Proc. Nat. Acad. Sci.* 72:1072–1076.
15. Nash, H. 1981. "Integration and Excision of Bacteriophage λ: The Mechanism of Conservative Site Specific Recombination." *Ann. Rev. Genetics* 15:143–167.
16. Better, M., C. Lu, R. C. Williams, and H. Echols. 1982. "Site-Specific DNA Condensation and Pairing Mediated by the Int Protein of Bacteriophage λ." *Proc. Nat. Acad. Sci.* 79:5837–5841.
17. Ross, W., and A. Landy. 1983. "Patterns of λ Int Recognition in the Regions of Strand Exchange." *Cell* 33:261–272.
18. Weisberg, R., and A. Landy. 1983. "Site-Specific Recombination in Phage Lambda." In *Lambda II*, ed. R. Hendrix, J. Roberts, F. Stahl, and R. Weisberg. Cold Spring Harbor, N.Y.: Cold Spring Harbor Laboratory, pp. 211–250.
19. Simon, M., J. Zieg, M. Silverman, G. Mandel, and R. Doolittle. 1980. "Phase Variation in the Evolution of a Controlling Element." *Science* 209:1370–1374.
20. Heffron, F., B. J. McCarthy, H. Ohtsubo, and E. Ohtsubo. 1979. "DNA Sequence Analysis of the Transposon Tn3: Three Genes and Three Sites Involved in Transposition of Tn3." *Cell* 18:1153–1163.
21. Cohen, S., and J. Shapiro. 1980. "Transposable Genetic Elements." *Sci. Amer.* 242:40–50.
22. Kleckner, N. 1981. "Transposable Elements in Prokaryotes." *Ann. Rev. Genetics* 15:341–404.
23. Craigie, R., and K. Mizuuchi. 1985. "Mechanism of Transposition of Bacteriophage Mu: Structure of a Transposition Intermediate" *Cell* 41:867–876.
24. Morisato, D., J. Way, H. Kim, and N. Kleckner. 1983. "Tn10 Transposase Acts Preferentially on Nearby Transposon Ends *in Vivo*." *Cell* 32:799–807.
25. Harshey, R. M. 1984. "Transposition Without Duplication of Infecting Bacteriophage Mu DNA." *Nature* 311:580–581.
26. Roberts, D., B. C. Hoopes, W. R. McClure, and N. Kleckner. 1985. "IS10 Transposition Is Regulated by DNA Adenine Methylation" *Cell* 43:117–130.
27. Boeke, J. D., D. J. Garfinkel, C. A. Styles, and G. R. Fink. 1985. "Ty Elements Transpose Through an RNA Intermediate." *Cell* 40:491–500.

CHAPTER 12

The Mutability and Repair of DNA

Since the developmental potential of an organism is determined by its genes, DNA must necessarily change, or mutate, as organisms evolve. But evolutionary changes occur only rarely, and it is the stability of DNA, as it is maintained and replicated over generations, that is most impressive. Since living cells require the correct functioning of thousands of proteins, each of which could be damaged by a mutation at many different sites in its gene, it is clear that DNA sequences usually must be passed on unchanged if progeny are to have a good chance of survival. The challenge for the cell is twofold. First, the enzymatic machinery that replicates DNA must be inherently accurate. Second, the cell must repair accidental damage to DNA that would destroy its function. DNA is a complex organic molecule of finite chemical stability: Not only does it suffer spontaneous damage like loss of bases, but it also is assaulted by natural (and unnatural) chemicals and radiation that break its backbone and chemically alter the bases. Since mutations result when damage changes the coding properties of bases, an organism could not survive the natural rate of damage to its DNA without specific enzymatic mechanisms to repair damaged sites. In fact, DNA repair is so important that a bacterium may devote several percent of its genome to specifying and controlling the enzymes involved.

The Nature of Mutations

Naturally occurring mutations include almost all conceivable changes in DNA sequence. Mutations that have only a subtle effect on a gene product, such as temperature-sensitive mutations, are often simple switches of one base for another. However, many natural mutations destroy the function of a gene completely. These more drastic changes, called **null mutations,** include not only base switches and insertions or deletions of a base, but also more extensive insertions and deletions and even gross rearrangements of chromosome structure. Such changes might be caused, for example, by the insertion of a transposon, which typically places many thousands of bases of foreign DNA in the coding sequence of a gene (Chapter 11), or by aberrant actions of cellular recombination processes.

Mutation by Single-Base Change[1, 2]

There are a number of reasons to consider first the simplest type of mutation, the exchange of one base for another. First, base switches reflect the basic accuracy of DNA replication. Second, many important mutagens act by making single-base changes. Finally, single-base mutations are critical to evolution, because they change genes in ways that are subtle enough to yield useful variants.

What sorts of single-base changes occur naturally? To determine this in a practical way, we have to be able to pick out quickly those mutants that result from only a single change. This was first accomplished by examining large numbers of spontaneous mutations that create a translation stop codon in the *E. coli* lactose operon *I* gene, which codes for the lactose repressor. It can be verified that these mutations are single-base changes in some codon originally identical to the stop codon in two out of three positions, by showing that active repressor is once again made from the mutant gene in a cell containing a suppressor (which allows the stop codon to be translated as an amino acid; see Chapter 14). This analysis shows that all types of base switches occur: **transitions,** which are changes from one purine to the other or from one pyrimidine to the other, and **transversions,** which are changes from purine to pyrimidine or from pyrimidine to purine (Figure 12-1). A second important fact is that not all possible sites mutate with the same efficiency; some are "hot spots" (first discovered by S. Benzer in his classic genetic analysis of the phage T4 *rII* gene), at which mutations occur much more often than at most sites.

Single-base switches are usually reversible, and often the rate of "back" mutation to the normal nucleotide arrangement is similar in order of magnitude to the rate of change to the mutant arrangement. This fact represents an important way to distinguish base switches from more drastic alterations like large deletions, for which the reverse reaction (called **reversion**) is impossible. An intermediate case is single-base insertions or deletions; these may revert, but much less often than single-base switches. Most spontaneous single-base exchanges are simply rare failures in the replication process, arising when a nucleotide is added to the growing chain even though it does not pair normally with the template base.

Error Levels per Incorporated Nucleotide Range from 10^{-7} to 10^{-11}

Since the *lacI* system detects all instances of a particular base change at 80 distinct sites, it is simple to calculate how often each mutation appears when a cell's DNA replicates and thus the rate at which the replication system makes an error at each site. The average rate for all 80 sites is about 10^{-9} mistakes for each replication. However, hot spots produce mutants at 25 times the average rate, and other sites are at least 25 times *less* active than the average. Many other measurements of base substitutions at particular sites in other genes show that the rate varies from 10^{-7} to 10^{-11} mistakes per replication event.

Control of Mutation Levels by the Relative Efficiencies of Forward Polymerizing and Backward Nuclease Activities[3, 4]

At first it might appear that the error frequencies directly reflect the inherent accuracy of AT and GC base-pair formation. However, strong theoretical arguments contradict this idea: There is not enough difference between right and wrong base pairs for a polymerase to make as few as one error for every 10^9 nucleotides it incorporates.

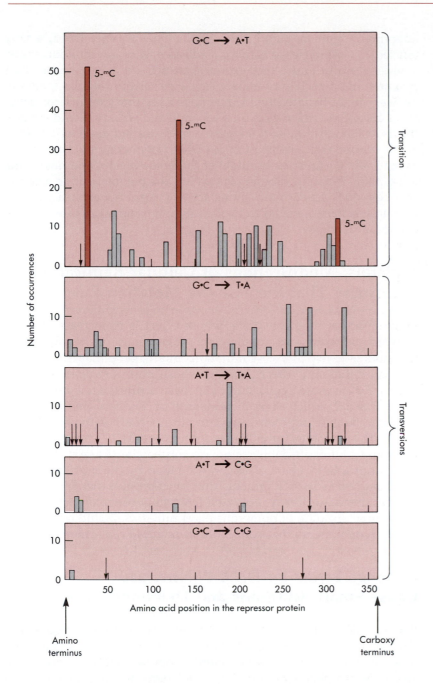

Figure 12-1

Nature and distribution of spontaneous mutations that create stop codons in the lactose repressor gene. This assay detects all four transversions and one of the two possible transitions. Other assays show that the other transition (AT→GC) is also found among spontaneous mutations. The vertical arrows indicate sites where mutations could have been detected but did not occur. The three cytosines methylated in the 5 position (5-me-C) are hot spots for mutation. [After J. H. Miller and K. B. Low, *Cell* 37 (1984):675–682.]

Furthermore, the lower limit of wrong base insertion should be set by the frequency that bases assume the "incorrect" tautomeric form (enol instead of keto, imino instead of amino), perhaps one in 10^4, much greater than the actual mutation frequency. This dilemma vanished with the discovery of the *proofreading* capability of DNA polymerases. Although the *initial* error frequencies during base-pair selection are consistent with measured tautomeric ratios, almost all the wrongly inserted bases are later removed by the 3'→5' exonuclease component of the respective DNA polymerase. At first, its association with every known bacterial DNA polymerase seemed most bizarre, but now we realize that the 3'→5' exonuclease activity is necessary for accurate DNA replication. In fact, studies with mutant DNA polymerase molecules show that mutation levels may be controlled by the ratio of backward (3'→5') nuclease activities to forward (5'→3') polymerizing capabilities. If the 3'→5' exonuclease is inefficient, then abnormally high mutation rates result. On the other hand, a very efficient 3'→5' nuclease activity leads to a very low mutation rate.

The failure to find 3'→5' exonuclease activity in eucaryotic DNA polymerases, which may make even fewer errors than bacterial enzymes, cannot be interpreted. The possibility that exonuclease activity of eucaryotic enzymes is on a separate and weakly bound subunit that is lost in purification is suggested by the recent discovery that the 3'→5' exonuclease activity of *E. coli* DNA polymerase III is carried on a small, 25,000-dalton subunit (ϵ) of the holoenzyme, and not on the large, 140,000-dalton subunit (α) that has the polymerization activity. In exceptional cases, proofreading activity might not exist, a possibility that has been invoked to explain the high mutation rate of mitochondrial genomes.

Mutators: Modifications of Cellular Replication Machinery That Increase the Mutation Rate[5, 6]

There exist bacterial mutations whose effect is to increase the rate at which mutations accumulate in other genes. Generally, these **mutators** give rise to excess mutations in any gene that is examined and must therefore affect proteins whose normal functions are required for accurate replication. The best characterized mutator in *E. coli* is *mutD*, which alters the ϵ subunit of DNA polymerase III and thus interferes with proofreading. In vitro, DNA polymerase III containing this altered subunit shows a decreased 3'→5' exonuclease activity and cannot efficiently remove a mismatched base at the growing end of a DNA chain. The genes defined by the mutators *mutH*, *mutL*, and *mutS* specify proteins that function in mismatch repair, a process that scans newly replicated DNA for wrongly inserted bases. The characterization of numerous other mutators should reveal other cellular components that contribute to accurate replication.

Slippage Errors by DNA Polymerase Cause Small Additions and Deletions[7]

Replication errors by DNA polymerase are not limited to single-base-pair changes. Deletions or additions of one or a few base pairs are thought to result from displacement ("looping out") of bases from either the template strand (giving a deletion) or the growing strand (giving an addition). These mistakes occur particularly at runs of identical bases in the DNA, where the looped-out structure can be stabilized by normal base pairing beyond the unpaired base (Figure 12-2). When the miscopied DNA strand acts as a template, the deletion or addition is usually copied accurately, fixing the mutation. An addition or deletion of one or a few base pairs that changes the translation reading frame within a protein coding sequence and thus completely disrupts its synthesis is often called a **frameshift mutation.**

Mutation by Large DNA Sequence Rearrangement[8, 9]

Many mutations are much less subtle than single-base changes, involving extensive rearrangements. These are often found among mutations selected to destroy the function of a gene. Large segments,

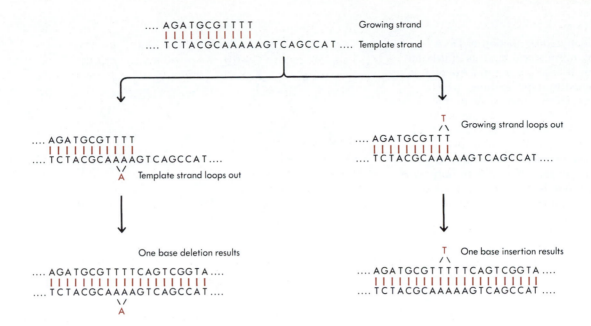

Figure 12-2
Small additions and deletions are caused by slippage at the replication growing point.

composed of hundreds or thousands of nucleotides, can be deleted by the aberrant operation of the cell's recombination enzymes (Chapter 11) or by errors in copying the template DNA strand. Similarly, segments can be inverted by recombination within a DNA molecule or exchanged between chromosomes of a higher cell, sometimes with the important consequence that genes near the site of exchange are no longer regulated correctly. Since chromosomes are generally stable, massive rearrangements must represent rare accidents that are preserved by selection or that happen to be harmless.

The Importance of Studying Chemical Mutagens[10, 11]

Since mutagenic chemicals greatly increase the probability that a mutation will occur, they must interfere in some important way with accurate replication of DNA. Mutagens are studied not only because they are useful for modifying genes and learning how accurate replication occurs, but also because naturally occurring mutagens affect all organisms on earth, including us. Bacteria and phages have allowed the mechanism of action of many mutagens to be worked out, and they also provide the most straightforward way to test if new chemicals may be dangerous mutagens that should be avoided.

Mutagens either act on DNA directly to change its template properties or in some way subvert replication so that a wrong base is inserted. We understand clearly how some mutagens work. **DNA-Reactive chemicals,** such as nitrous acid (HNO_2) or alkylating agents like methyl-nitrosoguanidine, act directly on DNA to change the bases into chemically distinct structures. These new bases often base-pair in a different way from the original bases, in which case the effect of the mutagen is to alter the genetic code directly (Figure 12-3). However, some modified bases cannot pair at all, and sometimes the damage completely removes the base from the DNA backbone; in these cases, mutations arise only through a special DNA repair process that introduces errors as part of the correction mechanism. (This process of error-prone repair is discussed later in this chapter.)

Figure 12-3
The oxidative deamination of DNA bases by nitrous acid, and its effects on subsequent base pairing. (a) Adenine is deaminated to hypoxanthine, which bonds to cytosine instead of to thymine. (b) Cytosine is deaminated to uracil, which bonds to adenine instead of to guanine. (c) Guanine is deaminated to xanthine, which continues to bond to cytosine, though with only two hydrogen bonds. Thymine and the uracil of RNA do not carry an amino group and so remain unaltered. [After W. Hayes, *The Genetics of Bacteria and Their Viruses* (Oxford, Eng.: Blackwell, 1964), p. 280, with permission.]

Figure 12-4
The base-pairing attributes of 5-bromouracil. (a) In the normal keto state, with a hydrogen atom in the N1 position, bromouracil bonds to adenine. (b) In the rare enol state, a tautomeric shift of this hydrogen atom determines specific pairing with guanine. [After W. Hayes, *The Genetics of Bacteria and Their Viruses* (Oxford, Eng.: Blackwell, 1964), p. 278, with permission.]

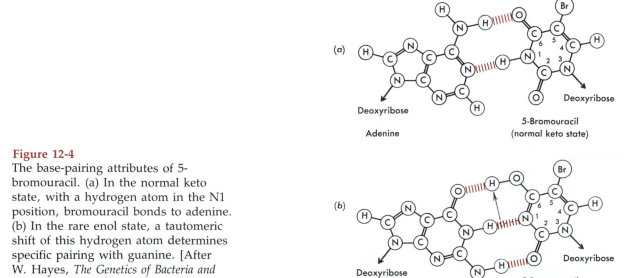

A second class of chemical mutagens is **base analogs,** which because of their similarity to the normal bases are incorporated (as nucleoside triphosphates) into DNA during normal replication. Their different structures, however, lead to less accurate base-pair formation than normal, causing frequent mistakes during the replication process. One of the most powerful base analog mutagens is 5-bromouracil, an analog of thymine. It is believed to cause mutations because its hydrogen atom at position 1 is not as firmly fixed as the corresponding hydrogen atom in thymine. Sometimes, this hydrogen atom is bonded to the oxygen atom attached to carbon atom 6 (Figure 12-4). When this happens, the 5-bromouracil can pair with guanine.

A third class of chemical mutagens consists of the **frameshift mutagens,** like proflavin, which cause the deletion or addition of a single base, or sometimes a few bases. Frameshift mutagens are flat molecules containing several connected rings (polycyclic) that bind to the equally flat purine or pyrimidine bases of DNA, just as the bases bind to or "stack" with each other in the double helix. These mutagens might act by binding to bases that loop out of either the template DNA strand or the growing strand during DNA synthesis (see Figure 12-2), thereby stabilizing the looped-out structure and greatly increasing the chance that it will lead to a mutation.

Cell Survival Depends on Continual Repair of DNA[12–15]

A mutant survives when its genetic change is not harmful, or, more rarely, is beneficial. But most mutations are damaging, and cells would not endure without enzymatic mechanisms to reverse the effects of natural mutagenic processes. As one example, the base cytosine suffers spontaneous deamination to the closely related base uracil, which pairs in DNA like thymine; thus, a GC base pair becomes an AT base pair when the DNA is replicated. Were this change not repaired, 1 in every 1000 cytosines in human DNA would become uracil during a lifetime, giving rise to an intolerable rate of mutation. Even when DNA repair is efficient, the rate of genetic change caused by natural damage may be much greater than that resulting from errors in DNA replication.

An even more immediate problem is posed by breaks in the DNA sugar-phosphate backbone and by bases so altered that they no longer even simulate natural bases and thus cannot pair at all. Clearly, such lesions prevent expression of any gene whose coding sequence they interrupt. Furthermore, they prevent DNA replication, because DNA polymerase almost always stops if it cannot make a correct base pair. Besides backbone breaks, which prevent a DNA strand from acting as a template, a typical lesion that prevents base pairing is the **pyrimidine dimer.** Ultraviolet radiation with a wavelength of about 260 nm is absorbed strongly by bases, and one consequence is the photochemical fusion of two adjacent pyrimidines into nonpairing structures like the cyclobutane-containing thymine dimer and the mutagenic "6-4" photoproduct (Figure 12-5). The DNA of a skin cell exposed to normal daylight, for example, would acquire thousands of dimers per day, were these not removed by repair enzymes. The human skin disease **xeroderma pigmentosum** is caused by a genetic defect in the enzymes that remove dimers and other ultraviolet-induced lesions. The UV component of sunlight results in abnormally high skin cell death and creates many skin cancer cells in people who have this disease.

Figure 12-5

Two important pyrimidine dimer photoproducts produced by ultraviolet light.

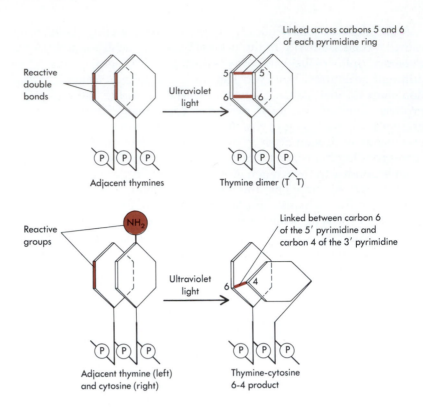

Some DNA Repair Enzymes Recognize and Reverse Specific Products of DNA Damage: Photolyase and O^6-Methylguanine Methyltransferase[14, 16–19]

Although most damage calls for more radical surgery, two common lesions in DNA can be repaired directly by enzymes that simply reverse the chemical change that engendered them. *Pyrimidine dimers* are the target of the universal enzyme **photolyase,** which binds a dimer and catalyzes a second photochemical reaction, this time using visible light, that changes the cyclobutane ring back into individual pyrimidine bases—a process called **photoreactivation.**

The second type of damage results from **alkylation,** the transfer of methyl or ethyl groups to reactive sites on the bases and to phosphates in the DNA backbone. Alkylating chemicals include nitrosamines and the very potent laboratory mutagen methyl-nitrosoguanidine (Figure 12-6). One of the most vulnerable sites of alkylation is the base guanine, which is particularly subject to methylation at the oxygen of carbon atom 6. The product O^6-methylguanine often mispairs with thymine, changing GC to AT when the DNA is replicated (Figure 12-7). This lesion is removed by the cellular enzyme O^6-**methylguanine methyltransferase,** encoded by the gene *ada*, which recognizes O^6-methylguanine in the DNA duplex and removes the methyl group by transferring it to an amino acid of the enzyme. The same enzyme also removes possibly disruptive methyl groups from phosphates of the DNA backbone. A bizarre and extravagant feature of this reaction is that there is no means to regenerate the unmethylated enzyme: A new enzyme molecule is spent for each methyl group removed.

If growing *E. coli* are exposed to a low and relatively harmless concentration of nitrosoguanidine, they are afterward much more resis-

Figure 12-6

Some alkylating mutagens. The reactive alkyl groups are shaded.

Figure 12-7
Reaction of guanine with an alkylating agent to yield O^6-methylguanine.

tant to the mutagenic and toxic effects of the chemical. This induced resistance, called the **adaptive response,** results from a hundredfold increase in the cell's content of O^6-methylguanine methyltransferase and a concurrent induction of the glycosylase that removes alkylated bases from the DNA backbone. The induction of both enzymes is blocked by mutations in the *ada* gene, whose product is therefore both an enzyme (the transferase) and a positive regulator of itself and the glycosylase. How the regulation works is unknown, but a reasonable guess is that the inducing signal is methylated DNA and that the reaction of the transferase activity with methyl groups initiates the important regulatory event.

Most DNA Repair Depends on Information in the Complementary DNA Strand[14]

It is the double-stranded nature of DNA that allows almost any damage to be repaired, no matter what the chemical change may be. Because a double helix holds two interconvertible copies of the genetic message, the loss of one is not serious: The information necessary to renew a damaged segment is present in the complementary strand. Of course, this is true only if damage is limited to one of the two complementary bases; but generally, only one base or two adjacent bases on a strand are damaged, so the chance of losing both bases of a complementary pair is slight. It is difficult to imagine how stable complex genomes ever could exist without the inherent duplication of information the double helix provides. Genomes consisting of single-stranded DNA must survive without this advantage, and, in fact, the largest single-stranded DNA genomes we know are those of phages like ϕX174, whose 5000-nucleotide DNA is only one-thousandth the size of the *E. coli* chromosome.

There are several pathways of **excision repair,** in which a damaged base or backbone segment of DNA is removed. All lead to the same intermediate product: either a single-strand break or a gap in the DNA at the site of damage that provides a 3'-hydroxyl end from which DNA polymerase can initiate synthesis to replace the damaged segment. Cells mutant in the DNA polymerase I gene are very deficient in excision repair; only this enzyme is abundant and mobile enough to fill small gaps efficiently when heavy damage requires many sites to be repaired at once. In contrast, the replicative enzyme DNA polymerase III is a large and complex enzyme that remains engaged at the growing fork for long periods and is present in only a few copies per cell. The other common essential enzyme is DNA ligase, which seals the nick left after DNA polymerase I acts.

The UvrABC Endonuclease Removes a Twelve-Base Segment Surrounding a Site of Damage[20, 21]

As for other biochemical pathways, the isolation of mutants has been essential to understanding the mechanism of DNA repair. Besides those in the photolyase gene *(phr)*, mutations in *E. coli* sensitive to ultraviolet light (and thus defective in repairing this damage) occur mostly in three genes: *uvrA*, *uvrB*, and *uvrC*. Unlike photoreactivation, the repair system these genes encode is not specific for ultraviolet damage, but instead senses any serious damage-induced distortion of the DNA helix. The three genes encode separate subunits of a single enzyme that removes a damaged DNA segment by cutting the strand that contains it. For many years, it was thought that the enzyme specified by these genes made one cut in DNA and that the nick translation activity of DNA polymerase acting on the cut end removed the damaged section; some phage-encoded enzymes do in fact work this way. However, biochemical studies of the *E. coli* enzyme were frustrated until cloning of the three *uvr* genes allowed their polypeptide products to be made abundantly and purified separately. When they were mixed and assayed in vitro on DNA containing pyrimidine dimers, a more complex activity dependent on all three products was revealed: the UvrABC enzyme cuts on *both* sides of the dimer (lesion), freeing a 12-base oligonucleotide containing the lesion and leaving a 12-base gap that is filled by DNA polymerase and sealed by DNA ligase (Figure 12-8).

It is clear from its activity on different types of damage that the UvrABC nuclease does not detect any particular lesion directly, but instead recognizes the *absence* of a normal DNA shape. How could this work? The two cuts it makes are on the same side of the helix, a little more than one turn apart, whereas the lesion is more nearly on the other side. We imagine that the enzyme "feels" the DNA shape in the 12-base segment between the sites where its nuclease active centers are aligned, perhaps wrapping around the helix to touch the other side. If the enzyme cannot conform to the natural shape of DNA because a bulky lesion is in its way, the nuclease domains are activated and the healing cuts are made.

Damaged Bases Can Be Removed by Glycosylases[14, 22]

Cells have a second way to excise a damaged base, begun by a glycosylase enzyme that detects an individual unnatural base and removes it from the deoxyribose sugar, initially leaving the sugar-phosphate backbone intact. The "hole" that results is called an **AP site,** an abbreviation for *apurinic* (lacking A or G) or *apyrimidinic* (lacking C or T). These also may result from natural loss of bases through breakage of

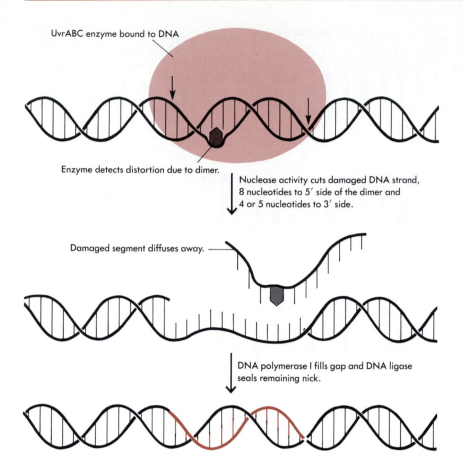

UvrABC enzyme bound to DNA

Enzyme detects distortion due to dimer.

Nuclease activity cuts damaged DNA strand, 8 nucleotides to 5′ side of the dimer and 4 or 5 nucleotides to 3′ side.

Damaged segment diffuses away.

DNA polymerase I fills gap and DNA ligase seals remaining nick.

Figure 12-8
Excision repair of dimer and other bulky lesion damage initiated by the UvrABC endonuclease.

the glycosidic bond. The "hole" is in turn recognized by an AP endonuclease that cuts the backbone, leaving a primer end from which DNA polymerase initiates synthesis to replace the missing nucleotide along with a few adjacent nucleotides (Figure 12-9).

There are many different glycosylases that remove specific lesions, in particular, aberrantly methylated bases. **Uracil-DNA glycosylase** corrects a problem caused by the natural instability of cytosine. Deamination of cytosine in DNA occurs spontaneously at a low rate to yield uracil. The glycosylase recognizes and removes uracil residues in DNA, allowing the C to be replaced by DNA polymerase. This repair system is foiled in a significant way if the cytosine is methylated on its 5 carbon, as occurs, for example, in sequences protected from restriction endonucleases. The 5-methylcytosine pairs normally in DNA and is not itself removed by a glycosylase. But deamination of 5-methylcytosine gives rise to thymine, not uracil. Thymine is a natural base and is not removed by uracil-DNA glycosylase or any other enzyme (Figure 12-10). Instead, it remains to pair with adenine in the next generation and to give rise to a mutation. Thus, 5-methyl-cytosines are hot spots for spontaneous mutation.

Mismatch Correction Removes Errors That Escape Proofreading[23, 24]

Some incorrectly paired bases escape even the proofreading activity of bacterial DNA polymerases. Like polymerization, proofreading has a substantial, but still finite, inherent accuracy. A good estimate is

Figure 12-9

Excision repair initiated by a glycosylase that recognizes and removes chemically damaged bases. As is shown, both the damaged nucleotide and a few adjacent nucleotides are replaced. When DNA polymerase 1 binds the free primer end, its 5'-3' exonuclease activity (Chapter 10) cuts a few nucleotides ahead of the missing base, and its polymerization activity fills the entire gap of several nucleotides. This activity (in which DNA polymerase 1 simultaneously polymerizes new DNA and degrades DNA ahead of the growing site, a process that can continue some distance beyond the initial primer end) is called **nick translation.**

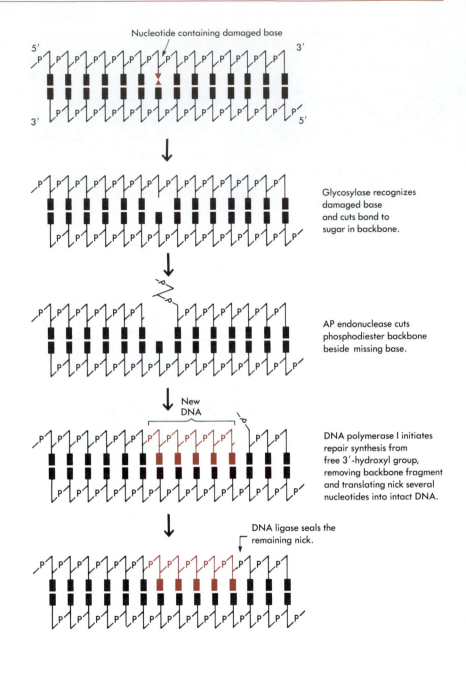

Glycosylase recognizes damaged base and cuts bond to sugar in backbone.

AP endonuclease cuts phosphodiester backbone beside missing base.

DNA polymerase I initiates repair synthesis from free 3'-hydroxyl group, removing backbone fragment and translating nick several nucleotides into intact DNA.

DNA ligase seals the remaining nick.

Figure 12-10

Molecular basis of 5-methylcytosine hot spots: different deamination products of cytosine and 5-methylcytosine.

Removed by uracil-DNA glycosylase and replaced with cytosine, so that no mutation occurs.

Not removed by uracil-DNA glycosylase; pairs with adenine at next replication to give a CG → TA transition.

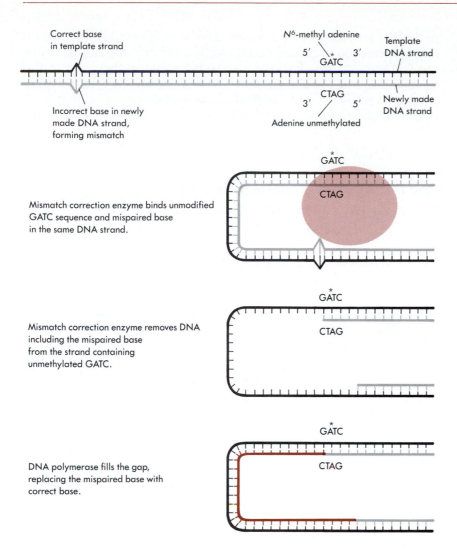

Correct base in template strand

N⁶-methyl adenine

Template DNA strand

5' GATC 3'

CTAG

3' 5'

Incorrect base in newly made DNA strand, forming mismatch

Adenine unmethylated

Newly made DNA strand

Mismatch correction enzyme binds unmodified GATC sequence and mispaired base in the same DNA strand.

Mismatch correction enzyme removes DNA including the mispaired base from the strand containing unmethylated GATC.

DNA polymerase fills the gap, replacing the mispaired base with correct base.

Figure 12-11

A model showing how the *E. coli* mismatch repair system could act to replace an incorrect and mispaired base in double-stranded DNA.

that DNA polymerase leaves about one mistake per 10^8 replicated base pairs. However, not even these usually appear as mutations, since the measured mutation rate can be as low as one mistake per 10^{10} or 10^{11} nucleotides. In *E. coli*, the final degree of accuracy is the responsibility of a **mismatch correction enzyme** encoded by genes *mutH, mutL,* and *mutS*. The enzyme scans newly replicated DNA for mismatched base pairs and removes a single-stranded segment containing the wrong nucleotide, thereby allowing a DNA polymerase to insert the correct base when it fills the resulting gap. The obvious problem that this entails is that of distinguishing which base of a mismatched pair is wrong, because both are natural components of DNA. If one of the pair were removed at random, the correct base would be taken about half of the time and a mutation would be ensured instead of prevented. Instead, a specific signal controlled by a timing device directs the mismatch excision system exclusively to the newly synthesized strand. The correction enzyme removes DNA containing a mismatched base only from a strand in which it also detects the DNA sequence GATC nearby (Figure 12-11). However, if a methylase encoded by the *dam* gene first modifies adenine in the GATC sequence to N^6-methyladenine, the correction enzyme cannot act and no excision occurs. Most GATC sequences in the cell become so modified after their synthesis, but only after a slight delay—

perhaps a few seconds or minutes. This is long enough for the mismatch correction enzyme to bind, so that the enzyme is directed to the DNA strand that has just been synthesized, rather than to the template strand. The correction enzyme detects not only singly mispaired bases but also small additions and deletions, so that it reduces the incidence of frameshift mutations as well as base substitutions.

Recombination Repairs DNA by Splicing into It Undamaged Strands Extracted from Duplex DNA[12, 25, 26]

Excision repair uses a complementary strand as a template to replace a damaged segment of DNA. But sometimes this template is not available, such as after a replication fork meets a lesion (e.g., a pyrimidine dimer) and separates the strands before excision repair can take place. Moreover, if both bases of a complementary pair are altered, perhaps crosslinked by a chemical like the carcinogen mitomycin C, then neither can act as a template for the other. Finally, a double helix that is broken straight across, particularly if a duplex segment is missing altogether, cannot be repaired directly with much chance of the connection being correctly remade. In all these cases, all information is lost at the site of damage, and it can be recovered only by taking a corresponding DNA segment from a separate but identical DNA molecule: This is called **recombinational repair.** An appropriate source of DNA almost always can be found in the cell. For example, when a lesion is left in a single-stranded segment by passage of the replication fork, the second daughter DNA carries the same sequence. Or if both strands are damaged in an unreplicated region, a vigorously growing bacterial cell often has one or more extra copies of its chromosome that can be used. Higher cells, of course, are generally diploid and carry the same, or nearly the same, sequence on each chromosome of a homologous pair. The key to recombinational repair is an enzyme that anneals the sequences on either side of a lesion to their complement in the undamaged DNA molecule, thus making the critical match that identifies the segment carrying the missing information. In *E. coli,* the RecA protein carries out this vital function (Chapter 11).

Inducibility of the SOS DNA Repair Genes[27]

Since a cell can often regulate the expression of genes according to a need for their products, it is not surprising that many DNA repair enzymes are induced by DNA damage. For example, methyltransferases are induced by abnormally alkylated DNA during the adaptive response (discussed earlier in this chapter). However, the most important and extensive group consists of the **SOS genes,** whose regulation is considered in detail in Chapter 16. The SOS genes are induced by damage that is severe enough to stop DNA synthesis altogether, rather than merely changing the pairing properties of bases. An example of an SOS-inducing lesion is a pyrimidine dimer, which cannot base-pair at all. When a replication fork meets a dimer,

Table 12-1 SOS Genes, Inducible by DNA Damage

Gene Name	Role in DNA Repair
Genes of known function	
uvrA *uvrB* *uvrC*	Encode excision endonuclease
umuD *umuC*	Encode proteins required for error-prone DNA repair and for most mutagenesis
sulA	Encodes protein that inhibits cell division, possibly to allow time for DNA repair
Genes involved in DNA metabolism whose specific role in DNA repair is unknown	
ssb	Encodes single-strand binding protein (SSB)
uvrD	Encodes helicase II (DNA unwinding protein)
himA	Encodes subunit of integration host factor, involved in site-specific recombination
recN	Involved in recombinational DNA repair
Genes of unknown function	
dinA *dinB* *dinD* *dinF*	

it stops and reinitiates some distance away, leaving a gap in the DNA to which the recombination enzyme RecA protein binds. Besides initiating DNA strand exchange (and thus recombinational repair), binding to single-stranded DNA activates RecA protein to carry out an enzymatic function entirely separate from recombination: to destroy by proteolytic cleavage the repressor of the SOS genes (the LexA repressor). In this way, RecA protein promotes DNA repair both by recombination and by mediating the induction of about 15 SOS genes (Table 12-1). Some of these genes have been discussed already: *uvrA*, *uvrB*, *uvrC*, and *recA* itself, all of which are expressed at a significant rate even without induction, but then much faster after the cell's DNA is damaged. Other SOS genes encode proteins involved in DNA synthesis and thus DNA repair as well. Meanwhile, the role of still other SOS genes is yet to be discovered.

Bases Too Damaged to Pair Give Rise to Mutations: Error-Prone Repair[28]

Although we know how base analogs and some modified bases cause mutations through mispairing, it is still unclear how mutations arise when bases are so damaged that they are unable to pair at all. Yet, this is the most significant process of all, since most natural mutagens and thus most **carcinogens** (cancer-causing agents; see below) act because cells can continue a DNA chain despite the apparently complete lack of template information at the site of a lesion. One important mutagenic structure of this sort is the ultraviolet-induced 6-4 photoproduct of two adjacent pyrimidine bases (see Figure 12-5). However, the extreme case is a DNA molecule lacking a base altogether (apurinic or apyrimidinic site). To show that such drastic damage can be repaired, we can use apurinic DNA made in vitro by brief treatment of single-stranded ϕX174 DNA with acid, which causes depurination of only one or a few nucleotides of the DNA. Since ϕX174 DNA is infectious to protoplasts (cells with part of their cell wall removed), it is possible to determine if the cell can repair the apurinic DNA molecule: Progeny phage are viable only if they have a nucleotide where the infecting DNA molecule had a hole. The cell does repair a fraction of the infecting apurinic DNA molecules, but it also usually makes a mutation in doing so, because there is no information by which it could choose the correct base to repair a single-stranded DNA molecule; the process is thus called **error-prone repair.** It seems likely that error-prone repair is an absolute last recourse for the cell, used only when there is no template available for accurate repair: It is better to incorporate a base that is probably wrong and let replication proceed than not to replicate at all. Naturally, the chance of mutation is very high, probably about three in four. Ironically, DNA repair is the *cause* of the mutation in this case.

Both UmuC and UmuD Proteins and RecA Protein Are Required for Error-Prone Repair[29, 30]

The major clue to the mechanism of error-prone repair is the involvement of two *E. coli* genes, *umuC* and *umuD* (for *UV* non*mu*table). These genes were originally identified as sites of mutations that prevent mutagenesis by ultraviolet irradiation. They are SOS genes, so that their products are abundant enough to act only after the cell is treated with an SOS inducer. (Error-prone repair is sometimes called **SOS repair**). Cells mutant in either gene are somewhat more sensitive than the wild type to killing by ultraviolet light, indicating that *umuC* and *umuD* promote survival by repairing damage. Because *umuC* and *umuD* are SOS genes, RecA protein is needed for error-prone repair, at least indirectly, to mediate induction of *umuC* and *umuD*. However, genetic experiments suggest that RecA protein also acts *directly* in error-prone repair.

How might error-prone repair work? We know that the mutations it results in are generally single-base changes at the site of the damage and not, for example, extensive replacements of DNA by some mechanism that splices in unrelated DNA sequences. Somehow, the UmuC and UmuD proteins cause a base to be inserted into the growing strand, even though no template base exists, a process that normally would be prohibited. This process has been named **bypass synthesis.** One model to explain bypass synthesis proposes that the UmuC and UmuD proteins might act to inhibit proofreading by the $3' \rightarrow 5'$ exonuclease activity of DNA polymerase. Another suggestion has been that UmuC and UmuD encode an entirely new DNA polymerase that can make mistakes. How RecA protein might be involved is also a mystery. One possibility follows from the fact that RecA protein attaches to DNA in the gap produced when a nonpairing lesion stops DNA synthesis: RecA protein might also bind the UmuC and UmuD proteins and thereby direct them to the site where error-prone DNA synthesis is required.

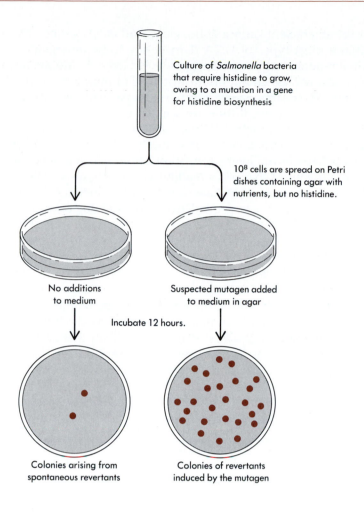

Culture of *Salmonella* bacteria that require histidine to grow, owing to a mutation in a gene for histidine biosynthesis

10^8 cells are spread on Petri dishes containing agar with nutrients, but no histidine.

No additions to medium

Suspected mutagen added to medium in agar

Incubate 12 hours.

Colonies arising from spontaneous revertants

Colonies of revertants induced by the mutagen

Figure 12-12
In the Ames test, the activity of a mutagen is detected as an increased incidence of wild-type revertants among auxotrophic bacteria grown in the presence of the chemical. Potent mutagens yield thousands of revertants from the 10^8 bacteria on a Petri dish.

Mutagens in Nature and Their Detection[31, 32]

It is important to identify mutagens and avoid human exposure to them for two reasons. First, random genetic change is more likely to be harmful than helpful in future generations. But even mutations that occur in somatic (nongerm) cells and thus are not passed to progeny threaten the organism. We now know, as was long suspected, that cancer is often the result of mutations in somatic cells that lead to uncontrolled cell growth. And a number of disorders associated with age, such as atherosclerosis, may be the expression of accumulated, unrepaired DNA damage. In fact, it has been suggested that the entire aging process may reflect irreversible damage to the DNA of somatic cells.

The clear demonstration that most mutagens are carcinogens (and vice versa) means that there is a simple way to determine if a substance is likely to be carcinogenic: We determine whether it is mutagenic to bacteria, whose short generation time allows a test to be made in a day or two (Figure 12-12). This method, now called the **Ames test,** was developed and refined by B. Ames, who used several tricks to make it very sensitive. Reversion to wild type of a mutation causing a growth defect, such as a requirement for the amino acid histidine, is measured, because even a few growing cells among many give rise to colonies that can easily be scored. The test bacteria bear mutations making their cell walls leaky, so that most compounds to be tested can permeate them, and they also carry a plasmid that expresses an efficient variant of the *umuC* and *umuD* genes, which are essential if most types of DNA damage are to be mutagenic.

A final requirement for carcinogens to be active as mutagens in the bacterial test reveals an important principle of mutagenesis in higher organisms. Most carcinogenic compounds, particularly natural products, are not mutagenic if they are applied directly to growing bacteria. First they must be *activated* by incubation with an extract of mammalian liver, a process that occurs naturally when foreign substances are metabolized in the liver. Activation converts carcinogens to electrophilic derivatives that react readily with the bases in DNA.

What are the important mutagens revealed by such tests as likely or certain carcinogens? It is proved from epidemiological studies that tobacco smoke is one of the most important carcinogens that humans encounter, and tobacco smoke is highly mutagenic in the Ames test. The potential danger of certain mutagenic industrial chemicals like ethylene dichloride is well known, and many other environmental contaminants may be significant carcinogens. But a more recent realization is that our greatest exposure to mutagens and carcinogens comes from eating—and not necessarily exotic foods, but even such staples as common vegetables. Plants make toxic chemicals as a defense against predators like insects, and many of these are metabolized in the liver into potent mutagens. The actual mutagens often are **oxidants** such as oxygen radicals that react with bases in DNA. Another dietary source of mutagens and a likely cause of cancer is lipids, which are metabolized to reactive oxidants by the liver. An optimistic note is that natural antioxidants like glutathione may counter many of the effects of these dietary hazards.

Summary

Organisms can survive only if their DNA is replicated faithfully and is protected from chemical and physical damage that would change its coding properties. The limits of accurate replication and repair of damage are revealed by the natural mutation rate. Thus, an average nucleotide is likely to be changed by mistake only about once every 10^9 times it is replicated, although error rates for individual bases vary by a hundredfold from this. Much of the accuracy of replication is inherent in the way DNA polymerase copies a template. The initial selection of the correct base is guided by complementary pairing. Accuracy is increased by proofreading, in which the $3' \rightarrow 5'$ exonuclease activity of DNA polymerase checks the last base inserted and removes a wrong one. Finally, in mismatch correction, the newly synthesized DNA strand is scanned by an enzyme that initiates replacement of DNA containing incorrectly paired bases. Cellular mutations called mutators that increase the natural mutation rate affect proteins involved in both proofreading and mismatch correction. Despite the safeguards, mistakes of all types occur: base substitutions, small and large additions and deletions, and gross rearrangements of DNA sequences.

Mutagens increase the natural rate of mutation. Base analogs simulate natural bases and are incorporated in their place, but pair less accurately and thus give rise to base switches. Reactive chemicals, including alkylating agents like nitrosamines and nitrosoguanidine, chemically change the bases so that pairing mistakes are made. Frameshift mutagens probably act by stabilizing looped-out regions in the template or growing strand during replication, ensuring that they are miscopied.

Cells have a large repertory of enzymes devoted to repairing DNA damage that would otherwise be lethal or that would modify DNA so as to engender damaging mutations. Photoreactivation is catalyzed by photolyase, an enzyme that reverses the photochemical fusion of adjacent pyrimidine bases into pyrimidine dimers caused by ultraviolet light. O^6-methylguanine methyltransferase enzymatically removes methyl groups from the dangerously reactive O^6 groups of guanine and probably from the phosphodiester backbone as well. The adaptive response, an induced resistance to alkylating agents that cause methylation, results from greatly increased synthesis of the transferase.

Although there are a few enzymes that simply reverse damage, a more universal process is excision repair, in which a damaged segment is removed and replaced through new DNA synthesis for which the undamaged strand serves as a template. In *E. coli*, excision repair is initiated in two ways. The UvrABC endonuclease cuts out a 12-base segment of a DNA strand that includes the lesion. Glycosylases remove only a damaged base, leaving a second step to an apurinic or apyrimidinic endonuclease, which cuts the backbone. Both pathways leave a primer end from which DNA polymerase can initiate nick translation to remove the damaged segment and renew the DNA. An alternative repair method, which is particularly important if excision repair cannot occur because no template for repair synthesis is available, is recombinational repair, in which an intact DNA strand is taken from a different but homologous duplex.

A central enzyme of DNA repair in *E. coli* is the RecA protein, which both catalyzes DNA strand exchange and thus promotes recombinational repair and also regulates a group of at least 15 DNA repair genes called the SOS genes. When cellular DNA is damaged, a protease function of RecA protein is activated to destroy the repressor (LexA repressor) of the SOS genes, leading to their induction. Two SOS genes, *umuC* and *umuD*, are required for error-prone repair, an important but still mysterious process by which mutations result from repair of DNA damage so severe that non-base-pairing lesions are created. The majority of all dangerous natural mutagens act through error-prone repair.

Mutagens are of concern to us because they permanently affect the genes that organisms inherit and because cancer is often caused by mutations in somatic cells that lead to uncontrolled growth. Thus, it is important to be able to identify and characterize mutagens to which organisms might be exposed. Bacterial tests allow this to be done conveniently because the short generation time of bacteria and our ability to handle many cells at once reveal the activity of a mutagen in a day. Natural mutagens are prevalent, but so are the biological defenses that organisms mount against them.

Bibliography

General References

Friedberg, E. C. 1985. *DNA Repair.* San Francisco: Freeman.

Kornberg, A. 1980. *DNA Replication.* San Francisco: Freeman. Also, *1982 Supplement to DNA Replication.* The essential textbook that describes the basic principles of replication and repair of DNA.

Cited References

1. Miller, J. H. 1983. "Mutational Specificity in Bacteria." *Ann. Rev. Genetics* 17:215–238.
2. Miller, J. H., and K. B. Low. 1984. "Specificity of Mutagenesis Resulting from the Induction of the SOS System in the Absence of Mutagenic Treatment." *Cell* 37:675–682.
3. Topal, M. D., and J. R. Fresco. 1976. "Complementary Base Pairing and the Origin of Substitution Mutations." *Nature* 263:285–293.
4. Loeb, L. A., and T. A. Kunkel. 1982. "Fidelity of DNA Synthesis." *Ann. Rev. Biochem.* 51:429–457.
5. Cox, E. C. 1976. "Bacterial Mutator Genes and the Control of Spontaneous Mutation." *Ann. Rev. Genetics* 10:135–156.
6. Echols, H., C. Lu, and P. M. J. Burgers. 1983. "Mutator Strains of *Escherichia coli, mutD* and *dnaQ,* with Defective Exonucleolytic Editing by DNA Polymerase III Holoenzyme." *Proc. Nat. Acad. Sci.* 80:2189–2192.
7. Ghosal, D., and H. Saedler. 1978. "DNA Sequence of the Mini-Insertion IS2-6 and Its Relation to the Sequence of IS2." *Nature* 275:611–617.
8. Albertini, A., M. Hofer, M. Calos, and J. Miller. 1982. "On the Formation of Spontaneous Deletions: The Importance of Short Sequence Homologies in the Generation of Large Deletions." *Cell* 29:319–328.
9. Ripley, L. S. 1982. "Model for the Participation of Quasi-Palindromic DNA Sequences in Frameshift Mutation." *Proc. Nat. Acad. Sci.* 79:4128–4132.
10. Roth, J. R. 1974. "Frameshift Mutations." *Ann. Rev. Genetics* 8:319–346.
11. Singer, B., and J. T. Kusmierek. 1982. "Chemical Mutagenesis." *Ann. Rev. Biochem.* 52:655–693.
12. Howard-Flanders, P. 1981. "Inducible Repair of DNA." *Sci. Amer.* 245:72–103.
13. Brash, D. E., and W. A. Haseltine. 1982. "UV-Induced Mutation Hotspots Occur at DNA Damage Hotspots." *Nature* 298:189–192.
14. Lindahl, T. 1982. "DNA Repair Enzymes." *Ann. Rev. Biochem.* 51:61–87.
15. Haseltine, W. 1983. "Ultraviolet Light Repair and Mutagenesis Revisited." *Cell* 33:13–17.
16. Samson, L., and J. Cairns. 1977. "A New Pathway for DNA Repair in *Escherichia coli.*" *Nature* 267:281–282.
17. Cairns, J., P. Robins, B. Sedgwick, and P. Talmud. 1981. "The Inducible Repair of Alkylated DNA." *Prog. Nucleic Acid Res. Mol. Biol.* 26:237–244.
18. Lindahl, T., B. Demple, and P. Robins. 1982. "Suicide Inactivation of the *E. coli* O⁶-Methylguanine-DNA Methyltransferase." *EMBO J.* 1:1359–1363.
19. Teo, I., B. Sedgwick, B. Demple, B. Li, and T. Lindahl. 1984. "Induction of Resistance to Alkylating Agents in *E. coli:* The *ada* Gene Product Serves Both as a Regulatory Protein and as an Enzyme for Repair of Mutagenic Damage." *EMBO J.* 3:2151–2157.
20. Sancar, G. B., and W. D. Rupp, 1983. "A Novel Repair Enzyme: UVRABC Excision Nuclease of *Escherichia coli* Cuts a DNA Strand on Both Sides of the Damaged Region." *Cell* 33:249–260.
21. Yeung, A. T., W. B. Mattes, E. Y. Oh, and L. Grossman. 1983. "Enzymatic Properties of Purified *Escherichia coli* uvrABC Proteins." *Proc. Nat. Acad. Sci.* 80:6157–6161.
22. Coulondre, C., J. Miller, P. Farabaugh, and W. Gilbert. 1978. "Molecular Basis of Base Substitution Hotspots in *Escherichia coli.*" *Nature* 274:775–780.
23. Wagner, R., and M. Meselson. 1976. "Repair Tracts in Mismatched DNA Heteroduplexes." *Proc. Nat. Acad. Sci.* 73:4135–4139.
24. Lu, A.-L., S. Clark, and P. Modrich. 1983. "Methyl-Directed Repair of DNA Base-Pair Mismatches *in vitro.*" *Proc. Nat. Acad. Sci.* 80:4639–4643.
25. Radding, C. 1982. "Homologous Pairing and Strand Exchange in Genetic Recombination." *Ann. Rev. Genetics* 16:405–437.
26. Szostak, J., T. Orr-Weaver, R. Rothstein, and F. Stahl. 1983. "The Double-Strand Break Repair Model for Recombination." *Cell* 33:25–35.
27. Walker, G. C. 1984. "Mutagenesis and Inducible Responses to Deoxyribonucleic Acid Damage in *Escherichia coli.*" *Microbiol. Revs.* 48:60–93.
28. Schaaper, R. M., T. A. Kunkel, and L. A. Loeb. 1983. "Infidelity of DNA Synthesis Associated with Bypass of Apurinic Sites." *Proc. Nat. Acad. Sci.* 80:487–491.
29. Kato, T., and Y. Shinoura. 1977. "Isolation and Characterization of Mutants of *Escherichia coli* Deficient in Induction of Mutations by Ultraviolet Light." *Mol. Gen. Genetics* 156:121–131.
30. Elledge, S., and G. Walker. 1983. "Proteins Required for Ultraviolet Light and Chemical Mutagenesis: Identification of the Products of the *umuC* Locus of *Escherichia coli.*" *J. Mol. Biol.* 164:175–192.
31. Ames, B. 1979. "Identifying Environmental Chemicals Causing Mutations and Cancer." *Science* 204:587–593.
32. Ames, B. 1983. "Dietary Carcinogens and Anticarcinogens." *Science* 221:1256–1264.

V

THE STEPS IN PROTEIN SYNTHESIS

CHAPTER 13

The Synthesis of RNA upon DNA Templates

Although the chemistry of DNA synthesis and RNA synthesis is very similar at the nucleotide level, the biological functions of the two processes are quite different. Whereas DNA synthesis must only replicate the genome accurately and is inherently unvarying, transcription of the genome into RNA reflects much of the complexity of genetic expression itself. Genes are usually regulated at the level of transcription, and all phases of RNA synthesis are subject to variation: where an RNA chain starts or stops on the DNA template, when it starts and stops, and how efficiently it does both. Much of the recent progress in understanding the molecular biology of gene expression has come from studying the DNA sequences and the proteins that control transcription, particularly the RNA polymerase enzymes themselves. The basic principles were first established for bacterial (mainly *E. coli*) RNA polymerase, but these principles are probably universal, despite a few important differences that eucaryotic enzymes reveal. In this chapter we will primarily consider transcription in bacteria and mention eucaryotic enzymes only briefly as a preview to Chapter 21.

The Structures of RNA

RNA is almost always single-stranded, in contrast to DNA, and is synthesized from the DNA template in a correspondingly different way. Before we consider their synthesis, it is important to point out that although they are single-stranded, RNA molecules nevertheless do have structures, some as definite and fixed as those of proteins and some much less specifically defined. Almost every RNA molecule has many short double-helical regions, which can form because the two sections of an RNA chain within a hairpin fold are in the correct antiparallel orientation to base-pair (Figure 13-1). In addition to the standard AU and GC base pairs, weaker GU base pairs (Chapter 14) also contribute to this secondary structure of single-stranded RNA. Even an RNA of random base sequence has much secondary structure, because short adjacent complementary sequences occur by chance. Furthermore, some RNAs (such as transfer RNA) have evolved to contain extensive internal complementary sequences that base-pair to give the whole molecule a specific shape, its tertiary structure. As we would expect, secondary structure also exists within

Figure 13-1
Schematic folding of an RNA chain, showing several double-helical regions held together by hydrogen bonds.

UGGCGUUCGUACUUAAAUAUGGAAU^U   A

GCCUCAAGCAUCGCUUUUAACCUUA_U   C

An actual example of an RNA loop from the
RNA of the virus R17

single chains of DNA when there is no complementary strand that allows a long duplex to form.

Two principles of chain folding that apply to polypeptides are probably equally true for RNA chains: The structure that each can form is encoded in the sequence of functional groups in the chain (whether amino acids or nucleotides), and this structure forms as the chain is being made. This is important to keep in mind, because as we shall see, the structure of a growing RNA chain can affect transcription of the remainder of the molecule.

Each cell contains many distinct RNA molecules, with lengths from less than 50 nucleotides to tens of thousands of nucleotides. These are almost always linear single strands, although a few circular RNAs are known to occur (Chapter 24).

Enzymatic Synthesis of RNA upon DNA Templates

The fact that RNA, like DNA, is a long, unbranched chain of four different nucleotides immediately suggests that the genetic information of DNA chains is transferred to a complementary sequence of RNA nucleotides. According to this hypothesis, the DNA strands at one or more stages in the cell cycle separate and function as templates onto which complementary ribonucleotides are attracted by DNA-like base pairing [adenine with thymine (uracil in RNA) and guanine with cytosine]. This hypothesis also implies that some control mechanism must dictate whether the separated DNA strands will function as templates for a complementary DNA strand or a complementary RNA strand.

Direct evidence for the hypothesis came from the discovery of the appropriate enzyme, RNA polymerase, which exists in virtually all cells; a cell of *E. coli* has about 3000 molecules of RNA polymerase. This enzyme links together ribonucleotides by catalyzing the forma-

Figure 13-2

Catalysis of 3′–5′ phosphodiester bond formation in RNA.

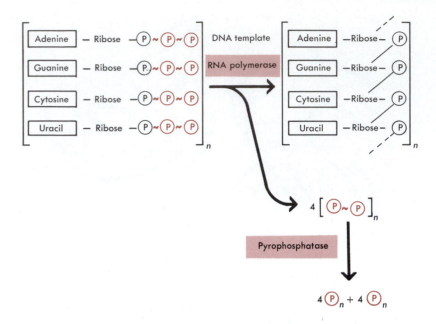

tion of the internucleotide 3′–5′ phosphodiester bonds that hold the RNA backbone together (Figure 13-2). It does so, however, only in the presence of DNA, a fact that suggests that DNA must line up the correct nucleotide precursors in order for RNA polymerase to work.

Although RNA polymerase normally uses double-stranded DNA in making RNA, the actual template is single-stranded DNA that the enzyme exposes by unwinding a bit of the helix as it passes. This is shown clearly in experiments with DNA from the virus ϕX174. This virus belongs to one of the viral classes that contain single-stranded DNA instead of the customary double-helical form. Only one of the two possible complementary DNA strands is present, and when it is used as the template for RNA polymerase, the enzymatic product has a complementary base sequence (Table 13-1). Moreover, in this special case, the RNA product remains attached to its DNA template, allowing the isolation of a hybrid DNA-RNA double helix. When the template is double-helical DNA, the RNA product quickly detaches from its template, and the two DNA strands again come together in specific register. This means that the front end of a messenger RNA becomes immediately available to bind to a ribosome (Chapter 14).

We thus see that the fundamental mechanism for the synthesis of RNA is very similar to that of DNA. In both cases, the immediate precursors are nucleoside triphosphates that use the energy in one of their pyrophosphate bonds to drive the reaction toward synthesis.

Table 13-1 Base Composition of Enzymatically Synthesized RNA Using Single-Stranded ϕX174 DNA as Template

	ϕX174 DNA	Observed Values of RNA Product	Predicted RNA Composition
Adenine	0.25	0.32	0.33
Uracil	0.33 (thymine)	0.25	0.25
Guanine	0.24	0.20	0.18
Cytosine	0.18	0.23	0.24
TOTAL	1.00	1.00	1.00

Also in both cases, a single enzyme works on all four possible nucleotides, whose correct selection is dictated by the obligatory need to base-pair with a polynucleotide template.

RNA synthesis is necessarily an extremely accurate process. There is no evidence, however, for any proofreading, and so the overall precision of transcription does not approach the almost perfect copying characteristic of DNA self-replication. But since cellular RNA is not self-replicating, the rare mistakes that do occur are not genetically perpetuated.

Along Each Gene Only One DNA Strand Acts as an RNA Template[1]

If each of the two DNA strands of a given gene served as an RNA template, each gene would produce two RNA products with complementary sequences. Since our genetic evidence tells us that each gene controls only one protein, we must assume that either only one of the two possible RNA strands is made or, if both are synthesized, only one is functional for some special reason. It appears that the former possibility is correct: In vivo, only one of the two possible RNAs is found. This can be seen by examining the RNA synthesized in vivo from the virus SP8, which multiplies in the bacterium *Bacillus subtilis*. The two complementary DNA strands of SP8, unlike those of most viruses, have quite different base compositions and can be relatively easily separated. It is thus possible to determine whether the RNA products have base sequences complementary to one or both of the DNA strands.

To make this determination, we use our ability to form artificial DNA-RNA hybrid molecules by mixing RNA molecules with single-stranded DNA molecules formed by heating double-helical DNA. Heating DNA molecules to temperatures just below 100°C breaks the hydrogen bonds holding the complementary strands together; they then quickly separate from each other (DNA denaturation). If the temperature is gradually lowered, the complementary strands again form the correct hydrogen bonds, and the double-helical form is regained (renaturation of DNA). If, however, this gentle cooling is done in the presence of single-stranded RNA that has been synthesized using the DNA as template, then DNA-RNA hybrids form as well as renatured DNA double helices (Fig. 13-3). These DNA-RNA hybrids are very specific and form only if stretches of the nucleotide sequence in the DNA are complementary to the RNA nucleotide sequences. This technique lets us see whether the in vivo RNA products will form hybrids with only one or with both of the complementary SP8 DNA strands. A clear result emerges: Only one DNA strand is copied.

The copying of only one DNA strand within a given gene explains why the base ratios of RNA need not be complementary, even though RNA is made on a DNA template. In a given DNA strand, there is no reason why the A should equal T, or the G equal C. These ratios must be 1:1 only when the corresponding bases on the two complementary chains are added together. Thus, we must assume that only in the exceptional DNA molecule will the single strands be found, on close inspection, to have even approximately complementary ratios. Correspondingly, only rarely will single-stranded RNA molecules be found in which A nearly equals U and G nearly equals C.

Figure 13-3

The use of DNA-RNA hybrids to show the complementarity in nucleotide sequence between an RNA molecule and one of the two strands of its DNA template. The left side of the diagram shows the formation of a hybrid molecule between an RNA molecule and one of the two strands of the template. The right side of the diagram shows why DNA-RNA hybrid formation allows complementarity to be detected: If the same RNA molecule is mixed with unrelated DNA, no hybrid molecules are formed.

Differential copying of the two strands of a DNA molecule by RNA polymerase can also be observed in vitro. This result depends on the use of DNA templates that have not been severely damaged. If, for example, T7 DNA has been denatured to produce single-stranded regions that RNA polymerase can bind, then both strands are copied. However, on its natural template, intact double-stranded DNA, RNA polymerase binds only to specific sequences called promoters, which direct it to copy the same strand that is copied during in vivo viral reproduction (Figure 13-4).

The transcription patterns along other chromosomes are often not as clear-cut as with T7 or SP8. Both strands of the T4 and λ chromosomes are transcribed, one strand serving as the template for some genes (usually contiguous groups) while the remaining genes are transcribed along the other strand (see Figure 17-15). The same picture holds for the *E. coli* chromosome, which can be transcribed both clockwise and counterclockwise.

Synthesis of RNA Chains Occurs in a Fixed Direction

Each RNA chain, like each DNA chain, has a direction defined by the orientation of the sugar-phosphate backbone. The chain end terminated by the 5' carbon atom is called the 5' end, while the end containing the 3' carbon atom is called the 3' end (see Figure 3-13). As with DNA synthesis, it always seemed safe to assume that RNA chains grow sequentially in only one of the two possible polarities, 5'→3' or 3'→5'. If they grow 5'→3', then we expect the begin-

Figure 13-4

Electron micrographs of in vitro transcription of T7 DNA. (a) Several growing RNA chains can be seen attached to the DNA template. All of the RNA chains are transcribed off the same T7 strand, initiating their growth at three clustered promoters near the end of the molecule. Total synthesis time is 5 minutes at 37°C. (b, c) Views of two attached RNA chains are shown at higher magnification. (d) Synthesis was terminated after 2½ minutes at 37°C. Twelve nascent chains all at one end can be seen. RNA polymerase is present where the RNA chains are attached, but it is not visible. (Courtesy of Dr. R. Davis, Stanford University.)

ning nucleotide to possess a ⓟ~ⓟ~ⓟ (triphosphate) group (Figure 13-5). But if the chain grows 3'→5', then the nucleotide at the growing end would contain the ⓟ~ⓟ~ⓟ group. Firm evidence shows that the direction of growth is 5'→3': Newly inserted nucleotides are found at the 3' ends, while the ⓟ~ⓟ~ⓟ groups are found attached to the nucleotides that commenced chain growth.

The 5'→3' direction is confirmed by work with the metabolic inhibitor 3'-deoxyadenosine, an analog of adenosine lacking the 3' oxygen that forms part of the internucleotide bond. When it is added to cells, 3'-deoxyadenosine is first phosphorylated to 3'-deoxyadenosine-ⓟ~ⓟ~ⓟ and then joined to the 3' growing end. Because it contains no 3'-OH group, further nucleoside triphosphates cannot attach, and RNA synthesis stops. But this means that synthesis must be 5'→3', because otherwise the inhibitor could not be incorporated at all.

(a) Chain growth in the 5'⟶3' direction

(b) Chain growth in the 3'⟶5' direction

(c) 3'-deoxyadenosine incorporation

Figure 13-5

Alternative directions for the synthesis of an RNA chain are shown in (a) and (b). Many types of experiments suggest that chains always grow in the 5'→3' direction. Part (c) shows the structure of 3'-deoxyadenosine triphosphate and its incorporation at the 3' end of an RNA chain growing 5'→3', halting RNA synthesis.

Construction of Bacterial RNA Polymerase from Subunits

Like DNA polymerase III, bacterial RNA polymerase has a very complicated subunit structure. The active form, the holoenzyme, sediments at about 15S and contains five different polypeptide chains, β', β, σ, α, and ω, with respective molecular weights (for *E. coli*) of 155,000, 151,000, 70,000, 36,500, and about 11,000 daltons. No covalent bonds join the various chains; aggregation results from the formation of secondary bonds. Each specific chain, with the exception of α, which appears twice, is present once within the active molecule, giving the complete holoenzyme ($\beta'\beta\alpha_2\omega\sigma$) a molecular weight of about 450,000 (Figure 13-6 and Table 13-2). If it were a sphere, the diameter of RNA polymerase would be about 100 Å, the length of a 30-base-pair segment of duplex DNA; however, the fact that it binds over 60 nucleotides suggests instead an elongated shape.

It is tempting to relate the complexity of bacterial RNA polymerase to the many individual enzymatic steps that it carries out to achieve initiation, elongation, and termination of RNA chains. Yet, phage T7 encodes an RNA polymerase containing only a single chain with a molecular weight of 98,000 daltons that manages all of this as well (Chapter 17). Perhaps much greater complexity is necessary for the elaborate and varied transcriptional controls that the bacterial cell requires. It would be of great interest to know the detailed three-dimensional structure of RNA polymerase. However, its large size makes X-ray crystallography a difficult undertaking, and furthermore, the requisite crystals have not yet been produced.

The attachment of σ to the other chains is not very firm, so it is relatively easy to isolate a $\beta'\beta\alpha_2\omega$ aggregate. This specific grouping is known as the **core enzyme,** for it catalyzes the formation of the internucleotide phosphodiester bonds equally well in the presence or absence of σ. The catalytic center is thought to be in the β subunit of the core. Subunit β is the binding site of the important antibiotic **rifampicin** (and the related antibiotic rifamycin), a potent inhibitor of the enzyme that prevents initiation of an RNA chain but does not affect its subsequent elongation. Mutations that make the cell resistant to rifampicin map in *rpoB*, the gene for β, as do many nonlethal mutations that subtly affect RNA polymerase function. The antibiotic **streptolydigin** inhibits elongation of an RNA chain, and mutations conferring resistance to it also are in the gene encoding β.

Figure 13-6
Schematic picture of RNA polymerase showing subunit construction.

Organization of RNA Polymerase Genes into Operons[2]

The genes for the *E. coli* RNA polymerase subunits (with the exception of ω, whose gene is unknown) are found in three operons that also contain essential genes for protein and DNA synthesis (Figure 13-7). Mostly, these genes encode some of the 50 or so ribosomal proteins (Chapter 14), although the σ operon also contains the gene for the replication protein DNA primase (Chapter 10). It is clear that

Table 13-2   Subunits of *E. coli* RNA Polymerase and Transcription Factors

	Gene Name	Map Position* (Minutes)	Polypeptide MW (Daltons)	Number in Enzyme	Function
RNA Polymerase Subunits					
β' (beta')	*rpoC*	89.5	155,000	1	DNA binding?
β (beta)	*rpoB*	89.5	151,000	1	Catalytic site for RNA polymerization
σ (sigma)	*rpoD*	66.5	70,000	1	Promoter recognition, initiation
α (alpha)	*rpoA*	72	36,500	2	?
ω (omega)	—	—	11,000	1	?
Transcription Factors					
ρ (rho)	*rho*	84.5	46,000	6	Termination
nusA	*nusA*	65	69,000	1	Elongation, termination

*The map position refers to the conventional (circular) *E. coli* genetic map (see Figure 7-22).
SOURCE: After G. Zubay, *Biochemistry* (Reading, Mass.: Addison-Wesley, 1983), p. 792. Used with permission.

Figure 13-7

The three operons that contain genes for the σ, β, β', and α subunits of RNA polymerase. The major promoters (P) are at the left, and RNA is synthesized rightward. Minor promoters (p), including one in the σ operon that is active in heat-shock conditions (p_{HS}), are found within the operons; they promote transcription only of genes to their right. The first two genes of the β operon, *L11* and *L1*, are sometimes considered to make up a separate operon, because there is a promoter before the *L10* gene (see Chapter 16). [After Z. F. Burton, C. Gross, K. Watanabe, and R. Burgess, *Cell* 32 (1983):335, with permission.]

this congregation must reflect common regulation of proteins needed in amounts that are closely related to the growth rate of the cell. How these operons are regulated is unknown, although the σ operon has two tandem promoters, as do the ribosomal RNA operons, which are thought to be controlled by the molecule ppGpp* in response to growth rate (Chapter 16). The genes in these operons are not transcribed or expressed equally: Ribosomal protein genes at the beginning of the σ and β operons are followed by **attenuators,** partially effective termination sites that allow only a fraction of the RNA polymerase to transcribe the remainder of the operon. Corresponding to the unequal transcription within each operon, fewer RNA polymerase subunits (3000) than ribosomal proteins (20,000) are needed in each cell. However, this agreeable relation is confounded by the DNA primase gene, which precedes the σ gene, even though the cell has 60 times as many molecules of σ as primase (3000 versus 50); the explanation is that the mRNA segment encoding DNA primase is much less stable than the rest. The occurrence of several minor promoters within these operons, including one in the σ operon that is activated by heat shock (Chapter 16), bespeaks a much more complex regulation than we so far understand.

RNA Polymerase Recognizes Specific Starting Sequences in DNA: Promoters[3–8]

The σ subunit by itself has no catalytic function; its role is the recognition of start signals along the DNA molecule. In vitro experiments show that in the absence of σ, the core enzyme will occasionally initiate RNA synthesis, but this is the result of starting mistakes. RNA chains made by the core enzyme alone are started randomly along both strands of a given gene. When σ is present, however, the correct site is selected, indicating that the holoenzyme has the capacity to bind specifically to a "starting" sequence, the *promoter*.

---

*This is the common abbreviation for guanosine with a pyrophosphate on its 5' and 3' carbons. In our previous nomenclature it is Ⓟ~Ⓟ—G—Ⓟ~Ⓟ.

Figure 13-8
Promoter consensus sequences and RNA polymerase contact sites.
(a) Below the −35 and −10 consensus sequences is the sequence of a particular promoter, named lactose UV5. RNA polymerase contacts bases (indicated by symbols above them) and phosphates in the helix backbone (indicated by symbols between the bases) throughout this promoter, but particularly near the −35 and −10 regions. In this particular experiment, it was found that at least the indicated 12-base segment is unwound, although a different measurement shows that a segment about 17 bases long is unwound. [After U. Siebenlist, R. Simpson, and W. Gilbert, *Cell* 20 (1980):270, with permission.] (b) "Footprint" analysis of the lactose UV5 promoter, revealing sequences protected by RNA polymerase from limited DNase digestion (see Chapter 16). This experiment shows that RNA polymerase covers 50 to 60 bases of DNA, blocking access by the relatively large nuclease to most phosphodiester bonds in this segment. [After A. Schmitz and D. Galas, *Nucleic Acid Res.* 6 (1979):124, with permission.]

What features of DNA structure direct bacterial RNA polymerase to bind only at promoters? Clearly, there must be nucleotide sequences giving DNA a specific local shape that fits RNA polymerase, just as any enzyme fits its substrate. A comparison of more than 100 promoters shows that two common sequences of six nucleotides occur about 10 base pairs and about 35 base pairs before the site where RNA synthesis starts. These are named the −35 (minus 35) and −10 (minus 10) sequences, according to the numbering scheme in which the DNA nucleotide encoding the beginning of the RNA chain is named +1. Few promoters have exactly the −10 and −35 **consensus sequences** shown in Figure 13-8, but most differ from them by only a few nucleotides. There is strong evidence that these sequences are the critical parts of the promoter. First, RNA polymerase contacts bases and phosphates of the DNA backbone in and alongside the −35 and −10 sequences. These contacts are revealed by "protection" experiments, in which the ability of RNA polymerase to shield regions of DNA from attack by chemicals and nucleases is measured, and by "chemical modification" experiments, in which chemical groups whose integrity is essential for RNA polymerase binding are identified (Chapter 16). Second, weaker promoters tend to differ in sequence somewhat from the consensus, whereas stronger promoters match it

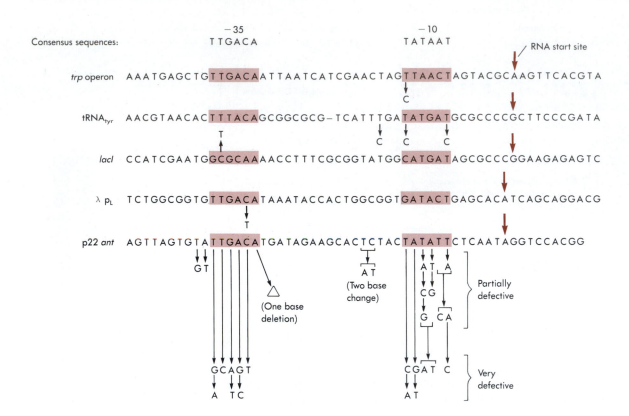

Figure 13-9

Mutations that decrease (downward arrows) and increase (upward arrow) initiation from five promoters. The strand identical in sequence to the RNA is shown. [After M. Rosenberg and D. Court, *Ann. Rev. Genetics* 13 (1979):319; and after P. Youderian, S. Bouvier, and M. Susskind, *Cell* 30 (1982): 843–853.]

more closely. Third, and most important, 75 percent of mutations that destroy promoter function change nucleotides in the common sequences (Figure 13-9); the other 25 percent are near the common sequences.

The two binding regions are separated by about 20 base pairs, or two complete turns of the double helix. Since both binding regions are therefore on one side of a DNA molecule, the enzyme must bind along one face of the helix, feeling in each region for the specific chemical shape made by base pairs in the bottoms of the grooves (Figure 13-10).

RNA Polymerase Unwinds DNA as It Binds the Promoter[9, 10]

E. coli RNA polymerase binds a promoter in two distinguishable steps (see Figure 13-10). First, it finds the promoter sequence and binds loosely in a "closed" complex. This initial recognition occurs mostly at the −35 region, since mutations that impair binding to the promoter change bases in the −35 region. Second, the closed complex is converted to an "open" complex, in which RNA polymerase is bound much more tightly. During this shift, about 17 base pairs beginning in the −10 region unwind, exposing the template strand against which RNA nucleotides are polymerized. It may be significant that the −10 region is rich in AT base pairs, because these come apart or "melt" much more easily than GC pairs. Not surprisingly, mutations that hinder formation of the open complex map in the −10 region. Many of these change the −10 sequence but preserve its AT content, showing that this region must have a specific shape for RNA polymerase to recognize, as well as being inherently susceptible to melting. Unwinding of the DNA as the open complex forms extends the stretch of

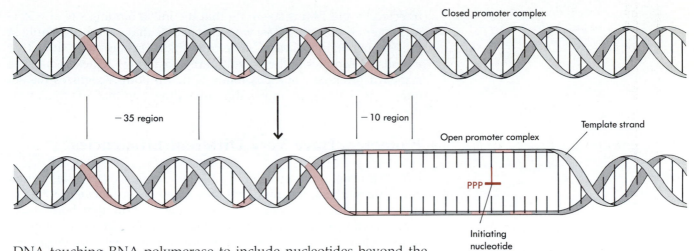

DNA touching RNA polymerase to include nucleotides beyond the −10 region that were initially on the other side of the DNA helix from the binding face (see Figure 13-10). Since these contacts can probably be made only if this DNA strand swings around as it unwinds, their formation may provide the energy for unwinding.

Different σ Factors Allow RNA Polymerase to Recognize Different Promoter Sequences[11-16]

The σ subunit must be present for RNA polymerase to engage a promoter, although the core enzyme alone can catalyze synthesis of an RNA chain. Can we distinguish whether it is σ or the core enzyme that recognizes the specific DNA sequences that constitute a promoter? A most important discovery shows that σ is responsible: there are other σ factors, distinct from the predominant σ that copurifies with core RNA polymerase (called σ^{70}), which direct RNA polymerase to initiate RNA synthesis at promoters whose sequences are quite different from the major bacterial promoters. These variant σ factors serve special functions in the cell, generally to redirect cellular RNA synthesis drastically when expression of a completely new set of genes is required. They were first discovered in the bacterium *Bacillus subtilis*, where their role is to begin expression of new proteins required for the process of sporulation (Chapter 22). In *E. coli*, a separate σ factor named σ^{32} promotes expression of the heat-shock genes, which probably serve special metabolic functions required when the temperature of cell growth is suddenly raised (Chapter 16). Still others are made by phages to invoke transcription of phage genes necessary at particular times of phage development (Chapter 17).

Since the promoter sequences that variant σ factors recognize are different from major bacterial promoters in both the −35 and −10 hexamers, it is possible either that σ binds both regions of DNA at once, even though it would have to be greatly elongated to span the distance required (about 70 Å), or that it binds the two regions in succession during the initiation process. However, a surprising recent discovery about the phage-encoded σ factor that directs synthesis of *E. coli* phage T4 "late" genes suggests that this σ factor may bind only to the −10 region: The −35 region of T4 "late" promoters can be changed (by genetic engineering) to other and unrelated sequences without harming promoter function, indicating that the −35 sequence is not involved in the specificity of RNA polymerase binding to T4 late promoters. Perhaps this means that promoter binding

Figure 13-10
RNA polymerase binding to DNA in the closed and open promoter complexes. This interpretation, based on the experiment summarized in Figure 13-8, shows where RNA polymerase may touch promoter DNA during the initial binding (closed complex) and after the strands are separated to allow initiation of RNA chain growth (open complex). Sites of contact are shown in color. The ATP that initiates the RNA chain is shown, although its presence is not required for the open complex to form. [After U. Siebenlist, R. Simpson, and W. Gilbert, *Cell* 20 (1980):269–281.]

directed by the T4 σ is so strong that the initial binding site at -35 is no longer necessary; at least we can conclude that σ binds primarily at the -10 region in this case. Since crosslinking experiments show that σ binds nucleotides in the unwound region at the -10 hexamer of the standard promoter, one possibility is that it both recognizes the sequence at -10 and promotes the unwinding step that lets RNA synthesis begin.

Promoters Have Very Different Efficiencies[8, 17]

It is clear from our discussion of promoter sequences that some work better than others, meaning that they allow faster repeated initiation of RNA synthesis. Biochemical experiments show that both steps, initial binding and open complex formation, differ among promoters. In the cell, different promoters function at rates that vary by several orders of magnitude—from only one initiation in ten or more minutes to one initiation every second or two. This is a fundamental way in which the rate of gene expression is set.

Regulatory Proteins Affect Transcription Initiation by Binding Near or In Promoters

The rate of transcription of a particular gene is not necessarily fixed, but instead may change according to the needs of the cell in different conditions of growth; such a gene is said to be "regulated." In *E. coli*, regulation of transcription is often mediated by proteins that, by binding to DNA near or within the promoter, increase or decrease the rate at which RNA polymerase initiates RNA synthesis. These proteins include repressors, which block binding of RNA polymerase and thus inhibit transcription, and **activators,** which bind with RNA polymerase and stimulate its activity. Promoters that require activators in order to function efficiently often match the -35 consensus sequence poorly, suggesting that the activator substitutes for this part of the binding site. A protein that is a repressor of one promoter may be an activator of another, and vice versa, depending on the precise location of its binding site; the name of a regulatory protein usually reflects the function that was discovered first. Some regulatory proteins are presented in Table 13-3, and their mechanism of action is considered in detail in Chapters 16 and 17.

Upstream DNA Sequences Can Increase Promoter Activity[18]

Although the -10 and -35 hexamers of a promoter are the essential sequences where RNA polymerase both binds and acts, other nearby DNA sequences can affect promoter function as well. For example, the natural sequences between 50 and 150 nucleotides upstream of the start site for ribosomal RNA synthesis are essential for full promoter activity. If these are deleted and replaced by foreign DNA, such as DNA of a plasmid into which the ribosomal RNA genes are cloned, the rate of initiation decreases about tenfold. In other cases, similarly distant AT-rich DNA is thought to stimulate the rate of initiation. Although upstream sequences sometimes may be binding sites for unidentified activators that stimulate RNA polymerase directly, they also could act in other ways. For example, an upstream sequence might attract a topoisomerase that induces a favorable state of local

Table 13-3 Regulatory Proteins That Affect RNA Polymerase
Function at Promoters

Protein	Effect on Initiation at Target Promoter	Function in Cell
Lactose repressor	Blocks	Prevents synthesis of enzymes that metabolize lactose when the sugar is absent.
Galactose repressor	Blocks	Prevents synthesis of enzymes that metabolize galactose when the sugar is absent.
LexA repressor	Blocks	Prevents synthesis of DNA repair enzymes until cellular DNA is damaged.
λ repressor	Blocks (promoters P_L and P_R)	Prevents synthesis of proteins required for lytic growth.
λ repressor	Activates (promoter P_{RM})	Increases synthesis of itself by stimulating its own promoter.
CAP protein	Activates	Increases synthesis of enzymes that metabolize other sugars when glucose is not available.
CAP protein	Blocks	Regulates its own gene and prevents synthesis of other proteins not needed in the absence of glucose.
AraC protein	Activates	Increases synthesis of proteins required to utilize arabinose when arabinose is available.
AraC protein	Blocks	Prevents synthesis of enzymes that metabolize arabinose when the sugar is not present.
λ cII protein	Activates	Stimulates synthesis of phage proteins required to establish lysogeny.

supercoiling. Or it could bring the DNA to a region where RNA polymerase has better access to it, perhaps by binding proteins fixed in the cell structure. Finally, it is possible that subtle differences in DNA structure imposed by specific sequences are transmitted a considerable distance along the helix, so that a particular upstream sequence might affect the detailed DNA structure in the -10 and -35 regions.

Superhelical Tension of the Template Affects Promoter Efficiency[19]

Since RNA polymerase must unwind DNA at the promoter before it initiates RNA synthesis, conditions that help or hinder unwinding should increase or decrease the rate of initiation. An important condition in the cell is the superhelical tension of the DNA template: The enzyme DNA gyrase keeps bacterial DNA negatively supertwisted, thus favoring unwinding. Simultaneously, negative supercoiling is kept in check by topoisomerase I, an enzyme that *relaxes* excess negative supercoils (see Chapter 9). Impairing the activity of these enzymes by mutation or by inhibitors generally affects promoter function in the expected way. Thus, the DNA gyrase inhibitor nalidixic acid inhibits expression of many genes. Mutations in the gene encoding topoisomerase I can have the opposite effect, increasing the expression of some genes. A mutation (*supX*) in the bacterium *Salmonella typhimurium* selected to reverse the effect of a defective leucine operon promoter turned out to inactivate topoisomerase I, providing direct evidence that the function of the leucine promoter is sensitive to supercoiling.

It is not always true that increased negative supercoiling stimulates promoter function; for some promoters, the opposite is true. In these cases, we guess that negative supercoiling changes the detailed structure of DNA at the promoter in a way that inhibits RNA polymerase function much more than easier melting of DNA stimulates it. The promoter from which mRNA for DNA gyrase is made is believed to have this property. Here there is a regulatory advantage: Synthesis of DNA gyrase decreases when there is enough in the cell to keep DNA negatively supercoiled to the proper extent.

RNA Chains Usually Initiate with pppA or pppG

Actual chain initiation occurs at one extreme of the region to which *E. coli* RNA polymerase binds, within the melted segment (Figures 13-8 and 13-10) and about 12 or 13 bases from the beginning of the −10 region. Thus, the catalytic center of the enzyme appears to be at its far end, toward the direction of RNA synthesis. The first nucleotide is usually pppA or pppG, but it is sometimes pppC and rarely pppU. Occasionally, initiation occurs at several adjacent nucleotides; thus, the enzyme is somewhat flexible, and it may scan part of the melted region to find a purine start, accepting a pyrimidine only as a second choice.

The nucleotide at the 5′ end of an RNA molecule isolated from a cell is not necessarily the one that initiated transcription. After synthesis, many RNA chains are cleaved by endonucleases to yield smaller fragments, many of which start with a pyrimidine. For example, although some tRNA chains start with U (Chapter 14), they never contain terminal triphosphate groups; instead, they are made by in vivo cleavage of longer chains. As virtually all cells contain a variety of nucleases, unambiguous data about initiating nucleotide sequences are likely to come from in vitro experiments, where, using highly purified enzymes, it is possible to exclude nuclease effects.

σ Dissociates After the First Nucleotides Are Joined[3, 20]

After chain elongation begins, σ dissociates from the core enzyme–DNA–nascent RNA complex and becomes free to attach to another core molecule (Figure 13-11). Conceivably, it is released merely because it no longer has any role to play. Alternatively, its continued presence might result in such tight binding to the promoter sequence

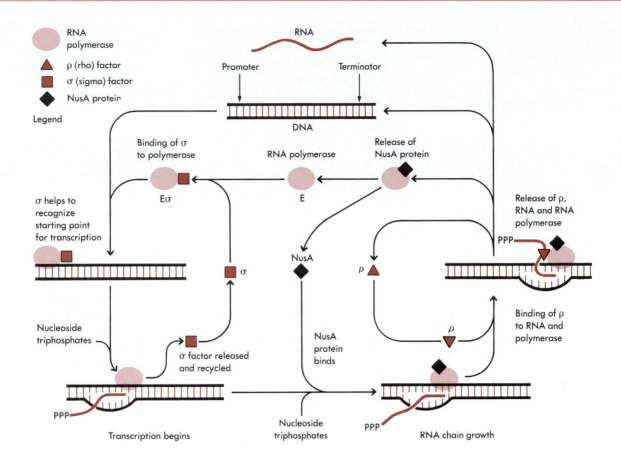

that subsequent enzyme movement along the template would be effectively impossible. Chain elongation may best occur when the nascent chain–enzyme complex is held to the DNA template by the much weaker and sequence-unspecific interactions that occur between the core enzyme and DNA.

Although holoenzyme binds much more tightly to the promoter than core enzyme does to any DNA, the opposite behavior characterizes binding to non-promotor DNA. Thus, σ suppresses the nonspecific DNA binding activity of RNA polymerase, allowing the enzyme to travel lightly along the helix until it finds a promoter sequence. This weaker binding is important to the way holoenzyme finds a promoter quickly. By "diffusing" along DNA in one dimension, the holoenzyme encounters a promoter much faster than it could by diffusing around the cell in three dimensions.

Since transcribing RNA polymerase molecules lack σ, and perhaps half of the enzyme molecules are active in a cell in normal growth conditions, there is no reason that the number of σ and core subunits in the cell should be equal. Nonetheless, the numbers are not too different; probably a pool of free σ is maintained so that holoenzyme is quickly reconstituted when core enzyme is released at a termination site.

DNA Is Unwound and Rewound by RNA Polymerase as the Enzyme Advances[9, 10, 21, 22]

RNA polymerase unwinds DNA continuously as the enzyme extends the growing RNA chain, exposing the template strand against which RNA precursors are polymerized. The growing point of an RNA

Figure 13-11
RNA synthesis. Here is a case where chain termination requires the ρ factor. In many other situations, RNA polymerase by itself can read the terminating signal.

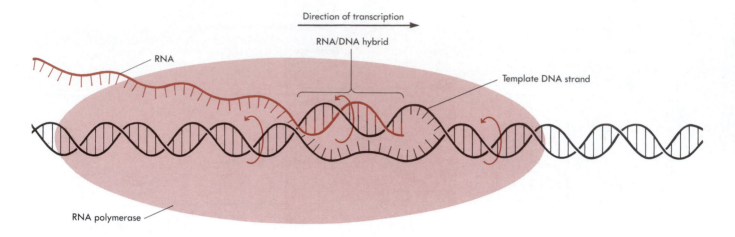

Direction of transcription

RNA/DNA hybrid

RNA

Template DNA strand

RNA polymerase

Figure 13-12
A model of the elongating complex of RNA polymerase, template DNA, and emerging RNA strand. Curved arrows indicate the rotation required for DNA to unwind as the template strand is exposed, for the RNA-DNA hybrid to form and unwind, and for the DNA strands to rewind again as the RNA chain is expelled. Although the enzyme is shown to be extended along nearly 60 nucleotides, as it is at the promoter, it is believed that only about 30 nucleotides of DNA actually are covered in the elongating complex. [After H. Gamper and J. Hearst, *Cell* 29 (1982):86, with permission.]

chain begins a region of RNA-DNA hybrid about 12 bases long, which then ends when the RNA chain leaves the template strand and the two DNA strands are rewound into double helix (Figure 13-12). The amount of DNA kept unwound by the elongating enzyme is a bit more than the length of the RNA-DNA hybrid, about 17 bases, the same amount that is unwound initially as the open complex forms at the promoter. DNA unwinding by RNA polymerase can be conveniently measured by allowing a topoisomerase to relax covalently closed circular DNA while it is being transcribed, then adding a detergent to remove the RNA polymerase and RNA, and finally measuring the number of supercoils in the DNA (Figure 13-13). One new negative supercoil appears for each 10.4 base pairs of DNA unwound at the time the topoisomerase acted; thus, by knowing the number of active enzymes, we can calculate the number of base pairs that each unwinds.

Although the elongating complex generally keeps about 17 bases of DNA unwound, this number can change dramatically when the RNA polymerase encounters special sequences. At terminators, for example, the DNA strands completely rewind and enzyme and RNA are released from the complex. The opposite occurs during transcription of the RNA primer for replication of plasmid ColE1, where rewinding of the DNA is inhibited at a specific site, and the remaining transcript remains as an RNA-DNA hybrid several hundred nucleotides long. It is not known why this occurs, although specific secondary structures of the preceding transcript seem to be involved.

The Transcription Factor NusA May Replace σ During Elongation and Termination of RNA Synthesis[23, 24, 25]

An RNA polymerase accessory factor named **NusA protein** was unrecognized for many years because the enzyme's basic activity does not require it and because NusA protein does not copurify with RNA polymerase. Its importance to RNA synthesis was realized when NusA protein was shown to bind tightly to the bacteriophage λ N gene protein, a regulatory factor whose function is to prevent termination of transcription during λ growth (Chapter 17). Like σ, NusA binds core RNA polymerase, but less tightly; thus, σ displaces NusA during purification. These properties suggest that in the cell, NusA

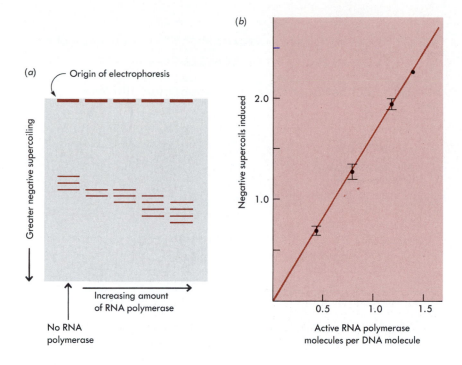

(a) Origin of electrophoresis

Greater negative supercoiling

Increasing amount
of RNA polymerase

No RNA
polymerase

(b)

Negative supercoils induced

2.0

1.0

0.5   1.0   1.5

Active RNA polymerase
molecules per DNA molecule

Figure 13-13

RNA polymerase–induced unwinding of DNA. (a) By allowing a topoisomerase to relax covalently closed, circular DNA as it is transcribed, one can trap the unwinding induced by an active RNA polymerase molecule. When the transcribing complex is disrupted by denaturation of RNA polymerase, the partially unwound DNA becomes negatively supercoiled and, being more compact, migrates faster than relaxed DNA in an agarose gel. (b) Graphing the number of supercoils induced against the number of active RNA polymerase molecules reveals the amount of unwinding per enzyme: 1.6 supercoils per RNA polymerase, or an unwinding of 17 base pairs per enzyme (1.6 turns of DNA multiplied by 10.5 base pairs per turn). [After H. Gamper and J. Hearst, *Cell* 29 (1982):84, with permission.]

binds to an RNA polymerase molecule that is making an RNA chain as soon as σ is released from it. However, NusA is not fixed irreversibly to one RNA polymerase molecule elongating an RNA chain, because a single NusA molecule can act on several RNA polymerase molecules during a transcription reaction. Therefore, even if there are no more NusA molecules in the cell than are recovered during purification (a few hundred per cell), these can be expected to act on all elongating RNA polymerase molecules. Presumably, σ again displaces NusA after RNA polymerase is released from the DNA through its greater affinity for free core enzyme (see Figure 13-11).

It is difficult to assign a specific role to NusA protein in transcription, although genetic experiments show that it is an essential protein. In vitro, NusA has subtle but distinct effects on the rate of RNA synthesis and on termination of RNA synthesis by RNA polymerase. It clearly has an important function in mediating interactions with regulatory factors such as the phage λ N gene protein.

Stop Signals Produce Chains of Finite Length[26–31]

There are signals in bacterial DNA called **terminators,** whose function is to stop RNA synthesis at specific points along the DNA template. At terminators, the enzyme stops polymerizing nucleotides into RNA, releases the RNA chain, and leaves the DNA to eventually reinitiate at another promoter. At some sites, termination requires an accessory protein, **ρ factor** (rho factor), whereas at other sites, the core enzyme alone carries out all these steps.

Much has been learned from sequence analysis of terminators in bacteria (and phages), particularly those active without ρ factor. Two distinct features of ρ-independent terminators are found: a dyad symmetry in the DNA, centered 15 to 20 nucleotides before the end of the

Figure 13-14

Sequence of a ρ-independent terminator and structure of the terminated RNA (the *trp* attenuator; Chapter 16). The mutations shown in gray partially or completely prevent termination. [After C. Yanofsky, *Nature* 289 (1981):751.]

RNA, and a run of about six As in the DNA template strand, which are transcribed into Us at the end of the RNA (Figure 13-14). Although a dyad symmetry often marks a DNA-binding site for an oligomeric protein, its importance here is that its transcript can form a hairpin loop (Figure 13-15). Experiments in vitro show that termination is prevented by incorporation of altered bases that do not pair normally, arguing that formation of the hairpin is important for termination. Furthermore, most mutations selected to destroy the function of terminators are single nucleotide changes that disrupt the regular double-helical stem (see Figure 13-14). Analysis of terminator mutations also reveals the importance of the poly dA template sequence; interrupting this sequence by a base change or removing part of it by deletion inactivates the terminator.

How could these two structural features cause termination? One model suggests that the first half of the hairpin stem anneals to and thus prematurely extracts the second half from the RNA-DNA hybrid region at the growing site, leaving only the poly U sequence annealed; since an RNA-DNA hybrid consisting of polyribo U and polydeoxy A is very unstable, the RNA chain will be quickly expelled from the DNA duplex (see Figure 13-15).

The ρ-dependent terminators lack the poly A sequence characteristic of ρ-independent ones, and not all are able to form strong hairpins. We do not know how ρ substitutes for these sequence features, but there are several hints. First, ρ hydrolyzes ATP in the presence of RNA, suggesting that it binds nascent RNA and uses the energy of ATP in its activity, perhaps to extract the RNA from the enzyme and template. Second, ρ probably binds RNA polymerase as it acts, since certain ρ proteins altered by mutations in the *rho* gene are active in terminating RNA synthesis only with RNA polymerase modified by specific mutations in the gene encoding its β subunit. Third, we know that RNA polymerase itself recognizes the ρ-dependent termination sequence in DNA, and ρ acts afterward to release the RNA: The evidence is that RNA polymerase by itself pauses at terminators that require ρ, but then moves on unless ρ is present in the reaction. One model of ρ function suggests that even a weak hairpin causes polymerase to pause, allowing ρ to attach and to dissociate the RNA and polymerase enzymatically.

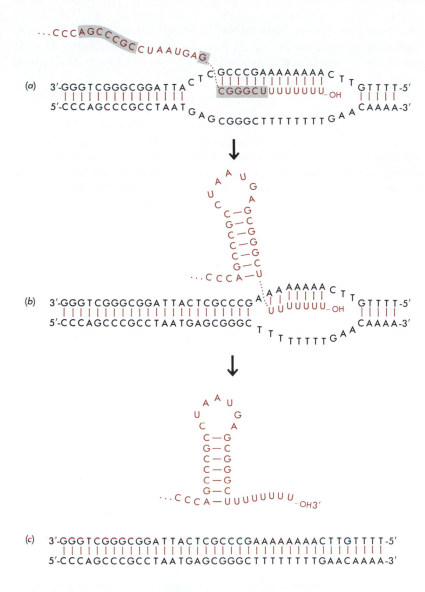

Figure 13-15
How a short duplex stem and a U-rich sequence in RNA might lead to termination and release. (a) The elongation complex has just finished synthesis of the U stretch. (b) RNA-RNA hybrid formation destroys part of the RNA-DNA elongation hybrid, leaving only the poly U stretch annealed. (c) The unstable poly U-poly dA hybrid dissociates, thus releasing the transcript. [After T. Platt, *Cell* 24 (1981):10.

There Are Three RNA Polymerases in Eucaryotic Cells

The synthesis of cellular RNA that is accomplished entirely by one RNA polymerase in bacteria is accomplished by three enzymes in all eucaryotic cells: RNA polymerases I, II, and III. Each has a specific role. RNA polymerase I (called pol I) makes ribosomal RNA; pol II makes mostly messenger RNA; and pol III makes transfer RNA, 5S ribosomal RNA, and a few other small RNA molecules. Like bacterial RNA polymerase, these are large and complex enzymes with a molecular weight of about a half million. They differ, however, in that they contain about ten, rather than five, subunits. There are some common subunits among the three enzymes, so it is likely that a more primitive enzyme became specialized in evolution for different functions in the complex eucaryotic cell. A further complexity of eucaryotic cells is that the genomes of mitochondria and chloroplasts are transcribed by their own organelle-specific RNA polymerases.

There exist for all three eucaryotic RNA polymerases, as for the bacterial enzyme, promoters and terminators that bound transcription units and regulate the rate at which they work. These signals are quite different for two, and probably for all three, of the enzymes. Promoters recognized by pol III have been well characterized, with the surprising result that the promoter is not in front of, but is instead within, the transcribed sequences (Chapter 21).

Since pol II makes messenger RNA and is thus the enzyme most directly involved in genetic regulation, there is great interest in understanding how it starts. Answers are just emerging, delayed by several difficulties. First, the enzyme as purified does not work on natural promoters; it shows only undirected transcription, reminiscent of bacterial core RNA polymerase unless other proteins, as yet not clearly defined, are added. These proteins might well include initiation factors like the bacterial σ, but the first indications are that not one, but several, different kinds of factors are missing. Second, analysis of mutants shows that the region of DNA that affects transcription is much longer than the bacterial promoter. Furthermore, certain DNA sequences important for the promoter's activity in the cell (named **enhancers**) are not active at all in the test tube. How enhancers work can only be guessed, but because they may be the key to understanding the basic features of regulation in eucaryotic cells (Chapter 21), enhancers are the subject of much current research.

Summary

RNA is made like DNA, in that an enzyme (RNA polymerase) links monomeric nucleoside triphosphate precursors into an RNA chain for which one DNA strand serves as a template. Although eucaryotic cells have three enzymes to make different types of RNA, bacterial cells have only one. *E. coli* RNA polymerase is a large enzyme of 450,000 molecular weight, constructed of five types of polypeptide chains: β, β', α, ω, and σ. The capacity to make RNA chains resides in the core enzyme, $\beta\beta'\alpha_2\omega$, whereas the role of the loosely bound σ subunit (σ factor) is to direct initiation to natural starting sites, called promoters. Besides the predominant σ factor, which is nearly as abundant as the core enzyme subunits, there are minor σ factors in *E. coli* and *B. subtilis* that direct initiation to distinct promoters.

Promoters have greatly varying efficiencies, corresponding to the different rates at which genes are expressed. Inspection of many promoters reveals a consensus sequence, a hypothetical ideal sequence that actual promoters approximate. The consensus sequence consists of two segments of six nucleotides each, located 35 and 10 bases before the RNA start site. Strong promoters match the consensus closely, whereas weak promoters match it less well. Most mutations affecting promoter function change bases in the consensus segments. Chemical and nuclease protection experiments show that RNA polymerase binds to both parts of the consensus sequence. Promoters recognized by variant σ factors have distinct consensus sequences, indicating that σ carries the specificity for promoter recognition. Promoters are the targets of many regulatory proteins, such as repressors, which prevent RNA polymerase binding, and activators, which stimulate RNA polymerase binding and thereby increase the rate of promoter function.

Initiation of RNA synthesis involves distinct steps. First, RNA polymerase binds the promoter in a closed complex. The closed complex is then converted to an open complex, in which a segment of DNA at the starting sequence is unwound. And finally, the first nucleotide bond is made. As RNA polymerase progresses along the DNA, a 17-base-pair region of DNA is kept unwound, continuously exposing the template DNA strand. The strands rewind as RNA polymerase passes, freeing the single-stranded RNA product. After initiation of RNA synthesis, σ is released from the enzyme and may be replaced by the transcription factor NusA protein during elongation of the RNA chain. NusA protein affects the behavior of RNA polymerase during elongation and termination of RNA synthesis, and it mediates the function of transcription antiterminators like the phage λ N gene protein.

Transcription ends at terminators, sequences that stop RNA polymerase and release both RNA and enzyme from the DNA. At many bacterial terminators, called ρ-independent terminators, RNA polymerase carries out these steps by itself. At other terminators, ρ-dependent terminators, RNA synthesis stops only if the termination factor ρ acts on the transcribing complex. The ρ-independent terminators consist of a symmetrical DNA sequence that forms a hairpin loop in the RNA, followed by an AT-rich segment encoding a stretch of uridines in the RNA, at the end of which the RNA is released. RNA polymerase also recognizes and pauses at ρ-dependent terminators, whereupon ρ binds the emerging RNA and catalyzes its release from the transcribing complex.

Bibliography

General References

Chamberlin, M. 1982. "Bacterial DNA-Dependent RNA Polymerases." In *The Enzymes*, vol. 15, part B, ed. P. Boyer. New York: Academic Press, p. 61.

Lewis, M. K., and R. R. Burgess. 1982. "Eukaryotic RNA Polymerases." In *The Enzymes*, vol. 15, ed. P. Boyer. New York: Academic Press, pp. 109–153.

Losick, R., and M. Chamberlin, eds. 1976. *RNA Polymerase*. Cold Spring Harbor, N.Y.: Cold Spring Harbor Laboratory. A collection of reviews and research articles about RNA polymerase and transcription, somewhat out of date but still very useful.

McClure, W. 1985. "Mechanism and Control of Transcription Initiation in Prokaryotes." *Ann. Rev. Biochem.* 54:171–204.

Platt, T. 1986. "Transcription Termination and the Regulation of Gene Expression." *Ann. Rev. Biochem.* 55, in press.

Von Hippel, P., D. Bear, W. Morgan, and J. McSwiggen. 1984. "Protein-Nucleic Acid Interactions in Transcription: A Molecular Analysis." *Ann. Rev. Biochem.* 53:389–446. An advanced review of the physical chemistry of transcription.

Cited References

1. Taylor, K., Z. Hradecna, and W. Szybalski. 1967. "Asymmetric Distribution of the Transcribing Regions on the Complementary Strands of Coliphage λ DNA." *Proc. Nat. Acad. Sci.* 57:1618–1625.

2. Burton, Z. F., C. Gross, K. Watanabe, and R. Burgess. 1983. "The Operon That Encodes the Sigma Subunit of RNA Polymerase Also Encodes Ribosomal Protein S21 and DNA Primase in *E. coli* K12." *Cell* 32:335–349.

3. Burgess, R., A. A. Travers, J. J. Dunn, and E. K. F. Bautz. 1969. "Factor Stimulating Transcription by RNA Polymerase." *Nature* 222:537–540.

4. Schmitz, A., and D. Galas. 1979. "The Interaction of RNA Polymerase and lac Repressor with the lac Control Region." *Nucleic Acid Res.* 6:111–137.

5. Siebenlist, U., R. Simpson, and W. Gilbert. 1980. "*E. coli* RNA Polymerase Interacts Homologously with Two Different Promoters." *Cell* 20:269–281.

6. Rodriguez, R., and M. Chamberlin, eds. 1982. *Promoters: Structure and Function.* New York: Praeger.

7. Youderian, P., S. Bouvier, and M. Susskind. 1982. "Sequence Determinants of Promoter Activity." *Cell* 30:843–853.

8. Hawley, D. K., and W. R. McClure. 1983. "Compilation and Analysis of *Escherichia coli* Promoter DNA Sequences." *Nucleic Acid Res.* 11:2237–2255.

9. Melnikova, A. F., R. Beabealashvilli, and A. D. Mirzabekov. 1978. "A Study of Unwinding of DNA and Shielding of the DNA Grooves by RNA Polymerase by Using Methylation with Dimethylsulphate." *Eur. J. Biochem.* 84:301–309.

10. Gamper, H., and J. Hearst. 1982. "A Topological Model for Transcription Based on Unwinding Angle Analysis of *E. coli* RNA Polymerase Binary, Initiation, and Ternary Complexes." *Cell* 29:81–90.

11. Losick, R., and J. Pero. 1981. "Cascades of Sigma Factors." *Cell* 25:582–584.

12. Johnson, W., C. Moran, and R. Losick. 1983. "Two RNA Polymerase Sigma Factors from *Bacillus subtilis* Discriminate Between Overlapping Promoters for a Developmentally Regulated Gene." *Nature* 302:800–804.

13. Elliot, T., and E. P. Geiduschek. 1984. "Defining a Bacteriophage T4 Late Promoter: Absence of a "-35" Region." *Cell* 36:211–219.

14. Grossman, A., J. Erickson, and C. Gross. 1984. "The *htpR* Gene Product of *E. coli* Is a Sigma Factor for Heat-Shock Promoters." *Cell* 38:383–390.

15. Kassavetis, G., and E. Geiduschek. 1984. "Defining a Bacteriophage T4 Late Promoter: The T4 Gene 55 Protein Suffices for Directing Late Promoter Recognition." *Proc. Nat. Acad. Sci.* 81:5101–5105.

16. Landick, R., V. Vaughn, E. Lau, R. Van Bogelen, J. Erickson, and F. Neidhardt. 1984. "Nucleotide Sequence of the Heat Shock Regulatory Gene of *E. coli* Suggests Its Protein Product May Be a Transcription Factor." *Cell* 38:175–182.

17. Hawley, D., and W. McClure. 1980. "*In vitro* Comparison of Initiation Properties of Bacteriophage λ Wild-Type P_R and X3 Mutant Promoters." *Proc. Nat. Acad. Sci.* 77:6381–6385.

18. Lamond, A. I. and A. A. Travers. 1983. "Requirement for an Upstream Element for Optimal Transcription of a Bacterial tRNA Gene." *Nature* 305:249–250.

19. Menzel, R., and M. Gellert. 1983. "Regulation of the Genes for *E. coli* DNA Gyrase: Homeostatic Control of DNA Supercoiling." *Cell* 34:105–113.

20. Travers, A. A., and R. R. Burgess. 1969. "Cyclic Reuse of the RNA Polymerase Sigma Factor." *Nature* 222:537–540.

21. Hanna, M., and C. Meares. 1983. "Topography of Transcription: Path of the Leading End of Nascent RNA Through the *Escherichia coli* Transcription Complex." *Proc. Nat. Acad. Sci.* 80:4238–4242.

22. Masukata, H., and J. Tomizawa. 1984. "Effects of Point Mutations on Formation and Structure of the mRNA Primer for ColE1 DNA Replication." *Cell* 36:513–522.

23. Greenblatt, J., and J. Li. 1981. "The *nusA* Gene Protein of *Escherichia coli*: Its Identification and a Demonstration That It Interacts with the Gene N Transcription Anti-Termination Protein of Bacteriophage Lambda." *J. Mol. Biol.* 147:11–23.

24. Greenblatt, J., and J. Li. 1981. "Interaction of the Sigma Factor and the *nusA* Gene Protein of *E. coli* with RNA Polymerase in the Initiation-Termination Cycle of Transcription." *Cell* 24:421–428.

25. Schmidt, M., and M. Chamberlin. 1984. "Amplification and Isolation of *E. coli nusA* Protein and Studies of Its Effects on in vitro RNA Chain Elongation." *Biochemistry* 23:197–203.

26. Roberts, J. W. 1969. "Termination Factor for RNA Synthesis." *Nature* 224:1168–1174.

27. Lowery-Goldhammer, C., and J. P. Richardson. 1974. "An RNA-Dependent Nucleoside Triphosphate Phosphohydrolase (ATPase) Associated with Rho Termination Factor." *Proc. Nat. Acad. Sci.* 71:2003–2007.

28. Rosenberg, M., D. Court, H. Shimatake, C. Brady, and D. Wulff. 1978. "The Relation Between Function and DNA Sequence in an Intercistronic Regulatory Region in Phage λ," *Nature* 272:414–422.

29. Fisher, R., and C. Yanofsky. 1983. "Mutations of the β Subunit of RNA Polymerase Alter Both Transcription Pausing and Transcription Termination in the *trp* Operon Leader Region *in vitro*." *J. Biol. Chem.* 258:8146–8150.

30. Platt, T., and D. Bear. 1983. "The Role of RNA Polymerase, Rho Factor, and Ribosomes in Transcription Termination." In *Gene Function in Procaryotes*, eds. J. Beckwith, J. Davies, and J. Gallant. Cold Spring Harbor, N.Y.: Cold Spring Harbor Laboratory, pp. 123–162.

31. Ryan, T., and M. Chamberlin. 1983. "Transcription Analyses with Heteroduplex *trp* Attenuator Templates Indicate That the Transcript Stem and Loop Structure Serves as the Termination Signal." *J. Biol. Chem.* 258:4690–4693.

CHAPTER 14

Involvement of RNA in Protein Synthesis

We will now begin to look at how single-stranded RNA molecules function during the process of **translation,** or protein synthesis. When the central dogma DNA → RNA → protein was gaining validity during the mid-1950s, there was general belief that all RNA was template RNA. In addition, it was hoped that when the RNA general structure was solved, mere inspection might tell us how RNA orders amino acid sequences. Now, however, we realize that these views were naive and that protein synthesis is a much more complicated process than the synthesis of nucleic acid. Moreover, not all RNA molecules are templates. Besides a template class, there are two other classes of RNA that play vital roles in protein synthesis.

Amino Acids Have No Specific Affinity for RNA

The fundamental reason behind the complexity of translation is that there is no specific affinity between the side groups of many amino acids and the purine and pyrimidine bases found in RNA. For example, the hydrocarbon side groups of the amino acids alanine, valine, leucine, and isoleucine do not form hydrogen bonds and would be actively repelled by the amino and keto groups of the various nucleotide bases. Likewise, it is hard to imagine specific RNA surfaces with unique affinities for the aromatic amino acids phenylalanine, tyrosine, and tryptophan. It is thus impossible for these amino acids, in unmodified form, to line up passively in a specific and accurate order against an RNA template prior to peptide bond formation.

Amino Acids Are Aligned on RNA Templates by Means of Adaptors[1]

Before the amino acids line up against the RNA template, they are chemically modified to possess a specific surface capable of combining with a specific number of the bases along the template. This chemical change consists of the addition of a specific adaptor molecule to each amino acid through a single covalent bond. It is this adaptor component that interacts with the template; at no time does the amino acid side group itself make contact with the template. Adding a specific adaptor residue to an amino acid is much more economical than chemically modifying the side group itself. The latter process might require many enzymes for just a single amino acid, and a similar number might be required to change the modified side group back to its original configuration after it became part of a polypeptide chain. On the other hand, only a single enzyme is needed either to attach or to detach an amino acid from its specific adaptor.

Specific Enzymes Recognize Specific Amino Acids[2, 3]

There need not be any obvious relation between the shape of the amino acid side group and the adaptor surface. Instead, the crucial selection of an amino acid is done by a specific enzyme. The enzyme that catalyzes the attachment of the amino acid to its adaptor must be able to bind specifically to both the amino acid side group and the adaptor. Proteins are extremely suitable for this task because their active regions can be rich in either hydrophilic or hydrophobic groups. There is no difficulty in folding a suitable polypeptide chain to produce a cavity that is specific for the side group of one particular amino acid. For example, tyrosine can be distinguished from phenyl-alanine by an enzyme having a specific cavity containing an atom that can form a hydrogen bond to the OH group on tyrosine. The formation of one hydrogen bond yields about 4 to 5 kcal/mol of energy, which would be lost if phenylalanine were chosen instead. Thus, with the help of a physical-chemical theory, we can predict that the probability that tyrosine is found in the "tyrosine cavity" is about a thousand times greater than the probability that phenylalanine is found in the tyrosine cavity.

There is more difficulty in immediately seeing how a similar accuracy can be achieved in distinguishing between amino acids differing only by one methyl residue, a group incapable of forming either salt linkages or hydrogen bonds. For example, glycine must be distinguished from alanine, and valine from isoleucine. There is, of course, no difficulty in understanding why the larger alanine side group cannot fit into the cavity designed for the smaller amino acid glycine. The problem arises when we ask why glycine will not sometimes fit into the alanine cavity, or valine into the isoleucine hole. If this should happen, there would be loss of the van der Waals forces arising out of a snug fit around a methyl group. These forces are now thought to generate about 2 to 3 kcal/mol, by themselves too small in value to account for the accuracy with which amino acids are ordered during protein synthesis. The maximum frequency with which a wrong amino acid is inserted into a growing polypeptide chain is between 1 in 1000 and 1 in 10,000. This means that the energy gained by selecting the correct amino acid must be at least 4 to 6 kcal, about twice the energy provided by the apparent van der Waals energy. Clearly, the process that discriminates between such amino acids is more complicated than initially conceived.

This long-troublesome paradox can be resolved if all such selection processes occur in multiple steps, providing several opportunities to reject each wrongly inserted amino acid. The combination of two steps, each of which utilizes 2 to 3 kcal of binding energy, would produce the 4 to 6 kcal necessary to explain the 10^{-3} error rate. Later we will see that many supposedly one-step discriminations do in fact involve several successive steps. This is true both of the chemical reactions involved in attaching amino acids to their adaptors and of the processes that align the adaptors on the template.

The Adaptor Molecules Are Themselves RNA Molecules[1, 4]

The molecules to which the amino acids attach are a group of relatively small RNA molecules called **transfer RNA** (**tRNA**). It is really not surprising that the adaptors are also RNA molecules, since a prime requirement for a useful adaptor is the ability to interact specifically with the free keto and amino groups on the single-stranded template RNA molecules. This interaction is ideally accomplished if the adaptor is also a single-stranded RNA molecule, since this allows the possibility of having very specific hydrogen bonds (perhaps of the base-pair variety) hold the template and adaptor together temporarily. The tRNA adaptors for the different amino acids all have different structures, each uniquely adapted for fitting a different nucleotide sequence on the template. Thus, there are many different types of tRNA.

Transfer RNA Molecules Contain Approximately Seventy-Five Nucleotides[5, 6]

By now, the sequences of several hundred tRNAs from a variety of different organisms have been determined. Each tRNA molecule con-

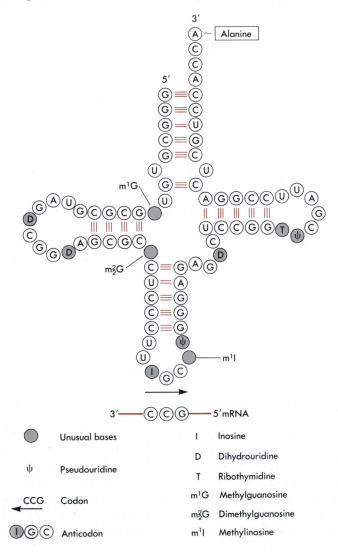

Figure 14-1

The complete nucleotide sequence of yeast alanine tRNA showing the unusual bases and anticodon position.

tains between 73 and 93 nucleotides (with a total molecular weight of 25 to 30 kdal) linked together in a single, covalently bonded chain (Figure 14-1). Although the exact sequence of residues varies, certain positions are conserved from one tRNA to another (Figure 14-2). For instance, one end of the chain, the 3' end, always terminates in a CCA sequence (cytidylic acid, cytidylic acid, adenylic acid). The other end, the 5' end, always carries a 5'-terminal monophosphate group and is often guanylic acid.

A striking aspect of all tRNA sequences is their high content of unusual bases (Figure 14-3). (By "unusual" we mean a base other than A, G, C, or U.) Many of these unusual bases differ from normal bases by the presence of one or more methyl (CH_3) groups. These methyl groups are added enzymatically after the nucleotides are linked together by 3'–5' phosphodiester bonds. Likewise, the other unusual bases arise by the enzymatic modification of a preexisting polynucleotide. Although the function of most of the unusual bases is not yet clear, some have been shown to play important regulatory roles in tRNA function.

Cloverleaf Representation of tRNA Molecules[6]

Although there is only one chain in a tRNA molecule, a majority of the bases are hydrogen-bonded to each other. Hairpin folds bring bases on the same chain into a double-helical arrangement where short stretches of nucleotides are complementary to one another. However, some nucleotides do not base-pair internally and are available to interact with other molecules during the functioning of tRNA in protein synthesis.

The exact nucleotide sequence of any one tRNA does not provide sufficient information for us to guess the way various bases hydrogen-bond to each other. A number of possible hairpin configurations can be imagined. But after the first few tRNA sequences had

Figure 14-2
A diagram of all tRNA sequences except for initiator tRNA. The positions of invariant and semi-invariant bases are shown. The numbering system is that of yeast tRNA^{Phe}. Y stands for pyrimidine, R for purine, and H for a hypermodified purine. R_{15}^+ and Y_{48}^+ are usually complementary. The dotted regions α and β in the D loop and in the variable loop contain different numbers of nucleotides in various tRNA sequences.

Figure 14-3
The structures of six common modified nucleotides found in yeast alanine or other tRNAs.

been determined, it was realized that there was only one way to fold the chain in two dimensions that would maximize the number of base pairs and lead to a common shape—the cloverleaf (see Figures 14-1 and 14-2).

Each cloverleaf contains four hydrogen-bonded stems and a number of non-hydrogen-bonded sections:

1. The 3' end consists of CCA$_{OH}$. This sequence plus a variable fourth nucleotide extend beyond the stem formed by base-pairing the 5' and 3' segments of the molecule. The amino acid always becomes attached to the 3'-terminal A.

2. As we move along the backbone away from the 3' end, we encounter TψC loop, the first loop of the cloverleaf. It is made up of 7 unpaired bases and nearly always contains the sequence 5'-TψCG-3'. Binding to the ribosomal surface may involve this region of the tRNA.

3. Next appears a loop of highly variable size, called the extra, or variable, loop.

4. The third loop, which consists of 7 unpaired bases, contains the **anticodon**—the 3 adjacent bases that bind (by base pairing) to the successive bases that make up a codon in messenger RNA. The anticodon is bracketed on its 3' end by a purine and on its 5' end by U. The purine is often alkylated.

5. The fourth loop contains 8 to 12 unpaired bases and characteristically contains the modified base dihydro-U (hence the name D loop).

Note that nearly all the invariant nucleotides in tRNA molecules occur in these non-hydrogen-bonded regions (see Figure 14-2).

Figure 14-4

Tertiary interactions in yeast phenylalanine tRNA. (a) The molecule is drawn in the conventional cloverleaf configuration with solid lines connecting bases that are hydrogen-bonded. (b) Sequence rearranged to show continuous stacking of the anticodon stem on the D stem and of the acceptor stem on the TψC stem. Note the close interaction between the TψC and D loops. [After A. Rich and S. H. Kim, *Ann. Rev. Biochem.* 45 (1976):805.]

(a)

(b)

The Three-Dimensional Structure of Yeast Phenylalanine tRNA[6-9]

Discovery of the cloverleaf pattern does not by itself tell us how the various tRNA loops are arranged in space. This information can only come from X-ray diffraction studies of tRNA crystals. Fortunately, very good crystals of yeast phenylalanine tRNA (tRNA^{Phe}) obtained in several laboratories allowed elucidation of its three-dimensional (tertiary) structure to atomic resolution. Much delight came from the finding that all the double-helical stems predicted by the cloverleaf model for tRNA^{Phe} do in fact exist (Figures 14-4 and 14-5). Moreover, these stem regions conform to the A-type helix structure expected for RNA-RNA duplexes (Chapter 9). Also, an actual GU base pair was visualized in the acceptor stem of the tRNA^{Phe} structure (see Figure 14-4).

In addition to the base-pairing interactions in the stems, there is a series of additional hydrogen bonds that bends the cloverleaf into a stable tertiary structure with a roughly L-shaped appearance (Figures 14-4, 14-5, 14-6, and 14-7). Examination of this structure reveals the following:

- The amino acid acceptor CCA group is located at one end of the L, some 70 Å from the anticodon, which occurs at the opposite end.

- The dihydrouracil-rich (D) and TψC loops form the corner of the L.

- Many tertiary hydrogen-bonding interactions involve bases that are invariant in all known tRNAs (see Figure 14-4a), strongly support-

(a)　(b)

Figure 14-5
Photographs of (a) a backbone and (b) a space-filling molecular model of yeast phenylalanine tRNA. The CCA acceptor stem is at the upper right, with the anticodon loop at the bottom. [Reproduced from Kim et al., *Science* 185 (1974):435, courtesy of S. H. Kim.]

Figure 14-6

Diagram illustrating the folding of the yeast tRNA^Phe molecule. The ribose-phosphate backbone is drawn as a continuous ribbon, and internal hydrogen bonding is indicated by crossbars. Positions of single bases are indicated by rods that are intentionally shortened. The anticodon is at the bottom of the figure, while the amino acid acceptor end is at the upper right. The anticodon and acceptor arms are shaded. The numbering of nucleotides is the same as in Figure 14-4. [After Kim et al., *Proc. Nat. Acad. Sci.* 71 (1974):4970.]

ing the belief that all tRNAs have basically the same tertiary configurations.

- Most of the hydrogen bonds that form the tertiary structure involve base pairs (or triplets) different from the conventional AT and GC pairs found in double-helical DNA (Figure 14-8). Other tertiary interactions involve groups of the ribose-phosphate backbone, including the 2' OH of the ribose sugars.

- Almost all the bases are so oriented that they can stack next to each other, thereby maximizing interaction between their hydrophobic flat faces (see Figure 14-7). Even the apparent unstabilized anti-

Figure 14-7

Diagram illustrating the hydrophobic stacking interactions between the nucleotides of yeast tRNA^Phe. Full stacking and partial stacking are indicated. Where adjacent stacking nucleotides are connected by a ribose-phosphate chain, the linkage is noted by a thin line. The heavy, solid lines attached to the nucleotide symbols represent in schematic fashion the purine or pyrimidine bases, while the hydrophobic interaction is indicated by the blocks between the bases. The connectivity of the molecule is indicated only by the numbering scheme. This accounts for the fact that many nucleotides appear to be unconnected. Bases not involved in stacking (bases 16, 17, 20, and 47) are omitted from the figure. [After Kim et al., *Proc. Nat. Acad. Sci.* 71 (1974):4970.]

codon region is held relatively rigid by stacking interactions. Stacking is probably as major a factor in stabilizing the tRNA configuration as are the tertiary hydrogen bonds.

- Only a few tertiary hydrogen bonds hold the anticodon stem to the remainder of the molecule, raising the possibility that the relative orientation of the anticodon region may change during protein synthesis.

Other tRNAs Have Similar Three-Dimensional Structures[10, 11, 12]

An important question raised by the three-dimensional structure of tRNA^{Phe} is whether all other tRNAs fold into the same L-shaped configuration. Indeed, subsequent studies of crystals of several other tRNAs show the same basic shape. Only the angle between the two helical arms of the L varies slightly. There are indications that this hinge may be somewhat flexible, perhaps allowing the angle to change during different steps of tRNA function. Thus, it is not surprising that all tRNAs can fit into the same cavities on the ribosome during protein synthesis or that some tRNA molecules can be activated by enzymes from quite unrelated organisms. On the other hand, proteins that discriminate between tRNAs must detect very subtle differences in the arrangement of chemical groups on the surface of the L-shaped molecule. Among the proteins that can select among tRNAs are the enzymes that activate tRNAs (aminoacyl-tRNA synthetases), those that trim and add modifications to tRNA precursor molecules during their synthesis, and those that modify and deliver a particular tRNA to the ribosome to start protein synthesis.

Attachment to tRNA by Aminoacyl-tRNA Synthetase Activates the Amino Acid[13–16]

The link between the 3'-terminal adenosine and the amino acid is a covalent bond between the amino acid carboxyl group and the terminal ribose component of the RNA (Figure 14-9). The use of the amino acid carboxyl group to attach to the adaptor has several interesting implications. In the first place, before the carboxyl group can form a peptide bond, the adaptor must be released. Thus, peptide bond formation and adaptor removal occur in a coordinated fashion. In the second place, the bond linking the tRNA to its specific amino acid is a high-energy bond, making these complexes "activated" precursors. The energy in the amino acid–tRNA bond (an aminoacyl bond) can be used in the formation of the lower-energy peptide bond.

The energy required for forming the aminoacyl bond comes from a high-energy pyrophosphate linkage ⓟ~ⓟ in ATP. Prior to formation of the AA~tRNA compounds, the amino acids are activated by enzymes called **aminoacyl-tRNA synthetases** to form amino acid adenylates (AA~AMP), in which the amino acid carboxyl group is attached by high-energy bonding to an adenylic acid (AMP) group (see Figure 14-9):

$$\text{AA} + \text{ATP} \xrightleftharpoons{\substack{\text{Aminoacyl-tRNA} \\ \text{synthetase}}} \text{AA} \sim \text{AMP} + ⓟ \sim ⓟ \qquad (14\text{-}1)$$

Figure 14-8
Various types of tertiary hydrogen bonding that are seen in yeast tRNA^{Phe}. The filled black circles represent the 1'-carbon of the ribose residues.

Figure 14-9
Activation of an amino acid by ATP and its transfer to the CCA end of its specific tRNA adaptor. Note that once attached, the amino acid can move rapidly between the 3'- and 2'-OH positions of the terminal adenosine.

The AA~AMP intermediate normally remains tightly bound to the activating enzyme until collision with a tRNA molecule specific for that amino acid. Then the same activating enzyme transfers the amino acid to the terminal adenylic acid residue of the tRNA:

$$\text{AA{\sim}AMP} + \text{tRNA} \underset{}{\overset{\substack{\text{Aminoacyl-tRNA}\\ \text{synthetase}}}{\rightleftharpoons}} \text{AA{\sim}tRNA} + \text{AMP} \quad (14\text{-}2)$$

A tRNA carrying its cognate amino acid is said to be "charged."

We therefore see that the activating enzymes are able to recognize and bind to both a specific amino acid and its tRNA adaptor. For this purpose, the enzymes must have two different combining sites: one that recognizes the side group of an amino acid and another that recognizes the tRNA specific for that amino acid. Similarly, the part of a tRNA molecule that is recognized by its activating enzyme can be totally different from the part that interacts with a specific group of template nucleotides. It follows that the amino acid side group itself never has to come into contact with a template molecule. It only needs to bind specifically to the correct activating enzyme.

Every cell needs at least 20 different kinds of activating enzymes and at least 20 kinds of tRNA molecules. There must be at least one of each for every amino acid. In fact, there are usually several types of tRNA for the same amino acid. Such tRNAs are called **isoacceptors.** Their existence is connected with the fact that the genetic code often uses more than one nucleotide sequence (codon) to specify a given amino acid, a characteristic called degeneracy (Chapter 15). Frequently, there is a unique tRNA molecule for each functional codon. There need not be, however, a separate activating enzyme for each of the several tRNAs corresponding to a given amino acid. In fact, isoacceptor tRNA molecules that bind to different codons are activated by the same aminoacyl-tRNA synthetase.

Many studies have indicated that aminoacyl-tRNA synthetases bind in and around the inside of the L shape assumed by tRNA molecules. The invariant uridine (or modified U) at position 8 of the tRNA chain (see Figure 14-2) is ideally positioned to interact (and may even form a transient covalent bond) with the synthetase. Recently, cocrystals of tRNA bound to an aminoacyl-tRNA synthetase have been obtained. It is hoped that these will tell us how activating enzymes can discriminate among tRNAs for different amino acids and yet bind the several isoacceptors for a particular amino acid.

AA~tRNA Formation Is Very Accurate[17-21]

How given aminoacyl-tRNA synthetases accurately select among very similar amino acids at first seemed very mysterious. So the discovery of how isoleucyl-tRNA synthetase precisely discriminates isoleucine from valine was particularly encouraging. These amino acids differ by only single methyl groups, and so the difference in their binding energy to isoleucyl-tRNA synthetase is only 2 to 3 kcal, an amount of energy that at first seems insufficient to prevent frequent mistakes. For example, when isoleucyl-tRNA synthetase is presented with a mixture of equimolar amounts of isoleucine and valine, approximately one valine~AMP activation occurs for every 100 correctly activated isoleucine~AMP complexes. If all these activated valine molecules were later transferred to isoleucine-tRNA molecules, an unacceptably high level of mistakes would result. Such transfers, however, are very infrequent, for when isoleucine tRNA binds to a valine~AMP–isoleucyl-tRNA synthetase complex, almost all valine~AMP molecules fall apart into their valine and AMP components (Figure 14-10). This form of proofreading demands continued binding of the amino acid to the active site of the synthetase through both enzymatic steps catalyzed by the synthetase. Discrimination between isoleucine and valine can thus occur twice, giving rise to an acceptable $10^{-2} \times 10^{-2} = 10^{-4}$ final error level.

It now appears that proofreading occurs only in those enzymes where it is not possible to achieve sufficient accuracy in a single recognition step. For instance, we have recently learned from the three-dimensional structure of tyrosyl-tRNA synthetase that this enzyme selects between the two very similar amino acids phenylalanine and

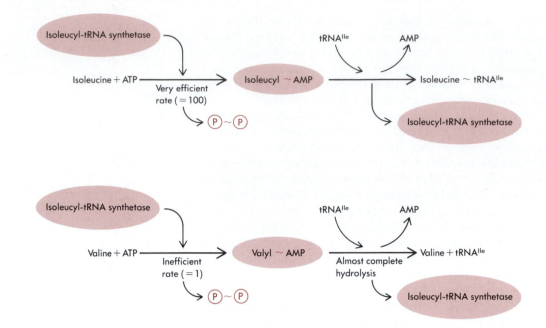

Figure 14-10

Discrimination between isoleucine and valine in the formation of isoleucyl-tRNA^{Ile}.

tyrosine by making specific hydrogen bonds to the OH group of tyrosine (Figure 14-11). Other amino acids that can be similarly discriminated without proofreading include asparagine, aspartic acid, cysteine, histidine, lysine, and tryptophan.

Once a particular amino acid is attached to its adaptor tRNA, it is not specifically recognized in any of the subsequent steps of protein synthesis. Evidence for this passive role of the amino acid came from an experiment where cysteine was first charged onto tRNA^{Cys} and then chemically converted to alanine (Figure 14-12). When this alanyl-tRNA^{Cys} was added to a hemoglobin-synthesizing reticulocyte lysate, all cysteine positions in the newly made hemoglobin were found to be occupied by alanine. Thus, the tRNA molecule is a true adaptor in that only it is selected by the template.

Aminoacyl-tRNA synthetases are an extremely diverse group of proteins with respect to both the size and number of their polypeptide subunits. Some synthetases attach the amino acid to the 3' OH, while others attach it to the 2' OH of the terminal ribose sugar of the tRNA. Once charged, however, the amino acid probably moves rap-

Figure 14-11

Hydrogen bonds made between tyrosyl-tRNA synthetase and tyrosyl adenylate. The three-dimensional structure, known from X-ray crystallography (ref. 18), reveals eleven possible hydrogen bonds that may be formed between the enzyme and the AA~AMP. In the synthetase, groups on the amino acid side chains as well as groups that are part of the main chain (MC) are involved. Not only does the tyrosine hydroxyl group make two specific hydrogen bonds, but there are van der Waals' interactions between the hydroxyl group and the enzyme that would not form if phenylalanine were bound instead. [After Fersht et al., *Nature* **314** (1985):235.]

Cys-tRNA^{Cys} Ala-tRNA^{Cys}

idly back and forth, since these two positions are chemically very similar (see Figure 14-9). Only later, when the amino acid is transferred to the growing polypeptide chain, is the position important. Then the amino acid must leave from the 3′ OH group.

Peptide Bond Formation Occurs on Ribosomes[22-25]

Once the amino acids have acquired their adaptors, they diffuse to the **ribosomes,** the spherical particles on which protein synthesis occurs. Protein synthesis never takes place free in solution, but only on the surfaces of the ribosomes, which might be regarded as miniature factories for making proteins. Their chief function is to orient the incoming AA~tRNA precursors and the template RNA so that the genetic code can be read accurately. Ribosomes thus contain specific surfaces that bind the template RNA, the AA~tRNA precursors, and the growing polypeptide chain in suitable stereochemical positions.

There are approximately 15,000 ribosomes in a rapidly growing *E. coli* cell. Each ribosome has a molecular weight of slightly less than 3 million daltons. Together the ribosomes account for about one-fourth the total bacterial cell mass, and hence a very sizable fraction of the total cellular synthesis is devoted to the task of making ribosomes. Only one polypeptide chain can be formed at a time on a single ribosome. Under optimal conditions, the production of a chain of 400 amino acids (molecular weight of about 40,000) requires about 10 seconds. The finished polypeptide chain is then released, and the free ribosome can be used immediately to make another protein.

All ribosomes are constructed from two subunits, the larger subunit being approximately twice the size of the smaller one (Figure 14-13). Both subunits contain both RNA and protein. In bacterial ribosomes, the RNA:protein ratio is about 2:1; in many other organisms, it is about 1:1. Both the large and small subunits contain a large number of different proteins. Intensive work has been done with the ribosomal proteins from *E. coli.* The 21 S proteins (labeled S1 to S21) from the smaller (30S) subunit (see Figure 14-13) show a variety of sizes (Figure 14-14). Recently, all have been sequenced, and it has become clear that each is present in only one copy per ribosome. Likewise, of the 34 L proteins (L1 to L34) in the larger (50S) subunit, most are present only once in a given ribosome. An exception is the protein called L7/L12, which is present in four copies, some of which possess acetyl groups at their N termini. L7/L12 plays a central role in the energy-utilizing steps of protein synthesis that occur on the ribosome.

Ribosomal RNA Does Not Carry Genetic Information

When ribosomes first became implicated in protein synthesis in 1953, it seemed natural to suppose that their tightly bound RNA component was the template that ordered the amino acids. In fact, it was

Figure 14-12
How the side chain of cysteine can be converted into alanine after attachment to tRNA^{Cys}. Alanines are subsequently incorporated into cysteine positions in hemoglobin molecules.

Figure 14-13

The structure of the *E. coli* ribosome. It is usually called the 70S ribosome, since 70S is a measure of how fast this ribosome sediments in the centrifuge. (S stands for the Svedberg, a measure of sedimentation rate.) Likewise, the designation 30S and 50S are the sedimentation constants of the smaller and larger ribosomal subunits; and 5S, 16S and 23S are the sedimentation constants of the smaller and larger ribosomal RNA molecules. All bacterial ribosomes have sizes similar to those of *E. coli*, possessing 30S and 50S subunits. In higher organisms, ribosomes are somewhat larger (80S), with 40S and 60S subunits.

Figure 14-14

Two-dimensional gel electropherograms of the 21 proteins from the small (left) and the 34 proteins from the large (right) *E. coli* ribosomal subunits. S20 and L26 have turned out to be the same protein; L8 represents a complex of L7/L12 with L10. In general, the size of the protein decreases with increasing number. That is, S1 is the largest protein (MW about 60,000) and S21 the smallest protein (MW about 8000) on the 30S ribosome, and L1 is the largest protein (MW about 25,000) and L34 the smallest protein (MW about 5000) on the 50S ribosome. [Courtesy of E. Kaldschmidt and H. G. Wittmann, *Proc. Nat. Acad. Sci.* 67 (1970):1276.]

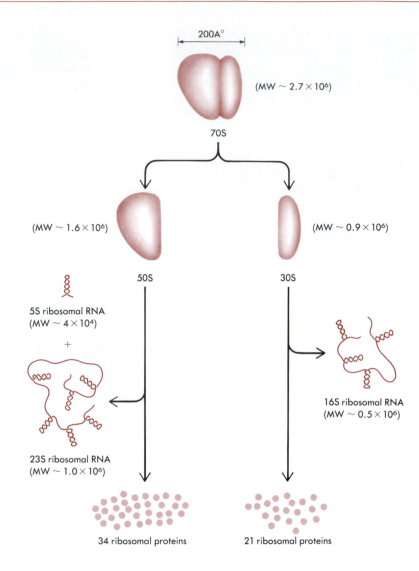

200A°

(MW ~ 2.7×10^6)

70S

(MW ~ 1.6×10^6)   50S

5S ribosomal RNA
(MW ~ 4×10^4)

+

23S ribosomal RNA
(MW ~ 1.0×10^6)

34 ribosomal proteins

(MW ~ 0.9×10^6)   30S

16S ribosomal RNA
(MW ~ 0.5×10^6)

21 ribosomal proteins

E. COLI C, 30 S

E. COLI C, 50 S

initially thought that all cellular RNA was located in the ribosomes and that the lighter, slowly sedimenting, soluble fraction (about 20 percent of total RNA) was a degradation product of the ribosomal RNA templates. The identification in 1956 of tRNA adaptor molecules in the soluble fraction corrected this latter assumption, but did not remove the belief that the remaining 80 percent of cellular RNA functioned as templates. In 1960, however, the RNA isolated from purified ribosomes (**ribosomal RNA,** or **rRNA**) was unambiguously shown not to have a template role.

Messenger RNA (mRNA) Reversibly Associates with Ribosomes[26, 27]

The active templates are an RNA fraction that account for only one to several percent of the total cellular RNA. This RNA reversibly binds to the surface of the smaller ribosomal subunit and in media of low magnesium ion concentration can be removed without affecting ribosome integrity. Because it carries the genetic message from the gene to the ribosomal factories, this particular kind of RNA is called **messenger RNA (mRNA)**. By moving across the ribosomal site of protein synthesis, mRNA brings successive codons into position to select the appropriate AA~tRNA precursors.

The existence of mRNA was first established in experiments with T2-infected *E. coli* cells. After T2 DNA enters a host cell, it must turn out RNA templates for the many virus-specific proteins needed for viral reproduction (Chapter 17). Many biochemists were therefore surprised by the 1959 finding that no new rRNA chains and hence no new ribosomes were synthesized following T2 infection. This result could only mean that T2-specific proteins are not synthesized on rRNA templates, also suggesting that in normal cells, rRNA chains are not the templates for protein synthesis. Subsequent research quickly revealed the presence of mRNA, first in virus-infected cells and soon after in normal *E. coli* cells.

Ribosomal RNA Comes in Three Basic Sizes[28]

Two large rRNA molecules and one small rRNA molecule are found in every bacterial ribosome. They are integral components, and unlike mRNA, cannot be removed without the complete collapse of the ribosome structure. The 16S rRNA molecule, found in the smaller ribosomal subunit, has a chain length of 1542 nucleotides, whereas the 23S molecule, a component of the larger ribosomal subunit, contains 2904 nucleotides. Each larger subunit contains, in addition, one very short rRNA molecule that sediments at 5S and has 120 nucleotides. All three rRNAs are single-stranded and have unequal amounts of guanine and cytosine and of adenine and uracil. Nonetheless, there is enough equivalence of base pairs so that many rRNA bases on the same chain are hydrogen-bonded into the hairpin-type turns found in tRNA.

Sequences Reveal Conserved Folding Patterns for rRNA Molecules[28, 29, 30]

We have recently arrived at a good picture of how the three ribosomal RNAs fold, despite the fact that ribosome crystal structures have not yet been worked out (see Figure 14-15a and b). The key to this under-

E. coli 16S rRNA

(a)

Bovine mitochondria 12S rRNA

(b)

Figure 14-15
A comparison of the structures of the small subunit rRNAs from (a) *E. coli* (16S rRNA) and (b) bovine mitochondria (12S rRNA). All stems are supported by comparative sequence evidence from different species. Long-range base-pairing interactions that define the four domains of this rRNA are indicated by I, II, III, and IV. Note that the domain structure is conserved despite the extreme difference in length between the two rRNAs (1542 nucleotides for 16S and 954 nucleotides for 12S). (Courtesy of R. Gutell and H. Noller.)

standing was the elucidation of complete rRNA sequences from a number of different organisms, some closely and some distantly related to *E. coli*. Although the chain lengths of the two large ribosomal RNAs vary significantly among organisms, the corresponding RNAs are often referred to as 16S-like or 23S-like.

When the 1542-nucleotide sequence of *E. coli* 16S rRNA was first determined in 1978, it was possible to propose a folding pattern based both on sequence complementarity and on the sensitivity of certain parts of the RNA sequence to ribonucleases and chemical reagents. (Single-stranded portions of the chain are more susceptible to such probes than are tightly base-paired regions.) Sequences of rRNAs from other bacterial species revealed that most of the proposed helical regions are conserved even if the exact nucleotide sequence is not. That is, if the order of residues on one side of a particular base-paired stem differs from that in *E. coli*, then a compensating change occurs on the other strand to restore complementarity. Now it appears that the 16S-like RNA from the small ribosomal subunit of all organisms folds into the same four-domain structure (Figure 14-15). These domains, each containing many stem and loop regions, are closed by long-range base-pairing interactions. Even where the chain length of the rRNA differs twofold from *E. coli*, the extra or missing nucleotides are simply accommodated by extending or truncating various loop structures (see Figure 14-15).

Similar conserved folding patterns can be formulated for the 23S-like and 5S rRNAs that reside in the large ribosomal subunit. These structures are shown in Figure 14-16. The striking conservation of each of the three ribosomal RNA folding patterns confirms our suspicion that all ribosomes are constructed in essentially the same way, despite extreme differences in size and protein composition.

The Function of Most rRNA Is Not Yet Known[28]

Until recently, there was no satisfactory hypothesis for why such a large fraction of the mass of ribosomes should be rRNA rather than protein. But now we believe that many of the unpaired bases in rRNA are involved in the binding of other RNA molecules to ribosomes. As we shall see later, several unpaired bases near the 3' end of 16S rRNA form temporary base pairs with the initiation sites (ribosome binding sites) of mRNA molecules. Likewise, rRNA sequences may interact with invariant tRNA sequences during tRNA binding. Also, rRNA-rRNA bonds may help hold the two ribosomal subunits together during translation. Other rRNA residues may form base pairs with small RNA-protein complexes involved in directing newly synthesized proteins to compartments beyond the cell membrane.

Despite all these RNA-RNA interactions, we are still very far from understanding what the hundreds of unpaired nucleotides of each rRNA component are doing. Even the ribosomal proteins that help to hold the rRNA in a well-defined three-dimensional shape seem to prefer to bind to the paired nucleotides that form the conserved stem structures of the folded rRNAs. The constancy of rRNA folding patterns across all species underscores the essential, albeit not yet well understood, role of RNA molecules in ribosome function.

Self-Assembly of Ribosomes in the Test Tube[23, 31–33]

The complete reassembly of the smaller *E. coli* subunit from its RNA and protein constituents was first accomplished in 1968. Such reconstituted particles are active in protein synthesis, showing identical behavior to normal 30S subunits. The fact that no additional components except for the 16S rRNA and the 21 proteins present in the final particle are required shows that the test-tube reconstitution of 30S ribosomes is a true "self-assembly" process. These studies further revealed that certain proteins can bind directly to the rRNA, while the assembly of others requires the prior association of other S proteins (Figure 14-17). This does not necessarily mean that one protein directly contacts another in the ribosome. Rather, the first proteins that assemble may induce conformational changes in the rRNA that create or expose binding sites for the remaining proteins. For instance, protein S4, which is required early in the reconstitution of 30S ribosomes, is believed to bind and stabilize the two helical stems that close domains I and II of the 16S folded structure shown in Figure 14-15.

(a)   16S-like rRNA

(b)   5S rRNA

(c)   23S-like rRNA

Figure 14-16

"Minimal" conserved folding patterns for the three ribosomal RNAs. Sequences from both kingdoms of bacteria (the Archebacteria and Eubacteria) as well as from higher cells and their organelles have been included in formulating these structures. Gaps indicate regions where non-conserved sequences of various lengths are present in individual molecules. Currently over 20 complete 16S-like, about 15 23S-like, and dozens of 5S structures are known. (Parts (a) and (c) courtesy of R. Gutell, C. Woese, and H. Noller.)

Test-tube reconstitution of functional 50S *E. coli* subunits has been much harder to achieve. Not only must a larger number of proteins be assembled in an orderly fashion, but two rRNAs (the 23S and 5S) must be built into the particle. Again, there is evidence that after about half the proteins have assembled with the two RNAs, a conformational change must occur before the remaining proteins can be properly positioned (see Figure 14-17). It is worth noting that the earliest proteins to assemble bind near the 5' end of the 23S RNA, while those that are added later interact more with the 3' portion of the molecule. Thus, ribosome assembly in the cell can occur coordinately with the 5'→3' transcription of the 23S rRNA itself.

Finding the right conditions for the test-tube assembly of the two ribosomal subparticles has been crucial in understanding the structure and function of the ribosome. For instance, partial reconstitution of particles lacking one or more specific proteins has enabled us to identify proteins involved in specific ribosomal functions (e.g., peptide bond formation and GTP hydrolysis). Similarly, mixed reconstitutions using proteins from mutant and normal ribosomes or from different bacterial species can reveal which protein carries the mutation or which proteins are functionally equivalent in ribosomes from different organisms. Finally, the ability to reconstitute ribo-

Figure 14-17
In vitro assembly pathways for the 30S and 50S *E. coli* ribosomal subparticles, illustrating the steps where energy input (conformational change) is required. In the case of 30S assembly, between 7 and 13 S proteins can be demonstrated to bind directly to the 16S rRNA, depending on the method of rRNA preparation. About 17 L proteins are thought to bind directly to the 23S rRNA and 3 to the 5S rRNA.

somes containing only a few components appropriately labeled with isotopes has been essential for the application of newer approaches to deciphering the overall structure of the ribosome.

Pre-tRNA and Pre-rRNA[34–37]

In bacteria, molecules of tRNA and rRNA are synthesized exactly as mRNA molecules are, by RNA polymerase using DNA as template. But even though rRNA and tRNA together make up over 98 percent of all RNA in the cell, less than 1 percent of the total DNA functions as their templates. This is because mature tRNAs and rRNAs are quite stable, whereas mRNAs break down rapidly.

However, neither tRNAs nor the three ribosomal RNAs are transcribed as such. Instead, they are derived from larger short-lived precursor molecules that nucleases break down into mature tRNA or rRNA molecules. The 16S, 23S, and 5S rRNAs are transcribed in that order into a single 30S pre-rRNA transcript of about 6500 nucleotides. The pre-rRNA contains additional sequences at both its 5′ and 3′ ends (called **leader** and **trailer regions**) as well as in the **spacer regions** between the three rRNA sections. All tRNAs are likewise transcribed in the form of larger precursors (Figure 14-18). Sometimes, only one tRNA with 5′ and 3′ flanking sequences is present; sometimes as many as seven different tRNAs are contained in a common pre-tRNA transcript.

In the case of pre-rRNA, it is clearly advantageous to include 16S, 23S, and 5S sequences in a single transcript to ensure that equivalent numbers of the three molecules will be available for ribosome assembly. It is not obvious why tRNAs should be made in precursor form. However, it is known that mutant pre-tRNAs that cannot be correctly converted to mature molecules often have defects that would interfere with the folding of the tRNA portion into the correct L-shaped conformation. Such mutant pre-tRNAs are rapidly degraded in the cell, providing a way of discarding defective molecules before they become mature tRNAs.

Pre-tRNAs Are Cleaved by an Enzyme That Contains a Catalytically Active RNA[34, 36, 38–41]

Since pre-tRNA molecules contain variable lengths of 5′ and 3′ flanking sequences, they must be acted upon by specific RNA processing enzymes (RNases) to create the exact termini found in mature tRNA. Most is known about an enzyme called **ribonuclease P (RNase P)**,

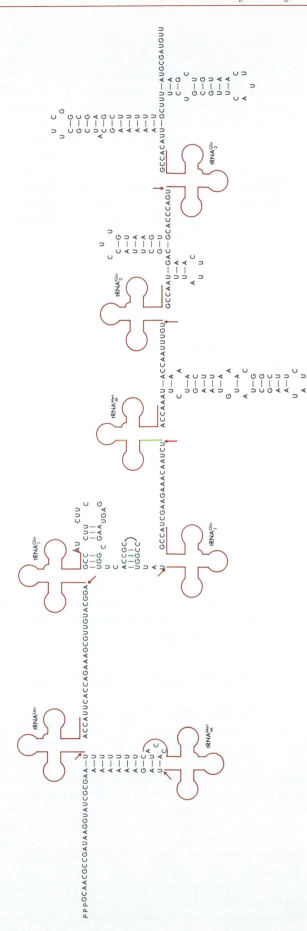

Figure 14-18

Sequence and proposed secondary structure for the pre-tRNA transcript of the *E. coli* operon called *supB-E*. The seven tRNA sequences are indicated schematically as cloverleaf structures. The duplicated genes for tRNA$^{Met}_M$, tRNA$^{Gln}_1$, and tRNA$^{Gln}_2$ have identical sequences. Note that the sequences and structures around the sites of RNase P cleavage (indicated by arrows) are all different. In *E. coli*, the CCA-3' ends of tRNAs are always encoded. The spacer regions are of different lengths and sequences. [After N. Nakajima, H. Ozeki, and Y. Shimura, *Cell* 23 (1981):239.]

which removes the extra 5′ nucleotides (see Figure 14-18). RNase P is an endonuclease that makes a single cut, simultaneously producing the 5′-P end of the tRNA and releasing the 5′ flanking sequence with a 3′-OH group. Since it acts on every *E. coli* pre-tRNA (whether containing one or several tRNAs), it must recognize some feature common to all these molecules. Yet, no particular nucleotide sequence appears either 5′ or 3′ to each cut site. Instead, RNase P probably detects certain aspects of the three-dimensional tRNA structure already correctly folded in its precursor molecule.

Purification of RNase P led to the realization that the enzyme is not a pure protein. Rather, it is a noncovalent complex of a small RNA molecule (377 residues long) and a small protein (MW about 20,000). Reconstitution studies, as well as the identification of mutants in the RNA and in the protein, have shown that both components contribute to RNase P activity under physiological conditions. A suggested secondary structure for the RNA component of *E. coli* RNase P is shown in Figure 14-19.

Most astounding is the recent discovery that the RNA component *alone* is sufficient to carry out pre-tRNA cleavage! In nonphysiological buffers containing high concentrations of magnesium ion or spermidine, the isolated RNA cuts as accurately and nearly as rapidly as the RNA-protein complex does in physiological buffers. This novel finding illustrates that proteins are not the only macromolecules that can bind substrates and catalyze reactions. Deletion and point mutation analyses of the RNase P RNA component are currently under way to pinpoint the regions that align the pre-tRNA substrate and carry out the cleavage reaction. The role of the RNase P protein also remains to be elucidated. RNA catalysis may provide an important clue to the early evolution of the gene expression apparatus.

Another tRNA processing enzyme called **RNase D** trims away the extra 3′ nucleotides from *E. coli* pre-tRNA molecules. It is an exonuclease that releases 5′ monophosphates and somehow knows to stop when it reaches the CCA end. Meanwhile, other specific enzymes introduce the many base and sugar modifications found in mature tRNA molecules. If the invariant CCA-3′ end of a mature tRNA becomes nibbled away, there also exists a specific CCA nucleotidyl transferase enzyme responsible for repairing this sequence so that an amino acid can be attached.

Pre-rRNA Molecules Also Contain tRNAs[42, 43]

Generally, only one or two specific regions on the *E. coli* chromosome encode each of the 30 to 40 different pre-tRNA molecules. In contrast, pre-rRNA chains are coded by seven different gene sets (called *rrn* operons) located at separate sites on the *E. coli* chromosome. Apparently, this number is required to provide the large amounts of ribosomal RNA needed for the rapid growth of the bacterium. The sections of the pre-rRNA that will become the mature 16S, 23S, and 5S ribosomal RNAs differ in only a very few positions (about 1 percent of the nucleotides). However the leader, trailer, and spacer regions can be strikingly different in both length and sequence from one *rrn* operon to another.

The extreme variation possible in pre-rRNA spacer and trailer regions has been underscored by the discovery of tRNA genes embedded in *rrn* operons. In the spacer between 16S and 23S sequences,

Figure 14-19
A possible secondary structure for
M1 RNA, the RNA subunit of RNase P
from *E. coli*. The structure shown was
derived from studies in solution of the
sensitivity of M1 RNA to digestion with
nucleases and comparison with the
analogous RNA from *S. typhimurium*.
The sequence of mature M1 RNA ends
at nucleotide 377. Δ indicates nucleotides deleted and the arrows point to
base differences in the M1 RNA of
S. typhimurium. (Courtesy of
S. Altman.)

either one or two tRNA genes are found; none, one, or two tRNAs are
coded in the trailer region beyond the 5S rRNA (Figure 14-20). Their
presence in pre-rRNA molecules requires a coupling of tRNA and
rRNA processing events. As in the processing of other pre-tRNAs,
RNase P makes the cleavage at the 5' ends of tRNAs in the pre-rRNA
transcripts. Why certain tRNAs have been selected for inclusion in
pre-rRNA transcripts is not clear. They are, as expected, among the
more abundant tRNA species in the *E. coli* cell.

Pre-rRNA Is First Cleaved in Double-Stranded Regions[43-47]

The 30S pre-rRNA transcript containing 16S, 23S, and 5S rRNA (as
well as one to three tRNA sequences) is normally very short-lived.
However, in mutant *E. coli* lacking an enzyme called **ribonuclease III
(RNase III)**, unusually high levels of the 30S pre-rRNA accumulate,
suggesting that RNase III plays a critical role in pre-rRNA processing.
RNase III was initially characterized as an endonuclease specific for
double-stranded RNA. Therefore, it was anticipated that analysis of
the leader, spacer, and trailer regions of pre-rRNA would reveal adjacent complementary sequences able to fold into long hairpin loop
structures that could be cut by RNase III. Instead, the complementary
sequences were unexpectedly discovered to lie thousands of nucleotides apart on the pre-rRNA transcript (see Figure 14-20). Leader sequences before the 16S coding region can base-pair with sequences
lying between 16S and the (first) spacer tRNA. Likewise, sequences

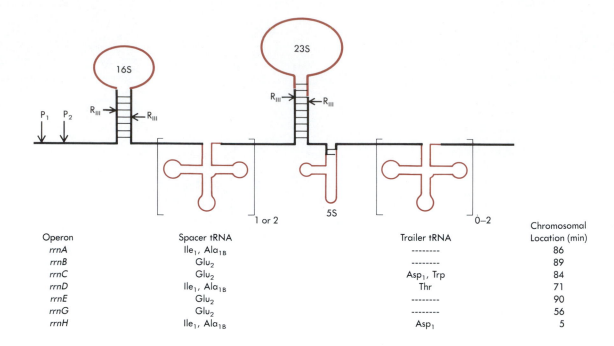

Operon	Spacer tRNA	Trailer tRNA	Chromosomal Location (min)
rrnA	Ile_1, Ala_{1B}	--------	86
rrnB	Glu_2	--------	89
rrnC	Glu_2	Asp_1, Trp	84
rrnD	Ile_1, Ala_{1B}	Thr	71
rrnE	Glu_2	--------	90
rrnG	Glu_2	--------	56
rrnH	Ile_1, Ala_{1B}	Asp_1	5

Figure 14-20

The seven ribosomal RNA (*rrn*) operons of *E. coli* K-12. Color indicates mature rRNA or tRNA sequences. R_{III} indicates sites of primary processing by RNase III. The discarded spacer regions vary in length and sequence between the different operons. P_1 and P_2 indicate the two transcription start sites located about 100 nucleotides apart. Although the *rrn* operons are distributed around the *E. coli* chromosome, the direction of their transcription is always the same as that of DNA replication. In some *E. coli* strains, *rrnH* is replaced by another operon, *rrnF*, located at 74 min; but all *E. coli* strains have a total of seven *rrn* operons. [Data from M. Ellwood and M. Nomura, *J. Bacteriol.* 149 (1982):458.]

between the (last) spacer tRNA and the beginning of 23S are complementary to sequences between the 23S and 5S rRNAs. These regions come together in the pre-rRNA transcript to form two stem structures, each of which closes a giant loop including the entire 16S or 23S rRNA molecule. RNase III then cuts at two precise sites within each stem to create separate pre-16S and pre-23S molecules (Figure 14-21).

Although the naked 30S pre-rRNA molecule contains recognition sites for RNase III cleavage, this is not the case for subsequent steps in rRNA processing. Ribosomal proteins begin to assemble on pre-rRNA soon after its transcription starts. Since complete synthesis of the 16S or the 23S sequence must precede RNase III cutting, many proteins will have already bound by the time the separate pre-16S and pre-23S rRNAs are created. Further 5' and 3' trimming of these intermediates requires the presence of ribosomal proteins. The same appears to be true of the processing steps that produce mature 5S rRNA from the 3' piece cleaved off by RNase III (see Figure 14-21). Thus, in the *E. coli* cell, the processing of the pre-rRNA and the assembly of ribosomal proteins occur concomitantly, probably via a precisely ordered series of steps, each corresponding to a slightly different ribosomal precursor particle.

Messenger RNA Molecules Come in a Large Variety of Sizes[48]

In contrast to tRNA molecules, which have molecular weights of about 2.5×10^4, and to rRNA molecules, which also have defined sizes (4×10^4, 5×10^5, and 10^6 for 5S, 16S, and 23S, respectively), mRNA molecules vary greatly in chain length and hence in molecular weight. Some of this heterogeneity reflects the large size spread in the length of polypeptide chain products. Most polypeptide chains contain 100 or more amino acids, and so most mRNA molecules must contain at least 100×3 nucleotides (because there are three nucleotides in a codon). Additional variations arise from unequal lengths of

Figure 14-21

Sequences around sites where pre-rRNA is cleaved by *E. coli* RNase III (R$_{III}$) or by *B. subtilis* RNase M5 (M5). The RNase III cuts are displaced by two base pairs within a perfect or nearly perfect helical region. Subsequently, other enzymes trim the ends down to mature 16S and 23S sequences (indicated by color type). RNase M5 makes two cuts one base pair apart in a *B. subtilis* 5S RNA precursor to create the 5' and 3' ends of mature 5S rRNA; its action requires the binding of other proteins. [After R. J. Bram, R. A. Young, and J. A. Steitz, *Cell* 19 (1980):393; and M. L. Sogin, N. R. Pace, M. Rosenberg, and S. Weissman, *J. Biol. Chem.* 251 (1976):3480.]

leader sequences at the 5′ ends and trailer sequences at the 3′ ends. These are sequences that do not code for any amino acids but which are essential for the proper synthesis and functioning of an mRNA. For example, the leader of *E. coli* β-galactosidase mRNA is 38 residues long, while that of *E. coli* tryptophan mRNA contains 162 bases. At its 3′ end, the *E. coli* tryptophan mRNA extends 35 nucleotides beyond the coding region. In *E. coli*, therefore, the mRNAs that code for average-size polypeptides of 300 to 500 amino acids usually contain between 1000 and 2000 nucleotides.

Still more heterogeneity in size comes from the fact that some mRNAs code for more than one polypeptide. Besides leader and trailer sequences, these polygenic messengers contain intergenic regions that are sometimes as long as leader sequences. In most, if not all, of these polygenic messengers, the polypeptide products have related functions. For example, a single mRNA molecule codes for the five specific enzymes needed to synthesize the amino acid tryptophan. It has recently been completely sequenced and contains about 6800 nucleotides, or an average of 1400 nucleotides coding for each enzyme and its adjacent intergenic regions.

Ribosomes Come Apart into Subunits During Protein Synthesis[22–24]

The construction of all ribosomes from easily dissociable subunits suggested a cycle during which the large and small subunits come apart at some stage in protein synthesis. This hunch was confirmed by experiments with *E. coli* and yeast that showed that most ribosomes are constantly dissociating into subunits and then reforming. Growth of cells in heavy isotopes followed by transfer to a light medium was the technique used to settle this point. Soon after transfer to the light medium, hybrid ribosomes (heavy 50S/light 30S and light 50S/heavy 30S) began to appear, the rate of their appearance suggesting that the subunits exchange once during every cycle of polypeptide synthesis. Confirmation comes from in vitro studies of protein synthesis (see Chapter 15) where the fate of heavy ribosomes can be followed in the presence of a great excess of light ribosomes. Within a minute or so (the time required to synthesize a complete polypeptide chain), almost all heavy ribosomes disappear, with the simultaneous appearance of hybrid ribosomes (Figure 14-22). The obligatory dependence of subunit exchange on protein synthesis is confirmed by in vitro experiments with the antibiotic sparsomycin. This compound, which blocks polypeptide chain elongation, also prevents any subunit exchange.

Polypeptide Chain Growth Begins at the Amino-Terminal End[49]

A completed polypeptide chain has at one end an amino acid bearing a free carboxyl group (the C terminus) and at the other end an amino acid bearing a free amino group (the N terminus). Protein synthesis by ribosomes always involves stepwise addition of single amino acids, starting with the amino terminus and ending with the carboxyl terminus (Figure 14-23). This point is clearly shown by brief exposure

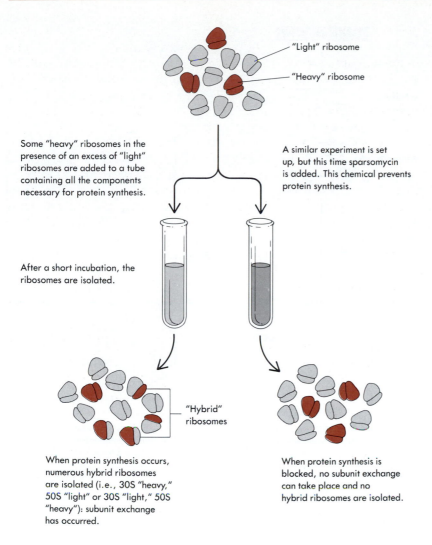

Figure 14-22
The obligatory dependence of subunit exchange on protein synthesis.

Some "heavy" ribosomes in the presence of an excess of "light" ribosomes are added to a tube containing all the components necessary for protein synthesis.

A similar experiment is set up, but this time sparsomycin is added. This chemical prevents protein synthesis.

After a short incubation, the ribosomes are isolated.

"Hybrid" ribosomes

When protein synthesis occurs, numerous hybrid ribosomes are isolated (i.e., 30S "heavy," 50S "light" or 30S "light," 50S "heavy"): subunit exchange has occurred.

When protein synthesis is blocked, no subunit exchange can take place and no hybrid ribosomes are isolated.

of hemoglobin-synthesizing reticulocytes to radioactively labeled amino acids, followed by immediate isolation of newly completed hemoglobin chains. If the time of labeling is short relative to the time required to complete a chain, very little radioactivity is found in the

Figure 14-23
Stepwise growth of a polypeptide chain. Initiation begins at the free NH_3^+ end, with the growing point of the chain (the carboxyl end) always terminated by a tRNA molecule. (Peptide bonds between amino acids are indicated by a heavy line.)

Figure 14-24

Experimental demonstration that hemo-globin chains grow in the $NH_3^+ \rightarrow COO^-$ direction. [After H. M. Dintzis, *Proc. Nat. Acad. Sci.* 47 (1961):247.]

(a) Only those chains already started have had time to be completed; they show a gradient of radioactivity, with more label at the COO^- end; no complete chains are found labeled at the NH_3^+ end.

(b) The completed chains are digested with protease, and the radioactivity in the resulting peptides is measured.

N-formylmethionine (fMet)

Methionine (Met)

Figure 14-25

The structure of *N*-formylmethionine compared with that of methionine.

amino acids of the amino-terminal end; most is found in the amino acids of the carboxyl end (Figure 14-24). Moreover, a clear gradient of increasing radioactivity is observed as one moves from the amino to the carboxyl end.

Bacterial Polypeptide Chains Start with *N*-Formylmethionine[22, 50]

The starting amino acid in the synthesis of virtually all bacterial poly-peptides is ***N*-formylmethionine** (**fMet**). This is a modified methio-nine that has a formyl group attached to its terminal amino group (Figure 14-25). A blocked amino acid like *N*-formylmethionine can be used only to start protein synthesis. The absence of a free amino group would prevent this amino acid from being inserted during chain elongation. The formyl group is enzymatically added onto methionine after the methionine has become attached to its tRNA adaptor.

Not all methionine tRNA molecules can be formylated. Instead, there are two types of tRNA^{Met}, only one of which (called tRNA$_F^{Met}$) permits the formylation reaction. Sequence analysis of tRNA$_F^{Met}$ and tRNA$_M^{Met}$ (for internal methionine) reveals that they both have the same anticodon sequence, posing the dilemma of how the same codon (AUG) can code both for the initiating amino acid (*N*-formyl-methionine) and for internal methionines in the chain (Figure 14-26).

The discovery that synthesis of bacterial proteins starts with a blocked amino acid was unexpected, because isolation of pure protein from growing bacteria revealed essentially no formylated end groups. This means that a **deformylase** enzyme exists that removes the formyl group from the growing chain very soon after synthesis commences (Figure 14-27). In addition, another enzyme (an **aminopeptidase**) subsequently removes the terminal methionine from many proteins. However, it does not act on all proteins, so a large number of bacterial proteins do have methionine at their N termini.

Messenger RNA Binding by the Smaller Ribosomal Subunit Involves rRNA-mRNA Pairing[50–53]

Initiation of protein synthesis in bacteria starts with the formation of a complex between the smaller 30S ribosomal subunit, fMet-tRNA$_F^{Met}$, and an mRNA molecule. Then a 50S subunit joins to form the functional 70S ribosome. Every bacterial mRNA molecule has one **ribosome binding site** for each of its independently synthesized polypeptide products. For example, three such sites are found on the RNA genome of phage R17, which codes for three major proteins. Within each site are nucleotide sequences whose function is the correct positioning of mRNA molecules on the ribosomal surface prior to the commencement of protein synthesis.

How does the ribosome distinguish between AUG codons at the start of a gene and those coding for internal methionines? The first known initiator sequences were established by binding ribosomes to

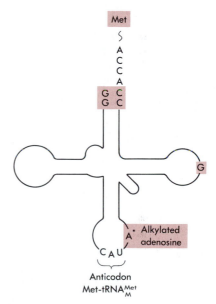

Figure 14-26
The three main points of difference between *N*-formyl-methionyl-tRNA$_F^{Met}$ and methionyl-tRNA$_M^{Met}$.

Figure 14-27
Enzymatic steps involving formylmethionine as the initiator of protein synthesis.

Phage Qβ A protein	CUG	AGU	AUA	AGA	GGA	CAU	AUG	CCU	AAA	UUA
Phage Qβ coat	CUU	UGG	GUC	AAU	UUG	AUC	AUG	GCA	AAA	UUA
Phage Qβ replicase	UUA	CUA	AGG	AUG	AAA	UGC	AUG	UCU	AAG	ACA
Phage λ Cro	AUG	UAC	UAA	GGA	GGU	UGU	AUG	GAA	CAA	CGC
Phage fl coat	UUU	AAU	GGA	AAC	UUC	CUC	AUG	AAA	AAG	UCU
Phage φX174 A	AAU	CUU	GGA	GGC	UUU	UUU	AUG	GUU	CGU	UCU
Phage φX174 A*	UUG	CUG	GAG	GCC	UCC	ACU	AUG	AAA	UCG	CGU
Phage φX174 B	AGG	UCU	AGG	AGC	UAA	AGA	AUG	GAA	CAA	CUC
Phage φX174 E	GCG	UUG	AGG	CUU	GCG	UUU	AUG	GUA	CGC	UGG
Lipoprotein	AUC	UAG	AGG	GUA	UUA	AUA	AUG	AAA	GCU	ACU
RecA	GGC	AUG	ACA	GGA	GUA	AAA	AUG	GCU	AUC	G
GalE	AGC	CUA	AUG	GAG	CGA	AUU	AUG	AGA	GUU	CUG
GalT	CCC	GAU	UAA	GGA	ACG	ACC	AUG	ACG	CAA	UUU
LacI	CAA	UUC	AGG	GUG	GUG	AAU	GUG	AAA	CCA	GUA
LacZ	UUC	ACA	CAG	GAA	ACA	GCU	AUG	ACC	AUG	AUU
Ribosomal L10	CAU	CAA	GGA	GCA	AAG	CUA	AUG	GCU	UUA	AAU
Ribosomal L7/L12	UAU	UCA	GGA	ACA	AUU	UAA	AUG	UCU	AUC	ACU
RNA polymerase β subunit	AGC	GAG	CUG	AGG	AAC	CCU	AUG	GUU	UAC	UCC

16S 3′ end $_{HO}$AUUCCUCCACUAG-5′

Figure 14-28

Examples of the several hundred known ribosome-binding sites (protein synthesis initiator regions) that function in *E. coli*. The nucleotides complementary to the 3′ end of 16S rRNA are in shaded boxes; lighter shading indicates GU base pairs. The initiator codon is colored. Note the variation in the length and positioning of the mRNA region that can base-pair and also the variation in exactly which 16S nucleotides interact.

specific mRNAs and then adding ribonuclease to trim away all mRNA sequences except those protected by ribosome attachment. When sequences of these 30-nucleotide-long protected fragments were analyzed, they were found to contain AUGs roughly near the middle, followed by nucleotides that coded for the first three or four amino acids of their respective polypeptide products. These are not the nucleotides specifically recognized by the 30S ribosome, for they vary from one gene to another. Instead, a group of 3 to 9 purine nucleotides appearing in the 5′ half of the protected fragment is critical for ribosome binding. Almost all ribosome binding sites have a sequence such as AGGA or GAGG centered some 8 to 13 nucleotides upstream from the initiator codon (Figure 14-28). These form base pairs with complementary residues found in a pyrimidine-rich sequence at the 3′ end of 16S rRNA chains (…GAUCACCUCCUUA$_{OH}$-3′). This pairing brings the initiating AUG (or GUG) codon into position so that it can bind to the anticodon of the initiator tRNA bound to the 30S subunit. Thus, the capacity to form two separate base-pairing interactions simultaneously (16S rRNA-mRNA and mRNA-tRNA) distinguishes a true initiator region from other sites containing the many internal AUGs in an mRNA.

As expected, mutations that change the purine-rich region of a ribosome binding site lower the efficiency of mRNA translation; translation can be restored by subsequent mutations that restore comple-

Figure 14-29

Sequence of the initiator region at the beginning of the mRNA of the phage T7 gene *0.3*. Its interaction with the 3′ end of 16S rRNA is abolished by a G to A transition, which creates a mutant phenotype. Function is restored by a second transition (A to G) at an adjacent position. A different set of base pairs with 16S rRNA are formed by the revertant. The initiator AUG codon is shown in color. [After J. J. Dunn, E. Buzash-Pollert, and F. W. Studier, *Proc. Nat. Acad. Sci. 75* (1978):2743.]

CACGAGGUAACAAG AUG   Wild-type sequence

3′ $_{HO}$-AUUCCUCCACUAG

16S rRNA

CACGAAGUAACAAG AUG   Mutant synthesizing <10% protein

CACGGAGUAACAAG AUG   Reversion by a second-site mutation

3′ $_{HO}$-AUUCCUCCACUAG

16S rRNA

mentarity to the 3' end of 16S rRNA, even if the sequence is different from that of the original mRNA (Figure 14-29). This raises the question of whether the frequency of initiation on different mRNAs can be directly correlated with the length or stability of the mRNA-rRNA interaction. Unfortunately, no simple relationship can be seen, even though ribosome binding sites do differ by factors of 10 to 100 in their ability to start protein synthesis. Rather, the sequence of the entire ribosome-bound region, as well as the folding of the mRNA molecule, appears to contribute in complex ways to the efficiency of initiation.

Once elongation of the polypeptide chain commences, the 16S rRNA-mRNA base pairs must somehow dissociate, leaving the mRNA free to move across the ribosomal surface. Such movement brings the ribosome into contact with mRNA regions to which it would not bind unless protein synthesis had been initiated upstream. As the ribosome moves along, it must temporarily disrupt the double-helical hairpin regions that characterize so many sections of free mRNA, thus creating single-stranded regions that can correctly select the anticodons of the AA~tRNA precursors.

Initiation Factors[50, 54]

A mixture of N-formyl-Met-tRNA$_F^{Met}$, mRNA, and the 30S and 50S subunits is not by itself sufficient for initiation. At least three separate proteins that are only loosely bound by ribosomes, called IF1, IF2, and IF3, must also be present (Figure 14-30). In the first step, these three **initiation factors** attach to the 30S subunit. GTP (guanosine triphosphate) stabilizes their binding and is probably directly bound by IF2. IF3 prevents the association of 30S subunits with 50S subunits and was therefore first identified as a ribosome dissociation factor. It is now believed that IF3 causes a subtle change in the shape of the 30S subunit that both prevents its association with 50S and aids mRNA binding. IF1 assists the binding of IF2 and IF3. Table 14-1 lists the molecular weights of the three initiation factors, as well as those of factors involved in the elongation and termination steps of protein synthesis.

In the second step, fMet-tRNA$_F^{Met}$ and the mRNA attach in either order to the IF-30S-GTP aggregate. In binding, the fMet-tRNA$_F^{Met}$ makes close contact with the IF2-GTP complex. Once the complete **30S initiation complex** is formed, IF3 is released. Then the joining of a 50S subunit leads to GTP hydrolysis as well as the release of the other two initiation factors (see Figure 14-30). The final complex is called a **70S initiation complex.**

The Direction of mRNA Reading is 5'→3'[22]

After an initiator region of an mRNA molecule has been correctly bound to a ribosome, it always moves in a fixed direction during protein synthesis. It does not have the option of moving either to the right or to the left, reflecting the fact that RNA molecules have a direction defined by the relative orientations of their 5' and 3' ends. The end that is read first, the 5' end, is also synthesized first. Thus, a ribosome can attach to an incomplete mRNA molecule in the process of being synthesized on its DNA template. If, on the other hand, polypeptide synthesis went 3'→5', then a length of mRNA corre-

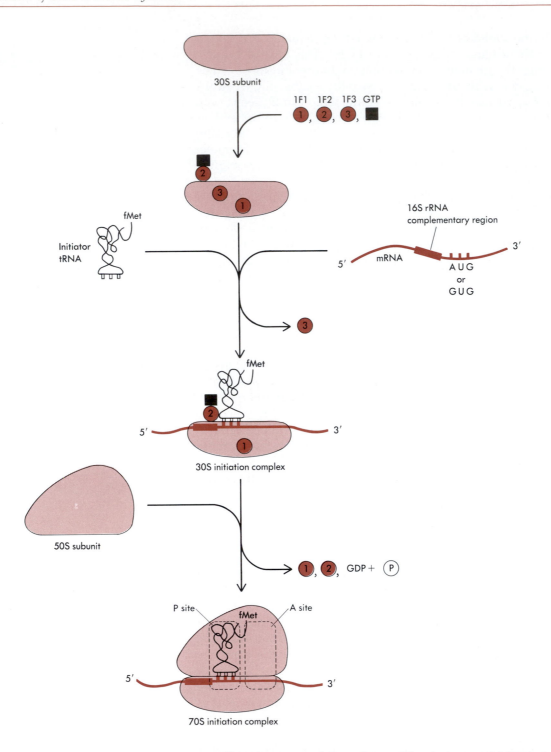

Figure 14-30
View of the initiation process in protein synthesis. Although the fMet-tRNA$_F^{Met}$ is shown as entering the P site directly, it is not known whether it initially contacts the A site. Once the 70S initiation complex is formed, the initiator is clearly bound in the P site.

sponding to a complete polypeptide chain would have to be completely synthesized before a ribosome could attach. The fact that protein synthesis goes in a 5'→3' direction means that long sections of mRNA unattached to ribosomes do not normally occur in rapidly growing bacterial cells.

Each Ribosome Has Two tRNA Binding Sites[22, 55]

Each 70S ribosome contains two cavities into which tRNA molecules can be inserted (Figure 14-31). They are the P (peptidyl) site and the

Table 14-1 Properties of Soluble Factors Involved in *E. coli* Protein Synthesis

	Approximate Molecular Weight	Ability to Bind GTP (GDP)	Approximate Abundance Relative to Number of Ribosomes
Initiation			
IF1	9,000	No	1/7
IF2	120,000	Yes	1/7
IF3	22,000	No	1/7
Elongation			
EF-Tu	45,000	Yes	10†
EF-Ts	30,000	Yes	1
EF-G	80,000	Yes	1
Termination			
RF1	36,000	No	1/20
RF2	38,000	No	1/20
RF3*	46,000	Yes	Not known

*The role of RF3 is not firmly established.
†Amounts of EF-Tu relative to ribosomes can vary from 3-fold to 14-fold, depending on the exact growth conditions. Because it is needed in such high amounts (up to 5 percent of the total cell protein), it is encoded by two different genes called *tufA* and *tufB*.

A (aminoacyl) site. Each tRNA binding site is formed partly by the 30S ribosome and partly by the 50S ribosome. The shape of the site is made specific for a given AA~tRNA by the presence of a particular mRNA codon. That is, although the ribosomal cavities can accept any of the AA~tRNAs, the mRNA-containing surface of the site makes it specific for a unique tRNA molecule.

We do not yet know whether the initiating fMet-tRNA$_F^{Met}$ enters the P site directly or whether it first goes into the A site and subsequently moves into the P site. It is clear, however, that it must end up in the P site and that IF2 must be released by GTP hydrolysis before the second AA~tRNA can bind to the ribosome. For all AA~tRNAs except fMet-tRNA$_F^{Met}$, it has been firmly established that entry to the ribosome occurs through the A site. After the second AA~tRNA is correctly placed in the A site, a peptide bond is formed to yield a dipeptide (two amino acids linked by a peptide bond) attached to the tRNA adaptor of the second amino acid. Then, in a step called **translocation,** the peptidyl-tRNA (tRNA with the growing peptide chain attached to its 3' end) and the mRNA codon to which it is bound are coordinately shifted to the P site. This process of amino acid addition repeats again and again, adding one amino acid at a time to form a complete chain (see Figure 14-31). In these events, the following steps should be emphasized:

1. The growing carboxyl end is always terminated by a tRNA molecule. The binding of this terminal tRNA to either the P or A site is the main force holding a growing polypeptide chain to the ribosome.

2. Formation of the peptide bond moves the attachment point of the growing chain from the P site to the A site. During this step, the carboxyl group of the amino acid in the P site is transferred from its

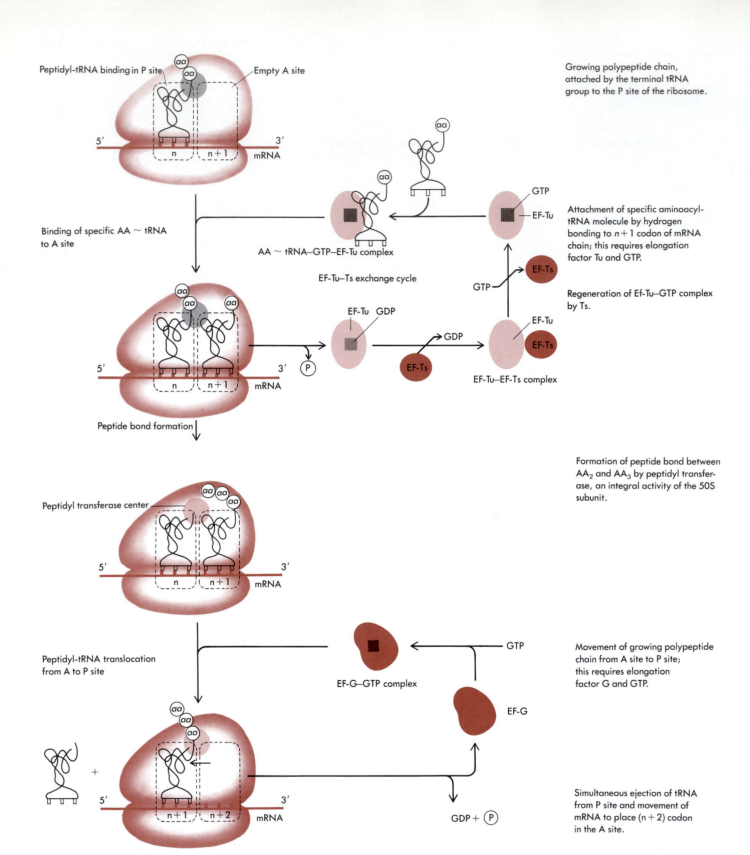

Peptidyl-tRNA binding in P site · Empty A site

Growing polypeptide chain, attached by the terminal tRNA group to the P site of the ribosome.

5′ — n — n+1 — 3′ mRNA

Binding of specific AA ∼ tRNA to A site

AA ∼ tRNA–GTP–EF-Tu complex

EF-Tu–Ts exchange cycle

GTP

EF-Tu

Attachment of specific aminoacyl-tRNA molecule by hydrogen bonding to n+1 codon of mRNA chain; this requires elongation factor Tu and GTP.

GTP

EF-Ts

Regeneration of Ef-Tu–GTP complex by Ts.

EF-Tu   GDP

GDP

EF-Tu

EF-Ts

5′ — n — n+1 — 3′ mRNA

P

EF-Ts

EF-Tu–EF-Ts complex

Peptide bond formation

Peptidyl transferase center

5′ — n — n+1 — 3′ mRNA

Formation of peptide bond between AA₂ and AA₃ by peptidyl transferase, an integral activity of the 50S subunit.

Peptidyl-tRNA translocation from A to P site

EF-G–GTP complex

GTP

EF-G

Movement of growing polypeptide chain from A site to P site; this requires elongation factor G and GTP.

5′ — n+1 — n+2 — 3′ mRNA

GDP + P

Simultaneous ejection of tRNA from P site and movement of mRNA to place (n+2) codon in the A site.

Figure 14-31

View of one cycle of peptide bond formation showing the role of the elongation factors, EF-Tu, EF-Ts, and EF-G, and the involvement of GTP hydrolysis.

Figure 14-32
The peptidyl transferase reaction. The elongation of the polypeptide chain by one amino acid is shown. Upon chain termination, the ester bond between the 3′ adenosine of the tRNA and the amino acid undergoes nucleophilic attack by water instead of by the amino group of the next amino acid.

tRNA to the amino group of the amino acid in the A site (Figure 14-32).

3. The new terminal tRNA molecule then moves (translocates) from the A site to the P site. At the same time, the mRNA template bound to the smaller ribosome subunit moves to place codon $n + 1$ in the position previously occupied by codon n.

4. Coincident with step 3, the tRNA molecule that was released from its amino acid during peptide bond formation is ejected from the P site.

5. The now vacant A site becomes free to accept a new AA~tRNA molecule whose specificity is determined by correct base pairing between its anticodon and the relevant mRNA codon.

Chain Elongation Requires GTP[22, 56]

Because the amino acid carboxyl groups are activated by attachment to their tRNA adaptors, it was initially thought that perhaps no more energy would be needed for making the peptide bond. This hunch, however, was wrong. In addition to the one molecule of GTP required for initiation, two molecules of GTP are hydrolyzed for each amino acid subsequently added to the growing polypeptide chain. GTP utilization can be directly traced to the participation of three proteins (the elongation factors) that are not normally part of ribosomes but which are necessary for the elongation steps in protein synthesis. Surprisingly, none of these proteins participates directly in the actual formation of peptide bonds.

Binding of AA~tRNA to the A Site Requires Elongation Factors EF-Tu and EF-Ts[22, 56]

The attachment of the AA~tRNA precursor to ribosomes was initially believed to be a nonenzymatic event occurring when the correct AA~tRNA randomly bumped an A site bounded by its particular mRNA codon. Further scrutiny, however, indicated that the binding reaction is far from simple. It starts when one of the **elongation factors (EF-Tu)** reacts with GTP and AA~tRNA to form an AA~tRNA–EF-Tu–GTP complex. EF-Tu is an exceedingly abundant protein in *E. coli*, present in approximately as many copies as there are tRNA molecules (see Table 14-1). It can bind every AA~tRNA except for fMet-tRNA$_F^{Met}$; the basis of this discrimination is not understood. After binding to EF-Tu, the AA~tRNA component is transferred to the ribosomal A site with release of a free EF-Tu–GDP complex and phosphate (see Figure 14-31).

The reutilization of EF-Tu to bring another AA~tRNA to the ribosomal A site requires the participation of another factor, **EF-Ts**. EF-Ts displaces GDP from EF-Tu by itself forming a complex with EF-Tu. When this complex encounters GTP, a GTP–EF-Tu complex is formed, releasing EF-Ts. EF-Tu–GTP can then bind another AA~tRNA and repeat the cycle (see Figure 14-31).

The Peptide-Bond-Forming Activity Is an Integral Component of the 50S Ribosome[23, 57]

Enzymatic catalysis of the peptide bond itself has been assigned to the larger ribosomal subunit. This activity is called **peptidyl transferase**, but all attempts so far to dissociate it from 50S particles have failed. Rather, reconstitution experiments indicate that about six L proteins plus the 23S rRNA are essential for peptidyl transferase activity; another six proteins and 5S rRNA contribute strongly. Thus, peptidyl transferase might best be pictured as an active *center* or site on the ribosome that simply aligns the two AA~tRNAs in precisely the correct way to allow peptide bond formation (see Figure 14-32). Since the AA~tRNAs represent activated forms of the amino acids, this reaction should occur spontaneously, without the input of additional energy.

Peptidyl-tRNA Translocation
Requires Elongation Factor G[22, 56]

The movement of peptidyl-tRNA from the A site to the P site is brought about by elongation factor G (**EF-G**), often called the **translocase.** In this process, an EF-G–GTP–ribosome complex first forms. Translocation then occurs coupled with ejection of the free tRNA from the P site (see Figure 14-31). Subsequent release of free reusable EF-G requires the hydrolysis of GTP to GDP and Ⓟ. The splitting of the high-energy bond is obligatory for entry of the ribosome into the next elongation cycle.

Movement of mRNA
Across the Ribosomal Surface[22, 58]

Normally, at each translocation step, the mRNA template is advanced precisely three nucleotides. One can imagine the existence of a machinery within the ribosome that obligatorily advances the mRNA chain three nucleotides at a time. Alternatively, the mRNA may not be moved by the ribosome per se, but instead may be pulled by the process that moves peptidyl-tRNA from the A site to the P site. According to this hypothesis, the movement of a codon triplet is a consequence of its binding to an anticodon in tRNA. Strongly favoring this latter proposal is the discovery of frameshift suppressor tRNAs (Chapter 15) that have anticodons containing four nucleotides and so bind to groups of four adjacent mRNA nucleotides. When such unusual tRNAs are translocated, the attached mRNA template is advanced by four bases, showing that mRNA movement is directly coupled to tRNA movement.

Inhibition of Specific Steps in
Protein Synthesis by Antibiotics[59, 60]

A number of antibiotics have proved very useful in delineating the steps by which proteins are built. For example, puromycin, a very powerful inhibitor of the growth of all cells, acts by interrupting chain elongation (Figure 14-33). Its structure resembles the 3' end of a charged tRNA molecule, and so it enters the ribosomal A site very efficiently, competitively inhibiting the normal entry of AA~tRNA precursors. More importantly, peptidyl transferase will use the antibiotic as a substitute, transferring nascent chains to puromycin acceptors. Since the puromycin residue is very small compared to a tRNA, it binds only weakly to the A site. Thus, the puromycin-terminated nascent chains fall off the ribosomes, producing incomplete polypeptides of varying lengths. Note that the incorporation of puromycin into the carboxy-terminal ends of growing chains proves that protein synthesis proceeds from the N terminus to the C terminus.

Other antibiotics act to inhibit other steps of protein synthesis. For instance, fusidic acid interferes with the release of EF-G after it has functioned in translocation. The stuck EF-G–GDP complex very effectively blocks the binding of the next AA~tRNA–EF-Tu–GTP to the ribosomal A site, which must precede a subsequent round of elongation. The peptidyl transferase reaction is inhibited by the antibiotics

Figure 14-33
Premature polypeptide chain termination by puromycin.

chloramphenicol and sparsomycin, both of which specifically bind to the 50S subunit. In contrast, streptomycin, which binds to the 30S subunit, is a powerful inhibitor of chain initiation.

Polypeptide Chains Fold Up Simultaneously with Synthesis

Under optimal conditions, the time required for the synthesis of an *E. coli* polypeptide chain containing 300 to 400 amino acids is 10 to 20 seconds. During this time, the elongating chain does not remain a random coil, but quickly assumes much of its final three-dimensional shape through the formation of many of its secondary bonds. Thus, with many proteins, most of the final shape may be assumed before the last few amino acids are added on. As a result, trace amounts of many enzyme activities can be found on ribosomes that have not yet released their polypeptide products. Likewise, antibodies, which recognize specific three-dimensional shapes on the surfaces of proteins (Chapter 23), can often bind to growing polypeptide chains still attached to ribosomes.

Chain Release Depends on Specific Release Factors That Read Chain-Terminating Codons[61]

Two conditions are necessary for the termination of protein synthesis. One is the presence of a codon that specifically signals that polypeptide elongation should stop. The other is the presence of a protein

release factor (RF) that reads the chain-terminating signal. Behind all this complexity is the fact that after a polypeptide chain has reached its full length, its carboxyl end is still bound to its tRNA adaptor. Termination must therefore involve the splitting off of the terminal tRNA. When this happens, the nascent chain quickly dissociates, since its binding to the ribosome occurred principally through its adaptor tRNA.

Elucidation of the genetic code (Chapter 15) revealed three codons specifically signifying "stop." Their existence initially led to the expectation of chain-terminating tRNA molecules; that is, tRNAs that would specifically bind to the stop codon but which had no amino acid attached to their 3' adenosine and which in some way promoted the release of the terminal tRNA group. Experiments, however, conclusively ruled out the existence of such molecules. The stop codons are instead read by proteins called release factors. In *E. coli*, there are two well-characterized RFs (see Table 14-1), one (RF1) that recognizes the stop codons UAG and UAA and another (RF2) that acts with UGA and UAA. They function by inducing the peptidyl transferase to transfer the growing chain to water instead of to an amino acid activated by attachment to tRNA (see Figure 14-32). Whether GTP hydrolysis is required for termination in *E. coli* is not yet firmly established, although a third factor (sometimes called RF3), which does bind GTP and GDP, appears to stimulate the reaction. After chain termination, the release of the 50S ribosomal subunit from the mRNA probably precedes that of the 30S particle; and still another protein factor may be involved in ribosome release. The ribosomal subunits are then available to begin synthesis of another polypeptide chain.

GTP May Function by Causing Conformational Changes[56, 62]

The repeated use of GTP at so many stages of the translation process (initiation, elongation, and possibly termination) calls for some explanation. There is no evidence for its use in the formation of any covalent bond, and so it does not appear to function as an ATP-like energy donor. Instead, it plays a key role in the noncovalent binding of the various translational factors (IF2, EF-Tu, and EF-G) to the ribosomal surface. Such binding processes require only the presence of GTP, while the GTP $\rightarrow$ GDP cleavage reaction is the signal for the release of the bound factor. An attractive hypothesis is that GTP attachment induces the translation factors to undergo shape changes that are necessary for their binding to ribosomes. Subsequent hydrolysis to GDP leads to a return of the original free configuration, resulting in an ejection of the respective factor from the ribosomal surface.

Reconstitution experiments have shown that the four copies of protein L7/L12 on the 50S ribosome are intimately involved in GTP hydrolysis. These ribosome components are essential for the binding of the various protein synthesis factors that utilize the energy stored in GTP during their functioning. However, there are reports that protein synthesis can proceed, albeit at a very slow rate, in the absence of GTP hydrolysis and elongation factors. Thus, ribosomes themselves must possess an innate capacity to bind AA~tRNA and translocate. The participation of factors and GTP increases the rate at which they do so to a biologically meaningful level.

GTP Hydrolysis Also Contributes to the Fidelity of Translation[2, 63, 64, 65]

Another extremely important way in which GTP functions on the ribosome is the role it plays in the accuracy of translation. Evidence has now accumulated for a proofreading step that occurs after AA~tRNA binding. When AA~tRNA is first brought to the A site complexed with EF-Tu–GTP, the initial mRNA-tRNA base-pairing interaction is not sufficiently specific to account for the low error rate (about 10^{-4}) of protein synthesis. Proofreading improves the error rate by discarding an incorrectly selected AA~tRNA. This consumes an AA~tRNA–EF-Tu–GTP complex (which involves GTP → GDP conversion) but does not extend the polypeptide chain. Thus, editing steps require the input of additional energy.

To study the proofreading mechanism, it has been necessary to devise in vitro systems that translate nearly as accurately as cells do. This has been achieved by carefully choosing ionic conditions that mimic those in vivo and by ensuring that a high GTP/GDP ratio is maintained throughout the experiment. Recent results suggest that the initial tRNA selection and subsequent proofreading steps contribute approximately equally to the fidelity of translation (each about 10^{-2}). It is worth noting that some mutations in ribosomal proteins can significantly affect the proofreading ability of the ribosome (Chapter 15).

Production of ppGpp on Ribosomes by an Idling Reaction in the Absence of Charged tRNA[66, 67, 68]

If, by accident, an uncharged tRNA molecule occupies an A site, not only does polypeptide growth temporarily cease, but an idling reaction occurs on the affected ribosome, which uses ATP as a ⓟ~ⓟ donor to convert pp5'-G-3'OH (GDP) into the unusual guanine nucleotide pp5'-G-3'pp, often called "magic spot." Production of ppGpp is stimulated only if the uncharged tRNA can correctly pair with the mRNA codon exposed in the A site. While during normal bacterial growth only very low levels of ppGpp are present, large amounts accumulate during amino acid starvation. These ppGpp molecules do not play a passive role, but act as signals that specifically stop the synthesis of rRNA and tRNA chains. They thus act as the control molecules that keep cells from wastefully producing more ribosomes than they can employ. This specific blockage of rRNA and tRNA synthesis under conditions of amino acid starvation is called the **stringent response**. Its existence is not obligatory, however, since there are *relaxed mutants* that continue to make rRNA and tRNA during amino acid starvation. Some relaxed cells lack a 75,000-dalton enzyme, the **stringent factor**, that can bind to ribosomes and carry out the conversion of ppG to ppGpp. Other relaxed mutants are altered in one of the L proteins (L11) of the 50S ribosome.

Several Ribosomes Translate an mRNA Molecule Simultaneously

The section of an mRNA molecule that is in contact with a single ribosome is relatively short (about 30 nucleotides). This allows a given

Free ribosomal subunits about to attach to mRNA

Completed chain

5'

3'

50S

30S   Free ribosomal subunits   ▬▬▬ mRNA

〜〜〜 Growing polypeptide chain   tRNA

Figure 14-34
Schematic picture of a polyribosome during protein synthesis. The mRNA molecule is moving from right to left.

mRNA molecule to move over the surfaces of several ribosomes at once, functioning simultaneously as a template for several identical polypeptide chains. The collection of ribosomes bound to a single mRNA chain is called a **polysome** or **polyribosome.** At a given time, the lengths of polypeptide chains attached to successive ribosomes in the polyribosome vary in direct proportion to the fraction of the messenger tape each ribosome has already read (Figure 14-34). This means that at any moment, the polypeptide chains being produced along the length of the mRNA are shortest at the 5' end and gradually lengthen toward the 3' end. There is great variation in polysome size, which depends both on the size of the mRNA chain and how closely the ribosomes are packed. The latter obviously depends on how frequently ribosomes can initiate at the beginning of a particular gene; this varies from one ribosome binding site to another. At maximum utilization of an mRNA chain, there is one ribosome for every 80 mRNA nucleotides. Thus, the polysomes that make hemoglobin molecules usually contain 4 to 6 ribosomes, while approximately 12 to 20 ribosomes are attached to the mRNA molecules concerned with the synthesis of proteins in the 30,000 to 50,000-dalton range (300 to 500 amino acids).

The ability of several ribosomes to function on a single mRNA simultaneously explains why a cell needs relatively little mRNA. Before polysomes were discovered, and when only one ribosome was thought to be attached to a given mRNA molecule, the fact that mRNA accounts for only 1 to 2 percent of the total cellular RNA seemed highly paradoxical. This followed from the calculation that given this amount of mRNA and an average chain length of about 1700 nucleotides, then at any instant, no more than 10 percent of the ribosomes in a cell could be making protein.

Proteins Designed for Export Are Made on Membrane-Bound Ribosomes and Are Cleaved After Synthesis[69–74]

Bacteria synthesize not only proteins that function inside the cell, but also proteins whose roles demand that they be properly localized in or beyond the cell membrane. In fact, the cell envelope of a Gram-negative bacterium such as *E. coli* is made up of three different compartments, each characterized by different populations of specific proteins. For instance, the inner membrane contains proteins in-

Table 14-2   Signal Sequences of Some Exported Bacterial Proteins*

Protein	Charged Segment	Hydrophobic Segment
Inner Membrane Proteins		
Phage fd, major coat protein	Met Lys Lys Ser Leu Val Leu Lys	Ala Ser Val Ala Val Ala Thr Leu Val Pro Met Leu Ser Phe Ala ↓ Ala Glu Gly
Phage fd, minor coat protein	Met Lys Lys	Leu Leu Phe Ala Ile Pro Leu Val Val Pro Phe Tyr Ser His Ser ↓ Ala Glu Thr
Periplasmic Proteins		
Alkaline phosphatase	Met Lys	Gln Ser Thr Ile Ala Leu Ala Leu Leu Pro Leu Leu Phe Thr Pro Val Thr Lys Ala ↓ Arg Thr Pro
Maltose binding protein	Met Lys Ile Lys Thr Gly Ala Arg	Ile Leu Ala Leu Ser Ala Leu Thr Thr Met Met Phe Ser Ala Ser Ala Leu Ala ↓ Lys Ile Glu
Leucine-specific binding protein	Met Lys Ala Asn Ala Lys	Thr Ile Ile Ala Gly Met Ile Ala Leu Ala Ile Ser His Thr Ala Met Ala ↓ Asp Asp Ile
β-lactamase of pBR322	Met Ser Ile Gln His Phe Arg	Val Ala Leu Ile Pro Phe Phe Ala Ala Phe Cys Leu Pro Val Phe Ala ↓ His Pro Glu
Outer Membrane Proteins		
Lipoprotein	Met Lys Ala Thr Lys	Leu Val Leu Gly Ala Val Ile Leu Gly Ser Thr Leu Leu Ala Gly ↓ Cys Ser Ser
LamB	Met Met Ile Thr Leu Arg Lys	Leu Pro Leu Ala Val Ala Val Ala Ala Gly Val Met Ser Ala Gln Ala Met Ala ↓ Val Asp Phe
OmpA	Met Lys Lys	Thr Ala Ile Ala Ile Ala Val Ala Leu Ala Gly Phe Ala Thr Val Ala Gln Ala ↓ Ala Pro Lys

*The hydrophobic portion (color) of the signal sequence is invariably preceded by a short charged segment. The arrows indicate the sites of cleavage by signal peptidase.

SOURCE: S. Michaelis and J. Beckwith, *Ann. Rev. Microbiol.* 36 (1982):435.

volved in energy metabolism and nutrient transport; the outer membrane has proteins that facilitate ion and nutrient entry and serve as bacteriophage receptors; the space in between (called the *periplasm*) contains hydrolytic enzymes such as nucleases and proteases, as well as nutrient-binding proteins.

How are these proteins conveyed across the inner membrane to their correct destination? Most exported proteins are synthesized in the form of longer precursors, with between 15 and 30 additional amino acids at their N termini. This N-terminal region, called the **signal sequence**, is invariably rich in uncharged, usually hydrophobic amino acids (Table 14-2). They function to initiate association of the nascent polypeptide with the membrane as soon as the signal sequence has emerged from the ribosome. It has been suggested that the signal sequence and the residues just beyond fold into two short helical segments that then bend into an antiparallel helical hairpin. The properties of such a hairpin are ideal for insertion into the hydrophobic lipid bilayer (Figure 14-35). Once the N terminus of a to-be-exported protein is anchored in the membrane, continued synthesis will extrude the remainder of the polypeptide through to the other side. Or, if another hydrophobic stretch is synthesized, this stretch may stick as it crosses the membrane and permanently fix the protein in the membrane. Meanwhile, the short N-terminal signal sequence is usually cleaved off by a specific **signal peptidase.** How proteins can be specifically targeted for the outer membrane rather than the periplasm is not yet known.

Ribosomes translating an mRNA whose product is to be exported are membrane-bound. This attachment is at least partially via the growing polypeptide chain. What other interactions are required between the ribosome and the membrane is not yet clear. However,

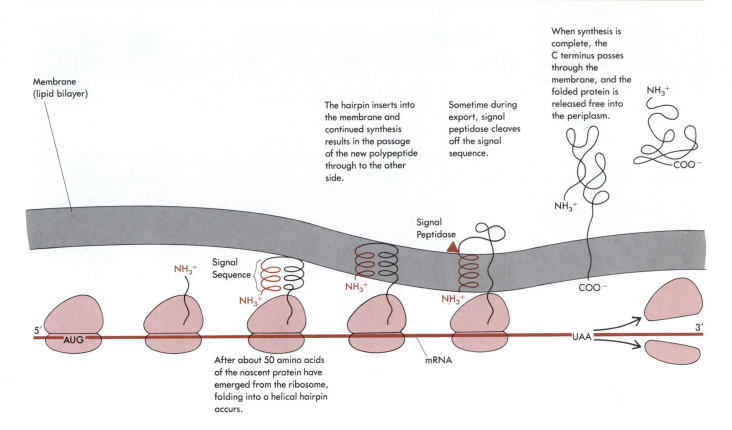

Membrane
(lipid bilayer)

The hairpin inserts into the membrane and continued synthesis results in the passage of the new polypeptide through to the other side.

Sometime during export, signal peptidase cleaves off the signal sequence.

When synthesis is complete, the C terminus passes through the membrane, and the folded protein is released free into the periplasm.

Signal Peptidase

Signal Sequence

After about 50 amino acids of the nascent protein have emerged from the ribosome, folding into a helical hairpin occurs.

mRNA

mutations in several genes that affect the export of many proteins in *E. coli* are currently under scrutiny. Their properties hint that the situation may be quite similar to that in mammalian cells, where firm evidence points to the participation of a small RNA-protein complex both in initiating secretion and in controlling the elongation of proteins designed for export (Chapter 21).

Figure 14-35
Steps in the process of cotranslational secretion. The diagram illustrates how polysomes engaged in synthesizing a protein designed for export become membrane-bound via the nascent polypeptide chain. [After D. M. Engelman and T. A. Steitz, *Cell* 23 (1981):411.]

Potent Methods for Studying Ribosome Structure[24, 28, 75–82]

It is very likely that the general outline of protein synthesis is now known. Meanwhile, progress on relating each of the many reactions to the structure of the ribosome itself has lagged far behind. The situation is clearly not hopeless, however, for data obtained using a number of techniques have recently begun to converge. These give us a good picture of the shapes of the two *E. coli* ribosomal subunits and where many of their important functional sites are located.

Two different kinds of electron microscopy have provided us with information concerning ribosome shape. One is direct visualization of fields containing isolated 70S ribosomes, 50S subunits, and 30S subunits (Figure 14-36). The other applies mathematical reconstruction methods to interpret images generated from naturally occurring ribosome microcrystals. There is general agreement that the 50S subunit resembles a hemisphere with three protuberances emerging from its flat face. The 30S subunit is more flattened and appears to be separated into an upper one-third and a lower two-thirds, with a platform protruding on one side just below the constriction. The deduced shapes for the two subunits and how they might fit together to form the 70S ribosome are shown in Figures 14-36 and 14-37.

Figure 14-36
Direct visualization of ribosome shape using electron microscopy. (a) A field of ribosomes from *E. coli* includes some that are oriented like A or B in the model in (b). A few ribosomal subunits are also present, identified by the letters s and l, which stand for small and large. The field has been negatively stained for electron microscopy; each ribosome is defined by an outline of the salt of a heavy metal, which in this case is uranium. [Adapted with the permission of the author from J. A. Lake, "The Ribosome," *Sci. Amer.* 245 (1981):86.]

(a)

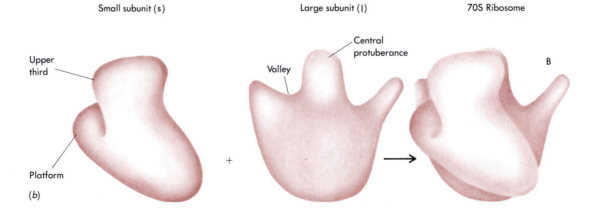

(b)

A much broader spectrum of methodologies has contributed to the assignment of functional sites to specific locations on the surfaces of these shapes. For instance, analysis of chemical crosslinks between proteins (Figure 14-38), between proteins and rRNA or tRNA sequences, and between different parts of an rRNA molecule have told us what is near what. The ability to reconstitute ribosomal subunits has allowed the labeling of certain proteins with chromophores for fluorescence energy transfer or with deuterium for neutron-scattering experiments. Such techniques yield measurements of the distances between proteins in or bound to the ribosome (Figure 14-39). Finally, antibodies have been raised against individual ribosomal proteins or

30S   50S   Scale Å   70S

(the rather rare) modified nucleotides in rRNA. Using the electron microscope, researchers can visualize these antibodies bound to ribosomes and can therefore localize particular sites on the two ribosomal subunits (Figure 14-40).

The picture that emerges is quite satisfying, since components that are known to interact occupy nearby positions (see Figure 14-37).

Figure 14-37
Shapes of the two subunits of *E. coli* ribosomes. Functional sites were localized by a combination of methods. The diagram of the 70S ribosome shows how a tRNA bound in the A site would be oriented. The A site is located between the 30S and 50S subunits. [After J. A. Lake, *Sci. Amer.* 245 (1981):84; H. M. Olson, P. G. Grant, B. S. Cooperman, and D. G. Glitz, *J. Biol. Chem.* 257 (1982):2649.]

Figure 14-38
Illustration of how chemical crosslinking can identify neighboring proteins in the ribosome. Introduction of reversible crosslinks allows analysis by a "diagonal" method employing a two-dimensional gel. Recall that mobility on a sodium dodecyl sulfate (SDS) gel is related to the molecular weight of the protein. [After Traut *et al.*, in *Ribosomes*, ed. G. Chamblis, G. R. Craven, J. Davis, L. Kahan, and M. Nomura (Baltimore, Md.: University Park Press, 1980), p. 89.]

Bacteria in H₂O

Bacteria in D₂O

Isolate ribosomal subunits: fractionate into 16S RNA and the 21 S proteins of the 30S subunit.

Normal 16S RNA   Normal S proteins   Deuterated S proteins   Deuterated 16S RNA

Mix together normal 16S RNA, 19 normal S proteins, and two deuterated S proteins.

Reconstituted 30S subunits containing two deuterated proteins.

(a)

Figure 14-39
How neutron-scattering studies can determine distances between specific proteins in the ribosome. [After D. M. Engelman and P. B. Moore, *Sci. Amer.* 235 (1976):44.]

1. The mRNA binding site and the 3' end of the 16S rRNA are located on the platform between the upper and lower portions of the 30S subunit. Initiation factors also bind in this region.

2. The two tRNA binding sites are in the cleft formed by the platform and upper third of the 30S ribosome.

3. The GTPase center (consisting of the four molecules of protein L7/L12) has been assigned to the most fingerlike of the three protuberances of the 50S subunit.

4. The peptidyl transferase center lies in the valley between the other two 50S protuberances.

5. The exit site for the growing polypeptide chain is almost on the exact opposite side of the 50S subunit from the transferase center. This suggests that the 30 amino acids of the nascent polypeptide chain passing through the interior of the ribosome must be in a nearly completely extended conformation.

6. The site where the 50S subunit interacts with membranes is located very near the polypeptide exit hole.

A neutron beam is passed through the deuterated 30S subunits and the pattern of scattered neutrons registered by the detector.

The interference pattern appears as a set of concentric rings in which the scattered neutron intensity is alternately higher and lower than the background level.

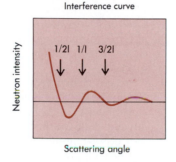

The neutron intensity is graphed as a function of the angular deflection of the beam; the distance between successive interference rings is inversely related to the distance, l, between the centers of mass of the two deuterated proteins in the 30S ribosome.

(b)

7. The two ribosomal subunits join to bring the 30S platform that binds mRNA and tRNA in close proximity to the 50S surface that contains the GTPase and peptidyl transferase activities.

Despite these advances in understanding the topography of the ribosome, the methods mentioned here cannot tell us how the ribosome works in atomic detail. Such answers must ultimately come from X-ray diffraction analysis. Only recently have we obtained good three-dimensional crystals of *E. coli* ribosomal subunits. This, however, is just the first step in a monstrous endeavor, for X-ray methods have not yet been used to solve structures of anywhere near this complexity. Meanwhile, crystals of individual ribosomal proteins, the 5S rRNA, and a 5S rRNA–protein complex have been prepared, and their structures are being solved. They will tell us much about RNA structure and RNA-protein interactions. Nonetheless, nagging questions, such as why the ribosome is two-thirds RNA, will probably be answered only after the complete three-dimensional structure of this complicated protein synthesis machine is known.

Figure 14-40

Localization of proteins on ribosomal subunits by immune electron microscopy. Ribosome and ribosomal subunit pairs are joined by antibody molecules (which have a Y shape) against specific ribosomal proteins: (a) anti-S14, (b) anti-L7/L12, (c) anti-L27. [Courtesy of J. A. Lake, *J. Mol. Biol.* 161 (1982):95.]

Protein S14 is located on the upper third of the small subunit.

Protein L7/L12 is located on the most fingerlike protuberance of the large subunit.

Protein L27 is located on the central protuberance of the large subunit.

Summary

Amino acids do not attach directly to RNA templates. They first combine with specific adaptor molecules to form aminoacyl (AA) adaptor complexes. The adaptor component has a strong chemical affinity for RNA nucleotides. All the adaptors are transfer RNA molecules (tRNA) with molecular weights of about 25,000. A given tRNA molecule is specific for a given amino acid.

All tRNA chains are folded into semirigid L-shaped molecules. About half the tRNA bases are found in double-helical stems, with the remainder forming loops stabilized by interloop tertiary bonds and stacking forces. One of the loops contains the anticodon, the group of three bases that base-pair to three successive template bases (the codon). Amino acids are attached through their carboxyl groups to the 3'-terminal adenosine of tRNA molecules by high-energy covalent bonds and thus become activated. There is a specific activating enzyme (aminoacyl-tRNA synthetase) for each amino acid; and so it is an enzyme, not tRNA, that recognizes the amino acid. The overall accuracy of protein synthesis can thus be no greater than the accuracy with which the activating enzymes can selectively recognize the various amino acids.

After activation, the AA~tRNA molecules diffuse to the ribosomes, which are roughly spherical particles on which the peptide bonds form. Ribosomes have molecular weights of about 3×10^6 daltons and in bacteria consist of about one-third protein and two-thirds ribosomal RNA (rRNA). They are always constructed from a large (50S) subunit and a small (30S) subunit. In a single bacterial ribosome there are 52 different proteins, in addition to three RNA chains (16S, 23S, and 5S rRNA). Ribosomal RNA contains no genetic information, and the function of most of its nucleotides remains a mystery. The template itself is a third form of RNA, messenger RNA (mRNA). Messenger RNA attaches to ribosomes and moves across them to bring successive codons into position to select the correct AA~tRNA precursors.

Protein chains grow stepwise, beginning at the amino-terminal end. This means that the growing end is always terminated by a tRNA molecule, which serves to hold the nascent chain on the ribosomal surface. Each ribosome has two tRNA-binding sites, the P (peptidyl) site and the A (aminoacyl) site. AA~tRNA molecules first enter the A site, allowing the subsequent formation of a peptide bond with the growing chain held in the P site. This transfers the nascent chain to the A site. A process called translocation then moves the tRNA carrying the nascent chain into the P site and simultaneously advances the mRNA to the next codon and ejects the free tRNA from the P site. Thus, another cycle can begin. Both the attachment of AA~tRNA to the A site and the translocation process require the splitting of GTP to GDP and the participation of protein factors. Initiation of synthesis requires separate factors from those contributing to chain elongation and also involves GTP hydrolysis. With bacterial ribosomes, initiation requires N-formylmethionyl-tRNA. Chain termination results when specific stop signals in the mRNA are read by protein release factors.

Completion of chain synthesis is followed by the dissociation of ribosomes into the large and small ribosomal subunits. Rejoining of subunits occurs only after the initiating aminoacyl-tRNA and the mRNA have first bound to the small ribosomal particle. A given mRNA molecule usually works simultaneously on many ribosomes (a polysome). Thus, at any given moment, many codons of the same template are "at work."

Bibliography

General References

Altman, S., ed. 1978. *Transfer RNA*. Cambridge, Mass: MIT Press. A collection of articles on tRNA structure and function.

Chambliss, G., G. R. Craven, J. Davies, K. Davis, L. Kahan, and M. Nomura, eds. 1980. *Ribosomes*. Baltimore, Md.: University Park Press. A compendium of excellent reviews and summaries, not only of ribosome structure but of all aspects of protein synthesis as of 1980.

Clark, B. and H. Petersen, eds. 1984. "Gene Expression: The Translational Step and Its Control." *Alfred Benzon Symposium*, vol. 19. Copenhagen: Munksgaard. A number of excellent reviews on aspects of translation currently under study.

"Protein Synthesis." 1969. *Cold Spring Harbor Symposium*. 34. Cold Spring Harbor, N.Y.: Cold Spring Harbor Laboratory. Summarizes the state of our understanding of protein synthesis three years after the genetic code was solved.

Nomura, M., A. Tissieres, and P. Lengyel, eds. 1974. *Ribosomes*. Cold Spring Harbor, N.Y.: Cold Spring Harbor Laboratory. An early but valuable treatise on ribosome structure and function.

P. R. Schimmel, D. Söll, and J. N. Abelson, eds. 1980. *Transfer RNA: Structure Properties and Recognition, and Biological Aspects*. Cold Spring Harbor, N.Y.: Cold Spring Harbor Laboratory. Two volumes with contributions on all aspects of tRNA.

Watson, J. D. 1963. "The Involvement of RNA in the Synthesis of Proteins." *Science* 140: 17–26. A history of work between 1953 and 1962 that established how RNA participates in protein synthesis.

Cited References

1. Crick, F. H. C. 1958. "On Protein Synthesis. Biological Replication of Macromolecules." *Symp. Exp. Biol.* 12:138–163.
2. Hopfield, J. J. 1974. "Kinetic Proofreading: A New Mechanism for Reducing Errors in Biosynthetic Processes Requiring High Specificity." *Proc. Nat. Acad. Sci.* 71:4135–4139.
3. Ninio, J. 1975. "Kinetic Amplification of Enzyme Discrimination." *Biochimie* 57:587–595.
4. Hoagland, M. B., M. L. Stephenson, J. F. Scott, L. I. Hecht, and P. C. Zamecnik. 1958. "A Soluble Ribonucleic Acid Intermediate in Protein Synthesis." *J. Biol. Chem.* 231:241–257.
5. Holley, R. W. 1968. "The Nucleotide Sequence of a Nucleic Acid." 1966. *Sci. Amer.* 214:30–39.
6. Rich, A., and U. L. RajBhandary. 1976. "Transfer RNA: Molecular Structure, Sequence, and Properties." *Ann. Rev. Biochem.* 45:805–860.
7. Kim, S. H., F. L. Suddath, F. L. Quigley, A. McPherson, J. L. Sussman, A. H. J. Wang, N. C. Seeman, and A. Rich. 1974. "Three-Dimensional Tertiary Structure of Yeast Phenylalanine Transfer RNA." *Science* 185:435–440.
8. Robertus, J. D., J. E. Ladner, J. T. Finch, D. Rhodes, R. S. Brown, B. F. C. Clark, and A. Klug. 1974. "Structure of Yeast Phenylalanine tRNA at 3 Å Resolution." *Nature* 250:546–551.
9. Rich, A., and S. H. Kim. 1978. "The Three-Dimensional Structure of Transfer RNA." *Sci. Amer.* 238:56–62.
10. Schevitz, R. W., A. D. Podjarny, N. Krishnamachari, J. J. Hughs, P. B. Sigler, and J. L. Sussman. 1979. "Crystal Structure of a Eukaryotic Initiator tRNA." *Nature* 278:188–190.
11. Woo, W., and A. Rich. 1980. "The Three Dimensional Structure of *E. coli* Initiator tRNA^{Met}." *Nature* 286:346–351.
12. Moras, D., M. B. Comarmond, J. Fisher, R. Weiss, J. C. Thierry, J. P. Ebel, and R. Geige. 1980. "Crystal Structure of Yeast tRNA^{Asp}." *Nature* 288:669–674.
13. Hoagland, M. B., E. B. Keller, and P. C. Zamecnik. 1956. "Enzymatic Carboxyl Activation of Amino Acids." *J. Biol. Chem.* 218:345–358.
14. Schimmel, P. R., and D. Soll. 1979. "Aminoacyl tRNA Synthetases: General Features and Recognition of Transfer RNAs." *Ann. Rev. Biochem.* 48:601–648.
15. Igloi, G. L., and F. Cramer. 1979. "Interaction of Aminoacyl tRNA Synthetases and Their Substrates with a View to Specificity." In *Transfer RNA*, ed. S. Altman. Cambridge, Mass.: MIT Press, pp. 294–349.
16. Schimmel, P., S. Putney, and R. Starzyk. 1982. "RNA and DNA Sequence Recognition and Structure-Function of Aminoacyl tRNA Synthetases." *Trends in Biochem. Sci.* 7:209–212.
17. Baldwin, A. N., and P. Berg. 1967. "Transfer Ribonucleic Acid-Induced Hydrolysis of Valyladenylate Bound to Isoleucyl Ribonucleic Acid Synthetase." *J. Biol. Chem.* 241:839–845.
18. Blow, D. M., and P. Brick. 1985. "Aminoacyl-tRNA Synthetases." In *The Structure of Biological Macromolecules and Assemblies*, vol. 2, *Nucleic Acid Binding Proteins*, ed. F. A. Jurnak and A. McPherson. New York: Wiley, pp. 442–469.
19. Fersht, A. R., J. P. Shi, J. Knill-Jones, D. M. Lowe, A. J. Wilkerson, D. M. Blow, P. Brick, P. Carter, M. M. Y. Waye, and G. Winter. 1985. "Hydrogen Bonding and Biological Specificity Analyzed by Protein Engineering." *Nature* 314:235–238.
20. Chapeville, F., F. Lipmann, G. V. Ehrenstein, B. Weisblum, W. J. Ray, Jr., and S. Benzer. 1962. "On the Role of Soluble Ribonucleic Acid in Coding for Amino Acids." *Proc. Nat. Acad. Sci.* 48:1086–1092.
21. Hecht, S. M., and A. C. Chinault. 1976. "Position of Aminoacylation of Individual *Escherichia coli* and Yeast tRNAs." *Biochem.* 73:405–409.
22. Hershey, J. W. B. 1980. "The Translational Machinery." In *Cell Biology: A Comprehensive Treatise*, vol. 4, ed. D. M. Prescott and L. Goldstein. New York: Academic Press.
23. Nierhaus, K. H. 1982. "Structure, Assembly, and Function of Ribosomes." *Current Topics in Microbiol. and Immun.* 97:82–155.
24. Wittmann, H. G. 1983. "Architecture of Prokaryotic Ribosomes." *Ann. Rev. Biochem.* 52:35–65.
25. Wittmann, H. G. 1982. "Components of Bacterial Ribosomes." *Ann. Rev. Biochem.* 51:155–183.
26. Brenner, S., F. Jacob, and M. Meselson. 1961. "An Unstable Intermediate Carrying Information from Genes to Ribosomes for Protein Synthesis." *Nature* 190:576–581.
27. Gros, F., H. Hiatt, W. Gilbert, C. G. Kurland, R. W. Risebrough, and J. D. Watson. 1961. "Unstable Ribonucleic Acid Revealed by Pulse Labelling of *Escherichia coli*." *Nature* 190:581–585.
28. Noller, H. F. 1984. "Structure of Ribosomal RNA." *Ann. Rev. Biochem.* 53:119–162.
29. Woese, C. R., R. Gutell, R. Gupta, and H. F. Noller. 1983. "Detailed Analysis of the Higher-Order Structure of 16S-Like Ribosomal Ribonucleic Acids." *Microbiol. Rev.* 47:621–669.
30. Brimacombe, R. 1984. "Conservation of Structure in Ribosomal RNA." *Trends in Biochem. Sci.* 9:273–277.
31. Traub, P., and M. Nomura. 1969. "Studies on the Assembly of Ribosomes *in Vitro*." *Cold Spring Harbor Symp. Quant. Biol.* 34:63–67.
32. Held, W. A., B. Ballou, S. Mizushima, and M. Nomura. 1974. "Assembly Mapping of 30S Ribosomal Proteins from *Escherichia coli*." *J. Biol. Chem.* 249:3103–3111.
33. Rohl, R., and K. H. Nierhaus. 1982. "Assembly Map of the Large Subunit (50S) of *Escherichia coli* Ribosomes." *Proc. Nat. Acad. Sci.* 79:729–733.
34. Altman, S., C. Guerrier-Takada, H. Frankfort, and H. Robertson. 1982. "RNA-Processing Nucleases." In *Nucleases*, ed. S. Linn and R. Roberts. Cold Spring Harbor, N.Y.: Cold Spring Harbor Laboratory, pp. 243–274.
35. Gegenheimer, P., and D. Apirion. 1981. "Processing of Procaryotic Ribonucleic Acid." *Microbiol. Rev.* 45:502–541.
36. Abelson, J. 1979. "RNA Processing and the Intervening Sequence Problem." *Ann. Rev. Biochem.* 48:1035–1069.
37. Nakajima, N., H. Ozeki, and Y. Shimura. 1981. "Organization and Structure of an *E. coli* tRNA Operon Containing Seven tRNA Genes." *Cell* 23:239–249.
38. Guerrier-Takada, C., K. Gardiner, T. Marsh, N. Pace, and

S. Altman. 1983. "The RNA Moiety of Ribonuclease P is the Catalytic Subunit of the Enzyme." *Cell* 35:849–857.

39. Deutscher, M. 1984. "Processing of tRNA in Prokaryotes and Eukaryotes." *Crit. Rev. Biochem.* 17:45–71.
40. Guthrie, C. 1980. "Folding Up a Transfer RNA Molecule Is Not Simple." *Quart. Rev. Biol.* 55:335–352.
41. Deutscher, M. P. 1985. "*E. coli* RNases: Making Sense of Alphabet Soup." *Cell* 40:731–732.
42. Lund, E., J. E. Dahlberg, L. Lindahl, R. Jaskunas, P. P. Dennis, and M. Nomura. 1976. "Transfer RNA Genes Between 16S and 23S rRNA Genes in rRNA Transcription Units of *E. coli.*" *Cell* 7:165–177.
43. Schlessinger, D. 1980. "Processing of Ribosomal RNA Transcripts in Bacteria." In *Ribosomes: Structure, Function and Genetics.*, ed. G. Chambliss, G. R. Craven, J. Davies, K. Davis, L. Kahan, and M. Nomura. Baltimore, Md.: University Park Press, pp. 767–780.
44. Dunn, J. J., and F. W. Studier. 1973. "T7 Early RNAs and *Escherichia coli* Ribosomal RNAs Are Cut from Large Precursor RNAs *in Vivo* by Ribonuclease III." *Proc. Nat. Acad. Sci.* 70:3296–3300.
45. Bram, R. J., R. A. Young, and J. A. Steitz. 1980. "The Ribonuclease III Site Flanking 23S Sequences in the 30S Ribosomal Precursor RNA of *E. coli.*" *Cell* 19:393–401.
46. Robertson, H. D. 1982. "*Escherichia coli* Ribonuclease III Cleavage Sites." *Cell* 30:669–672.
47. Sogin, M. L., N. R. Pace, M. Rosenberg, and S. Weissman. 1976. "Nucleotide Sequence of a 5S Ribosomal RNA Precursor from *Bacillus subtilis.*" *J. Biol. Chem.* 251:3480–3488.
48. Yanofsky, C., T. Platt, I. P. Crawford, B. P. Nicholas, G. E. Christie, H. Horowitz, M. VanCleemput, and A. M. Wu. 1981. "The Complete Nucleotide Sequence of the Tryptophan Operon of *Escherichia coli.*" *Nucleic Acid Res.* 9:6647–6668.
49. Dintzis, H. M. 1961. "Assembly of the Peptide Chain of Hemoglobin." *Proc. Nat. Acad. Sci.* 47:247–261.
50. Kozak, M. 1983. "Comparison of Initiation of Protein Synthesis in Procaryotes, Eucaryotes, and Organelles." *Microbiol. Rev.* 47:1–45.
51. Shine, J., and L. Dalgarno. 1974. "The 3'-Terminal Sequence of *E. coli* 16S rRNA: Complementarity to Nonsense Triplets and Ribosome Binding Sites." *Proc. Nat. Acad. Sci.* 71:1342–1346.
52. Steitz, J. A. 1980. "RNA-RNA Interactions During Polypeptide Chain Initiation." In *Ribosomes: Structure, Function and Genetics*, ed. G. Chambliss, G. R. Craven, J. Davies, K. Davis, L. Kahan, and M. Nomura. Baltimore, Md.: University Park Press, pp. 479–495.
53. Gold, L., D. Pribnow, T. Schneider, S. Shinedling, B. S. Singer, and G. Stormo. 1981. "Translational Initiation in Prokaryotes." *Ann. Rev. Microbiol.* 35:365–403.
54. Grunberg-Manago, M. 1980. "Initiation of Protein Synthesis as Seen in 1979." In *Ribosomes: Structure, Function and Genetics*, ed. G. Chambliss, G. R. Craven, J. Davies, K. Davis, L. Kahan, and M. Nomura. Baltimore, Md.: University Park Press, pp. 445–478.
55. Watson, J. D. 1964. "The Synthesis of Proteins Upon Ribosomes." *Bull. Soc. Chim. Biol.* 46:1399–1425.
56. Weissbach, H. 1980. "Soluble Factors in Protein Synthesis." In *Ribosomes: Structure, Function and Genetics*, ed. G. Chambliss, G. R. Craven, J. Davies, K. Davis, L. Kahan, and M. Nomura. Baltimore, Md.: University Park Press, pp. 377–412.
57. Garrett, R. A., and P. Woolley. 1982. "Identifying the Peptidyl Transferase Centre." *Trends in Biochem. Sci.* 7:385–386.
58. Johnson, A., H. Adkins, E. Matthews, and C. Cantor. 1982. "Distance Moved by Transfer RNA During Translocation from the A Site to the P Site on the Ribosome." *J. Mol. Biol.* 156:113–140.
59. Traut, R. R., and R. E. Monro. 1964. "The Puromycin Reaction and Its Relation to Protein Synthesis." *J. Mol. Biol.* 10:63–72.
60. Cundliffe, E. 1980. "Antibiotics and Prokaryotic Ribosomes: Action, Interaction, and Resistance." In *Ribosomes: Structure, Function and Genetics*, ed. G. Chambliss, G. R. Craven, J. Davies, K. Davis, L. Kahan, and M. Nomura. Baltimore, Md.: University Park Press, pp. 377–412.

61. Caskey, C. T., W. C. Forrester, and W. Tate. 1984. "Peptide Chain Termination." In *Alfred Benzon Symposium*, vol. 19, ed. B. Clark and H. Petersen. Copenhagen: Munksgaard, pp. 457–466.
62. Kaziro, Y. 1978. "The Role of Guanosine 5'-Triphosphate in Polypeptide Chain Elongation." *Biochim. Biophys. Acta* 505:95–127.
63. Kurland, C. G. 1982. "Translational Accuracy *in Vitro.*" *Cell* 28:201–202.
64. Yarus, M., and R. Thompson. 1983. "Precision of Protein Biosynthesis." In *Gene Function in Prokaryotes*, ed. J. Beckwith, J. Davies, J. A. Gallant. Cold Spring Harbor, N.Y.: Cold Spring Harbor Laboratory, pp. 23–63.
65. Kurland, C. G., and J. Gallant. 1985. "The Secret Life of the Ribosome." In *Accuracy in Molecular Processing; Its Control and Relevance to Living Systems*, ed. D. Galas, R. Rosenburger, T. Kirkwood. London: Chapman and Hall, 1985.
66. Gallant, J. A. 1979. "Stringent Control in *E. coli.*" *Ann. Rev. Genetics* 13:393–415.
67. Nomura, M., R. Gourse, and G. Baughman. 1984. "Regulation of the Synthesis of Ribosomes and Ribosomal Components." *Ann. Rev. Biochem.* 53:75–117.
68. Lamond, A. I., and A. A. Travers. 1985. "Stringent Control of Bacterial Transcription." *Cell* 41:6–8.
69. Silhavy, T. J., S. A. Benson, and S. D. Emr. 1983. "Mechanisms of Protein Localization." *Microbiol. Rev.* 47:313–344.
70. Michaelis, S., and J. Beckwith. 1982. "Mechanism of Incorporation of Cell Envelope Proteins in *Escherichia coli.*" *Ann. Rev. Microbiol.* 36:435–465.
71. Wickner, W. 1983. "M13 Coat Protein as a Model of Membrane Assembly." *Trends in Biochem. Sci.* 8:90–94.
72. Randall, L. L., and S. J. S. Hardy. 1984. "Export of Protein in Bacteria: Dogma and Data." In *Modern Cell Biology*, vol. 3, ed. B. Satir. New York: Liss, pp. 1–20.
73. Davis, N. G., and P. Model. 1985. "An Artificial Anchor Domain: Hydrophobicity Suffices To Stop Transfer." *Cell* 41:607–614.
74. Engelman, D. M., and T. A. Steitz. 1981. "The Spontaneous Insertion of Proteins into and Across Membranes: The Helical Hairpin Hypothesis." *Cell* 23:411–422.
75. Lake, J. A. 1981. "The Ribosome." *Sci. Amer.* 245:84–97.
76. Engelman, D. M., and P. B. Moore. 1976. "Neutron Scattering Studies of the Ribosome." *Sci. Amer.* 235:43–54.
77. Ofengand, J. 1980. "The Topography of tRNA Binding Sites on the Ribosome." In *Ribosomes: Structure, Function and Genetics*, ed. G. Chambliss, G. R. Craven, J. Davies, K. Davis, L. Kahan, and M. Nomura. Baltimore, Md.: University Park Press, pp. 497–530.
78. Cooperman, B. S. 1980. "Functional Sites on the *E. coli* Ribosome as Defined by Affinity Labeling." In *Ribosomes: Structure, Function and Genetics*, ed. G. Chambliss, G. R. Craven, J. Davies, K. Davis, L. Kahan, and M. Nomura. Baltimore, Md.: University Park Press, pp. 531–554.
79. Traut, R. R., J. M. Lambert, G. Boileau, and J. W. Kenny. 1980. "Protein Topography of *Escherichia coli* Ribosomal Subunits as Inferred from Protein Crosslinking." In *Ribosomes: Structure, Function and Genetics*, ed. G. Chambliss, G. R. Craven, J. Davies, K. Davis, L. Kahan, and M. Nomura. Baltimore, Md.: University Park Press, pp. 89–110.
80. Stoffler, G., R. Bald, B. Kastner, R. Luhrmann, M. Stoffler-Meilicke, and G. Tischendorf. 1980. In *Ribosomes: Structure, Function and Genetics*, ed. G. Chambliss, G. R. Craven, J. Davies, K. Davis, L. Kahan, and M. Nomura. Baltimore, Md.: University Park Press, pp. 171–206.
81. Cantor, C. R. and P. R. Schimmel. 1980. *Biophysical Chemistry*, Part II. San Francisco: Freeman, pp. 448–454.
82. Zimmermann, R. A. 1980. "Interactions Among Protein and RNA Components of the Ribosome." In *Ribosomes: Structure, Function and Genetics*, ed. G. Chambliss, G. R. Craven, J. Davies, K. Davis, L. Kahan, and M. Nomura. Baltimore, Md.: University Park Press, pp. 135–170.

CHAPTER 15

The Genetic Code

By 1960, the general outline of how RNA participates in protein synthesis had been established. Nevertheless, there was little optimism that we would soon have a detailed understanding of the genetic code itself. It was believed that identification of the codons for a given amino acid would require exact knowledge of both the nucleotide sequences of a gene and the corresponding amino acid order in its protein product. At that time, the elucidation of amino acid sequences, although a laborious process, was already a very practical one. On the other hand, the then-current methods for determining DNA sequences were very primitive. Fortunately, this apparent roadblock did not hold up progress. In 1961, just one year after the discovery of mRNA, the use of artificial messenger RNAs partially cracked the genetic code, with the unambiguous demonstration of a codon for the amino acid phenylalanine. To explain how this discovery was made, we must first describe how biochemists study protein synthesis in cell-free systems.

Addition of mRNA Stimulates in Vitro Protein Synthesis[1]

The need for three forms of RNA—tRNA, rRNA, and mRNA—for protein synthesis was demonstrated largely by experiments using cell-free extracts prepared from cells actively engaged in protein synthesis. In these experiments, carefully disrupted cells were fractionated to see which cell components were necessary for incorporation of amino acids into proteins. All these experiments utilized radioactively labeled (^{3}H, ^{14}C, or ^{35}S) amino acids because in vitro protein synthesis is quite inefficient; that is, there is no detectable net synthesis compared to the amount of protein present in the extract. Only by using labeled precursors can the incorporation of amino acids into new proteins be convincingly demonstrated (Figure 15-1).

A typical time course of the in vitro incorporation of radioactive amino acids into proteins is illustrated by the graph in Figure 15-2. It shows that synthesis in an *E. coli* extract proceeds rapidly for several minutes and then gradually stops. During this interval, there is a corresponding loss of mRNA, owing to the action of degradative enzymes present in the extract. This suggests that one cause of the inefficiency of cell-free protein synthesis is loss of the template component. The correctness of this supposition is shown by adding new mRNA to extracts that have stopped making protein. Such an addition causes an immediate resumption of synthesis. Extracts depleted of mRNA are very valuable in testing mRNA activity. Because they contain very little functional mRNA, they can be used to detect small amounts of template activity in externally added mRNA.

Figure 15-1
Experimental details of in vitro studies of protein synthesis.

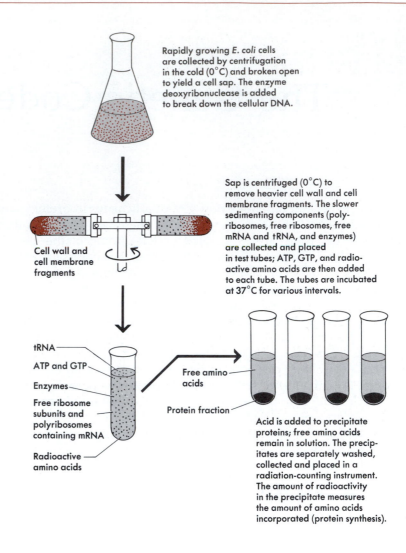

Rapidly growing *E. coli* cells are collected by centrifugation in the cold (0°C) and broken open to yield a cell sap. The enzyme deoxyribonuclease is added to break down the cellular DNA.

Sap is centrifuged (0°C) to remove heavier cell wall and cell membrane fragments. The slower sedimenting components (poly-ribosomes, free ribosomes, free mRNA and tRNA, and enzymes) are collected and placed in test tubes; ATP, GTP, and radio-active amino acids are then added to each tube. The tubes are incubated at 37°C for various intervals.

Cell wall and cell membrane fragments

tRNA
ATP and GTP
Enzymes
Free ribosome subunits and polyribosomes containing mRNA
Radioactive amino acids

Free amino acids
Protein fraction

Acid is added to precipitate proteins; free amino acids remain in solution. The precip-itates are separately washed, collected and placed in a radiation-counting instrument. The amount of radioactivity in the precipitate measures the amount of amino acids incorporated (protein synthesis).

Specific Proteins Can Be Made in Cell-Free Systems[2, 3, 4]

Even with the addition of excess mRNA, there is still only a small amount of amino acid incorporation into polypeptide chains in the in vitro systems studied. Therefore, doubts were initially raised about whether the newly made proteins had structures at all similar to those of natural proteins. Fortunately, the use of specific viral RNA showed that these doubts were unfounded and that the genetic code can be accurately read in cell-free systems. In the first successful experiments, which took place in 1962, RNA isolated from the bacterial virus f2 was added to preincubated *E. coli* extracts. The RNA acted as a template and promoted the incorporation of amino acids into protein. Some complete polypeptide chains were made and released from the ribosomes. These newly made protein products were then compared with the f2 coat protein (MW about 14,000) and were found to have amino acid sequences identical to that of the coat protein synthesized in vivo. Thus, under cell-free conditions, mRNA molecules on the ribosomal surface can select the appropriate AA~tRNA precursors.

Figure 15-2
Ⓐ indicates the breakdown of endoge-nous mRNA, while Ⓑ shows the breakdown of exogenously added mRNA whose time of addition is indi-cated by the arrow.

Over the ensuing decade, the cell-free synthesis of a variety of other viral and bacterial proteins was achieved. One of the earliest of these successes was that of the T4 lysozyme, a 20,000-dalton enzyme that helps digest bacterial cell walls. Its cell-free synthesis in enzymatically active form further showed the great fidelity of translation possible in vitro. Although at first there was some difficulty in making very long polypeptide chains, by 1969 the synthesis of enzymatically active β-galactosidase (a protein of 1300 amino acids) was convincingly demonstrated. Subsequently, a number of important mammalian proteins (e.g., hemoglobin, light and heavy immunoglobulin chains, actin, collagen, and myosin) have been made in the test tube, proving that the conditions for in vitro synthesis of higher cell proteins are now as well understood as those of their bacterial counterparts.

Stimulation of Amino Acid Incorporation by Synthetic mRNA[5]

While the foregoing experiments with natural mRNAs told us that reading specificity is preserved in cell-free systems, by themselves they said nothing about the exact form of the genetic code. They could not tell us which three-letter words (out of the possible 64) code for any particular amino acid. To obtain such data, use was made of synthetic polyribonucleotides, whose formation involves the enzyme polynucleotide phosphorylase. This enzyme, found in all bacteria, catalyzes the reaction

$$\text{RNA} + \textcircled{P} \rightleftharpoons \text{ribonucleoside—}\textcircled{P}\sim\textcircled{P} \qquad (15\text{-}1)$$

With normal cell metabolite concentrations, the equilibrium conditions favor RNA degradation to nucleoside diphosphates, so that the main cellular function of polynucleotide phosphorylase may be to control mRNA lifetime (Chapter 16). By use of high initial nucleoside diphosphate concentrations, however, this enzyme can be made to catalyze the formation of internucleotide 3'–5' phosphodiester bonds and thus make RNA molecules (Figure 15-3). No template DNA or RNA is required for RNA synthesis with this enzyme; the base composition of the synthetic product depends entirely on the ratio of the various ribonucleoside diphosphates added to the reaction mixture. For example, when only adenosine diphosphate is used, the resulting RNA contains only adenylic acid and is thus called polyadenylic acid or poly A. It is likewise possible to make poly U, poly C, and poly G. Addition of two or more different diphosphates produces mixed copolymers such as poly AU, poly AC, poly CU, and poly AGCU. In all these mixed polymers, the base sequences are approximately random, with the nearest-neighbor frequencies determined solely by the relative concentrations of the reactants. For example, poly AU molecules with two times as much A as U have sequences like UAAUAUAAAUAAUAAAAUAUU. . . .

$$5' \ldots \text{ApApA}_{\text{OH}}{}^{3'} + \text{ppA} \xrightleftharpoons[\text{phosphorylase}]{\text{Polynucleotide}} \ldots \text{ApApApA}_{\text{OH}} + \textcircled{P}$$

$$\text{Poly A}_n \qquad\qquad\qquad\qquad \text{Poly A}_{n+1}$$

Figure 15-3
Synthesis (degradation) of RNA molecules using the enzyme polynucleotide phosphorylase.

Poly U Codes for Polyphenylalanine[6]

Under the right conditions in vitro, almost all synthetic polymers will attach to ribosomes and function as templates. Luckily, high concentrations of magnesium were used in the early experiments. A high magnesium concentration circumvents the need for initiation factors and the special initiator fMet-tRNA$_F^{Met}$, allowing chain initiation to take place without the proper signals in the mRNA (Chapter 14).

Poly U was the first synthetic polyribonucleotide discovered to have mRNA activity. It selects phenylalanyl tRNA molecules exclusively, thereby forming a polypeptide chain containing only phenylalanine (polyphenylalanine). Thus, we know that a codon for phenylalanine is composed of a group of three uridylic acid residues, UUU. (That a codon has three nucleotides was known from the genetic experiments described in Chapter 8.)

On the basis of analogous experiments with poly C and poly A, CCC was assigned as a proline codon and AAA as a lysine codon. Unfortunately, this type of experiment did not tell us what amino acid GGG specifies. The guanine residues in poly G firmly hydrogen-bond to each other and form multistranded triple helices that do not bind to ribosomes.

Mixed Copolymers Allowed Additional Codon Assignments[7]

Poly AC molecules can contain eight different codons, CCC, CCA, CAC, ACC, CAA, ACA, AAC, and AAA, whose proportions vary with the copolymer A/C ratio. When AC copolymers attach to ribosomes, they cause the incorporation of asparagine, glutamine, histidine, and threonine, in addition to the proline previously assigned to CCC codons and the lysine previously assigned to AAA codons. The proportions of these amino acids incorporated into polypeptide products depend on the A/C ratio. Thus, since an AC copolymer containing much more A than C promotes the incorporation of many more asparagine than histidine residues, we conclude that asparagine is coded by two As and one C and that histidine is coded by two Cs and one A (Table 15-1). Similar experiments with other copolymers allowed a number of additional assignments. Such experiments, however, did not reveal the order of the different nucleotides within a codon. There is no way of knowing from random copolymers whether the histidine codon containing two Cs and one A is ordered CCA, CAC, or ACC.

Transfer RNA Binding to Defined Trinucleotide Codons[8]

A direct way of ordering the nucleotides within some of the codons was developed in 1964. This method utilized the fact that even in the absence of all the factors required for protein synthesis, specific aminoacyl-tRNA molecules can bind to ribosome-mRNA complexes. For example, when poly U is mixed with ribosomes, only phenylalanyl-tRNA will attach. Correspondingly, poly C promotes the binding of prolyl-tRNA. Most important, this specific binding does not demand

Table 15-1 Amino Acid Incorporation into Proteins*

Amino Acid	Observed Amino Acid Incorporation	Tentative Codon Assignments	Calculated Triplet Frequency				Sum of Calculated Triplet Frequencies
			3A	2A1C	1A2C	3C	
Poly AC (5:1)							
Asparagine	24	2A1C		20			20
Glutamine	24	2A1C		20			20
Histidine	6	1A2C			4.0		4
Lysine	100	3A	100				100
Proline	7	1A2C, 3C			4.0	0.8	4.8
Threonine	26	2A1C, 1A2C		20	4.0		24
Poly AC (1:5)							
Asparagine	5	2A1C		3.3			3.3
Glutamine	5	2A1C		3.3			3.3
Histidine	23	1A2C			16.7		16.7
Lysine	1	3A	0.7				0.7
Proline	100	1A2C, 3C			16.7	83.3	100
Threonine	21	2A1C, 1A2C		3.3	16.7		20

*The amino acid incorporation into proteins was observed after adding random copolymers of A and C to a cell-free extract similar to that described in Figure 15-1. The incorporation is given as a percentage of the maximal incorporation of a single amino acid. The copolymer ratio was then used to calculate the frequency with which a given codon would appear in the polynucleotide product. The relative frequencies of the codons are a function of the probability that a particular nucleotide will occur in a given position of a codon. For example, when the A/C ratio is 5:1, the ratio of AAA/AAC = 5 × 5 × 5:5 × 5 × 1 = 125:25. If we thus assign to the 3A codon a frequency of 100, then the 2A and 1C codon is assigned a frequency of 20. By correlating the relative frequencies of amino acid incorporation with the calculated frequencies with which given codons appear, tentative codon assignments can be made.

the presence of long mRNA molecules. In fact, the binding of a *trinucleotide* to a ribosome is sufficient. The addition of the trinucleotide UUU results in phenylalanyl-tRNA attachment, whereas if AAA is added, lysyl-tRNA specifically binds to ribosomes. The discovery of this trinucleotide effect provided a relatively easy way of determining the order of nucleotides within many codons. For example, the trinucleotide 5'-GUU-3' promotes valyl-tRNA binding, 5'-UGU-3' stimulates cysteinyl-tRNA binding, and 5'-UUG-3' causes leucyl-tRNA binding (Table 15-2). Although all 64 possible trinucleotides were syn-

Table 15-2 Binding of Aminoacyl tRNA Molecules to Trinucleotide-Ribosome Complexes

Trinucleotide						AA~tRNA Bound
5'-UUU-3'	UUC					Phenylalanine
UUA	UUG	CUU	CUC	CUA	CUG	Leucine
AAU	AUC	AUA				Isoleucine
AUG						Methionine
GUU	GUC	GUA	GUG	UCU*		Valine
UCU	UCC	UCA	UCG			Serine
CCU	CCC	CCA	CCG			Proline
AAA	AAG					Lysine
UGU	UGC					Cysteine
GAA	GAG					Glutamic acid

*Note that this codon was misassigned by this method.

Figure 15-4
Using a combination of organic synthesis and copying by DNA polymerase I, double-stranded DNA with simple repeating sequences can be generated. RNA polymerase will then synthesize long polyribonucleotides corresponding to one or the other DNA strand, depending on the choice of ribonucleoside triphosphates added to the reaction mixture.

thesized with the hope of definitely assigning the order of every codon, not all codons were determined in this way. Some trinucleotides bind to ribosomes much less efficiently than UUU or GUU, making it impossible to know whether they code for a specific amino acid.

Codon Assignments from Repeating Copolymers[9]

At the same time that the trinucleotide binding technique became available, a combination of organic chemical and enzymatic techniques were being used to prepare synthetic polyribonucleotides with known repeating sequences (Figure 15-4). Ribosomes start protein synthesis at random points along these regular copolymers; yet they incorporate specific amino acids into polypeptides. For example, the repeating sequence CUCUCUCU . . . is the messenger for a regular polypeptide in which leucine and serine alternate. Similarly, UGUGUG . . . promotes the synthesis of a polypeptide containing two amino acids, cysteine and valine. And ACACAC . . . directs the synthesis of a polypeptide alternating threonine and histidine (Table 15-3). Use of the copolymer built up from repetition of the three-nucleotide sequence AAG (AAGAAGAAG) directs the synthesis of three types of polypeptides: polylysine, polyarginine, and

Table 15-3 Assignment of Codons Using Repeating Copolymers Built from Two or Three Nucleotides

Copolymer	Codons Recognized	Amino Acids Incorporated or Polypeptide Made	Codon Assignment
$(CU)_n$	CUC|UCU|CUC . . .	Leucine	5'-CUC-3'
		Serine	UCU
$(UG)_n$	UGU|GUG|UGU . . .	Cysteine	UGU
		Valine	GUG
$(AC)_n$	ACA|CAC|ACA . . .	Threonine	ACA
		Histidine	CAC
$(AG)_n$	AGA|GAG|AGA . . .	Arginine	AGA
		Glutamine	GAG
$(AUC)_n$	AUC|AUC|AUC . . .	Polyisoleucine	5'-AUC-3'
	UCA|UCA|UCA . . .	Polyserine	UCA
	CAU|CAU|CAU . . .	Polyhistidine	CAU

Table 15-4   The Genetic Code

		Second Position					
		U	C	A	G		
First Position (5' End)	U	UUU ⎤ Phe UUC ⎦ UUA ⎤ Leu UUG ⎦	UCU ⎤ UCC ⎥ Ser UCA ⎥ UCG ⎦	UAU ⎤ Tyr UAC ⎦ UAA* *Stop* UAG* *Stop*	UGU ⎤ Cys UGC ⎦ UGA* *Stop* UGG Trp	U C A G	Third Position (3' End)
	C	CUU ⎤ CUC ⎥ Leu CUA ⎥ CUG ⎦	CCU ⎤ CCC ⎥ Pro CCA ⎥ CCG ⎦	CAU ⎤ His CAC ⎦ CAA ⎤ Gln CAG ⎦	CGU ⎤ CGC ⎥ Arg CGA ⎥ CGG ⎦	U C A G	
	A	AUU ⎤ AUC ⎥ Ile AUA ⎦ AUG† Met	ACU ⎤ ACC ⎥ Thr ACA ⎥ ACG ⎦	AAU ⎤ Asn AAC ⎦ AAA ⎤ Lys AAG ⎦	AGU ⎤ Ser AGC ⎦ AGA ⎤ Arg AGG ⎦	U C A G	
	G	GUU ⎤ GUC ⎥ Val GUA ⎥ GUG† ⎦	GCU ⎤ GCC ⎥ Ala GCA ⎥ GCG ⎦	GAU ⎤ Asp GAC ⎦ GAA ⎤ Glu GAG ⎦	GGU ⎤ GGC ⎥ Gly GGA ⎥ GGG ⎦	U C A G	

*Chain-terminating, or "nonsense," codons.
†Also used to specify the initiator formyl-Met-tRNA$_F^{Met}$. The Val triplet GUG is therefore "ambiguous" in that it codes both valine and methionine.

polyglutamic acid. Poly $(AUC)_n$ behaves the same way, being a template for polyisoleucine, polyserine, and polyhistidine (Table 15-3). Further codon assignments were obtained from repeating tetranucleotide sequences. The sum of all these observations permitted the definite assignments of specific amino acids to 61 out of the possible 64 codons (Table 15-4). The remaining three codons, as shown in Table 15-4, code for chain termination.

The Code Is Degenerate

Many amino acids are specified by more than one codon, a phenomenon called **degeneracy.** For example, both UUU and UUC code for phenylalanine, while serine is encoded by UCU, UCC, UCA, UCG, AGU, and AGC. In fact, when the first two nucleotides are identical, the third nucleotide can be either cytosine or uracil and the codon will still code for the same amino acid. Often, adenine and guanine are similarly interchangeable. However, not all degeneracy is based on equivalence of the first two nucleotides. Leucine, for example, is coded by UUA and UUG, as well as by CUU, CUC, CUA, and CUG (Figure 15-5). Codon degeneracy, especially the frequent third-place equivalence of cytosine and uracil or guanine and adenine, explains how there can be great variation in the AT/GC ratios in the DNA of various organisms (Chapter 9) without correspondingly large changes in the relative proportion of amino acids in their proteins.

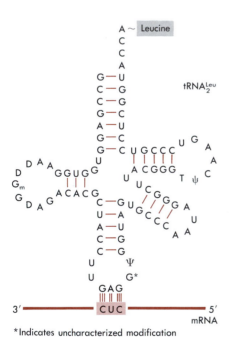

*Indicates uncharacterized modification

Figure 15-5

Two different *E. coli* tRNA molecules that accept leucine residues. They recognize different code words.

Perceiving Order in the Makeup of the Code[10]

Inspection of the distribution of codons in the genetic code suggests that the code evolved in such a way as to minimize the deleterious effects of mutations. For instance, mutations in the first position of a code word will usually give a similar (if not the same) amino acid. Furthermore, code words with pyrimidines in the second position specify mostly hydrophobic amino acids, while those with purines in the second position correspond mostly to polar amino acids (see Table 15-4). Hence, since transitions are the most common type of point mutations (Chapter 12), a change in the second position of a code word will usually replace one amino acid with a very similar one. Finally, if a code word suffers a transition mutation in the third position, almost never will a different amino acid be specified. Even a transversion mutation in this position will have no consequence about half the time.

Another consistency noticeable in the code is that whenever the first two positions of a codon are both occupied by G or C, each of the four nucleotides in the third position specifies the same amino acid (e.g., proline, alanine, arginine, or glycine). On the other hand, whenever the first two positions of the codon are both occupied by A or U, the identity of the third nucleotide does make a difference. Since GC base pairs are stronger than AU base pairs, mismatches in pairing the third codon base are often tolerated if the first two positions make strong GC base pairs. Thus, having all four nucleotides in the third position specify the same amino acid may have evolved as a safety mechanism to minimize errors in the reading of such codons.

Wobble in the Anticodon[11]

It was first proposed that a specific tRNA anticodon would exist for every codon. If that were the case, at least 61 different tRNAs, possibly with an additional three for the chain-terminating codons, would be present. Evidence began to appear, however, that highly purified tRNA species of known sequence (e.g., alanyl-tRNA^Ala) could recognize several different codons. Cases were also discovered in which an anticodon base was not one of the four regular ones, but a fifth base, inosine. Like all the other minor tRNA bases, inosine arises through enzymatic modification of a base present in an otherwise completed tRNA chain. The base from which it is derived is adenine, whose carbon 6 is deaminated to give the 6-keto group of inosine.

To explain these observations, the **wobble concept** was devised in 1966. It states that the base at the 5' end of the anticodon is not as spatially confined as the other two, allowing it to form hydrogen bonds with any of several bases located at the 3' end of a codon. Not all combinations are possible, with pairing restricted to those shown in Table 15-5. For example, U at the wobble position can pair with either adenine or guanine, while I can pair with U, C, or A (Figure 15-6). The pairings permitted by the wobble rules are those that give ribose-ribose distances close to that of the standard AU or GC base pairs. Purine-purine or pyrimidine-pyrimidine pairs would give ribose-ribose distances that are too long or too short, respectively.

The wobble rules do not permit any single tRNA molecule to recognize four different codons. Three codons can be recognized only when inosine occupies the first (5') position of the anticodon.

(a) "Wobble" enables the anticodon base to form hydrogen bonds with bases other than those in standard base pairs.

Anticodon arm of tRNA

Anticodon arm of tRNA

(b) U in the first anticodon position can pair with A or G.

(c) I in the first anticodon position can pair with U, C, or A.

Figure 15-6
Examples of wobble pairing. Note that the ribose-ribose distances for all the wobble pairs are close to that of the standard AU or GC base pair.

Almost all the evidence gathered since 1966 supports the wobble concept. For example, the concept correctly predicted that at least three tRNAs exist for the six serine code words (UCU, UCC, UCA, UCG, AGU, and AGC). The other two amino acids (leucine and arginine) that are encoded by six codons also have different tRNAs for the sets of codons that differ in the first or second position.

In the recently established three-dimensional tRNA structures, the three anticodon bases, as well as the two following (3') bases in the anticodon loop, all point in roughly the same direction, with their exact conformations largely determined by stacking interactions between the flat surfaces of the bases (see Figures 14-6 and 14-7). Thus, the first (5') anticodon base is at the end of the stack and is perhaps less restricted in its movements than the other two anticodon bases—hence, wobble in the third (3') position of the codon. By contrast, not only does the third (3') anticodon base appear in the middle of the

Table 15-5 Pairing Combinations with the Wobble Concept

Base in Anticodon	Base in Codon
G	U or C
C	G
A	U
U	A or G
I	A, U, or C

Table 15-6 Codon Usage Observed for Some *E. coli* Ribosomal Proteins*

		Second Position				
		U	C	A	G	Third Position (3' End)
First Position (5' End)	U	10 UUU ⎤ Phe 23 UUC ⎦ 1 UUA ⎤ Leu 2 UUG ⎦	18 UCU ⎤ 18 UCC ⎤ Ser 1 UCA ⎥ 1 UCG ⎦	3 UAU ⎤ Tyr 13 UAC ⎦ UAA *Stop* UAG *Stop*	1 UGU ⎤ Cys 6 UGC ⎦ UGA *Stop* 3 UGG Trp	U C A G
	C	4 CUU ⎤ 3 CUC ⎤ Leu 0 CUA ⎥ 79 CUG ⎦	3 CCU ⎤ 0 CCC ⎤ Pro 4 CCA ⎥ 36 CCG ⎦	3 CAU ⎤ His 15 CAC ⎦ 9 CAA ⎤ Gln 33 CAG ⎦	48 CGU ⎤ 26 CGC ⎤ Arg 0 CGA ⎥ 0 CGG ⎦	U C A G
	A	13 AUU ⎤ 51 AUC ⎥ Ile 0 AUA ⎦ 30 AUG Met	36 ACU ⎤ 26 ACC ⎤ Thr 3 ACA ⎥ 0 ACG ⎦	3 AAU ⎤ Asn 42 AAC ⎦ 90 AAA ⎤ Lys 24 AAG ⎦	1 AGU ⎤ Ser 12 AGC ⎦ 1 AGA ⎤ Arg 0 AGG ⎦	U C A G
	G	54 GUU ⎤ 6 GUC ⎤ Val 40 GUA ⎥ 16 GUG ⎦	93 GCU ⎤ 10 GCC ⎤ Ala 45 GCA ⎥ 28 GCG ⎦	17 GAU ⎤ Asp 45 GAC ⎦ 61 GAA ⎤ Glu 16 GAG ⎦	49 GGU ⎤ 34 GGC ⎤ Gly 0 GGA ⎥ 0 GGG ⎦	U C A G

*1209 codons are tabulated.
SOURCE: T. Ikemura, *J. Mol. Biol.* 158 (1982):573.

stack, but the adjacent base is always a bulky modified purine residue (called H in Figure 14-7). Thus, restriction of its movements may explain why wobble is not seen in the first (5') position of the code.

Transfer RNAs Are Synthesized in Varying Amounts[12, 13]

Sometimes, several tRNA molecules with distinctly different sequences have the same anticodon. Often, one of these species is present in very large amounts compared to the other species (as much as 11 times greater). In *E. coli*, for example, there are a major tyrosine tRNA and a minor tyrosine tRNA, both with the anticodon 3'-AUG-5'. Origin of the minor species by enzymatic modification of the major species following gene transcription is ruled out by the nature of the sequence differences. Rather, the two tRNAs are encoded by different genes, as is the case for the other major-minor tRNA pairs. Different promoter strengths or differences in the survival of various processing intermediates account for the different levels of mature tRNA produced. The normal function played by minor tRNA species is not yet known. Conceivably these tRNAs play regulatory roles, explaining why only small numbers of molecules need to be present. And as we will soon see, minor tRNAs are frequently involved in suppressor gene action.

Table 15-7   Codon Usage in the Genes of Animals*

		Second Position					
		U	C	A	G		
First Position (5' End)	U	13 UUU ⎤ Phe 28 UUC ⎦ 2 UUA ⎤ Leu 9 UUG ⎦	16 UCU ⎤ 18 UCC ⎥ Ser 9 UCA ⎥ 2 UCG ⎦	10 UAU ⎤ Tyr 23 UAC ⎦ UAA *Stop* UAG *Stop*	10 UGU ⎤ Cys 13 UGC ⎦ UGA *Stop* 12 UGG Trp	U C A G	Third Position (3' End)
	C	9 CUU ⎤ 27 CUC ⎥ Leu 7 CUA ⎥ 47 CUG ⎦	14 CCU ⎤ 17 CCC ⎥ Pro 10 CCA ⎥ 5 CCG ⎦	10 CAU ⎤ His 21 CAC ⎦ 10 CAA ⎤ Gln 28 CAG ⎦	8 CGU ⎤ 11 CGC ⎥ Arg 4 CGA ⎥ 5 CGG ⎦	U C A G	
	A	11 AUU ⎤ 24 AUC ⎥ Ile 4 AUA ⎦ 16 AUG Met	15 ACU ⎤ 28 ACC ⎥ Thr 11 ACA ⎥ 6 ACG ⎦	8 AAU ⎤ Asn 28 AAC ⎦ 19 AAA ⎤ Lys 49 AAG ⎦	12 AGU ⎤ Ser 21 AGC ⎦ 8 AGA ⎤ Arg 10 AGG ⎦	U C A G	
	G	9 GUU ⎤ 21 GUC ⎥ Val 5 GUA ⎥ 33 GUG ⎦	28 GCU ⎤ 38 GCC ⎥ Ala 14 GCA ⎥ 6 GCG ⎦	16 GAU ⎤ Asp 24 GAC ⎦ 21 GAA ⎤ Glu 34 GAG ⎦	22 GGU ⎤ 32 GGC ⎥ Gly 16 GGA ⎥ 11 GGG ⎦	U C A G	

*2244 codons are tabulated.
SOURCE: R. Grantham, C. Gautier, M. Gouy, M. Jacobzone, and R. Mercier, "Codon Catalog Usage Is a Genome Strategy Modulated for Gene Expressivity," *Nucleic Acid Res.* 9 (1981):r43–r74.

Correlation of tRNA Abundance with Codon Usage in Natural mRNAs[13, 14]

The fact that most amino acids are specified by more than one codon poses the question of whether the alternative codons are used with approximately equal frequencies or whether some are preferentially employed. For example, in organisms containing AT-rich DNA, we might expect that most codons have U or A at the third position. In fact, the complete sequence of the single-stranded DNA genome of bacteriophage ϕX174, which consists of 24 percent A, 22 percent C, 23 percent G, and 31 percent T, does reveal that U is the preferred base utilized in the third position of codons.

Even in organisms where there is no bias in base composition, codon usage is not random. Now that the sequence of many *E. coli* genes are known, it has become apparent that some codons are utilized repeatedly, while others almost never appear. This is illustrated in Table 15-6, which summarizes the codon frequency in a number of *E. coli* ribosomal protein genes. In the genes of higher organisms, codons are likewise used nonrandomly, but those that are favored or discriminated against are not necessarily the same as in *E. coli* (Table 15-7).

Careful analysis has revealed that the occurrence of codons in *E. coli* mRNAs is strongly correlated with the relative abundance of

Figure 15-7

Correlation between tRNA amount and the frequency of tRNA usage in genes expressed in *E. coli*. The amounts of 26 different *E. coli* tRNA species were measured relative to the amount of $tRNA_1^{Leu}$, arbitrarily set at 1.0. Based on the known anticodon sequences of these tRNAs, the frequency of appearance of the codons they decode in several sequenced genes (*recA*, β-lactamase) or sets of genes (ribosomal proteins, φX174) was calculated. Note that there is a very good correspondence between tRNA abundance and usage for the highly expressed genes-(*recA* and ribosomal protein genes) and a much more scattered distribution for the transposon gene (β-lactamase) and the phage genes (φX174). If every tRNA species were used equally, a horizontal distribution would be observed. [After T. Ikemura, *J. Mol. Biol.* 146 (1981):1.]

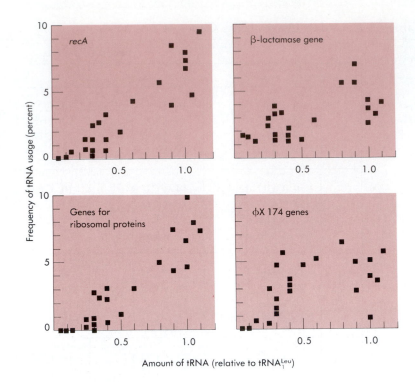

their respective tRNA species (Figure 15-7). This is especially true of the mRNAs for proteins that are synthesized in very large amounts (for instance, RecA protein and ribosomal proteins). Such mRNAs contain codons read by the most abundant isoacceptor tRNAs and lack codons corresponding to minor tRNA species. The same correlation is seen, but to a lesser degree, with other *E. coli* proteins, like those encoded by transposons or phage genomes. In the cells of higher organisms also, codons corresponding to the major tRNA isoacceptors are almost always used with the highest frequency. Two especially good examples are the abundant yeast proteins alcohol dehydrogenase isozyme I and glyceraldehyde phosphate dehydrogenase, for which more than 96 percent of the amino acids are coded by only 25 triplets, whose corresponding tRNAs are highly abundant. Hence, in strongly expressed genes, codon choice is constrained by tRNA availability and seems to be designed to optimize translational efficiency. This hints that codons corresponding to rare tRNAs may in fact slow down translation.

Even when a single tRNA can wobble to read several different codons, one codon is usually preferred. Analysis of such preferences has led to several interesting realizations regarding codon-anticodon interaction:

- Certain modifications of U in the 5′ (wobble) position of the anticodon give a preference for A over G in the third position of the codon.

- I at the 5′ (wobble) position of the anticodon preferentially pairs with U or C, rather than with A.

- When the first two letters of a codon form AU base pairs with the anticodon, the third base is preferentially C (giving a GC base pair) rather than U (giving a third AU pair). In other words, the more stable codon-anticodon interaction is preferred.

Knowing these rules in addition to the relative abundances of the different types of tRNA in any particular cell now permits fairly good prediction of codon usage in the mRNAs frequently translated in that cell.

AUG and GUG as Initiation Codons[15]

The discovery of a tRNA specific for *N*-formylmethionine (tRNA$_F^{Met}$) initially suggested that it would recognize a codon different from the AUG methionine codon. But as shown in Figure 14-26, elucidation of the tRNA$_F^{Met}$ sequence revealed that its anticodon 3'-UAC-5' was identical to that of tRNA$_M^{Met}$. Both fMet and Met must therefore be directly coded by AUG. Discrimination between the two forms occurs through their differential binding by protein synthesis factors. Only fMet-tRNA$_F^{Met}$ can bind to the initiation factor IF2 to form the 30S initiation complex; and only Met-tRNA$_M^{Met}$ is able to bind to the elongation factor EF-Tu during polypeptide chain elongation.

A further complication is the realization, coming first from in vitro studies and later from mRNA sequences, that fMet-tRNA$_F^{Met}$ can bind to and initiate synthesis at GUG as well as AUG codons. GUG is used as an initiator codon only about one-thirtieth as frequently as AUG initiators in *E. coli*. Normally, GUG is a valine codon and its recognition by tRNA$_F^{Met}$ would require an unusual sort of wobble in the codon-anticodon interaction. The ambiguity is at the first (5') as opposed to the third (3') position of the codon and would therefore involve the 3' base of the anticodon. A possible explanation for the unexpected ability of the initiator tRNA to engage in first-position wobble lies in the tRNA$_F^{Met}$ sequence. The nucleotide adjacent to the 3' end of the anticodon is an unmodified adenine, not the bulky alkylated derivative found in virtually all other tRNAs (H in Figure 14-2). In fact, UUG and CUG codons also specify initiation, although even more rarely than GUG. Thus, the unique wobble capability of the third (3') position of the tRNA$_F^{Met}$ anticodon seems to extend to all possible partners.

Three Different Codons Direct Chain Termination[16]

The three codons UAA, UAG, and UGA do not correspond to any amino acid. Instead, they signify chain termination. As mentioned in Chapter 14, these codons are read not by special chain-terminating tRNAs but by specific proteins, the release factors (see Chapter 14). Two release factors, RF1 and RF2, have been identified, each of which recognizes two codons. One is specific for UAA and UAG and the other for UAA and UGA. Knowledge of how two different codons can be recognized by each release factor must await detailed knowledge of the release factors' three-dimensional structures. The use of proteins to read the stop signals emphasizes the fact that not only polynucleotides can specifically interact with other polynucleotides. The specific hydrogen-bond-forming groups of the bases can also be recognized by proteins containing specific constellations of amino acids that have groups prone to hydrogen bonding.

Why three different stop codons exist is not at all obvious, and initially much speculation arose as to which codon would be prefer-

Figure 15-8
Nucleotide sequences at the 3′ ends of coding regions of mRNAs translated in *E. coli.* The stop codons are indicated by color boxes. Tandem stop codons do occur, but are rare. Likewise, when many mRNAs are compared, the average distance from the terminator to the next in-frame stop codon is about as expected on a random basis. [After J. Kohli and H. Grosjean, *Mol. Gen. Genetics* 182 (1981):430.]

	5′				3′
MS2 A protein	— AGA	UAG	AGC		—
MS2 coat protein	— UAC	UAA	UAG	ACG	—
β-galactosidase	— AAA	UAA	UAA	UAA	—
IlvG protein	— GUU	UAG	UCA		—
Ribosomal protein S13	— AAA	UAA	UCG GGG	UGA	—
Ribosomal protein S7	— AAU	UGA	ACG		—

entially used to end polypeptide coding regions. Now that the nucleotide sequences at the ends of a number of *E. coli* genes have been elucidated, it is clear that UAA is the preferred, but by no means exclusive, signal. In most cases, only a single stop codon appears. But some genes end with two or even three successive stop signals (Figure 15-8). The presence of more than one stop codon may be a precaution against the rare case in which the first codon fails; but why this device is used only occasionally is unclear. For example, in the RNA phages, homologous coat protein genes terminate with either one stop codon (in phage Qβ) or two stop codons (in phages R17 and MS2).

Nonsense Versus Missense Mutations[17, 18]

An alteration that changes a codon specific for one amino acid to a codon specific for another amino acid is called a **missense mutation**. The change to a chain-termination codon is known as a **nonsense mutation**. Given the existence of only three chain-termination codons, most mutations involving single-base replacements (**point mutations**) are likely to result in missense rather than nonsense. Since a new protein arising by missense mutation contains only a single amino acid replacement, it frequently possesses some of the biological activity of the original protein. Often, missense proteins fail to function only at higher than normal temperatures and are therefore known as **temperature-sensitive mutations** (Chapter 7). Many of the abnormal hemoglobins (see Figure 3-12) are the result of missense mutations. Amino acid replacement data obtained from these changed hemoglobin molecules strongly support the idea that these mutations result from the substitution of single nucleotides.

Nonsense Mutations Produce Incomplete Polypeptide Chains[17, 18]

When a nonsense mutation occurs in the middle of a genetic message, an incomplete polypeptide is released from the ribosome owing to premature chain termination. The size of the incomplete polypeptide chain depends on the location of the nonsense mutation. Mutations occurring near the beginning of a gene result in very short fragments, while mutations near the end produce fragments of almost normal length. Most incomplete chains have no biological activity, making most nonsense mutations in vital genes easily detectable. In contrast, the majority of missense mutations have some biological activity and can be easily overlooked. Thus, after treating *E. coli* with a mutagen, a sizable fraction of the *detectable* mutations is of the nonsense variety.

Suppressor Mutations Can Reside in the Same or a Different Gene[17–21]

Often, the effects of harmful mutations can be reversed by a second genetic change. Some of these subsequent mutations are very easy to understand, being simple **reverse** (back) **mutations,** which change an altered nucleotide sequence back to its original arrangement. Much more difficult to understand are the mutations occurring at different locations on the chromosome that suppress the change due to a mutation at site A by producing an additional genetic change at site B. Such **suppressor mutations** fall into two main categories: those occurring within the same gene as the original mutation but at a different site in this gene (**intragenic suppression**) and those occurring in another gene (**intergenic suppression**). Genes that cause suppression of mutations in other genes are called **suppressor genes**.

Now we realize that both types of suppression work by causing the production of good (or partially good) copies of the protein made inactive by the original harmful mutation. For example, if the first mutation caused the production of inactive copies of one of the enzymes involved in making arginine, then the suppressor mutation allows arginine to be made by restoring the synthesis of some good copies of this same enzyme. However, the mechanisms by which intergenic and intragenic suppressor mutations cause the resumption of the synthesis of good proteins are completely different.

Those mutations that can be reversed through additional changes in the same gene often involve insertions or deletions of single nucleotides. These shift the reading frame (Chapter 8) so that all the codons following the insertion (or deletion) are completely changed, thereby generating new amino acid sequences. Often, the shifted reading frame will contain nonsense codons, and as a result, prematurely terminated polypeptides will be produced by the mutant cell. Intragenic suppression may occur when a second mutation deletes (or inserts) a new nucleotide near the original change and thus restores the original codon arrangement beyond the second change (Figure 15-9). Even though there are still scrambled codons between

Figure 15-9
Intragenic suppression of a nucleotide deletion or insertion mutation. (a) The effects of a single-nucleotide deletion mutation upon the reading of the genetic message. (b) The mechanism by which a nucleotide addition mutation can suppress the havoc caused by the previous deletion mutation. In a similar way, the effect of a nucleotide addition could be overcome by a subsequent nucleotide deletion.

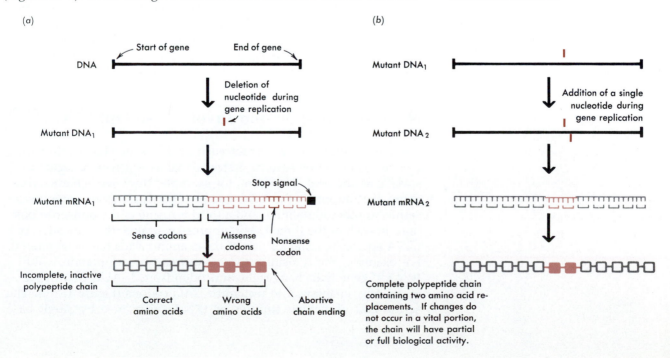

the two changes, there is a good probability, because of degeneracy, that the scrambled codons all code for some amino acid. If so, full-length, often functional proteins may be produced.

When the original mutation is a missense mutation, intragenic suppression can also result from a second missense mutation. In these cases, the original loss of enzymatic activity is due to an altered three-dimensional configuration resulting from the presence of a wrong amino acid. A second missense mutation in the same gene brings back biological activity if it somehow restores the original configuration around the functional part of the molecule. An example of this type of suppression in the tryptophan synthetase system was shown in Chapter 8 (see Figure 8-17).

Suppressor Genes Upset the Reading of the Genetic Code[17, 18, 20, 21]

Suppressor genes do not act by changing the nucleotide sequence of a mutant gene. Instead, they change the way the mRNA template is read. There are a number of different suppressor genes in *E. coli*. Since each causes the misreading of a specific nonsense or missense codon, suppressor genes can reverse the effects of only a small fraction of the point mutations that might arise within a given gene. For example, if we collect a large number of mutations blocking the synthesis of the enzyme β-galactosidase (Chapter 16), only several percent of these mutations will be suppressed by suppressor gene *a*. These few mutations will have nucleotide replacements in codons whose reading is specifically altered by gene *a*. Similarly, a completely different small fraction of β-galactosidase mutations can be suppressed by suppressor gene *b*. Thus, we see that specific codons are misread by specific suppressor genes.

On the other hand, since each suppressor gene causes the misreading of a specific codon, it is easy to understand how a given suppressor gene can suppress mutations in a number of different protein-coding genes. For example, the ability to synthesize both arginine and tryptophan in certain double mutants unable to make either amino acid can be restored by the presence of a single suppressor gene. We merely need to postulate that both these growth requirements are caused by the same specific codon change to missense or nonsense.

Nonsense Suppression Involves Mutant tRNAs[21–27]

There are suppressor genes for each of the three chain-terminating codons. They act by reading a stop signal as if it were a signal for a specific amino acid. There are, for example, three well-characterized genes that suppress the UAG codon. One suppressor gene inserts serine, another glutamine, and a third tyrosine at the nonsense position. In each of the three UAG suppressor strains, the anticodon of a tRNA species specific for one of these amino acids has been altered. For example, the tyrosine suppressor arises by a mutation within a tRNA^{Tyr} gene that changes the anticodon from 3'-AUG-5' to 3'-AUC-5', thereby enabling it to recognize UAG codons (Figure 15-10). The serine and glutamine suppressor tRNAs also arise by single-base

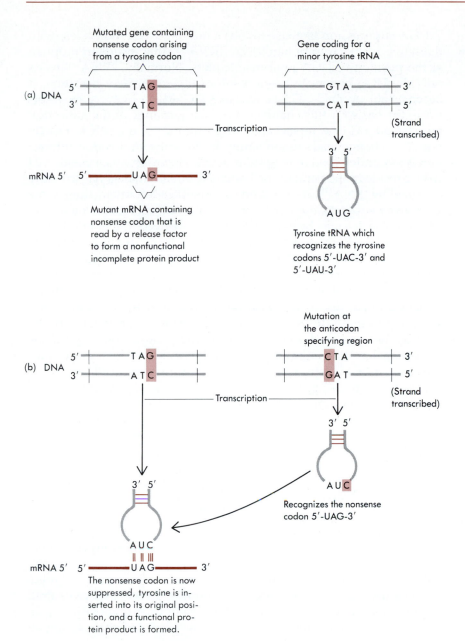

changes in their anticodons. An example of tRNA-mediated UAG suppression was first demonstrated with a nonsense mutation that blocked synthesis of the coat protein of the RNA phage R17. When mutant RNA containing this nonsense codon was used in an in vitro system, no coat protein was produced until purified tRNA from the correct suppressor strain was added.

The suppression of UAA nonsense codons is also mediated by mutant tRNAs. Tyrosine and lysine tRNAs, which normally read codons only one base different from UAA, are examples of tRNAs that can easily mutate to nonsense-suppressor tRNAs (see Table 15-4). Note that wobble in the third position of the codon enables these UAA suppressor tRNAs also to read (suppress) the UAG codon. The converse is not true.

UGA suppression likewise involves mutant tRNA molecules, with the active agent a tryptophan tRNA (tRNA^{Trp}) that inserts tryptophan at the position of UGA nonsense triplets. Normally, tRNA^{Trp} reads only the UGG codon, but as a result of the suppressor mutation, it becomes able to read UGA as well as UGG. Much to everyone's surprise, the basis for this change is not a base change at the anticodon but instead a G to A replacement at position 24 of the tRNA. Exactly how this change leads to ambiguity in the anticodon-codon interaction is not understood in molecular detail. However, since position 24 would reside in the bend of the L-shaped three-dimensional structure assumed by tRNA^{Trp}, its effect on codon-anticodon interactions some 15 Å away supports the notion that small conformational changes in tRNA molecules may be essential for accurate decoding of mRNAs on the ribosome. Perhaps not surprisingly, it turns out that even the normal tRNA^{Trp} can read the UGA terminator codon with low but detectable frequency. UGA stop signals are therefore termed "leaky."

The discovery that cells with nonsense suppressors contain mutationally altered tRNAs raised the question of how their codons corresponding to these tRNAs could continue to be read normally. In the case of the tyrosine UAG suppressor, the answer comes from the discovery that three separate genes code for tRNA^{Tyr}. One codes for the major tRNA^{Tyr} species, while the other two are duplicate genes coding for a species present in smaller amounts. One or the other of the two duplicate genes, which reside only 200 base pairs apart on the *E. coli* chromosome, is always the site of the suppressor mutation. No such dilemma exists for UGA suppression, since the suppressing tRNA^{Trp} retains its capacity to read UGG (tryptophan) codons.

Nonsense Suppressors Also Read Normal Termination Signals[16, 21, 28]

The act of nonsense suppression can be viewed as a competition between the suppressor tRNA and the release factor. When a stop signal comes into the ribosomal A site, either read-through or polypeptide chain termination will occur, depending on which arrives first. Suppression of both the UAG and UGA codons is quite efficient. In the presence of the suppressor tRNAs, over half the chain-terminating signals are read as specific amino acid codons. In contrast, suppression of the UAA codon usually averages between 1 and 5 percent. At first it was believed that the efficient suppression of UAG and UGA meant that they seldom, if ever, served as normal terminator signals. If they were frequently used, the presence of their specific suppressors might prevent much normal chain termination, leading to the production of aberrantly long polypeptides and cessation of cell growth. Now, however, the accumulation of sequence data tells us that UAG and UGA codons appear quite frequently at the ends of genes. Moreover, the use of a double stop signal is rather rare. Thus, it remains a mystery why UAA suppression is so inefficient and why UAA suppressor strains are particularly slow growers. Equally unresolved is why the UAG and UGA suppressor tRNAs do not generate an unacceptable number of abnormally long proteins.

Another aspect of nonsense suppression that is not yet understood in molecular terms is the contribution that the nucleotide sequence surrounding a stop signal makes to the efficiency of suppression (read-through). The response of a particular suppressor tRNA to a

Figure 15-11
Schematic diagram of the portion of the phage Qβ genome where read-through (natural suppression) leads to production of a minor coat protein (A₁ protein). After tryptophan is inserted at the normal UGA termination signal for the major coat protein, another 195 codons are read before tandem stop codons are encountered.

particular nonsense codon can vary as much as tenfold, depending on its location in an mRNA. Moreover, such "context effects" on the reading of stop codons are not confined to cells containing suppressor genes. In fact, several cases are known where low-level read-through of a normal termination signal is biologically important. The best documented example occurs during translation of the RNA genome of bacteriophage Qβ. This phage has two coat proteins, a major component (MW about 14,000) and a minor component (MW about 38,000), with identical N-terminal amino acid sequences. The minor protein is derived by the occasional insertion of tryptophan at the UGA termination codon for the major protein (Figure 15-11). Thus, about 5 percent of the time, a much-extended polypeptide is produced that is essential for infectivity of the virus particle.

Mutations in Normal Stop Signals[29]

The frequent occurrence of single stop signals at the ends of genes means that mutations should occasionally be detected that convert stop codons to codons that specify amino acids. Since the 3' ends of normal messages contain untranslated nucleotide sequences, translation of such mutant messages should lead to abnormally long proteins. Indeed, several such mutations have been detected, one of the first being in the human gene for the α chain of the hemoglobin molecule. Normally, the α chain is 141 amino acids long, with translation ceasing upon the reading of a single UAA stop codon. But when the stop codon is eliminated by a U to C change, glutamine becomes inserted at position 142 and translation proceeds to produce a chain of 172 amino acids (Figure 15-12). Thus, even before the sequence of the α chain gene was determined, we knew that the 3' end of its mRNA molecule contains at least 93 nucleotides normally never translated.

α Wild type   — U C C A A A U A C C G U U A A —

                            140

                Ser Lys Tyr Arg TER

                 140   148   172
               Ser Lys Tyr Arg Gln Ala Gly Ala Ser Val Ala......Val Phe Glu

α Constant Spring — U C C A A A U A C C G U C A A G C U G G A G C C U C G G U A G C ————

                 140
               Ser Asn Thr Val Lys Leu Glu Pro Arg TER

α Wayne   — U C C A A U A C C G U U A A G C U G G A G C C U C G G U A G —

                   Deletion of A

Figure 15-12
The origin of an extralong human hemoglobin α chain (hemoglobin "Constant Spring") through a U to C change in its UAA stop codon. An abnormally long α chain also characterizes hemoglobin "Wayne," which has a deletion of an A in the 139 lysine codon. The deletion shifts the reading frame so as to abolish the normal UAA stop codon. A new UAG stop codon is created at position 147, leading to a "Wayne" α chain of 146 residues.

Transfer RNA-Mediated Missense Suppression[20, 30]

Suppression of missense as well as nonsense mutations can be mediated by mutant tRNAs. For instance, there is a mutation in the tryptophan synthetase A gene in which a replacement of glycine with arginine gives rise to an inactive enzyme (Chapter 8). But there are also suppressor mutations that cause the insertion of glycine at the mutant arginine site and thus restore enzyme function. The efficiency of suppression is low, however, so even in the presence of the suppressor gene, both active and inactive forms of the enzyme are made.

The suppressor activity was traced to tRNAs that cochromatographed with glycyl-tRNA fractions, suggesting that the mutations had altered a tRNA^{Gly} anticodon. In *E. coli*, there are three different tRNA^{Gly} isoacceptors, each containing a different anticodon and thus recognizing different subsets of the four glycine codons (Table 15-8). One, called tRNA$_1^{Gly}$, recognizes only GGG. But this codon is also recognized by tRNA$_2^{Gly}$, whose anticodon can wobble to pair with either GGG or GGA. Thus, tRNA$_1^{Gly}$ is not essential for protein synthesis, and so a change in its anticodon from 3'-CCC-5' to 3'-UCC-5' allows it to decode the AGG arginine codon without simultaneously depriving the cell of the ability to read the GGG glycine codon.

In other cases, suppression arises because a mutation in a tRNA leads to its mischarging. Such changes, which can occur at various positions within the molecule, cause the tRNA to be recognized by the wrong aminoacyl-tRNA synthetase.

Frameshift Suppression[31–37]

There are also suppressor genes that mask the effects of certain frameshift mutations created by the insertion of nucleotides. Again, mutant tRNAs, such as glycine and proline tRNAs, are responsible. Complete sequence analysis of one of the frameshift glycine tRNAs (called tRNA$_{SufD}^{Gly}$, or tRNA$_{GGGG}^{Gly}$) reveals the presence of an extra base in its anticodon, with a CCC sequence being replaced by CCCC (Figure 15-13). The remainder of its sequence is identical with that of tRNA$_1^{Gly}$, indicating that it arose from an insertion mutation in *glyU*, the gene that specifies this tRNA (see Table 15-7). Possession of this extra base not only allows tRNA$_1^{Gly}$ to pair with a four-base codon, but also causes the subsequent translocation of a four-nucleotide length of mRNA from the A site to the P site of the ribosome. Thus, the reading frame is restored to the correct position.

Table 15-8 The Glycine-tRNAs of *E. coli*, Showing the Origin of Missense Suppressor tRNAs That Insert Glycine at Arginine Codons

Gene	tRNA Product	Anticodon	Glycine Codons Recognized	Derived Missense Suppressor Gene	Anticodon in tRNA Product	Arginine Codons Recognized
glyU	tRNA$_1^{Gly}$	3'-CCC-5'	GGG	*glyUsu*	3'-UCU-5'	AGG or AGA
glyT	tRNA$_2^{Gly}$	3'-CCU-5'	GGA	*glyTsu**	3'-UCU-5'	AGA
			GGG			AGG
glyV	tRNA$_3^{Gly}$	3'-CCA-5'	GGU			
			GGC			

*Strains with this mutation grow very badly (the mutation is semilethal). The origin of the residual ability to read GGA codons is unclear.

More recently, additional insertion suppressor strains of both bacteria and yeast have been characterized; some of these have tRNAs that appear to read four bases at a time, even if proper base pairing cannot occur in the fourth position. Other frameshift suppressor tRNAs act on deletions, suggesting that there is also a way of reading only two bases at a given time. Finally, frameshifting can be induced both in vivo and in vitro by a lack or overabundance of a normal charged tRNA. In other words, any condition that gives a wrong AA~tRNA a competitive advantage for binding to the ribosomal A site can potentially bring about a frameshift or misreading of the codon.

Ribosomal Mutations Also Affect the Reading Accuracy[38, 39]

Specific mutations in several of the 30S ribosomal proteins greatly influence the accuracy of reading. These effects of exact ribosome structure can be seen both in normal cells and in cells containing nonsense or missense suppressor tRNAs. A number of different amino acid replacements each must distort the ribosome so that an incorrect aminoacyl-tRNA molecule is more frequently used to elongate the polypeptide chain (Figure 15-14). One such mutation, *ram* (ribosomal ambiguity), will weakly suppress all three nonsense codons as well as missense mutations in the absence of any suppressing tRNAs. Recent evidence suggests that the proofreading step in polypeptide chain elongation, rather than the initial tRNA selection, is affected in *ram* mutants.

Streptomycin Causes Misreading[38, 39]

The finding that distorted ribosomes can misread the genetic code is strongly supported by experiments showing that the addition of the antibiotic streptomycin to either in vitro systems or living cells pro-

Figure 15-13
The nucleotide sequence of the frameshift suppressor tRNA$_\text{SufD}^\text{Gly}$ drawn in the cloverleaf configuration. Note the additional nucleotide in the anticodon loop.

Figure 15-14
How a missense mutation in a gene coding for one of the ribosomal proteins acts as a suppressor mutation.

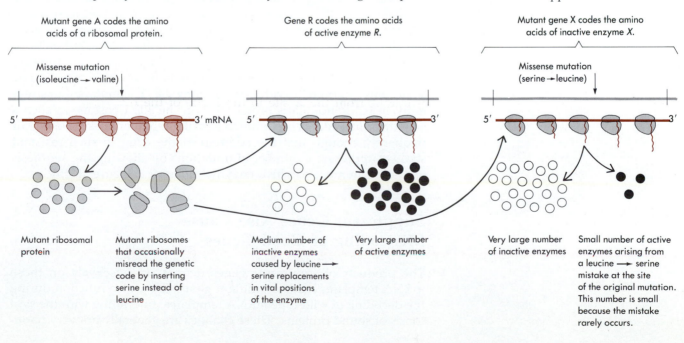

Mutant gene A codes the amino acids of a ribosomal protein.

Missense mutation (isoleucine → valine)

5' 3' mRNA

Mutant ribosomal protein

Mutant ribosomes that occasionally misread the genetic code by inserting serine instead of leucine

Gene R codes the amino acids of active enzyme R.

5' 3'

Medium number of inactive enzymes caused by leucine → serine replacements in vital positions of the enzyme

Very large number of active enzymes

Mutant gene X codes the amino acids of inactive enzyme X.

Missense mutation (serine → leucine)

5' 3'

Very large number of inactive enzymes

Small number of active enzymes arising from a leucine → serine mistake at the site of the original mutation. This number is small because the mistake rarely occurs.

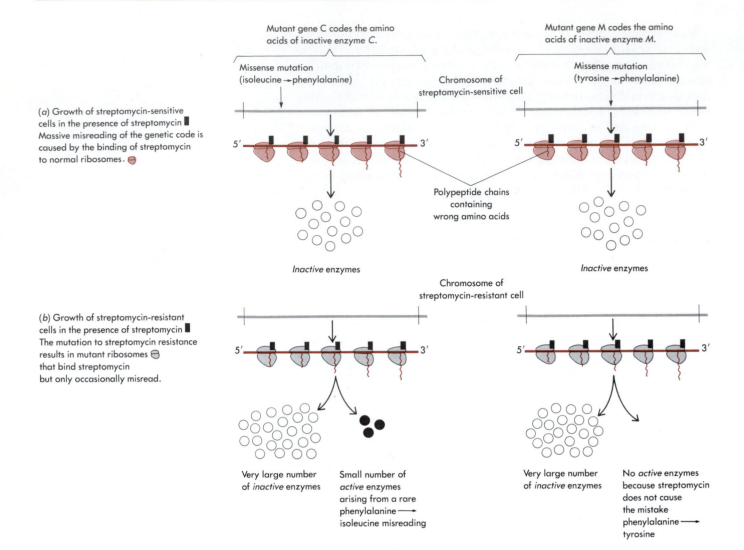

Mutant gene C codes the amino acids of inactive enzyme C.

Mutant gene M codes the amino acids of inactive enzyme M.

Missense mutation (isoleucine → phenylalanine)

Missense mutation (tyrosine → phenylalanine)

Chromosome of streptomycin-sensitive cell

(a) Growth of streptomycin-sensitive cells in the presence of streptomycin ▪ Massive misreading of the genetic code is caused by the binding of streptomycin to normal ribosomes.

5′ 3′ 5′ 3′

Polypeptide chains containing wrong amino acids

Inactive enzymes

Inactive enzymes

Chromosome of streptomycin-resistant cell

(b) Growth of streptomycin-resistant cells in the presence of streptomycin ▪ The mutation to streptomycin resistance results in mutant ribosomes that bind streptomycin but only occasionally misread.

5′ 3′ 5′ 3′

Very large number of *inactive* enzymes

Small number of *active* enzymes arising from a rare phenylalanine → isoleucine misreading

Very large number of *inactive* enzymes

No *active* enzymes because streptomycin does not cause the mistake phenylalanine → tyrosine

Figure 15-15
The action of streptomycin on strepto-mycin-sensitive and streptomycin-resis-tant *E. coli* cells.

motes mistakes in the translation of the genetic code. Streptomycin does this by binding to the 30S ribosome and somehow disturbing the normal mRNA-tRNA-ribosome interaction. The extent of misreading depends on whether the streptomycin is added to streptomycin-sensitive or streptomycin-resistant cells. Addition of the antibiotic to sensitive cells results in large-scale misreading. Since streptomycin also acts to inhibit polypeptide chain initiation, sensitive cells become so defective in protein synthesis that they die. The mutation to strep-tomycin resistance alters protein S12 on the 30S ribosome in such a way that misreadings occur much less often. Nonetheless, they occur frequently enough that the addition of the drug to such resistant strains suppresses a number of mutations by allowing the synthesis of a small number of active enzyme molecules (Figure 15-15).

Suppressor Genes Also Cause Misreading of Good Genes[21, 40, 41]

The products of suppressor genes do not act selectively on those mRNA templates made on mutant genes. They also function during the decoding of wild-type mRNA templates, interfering with the synthesis of sound proteins. These changes are generally not very harmful to the growing cell, since many more good copies of each protein than bad ones are produced. There is, however, no advantage for a normal cell to harbor suppressor mutations that cause it to produce even a small fraction of bad proteins. Suppressors tend to be selected against in evolution unless a harmful mutation is present for whose effect they must compensate.

It is now possible to make a general prediction about the function of suppressor genes. A gene becomes a suppressor gene by mutation. Before this mutation occurs, the gene is normal and active. It may code for a specific tRNA or for one of the ribosomal proteins or for an aminoacyl-tRNA synthetase. Even a tRNA-modifying enzyme may be involved; there is now evidence that certain modifications of bases in or adjacent to anticodons can affect suppression (Figure 15-16). Presumably, all such genes have evolved so that their products have the optimal configuration for ensuring accurate reading of the genetic code. If a mutation yields an altered product that increases the misreading level, the gene becomes a suppressor gene. Only when an increased mistake level is necessary for cellular existence do such mutant genes have a selective advantage over their normal counterparts.

The Code Is Nearly Universal

Poly U stimulates phenylalanine incorporation in cell-free extracts from a variety of different organisms ranging from bacteria to mammals. Likewise, poly C promotes proline incorporation and poly A causes lysine incorporation in all extracts tested, regardless of their cellular source. Such indications of the universality of the code among contemporary organisms have recently been confirmed by extensive DNA sequence data on protein-coding genes from many different species. Thus, the genetic code appears to have remained constant over a long evolutionary period.

Invariability in most of the code is expected. Consider what might happen if a mutation changed the genetic code. Such a mutation might, for example, alter the sequence of the serine tRNA molecule of the class that corresponds to UCU, causing them to recognize UUU sequences instead. This would be a lethal mutation in haploid cells containing only one gene directing the production of tRNA^{Ser}, for serine would not be inserted into many of its normal positions in proteins. Even if there were more than one gene for tRNA$^{Ser}_{UCU}$ (e.g., in a diploid cell), this type of mutation would still be lethal, since it would cause the simultaneous replacement of many phenylalanine residues by serine in cell proteins.

An Altered Code in Mammalian Mitochondria[42, 43, 44]

In view of what we have just said, it was completely unexpected to find that in certain subcellular organelles, the genetic code is in fact slightly different from the "universal" code. This realization came during the elucidation of the entire DNA sequence of the 16,569-base-pair human mitochondrial genome. Mitochondria are the cellular sites of oxidative phosphorylation; they generate ATP using enzymes bound to the inner side of their double membrane exterior. Most proteins in mitochondria, including those necessary for RNA and protein

Figure 15-16
Structure of *Drosophila* tRNA^{Tyr}, in which the modification state of an anticodon base affects suppression. Whereas the tRNA containing an unmodified G in the first position of the anticodon can occasionally read (suppress) the UAG stop codon, introduction of a Q base modification at this position abolishes suppressor activity. The structure of the Q base is shown in the inset. In another system, tRNA undermethylation is responsible for the acquisition of UGA suppressor activity. (Courtesy of E. Kubli.)

Figure 15-17

Map of the 16,500-base-pair mammalian mitochondrial genome. The map is split into two parts, the outer circle indicating those genes transcribed from the H (heavy)-strand DNA and the inner circle indicating those genes transcribed from the L (light)-strand DNA. RNA genes (for 16S and 12S rRNA or for tRNAs) are indicated in black, whereas protein-coding genes are colored. Note that whenever a space exists on one strand, it is filled by a gene on the other strand. In the designation of codons recognized by certain tRNAs, Y stands for pyrimidine, R stands for purine, and N stands for any base. ATPase 6 and 8 are components of the mitochondrial ATPase complex. CO I, CO II and CO III are cytochrome oxidase subunits. Cyt *b* stands for cytochrome *b*. ND1-5 code for components of the respiratory chain NADH dehydrogenase and were previously designated URF1-5 (for *unidentified* open *reading frame*). URF6 remains an open reading frame of unknown function. O_H and O_L designate the origins of DNA replication for the H (heavy) strand (which is initiated first) and for the L (light) strand, respectively.) (Courtesy of G. Attardi.)

synthesis, are encoded by nuclear genes and are somehow specifically transported across the mitochondrial membrane. Mitochondria contain a double-stranded circular DNA genome, which, oddly enough, gets smaller as one proceeds from lower to higher eucaryotes. (The yeast mitochondrial DNA is five times larger than that of human mitochondria.) Mitochondrial genomes encode two rRNA molecules (which are exceptionally small in mammalian cells), a few proteins, and all the tRNAs necessary to synthesize these few proteins.

In the DNA of human (and mouse) mitochondria, the genes are extremely closely packed, with one or more tRNAs serving as spacers between the regions that code for protein or rRNA (Figure 15-17). Sequences of the regions known to specify proteins have revealed the following differences between the "universal" and the mammalian mitochondrial genetic code (Table 15-9):

- UGA is not a stop signal, but codes for tryptophan. Hence, the anticodon of mitochondrial tRNA^{Trp} recognizes both UGG and UGA, as if obeying the traditional wobble rules.

- Internal methionine is encoded by both AUG and AUA; initiating methionines are specified by AUG, AUA, AUU, and AUC.

- AGA and AGG are not arginine codons (of which there are six in the "universal" code), but specify chain termination. Thus, there are four stop codons (UAA, UAG, AGA, and AGG) in the mitochondrial code.

Perhaps not surprisingly, mitochondrial tRNAs are likewise unusual both in their structure and with respect to the rules by which they decode mitochondrial messages. Only 22 tRNAs are present in mammalian mitochondria, whereas a minimum of 32 tRNA molecules are required to decode the "universal" code according to the wobble rules. Consequently, when an amino acid is specified by four codons (with the same first and second positions), only a single mitochondrial tRNA is involved. (Recall that a minimum of two tRNAs

Table 15-9 Genetic Code of Mammalian Mitochondria*

		Second Position					
		U	C	A	G		
First Position (5' End)	U	UUU ⎤ UUC ⎦ Phe (GAA)† UUA ⎤ UUG ⎦ Leu (UAA)	UCU ⎤ UCC UCA UCG ⎦ Ser (UGA)	UAU ⎤ UAC ⎦ Tyr (GUA) UAA *Stop* UAG *Stop*	UGU ⎤ UGC ⎦ Cys (GCA) UGA ⎤ UGG ⎦ Trp (UCA)	U C A G	Third Position (3' End)
	C	CUU ⎤ CUC CUA CUG ⎦ Leu (UAG)	CCU ⎤ CCC CCA CCG ⎦ Pro (UGG)	CAU ⎤ CAC ⎦ His (GUG) CAA ⎤ CAG ⎦ Gln (UUG)	CGU ⎤ CGC CGA CGG ⎦ Arg (UCG)	U C A G	
	A	AUU ⎤ AUC ⎦ Ile (GAU) AUA ⎤ AUG ⎦ Met (CAU)‡	ACU ⎤ ACC ACA ACG ⎦ Thr (UGU)	AAU ⎤ AAC ⎦ Asn (GUU) AAA ⎤ AAG ⎦ Lys (UUU)	AGU ⎤ AGC ⎦ Ser (GCU) AGA *Stop* AGG *Stop*	U C A G	
	G	GUU ⎤ GUC GUA GUG ⎦ Val (UAC)	GCU ⎤ GCC GCA GCG ⎦ Ala (UGC)	GAU ⎤ GAC ⎦ Asp (GUC) GAA ⎤ GAG ⎦ Glu (UUC)	GGU ⎤ GGC GGA GGG ⎦ Gly (UCC)	U C A G	

*Differences between the mitochrondial and "universal" genetic code (Table 15-4) are shown by shading.
†Each bracketed group of codons is read by a single tRNA whose anticodon, written 5'→3', is in parentheses. Each four-codon group is read by a tRNA with a U in the first (5') position of the anticodon. Two-codon groups with codons ending in either U/C or A/G are read with GU wobble by tRNAs, with G or U, respectively, in the first position of the anticodon. The anticodons often contain modified bases.
‡Note that the C in the first anticodon position engages in unusual pairing.

would be required by nonmitochondrial systems.) Such mitochondrial tRNAs all have in the 5' (wobble) position of their anticodons a U residue, which somehow can engage in pairing with any of the four nucleotides in the third codon position. In cases where purines in the third position of the codon correspond to different amino acids from pyrimidines in that position, a modified U in the first position of the anticodon of the mitochondrial tRNA restricts wobble to pairing with the two purines only.

In addition to obeying unusual wobble rules, mitochondrial tRNAs often have bizarre structures. For instance, the invariant GTψCRA sequence is in most cases missing (see Figure 14-2). Other invariant nucleotides located in the D and T loops are altered. The TψC loop can have from three to nine residues, rather than the seven nucleotides found consistently in the cloverleaf structure of nonmitochondrial tRNAs. An extreme case is one of the serine tRNAs, which completely lacks the D arm of the usual cloverleaf structure (Figure 15-18). Such irregularities hint that mitochondrial tRNAs differ significantly from other tRNAs both in their three-dimensional folding and in the way they interact with the mitochondrial ribosome.

Finally, the exceedingly economic usage of the genetic material in mammalian mitochondria creates still other unexpected features of mRNA synthesis and function:

• Both strands of the mitochondrial DNA are transcribed from promoters residing near the origin of DNA replication (see Figure

Figure 15-18
Gene sequences for two human mitochondrial tRNAs folded into proposed cloverleaf forms. CCA is added posttranscriptionally to the 3' ends.

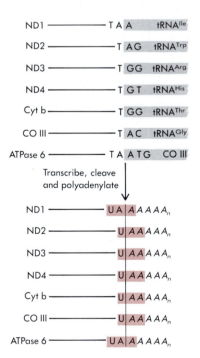

Figure 15-19
RNA processing and polyadenylation model for generating translational stop signals at the ends of the coding regions for mitochondrial proteins. The top presents the DNA sequences, showing how the coding regions abut tRNA or other protein-coding sequences. Below are the RNA sequences that would arise by transcription, site-specific cleavage at the vertical line, and addition of A residues. [After S. Anderson et al., *Nature* 290 (1981):457.]

15-17) into very long RNA molecules. Separate mRNAs are then generated by specific processing at the 5' and 3' ends of the spacer tRNA sequences.

- Initiator codons lie directly at or very close to the 5' ends of these processed mRNAs. No upstream sequences that might base-pair with the 3' end of the mitochondrial small subunit rRNA are apparent.

- The coding regions run all the way to the 3' processing sites of the mRNAs. Usually, a complete translational stop codon is not even encoded in the DNA. Rather, after the 3' end of the mRNA has been freed by cleaving off the adjacent tRNA, a string of A residues is added by a poly A–adding enzyme. This creates a UAA stop signal at the junction between the mRNA and its poly A tail (Figure 15-19).

More recently, the mitochondria of other organisms (e.g., yeast) have also been found to have slightly altered genetic codes (see Chapter 18). Moreover, they differ in detail even from the code of mammalian mitochondria (see Table 15-9). Mitochondrial DNA is known to evolve exceedingly rapidly. Hence, it is not clear whether the apparently simpler mitochondrial codes represent remnants of a primitive genetic code or whether they are more streamlined versions of the widely adopted "universal" code.

The Unexpected Existence of Overlapping Genes and Signals[45, 46, 47]

In most of the phage and bacterial genes that were sequenced earliest, the genome showed a neat sequential ordering of signals specifying the different steps in gene expression. That is, the promoter signal was followed by a 5' leader region containing the bases complementary to the 16S rRNA needed for the initiation of protein synthesis; then there appeared the polypeptide coding region; and beyond the

3' stop codon(s) were trailer nucleotides containing sequences specifying termination of RNA synthesis.

The discovery in 1977 of a gene embedded completely within another gene therefore came as a total surprise. It clearly demonstrated that a given DNA sequence can be multifunctional. Work leading to the elucidation of the complete sequence of the single-stranded DNA genome of phage φX174 first revealed that gene E, specifying a lysis protein, is translated in the +1 frame relative to the D gene, within which it resides entirely. Subsequently, it was realized that another φX174 gene, called B (involved in single-stranded DNA synthesis), is encoded in the −1 frame within the A gene. And still a third gene, K (also involved in lysis), overlaps both the A and C genes (Figure 15-20). Preceding the coding regions for each of the three overlapping genes there appears the expected short complementarity to 16S rRNA, always appropriately positioned relative to the initiating AUG (see Figure 14-28). Thus, the sequence of the "parent" gene not only encodes its own protein, but also provides the ribosome binding signal for initiating translation of the overlapping gene. Similarly, termination of translation is specified by an out-of-frame nonsense codon in the "parent" gene. Most remarkable is a short stretch of DNA found in both φX174 and its close relative G4 where the K gene is read in a third frame relative to the A and C genes, which already overlap slightly at their boundary (Figure 15-21). The N terminus of K, like all of E, is read in the +1 frame relative to its "parent" gene. Since φX174 DNA preferentially uses Ts in the third positions of its codons, both K and E polypeptides have a high proportion of hydrophobic amino acids (codons with U in the second position) and associate with the cell membrane during their function.

In addition to extensive overlapping protein-coding regions, other types of signals residing within protein-coding regions were discovered during the DNA sequence analyses of φX174 and G4 (see Figure 15-20). For example, promoters for phage mRNA transcription have been identified in the A and C protein-coding regions. The origin of single-stranded DNA synthesis is located early in the A gene. The A protein itself is translated from two different AUGs in the same reading frame, giving rise to two polypeptides (called A and A*) of different size and function. Again, the ribosome binding signal for the shorter (A*) protein is simply buried within the coding region of the larger protein.

Subsequent work has shown that the small, single-stranded DNA phages are not alone in indulging in extremely economical utilization of their DNA sequences. Completely overlapping genes have so far been identified only in viruses. But in bacterial mRNAs that encode more than one protein (polycistronic mRNAs), it is quite common to find overlap of a few nucleotides where one coding region stops and another starts (Figure 15-22). Likewise, minor promoters for RNA transcription are often buried within protein-coding regions of *E. coli* genes.

Figure 15-20
Map of the 5375-nucleotide genome of phage φX174. The reading frames of the overlapping genes relative to their "parent" genes are indicated by colored shading. Light color means translation in the −1 frame; dark color indicates translation in the +1 frame; gray means translation in the same frame. [Adapted from Sanger et al. *Nature* 265 (1977):687.]

Figure 15-21
Region of the phage G4 genome where the C, A(A*), and K genes all overlap. Although this situation was first discovered in G4, an analogous region is present in φX174 DNA (see Figure 15-20). [After G. N. Godson, B. G. Barrell, R. Staden, and J. C. Fiddes, *Nature* 276 (1978):236.]

trpE-trpD	CAGGA GACTTTCTGATGGCT
trpD-trpC	CACGAGG GTAAATGATGCAA
trpC-trpB	TAAGGA AAGGAACAATGACA
trpB-trpA	CGAGG GGAAATCTGATGGAA

Figure 15-22

Gene overlap at several intercistronic junctions in the tryptophan operon of *E. coli.* Initiator and stop codons are boxed in color and gray, respectively; nucleotides complementary to 16S rRNA are shown in color. [After C. Yanofsky et al., *Nucleic Acid Res.* 9 (1981):6647–6669.]

Natural Frameshifting Allows Translation of Some Overlapping Genes[48, 49, 50]

The very existence of overlapping genes poses the dilemma of how they are translated. Does a polysome formed on a single mRNA have ribosomes decoding the same gene in two different reading frames? Or are separate mRNAs utilized to generate the alternative protein products? As of yet, we do not have the complete answer. However, studies of the translation of an overlapping gene present on the RNA bacteriophage MS2 messenger suggest that decoding errors leading to termination of synthesis of the "parent" polypeptide may be essential for initiating translation at the beginning of the overlapping gene.

The MS2 phage RNA contains three major genes separated by intercistronic spaces (Figure 15-23). A fourth protein, involved in host cell lysis, is specified by an overlapping gene whose translation begins in the second gene (the coat protein gene), extends through the intercistronic space, and ends early in the third (replicase) gene. The lysis protein is translated in the +1 frame relative to both the coat and replicase proteins. It was noticed that expression of the lysis gene is obligatorily coupled to the production of coat protein. Mutant analysis employing recombinant DNA techniques (Chapter 19) led to the conclusion that a small fraction (several percent) of the ribosomes engaged in translating the coat protein make a translocation (frameshift) error somewhere before the beginning of the lysis gene. As a consequence, they encounter out-of-phase nonsense codons preceding the lysis gene and are released. Translation can then be initiated at the AUG beginning the lysis gene subsequent to termination of the frameshifted coat polypeptide chain. Since the initiation of protein synthesis is slow compared to elongation, such a mechanism circumvents the problem that frequent readout of the parent gene is bound to interfere with initiation complex formation at the start of an overlapping gene.

It is not yet known whether the translation of all overlapping genes requires that a frameshifting error be made in translating the "parent" gene. Alternatively, ribosomes may bind de novo, as they do at the beginnings of most genes in polycistronic mRNAs.

Ribosomal frameshifting has also been shown to be utilized by other small genomes, multiplying their coding potential by allowing the synthesis of two different polypeptide chains from the same DNA sequence. For instance, bacteriophage T7, which has a linear, double-stranded DNA genome containing about 30 genes, makes a major and a minor coat protein from the same gene. The two proteins have identical sequences at their N termini but differ in both sequence and length at their C termini. The longer, minor coat protein arises because about 10 percent of the ribosomes shift into the −1 frame near the end of the coding region for the major protein. They then continue beyond the normal stop and encounter a nonsense codon in the −1 frame some 20 amino acids downstream. In this case, the position where the ribosomes make the frameshifting error has been localized

Figure 15-23

Genetic map of the RNA phage MS2. [Nucleotide numbers after W. Fiers et al., *Nature* 260 (1976):500–507.]

to the sequence U UUC AAA. A possible scenario is that either the tRNA^Phe or the tRNA^Lys decoding this region occasionally moves the mRNA only two residues along during translocation, resulting in a shift to the −1 reading frame for continued translation.

Evolution of the Code[51–56]

The question of how the present-day genetic code evolved has commanded much attention ever since the structure of DNA was elucidated in 1953. Exactly how each of the 20 amino acids came to be paired with its codon(s) remains a matter of speculation. However, by the time some of these pairings became fixed, a primitive coding system must have been in operation. Presumably, the system continued to evolve through many stages until the current level of sophistication was reached.

Recent analyses of DNA sequences from many genomes provide hints of how such an early coding system may have been structured. The common characteristic found in today's DNA is a constancy of the relative positions of purines and pyrimidines within protein-coding regions. Specifically, codons of the form RNY predominate (where R is a purine, Y is a pyrimidine, and N can be either), suggesting that all codons may have been of this type originally. A primitive message composed exclusively of RNY codons could have been translated in only one of the three possible frames, circumventing the need for special start signals to fix the reading frame. Interestingly, among the eight amino acids specified by RNY in today's code (glycine, isoleucine, threonine, asparagine, serine, valine, alanine, and aspartic acid) are amino acids that are most likely to have been generated by prebiotic synthesis, as well as those that often appear in meteorites.

If today's genomes are searched for RNY periodicity, the extent of coding regions and their correct reading frames can usually be identified. As shown in Figure 15-24, genes that are well expressed seem to have best preserved the RNY pattern over their entire length (e.g., yeast glyceraldehyde-3-phosphate dehydrogenase [G3PDH]). In other genes, the original RNY message can still be detected but appears badly mutated (e.g., the beginning of chicken lysozyme) or is shifted over parts of the coding region into other reading frames by appropriate deletion-insertion pairs (e.g., *E. coli* TufA). The overlapping genes of bacteriophages are most interesting in this regard. Either they correspond to a shifted region in the "parent" gene (and are themselves read in the RNY frame), or they arise by translation of the NYR frame, producing proteins with a high proportion of hydrophobic amino acids. When *E. coli* 16S and 23S rRNA sequences are subjected to a search for RNY periodicity, their purine-pyrimidine distributions appear random, suggesting either that they never carried a coding region or that any primeval message has been mutated to randomness.

Why might the RNY pattern have been favored during early evolution over other possible codon constellations? RNY is a self-complementary sequence. Moreover, a repeating RNY pattern can be perceived in a master tRNA sequence compiled from the several hundred tRNAs analyzed so far. Thus, it has been suggested that primitive tRNAs may have served dual roles as both adaptors and mRNAs. According to this scheme, before ribosomes appeared, the entire process of gene expression could have been carried out by a single class of RNA molecules.

Figure 15-24

Search for the primeval RNY message in three gene-containing DNA sequences. The degree of mutation (m%) away from RNY periodicity is plotted against the base sequence number for the three reading frames in each case. The reading frame f that best fits the RNY message (has the smallest m%) will give the line shown in color. Below, f is plotted and can be compared with the actual coding region, designated by an arrowed line. In each of the three cases illustrated, the frame and extent of the coding region are well predicted. The chicken lysozyme gene appears badly mutated at its beginning, whereas *E. coli* TufA (one of the two genes for EF-Tu) has a short region in its middle where the reading appears to have shifted from one frame to another. [After J. C. W. Shepherd, *Cold Spring Harbor Symp. Quant. Biol.* 47 (1983):1099.]

(a)

(b)

(c)

Summary

The genetic code was determined through the study of protein synthesis in cell-free extracts. Addition of new mRNA to an extract depleted of its original messenger component results in the production of new proteins whose amino acid sequences are determined by the externally added mRNA. The first (and probably most important) step in cracking the genetic code occurred when the synthetic polyribonucleotide poly U was found to code specifically for polyphenylalanine. Use of other synthetic polyribonucleotides, both homogeneous (poly C, and so on) and mixed (poly AU, and so on), then assigned the base composition of codons for the various amino acids. Determination of the exact order of nucleotides in codons subsequently came from a study of specific trinucleotide-tRNA-ribosome interactions and the use of regular copolymers as messengers.

In the "universal" genetic code used by every organism from bacteria to humans, 61 codons signify specific amino acids; the remaining three are stop signals. The code is highly degenerate, with several codons usually corresponding to a single amino acid. A given tRNA can sometimes specifically recognize several codons. This ability arises from wobble in the base at the 5' end of the anticodon. The codon for the starting amino acid, *N*-formylmethionine, is usually AUG, which also specifies internal methionine, or more rarely GUG, which normally codes for valine. The stop codons UAA, UAG, and UGA are read by specific proteins, not specialized tRNA molecules.

Certain mutations, called intergenic suppressor mutations, appear to increase the frequency of mistakes in reading the genetic code. As a result of this increase in the mistake level, a mutant gene may occasionally produce a normal product. Suppressor genes exist for both missense (amino acid replacement) and nonsense (chain-terminating) codons, as well as for frameshift mutations caused by single-nucleotide insertions. Transfer RNAs with altered-anticodons are the molecular basis of most suppressor gene actions, although mutant ribosomal proteins can also affect reading accuracy.

A slightly different genetic code is utilized in certain rapidly evolving cellular organelles such as mitochondria. Fewer tRNAs are required to read this altered code. Many mitochondrial tRNAs have bizarre structures.

Overlapping genes have been recognized in many viral genomes. Other economies in genome utilization involve occasional read-through of termination signals or ribosome frameshifting to produce more than one polypeptide from the same coding region.

Bibliography

General References

Celis, J. E., and J. D. Smith, eds. 1979. *Nonsense Mutations and tRNA Suppressors.* New York: Academic Press. A most impressive collection of reviews on termination and suppression.

Clark, B., and H. Petersen, eds. 1984. "Gene Expression: The Translational Step and Its Control." *Alfred Benzon Symposium,* vol. 19. Copenhagen: Munksgaard. A number of current reviews on frameshifting and accuracy in translation are included.

Crick, F. H. C. 1963. "The Recent Excitement in the Coding Problem." *Prog. Nucleic Acid Res.* 1:164. A superb analysis of the state of the coding problem as of late 1962.

"The Genetic Code." 1966. *Cold Spring Harbor Symp. Quant. Biol.* 31. A historic collection of papers, presented in June 1966, just after the general features of the code became clear.

Söll, D., J. N. Abelson, and P. R. Schimmel, eds. 1980. *Transfer RNA, Biological Aspects.* Cold Spring Harbor, N.Y.: Cold Spring Harbor Laboratory. This volume of a two-volume compendium on tRNA contains many valuable articles on the role of tRNA in suppression.

Ycas, M. 1969. *The Biological Code.* New York: Wiley (Interscience). The most complete monograph on the universal code, written by one of the early workers in the field.

Cited References

1. Zamecnik, P. C. 1969. "An Historical Account of Protein Synthesis, with Current Overtones—A Personalized View." *Cold Spring Harbor Symp. Quant. Biol.* 34:1–16.
2. Nathans, D., G. Notani, J. H. Schwartz, and N. D. Zinder. 1962. "Biosynthesis of the Coat Protein of Coliphage f2 by *E. coli* Extracts." *Proc. Nat. Acad. Sci.* 48:1424–1431.
3. Salser, W., R. F. Gesteland, and B. Ricard. 1969. "Characterization of Lysozyme Messenger and Lysozyme Synthesized *in-Vitro.*" *Cold Spring Harbor Symp. Quant. Biol.* 34:771–780.
4. Pelham, H. R. B., and R. J. Jackson. 1976. "An Efficient mRNA-Dependent Translation System from Reticulocyte Lysates." *Eur. J. Biochem.* 67:247–256.
5. Grunberg-Manago, M., and S. Ochoa. 1955. "Enzymatic Synthesis and Breakdown of Polynucleotides: Polynucleotide Phosphorylase." *J. Amer. Chem. Soc.* 77:3165–3166.
6. Nirenberg, M. W., and J. H. Matthaei. 1961. "The Dependence of Cell-Free Protein Synthesis in *E. coli* upon Naturally Occurring or Synthetic Polyribonucleotides." *Proc. Nat. Acad. Sci.* 47:1588–1602.
7. Speyer, J. F., P. Lengyel, C. Basilio, A. J. Wahba, R. S. Gardner, and S. Ochoa. 1963. "Synthetic Polynucleotides and the Amino Acid Code." *Cold Spring Harbor Symp. Quant. Biol.* 28:559–568.
8. Nirenberg, M., and P. Leder. 1964. "The Effect of Trinucleotides upon the Binding of sRNA to Ribosomes." *Science* 145:1399–1407.
9. Nishimura, S., D. S. Jones, and H. G. Khorana. 1965. "The *in Vitro* Synthesis of a Copolypeptide Containing Two Amino Acids in Alternating Sequence Dependent upon a DNA-Like Polymer Containing Two Nucleotides in Alternating Sequence." *J. Mol. Biol.* 13:302–324.
10. Lagerkvist, U. 1981. "Unorthodox Codon Reading and the Evolution of the Genetic Code." *Cell* 23:305–306.
11. Crick, F. H. C. 1966. "Codon-Anticodon Pairing: The Wobble Hypothesis." *J. Mol. Biol.* 19:548–555.
12. Smith, J. D. 1976. "Transcription and Processing of Transfer RNA Precursors." *Prog. Nucleic Acid Res.* 16:25–73.
13. Ikemura, T., and H. Ozeki. 1982. "Codon Usage and Transfer RNA Contents: Organism-Specific Codon-Choice Patterns in Reference to the Isoacceptor Contents." *Cold Spring Harbor Symp. Quant. Biol.* 47:1087–1096.
14. Grosjean, F., and W. Fiers. 1982. "Preferential Codon Usage in Prokaryotic Genes: The Optimal Codon-Anticodon Interaction Energy and the Selective Codon Usage in Efficiently Expressed Genes." *Gene* 18:199–209.
15. Kozak, M. 1983. "Comparison of Initiation of Protein Synthesis

in Procaryotes, Eucaryotes, and Organelles." *Microbiol. Revs.* 47:1–45.

16. Kohli, J., and H. Grosjean. 1981. "Usage of the Three Termination Codons: Compilation and Analysis of the Known Eukaryotic and Prokaryotic Translation Termination Sequences." *Mol. Gen. Genetics* 182:430–439.

17. Garen, A. 1968. "Sense and Nonsense in the Genetic Code." *Science* 160:149–159.

18. Brenner, S., A. O. W. Stretton, and S. Kaplan. 1965. "Genetic Code: The Nonsense Triplets for Chain Termination and Their Suppression." *Nature* 206:994–998.

19. Crick, F. H. C., L. Barnett, S. Brenner, and R. J. Watts-Tobin. 1961. "General Nature of the Genetic Code for Proteins." *Nature* 192:1227–1232.

20. Hill, C. W. 1975. "Informational Suppression of Missense Mutations." *Cell* 6:419–427.

21. Steege, D. A., and D. G. Soll. 1979. "Suppression." In *Biological Regulation and Development I,* ed. R. F. Goldberger. New York: Plenum, pp. 433–486.

22. Capecchi, M. R., and G. Gussin. 1965. "Suppression *in Vitro*: Identification of Serine-sRNA as a 'Nonsense' Suppressor." *Science* 149: 417–422.

23. Engelhardt, D. L., R. Webster, R. Wilhelm, and N. Zinder. 1965. "*In Vitro* Studies on the Mechanism of Suppression of a Nonsense Mutation." *Proc. Nat. Acad. Sci.* 54:1791–1797.

24. Goodman, H. M., J. Abelson, A. Landy, S. Brenner, and J. D. Smith. 1968. "Amber-Suppression: A Nucleotide Change in the Anticodon of a Tyrosine Transfer RNA." *Nature* 217:1019–1024.

25. Hirsh, D. 1971. "Tryptophan Transfer RNA as the UGA Suppressor." *J. Mol. Biol.* 58:439–458.

26. Buckingham, R. H., and C. G. Kurland. 1980. "Interactions Between UGA-Suppressor tRNA^{Trp} and the Ribosome: Mechanisms of tRNA Selection." In *Transfer RNA, Biological Aspects,* ed. D. Soll, J. N. Abelson, and P. R. Schimmel. Cold Spring Harbor, N.Y.: Cold Spring Harbor Laboratory, pp. 421–426.

27. Ozeki, H., H. Inokuchi, F. Yamao, M. Kodaira, H. Sakano, T. Ikemura, and Y. Shimura. 1980. "Genetics of Nonsense Suppressor of tRNAs in *Escherichia coli*." In *Transfer RNA, Biological Aspects,* ed. D. Soll, J. N. Abelson, and P. R. Schimmel. Cold Spring Harbor, N.Y.: Cold Spring Harbor Laboratory, pp. 341–349.

28. Weiner, A. M., and K. Weber. 1973. A single UGA Codon Functions as a Natural Termination Signal in the Coliphage Qβ Coat Protein Cistron." *J. Mol. Biol.* 80:837–855.

29. Bunn, H. F., and B. G. Forget. 1986. *Hemoglobin: Molecular, Genetic and Clinical Aspects.* Philadelphia: Saunders, pp. 407–414.

30. Carbon, J., P. Berg, and C. Yanofsky. 1966. "Missense Suppression Due to a Genetically Altered tRNA." *Cold Spring Harbor Symp. Quant. Biol.* 31:487–497.

31. Riddle, D. L., and J. Carbon. 1973. "Frameshift Suppression: A Nucleotide Addition in the Anticodon of a Glycine Transfer RNA." *Nature New Biol.* 242:230–234.

32. Roth, J. R. 1981. "Frameshift Suppression." *Cell* 24:601–602.

33. Bossi, L., and D. Smith. 1984. "Suppressor *sufj*: A Novel Type of tRNA Mutant That Induces Translational Frameshifting." *Proc. Nat. Acad. Sci.* 81:6105–6109.

34. Atkins, J., R. Gesteland, B. Reid, and C. Anderson. 1979. "Normal tRNAs Promote Ribosomal Frameshifting." *Cell* 18:1119–1131.

35. Weiss, R. 1984. "Molecular Model of Ribosome Frameshifting." *Proc. Nat. Acad. Sci.* 81:5797–5801.

36. Weiss, R., J. Murphy, G. Wagner, and J. Gallant. 1984. "The Ribosome's Frame of Mind." *Alfred Benzon Symposium,* vol. 19, ed. B. Clark and H. Petersen. Copenhagen: Munksgaard, pp. 208–220.

37. Kurland, C. G. 1979. "Reading Frame Errors on Ribosomes." In *Nonsense Mutations and tRNA Suppressors,* ed. J. E. Celis and J. D. Smith. San Francisco: Academic Press, pp. 97–108.

38. Gorini, L. 1974. "Streptomycin and Misreading of the Genetic Code." In *Ribosomes.* Cold Spring Harbor, N.Y.: Cold Spring Harbor Laboratory, pp. 791–803.

39. Yarus, M., and R. Thompson. 1983. "Precision of Protein Biosynthesis." In *Gene Function in Prokaryotes,* ed. J. Beckwith, J. Davies, J. A. Gallant. Cold Spring Harbor, N.Y.: Cold Spring Harbor Laboratory, pp. 23–63.

40. Glass, R. E., V. Nene, and M. G. Hunter. 1982. "Informational Suppression as a Tool for the Investigation of Gene Structure and Function." *Biochem. J.* 203:1–13.

41. Bienz, M., and E. Kubli. 1981. "Wild-Type tRNA$^{Tyr}_G$ Reads the TMV RNA Stop Codon, but Q Base-Modified tRNA$^{Tyr}_Q$ Does Not." *Nature* 294:188–190.

42. Anderson, S., A. T. Bankier, B. G. Barrell, M. H. L. deBruijn, A. R. Coulson, J. Drouin, I. C. Eperon, D. P. Nierlich, B. A. Roe, F. Sanger, P. H. Schreier, A. J. H. Smith, R. Staden, and I. G. Young. 1981. "Sequence and Organization of the Human Mitochondrial Genome." *Nature* 290:457–465.

43. Attardi, G. 1985. "Animal Mitochondrial DNA: An Extreme Example of Genetic Economy." *Internatl. Rev. Cytol.* 93:93–145.

44. Battey, J., and D. A. Clayton. 1980. "The Transcription Map of Human Mitochondrial DNA Implicates Transfer RNA Excision as a Major Processing Event." *J. Biol. Chem.* 255:11599–11606.

45. Sanger, F., G. M. Air, B. G. Barrell, N. L. Brown, A. R. Coulson, J. C. Fiddes, C. A. Hutchinson, P. M. Slocombe, and M. Smith. 1977. *Nature* 265:687–695.

46. Godson, G. N., B. G. Barrell, R. Staden, and J. C. Fiddes. 1978. "Nucleotide Sequence of Bacteriophage G4 DNA." *Nature* 276:236–247.

47. Yanofsky, C., T. Platt, I. P. Crawford, B. P. Nichols, G. E. Christie, H. Horowitz, M. Van Cleemput, and A. M. Wu. 1981. *Nucleic Acid Res.* 9:6647–6668.

48. Atkins, J. F., J. A. Steitz, C. W. Anderson, and P. Model. 1979. "Binding of Mammalian Ribosomes to MS2 Phage RNA Reveals an Overlapping Gene Encoding a Lysis Function." *Cell* 18:247–256.

49. Kastelein, R. A., E. Remaut, W. Fiers, and J. van Duin. 1982. "Lysis Gene Expression of RNA Phage MS2 Depends on a Frameshift During Translation of the Overlapping Coat Protein Gene." *Nature* 295:35–41.

50. Dunn, J. J., and F. W. Studier. 1983. "Complete Nucleotide Sequence of Bacteriophage T7 DNA and the Locations of T7 Genetic Elements." *J. Mol. Biol.* 166:477–535.

51. Shepherd, J. C. W. 1982. "From Primeval Message to Present-Day Gene." *Cold Spring Harbor Symp. Quant. Biol.* 47:1099–1108.

52. Crick, F. H. C., S. Brenner, A. Klug, and G. Pieczenik. 1976. "A Speculation on the Origin of Protein Synthesis." *Origins Life* 7:389–397.

53. Woese, C. 1980. "Just So Stories and Rube Goldberg Machines: Speculations on the Origin of the Protein Synthetic Machinery." In *Ribosomes: Structure, Function and Genetics,* ed. G. Chambliss et al. Baltimore, Md.: University Park Press, pp. 357–376.

54. Eigen, M., and R. Winkler-Oswatitsch. 1981. "Transfer-RNA: The Early Adaptor." *Naturwissenschaften* 68:217–228.

55. Eigen, M., and R. Winkler-Oswatitsch. 1981. "Transfer-RNA: An Early Gene?" *Naturwissenschaften* 68:282–292.

56. Crothers, D. M. 1982. "Nucleic Acid Aggregation Geometry and the Possible Evolutionary Origin of Ribosomes and the Genetic Code." *J. Mol. Biol.* 162:379–391.

VI

REGULATION OF GENE FUNCTION IN BACTERIAL CELLS

CHAPTER 16

Regulation of Protein Synthesis and Function in Bacteria

Having described the general features of DNA and RNA function in protein synthesis, we are now ready to examine the cellular mechanisms that control the rate of synthesis of proteins. Since under any given environmental condition there is great variation in the number of different protein molecules in any cell, there must be devices to ensure the selective synthesis of proteins required in large numbers. There also must be ways to change the relative numbers of protein molecules, so that cells can respond efficiently to varying nutritional and other environmental conditions. We now realize that the rate at which each step in the synthesis of a protein occurs, from the binding of RNA polymerase as it initiates mRNA synthesis to the completion of the polypeptide on a ribosome, is partially under internal genetic control and partially determined by the external chemical environment. To show how these factors can interact, we shall focus our attention on bacterial systems, since they have been the basis of most of the important concepts formulated up to now.

Different Proteins Are
Produced in Different Numbers

We have already estimated from its length that the *E. coli* chromosome codes for 2000 to 4000 different polypeptide chains. Exactly how many different proteins are simultaneously present in a given cell is not yet known. Based on the probable number of enzymes needed to make the necessary metabolites, general estimates argue for the presence of at least 600 to 800 different enzymes in a cell growing with glucose as its sole carbon source. Some of these enzymes, particularly those connected with the first steps in glucose degradation and with the reactions that make the common amino acids and nucleotides, are present in relatively large amounts. Also required in large amounts are the enzymes needed to produce ATP. In contrast, other enzymes, particularly those involved in making the much smaller amounts of the necessary coenzymes, are present only in trace quantities. There must also be relatively large amounts of the various structural proteins used to construct the cell wall, the cell membrane, and the ribosomes.

It is useful to keep in mind how many copies of both scarce and abundant proteins there are in an *E. coli* cell. One well-studied protein present in moderately high amounts is the enzyme β-galactosidase (MW 4.6×10^5 daltons), which splits the sugar lactose

Figure 16-1

The sugar lactose can be hydrolytically cleaved to galactose and glucose by the enzyme β-galactosidase. Mutants that fail to make this protein cannot utilize lactose as a carbon source.

into its glucose and galactose moieties (Figure 16-1). Each active molecule has a tetrameric structure, being composed of four identical polypeptide chains of molecular weight 116,000. This is a very important enzyme, because lactose cannot be used as either a carbon or energy source unless it is first broken down to the simpler sugars glucose and galactose. An *E. coli* cell growing with lactose as its exclusive carbon source generally contains about 3000 molecules of β-galactosidase, which represents about 1.5 percent of the total protein. This is the maximum quantity that can be synthesized if just one gene coding for the β-galactosidase amino acid sequence is present on each *E. coli* chromosome. If this gene is present in two copies, 3 percent of the total protein produced by the cell can be β-galactosidase.

Now we know it is possible for a single gene that is transcribed and translated at an optimal rate to be expressed many times faster than β-galactosidase and to account for perhaps half of the protein synthesized by the cell (a situation contrived by genetic engineering). However, the cell is unhealthy with so much of its resource directed to making a single protein, and no cell does this naturally.

Other abundant proteins that have been accurately counted and cataloged include components of the protein synthesis machinery. There are 53 different ribosomal proteins (average MW 20,000) that collectively account for about 20 percent of the total protein in cells using glucose as a carbon source. Thus, the average ribosomal protein makes up 0.3 percent of the total *E. coli* protein, and there are about 20,000 copies of each. One of the most abundant *E. coli* proteins is the protein synthesis elongation factor EF-Tu (see Chapter 14), present in around 100,000 molecules per cell.

In contrast to these prominent molecules, regulatory proteins and enzymes that make minor cellular components are present in only a few to a few hundred copies per cell. Thus, they are more difficult to detect and purify, and we often have less precise information about their amounts.

Relation Between Amount of and Need for Specific Proteins

When a protein is needed, it can be present in a vastly greater amount than when environmental conditions do not require it. For example, there are approximately 3000 β-galactosidase molecules in each normal *E. coli* cell growing in the presence of β-galactosides, such as lactose, and less than one-thousandth of this number in cells growing on other carbon sources. Substrates like lactose, whose introduction into a growth medium specifically increases the amount of an enzyme, are known as **inducers;** their enzymes are called **inducible enzymes.** The opposite response is shown by many enzymes involved in cellular biosynthesis. For example, *E. coli* cells growing in a medium without any amino acids contain all the enzymes necessary for the biosynthesis of the 20 necessary amino acids. When, on the other hand, the growth medium contains these amino acids, their biosynthetic enzymes are almost entirely missing. Biosynthetic enzymes whose amount is reduced by the presence of their *end products* (e.g., histidine is the end product of the histidine biosynthetic enzymes) are called **repressible enzymes.** Those end-product metabolites whose introduction into a growth medium specifically decreases the amount of a specific enzyme are known as **corepressors.** The inductive and repressive responses are equally useful to bacteria: When enzymes are needed to transform a specific food molecule or to synthesize a necessary cell constituent, they are present; when they are unnecessary, they are effectively absent.

Adaptation is not, however, an all-or-nothing response. Under conditions of intermediate need, there may be an intermediate enzyme level (Figure 16-2). Similar variation is found in the quantities of structural proteins, best shown by the variation in the number of ribosomes themselves. When bacteria are growing at their maximum rate, ribosomes amount to 25 to 30 percent of the cell mass. If, however, their growth rate is cut down by unfavorable nutritional conditions, the bacteria need fewer ribosomes to maintain their slower rate of protein synthesis, and the ribosome content can drop to as little as one-fifth of its maximum value.

Variation in Protein Amount Can Reflect the Number of Specific mRNA Molecules

In actively dividing bacteria, most individual protein molecules, once synthesized, are quite stable. Thus, variation in the amounts of proteins usually reflects rates of synthesis, not relative stability, although there are important exceptions. As the cell adapts to new conditions of growth, the change in the amount of a protein is often the result of a change in the number of available mRNA templates. The number of β-galactosidase templates in cells actively making β-galactosidase, for example, greatly exceeds the number found in cells not engaged in synthesizing this enzyme. Now our best estimate is that during maximum β-galactosidase synthesis, about fifty β-galactosidase mRNA molecules are present in each cell. In contrast, when no lactose is present, the average cell contains fewer than one mRNA molecule specific for β-galactosidase synthesis.

Figure 16-2
Variation in the amount of enzyme per cell as a function of the amount of inducer (end product) present in the growth medium.

Constitutive Proteins Are Not Under Direct Environmental Control

There are a number of proteins within the cell whose amounts do not seem to be influenced by the external environment. Such proteins are called **constitutive proteins.** In *E. coli*, for example, the amounts of the enzymes controlling the degradation of glucose do not radically change when glucose is either removed from or added to the growth medium. Perhaps this means that *E. coli* so often finds glucose in nature that there is no advantage in evolving a means to repress the production of these enzymes. Similarly, it appears to be false economy for the cell to control proteins made in very small amounts; the very low copy number lactose repressor, for example, is made constitutively. The rate of synthesis of a constitutive protein is set genetically by: 1) the natural efficiency of the promoter from which its messenger is made; 2) the rate at which ribosomes read the messenger; and 3) the sensitivity of its messenger to degradation by nucleases.

Gene-Specific Regulatory Proteins Control Much RNA Synthesis

Special proteins called *regulatory proteins* determine when many inducible and repressible enzymes are to be made. Each regulatory protein affects the expression of one or more specific genes. They are of two fundamental types, *positive regulators* and *negative regulators,* distinguished by the contrasting effect of mutations in the genes that encode them (Figure 16-3). Thus, mutational inactivation of a negative regulator, or *repressor,* leads to synthesis of its target proteins irrespective of need. Such mutants are called *constitutive mutants.* In contrast, mutational loss of a positive regulator, sometimes called an *activator,* abolishes any substantial synthesis of the proteins it regulates, even when they are required in greater amounts by a change in cellular conditions.

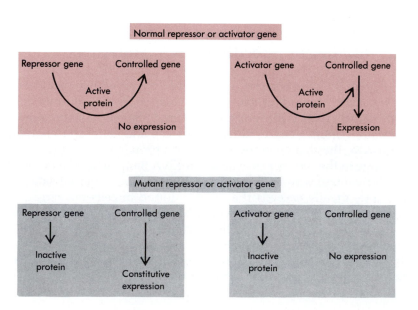

Figure 16-3

How repressors and activators are genetically distinguished.

Note that the terms *repressible* and *inducible* refer to the action of an external controlling molecule, like an amino acid or the sugar lactose, whereas the names *repressor* and *activator* reflect the action of a controlling gene in the cell. Therefore, both repressible and inducible genes could be controlled by either a repressor or an activator.

Repressors Prevent Initiation of mRNA Synthesis by Binding to DNA[1, 2]

Our first real insight into how genes can be regulated came from the isolation of two repressors: the lactose operon repressor (*lac* repressor) and the bacteriophage λ repressor (Chapter 17). Since a cell contains only 10 to 20 copies of lactose repressor and 100 to 200 copies of λ repressor, their detection and isolation in 1967 was a remarkable feat. Now about ten repressors are known, but this original pair is by far the best studied and understood.

Like most that have been studied, the *lac* and λ repressors are *transcriptional* repressors: They act by binding at specific sites on their respective DNA molecules, blocking the initiation of transcription of the corresponding mRNA molecules. The specific nucleotide sequences that bind the repressors are called *operators*. In general, an operator must be at least 10 to 12 bases long in order to interact specifically with the appropriate recognition site on a repressor. A base number this large greatly lessens the chance that a similar sequence will exist somewhere else along the same chromosome. If a smaller number of bases were used, too many false bindings would occur. The large number of specific interactions that a repressor can make with a DNA segment this long has the added benefit that the binding can be very strong. Once a *lac* repressor has bound to DNA, it effectively remains attached until subsequent interaction with its inducer.

Powerful Methods to Identify Protein Binding Sites on DNA[3, 4]

How can a protein binding site in DNA, such as an operator, be identified? Originally, operators were characterized through isolation and sequencing of DNA fragments protected from nuclease attack by specific binding of repressors. But this is a very laborious procedure. Fortunately, our ability to make and sequence specific restriction nuclease fragments has given rise to much quicker methods to outline the sites where proteins act and to learn exactly which chemical groups in DNA (methyl, amino, or phosphate) a protein contacts.

One basic principle, identical to that which the chemical method of DNA sequencing uses, underlies these methods: If a DNA fragment is labeled with a radioactive atom only at one end of one strand, the location of any break in this strand can be deduced merely from the size of the labeled fragment that results. The size, in turn, can easily be determined by high-resolution electrophoresis in a polyacrylamide gel. In the **footprinting** method (named facetiously after fingerprinting methods of chromatographic identification), the binding site is marked by internucleotide bonds that are shielded from the cutting action of a nuclease by the binding protein (Figure 16-4). The related **chemical protection** method relies on the ability of a bound protein to modify the reactivity of bases in the binding site to those base-specific

Figure 16-4

The footprinting method to map protein binding sites in DNA. (a) How footprinting reveals the DNA site to which a protein binds. Often, the same DNA fragment is subjected to chemical DNA sequencing and electrophoresed alongside the footprinted DNA samples. In this way, the exact position in the DNA sequence of a nuclease cut can be determined. (b) Photograph of an actual footprint, showing binding of lactose repressor to its operator. The DNA sequence is aligned correctly to the gel, so that radioactivity in a band indicates a nuclease cut at the nucleotide written alongside. The DNA strand labeled is the lower strand shown in Figure 16-5. In this case, the DNA fragment used was 48 nucleotides long. [Courtesy of D. Galas and A. Schmitz, *Nucleic Acid Res.* 5 (1978):3161. Used with permission.]

reagents that (after a further reaction) give rise to backbone cuts in the standard chemical DNA-sequencing method.

By changing the order of the first two steps, a third method determines which features of the DNA structure are *necessary* for the pro-

Figure 16-5
The *lac* operator sequence, showing its relation to the starting point for transcription of *lac* mRNA (nucleotide 1). Base-pair changes that lead to poor binding of the *lac* repressor are shown below their wild-type equivalents. The horizontal arrows designate bases having identical counterparts located symmetrically in the other half of the operator. The center of symmetry is at nucleotide 11.

tein to bind. An average of one chemical change per DNA is made, and then protein-DNA complexes are isolated. If a modification at a particular site does not prevent binding, DNA isolated from the complex will contain the modified chemical group, and the harmless modification allows the DNA to be broken at this site by further chemical treatment. If, on the other hand, binding is blocked, then no DNA modified at the site will be found complexed to the binding protein and the isolated fragments will not be broken at this site by subsequent chemical treatment. By all three methods, we can learn where a protein makes specific contacts both with bases and with the phosphates in the sugar-phosphate backbone of DNA.

The Structure of Operators

Many operators have now been characterized through footprinting and chemical protection experiments, as well as DNA sequence analysis of operator mutations that affect repressor binding. Their most important feature is an inverted repeat (or near repeat) that gives most operators a symmetrical shape. For example, of 24 base pairs in the lactose operator fragment isolated after nuclease digestion of DNA complexed with *lac* repressor, some 16 are related by a twofold axis of symmetry centered at the base pair in position 11 (Figure 16-5). It was realized as soon as operator sequences were determined that such symmetry could allow two subunits of a multimeric repressor to bind simultaneously to the operator. That the symmetry is not perfect was originally interpreted as meaning that not all base pairs on a given side of the operator bind to the repressor. This idea led to the speculation that only the symmetrical pairs are involved in the protein–nucleic acid interactions. However, analysis of mutant operators reveals that base-pair changes that directly reduce the strength of repressor-operator binding occur in the nonsymmetrical as well as symmetrical regions.

How Repressors Bind Operators[5–12]

Recent success in determining by X-ray crystallography the three-dimensional structure of several specific DNA binding proteins, provided in abundant amounts by genetically engineered plasmids, has given a striking view of how regulatory proteins like repressors bind specific DNA sites. The first structure determined was that of *E. coli* CAP protein, followed soon by λ Cro protein, λ repressor, and the repressor of phage 434, a close relative of λ. As protein crystallographers had expected, the active binding unit is a dimer of two identical globular polypeptide chains oriented oppositely in space to give

Figure 16-6

(a) A drawing of λ repressor bound to operator, showing both the NH₂-terminal DNA binding domains and the associated COOH-terminal domains (as they might appear). (b) How two λ repressor monomers fit onto the operator. The diagram on top shows the portion of each polypeptide chain at the operator binding surface. There are two polypeptides, placed symmetrically on the halves of the operator. Cylinders are α helices, and they are connected by less structured segments of polypeptide. The amino terminus of the polypeptide is behind the DNA and connects to α helix 1. The bottom diagram shows which bases and backbone phosphates contact the repressor, as determined by chemical protection experiments. [After C. Pabo and R. Sauer, *Ann. Rev. Biochem.* 53 (1984):300, with permission.] (c) The DNA binding faces of three DNA binding regulatory proteins. In each case, the number 3 α helices (named "F" in CAP protein) in the dimer are separated by 34 Å, the length of one turn of B-form DNA. (Courtesy of T. A. Steitz and I. T. Weber.)

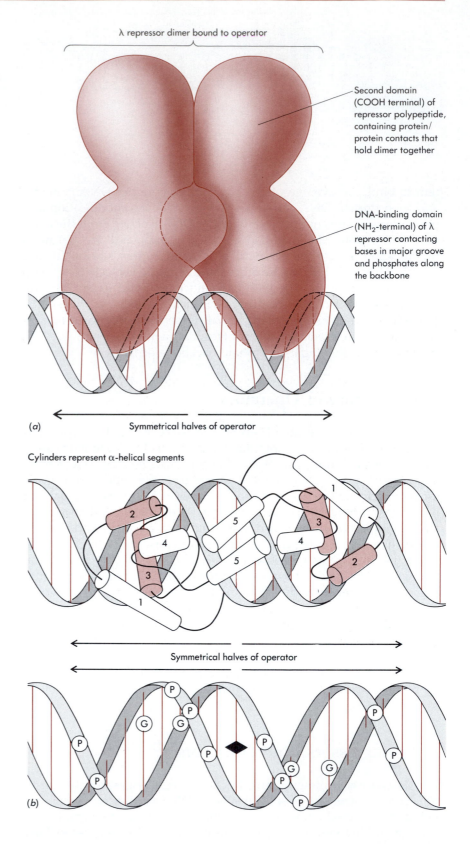

λ repressor dimer bound to operator

Second domain (COOH terminal) of repressor polypeptide, containing protein/protein contacts that hold dimer together

DNA-binding domain (NH₂-terminal) of λ repressor contacting bases in major groove and phosphates along the backbone

(a) Symmetrical halves of operator

Cylinders represent α-helical segments

Symmetrical halves of operator

(b)

an overall shape with a twofold axis of symmetry matching that of the binding site in DNA. In the case of λ repressor, for example, the amino terminus of each chain folds into a domain that binds to one-half of the operator (Figure 16-6a).

Figure 16-6 *(Continued)*

λ Cro   λ Repressor   Cap

(c)

The most critical DNA contacts are made by two short, adjacent α helices located at the binding face of each protein monomer; these are α helices numbers 2 and 3 in the λ repressor structure of Figure 16-6b and are shown for CAP, Cro, and λ repressor together in Figure 16-6c. In λ repressor, helix 3 protrudes from the protein surface, and it is separated from its counterpart in the second subunit of a λ repressor dimer by exactly the helix repeat distance of B-DNA (34 Å). Model building suggests that helix 3 fits snugly into the major groove, so that the repressor dimer binds one side of the DNA through the contacts made by the helix 3 of each monomer with half of the operator (Figure 16-6b). Most of the contacts that provide specific recognition of the λ operator result from the hydrogen-bonding complementarity between the edges of the base pairs in the major grooves and amino acid side chains of helix 3 (Figure 16-7). Strong evidence for this model is the fact that most mutations affecting λ repressor binding to the operator change amino acids in the two adjacent α helices. Very recently, the model has been confirmed directly by X-ray analysis of the repressor of phage 434 (a close relative of λ) bound to a synthetic DNA oligonucleotide containing the 434 operator sequence.

A knowledge of the actual structure and amino acid sequence of the critical pair of α helices in these several cases allows us now to predict whether this structure is likely to form simply by inspecting the amino acid sequence of a protein. Examination of more than ten DNA binding proteins indicates that a pair of α helices is a general feature

Figure 16-7
Postulated hydrogen-bond contacts between the side chains of two amino acids of CAP protein, glutamic acid at position 181 and lysine at position 188, and two base pairs in the CAP binding site. (Courtesy of T. A. Steitz and I. T. Weber.)

Figure 16-8
Overlap between the lactose operon promoter and operator sites. The DNA strand identical in sequence to the mRNA is shown.

of the active sites of regulatory proteins that bind specific DNA sequences. The discovery of this "helix-turn-helix" motif in DNA binding regulatory proteins is a most important advance in our understanding of the structural basis of genetic regulation.

Repressors Prevent RNA Polymerase Binding

The identification of DNA binding sites for both RNA polymerase (which bind to promoters) and repressors (which bind to operators) immediately suggested how repressors work: They physically prevent RNA polymerase from binding and initiating RNA synthesis at the promoter. The lactose operator (Figure 16-8) is made up of the same 21 base pairs of DNA that is the template for the first 21 nucleotides of lactose operon mRNA. Thus, *lac* repressor covers the template just where the catalytic center of RNA polymerase would lie, thereby denying it access to the template. Other operators are found to overlap different parts of the promoter, but all interfere with RNA polymerase binding. As we would expect, the repressor prevents only initiation, having no effect on chain growth once elongation has commenced.

Corepressors and Inducers Determine the Functional State of Repressors

Repressors are not always able to prevent specific mRNA synthesis. Otherwise, they would permanently inhibit the synthesis of their specific proteins. Instead, many repressor molecules can exist in both an active and an inactive form, depending on whether they are combined with their appropriate inducers (or corepressors). The attachment of an inducer inactivates the repressor. For example, when combined with a β-galactoside sugar like lactose or allolactose (a metabolite of lactose that is the natural inducer), the *lac* repressor cannot bind to its specific operator. Thus, the addition of β-galactosides to growing cells permits β-galactosidase synthesis by decreasing the concentration of active *lac* repressor molecules. In contrast, the binding of a corepressor changes an inactive repressor into an active form. For example, the addition of the amino acid tryptophan to cells activates the repressor that controls the synthesis of enzymes involved in tryptophan biosynthesis. This quickly shuts off synthesis of their specific mRNA molecules (Figure 16-9).

No covalent bond is formed between repressors and their specific inducers or corepressors. Instead, a portion of each repressor mole-

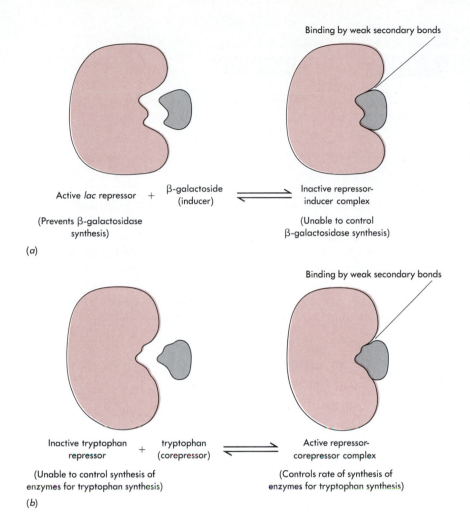

Binding by weak secondary bonds

Active *lac* repressor + β-galactoside (inducer) ⇌ Inactive repressor-inducer complex

(Prevents β-galactosidase synthesis)

(Unable to control β-galactosidase synthesis)

(a)

Binding by weak secondary bonds

Inactive tryptophan repressor + tryptophan (corepressor) ⇌ Active repressor-corepressor complex

(Unable to control synthesis of enzymes for tryptophan synthesis)

(Controls rate of synthesis of enzymes for tryptophan synthesis)

(b)

Figure 16-9
Drawing illustrating the opposite effects of corepressors and inducers upon the activity of repressors. Depending on whether the enzymes are inducible or repressible, the free repressors are either active or inactive.

cule is complementary in shape to a specific portion of its inducer or corepressor, and the two are joined by weak secondary bonds (hydrogen bonds, salt linkages, or van der Waals forces). Since these bonds are weak, they are rapidly made and broken, allowing the repressor state (active or inactive) to adjust quickly to the physiological need. For example, the synthesis of β-galactosidase mRNA ceases almost immediately after the removal of lactose. The reversible change in shape that a repressor undergoes upon binding an inducer is another example of an allosteric transformation (Chapter 4).

Repressors Can Control More Than One Protein

In some cases, repressors control the synthesis of only one protein. Often, however, a single repressor affects the synthesis of several enzymes. The *lac* repressor of *E. coli*, for example, controls at least three enzymes from its single operator: β-galactosidase itself; galactoside permease, which controls the rate of entry of β-galactosides into the bacteria; and galactoside acetylase, whose function is not yet known. When active β-galactoside permease is absent, *E. coli* cells are unable to concentrate β-galactosides within themselves. Since β-galactosidase, β-galactoside permease, and galactoside acetylase (?) are ordinarily needed to metabolize β-galactosides, their *coordinated*

Figure 16-10
The lactose operon and its associated regulatory genes, drawn to scale based on known sizes of their gene products. The number of base pairs in each gene is shown.

synthesis is clearly desirable. Coordinated synthesis is brought about by having the enzymes coded by adjacent genes, which are transcribed into a single mRNA molecule carrying all three genetic messages (Figures 16-10 and 16-11). An even larger number of genes is coordinately repressed by the repressor of tryptophan biosynthesis. Again, this is achieved by having a single mRNA molecule carry the messages of all these genes. Such a collection of adjacent genes controlled from a single operator is called an *operon*.

Absence of an Operator Leads to Constitutive Synthesis

Like repressors, the operators to which they bind have *negative* functions: If a functional operator is absent, the corresponding repressor cannot inhibit the synthesis of the specific mRNA, and as a result, there is constitutive synthesis of its corresponding protein product(s).

The existence of operators was first revealed by genetic analysis. The structure of the operator can mutate to an inactive form, preventing the binding of its repressor. When this happens, constitutive enzyme synthesis results. These mutants are therefore called O^c (constitutive) mutants. O^c mutations can easily be distinguished from mutations in the repressor genes by measuring enzyme synthesis in special, partially diploid cells containing two copies of the relevant chromosomal regions. The lactose operons of cells containing one nonfunctional and one functional repressor gene are still repressible,

Figure 16-11
How the interaction of repressor, inducer, and operator controls the synthesis of the *E. coli* proteins β-galactosidase, β-galactoside permease, and galactoside acetylase.

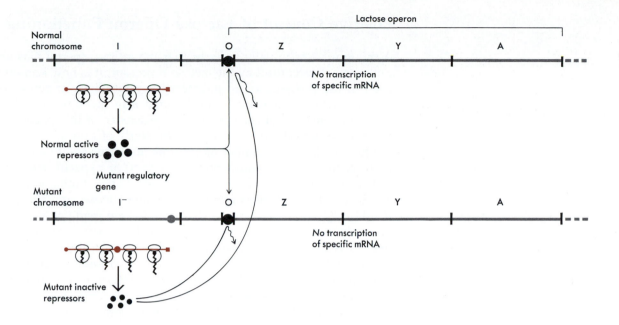

since good repressor molecules can act on both operators (Figure 16-12). In contrast, cells containing only one bad operator will express lactose operon enzymes constitutively, no matter what the condition of the repressor gene (Figure 16-13).

Figure 16-12
The use of partially diploid cells to show that the presence of functional repressors is dominant over the presence of inactive repressors. No significant amounts of β-galactosidase molecules will be produced in these cells in the absence of externally added β-galactosides.

Figure 16-13
The control of specific mRNA synthesis by normal and mutant operators.

Positive Control of Lactose Operon Functioning

Although it was initially thought that the lactose operon was solely under negative control by the lactose repressor, it is now known that even in the presence of an inducer that neutralizes its repressor, a protein-mediated positive control signal also must be received for normal expression of the operon. The existence of this positive control protein was discovered during the study of how glucose blocks the functioning of a number of operons (**glucose-sensitive operons**), each controlling the breakdown (catabolism) of a specific sugar (e.g., lactose, galactose, arabinose, and maltose). For example, when *E. coli* is grown in the presence of both glucose and lactose, only the glucose is utilized, and the lactose operon proteins are made at a low level. Similarly, when both glucose and galactose are present, the galactose operon is inactive. The reason that glucose is preferred over other sugars can only be guessed. Perhaps, during evolution, bacteria more often found themselves in a glucose-rich environment than in one dominated by other sugars.

In any case, glucose inhibition does not operate by influencing the rate of entry of the various sugars into the cells. Instead, it acts by influencing the transcription pattern. This was first suggested by the discovery of mutations in the lactose operon control region that make the lactose operon insensitive to glucose effects. With such mutants, the lactose operon can be maximally induced, even in the presence of large amounts of glucose.

Glucose Catabolism Affects the Cyclic AMP Level

The action of glucose on transcription is not direct. Instead, one of its breakdown products (catabolites) acts by lowering the intracellular amount of *cyclic AMP* (cAMP). This key metabolite is required for the transcription of all the operons that are inhibited by glucose catabolism. The manner in which a glucose catabolite controls the cyclic AMP level within a cell is still unknown. ATP is the direct metabolic precursor of cAMP, and the enzyme responsible for this transformation, **adenylcyclase**, may be directly inhibited by a specific catabolite (Figure 16-14). On the other hand, since there is an enzyme (**phosphodiesterase**) that specifically converts cAMP to AMP, inhibition might be controlled by the rate of cAMP breakdown.

Activation of the Catabolite Activator Protein (CAP) by cAMP Binding[13]

Cyclic AMP does not directly promote *lac* mRNA synthesis, but works instead by binding to the **catabolite gene activator protein** (**CAP**), a dimeric molecule with a molecular weight of 45,000 daltons. CAP has no influence on transcription until cAMP has bound to it. It then acquires the ability to bind to very specific sites on DNA, and by so doing increases the rate of transcription of adjacent operons. CAP is the positive control element for all the glucose-sensitive operons; so cells with mutations that block its functioning are simultaneously unable to utilize a large number of different sugars.

Figure 16-14
Control of catabolite-sensitive transcription through cyclic AMP.

CAP Controls RNA Polymerase Binding at the Lactose Promoter[13, 14]

The binding of neither CAP nor a specific repressor has any influence on the rate at which mRNA chains grow. Instead, both act—one positively, the other negatively—by controlling the rate at which RNA polymerase molecules attach to promoters. Whereas repressor blocks binding of RNA polymerase, CAP (when complexed with cyclic AMP) *assists* RNA polymerase in binding effectively to the lactose promoter, allowing much more frequent initiation of RNA synthesis. Like the operator, the site where CAP binds to lactose operon DNA is marked both by mutations that prevent its function (no stimulation of lactose enzymes in conditions of high cellular cyclic AMP) and by footprinting and chemical protection experiments, which identify a specific binding sequence. The important result is that CAP complexed with cyclic AMP binds next to RNA polymerase at the promoter (Figures 16-15 and 16-16).

How CAP enhances RNA polymerase binding at the promoter is a question of current interest and much importance for understanding the way regulatory proteins work. There are two main possibilities. First, RNA polymerase is close enough that it might directly complex with CAP as it binds. Such a protein-protein contact would provide extra energy to greatly strengthen RNA polymerase binding. Second, CAP binding could change the detailed structure of the DNA helix surrounding its binding site into a shape that RNA polymerase recognizes much better than unbound DNA. We know that DNA is flexible enough to allow such local distortions from the normal double-helical structure.

Figure 16-15
The nucleotide sequence and organization of the *E. coli* lactose operon control region.

In Vitro Analysis of Promoter Functioning

An important step in establishing our concepts about how promoters and operators function has come from in vitro transcription experiments. For example, when *lac* DNA is used as the in vitro template, the binding of RNA polymerase and the initiation of *lac* mRNA synthesis requires the simultaneous presence of cAMP and CAP. And when the DNA templates have been isolated from mutant cells whose *lac* operon is transcribed even when glucose is present, no requirement for CAP is found in the in vitro systems. Such experiments provide the clearest demonstrations that cAMP, CAP, and repressors act at the transcriptional level and not through effects on membrane permeability.

DNA Repair Genes at Many Different Sites on the Chromosome Are Regulated by a Single Repressor[15, 16, 17]

The *lac* repressor binds to only a single operator, and this operator happens to be near the repressor gene itself. But neither feature is essential to the function of the repressor. Because a repressor diffuses throughout the cell, it can regulate any number of genes, including some that are distant from its own. By inactivating such a repressor, the cell can invoke expression of a variety of cellular genes to serve a common function. A striking example is provided by the set of about fifteen *SOS genes*, which encode proteins that help the cell repair damage to its DNA, such as that caused by ultraviolet light (Figure 16-17). Each SOS gene has an operator recognized by the LexA repressor (encoded by the *lexA* gene). DNA damage is detected by the recombination enzyme RecA protein, which binds the single-stranded DNA left by replication of damaged regions (and begins to repair this damage by recombination; Chapter 11). It was a surprising discovery that the inducer of the SOS genes is not a small molecule like lactose or tryptophan, but is instead RecA protein itself. Binding to single-stranded DNA activates a site on the RecA polypeptide, distinct from those involved in strand pairing, that has a protease activity specific for LexA repressor (and phage repressors; Chapter 17). LexA molecules are cut into two fragments, so that their operator binding activity is lost and mRNA for the SOS genes can be made. Unlike the case of the lactose repressor, inactivation of LexA is irreversible: Repression can be reestablished only when more LexA is made.

The details of SOS gene induction show how flexibly and subtly a repressor control can serve the needs of the cell. Besides the SOS genes, LexA also represses the *recA* and *lexA* genes themselves. Here

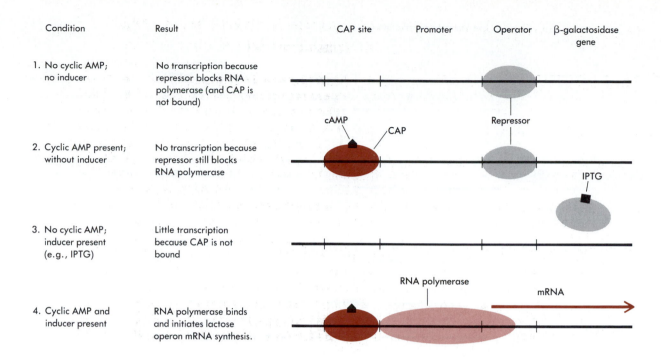

Condition	Result	CAP site	Promoter	Operator	β-galactosidase gene
1. No cyclic AMP; no inducer	No transcription because repressor blocks RNA polymerase (and CAP is not bound)				

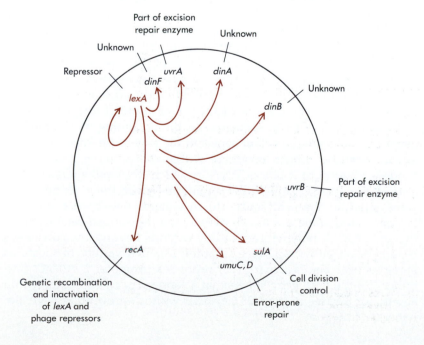

repression is not complete: A substantial basal level of both proteins is made, even when a normal level of LexA repressor is present. Destruction of LexA then greatly increases the rate of synthesis of RecA and LexA proteins, an increase that serves a different purpose in each case. More RecA protein accumulates to satisfy the need for a high rate of DNA repair by recombination. In contrast, the increased rate of LexA synthesis is not sufficient to allow it to accumulate in conditions of DNA damage, because the abundant RecA protease activity easily destroys all new LexA made at the higher rate. But this faster synthesis is important when damage is repaired and the protease activity disappears: LexA then is made rapidly so that repression can

Figure 16-16
Control of initiation of lactose operon transcription by CAP, repressor, cyclic AMP, and an inducer (like allolactose, the natural inducer, or isopropyl-thiogalactoside [IPTG], a useful synthetic inducer).

Figure 16-17
Location of SOS DNA repair genes on the circular chromosome of *E. coli*. All the SOS genes are regulated by the LexA repressor. The names of the genes are given inside the circle, and their functions (where known) outside. (After G. Walker, *Microbiological Reviews* 48 (1984):67, with permission.)

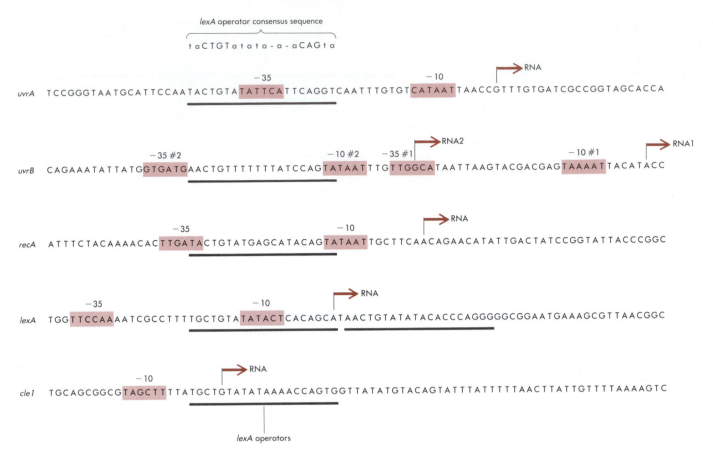

Figure 16-18
DNA sequences of five SOS gene promoter regions, showing the relative positions of promoters and operators. Promoter 1 of *uvrB* is repressed by LexA, whereas promoter 2 is not. The *lexA* gene has two adjacent operators to which its own product binds. The *cle1* gene of plasmid ColE1 encodes colicin E₁, an antibacterial toxin. In the *lexA* operator consensus sequence, highly conserved bases are shown in black, and less highly conserved bases are in color. (After A. Sancar, G. Sancar, W. Rupp, J. Little, and D. Mount, *Nature* 298 (1982):98, with permission.)

be reestablished efficiently. Another regulatory variation is found in the expression of *uvrB*, an SOS gene that encodes part of the major enzyme of excision repair. UvrB protein also is made at a high basal level in the presence of active LexA repressor, but for a different reason. Its mRNA is initiated from two adjacent but separate promoters, only one of which is repressed by LexA; the second is always active, providing the cell a substantial level of the excision enzyme at all times so that it can begin this type of repair as soon as the damage occurs (Figure 16-18). Still a further refinement is that operators of the various SOS genes bind LexA repressor with different affinities. Some bind it weakly, allowing their genes to be expressed after mild DNA damage that yields moderate levels of protease activity and reduces the LexA concentration only perhaps two- or three-fold. Other operators bind LexA more avidly and are induced only by extensive damage that reduces the LexA concentration much further.

A comparison of DNA sequences at the beginning of several SOS genes illustrates an important aspect of the way repressors work: The operator can be almost anywhere within the promoter, as long as repressor bound to it blocks DNA to which RNA polymerase binds (see Figure 16-18). Thus, some operators cover the −35 hexamer, some straddle the −35 and −10 hexamers, one covers the −10 hexamer, and some are mostly within the transcribed region. Since the site of RNA polymerase binding extends beyond the consensus hexamers on either side (see Chapter 13), this variety of locations makes sense. The fact that LexA represses promoter 2 but not promoter 1 of the *uvrB* gene shows that a repressor can approach within about 7 or 8 bases of the −35 hexamer without preventing RNA polymerase function.

Arabinose C Protein Is Both an Activator and a Repressor[14, 18–22]

Whereas the catabolite gene activator protein (CAP) passes the cellular signal for glucose starvation to many operons and genes, some positive regulators (activators) deliver specific signals to only one or a few genes. The arabinose *C* gene encodes one such protein, which activates transcription of the arabinose *BAD* operon, consisting of genes *araB, araA,* and *araD* (Figure 16-19). C protein must be complexed with arabinose to do this, so that the enzymes are made only when their substrate is present. Like the lactose operon, the arabinose operon is sensitive to glucose and thus is controlled by cyclic AMP and CAP in addition to arabinose and C protein. How can *two* activators, C protein and CAP, regulate one operon?

Perhaps not surprisingly, the mechanism is a little complicated. C protein is not only an activator but also a repressor, and in fact, it acts as a repressor in two different ways. C protein binds three separate sites in the arabinose operon control region, *araI, araO$_1$,* and *araO$_2$* (Figure 16-20). The site in the middle, *araO$_1$,* overlaps the promoter for gene *C* and serves only to repress synthesis of gene *C* mRNA when sufficient C protein has accumulated in the cell; thus, the *C* gene is *autoregulated.* The other two C protein binding sites, *araI* and *araO$_2$,* act along with the CAP binding site to mediate repression and activation of the *BAD* operon. It is thought that when the CAP binding site is empty because the cyclic AMP concentration is low, one C protein molecule (perhaps a dimer or tetramer) binds both *araI* and the distant site *araO$_2$,* thereby drawing the DNA into a loop (see Figure 16-20). In this *repressing* configuration, C protein cannot activate RNA polymerase at the *BAD* promoter, whether or not arabinose is bound to it. However, the binding of CAP-cAMP next to *araI* changes the activity of C protein at *araI,* perhaps by touching it directly. C protein now releases the DNA site *araO$_2$,* thus undoing the loop; moreover, it can also activate RNA polymerase to initiate synthesis at the adjacent *BAD* promoter. For C protein to activate *BAD*

Figure 16-19
The arabinose operon of *E. coli:* an example of positive control.

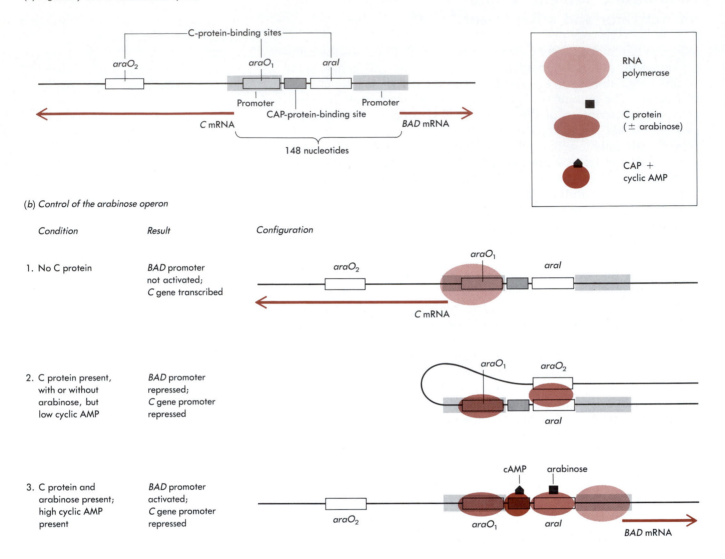

(a) *Regulatory sites of the arabinose operon*

(b) *Control of the arabinose operon*

Figure 16-20
Regulation of the *E. coli* arabinose operon. (a) Regulatory sites of the arabinose operon. (b) Control of the arabinose operon.

transcription, it first must bind arabinose; therefore, two small molecules, cyclic AMP and arabinose, are required for efficient *BAD* expression. Exactly why a distant site should be required for repression is not clear, but evidence for this mechanism is quite persuasive. In particular, the function of CAP-cAMP in breaking the looped repressing configuration is shown by the fact that deletion of site *araO₂* allows the *BAD* promoter to work in the absence of CAP and cyclic AMP, although it still needs C protein and arabinose.

As with the lactose operon, the cell uses an operon-specific regulator (*lac* repressor or C protein) along with CAP protein to detect the presence of a specific sugar and the absence of glucose, although it has evolved a very different mechanism for the arabinose operon. It will be surprising if many other variations do not appear when more regulatory schemes are known in this detail. Like arabinose C protein, any DNA binding protein could be either an activator or a repressor, depending on whether it blocks or stimulates binding of RNA polymerase to a promoter. Indeed, the phage λ repressor does both (Chapter 17). Sometimes, CAP protein also acts as a repressor, preventing expression of genes not needed when glucose is limiting. For example, the complex control of the galactose operon of *E. coli* involves CAP protein acting both as activator and repressor. In addition, there is a separate galactose-regulated repressor that binds two operators and might thereby draw the DNA into a loop like that envisioned for the arabinose operon. Very recent evidence suggests that CAP may have a secondary role as a repressor even in the lactose operon. Besides the major promoter of the lactose operon, there is a second, upstream promoter where RNA polymerase binds strongly but initiates RNA synthesis very inefficiently. Because the two promoters overlap, this binding blocks productive interaction of RNA polymerase with the major promoter. The CAP binding site overlaps the second promoter, so that CAP protein prevents obstructive binding of RNA polymerase to the second promoter at the same time it directly stimulates RNA polymerase binding to the major promoter. Both functions of CAP are important for *lac* operon expression.

The CAP and arabinose C proteins represent two classes of activators found in bacteria: general activators like CAP, which each control many genes whose functions are required in certain conditions (e.g., glucose, nitrogen, or phosphate limitation), and specific activators like arabinose C, which control a few genes whose products metabolize particular compounds (e.g., arabinose, maltose, or serine). In Chapter 17, we describe another transcription activator whose mechanism is well understood: the phage λ cII protein.

Genes for Heat-Shock Proteins of *E. coli* Are Recognized by a σ Factor (σ³²) That Specifically Recognizes Their Promoters[23-29]

Repressors and operon-specific activators like the arabinose C protein that control relatively few genes usually act by changing the availability or initiation efficiency of promoters. However, they do not change the activity of RNA polymerase, which remains available to any receptive promoter. A different strategy is used when conditions require a massive redirection of gene expression. Such is the case, for example, after a sudden rise in the growth temperature, called **heat shock**. In almost all kinds of cells subjected to heat shock, certain proteins (about 17 in *E. coli*) begin to be made much faster than usual. If the new temperature is not too high (e.g., 42°C for *E. coli*), the rate of heat shock protein synthesis soon declines and the normal pattern of protein synthesis soon resumes (in about 20 minutes for *E. coli*). But at temperatures too high for growth (50°C for *E. coli*), synthesis of only heat-shock proteins continues. Remarkably, some heat-shock proteins of widely different species are closely related; in fact, there are even similarities between those of bacteria and those of eucaryotes. For example, the gene for the 66,000-dalton *E. coli* heat-shock protein DnaK is identical in almost half its nucleotides to the gene for the 70,000-dalton *Drosophila* heat-shock protein Hsp70. Although we know that some heat-shock proteins are required in *E. coli* for phage growth—the proteins GroEL and GroES are required for phage λ particle assembly, and DnaK for λ DNA synthesis—we have no idea what function they have in common that is essential to save cells from the rigors of a sudden temperature rise or certain other stressful treatments such as the addition of ethanol to the culture that also induce the heat shock genes. (Perhaps the function of heat-shock proteins in phage assembly somehow reflects the fact that phage infection is itself a shock.) What makes heat shock particularly interest-

Table 16-1 Sequences of Heat Shock Promoters*

Promoter	−35 Region		−10 Region	+1
groE	TTTCCCCCTTGAA	GGGGCGAAGCCAT	CCCCATTTCTCTGGTCAC	
dnaK promoter 1	TCTCCCCCTTGAT	GACGTGGTTTACGA	CCCCATTTAGTAG TCAA	
dnaK promoter 2	TTGGGCAGTTGAA	ACCAGACGTTTCG	CCCCTATTACAGACTCAC	
C62.5 gene	GCTCTCGCTTGAA	ATTATTCTCCCTTGT	CCCCATCTCTCCCACATC	
σ^{70}	TGCCACCCTTGAA	AAACTGTCGATGTGG	GACGATATAGCAG ATAA	
Heat-shock promoter (σ^{32}) consensus	T - - C - C - CTTGAA	13–15 bp	CCCCAT- T	
Standard promoter (σ^{70}) consensus	TTGACA		TATAAT	

*DNA sequences of the RNA-like strand of *E. coli* heat-shock promoters. Note that the "standard" σ, σ^{70}, is itself a heat-shock protein. Bases matching the σ^{32} consensus sequence are underscored.
SOURCE: Adapted from D. W. Cowing, J. C. A. Bardwell, E. A. Craig, C. Woolford, R. W. Hendrix, and C. A. Gross, "Consensus for *Escherichia coli* Heat Shock Gene Promoters," *Proc. Nat. Acad. Sci.* 82 (1985):2679–2683, with permission.

ing is the fact that the entire pattern of gene expression can be so abruptly shifted.

A critical discovery about the mechanism of this shift is that promoters for *E. coli* heat-shock proteins are recognized by a unique form of RNA polymerase holoenzyme containing a sigma (σ) factor with a molecular weight of 32,000 daltons (σ^{32}), encoded by gene *rpoH* (previously called *htpR*). Factor σ^{32} is a minor protein, much less abundant than the primary σ factor, σ^{70}; it binds tightly to core enzyme to form a holoenzyme that can be purified separately from holoenzyme containing σ^{70}. This minor holoenzyme acts exclusively on promoters of heat-shock proteins, and not on the "standard" promoters of other genes. Correspondingly, heat-shock promoters have different sequences from the standard promoter (Table 16-1), particularly in the −10 segment that σ factors may recognize (Chapter 13).

Although it is clear that σ^{32} is responsible for recognizing heat-shock promoters, the critical regulatory event that causes it to function after a temperature shift is still a mystery. Also completely unexplained is why σ^{32} stops acting on promoters of heat shock genes when cells adapt to a moderately higher temperature of growth.

Separate σ factors also control sporulation in *B. subtilis* (Chapter 22) and phage growth in both *B. subtilis* and *E. coli* (Chapter 17). *B. subtilis* contains at least four σ factors, possibly because sporulation is a radical and prolonged transformation of cell structure that implies a correspondingly large change in gene expression. There is recent evidence that the control of genes required for nitrogen metabolism in *E. coli* and other bacteria, including genes for nitrogen fixation, involves a distinct σ factor, which is encoded by gene *ntrA* in *E. coli*.

Amino Acid Biosynthetic Operons Are Regulated by Varying Termination of mRNA Synthesis Before the Structural Genes[30, 31]

Initially, it seemed likely that the synthesis of five enzymes of the tryptophan *(trp)* operon (see Figure 8-10) would also be regulated primarily by a repressor. Here the important feature is that transcription of the five contiguous *trp* genes is needed only when tryptophan is limiting (Figure 16-21). Thus, tryptophan acts not as an inducer but as a corepressor that activates its specific repressor so that it can bind to the *trp* operator and prevent transcription of its respective operon. When the tryptophan concentration is low, the *trp* operator is free and the synthesis of *trp* mRNA commences at the adjacent promoter. Surprisingly, however, a *trp* mRNA molecule, once so initiated, does not automatically grow to a full length. Instead, most *trp* mRNA molecules stop growing before the transcription of even the first *trp* gene (*trpE*) begins, unless a second and novel device confirms that little tryptophan is available.

The key to understanding this unexpected result came from sequence analysis of the 5′ end of tryptophan operon mRNA, which revealed that 161 nucleotides of RNA are made from the tryptophan promoter before RNA polymerase encounters the first codon of *trpE* (Figure 16-22). Near the end of the sequence and before *trpE* is a transcription terminator (Chapter 13), composed of a characteristic hairpin loop in the RNA (made from sequences in regions 3 and 4 of Figure 16-22), followed by 8 uridine residues. At this so-called **attenuator**, RNA synthesis usually stops (and, we might have thought, should always stop), yielding a leader RNA 139 nucleotides long. How, then, can mRNA for the whole operon ever be made? Three features of the leader sequence allow the attenuator to be passed by RNA polymerase when the cellular concentration of tryptophan is

Figure 16-21
The tryptophan operon of *E. coli,* showing the relation of the leader to the structural genes that code for the *trp* enzymes.

Figure 16-22
Features of the nucleotide sequence of the *trp* operon leader RNA.

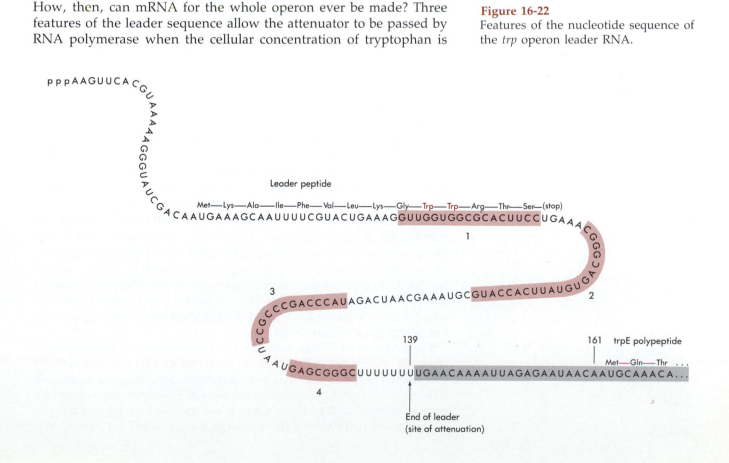

Figure 16-23

How transcription termination at the *trp* operon attenuator is controlled by the availability of tryptophan.

1. High tryptophan

Translation termination codon

Leader RNA

1   2   3   4

Leader peptide coding region

Sequence 3 can pair with sequence 4; the transcription termination hairpin forms and RNA synthesis stops; trp enzymes are not expressed.

Ribosome proceeds to leader peptide termination codon, blocking sequence 2

2. Low tryptophan

Leader peptide

Tandem tryptophan codons

2   1   3   4

End of leader

trp operon mRNA

Ribosome stalls at adjacent tryptophan codons, leaving sequence 2 free to pair with sequence 3, thus preventing formation of the 3,4 termination hairpin; RNA polymerase proceeds into the operon, making mRNA for tryptophan enzymes.

3. No protein synthesis

1   2   3   4

If no ribosome begins translation of the leader peptide AUG, hairpin 1,2 forms, preventing formation of 2,3 and thus allowing the terminator hairpin 3,4 to form; tryptophan enzymes are not expressed.

low. First, there is a second hairpin besides the terminator hairpin that can form, made of regions 1 and 2 of the leader (Figure 16-23). Second, region 2 also is complementary to region 3; thus, yet another hairpin, consisting of regions 2 and 3, can form, thereby preventing the terminator hairpin (3, 4) from forming. Third, the leader RNA codes for a short leader peptide of 14 amino acids that is preceded by a strong ribosome binding site. The leader peptide has a striking sequence feature: two tryptophan codons in a row. Their importance is underscored by corresponding sequences found in similar leader peptides of other operons encoding enzymes that make amino acids (Table 16-2). Thus, the leucine operon leader peptide has four adjacent leucine codons in a row, and the histidine operon leader peptide has seven histidines in a row!

The function of these codons is to stop a ribosome attempting to translate the leader peptide; thus, when tryptophan is scarce, little charged tryptophan tRNA is available, and the ribosome must stall when it reaches the tryptophan codons. RNA around the tryptophan codons stays bound to the ribosome and cannot be part of a hairpin loop. Figure 16-23 shows the consequence. A ribosome caught at a

Table 16-2 Leader Peptides of Attenuator-Controlled Operons Containing Genes for Amino Acid Biosynthesis*

Operon	Amino Acid Sequence of Leader Peptides
Tryptophan	Met Lys Ala Ile Phe Val Leu Lys Gly Trp Trp Arg Thr Ser
Threonine	Met Lys Arg Ile Ser Thr Thr Ile Thr Thr Thr Ile Thr Ile Thr Thr Gly Asn Gly Ala Gly
Histidine	Met Thr Arg Val Gln Phe Lys His His His His His His His Pro Asp
Isoleucine-valine GEDA	Met Thr Ala Leu Leu Arg Val Ile Ser Leu Val Val Ile Ser Val Val Val Ile Ile Ile . . . . . . Pro Pro Cys Gly Ala Ala Leu Gly Arg Gly Lys Ala
Leucine	Met Ser His Ile Val Arg Phe Thr Gly Leu Leu Leu Leu Asn Ala Phe Ile Val Arg Gly Arg Pro . . . . . . Val Gly Gly Ile Gln His
Phenylalanine	Met Lys His Ile Pro Phe Phe Phe Ala Phe Phe Phe Thr Phe Pro
Isoleucine-valine B	Met Thr Thr Ser Met Leu Asn Ala Lys Leu Leu Pro Thr Ala Pro Ser Ala Ala Val Val Val . . . . . . Val Arg Val Val Val Val Val Gly Asn Ala Pro

*The biosynthesis of isoleucine and valine is complex: The genes are encoded in several operons, and the pathway to leucine synthesis is a branch of the valine pathway. Thus, isoleucine, valine, and leucine are all involved in attenuation of the isoleucine-valine operons. (After C. Bauer, J. Carey, L. Kasper, S. Lynn, D. Waechter, and J. Gardner in *Gene Function in Prokaryotes*, Cold Spring Harbor, 1983, p. 68, with permission.)

tryptophan codon is placed so that region 2 is left free to pair with region 3; then the terminator hairpin (3, 4) cannot be made, and RNA polymerase passes the attenuator into the operon, allowing *trp* enzyme expression. If, on the other hand, there is enough tryptophan (and therefore enough charged *trp* tRNA) for the ribosome to proceed through the tryptophan codons, the ribosome blocks sequence 2, allowing the 3, 4 terminator hairpin to form and aborting transcription at the end of the leader RNA. A third possibility is that translation stops before the tryptophan codons (because some other amino acid is lacking) or fails to initiate altogether; then, sequence 1 is free to pair with sequence 2, and the terminator structure again forms to stop *trp* operon expression. The leader peptide is thought to be important only for positioning ribosomes during peptide synthesis. In fact, it is immediately destroyed by cellular proteases.

It is possible that the coexistence of repression and attenuation controls allows a finer tuning to the level of intracellular tryptophan, resulting in a two-stage response to progressively more stringent tryptophan starvation—the initial response being the cessation of repressor binding, with greater starvation leading to relaxation of attenuation. Other amino acid operons like *his* and *leu* have no repressors; instead, they rely entirely on attenuation for their control.

RNA Synthesis and Protein Synthesis Are Coupled in Bacteria[32, 33]

A polypeptide chain-terminating codon in a gene can have a surprising effect on genes transcribed later in the same operon: These genes may be expressed much less efficiently than normal, even though their own coding sequences are not interrupted by any terminating codons. This interference in downstream gene expression by an upstream translation stop is called **genetic polarity.** Its cause is termination of RNA synthesis induced by the failure of translation, and the cellular agent responsible is the rho (ρ) transcription termination factor (Chapter 13). This is shown by the fact that mutations isolated to prevent polarity, that is, to allow downstream gene expression after a normally polar mutation, map in the *rho* gene and reduce the activity of ρ protein. Rho must bind the RNA being synthesized in order to stop RNA polymerase, and usually this is prevented by ribosomes bound to the RNA. But discharge of ribosomes at a nonsense codon leaves the RNA free to bind ρ factor. Not surprisingly, the greater the distance between a nonsense codon and a new ribosome binding site, the more likely ρ is to act and the more severe is polarity. Polarity is thought to represent a cellular mechanism to prevent futile transcription, as might happen, for example, if the cell were starved and unable to support efficient protein synthesis.

Although ρ may be able to bind any substantial stretch of newly transcribed RNA that is not covered by ribosomes or enclosed in secondary structure, it cannot stop RNA polymerase just anywhere on the DNA. Instead, there are distinct sites where RNA polymerase pauses and allows ρ to release the RNA. Thus, RNA synthesis stops not where the nonsense codon that interrupts translation happens to be, but further downstream where the first RNA terminator is encountered.

The likelihood that most genes of substantial size have ρ-dependent terminators has suggested that some means exist to prevent ρ-dependent termination in genes encoding RNA that is not a messenger, such as ribosomal RNA genes. There is evidence that RNA polymerase initiating at ribosomal RNA promoters does transcribe through terminators, suggesting that the enzyme has been modified in some way. A precedent for operon-specific factors that regulate termination is provided by bacteriophage λ: It encodes two **antitermination** proteins, which modify RNA polymerase transcribing phage genes to allow it to ignore terminators (Chapter 17). Perhaps λ needs antiterminators because its genes are gathered in a few very long operons in which natural polarity caused by untranslated or sparsely translated segments would stop RNA synthesis prematurely without a specific device to defeat it.

Unequal Production of Proteins Coded by a Single mRNA Molecule

Variation in the number of molecules of different proteins arises also from the fact that the proteins coded by a single mRNA molecule need not be produced in similar numbers. This point is demonstrated by the study of the lactose operon proteins: Many more copies of β-galactosidase are synthesized than of galactoside permease or galactoside acetylase. The ratios in which these proteins appear are $1 : \frac{1}{2} : \frac{1}{5}$, respectively. This may mean that ribosomes attach to the different starting points along a given mRNA molecule at different rates, depending on the critical binding sequence preceding the AUG initiation codon (Chapter 14). Alternatively, the ribosomes may attach only at the β-galactosidase gene, with translation of subsequent genes depending on the frequency of ribosome detachment following the reading of a chain-terminating signal. This hypothesis fits nicely with the observation that the β-galactosidase gene is translated the most often and the acetylase the least often. Another factor that affects translation is the existence of codons whose corresponding tRNA species are present in very limiting amounts. Ribosomes must pause longer at such codons than at more common ones, waiting while scarce tRNAs diffuse to them.

It seems reasonable that there should be mechanisms to permit differential reading rates along single mRNA molecules. Although the coordinated appearance of related enzymes is obviously of great advantage to a cell, there is no reason why equal amounts should be produced. Equal numbers of molecules would be useful only if the specific catalytic activity rates (turnover numbers) of related proteins were equal. In general, however, there are great variations in individual turnover numbers.

Bacterial mRNA Is Often Metabolically Unstable

When corepressor or inducer molecules are added to or removed from growing bacteria, the rate of synthesis of the respective proteins is rapidly altered. This rapid adaptation to a changing environment is possible not only because growth requires continual synthesis of new mRNA molecules, but, even more significantly, because many bacterial mRNA molecules are metabolically unstable. The average lifetime of many *E. coli* mRNA molecules at 37°C is about 2 minutes, after which they are enzymatically broken down. The resulting free nucleotides are then phosphorylated to the high-energy triphosphate level and reutilized in the synthesis of new mRNA molecules.

Thus, there is almost complete replacement of the templates for many proteins every several minutes. For example, within several minutes after addition of suitable β-galactosides, *E. coli* cells synthesize β-galactosidase at the maximum rate possible for that particular inducer level. This is true because the rate of destruction soon matches the rate of synthesis of lactose mRNA, and the *amount* of mRNA therefore stays constant. If, on the other hand, all mRNA molecules were metabolically stable, the maximum rate of β-galactosidase synthesis would not be reached until a generation or so had passed. Moreover, the existence of unstable lactose mRNA also means that once β-galactosides are removed, synthesis of β-galactosidase quickly halts and does not resume until it is again needed (Figure 16-24).

The average lifetimes of mRNA molecules for different operons vary considerably; some, which perhaps encode proteins needed constantly, are very stable. This means that mRNA lifetime is itself genetically programmed by a nucleotide sequence (or sequences) that determines the chance that a nuclease attacks the molecule. Exactly what features of mRNA structure are important, and what nucleases are responsible, are only partly understood. One signal for nuclease attack is a specific type of hairpin loop that is a substrate for ribonuclease III, the enzyme responsible for processing both stable (e.g., ribosomal) and messenger RNA in *E. coli* (Chapter 14). Cleavage at these loops by RNase III opens the RNA to degradation by other nucleases, thus accelerating its destruction and decreasing its lifetime. The use of mRNA as a messenger also may affect its stability, since there are suggestions that bound ribosomes protect mRNA from degradation. Perhaps degradative enzymes must compete with ribosomes for sites where breakdown can begin. It is important to realize that the half-life of a messenger may be comparable to the time needed by RNA polymerase to transcribe a large operon. Thus, deg-

Figure 16-24
Rapid rise (or fall) of β-galactosidase mRNA upon the addition (or removal) of β-galactosidase inducers. The *E. coli* cells in this experiment were grown at 37°C under conditions where the cells divided every 40 minutes.

radation may begin before transcription finishes, in which case an entire mRNA molecule never exists.

Ribosome Attachment to mRNA Can Be Regulated[34]

From our first understanding that information is passed from DNA through RNA to proteins, it has been clear that synthesis of a particular protein could be switched on or off at two stages: when its messenger is made (transcriptional regulation), or when the messenger is read on ribosomes (translational regulation). Genetic studies led first to the discovery of transcriptional control, the mode we believe to predominate in both bacterial and higher cells. But we now know that ribosome binding to initiation sites on mRNA can be regulated, and we suspect that this will turn out to be a common mode of regulation, especially where quick and subtle changes in gene expression are required.

Messenger RNA Structure Determines the Availability of the Translation Initiation Site[35, 36]

Unlike promoters, which we believe to be uniformly accessible to RNA polymerase because they exist in the regular structure of duplex DNA, ribosome binding sites may be trapped in the secondary structure into which most single-stranded RNA molecules fold (Chapter 14). If this happens, ribosomes cannot initiate protein synthesis until the structure is broken, either by thermodynamic fluctuations or by a more directed mechanism. Sometimes, ribosome progression along the RNA from an upstream protein-coding sequence is required to open the RNA at an initiation site—a fact that explains the interdependence of translation of different genes in RNA phage (Chapter 17).

Such translation-dependent exposure of an enclosed ribosome binding site can be a specific regulatory device, linked to a controlling signal. This type of mechanism allows cells to express a gene specifying resistance to erythromycin, an antibiotic that binds the ribosomes of sensitive cells and inhibits protein synthesis. The gene for erythromycin resistance, carried on a plasmid that confers bacterial resistance to the antibiotic (Chapter 11), encodes an enzyme that methylates an adenine residue in 23S ribosomal RNA, thereby preventing the antibiotic from binding. Sequence analysis reveals a regulatory structure quite similar to that of the tryptophan operon, in that a leader polypeptide precedes the methylase gene. The controlling mechanism also involves RNA folding into alternative hairpins, but these RNA structures serve a different end. Here the "repressed" conformation prevents initiation of protein synthesis rather than causing termination of RNA synthesis. When a ribosome translates the leader polypeptide freely and is released, an RNA hairpin can form that buries the initiation sequence for methylase, preventing its translation (Figure 16-25). But when erythromycin binds and slows the progress of a ribosome making leader polypeptide, a different secondary structure forms that exposes the methylase initiation region, allowing methylase expression. Presumably, there is always a low level of methylase expression and thus a few erythromycin-

1. *Off conformation* (no erythromycin present)

Initiation sequence of methylase gene

Ribosome translating leader proceeds to the end of the leader and releases the RNA, allowing the 1,2 hairpin to form; this allows the 3,4 hairpin to form, blocking access of a ribosome to the methylase initiation sequence.

mRNA

Leader polypeptide coding region

Methylase gene coding region

2. *On conformation* (erythromycin present)

Ribosome inhibited by erythromycin stalls while translating the leader polypeptide, thereby covering sequence 1; this allows hairpin 2,3 to form, leaving the methylase initiation sequence in 4 free to bind a second ribosome.

GAG

Figure 16-25
Control of erythromycin methylase gene expression by a leader peptide and by alternate RNA structures that determine whether ribosomes can initiate translation of its gene. (After S. Horinouchi and B. Weisblum, *Mol. Gen. Genet.* 182 (1981):343, with permission.)

resistant ribosomes that can complete translation of the methylase gene when its initiation site is available. Induction is autocatalytic, since more methylase protects more ribosomes from erythromycin, thus allowing more methylase translation. Eventually, enough ribosomes are protected that none stall in the leader, and methylase synthesis is shut down.

A Small RNA Is a Translational Repressor of Tn10 Transposase[37]

Just as a ribosome binding site can be blocked if it is joined in a secondary structure with another part of its own mRNA, it also can be base-paired to a separate complementary RNA that prevents ribosome attachment by covering the critical sequences. Transposon Tn10 encodes such an RNA, which serves to inhibit the expression of the Tn10 transposase enzyme and in fact any related transposase in the cell that has the same sequences at its ribosome binding site. Two adjacent promoters near the beginning of the transposase gene of Tn10, designated pIN and pOUT, are important to this regulation (Figure 16-26a). Promoter pIN directs synthesis of transposase mRNA. RNA is made from pOUT in the opposite direction, but it overlaps the region transcribed from pIN by 35 nucleotides. Thus, the first 35 nucleotides of the pOUT transcript are complementary in sequence to the beginning of the transposase mRNA, a region that includes its translation initiation site. These overlapping sequences are believed to base-pair in the cell, covering the transposase initiation site and preventing ribosome binding (Figure 16-26b). The consequence is that expression of transposase is very low, and transposition happens only rarely (Chapter 11).

It might seem that the same end would be achieved if pIN were simply a less efficient promoter. However, one advantage of regulation by translational repression is that the cell is protected from acci-

Figure 16-26
Transposition of Tn10 is controlled by an RNA that prevents translation of transposase mRNA. (a) Map of Tn10, including the overlapping, opposed promoters pIN and pOUT. (b) How pOUT RNA is thought to cover the ribosome binding site of transposase mRNA made from pIN, thereby preventing synthesis of transposase.

(a)

Transposon Tn10 (9.3 kbp)

IS10 insertion element making up left end of Tn10 and encoding transposase

Tetracycline resistance gene

IS10 at other end of Tn10

IS10R (1.3 kbp)   Tet^R   IS10L

Transposase mRNA made from promoter pIN

Transposase gene

Regulatory mRNA made from promoter pOUT

(b)

Ribosome-binding sequence and initiating AUG of transposase

Transposase mRNA

5'

5'

Regulatory RNA annealed to beginning of transposase mRNA, preventing binding and initiation by a ribosome

dental transcription passing through the transposase gene from outside the transposon; even if too much transposon mRNA is suddenly made, the RNA repressor will be available in some excess to inhibit its translation. Another way that Tn10 transposase is regulated is discussed in Chapter 11.

It is clear that an RNA repressor like this could be encoded anywhere in the genome and thus could be a repressor of distant as well as adjacent genes. In fact, it is possible to construct, by genetic engineering, a plasmid that makes an RNA complementary to the translation initiation region of a known gene. This strategy has been used to prevent expression of particular genes, and it may become a powerful way to artificially manipulate genetic expression in cells.

Ribosomal Proteins Are Translational Repressors of Their Own Synthesis[38–41]

Correct expression of ribosomal protein genes poses an interesting regulatory problem for the cell. Each ribosome contains some 50 distinct proteins that must be made at exactly the same rate. Furthermore, the rate at which a cell makes protein, and thus the number of ribosomes it needs, is tied closely to the cell's growth rate; a change in growth conditions quickly leads to an increase or decrease in the rate of synthesis of all ribosomal components. How is all this coordinated regulation accomplished?

Figure 16-27
Ribosomal protein operons of *E. coli.* The protein that acts as repressor of the other proteins is shaded gray, and its site of action (on the mRNA) is shown by an arrow. (After M. Nomura, R. Gourse, and G. Baughman, *Ann. Rev. Biochem.* 53 (1984):82, with permission.)

Control of ribosomal protein genes is simplified by their organization into several different operons, each containing genes for up to 11 ribosomal proteins (Figure 16-27). Some nonribosomal proteins that also are required according to growth rate are contained in these operons, including subunits α, β, and β' of RNA polymerase. Like other operons, these may sometimes be regulated at the level of RNA synthesis. However, the primary control of ribosomal protein synthesis is on *translation* of the mRNA, not on its synthesis. This is shown by a simple experiment in which extra copies of a ribosomal protein operon are put into the cell. The amount of mRNA increases correspondingly, but synthesis of the proteins stays nearly the same; thus, the cell compensates for extra mRNA by curtailing its activity as a template. This happens because ribosomal proteins are repressors of their own translation (and thus are *autogenous* regulators). For each operon, one (or a complex of two) of the proteins binds the messenger near the translation initiation sequence of one of the first genes of the operon, preventing ribosomes from binding and initiating translation (see Figure 16-27). Repression of translation of the first gene also prevents expression of some or all of the rest. We believe that this happens because translation of several successive genes in the messenger is linked, each ribosome binding site being opened by a ribosome passing from the preceding gene.

The advantage to the cell in having a protein repress its own translation is to allow quick and precise adjustments of expression. A few unused molecules of protein L4, for example, will shut down its synthesis, as well as synthesis of the other ten ribosomal proteins in its operon. In this way, these proteins are made just at the rate they are needed for assembly into ribosomes (Chapter 14).

How one protein can function both as a ribosomal component and as a regulator of its own translation is shown by comparing the sites

Protein S7 can be crosslinked to this base and therefore must be close to it in the ribosome.

Protein S7 binding site in 16S ribosomal RNA

Translation initiation region in the messenger RNA for protein S7

Figure 16-28

A comparison of the region where ribosomal protein S7 (encoded by the *str* operon; Figure 16-27) binds 16S RNA in the ribosome, with the translation initiation site in its mRNA. Similar sequences, presumably parts of the binding site, are shaded. [After M. Nomura, J. Yates, D. Dean, and L. Post, *Proc. Nat. Acad. Sci.* 77 (1980):7086, with permission.]

where it binds in ribosomal and in messenger RNA. They are similar in sequence and in secondary structure, indicating that the same binding specificity accounts for both activities (Figure 16-28). The comparison also suggests a precise mechanism of regulation. Since the binding site in the messenger includes the initiating AUG, mRNA bound by excess protein S7 (in this example) clearly cannot attach to ribosomes to initiate translation. However, binding is stronger to ribosomal RNA than to mRNA, so that translation is repressed only when all need for the protein in ribosome assembly is satisfied.

Protein Synthesis Termination Factor RF2 Regulates Its Own Translation at the Termination Step[42]

Although most well-documented translational control mechanisms operate at the stage of polypeptide chain initiation, a remarkable exception has been recently discovered. The nucleotide sequence of the gene for RF2, one of the two *E. coli* factors necessary for polypeptide chain termination, suggests that RF2 regulates its own synthesis not at the first, but instead at the final, step of translation.

RF2 is the protein synthesis termination factor that catalyzes release of nascent chains at UGA and UAA stop codons, while the factor RF1 acts at UAG and UAA triplets (Chapter 14). Surprisingly, the carefully sequenced gene for RF2 does not contain a continuous open reading frame. Instead, the first 25 amino acids are encoded by one stretch of DNA, and the remaining 315 amino acids are specified by a sequence occurring in the +1 reading frame (Figure 16-29). Between these two coding regions lie a UGA stop codon and a C nucleotide, which somehow are read as aspartic acid whose codon is GAU or GAC, during synthesis of the full-length (38 kdal) RF2 protein.

NH$_2$ ... Gly - Tyr - Leu - Asp - Tyr - Asp ... COOH
23   24   25   26   27   28

Figure 16-29
Translation of mRNA for the protein synthesis release factor RF2 involves a shift in reading frame. The occurrence of this frameshift is thought to provide regulation of RF2 gene expression because the RF2 protein is itself required for release of the polypeptide at UGA; if RF2 protein is scarce, release is slow and there is time for the relatively inefficient frameshifting step to occur. (After W. Craigen, R. Cook, W. Tate, and C. Caskey, *Proc. Nat. Acad. Sci.*, 82 (1985):3619.)

We suppose that this translation frameshift regulates expression of the gene for RF2 in the following way. If ample RF2 is available in the cell, the 25-amino-acid polypeptide is released at the UGA codon and no functional RF2 is made. If RF2 is scarce, on the other hand, release of the polypeptide is delayed and there is time for the (presumably) slow frameshift to occur: Aspartic acid is inserted in position 26 when the ribosome shifts to GAC, and translation continues to the final UAG stop codon. Conveniently, UAG is recognized by the other release factor, RF1. Why this elegant translational self-control circuit is utilized only by the RF2 gene and not by the RF1 gene is not known; RF1 is encoded by a conventional continuous open reading frame terminated by UGA (which, as we have seen, is recognized by RF2).

ppGpp Is a Cellular Signal of Amino Acid Starvation[41, 43–46]

If ribosomal proteins are made from their mRNA only when ribosomal RNA is available for their incorporation into ribosomes, then what really determines how fast ribosomes appear is the rate of ribosomal RNA synthesis. How, then, does ribosomal RNA synthesis change according to the cell's need? Here we have no clear answers, although fairly strong circumstantial evidence says that the nucleotide ppGpp (Chapter 14) is an important agent of the regulation. Thus, starving a cell for an amino acid both causes the cellular level of ppGpp to rise abruptly and turns down rRNA synthesis and tRNA synthesis. Similarly, more subtle changes in ppGpp concentration that occur as growth conditions improve or degenerate may lead to adjustments in the rate of ribosomal RNA synthesis. Whether ppGpp acts directly on RNA polymerase to change its rate of initiation at rRNA and tRNA promoters or whether it acts through some activator or repressor is not yet known. However, it is likely that the ability of ribosomal RNA promoters to be controlled is conferred by sequences within the promoters themselves, since in vitro fusion of an rRNA promoter (those sequences between the RNA start site and −50; Chapter 13) allows any gene to be controlled like the rRNA genes.

The effects of ppGpp are not limited to ribosomal RNA synthesis. High concentrations of the nucleotide not only reduce expression of many genes besides those for ribosomal RNA, but also stimulate expression of many others. Presumably, these other genes encode proteins needed at higher levels when the cell is starved. Furthermore, ppGpp directly affects the activity of many enzymes, including those of the ribosome that catalyze peptide bond formation; here its function is thought to be in slowing the rate of protein synthesis, thus reducing translation errors by allowing more time for a correct (but scarce) aminoacyl-tRNA to find the ribosome. Therefore, ppGpp is probably a general messenger molecule that integrates a complex cellular response to amino acid starvation, just as the cellular level of cyclic AMP (acting through CAP protein) signals to the cell a sufficiency of glucose.

Alarmones: Possible Cellular Signals of Distress[45, 47, 48]

The discovery of ppGpp as a signal of amino acid starvation suggested that other unusual metabolic products made as a direct result of some specific deficiency could have similar roles. Named **alarmones** (*alar*m + hor*mones*), they would affect cellular enzymes or proteins so as to counteract the problem whose metabolic consequence they represent, moving to different sites in the cell, just as hormones move to different tissues of more complex organisms. Clearly, both ppGpp and cyclic AMP, the signal of carbon source starvation that affects expression of many genes through binding CAP protein, act in this way. Two other compounds have been considered as possible alarmones (Table 16-3). Diadenosine tetraphosphate *(AppppA)* is made by bacterial aminoacyl-tRNA synthetases in the absence of tRNA, as an aberration of the tRNA-charging reaction; it thus might accumulate when tRNA is unavailable in the cell, thereby signaling this condition. In eucaryotic cells, AppppA stimulates the activity of DNA polymerase α and has been suggested as a signal of disrupted replication that promotes new DNA synthesis. Folate starvation of bacteria increases synthesis of *ZTP*, a possible alarmone for this serious deficiency.

The Concentration of a Protein Can Be Determined by Its Sensitivity to Proteolytic Degradation[49–52]

Most bacterial proteins are stable during normal growth. That is, if a protein's synthesis stops, such as by addition of a corepressor to a growing culture, then its concentration in the cell is reduced only through dilution by other cell components that continue to be made. Many generations may be required to remove the protein effectively from the cells. For an enzyme like β-galactosidase, this is of little consequence. But some proteins must be removed more quickly if cell growth is to succeed. Among these are the LexA repressor of the SOS genes (see page 480) and the phage λ repressor (Chapter 17), which are cleaved into two segments and inactivated by RecA protein when the cell's DNA is damaged. Another such protein is an inhibitor of cell division in *E. coli* encoded by the *sulA* gene. This is itself an SOS gene whose function is to stop further cell division until the chromo-

Table 16-3 Metabolites That May Be Alarmones

Compound	Conditions that Induce Its Synthesis	Known or Suggested Function
1. cAMP (3'-5' cyclic AMP)	Absence of carbon source	To stimulate synthesis of enzymes that metabolize other sugars; to control expression of other proteins required when carbon is limiting.
2. ppGpp, pppGpp (guanosine tetraphosphate, guanosine pentaphosphate)	Amino acid starvation	To inhibit stable RNA synthesis; to control other cellular processes so as to favor adaptation to condition of limiting amino acids
3. AppppA (diadenosine tetraphosphate)	Bacteria: shortage of tRNA (?), oxidative damage (?)	?
	Eucaryotes: replication fork arrest; normal transition to S phase of cell cycle	To stimulate DNA synthesis, cell proliferation
4. ZTP (5-amino 4-imidazole carboxamide riboside 5'-triphosphate)	Folate starvation	?

some is repaired. The SulA protein is especially sensitive to attack by the cellular *Lon protease;* only its high rate of synthesis allows SulA to accumulate when the SOS genes are induced. As soon as SOS genes are no longer needed and are once again repressed, remaining SulA protein is quickly destroyed by Lon protease and cell division and growth resume. It is therefore clear why a mutant cell lacking Lon protease is killed by very small doses of ultraviolet light: The accumulated SulA division inhibitor is not destroyed, and the cell can never divide again, even though induction of other SOS genes allows repair of the cell's DNA.

What structural feature makes some proteins very sensitive to proteases, whereas others are completely resistant? Most proteins have probably evolved a tight tertiary structure that prevents the active site of a protease molecule, which is of comparable size, from approaching its peptide bonds. In contrast, proteins like SulA may possess exposed segments that a protease can reach. A second role of Lon protease supports this idea: It is responsible for destroying polypeptide fragments made when translation stops prematurely at a nonsense codon or aborts by accident. Such pieces of protein could be loosely or incorrectly folded, allowing regions to protrude and to be attacked by the protease.

Summary

Cells have control mechanisms to ensure that proteins are synthesized in the required amounts. Most of our detailed understanding of these mechanisms at a molecular level is limited to bacterial cells, in particular *E. coli*. Bacteria contain many enzymes whose rate of synthesis depends on the availability of external food molecules. These external molecules, called inducers and corepressors, usually determine the rate of synthesis of enzymes by controlling the synthesis of their mRNA templates. Inducers and corepressors act by binding to regulatory proteins, called activators and repressors. Activators are positive regulators, because their presence is required for the regulated enzyme to be made; on the other hand, repressors act negatively, because their regulatory activity is to prevent synthesis of proteins. Thus, when lactose is absent, the lactose operon repressor prevents synthesis of enzymes that metabolize lactose; however, upon binding an inducer (a molecule related to lactose), the repressor loses this ability and allows the enzymes to be made. The arabinose operon C protein, an activator, causes arabinose enzymes to be made when it binds the inducer arabinose. In contrast to inducers, corepressors prevent enzyme synthesis; tryptophan is a corepressor that gives the tryptophan repressor an active configuration by binding to it, so that enzymes that synthesize tryptophan are no longer made.

Regulatory proteins that control mRNA synthesis usually act by binding DNA near or in the promoters of genes they regulate. Thus, operators, the binding sites of repressors, overlap promoters and directly prevent RNA polymerase binding. Many activators, on the other hand, bind beside RNA polymerase so as to stimulate its binding to the promoter. One class of activators fundamentally changes RNA polymerase itself: new σ factors, present in place of the "standard" σ factors (σ^{70} in *E. coli*) in a small fraction of the holoenzyme, recognize promoters distinct from the "standard" promoter recognized by σ^{70}. For example, σ^{32} directs RNA polymerase to bind heat-shock promoters when the cell's temperature is abruptly raised, thereby greatly increasing expression of the heat-shock proteins.

The length of DNA controlled by a specific promoter, the operon, often consists of several genes with related metabolic functions (e.g., the production of successive enzymes in the synthesis of an amino acid or nucleotide). Enzymes in an operon are not necessarily (or even usually) made at the same rate, since ribosomes may read different parts of the messenger unequally, or some segments of the messenger may be more stable than others. One repressor or activator can control more than one promoter; the LexA repressor of *E. coli*, for example, controls about 15 operons encoding DNA repair enzymes. Messenger RNA is destroyed by nuclease attack, with a half-life that is typically a few minutes, so that the template disappears quickly when a regulatory signal changes.

The rate of synthesis of many proteins is not controlled, but is instead constitutive, fixed at a rate that may be very low or very high. This rate is determined by the inherent strength of the promoter, the lifetime of its messenger, and the rate at which ribosomes read the messenger.

A distinct class of regulation acts by determining the efficiency of RNA polymerase termination rather than the rate of RNA chain initiation. Operons encoding enzymes of amino acid biosynthesis are controlled at termination sites (in this case called attenuators) that separate a leader RNA from mRNA for the enzyme structural genes; the signal of amino acid deficiency is transmitted by ribosomes that, in translating the leader RNA, prevent formation of an RNA hairpin that is essential for termination. Although they are best characterized for the *E. coli* bacteriophage λ, operon-specific antiterminator proteins also are thought to exist in *E. coli*.

Translational control is mediated by devices that determine whether the translation initiation site of an mRNA is available to ribosomes. For example, erythromycin resistance is produced by a methylase whose synthesis depends on its ribosome binding site being uncovered from RNA secondary structure when the antibiotic is present. Expression of the Tn10 transposase is repressed by a complementary RNA that covers and thus inactivates its ribosome binding site. Ribosomal proteins are their own translational repressors, acting by blocking the ribosome initiation sites of their own mRNAs when they accumulate in excess.

A small molecule can signal important cellular conditions to many different genes or operons at the same time. Carbon source deficiency causes the enzyme adenylcyclase to make cyclic AMP, which in turn stimulates CAP protein to act as a transcriptional activator of many promoters, including promoters of the arabinose and lactose operons (the *lac* operon, like the *ara* operon, is therefore regulated both negatively and positively). Amino acid starvation leads to synthesis of ppGpp, which affects both the transcription of various genes and the activity of many enzymes whose functions are required when an amino acid is scarce. Small molecules such as these have been called alarmones, because they integrate the cellular response to particular conditions of stress.

Although proteins are generally stable in bacteria, some are destroyed specifically and quickly when regulatory change subjects them to proteolytic degradation; in this way, the concentration of a protein can be decreased abruptly, whereas a stable protein disappears only through dilution during cell growth. The LexA repressor and phage λ repressor are cleaved by RecA protein activated when cellular DNA is damaged, and the SulA cell division inhibitor is attacked by the Lon protease, so that it disappears quickly when its synthesis stops.

Bibliography

General References

Beckwith, J., J. Davies, and J. Gallant, eds. 1983. *Gene Function in Prokaryotes.* Cold Spring Harbor, N.Y.: Cold Spring Harbor Laboratory. Current reviews of many subjects discussed in the chapter, including attenuation, translational repression, termination and polarity, and global control systems.

Gottesman, S. 1984. "Bacterial Regulation: Global Regulatory Networks." *Ann. Rev. Genetics* 18:415–441.

Ingraham, J., O. Maaløe, and F. Neidhardt. 1983. *Growth of the Bacterial Cell.* Sunderland, Mass.: Sinauer. Considers physiology, biochemistry, and genetics and emphasizes overall regulation of cell growth.

Jacob, F., and J. Monod. 1961. "Genetic Regulatory Mechanisms in the Synthesis of Proteins." *J. Mol. Biol.* 3:318–356. A classic review of great historical interest that ties together the concept of messenger RNA with the problem of the control of protein synthesis.

Miller, J., and W. Reznikoff, eds. 1980. *The Operon.* Cold Spring Harbor, N.Y.: Cold Spring Harbor Laboratory. Everything about the lactose operon, as well as chapters about several others (tryptophan, arabinose, and phage λ early operons).

Raibaud, O., and M. Schwartz. 1984. "Positive Control of Transcription Initiation in Bacteria." *Ann. Rev. Genetics* 18:173–206.

Cited References

1. Gilbert, W., and B. Müller-Hill. 1966. "Isolation of the Lac Repressor." *Proc. Nat. Acad. Sci.* 56:1891–1898.
2. Ptashne, M. 1967. "Specific Binding of the λ Phage Repressor to λ DNA." *Nature* 214:232–234.
3. Galas, D., and A. Schmitz. 1978. "DNase Footprinting: A Simple Method for the Detection of Protein-DNA Binding Specificity." *Nucleic Acid Res.* 5:3157–3170.
4. Siebenlist, U., R. Simpson, and W. Gilbert. 1980. "E. coli RNA Polymerase Interacts Homologously with Two Different Promoters." *Cell* 20:269–281.
5. Pabo, C., and M. Lewis. 1982. "The Operator-Binding Domain of Lambda Repressor: Structure and DNA Recognition." *Nature* 298:443–447.
6. Hecht, M., H. Nelson, and R. Sauer. 1983. "Mutations in λ Repressor's Amino-Terminal Domain: Implications for Protein Stability and DNA Binding." *Proc. Nat. Acad. Sci.* 80:2676–2680.
7. Steitz, T., D. Ohlendorf, D. McKay, W. Anderson, and B. Matthews. 1982. "Structural Similarity in the DNA-Binding Domains of Catabolite Gene Activator and *cro* Repressor Proteins." *Proc. Nat. Acad. Sci.* 79:3097–3100.
8. Anderson, W., Y. Takeda, D. Ohlendorf, and B. Matthews. 1982. "Proposed α-Helical Super-Secondary Structure Associated with Protein: DNA Recognition." *J. Mol. Biol.* 159:745–751.
9. Sauer, R., R. Yocum, R. Doolittle, M. Lewis, and C. Pabo. 1982. "Homology Among DNA-Binding Proteins Suggests Use of a Conserved Super-Secondary Structure." *Nature* 298:447–451.
10. Pabo, C., and R. Sauer. 1984. "Protein-DNA Recognition." *Ann. Rev. Biochem.* 58:293–321.
11. Wharton, R., and M. Ptashne. 1985. "Changing the Binding Specificity of a Repressor by Redesigning an α-Helix." *Nature* 316:601–605.
12. Anderson, J., M. Ptashne, and S. Harrison. 1985. "A Phage Repressor-Operator Complex at 7Å Resolution." *Nature* 316:596–601.
13. DeCrombrugghe, B., S. Busby, and H. Buc. 1984. "Cyclic AMP Receptor Protein: Role in Transcription Activation." *Science* 224:831–838.
14. Malan, T. P., and W. R. McClure. 1984. "Dual Promoter Control of the *Escherichia coli* Lactose Operon." *Cell* 39:173–180.
15. Walker, G. 1984. "Mutagenesis and Inducible Responses to Deoxyribonucleic Acid Damage in *Escherichia coli.*" *Microbiol. Revs.* 48:60–93.
16. Little, J. W., S. H. Edmiston, L. Z. Pacelli, and D. W. Mount. 1980. "Cleavage of the *Escherichia coli lexA* Protein by the *recA* Protease." *Proc. Nat. Acad. Sci.* 77:3225–3229.
17. Sancar, G., A. Sancar, J. Little, and W. Rupp. 1982. "The *uvrB* Gene of *Escherichia coli* Has Both *lexA*-Repressed and *lexA*-Independent Promoters." *Cell* 28:523–530.
18. Dunn, T., S. Hahn, S. Ogden, and R. Schleif. 1984. "An Operator at −280 Base Pairs That Is Required for Repression of *araBAD* Operon Promoter: Addition of DNA Helical Turns Between the Operator and Promoter Cyclically Hinders Repression." *Proc. Nat. Acad. Sci.* 81:5017–5020.
19. Lee, N., W. Gielow, and R. Wallace. 1981. "Mechanism of *araC* Autoregulation and the Domains of Two Overlapping Promoters P_C and P_{BAD}, in the L-Arabinose Regulatory Region of *Escherichia coli.*" *Proc. Nat. Acad. Sci.* 78:752–756.
20. Raibaud, O., and M. Schwartz. 1984. "Positive Control of Transcription Initiation in Bacteria," *Ann. Rev. Genetics* 18:415–441.
21. Majumdar, A., and S. Adhya. 1984. "Demonstration of Two Operator Elements in *gal; In Vitro* Repressor Binding Studies." *Proc. Nat. Acad. Sci.* 81:6100–6104.
22. Busby, S., H. Aiba, and B. de Crombrugghe. 1982. "Mutations in the *Escherichia coli* Galactose Operon That Define Two Promoters and the Binding Site of the Cyclic AMP Receptor Protein." *J. Mol. Biol.* 154:211–227.
23. Neidhardt, F. C., R. A. VanBogelen, and V. Vaughn. 1984. "Genetics and Regulation of Heat Shock Proteins." *Ann. Rev. Genetics* 18:295–329.
24. Grossman, A., J. Erickson, and C. Gross. 1984. "The *htpR* Gene Product of *E. coli* is a Sigma Factor for Heat-Shock Promoters." *Cell* 38:383–390.
25. Landick, R., V. Vaughn, E. T. Lau, R. A. VanBogelen, J. W. Erickson, and F. C. Neidhardt. 1984. "Nucleotide Sequence of the Heat Shock Regulatory Gene of *E. coli* Suggests Its Protein Product May Be a Transcription Factor." *Cell* 38:175–182.
26. Yura, T., T. Tobe, K. Ito, and T. Osawa. 1984. "Heat Shock Regulatory Gene (*htpR*) of *Escherichia coli* is Required for Growth at High Temperature but is Dispensable at Low Temperature." *Proc. Nat. Acad. Sci.* 81:6803–6807.
27. Cowing, D. W., J. C. A. Bardwell, E. A. Craig, C. Woolford, R. W. Hendrix, and C. A. Gross. 1985. "Consensus Sequence for *Escherichia coli* Heat Shock Gene Promoters." *Proc. Nat. Acad. Sci.* 82:2679–2683.
28. Bardwell, J. C. A., and E. A. Craig. 1984. "Major Heat Shock Gene of *Drosophila* and the *E. coli* Heat Inducible *dnaK* Gene Are Homologous." *Proc. Nat. Acad. Sci.* 81:848–852.
29. Hirschman, J., P. K. Wong, K. Sei, J. Keener, and S. Kustu. 1985. "Products of Nitrogen Regulatory Genes *ntrA* and *ntrC* of Enteric Bacteria Activate *glnA* Transcription in Vitro; Evidence that the *ntrA* Product is a σ Factor." *Proc. Nat. Acad. Sci.* 82:7525–7529.
30. Yanofsky, C., and R. Kolter. 1982. "Attenuation in Amino Acid Biosynthetic Operons." *Ann. Rev. Genetics* 16:113–134.
31. Bauer, C., J. Carey, L. Kasper, S. Lynn, D. Waechter, and J. Gardner. 1983. "Attenuation in Bacterial Operons." In *Gene Function in Prokaryotes*, ed. J. Beckwith, J. Davies, and J. Gallant. Cold Spring Harbor, N.Y.: Cold Spring Harbor Laboratory, pp. 65–89.
32. Adhya, S., and M. Gottesman. 1978. "Control of Transcription Termination." *Ann. Rev. Biochem.* 47:967–996.
33. Li, S., C. L. Squires, and C. Squires. 1984. "Antitermination of *E. coli* rRNA Transcription Is Caused by a Control Region Segment Containing Lambda *nut*-like Sequences." *Cell* 38:851–861.
34. Campbell, K., G. Stormo, and L. Gold. 1983. "Protein-Mediated Translational Repression." In *Gene Function in Prokaryotes*, ed. J. Beckwith, J. Davies, and J. Gallant. Cold Spring Harbor, N.Y.: Cold Spring Harbor Laboratory, pp. 185–210.

35. Horinouchi, S., and B. Weisblum. 1980. "Posttranscriptional Modification of mRNA Conformation: Mechanism That Regulates Erythromycin-Induced Resistance." *Proc. Nat. Acad. Sci.* 77:7079–7083.

36. Horinouchi, S., and B. Weisblum. 1981. "The Control Region for Erythromycin Resistance: Free Energy Changes Related to Induction and Mutation to Constitutive Expression." *Mol. Gen. Genet.* 182:341–348.

37. Simons, R., and N. Kleckner. 1983. "Translational Control of IS10 Transposition." *Cell* 34:683–691.

38. Nomura, M., J. Yates, D. Dean, and L. Post. 1980. "Feedback Regulation of Ribosomal Protein Gene Expression in *Escherichia coli*: Structural Homology of Ribosomal RNA and Ribosomal Protein mRNA." *Proc. Nat. Acad. Sci.* 77:7084–7088.

39. Baughman, G., and M. Nomura. 1983. "Localization of the Target Site for Translational Regulation of the L11 Operon and Direct Evidence for Translational Coupling in *Escherichia coli*." *Cell* 34:979–988.

40. Baughman, G., and M. Nomura. 1984. "Translational Regulation of the Lll Ribosomal Protein Operon of *Escherichia coli*: Analysis of the mRNA Target Site Using Oligonucleotide-Directed Mutagenesis." *Proc. Nat. Acad. Sci.* 81:5389–5393.

41. Nomura, M., R. Gourse, and G. Baughman. 1984. "Regulation of the Synthesis of Ribosomes and Ribosomal Components." *Ann. Rev. Biochem.* 53:75–117.

42. Craigen, W. J., R. G. Cook, W. P. Tate, and C. T. Caskey. 1985. "Bacterial Peptide Chain Release Factors: Conserved Primary Structure and Possible Frameshift Regulation of Release Factor 2." *Proc. Nat. Acad. Sci.* 82:3616–3620.

43. Sarmientos, P., and M. Cashel. 1983. "Carbon Starvation and Growth Rate-Dependent Regulation of the *Escherichia coli* Ribosomal RNA Promoters: Differential Control of Dual Promoters." *Proc. Nat. Acad. Sci.* 80:7010–7013.

44. Ryals, J., R. Little, and H. Bremer. 1982. "Control of rRNA and tRNA Synthesis in *Escherichia coli* by Guanosine Tetraphosphate." *J. Bacteriol.* 151:1261–1268.

45. Stephens, J., S. Artz, and B. Ames. 1975. "Guanosine 5'-Diphosphate 3'-Diphosphate (ppGpp): Positive Effector for Histidine Operon-Transcription and General Signal for Amino-Acid Deficiency." *Proc. Nat. Acad. Sci.* 72:4389–4393.

46. Lamond, A. I., and A. A. Travers. 1984. "Genetically Separable Functional Elements Mediate the Optimal Expression and Stringent Regulation of a Bacterial tRNA Gene." *Cell* 40:319–326.

47. Lee, P., B. Bochner, and B. Ames. 1983. "Diadenosine 5',5'''-P^1,P^4-Tetraphosphate and Related Adenylylated Nucleotides in *Salmonella typhimurium*." *J. Biol. Chem.* 258:6827–6834.

48. Bochner, B., and B. Ames. 1982. "ZTP (5-Amino 4-Imidozole Carboxamide Riboside 5'-Triphosphate): A Proposed Alarmone for 10-Formyl-Tetrahydrofolate Deficiency." *Cell* 29:929–937.

49. Mizusawa, S., and S. Gottesman. 1983. "Protein Degradation in *Escherichia coli*: The *lon* Gene Controls the Stability of sulA Protein." *Proc. Nat. Acad. Sci.* 80:358–362.

50. Charette, M. R., G. W. Henderson, and A. Markovitz. 1981. "ATP Hydrolysis-Dependent Protease Activity of the *lon(capR)* Protein of *Escherichia coli* K-12." *Proc. Nat. Acad. Sci.* 78:4728–4732.

51. Chung, C. H., and A. L. Goldberg. 1981. "The Product of the *lon(capR)* Gene in *E. coli* Is the ATP-Dependent Protease La." *Proc. Nat. Acad. Sci.* 78:4931–4935.

52. Baker, T. A., A. D. Grossman, and C. A. Gross. 1984. "A Gene Regulating the Heat Shock Response in *Escherichia coli* Also Affects Proteolysis." *Proc. Nat. Acad. Sci.* 81:6779–6783.

CHAPTER 17

The Replication
of Bacterial Viruses

Molecular geneticists have often focused their study on viruses because of the apparent simplicity of the virus. In the beginning, viruses were thought of as naked genes; but gradually it became clear that the correct analogy is the naked chromosome, since many of the viruses first thought to be so uncomplicated are now known to contain several hundred genes. And when we began to discover the various ways bacteria can control gene expression, it seemed probable that the essence of a virus's existence was the lack of any such regulatory devices. Being so constituted, they would be able to multiply rapidly at the expense of their host cells' metabolism. Again the first hunches proved wrong. The replication of even the smallest of viruses is a very complicated affair, achieved only with the aid of highly evolved regulatory systems designed to ensure that the right molecules are synthesized at just the right time in the life cycle of a virus. We now understand the molecular interactions that underlie many of the basic regulatory mechanisms that direct growth of bacterial viruses (bacteriophages), and comparable descriptions of eucaryotic virus development are in sight (Chapter 24). In this chapter, we first consider some general principles of virus structure and multiplication and then describe in detail the growth of a few important bacteriophages that have provided our most advanced knowledge of molecular genetics.

The Core and Coating of Viruses[1-4]

Both the size and structural complexity of viruses show great variation. Some viruses have molecular weights as small as several million, whereas others approach the size of very small bacteria. However, all viruses differ fundamentally from cells, which have both DNA and RNA, in that viruses contain only one type of nucleic acid, which may be either DNA or RNA. The genetic nucleic acid component is always present in the center of the virus particle, surrounded by a protective coat, or shell, called a capsid. Some of the shells are quite complex; they contain many proteins arranged in several layers and have complex appendages like phage λ tails (Figure 17-1). Other viruses have lipid or carbohydrate molecules as well as protein (Figure 17-2). In still other cases, such as the small RNA bacterial viruses MS2, f2, R17, and Qβ, the shell contains only one type of protein and no lipid or carbohydrate (Figure 17-3).

All capsids contain many copies of the protein components, often arranged either in an elongated particle with helical symmetry or in a

Figure 17-1
The structure of the λ phage particle. (a) An electron micrograph of several negatively stained λ particles. (b) A schematic drawing showing detailed features revealed by electron microscopy and biochemical analysis. [Courtesy R. W. Hendrix, University of Pittsburgh.]

(a)

(b)

Head

Tail tube

1350 Å

Conical part

150 Å

Tail fiber

230 Å

nearly spherical particle, which is actually an icosahedron (a polyhedron with 20 faces) and has cubical (or quasi-cubical) symmetry. Thus, tobacco mosaic virus, TMV (Chapter 5) has about 2150 identical protein molecules (MW about 17,000 daltons) helically arranged around a central RNA molecule containing approximately 6000 nucleotides, whereas the icosahedral bacteriophage MS2 (as well as f2, R17, and Qβ) has 180 identical proteins (MW about 14,000 daltons) cubically arranged about an inner RNA molecule with 3569 nucleotides (Figure 17-3).

The use of a large number of identical protein molecules in the construction of the capsid is an obligatory feature of the structures of all viruses. This follows from their limited nucleic acid content, which in turn places a restriction on the maximum number of amino acids in the proteins coded by the viral chromosome. For example, the 3569 nucleotides in an MS2 RNA chain could code for about 1200 amino acids, corresponding to a protein molecular weight of about 1.3×10^5 daltons. This is much smaller than the molecular weight of the MS2 protein shell (2.5×10^6). Thus, even if the entire MS2 RNA chain coded for its coat protein (which we know it does not), approximately 20 identical protein molecules would be needed. This use of a large number of identical protein subunits is why the simpler viruses, which often contain only one major type of protein molecule in their coat, have either helical or cubical (or quasi-cubical) symmetry. Only these two types of symmetry permit the identical protein subunits to be packed together in a regular (or quasi-regular) fashion and thus to have virtually identical chemical environments, except for their contacts with the nucleic acid core.

Nucleic Acid: The Genetic Component of All Viruses

Viruses originally provided the clearest demonstration that genetic specificity is carried by nucleic acid molecules. Many viral nucleic acids, both DNA and RNA, are easily isolated from their protein shell and prepared in highly purified form. When they are added to host cells, new infective virus particles are produced, identical to those from which the nucleic acid was isolated. Such experiments showed definitively that the viral nucleic acid carries the genetic specificity to

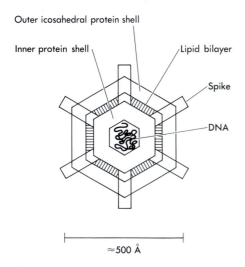

Outer icosahedral protein shell

Inner protein shell

Lipid bilayer

Spike

DNA

≈500 Å

Figure 17-2
The structure of PM2, a lipid-containing bacteriophage that infects the bacterium *Pseudomonas*. [After R. Schaefer and R. M. Franklin, *J. Mol. Biol.* 97 (1975):32, with permission].

code both for its own replication and for the amino acid sequences in its specific coat proteins.

Viral Nucleic Acid May Be Either Single- or Double-Stranded

The nucleic acid of most viruses has the form of its cellular counterparts. Thus, all the best-known DNA viruses, such as smallpox (or its harmless relative *Vaccinia*), SV40, bacteriophage λ, and the T2, T4, and T6 group of bacteriophage, have DNA in the double-stranded helical form. Correspondingly, the RNA of TMV, influenza virus, poliomyelitis virus, and the bacteriophages MS2 and R17 is single-stranded. There are, however, several groups of bacterial viruses in which the DNA is single-stranded, as well as at least two groups of RNA viruses (e.g., the reoviruses) in which the RNA is double-stranded.

It does not really matter whether the genetic message is initially present as a single strand or as the double helix, for the single strand can quickly be used to form a complementary DNA or RNA chain soon after it enters a suitable host cell. The really important fact is that the genetic information is present as a sequence of nucleotide bases.

Viral Nucleic Acid and Protein Synthesis Occur Independently

Exactly what happens after a viral nucleic acid molecule enters a susceptible host cell depends on the specific viral system. Particularly important is whether the virus contains DNA or RNA. If the genetic component is DNA, then the DNA serves as a template both for its own replication and for the virus-specific RNA necessary for the synthesis of its specific proteins. Similarly, if RNA is the genetic component, the RNA molecules have two template roles, the first to make more RNA molecules and the second to make the virus-specific proteins. In both cases, the end result of virus infection is the same: the production of many new copies of both the viral nucleic acids and the coat proteins.

The new progeny molecules then spontaneously aggregate to form mature virus particles. For most viruses, enzymes play no role in these final aggregation events, since the formation of new covalent bonds is not needed either to build a stable protein coat or to affix it firmly to the nucleic acid core. Only weak secondary bonds (salt linkages, van der Waals forces, and hydrogen bonds) need be involved. This last point is shown clearly with TMV. Here the rod-shaped particles can be gently broken down and their free RNA and coat protein components separated. When the components are again mixed together, new infectious particles, identical to the original rods, quickly form. We thus see that the essential aspects of viral multiplication often are known once we understand the principles by which viral nucleic acid and protein components are individually synthesized.

Viral Nucleic Acids Code for Both Enzymes and Coat Protein

The viral nucleic acid genetic component must code the amino acid sequences in the protein(s) that make(s) up the protective coat. These coat proteins are never found in normal uninfected cells, and are

250 Å

Figure 17-3
Capsid structure of a small icosahedral virus (e.g., R17, f2, MS2, or Qβ) that contains 180 subunits of a single type of coat protein.

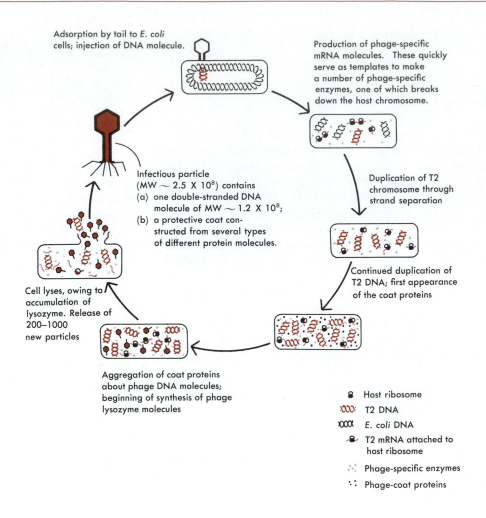

Figure 17-4
Chemical details in the life cycle of the double-stranded DNA virus T2 (T4).

completely specific to a given virus. In addition, one or more new enzymes are usually synthesized to permit successful viral multiplication.

For example, some DNA viruses carry information for the amino acid sequences of enzymes that synthesize precursor nucleotides for viral DNA. One of the most striking cases involves bacteriophage T4 multiplication (Figure 17-4). No cytosine is present in its DNA, a fact that initially suggested that T4 DNA might be very different from normal DNA. Instead, there is always present the closely related base 5-OH-methylcytosine, which, like cytosine, forms base pairs with guanine (Figure 17-5). The three-dimensional structure of T4 DNA is thus basically the same as that of normal double-helical DNA.

No 5-OH-methylcytosine is found in uninfected *E. coli* cells, and so the several new enzymes required for its biosynthesis must be coded by the T4 DNA. In addition, the rate of DNA synthesis in T4-infected cells is several times faster than in normal cells. This faster rate is achieved by having other T4 genes code for many of the enzymes involved in normal nucleotide metabolism, as well as for a new DNA polymerase (Figure 17-6).

New virus-specific enzymes are also frequently needed to ensure the release of progeny virus particles from the host cell. This is a vital need for those viruses multiplying in bacteria with rigid cell walls. Since these cell walls do not spontaneously disintegrate, they could effectively inactivate progeny particles by preventing their release

(a)

(b)

Figure 17-5
Phages of the T-even group do not contain cytosine in their DNA. Instead they contain the related base 5-OH-methylcytosine (a), which base-pairs exactly like cytosine. One or more glucose residues are attached to some of their 5-CH_2-OH groups. (b) The base 5-OH-methylcytosine with one glucose molecule attached. The biological significance of these unusual bases has not yet been clearly established. One hypothesis asserts that their function is to protect T-even DNA from a phage-specific enzyme that only breaks down unmodified DNA. This hypothesis would explain how the *E. coli* DNA is selectively broken down during phage growth.

and transfer to new host cells. To solve this problem, many phages have a gene that codes for the amino acid sequence of lysozyme, a cell wall–destroying enzyme. This enzyme begins to be synthesized when the coat proteins appear and causes the rupture of the cell wall at about the time virus maturation is complete.

One of these enzymes specifically breaks down the normal DNA precursor.

A second enzyme converts.

A third enzyme adds a (P) ~ (P) group to

This explains why no cytosine is found in DNA synthesized after infection.

This is a substrate for DNA polymerase and is incorporated in DNA.

A still further enzyme adds glucose to otherwise complete DNA molecules.

Uridine-(P)~(P)~glucose

Uridine-(P)~(P)

Glucose Glucose Glucose

Glucose Glucose Glucose

Glucosylated DNA

Figure 17-6
The biochemical mechanism that brings about the synthesis of DNA lacking cytosine and containing instead 5-OH-methylcytosine and its glucose derivatives. Immediately after infection, a number of specific enzymes are synthesized. These are coded for by the viral DNA, and each has a specific role in ensuring the successful multiplication of the phage.

Figure 17-7
The morphogenetic pathway leading to the formation of the T4 particle. The numbers refer to the various genes involved in each step. [After W. B. Wood and R. A. Crowther, in *Bacteriophage T4*, ed. C. K. Mathews, E. M. Kutter, G. Mosig, and P. B. Berget (Washington, D.C.: American Society for Microbiology, 1983), p. 263, with permission.]

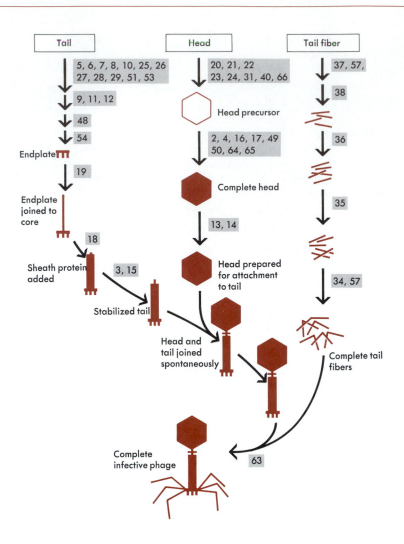

Virus Assembly Follows Morphogenetic Pathways

The assembly of structurally complex viruses like λ and T4 is much more involved than the simple aggregation process needed for simple viruses like TMV. Some 40 different T4 gene products, each coded by a specific T4 gene, interact to produce the mature virus particle. Many of the genes code for the various structural proteins, while at least one other codes for an enzyme that probably converts a precursor protein into the form found in mature T4 particles. In the assembly process, the various components do not associate with one another at random. Instead, the assembly occurs in a definite sequence (a **morphogenetic pattern**), possibly because each step provides the essential substrate for the next. Three different branches, the first concerned with the head, the second with the tail, and the third with the tail fibers, come together as shown in Figure 17-7 to produce the final infectious particles.

Viral Infection Often Radically Changes Host Cell Metabolism

Sometimes, the synthesis of virus-specific nucleic acids and proteins coincides with normal cell synthesis, and the host genome functions more or less normally during the viral life cycle. This is true for the filamentous single-stranded DNA phages fd and M13, cellular parasites that are continuously replicated and extruded through the membrane of the bacterium as it grows. In many cases, however, soon after infection, most cellular metabolism is redirected to the exclusive synthesis of viral components. In the most extreme cases, all DNA and RNA synthesis on the host chromosome ceases, the preexisting RNA templates are degraded, and all subsequent protein synthesis occurs on new RNA templates coded by the viral nucleic acid.

The extent to which a virus is able to control its host's synthetic facilities varies greatly, depending both on the nature of the infecting virus and on the type of host cell. For bacteriophage in particular, the larger the viral nucleic acid content, the larger the number of viral genes directed toward stopping host cell functions unnecessary for the production of new virus particles and replacing bacterial enzymes with phage-encoded enzymes more suited to the needs of the virus. For example, during T4 multiplication, the host *E. coli* chromosome is enzymatically broken down by enzymes coded for by T4 DNA. However, the T4 chromosome is not attacked enzymatically, probably because it contains the unusual base 5-OH-methylcytosine. Furthermore, T4 provides its own enzymes for DNA replication, repair, and recombination and encodes proteins that change the specificity of *E. coli* RNA polymerase to favor T4 promoters over those of *E. coli*.

Synthesis of Virus-Specific Proteins

Although many viruses radically redirect replication and transcription to their own DNA in infected cells, virus-specific proteins are synthesized in the same basic way as normal cellular proteins. Viral messenger RNA molecules attach to host ribosomes, forming polysomes to which the AA ~ tRNA precursors are attached. Viral proteins, like host proteins, start with fMet-tRNA (bacteria) or Met-tRNA (higher cells), and the termination codons for viral and host protein synthesis are the same. For bacteriophage, the only virus-encoded change that we know in the cellular machinery of protein synthesis is in tRNAs. Some new tRNA species arise by enzymatic modification of preexisting tRNAs, and other completely new tRNAs are encoded by phage DNA. Their significance is still unclear. Perhaps they are connected with differences in base composition between the phage chromosome and host chromosome. For example, T4 DNA has twice as many AT base pairs as GC base pairs, while *E. coli* DNA has roughly similar amounts of both types. As not all potential wobble pairs are equally strong, growth of an AT-rich virus using tRNAs adapted for a chromosome with equal amounts of the four bases might slow down the translation of many virus-specific proteins. Hence, there may be a selective advantage of new tRNA species with a different set of favored anticodons.

The Distinction Between Early and Late Proteins

After a viral chromosome enters a cell, its genes do not usually begin working all at the same time. Instead there is a time schedule by which they function. Some viral proteins appear immediately after infection, whereas the synthesis of others may not begin until more than halfway through a viral growth cycle. Also, some genes start working early and continue to do so throughout viral growth, while others function for only several minutes and then shut off (Figure

Figure 17-8
Time of synthesis of various T4-specific proteins and DNA at 37°C.

17-8). There are obvious advantages to such sequential appearance. Some of the early gene products are enzymes necessary to redirect cellular synthesis from host to viral genes (they clearly must be turned on early), while others, such as the viral coat proteins or the lysozyme for bacterial cell lysis, must only appear later. Hence, even the most cursory knowledge of a viral life cycle suggests that quite intricate control systems must exist in order to guarantee that the right proteins appear just when they are needed. As far as we know, these control systems usually operate at the transcriptional level. Thus, some specific mRNA molecules appear early, while others are found only late in the growth cycle. At first it was convenient to subdivide the various mRNAs into just two categories—early and late mRNAs—and likewise to classify a particular protein as either early or late. But often the situation is more complex, and, particularly for bacteriophage, the literature is filled with terms like "immediate early," "delayed early," "middle," "quasi-late," and so on.

Controlling the Time of Bacteriophage Gene Expression Through Gene Order

Under optimal growth conditions, the time needed to synthesize a bacterial mRNA molecule with a molecular weight of about 2 million daltons is of the order of 3 or 4 minutes. Molecules of this size (6000 nucleotides) can code for four or five average-size proteins. But synthesis of these proteins will not begin at the same time; the proteins coded at the 5' end can be completely translated before the transcription of the gene at the 3' end has even started. Genes that need to function first thus tend to occur at the start of operons. Likewise, when the optimal time for the beginning of a gene's function is 3 to 4 minutes after phage infection, it is likely to be placed toward the 3' end of a fairly large operon.

The delayed appearance of the "late" bacteriophage proteins, however, is not completely explained by the existence of very long operons. In T4, for example, most operons are relatively short, and few have more than several thousand base pairs. Yet, perhaps one-third of its 200 or so genes do not begin to function until about 10 minutes after its chromosome has entered the host cell. Part of the delay must be due to the time required for the accumulation and function of phage-specific proteins that direct the expression of particular sets of phage genes.

New σ Factors Direct RNA Polymerase to New Promoters During Phage Development[5-11]

The central strategy that most complex bacteriophages (e.g., T4, λ, T7, Mu and the *B. Subtilis* phage SPO1) use to ensure efficient and correctly timed expression of their genes is to modify the host RNA polymerase with phage proteins that direct it to specific phage genes, or in some cases to make an entirely new phage-specific RNA polymerase. Since the σ subunit endows RNA polymerase with the specificity to recognize a particular promoter sequence (Chapter 13), a phage can program the enzyme by providing new and distinct σ factors that recognize only phage promoters of a particular sequence. As an example, the proteins encoded by genes *33* and *55* of phage T4 bind core RNA polymerase in place of the predominant cellular σ^{70}, causing it to transcribe only phage late genes. Correspondingly, the T4 late promoters have a consensus sequence different from that recognized by σ^{70}.

Orderly development of the *B. subtilis* phage SPO1 is provided by two or three σ factors that are made in sequence. Several genes of SPO1, identified by suppressible nonsense mutations, encode positive regulatory factors required for the expression of other SPO1 genes. The 26 kdal gene *28* product is a phage-specific σ factor (σ^{gp28}) required for synthesis of mRNA for SPO1 middle genes (those expressed beginning about midway into the infection cycle), whereas the gene *33* and *34* proteins act (possibly together) as a σ factor for SPO1 late genes. Gene *28* is itself an early gene, and its promoter is recognized by the standard *B. subtilis* σ factor σ^{55} (the counterpart of the *E. coli* standard σ factor σ^{70}). When the gene *28* protein appears early in infection, it displaces σ^{55} from the RNA polymerase core, possibly by binding the core more tightly; this new holoenzyme then recognizes SPO1 middle promoters. After σ^{gp28} is made, genes *33* and *34* are expressed to provide yet another holoenzyme, which is specific for late gene promoters. The timing of their expression suggests that genes *33* and *34* are controlled by middle promoters, although this has not yet been shown directly. Thus the program of gene expression in phage SPO1 infection is determined by a cascade of σ factors, each controlled by the σ factor active in the previous stage of phage growth.

DNA sequence analysis of SPO1 middle and late promoters shows that they differ not only from the "standard" consensus promoter recognized by the predominant holoenzyme containing σ^{55}, but also from each other (Table 17-1). Distinct consensus sequences can be derived for middle and for late promoters, and it is thought that each sequence defines those bases that the corresponding σ factor (σ^{gp28} or $\sigma^{gp33-34}$) recognizes to direct specific promoter recognition. In fact, phage SPO1 provided the first evidence that σ factors confer DNA sequence specificity to RNA polymerase in recognizing promoters (Chapter 13).

Table 17-1 Consensus Sequences for Phage SPO1 Middle and Late Promoters, Compared with the "Standard" Bacterial Consensus Sequence*

Promoter Class	−35 Consensus Sequence	−10 Consensus Sequence	σ Factor Recognizing Promoter
Major bacterial; SPO1 early	TTGACA	TATAAT	σ^{55} (bacterial σ factor)
SPO1 middle	AGGAGA	TTTNTTT	σ^{gp28} (SPO1 encoded middle gene σ factor)
SPO1 late	CGTTAGA	GATATT	$\sigma^{gp33-34}$ (SPO1 encoded [putative] late gene σ factor)

*In the phage promoters, T stands for hydroxymethyl uracil, which replaces thymine in phage DNA. Adapted from M. Costanzo, "Bacteriophage SP01 Regulatory Genes." 1983. Ph.D. thesis, Harvard University.

The Complete 39,936-Nucleotide Sequence of Phage T7 DNA Is Known[12]

Despite their importance in the history of molecular genetics, the *E. coli* phages T2 and T4 have lost some prominence to smaller but still complex phages that are easier to study. Arguably the best-understood complex phage is T7, which also infects *E. coli*. Its entire 39,936-nucleotide DNA sequence is known, and the functions of enough of its genes are understood that we can give a satisfying description of its life cycle at the molecular level (Figure 17-9). The T7 DNA sequence reveals 55 possible genes; the proteins encoded by 44 of these have been identified, although a function is known for only about 30 of them. Like other relatively small genomes that have been sequenced, T7 has no waste, virtually all of its DNA occurring either as genes or regulatory sites such as promoters and ribosome binding sites. As an example of this economy, the DNA sequence in Figure 17-10, about 0.5 percent of T7, contains the longest segment between two genes in the entire DNA. These 193 nucleotides include a promoter for the T7 RNA polymerase, sequences encoding mRNA hairpins recognized by the RNA processing enzyme RNase III (Chapter

Figure 17-9

An accurately scaled map of bacteriophage T7, with a detailed view of the first third.

Figure 17-10
Regulatory signals in the DNA sequence between T7 genes *1* and *1.1*.

14), a second T7 RNA polymerase promoter, an essential AT-rich section of the origin of DNA replication (Chapter 10), an initiation site for T7 DNA primase, and sequences encoding the ribosome binding site in mRNA for gene *1.1*. At most, five or six nucleotides separate these distinct sites.

T7 Encodes Most Enzymes Required for Its Replication and Recombination[12]

T7 provides its own replication proteins to speed and direct the multiplication of its DNA; 200 copies of the infecting DNA are made within 15 minutes after T7 enters the bacterial cell. These proteins have all the essential activities to replicate double-stranded DNA, but they are a condensed, minimal set, as compared to the complex group of 30 or so polypeptides apparently required to replicate *E. coli* DNA (Chapter 10). The DNA polymerase polypeptide encoded by gene *5* is active when it is complexed with only one other protein, the bacterial protein thioredoxin, whose function in this role is unknown. A primase and a helicase (DNA-unwinding enzyme) combined in one protein encoded by gene *4*, a DNA ligase (encoded by gene *1.3*), a single-stranded DNA binding protein (encoded by gene *2.5*), and a nuclease to remove RNA primers (encoded by gene *6*) complete the set.

Crossing over between different T7 DNA molecules that infect one cell also depends on proteins made by T7. In fact, the gene *6* nuclease again is required, along with a second nuclease encoded by gene *3*. These probably act by partially degrading replicating molecules, exposing single-stranded regions in different molecules that can anneal to initiate heteroduplex formation (Chapter 11).

A New RNA Polymerase Coded by T7 Recognizes Phage-Specific Promoter Sequences[13,14,15]

Like SPO1, T7 makes proteins that redirect cellular transcription to viral DNA. Its principal strategy is different, however. Instead of encoding new σ factors for the host core RNA polymerase, T7 makes an entirely new RNA polymerase (encoded by T7 gene *1*) that is highly specific for a set of distinct phage promoter sequences, the T7 late promoters. Of course, *E. coli* RNA polymerase must make the messenger RNA for gene *1* (as well as a few other early genes), but the new enzyme can then take over transcription of phage late genes. Compared to the *E. coli* enzyme which has a molecular weight of about 450,000 daltons, the T7 RNA polymerase is small, containing a single polypeptide chain with a molecular weight of 98,000 daltons. The T7 RNA polymerase is a very active enzyme and is highly specific for the T7 late promoters. Corresponding to the distinct structure of the enzyme, the consensus sequence derived from comparison of the 17 T7 late promoters is quite different from the *E. coli* RNA polymerase consensus; instead of separate -35 and -10 regions, the T7 late promoters consist of a single highly conserved sequence from -17 to $+6$ relative to the RNA start site (Figure 17-11). Only one T7 promoter differs from the 23-nucleotide consensus by as many as 7 nucleotides, and the five strongest promoters (class III, to be discussed shortly) have exactly the consensus sequence. Its great specificity and activity make the T7 RNA polymerase (and a related enzyme from the *Salmonella typhimurium* phage SP6) very useful for in vitro synthesis of RNA specific to any DNA segment that can be cloned next to a T7 or SP6 promoter.

Figure 17-11

Nucleotide sequence of promoters for T7 RNA polymerase. The replication promoters are found in the origin of DNA replication and are thought to be required for transcriptional activation of replication by T7 RNA polymerase. Bases that differ from the consensus sequence are shaded.

```
                                          −10              +1 (start of RNA)
                                           |                |
 Conserved sequence              TAATACGACTCACTATAGGGAGA

 Class II promoters                              +1
                                                  |
     φ1.1A      AACGCCAAAT   CAATACGACTCACTATAGAGGGA  CA
     φ1.1B      TTCTTCCGGT   TAATACGACTCACTATAGGAGAA  CC
     φ1.3       GGACTGGAAG   TAATACGACTCAGTATAGGGACA  AT
     φ1.5       AGTTAACTGG   TAATACGACTCACTAAAGGAGGT  AC
     φ1.6       TGGTCACGCT   TAATACGACTCACTAAAGGAGAC  AC
     φ2.5       AGCACCGAAG   TAATACGACTCACTATTAGGGAA  GA
     φ3.8       CGTGGATAAT   TAATTGAACTCACTAAAGGGAGA  CC
     φ4c        CCGACTGAGA   CAATCCGACTCACTAAAGAGAGA  GA
     φ4.3       AGTCCCATTC   TAATACGACTCACTAAAGGAGAC  AC
     φ4.7       TTCATGAATA   CTATTCGACTCACTATAGGAGAT  AT

 Class III promoters

     φ6.5       GTCCCTAAAT   TAATACGACTCACTATAGGGAGA  TA
     φ9         GCCGGGAATT   TAATACGACTCACTATAGGGAGA  CC
     φ10        ACTTCGAAAT   TAATACGACTCACTATAGGGAGA  CC
     φ13        GGCTCGAAAT   TAATACGACTCACTATAGGGAGA  AC
     φ17        GCGTAGGAAA   TAATACGACTCACTATAGGGAGA  GG

 Replication promoters

     φOL        TTGTCTTTAT   TAATACAACTCACTATAAGGAGA  GA
     φOR        CACGATAAAT   TAATACGACTCACTATAGGGAGA  GG
```

T7 Genes Are Expressed in Sequence[12]

T7 illustrates nicely the way that timing of phage gene expression can be controlled through ordered and sequential transcription. T7 genes are all transcribed from one strand; RNA polymerase travels rightward on the DNA, according to the conventional genetic map (Figure 17-12). When T7 infects the cell, *E. coli* RNA polymerase initiates synthesis of T7 early gene mRNA from the extreme left end; here are situated three closely spaced promoters, which have the standard *E. coli* consensus sequence at -35 and -10. A terminator between genes 1.3 and 1.4 stops the enzyme to yield an mRNA about 7000 nucleotides long, which is cut between gene-coding segments by the processing nuclease RNaseIII to form several smaller RNAs. This early mRNA encodes the class I proteins: T7 RNA polymerase, some T7 proteins that inactivate host enzymes, T7 DNA ligase, and other proteins of unknown function. Synthesis of class I proteins begins and ends in the first 6 minutes, as can be seen easily by examining radioactively labeled proteins displayed on a gel (Figure 17-13).

The late mRNA of T7 is made by the class I gene product T7 RNA polymerase. Two groups of late mRNAs and their corresponding proteins are made: those for class II genes, encoding mostly enzymes required for DNA metabolism, and those for class III genes, which encode mostly structural proteins of the phage particle. Class II proteins are made between about 6 and 12 minutes after infection, whereas synthesis of class III mRNA and proteins begins at 8 minutes and then continues vigorously until the cell lyses. Since both are transcribed from promoters recognized by T7 RNA polymerase, how can class II and class III mRNAs be made at different times? It is believed

Figure 17-12
The pattern of mRNA synthesis from T7 DNA by *E. coli* RNA polymerase and T7 RNA polymerase. Upward vertical lines designate promoters, and downward lines, terminators. Vertical bars through the RNAs are sites of processing by RNase III. The readthrough RNA, made by T7 RNA polymerase that initiates at a class II promoter and proceeds to the end of the DNA, is minor but important, because it is the only source of mRNA for genes *11* and *12*. [From J. J. Dunn and F. W. Studier, *J. Mol. Biol.* 166 (1983):477, with permission.]

Figure 17-13

The appearance of T7-specific proteins during infection. Proteins being synthesized at different times during infection were made radioactive with a short pulse of radioactive amino acid and were separated in a denaturing polyacrylamide gel and detected by autoradiography. Numbers at the side designate the T7 genes encoding the indicated proteins. Many bacterial proteins are made at the beginning of infection, but after a few minutes, host synthesis is shut down and only T7 proteins are labeled. Inspection shows that proteins of different classes are made at the indicated times (see Figure 17-9); however, not all proteins are visible here, because some are made in very small amounts. [From J. J. Dunn and F. W. Studier, *J. Mol. Biol.* 166 (1983):477, with permission.]

that the T7 DNA molecule enters the cell slowly, and always from its left end, so that for a few minutes only the weaker class II promoters are inside; when more DNA is injected, the stronger class III promoters are available and they begin to dominate late transcription. The nearly exclusive synthesis of class III proteins at later times may result simply from the high concentration of class III mRNA, which saturates the available ribosomes.

It is clear that the major T7 proteins are made when they are needed: RNA polymerase early, to begin expression of the remainder of T7 genes; replication proteins next, to begin the production of new DNA; and structural proteins at the end when they are needed for the final step of phage growth, the assembly of new particles.

Viral Proteins Attack Specific Cellular Components and Help T7 Take Over the Cell[16-19]

T7 encodes proteins that disable both host RNA and host DNA synthesis. The small (7-kilodalton) gene 2 protein binds and inactivates *E. coli* RNA polymerase, although this does not happen until enough early phage RNA has accumulated to ensure successful phage growth. A protein kinase (encoded by gene *0.7*) may contribute to this inactivation by phosphorylating the β and β' subunits of RNA polymerase; the kinase also phosphorylates many other cellular proteins, but it is not known what effect this has.

A drastic attack is reserved for components of cellular DNA synthesis. Here, T7 destroys the bacterial DNA by the combined effects of an endonuclease (encoded by gene *3*) and an exonuclease (encoded by gene *6*). In fact, the bacterial DNA is degraded to its nucleotide components, which provide the essential precursors for synthesis of T7 DNA. It is not clear why T7 DNA is not also destroyed by the nucleases, but one possibility is that phage DNA is repaired more efficiently after nuclease attack.

Bacteria defend themselves against phages and other foreign DNA with DNA restriction enzymes, against which cellular DNA is protected by methylation (Chapter 9); T7 evades the cellular restriction nucleases *Eco* B and *Eco* K with its gene *0.3* protein, a specific inhibitor that binds and inactivates the nucleases. Curiously, the bacteriophage T3, which is closely related in structure and gene function to T7, has a gene at the site corresponding to that of T7 *0.3* that achieves the same goal by a totally different enzymatic mechanism: It hydrolyzes the compound *S*-adenosylmethionine (SAM), an essential cofactor of the restriction enzymes.

Finally, T7 lysozyme destroys the bacterial cell wall and allows newly formed phage particles to escape and renew their growth cycle in other bacteria.

Bacteriophage N4 Virion RNA Polymerase Recognizes Promoters in Single-Stranded DNA[20,21]

The relatively unknown *E. coli* phage N4 makes an unusual and interesting RNA polymerase that selects initiation sites differently from other known RNA polymerases. N4 virion RNA polymerase is a single large polypeptide chain of 300,000 daltons that is packaged in the phage particle at the end of one growth cycle and then injected along with phage DNA to act in the next generation. Its function is to transcribe the earliest phage genes from three promoters at one end of the phage DNA, and it recognizes only these promoters. Most surprisingly, N4 virion RNA polymerase is inactive on native double-helical DNA, but it binds and initiates RNA synthesis at specific promoter sites on single-stranded N4 DNA in vitro. The promoters include the 18 nucleotides before each RNA start site, a region that is nearly identical in sequence in all three. Early N4 transcription requires the cellular activities of DNA gyrase and SSB (single-stranded DNA binding protein), suggesting that the promoter region denatures when gyrase introduces negative supercoils in the DNA and that a melted region is stabilized by binding to SSB. Since the promoter includes one of two hairpin loops that could form in single-stranded DNA around the RNA start sites, the DNA structure recognized by virion RNA polymerase might be at least partly helical.

Phage λ: A Simple Model of Developmental Regulation[22]

Phage λ is rarely regarded as simple, but only because so many facts are known about it, including its complete DNA sequence. However, the important regulatory events in its life cycle can be summarized

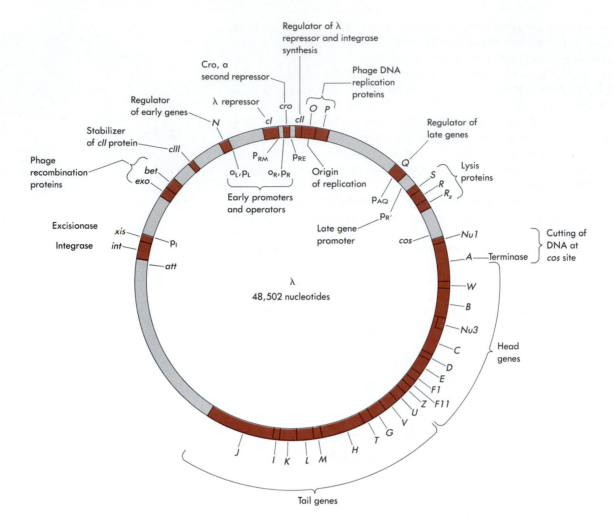

Figure 17-14
A map of phage λ in the intracellular circular form, showing the major λ genes. Before it is packaged into the phage coat, the DNA is cut at the site named *cos*; this means that the genetic map is linear, with one end near gene *A* and the other end near gene *R*.

quite easily, as we do in the following section. Besides providing the best-understood example of repressors and operators, λ is important because it is a **temperate phage** that follows either of two distinct developmental pathways. First, λ can undergo **lytic growth** like T7, reproducing itself manyfold while it kills the cell. Second, it can enter **lysogenic growth,** in which its DNA is combined with the bacterial chromosome and carried passively by the growing bacterium until it is roused by a special signal. We can give an accurate description of the basic regulatory mechanics that drive each developmental pathway, although understanding exactly how the choice between them is made is still an important challenge.

An Outline of λ Growth[23]

To understand how each mode of growth is achieved, we first should know the endpoints of each. Lytic growth culminates with release of newly made phage from the dead cell. Lysogenic growth is established when λ repressor binds the operators that control the two early promoters p_L and p_R (Figure 17-14), and when phage DNA is inserted into the host DNA. Because expression of all λ genes (except that for repressor itself and another minor gene) ultimately requires that RNA polymerase start at p_L and p_R, phage genes are inactive during lysogenic growth. Now the questions are: What happens in each pathway of λ growth and why does a particular phage DNA molecule take one pathway or the other?

As we might expect, the two directions of development reflect the expression of different genes. However, the first stages of growth when λ infects a cell are the same, regardless of the eventual outcome. The phage DNA ends meet and become connected through the annealing of complementary single-stranded tails 12 nucleotides long ("sticky ends"), much like the joining of restriction fragments; the remaining nicks are sealed by DNA ligase, forming a covalent circle. Because there is no repressor in the beginning, RNA synthesis starts immediately from p_L and p_R (Figure 17-15). Gene N, the first in the leftward operon, encodes a transcription antiterminator that allows RNA synthesis to extend leftward of gene N and rightward of gene cro, to include the remainder of the early genes. Early genes on the right include O and P, whose products direct phage DNA replication, and Q, whose product turns on the late genes encoding lysis and virion proteins. Only if the lytic pathway predominates and RNA synthesis continues from p_R for some time does enough Q protein accumulate to act effectively.

Two early λ genes, cro and cII, are critical to determining the direction of phage development. Both Cro and cII proteins accumulate during early infection, but they have opposite effects. The **cII protein** activates transcription from a special promoter to provide a burst of phage repressor (encoded by the cI gene), thereby favoring lysogenic growth. If enough repressor is made, it shuts off the early promoters, thus repressing all phage genes required for lytic growth; in particular, the late gene regulator Q is unavailable to provide lysis proteins and phage coats. **Cro protein,** on the other hand, favors lytic growth, although its mechanism is a little more subtle. First, it turns down (but not off) RNA synthesis from p_L and p_R, reducing synthesis of cII protein and disfavoring repressor synthesis. Cro also turns down synthesis of the O, P, and Q proteins, but enough of them is made to complete lytic growth. The second important effect of Cro is to inhibit repressor synthesis directly by blocking the promoter p_{RM} from which repressor is made during lysogenic growth (Figures 17-15 and 17-16).

Thus, Cro acts in opposition to repressor (and to cII, which stimulates repressor synthesis). In any given infected cell, only one will prevail; in normal conditions of cell growth, the odds are about equal for each. We will consider how the outcome is decided.

Repressor and Cro Both Act by Binding the Early Operators O_L and O_R [24–30]

As described in Chapter 16, λ and other repressors are dimers that bind a symmetrical operator structure in DNA. Cro is also a dimeric repressor, although for historical reasons only the cI gene product is called the λ repressor. Both proteins control RNA synthesis from promoters p_L and p_R by binding the operators o_L and o_R. How can they bind the same sites on DNA, yet have such different biological effects?

The answer lies in the complex structure of the operators o_L and o_R. We will describe how repressor and Cro work at o_R, keeping in mind that the structure and function of o_L is quite similar. Operators o_L and o_R are each made up of three adjacent suboperators; in o_R, these are

Both pathways of phage growth begin when RNA polymerase binds the early promoters p_L and p_R, and makes mRNA for the *N* and *cro* genes.

N protein then acts at three sites to extend RNA synthesis to other early genes; O and P proteins allow DNA synthesis to begin; cII protein is made and begins to act as described below.

The following two pathways then occur in competition

Toward lysogenic development: dominance of repressor

Toward lytic development: dominance of Cro

cII protein stimulates synthesis of mRNA for repressor from promoter p_{RE} and mRNA for integrase from promoter p_I.

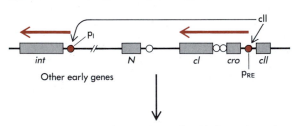

If enough repressor is made, it binds o_R and o_L, blocking RNA synthesis from the early promoters; furthermore, repressor bound to o_R stimulates synthesis of more repressor mRNA from p_{RM}, so that its concentration is maintained in the cell; integrase promotes integration of phage DNA, and lysogeny is established.

Cro protein occupies the operators o_R and o_L, preventing synthesis of repressor mRNA from promoter p_{RM}, but allowing enough rightward transcription for Q protein to accumulate.

Q protein stimulates transcription of phage late genes, which encode structural proteins that package DNA into new phage particles.

Figure 17-15

Gene expression during phage λ development along two distinct pathways.

named o_{R1}, o_{R2}, and o_{R3}. Both repressor and Cro bind to all three suboperators (Figures 17-16 and 17-17) and thereby control the overlapping promoters p_R and p_{RM}. Since each suboperator has two symmetrical elements, there are a total of six sequences related, *but not identical*, to a consensus binding sequence (Figure 17-17). It is this nonidentity of the binding sites that allows repressor and Cro to work

The earliest stage of infection

Lysogeny is established. Repressor bound to o_{R1} and o_{R2} blocks expression of phage genes to the right, and repressor bound to o_{R2} stimulates binding of RNA polymerase to p_{RM}, thereby provoking its own further synthesis.

Lytic growth is established. Cro bound to o_{R3} prevents repressor synthesis from p_{RM}; the concentration of repressor is low enough that o_{R1} and o_{R2} are usually free, so that early rightward phage genes are transcribed.

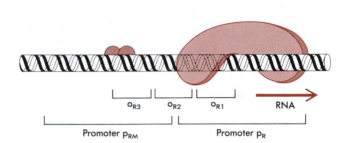

If its concentration becomes excessive, repressor binds o_{R3} and blocks its own synthesis.

Cro binds o_{R1} and o_{R2} when its concentration rises, blocking promoter p_R; this prevents further synthesis of Cro and also limits expression of other rightward early genes.

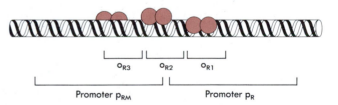

differently, because the sites do not bind the proteins with equal strength. Therefore, if the concentration of repressor or Cro is low, the tightest binding sites are filled but the others are empty. Furthermore, the nonidentity of binding sequences has opposite effects on the binding of repressor and Cro to o_R. The consequence of these different binding strengths for the control of promoters p_R and p_{RM} is shown in Figure 17-16, and the corresponding biological effects are described in Figure 17-15.

When neither repressor nor Cro is present, RNA polymerase binds p_R and makes mRNA for genes to the right. In this condition, RNA polymerase does *not* bind p_{RM}, for the reason discussed shortly. When repressor is made in abundance, it binds o_{R1} and o_{R2}, thereby blocking p_R (Figure 17-16). The repressor dimer at o_{R2} contacts not

Figure 17-16
The occupancy of the λ rightward operator and promoters p_R and p_{RM} by repressor, Cro, and RNA polymerase during successive stages of λ development along two pathways. [Adapted from M. Ptashne, A. Jeffrey, A. D. Johnson, R. Maurer, B. J. Meyer, C. O. Pabo, T. M. Roberts, and R. T. Sauer, *Cell* 19 (1980):1, with permission.]

Consensus sequences:

Operator (half-site):   5′—TATCACCG—3′

Promoter −35 region: 5′—TTGACA—3′
Promoter −10 region: 5′—TATAAT—3′

Figure 17-17

DNA sequence of the λ rightward operator and its associated promoters, and the consensus nucleotide sequences of each.

only DNA, but also the repressor dimer at o_{R1}; it therefore binds more tightly than it would if o_{R1} were not occupied. This interaction between repressor dimers is called cooperativity, and it is important for effective repression.

Repressor binding to o_{R2} has a second important effect besides ensuring repression of the rightward promoter. Repressor also acts as an *activator* for RNA polymerase binding to the promoter p_{RM}, which is the promoter from which mRNA for repressor itself is made in a lysogen; thus, repressor stimulates its own synthesis. This requirement for self-activation explains why only p_R, and not p_{RM}, is active during early infection when no repressor is present.

Finally, at very high concentrations, repressor binds o_{R3} and blocks p_{RM}, thereby preventing its own overproduction. Synthesis of the λ repressor is therefore self-regulated both positively and negatively.

If Cro also binds o_{R1} o_{R2}, and o_{R3}, then why does it act differently from repressor? The reason is that Cro binds most tightly to o_{R3}, not o_{R1} and o_{R2}. Therefore, its first effect is to block p_{RM}, preventing the synthesis of repressor that is necessary for lysogeny. At higher concentrations, Cro also fills o_{R1} and o_{R2}, but this only happens part of the time, so that the rightward promoter stays active enough for the essential genes on the right to be expressed. There is no cooperativity between Cro dimers in binding to the operators, nor does Cro act like λ repressor as an activator of either promoter at o_R.

The λ cII Protein Directs Phage Development According to Cellular Growth Conditions[31-35]

Our description of the competition between repressor and Cro left unanswered why one or the other prevails in a particular infected cell. Clearly, the amount of each protein must be critical to the balance. As the system has evolved, it is the extra burst of repressor synthesis provided by cII protein that gives the lysogenic pathway a good chance to succeed. The cII protein is an activator that causes RNA polymerase to initiate synthesis at the promoter p_{RE}, located between genes *cro* and *cII* itself (see Figure 17-15). Transcription from p_{RE} goes leftward (and through the *cro* gene backward) and makes mRNA for repressor. Once repressor is made from this mRNA, it then can bind o_{R1} and o_{R2}, thus activating the promoter p_{RM} which functions thereafter to provide repressor during lysogenic growth. But the initial function of cII protein is necessary for lysogeny to be established: A *cII* mutant phage almost always grows lytically.

The activity of cII protein is not firmly programmed, but instead changes with certain conditions of cell growth. For example, when

5'—ATCTAAGGAAATACTTACATATGGTTCGTGCAAACAAACGCAACGAGGCT—3'
3'—TAGATTCCTTTATGAATGTATACCAAGCACGTTTGTTTGCGTTGCTCCGA—5'

(TAATAT)
−10

(ACAGTT)
−35

Standard promoter
consensus sequences

Beginning of
pRE mRNA

cII protein binding site

glucose is plentiful and the cyclic AMP level is low (Chapter 16), there is less cII activity, so that lytic growth is favored over lysogenic growth. The cII protein is very unstable in the cell; its concentration, and thus the total cII activity, is determined by the rate at which it is degraded by cellular proteases. The effect of cyclic AMP is thought to be mediated by these proteases, although the details of this regulation are still unknown. Because its level is controlled by such environmental factors as the presence of glucose, it is possible that cII protein acts primarily as a sensor of cellular growth conditions for the phage, measuring how favorable it would be to join the cell in lysogenic growth rather than to multiply lytically.

The reason why promoter p_{RE} is essentially inactive unless cII is present is apparent in the nucleotide sequence of the promoter (Figure 17-18). Although its −10 region matches the promoter consensus sequence in three out of six positions, including two of the most strongly conserved, there is no match to the −35 consensus sequence at all. However, evidence from mutations preventing cII function, as well as DNA protection experiments, shows that cII protein binds DNA at two 5'-TTGC-3' sequences that flank the site where RNA polymerase contacts DNA at −35. Thus, cII protein might bind both DNA and RNA polymerase here, substituting for the attraction that a good −35 sequence otherwise provides. The cII binding regions are separated by 10 nucleotides (one turn of DNA helix), and each is about 5 nucleotides (one-half turn of DNA helix) from the center of the contact region at −35. Thus, cII protein binds the other side of the DNA from RNA polymerase, so that it could reach around to touch the RNA polymerase, but still allow the enzyme to contact the DNA backbone in the −35 region.

A second cII protein-dependent promoter, p_I, has a nucleotide sequence similar to that of p_{RE} and is located before gene *int* (see Figure 17-16); it provides a high concentration of the integrase enzyme that catalyzes site-specific recombination of λ DNA into the bacterial chromosome to form the prophage (Chapter 11). A third cII-dependent promoter, p_{AQ}, located in the middle of gene *Q*, acts in an unknown way to retard lytic development and thus to promote lysogenic development.

Positive Control by the λ Gene *N* and Gene *Q* Antiterminator Proteins[36–43]

Phage λ provided the first example of a type of positive transcriptional regulation entirely distinct from promoter activation by regulators like cII protein, CAP, and the arabinose operon C protein (Chapter 16). The transcripts that λ N and Q proteins affect are initiated perfectly well in their absence, but then terminate a few hundred to a thousand nucleotides downstream unless RNA polymerase has been modified by the regulator; λ N and Q proteins are therefore called antiterminators.

Figure 17-18
DNA sequence of λ promoter p_{RE}, including the binding site for the activator cII. The promoter is directed leftward so that the consensus sequences are on the bottom strand in order that the direction agree with the conventional genetic map of λ.

N protein recognition site in the λ early rightward operon, and the first terminator at which it acts

Q protein recognition site in the λ late gene operon, and the terminator at which it acts

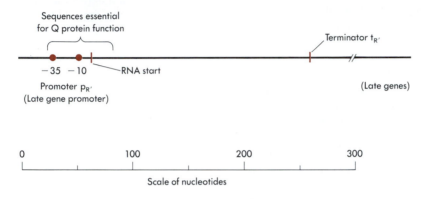

Figure 17-19
Recognition sites and sites of action of the λ N and Q protein transcription antiterminators. Box A is a seven nucleotide sequence associated with *nut* sites that is probably required for efficient N protein function. The sequence of the RNA-like strand of *nutR* is shown above *nutR*. Complementary regions of *nutR* that might pair to form a hairpin loop are indicated in color.

N protein regulates early gene expression by acting at three terminators: one to the left of gene *N* itself, one to the right of gene *cro*, and one between genes *P* and *Q* (Figures 17-15 and 17-19). Q protein has one known target, a terminator that is 200 nucleotides downstream from the single late gene promoter, $p_{R'}$, located between genes *Q* and *S* (see Figures 17-14, 17-15, and 17-19). This late gene operon of λ is remarkably large for a procaryotic transcription unit: about 26 kilobases, a distance that takes about 10 minutes for RNA polymerase to traverse. Possibly in this great expanse of DNA there are other, unidentified terminators at which Q protein activity is also required.

Although Q protein–modified RNA polymerase transcribes the whole late region evenly, and all late mRNA has about the same stability, the λ late proteins are made at rates that differ vastly by up to a thousandfold. This illustrates how important the efficiency of mRNA translation by ribosomes can be in determining the rate of gene expression.

We are just beginning to understand how these antiterminators act. Like other regulatory proteins, N and Q proteins recognize specific nucleotide sequences, and they affect only those operons that contain these sequences. Thus, N protein prevents termination in the early operons of λ, but not in other bacterial or phage operons. The λ N protein does not even work in the early operons of phages closely related to λ, like φ80 or the *Salmonella typhimurium* phage P22, which have their own specific "N" proteins.

Genetic experiments first showed that the specific recognition sequences for antiterminators do not occur in the terminators where they act, but instead occur in the operons well before the terminators. N protein recognizes sites named *nut* (for *N utilization*) that are 60 and 200 nucleotides downstream from the early leftward and rightward RNA start sites (Figure 17-19), but it then prevents transcription termination at terminators hundreds to thousands of nucleotides further downstream. This means that N protein must modify RNA polymerase as it passes the *nut* site, perhaps becoming a firmly attached subunit of the enzyme and thereby changing its ability to recognize terminators. Genetic experiments have shown that the products of the bacterial genes *nusA, nusB,* and *nusE* act along with λ N protein. The NusA protein is an important cellular transcription factor (Chapter 13). NusE is the small ribosomal subunit protein S10, but its role in N protein function is unknown. No cellular function of NusB protein is known.

Unlike N protein, the λ Q protein recognizes sequences adjacent to and partly within the late gene promoter (see Figure 17-19). Fortunately, Q protein is active in vitro during RNA synthesis by purified RNA polymerase, so that progress has been made in understanding how Q protein prevents termination: It changes the ability of the enzyme to be stopped temporarily (to "pause") at certain nucleotide sequences in DNA that are believed to be essential components of terminators (Chapter 13).

Retroregulation: An Interplay of Controls on RNA Synthesis and Stability Determines *int* Gene Expression[44]

The cII protein activates the promoter p_I that serves the *int* gene, as well as the promoter p_{RE} responsible for repressor synthesis (see Figure 17-15). Therefore, conditions favoring cII protein activity give rise to a burst of both repressor and integrase enzyme. However, the *int* gene is transcribed from p_L as well as from p_I, so that we would expect integrase to be made even in the absence of cII protein. This does not happen. The reason is that *int* messenger RNA initiated at p_L is degraded by cellular nucleases, whereas messenger RNA initiated at p_I is stable and can be translated into integrase protein. This occurs because the two messengers have different structures at their 3' ends. RNA initiated at p_I stops at a terminator about 300 nucleotides after the end of the *int* gene; it has a typical stem-and-loop structure followed by six uridine nucleotides at the end (Figure 17-20). When RNA synthesis is initiated at p_L, on the other hand, RNA polymerase is modified by the N protein antiterminator and thus goes through and beyond the terminator. Then a more extended RNA stem can form that is a substrate for nucleases (Figure 17-20). It is attacked first by the RNA processing nuclease RNaseIII, which leaves an end from which exonucleases can degrade the mRNA and thus destroy it. Because the site responsible for this negative regulation is downstream of the gene it affects, and because degradation proceeds backward through the gene, this process is called **retroregulation.**

The biological function of retroregulation is clear. When cII activity is low and lytic development is favored, there is no need for integrase enzyme; thus, its template is destroyed. However, when cII activity is high and lysogeny is favored, the *int* gene is expressed to promote recombination of the repressed phage DNA into the bacterial chromosome.

There is yet a further subtlety in this regulatory device. When a prophage is induced, it needs to make integrase (as well as excisionase; Chapter 11) to catalyze reformation of free phage DNA by recombination out of the bacterial DNA, and it must do this whether or not

DNA that encodes the site of retroregulation after the λ *int* gene

5′— TGATGACAAAAAATTAGCGCAAGAAGACAAAAATCACCTTGCGCTAATGCTCTGT ···
3′— ACTACTGTTTTTTAATCGCGTTCTTCTGTTTTTAGTGGAACGCGATTACGAGACA ···

Site of termination
without antitermination
activity of N protein

Symmetry forming
hairpin loops in RNA

Direction of transcription

Symmetry forming
hairpin loops in RNA

int gene

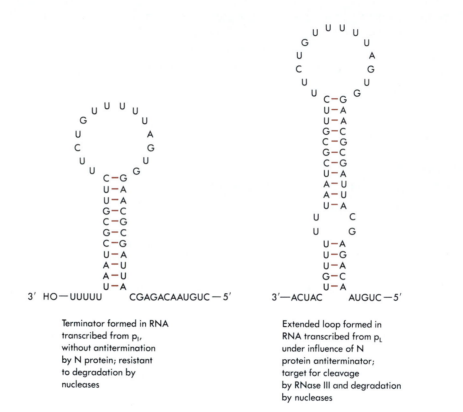

Terminator formed in RNA
transcribed from p_I,
without antitermination
by N protein; resistant
to degradation by
nucleases

Extended loop formed in
RNA transcribed from p_L
under influence of N
protein antiterminator;
target for cleavage
by RNase III and degradation
by nucleases

Figure 17-20
DNA coding site and transcribed RNA
structures active in retroregulation of
Int protein synthesis.

cII activity is high. In this case, the phage must make stable integrase mRNA from p_L despite the antitermination activity of N protein. Stable integrase mRNA is made, because the phage attachment site at which recombination occurs is *between* the end of the *int* gene and those sequences encoding the extended stem from which mRNA degradation is begun. Thus, the site causing degradation is removed from the end of the *int* gene, and *int* mRNA made from p_L in the prophage is stable; ample integrase is made to help catalyze excision.

Specific Initiation Factors for Viral DNA Replication: The λ O and P Proteins[45, 46]

Like ϕX174 (Chapter 10), phage λ encodes initiation factors, the products of its *O* and *P* genes, that direct the bacterial replication complex to a specific origin of replication on λ DNA. O protein recognizes the λ origin and is therefore the counterpart of the *E. coli* dnaA protein.

For this reason, λ does not require dnaA protein for its replication. O protein also binds P protein, which in turn interacts with the E. *coli* replication protein dnaB; thus, P protein links λ to the bacterial replication enzymes. Except for dnaA and dnaC (possibly the counterpart of the λ P protein), λ is thought to utilize most of the normal E. *coli* replication proteins.

Remarkably, the λ origin of replication is *within* the O gene, woven into the coding sequence of the O protein (Figure 17-21). The flexibility of the genetic code, due largely to the multiplicity of codons available for most amino acids, allows this DNA to be at the same time a complex protein binding sequence and a continuous protein coding segment. The origin includes four adjacent and nearly identical 20 nucleotide repeated sequences each containing two symmetrically related half-sites, like the λ operators (Figure 17-21). Four O protein dimers bind to the origin, as shown by the fact that purified O protein protects the four sites from DNase digestion in vitro. Electron microscopy reveals the O protein–origin complex to be a large globular structure similar to that made by integrase protein binding to the attachment site (Chapter 11).

How O protein binding causes DNA replication to initiate at the λ origin is now beginning to be understood. Even when O and P proteins are already present, replication starts only if RNA polymerase is actively transcribing DNA near or through the origin. One guess is that transcription melts the DNA, exposing a template to which the initiation complex can direct DNA primase, so that primers for DNA polymerase are made. An AT-rich region in the origin, to the right of the O protein binding sites, might be involved in this melting. Recent success in showing replication of λ DNA in vitro, dependent on O and P proteins, RNA polymerase, and other cellular replication proteins, should lead to rapid progress in understanding the biochemical details.

Sticky Ends Are Made as Unit-Length λ DNA Molecules Are Cut from Concatemers During Packaging into Phage Heads[47]

The nature of λ DNA replication changes during the growth cycle. Early in infection, the circular infecting DNA is replicated bidirectionally from its origin in gene *O*, like the DNAs of SV40 or E. *coli* itself (see Figure 10-4), yielding more circular molecules. Later, there is a switch to rolling circle replication, similar to that of φX174 (see Figure 10-22) but with two differences. First, there is no enzyme like φX174 A protein to circularize each DNA as it is made, so that a continuous long tail of tandemly duplicated λ DNA molecules (a concatemer) forms. Second, DNA primase acts continually to initiate lagging strand synthesis as the circle rolls, so that the product is double-stranded. As they are packaged into phage heads, unit-length DNA molecules are cut from the concatemer by the enzyme **terminase,** the

Figure 17-21

Binding sites for λ O protein in the origin of replication, showing their overlap with the coding sequence for the O protein. Each half-site in the four symmetrical binding sequences is indicated by an arrow.

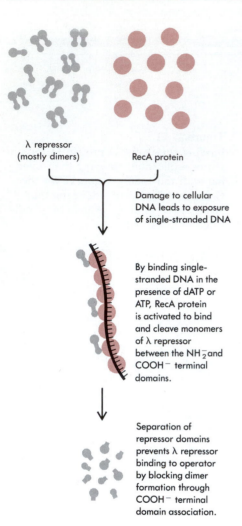

λ repressor
(mostly dimers)

RecA protein

Damage to cellular DNA leads to exposure of single-stranded DNA

By binding single-stranded DNA in the presence of dATP or ATP, RecA protein is activated to bind and cleave monomers of λ repressor between the NH_2^- and $COOH^-$ terminal domains.

Separation of repressor domains prevents λ repressor binding to operator by blocking dimer formation through $COOH^-$ terminal domain association.

Figure 17-22
Activation of *E. coli* RecA protein by DNA to bind and destroy λ repressor by proteolytic cleavage.

product of λ genes *A* and *Nu1*. Terminase makes 12-nucleotide staggered cuts in the two DNA strands at a site named *cos* (see Figure 17-14), thereby generating the "sticky ends" that allow an infecting molecule to circularize in the cell. The concerted reactions of cutting and packaging occur only on concatemers, in which each unit length of DNA is bounded by a *cos* site on either side. It is not clear how or why two *cos* sites separated by this considerable length of DNA might be required for DNA packaging. However, one consequence is that the monomer circles made early in infection cannot be packaged.

Destruction of the λ Repressor by Proteolytic Cleavage Leads to Prophage Induction[48,49]

Although the state of lysogeny is stable, it is not irreversible. Abrupt inactivation of λ repressor in a lysogen leads to induction of lytic growth of the prophage, just as though it were a newly infecting phage. Prophage DNA still must be freed from host DNA through reversal of the integration event, but this is taken care of by the enzymes integrase and excisionase (Chapter 11). Although a λ prophage can be induced to grow in trivial ways—heat inactivation of a temperature-sensitive repressor, for example—prophage induction is also part of an important physiological process: induction of the SOS DNA repair genes (Chapters 12 and 16). Single-stranded DNA exposed by damage to cellular DNA binds the *E. coli* RecA protein, activating in it a protease function that destroys LexA protein, the repressor of the SOS genes. A likely source of single-stranded DNA is the gap left in cellular DNA when replication is stopped by damaged bases (Chapter 11). The λ repressor is also cleaved by this activated RecA protein (Figure 17-22); in fact, it was the study of λ repressor that led to the discovery of the mechanism of SOS gene induction. Treatment of a lysogen with agents that damage DNA, such as ultraviolet irradiation or the crosslinking chemical mitomycin C, causes induction of the prophage and eventual lysis of the cell. It is thought that this gives the prophage a means of escaping a cell subjected to damage so severe that its survival is threatened.

Phage P22 Antirepressor Inactivates Phage Repressors by Noncovalent Binding[50,51]

The *Salmonella* phage P22, one of the related lambdoid phages (including λ, φ80, 82, and 434), makes a unique protein that inactivates repressors in a way entirely different from *recA*-dependent cleavage. P22 **antirepressor** binds specifically to repressors of lambdoid phages, including its own, thereby preventing them from complexing with phage operators. Thus, P22 induces the prophage in a lysogen it infects; furthermore, P22 can infect a lysogen of P22 itself, because antirepressor inactivates the P22 repressor made by the prophage. In contrast, lysogens of other lambdoid phages, which have no antirepressor genes, are immune to infection by the same phage, because repressor made by the prophage shuts down the infecting phage just as it does the prophage. It may seem paradoxical that P22 itself can form a lysogen if it makes antirepressor; but in conditions favoring lysogenic growth, enough P22 repressor is made from the entering phage to overwhelm the antirepressor, and a separate mechanism prevents further synthesis of antirepressor from the prophage.

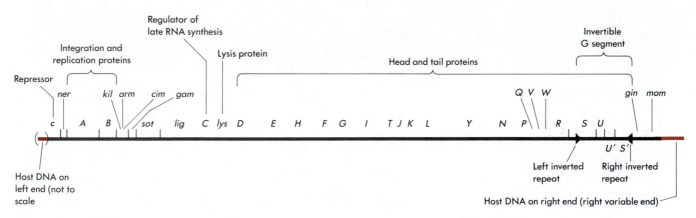

Figure 17-23
Genetic and physical map of bacterio-phage Mu. [Adapted from F. J. Grundy and M. M. Howe, *Virology* 134 (1984):296, with permission.]

Phage Mu Is a Transposon[52]

Phage Mu is a temperate bacteriophage, like λ in many ways, but distinct in one important respect: In the major transactions of its DNA—integration into bacterial DNA and replication—Mu acts as a transposon and not as a typical phage. In effect, Mu is a transposon that happens to have not only genes that regulate its transposition, but also genes that encode structural proteins that package its DNA. It is unique also in transposing much more often than other transposons, providing about a hundred new copies of Mu during its hour-long life cycle.

Since Mu is a transposon, the ends of its DNA are never free but instead are embedded in other DNA, namely, that of its bacterial host. This is true not only of the integrated Mu prophage, but also of packaged phage DNA, which has short (50 to 2000 base pairs) segments of bacterial DNA at either end (Figure 17-23). Like other transposons, Mu has inverted repeats at and near the ends of its DNA, and it makes a short (5-base-pair) duplication of the bacterial DNA into which it is inserted.

Mutagenesis by Nearly Random Insertion of Mu Prophage into Cellular Genes[52–57]

As is true for many other transposons (and completely different from phage λ), the transpositional integration of Mu into bacterial DNA to form a prophage is almost random. Thus, each lysogen made when Mu infects a bacterial cell is likely to have the prophage inserted at a different site. Since a bacterial gene interrupted by Mu DNA is usually inactive, Mu makes mutations in the bacterial host as it integrates. This is particularly useful, since once a lysogen has been isolated in which the prophage has interrupted some gene of interest, it is a straightforward process to clone either of the separated pieces of this gene; one simply screens among randomly cloned segments of lysogen DNA for those containing one or the other end of the Mu phage DNA.

Specially constructed variants of Mu named *Mud* (shortened from Mu d*lac*, meaning Mu defective *lac*tose operon) allow genes that can be induced by some treatment of interest to be isolated. At one end, *Mud* phages have part of the lactose operon containing the β-galactosidase gene but lacking the lactose operon promoter. Integra-

Figure 17-24
Prophage integration and replication of
phage Mu by transposition.

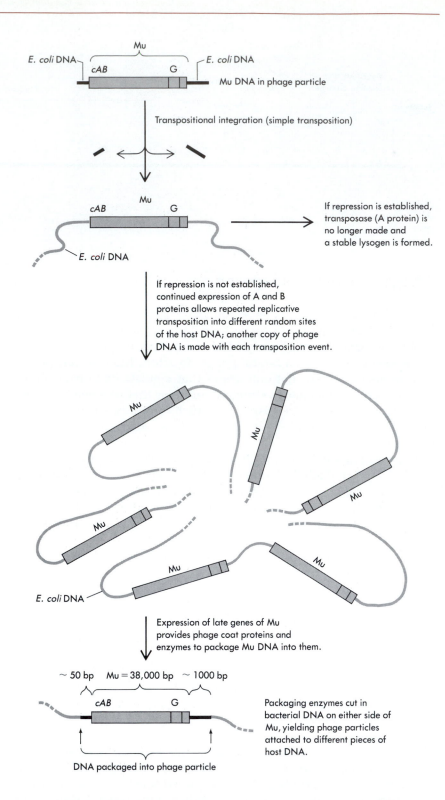

tion of *Mud* into a bacterial gene in the correct direction puts β-
galactosidase expression under control of the promoter of the bacterial gene in place of the unknown gene product, thereby revealing
and marking that gene. As an example, the characterization of *Mud*
lysogens in which β-galactosidase is induced by DNA-damaging
agents like ultraviolet light led to identification of many of the SOS
genes (Chapters 12 and 16).

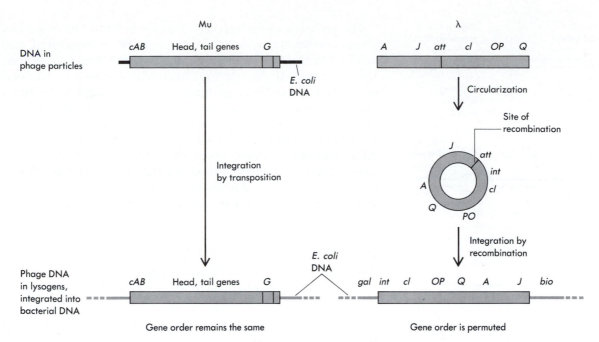

Phage Mu Replicates by Repeated Transposition into Cellular DNA[58–62]

We can understand the life cycle of Mu by remembering how transposons move and replicate (Chapter 11). Integration and replication of Mu are examples, respectively, of *simple transposition* and *replicative transposition*. As for most other transposons, the essential elements are a transposase, here the Mu gene *A* protein, and the ends of the transposon, which are at the junction between Mu and bacterial DNA and contain binding sites for the gene *A* protein. When Mu infects a cell, the gene *A* protein directs transposition of Mu from the fragments of bacterial DNA that bounded it into a new site in the bacterial host's DNA (Figure 17-24). The old end pieces of bacterial DNA are probably degraded by nucleases in the cell. Since Mu moves precisely from its ends, its gene order is the same as an infecting particle or prophage; in contrast, λ integrates as a circle by crossing over from a site near the middle of the virion DNA molecule, so that its gene order becomes permuted (Figure 17-25).

If enough Mu repressor is made, Mu genes are repressed and this newly integrated copy remains as a prophage. If, on the other hand, Mu repressor is not made efficiently, then the lytic cycle continues.

Phage Mu replicates during its lytic cycle by continual replicative transposition into different (nearly) random sites of the bacterial DNA. Since one copy remains at the old site and one appears at the new, each transposition event yields another copy of Mu. In addition to transposase (A protein), the protein encoded by gene *B* is required for Mu replication, although its function is not known. No resolvase (Chapter 11) is made by Mu to divide the cointegrates that are formed. Instead, each progeny Mu is packaged into a phage coat along with some of the bacterial DNA that happens to be adjacent— 50 to 100 base pairs on the left, and 1000 to 2000 base pairs on the right. Since Mu inserts at random sites during replication, almost all phage particles are bounded by different bacterial DNA sequences.

Figure 17-25

A comparison of prophage integration by phages Mu and λ. Phage Mu integrates by moving from its ends, so that its gene order always remains the same. The linear λ DNA molecule circularizes when it enters the cell and then integrates by reciprocal recombination between specific sites on phage and bacterial DNA; the λ gene order is thereby permuted.

Figure 17-26
DNA structures resulting from denaturing and reannealing Mu virion DNA molecules. Straight double lines indicate double-stranded (i.e., homologous) DNA, and wavy lines indicate single-stranded DNA.

The presence of bacterial DNA at the ends of Mu virion DNA was shown originally by electron microscopy of denatured and reannealed Mu DNA (Chapter 9). For phage like λ and T7, whose virion DNAs consist of discrete and invariable sequences, reannealing of denatured virion DNA produces only homoduplexes—that is, continuous duplex DNA molecules. However, since Mu DNA molecules have different bacterial sequences at their ends, and since the two strands that meet to reanneal almost certainly were not partners originally, reannealed molecules of Mu phage DNA are partial heteroduplexes bearing single-stranded segments of bacterial DNA at their ends—the so-called split ends (Figure 17-26a). (The shorter single-stranded pieces at the left end are not visible in the electron microscope.)

A Surprise in the Structure of Mu DNA: An Invertible DNA Segment That Controls Mu Host Range[63, 64]

Some reannealed Mu DNA molecules have a second region of nonhomology: a bubble formed by two 3000-base-pair segments of single-stranded DNA near the right end of the DNA (Figure 17-27b). Furthermore, in a few molecules, only this segment of DNA has reannealed, and all the rest is left as long, single-stranded tails (Figure 17-26c). These structures arise because the 3000-base-pair **G segment** of Mu has been inverted in about half of the molecules; thus, about half of the reannealing structures cannot pair across the G segment (Figure 17-26b), or instead pair only in the G segment and then cannot pair in the rest (Figure 17-26c). Inversion occurs by a mechanism like that of phase variation in *Salmonella* (p. 331), by recombination across inverted repeats (of 34 base pairs in this case), and it is catalyzed by a similar enzyme, the product of the Mu *gin* gene.

The consequence of G segment inversion is to substitute one pair of phage proteins involved in tail fiber formation (S and U) for a second pair (S′ and U′) that is equivalent in forming tails, but different in detailed structure. The same segment of DNA, situated to the left of the G segment, encodes the beginning of tail fiber genes *S* and *S′* (Figure 17-27). However, the ends of genes *S* and *S′*, and all of genes *U* and *U′*, are different and are on the G segment oriented oppositely. When genes *S* and *U* are pointed rightward [the G(+) orientation] and are thus expressed along with the other late genes, tail fibers contain the S (and maybe U) protein; then Mu can bind surface receptors of *E. coli* strain K12 and infect it, but cannot bind certain other bacteria. On the other hand, when S′ and U′ proteins are made [the G(−) orientation], Mu tail fibers contain S′ (and maybe U′), and Mu can adsorb to and infect *E. coli* strain C, as well as the bacteria *Shigella sonnei* and *Serratia marcescens*. Thus, inverting its G segment allows Mu to expand or control its host range. G segment inversion does not occur frequently during lytic growth of the phage, but it does occur during growth of lysogens. Thus, a population of lysogens grown for many generations contains about half G(+) and half G(−) prophage.

Expression of the Phage Mu *mom* Gene Is Determined by the State of DNA Methylation[65–69]

Like T7, phage Mu makes an enzyme that protects its DNA against attack by cellular restriction endonucleases. After a Mu lysogen is

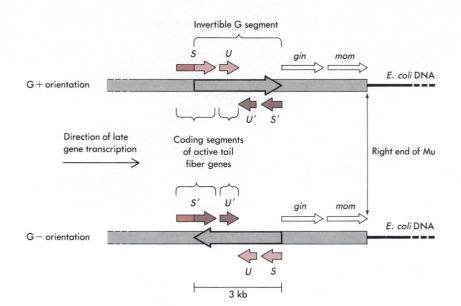

Direction of late gene transcription

Coding segments of active tail fiber genes

Right end of Mu

Figure 17-27
Structure and coding segments of the invertible G segment of phage Mu in its two orientations. [Adapted from F. J. Grundy and M. M. Howe, *Virology* 134 (1984):296, with permission.]

induced, the enzyme encoded by the Mu *mom* gene modifies about 15 percent of the adenine in phage and bacterial DNA to N^6-(1-acetamido)-adenine; this happens wherever adenine exists in the sequence 5'-C/G-A-C/G-N-Py-3' (where N is any nucleotide and Py is a pyrimidine). The recognition sites of many restriction endonucleases include this sequence, and the modification spoils them as substrates. As is true of other late genes, *mom* expression requires the Mu gene C protein, the activator of late gene transcription. But the *mom* gene is also controlled by the bacterial Dam methylase, the enzyme that methylates adenine in DNA wherever it occurs in the sequence 5'-GATC-3' (Chapter 12). Three GATC sequences in the hundred nucleotides before the *mom* gene promoter must be methylated by the Dam methylase for the *mom* gene to be transcribed. So far, it is not clear how the regulation works or what its function may be, but it will be important to understand this kind of transcriptional regulation by methylation of DNA.

The Single-Stranded DNA Phages: The Smallest DNA Viruses[70,71,72]

Single-stranded DNA (ssDNA) phages like φX174 and M13 have been important for studies of both DNA replication (Chapter 10) and genome structure at the DNA sequence level (Chapter 15). Phage φX174 was the first DNA phage to be sequenced completely, followed soon by its relatives G4, f1, fd, and M13. All are relatively simple, containing 10 or 11 genes that mediate DNA replication and assembly of progeny virions (Figure 17-28). The level of gene expression is determined mainly by the efficiencies of several promoters and terminators on the double-stranded replicative form DNA (Chapter 10), which is the template for synthesis of phage mRNA by the host RNA polymerase.

There are two classes of ssDNA phages, distinct in virion structure and in their effect on the cell in which they grow. The **spherical** (actually polyhedral) **ssDNA phages**, including φX174, G4, and S13, are packaged in an icosahedral shell made of one major and several minor proteins (Figure 17-29a). Like typical phages, they infect by

Figure 17-28
Genetic maps of typical spherical
(ϕX174) and filamentous (fd) ssDNA
phages. [Map of ϕX174 after
A. Kornberg, *1982 Supplement to DNA
Replication*, W. H. Freeman, San Fran-
cisco, p. 162; map of fd after N. D.
Zinder and K. Horiuchi, *Microbiol. Revs.*
49 (1985):102.]

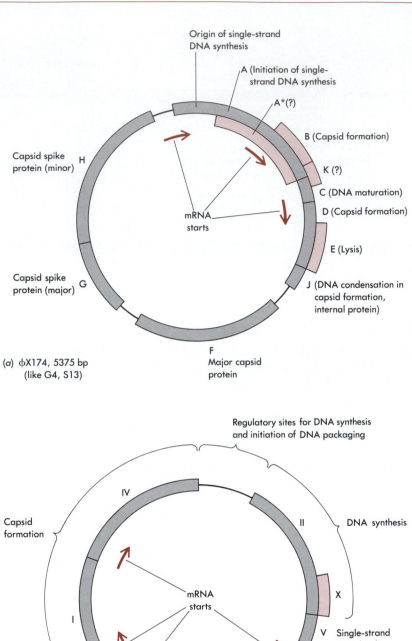

(a) ϕX174, 5375 bp
(like G4, S13)

(b) fd, 6407 bp
(like f1, M13)

adsorbing to receptors on the outside of the cell, replicate within the
cell, and destroy the cell by lysis to escape. The **filamentous ssDNA
phages**, including M13, f1, and fd, are thin particles as long as the cell
(1 μm), within which phage DNA forms an extended loop (Figure
17-29b). They also infect by adsorbing an external receptor, in this
case the F pilus carried by male bacteria (Chapter 7). Although their
DNA is replicated and transcribed like that of spherical phages, fila-

Spike, containing proteins encoded by gene *G* (60 copies per phage) and gene *H* (12 copies per phage)

Major capsid protein, product of gene *F* (60 copies per phage)

250 Å

(a) φX174 (spherical phage)

— A protein (pIII)
— B protein (pVIII)
≈ C protein (pVII-pIX)
⌣ D protein (pVI)

60 Å

1μ

(b) f1 (filamentous phage)

Figure 17-29

Structures and electron micrographs of (a) spherical and (b) filamentous ssDNA phages. The orientation that the DNA molecule is thought to have within the elongated filamentous particle is indicated by gene numbers; the morphogenetic signal sequence is believed to be in the intergenic (IG) region at the right, possibly bound to gene *VII* and *IX* proteins. The diagram also shows how the major coat protein is polymerized in helical array around the DNA. [Electron micrograph in (a) courtesy of Professor R. C. Williams; electron micrograph in (b) courtesy of Professor R. Webster; structures in (a) after A. Kornberg, *DNA Replication*, W. H. Freeman, San Francisco, 1980, p. 499; structure in (b) courtesy of Professor R. Webster.

mentous ssDNA phages do not kill the cell or even make a lysis protein. Instead, phage particles are extruded continuously as the cell continues to grow (albeit more slowly than normal), and the synthesis of phage proteins and DNA continues at a moderate rate that allows cellular metabolism to continue.

Filamentous Single-Stranded DNA Phages Are Secreted Through the Cell Membrane[73-78]

Most phages kill cells because infection destroys the vital cell membrane, equilibrating the cell with its surroundings. Filamentous ssDNA phages, on the other hand, preserve the cell's integrity by adapting its natural protein secretion mechanism for their escape. The major coat protein of phage fd, encoded by gene *VIII*, does not complex with phage DNA as it is made, but instead inserts into the cell membrane—the first step of protein secretion (Chapter 14). As is true of many proteins that go into or through membranes, insertion of coat protein is guided by a signal sequence (of 23 amino acids in this case), which is removed by a protease as insertion occurs. The mature coat protein of 50 amino acids has a basic portion within the cell, a hydrophobic segment that spans the membrane, and an acidic portion on the outside.

Newly synthesized phage DNA is covered first not by the gene *VIII* coat protein in the membrane, but by the abundant, cytoplasmic phage gene *V* protein; about 1500 molecules of gene *V* protein bind to each newly made phage DNA strand. This complex of DNA and gene *V* protein is then drawn to the inner surface of the membrane, where phage assembly begins. The phage particle is formed in the mem-

Figure 17-30

Pathway of filamentous phage particle morphogenesis. The particle is secreted from the cell as it is formed through replacement of gene *V* protein bound to the newly synthesized single-stranded DNA by the membrane-bound gene *VIII* major capsid protein. [After R. E. Webster and J. S. Cashman, in *The Single-Stranded DNA Phages*, ed. D. T. Denhardt, D. Dressler, and D. S. Ray (Cold Spring Harbor; N.Y.: Cold Spring Harbor Laboratory, 1978), p. 558.]

Rolling circle replicating structure

Viral strand is covered by gene *V* protein as it is made.

Gene *VIII* coat-protein inserted into inner membrane

Inner membrane

Cell wall

Outer membrane

Intracellular fd progeny DNA bound to gene *V* protein

Gene *V* protein is replaced by gene *VIII* capsid protein as particle is extruded through inner membrane.

Completed phage particle passes through cell wall and outer membrane to outside of cell.

brane through replacement of gene *V* protein by the gene *VIII* coat protein, a process that occurs as the growing filament emerges from the cell (Figure 17-30). Attraction between the basic end of the coat protein within the cell and the acidic DNA probably helps drive the assembly process. Although no phage-specific DNA sequences are required for gene *V* protein to bind, only phage DNA containing a **morphogenetic signal sequence** located between genes *IV* and *II* is assembled into particles at the membrane. It is thought that the minor phage proteins encoded by genes *VII* and *IX* bind to the morphogenetic sequence and are responsible for initiating the assembly process.

Unlike polyhedral phages, filamentous phages have no definite length and thus can accommodate variable amounts of DNA. This flexibility, as well as the fact that phage DNA can be prepared either double-stranded (as the circular replicative form isolated from the cell) or single-stranded (as DNA from phage particles), makes them very useful as cloning vectors. Isolated M13 or fd replicative form DNA can be used like plasmid DNA to make recombinants in vitro; the recombinant DNAs are put into the cell by transformation, again like plasmid DNA, whereupon they give rise to phage particles whose DNA carries the inserted DNA sequence. Even phage λ DNA can be cloned into M13 DNA, although it is eight times as long as M13 itself!

Gene *V* protein of fd has a regulatory function in addition to its role in phage assembly: It represses translation of mRNA for phage genes *II* and *X*. The fd gene *II* protein, like the φX174 A protein, is responsible for initiating phage DNA synthesis by nicking the replicative form DNA, and the control of its expression is believed to limit phage DNA synthesis to a level that the cell can live with.

Viral RNA Self-Replication: Requirements for a New Virus-Specific Enzyme

Cellular RNA molecules never serve as templates for the formation of new RNA strands. Therefore, the replication of most RNA viruses demands the participation of a completely new enzyme capable of

Figure 17-31
Electron micrograph of a collection of f2 RNA phage particles negatively stained so that they appear light against a dark background. (Courtesy of Dr. Norton Zinder of Rockefeller University.)

forming new RNA strands upon parental RNA templates. This enzyme, called **RNA replicase** (or **synthetase**), is usually formed just after the viral RNA enters the cell and attaches to host ribosomes. Like both DNA polymerase and RNA polymerase, RNA replicase catalyzes the formation of a complementary strand upon a single-stranded template. The fundamental mechanism for the copying of all nucleic acid base sequences is thus the same. Pairing of complementary bases is always used to achieve accurate replication of specific nucleotide sequences.

After their formation, the complementary RNA strands in turn serve as templates for new rounds of RNA synthesis. In this process, some of the progeny strands become part of new virus particles, while others are used as templates for specific protein formation. Exactly how this happens depends on whether the mature virus contains single-stranded or double-stranded RNA. In both cases, however, only one of the two complementary strands serves as mRNA for a given gene. Thus, there is a similarity in the formation of mRNA in DNA and RNA viral systems.

During the reproduction of most single-stranded RNA viruses (e.g., polio and the various RNA phages), the strand with the template function is the same strand found in the mature particles (i.e., +). This allows the infectious strand to code for the RNA replicase molecules necessary to initiate viral RNA replication. The single-stranded RNA animal virus vesicular stomatitis virus (VSV) behaves quite differently, however. The VSV particle contains the − strand, so that the infecting RNA cannot encode the replicase. Instead, each VSV particle has packaged within it a replicase molecule, made in the previous cycle of viral infection, that enters the cell along with the viral RNA molecule and immediately begins to copy it into a + strand that acts as the mRNA.

RNA Phages Are Very Simple[79–86]

The well-characterized RNA phages of *E. coli*—MS2, R17, f2, and Qβ—are very small (about 250 Å in diameter) and are the simplest viruses known, containing only four genes in their 3600- to 4200-nucleotide-long chromosomes (Figures 17-31 and 17-32). They are prolific, each infecting particle giving 5000 to 10,000 progeny after 30 to 60 minutes' growth at 37°C (Figure 17-33). For the nearly identical phages of the MS2 family (which also includes R17 and f2), two of the

Figure 17-32
Genetic maps of RNA phages MS2 (f2, R17) and Qβ. Gene boundaries are given in nucleotides from the 5' end of the RNA. [After R. B. Winter and L. Gold, *Cell* 33 (1983):878, with permission.]

four genes code for their two types of structural polypeptide chains. One, the 129-amino-acid coat protein, has a molecular weight of 13,700 daltons and is present 180 times in each virus particle. The other structural protein is the A (attachment) protein needed for adsorption of these phages to their host bacteria, as well as for the subsequent penetration of the viral RNA through the bacterial cell wall and membrane. One copy is found in each virus particle. It has a molecular weight of 44,000 and is made from 393 amino acids. Thus, almost half of RNA phage genetic material is used to code for its structural proteins. Most of the remainder encodes the virally specified component of the replicase, a 544-amino-acid polypeptide of molecular weight 61,000. Overlapping the coat protein and replicase-coding segments is the 75-amino-acid lysis protein, which is encoded in a different reading frame from both and is made by a small fraction of the ribosomes that shift reading frames while they are translating the coat protein–coding segment (Chapter 15). The slightly larger genome of Qβ (see Figure 17-32) is different in two ways: It has no separate lysis gene, the maturation protein A_2 having a second role as a lysis protein, and it encodes a second minor virion protein (A_1) made by ribosomes that occasionally read through the UGA termination codon of the coat protein.

The Nucleotide Sequence of MS2 Suggests Extensive Secondary Structure[87,88]

Our detailed understanding of RNA phage structure required learning the 3569-nucleotide-long sequence of MS2 RNA, an impressive achievement requiring methods much more difficult than those required for modern DNA sequencing (Figure 17-34). The sequence confirms earlier evidence from physical experiments that phage RNA has much internal structure. Although the base-paired structure shown in Figure 17-34 is hypothetical, the existence of some of the base-paired segments is shown by their resistance to digestion by the nuclease T1. Besides the predominantly local pairing shown in the figure, it is likely that distant segments also join to give the RNA an overall tight structure, accounting for its distinctive behavior as a messenger.

The sequence shows that translation does not start at the 5' end; instead, 129 nucleotides precede the GUG codon specifying the initiating formyl-methionyl group of A protein. Furthermore, there is no termination codon at the 3' end: The UAG termination signal for the

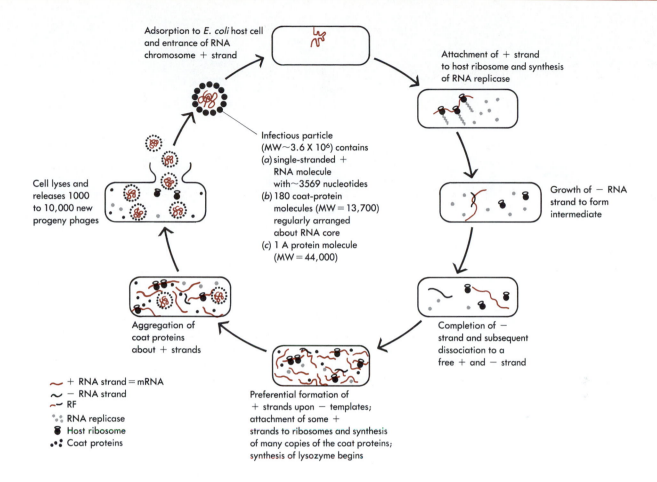

Adsorption to *E. coli* host cell and entrance of RNA chromosome + strand

Attachment of + strand to host ribosome and synthesis of RNA replicase

Growth of − RNA strand to form intermediate

Completion of − strand and subsequent dissociation to a free + and − strand

Preferential formation of + strands upon − templates; attachment of some + strands to ribosomes and synthesis of many copies of the coat proteins; synthesis of lysozyme begins

Aggregation of coat proteins about + strands

Cell lyses and releases 1000 to 10,000 new progeny phages

Infectious particle (MW~3.6 X 10⁶) contains
(a) single-stranded + RNA molecule with ~3569 nucleotides
(b) 180 coat-protein molecules (MW = 13,700) regularly arranged about RNA core
(c) 1 A protein molecule (MW = 44,000)

~ + RNA strand = mRNA
~ − RNA strand
~ RF
RNA replicase
Host ribosome
Coat proteins

replicase gene is followed by an untranslated sequence of 174 nucleotides. It is common for mRNA to have untranslated leading and trailing sequences, which may act in part to determine its resistance to nucleases and thus its stability; in this case, they may also be important for replicase function or in forming the structure of the phage particle.

Figure 17-33
The life cycle of a single-stranded RNA phage of the MS2 (R17, f2) family. Growth of phage Qβ is essentially the same, except that the particle contains both A_1 and A_2 proteins, and the A_1 protein is also the phage lysis protein.

Initial Binding of Ribosomes to One Site on the Phage RNA

After entry of an RNA phage genome into a host cell, ribosomes attach only to the ribosome binding site at the beginning of the coat protein gene (Figure 17-35a). No attachment occurs at the beginning of either the A protein gene or the replicase gene, because their ribosome binding sites are blocked by the secondary structure of the viral RNA. In contrast, the initiating AUG for coat protein appears to be accessible to ribosomes because it is exposed at the end of a stem (see Figure 17-34).

The first of the two blocked sites to become available is the replicase site, which opens up when reading of the coat protein gene temporarily disrupts the inhibitory hydrogen-bonded hairpin loops (Figure 17-35b). Thus, functioning of the replicase gene always depends on prior ribosome attachment at the coat protein binding site. This dependence was first suggested by the discovery that a nonsense mutation early in the coat protein gene, at the sixth amino acid, is "polar"—

Figure 17-34

A portion of the sequence of MS2 RNA and its hypothetical secondary structure. Positions are given as nucleotide number from the 5' end of the RNA. [Adapted from W. Fiers, *Comp. Virology* 13 (1979):69, with permission.]

meaning that it prevents replicase expression even though the replicase-coding sequence is intact. It is clear that according to the structural model of Figure 17-34, a ribosome stopping after six amino acids would not open enough RNA to expose the replicase initiation site. The fact that nonsense mutations at positions 50, 54, or 70 are *not* polar also is consistent with the model, since the replicase initiation sequence would already be open when a ribosome had gone this far.

Coat Protein Chains Can Repress the Translation of the Replicase Gene[89, 90]

The RNA phage coat protein polypeptide chains are made in much greater numbers than the replicase polypeptide chains, even though the replicase ribosome binding site is opened each time the coat cistron is translated. This reflects an in vivo control device: Newly made coat protein subunits specifically attach to the replicase gene ribosome binding site and prevent ribosome attachment (Figure 17-36). Thus, the coat protein is a specific **translational repressor** of the replicase gene. Within 10 minutes after infection, enough coat protein has accumulated to block all further replicase synthesis.

Stopping replicase synthesis less than halfway through the life cycle makes obvious sense. By this time, enough enzyme has accumulated to support viral RNA replication, and the primary process needed in the latter phases of the life cycle is the production of the structural proteins of the virus.

The RNA segment that coat protein binds to in order to repress replicase initiation has a structure different from that believed to exist in the virion RNA molecule (see Figure 17-34). It is a 20-nucleotide hairpin loop, a structure about the size of one coat protein molecule, which may be able to form in vivo only when the original base pairing is opened by ribosomes reading the coat protein gene (Figure 17-37). Although this hairpin apparently is not tight enough by itself to prevent ribosomes from initiating translation of replicase, binding of coat protein is thought to stabilize it enough to block ribosome entry and thereby repress replicase synthesis.

This RNA hairpin has been an important tool for understanding the specific interaction of a protein with a polynucleotide different from double-stranded DNA, which is the target of most known regulatory proteins. Measurements of coat protein binding to variants of the hairpin, synthesized in vitro with the enzymes polynucleotide phosphorylase and T4 RNA ligase, show that both the single-stranded loop and the helical stem are recognized by coat protein. Only four of the particular bases present in the hairpin are necessary for coat protein binding, although appropriate bases must exist in other positions so that the structure can form (see Figure 17-37). Unlike many other DNA or RNA binding proteins, coat protein contacts mostly the bases and not the phosphates in the backbone chain.

Functional Complexes of Virally Coded Replicase Chains and Host Proteins[91, 92, 93]

The 61,000-dalton polypeptide product of the replicase gene does not function by itself but only after it has combined with three different host polypeptide chains (Figure 17-38). Two of the host polypeptide chains (with molecular weights of 30,000 and 45,000) are the Tu and

(a) In the absence of protein synthesis, the A and replicase sites are blocked.

(b) Passage of ribosomes over the coat protein gene opens up the secondary structure normally blocking the replicase site, thereby making the replicase gene available for translation.

Figure 17-35
Schematic diagram showing how secondary structure affects the accessibility of ribosome binding sites of RNA phage RNA.

Figure 17-36
Repression of replicase synthesis by binding of a coat protein molecule to the initiation site of the replicase gene.

Original structure
in phage R17 RNA

(a)

(b)

(c)

N   any base
N---N any base pair
Py pyrimidine (U or C)

Figure 17-37
Postulated secondary structure of MS2
RNA containing the initiation site for
replicase (a) before ribosomes bind and
(b) after the original structure is opened
by translation of the coat protein gene,
allowing a hairpin to form. Essential
elements of the coat protein binding
RNA hairpin are shown in (c). [After
O. C. Uhlenbeck, J. Carey, P. J.
Romaniuk, P. T. Lowary, and D. Beckett, *J. Biomolecular Structure and Dynamics* 1 (1983):539, with permission.]

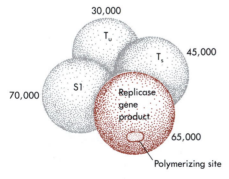

Figure 17-38
Schematic representation of Qβ replicase, showing its construction from one
phage coded polypeptide chain and
three host components.

the Ts components of the T factor, the *E. coli* protein complex involved in the placement of incoming AA ~ tRNA molecules into the
A site of ribosomes (Chapter 14). The remaining host factor is S1, a
70,000-dalton protein that is usually found tightly bound to the 30S
ribosomal subunit. We are certain only of the role of the virally coded
component. It contains the catalytic site responsible for making the
internucleotide 3'–5' phosphodiester bonds between a growing chain
and ribonucleoside triphosphate precursors. The T factor probably
functions in chain initiation, helping to recognize the 3' ends of the
RNA templates. Favoring this hypothesis is the presence of tRNA-
like sequences (e.g., CCA–OH) at the 3' ends of both the viral +
and − strands. As T factors are adapted specifically to recognize the
tRNA molecules, their presence within the complete replicase complex may have evolved to help replicase molecules preferentially attach to 3' ends of viral as opposed to cellular RNA molecules. The S1
component of replicase is required only for the initiation stage of replication, and only on + strand templates; thus, S1 may help replicase
to recognize a specific binding site on the viral RNA molecule.

RNA Phage RNA Self-Replication Does Not Involve Fully Double-Helical Intermediates[94, 95, 96]

Replication of viral RNA starts with the attachment of replicase to the
infecting + strand. Surprisingly, the enzyme binds not only to the 3'
end, where 5' → 3' synthesis of the product − strand must begin,
but also to two internal sites, one that includes the ribosome binding
site of the coat gene and one that is within the replicase gene. It is
likely that these sites are close to one another in the complex three-
dimensional structure of the viral RNA molecule. Binding of replicase
to the coat gene has regulatory function: It prevents ribosomes from
attaching and translating coat protein, a process that might interfere
with replication.

Although base pairing determines the − strand sequences, no
double-helical **replicative intermediate** is generated, and a free −
strand is the product of the first round of synthesis (Figure 17-39).
Only at the growing end of the − strand is a short double-helical

stretch present within the replicative intermediate. However, if the replicase is removed by treatments with protein-denaturing agents, all the nascent strands base-pair with their parental complements to form double-helical sections. Therefore, replicase itself is responsible for keeping the strands apart, despite their potential for full annealing if the protein is removed.

Both the complete progeny − strand and the original + strand in turn serve as templates to make more + and − strands. In this process, many more + strands are made than − strands, indicating that replicase molecules preferentially bind to the 3′ end of − strands, as opposed to 3′ ends of + strands.

Laboratory-made + strands have the same length as those made in vivo and have equal infectivity when added to growing bacteria. Moreover, the frequency of mutants is similar to that in nature. This means that the copying achieved in vitro is just as accurate as that occurring within cells.

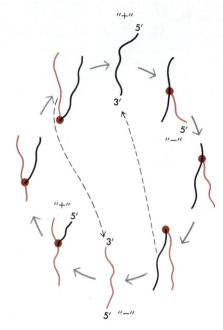

Figure 17-39
Schematic model for the replication of RNA phage (Qβ, MS2, R17, f2) RNA directed by the phage replicase. Here, for simplicity of visualization, the single-stranded regions are shown in an extended rather than tightly collapsed form.

Only Nascent Stretches of + RNA Serve as Templates for the A Protein

Reading of the A gene is limited to the periods of viral RNA replication that are characterized by incomplete stretches of nascent RNA. Just after a viral + strand has started to grow, the newly formed A binding site is accessible for ribosome attachment (Figure 17-40). But it quickly becomes closed as extensive patches of secondary structure form upon further chain elongation. Open A sites bind ribosomes even better than open coat protein sites, so that their inherent affinity for ribosomes by itself does not determine how often they will be translated; the degree to which they are not blocked by secondary structure is more important.

Self-Assembly of Progeny Particles and the Formation of Intracellular Viral Crystals

Assembly of progeny phages spontaneously starts as soon as the A protein, coat protein, and progeny + strands begin to accumulate. Only progeny + strands are ever encapsulated; thus the newly made − strands are left free to produce still more progeny + strands. In the aggregation process, the A protein is thought to bind early to the phage RNA, forming a tight complex that persists through the next cycle of phage growth when the A protein enters the bacterial cell along with its respective RNA molecule. Conceivably, binding of the A protein converts progeny + strands into forms that hasten the aggregation of coat protein around them.

As infection progresses, the progeny phage often form such a substantial fraction of the internal bacterial contents that they begin to form tiny crystals that are easily visualized by the electron microscope (Figure 17-41). Since RNA phage crystals have also been made in vitro, there is promise that the molecular structure of RNA phage will soon be known from x-ray diffraction analysis.

Figure 17-40
Translation of the RNA phage A protein gene from nascent + strands.

Figure 17-41
Electron micrograph of a thin section of
E. coli cells infected with the RNA
phage f2. Most of the progeny particles
lie next to each other in crystalline
array. (Courtesy of Dr. Norton Zinder
of Rockefeller University.)

Figure 17-41
Electron micrograph of a thin section of
E. coli cells infected with the RNA
phage f2. Most of the progeny particles
lie next to each other in crystalline
array. (Courtesy of Dr. Norton Zinder
of Rockefeller University.)

The Smallest Known Viruses Are Almost at the Lower Limits for a Virus

The need ordinarily to code for both virus-specific RNA replicase and viral coat protein effectively places a lower limit on how small an RNA-containing virus can be. First, it would be surprising if a replicase much smaller than, say, 30,000 daltons could carry out all the tasks involved in RNA self-replication. Second, it is hard to imagine stable coat protein subunits made up of less than 50 to 75 amino acids. Thus, the smallest complete RNA virus likely to exist probably contains at least 1500 nucleotides (half the MS2 number). The expectation of anything much smaller seems highly unrealistic.

Analogous reasoning for the minimal size of a DNA virus is not yet possible, since it is possible that a circular viral DNA molecule might be replicated using only host enzymes. If so, then someday we might find a DNA virus whose only gene codes for its coat protein. Conceivably, such a DNA chain might contain as few as 300 to 500 base pairs.

Making intelligent guesses of the upper size for a virus is much more difficult. The largest viruses now known code for some 200 to 300 genes, but there seems to be no reason why a much larger number could not be coded for. Perhaps, above a certain size, there is no substantial need for new genes to manipulate the host cell in favor of viral products.

Always very clear, however, is the absolute difference between a virus and a cell. As little as 20 years ago, there was much confusion. At that time, a key criterion for a virus was still its ability to pass through membrane filters that hold back bacteria. Under this definition, the very small disease-causing bodies known as *Rickettsiae* were generally treated as viruses. But they since have been found to contain DNA and RNA, as well as a protein-synthesizing system, and so are realized to be very small bacteria. The essence of viruses is thus the absence of protein-synthesizing systems. Hence, their multiplication must always involve the breakdown of their surrounding protective coats, thereby letting their chromosomes come into contact with cellular enzymes.

Summary

All known viruses contain a core of either DNA or RNA, surrounded by a protective capsid of protein. More complicated viruses also contain lipids or carbohydrates, or both. The bulk of the viral coat is invariably constructed of one or a few types of polypeptides that are present in many copies. Viral DNA or RNA encodes both viral coat proteins and enzymes that are required to replicate the genome. After coat proteins and copies of the viral nucleic acid are made in an infected cell, these components specifically aggregate to form new infectious viruses identical to the infecting parental particle. When RNA is the genetic component, it is always replicated by a virus-specific enzyme, whereas DNA viruses often are replicated by cellular DNA polymerases. The viral genome may be either single-stranded or double-stranded. Single-stranded viral DNA or RNA first is copied into a complementary strand, which in turn serves as a template for synthesis of more viral nucleic acid; thus, the basic rules by which nucleic acids are duplicated through complementary base pairing are always the same.

Viruses often specify proteins that subvert cellular metabolism to promote selective expression of the viral genome. The more complex DNA bacteriophages (bacterial viruses) either alter the host RNA polymerase so that it initiates only at phage promoters or provide their own virus-specific RNA polymerase. Furthermore, they may encode nucleases that destroy the bacterial DNA and lysozyme to disrupt the bacterial cell wall and liberate progeny phage particles. Virus-encoded regulatory proteins act in a defined sequence to establish a correctly timed program of gene expression during viral growth.

Bacteriophages have provided our best understanding of the molecular basis of gene regulation, particularly the regulation of transcription. Phages SPO1 and T4 encode new σ factors that direct the host RNA polymerase to distinct phage-specific promoter sequences. In contrast, phages T7 and N4 encode entirely new RNA polymerases specific for phage promoter sequences.

The temperate phage λ can either grow lytically upon infection, expressing all genes required for its replication, or instead join the host chromosome by recombination and grow as a repressed prophage that is replicated passively by the host cell. The opposing activities of λ Cro protein and λ repressor direct these alternate developmental pathways; the outcome of their competition is influenced by the separate regulator cII, which senses the environmental condition of the cell and activates transcription of λ repressor if conditions are favorable for lysogeny. The positive regulators of early and late gene transcription of λ are antiterminators that modify RNA polymerase and thereby allow it to pass transcription terminators that precede regulated genes in specific operons.

Phages encode proteins that direct their own replication. The λ O and P proteins, like the ϕX174 A protein, initiate replication at a specific phage origin. Phage Mu replicates by transposition, so that its specific replication protein is a transposase. In lytic growth, Mu transposes repeatedly into different sites of the host genome, and it is cut from the chromosome for packaging into phage coats with random bits of bacterial DNA still attached to its ends.

Phage Mu is a temperate phage that also integrates into host DNA by transposition, entering nearly random sites of the bacterial DNA and making mutations in the host as it does. The invertible G segment of Mu specifies different tail fiber proteins in its two orientations, thereby allowing the host adsorption specificity of Mu to change.

The small, single-stranded DNA phages like ϕX174 and fd encode mostly structural proteins, except for the initiator of single-stranded DNA synthesis. DNA of filamentous single-stranded DNA phages is packaged through a process of secretion, in which coat protein waiting in the cell membrane coalesces with newly replicated phage DNA to form a particle that is extruded from the cell as it is made.

RNA phages encode a replicase that first synthesizes a − strand copy of the infecting + strand and then more viral + copies from the − strand template. The efficiency with which RNA phage genes are translated is determined by the complex secondary structure of the phage RNA; at first only its coat protein initiation codon is accessible to ribosomes, until the act of translating the coat gene opens the RNA for translation of the replicase. RNA phage coat protein is a specific translational repressor of replicase synthesis, acting by binding a hairpin loop in phage RNA that includes the replicase translation initiation site.

Bibliography

General References

Denhardt, D. T., D. Dressler, and D. S. Ray, eds. 1978. *The Single-Stranded DNA Phages.* Cold Spring Harbor, N.Y.: Cold Spring Harbor Laboratory.

Hendrix, R. W., J. W. Roberts, F. W. Stahl, and R. A. Weisberg, eds. 1983. *Lambda II.* Cold Spring Harbor, N.Y.: Cold Spring Harbor Laboratory.

Luria, S. E., J. E. Darnell, Jr., D. Baltimore, and A. Campbell. 1978. *General Virology.* 3rd ed. New York: Wiley.

Mathews, C. K., E. M. Kutter, G. Mosig, and P. B. Berget, eds. 1983. *Bacteriophage T4.* Washington, D.C.: American Society for Microbiology.

Stent, G. S. and R. Calendar. 1978. *Molecular Genetics: An Introductory Narrative.* 2nd ed. San Francisco: W. H. Freeman.

Zinder, N. D., ed. 1975. *RNA Phages.* Cold Spring Harbor, N.Y.: Cold Spring Harbor Laboratory.

Cited References

1. Crick, F. H. C., and J. D. Watson, 1957. "Virus Structure: General Principles." *CIBA Found. Symp. Nature Viruses.* 5–13.

2. Caspar, D. L. D., and A. Klug. 1962. "Physical Principles in the Construction of Regular Viruses." *Cold Spring Harbor Symp. Quant. Biol.* 27:1–241.

3. Wood, W. B., and R. S. Edgar. 1973. "Building a Bacterial Virus." *Scientific American* 1967. In *The Chemical Basis of Life*, ed. P. C. Hanawalt and R. H. Haynes. San Francisco: Freeman.

4. Harrison, S. C. 1983. "Virus Structure: High-Resolution Perspectives." *Adv. Virus Res.* 28:175–240.

5. Losick, R. and J. Pero. 1981. "Cascades of Sigma Factors." *Cell* 25:582–584.

6. Chelm, B. K., J. R. Greene, C. Beard, and E. P. Geiduschek. 1982. "The Transition of Early and Middle Gene Expression in the Development of Phage SPO1: Physiological and Biochemical Aspects." In *Molecular Cloning and Gene Regulation in* Bacilli, ed. A. T. Ganesan, S. Chang, and J. A. Hoch. San Diego: Academic Press, pp. 345–359.

7. Fox, T. D., R. Losick, and J. Pero. 1976. "Regulatory Gene *28* of Bacteriophage SPO1 Codes for a Phage-Induced Subunit of RNA Polymerase." *J. Mol. Biol.* 101:427–433.

8. Duffy, J. J. and E. P. Geiduschek. 1977. "Purification of a Positive Regulatory Subunit from Phage SPO1-Modified RNA Polymerase." *Nature* 270:28–32.

9. Costanzo, M., L. Brzustowicz, N. Hannett, and J. Pero. 1984. "Bacteriophage SPO1 Genes *33* and *34*: Location and Primary Structure of Genes Encoding Regulatory Subunits of *Bacillus Subtilis* RNA Polymerase." *J. Mol. Biol.* 180:533–547.

10. Costanzo, M. 1983. "Bacteriophage SPO1 Regulatory Genes." Ph.D thesis, Harvard University.

11. Lee, G., and J. Pero. 1981. "Conserved Nucleotide Sequences in Temporally Controlled Bacteriophage Promoters." *J. Mol. Biol.* 152:247–265.

12. Dunn, J. J., and F. W. Studier. 1983. "Complete Nucleotide Sequence of Bacteriophage T7 DNA and the Locations of T7 Genetic Elements." *J. Mol. Biol.* 166:477–535.

13. Chamberlin, M., J. McGrath, and L. Waskell. 1970. "New RNA Polymerase from *E. coli* Infected with Bacteriophage T7." *Nature* 228:227–231.

14. Bailey, J. N., J. R. Klement, and W. T. McAllister. 1983. "Relationship Between Promoter Structure and Template Specificities Exhibited by the Bacteriophage T3 and T7 RNA Polymerases." *Proc. Nat. Acad. Sci.* 80:2814–2818.

15. Davanloo, P., A. H. Rosenberg, J. J. Dunn, and F. W. Studier. 1984. "Cloning and Expression of the Gene for Bacteriophage T7 RNA Polymerase." *Proc. Nat. Acad. Sci.* 81:2035–2039.

16. Mark, K., and F. W. Studier. 1981. "Purification of the Gene *0.3* Protein of Bacteriophage T7, an Inhibitor of the DNA Restriction System of *Escherichia coli*." *J. Biol. Chem.* 256:2573–2578.

17. Hesselbach, B. and D. Nakada. 1977. "'Host Shutoff' Function by Bacteriophage T7: Involvement of T7 Gene 2 and Gene *0.3* in the Inactivation of *Escherichia coli* RNA Polymerase." *J. Virology* 24:736–745.

18. Zillig, W., H. Fujiki, W. Blum, D. Janeković, M. Schweiger, H. Rahmsdorf, H. Ponta, and M. Hirsch-Kauffman. 1975. "*In Vivo* and *in Vitro* Phosphorylation of DNA-Dependent RNA Polymerase of *Escherichia coli* by Bacteriophage-T7-Induced Protein Kinase." *Proc. Nat. Acad. Sci.* 72:2506–2510.

19. Center, M. S., F. W. Studier, and C. C. Richardson. 1970. "The Structural Gene for a T7 Endonuclease Essential for Phage DNA Synthesis." *Proc. Nat. Acad. Sci.* 65:242–248.

20. Haynes, L. L., and L. B. Rothman-Denes. 1985. "N4 Virion RNA Polymerase Sites of Transcription Initiation." *Cell* 41:597–605.

21. Falco, S. C., W. Zehring, and L. B. Rothman-Denes. 1980. "DNA-Dependent RNA Polymerase from Bacteriophage N4 Virions, Purification and Characterization." *J. Biol. Chem.* 255:4339–4347.

22. Hendrix, R. W., J. W. Roberts, F. W. Stahl, and R. A. Weisberg, eds. 1983. *Lambda II.* Cold Spring Harbor, N.Y.: Cold Spring Harbor Laboratory.

23. Friedman, D., and M. Gottesman. 1983. "Lytic Mode of Lambda Development." In *Lambda II*, ed. R. W. Hendrix, J. W. Roberts, F. W. Stahl, and R. A. Weisberg. Cold Spring Harbor, N.Y.: Cold Spring Harbor Laboratory, pp. 21–51.

24. Gussin, G., A. Johnson, C. Pabo, and R. Sauer. 1983. "Repressor and Cro Protein: Structure, Function, and Role in Lysogenization." In *Lambda II*, ed. R. W. Hendrix, J. W. Roberts, F. W. Stahl, and R. A. Weisberg. Cold Spring Harbor, N.Y.: Cold Spring Harbor Laboratory, pp. 93–121.

25. Ptashne, M. 1986. *A Genetic Switch.* Palo Alto: Blackwell Scientific Publications and Cell Press.

26. Johnson, A. D., A. R. Poteete, G. Lauer, R. T. Sauer, G. K. Ackers, and M. Ptashne. 1981. "λ Repressor and Cro—Components of an Efficient Molecular Switch." *Nature* 294:217–223.

27. Meyer, B. J., R. Maurer, and M. Ptashne. 1980. "Gene Regulation at the Right Operator (O_R) of Bacteriophage λ. II. O_R1, O_R2, and O_R3: Their Roles in Mediating the Effects of Repressor and cro." *J. Mol. Biol.* 139:163–194.

28. Hochschild, A., N. Irwin, and M. Ptashne. 1983. "Repressor Structure and the Mechanism of Positive Control." *Cell* 32:319–325.

29. Eisen, H., P. Brachet, L. Pereira da Silva, and F. Jacob. 1970. "Regulation of Repressor Expression in λ." *Proc. Nat. Acad. Sci.* 66:855–862.

30. Reichardt, L., and A. D. Kaiser. 1971. "Control of λ Repressor Synthesis." *Proc. Nat. Acad. Sci.* 68:2185–2189.

31. Wulff, D. L., and M. Rosenberg. 1983. "Establishment of Repressor Synthesis." In *Lambda II*, ed. R. W. Hendrix, J. W. Roberts, F. W. Stahl, and R. A. Weisberg. Cold Spring Harbor, N.Y.: Cold Spring Harbor Laboratory, pp. 53–73.

32. Herskowitz, I., and D. Hagen. 1980. "The Lysis-Lysogeny Decision of Phage λ; Explicit Programming and Responsiveness." *Ann. Rev. Genetics* 14:399–445.

33. Shimatake, H., and M. Rosenberg. 1981. "Purified λ Regulatory Protein cII Positively Activates Promoters for Lysogenic Development." *Nature* 292:128–132.

34. Hoyt, M. A., D. M. Knight, A. Das, H. I. Miller, and H. Echols. 1982. "Control of Phage λ Development by Stability and Synthesis of cII Protein: Role of the Viral *cIII* and Host *hflA*, *himA* and *himD* Genes." *Cell* 31:565–573.

35. Hoopes, B. C., and W. R. McClure. 1985. "A cII-Dependent Promoter Is Located Within the *Q* Gene of Bacteriophage λ." *Proc. Nat. Acad. Sci.* 82:3134–3138.

36. Roberts, J. W. 1969. "Termination Factor for RNA Synthesis." *Nature* 224:1168–1174.

37. Adyha, S., M. Gottesman, and B. de Crombrugghe. 1974. "Release of Polarity in *Escherichia coli* by Gene N of Phage λ: Termination and Antitermination of Transcription." *Proc. Nat. Acad. Sci.* 71:2534–2538.

38. Franklin, N. 1974. "Altered Reading of Genetic Signals Fused to the N Operon of Bacteriophage λ: Genetic Evidence for Modification of Polymerase by the Protein Product of the N Gene." *J. Mol. Biol.* 89:33–48.

39. Friedman, D. I., G. S. Wilgus, and R. J. Mural. 1973. "Gene N Regulator Function of Phage λimm21: Evidence That a Site of N Action Differs from a Site of N Recognition." *J. Mol. Biol.* 81:505–516.

40. Salstrom. J. S., and W. Szybalski. 1978. "Coliphage λ *nut* L⁻: A Unique Class of Mutants Defective in the Site of Gene N Product Utilization for Antitermination of Leftward Transcription." *J. Mol. Biol.* 124:195–221.

41. Herskowitz, I., and E. R. Signer. 1970. "A Site Essential for Expression of All Late Genes in Bacteriophage λ." *J. Mol. Biol.* 47:545–556.

42. Grayhack, E. J., X. Yang, L. F. Lau, and J. W. Roberts. 1985. "Phage Lambda Gene Q Antiterminator Recognizes RNA Polymerase Near the Promoter and Accelerates It Through a Pause Site." *Cell* 42:259–269.

43. Ray, P. N. and M. L. Pearson. 1974. "Evidence for Posttranscriptional Control of the Morphogenetic Genes of Bacteriophage Lambda." *J. Mol. Biol.* 85:163–175.

44. Gottesman, M., A. Oppenheim, and D. Court. 1982. "Retroregulation: Control of Gene Expression from Sites Distal to the Gene." *Cell* 29:727–728.

45. Tsurimoto, T., and K. Matsubara. 1981. "Purified Bacteriophage λ O Protein Binds to Four Repeating Sequences at the λ Replication Origin." *Nucleic Acid Res.* 9:1789–1799.

46. LeBowitz, J. H., M. Zylicz, C. Georgopoulos, and R. McMacken. 1985. "Initiation of DNA Replication on Single-Stranded DNA Templates Catalyzed by Purified Replication Proteins of Bacteriophage λ and *Escherichia coli.*" *Proc. Nat. Acad. Sci.* 82:3988–3992.

47. Feiss, M. and A. Becker. 1983. "DNA Packaging and Cutting." In *Lambda II,* ed. R. W. Hendrix, J. W. Roberts, F. W. Stahl, and R. A. Weisberg. Cold Spring Harbor, N.Y.: Cold Spring Harbor Laboratory, pp. 305–330.

48. Roberts, J. W., and C. W. Roberts. 1975. "Proteolytic Cleavage of Bacteriophage Lambda Repressor in Induction." *Proc. Nat. Acad. Sci.* 72:147–151.

49. Craig, N. L., and J. W. Roberts. 1980. "*E. coli* recA Protein-Directed Cleavage of Phage λ Repressor Requires Polynucleotide." *Nature* 283:26–30.

50. Susskind, M. M., and P. Youderian. 1983. "P22 Antirepressor and Its Control." In *Lambda II,* ed. R. W. Hendrix, J. W. Roberts, F. W. Stahl, and R. A. Weisberg. Cold Spring Harbor, N.Y.: Cold Spring Harbor Laboratory, pp. 347–363.

51. Susskind, M. M., and D. Botstein. 1975. "Mechanism of Action of *Salmonella* Phage P22 Antirepressor." *J. Mol. Biol.* 98:413–424.

52. Shapiro, J. A., ed. 1983. *Mobile Genetic Elements.* New York: Academic Press.

53. Taylor, A. L. 1963. "Bacteriophage-Induced Mutation in *Escherichia coli.*" *Proc. Nat. Acad. Sci.* 50:1043–1051.

54. Bukhari, A., and D. Zipser. 1972. "Random Insertion of Mu-1 DNA Within a Single Gene." *Nature New Biol.* 236:240–243.

55. Daniell, E., R. Roberts, and J. Abelson. 1972. "Mutations in the Lactose Operon Caused by Bacteriophage Mu." *J. Mol. Biol.* 69:1–8.

56. Daniell, E., D. E. Kohne, and J. Abelson. 1975. "Characterization of the Inhomogeneous DNA in Virions of Bacteriophage Mu by DNA Reannealing Kinetics." *J. Virology* 15:739–743.

57. Castilho, B. A., P. Olfson, and M. J. Casadaban. 1984. "Plasmid Insertion Mutagenesis and *lac* Gene Fusion with Mini-Mu Bacteriophage Transposons." *J. Bacteriol.* 158:488–495.

58. Faelen, M., A. Toussaint, and M. Couturier. 1971. "Mu-1 Promoted Integration of a λ-*gal* Phage in the Chromosome of *E. coli* K12." *Mol. Gen. Genetics* 113:367–370.

59. Van de Putte, P., and M. Gruijthuijsen. 1972. "Chromosomal Mobilization and Integration of F Factors in the Chromosome of a *recA* Strain of *E. coli.*" *Mol. Gen. Genetics* 118:173–183.

60. Ljungquist, E., and A. I. Bukhari. 1977. "State of Prophage Mu Upon Induction." *Proc. Nat. Acad. Sci.* 74:3143–3147.

61. Harshey, R. M. 1984. "Transposition Without Duplication of Infecting Bacteriophage Mu DNA." *Nature* 311:580–581.

62. Mizuuchi, K. 1983. "*In vitro* Transposition of Bacteriophage Mu: A Biochemical Approach to a Novel Replication Reaction." *Cell* 35:785–794.

63. Van de Putte, P., S. Cramer, and M. Giphart-Gassler. 1980. "Invertible DNA Determines Host Specificity of Bacteriophage Mu." *Nature* 286:218–222.

64. Grundy, F. J., and M. M. Howe. 1984. "Involvement of the Invertible G Segment in Bacteriophage Mu Tail Fiber Biosynthesis." *Virology* 134:296–317.

65. Toussaint, A. 1976. "The DNA Modification Function of Temperate Phage Mu-1." *Virology* 70:17–27.

66. Hattman, S. 1979. "Unusual Modification of Bacteriophage Mu DNA." *J. Virology* 34:277–279.

67. Hattman, S., M. Goradia, C. Monaghan, and A. I. Bukhari. 1983. "Regulation of the DNA Modification Function of Bacteriophage Mu." *Cold Spring Harbor Symp. Quant. Biol.* 47:647–654.

68. Plasterk, R. H. A., M. Vollering, A. Brinkman, and P. Van de Putte. 1984. "Analysis of the Methylation-Regulated Mu *mom* Transcript." *Cell* 36:189–196.

69. Swinton, D., S. Hattman, P. F. Crain, C.-S. Cheng, D. L. Smith, and J. A. McCloskey. 1983. "Purification and Characterization of the Unusual Deoxynucleoside, α-N-(9-β-D-2'-Deoxyribofuranosylpurin-6-yl) Glycinamide, Specified by the Phage Mu Modification Function." *Proc. Nat. Acad. Sci.* 80:7400–7404.

70. Denhardt, D. T., D. Dressler, and D. S. Ray, eds. 1978. *The Single-Stranded DNA Phages.* Cold Spring Harbor, N.Y.: Cold Spring Harbor Laboratory.

71. Fiers, W., and R. L. Sinsheimer. 1962. "The Structure of the DNA of Bacteriophage φX174. I. The Action of Exopolynucleotidases." *J. Mol. Biol.* 5:408–419.

72. Sanger, F., G. M. Air, B. G. Barrell, N. L. Brown, A. R. Coulson, J. C. Fiddes, C. A. Hutchison III, P. M. Slocombe, and B. Smith. 1977. "Nucleotide Sequence of Bacteriophage φX174 DNA." *Nature* 265:687–695.

73. Webster, R. E., and J. S. Cashman. 1978. "Morphogenesis of the Filamentous Single-Stranded DNA Phages." In *The Single-Stranded DNA Phages,* ed. D. T. Denhardt, D. Dressler, and D. S. Ray. Cold Spring Harbor, N.Y.: Cold Spring Harbor Laboratory, pp. 557–569.

74. Zinder, N. D., and K. Horiuchi. 1985. "Multiregulatory Element of Filamentous Bacteriophages." *Microbiol. Revs.* 49:101–106.

75. Webster, R. E., R. A. Grant, and L. W. Hamilton. 1983. "Orientation of the DNA in the Filamentous Bacteriophage f1." *J. Mol. Biol.* 152:357–374.

76. Messing, J., B. Gronenborn, B. Müller-Hill, and P. H. Hofschneider. 1977. "Filamentous Coliphage M13 as a Cloning Vehicle: Insertion of a *Hind*II Fragment of the *lac* Regulatory Region in M13 Replicative Form *in Vitro.*" *Proc. Nat. Acad. Sci.* 74:3642–3646.

77. Model, P., C. McGill, B. Mazur, and W. D. Fulford. 1982. "The Replication of Bacteriophage f1: Gene V Protein Regulates the Synthesis of Gene II Protein." *Cell* 29:329–335.

78. Benedict Yen, T. S., and R. E. Webster. 1982. "Translational Control of Bacteriophage f1 Gene II and Gene X Proteins by Gene V Protein." *Cell* 29:337–345.

79. Zinder, N. D., ed. 1975. *RNA Phages.* Cold Spring Harbor, N.Y.: Cold Spring Harbor Laboratory.

80. Fiers, W. 1979. "Structure and Function of RNA Bacteriophages." *Comp. Virology* 13:69–204.

81. Atkins, J. G., J. A. Steitz, C. W. Anderson, and P. Model. 1979. "Binding of Mammalian Ribosomes to MS2 Phage RNA Reveals an Overlapping Gene Encoding a Lysis Function." *Cell* 18:247–256.

82. Beremand, M. N., and T. Blumenthal. 1979. "Overlapping Genes in RNA Phage; a New Protein Implicated in Lysis." *Cell* 18:257–266.

83. Model, P., R. E. Webster, and N. D. Zinder. 1979. "Characterization of Op3, a Lysis-Deficient Mutant of Bacteriophage f2." *Cell* 18:235–246.

84. Blumenthal, T., and G. C. Carmichael. 1979. "RNA Replication: Function and Structure of Qβ-Replicase." *Ann. Rev. Biochem.* 48:525–548.

85. Winter, R. B., and L. Gold. 1983. "Overproduction of Bacteriophage Qβ Maturation (A₂) Protein Leads to Cell Lysis." *Cell* 33:877–885.

86. Kastelein, R. A., E. Remaut, W. Fiers, and J. van Duin. 1982. "Lysis Gene Expression of RNA Phage MS2 Depends on a Frameshift During Translation of the Overlapping Coat Protein Gene." *Nature* 295:35–41.

87. W. Fiers, R. Contreras, F. Duerinck, G. Haegemeean, D. Iserentant, J. Merregaert, W. Min Jou, F. Molemans, A. Raeymaekers, A. Van den Berghe, G. Volckaert, Y. M. Ysebaert. 1976. "Complete Nucleotide Sequence of Bacteriophage MS2 RNA: Primary and Secondary Structure of the Replicase Gene." *Nature* 260:500–507.

88. Gussin, G. N., M. R. Capecchi, J. M. Adams, J. E. Argetsinger, J. Tooze, K. Weber, and J. D. Watson. 1966. "Protein Synthesis Directed by RNA Phage Messengers." *Cold Spring Harbor Symp. Quant. Biol.* 31:257–272.

89. Campbell, K. M., G. D. Stormo, and L. Gold. 1983. "Protein-Mediated Translational Repression." In *Gene Function in Prokaryotes,* ed. J. Beckwith, J. Davies, and J. A. Gallant. Cold Spring Harbor, N.Y.: Cold Spring Harbor Laboratory, pp. 185–187.

90. Uhlenbeck, O. C., J. Carey, P. J. Romaniuk, P. T. Lowary, and D. Beckett. 1983. "Interaction of R17 Coat Protein with Its RNA

Binding Site for Translational Repression." *J. Biomol. Structure and Dynamics* 1:539–552.

91. Spiegelman, S., I. Haruna, I. B. Holland, G. Beaudreau, and D. Mills. 1965. "The Synthesis of a Self-Propagating and Infectious Nucleic Acid with a Purified Enzyme." *Proc. Nat. Acad. Sci.* 54:919–927.

92. Blumenthal, T., T. A. Landers, and K. Weber. 1972. "Bacteriophage Qβ Replicase Contains the Protein Biosynthesis Elongation Factors EFTu and EFTs." *Proc. Nat. Acad. Sci.* 69:1313–1317.

93. Kamen, R. I. 1975. "Structure and Function of the Qβ RNA Replicase." In *RNA Phages,* ed. N. D. Zinder. Cold Spring Harbor, N.Y.: Cold Spring Harbor Laboratory, pp. 203–234.

94. Kolakofsky, D., and C. Weissmann. 1971. "Possible Mechanism for Transition of Viral RNA from Polysome to Replication Complex." *Nature New Biol.* 231:42–46.

95. Vollenweider, H. G., T. H. Koller, H. Weber, and C. Weissmann. 1976. "Physical Mapping of Qβ RNA." *J. Mol. Biol.* 101:367–377.

96. Meyers, F., H. Weber, and C. Weissmann. 1981. "Interactions of Qβ Replicase with Qβ RNA." *J. Mol. Biol.* 153:631–660.

VII

FACING UP TO EUCARYOTIC CELLS

CHAPTER 18

Yeasts as the *E. coli* of Eucaryotic Cells

The very concentrated effort that has gone into the study of all aspects of *E. coli* is one of the major reasons why molecular biology has advanced so rapidly over the past 40 years. As a procaryotic organism, however, *E. coli* possesses a structural organization that is fundamentally different from the more complex eucaryotic organisms. In possessing chromosome-containing nuclei, eucaryotic cells are so arranged that the fundamental processes of transcription and translation occur in physically separate compartments. Thus, by the time the central dogma was an established fact, serious opinions were already voiced that the molecular devices used to control the patterns of gene expression would be radically different between procaryotes and eucaryotes. By now, these suspicions have matured into hard facts. Therefore, no matter how well we understand *E. coli*, it cannot serve as the intellectual scaffold from which to approach the organization and functioning of higher cells. Instead, we need to understand one or more of the simplest eucaryotic organisms at the same level of detail that we presently comprehend *E. coli*.

The organisms that stand out as the obvious subjects for study are the eucaryotic microorganisms that superficially grow and divide like bacteria. Of these, the best understood by far are several of the yeasts. Their well-dissected genetic systems provide a virtually autocatalytic stimulus for their further study by the community of molecular biologists. Most importantly, these yeasts have only 3½ times more DNA in their haploid genetic complements than *E. coli* cells, while *Drosophila* contains 25 times more DNA and higher vertebrate cells contain at least 1000 times more DNA. Still another factor favoring the emphasis on yeasts is economic. Work with higher vertebrate cells is far more expensive because of the high cost of the media required to grow them in culture. If we can solve the same problem equally well with yeasts or with human cells, common sense tells us to stick with the simpler and less expensive system.

For these reasons, an ever-increasing number of molecular biologists have taken up yeasts as their primary research system over the last decade. At this rate, the interest expressed in yeasts will quickly match the maximal interest ever generated by *E. coli*.

New Cells Are Produced Through Either Budding or Fission

The best-characterized yeast, genetically speaking, is bakers' yeast, *Saccharomyces cerevisiae*. This species has oblately spheroid or ovoid shaped cells some 3 μm in diameter, which under optimal nutritional

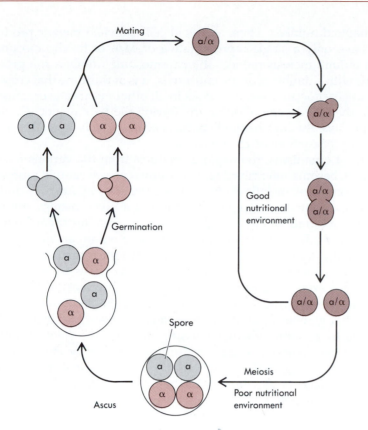

Mating

Germination

Good
nutritional
environment

Spore

Ascus

Meiosis

Poor nutritional
environment

Figure 18-1
Life cycle of *Saccharomyces cerevisiae*. The two mating types are **a** and α. In a rich nutritional environment, both haploid and diploid cells will reproduce by mitosis; in a poor environment, diploid **a**/α cells (formed by mating between *a* and α haploid cells) will undergo meiosis. The life cycle of *Schizosaccharomyces pombe* is very similar, the main difference being that these cells divide by fission rather than budding.

conditions double their mass every 90 minutes. Each cell does this by forming a single budlike appendage into which a full complement of daughter chromosomes moves during a mitotic division (Figures 18-1 and 18-2). The bud is then pinched off to yield a smaller daughter cell, which increases in size until it becomes as large as its parent cell. Then it, in turn, begins to grow a bud. Like all other yeasts, *S. cerevisiae* can exist in either the haploid or diploid state, with the haploid cells being either of two sexes (mating types), designated **a** and α. Fusion of **a** and α cells yields **a**/α diploid cells, which under satisfactory nutritional conditions will grow and divide, maintaining the diploid state. Under conditions of nutritional deficiency, however, the diploid cells undergo meiosis to yield four progeny haploid cells, which become encapsulated as spores within a thick, cell-walled, saclike body called an **ascus** (plural **asci**). Upon rupture of the ascus, the spores are released and subsequently germinate to commence new rounds of haploid existence (see Figure 18-1).

Also very well studied is the fission yeast *Schizosaccharomyces pombe*, whose two haploid sexes are called h^+ and h^-. This particular species does not form buds, but grows in size until an equatorial cell division process produces two equally sized daughter cells. Otherwise, its life cycle is similar to *S. cerevisiae*.

The Primary Products of Meiosis Are Constrained Within Tetrads[1-4]

Yeasts were initially only of applied interest to most biologists because of their use in the baking and brewing industries. However, when their genetic analysis began in the 1930s, yeasts became valued as tools for studying recombination, in particular, the details of crossing over. In meiosis, the diploid number of chromosomes is reduced

Figure 18-2
Budding yeast cell. (Courtesy of Barbara J. Stevens.)

to a haploid number. Thus, a diploid cell will yield meiotic products that have only a single representative of each of its chromosomes. Even though meiosis reduces the chromosome number, the process begins with a doubling of chromosomes. It is at this time that crossing over in yeast occurs, just as it does in all other eucaryotic organisms. No further DNA replication occurs during the two meiotic divisions that produce the four haploid spores, each of which therefore carries one copy of each yeast chromosome.

In most organisms, the haploid products from the various meiotic divisions become intermixed (e.g., a human sperm sample randomly arises from many meiotic events). However, in many fungi, including yeasts, the four primary spore products of a given meiosis are contained as a **tetrad** within a single ascus that separates them from the products of other meiotic events. Each spore within an ascus can be individually dissected away with the help of a micromanipulator, and its genetic constitution can then be analyzed. Yeasts thus provide a system for examining the two primary DNA products of individual crossing over events, something that cannot be accomplished with most other genetically studied organisms. For example, they made it possible to examine whether crossing over always yields daughter chromosomes whose genetic constitutions are reciprocally related. In the early 1940s, analysis of such yeast tetrads revealed that in a certain percentage of asci (approximately 4%), a given locus did not show the anticipated 2:2 segregation pattern but instead displayed occasional 3:1 patterns (see Figure 11-18). This phenomenon became known as **gene conversion** because one of the two original genes converted to its partner allele. An even rarer event—occurring in approximately 0.2% of asci—was also observed. In this case, spores— which are expected to contain only one form of the gene or the other— were not genetically pure; when they divided, they produced some progeny with one type of gene and some progeny with the other type of gene.

Today, these once baffling results are seen as valuable clues as to how crossing over occurs at the DNA level. First of all, studies of yeast tetrads show that although the number of chromosomes that are present at the time of crossing over is four, only two of these chromosomes engage in a single crossing over event. It is further important to remember that each of the participating chromosomes is a DNA duplex. Hence a single crossing over involves interactions among four strands of DNA. Those loci that segregate aberrantly are located at or very near to the physical sites of crossing over. Analysis of these aberrant types of tetrads has shown that a crossing over event often involves formation of joint DNA molecules containing DNA heteroduplexes (Chapter 11). Failure to repair the heteroduplex leads to formation of a spore that contains this heteroduplex and that will yield a mixed colony upon germination and subsequent mitotic divisions. In contrast, if the heteroduplex is removed by DNA repair a gene conversion event may occur.

Establishing Chromosomal Assignments and Locations Through Tetrad Analysis[5, 6, 7]

Tetrad analysis is now routinely employed to assign newly found mutant genes to linkage groups (chromosomes) and then to map each gene in relation both to the chromosomal centromere and to the other genes on the same chromosome. This is accomplished by analyzing the segregation patterns of different heterozygous markers among the four haploid spores. Initially, such mapping was much more time-consuming than mapping *E. coli* genes. The intensively studied yeast *S. cerevisiae* has 16 established chromosomes, while *S. pombe* has 3 chromosomes. Until they were all found and each characterized by suitable markers, the mapping of a new mutant gene could consume many months. Now, however, there are several classes of genetically identified "tester" strains that greatly speed up mapping. Especially helpful are tester strains containing extra copies of a given chromosome, which lead to aberrant segregation patterns when tested against strains carrying mutations on this chromosome. Also very useful are strains in which specific, marked chromosomes are unstable and whose loss after an appropriate genetic cross can immediately tell us that a given gene is on that chromosome.

Through the use of such "tricks," the number of mapped genes in *cerevisiae* is now 568, roughly half the number of mapped *E. coli* genes, and is increasing rapidly (Figure 18-3). However, since yeasts contain some four times more DNA than *E. coli*, the mapping of the yeast genome still remains in relative infancy.

Separation of Daughter Chromosomes on Microtubule-Containing Spindles[8, 9, 10]

Cell division in yeasts, as in all other eucaryotes, involves the participation of spindles upon which the chromosomes line up prior to the regular segregation processes that pull daughter sets of chromosomes in opposite directions. A major constituent of those spindles are **microtubules,** hollow 240 Å diameter filaments composed of the monomeric building blocks α and β tubulins (molecular weights about 55,000 daltons). Unlike that of most eucaryotic cells, the nuclear membrane of yeasts does not break down prior to formation of the spindle. Instead, the **spindle pole bodies (SPBs),** microtubule-organizing centers analogous to the centrosomes of animal cells, are integrated constituents of the nuclear membrane; microtubules emanate from the SPBs into the nucleoplasm. Each yeast cell contains one SPB that duplicates prior to the initiation of mitosis to generate a daughter SPB immediately adjacent to its parent on the nuclear membrane. As these become positioned to the opposite side of the nuclear membrane, the microtubules elongate, extending into the nucleus and eventually interdigitating to form a complete spindle (Figures 18-4 and 18-5).

Two types of spindle microtubules span the nucleus. One type runs continuously across the entire spindle, while the other, more common form consists of shorter microtubules attached to only one SPB. It is this second group of tubules that attaches the chromosomes to the spindle and whose subsequent shortening pulls the two sets of homologous chromosomes to opposite poles. At the same time, the longer pole-to-pole microtubules grow even further, thereby elongating the spindle greatly and positioning the daughter sets of chromosomes into positions that favor the subsequent cleavage of the nucleus into similar-size daughter nuclei.

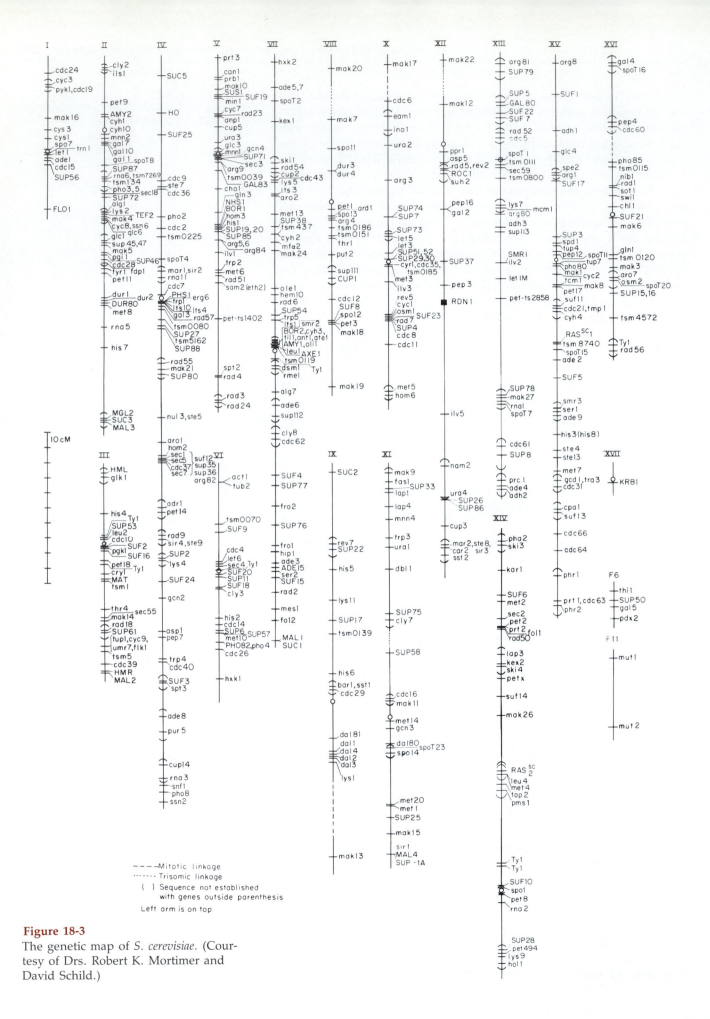

Figure 18-3

The genetic map of *S. cerevisiae.* (Courtesy of Drs. Robert K. Mortimer and David Schild.)

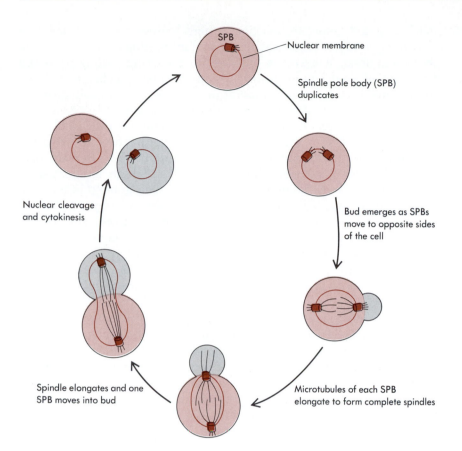

Figure 18-4
Mitotic spindle formation and movement in yeast.

Within the diagram:
- SPB
- Nuclear membrane
- Spindle pole body (SPB) duplicates
- Bud emerges as SPBs move to opposite sides of the cell
- Microtubules of each SPB elongate to form complete spindles
- Spindle elongates and one SPB moves into bud
- Nuclear cleavage and cytokinesis

While in higher eucaryotic cells the microtubule attachment sites at the centromeric regions on chromosomes are morphologically conspicuous and are called **kinetochores,** no recognizable equivalents occur on yeast chromosomes. This difference probably reflects the much smaller size of individual yeast chromosomes, each of whose attachment to the spindle at the centromere is mediated by just single microtubules rather than the microtubule bundles that emanate from the kinetochores of higher cells.

The cell division cycle terminates with the division of the cytoplasm into two daughter cells (cytokinesis) by the formation of an ever-constricting ring of circumferential filaments containing the proteins *actin* and *myosin,* the same proteins involved in muscle contraction. The relative motion of these two molecules along each other during cytokinesis may underlie the contractive and constrictive movements of the filaments.

Mutations That Help Define the Cell Cycle[11–14]

During a cell cycle (Figure 18-6), the continuous processes that duplicate most cellular molecules (e.g., glycolytic enzymes) are punctuated by stage-specific events or landmarks. Mutations causing stage-specific blockage of the cell cycle are an increasingly important tool in elucidating these landmarks (Table 18-1). The most useful cell division cycle (*cdc*) mutations are temperature-sensitive mutations in genes essential for completing the cell cycle. Many gene products of

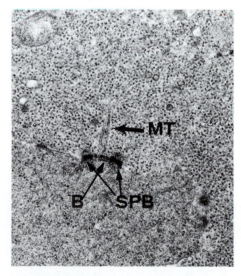

Figure 18-5
Formation of spindles in yeast. Here, the bridge (B) of the duplicated SPBs bears cytoplasmic microtubules (MT). (Courtesy of Breck Byers.)

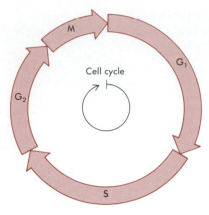

Figure 18-6

Phases in the life cycle of a yeast cell. M = mitosis, G_1 = pre-DNA synthesis, S = DNA synthesis, G_2 = period between DNA synthesis and mitosis.

such mutated genes are inactive at high temperatures but active at low temperatures. When such a *cdc* mutant is subjected to restrictive hot conditions, it becomes blocked from moving through the cell cycle at that stage where its normal equivalent product begins to function. Thus, if we raise the temperature of an asynchronously growing yeast population containing a heat-sensitive *cdc* gene, the population becomes converted into a synchronous population of cells all blocked at the same stage of the cell cycle (Figure 18-7). There are also the opposite type of temperature-sensitive *cdc* mutant genes that are inactive at low temperatures while being active at high temperatures.

By now, some 50 different *cdc* genes have been defined in terms of the landmark events beyond which they do not pass in the cell cycle. They all function during the growth of **a** and α haploid cells, as well as during **a**/α diploid cell growth, with many also playing vital roles during meiosis and spore formation. A particularly important problem is to determine the roles of these proteins. As of now, only a few are known. One such gene codes for the enzyme thymidylate synthetase, which provides thymine residues needed for DNA synthesis. Another gene codes for one of the two forms of β tubulin. Still another codes for DNA ligase, an enzyme that, among other functions, joins together the fragments forming the lagging strand during DNA synthesis. How all these *cdc* genes function harmoniously and in the right order is only beginning to be understood.

The beginning event in the cell cycle, called *start*, takes place in the GI stage and has a unique function. The decision or commitment to undergo another cell cycle is made at start. Once a cell has passed

Table 18-1   Various *cdc* Genes

Gene	Function and Remarks
cdc4	Required for initiation of DNA synthesis in the mitotic cell division cycle and for premeiotic DNA synthesis in meiosis; required for replication of nuclear chromosomes and the 2 micron plasmid but not mitochondrial DNA.
cdc7	Similar to *cdc4* but not required for premeiotic DNA synthesis.
cdc8	Structural gene for thymidylate kinase; required for DNA precursor synthesis and hence replication of all DNA.
cdc9	Structural gene for DNA ligase.
cdc16	Required for completion of chromosome distribution; possibly codes for a microtubule-associated protein.
cdc21	Structural gene for thymidylate synthetase; required for DNA precursor synthesis and hence replication of all DNA.
cdc25	Necessary for completion of synthesis of the spindle pole body and exit from G1; involved in cyclic AMP metabolism.
cdc28	Necessary for duplication of the spindle pole body and exit from G1; homologous to mammalian protein kinases and has protein kinase activity.
cdc31	Necessary for duplication of the spindle pole body; Ca^{++} binding protein.
tub2	Structural gene for β-tubulin; necessary for nuclear migration and chromosome distribution.

SOURCES: Sclafani, R. A. and W. L. Fangman. 1984. "Yeast Gene *cdc8* Encodes Thymidylate Kinase and Is Complemented by Herpes Thymidine Kinase Gene *TK*." *Proc. Nat. Acad. Sci.* 81:5821–5825. S. I. Reed, J. A. Hadwiger, and A. T. Lorincz. 1985. "Protein Kinase Activity Associated with the Product of the Yeast Cell Division Cycle Gene *cdc28*." *Proc. Nat. Acad. Sci.* 82:4055–4059.

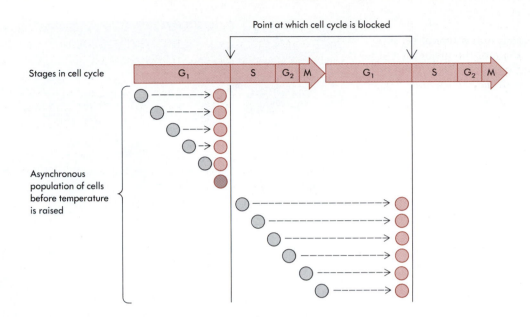

through start, it will commence DNA synthesis and pass through all the successive stages in the cell cycle until its daughters reach start. These cells will then initiate another cell cycle (passing through start), unless they become arrested in the cell division cycle by unfavorable nutritional conditions or by an appropriate sex hormone. The completion of start is marked by the duplication of the spindle pole body. Once that event occurs in *S. cerevisiae*, DNA synthesis commences, the bud begins to grow, and the normal sequential passage of cell cycle events is ensured.

Yeasts Display All the Characteristic Internal Compartments of Eucaryotic Cells[15–19]

Yeast cells are organized into the same major membrane-bounded compartments as all other nonphotosynthetic eucaryotic cells (Figure 18-8). In addition to the *nucleus*, yeast cells contain:

- The *endoplasmic reticulum (ER)* with its attached ribosomes that synthesize the integral membrane proteins, as well as proteins destined for secretion, largely into the outer cell wall.

- The *Golgi apparatus*, to which many proteins made by the endoplasmic reticulum are first transported and where the initial stages of protein modification (e.g., addition of sugar groups) occur; somehow in the Golgi sacs, newly synthesized proteins are tagged with markers that direct them to their correct cellular destinations.

- The energy-generating *mitochondria*.

- The *cytosol*, which in rapidly growing cells is tightly packed with essential enzymes as well as with very large numbers of free polyribosomes engaged in protein synthesis.

- The *vacuole* (analogous in several respects to the *lysosome* of mammalian cells), which contains a variety of degradative enzymes and is a storage area for certain cellular nutrients.

- The *peroxisomes*, which sequester enzymes (e.g., catalase) that destroy the free radicals derived from oxidative reactions.

Figure 18-7
Heat-sensitive *CDC* mutants become blocked at a specific stage in the cell cycle as the temperature is raised. Therefore, an asynchronous population of these cells becomes converted to a culture all blocked at the same stage. When the temperature is lowered, the cells in the culture will grow synchronously for a short time. (Redrawn from B. Alberts et al, *Molecular Biology of the Cell*, p. 641.)

Figure 18-8

Electron micrograph of a typical yeast cell, showing some of the major membrane-bounded organelles. (Courtesy of Drs. R. Schekman, P. Novick, and P. C. Esmon.)

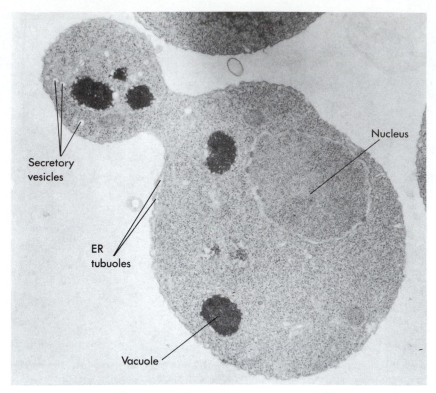

These last two forms of organelles allow cells to segregate potentially harmful hydrolytic enzymes or metabolic intermediates from most of the cellular cytoplasm. For example, cellular function would rapidly cease if the proteolytic enzymes found within vacuoles came into contact with most cellular proteins. Likewise, free peroxide (H_2O_2) would oxidatively destroy many cellular components if it had access to them. Because the generation of H_2O_2 is limited to the interior of peroxisomes, where catalase almost instantly breaks it down, the cellular level of H_2O_2 never reaches dangerous levels.

In yeast, most of the proteins made in the endoplasmic reticulum are rapidly passed through the Golgi apparatus to other intracellular membranes, to the periplasmic space, and to the outermost layer of the cell wall. Consequently, the endoplasmic reticulum and Golgi compartments are not swollen with protein products and thus are not morphologically conspicuous, as they are in liver cells, for example. In fact, only after yeast cells containing mutations in genes involved in protein transport were examined, did the Golgi compartments of yeast become unambiguously identified (Figure 18-9). It is hard to believe, however, that the Golgi would have remained unidentified for long. They allow the highly complex glycoproteins of the surface layers to be assembled in several discrete steps at internal locations, transferred to secretory granules, and finally delivered to their cellular destinations.

Like all other eucaryotic cells, yeasts contain the key protein *clathrin*, a major outer component of the *coated vesicles* that transport proteins from one membrane site to another (e.g., from the plasma membrane to lysosomes). Through possession of clathrin, yeasts have the capability to selectively endocytose those proteins of their external membranes to which extracellular molecules have attached. Which extracellular molecules are taken up in such a manner remains to be discovered. However, now that we know that clathrin is present in yeasts, it is difficult to imagine that it does not sometimes function as

Figure 18-9
Electron micrograph showing an accumulation of Golgi in a yeast cell with a mutation affecting protein transport. (Courtesy of Drs. R. Schekman, P. Novick, and P. C. Edmon.)

it does in higher eucaryotic cells (e.g., in the uptake of growth or antigrowth factors).

Signals on Proteins Determine Proper Cellular Location

Yeast proteins contain signals that are responsible for their proper localization within the cell. These signals determine whether the protein should go to the vacuole or the mitochondria, be secreted from the cell, or return to the nucleus. One way in which these localization determinants have been identified is by producing hybrid proteins, through *in vitro* methods, which consist of joining one part of the protein that is localized to one cellular compartment to a test protein that is ordinarily found in a different cellular compartment. Test proteins are, for example, *E. coli* β-galactosidase or dihydrofolate reductase. A segment is considered to have localization information if it determines who wins this molecular tug-of-war. Proteins destined to the vacuole or to be secreted from the cell both proceed first to the endoplasmic reticulum, under direction of a hydrophobic signal at their amino terminus. These proteins then must contain further routing information, since their intracellular paths diverge after the Golgi. Mitochondrial proteins do not pass through the ER, but rather go directly to the mitochondria. In all of these cases, the signal is likely to be a binding site for guiding the protein species to its proper place in the cell. Finally, proteins that concentrate in the nucleus (regulatory proteins such as α2 and *GAL4*, and ribosomal proteins) all contain short amino-terminal segments that can drag β-galactosidase into the nucleus. In this case, it is not known whether the localization determinant on these nuclear proteins is responsible for entry of the protein into the nucleus, or for retaining the protein after it has diffused into the nucleus. In the one case in which this distinction has been made (nucleoplasmin in *Xenopus*), the signal is responsible for uptake rather than retention. Whatever the molecular mechanism, it is remarkable that a protein the size of β-galactosidase is able to enter the nucleus, presumably through the nuclear pores.

Cytoskeletal Elements Within Yeast Cells[20, 21]

Because yeast cells are small and appear to be morphologically simple, there initially was no reason to presume that the cytosol of yeast would contain the multiplicity of filamentous elements contained in higher eucaryotic cells. Recall that the **cytoskeleton** of higher eucaryotic cells is composed of actin-containing microfilaments, intermediate (100 Å) filaments, and tubulin-containing microtubules. However, the finding that some microtubules are present in the yeast cytosol, as well as within the spindles of nuclei, prompted the question of whether they singly or in conjunction with other filamentous elements help direct the intracellular flow of molecules and membrane-bounded organelles. The question has become even more significant with the recent finding that actin- and myosin-containing filaments, as well as intermediate (100 Å) filaments, are components of the circumferential rings whose formation leads to the final separation of the daughter cytoplasm to conclude a cell division cycle. Such filaments may also act earlier in the cell cycle, when they are directed into growing *S. cerevisiae* buds. There they may serve to direct the secretory granules bringing newly made molecules for insertion into the expanding outer components of the buds.

The Thick and Tough Cell Walls of All Yeasts[22]

Surrounding the plasma membrane of yeast cells is the yeast cell wall, a thick, complex mixture of secreted proteins and elongated, intertwined polysaccharides. On its outer surface is a tight aggregation of highly immunogenic *mannoproteins*, which consist of large numbers of mannose groups covalently bound to N-acetylglucosamine groups that in turn are attached to specific amino acid side groups of a number of cell wall–specific polypeptide chains. Between this mannoprotein layer and the plasma membrane is the more open periplasmic region. In this region, a number of degradative enzymes (e.g., the sucrose-hydrolyzing agent invertase and the phosphate-releasing agent alkaline phosphatase) lie among the long, thin polysaccharides glucan (a polymer of D-glycopyranose groups) and chitin (a polymer of N-acetylglucosylamine), which in some way are anchored to both the plasma membrane and the mannoprotein layer (Figure 18-10).

For years, their thick and largely impenetrable outer surfaces prevented much research on yeasts. Cytological work, for example, was difficult because embedding substances could not penetrate. Nor was there any way to introduce externally added DNA into intact cells or to isolate from ruptured cells even semi-intact organelles. Many of these problems can now be circumvented by simultaneously employing several cell wall–degrading enzymes that sever the restraining glucan chains many times and make many holes in the external man-

Figure 18-10
The yeast cell surface.

noprotein layer to produce much less rigid **spheroplasts** (proto-plasts).

Partition of the Yeast Genome into Linear Chromosomes[23–28]

S. cerevisiae nuclear DNA (14×10^3 kilobase pairs) is distributed among 16 discrete chromosomes, each of which contains a single DNA molecule (Table 18-2). Until recently, even the smallest of these DNA molecules was too large to be effectively isolated in pure form. Now, however, the development of pulse field electrophoretic

Table 18-2 Genetic and Physical Sizes of Yeast Chromosomes

Band	Chromosome	Recombinant length (cM)	DNA (kb)*
1	1	98	198
2	6	137	225
3	3	137	311
4	9	198	495
5a	8	183	635
5b	5	233	709
6	11	242	763
7	10	200	835
8	14	283	864
9	2	244	948
10	13	229	991
10b	16	176	1047
11a	15	361	1146
11b	7	391	1285
12	4	485	1571
13	12	294	2194

*DNA content/chromosome, assuming 14,000 kb per haploid genome (96).
SOURCE: Data from R. K. Mortimer and D. Schild, *Microbiol. Revs.* (1985) 49:206.

Figure 18-11
Identification of the bands corresponding to chromosomes II, XIII, XV, and
XVI by DNA·DNA hybridization (strain
AB972). DNA from the gel whose
ethidium bromide staining pattern is
shown on the left was transferred to a
single sheet of nitrocellulose. This sheet
was then cut into four strips that were
separately hybridized to four chromo-
some-specific probes, and the filter
strips were then positioned in their
original alignment before the
autoradiogram on the right was exposed. The ethidium bromide staining
pattern and the autoradiogram are
printed to the same scale and are
aligned appropriately with one another.
[Courtesy of G. F. Carle and M. V.
Olson. Printed from *Proc. Nat. Acad.
Sci.* (1985) 82:3758, with permission.]

methods allows the almost routine separation on agarose gels of 15 of
their chromosomes. Only chromosome 12, over half of which contains an average of 125 tandemly arranged ribosomal genes (rDNA),
does not cleanly separate. This is most likely due to the wide variation
in the average number of the 8.6 kbp rDNA genes found per chromosome. Variation of this magnitude is to be expected because of unequal crossing over events (Chapter 7) occurring within the rDNA
region. Given the existence of a DNA probe for a given yeast gene, it
is now a simple matter to map this gene to a precise chromosome
using DNA-DNA hybridization techniques (Chapter 19) (Figure
18-11). Interestingly, there is significant variation in the lengths of
individual chromosomes from one strain of *S. cerevisiae* to another
(chromosomal length polymorphisms). The molecular basis for these
polymorphisms remains to be established. One possibility is variation
in the length of the telomeric sequences found at the ends of all chromosomes (see below).

Only three chromosomes carry the genetic information of *S. pombe;*
all its chromosomes are also linear. Many yeast cells also possess
multiple copies of autonomously replicating small, circular plasmid
DNAs, 2μ circles (see Chapter 19), that normally do not carry any
essential genes. Because they lack centromeres, they are not regularly
partitioned on the mitotic spindle during cell division. Thus, still to be
explained is why these yeast plasmids tend to be maintained rather
than lost in a random fashion from a small fraction of all daughter
cells. The function of these plasmids is not known.

Yeast DNA, like that of all other eucaryotic cells, is complexed with
the four main histones (H2A, H2B, H3, and H4) used to construct the
146-base-pair nucleosomal cores of chromatin (a nucleosome is a globular subunit of chromosomal DNA complexed with histones; see
Chapter 21). Apparently absent in *S. cerevisiae* is the H1 histone, which
in higher eucaryotes plays a key role in the compaction of chromatin
to the highly supercoiled state that exists during mitosis (Chapter 21).
The lack of any yeast equivalent to H1 may explain why, with the
exception of *S. pombe,* individual yeast chromosomes cannot be visualized during mitosis. Yeast chromatin, however, does contain major
amounts of an acidic, nonhistonelike group of proteins whose amino
acid composition and charge closely resemble those of similar proteins found in the chromosomes of higher eucaryotes. What their
function is remains a mystery despite the several decades in which
their existence has been known.

Also still to be established is why, despite the linear nature of their underlying DNA molecules, sections of yeast chromatin appear to be rotationally restrained into superhelical regions. Conceivably, structures similar to those that maintain the circular *E. coli* DNA in some 40 to 50 independently negatively supercoiled domains are also essential ingredients in the functioning of yeast chromatin.

Absence of Conventional Nucleosomes at the Centromeric Regions[29]

The centromeres of mitotic chromosomes display such morphological discontinuities with the chromosome arms that it was originally suspected that they represented regions separating two (or more) discrete DNA molecules. However, the discovery that a single DNA molecule runs through an entire yeast chromosome made it clear that chromosomal DNA was continuous through the centromere and that there were specialized DNA sequences to promote interaction with the mitotic spindle. The first of such *CEN* (centromeric) **sequences** was isolated from a shortened aberrant circular form of chromosome III of *S. cerevisiae*, formed by accidental crossing over between two of the mating-type genes that reside on this chromosome. Because of its smaller circular form, this DNA can easily be separated from other yeast DNA and used as an enriched source of restriction fragments for the "chromosome walks" (discussed later in the next chapter) used to find the *CEN3* sequence. When this DNA is inserted into yeast plasmid DNAs, it promotes their regular segregation during mitosis and meiosis in the manner expected for centromere-linked genes. Subsequently, *CEN* sequences have been isolated from several other yeast chromosomes, with each example showing sequences homologous to *CEN3*. All the essential *CEN* sequences are contained within a 130-base-pairs long, very AT-rich region encompassing two shorter blocks of high homology (Figure 18-12).

The *CEN* sequences reside within a region 220 to 250 base pairs long that is more resistant to nuclease digestion than the vast bulk of yeast chromatin. On both sides of this region, the nucleosomes are singly placed along the DNA. The resistance to nuclease digestion suggests that at the *CEN* sequences there is either a modified nucleosome or a complex of proteins, other than the normal histone components of nucleosomes, replacing the *CEN* sequences. In any case, the protein-bound region of the centromere (the kinetochore) mediates attachment to the microtubules of the mitotic spindle (Figure 18-13). Several yeast proteins that bind specifically to *CEN* sequences have already been isolated, and it should soon become clear whether their presence prevents the simultaneous association with any of the four main histone classes. Eventually, we will succeed with the in vitro reconstitution of the centromere-microtubule complexes and gain a more detailed understanding of how chromosomes attach to the mitotic spindle.

	Element I		Element II		Element III
CEN3	ATAAGTCACATGAT	←	88 bp (93% AT)	→	TGATTTCCGAA
CEN11	ATAAGTCACATGAT	←	89 bp (94% AT)	→	TGATTTCCGAA
CEN4	'AAAGGTCACATGCT	←	82 bp (93% AT)	→	TGATTACCGAA

Figure 18-12
Sequence elements common to three yeast centromeres. Elements I and III are regions of very high homology flanking a longer central element of high AT content.

Core histones
of nucleosome

160 bp DNA
11 nm diameter

Kinetochore protein (s)

Chromosome

220 bp centromeric DNA
≅15–20 nm diameter

24 nm
diameter

Figure 18-13
Diagram of *S. cerevisiae* chromosome, showing the nuclease-resistant centromere core (shaded box) and the nucleosomes that are placed along the DNA on either side of the centromere core. At the *CEN* region, the nucleosomes are replaced with other proteins that attach to microtubules of the mitotic spindle. This proteinaceous region of the centromere is called the *kinetochore*. [After J. S. Bloom and J. Carbon, *Cell* 29 (1982):305.]

Common Telomeric Sequences Are at the Ends of All Yeast Chromosomes[30–34]

The ends of all yeast chromosomes, like those of all other linear eucaryotic chromosomes, have unique structures that are called **telomeres.** Each yeast telomere ends with approximately 100 base pairs of irregularly repeated sequences, of the form

$$5'\text{-}C_{1-3}A \ldots$$
$$3'\text{-}G_{1-3}T, \ldots$$

attached to specific X and Y sequences.

Within these specific terminal X and Y sequences are additional internal sections of the repetitive elements (Figure 18-14). All known telomeres contain similar tandemly repeated sequences. As we pointed out when we discussed DNA replication (Chapter 10), these repetitive elements have evolved as a device to preserve the integrity of the ends of DNA molecules, which often cannot be finished by the conventional mechanisms of DNA replication. The ends of yeast chromosomes are maintained by enzymatically adding $3'\text{-}G_{1-3}T\text{-}5'$ groups to the 3' ends of DNA chains as well as by one or more primaselike enzymes that make the complementary $5'\text{-}C_{1-3}A\text{-}3'$ polyribonucleotide primers.

What determines the length of such terminal additions is not known. But the process must be genetically controlled, since the average number of $C_{1-3}A$ repeats is fixed for a given yeast strain. This number varies between different yeast strains, and more than one yeast gene is involved in determining the number.

Broken chromosome ends that have completely lost their telomeres can often be repaired. This long-puzzling observation can now be explained by the internal copies of the repetitive sequences. When a yeast chromosome becomes broken in or near a poly $C_{1-3}A$ tract, it can acquire new telomeric sequences through a repair process (Chapter 11) in which a broken end invades an intact double helix at a

Figure 18-14
Structure of yeast telomeres and adjacent sequences. Not all yeast telomeres contain a Y' element.

region of sequence homology. The genetic information of the intact strand is then used to replace that missing from the broken end (Figure 18-15). Such events lead to constantly changing telomeric structures, giving the sequences at the end the potential for continued dispersal and amplification.

Artificial Yeast Chromosomes Can Be Made by Ligation of Telomeres, Origins of Replication, and Centromeres[35-38]

The isolation and characterization of the three elements essential for eucaryotic chromosome replication—origins of DNA replication, centromeres, and telomeres (see Chapter 19 for more details)—opened up the possibility of linking together single copies of each of these units with appropriate specific yeast DNA segments to form completely new, artificial chromosomes. Such chromosomes can be tested for their functioning by inserting them into appropriate yeast strains.

Very short, linear, artificial chromosomes segregate incorrectly much more frequently than larger (300 kilobase-pair) artificial chromosomes, which segregate with the same degree of accuracy as their similarly sized normal counterparts. Increasing their lengths may lead to more accurate chromosome partitioning by increasing the time the newly duplicated chromosomes remain attached to each other. Conceivably, if daughter chromosomes separate from each other too soon, they cannot line up correctly on the spindle. Favoring this possibility is the fact that very small circular chromosomes (e.g., plasmids that replicate in yeasts) segregate more accurately than their linear counterparts. Replication of circular DNA molecules initially leads to intertwined daughter double helices whose eventual separation depends on topoisomerase II action, allowing them to segregate accurately (only 1 in 100 mistakes).

DNA Synthesis Does Not Occur Throughout the Cell Cycle[39]

In contrast to bacterial DNA synthesis, DNA synthesis in yeasts does not normally occur over the entire mitotic cycle. Upon separation of a daughter cell (bud) from its parent, an interval (designated the G1 phase) of some 10 to 30 minutes passes before DNA synthesis commences. The **S phase** (the period of DNA synthesis) in turn occupies some 20 to 30 minutes. The final 30 minutes occupied by the G2 and

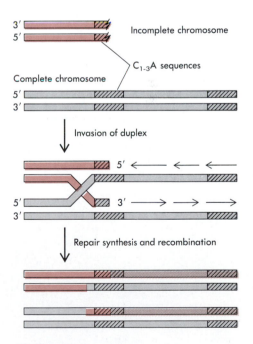

Figure 18-15
Chromosomes broken in or near internal $C_{1-3}A$ repetitive sequences (///) can complete their ends by a repair process in which they recombine with an intact chromosome at homologous sequences.

M phases is the time needed for growth of the spindle, chromosome segregation, and the terminal cytokinesis events that split off the daughter cell (see Figure 18-6).

DNA Replication Commences at Many Sites Along Each Yeast Chromosome[40]

Yeast DNA molecules, like DNA molecules of all eucaryotic organisms, contain multiple origins of replication. Electron microscopic examination reveals that origins are generally spaced about 35 μm apart, suggesting the existence of some 400 origins in the entire yeast genome. Replication is normally bidirectional, proceeding 2 kilobase pairs per minute (at 30°C) or faster at each replicating fork. It has been suggested that synthesis starts at approximately the same time at all origins. Assuming that this were true, then a fork movement of 2 kilobase pairs per minute might mean that the interval for DNA replication would occupy no more than 10 minutes—an interval considerably less than 25 to 30 minutes usually taken for DNA replication within cells doubling their masses every 90 to 100 minutes. It is possible that some replication origins are much more widely spaced from each other, with a correspondingly longer time needed to complete replication between them; or it is possible that the times at which DNA replication starts are staggered.

In *S. cerevisiae*, DNA replication often begins just as budding becomes observable; but there is no strict correlation between these two events, as proved by cell cycle mutants that dissociate them. The enzymes involved in yeast DNA synthesis are still largely unknown. It is hoped that this situation will soon be corrected, as yeast should provide ideal mutants for mastering eucaryotic DNA synthesis.

Yeast Genomes Are Transcribed by Three Different Forms of RNA Polymerase[41]

Yeast DNA, like the DNA of all eucaryotic cells, is transcribed by three different and easily separated forms of RNA polymerase, each of which transcribes different sets of genes. The function of RNA polymerase I is to transcribe the ribosomal genes (rDNA) into a precursor rRNA that eventually yields the two large rRNAs (25S and 17S) and one of the two small RNAs (5.8S) in ribosomes. RNA polymerase II makes all the RNA that is to serve as mRNA and some smaller-sized RNAs whose functions are just recently being elucidated (Chapter 20). RNA polymerase III transcribes the genes coding for tRNA, 5S ribosomal RNA, and a number of other small RNAs. The three polymerases can conveniently be distinguished from one another by their different sensitivities to the fungal toxin α-amanitin. Each of the yeast RNA polymerases, like the more multipurpose bacterial RNA polymerase (which transcribes all the genes of a given bacterial genome), is large (MW 4 to 6 $\times$ 10^5) and is composed of a number of different polypeptide chains. The exact roles of most of these polypeptides are still unclear, but some subunits may be shared between the three polymerases.

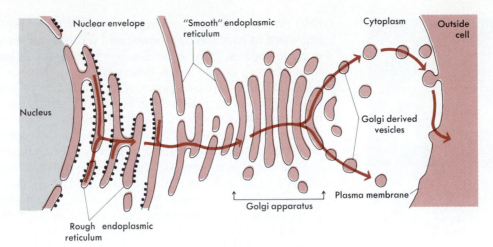

Newly Made RNA Molecules Move Through the Pores of the Nuclear Membrane[42]

The chromosomes of yeast function within a nuclear envelope identical in appearance to that of higher eucaryotic cells. The nuclear envelope can be regarded as a specialized outpocketing of the endoplasmic reticulum, since its external surface is covered with ribosomes that secrete proteins into the cavity separating the two unit membranes (Figure 18-16). Moreover, electron micrographs often show this cavity to be connected to the main cavities of the endoplasmic reticulum. Distributed over the surface of the nuclear envelope of most eucaryotic cells are very large numbers of uniformly sized (100 nm) nuclear pores. These pores have eight-sided polygonal shapes and usually appear to be filled with plugs made up of an amorphous material (Figure 18-17).

Through these pores must pass all the RNA that moves from the nucleus to the cytosol to function during protein synthesis. The forces that direct this unidirectional RNA transport have until recently been completely mysterious. An important new clue has arisen from experiments showing that tRNA molecules are exported not by random diffusion but by some carrier-mediated process. It is conceivable that ribosome-size granules that line the nuclear pores are involved in the transport of both tRNAs and mRNAs across the nuclear membrane (Figure 18-17c).

Molecular Specialization for Eucaryotic Protein Synthesis[43-49]

The protein-synthesizing apparatus in all eucaryotic cells, including yeast cells, shows striking differences when compared with the protein-synthesizing apparatus of bacterial cells. In the first place, the ribosomes of all eucaryotic cells are larger, sedimenting roughly at 80S in comparison to the 70S value found for procaryotes. Correspondingly, eucaryotic ribosomes have bigger subunits that sediment at 60S and 40S. The 60S subunit contains a 28S (25S in yeast) rRNA chain, while the 40S particle has an 18S (17S in yeast) rRNA chain. Two small RNAs are present in the large ribosomal subunit: one 5S rRNA, analogous to 5S RNA in *E. coli* ribosomes, and an additional 5.8S molecule. Eucaryotic ribosomes are also somewhat more protein-

Figure 18-16
Schematic representation of the various membrane-bounded cavities in the eucaryotic cell and their possible role in synthesis and excretion. The rough endoplasmic reticulum bears ribosomes on its outer surfaces. The Golgi apparatus, which secretes vesicles from one face, is formed on the other face by elements of the smooth endoplasmic reticulum. The solid-colored arrow indicates both the general membrane flow within the cell and also the probable route of proteins that are to be secreted. Areas cut off from the cytoplasm by membranes are shaded in color.

Figure 18-17
Nuclear pores. (a) Electron micrograph of part of a nuclear envelope with a number of pores. (b) Cross section through two pores (n = nucleus, e = envelope, c = cytoplasm). (c) Diagram of a pore, showing the eight pairs of granules that line it and a central granule. Although yeast does have nuclear pores, detailed electron microscopic studies have not yet been carried out. (Courtesy of K. Roberts, John Innis Institute, Norwich, England.)

rich; proteins account for about one-half their mass (as compared to one-third in bacterial ribosomes) and represent more individual polypeptide species. Why these differences exist is not known. The thing to remember is that on a structural level, there is a larger gap in ribosome size between procaryotic and eucaryotic microorganisms (e.g., between bacteria and the yeasts, *Neurospora,* or *Aspergillus*) than between the most evolutionarily divergent eucaryotic cells.

Each mature yeast rRNA chain, with the exception of that of 5S rRNA, is initially synthesized as part of a much longer rRNA precursor (37S) that is processed to give rise to the two large (25S and 17S) rRNA species and one of the small (5.8S) rRNA species. In yeast, but not in higher eucaryotes, the 5S gene is located close to the rDNA units and is part of the same repeating unit, even though it is transcribed by a different RNA polymerase and in the opposite direction. Because the rDNA units are tandemly located next to each other, intramolecular crossing over between them occasionally generates yeast cells containing small circles of rDNA, which are rapidly lost since they do not possess centromeres. Tandem repetition, however, is important for maintaining the sequence homogeneity of the individual repeats within the larger unit by mechanisms that will be discussed in Chapter 20.

Within all eucaryotic cells, the actively transcribed rDNA units are somehow compacted into dense-appearing nucleolar bodies. In yeasts, these nucleoli are crescent shaped and closely associated with the nuclear membrane.

Caps at the 5′ Ends of Eucaryotic mRNAs[50–53]

Eucaryotic mRNAs, including those of yeast, have several important features that differentiate them from procaryotic mRNAs. First, they are modified at their 5′ ends by the addition of a guanine nucleotide. This alone would not be considered unusual if the linkage were a conventional 3′–5′ phosphodiester bond. Instead, GTP reacts with the triphosphate at the 5′ end of an mRNA chain in a 5′–5′ condensation to form the structure 3′-G-5′ppp5′-N-3′p . . . , generally known as a **cap.** Thus, the 5′ and 3′ ends of most eucaryotic mRNAs are terminated by ribose moieties with free 2′- and 3′-OH groups. Subsequent to cap formation, a methyl group is added to the backward guanine residue (at the 7 position of the purine ring) and often also to the 2′-OH groups of the first and/or second adjacent nucleotides (Figure 18-18).

Why do the 5′ ends of eucaryotic mRNAs need to be so blocked? One possible reason is that these caps (and apparently specific proteins that bind to them) help ribosomes attach to mRNA chains so that they start translation at the correct AUG codon. Favoring this hypothesis is the absence of sequences complementary to 18S rRNA preceding coding regions in eucaryotic mRNA molecules. Specific ribosome binding sequences analogous to those of procaryotic mRNAs thus do not exist on eucaryotic mRNA molecules. Instead, eucaryotic ribosomes search out AUG initiator codons by binding to the caps and then migrating to the closest (5′-most proximal) AUG codon to start translation. Apparent exceptions to this rule are several viral mRNAs (e.g., polio RNA) that function perfectly normally in eucaryotic cells but lack the 5′ cap structure. Their ends are blocked instead by specific proteins that perhaps substitute as positioning agents.

Figure 18-18
A generalized structure for eucaryotic mRNA, showing the posttranscriptional modifications at the 5' and 3' ends.

Yeast mRNA Molecules Code for Single (Never Multiple) Polypeptide Chains[54, 55]

A given yeast mRNA molecule, like most other eucaryotic mRNAs, carries the genetic message for only a single polypeptide chain. Unlike many bacterial mRNAs, where translation can begin and end at several sites to give rise to several independent polypeptide chains, eucaryotic mRNAs are constructed so that translation usually commences at the first AUG codon following the capped 5' end. (Exceptions will be discussed in Chapter 24.) Thus, eucaryotic mRNAs are generally **monocistronic,** as opposed to the frequently **polycistronic** mRNAs of bacteria. Like bacterial mRNAs, however, eucaryotic mRNA molecules usually have extensive untranslated sequences before and after their protein-coding regions (leader and trailer sequences).

The inability of eucaryotic mRNAs to encode multiple proteins and thereby ensure their coordinate regulation probably explains why many eucaryotic polypeptides consist of several independent domains having related enzymatic functions. The yeast *HIS4* mRNA, for example, codes for a single protein that carries out three different enzymatic steps in histidine biosynthesis. Yet, it is rare that a single polypeptide carries out all the steps involved in a given metabolic pathway. Thus, regulatory molecules that coordinate the transcription of several functionally related mRNAs must act at multiple chromosomal sites. It is important to note that this latter feature is not unique to eucaryotes. For example, the enzymes that carry out the biosynthesis of arginine in *E. coli* are encoded by several different mRNA molecules whose expression is coordinately regulated, rather than being regulated by a single polycistronic mRNA.

Poly A at the 3' Ends of Eucaryotic mRNAs[56–60]

Still very mysterious is the observation that most yeast mRNA molecules, like those of all other eucaryotic cells, contain relatively long stretches (about 200 residues) of poly A at their 3' ends. These poly A

DNA

mRNA

Transcription

Endonuclease recognizes
AAUAAA and cuts the
mRNA 11 to 30 nucleotides
away

Cut

Poly-A tail added by
poly-A polymerase

Poly-A tail
(about 200 residues)

Figure 18-19
An AATAAA sequence appears toward
the 3′ end of many eucaryotic genes,
which encodes an AAUAAA sequence
that instructs an endonuclease to cleave
the strand at a site 11 to 30 bases
downstream; a poly A tail is then
added.

Table 18-3 Wobble Rules for Yeast
tRNAs

First Position of Anticodon		
Third Position of Codon	*E. coli*	Yeast
U	A, G, or I	G or I
C	G or I	G or I
A	U or I	U*
G	C or U	C

*U indicates a modified uridine residue.
SOURCE: Data from C. Guthrie and J. Abelson
in J. N. Strathern, E. W. Jones, and J. R.
Broach, eds., *The Molecular Biology of the Yeast
Saccharomyces, Metabolism and Gene Expression*
(Cold Spring Harbor, N.Y.: Cold Spring Har-
bor Laboratory, 1982), p. 492.

tracts (also called *tails*) are not specified by the genomic DNA, but are added after transcription to the newly made mRNA molecules released from the DNA into the nucleus. The enzyme responsible, poly A polymerase, specifically recognizes mRNA, as evidenced by the fact that poly A is not found at the 3′ ends of rRNA or tRNA. However, the poly A does not add onto the 3′ ends of completed mRNA chains.

Instead the poly A is added upstream from actual transcription termination sites onto 3′ ends created by a very specific endonuclease that, after recognizing certain specific AAUAAA sequences, cleaves the bound mRNA chains some 11 to 30 bases downstream (3′) to the AAUAAA recognition sequence (Figure 18-19). Still to be worked out are the signals (DNA sequences) used to specify actual termination of yeast mRNA synthesis.

For reasons that are still unknown, higher-cell histone mRNAs (not, however, those in *S. cerevisiae*) lack poly A, as do a few other mRNAs. Thus, whatever the function of poly A, it is not essential for mRNA synthesis or export from the nucleus. Poly A remains attached to the 3′ end after an mRNA binds to ribosomes and initiates translation. There is evidence, that as polypeptides are repeatedly made off the same mRNA molecule, the length of its poly A tract may shorten.

S. cerevisiae mRNA Molecules Are Translated by Only Forty-Six Different tRNA Molecules[61, 62]

Given the existence of 61 different sense codons in the genetic code, it was initially thought that an equivalent number of tRNA molecules would be required for their reading. But with the realization that *wobble* (a degree of allowed nonspecificity) occurred in the pairing between codons and anticodons at the third position, the possibility arose that many fewer than 61 distinct tRNAs might suffice (Chapter 15). The yeast *S. cerevisiae* now provides the cleanest system for understanding wobble, because (1) a majority of its tRNAs have been isolated, sequenced, and their anticodons determined; (2) the cloning of many of its tRNA genes has already occurred; and (3) the cloning of many protein-coding genes has revealed the precise codons used to signify their various amino acids.

There are some 20 species of *S. cerevisiae* tRNAs present in relatively large amounts (one for every amino acid) and another 26 species present in lesser amounts. Collectively, they are coded by approximately 360 separate tRNA genes. So on the average, each tRNA is coded by eight separate genes of virtually identical sequence. However, each amino acid is not represented by equal numbers of tRNA genes. Those amino acids that are relatively more abundant in proteins tend to be represented by more tRNA genes. There is almost no tandem clustering of the tRNA genes; members of a given set are usually located on different chromosomes. Thus, the number of copies of a given tRNA gene is effectively fixed, unlike the rRNA genes, which constantly fluctuate in number because of unequal crossing over within the rDNA tandem array.

Examination of established tRNA anticodons shows that wobble in yeast is more restricted than in *E. coli* (Table 18-3), and so the minimum number of tRNA species needed to translate the yeast mRNAs must be correspondingly larger. Single tRNAs recognize the pairs of related codons ending in either U or C, while two different tRNAs are needed to recognize the related codons ending in A or G. Thus, three different tRNAs are required for those amino acids like proline or threonine that are specified by blocks of four codons. Since two different tRNAs are needed for the initiating methionines and the internally located methionines, 46 different yeast tRNAs (Table 18-3) can be predicted to exist, given the caveat that still to be discovered modified bases within certain tRNA anticodons might upset the simple rules outlined here.

As the exact codons used to specify yeast proteins began to be known, it became clear that they do not represent a random collection. Instead, only 26 different codons typically specify the amino acids of yeast proteins, these being the codons recognized by the 20 more abundant tRNAs (see Table 18-4). The choice of these common codons follows three general rules:

1. When an amino acid is specified by a block of three or four related codons, C and U appear in the wobble position with similar frequencies, with A or G almost never used (with the exception of proline).
2. In cases of pyrimidine degeneracy (UUU and UUC specifying the same amino acid), either U or C can occupy the wobble site, with C being the preferred base when the other codon members are U and/or A.
3. In twofold purine degeneracy, either A or G in the wobble position is used for any given amino acid.

In each situation, the preferred codon-anticodon pairings are those which use conventional base pairing and which maximize the hydrogen-bonding strengths (e.g., UAC is preferred over UAA to specify tyrosine, since it avoids the use of the weaker AU hydrogen bond).

When less favored codons are used in yeast mRNAs, they must be recognized by one of the less abundant tRNA species. The more than occasional use of these less favored codons within a given coding sequence might thus be expected to slow down the rate of translation and thus be a characteristic of a protein made in small amounts. Conceivably, the most abundant proteins make almost exclusive use of the favored codons. Also important is the apparent inability of a cell to mutate away one of its minor tRNA species and still remain viable. Loss of the tRNA species would lead to untranslatability of the corresponding codon, and so many vital cell proteins would not be made.

Control Elements at the Ends of Yeast Genes[63, 64]

Because of the absence of operons in yeasts, there is a one-to-one correspondence between the transcriptional units along chromosomes and the genes that yeast geneticists have defined in terms of cellular function. A yeast gene can thus be defined as a DNA segment that is transcribed into a single mRNA product together with the adjacent *control elements* at its two ends (Figure 18-20). These control elements are specific groups of base pairs that by binding to specific cellular proteins, direct the starting and stopping of RNA chains at correct positions along the chromosomes (see also Chapter 21). Gene boundaries are thus effectively set by the location of these control elements.

Those control elements used to start yeast RNA chains are *promoters* (using the terminology first used to describe bacterial genes in Chap-

Table 18-4 The 46 Different Predicted Yeast tRNAs and Their Corresponding Amino Acids

Amino acid	Codon	Codon Usage	tRNAs Anticodon Predicted	tRNAs Anticodon Found	Amino acid	Codon	Codon Usage	tRNAs Anticodon Predicted	tRNAs Anticodon Found
Ala	GCU	208	IGC	IGC	Leu	CUU	3	IAG	
	GCC	76				CUC	1		
	GCA	3	U*GC			CUA	7	U*AG	UAG
	GCG	2	CGC			CUG	0	CAG	
Arg	CGU	7	ICG	ICG	Lys	AAA	15	U*UU	U*UU
	CGC	0				AAG	184	CUU	CUU
	CGA	0	U*CG		Met_m	AUG		CAU	CAU
	CGG	0	CCG				56		
	AGA	85	U*CU	U*CU	Met_i	AUG		CAU	CAU
	AGG	2	CCU		Phe	UUU	12	GAA	G_mAA
Asn	AAU	5	GUU			UUC	74		
	AAC	101			Pro	CCU	16	IGG	
Asp	GAU	51	GUC	GUC		CCC	1		
	GAC	106				CCA	85	U*GG	
Cys	UGU	22	GCA	GCA		CCG	0	CGG	
	UGC	1			Ser	UCU	103	IGA	IGA
Gln	CAA	61	U*UG			UCC	94		
	CAG	4	CUG			UCA	6	U*GA	U*GA
Glu	GAA	150	U*UC	U*UC		UCG	1	CGA	CGA
	GAG	7	CUC			AGU	5	GCU	
Gly	GGU	214	ICC	GCC		AGC	1		
	GGC	6			Thr	ACU	75	IGU	IGU
	GGA	3	U*CC			ACC	75		
	GGG	2	CCC			ACA	8	U*GU	
His	CAU	11	GUG	GUG		ACG	4	CGU	
	CAC	57			Trp	UGG	27	CCA	C_mCA
Ile	AUU	75	IAU		Tyr	UAU	10	GUG	GΨA
	AUC	71				UAC	76		
	AUA	1	U*AU		Val	GUU	121		
	UUA	14	U*AA	ZAA		GUC	97	IAC	IAC
	UUG	170	CAA	CAA		GUA	1	U*AC	N*AC
						GUG	2	CAC	CAC

m = methylated; *indicates modified; underscoring denotes the more abundant tRNAs.
SOURCE: Data from C. Guthrie and J. Abelson in J. N. Strathern. E. W. Jones, and J. R. Broach, eds., *The Molecular Biology of the Yeast Saccharomyces, Metabolism and Gene Expression* (Cold Spring Harbor, N.Y.: Cold Spring Harbor Laboratory, 1982), p. 490.

ter 8). But yeast promoters differ in a fundamental way from bacterial promoters through their possession of **upstream activating sequences (UASs)** and **upstream repressing sequences (URSs),** which can be several hundred base pairs 5′ to the start of transcription. Binding of positive control proteins to UAS sequences turns up the rate of transcription of their respective genes, while binding of negative control proteins to URS sequences turns down the transcription rate of those genes that need to be negatively regulated. Given the very subtle devices that we know are utilized in the regulation of bacterial DNA transcription, we must anticipate that many yeast

Figure 18-20
A schematic yeast gene showing the relative positions of its control elements.

genes will contain both UAS and URS elements, thereby allowing their expression to be complexly tuned to reflect a multiplicity of external environmental conditions.

In addition to this obligatorily upstream element, yeast promoters have two other essential components: (1) TATAAA-like blocks (called **TATA** sequences) usually placed some 20 to 40 base pairs 5′ to the transcription start site (TATA sequences may play similar roles to the block at −35 base pairs of the *lac* promoter) and (2) an initiator element encompassing the start site, which precisely determines where transcription begins.

Many yeast genes are transcribed at constant rates throughout the cell cycle. The speed of such **constitutive** synthesis is a function of the specific base pairs at the yeast promoters. Depending on the promoter sequence, the rate of constitutive transcription can vary from very slow to very fast. Other promoters work at variable rates that depend on the extent of their binding to specific control proteins, many of whose amounts are environmentally controlled. Such promoters control the speed at which "regulated" genes function.

Regulated Expression of the Galactose-Utilizing Genes[65–68]

Gene regulation in yeasts, as in bacteria, was first discovered when mutations were found in genetic elements that block the functioning of still other genes. The easiest such systems to analyze are those whose functioning is increased or decreased by externally added nutrients (e.g., sugars, amino acids, and phosphate ion). Already very well investigated are several regulated genes of *S. cerevisiae* that code for three very well characterized enzymes involved in the utilization of the sugar galactose (yeasts do not contain the lactose-hydrolyzing enzyme β-galactosidase). They are galactokinase, a product of the *GAL1* gene; galactose transferase, the product of the *GAL7* gene; and galactose epimerase, encoded by the *GAL10* gene (Figure 18-21). Together they transform galactose into glucose-1-phosphate, which becomes converted to glucose-6-phosphate and enters the glycolytic pathway using enzymes that are always present. Yeast cells produce these three enzymes only when they are grown in the presence of galactose; these genes are not transcribed when yeast is grown in other carbon sources, such as glycerol or glucose.

Although these three *GAL* genes are located close together, they do not comprise an operon; they are each transcribed into separate

Figure 18-21
Galactose metabolism in yeast.

Figure 18-22
Sequence organization of the *GAL1–GAL10* promoter region (color) and the *GAL7* promoter region (gray boxes). *GAL1* and *GAL10* are transcribed in opposite directions away from a common promoter element, while a separate promoter functions for the adjacent *GAL7* gene.

mRNA molecules. The *GAL1* and *GAL10* mRNAs are divergently transcribed away from a common promoter element, while a separate promoter functions for the immediately adjacent *GAL7* gene (Figure 18-22). Regulating these three genes are two other genes, *GAL4* and *GAL80*. Genetic analysis first suggested that *GAL4* is a positive regulator, which activates the transcription of *GAL1*, *GAL7*, and *GAL10* genes. In contrast, *GAL80* is thought to act as a negative regulator.

The function of *GAL4* as a positive regulator has been directly confirmed by isolation of the *GAL4* product; it is a 99 kdal protein that binds to specific "upstream" sequences within the *GAL1–GAL10* and *GAL7* promoters. Each promoter contains two 30-base-pair blocks that are protected from nuclease digestion when bound to the *GAL4* protein. *GAL4* binding is mediated by a stretch of 74 amino acids at its amino-terminal end. The sequences within the 30-base-pair blocks are closely homologous, showing 18 to 22 identical base pairs within a consensus sequence of 23 residues. As expected from its DNA binding function, the *GAL4* protein is normally found within the yeast nucleus.

How the regulatory protein encoded by *GAL80* works is not known. It is argued to be a negative regulator of transcription of the *GAL* genes because yeast mutants that lack *GAL80* function express the *GAL* genes constitutively, that is, even in the absence of galactose. Somehow, in the absence of galactose, the product of *GAL80* prevents *GAL4* from functioning. It might code for a repressorlike protein that binds to the *GAL1-GAL10* promoter region. The best guess now is that the *GAL80* protein is not a DNA binding protein but somehow directly inactivates the *GAL4* product when galactose is absent (but not when it is present!). In this way, the presence of galactose would turn on galactose-utilizing genes. If this scheme is correct, then we should expect the *GAL80* protein to inhibit the binding of *GAL4* to DNA. This proposal will be experimentally tested in the near future.

Different Promoter Elements for Constitutive and Regulated Expression[69–73]

Many highly regulated genes that are turned on by positive regulatory proteins also maintain finite basal levels of constitutive expression that is not under environmental control. Examination of the promoter elements of one of these genes, the well-studied *HIS3* gene, suggests that the two types of expression involve quite distinct promoter elements. Constitutive *HIS3* expression uses upstream, TATA, and initiation elements that are different from those involved in its regulated expression (Figure 18-23). Most interesting is the upstream element for constitutive expression, 17AT base pairs, which can act bidirectionally. Such poly dA:dT blocks are also found upstream of several other constitutive start sites, suggesting that they may be a general signal for much, if not all, constitutive expression. Favoring this idea is the fact that the level of constitutive expression rises with the length of the poly dA:dT tract. Since poly dA:dT tracts cannot

Figure 18-23
Constitutive expression of *HIS3* differs from regulated expression in the control elements involved: poly dA:dT versus TGACTC sequences, TATA elements (T_c and T_R), and initiation sites (1 and 12). In regulated expression, the protein coded by the gene *GCN4* binds to pairs of TGACTC control blocks to turn on the synthesis of amino acids in conditions of amino acid starvation.

form nucleosomes, the obvious hypothesis is that they provide a region of DNA free of nucleosomes for interacting with the proteins involved in initiating RNA chains.

High-level regulated expression of many genes coding for enzymes involved in amino acid biosynthesis requires the protein product of the *GCN4* gene. The regulation of this large set of genes is termed "general amino acid control." Activation of these genes occurs when cells are starved for one of the amino acids, for example, histidine in the case of *HIS3*. The upstream elements for regulated expression of *HIS3* are pairs of TGACTC blocks that bind to the *GCN4* protein. Similar TGACTC blocks are found upstream of all the promoters under *GCN4* control. It is interesting to note that the transcription of the *GCN4* gene itself is not regulated by amino acid starvation; its mRNA product is made at a constant rate. Instead, the expression of the *GCN4* gene is regulated at the translational level, with the efficiency of translation increasing with amino acid starvation. The *GCN4* mRNA has an unusually long 5′ untranslated region that contains four AUG start codons. If these codons are removed, the efficiency of translation dramatically increases. It is possible that these untranslated AUG sites bind ribosomes to prevent other ribosomes from initiating at the correct *GCN4* start sites. How amino acid starvation might effectively cover up these inhibitory AUG codons remains a mystery.

Only a Minority of the S. cerevisiae Genes Are Split[74, 75]

While the vast majority of the genes of higher eucaryotes are split into exons and introns (see Chapter 20), only a small minority of the genes of *S. cerevisiae* contain introns. Among those that do are the genes coding for actin and for many tRNA molecules. However, the *S. pombe* genome has a much larger number of split genes. This strongly hints that the two forms of yeast may not be as closely related as first believed. That *S. cerevisiae* and *S. pombe* may be evolutionarily very distant is also suggested by comparisons of their rDNAs. Conceivably, these two examples of unicellular fungi are as distant from each other as from the genomes of higher vertebrates.

Insertion of Transposable Genetic Elements Affects Gene Expression[76–81]

Mutated control elements frequently arise through the insertion of large blocks of base pairs that are present in virtually identical form at

Figure 18-24

A map of a Ty1-like genome. ENH = an enhancer element promoting gene expression; ORF = open reading frame. LTR = long terminal repeat, found in retroviruses and related elements such as Ty1. The LTRs of Ty1 are called "delta." (Courtesy of Dr. Gerry Fink.)

a number of other sites on the yeast genome. These *repetitive sequences* are usually movable genetic elements (Chapters 7 and 11) that carry internal genes, some of which code for the enzymes that transpose (move) them to other chromosomal locations.

Why the placement of a movable element within a promoter might destroy promoter activity is easy to comprehend. Initially, however, it was hard to understand how such an insertion event radically increased the expression of previously dormant genes. The explanation came with the finding that the flanking sequences of movable elements often contain strong promoter elements. They increase gene expression by providing much more effective (and sometimes less strongly regulated) start signals for transcription.

The most common movable element in yeast genomes is called Ty1 (Figure 18-24); structurally, it is closely related to the copia movable element of *Drosophila* and to the genomes of retroviruses that multiply in vertebrate cells (Chapter 24). Some 35 copies of complete 6-kilobase-pair Ty1 elements are present within the genomes of most yeast strains. Each contains a large central region with two large, open reading frames that code for two major protein products, one of which is a reverse transcriptase. It is flanked by two identical terminal 334-base-pair repeats called delta (δ) elements, each of which contains a promoter element as well as sequences recognized by the transposing enzymes. Transcribed from each Ty1 element is a single, 6000-nucleotide mRNA molecule that begins at the promoter located at the 5' end of Ty1. Usually, these Ty1 elements are frequently transcribed within host cells; the total Ty1 RNA comprises 5 to 10 percent of the total yeast mRNA.

The close similarity in genome structure and function between the Ty1 element and the retrovirus genome strongly suggests not only that these two elements have a common evolutionary origin, but also that they move to new chromosomal locations by essentially the same molecular mechanism. It seemed very likely, from the moment its genome structure was worked out, that the Ty1 element would not move like bacterial transposons (by a transfer of information from DNA to DNA). Instead, it seemed highly probable that Ty1 would follow the retroviral pattern and transpose by a reverse transcription of its RNA chain to yield a complementary DNA chain. After conversion of this DNA chain to a double helix, it would then insert into a new chromosomal location.

Proof that such reverse transcription does in fact occur comes from the discovery that one of the two Ty1 genes codes for a reverse transcriptase, as well as from the finding that when a Ty1 is genetically modified to contain an intron (see Chapter 20) and subsequently moves to a new site, the intron is lost. This is to be expected if transposition involves an RNA intermediate, since RNA molecules with an intron will have the intron spliced out following transcription. Thus, Ty1 elements can appropriately be referred to as **retrotransposons.**

Rapid Reversals of Sex That Do Not Obey Conventional Genetic Rules[82]

Very important information about the expression of yeast genes also comes from the study of the mating types (sexes) of *S. cerevisiae*. An allele present in the mating type (*MAT*) locus on chromosome 3 determines whether a haploid cell is of the **a** or α mating type (sex). The *MAT***a** allele specifies the **a** mating type, while the α mating type is manifested when the *MAT*α allele occupies the *MAT* locus. These two alternative alleles set into motion two different pathways of gene expression that are beginning to help us understand how differentiated cells choose to produce one of several mutually incompatible sets of proteins. Such studies by themselves are bound to attract much future attention to the *MAT* locus.

Already, the *MAT* locus has been the focus of a great deal of interest through studies on why many strains of yeast do not have stable sexes but instead rapidly alternate between the **a** and α sexes. In **homothallic yeasts,** the mating-type genes of haploid cells interconvert much more rapidly (almost every other cell division) than could be anticipated by any mechanism involving conventional spontaneous mutations. Such rapid switches, however, are absent in **heterothallic strains,** where the *MAT* locus behaves like a conventional genetic element. An allele present at a second locus, *HO,* located on chromosome 4, determines whether a yeast strain is homothallic or heterothallic.

Most *S. cerevisiae* strains found in nature are homothallic, with heterothallic strains usually restricted to laboratory variants that have been selected for this trait. The prevalence of the homothallic state is not limited to *S. cerevisiae,* but is characteristic of virtually all other yeasts, as well as many other classes of fungi whose life cycles also alternate between the haploid and diploid states. Why homothallism is so widespread is still unclear. Perhaps it is favored because it promotes the early diploidization of the descendants of all spores, thereby sheltering the yeasts' own progeny from the burden of harmful mutations that would almost immediately be expressed in the haploid state. Moreover, since diploidy is a prerequisite for the meiotic cyclic generation of spores, sex reversals that lead to mating and diploidization have the further advantage of leading to a more rapid sporulation response to unfavorable environmental circumstances.

The Cassette Model for Mating-Type Switches[83–87]

Initially, it was guessed that the rapid reversals of sex might be due to flip-flop inversions of the *MAT* locus, like those that lead to the alternative expression of the two different proteins found within the flagella of *Salmonella* bacteria (see Figure 11-25). Genetic analysis, however, ruled out such explanations and instead led to an unorthodox proposal for which there was no precedence. The key came from genetic crosses that showed that chromosome 3 possesses both an active *MAT* gene and two unexpressed mating-type loci. One unexpressed locus, *HML,* is situated 200 kilobase pairs to the left of the *MAT* locus and contains a silent copy of the α information; the other silent gene, to the right of *MAT,* is *HMR,* and it contains **a** specificity. The switching of these genes involves the donation of genetic information (called a **cassette**) from one of the unexpressed genes to the

Figure 18-25
The cassette model of mating-type switching. Here the **a** cassette replaces the α cassette in the mating-type locus that is expressed (*MAT*). Parenthesis shading indicates that a gene is silent. (Redrawn from I. Herskowitz and Y. Oshima, in *The Molecular Biology of the Yeast Saccheromyces, Life Cycle and Inheritance*, p. 194.)

MAT locus (Figure 18-25). In this process, the original *MAT* gene is somehow displaced. However, transposition of a cassette from a silent gene does not lead to the loss of any genetic information from that site. Thus, the genetic information needed to promote subsequent mating-type switches is preserved.

The cassette model, bizarre as it first seemed, is supported by the cloning of the two forms (α and **a**) of active *MAT* genes. *MAT*α and *MAT***a** have easily distinguishable structures; there is no sequence homology within their long (about 600 base pairs) internal Y**a** and Yα segments, which are bounded by common X and Z segments (Figure 18-26). These differences in the Y segments of the alternative forms of *MAT* are faithfully reproduced in the two silent genes; *HML*α has a Y segment identical to that of *MAT*, and *HMR***a** has the Y sequence found in MAT**a**. The silent loci thus possess exactly the information needed to generate the postulated cassettes.

SIR Genes Turn Off the *HML*α and *HMR***a** Genes[88–91]

The *HML*α and *HMR***a** genes were originally thought to be silent because they lacked essential promoter sites required for their expression. But this hypothesis became untenable when the transcription of the specific mRNAs of *MAT*α and *HMR***a** was shown to start within the internally located Y segments that have identical structures both in *MAT* loci and in the silent *HM* loci. The silent genes must therefore possess promoters identical to those found within their active equivalents. At around the same time, geneticists discovered four unlinked genes—*SIR1, 2, 3,* and *4,*—whose gene products act together in *trans* (they do not have to be on the same chromosome) to prevent the expression of the silent mating-type genes. If any one of these *SIR* genes fails to act, the *HML*α and *HMR***a** genes are transcribed at rates equal to that of the gene at the *MAT* locus. To find the sites where the *SIR* products act, deletions were first created within *HML*α and *HMR***a** by in vitro mutagenesis. Then the genetically altered DNAs were inserted into appropriate yeast strains using DNA transformation procedures (Chapter 19).

Such studies reveal essential (E) regions that are adjacent to both the *HML* and *HMR* genes and that control the expression of those genes, making them silent (Figure 18-27). E regions act in *cis,* only affecting genes on the same chromosome. Within E is a block of base pairs (common to both *HML* and *HMR* E regions) that is the likely site where the *SIR* products act. However, these blocks are located some 1500 base pairs away from the promoters they control, raising the question of how the *SIR* proteins binding to the E region can inhibit (silence) a promoter as far as 2 kilobase pairs away. Equally puzzling is the fact that an E region can work in either orientation and that its

Figure 18-26
Sequence homology in the *MAT*, *HML*α, and *HMR***a** loci as determined by heteroduplex analysis. Note that the X and Z segments are common to all, while *HML* has the same Y segment as *MAT*α, and *HMR***a** has the same Y segment as *MAT***a**. Thus, the silent loci contain the exact information needed to generate the postulated cassettes. (Redrawn from I. Herskowitz and Y. Oshima, in *The Molecular Biology of the Yeast Saccheromyces, Life Cycle and Inheritance*, p. 196.)

Figure 18-27
SIR proteins act at essential E regions to turn off the expression of the *HML* and *HMR* genes. The E regions are located some 1500 base pairs away from the promoters they control to silence the *HML* and *HMR* genes. Transcription from *MAT* starts at a point within Y and proceeds in both directions. (Redrawn from A. J. S. Klar, J. N. Strathern, and J. B. Hicks, R. Losick and L. Shapiro, eds., *Microbial Development*, p. 161.)

relative location (3' or 5') with regard to the promoters that it silences does not matter. Perhaps the *SIR* proteins work by influencing the structure of chromatin within the *HML* and *HMR* genes. Favoring this proposal are experiments showing that when *SIR* control is absent, the chromatin within *HML* and *HMR* is more susceptible to nuclease attack, perhaps because of a failure of nucleosomes to form within these regions.

Double-Stranded Breaks at the MAT Locus Initiate Cassette Movements[92–95]

It was initially unclear how a copy of a silent *HM* locus replaces the resident allele of the *MAT* locus. Then the seminal observation was made of double-stranded cuts at the junction between the Y and Z segments, events that have now been proved to initiate cassette switching. Such breaks are made by a site-specific endonuclease that recognizes bases within the sequence CAGCTTTCCGCAACAG-TAAA that is present at the Y/Z junction. This endonuclease is coded by the *HO* gene and is termed the Y/Z endonuclease. Such breaks are made only in *HO*+ (homothallic) strains, a finding that quickly led to experiments showing that *HO*− (heterothallic) strains do not switch mating types because they fail to make this enzyme. The discovery that a double-stranded break initiated switching events came soon after the discovery that double-stranded DNA scissions, and not just single-chain cuts, can stimulate crossing over in yeasts. Many of the fundamental steps involved in cassette movement may thus be identical to those that carry out the crossing events that exchange blocks of genes between homologous chromosomes. The *RAD52* gene is one such example; it is needed not only for mating-type interconversion, but also for repair of X-ray induced damages and for recombination.

Following an initiating double-stranded cut, the *MAT* Y region becomes degraded, which explains the loss of the parental *MAT* allele. To replace the missing Y segment, the two DNA free ends at the Z end of the cut together with the free ends generated by the removal of the Y segment invade the homologous regions of the appropriate *HM* gene. DNA replication onto the 3' ends of the invading double helices then repair the missing Y segment; in the process, the *MAT* locus acquires the genetic specificity of the invaded *HM* locus (Figure 18-28). By this stage, the two interlocked mating-type loci are held

Figure 18-28
Detailed mechanism of cassette switching. A double-stranded cut is made, and the *MAT* Y segment is degraded. The free ends of the Z and X segments invade homologous regions of the appropriate *HM* gene, creating short heteroduplexes. The 3′ extension during DNA repair replaces the missing Y segment with that of the invading *HM* gene. As shown in other figures, the *HM* and *MAT* genes involved in cassette switching are located on the same chromosome. [After A. J. S. Klar, J. N. Strathern, and J. B. Hicks, R. Losick and L. Shapiro, eds., *Microbiological Development* (Cold Spring Harbor, N.Y.: Cold Spring Harbor Laboratory), p. 175.]

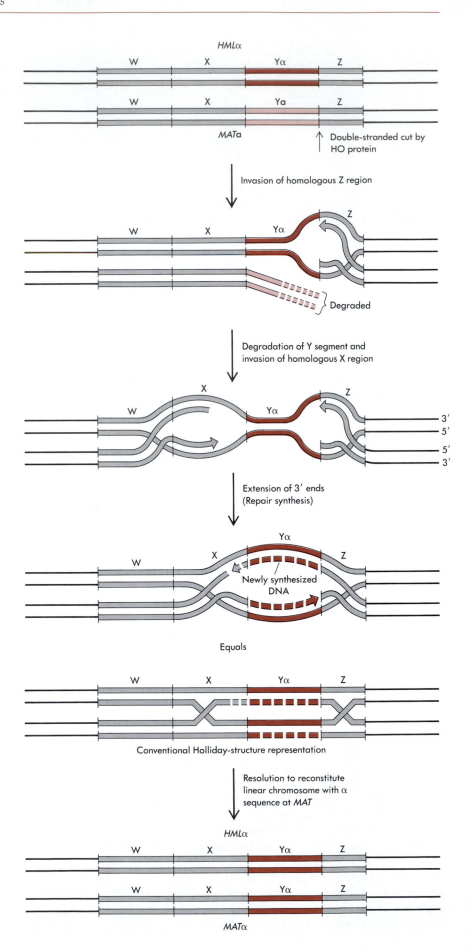

together by two regions of strand exchanges that must be resolved by subsequent single-stranded breaks similar to those that separate the joined intermediates of conventional crossing over (Chapter 11). These final resolving breaks must occur in a way that does not allow parts of the *MAT* locus to become exchanged with segments adjacent to the *HML* or *HMR* loci. Thus, isomerization processes that would lead to the exchange of flanking markers (Chapter 11) and that would produce a defective chromosome, do not occur.

The site that is recognized and cut by the *HO*-coded endonuclease is present not only at *MAT* but also at *HML* and *HMR*. However, only the site at *MAT* is cut *in vivo*. The *SIR* gene products are responsible for distinguishing between the cutting sites at these different locations: in mutants defective in *SIR*, cutting can occur at *HML* and *HMR*. Thus *SIR*, which was suggested above to be responsible for condensing the DNA in the *HML* and *HMR* regions to turn off transcription from these loci, also has the effect of preventing these loci from being cut by an endonuclease.

Programming the Direction of Cassette Movements[96, 97]

Switching is a highly programmed phenomenon that is limited to haploid cells and that only occurs in the parent cells, as opposed to the daughter, or bud, cells. Bud cells must divide once before they become capable of switching. Any "experienced" cells (those that have divided once) have a very high probability (80 percent) of switching. A switch is never observed in only one of a pair, suggesting that the cassettes move before the time of *MAT* DNA synthesis, with the switched cassette being subsequently replicated and distributed to both daughter cells (Figure 18-29). That switching is restricted to a predetermined point in the cell cycle is confirmed by recent experiments demonstrating that Y/Z endonuclease is made in the G1 phase and acts only during the next S phase of the cell cycle. The activity of *HO* Y/Z endonuclease must thus be inherently unstable, perhaps being inactivated while executing its function.

The switching of almost all experienced cells indicates that the direction of switching has been programmed to move from **a** to α and from α to **a** mating types. If the choice existed for the gapped *MAT* gene to randomly invade either the *HML* or *HMR* loci with equal probabilities, then the maximum frequency of observable switches would be 50 percent. This point was revealed through observing the switching behavior of strains in which *a* sequences occur at the *HML* loci and α sequences at the *HMR* locus. In such strains, observable switches occur much less often (about 5 percent). The lower values do not represent a less frequent switching, but instead indicate that a given cassette has been replaced by one of identical specificity (Figure 18-30). Thus, the two silent *HM* loci compete with each other as donors of the genetic information that will fill in the gapped *MAT* locus. Apparently, the *MAT* α locus usually preferentially associates with and then invades the *HMR* locus, while the *MAT***a** locus preferentially associates with and invades the *HML* locus. Such pairings are brought about by DNA binding proteins directly or indirectly controlled by the mating-type loci.

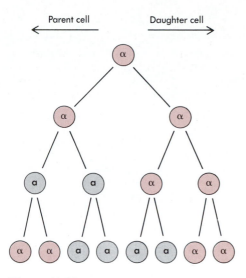

Figure 18-29
Division of a cell containing the *HO* gene results in either two cells with the same mating-type phenotype as the parental cell or two cells of switched phenotype. The switches always occur in pairs.

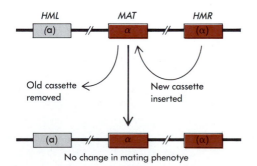

Figure 18-30
In yeast strains in which *HML* has **a** cassette and *HMR* has α cassette, the frequency of phenotypic switches from α to **a** is much lower than when HML is α and HMR is **a**. Although the actual switches occur at the normal frequency, the *MAT*α gene is most often replaced with one of identical specificity.

MATα and *MATa* Genes
Code for Control Proteins[98-103]

α cells contain a number of proteins, "α-specific proteins," that are not produced by **a** cells. Correspondingly, **a** cells contain a set of **a**-specific proteins that are not produced by α cells. The genes for the α-specific proteins are transcribed only in α cells, and the genes for the **a**-specific proteins are transcribed only in **a** cells. The mating-type locus is responsible for the expression of the appropriate set of cell-type-specific genes: *MATa* and *MATα* code for proteins that either activate or inhibit expression of genes that are scattered throughout the yeast genome. The mating-type locus is thus the master control locus for cell specialization. The precise manner in which the products of the mating-type locus govern expression of the specialized gene sets is shown in Figure 18-31. The α-specific genes are turned on in α cells by a protein coded by *MATα*, the α1 protein. In contrast, the **a**-specific genes are synthesized constitutively in **a** cells—there is no activator protein encoded by *MATa* that is needed. Therefore, the set of **a**-specific genes needs to be turned off in α cells. This task is performed by a protein coded by the second gene within *MATα*, the α2 protein. α2 protein is repressor, a site-specific DNA binding protein that recognizes a target site (an **a**-specific gene operator) located upstream of each of the members of the **a**-specific gene set.

This scheme leaves no obvious function for the *MATa* gene in haploid cells, and, in fact, it has no known function in these cells. In diploids, however, the α-specific genes must be turned off and the pathway of meiosis and sporulation must be turned on. This is accomplished by a third regulatory activity coded by *MAT,* which requires both the α2 protein and a product coded by *MATa,* the **a**l protein. This **a**1-α2 activity is responsible for many of the properties of **a**/α cells and turns off expression of a wide variety of genes. For example, it turns off expression of the α-specific genes by blocking transcription of the α1 protein. Thus **a**/α cells do not express the α-specific gene set. Furthermore, **a**/α cells do not undergo mating-type interconversion because **a**1-α2 turns off transcription of the *HO* gene. In addition, **a**1-α2 turns on meiosis and sporulation by repressing transcription of yet another gene (*RME1*). The *RME1* product is an inhibitor of meiosis and sporulation and is produced in **a** and in α cells; hence, they are unable to enter the pathway of meiosis and sporulation. In **a**/α cells, however, *RME1* synthesis does not occur, and cells can undergo meiosis and sporulation. Yet another example of repression by **a**1-α2 is the Ty1 element, whose transcript is repressed (for unknown physiological reasons) by **a**1-α2. The molecular nature of **a**1-α2 is not presently known. A plausible hypothesis is that the **a**1 polypeptide chain associates with the α2 polypeptide chain to form a repressor with novel recognition properties. Possible binding sites for **a**1-α2 are related sequences that are found in the upstream regions of genes repressed by **a**1-α2.

The three regulatory activities encoded by the yeast mating-type locus—α1, α2, and **a**1-α2—provide the first opportunity to study a class of regulatory proteins that do not respond to the environment (for example, to nutritional signals such as the presence of sugars or amino acids) but rather confer on a cell the proper expression of sets of genes responsible for differentiation. It is of course of interest to know whether higher eucaryotes have master regulatory loci and master regulatory proteins similar to those of yeast that are responsible for maintaining cellular differentiation.

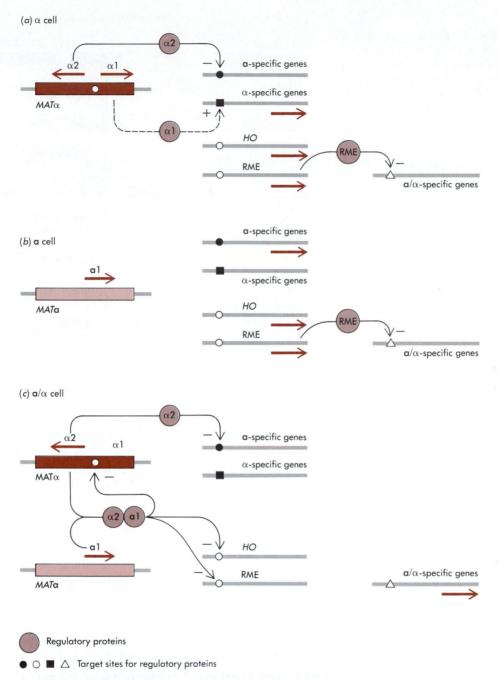

Regulatory proteins

● ○ ■ △ Target sites for regulatory proteins

Pheromones Make Haploid Cells Competent to Mate[104-107]

When mixed, α and **a** cells do not automatically bind strongly to each other and fuse to form α/**a** diploid cells. They must first convert each other into cells that are competent for mating by releasing specific hormonelike molecules called **pheromones,** which bind to receptors on cells of the opposite mating type. α cells secrete into the medium α-factor, which is 13 amino acids long. α-factor binds to a specific receptor on the **a** cell and induces the **a** cell to produce **a**-factor, which is 11 amino acids long. **a**-factor binds to a specific receptor on the α cell. A series of biochemical reactions follows that promotes the clumping of the cells to each other in α-a pairs and stops their movement through the cell cycle at the G1 stage. In this stage, the cells

Figure 18-31
The function of *MAT* genes as master control loci in (a) α haploid, (b) **a** haploid, and (c) **a**/α diploid cells. Positive control functions are denoted by dotted arrows, and negative control functions are denoted by solid arrows. **a**/α specific genes include those necessary for meosis and spore formation.

Figure 18-32
Processing of initial *MFα1* protein product into biologically active pheromones. S1, S2, S3, and S4 are spacers, which are cleaved as the polypeptide is processed. αF1, αF2, αF3, and αF4 are four identical mature α-factor pheromone molecules. Several different proteolytic enzymes catalyze the processing. [Courtesy of J. Kurjan and I. Herskowitz from *Cell* 30 (1982):940.]

cannot begin to replicate their DNAs, but they retain the capacity to synthesize the other major classes of cellular molecules. Finally, these biochemical reactions lead to specific changes in the cell walls and membranes that permit fusion between the paired cells.

The α pheromone is coded by the *MFα* gene, while the **a** pheromone is coded by the *MF***a** gene. The initial protein products of the *MF* genes are not yet active, but become so upon their degradation by specific proteolytic enzymes to produce the relatively short and biologically active oligopeptide pheromones (Figure 18-32). Conceivably, the relatively small size of both these pheromones permits their penetration through the cell wall to reach the outer surface of the plasma membrane of their target cells. If they function like other known polypeptide hormones, they bind to specific receptor molecules inserted into the plasma membrane of the target cell.

Meiosis Involves Proteins from the Mitotic Cell Cycle and Others That Are Unique to Meiosis[108–112]

Yeast cells undergo the process of meiosis in a manner very much like the meiosis of higher eucaryotes: diploid cells give rise to haploid meiotic products in a characteristic pathway in which the chromosome number is duplicated and then reduced in two successive meiotic segregations. In yeast, the meiotic products are then encased in spore coats to form spores. The entire process of meiosis and ascospore formation in yeast is termed "sporulation." Studies of yeast are providing considerable information about how cells make a decision to enter the pathway of meiosis, and how the numerous events of meiosis are orchestrated.

For **a**/α diploid yeast cells to enter the meiotic pathway, the cells first must receive a nutritional signal (nitrogen starvation and a poor carbon source), which causes cyclic AMP levels to drop. In subse-

Figure 18-33
Electron micrograph of a yeast cell, showing multiple mitochondria distributed throughout its cytoplasm. (Courtesy of Dr. Barbara J. Stevens.)

quently initiating meiosis, they exit the mitotic cycle at the GI phase carrying out a round of DNA replication. This is followed by recombination, which is especially high in meiotic cells (compared to mitotic cells), and then the two meiotic chromosome distributions. Many of the same proteins that are used for the mitotic cycle are also needed in meiosis. For example, there are several *CDC* genes needed for both meiotic and mitotic DNA replication and for nuclear division (chromosome distribution), which presumably make up the DNA replication apparatus and the nuclear spindle. Likewise, there are at least two genes needed for both mitotic and meiotic recombination. There are, however, also many genes whose roles are specific to the meiotic process. Notable examples are genes needed for initiation of meiotic DNA replication, for meiotic recombination, and for meiotic chromosome distribution. There are estimated to be at least 50 genes whose expression is specifically triggered during meiosis.

Mitochondria Have Their Own Chromosomes[113, 114]

Each haploid yeast cell contains about 10 to 50 mitochondria distributed throughout its cytoplasm (Figure 18-33). This variation in number does not usually represent a variation in the amount of mitochondrial matter per cell, which ordinarily accounts for some 10 percent of the cellular mass. Instead, it reflects a variation in the size of individual mitochondria, which can both divide to form smaller mitochondria or fuse together to create the very long and often branched mitochondria found in many rapidly growing cells.

Their tiny, sometimes bacterium-like shapes and the fact that they grow in length and then divide suggested to early-twentieth-century cytologists that mitochondria might have their own genetic elements and might have evolved during the early days of eucaryotic life from a procaryotic-like progenitor. Then, around 1950, evidence appeared for a special class of yeast mutation that did not show Mendelian segregation during mitosis that is characteristic of nuclear genes. These unusual mutants instead segregated as if they were genes located in the cytoplasm. These so-called *"petite"* mutations create mitochondria that are missing several vital enzymes, rendering them unable to generate ATP. As a result, the mutant cells grow slowly and form smaller ("petite") colonies on solid agar surfaces. The simplest way to explain these abnormal mitochondria was to postulate the existence of genetic molecules within normal mitochondria that code for the missing proteins. This conjecture became fact with the discovery that all eucaryotic mitochondria possess relatively tiny DNA molecules (mitochondrial DNA, or mtDNA) that in almost all organisms have circular shapes. Even though the individual molecules are quite small, the amount of these mtDNA molecules per cell is quite appreciable. Mitochondrial DNA is present in 50 to 100 copies per diploid cell and amounts to about 20 percent of the total DNA in a yeast cell. At around the same time, it was shown that chloroplasts also contain DNA molecules that code for many of their constituent proteins.

Synthesis of mtDNA Occurs Throughout the Yeast Life Cycle[115, 116]

Replication of circular yeast mtDNA is not limited to the S phase of the cell cycle. This may reflect the fact that mtDNA synthesis is carried out, at least in part, by different proteins than those that replicate nuclear DNA. Once started, mitochondrial DNA elongation occurs at a slower rate than that of its chromosomal equivalents. Single molecules of normal mtDNA require replication times equal to that of the entire cell cycle. Conceivably, the minimal length of the yeast cell cycle is dictated by the size of its mtDNA. Alternatively, there may have been no strong selective advantage for yeasts to reduce the mtDNA genome below the minimum size that can be replicated during the cell cycle.

Another interesting difference between nuclear and mitochondrial DNA replication is the absence of any proofreading (exonuclease) activity of the mitochondrial DNA polymerase. This leads to a much higher mutation rate within the mitochondrial genome than within chromosomal genes, and so mtDNA can evolve extremely rapidly. Perhaps this lack of an error repair mechanism is partly compensated by the large number of mitochondria within a single cell.

Much less is known about the replication of the DNA from yeast mitochondria than about the much smaller vertebrate mitochondrial DNA. The much larger size of yeast mtDNA has so far prevented the isolation of intact replicating molecules. Therefore, it is not known whether yeast mtDNA employ displacement-type replication like that employed by human mtDNA (Chapter 10).

The mitochondrial genome can undergo many different types of gross chromosomal rearrangements. A particularly common one is intramolecular recombination between repeated sequences. This generates deletion mutations of mtDNA and thus leads to mutants with a petite growth behavior. In some cases, the defective mtDNA can outgrow normal mitochondria in the same cell and create a cellular defect in mitochondrial function. A remarkable example of this type of deletion is one deletion mutant that lacks virtually all the normal mitochondrial genome, containing only many repetitions of those DNA sequences that make up the origin of DNA replication. A second type of gross chromosomal rearrangement is a specially programmed transposition of a segment of DNA throughout the population of mtDNA molecules. Some mtDNAs contain a 1.1 kbp insert ("omega") in the ribosomal RNA genes. When mtDNAs carrying omega are introduced into a cell that contains mtDNAs lacking omega, the omega element is rapidly transferred as if it were a movable genetic element to the omega-deficient mtDNAs. This occurs because the omega element codes for a site-specific endonuclease that produces a double-stranded break at a site within the ribosomal RNA gene of omega-deficient strains. Just as in mating-type interconversion, an efficient repair process ensues after a double-stranded break is produced. Repair of this break results in a copying in of the omega element as part of the repair process.

Noncoding Segments Along the Mitochondrial Genome[117, 118, 119]

When mitochondria from two different haploid cells are placed together in a diploid cell as a result of mating, they become physically intermixed, allowing genetic recombination to occur between them. Because such crossing-over events are very frequent, genetic crosses can be used only to show linkage between very closely spaced genetic units. The only way to assign the location of the genes on an mtDNA molecule is to isolate and sequence the relevant DNA segments. Such sequence analyses reveal a most unexpected structure for yeast mtDNA. Not only are the segments corresponding to well-defined genes frequently separated by long segments of apparently useless DNA (Figure 18-34), but some genes themselves are interrupted by

Figure 18-34
Physical map of a long mitochondrial genome unit of wild-type *S. cerevisiae*. Black areas correspond to mitochondrial genes or their exons, colored areas to intervening sequences and intergenic open reading frames (ORF1–ORF5). Radial lines indicate tRNA genes (the *thr*1 gene is the only one to have an anticlockwise orientation). Among mitochondrial genes, OXI1, 2 and 3 encode sub-units II, III, and I, respectively, of cytochrome c oxidase; cob, cytochrome b; *aap*1, *oli*2 and *oli*1, sub-units 8, 6, and 9 of ATPase; *var*1, a protein associated with the small mitochondrial ribosome sub-unit; 9S corresponds to the central part of tRNA synthesis locus; 15S and 21S are the genes for the small and large ribosomal RNAs, respectively. Triangles indicate the location of ori sequences 1–8. Mitochondrial transcripts are represented on an inner circle. They have a clockwise orientation except for two short transcripts in the ORF1-*thr*1 region. Dots correspond to initiation sites, arrows to termination sites, crosslines to processing sites. For some transcripts, initiation and/or termination sites are not known so far. [Adapted from M. de Zamaroczy and G. Bernardi, *Gene* 37 (1985):2, with permission.]

Table 18-5 Genetic Code for Yeast Mitochondrial DNA

Second Position

First Position		U	C	A	G	Third Position
U		UUU ⎤ Phe UUC ⎦ UUA ⎤ Leu UUG ⎦	UCU ⎤ UCC UCA Ser UCG ⎦	UAU ⎤ Tyr UAC ⎦ UAA ⎤ Ter UAG ⎦	UGU ⎤ Cys UGC ⎦ UGA⎤ Trp UGG⎦	U C A G
C		CUU ⎤ CUC Leu CUA (Thr) CUG ⎦	CCU ⎤ CCC CCA Pro CCG ⎦	CAU ⎤ His CAC ⎦ CAA ⎤ Gln CAG ⎦	CGU ⎤ CGC CGA Arg CGG ⎦	U C A G
A		AUU ⎤ AUC Ile AUA ⎦ AUG Met	ACU ⎤ ACC ACA Thr ACG ⎦	AAU ⎤ Asn AAC ⎦ AAA ⎤ Lys AAG ⎦	AGU ⎤ Ser AGC ⎦ AGA ⎤ Arg AGG ⎦	U C A G
G		GUU ⎤ GUC Val GUA GUG ⎦	GCU ⎤ GCC GCA Ala GCG ⎦	GAU ⎤ Asp GAC ⎦ GAA ⎤ Glu GAG ⎦	GGU ⎤ GGC GGA Gly GGG ⎦	U C A G

Grey shading denotes those codons that are read differently from the universal genetic code.

long, noncoding regions (introns) more reminiscent of the organization of eucaryotic DNA than of procaryotic DNA (Chapter 20). This observation suggests that the original procaryotic DNA that gave rise to mtDNA may have been more like today's eucaryotic DNA than the DNA in currently existing procaryotes.

Mitochondria Contain Their Own Specific Protein-Synthesizing Systems[120-124]

The proteins encoded by mitochondrial DNA are synthesized within the mitochondria themselves on a special class of small ribosomes called **mitoribosomes,** whose rRNA components are transcribed from two separate stretches of mtDNA. The tRNAs involved have unique structures that are also specified by mtDNA. Only 24 or 25 different tRNA molecules are used to translate yeast mitochondrial mRNA. Most of the 64 different codons specify the same amino acids and stop signals as specified by the genetic code for chromosomal DNA (Table 18-5). Only one codon is always read differently by the yeast tRNAs; UGA specifies tryptophan rather than a stop signal.

Also, the codon CUN, normally read as leucine, can sometimes be read as threonine. Moreover, as yet, all yeast mitochondrial genes have been found to terminate with UAA. The fact that only 25 tRNAs exist means that more *wobble* is possible than with the nuclearly coded tRNAs. When a single amino acid is specified by four codons with the same first and second positions (e.g., the four proline codons CCU, CCC, CCA, CCG), a single tRNA is sufficient; wobble permits pairing of any third codon base with the U that invariably resides in the first anticodon position of the tRNA. Furthermore, pairs of codons with pyrimidines in their third position are read by G in the tRNA anticodon, while pairs with purines in the third position are read by U in the tRNA.

Yeast mitoribosomes, like all other ribosomes, have two separable subunits (37S and 50S), each of which contains a single rRNA chain (15S and 21S, respectively) and a large number of different ribosomal proteins, each specific for the mitochondrial ribosomes.

The Genomes of Mitochondria Code Only for a Small Fraction of Their Proteins[125]

Given the almost bacterial size of mitochondria, it would not be surprising if mtDNAs were as large as the genomes of small bacteria. However, even the largest of mitochondrial DNAs can encode only a small percentage of mitochondrial proteins. Yeasts have among the largest mtDNAs (about 50 kilobase pairs) of any organism, and yet they code for only a small percentage of the proteins present in their mitochondria. Not even the majority of the enzymes involved in the generation of ATP are coded by mtDNA. Only seven well-defined proteins, in addition to the 15S and 21S rRNA molecules and the 25 different tRNAs, are known to be encoded by the yeast mtDNA. In contrast, all of the 60 or so proteins found in mitoribosomes, 20 different aminoacyl synthetases, and all the enzymes needed to replicate and transcribe the mtDNA are coded by chromosomal genes located in the nucleus. The mRNA molecules transcribed off these nuclear genes are translated by cytoplasmic ribosomes, and their protein products are later transported to their sites of function within the mitochondria.

The functioning of its protein-synthesizing system is not even necessary for mitochondrial growth and division. There are mutants that fail to make functional mitoribosomes and so cannot translate any of the genetic messages transcribed off mtDNA. Yet, the resulting mitochondria contain all of their usual chromosomally coded components and have essentially normal appearance. So it is still unknown why mitochondria need to contain any genetically active DNA at all.

We have seen that the yeast mitochondrial genome presents somewhat of a puzzle. It does some things, but not all that many. And it carries out these activities with a genome that is vastly larger than necessary. Why has the large mitochondrial genome persisted? One way in which yeast differs from higher eucaryotes is that the nuclear genome in higher eucaryotes contains vast amounts of noncoding DNA—satellites and introns. Yeast contains only modest amounts of such sequences. In contrast, the yeast mtDNA is ten times larger than the mammalian mtDNA. Perhaps the mitochondrial genome plays the additional role in yeast as a reservoir of genetic diversity, capable of serving the nuclear genome by contributing evolved sequences.

Figure 18-35

A cluster of viruslike killer particles (a) and retroviruslike particles (b) in the cytoplasm of yeast cells. [(a) Courtesy of J. D. Welsh. (b) Courtesy of D. J. Garfinkel.]

No Infectious Viruses Have Been Found for Yeasts[126, 127]

No infectious viruses are known to exist for any form of yeast. However, two different viruslike particles that contain RNA are found within many yeasts and may represent descendants of true viruses of past evolutionary epochs that today have lost their ability to effectively deliver their RNA genomes through the cell walls of yeast. The most commonly observed particles contain double-stranded RNA, are surrounded by a protein-containing capsid, and exist in multiple copies in the cytoplasm of most yeast strains. This viruslike RNA encodes a "killer" protein toxin that is secreted through the cell wall and is lethal to yeast cells that lack these viruslike particles. The killer toxin apparently acts by binding to the cell membrane of non-killer cells, inducing the release of ATP and K^+ into the surrounding medium.

Also in many yeast strains are particles morphologically similar to the retroviruses of higher cells (Figure 18-35). These particles have recently been shown to contain the RNA transcripts of the *Ty*-type movable DNA elements (transposons) that exist within virtually all yeast strains. The number of copies of these retrovirus-like particles depends on the rate of transcription of the yeast genome and varies with its surrounding environment. Why yeast cells produce such retroviruslike particles if they are never used as infectious agents remains a mystery. The same dilemma exists for the encapsulation of the "killer" RNA genomes.

Summary

Saccharomyces cerevisiae and *Schizosaccharomyces pombe* are now being extensively studied at the molecular level for the fundamental insights they can reveal about the structure and function of eucaryotic genes. Both these yeasts have well-studied, simple life cycles that involve both haploid and diploid phases, with the diploid cells undergoing conventional meiotic divisions to yield the haploid spores. All four products of a given meiotic division are encased within a thick-walled ascus. This allows the examination of the reciprocal chromosome partners of individual crossing-over events. Such tetrad analysis reveals the phenomenon of gene conversion, making yeasts among the preferred organisms for studying crossing over at the DNA level. *S. cerevisiae* cells divide through a budding process that separates parental cells from their daughter (bud) cells. In contrast, *S. pombe* is a fission yeast in which parental cells grow and then divide to form two morphologically identical daughter cells.

The average haploid yeast cell contains 14×10^6 base pairs; about 3½ times more DNA than is found in the *E. coli* chromosome. In *S. cerevisiae*, the DNA is distributed among 16 different chromosomes, while *S. pombe* has three different chromosomes. Already, more than 560 different *S. cerevisiae* genes have been mapped. During mitosis and meiosis, the yeast chromosomes are separated on microtubule-containing spindles, which remain completely encased by intact nuclear membranes and which, unlike those of higher eucaryotic cells, do not break down at the time of spindle formation. The microtubules of the spindle emanate from spindle pole bodies that are inserted into the nuclear membranes and which themselves divide at the time the decision for cell division is made. How the yeast cell cycle operates is beginning to be revealed through the analysis of *CDC* (cell division cycle) mutants blocked at specific stages of the cell cycle.

Like all other eucaryotic cells, yeasts contain internally located structural proteins that come together to form organized cytoskeletal elements. The contractile proteins actin and myosin already have been identified, as well as the α and β tubulins that aggregate to form the microtubules. Because of their small size, it was only recently discovered that yeasts contain all the major membrane-bounded internal compartments found in higher eucaryotic cells. Characterization of several of these organelles (e.g., the endoplasmic reticulum) was not straightforward until the discovery of mutants that block their functioning. Outside of all yeast cells is a thick cell wall impermeable to large molecules. Enzymes that specifically digest away much of the cell wall enable the generation of osmotically unstable protoplasts, now particularly useful because of their ability to take up externally added DNA. The DNA of the yeast chromosome is largely organized into nucleosomes through the coiling of DNA around histone cores made of H2A, H2B, H3, and H4 units. So far, there is no evidence for H1-like histones. Each chromosome contains a single centromeric DNA sequence that does not bind histones, but instead binds specialized proteins that in turn have the potential to attach to spindle microtubules. At the ends (telomeres) of all yeast chromosomes are some 100 base pairs of tandemly repeated $C_{1-3}A$ sequences that serve to maintain the ends of DNA molecules during DNA replication. DNA synthesis is initiated at many separate sites (origins of replication) along each chromosome. Synthesis occurs at a limited phase (S) during the cell cycle and proceeds bidirectionally at some 2 to 4 μm per minute. Nothing is known about the precise signals used to initiate DNA synthesis.

Three different forms of RNA polymerase are used to transcribe the yeast genome. RNA polymerase I transcribes the ribosomal DNA; RNA polymerase II transcribes the genes used to make the mRNA chains that code for proteins; and RNA polymerase III makes tRNA, 5S RNA, and several still functionally obscure forms of small RNA chains. Some primary RNA yeast transcripts have introns that become spliced out within the nucleus to yield smaller products. All mRNA molecules acquire caps at their 5' ends and poly A tails at their 3' ends within the nucleus before movement of their final functional forms through the nuclear pores into the cytoplasm. Each yeast mRNA molecule is monocistronic, coding for only a single polypeptide chain. A number of yeast polypeptides become folded into several enzymatic sites, each of which catalyzes a different chemical reaction.

Translation of mRNA chains on ribosomes requires the participation of 46 different tRNA molecules. Those tRNA molecules used to translate the more abundantly used codons are present in many more copies than those that translate infrequently utilized codons.

The rate at which a given gene is expressed is controlled at the level of both transcription and translation. Transcription levels are determined in part by the promoter sequences located 5' to the initiation sites for transcription. Each promoter contains sequences near the initiation site (initiation sequences and TATA blocks) as well as upstream activating sequences (UASs) or URSs (upstream repressing sequences). Some upstream elements act independently of the particular intracellular environment (constitutive expression), while others are regulated by specific control proteins whose activity is metabolically controlled (regulated expression). A given gene can contain both constitutive and regulatable upstream activating elements. The proteins that regulate these upstream elements can function either positively (turn on gene expression) or negatively (turn it down). Promoters also can have their activity either increased or decreased through the insertion of movable genetic elements. The main movable element in *S. cerevisiae* is Ty1, an element that moves by reverse transcription of its RNA product into DNA chains that can reinsert themselves back into many other chromosomal sites. Thus, Ty1 is an example of a retrotransposon.

The two alternative forms (*MATα* and *MATa*) of the genes determining mating type in *S. cerevisiae* code for control proteins, some acting positively and others acting negatively. The two forms of *MAT* are inherently unstable, being rapidly interchanged through the insertion of genetic cassettes emanating from normally unexpressed loci. These silent genes are turned off by products of the *SIR* genes, which act on sequences far removed from the promoter elements that they control. How this occurs remains unknown. The instability of *MAT* cassettes results from the action of the *HO* gene, which codes for the specific endonuclease that makes the double-stranded break at the *MATα* (or *MATa*) allele that initiates cassette movement.

Yeast mitochondria, like all mitochondria, contain genetically active DNA that replicates within the mitochondria, using enzymes imported into the mitochondria from the cytoplasm. Mitochondrial DNA (mtDNA) codes for the rRNA of its ribosomes, the tRNAs it needs to translate its own mRNA, and a small number of the enzymes involved in the mitochondrial generation of ATP. Mutations occur very frequently during mtDNA replication because the DNA polymerase involved has no proofreading capability. Many mtDNA mutants have lost most of their functional base sequences and in extreme cases contain only hundreds of tandemly repeated copies of their origin of replication. Such mitochondria have no protein-synthesizing capability, yet divide much more quickly and efficiently than normal mitochondria under conditions where oxidative ATP generation is not necessary. Why mitochondria have maintained genetically active DNA remains a puzzle.

Bibliography

General References

Egel, R., J. Kohli, P. Thuriaux, and K. Wolf. 1980. "Genetics of the Fission Yeast *Schizosaccharomyces pombe." Ann. Rev. Genetics* 14:77–108.

Fincham, J. R. S., and P. R. Day. 1971. *Fungal Genetics.* Philadelphia: Davis.

Klar, A. J. S., J. N. Strathern, and J. B. Hicks. 1984. "Developmental Pathways in Yeast." In *Microbial Development*, ed. R. Losick and L. Shapiro. Cold Spring Harbor, N.Y.: Cold Spring Harbor Laboratory, pp. 151–196.

Strathern, J. N., E. W. Jones, and J. R. Broach. 1981. *The Molecular Biology of the Yeast* Saccharomyces, *Life Cycle and Inheritance.* Cold Spring Harbor, N.Y.: Cold Spring Harbor Laboratory.

Strathern, J. N., E. W. Jones, and J. R. Broach, eds. 1982. *The Molecular Biology of the Yeast* Saccharomyces. *Metabolism and Gene Expression.* Cold Spring Harbor, N.Y.: Cold Spring Harbor Laboratory.

Cited References

1. Fogel, S., and D. D. Hurst. 1967. "Meiotic Gene Conversion in Yeast Tetrads and the Theory of Recombination." *Genetics* 57:455–481.
2. Esposito, M. S. 1971. "Post Meiotic Segregation in *Saccharomyces." Mol. Gen. Genetics* 111:297–299.
3. Fogel, S., R. K. Mortimer, and K. Lusnak. 1983. "Meiotic Gene Conversion in Yeast: Molecular and Experimental Perspectives." In *Yeast Genetics: Fundamental and Applied Aspects*, ed. J. F. T. Spencer, D. M. Spencer, and A. R. W. Smith. New York: Springer-Verlag, pp. 65–107.
4. Davidow, L. S., and B. Byers. 1984. "Enhanced Gene Conversion and Post Meiotic Segregation in Pachytene-Arrested *Saccharomyces cerevisiae." Genetics* 106:165–183.
5. Mortimer, R. K., and D. Schild. 1981. "Genetic Mapping in *Saccharomyces cerevisiae*." In *The Molecular Biology of the Yeast* Saccharomyces, *Life Cycle and Inheritance*, ed. J. N. Strathern, E. W. Jones, and J. R. Broach. Cold Spring Harbor, N.Y.: Cold Spring Harbor, Laboratory, pp. 11–26. Contains an excellent explanation of tetrad analysis and other mapping techniques.
6. Kuroiwa, T., H. Kojima, I. Miyakawa, and N. Sando. 1984. "Meiotic Karyotypes of the Yeast *Saccharomyces cerevisiae." Exp. Cell Res.* 153:259–265.
7. Mortimer, R. K., and D. Schild. 1985. "Genetic Map of *Saccharomyces cerevisiae*, Edition 9." *Microbial Rev.* 49:181–213.
8. Moens, P. B., and E. Rapport. 1971. "Spindles, Spindle Plaques and Meiosis in the Yeast *Saccharomyces cerevisiae." J. Cell Biol.* 50:344–361.
9. Byers, B. 1975. "The Behavior of Spindles and Spindle Plaques in the Cell Cycle and Conjugation of *Saccharomyces cerevisiae." J. Bacteriol.* 124:511–523.
10. Byers, B., K. Shriver, and L. Goetsch. 1978. "The Role of Spindle Pole Bodies and Modified Microtubule Ends in the Initiation of Microtubule Assembly in *Saccharomyces cerevisiae." J. Cell Sci.* 30:331–352.
11. Hartwell, L. H. 1974. "*Saccharomyces cerevisiae* Cell Cycle." *Bacteriol. Revs.* 38:164–198.
12. Game, J. C. 1976. "Yeast Cell Cycle Mutant cdc21 Is a Temperature Sensitive Thymidylate Auxotroph." *Mol. Gen. Genetics* 146:313–315.
13. Johnston, L. H., and K. A. Nasmyth. 1978. "*Saccharomyces cerevisiae* Cell Cycle Mutant cdc9 Is Defective in DNA Ligase." *Nature* 274:891–893.
14. Hiroaka, Y., T. Toda, and M. Yanagida. 1984. "The *NDA3* Gene of Fission Yeast Encodes β-Tubulin: A Cold-Sensitive *nda3* Mutation Reversibly Blocks Spindle Formation and Chromosome Movement in Mitosis." *Cell* 39:349–358.
15. Schekman, R., and P. Novick. 1982. "The Secretory Process and Yeast Cell Surface Assembly." In *The Molecular Biology of the Yeast* Saccharomyces, *Metabolism and Gene Expression*, ed. J. N. Strathern, E. W. Jones, and J. R. Broach. Cold Spring Harbor, N.Y.: Cold Spring Harbor Laboratory, pp. 361–398.

16. Bitter, G. A., K. K. Chen, A. R. Banks, and P.-H. Lai. 1984. "Secretion of Foreign Proteins from *Saccharomyces cerevisiae* Directed by α-Factor Gene Fusions." *Proc. Nat. Acad. Sci.* 81:5330–5334.
17. Brake, A. J., J. P. Merryweather, D. G. Coit, U. A. Heberlein, F. R. Masiarz, G. T. Mullenbach, M. S. Urdea, P. Valenzuela, and P. J. Barr. 1984. "α-Factor-Directed Synthesis and Secretion of Mature Foreign Proteins in *Saccharomyces cerevisiae." Proc. Nat. Acad. Sci.* 81:4642–4646.
18. Hall, M. N., L. Hereford, and I. Herskowitz. 1984. "Targeting of *E. coli* Beta-Galactosidase to the Nucleus in Yeast." *Cell* 36:1057–1065.
19. Hurt, E. C., B. Pesold-Hurt, K. Suda, W. Oppliger, and G. Schatz. 1985. "The First Twelve Amino Acids (Less Than Half of the Presequence) of An Imported Mitochondrial Protein Can Direct Cytosolic Dihydrofolate Reductase into the Yeast Mitochondrial Matrix." *EMBO J.* 4:2061–2068.
20. Adams, A. E. M., and J. R. Pringle. 1984. "Relationship of Actin and Tubulin Distribution to Bud Growth in Wild-Type and Morphogenic-Mutant *Saccharomyces cerevisiae." J. Cell Biol.* 98:934–945.
21. Kilmartin, J. V., and A. E. M. Adams. 1984. "Structural Rearrangements of Tubulin and Actin During the Cell Cycle of the Yeast *Saccharomyces." J. Cell Biol.* 98:922–933.
22. Ballou, C. E. 1982. "The Yeast Cell Wall and Cell Surface." In *The Molecular Biology of the Yeast* Saccharomyces, *Metabolism and Gene Expression*, ed. J. N. Strathern, E. W. Jones, and J. R. Broach. Cold Spring Harbor, N.Y.: Cold Spring Harbor Laboratory.
23. Lohr, D., and K. E. Van Holde. 1975. "Yeast Chromatin Subunit Structure." *Science* 188:165–166.
24. Moll, R., and E. Wintersberger. 1976. "Synthesis of Yeast Histones in the Cell Cycle." *Proc. Nat. Acad. Sci.* 73:1863–1867.
25. Carle, G. F., and M. V. Olson. 1984. "Separation of Chromosomal DNA Molecules by Orthogonal-Field-Alteration Electrophoresis." *Nucleic Acid Res.* 12:5647–5664.
26. Carle, G. F., and M. V. Olson. 1985. "An Electrophoretic Karyotype for Yeast." *Proc. Nat. Acad. Sci.* 82:3756–3760.
27. Schwartz, D. C., and C. R. Cantor. 1984. "Separation of Yeast Chromosome-Sized DNAs by Pulsed Field Gradient Gel Electrophoresis." *Cell* 37:67–75.
28. Strathern, J. S., C. S. Newlon, I. Herskowitz, and J. B. Hicks. 1979. "Isolation of a Circular Derivative of Yeast Chromosome III: Implications for the Mechanism of Mating Type Interconversion." *Cell* 18:309–319.
29. Carbon, J. 1984. "Yeast Centromeres: Structure and Function." *Cell* 37:351–353.
30. Blackburn, E. 1984. "Telomeres: Do the Ends Justify the Means?" *Cell* 37: 7–8.
31. Dunn, B., P. Szauter, M. L. Pardue, and J. W. Szostak. 1984. "Transfer of Yeast Telomeres to Linear Plasmids by Recombination." *Cell* 39:191–201.
32. Shampay, J., J. W. Szostak, and E. H. Blackburn. 1984. "DNA Sequences of Telomeres Maintained in Yeast." *Nature* 310:154–157.
33. Walmsley, R. M., C. S. M. Chan, B.-K. Tye, and T. D. Petes. 1984. "Unusual DNA Sequences Associated with the Ends of Yeast Chromosomes." *Nature* 310:157–160.
34. Walmsley, R. M., and T. D. Petes. 1985. "Genetic Control of Chromosome Length in Yeast." *Proc. Nat. Acad. Sci.* 82:506–510.
35. Murray, A. W., and J. W. Szostak. 1983. "Pedigree Analysis of Plasmid Segregation in Yeast." *Cell* 34:961–970.
36. Murray, A., and J. W. Szostak. 1983. "Construction of Artificial Chromosomes in Yeast." *Nature* 305:189–193.
37. Hieter, P., C. Mann, M. Snyder, and R. W. Davis. 1985. "Mitotic Stability of Yeast Chromosomes: A Colony Color Assay That Measures Nondisjunction and Chromosome Loss." *Cell* 40:381–392.
38. Koshland, D., J. C. Kent, and L. H. Hartwell. 1985. "Genetic Analysis of the Mitotic Transmission of Minichromosomes." *Cell* 40:393–403.

39. Williamson, D. H. 1965. "The Timing of Deoxyribonucleic Acid Synthesis in the Cell Cycle of *Saccharomyces cerevisiae*." *J. Cell Biol.* 25:517–510.

40. Burke, W., and W. L. Fangman. 1975. "Temporal Order in Yeast Chromosome Replication." *Cell* 5:263–269.

41. Sentenac, A., and B. Hall. 1982. "Yeast Nuclear RNA Polymerases and Their Role in Transcription." In *The Molecular Biology of the Yeast* Saccharomyces, *Metabolism and Gene Expression*, ed. J. N. Strathern, E. W. Jones, and J. R. Broach. Cold Spring Harbor, N.Y.: Cold Spring Harbor Laboratory, pp. 561–606.

42. Zazloff, Michael. 1983. "tRNA Transport from the Nucleus in a Eucaryotic Cell: Carrier-Mediated Translocation Process." *Proc. Nat. Acad. Sci.* 80:6436–6440.

43. Bell, G. I., L. J. DeGennaro, D. H. Gelfand, R. J. Bishop, P. Valenzeula, and W. J. Rutter. 1977. "Ribosomal RNA Genes of *Saccharomyces cerevisiae*. I. Physical Map of the Repeating Unit and Location of the Regions Coding for 5S, 5.8S, 18S Ribosomal RNAs." *J. Biol. Chem.* 252:8118–8125.

44. Petes, T. D., and D. B. Botstein. 1977. "Simple Mendelian Inheritance of the Reiterated Ribosomal DNA of Yeast." *Proc. Nat. Acad. Sci.* 74:5091–5095.

45. Philippsen, P., M. J. Thomas, R. A. Kramer, and R. W. Davis. 1978. "Unique Arrangement of Coding Sequences for 5S, 5.8S, 18S and 25S Ribosomal RNA in *Saccharomyces cerevisiae* as Determined by R-Loop and Hybridization Analysis." *J. Mol. Biol.* 123:387–404.

46. Petes, T. D. 1979. "Yeast Ribosomal DNA Genes Are Located on Chromosome XII." *Proc. Nat. Acad. Sci.* 76:410–414.

47. Egel, R. 1981. "Intergenic Conversions and Reiterated Genes." *Nature* 290: 191–192.

48. Jackson, J. A., and G. R. Fink. 1981. "Gene Conversion Between Duplicated Genetic Elements in Yeast." *Nature* 292:306–311.

49. Pringle, J. R., and L. H. Hartwell. 1981. "The *Saccharomyces cerevisiae* Cell Cycle." In *The Molecular Biology of the Yeast* Saccharomyces, *Life Cycle and Inheritance*, ed. J. N. Strathern, E. W. Jones, and J. R. Broach. Cold Spring Harbor, N.Y.: Cold Spring Harbor Laboratory.

50. Perry, R. P., and D. E. Kelly. 1976. "Kinetics of the Formation of 5′ Terminal Caps in mRNA." *Cell* 8:433–442.

51. Adams, J. M., and S. Cory. 1975. "Modified Nucleosides and Bizarre 5′ Termini in Mouse Melanoma Messenger RNA." *Nature* 255:28–31.

52. Furuichi, Y., M. Morgan, S. Muthukrishman, and A. J. Shatkin. 1975. "Reovirus Messenger RNA Contains a Methylated Blocked 5′-Termini Structure: m⁷G(5′)ppp(5′)-GᵐpCp." *Proc. Nat. Acad. Sci.* 72:362–356.

53. Kozak, M. 1978. "How Do Eucaryotic Ribosomes Select Initiation Regions in Messenger RNA?" *Cell* 15:1109–1123.

54. Keesey, J. K., Jr., R. Bigelis, and G. R. Fink. 1979. "The Product of the *his4* Gene Cluster in *Saccharomyces cerevisiae*; A Trifunctional Polypeptide." *J. Biol. Chem.* 254:7427–7433.

55. Sherman, F., and J. W. Stewart. 1982. "Mutations Altering Initiation of Translation of Yeast Iso-1-Cytochrome c; Contrasts Between the Eucaryotic and Procaryotic Initiation Process." In *The Molecular Biology of the Yeast* Saccharomyces, *Metabolism and Gene Expression*, ed. J. N. Strathern, E. W. Jones, and J. R. Broach. Cold Spring Harbor, N.Y.: Cold Spring Harbor Laboratory, pp. 301–333.

56. McLaughlin, C. S., J. R. Warner, M. Edmonds, H. Nakazato, and M. H. Vaughn. 1973. "Polyadenylic Acid Sequences in Yeast Messenger Ribonucleic Acid." *J. Biol. Chem.* 248:1466–1471.

57. Hereford, L. M., and M. Roshbash. 1977. "Number and Distribution of Polyadenylated RNA Sequences in Yeast." *Cell* 10:453–462.

58. Fahrner, K., J. Yarger, and L. Hereford. 1980. "Yeast Histone mRNA Is Polyadenylated." *Nucleic Acid Res.* 8:5725–5737.

59. Bennetzen, J. L., and B. D. Hall. 1982. "The Primary Structure of the *Saccharomyces cerevisiae* Gene for Alcohol Dehydrogenase I." *J. Biol. Chem.* 257:3018–3025.

60. Zaret, K. S., and F. Sherman. 1982. "DNA Sequence Required for Efficient Transcription Termination in Yeast." *Cell* 28:563–573.

61. Crick, F. H. C. 1966. "Codon-Anticodon Pairing: The Wobble Hypothesis." *J. Mol. Biol.* 19:548–555.

62. Guthrie, C., and J. Abelson. 1982. "Organization and Expression of tRNA Genes in *Saccharomyces cerevisiae*." In *The Molecular Biology of the Yeast* Saccharomyces, *Metabolism and Gene Expression*, ed. J. N. Strathern, E. W. Jones, and J. R. Broach. Cold Spring Harbor, N.Y.: Cold Spring Harbor Laboratory, pp. 487–528.

63. Guarente, L. 1984. "Yeast Promoters: Positive and Negative Elements." *Cell* 36:799–800.

64. Brent, R. 1985. "Repression of Transcription in Yeast." *Cell* 42:3–4.

65. St. John, T. P., and R. W. Davis. 1981. "The Organization and Transcription of the Galactose Gene Cluster of *Saccharomyces*." *J. Mol. Biol.* 152:285–315.

66. Oshima, Y. 1982. "Regulatory Circuits for Gene Expression: The Metabolism of Galactose and Phosphate." In *The Molecular Biology of the Yeast* Saccharomyces, *Metabolism and Gene Expression*, ed. J. N. Strathern, E. W. Jones, and J. R. Broach. Cold Spring Harbor, N.Y.: Cold Spring Harbor Laboratory, pp. 159–180.

67. Silver, P. A., L. P. Keegan, and M. Ptashne. 1984. "Amino Terminus of the Yeast *GAL4* Gene Product Is Sufficient for Nuclear Localization." *Proc. Nat. Acad. Sci.* 81:5951–5955.

68. Bram, R. J., and R. D. Kornberg. 1985. "Specific Protein Binding to Far Upstream Activating Sequences in Polymerase II Promoters." *Proc. Nat. Acad. Sci.* 82:43–47.

69. Donahue, T. F., R. S. Daves, G. Lucchini, and G. R. Fink. 1983. "A Short Nucleotide Sequence Required for Regulation of *his4* by the General Control Systems of Yeast." *Cell* 32.89–98.

70. Thireos, G., M. D. Penn, and H. Greer. 1984. "5′ Untranslated Sequences Are Required for the Translational Control of a Yeast Regulatory Gene." *Proc. Nat. Acad. Sci.* 81:5096–5100.

71. Struhl, K. 1984. "Genetic Properties and Chromatin Structure of the Yeast *gal* Regulatory Element: An Enhancer-Like Sequence." *Proc. Nat. Acad. Sci.* 81:7865–7869.

72. Hinnebusch, A. G., G. Lucchini, and G. R. Fink. 1985. "A Synthetic *HIS4* Regulatory Element Confers General Amino Acid Control on the Cytochrome c Gene (*CYC1*) of Yeast." *Proc. Nat. Acad. Sci.* 82:498–502.

73. Struhl, K., W. Chen, D. E. Hill, I. A. Hope, and M. A. Dettinger. 1985. "Constitutive and Coordinately Regulated Transcription of Yeast Genes: Promoter Elements, Positive and Negative Regulatory Sites, and DNA Binding Proteins." *Cold Spring Harbor Symp. Quant. Biol.* In press.

74. Mao, J., B. Appel, J. Schaack, S. Sharp, H. Yamada, and D. Soll. 1982. "The 5S RNA Genes of *Schizosaccharomyces pombe*." *Nucleic Acids Res.* 10:487–498.

75. Hindley, I., and G. A. Phear. 1984. "Sequence of the Cell Division Gene *CDC2* from *Schizosaccharomyces pombe*; Patterns of Splicing and Homology to Protein Kinases." *Gene* 31:129–134.

76. Roeder, G. S., P. J. Farabaugh, D. T. Chaleff, and G. R. Fink. 1980. "The Origins of Gene Instability in Yeast." *Science* 209:1375–1380.

77. Scherer, S., C. Mann, and R. W. Davis. 1982. "Reversion of a Promoter Deletion in Yeast." *Nature* 298:815–819.

78. Boeke, J. D., D. J. Garfinkel, C. A. Styles, and G. R. Fink. 1985. "Ty Elements Transpose Through an RNA Intermediate." *Cell* 40:491–500.

79. Clare, J., and P. Farabaugh. 1985. "Nucleotide Sequence of a Yeast Ty Element: Evidence for an Unusual Mechanism of Gene Expression." *Proc. Nat. Acad. Sci.* 82:2829–2833.

80. Kostriken, R., J. N. Strathern, A. J. S. Klar, J. B. Hicks, and F. Heffron. 1983. "A Site-Specific Endonuclease Essential for Mating-Type Switching in *Saccharomyces cerevisiae*." *Cell* 35:167–174.

81. Mellor, J., S. M. Fulton, M. J. Dobson, W. Wilson, S. M. Kingsman, and A. J. Kingsman. 1985. "A Retrovirus-Like Strategy for Expression of a Fusion Protein Encoded by Yeast Transposon Ty1." *Nature* 313:243–246.

82. Herskowitz, I., and Y. Oshima. 1981. "Control of Cell Type in *Saccharomyces cerevisiae*: Mating Type and Mating-Type Interconversion." In *The Molecular Biology of the Yeast* Saccharomyces, *Life Cycle and Inheritance*, ed. J. N. Strathern, E. W. Jones, and J. R. Broach. Cold Spring Harbor, N.Y.:, Cold Spring Harbor Laboratory, pp. 181–210.

83. Hicks, J. B., J. N. Strathern, and I. Herskowitz. 1977. "The

Cassette Model of Mating Type Interconversion." In *DNA Insertion Elements, Plasmids and Episomes,* ed. A. I. Bukhari, J. A. Shapiro, and S. L. Adhya. Cold Spring Harbor Laboratory Cold Spring Harbor, N.Y.

84. Strathern, J. N., E. Spatola, C. McGill, and J. B. Hicks. 1980. "The Structure and Organization of the Transposable Mating Type Cassettes in *Saccharomyces.*" *Proc. Nat. Acad. Sci.* 77:2839–2843.

85. Astell, C., L. Ahlstrom-Jonasson, M. Smith, K. Tatchell, K. Nasmyth, and B. D. Hall. 1981. "The Sequence of the DNAs Coding for the Mating-Type Loci of *Saccharomyces cerevisiae.*" *Cell* 27:15–23.

86. Beach, D. H. 1983. "Cell Type-Switching by DNA Transposition in Fission Yeast." *Nature* 305:682–688.

87. Beach, D. H., and A. J. S. Klar. 1984. "Rearrangements of the Transposable Mating-Type Cassettes of Fission Yeast." *EMBO J.* 3:603–610.

88. Nasmyth, K. 1982. "Regulation of Yeast Mating-Type Chromatin Structure by *SIR:* An Action at a Distance Affecting Both Transcription and Transposition." *Cell* 30:567–578.

89. Abraham, J., K. A. Nasmyth, J. N. Strathern, A. J. S. Klar, and J. B. Hicks. 1984. "Regulation of Mating Type Information in Yeast: Negative Control Requiring Sequences Both 5' and 3' to the Regulated Region." *J. Mol. Biol.* 176:307–331.

90. Miller, A. M., and K. A. Nasmyth. 1984. "Role of DNA Replication in Repression of Silent Mating Type Loci in Yeast." *Nature* 312:247–250.

91. Brand, A. H., L. Breeden, J. Abraham, R. Sternglanz, and K. Nasmyth. 1985. "Characterization of a "Silencer" in Yeast: A DNA Sequence with Properties Opposite to Those of a Transcriptional Enhancer." *Cell* 41:41–48.

92. Strathern, J. N., A. J. S. Klar, J. B. Hicks, J. A. Abraham, J. M. Ivy, K. A. Nasmyth, and C. McGill. 1982. "Homothallic Switching of Yeast Mating Type Cassettes Is Initiated by a Double-Stranded Cut in the MAT Locus." *Cell* 31:183–192.

93. Kostriken, R., J. N. Strathern, A. J. S. Klar, J. B. Hicks, and F. Heffron. 1983. "A Site-Specific Endonuclease Essential for Mating-Type Switching in *Saccharomyces cerevisiae.*" *Cell* 35:167–174.

94. Klar, A. J. S., and J. N. Strathern. 1984. "Resolution of Recombination Intermediates Generated During Yeast Mating Type Switching." *Nature* 310:744–747.

95. Klar, A. J. S., J. N. Strathern, and J. A. Abraham. 1984. "The Involvement of Double-Strand Chromosomal Breaks for Mating-Type Switching in *Saccharomyces cerevisiae.*" *Cold Spring Harbor Symp. Quant. Biol.* 49:77–88.

96. Rine, J., R. Jensen, D. Hagen, L. Blair, and I. Herskowitz. 1981. "Pattern of Switching and Fate of the Replaced Cassette in Yeast Mating-Type Interconversion." *Cold Spring Harbor Symp. Quant. Biol.* 45:951–960.

97. Klar, A. J. S., J. B. Hicks, and J. N. Strathern. 1982. "Directionality of Yeast Mating-Type Interconversion." *Cell* 28:551–561.

98. Strathern, J. N., J. B. Hicks, and I. Herskowitz. 1980. "Control of Cell Type in Yeast by the Mating Type Locus: The $\alpha 1$–$\alpha 2$ Hypothesis." *J. Mol. Biol.* 147:357–372.

99. Klar, A. J. S., J. N. Strathern, J. R. Broach, and J. B. Hicks. 1981. "Regulation of Transcription in Expressed and Unexpressed Mating Type Cassettes of Yeast." *Nature* 289:239–244.

100. Nasmyth, K. A. 1982. "Molecular Genetics of Yeast Mating Type." *Ann. Rev. Genetics* 16:439–500.

101. Sprague, Jr., G. F., L. C. Blair, and J. Thorner. 1983. "Cell Interactions and Regulation of Cell Type in the Yeast *Saccharomyces cerevisiae.*" *Ann. Rev. Microbiol.* 37:623–660.

102. Johnson, A. D., and I. Herskowitz. 1985. "A Repressor (MAT$\alpha 2$ Product) and its Operator Control Expression of a Set of Cell-Type Specific Genes in Yeast." *Cell* 42:237.

103. Miller, A. M., V. L. MacKay, and K. A. Nasmyth. 1985. "Identification and Comparison of Two Sequence Elements That Confer Cell-Type Specific Transcription in Yeast." *Nature* 314:598–602.

104. Thorner, J. 1981. "Pheromonal Regulation of Development in *Saccharomyces cerevisiae.*" In *The Molecular Biology of the Yeast* Saccharomyces, *Life Cycle and Inheritance,* ed. J. N. Strathern, E. W. Jones, and J. R. Broach. Cold Spring Harbor, N.Y.: Cold Spring Harbor Laboratory, p. 143.

105. Kurjan, J., and I. Herskowitz. 1982. "Structure of a Yeast Pheromone Gene (MFα): A Putative α-Factor Precursor Contains Four Tandem Copies of Mature α-Factor." *Cell* 30:933–943.

106. Jenness, D. D., A. C. Burkholder, and L. Hartwell. 1983. "Binding of α-Factor Pheromone to Yeast **a** Cells: Chemical and Genetic Evidence for an α-Factor Receptor." *Cell* 35:521–529.

107. Hagen, D. C., and G. F. Sprague, Jr. 1984. "Induction of the Yeast α-Specific *STE3* Gene by the Peptide Pheromone **a**-Factor." *J. Mol. Biol.* 178:835–852.

108. Wagstaff, J. E., S. Klapholz, and R. E. Esposito. 1982. "Meiosis in Haploid Yeast." *Proc. Nat. Acad. Sci.* 79:2986–2990.

109. Clancy, M. J., B. Buten-Magee, D. J. Straight, A. L. Kennedy, R. M. Partridge, and P. T. Magee. 1983. "Isolation of Genes Expressed Preferentially During Sporulation in the Yeast *Saccharomyces cerevisiae.*" *Proc. Nat. Acad. Sci.* 80:3000–3004.

110. Matsumoto, K., I. Uno, and T. Ishikawa. 1983. "Initiation of Meiosis in Yeast Mutants Defective in Adenylate Cyclase and Cyclic AMP-Dependent Protein Kinase." *Cell* 32:417–423.

111. Percival-Smith, A., and J. Segall. 1984. "Isolation of DNA Sequences Preferentially Expressed During Sporulation in *Saccharomyces cerevisiae.*" *Mol. Cell. Biol.* 4:142–150.

112. Mitchell, A. P., and I. Herskowitz. 1986. "Meiosis and Sporulation Are Activated by Repression of the *RME1* Product in Yeast." *Nature,* 319:738-742.

113. Stevens, B. 1981. "Mitochondrial Structure." In *The Molecular Biology of the Yeast* Saccharomyces, *Life Cycle and Inheritance,* ed. J. N. Strathern, E. W. Jones, and J. R. Broach. Cold Spring Harbor, N.Y.: Cold Spring Harbor Laboratory, pp. 471–504.

114. Slonimski, P., P. Borst, and G. Attardi, eds. 1982. *Mitochondrial Genetics.* Cold Spring Harbor, N.Y.: Cold Spring Harbor Laboratory.

115. Williamson, D. H. 1965. "The Timing of Deoxyribonucleic Acid Synthesis in the Cell Cycle of *Saccharomyces cerevisiae.*" *J. Cell Biol.* 25:517–528.

116. Zinn, A. R., and R. A. Butow. 1985. "Nonreciprocal Exchange Between Alleles of the Yeast Mitochondrial 21S rRNA Gene: Kinetics and the Involvement of a Double-Strand Break." *Cell* 40:887–895.

117. Dujon, B., P. P. Slonimski, and L. Weill. 1974. "Mitochondrial Genetics. IX. A Model for Recombination and Segregation of Mitochondrial Genomes in *Saccharomyces cerevisiae.*" *Genetics* 78:415–437.

118. Bernardi, G. 1976. "Organization and Evolution of the Mitochondrial Genome of Yeast." *J. Mol. Evol.* 9:25–35.

119. Borst, P., and L. A. Grivell. 1978. "The Mitochondrial Genome of Yeast." *Cell* 15:705–723.

120. O'Brien, T. W., and D. E. Matthews. 1976. "Mitochondrial Ribosomes." In *Handbook of Genetics,* vol. 15, ed. R. C. King. New York: Plenum, pp. 535–550.

121. Martin, N., and M. Rabinowitz. 1978. "Mitochondrial tRNAs in Yeast: Identity of Isoaccepting tRNAs." *Biochemistry* 17:1628–1634.

122. Faye, G., N. Dennebouy, C. Kujawa, and C. Jacq. 1979. "Inserted Sequences in the Mitochondrial 23S Ribosomal RNA Gene of the Yeast *Saccharomyces cerevisiae.*" *Mol. Gen. Genetics* 168:101–109.

123. Tabak, H. F., N. B. Hecht, H. Menke, and C. P. Hollenberg. 1979. "The Gene for the Small Ribosomal RNA on Yeast Mitochondrial DNA: Physical Map, Direction of Transcription and Absence of an Intervening Sequence." *Curr. Genetics* 1:33–43.

124. Bonitz, S. G., R. Berlani, G. Coruzzi, M. Li, G. Macino, F. G. Nobrega, M. P. Nobrega, B. E. Thalenfeld, and A. Tzagoloff. 1980. "Codon Recognition Rules in Yeast Mitochondria." *Proc. Nat. Acad. Sci.* 77:3167–3170.

125. Schatz, G. 1979. "How Mitochondria Import Proteins from the Cytoplasm." *FEBS Lett.* 103:203–211.

126. Garfinkel, D. J., J. D. Boeke, and G. R. Fink. 1985. "Ty Element Transposition: Reverse Transcriptase and Virus-like Particles." *Cell* 42:507–517.

127. Wickner, R. B. 1981. "Killer Systems in *Saccharomyces cerevisiae.*" In: J. N. Strathern, E. W. Jones, and J. R. Broach, eds. *The Molecular Biology of the Yeast* Saccharomyces, *Life Cycle and Inheritance,* pp. 415–444.

CHAPTER 19

Recombinant DNA at Work

Only through the employment of recombinant DNA procedures have the molecular features of eucaryotic genomes become effectively open to experimental analysis. The more than 14,000 kilobase pairs of DNA found within even the simplest of eucaryotic genomes intimidated the most motivated of molecular geneticists as late as the early 1970s. And as we mentioned in earlier chapters, the 4,000 kilobase-pair genome of *E. coli* would have frightened away even the most determined adventurer had it not been for its viruses, which were powerful tools for genetic analysis.

From the very start, the key attraction of *E. coli*, as opposed to any small eucaryotic microorganism, was the great abundance of phages that multiplied within it and which happily turned out to include those with genomes of RNA as well as of DNA. Moreover, it turned out not to be too hard to devise methods for isolating many of these viral chromosomes as intact molecules whose sizes could be exactly determined. Thus, no radical surprises were expected, and as we have related earlier, none were found when the chromosome of several *E. coli* phages and then the chromosome of *E. coli* itself were exposed to the ultimate examination of DNA sequencing.

Because of their lack of phagelike genetic parasites, yeasts were relatively ignored until about 10 years ago; until then, only a few molecular biologists, upon deciding to move up from *E. coli* to eucaryotic cells, chose to work with either *S. cerevisiae* or *S. pombe*. Most molecular biologists settled upon the cells of higher vertebrates, within which a large variety of different classes of viruses were known to multiply. By concentrating on these vertebrate viruses, they hoped to reveal the essential tricks that govern the expression of all eucaryotic genomes. We now realize that such a restriction to viral genomes would have provided only limited knowledge had not the vast powers of recombinant DNA appeared on the scene.

When in 1973 the first practical procedure became available for making recombinant DNA molecules using *E. coli* vectors, the then only slowly expanding world of yeast genetics immediately saw the multiple advantages that would accrue if similar systems could be developed for cloning molecules within yeast cells. By 1977, a practical method for introducing genetically active DNA into yeast cells had been worked out, and the extraordinary effort to understand the structure and functioning of the yeast genome had begun. The procedures now available for genetically engineering yeast cells are the most powerful so far developed for any eucaryotic organism. In discussing these recombinant DNA methods, we shall see that some remain restricted to work with yeasts. Others, often those developed

595

Figure 19-1

Cloning a yeast restriction fragment into plasmid pBR322. Fragments of at least 5 kilobase pairs can be stably introduced and are retained throughout plasmid replication. The boxes indicate the restriction endonuclease recognition sites.

for use with other life forms, are broadly applicable to all known organisms.

Libraries of Yeast Genes Within *E. coli* Vectors[1, 2]

Recombinant DNA research often starts with the preparation of a **gene library** containing all the DNA of a given organism inserted as discrete restriction enzyme–generated fragments into many thousands of different **plasmid, λ, or cosmid vectors**. The most useful libraries are those containing genome insertions of the sizes needed to encompass complete genes. The construction of a gene library depends on the use of restriction enzymes like *Eco*RI that do not make too frequent cuts within DNA (since they recognize only groups of six consecutive base pairs). Even then, it is usually necessary to use incompletely digested DNA when the desired genes are suspected of being extremely large (the situation with many of the genes of vertebrates; Chapter 20).

Most gene libraries are made using *E. coli* plasmids or phage components as vectors. Not only were they the first vectors to be developed, but they are constantly being improved. Moreover, they are particularly useful when the genome inserts come from other organisms. Once a yeast DNA fragment is cloned and grown as part of an *E. coli* vector, it becomes isolated from all other yeast sequences, and much less care is needed to purify it than if it were present on a yeast plasmid in a yeast cell.

Genomic fragments of at least 5 kilobase pairs are very stably reproduced when inserted into plasmids like pBR322 (Figure 19-1). In contrast, larger inserts, which slow down the multiplication rates of their host plasmids, tend to be replaced by smaller derivative plasmids that have lost sections of their DNA inserts. Larger fragments, however, can be stably grown within specially tailored phage λ systems that require DNA inserts of some 22 kilobase pairs in order to be packaged within mature particles. Even larger fragments can be stably inserted into **cosmid vectors** (Chapter 8) that can hold some 40 to 50 kilobase pairs of foreign DNA.

Because the yeast genome is approximately 14,000 kilobase pairs, fragments representing the entire yeast genome can thus be contained in either 3000 pBR322 plasmids, 1000 λ particles, or 300 cosmids. The very much larger sizes of higher plant and animal genomes means that the complete libraries of their genomes are also larger. A λ library requires some 250,000 particles to contain all the human genome. Constructing a genomic library of even this size, however, is now essentially a trivial problem.

Ingenuity, while not required in the making of a gene library, is still often needed for screening libraries for desired genes. Such searches are usually routine only for those genes in which mutations have been found that can be complemented by the introduction of the appropriate library member (DNA sequence) into the mutant cell (see Chapter 8's discussion of the complementation test). In the absence of such mutants, the usual way to pinpoint the appropriate library member is by means of its DNA sequence. A number of genes, for example, have been detected by their ability to hybridize to specific cellular mRNA molecules, known to promote the synthesis of a specific protein product, such as a histone (Figure 19-2). Such techniques are generally very laborious, and on many occasions success in employing recombinant DNA requires an elegant trick (rather than brute force) to pick out the desired gene.

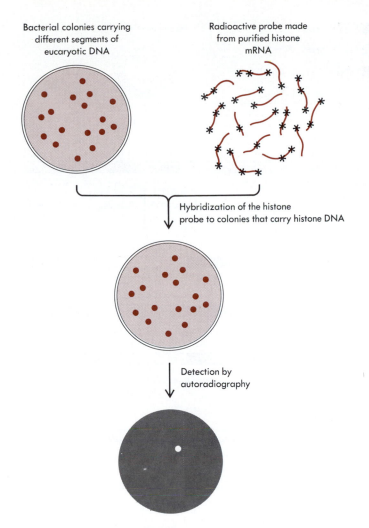

Bacterial colonies carrying
different segments of
eucaryotic DNA

Radioactive probe made
from purified histone
mRNA

Hybridization of the histone
probe to colonies that carry histone DNA

Detection by
autoradiography

Figure 19-2
Identification of a specific DNA library member by its ability to hybridize to specific mRNA molecules whose protein product can be identified.

The Unexpected Expression of Many Yeast Genes in *E. coli*[3, 4]

It initially seemed obvious that intact eucaryotic genes would not function properly, if at all, in procaryotic *(E. coli)* cells. There was no reason why the promoters recognized by the three types of eucaryotic RNA polymerases would be recognized by the single bacterial *(E. coli)* RNA polymerase. Nor did any reason exist for the eucaryotic RNA termination sites to be recognized by the proteins that bring about RNA chain termination in *E. coli*. Even if synthesized, eucaryotic mRNA molecules may not possess the correct ribosome binding site sequences to allow them to associate with bacterial ribosomes.

For these reasons, much surprise was generated when the *E. coli* mutant strain *leuB,* which has lost the ability to make the enzyme β-isopropylate dehydrogenase and so cannot make leucine, was observed to lose its leucine requirement after the introduction of a plasmid carrying the yeast gene *(LEU2)* that codes for the homologous enzyme in yeast. Apparently, the yeast genome contains many sequences that mimic functional *E. coli* promoters, with one of them properly situated upstream of the *LEU2* gene. These sequences also occur in front of many other yeast genes, and it is possible that as many as 30 percent of the yeast genes are transcribed into RNA when they are inserted into the *E. coli* chromosome. For such RNAs to function as messengers, they must also possess ribosome binding-site like sequences immediately adjacent to their AUG translational initiation

Figure 19-3

Integrative transformation of yeast by homologous recombination between a recombinant plasmid carrying a yeast gene (in this case *LEU2⁺*) and a yeast chromosome that has a mutation in the *LEU2* gene (*LEU2⁻*) (arrow indicates position of crossover). Single crossover events are most common, although double crossovers occasionally create transformants. [After D. Botstein and R. W. Davies.]

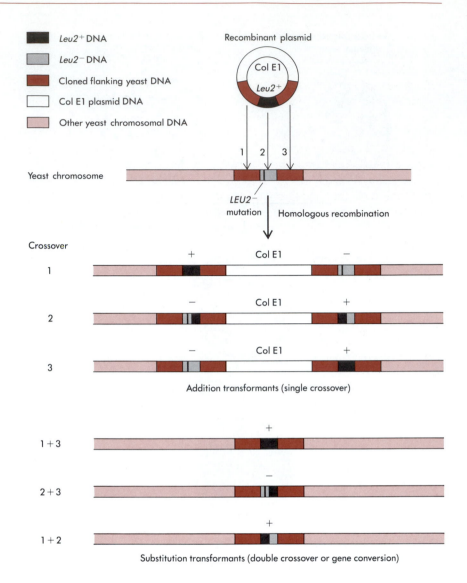

sites. In this way, mutant *E. coli* genes involved in tryptophan, histidine, arginine, and uracil synthesis have been complemented by yeast genes encoding equivalent proteins. The functioning of such a complementary yeast gene in *E. coli* can quickly lead to its isolation (cloning) and amplification. By so purifying the newly resident *E. coli* plasmid, complete copies of the yeast genes are automatically generated.

The Return of Cloned Genes to Yeast Using DNA Transformation[5, 6]

Only limited information can be obtained by cloning a eucaryotic gene within *E. coli*. For more complete answers as to how their functioning is controlled, eucaryotic genes must be reintroduced into the cells where they normally function. Free DNA molecules, however, are not taken up by normal yeast cells; their entry requires exposure of potential host cells to a set of enzymes that collectively degrade away much of the cell wall to produce the more permeable **spheroplasts.** DNA is then added in the presence of calcium ions (Ca^{2+}) and polyethylene glycol, which somehow makes the plasma membrane permeable and further encourages the passage of DNA through it.

The first demonstration that externally added DNA could be introduced into yeast cells in a genetically active form utilized the yeast gene *LEU2*[+]. Because it had been cloned away from the remainder of the yeast genome on a bacterial plasmid (ColE1), very high copy numbers of the gene could be added to mutant *LEU2*[−] haploid cells. All the resulting transformed *LEU2*[+] cells that subsequently grew represented the descendants of primary transformants in which the added DNA became incorporated into the host cell chromosome carrying the *LEU2*[−] gene by recombination enzymes that mediate homologous crossing-over events. On no occasion was the added *LEU2*[+] gene inserted at a nonhomologous region.

The most common event that inserts the *LEU2*[+] gene is a single reciprocal crossover event involving homologous yeast sequences on the plasmid and respective chromosomes (Figure 19-3). Such events lead to the placement of the entire plasmid with its *LEU2*[+] gene insert into the homologous chromosome (**integrative transformation**). No genetic information is lost from either the chromosome or the plasmid, and the resulting chromosome contains two copies of the yeast gene: one usually the original unmodified *LEU2*[−] gene, the other the wild-type *LEU2*[+] gene that came along with the plasmid DNA. By so acquiring an intact copy of the wild-type gene, the cells acquire the ability to make leucine and can easily be selected because they no longer have a leucine growth factor requirement. Less frequently, two crossovers occur in the homologous regions leading to substitution events in which the mutant *LEU2*[−] gene is replaced by DNA sequences originating from the added plasmid. When the substituted sequences overlap those sequences causing loss of function, the ability to make leucine is restored.

After these original positive results, a number of other genes were successfully reintroduced into yeast using genes available in cloned form on bacterial plasmids. In all cases, the transformation frequencies are low, with the most common of these rare events being the insertion of the entire donor plasmid sequence into the appropriate homologous chromosome.

Increasing Transformation Frequencies Using the 2μ DNA Plasmid[6, 7, 8, 9]

Much higher transformation frequencies (one hundred to a thousand times higher) result when genes like *LEU2* are introduced into yeast cells on plasmids containing replication origins that let them multiply autonomously in yeast. The added genes (e.g., *LEU2*) do not have to be inserted into host chromosomes by homologous crossing-over events in order for their replication to be ensured once every cell generation. Thus, the very low transformation rates observed when genes are added on nonreplicating plasmids does not reflect only inefficient uptake of added DNA. Instead, another limiting factor is the naturally very low frequency of homologous crossing over in mitotically growing yeast cells.

The first sequences used to make a functional yeast vector were those of the so-called **2μ plasmid,** the only plasmid that has so far been found to occur in yeast (Figure 19-4). While most strains of *Saccharomyces cerevisiae* contain this circular, 6300-base-pair plasmid, its function has not yet been discovered. No consistent difference in properties has been observed in those relatively rare strains that lack the 2μ plasmid. When present, the plasmids number some 30 copies

(a) No origin of replication; stable if integrates into yeast chromosome.

(b) 2μ contains origin of replication and *rep* genes; high copy number; unstable.

(c) ARS allows autonomous replication; unstable; high copy number.

(d) Contains centromere; stable; single copy.

Figure 19-4
The four different plasmids found in yeast. All contain a bacterial origin of replication and selectable marker such as Amp^R for selection of bacteria.

per haploid cell and are located in the nucleus, where their replication is largely accomplished by the same cellular enzymes used to replicate the linear yeast chromosomes. For this reason, they duplicate only during the S phase, the time of replication for the cellular chromosomes. When a single 2μ circle is introduced into a plasmid-free yeast, however, its replication, as well as that of its immediate descendants, occurs more than once every cell cycle until the normal complement of some 30 copies per cell is reached. Allowing this override of the normal cell cycle control of replication are two *Rep* proteins coded by two specific genes found on the 2μ DNA. In the absence of functional *Rep* genes, newly acquired 2μ circles can be present at a maximum of 1 to 2 copies per cell. And because they are not regularly segregated on mitotic (or meiotic) spindles, they are randomly lost from many daughter cells. Only when they carry a gene (e.g., *LEU2$^+$*) whose presence is needed for the growth of their host *(LEU2$^-$)* cell is their stability within a given strain ensured. This is done by growing the transformed yeast strain in minimal medium lacking leucine.

Plasmids carrying 2μ DNA sequences are called **YEp** or **yeast episomal plasmids,** while those which transform by integration (crossing over) into a yeast chromosome are called **YIp** or **yeast integrative plasmids** (see Figure 19-4).

The Construction of New Yeast Plasmids Using ARS Sequences[10]

Another source of yeast sequences for making new plasmids that multiply in yeast cells are the many genomic segments at which DNA replication is ordinarily initiated along the yeast chromosome. Each of the 200 to 400 such regions have the potential to act as the origin of replication for a circular plasmid. Because these regions are present in so many copies per genome, they frequently lie very close to genes we wish to study. In fact, the first **autonomously replicating sequence (ARS)** was discovered when the bacterial plasmids into which the *TRP1* gene was inserted displayed the capacity to replicate in yeast (see Figure 19-4). These yeast replicating plasmids **(YRp),** however, are not stably maintained in transformed cells. For unknown reasons, YRp plasmids tend to remain associated with the mother cell and are not efficiently distributed to the daughter cell. The replication of YRp plasmids, like that of any chromosomal DNA segment, occurs but once every cell generation. Rarely, stably transformed cells emerge through insertion of the intact YRp into one of the yeast chromosomes through crossing over within autonomously replicating sequences.

Further Stabilizing Plasmids Using *CEN* Sequences[11, 12, 13]

Highly stable yeast plasmid vectors can be made by including, in addition to autonomously replicating sequences, certain yeast DNA sequences that function as centromeres. These sequences are necessary for the attachment of the chromosomes to the microtubules of the mitotic spindle, presumably through some specific connector proteins that join the microtubules to the centromere (see Figure 19-4). Such *CEN*-containing plasmids (YCp) function as true chromosomes and segregate accurately during both mitosis and meiosis. *CEN* sequences can also be used to construct linear chromosomes if circularization is prevented by the addition of the telomeric sequences that terminate chromosomes.

Shuttling DNA Between Yeasts and Bacteria

All versatile vectors for use in yeast cells contain origins of replication active in both yeast and *E. coli* as well as one or more markers (e.g., for antibiotic resistance) that can be used to select those *E. coli* cells into which the vectors have entered. The vectors can thus be amplified (purified) in *E. coli* before being used to transform yeast cells. Furthermore, such **shuttle vectors** allow genetic sequences to be routinely transferred back and forth between yeast and *E. coli* (Figure 19-5).

YEp, YRp, and YCp vectors now provide the ideal tools to isolate yeast genes by virtue of their ability to complement mutations in yeast. Ordinarily, there is no longer any reason to clone a yeast gene through its ability to function in *E. coli*. Instead, fragments of the yeast genome are used to make libraries in yeast vectors that carry a marker (such as *LEU2*); the marker can then be used to select for the vector's transfer into a corresponding yeast mutant. The resulting library can then be amplified in *E. coli* or directly used to transform yeast strains that have defective genes of interest. Once the appropriate transformed yeast cell has been found, its descendants can be used directly as a source of the appropriate plasmid or as a source of DNA for amplification and subsequent isolation from *E. coli* cells.

Double-Stranded Cuts Promote High-Level Directed Integration into the Yeast Genome[14]

Initially, it seemed that the best way to optimize the insertion of yeast genes into their homologous chromosomes was to add them to appropriate cells in the form of intact, circular, integrative plasmids. In this way, the possibility of unwanted digestion of the free ends of linear DNA molecules by cellular nucleases would be avoided. Then, most unexpectedly, it was discovered that the frequency of crossing over–mediated integrative transformation into a specific chromosomal location was enhanced by a single double-stranded cut in the plasmid DNA. In such cases, integration occurs at frequencies some hundredfold higher and always into the chromosomal site homologous to the cut site (Figure 19-6). The vectors of choice for many transformation experiments have thus become linearized, yeast integrative plasmids, which only transform by homologous crossing over.

The observation that a double-stranded cut in DNA promotes homologous crossing over initiated a profound rethinking of the way crossing over occurs in eucaryotic cells. The free ends so made are highly recombinogenic, invading homologous genetic segments to form the heteroduplex intermediates. These intermediates are then resolved when appropriate nucleases cut single-stranded DNA to yield recombined DNA molecules.

Gene Retrievals Brought About by Gapped Shuttle Vectors[15]

Specific genes can be retrieved from their normal chromosomal locations through homologous crossing-over events that transfer them to specially constructed classes of **retriever vectors.** The most versatile

Figure 19-5
The possession of both bacterial and yeast origins of replication allows shuttle vectors to transform both bacteria and yeast and be transferred back and forth between them.

Figure 19-6
When double-stranded cuts are made in yeast integrating plasmids (YIp), integration into the yeast chromosomes always occurs at the homologous chromosomal site and at frequencies 100 times greater than with circular YIp.

retriever vectors contain origins of replication that let them multiply in *E. coli* as well as yeast, two selectable markers that favor their selective growth in either yeast or *E. coli*, and a homolog to the gene to be recovered from the yeast chromosome. Once such a retriever vector has been constructed, the gene in question is removed by appropriate restriction enzyme cuts, leaving behind the flanking yeast sequences on its two sides. The now "gapped vector" is added to the cell possessing the mutant gene to be recovered. Once it enters the nucleus, its two highly recombinogenic ends invade the gene's flanking segments and so position their 3′ ends to begin copying homologous sequences and thereby replace the lost (gapped) sequences with those of the chromosomal gene (Figure 19-7). These repair events generate two points of strand exchange (crossover) between the plasmid and the chromosome, which subsequently will be resolved by nuclease cuts to yield an intact plasmid as well as a complete chromosome. Only those plasmids so repaired are able to multiply and to complement their host yeast cell's deficiency with their selectable markers (e.g., $LEU2^+$). The stage is thus set for the mutant chromosomal gene to be shuttled to *E. coli* for amplification and sequencing.

Retriever vectors are a powerful weapon in the armory of the yeast geneticist because they greatly simplify the recovery of mutations produced *in vivo* onto plasmids so that they can be subjected to molecular scrutiny—e.g., sequencing. In particular, they provide a means to establish the precise structures of the mating-type genes, whose unusual behavior implies the existence of novel control elements (Chapter 18).

The Test-Tube Production of Mutated Control Elements[16–24]

Once a gene has been cloned and can be returned in a functional form to its normal cellular environment, the DNA sequences that control its expression can be directly worked out using a broad array of in vitro mutagenesis techniques. Until several years ago, the only way

Plasmid containing wild-type allele

Yeast origin

Bacterial origin

URA3

Tet^R

LEU2

Cleavage with restriction enzyme

Gapped plasmid

Tet^R

URA3^−

LEU2

Mutant allele

Yeast chromosome

Invasion of intact duplex and copy to fill gap

3′

5′

5′

3′

Plasmid carrying URA3^− can now replicate in yeast

URA3^−

URA3^−

Figure 19-7
A retriever vector. A specific restriction enzyme is used to splice out the desired gene. This "gapped vector" is added to the cell possessing the mutant gene desired. Once inside the yeast cell, it invades the gene's flanking segments and copies the sequence of the gene homologous to the one it lost.

to identify control sequences was to find mutations that affected their functioning. This approach, as indicated earlier (Chapters 16 and 17), has been most successful with *E. coli* and its phages. Naturally occurring mutations have also led to most of our current understanding of gene expression in yeast, and the discovery of additional key mutants will undoubtedly yield much new information. On the other hand, it is now becoming so easy to do *in vitro* mutagenesis procedures that we must anticipate a growing use of such procedures, even with organisms whose genetic analysis is relatively quick and inexpensive. A precise definition of the blocks of base pairs that make up control

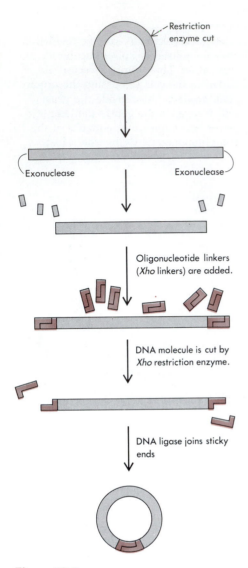

Restriction enzyme cut

Exonuclease   Exonuclease

Oligonucleotide linkers (*Xho* linkers) are added.

DNA molecule is cut by *Xho* restriction enzyme.

DNA ligase joins sticky ends

Figure 19-8
The creation of short deletions by the use of oligonucleotide linkers. A plasmid is first cut with a restriction enzyme at one site and then treated with a nuclease to create a slightly shorter linear DNA. Oligonucleotide linkers are added to the ends, and the plasmid is cut by the appropriate restriction enzyme to yield sticky ends that are then joined by DNA ligase.

elements will come from an almost routine creation of deletions, insertions, and base-pair replacements at predetermined locations or within a defined region in a given DNA.

Of all the changes, long deletions between two restriction enzyme sites are the easiest to make. By light digestion of a plasmid with a restriction enzyme, large numbers of partially digested molecules lacking just one fragment can be isolated. Their subsequent circularization yields smaller plasmids that frequently have biological activity. Much shorter deletions can be made by first cutting a plasmid at one restriction site to generate linear DNAs that can then be treated with nucleases to remove small groups of bases at each end. Ligating these slightly shorter linear fragments back into circular forms is often best accomplished by adding specific oligonucleotide linkers to the ends (see Figure 9-44). When subsequently cut by the appropriate restriction enzyme, these yield sticky ends that routinely can be joined together by DNA ligase (Figure 19-8). The making of precise insertion mutations also benefits from the use of such oligonucleotide linkers. These linkers can be added at cut sites without using nuclease to yield insertions of known sizes.

Single-base-pair changes at cytosine residues can be created using chemicals such as bisulfite, which deaminates cytosine to uracil, which base-pairs like thymine. Such chemical reactions thereby convert GC base pairs into AT base pairs. These reagents, however, only work on single-stranded DNA, where the base-pairing groups on the bases are freely accessible to the chemical. Thus, the ends of a DNA molecule cut by a restriction enzyme must first be exposed to exonucleases, which degrade from either the 3' end or the 5' end. Desired single-base-pair replacements can also be made by allowing DNA repair or DNA synthesis to occur either in the presence of base analogs that frequently mispair or in the total absence of one of the four nucleotides, which leads to incorporation of wrong bases at predetermined sites (Figure 19-9).

Now, however, these essentially hit-and-miss procedures for making single-base-pair substitutions are rapidly being replaced by much more powerful procedures that use 15- to 20-base-long oligonucleotide chains to make all possible classes of base-pair changes at any determined site along a DNA molecule. Oligonucleotides of this length form very stable DNA-DNA double helices with complementary segments of DNA. Less stable but nonetheless firm unions can also be made with DNA sequences that contain one to two mismatches. In the presence of DNA polymerase and the four nucleotide precursors of DNA, these slightly mismatched short DNA segments can function as primers for DNA extension when added to the circular, single-stranded chains of phages like M13. The resulting complete double helices will be perfectly complementary except for mismatched bases at the primer site. After their entry into *E. coli*, DNA replication will generate two types of daughter DNA strands: wild-type strands and mutant strands, the latter containing the changes imposed by the mismatched primer (Figure 19-10). The strands can easily be distinguished by the way they bind the corresponding oligonucleotides. The wild-type daughter DNAs will bind more strongly to wild-type oligonucleotide, while the short DNA segment that primed the mutant sequences will bind best to the mutated progeny chains.

Until recently, such oligonucleotide-directed mutagenesis was restricted to the laboratories of organic chemists capable of synthesizing the required probes. Now, however, such oligonucleotides are easily

```
          ...ATGGCGGTA...
          ...TACCGCCAT...

               │  Nick and exonuclease
               │  digestion
               ▼
          ...ATGG
          ...TACCGCCAT...

               │  Polymerase +
               │  dATP, dGTP, dCTP
               ▼
          ...ATGGCGGXA...
          ...TACCGCCAT...

               │  Replication
               ▼
   ...ATGGCGGXA...        ...ATGGCGTTA...
   ...TACCGCCXT...   and  ...TACCGCAAT...
       Mutant                Wild type
```

Figure 19-9
Production of single-base-pair replacements by creating a single-stranded section in the DNA and allowing repair to occur in the absence of one of the nucleotide bases. An incorrectly inserted base (in this case either A, G, or C) is represented by X, and X' represents its complement.

synthesized by the nonprofessional chemist using commercially available "DNA synthesizers" to polymerize the blocked nucleotide precursors that are also widely available to all who can afford their increasingly modest prices.

Using Integrative Transformation to Replace Chromosomal Genes with Mutated Derivatives[25, 26, 27]

Not only can inactive mutant genes be replaced by active wild-type genes, but integrative transformation provides a powerful way to replace active genes with inactive mutant genes. This genetic tool was first employed using a YIp containing both a deleted, nonfunctional *HIS3⁻* gene and an active *URA3* gene together with their immediately flanking sequences. Transformation of yeast with this plasmid results in 50 percent of the plasmids being inserted at the *HIS3* locus, while the other 50 percent of the plasmids insert into the *URA3* gene, which is located on a different chromosome. Insertions adjacent to the *HIS3* locus result in both the wild-type and nonfunctional forms of the *HIS3* locus in tandem on the yeast chromosome (Figure 19-11). Subsequent growth of these transformants leads to the occasional crossover between mutant and wild-type genes. Such crossovers remove the intervening vector and *URA3⁺* sequences, events that can be detected in the cells by the resumption of the *URA3⁻* phenotype. Of these *URA3⁻* revertants, some will retain the *HIS3⁺* gene, while others will possess the mutant form (see Figure 19-11). Using this general procedure, we can insert any mutant allele into yeast cells, as long as the gene either is not essential or (like *HIS3*) has the possibility of its product being supplied externally.

Integrative transformation can be used in an even simpler procedure to totally inactivate any chromosomally located gene whose active form has already been cloned on a plasmid. The trick is to transform using a segment of the gene that lacks both ends of the gene; in other words, an internal fragment. When crossing over occurs be-

Figure 19-10
Creation of base-pair mutants by using oligonucleotide DNA primers containing mismatched base pairs. These primers are added together with the four nucleotide precursors, DNA polymerase, and DNA ligase to a single-stranded phage such as M13 to create double-stranded M13 molecules. These are then introduced into *E. coli*, where replication of the double-stranded DNA creates one mutant and one wild-type daughter molecule.

Figure 19-11
Replacement of the active *HIS3* gene with the mutant *HIS3*⁻ gene by integrative transformation. A YIp containing mutant *HIS* and a functional *URA* gene may transform by separating at either the *HIS* or *URA* locus on the yeast chromosome. (a) Transformation near the *HIS* locus results in either transformed or wild-type yeast chromosomes, depending on where the plasmid cuts out. (b) Integration of the plasmid near the chromosomal *HIS3*⁺ locus leads to production of two *HIS3* genes in tandem. Excision of the plasmid (with loss of the plasmid *URA3* marker) can occur by the reverse process—recombination between the two copies of the *HIS3* gene. If recombination occurs in interval 1, then the *HIS3*⁻ mutation originally on the plasmid is left behind in the genome. If recombination occurs in interval 2, then the functional *HIS3* gene is left in the genome.

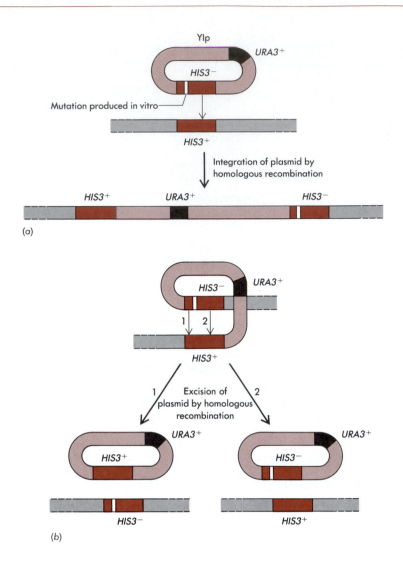

Figure 19-12
Using integrative transformation to determine whether a given chromosomal gene (in this case the yeast *actin* gene) is essential. A YIp was created in which the two ends of the yeast *actin* gene were removed and which contained the *URA3*⁺ gene. Integration of the plasmid by homologous recombination between the *actin* fragment of the plasmid and the chromosomal *actin* gene leads to a *URA3*⁻ transformant. The integration event leads to production of two versions of the *actin* gene, both of them defective. Yeast lacking a functional *actin* gene do not survive. Consequently, the type of insertional inactivation described here was done into a diploid strain, so that one active *actin* gene was present. Half of the haploid meiotic products—those carrying the inactivated *actin* gene—were inviable. Thus, the *actin* gene is shown to be essential in yeast.

tween the fragment and the chromosome, both of the resulting duplicated gene copies will be defective, one because it lacks the beginning of the gene, the other because it lacks the end. This procedure was first used to show that the gene coding for yeast actin is essential. In these experiments, a diploid *URA3⁻* cell was transformed to the *URA3⁺* phenotype by a plasmid carrying both a *URA3⁺* gene and the incomplete internal fragment of the actin gene. When the 50 percent of the *URA3⁺* cells in which integration had occurred within the actin gene (as opposed to within the *URA3* gene) sporulated, only *URA3⁻* spores were produced. No *URA3⁺*-containing chromosomes survived in the haploid state, since they lacked any functioning actin genes (Figure 19-12). This procedure provides a powerful way to test whether genes of still unknown function carry out essential roles.

Measuring the Activity of Fused Genes²⁸, ²⁹

Gene expression is best studied when the genes concerned code for products that can be assayed quickly and accurately with minimal expense. That the lactose operon of *E. coli* is still the best understood of all genetic segments is no accident. There is a simple colorimetric test for its principal product, the enzyme β-galactosidase, that unambiguously detects the enzyme's presence either in permeabilized cells or in cell-free extracts. Were β-galactosidase with its 1021 amino acids an unstable, hard-to-measure protein, we would never have concentrated on its underlying gene.

Yeasts, however, do not metabolize lactose and so normally do not produce any β-galactosidase-like enzyme. However, a close derivative of β-galactosidase can be made in yeasts using genetic engineering tricks to fuse the 5' upstream control elements and short amino-terminal sequences of selected yeast genes to the very long DNA fragments coding the carboxyl-terminal 990 amino acids of β-galactosidase (Figure 19-13). Since the first 30 amino acids at the amino-terminal end of β-galactosidase are not required for enzyme activity, the resulting fusion proteins have full β-galactosidase activity and effective stability. Such fused genomes allow the activity of mutated yeast 5' control elements to be monitored through the amount of β-galactosidase activity that is synthesized, rather than by the much more laborious methods needed, say, to quantitate the amount of the uracil biosynthetic enzyme coded by the *URA3* gene or the amount of cytochrome c1, the product of the *CYC1* gene.

Figure 19-13
Assaying for unstable, hard-to-measure proteins using gene fusion techniques. Here the *GAL4* promoter is assayed by fusing the promoter to segments of the β-galactosidase gene that code for the carboxy-terminal 990 base pairs of enzyme. A simple colorimetric test is available to measure the amount of β-galactosidase activity, and since the genes are fused, this is also a measure of the expression of the *GAL4* gene.

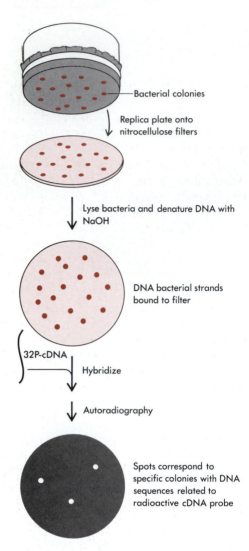

Bacterial colonies

Replica plate onto
nitrocellulose filters

Lyse bacteria and denature DNA with
NaOH

DNA bacterial strands
bound to filter

32P-cDNA

Hybridize

Autoradiography

Spots correspond to
specific colonies with DNA
sequences related to
radioactive cDNA probe

Figure 19-14
Detection of related DNA sequences by
hybridization of single-stranded DNA
on nitrocellulose filters. Bacterial colonies are replica plated onto nitrocellulose filters and treated with sodium
hydroxide (NaOH) to lyse the bacteria
and denature the DNA, which remains
attached to the filter after baking. The
filter is then exposed to specific radioactively labeled cDNA probes. Colonies
with DNA sequences related to the
cDNA probe hybridize to the probe and
can be detected by their radioactivity.

Detecting Related DNA Sequences Through Hybridization Procedures[30, 31, 32]

Once a gene has been cloned, its DNA can be used to make a restriction map and to determine whether similar DNA sequences exist elsewhere. The degree of relatedness between two DNA segments is best
approached by measuring the ability of their component single
strands to hybridize under appropriate conditions to form complementary double helices. A given cloned gene (restriction fragment)
actually becomes a "probe" to search out other DNA segments of
identical or nearly identical sequences. Now there are simple procedures in which such probes are used for the mass screening of many
thousands of plaques or bacterial colonies. Such procedures always
involve the production of replicas of the plaques or bacterial colonies
on nitrocellulose paper. The DNA within these replicas is first fixed
on the paper and then exposed to radioactively labeled DNA probes
to pinpoint those plaques or colonies that contain related DNA sequences (Figure 19-14).

Probes can also be made of RNA. The addition of labeled rRNA or
tRNA chains to appropriate library replicas quickly reveals those
members carrying the respective rDNA or tDNA genes. Messenger
RNA molecules also can be used as probes when they are available in
partially purified form. But since most cells contain more than a thousand different mRNA molecules, it is seldom possible to obtain
worthwhile probes from the ordinary cell. Only when a single cell
devotes much of its protein-synthesizing capacity to making single
proteins (e.g., hemoglobin and immunoglobin) can the respective
mRNA probes pick out their corresponding genes from a genomic
library. To obtain pure mRNA probes, the mRNA molecules themselves must be cloned through their reverse transcription into complementary DNA (cDNA) and subsequent insertion into appropriate
vectors.

Analyzing Related Genes Through "Southern" Blotting[33]

Given the availability of a suitable mRNA, cDNA, or genomic probe,
important preliminary data on the structure of related genes are generated through a technique developed in Edinburgh by the molecular
biologist E. M. Southern and which is now called **Southern blotting.**
It starts with the digestion of a DNA population by one or several
restriction enzymes. The resulting fragments, after separation on gels
according to their size, are transferred to nitrocellulose sheets so that
an exact replica of the DNA fragments in the gel is present on the
nitrocellulose filter. A specific radioactive probe is then applied to the
filter under hybridizing conditions. Subsequent autoradiography of
the filter results in a specific pattern of bands corresponding to the
discrete restriction fragments that are complementary to the DNA or
RNA probe (Figure 19-15).

In conjunction with recombinant DNA technology, Southern blots
can easily provide a physical map of restriction sites within a gene in
its normal chromosomal location, as well as provide information on
the number of copies of the gene in the genome, and the degree of
similarity of the gene under study to other known homologs.

DNA molecule

Cleavage with one or more restriction enzymes

Restriction fragments

Agarose gel electrophoresis

Gel with fragments fractionated by size

Flow buffer used to transfer DNA

Transfer to nitrocellulose filter

Gel

Filter

Hybridization with radioactively labeled DNA probe

Nitrocellulose filter with DNA fragments positioned identically to that in the gel

Autoradiograph showing hybrid DNA

Figure 19-15
Southern blotting method of analyzing DNA segments that share homology with a nucleic acid probe.

Complementary DNA
Cloning Using Poly A Tails[34]

There are many important uses for the gene libraries made by cloning the population of mRNA molecules present within specialized cells. In particular, they afford a way to isolate those gene segments that

Figure 19-16
Poly A tails on mRNA may be used to form oligo dT primers for reverse transcription to create double-stranded DNA. This DNA can then be inserted into a plasmid, which is then inserted into *E. coli*. In this way, a desired gene or gene segment can be isolated and amplified.

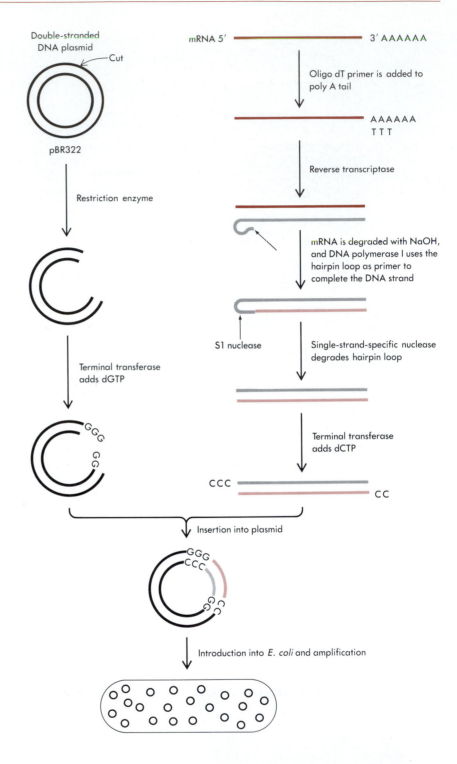

code for exons away from their noncoding, intron components. In this procedure, the information in mRNA is copied into cDNA (complementary DNA) replicas using the retrovirus enzyme *reverse transcriptase*, which normally functions to convert infecting retroviral RNA chains into DNA copies (Chapter 24). Reverse transcriptase, like all other DNA polymerases, can initiate DNA synthesis only by adding onto primers. Fortunately, it is easy to prime cDNA synthesis by utilizing the stretches of poly A (called a **poly A tail**) at the 3′ ends of most mRNA molecules. Added oligo dT binds to these poly A tails and serves as the necessary primer. Synthesis of a single-stranded complementary DNA chain can run to the 5' end of its template, where it often makes a hairpin loop turn, which serves as a primer for the synthesis of a complementary DNA strand. The loop of the hairpin can subsequently be dissected away by a nuclease specific for single-stranded DNA.

The resulting double-stranded cDNA can then be inserted into a plasmid like pBR322 either by adding complementary tails (e.g., poly G and poly C) using the enzyme terminal transferase or by attaching chemically synthesized restriction enzyme sites (linkers) using the enzyme DNA ligase (Figure 19-16). The cDNA-containing plasmids so constructed can then be introduced into *E. coli* and amplified.

Complementary DNA libraries will of necessity be heavily biased toward members representing the more abundant classes of mRNA found in given cells. It should not be taken for granted, however, that they must always contain the member we expect. The reverse transcription of long mRNA molecules into complete cDNA copies frequently aborts, producing short fragments representing only part of the gene. The making of a highly representative large-copy-number **cDNA library** is thus a much harder task than the construction of its equivalent **genomic library**.

Identifying the Products of cDNA Clones[35, 36, 37]

One of the most direct ways to screen a cDNA library is to use individual members as traps for the collection of specific mRNA molecules that in turn are added to *in vitro* systems for protein synthesis. With luck, the mRNA class selected out by its hybridization to the denatured strands of a given cDNA library member will direct the synthesis of a protein that can be identified by its position on a two-dimensional protein gel (Figure 19-17). Such procedures work best when the proteins concerned have been previously well characterized and are major cellular constituents (e.g., the hemoglobin molecules synthesized by red blood cells, or one of four major histones). But with organisms like yeast, whose proteins have not yet been well characterized, this form of screening is not a practical way to find most desired cDNA clones. Once a cDNA clone has been so identified, however, it is very straightforward to pick out the corresponding genomic clone using DNA-DNA hybridization procedures.

Those cDNAs made off less abundant mRNAs will obviously be much harder to identify. Directly looking for the protein products necessarily involves the testing of many hundreds or thousands of clones, often a costly task necessarily limited to industries. If, however, the proteins of interest have been characterized by partial or full amino acid sequencing, then an important shortcut may be used. Given an amino acid sequence, intelligent guesses can be made as to its corresponding mRNA (DNA) sequence. Because all amino acids but one are specified by more than one codon (Chapter 15), it is not possible to go from an amino acid sequence to a DNA sequence unambiguously. By focusing, however, on sequences that mainly contain the less common amino acids, it is usually possible to define a small collection of oligonucleotides, one of which should be exactly complementary to the segment of interest (Figure 19-18). Such a restricted collection can then be used as probes to identify the complementary cDNA clones by hybridization. Already this approach has

Figure 19-17
Identification of a cDNA-containing plasmid by translation of complementary mRNA selected from cellular mRNA by hybridization to the cDNA insert. The protein product is detected by gel chromatography.

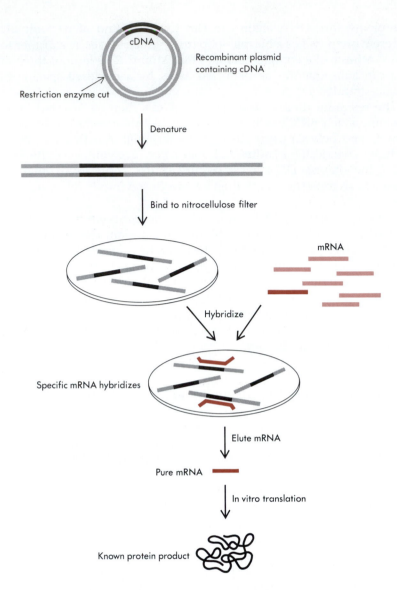

been used to isolate a number of important cDNA clones, and while it never is as simple as described here, it nevertheless is a practical technique that is bound to be increasingly employed.

Identifying cDNA Clones Through Their Expression in *E. coli*[38–43]

It is increasingly possible to find desired eucaryotic cDNA clones through techniques that allow their coding regions to function in *E. coli* strains that have been especially tailored to express foreign genes. Such strains are in fact called **expression strains.** In these techniques, the eucaryotic sequences are usually attached just downstream from a strong bacterial (or phage) promoter-ribosome binding site complex like those that govern the tryptophan and lactose operons or the expression of the phage λ genome. These gene fusions can easily be made to yield fusion proteins containing short bacterial amino-terminal segments linked to virtually all the amino acids of the eucaryotic protein (Figure 19-19). Often these hybrid proteins can be detected because they possess almost the total functional potential

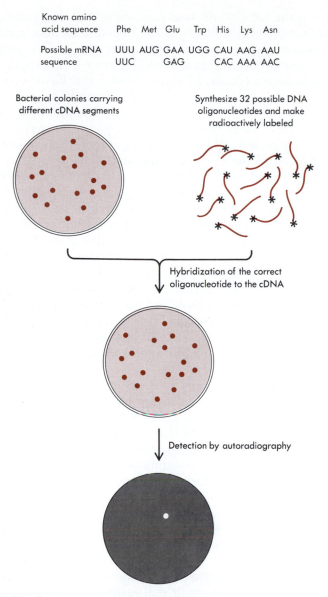

Known amino acid sequence	Phe	Met	Glu	Trp	His	Lys	Asn
Possible mRNA sequence	UUU UUC	AUG	GAA GAG	UGG	CAU CAC	AAG AAA	AAU AAC

Bacterial colonies carrying different cDNA segments

Synthesize 32 possible DNA oligonucleotides and make radioactively labeled

Hybridization of the correct oligonucleotide to the cDNA

Detection by autoradiography

Eucaryotic DNA

Bacterial promoter and ribosome binding site

Bacterial plasmid

Gene fusion

Introduction into *E. coli*; transcription of eucaryotic DNA into mRNA and translation of mRNA to produce hybrid protein

mRNA

Retrieval of eucaryotic protein product

H$_2$N

Protein product with bacterial segment attached to amino terminus

Figure 19-18, above

Cloning genes by use of a probe designed from a known amino acid sequence. Given the amino acid sequence, it is possible to synthesize DNA oligonucleotides for the possible mRNA sequences. The correct oligonucleotide will hybridize to the appropriate cDNA that we wish to identify that is carried in the plasmid. Since the oligonucleotide probe is radioactively labelled, the hybrid DNA can be detected by autoradiography.

Figure 19-19, right

Expression of eucaryotic DNA in *E. coli* allows high-level expression and identification of its protein product. Eucaryotic DNA can be introduced into a plasmid just after a strong bacterial promoter. This allows its subsequent translation into mRNA when introduced into *E. coli*. The mRNA with eucaryotic sequences will be translated into functional proteins, since *E. coli* ribosome binding sites are also present on the plasmid. An amino-terminal bacterial segment is usually attached to the protein product, but this often can be enzymatically cleaved away.

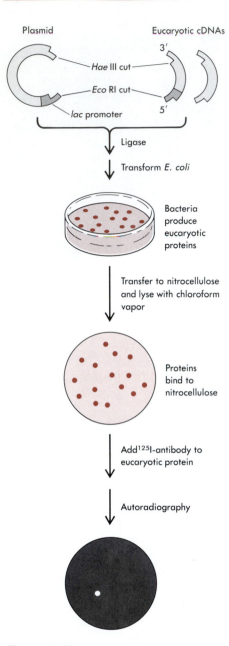

Figure 19-20
Immunological screening of expression vectors to detect desired eucaryotic-bacterial hybrid proteins. Proteins produced by bacteria are transferred to nitrocellulose filters and treated with chloroform to make them permeable to antibodies. An antibody specific for the hybrid protein we want to isolate is radioactively labeled and added to the filter, specifically binding to the correct protein. The protein can then be detected by autoradiography.

(e.g., enzymatic activity) of their pure eucaryotic equivalent. However, the hybrid proteins can usually be found more quickly by screening very large numbers of **expression vectors** for their possession of fusion products that specifically bind to appropriate antibodies. In such procedures, bacterial colonies that have been treated with chloroform to make them permeable to antibodies are exposed to radioactively labeled antibodies of the correct specificity. The usually rare colonies that are making immunologically cross-reacting proteins are easily picked out by autoradiography (Figure 19-20). To replace cleanly all of the *E. coli* coding sequences with those coding for an eucaryotic protein, appropriate restriction enzyme sites must occur within a very short and specific sequence of base pairs. It is now possible to produce such sites by in vitro synthesis. Consequently, new expression vectors have been constructed that greatly simplify the task of producing eucaryotic proteins without the bacterial peptides, and thus often make possible the synthesis in *E. coli* of perfect replicas of desired eucaryotic proteins. These vectors are often required when the primary purpose is not the detection of a given eucaryotic gene but the synthesis of its effectively pure product. Expression in *E. coli* is proving especially valuable for the purpose of making large amounts of normally rare eucaryotic proteins (e.g. the various interferons and growth factors) for their subsequent molecular (functional) characterization.

The bacterial promoters used in *E. coli* expression vectors usually have their expression controlled either like the lactose promoter, by an externally added molecule (e.g., an inducer), or like mutant λ, by a temperature-sensitive repressor that functions at 31°C but not at 37°C. In this way, an expression vector can be instructed at appropriate moments to make very large amounts of a protein whose continuous presence in large amounts during the growth of a bacterial population would be deleterious.

The routine use of **cDNA inserts** as opposed to genomic inserts of eucaryotic DNA into bacterial expression vectors is the consequence of the split nature of so many eucaryotic genes. As we explained in the previous chapter, while only a small fraction of the genes of *S. cerevisiae* are split, in higher eucaryotic cells only the exceptional gene is not split. Thus, many (if not most) of the genes of a higher eucaryotic organism will produce abnormally large, usually nonfunctional protein products when placed in an *E. coli* expression vector. Of course, the hassle of cDNA cloning can be avoided when the overexpression of an *E. coli* gene (none of which are split) itself is desired. Then the much more routinely cloned genomic segments can be inserted into the appropriate expression vector.

Eucaryotic Expression Vectors[44]

The development of multiple transforming systems (e.g., YEp, YRp, and YCp) for yeasts automatically creates the potential for using yeasts as hosts for expression vectors. The more we learn about how gene expression is controlled in eucaryotes, the more intelligently we can develop expression vector systems that can rival those already in use in *E. coli*. Several factors already encourage the development of eucaryotic systems for the expression of eucaryotic genes. First, such systems may more readily allow the passage through their external

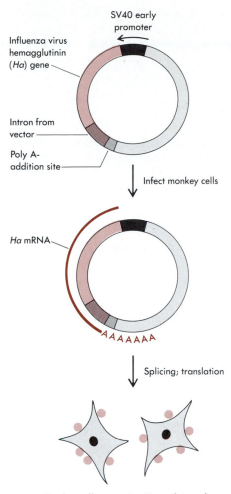

SV40 early
promoter

Influenza virus
hemagglutinin
(*Ha*) gene

Intron from
vector

Poly A-
addition site

Infect monkey cells

Ha mRNA

AAAAAAA

Splicing; translation

Monkey cells expressing *Ha* on their surface

Figure 19-21
Many eucaryotic proteins cannot pass through procaryotic membranes. Therefore, expression vectors have been created that can be expressed in eucaryotic cells. Here, the SV40 early promoter is used to express flu virus hemagglutinin in monkey cells.

plasma membrane of proteins that normally are exported outside the cells in which they are translated. Such secreted proteins are frequently held together by large numbers of internal disulfide bonds (S–S) that are formed only after the passage of their polypeptide chains through cell membranes. Many of these exported eucaryotic proteins, for reasons that are still unclear, cannot pass through the *E. coli* membrane and, as a consequence, never form the correct sets of S–S bonds. The functional expression of many eucaryotic proteins thus demands their synthesis within the proper eucaryotic (as opposed to procaryotic) cells (Figure 19-21).

Also favoring the use of eucaryotic cells for the expression of eucaryotic genes is the ability of eucaryotic cells to process (splice) the newly synthesized pre-mRNA products of split genes. There is no reason why a yeast gene on a yeast vector ever needs to be converted into cDNA. In contrast, when such genomic insertions are placed in *E. coli*, only those genes lacking noncoding inserts (called introns) are likely to produce functional products. Thus, it is hoped that expression vectors for use in all genetically well-characterized organisms will soon be developed. When it comes to choosing which vector to use for a stated purpose, the fact that eucaryotic microorganisms like yeast can be grown both faster and cheaper will favor their use whenever possible.

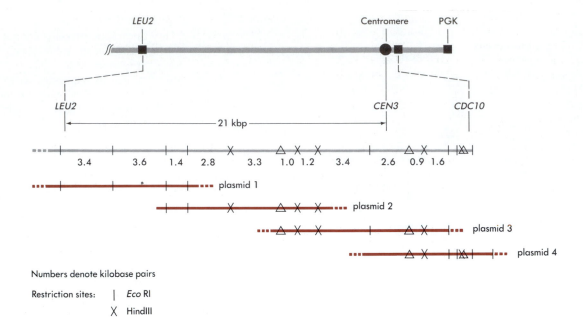

Numbers denote kilobase pairs

Restriction sites: | *Eco* RI
X HindIII
△ BamHI

Figure 19-22
The centromere region of *S. cerevisiae* chromosome 3 was cloned by the method of chromosome walking. Starting with plasmids that contain genes flanking the centromere (plasmid 1 and plasmid 4), the region in between was cloned by overlap hybridization. The four yeast DNA inserts obtained in this chromosome walk and the regions of overlap are indicated as colored lines. [Redrawn from L. Clarke and J. Carbon, *Nature* 287 (1980):505, with permission.]

Analyzing Long Stretches of Eucaryotic DNA Using Chromosome Walks[45, 46, 47]

Eventually, the structure of very long stretches of eucaryotic DNA, if not complete eucaryotic chromosomes, will be worked out. In accomplishing this task, DNA hybridization techniques will provide the easiest way to pick out clones from gene libraries that have overlapping sequences. Thanks to such "chromosome walking," (also called "overlap hybridization") the first centromeric sequences on a yeast chromosome has already been isolated. Two cloned genes *CDC*10 and *LEU2*, that had been mapped to opposite sides of the centromere were used as starting points for two walks that finally reached each other in a total of 4 steps (Figure 19-22). Each of the clones so identified were then tested to see which one(s) ensured regular plasmid segregation during mitosis and meiosis. Walks have also been used to establish the structure of some very large *Drosophila* genes that are much too long even to be encompassed on cosmid clones.

Unfortunately, chromosome walks are almost impossible to carry out in higher plants and animals, since their chromosomes contain large numbers of nearly identical repetitive sequences (Chapter 20). Their presence adjacent to so many genes leads to large numbers of cross-hybridizing clones, whose related sequences reflect the presence of common repetitive elements rather than overlapping unique chromosomal DNA. Hence, an easy and effective method to link up restriction enzyme–generated fragments along the chromosomes of higher organisms is not yet available.

Summary

Effective experimentation on the structure and functioning of the eucaryotic genome is only possible using recombinant DNA techniques. To understand how eucaryotic genes work, we must first isolate them in pure form through gene-cloning procedures and then reintroduce them into cells where their expression can be studied. The first step in the cloning of a gene is usually its insertion into a gene library made with an appropriate plasmid, phage λ, or cosmid vector. The members of these libraries are then individually propagated in cells where the function of the respective library member can be tested.

A simple way to screen a library is to see whether it contains members that can complement the functions of genetic defects present in mutant cells. In this way, it is relatively easy to isolate intact copies of any *E. coli* gene for which a functionally deficient mutant already exists. It was surprising to find that many yeast genes complement their functionally equivalent mutated *E. coli* genes. This shows not only that some transcriptional start signals on the yeast genome can be read by *E. coli* RNA polymerase, but also that the ribosomes and tRNAs of *E. coli* can translate, albeit sometimes poorly, yeast mRNAs.

The identification of the first cloned yeast gene allowed transformation procedures to be developed for the introduction of yeast DNA into yeast protoplasts. When such yeast DNA is present on an *E. coli* plasmid, the reintroduced DNA becomes stabilized within the yeast genome by crossing over with a homologous section of yeast DNA (integrative recombination). Yeast DNA can also be functionally introduced into cells using any of three different plasmids: a derivative of the resident 2μ yeast episomal plasmid (YEp); the yeast replicating plasmid (YRp), which contains a yeast origin of DNA replication called an autonomously replicating sequence (ARS); and a plasmid (YCp) containing both a centromere and a yeast origin of replication. Vectors containing two different origins of replication that allow them to replicate in two different organisms (e.g., in both *E. coli* and yeast) are called shuttle vectors.

The frequency of integrative recombination is greatly increased by making a double-stranded cut in the DNA of the gene we wish to integrate. The resulting free ends promote recombination by invading homologous DNA segments to form heteroduplex intermediates. These subsequently become resolved through appropriate nuclease cuts to yield genetically recombined DNA. By using shuttle vectors containing gaps in the gene being studied, any desired normal or mutant gene can be retrieved and its structure subsequently worked out. Integrative recombination now allows any yeast gene to be replaced with its mutated derivative and so provides a powerful way to probe the physiological role of genes whose function is not known.

After a gene is cloned, the manner in which its expression is controlled is studied using in vitro mutagenesis procedures. The most powerful of these procedures allows specified base changes to be made at precise locations along already sequenced DNA molecules. Also very useful are gene fusion procedures in which the promoter for one gene is attached to the amino acid coding element of another gene. In this way, the activity of mutated promoters can be measured through the presence of an easily assayable gene product (e.g., the color test for *E. coli* galactosidase activity) instead of by laborious measurements of enzymatic activity.

Many of the elegant recombinant DNA techniques now available for work with yeast genes remain restricted to this system. In most higher eucaryotic organisms, the recombinant DNA procedures used both for initial gene cloning and for later understanding of the rules by which eucaryotic gene expression is controlled are much more laborious. When practical, hybridization procedures (DNA-DNA or DNA-RNA) are used not only to search for genes with homologous sequence, but also to work out the sequences of long DNA stretches of a given chromosome (chromosome walks). When a gene to be cloned is suspected of being split (containing introns), then the most efficient way to isolate it usually involves reverse transcription of its mRNA product into cDNA molecules. These cDNAs are then placed into appropriate expression vectors, which allow expression of the DNA. Although the first good expression vectors involved *E. coli* systems, there are now available equally useful expression vectors for a variety of eucaryotic organisms.

Bibliography

General References

Botstein, D., and R. W. Davis. 1982. "Principles and Practice of Recombinant DNA Research with Yeast." In *The Molecular Biology of the Yeast* Saccharomyces, *Metabolism and Gene Expression*, ed. J. N. Strathern, E. W. Jones, and J. R. Broach. Cold Spring Harbor, N.Y.: Cold Spring Harbor Laboratory, pp. 607–638.

Davis, R. W., D. Botstein, and J. R. Roth. 1980. *Advanced Bacterial Genetics*. Cold Spring Harbor, N.Y.: Cold Spring Harbor Laboratory.

Maniatis, T., E. F. Fritsch, and J. Sambrook. 1982. *Molecular Cloning*. Cold Spring Harbor, N.Y.: Cold Spring Harbor Laboratory.

Silhavy, T. J., M. L. Berman, and L. W. Enquist. 1984. *Experiments with Gene Fusions*. Cold Spring Harbor, N.Y.: Cold Spring Harbor Laboratory.

Watson, J. D., J. Tooze, and D. T. Kurtz. 1983. *Recombinant DNA: A Short Course*. San Francisco: Freeman.

Cited References

1. Collins, J., and B. Hohn. 1978. "Cosmids: A Type of Plasmid Gene Cloning Vector That Is Packageable *in Vitro* in Bacteriophage Heads." *Proc. Nat. Acad. Sci.* 75:4242–4246.

2. Maniatis, T., R. C. Hardison, E. Lacy, J. Lauer, C. O'Connell, D. Quon, G. K. Sim, and A. Efstratiadis. 1978. "The Isolation of Structural Genes from Libraries of Eucaryotic DNA." *Cell* 15:687–701.

3. Struhl, K., J. R. Cameron, and R. W. Davies. 1976. "Functional Genetic Expression of Eucaryotic DNA in *E. coli*." *Proc. Nat. Acad. Sci.* 73:1471–1475.

4. Ratzkin, B., and J. Carbon. 1977. "Functional Expression of Cloned Yeast DNA in *E. coli*." *Proc. Nat. Acad. Sci.* 74:487–491.

5. Hinnen, A., J. B. Hicks, and G. R. Fink. 1978. "Transformation of Yeast." *Proc. Nat. Acad. Sci.* 75:1929–1933.

6. Beggs, J. D. 1978. "Transformation of Yeast by a Relicating Hybrid Plasmid." *Nature* 275:104–109.

7. Gerbaud, C., P. Fournier, H. Blanc, M. Aigle, H. Heslot, and M. Guerineau. 1979. "High Frequency Yeast Transformation by Plasmids Carrying Part or Entire 2μm Yeast Plasmid." *Gene* 5:233–237.

8. Broach, J. R. 1981. "The Yeast Plasmid 2μ Circle." In *The Molecular Biology of the Yeast* Saccharomyces, *Life Cycle and Inheritance*, eds. J. N. Strathern, E. W. Jones, and J. R. Broach. Cold Spring Harbor, N.Y.: Cold Spring Harbor Laboratory, pp. 445–470.

9. Struhl, K., D. T. Stinchcomb, S. Scherer, and R. W. Davies. 1979. "High-Frequency Transformation of Yeast: Autonomous Replication of Hybrid DNA Molecules." *Proc. Nat. Acad. Sci.* 76:1035–1039.

10. Stinchcomb, D. T., K. Struhl, and R. W. Davies. 1979. "Isolation and Characterization of a Yeast Chromosomal Replicator." *Nature* 282:39–43.

11. Hsiao, C.-L., and J. Carbon. 1981. "Characterization of a Yeast Replication Origin (*ARS2*) and Construction of Stable Minichromosomes Containing Cloned Yeast Centromere DNA (*CEN3*)." *Gene* 15:157–166.

12. Clarke, L., and J. Carbon. 1980. "Isolation of a Yeast Centromere and Construction of Functional Small Circular Chromosomes." *Nature* 287:504–509.

13. Clarke, L., and J. Carbon. 1985. "The Structure and Function of Yeast Centromeres." *Annual Review of Genetics* 19:29–57.

14. Orr-Weaver, T. L., J. W. Szostak, and R. L. Rothstein. 1981. "Yeast Transformation: A Model System for the Study of Recombination." *Proc. Nat. Acad. Sci.* 78:6354–6358.

15. Hicks, J. B., J. N. Strathern. A. J. S. Klar, and S. L. Dellaporta. 1982. "Cloning by Complementation in Yeast: The Mating Type Genes." In *Genetic Engineering—Principles and Methods*, vol. 4, ed. J. K. Setlow and A. Hollaender. New York: Plenum, pp. 219–248.

16. Shortle, D., and D. Nathans. 1979. "Regulatory Mutants of Simian Virus 40: Constructed Mutants with Base Substitutions at the Origin of Viral DNA Replication." *J. Mol. Biol.* 131:801–817.

17. Shortle, D., D. DiMaio, and D. Nathans. 1981. "Directed Mutagenesis." *Ann. Rev. Genetics* 15:265–294.

18. McKnight, S. L., and R. Kingsbury. 1982. "Transcriptional Control Signals of a Eukaryotic Protein-Coding Gene." *Science* 217:316–324.

19. Winter, G., A. R. Fersht, A. J. Wilkinson, M. Zoller, and M. Smith. 1982. "Redesigning Enzyme Structure by Site-Directed Mutagenesis." *Nature* 296:756–758.

20. Zakour, R. A., and L. A. Loeb. 1982. "Site-Specific Mutagenesis by Error-Directed DNA Synthesis." *Nature* 195:708–710.

21. Smith, M. (1985). "In vitro Mutagenesis." *Annual Review of Genetics* 19:423–463.

22. Heffron, F., M. So, and B. J. McCarthy. 1978. "*In vitro* mutagenesis of a circular DNA molecule by using synthetic restriction sites." *Proc. Nat. Acad. Sci.* 75:6012–6016.

23. Totchell, K., K. A. Nasmyth, B. D. Hall, C. Astell, and M. Smith. 1981. "*In vitro* mutation analysis of the mating-type lows in yeast." *Cell* 27:25–35.

24. McKnight, S. L., and R. Kingsbury. 1982. "Transcriptional Control Signals of a Eukaryotic Protein-Coding Gene." *Science* 217:316.

25. Scherer, S., and R. W. Davies. 1979. "Replacement of Chromosome Segments with Altered DNA Sequences Constructed *in Vitro*." *Proc. Nat. Acad. Sci.* 76:4951–4955.

26. Shortle, D., J. Haber, and D. Botstein. 1982. "Lethal Disruption of the Yeast Actin Gene by Integrative DNA Transformation." *Science* 217:371–373.

27. Rothstein, R. J. 1983. "One-step Gene Disruption in Yeast." *Methods in Enzymology* 101:202–211.

28. Guarente, L., and M. Ptashne. 1981. "Fusion of *E. coli* lacZ to the Cytochrome *c* Gene of *Saccharomyces cerevisae*." *Proc. Nat. Acad. Sci.* 78:2199–2203.

29. Rose, M., M. J. Casadaban, and D. Botstein. 1981. "Yeast Genes Fused to Beta-Galactosidase in *E. coli* Can Be Expressed Normally in Yeast." *Proc. Nat. Acad. Sci.* 78:2460–2464.

30. Grunstein, M., and D. S. Hogness. 1975. "Colony Hybridization: A Method for the Isolation of Cloned DNAs That Contain a Specific Gene." *Proc. Nat. Acad. Sci.* 72:3961–3965.

31. Efstratiadis, A., F. C. Kafatos, A. M. Maxam, and T. Maniatis. 1976. "Enzymatic *in Vitro* Synthesis of Globin Genes." *Cell* 7:279–288.

32. Benton, W. D., and R. W. Davis. 1977. "Screening λgt Recombinant Clones by Hybridization to Single Plaques *in Situ*." *Science* 196:180–182.

33. Southern, E. M. 1975. "Detection of Specific Sequences Among DNA Fragments Separated by Gel Electrophoresis." *J. Mol. Biol.* 98:503–517.

34. Efstratiadis, A., and L. Villa-Komaroff. 1979. "Cloning of Double-Stranded DNA." In *Genetic Engineering*, vol. 1, ed. J. K. Setlow and A. Hollaender. New York: Plenum, pp. 15–27.

35. Paterson, B. M., B. E. Roberts, and E. L. Kuff. 1977. "Structural Gene Identification and Mapping by DNA-mRNA Hybrid-Arrested Cell-Free Translation." *Proc. Nat. Acad. Sci.* 74:4370–4374.

36. Noyes, B. E., M. Mevarech, R. Stein, and K. L. Agarwal. 1979. "Detection and Partial Sequence Analysis of Gastrin mRNA by Using an Oligodeoxynucleotide Probe." *Proc. Nat. Acad. Sci.* 76:1770–1774.

37. Wallace, R. B., J. Schaffer, R. F. Murphy, J. Bonner, T. Hirose, and K. Itakura. 1979. "Hybridization of Synthetic Oligodeoxynucleotides to 0X174 DNA: The Effect of Single Base Pair Mismatch." *Nucleic Acid Res.* 6:3543–3557.

38. Backman, K., and M. Ptashne. 1978. "Maximizing Gene Expression on a Plasmid Using Recombinant DNA *in Vitro*." *Cell* 13:65–71.

39. Goeddel, D. V., H. L. Heyneker, T. Hozumi, R. Arentzen, K. Itakura, D. G. Yansura, M. J. Ross, G. Miozzari, R. Crea, and P. H. Seeburg. 1979. "Direct Expression in *E. coli* of a DNA Sequence Coding for Human Growth Hormone." *Nature* 281:544–548.

40. Guarente, L., G. Lauer, T. M. Roberts, and M. Ptashne. 1980. "Improved Methods for Maximizing Expression of a Cloned Gene: A Bacterium That Synthesizes Rabbit β-Globin." *Cell* 20:543–553.

41. Remaut, E., P. Stanssens, and W. Fiers. 1981. "Plasmid Vectors for High-Efficiency Expression Controlled by the p_L Promoter of Coliphage Lambda." *Gene* 15:81–93.

42. Silhavy, T. J., M. L. Berman, and L. W. Enquist. 1984. *Experiments with Gene Fusions*. Cold Spring Harbor, N.Y.: Cold Spring Harbor Laboratory.

43. Young, R. A., and R. W. Davies. 1983. "Efficient Isolation of Genes Using Antibody Probes." *Proc. Nat. Acad. Sci.* 80:1194–1198.

44. Gluzman, Y. 1982. *Eukaryotic Viral Vectors*. Cold Spring Harbor, N.Y.: Cold Spring Harbor Laboratory.

45. Bender, W., P. Spierer, and D. Hogness. 1979. "Gene Isolation by Chromosomal Walking." *J. Supra. Mol. Struc.* 10 (suppl.):32.

46. Chinault, A. C., and J. Carbon. 1979. "Overlap Hybridization Screening: Isolation and Characterization of Overlapping DNA Fragments Surrounding the *LEU2* Gene on Yeast Chromosome III." *Gene* 5:111–126.

47. Clarke, L., and J. Carbon. 1980. "Isolation of a Yeast Centromere and Construction of Small Circular Chromosomes." *Nature* 287:504–509.

VIII

THE FUNCTIONING
OF EUCARYOTIC
CHROMOSOMES

CHAPTER 20

The Unexpected Structures of Eucaryotic Genomes

After many years of work on bacteria and their viruses, even the most skeptical scientists were certain that our understanding of how proteins are encoded in DNA would be automatically transferable to higher plants and animals. Early studies had established that the central dogma (DNA makes RNA and RNA makes protein) holds for mammalian cells as it does for bacteria. Moreover, the genetic code appeared to be universal. Hence, no one would have expected that another fundamental tenet of gene expression might not apply to eucaryotic cells. This was the concept of colinearity (Chapter 8), first deduced from genetic studies and later made absolutely firm when DNA, RNA, and protein sequences demonstrated a direct correspondence between the sequence of a gene, its mRNA, and its protein product.

The discovery of interruptions in eucaryotic genes therefore came as a complete shock to the scientific world in 1977. It was a unique sort of discovery in that experimental data accumulated in several laboratories forced a conclusion that no one could have foreseen. During the early years of the same decade, intensified studies had correctly indicated that every aspect of gene expression and its regulation in higher cells would be more complex than in bacteria. But only with the advent of cloning and rapid DNA-sequencing technology did the revolution in our understanding of the structure and functioning of eucaryotic genes begin.

The Amount of DNA per Cell Increases About Eight Hundredfold from *E. coli* to Mammals[1, 2]

For years, biologists had been puzzled about the very large DNA content found in most eucaryotic cells. For instance, the amount of DNA present per mammalian cell is approximately 800 times that in *E. coli*. Since there was no reason to believe that gene size increased from the lower to the higher forms of life, this number was used to calculate an upper limit of the number of different genes. Using the assumption that each gene is present in only one copy per haploid genome, a mammalian cell was predicted to be capable of synthesizing over two million different proteins. Thus, the task of relating a given protein to its gene was viewed as much, much more formidable than the corresponding job with bacteria.

Figure 20-1

Haploid amounts of DNA in various cells, expressed as multiples of the amounts found in *E. coli* (4×10^{-12} mg = 2.4×10^9 MW). Many of these values should be taken as approximate. [After R. Holliday, *Symp. Soc. Gen. Microbiol.* 20 (1970):362.]

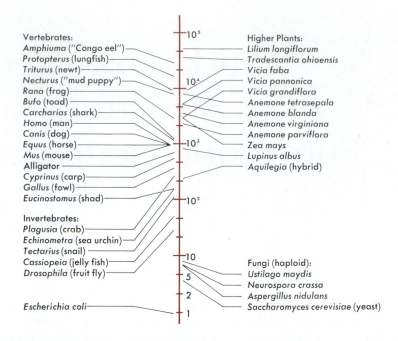

In some cases, however, the amount of DNA appeared misleading. There are groups of amphibians that contain 25 times more DNA in their cells than is present in mammalian cells (Figure 20-1). Here, there is no reason to believe that greater biological complexity is involved. Rather, this situation hints that we should be very cautious about relating DNA content directly to the number of different proteins that may be synthesized by a given cell. Morphological examination with the electron microscope tells us that a much larger variety of subcellular structures do exist in the vertebrate cell than in *E. coli* (compare Figure 20-2 with Figure 4-7). Correspondingly, a larger number of structural proteins and enzymes should be necessary for their construction and function. This expectation is borne out by two-dimensional gel displays of proteins from mammalian cells, which do reveal many times more polypeptide spots than those of *E. coli* proteins (see Figure 4-6). Nonetheless, it would be surprising if the vertebrate cell were more than 50 to 100 times more complex genetically than *E. coli*.

The Surprisingly Long Length of Pre-mRNA Molecules[3, 4]

Other mystifying observations were made when scientists began to examine mammalian RNA synthesis in the late 1960s. When labeled RNA precursors were used and nuclear RNA was analyzed in sucrose gradients, the nonribosomal, non-tRNA molecules sedimented as if they were very large and very heterogeneous in size. This gave rise to the term **heterogeneous nuclear RNA (hnRNA)**. Heterogeneous nuclear RNA possesses many properties expected of lengthy messenger precursors. Its average size is about ten times the length of cytoplasmic mRNAs. It has a very short lifetime; about half the labeled nucleotides incorporated during a short pulse into hnRNA disappear within a 3-minute chase period ($t_{1/2} \cong 3$ minutes). And somewhat less than 10 percent of the label originally in hnRNA eventually appears in the cytoplasm in the form of mRNA-like molecules.

(a)

0.5 μm

Figure 20-2
The complexity of a vertebrate cell.
(a) An electron micrograph of part of a mammalian liver cell is seen in cross section. The major cellular compartments and structures are indicated. (Courtesy of Daniel Friend.) (b) A schematized version. [After B. Alberts et al., *Molecular Biology of the Cell* (New York: Garland, 1983), 322.] Also see Figures 1-1 and 1-2, demonstrating similar structures in a higher plant cell.

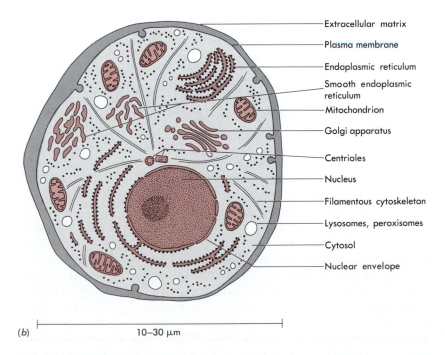

(b)   10–30 μm

Once poly A tails were identified on eucaryotic mRNAs in the early 1970s, hnRNA was also immediately checked. It, too, possessed 100- to 200-nucleotide-long stretches of A residues at its 3' termini. This served to strengthen further the belief that hnRNA is indeed a nuclear precursor form of cytoplasmic message. Thus, it seemed that extraordinarily long 5' leaders must be trimmed away by RNA processing events, sending 3' coding sequences with their poly A tails already attached out to the cytoplasm to be translated by the ribosomes.

This neat picture, however, did not survive long. The 5' cap structures on eucaryotic mRNAs were discovered shortly thereafter. Careful analyses revealed that the number of 5' caps on hnRNA, like the number of poly A tails, was approximately equal to the number of molecules present. How could both the 5' and 3' ends of hnRNA be preserved during processing into smaller mRNA molecules? The facts were thus unreconcilable with any simple model for the generation of mRNA from hnRNA. There the matter rested until 1977, when the discovery of interruptions in genes finally provided a rationale for the existence of extraordinarily long pre-mRNA molecules.

Amazing Sequence Arrangement at the 5' Ends of Adenovirus mRNAs[5–9]

Studies of the abundant messenger RNAs produced late in infection of human cells by adenovirus 2 were among the first to provide incontrovertible evidence that parts of the same mRNA molecule can be specified by noncontiguous segments of eucaryotic DNA. The genome of adenovirus is a double-stranded linear molecule of about 36 kilobase pairs (Chapter 24). Coding regions for the late (structural) proteins are located in the rightmost two-thirds of the viral DNA. RNA sequencing had revealed that several different late mRNAs possess not only the expected cap structures, but also an identical 11-nucleotide sequence at their 5' ends. Surprisingly, these 5' regions were not resistant to ribonuclease digestion in hybrids formed between the mRNAs and restriction fragments from the late region of the adenovirus genome, as would be expected if the DNA and RNA sequences were complementary along their entire length.

These hints that the 5' ends might not be encoded adjacent to the main bodies of the adenovirus late mRNAs were followed up in a series of elegant **R-looping experiments.** If an RNA molecule is incubated with a complementary double-stranded DNA under conditions where RNA-DNA hybrids form in preference to DNA duplexes, the DNA strand that possesses the same sequence as the RNA will be displaced. Such displacement loops (R loops) can be easily visualized using the electron microscope and their ends mapped relative to restriction endonuclease cut sites in the DNA (Figure 20-3a). When the late adenovirus mRNAs were hybridized under R-loop conditions with DNA fragments known to contain their protein-coding regions, projections were indeed formed by the 5' ends (as well as by the 3' poly A tails) of the mRNAs (Figure 20-3b). By preparing single-stranded restriction fragments from other parts of the adenovirus genome and adding them each in turn during the preparation of R loops, researchers were able to locate the DNA regions specifying the 5' projections (Figure 20-3c). Amazingly, several different late mRNAs had 5' ends derived from the same three separate locations in

Figure 20-3
How R-loop mapping showed the origins of 5′ leader sequences on adenovirus-2 late mRNAs. (a) The formation of R-loop structures in which the DNA strand with the sequence corresponding to the mRNA is displaced. (b) Electron micrograph and schematic diagram of an R loop observed after incubating hexon mRNA with a complementary DNA restriction fragment from the late region of the adenovirus-2 genome. (The black lines are DNA; the color line is mRNA.) Note that both 5′ and 3′ extensions are seen. [Electron micrograph courtesy of S. M. Berget, C. Moore, and P. A. Sharp, *Proc. Nat. Acad. Sci.* 74 (1977):3171–3175.] (c) Electron micrograph and schematic diagram of an R loop observed after incubating fiber mRNA with two DNAs: the adenovirus genome and a restriction fragment from the early region. (The Ad2 DNA and the restriction fragment are shown in black; the color line is mRNA.) The structure formed shows that part of the 5′ extension is specified by the DNA between positions 11.3 and 18.1 on the viral genome. [Electron micrograph courtesy of L. T. Chow, R. E. Gelinas, T. R. Broker, and R. J. Roberts, *Cell* 12 (1977):1–8.] (d) Map of the adenovirus-2 genome showing the location of DNA specifying the late transcripts and the tripartite leader (arising from positions 16.6, 19.6, and 26.6) found on each of the late mRNAs.

the leftmost one-third of the adenovirus genome. Short sequences occurring at map positions 16.6, 19.6, and 26.6 seemed to be joined to form a leader of 150 to 200 nucleotides common to many of the adenovirus late mRNAs! (Figure 20-3d.) Clearly, an unprecedented mechanism had to underlie the biosynthesis of these particular mRNAs. Was the phenomenon peculiar to viruses or a manifestation of some general mechanism for the expression of genes in mammalian cells?

Discovery of Interruptions in Ovalbumin, Immunoglobulin, and β-Globin Genes[10–17]

At the same time that the preceding data were being collected on adenovirus mRNAs, comparably bizarre observations were being made concerning several cellular genes, including immunoglobulin (see Chapter 23), β-globin, and the chicken gene for ovalbumin. Ovalbumin, the major component of egg white, is synthesized only by specialized cells of the oviduct. The ovalbumin gene was being investigated because its expression is regulated by female sex hormones. To understand this control process at the molecular level, several groups had set out to clone the ovalbumin gene using the newly available recombinant DNA techniques. The first step was isolation of the mRNA, only a moderately formidable task because of its abundance in the oviduct cells of laying hens. From the mRNA, a complementary DNA (cDNA) was produced (Chapter 19), cloned, and sequenced. The structure deduced for the 1872-nucleotide mRNA provided no real surprises: 1158 coding nucleotides (corresponding to the 386 amino acids of the protein) were preceded by a 64-nucleotide 5′ leader sequence and followed by a nontranslated stretch of 650 nucleotides at the 3′ end. The next step was to use the cloned cDNA to investigate possible differences in the structure of the ovalbumin gene between the cells of the laying-hen oviduct and the cells of tissues not engaged in synthesizing ovalbumin. Here the blot hybridizations (see Chapter 19) revealed a totally unanticipated situation. It was expected that analysis of genomic restriction fragments generated by enzymes that did not cut the ovalbumin cDNA would show only a single radioactive band containing the entire ovalbumin gene. Instead, several bands appeared in the autoradiograph of the hybridized blot! (Figure 20-4.) Moreover, the multiple-band pattern was the same whether the chromosomal DNA was derived from oviduct cells or erythrocytes (red blood cells, which do not make ovalbumin). Initially, such puzzling results were ascribed to experimental artifacts.

Meanwhile, similar experiments mapping restriction sites in the β-globin genes of the rabbit and mouse were also generating unanticipated results. The data could be explained only by assuming the existence of an interruption of about 600 base pairs in the middle of the protein-coding sequence of the gene from both animals. Again, the insertion did not appear to be involved in inactivating the β-globin gene; it was present in DNA from both erythroid (globin-producing) and nonerythroid tissue.

Visual evidence for split gene organization soon came from the examination of R loops formed between ovalbumin or globin mRNAs and cloned DNA segments derived from these genes. In the ovalbumin case, seven unhybridized DNA regions looped out of the DNA-RNA hybrid molecule, having failed to find complementary RNA sequences with which to base-pair (Figure 20-5). Thus, at least

Chicken oviduct or erythrocyte DNA

Digest with *Eco*RI

DNA Fragments

Separate by electrophoresis

DNA fragments moving toward the positive pole

Electrophoresis gel

DNA fragments generated by digestion with *Eco*RI are separated by gel electrophoresis.

Cellulose nitrate paper

Denature

Gel

Double-stranded DNA is transferred by blotting to filter paper, fixed to it, and denatured (the two strands are separated).

Labeled, cloned complementary DNA

Hybridize with labeled complementary DNA

Wash

X-Ray film

Filter paper

Probe fragments reassociate with complementary DNA and bind to bands. Unbound DNA probe is washed away.

Autoradiograph

Eco *a* — 9,200
Eco *b* — 2,350
Eco *c* — 1,750
Eco *d* — 1,250

Size of DNA in base pairs

Erythrocyte   Oviduct

DNA bands are made visible by autoradiography.

Figure 20-4

Demonstration of the interrupted structure of the ovalbumin gene by a blotting experiment in which chromosomal DNA from chicken erythrocytes or oviduct cells was cleaved with *Eco* RI. This restriction enzyme does not cleave the ovalbumin cDNA. Thus, the four bands seen in the autoradiograph reveal a split gene structure that appears to be the same in the two tissues. [After P. Chambon, *Sci. Amer.* 244 (1981):60–71.]

eight discontinuous segments of genomic DNA specify the ovalbumin mRNA. Electron micrographs of R loops between β-globin mRNA and its cloned gene revealed a single, large, looped-out segment (corresponding to the 600-base-pair insert deduced from mapping restriction sites). Careful measurements suggested the additional presence of a second much smaller interruption closer to the beginning of the β-globin gene.

Many Genes Are More Intron Than Exon[18, 19, 20]

Sequence analysis subsequently defined the exact extent and nature of the discontinuities in the ovalbumin and β-globin genes. Regions not represented in the mRNA are called **introns** or **intervening**

Figure 20-5

The organization of the chicken ovalbumin gene. DNA containing the gene was allowed to hybridize with ovalbumin mRNA under R-loop conditions. The eight exons (L, 1–7) of the gene pair with the complementary regions of RNA, and the seven introns (A–G) loop out from the DNA-RNA hybrid. The 5′ and 3′ ends of the mRNA are indicated, as is the poly A tail. [After P. Chambon, *Sci. Amer.* 244 (1981):60–71.]

sequences. Segments that do specify the mRNA are called **exons** (because they are expressed). Exons include not only the protein-coding sequences of a gene, but also the untranslated 5′ and 3′ residues (leader and trailer sequences) present in the mRNA. Recall that the poly A tail of mRNA is not encoded in the genome, but added post-transcriptionally (Chapter 18).

Over three-quarters of the 7700 base pairs of the chicken ovalbumin gene are contained in introns (see Figure 20-5). As indicated by the R-loop patterns, the introns are seven in number. They range in size from about 250 to 1600 nucleotides. Whereas one intron resides in the untranslated 5′ leader sequence, the other six interrupt the protein-coding portion of the gene. Like the introns, the eight ovalbumin exon segments vary considerably in length. The shortest is 47 base pairs and the longest is 1043 base pairs.

A simpler distribution of introns and exons was revealed by sequencing several mammalian β-globin genes (Figure 20-6). They all contain one large intron (about 600 base pairs) and one small intron (about 100 base pairs), as had been predicted by the examination of R loops. This gene is therefore approximately equally divided between introns and exons. If the nucleotide sequences of the mouse, rabbit, and human β-globin genes are compared in detail, little divergence is seen in their protein-coding regions, reflecting the evolutionary conservation of the β-globin polypeptide. The introns, by contrast, differ between species in both length and exact sequence. Yet, they occur at precisely the same positions within the β-globin-coding region of each species, with the small intron between the codons specifying amino acids 30 and 31, and the large intron between the codons for residues 104 and 105.

Now that many more genes have been analyzed, it is clear that the split gene patterns exemplified by chicken ovalbumin and the mammalian β-globins are typical of the majority of vertebrate protein-coding genes. In a few cases, the distribution of residues between introns and exons is quite extraordinary. For instance, the chromosomal gene for mouse dihydrofolate reductase extends over about 31,000 base pairs, but gives rise to an mRNA of only about 1600 nucleotides, which in turn possesses only 558 coding residues. The chicken collagen gene contains at least 50 exons, which divide its coding region into regular repeating units that reflect comparable repeats of the amino acid sequence along the polypeptide chain (Figure 20-7).

Introns can be of many different lengths and sequences. They appear in what will become the 3' trailer region as well as the protein-coding and 5' leader regions of various mRNAs. They need not reside neatly between codons, as they do in the β-globin or collagen genes. Often they split a triplet, so that the information needed to specify a particular amino acid lies thousands of base pairs apart on the chromosomal DNA.

Figure 20-6
Structure of the mammalian β-globin gene, showing how introns are removed and exons preserved during the maturation of the mRNA. Exon 1 specifies the 5' nontranslated region and amino acids 1–30; exon 2, amino acids 31–104; and exon 3, amino acids 105 to the terminus plus the 3' nontranslated region.

Not All Genes Are Split

Even as the roster of split genes grows daily, firm evidence accumulates that some protein-coding genes are not interrupted. For example, the multiple genes encoding the histone proteins that package eucaryotic DNA into nucleosomes (Chapter 21) only rarely have introns. The interferons, proteins produced by vertebrates in response to viral infections (Chapter 24), are also often specified by uninterrupted genes. Many mRNAs synthesized by viruses that infect mammalian cells are colinear with their respective DNA genomes. Thus, the presence of introns cannot be essential for gene expression, nor does there appear to be any obvious correlation with how the expression of a gene is regulated.

Although the best initial guess for any uncharacterized mammalian protein-coding gene is that it contains introns, this is not true of genes from all eucaryotic organisms. Generally, as one proceeds down the evolutionary ladder, fewer and fewer protein-coding genes are interrupted, and the introns that do exist tend to be shorter. For instance, whereas the human and rat cytochrome *c* genes are interrupted, the yeast cytochrome *c* gene lacks introns. The *Drosophila* alcohol

(a)

Figure 20-7

The chick type 1α2 collagen gene has over 50 introns. (a) In the schematic representation, exons are indicated by bars and introns by the connecting line. The 5' and 3' ends of the 38-kilobase-pair gene are located. Much of the collagen polypeptide is arranged in a triple helical structure with glycine appearing every third residue. The exons encoding this portion of the chain (colored) always contain a multiple of 9 base pairs; most are 54 base pairs, but others are as short as 45 or as long as 108 base pairs. (b) Electron micrograph and schematic diagram of R-loop structures formed between α2 collagen mRNA and cloned DNA derived from the central portion of the gene. For historical reasons, the exons are numbered beginning at the 3' end of the gene. When two numbers appear, the segment contains two exons separated by a small intron often seen as a pimple. [Courtesy of H. Ohkubo et al., *Proc. Nat. Acad. Sci.* 77 (1980):7059–7063. Used with permission.]

(b)

dehydrogenase gene has introns, while the comparable yeast gene does not. Thus, in the yeast *Saccharomyces cerevisiae*, a gene containing introns is the exception rather than the rule. It now appears that this is the case for many other lower eucaryotes as well.

tRNA and rRNA Genes Also Contain Introns[21, 22, 23]

At the same time that interruptions in protein-coding genes were being analyzed in 1977, the first tRNA genes to have introns were discovered. The seemingly dull task of sequencing the gene for a yeast suppressor tRNA^{Tyr}, whose RNA sequence was already known, turned out to be quite interesting when 14 extra base pairs (not present in the RNA) appeared in the DNA sequence specifying the anticodon loop. Since the suppressor strain had arisen by spontaneous mutation and the anticodon sequence showed the expected single-base change, it seemed highly likely that this must be an active gene. Equally surprising was the finding that three other distinct genes for tRNA^{Tyr} all had the identical sequence (except in the anticodon), including the 14-base-pair insertion. Soon after, 19-base-pair insertions were identified in yeast tRNA^{Phe} genes, also located just 3' to the anticodon sequence.

We now know that (nuclear) tRNA genes from many eucaryotic organisms contain introns. Transfer RNA introns come in lengths of 14 to 60 base pairs. They almost always reside between the first and second nucleotides 3' to the anticodon. Curiously, they appear confined to a particular subset of tRNAs: those accepting tyrosine, phenylalanine, serine, leucine, isoleucine, tryptophan, proline, and lysine. The introns present in any family of tRNA genes from one organism (e.g., the eight yeast tRNA^{Tyr} genes) are identical or nearly identical. But intron sequences bear no resemblance between genes specifying different tRNAs (Figure 20-8).

Figure 20-8
Three yeast tRNA precursors containing introns. The anticodons (boxed in gray) often (but not always) can be drawn to pair with intron sequences (shown in color). Arrows indicate the sites where the tRNA-splicing machinery must cut and rejoin the pre-tRNA to form the mature tRNA molecule. Positions of sequence variation among the several genes specifying each tRNA species are shown. Such tRNA precursors can be isolated from yeast strain ts36 and have been used to decipher the biochemistry of yeast tRNA splicing.

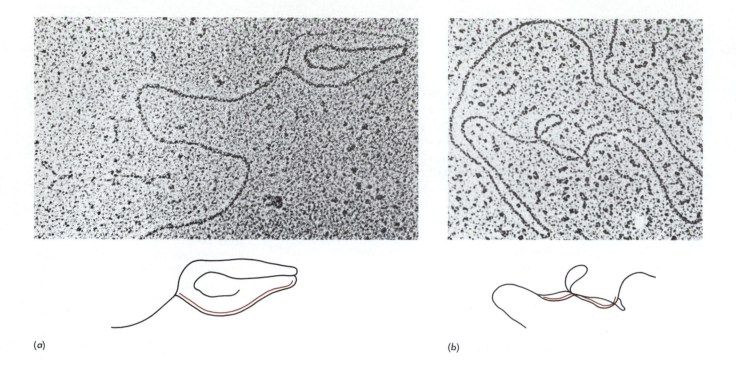

(a)   (b)

Figure 20-9

Demonstration that the large intron of mouse β-globin gene is transcribed. Electron micrographs and schematic diagrams of the R-loop structures formed by incubating the cloned gene with either (a) 15S (precursor) RNA or (b) 10S (messenger) RNA are shown. (Black lines indicate DNA; colored lines indicate RNA.) The 15S RNA had previously been characterized as having many of the features expected of a precursor to the mature mRNA, including a 5' cap and a 3' poly A tail. In (b), only the larger of the two β-globin introns is big enough that the nonhybridized DNA can be seen buckling out from the R loop. [Electron micrographs courtesy of S. M. Tilghman et al., *Nat. Acad. Sci.* 75 (1978):1309–1313. Used with permission.]

Some ribosomal RNA genes from eucaryotic organisms also contain introns. So far, known active (nuclear) rRNA genes with introns all specify the large (23S-like) ribosomal RNA and are from lower eucaryotic species.

Thus, we see that introns are not restricted to any particular type of eucaryotic gene. Such interruptions can appear in genes whose terminal products are RNA as well as in protein-coding genes. Even mitochondrial and chloroplast genomes sometimes contain introns in regions producing mRNA, rRNA, or tRNA.

Introns Are Removed at the RNA Level[23–27]

Once it became clear that many chromosomal genes contain interruptions, much speculation arose as to how the corresponding intronless RNAs might be generated. An initial notion that the creation of such interruptions might be a mechanism for gene inactivation was quickly dispelled when genes from tissues active in the production of particular proteins were shown to have exactly the same discontinuities as DNA from inactive tissues. The possibility of "hopping polymerases" that somehow knew to transcribe only the exon portions of genes was also considered. A third possibility was RNA splicing, whereby the introns would be removed by RNA processing after the entire gene had been transcribed into RNA. Whatever the mechanism, it clearly had to be exquisitely precise. If intron removal were off by even one nucleotide, then missense or frameshifted proteins would result.

Many lines of evidence quickly converged to establish that **RNA splicing** must be the mechanism of intron removal for both protein-coding and RNA-specifying genes. For example, when R loops were prepared using nuclear RNA (hnRNA) and an interrupted gene like β-globin or ovalbumin, occasional hybrid molecules appeared where the DNA regions corresponding to introns were not looped out (Fig-

Cytoplasmic mRNA   Oviduct nuclear RNA

Position of size markers

Direction of electrophoresis

28S

23S

18S

16S

Figure 20-10
Detection of high-molecular-weight ovalbumin mRNA precursors by **Northern blotting.** In this procedure an RNA sample is first fractionated on a gel and then transferred to a specially modified paper with which it makes covalent crosslinks. [See J. C. Alwine, D. J. Kemps, and G. R. Stark, *Proc. Nat. Acad. Sci.* 74 (1977):5350–5354.] Then, as in Southern blotting (Figure 19-15), bands of complementary RNA are detected by hybridization with labeled DNA probes. Here, note the presence of many discrete nuclear RNA species up to four times longer than the mature ovalbumin mRNA. These have been detected with a cDNA probe (containing only mRNA sequences). Hybridization with probes containing only intron sequences from the cloned gene can reveal the sizes of nuclear RNAs that still retain various introns. (The asterisk indicates an autoradiographic artifact, not a reproducibly observed RNA.) [Courtesy of D. R. Roop, J. L. Nordstrom, S. Y. Tsai, M. J. Tsai, and B. W. O'Malley, *Cell* 15 (1978):671–685. Used with permission.]

ure 20-9). Rather, a complete R loop was formed, indicating that a continuous transcript of the entire gene was being detected. Likewise, hybridization experiments revealed the existence of tiny amounts of nuclear poly A–containing RNA that was not only significantly longer than the corresponding cytoplasmic mRNA but also contained both intron and exon sequences (Figure 20-10). Usually, such longer nuclear molecules are detected in several different sizes. These apparent processing intermediates occur in varying amounts, suggesting that some introns in a particular pre-mRNA are removed more quickly than others. Yet, the deduced order of intron removal for several closely scrutinized genes correlates neither with intron size nor with intron position in the transcript.

At long last, the giant hnRNA molecules of mammalian cell nuclei had been explained. At least many of them must be mRNA precursors undergoing progressive RNA splicing before export from the nucleus. Their short lifetimes, large sizes, and equivalent contents of 5' caps and 3' poly A tails are totally consistent with a scenario whereby long internal stretches of nucleotides are excised from pre-mRNAs as they are fashioned into mature messenger molecules.

Once it was discovered that some tRNA genes are interrupted, tRNA precursors containing introns were also immediately detected. A yeast strain that had long been known to be defective in the production of functional RNAs was found to contain unspliced tRNAs. These pre-tRNA molecules, which are matured at their 5' and 3' ends but still retain intron sequences, proved to be ideal substrates for characterizing tRNA-splicing enzymes.

Similarly, after analysis of rRNA gene structure had revealed extra sequences in the region specifying the large rRNA, rRNA precursors containing introns were detected, most notably in the ciliated protozoan *Tetrahymena* (Figure 20-11). This organism provided experimental material that allowed the discovery of self-splicing RNA, which will be discussed shortly.

Intron

Exon   Exon

Tetrahymena rDNA

Transcription

5'   3'

pre-rRNA

Splicing

+

9S

Further processing

17S   5.8S   26S

Mature rRNAs

Figure 20-11
Synthesis of *Tetrahymena* rRNAs. A 9S fragment (about 0.4 kilobase) is excised from the 35S pre-rRNA as the first step in rRNA processing. The caret in the 26S rRNA represents the region from which the intron has been removed. [After A. J. Zaug and T. R. Cech, *Cell* 19 (1980):331–338.]

Deciphering Splicing Systems[28]

The availability of RNA precursor molecules containing introns challenged biochemists to ask whether accurate splicing could be carried out in the test tube. Indeed, systems now exist in which introns can be excised from nuclear pre-mRNA as well as from pre-tRNA and pre-rRNA molecules in vitro. From such studies, much has been learned about the biochemical mechanisms that achieve precise cutting at the **splice junctions** (the boundaries between intron and exon segments) and the subsequent ligation (joining) of exon sequences.

Meanwhile, an understanding of the structural features required for splicing RNA precursor molecules has also evolved. DNA sequences of split genes from many eucaryotic organisms have rapidly accumulated, allowing the identification of common sequence elements. Both naturally occurring and artificially induced mutations that affect splicing have been systematically scrutinized. These mutant analyses have confirmed our original guesses about which sequences are necessary, have uncovered other sequences, and have raised additional questions about splicing mechanisms.

We now believe that splicing reactions can be grouped into three distinct classes. One mechanism applies to rRNA molecules, another to mRNA molecules, and a third to tRNA molecules. This division corresponds exactly to the synthesis of these three kinds of transcripts by the three different eucaryotic RNA polymerases I, II, and III, respectively. Although these three types of splicing reactions all occur in eucaryotic cell nuclei, splicing also sometimes takes place in the other DNA-containing cellular compartments—the mitochondria and chloroplasts. The splicing of some mitochondrial mRNA and rRNA precursors closely resembles that of nuclear pre-rRNA.

The three splicing systems can be readily distinguished by examining the nature of the signals recognized by the splicing apparatus and their location in the precursor RNA. In pre-mRNAs, short sequences located directly at (or near) the splice junctions seem to be the principal determinants of where and how splicing occurs. In pre-tRNAs, splice junction sequences are not conserved. Rather, the common feature of tRNA introns is their location 3' to the anticodon. This placement would not disrupt the highly conserved three-dimensional folding of tRNA molecules (see Figures 14-5 and 14-6), arguing strongly that exon sequences and structures are major recognition elements for tRNA splicing. In sharp contrast, certain sequences internal to the intron are essential for pre-rRNA splicing (and for the excision of introns from at least some mitochondrial transcripts). These segments apparently interact to stabilize a precise folding of the RNA that juxtaposes the two splice junctions in such a way that a defined series of cleavage-ligation reactions occurs. The contrasting features of the three nuclear splicing systems are summarized in Table 20-1.

A Separate Nuclease and Ligase Act in tRNA Splicing[29, 30, 31]

The biochemistry of tRNA splicing has been best described for yeast, where the enzymes involved have been purified (Figure 20-12). In the first stage, two cuts precisely at the ends of the intron are executed by a splicing endonuclease associated with the nuclear membrane. Using a mechanism common to many ribonucleases, the endonucle-

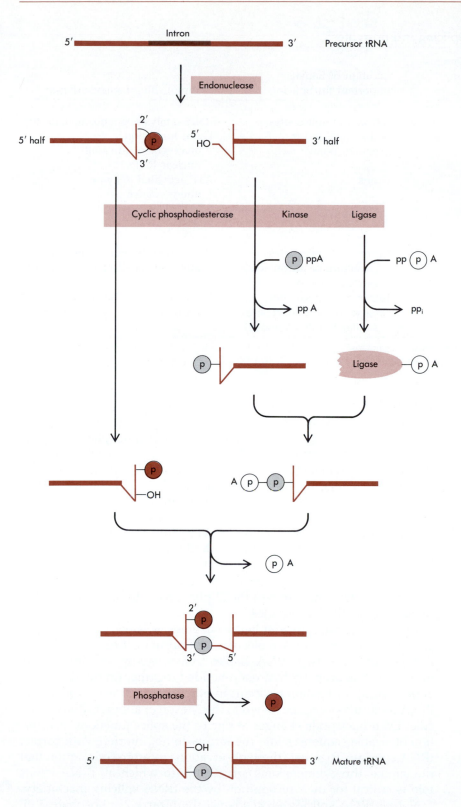

Figure 20-12
The splicing of yeast tRNA. Although the splicing of mammalian pre-tRNAs proceeds via a distinctly different pathway, adenylylated enzyme intermediates also appear to be involved. [After C. L. Greer, C. L. Peebles, P. Gegenheimer, and J. Abelson, *Cell* 32 (1983):537–546.]

ase leaves a 2'-3' cyclic phosphate on the two newly generated 3' ends and a hydroxyl group at the new 5' ends. In the second stage of splicing, the tRNA halves are joined by a splicing ligase in a series of reactions requiring ATP. The process begins when a phosphate group is added to the 5'-OH group of the 3' tRNA exon with the concomitant conversion of ATP to ADP. Next, the 5' phosphate end is activated by the transfer of an AMP group from the splicing ligase itself.

Table 20-1 Major Features of RNA Splicing in Eucaryotes[†]

Final Product	RNA Polymerase Synthesizing Pre-RNA	Location of Signals Important for Splicing	Distinguishing Characteristics
rRNA	I	Intron and splice sites	RNA catalysis, guanosine cofactor
mRNA	II	Splice sites, intron branch point	Lariat formation, involvement of small nuclear RNPs
tRNA	III	Exons	ATP-activated enzyme intermediates
Organelle RNA (mitochondrial, chloroplast)	Organellar		
Group I Introns* (rRNA, tRNA, mRNA)		Intron and splice sites (structure of intron same as rRNA above)	RNA catalysis, guanosine cofactor, intron-encoded splicing proteins
Group II Introns* (mRNA)		Intron and splice sites (sequences at splice sites similar to mRNA above)	RNA catalysis, lariat formation, intron-encoded splicing proteins

[†]See Chapter 28 for a discussion of splicing in procaryotic organisms.
*For a discussion of the generality of the Group I and Group II designations, see T. R. Cech, *Cell*, 44 (1986):207.

This AMP was originally derived from a second ATP, but first appears in the form of the adenylylated enzyme. Meanwhile, a cyclic phosphodiesterase activity has opened the 2′-3′ cyclic phosphate on the 5′ exon to form a 2′ phosphate. The final reaction that joins the two tRNA exons is remarkable in that it creates a 2′ phosphate 3′-5′ phosphodiester linkage in the newly spliced tRNA. A hallmark of this unusual mechanism is that the phosphate group at the spliced position is not one that was present in the pre-tRNA. Rather, it was donated during the kinase reaction by ATP. Both the kinase and ligase activities, as well as the cyclic phosphodiesterase activity, appear to be carried by the single polypeptide chain of the splicing ligase (MW about 90,000). After the two halves of the tRNA are joined, a separate phosphatase enzyme removes the 2′ phosphate, for it is not found in mature yeast tRNA molecules.

The separate nuclease and ligase enzymes isolated from yeast cells are specific for pre-tRNA molecules and do not act on other RNAs. Furthermore, the two tRNA halves generated by excision of the intron must be properly hydrogen-bonded together (in the tRNA cloverleaf form) for ligation to occur; in other words, the 5′ exon of one tRNA molecule cannot be joined to the 3′ exon of another tRNA molecule. Even nucleotide changes at one of the splice junctions or insertion of sizable sequences into the intron do not interfere with correct tRNA splicing. Together, these observations reinforce our notion that the similar three-dimensional structure into which all tRNA exons fold is critical for their recognition by the tRNA-splicing machinery.

Not all tRNA splicing proceeds via the bizarre 2′ phosphate 3′-5′ phosphodiester intermediate identified in yeast. Although this intermediate probably also occurs in other lower eucaryotes and plants, in vertebrates the phosphate group that joins the two exons in the final spliced tRNA is not donated by ATP. Instead, the phosphate that originally linked the 5′ exon and the intron in the pre-tRNA is retained at the spliced junction. The details of the biochemical mechanism must therefore be different from that diagrammed in Figure 20-12. Yet, the three-dimensional structural features of tRNAs that allow pre-tRNA recognition by splicing enzymes appear to be conserved in all organisms, as yeast pre-tRNAs containing introns are readily spliced by the vertebrate machinery.

The Self-Splicing of rRNA: G Nucleoside Involvement and the Creation of a Circular Intron[32, 33, 34]

Ribosomal RNA processing in several lower eucaryotes involves excision of intron sequences from a site near the 3' end of the region that becomes the longest (23S-like) mature rRNA. Dissection of this splicing reaction in *Tetrahymena thermophila* has revealed that proteins are not essential. Instead, the splicing activity is intrinsic to the precursor rRNA molecule itself.

In this fascinating mechanism, a 413-nucleotide intron is excised and circularized by a series of **phosphoester transfer** reactions (Figure 20-13). The train of events begins when the 3'-OH group of a G nucleoside attacks the phosphodiester bond at the 5' end of the intron and adds itself to the intron via a normal 3'-5' phosphodiester bond. Next, the newly generated 3'-OH group on the 5' exon similarly attacks the phosphodiester bond at the other end of the intron, creating a spliced joint and releasing the intron (terminated again by a 3'-OH). Finally, the linear intron undergoes autocyclization: Its 3'-OH group attacks a bond near its 5' end, releasing a 15-nucleotide-long fragment that starts with the G residue added in the initial step.

Elegant proof that the pre-rRNA can splice itself has been obtained by cloning DNA sequences specifying the intron and its flanking regions downstream from the *E. coli lac* promoter, amplifying this recombinant DNA plasmid in *E. coli*, and then transcribing with *E. coli* RNA polymerase in vitro. The RNA synthesized readily undergoes self-splicing in the test tube, despite the fact that it never contacted any *Tetrahymena* proteins.

The requirements for this amazing RNA-catalyzed reaction are minimal. Because the 5' end of the G nucleoside cofactor does not participate directly in any of the phosphoester transfer reactions, it can be modified in a number of ways; for example, 5'-GMP, GDP, or GTP work as well as a simple G nucleoside. By contrast, nucleosides modified at the 2'- or 3'-OH groups are inactive as cofactors. G analogs in which those groups on the purine ring that normally hydrogen-bond to C have been altered do not work as well, suggesting that standard GC base-pairing interactions may be important for positioning the G cofactor properly on the RNA surface. Both monovalent and divalent (usually Mg^{2+}) cations are required, but no GTP or ATP hydrolysis is involved. The lack of an energy requirement may be explained by the linking of each ligation step to a cleavage step: In the overall splicing reaction, phosphodiester bonds are neither lost nor gained. Since every one of the phosphoester transfer reactions is in theory reversible, some explanation is needed for why splicing proceeds in the forward direction. In vivo, rapid degradation of the intron (observed for both its linear and circular form) or export of the spliced rRNA to the cytoplasm may be responsible.

Figure 20-13
The self-splicing of *Tetrahymena thermophila* rRNA. Here the initiating G nucleoside is GMP (boxed). Intron sequences are in color. Square brackets indicate a postulated intermediate that has not been isolated. A possible secondary structure for the 15-mer released by the final reaction is shown below, with sequence differences in the intron of *Tetrahymena pigmentosum* indicated. [After A. J. Zaug, P. J. Grabowski, and T. R. Cech, *Nature* 30 (1983):578–583.]

Figure 20-14

Similar sequences and secondary structures found in the *Tetrahymena thermophila* rRNA intron (a) and a yeast mitochondrial protein intron (the fourth in the pre-mRNA for cytochrome oxidase) (b) Large arrows designate the splice sites. Boxed sequences are invariant between these and some other mitochondrial introns. Triangles point to the sites of splicing defective mutants. Sequence elements A and B, 9R and 9R', and 9L and 2 are believed to base pair, establishing a defined intron tertiary structure. [Courtesy of F. Michel and B. Dujon, *EMBO J.* 2 (1983):33–38. Used with permission as modified by T. R. Cech and B. L. Bass.]

The observation that the excised intron is capable of undergoing an internal splicing reaction to produce a covalent circle tells us that the cleavage-ligation activity of the pre-rRNA must reside within the intron itself. Indeed, certain small deletions of the intron far from the splice junctions totally prevent splicing. A secondary structure model for the *Tetrahymena thermophila* intron that brings its 5' and 3' ends into close proximity is illustrated in Figure 20-14. Also indicated are certain short sequences found in common with some mitochondrial introns. In the mitochondrial introns, these sequences have been pinpointed by mutational analyses as essential elements in splicing.

Mitochondrial Introns Encode Splicing Proteins[35-40]

Whereas the mitochondrial DNA of mammalian cells does not contain introns, many mitochondrial genomes in lower eucaryotes do. Interruptions have been detected in both protein-coding and rRNA-specifying genes. Analyses of numerous mutations affecting expression of such mitochondrial genes suggest that the splicing of their transcripts may be carried out in various ways; even intron removal from two different pre-mRNAs can have different requirements. Among the components identified as essential for splicing are nuclear-encoded proteins, specific sequences within introns, and polypeptides encoded by the intron sequence itself (or by the intron of another gene).

Participation of intron-encoded proteins is a unique aspect of the mitochondrial splicing process. In some interrupted genes in yeast mitochondria, an open reading frame that begins in the first exon simply continues through the splice junction into intron sequences (Figure 20-15). Nonsense mutations introduced into these lengthy intron-coding regions not only lead to the accumulation of prematurely terminated polypeptide chains, but also block the splicing of that intron. Hence, such intron-encoded polypeptides are agents in the destruction of their own mRNA, which unavoidably occurs upon intron excision. So far, this scenario seems limited to situations where transcription and translation occur in the same cellular compartment, such as the mitochondrion. Recall that the other splicing systems that act on nuclear pre-mRNAs, tRNA, or rRNA precursors operate in the ribosome-free environment of the nucleus.

Yet, it is not intron-encoded proteins but another intron feature that ties together nuclear rRNA splicing in *Tetrahymena* and the removal of some mitochondrial mRNA and rRNA introns in yeast and other fungi. Four short regions (each 10 to 12 nucleotides long) of sequence homology have been recognized in all these introns, despite their diverse origins (see Figure 20-12). The four conserved sequences can be base-paired to form two stems that appear crucial in defining a particular three-dimensional structure for the intron; splicing mutations occurring in any one of these regions can be suppressed by changing the complementary region to restore complementarity. Two additional intron regions that are not conserved in exact sequence but are always complementary to each other are also sites of splicing mutations. The remarkable phylogenetic conservation of these three pairs of short complementary sequences argues strongly that precise intron folding is essential for splicing. Not surprisingly, the excision of (at least some) mitochondrial introns also appears to involve RNA catalysis. Additional requirements for intron- and nuclear-encoded proteins may be explained by their role in stabilizing the proper intron conformation for self-splicing.

Conserved Sequences at Exon-Intron Boundaries in Pre-mRNA[16, 41–44]

In contrast to the importance of exon sequences for tRNA splicing or intron sequences for rRNA and mitochondrial splicing, signals crucial for the splicing of mammalian pre-mRNAs reside directly at the splice junctions themselves. (The exception to this rule is a short, conserved intron sequence in yeast.) Even when only a few sequences of split protein-coding genes had accumulated, it became obvious that the nucleotides at exon-intron boundaries are not random. Introns always seemed to begin with GT and end with AG.

Now that the sequences of several hundred interrupted genes have been compiled, it is clear that the **GT-AG rule** is nearly always obeyed and that somewhat longer consensus sequences can be written for both the 5′ and 3′ splice junctions. The nine-nucleotide consensus for 5′ junctions extends three residues upstream and six residues downstream from the splice point (Figure 20-16). The average 5′ junction matches this consensus in at least seven positions; only rarely does a 5′ splice junction with as little as a five-residue match

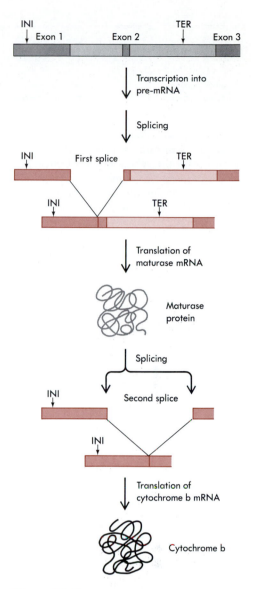

Figure 20-15

The action of an intron-encoded protein in yeast mitochondrial splicing. Only the first third of the cytochrome *b* gene is shown. After the first intron is excised, the RNA can be translated into maturase. Action of the maturase in excising the second intron creates the cytochrome *b* mRNA, but concomitantly destroys the maturase mRNA. Thus, no more synthesis of maturase takes place. INI and TER indicate translational start and stop codons. [After J. Lazowska, C. Jacq, and P. P. Slonimski, *Cell* 22 (1980):333–348.]

Figure 20-16
Consensus sequences for 5′ and 3′ splice sites in nuclear protein-coding genes from the whole range of eucaryotic organisms. The nearly invariant GT and AG dinucleotides are indicated. [For details, see S. M. Mount, *Nucleic Acid Res.* 10 (1982):459–472.]

occur. The consensus sequence for 3′ splice junctions is different from the 5′ consensus (Figure 20-16). It is composed of a pyrimidine-rich region of variable length (but always greater than ten nucleotides) followed by a short consensus sequence extending only three residues upstream and one residue downstream from the splice point. The pyrimidine-rich region is characteristically devoid of the sequence AG, the invariant dinucleotide that occurs just preceding the 3′ intron-exon boundary. Note that at both splice junctions, the conserved sequences extend farther into the intron than the exon, thereby placing only minimal constraints on the sequence of the mature mRNA. Quite remarkably, splice junction sequences from genes representing the whole range of eucaryotic species (yeast to human) conform well to the same 5′ and 3′ consensus sequences. This suggests that whatever recognizes these splicing signals must likewise have been highly conserved over evolutionary time.

Some of the most potent evidence supporting the importance of conserved junction sequences for pre-mRNA splicing has come from analyses of mutant human genes. **Thalassemias** are genetic diseases in which the synthesis of hemoglobin polypeptides is defective. Recently, a number of thalassemias have been traced to mutations that interfere with the correct removal of introns from globin pre-mRNAs (Figure 20-17). Use of a 5′ splice site is abolished by changes in the invariant GT sequence and is decreased by mutations elsewhere in the nine-nucleotide consensus. The surprising result, observed with both the thalassemias and other 5′ splice site mutants, is that splicing is not simply prevented. Rather, new sites called **cryptic splice sites,** which match the 5′ consensus but are normally not used, become activated. Also leading to aberrant splicing of the pre-mRNA molecule in other thalassemias are mutations that create a new 5′ or 3′ splice junction consensus sequence. While the bizarre results of splice site mutations do not tell us how the splicing machinery discriminates between true splice junctions and other regions that fit the consensus sequences, they do emphasize that junction recognition is basic to pre-mRNA splicing.

It is therefore not surprising that huge blocks of sequence can be deleted from the interior of most introns with little noticeable effect on splicing. Only when deletions begin to impinge on the conserved splice junction sequences do aberrations in splicing occur. Moreover, chimeric introns—created by cutting the DNA within two different introns and joining the 5′ and 3′ halves one to the other—function quite nicely, showing that splice junction signals are interchangeable. All these observations underscore the versatility of the pre-mRNA splicing machinery and are entirely compatible with the huge variations known to occur in intron size and sequence.

Small Nuclear RNA-Protein Complexes in Search of Functions[45, 46, 47]

A distinctive feature of mammalian cells, as compared to bacteria, is their great abundance of small RNA species that are neither transfer RNAs nor ribosomal RNAs. Many of these species reside in the nucleus rather than the cytoplasm. Cells of all vertebrate species contain six different **small nuclear RNAs (snRNAs)** called U1, U2, U3, U4, U5, and U6, which range in size from about 100 (U6) to 215 (U3) nucleotides (Figure 20-18). These RNAs are relatively stable (having lifetimes at least as long as a cell generation), possess unusual

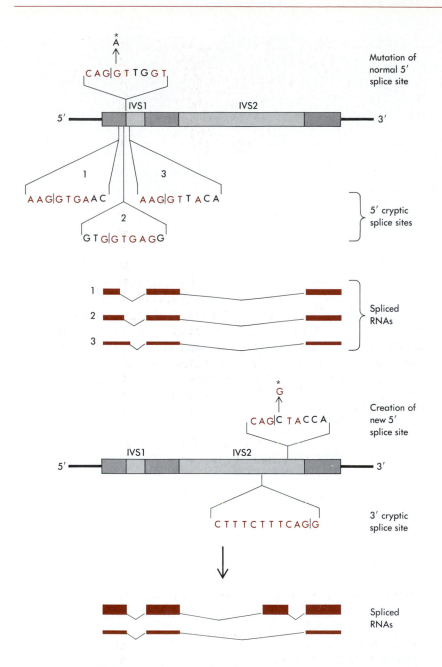

Mutation of
normal 5'
splice site

IVS1   IVS2

5'   3'

1   3

AAG|GTGAAC   AAG|GTTACA

2

GTG|GTGAGG

5' cryptic
splice sites

1

2

3

Spliced
RNAs

CAG|C TACCA

Creation of
new 5'
splice site

IVS1   IVS2

5'   3'

CTTTCTTTCAG|G

3' cryptic
splice site

Spliced
RNAs

Figure 20-17
Human β-thalassaemia genes that produce wrongly spliced β-globin RNAs. Mutation of the first nucleotide (G to A) of the first intron (IVS 1) leads to no correct splicing but instead utilization of three cryptic 5' splice sites. Mutation of a single base pair within the second intron (IVS 2) creates a new 5' splice site and uncovers a cryptic 3' splice site, leading to a predominance of misspliced over correctly spliced RNA. The more abundant products are indicated by thicker bars. Nucleotides conforming to the 5' or 3' splice junction consensus sequences (Figure 20-16) are in color. Vertical lines indicate the splice point. None of the misspliced RNAs can be translated into functional β-globin chains. [After R. Treisman, S. H. Orkin, and T. Maniatis, *Nature* 302 (1983):591–596.]

trimethylated ($m_3^{2,2,7}$G) cap structures at their 5' ends (except for U6), and contain some other modified nucleotides (as do tRNAs and rRNAs). They all exist not as naked RNA molecules but stably complexed with certain abundant nuclear proteins. The resulting particles are referred to as **small nuclear ribonucleoproteins (snRNPs)**. Although snRNPs are only about one-fifth the mass of small ribosomal subunits (10 to 12S) and have higher protein/RNA ratios, like ribosomes, each snRNP contains several different protein species. Some of these polypeptides are shared by those snRNPs containing U1, U2, U5, and U4/U6 RNAs (only in the last case do two RNAs coexist in the same snRNP particle), whereas other polypeptides are unique to particular snRNPs. The existence of common and unique protein components was first recognized with the discovery that some patients with the disease systemic lupus erythematosus make **autoantibodies** (normally forbidden antibodies that bind one's own cellular components) directed against snRNP proteins. For instance, some autoantibodies precipitate U1, U2, U5, and U4/U6 snRNPs and therefore recognize

Figure 20-18

Sequences and probable secondary structures for human U1 and U2 snRNAs. U1 nucleotides complementary to the 5' splice junction consensus sequence (Figure 20-16) are boxed in color; those complementary to the 3' splice junction consensus sequence are boxed in gray. This region of U1 snRNA is not only available for interaction with other RNAs but is completely conserved in sequence from mammals to sea urchins. U1 and U2 are the most abundant of all snRNAs in mammalian cells. Both appear to be involved in pre-mRNA splicing. Note the presence of many 2'-O-methyl modifications (m) as well as pseudouridines (ψ), 6-methyladenosine (6m), and the trimethyl cap structure at the 5' ends. [See H. Busch, R. Reddy, and L. Rothblum, *Ann. Rev. Biochem.* 5 (1982):617–654.]

shared polypeptides; other autoantibodies precipitate U1, U2, or U3 snRNPs singly because they bind polypeptides unique to each of these particles.

Several properties of snRNPs argue that they play important roles in nuclear function. First, both their protein components and the sequences of their RNAs are highly conserved across all higher eucaryotic species (e.g., mammals, amphibians, and insects). Even in lower eucaryotes like yeast or the slime mold *Dictyostelium*, small nuclear RNAs with homologous sequences have been characterized. Second, snRNPs are very plentiful, particularly in mammalian cells, where the number of each type is equal to 1 to 10 percent the number of ribosomes. Since the U3 snRNP has been localized to the nucleolus, it is expected to participate somehow in ribosome biogenesis. The remaining snRNPs reside in the nucleoplasm (the remainder of the nucleus) and are often found associated with hnRNA. Hence, it is suspected that they contribute to various aspects of pre-mRNA processing.

Involvement of the U1 snRNP in Pre-mRNA Splicing[48–51]

U1 RNA is the most abundant of all the small nuclear RNAs in vertebrate cells. It has at its 5' end a sequence that is precisely complementary to the nine-nucleotide consensus for 5' splice junctions (see Figure 20-18). This generated the idea that the U1-containing small RNA–protein complex (U1 snRNP) might use base pairing to interact with splice sites and thus be an essential component of the pre-mRNA splicing apparatus. That this is the case has been verified by the use of patient autoantibodies directed against protein components of the U1 snRNP. When added to in vitro splicing systems, such antibodies effectively block intron removal. Moreover, digesting away only the 5' end of U1 RNA (just the region complementary to the 5' splice junction consensus sequence) prevents in vitro splicing. Studies performed in a more purified system demonstrate that U1 snRNPs specifically bind and protect from nuclease digestion a stretch of pre-mRNA sequences just at the 5' end of the intron. Both

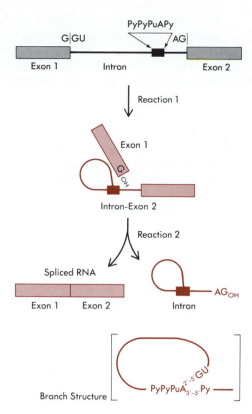

Figure 20-19
Formation of a branched RNA intermediate during pre-mRNA splicing. The
reaction sequence applies to both mammalian and yeast in vitro splicing, but
the sequence at the branch point is
more rigorously defined in yeast. The
phosphate group at the spliced junction
in the newly matured mRNA is the one
present at the 5' end of exon 2 in the
pre-mRNA. [After E. Brody and J.
Abelson, *Science* 228 (1985):963–967.]

the RNA and protein components of the snRNP are essential for this
selective binding, reminiscent of the requirements for 30S ribosomes
to interact with start sites on procaryotic mRNAs during polypeptide
chain initiation (Chapter 14).

What recognizes the 3' ends of introns during pre-mRNA splicing
is not yet clear. Another most intriguing unsolved problem is how the
splicing machinery faithfully joins one end of an intron to the other
and avoids mistakenly excising exons from a transcript that contains
multiple introns. One possibility is that some splicing component
tracks along the intron, ensuring that neighboring splice sites are
paired. Yet, even if such a mechanism exists, it is difficult to explain
why cryptic sites in nonmutant pre-mRNAs are ignored and how the
occasional intron that utilizes multiple 5' or 3' splice sites functions
(Chapter 21).

Lariat Structures Are Formed During Pre-mRNA Splicing[52–58]

As was the case for tRNA and rRNA splicing, probing the biochemical mechanism of pre-mRNA splicing has uncovered the existence of
unusual intermediates leading to intron removal. An early step,
which requires prior U1 snRNP binding and ATP hydrolysis, is a cut
at the 5' splice junction accompanied by formation of an unusual 2'-5'
phosphodiester bond between the 5'-terminal G residue of the intron
and a site about 30 nucleotides upstream from the 3' end of the
intron. This creates an RNA lariat structure in which an A residue
invariably forms the branch point (Figure 20-19). The branched A
nucleotide engages in three phosphodiester bonds: the two ordinary
3'-5' bonds plus a novel 2'-5' bond. In mammalian pre-mRNAs, the
sequence surrounding this A residue is only very loosely defined (see
Figure 20-19), and intron sequences can be deleted nearly at random.

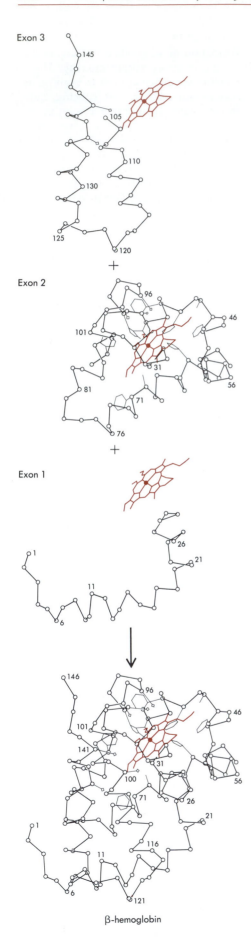

Exon 3

Exon 2

Exon 1

β-hemoglobin

Figure 20-20
A backbone protein model showing how the three exons of the β-hemoglobin gene specify different domains of the protein. The heme group is shown in color. The amino acid side chains contacting the heme are drawn symbolically. [After M. Go, *Nature* 291 (1981):90–92.]

In yeast, on the other hand, the branch always occurs at the last A within a particular sequence (TACTAAC), which appears in every intron 20 to 55 nucleotides upstream of the 3′ splice site. Deletion of this yeast sequence or certain mutations within it abolish excision of that intron. Apparently, the yeast splicing machinery is more demanding than the mammalian apparatus in requiring that a defined sequence occur at the branch point.

Many components must be required to assemble an active splicing complex, for lariat intermediates formed in in vitro splicing reactions sediment very rapidly—in the range of 40S (for yeast) to 60S (for mammalian systems). U2 as well as U1 snRNPs appear to be essential components of this complex (**spliceosome**). During the splicing reaction, U2 snRNPs associate with the intron branch point while U1 snRNPs interact with the 5′ splice site and some other component (another snRNP?) binds the 3′ splice junction in the pre-mRNA. All these participants seem to assemble along an ordered pathway, just as ribosomal subunits do during the initiation of protein synthesis (Chapters 14 and 21). Very soon, a complete description of the many components required for the pre-mRNA splicing process should become available.

Exons Sometimes Correspond to Protein-Folding Domains[59, 60]

In many instances where we know both the three-dimensional structure of a protein and the distribution of introns and exons in its gene, an intriguing correlation can be perceived; namely, exons often correspond to distinct **protein-folding** (or functional) **domains** of the encoded polypeptide (Chapter 5). A vivid example is provided by the immunoglobulins (Chapter 23), where exon boundaries precisely coincide with junctions between the separate folded domains of the protein molecule.

Another interesting case is the β-globin gene, whose three exons can be identified with particular functional as well as structural regions of the protein (Figure 20-20). First, it was shown that the amino acid residues specified by the central exon alone are capable of binding heme tightly and specifically. Second, a detailed analysis of the course of the polypeptide chain in the β-globin three-dimensional structure suggested that the molecule possesses four distinct subdomains or compact structures, each composed of continuous stretches of amino acid residues (Chapter 5). One subdomain corresponds to the first exon of the gene and one to the third exon, but two subdomains are specified by the central exon. Quite remarkably, when the sequence of the gene for soybean leghemoglobin (a related oxygen-binding protein in leguminous plants) was determined, a third intron was found precisely at the division between the two subdomains of what is the central exon of animal globins.

Likewise in the enzyme alcohol dehydrogenase, introns neatly divide structural and functional domains of the protein. Alcohol dehy-

(a)

(b)

Figure 20-21
Division of the alcohol dehydrogenase gene into structural domains. (a) The three-dimensional structure of horse liver alcohol dehydrogenase. One of the two identical subunits is shaded, and the zinc ligands are shown as balls. (b) The amino acid sequence of the liver enzyme is aligned with intron positions (gaps) in the maize alcohol dehydrogenase gene. The catalytic domain is shaded gray and its contacts with one of the zinc atoms are shown. The α-helical regions are coiled and strands of β-pleated sheet are indicated by colored lines. The introns are numbered and their positions in the three-dimensional structure are indicated by ▲ in (a). [After C.-I. Branden, H. Eklund, C. Cambillau, and A. J. Pryor, *EMBO J.* 3 (1984):1307–1310.]

drogenase utilizes the coenzyme NAD⁺, which binds to a domain of the folded polypeptide chain similar in its three-dimensional structure to the NAD⁺ binding domains of two other dehydrogenases (lactate dehydrogenase and glyceraldehyde-3-phosphate dehydrogenase). Exon-intron junctions in the gene mark the edges of the coenzyme binding domain: this portion of the protein is encoded by five of the ten exons (numbers 5 to 9), whereas amino acid residues involved in catalysis are specified by exons 1 to 4 and 10 (Figure 20-21). Moreover, within the part of the gene encoding the coenzyme binding domain itself, introns are distributed in an interesting way: Intron 7 separates two structurally similar mononucleotide binding units (recall from Chapter 6 that NAD⁺ is a dinucleotide), and introns 5 and 6 in turn subdivide the first of these units into three repeating sections, each composed of a short stretch (about 20 residues) of α helix followed by β strand.

Split Genes May Enhance the Rate of Evolution[61–66]

The existence of introns immediately raises evolutionary questions. Were these interruptions in genes acquired by eucaryotic cells because they fulfilled some special purpose? Or did they exist in primordial messages, to be later discarded by bacteria under pressure to streamline their genomes for maximal reproduction rates? As yet, the evidence in favor of one or the other of these alternatives is not conclusive. However, several aspects of the structures of interrupted genes hint that the presence of introns (coupled with the ability to excise them at the RNA level) could have been used to advantage during the evolution of higher eucaryotic genomes.

Since exons can correspond to separate structural or functional domains of polypeptide chains, mixing strategies may have been employed during evolution to create new proteins. Having individual protein-coding segments separated by lengthy intron sequences spreads a eucaryotic gene out over a much longer stretch of DNA than the comparable bacterial gene. This will simultaneously increase the rate of recombination between one gene and another and also lower the probability that the recombinant joint will fall within an exon and create some unacceptable aberrant structure in the new protein. Instead, recombination will most likely take place between intron sequences, thereby generating new combinations of independently folded protein domains. The transcript of such a new gene would have at least one hybrid intron, but this should pose no problem for the splicing machinery, since internal intron sequences are largely dispensable and splice junction signals interchangeable.

As yet, there are no clear-cut examples where different exons of one gene can be traced to exons present in two other genes. However, a number of protein-coding genes are known to contain two or more exons that are obviously closely related to one another. These presumably were generated by unequal crossing over between intron sequences followed by divergence of the duplicated exons to provide new functions. Both the immunoglobulins (Chapter 23) and collagen (see Figure 20-7) are conspicuous examples of genes that contain multiple copies of related exons. The chicken ovomucoid gene appears to have arisen by triplication of an ancestral protein-coding DNA segment split by one intron. Homologies between the three domains (which are themselves separated by introns) are quite striking (Figure 20-22).

Perhaps even more important for understanding the evolution of higher organisms is the realization that introns allow a new protein to be tested while enabling the organism to retain expression of the original (perhaps essential) product. We know that even a point mutation can occasionally create a new 5' or 3' splice junction signal where none previously existed. Because the pre-mRNA splicing machinery has the capacity to join a single 3' splice site to either of two (or more) 5' splice junctions (or vice versa), it is conceivable that a mutated gene might then be spliced to yield two different mRNAs. Thus, both the old and a new protein could be synthesized simultaneously. Depending on the position of the mutation, coding regions could be either expanded or contracted in this manner. Eucaryotic cells thus enjoy an evolutionary advantage, for in procaryotes a mutation necessarily destroys the old gene product when creating a new product, unless a duplication of that particular gene has already taken place (Chapter 7).

Examples of genes that have added coding sequences at splice junctions abound if one compares homologous proteins from different organisms. Interestingly, where three-dimensional protein structures are known, the added sequences often correspond to loops on the surface of the folded polypeptide chain and thereby do not perturb

Domain I

Intron B

Intron C

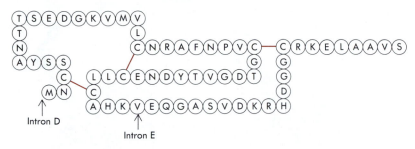

Domain II

Intron D

Intron E

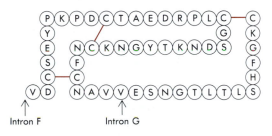

Domain III

Intron F

Intron G

Figure 20-22
Alignment of the three domains of chicken ovomucoid, showing the similar placement of disulfide bonds (indicated by colored lines) and positions of introns in its gene. The amino acid (A.A.) and nucleic acid (N.A.) sequence homologies are listed below. [After J. P. Stein, J. F. Catterall, P. Kristo, A. R. Means, and B. W. O'Malley, *Cell* 21 (1980):681–687.]

% Homology

Domains	A.A.	N.A.
I/II	46	66
II/III	30	42
I/III	33	50

the internal structure of the molecule (Figure 20-23). Recently, the α-crystallin gene of the mouse has been shown to specify two polypeptide products that differ by a 23-amino-acid insertion; two mRNAs, one with an additional internal exon, are responsible. This is clearly an alternative use of splice junctions (Chapter 21) and could represent a gene caught in the act of testing a new product.

Single Copies of Most Protein-Coding Genes per Haploid Complement[67]

The presence of lengthy intron sequences within most protein-coding genes multiplies our estimate of the amount of DNA required to encode a vertebrate protein by about tenfold. Yet, this factor can account for only a fraction of the excess DNA known to reside in higher eucaryotic genomes. Another way of expanding the DNA content would be to have multiple copies of some or all genes.

Figure 20-23
How the creation of new splice junctions can account for length differences in related proteins. On the left, a hypothetical gene containing two introns labeled A and B is transcribed; the introns are excised precisely, and the RNA is spliced to form the functional mRNA. The mRNA is then translated into protein. The letters A and B on the protein mark the splice junctions. On the right, a mutation alters the point of excision of intron B. This results in an extension of the exon and hence an added segment in the functional mRNA. On translation, this produces a protein that has an additional surface loop at the splice junction B. Such a situation is found in the family of serine proteases. [After C. S. Craik; W. J. Rutter, and R. Fletterick, *Science* 220 (1983):1125–1129.]

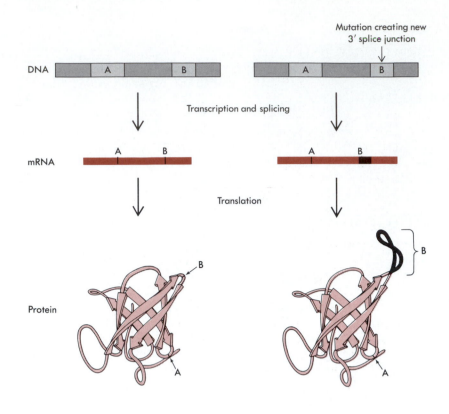

Figure 20-24
Arrangement of human α- and β-globin gene clusters and temporal expression of the individual genes. The colored boxes are functional genes, while those designated ψ are pseudogene regions. At all times in development, two α-like and two β-like polypeptide chains assemble to form functional hemoglobin tetramers. [After T. Maniatis, P. Melon, V. Parker, N. Proudfoot, and B. Seed, in *Molecular Genetics of Neuroscience*, ed. S. J. Bird and F. E. Bloom (New York: Raven Press, 1982), pp. 87–101.]

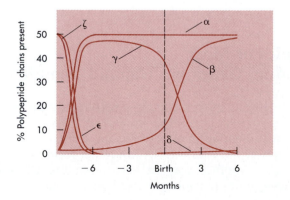

The number of DNA-RNA hybrid molecules that are formed when purified mRNA molecules are mixed with cellular DNA provides a method for measuring the number of times a given gene appears within a haploid genetic complement. Such measurements became possible when the first specific mRNA molecules were isolated from cells engaged in the selective synthesis of specific proteins. For example, studies using the 10S mRNA molecules that code for the α or β polypeptide chains of hemoglobin revealed that only one copy of each of these genes is present per haploid genetic complement. This conclusion holds not only for the egg and sperm but also for the cells (erythroblasts) differentiated for the almost exclusive synthesis of hemoglobin. Likewise, the gene encoding the silkworm protein fibroin is present in only one copy per haploid complement, even in the silk gland cells actively making silk. These examples are not exceptional; now that cDNAs corresponding to many mRNAs have been isolated and used to probe genomic DNA, we repeatedly find that only one hybridizing region is detected. These solitary numbers reflect the fact that the final level of protein synthesis is usually the result of two levels of amplification: the first occurring when a gene is transcribed many times to produce a large number of messenger RNA molecules, and the second occurring when these mRNA molecules in turn act as templates for many, many rounds of protein synthesis. The one fibroin gene of the silkworm, for example, can give rise in four days to some 10^5 fibroin mRNA molecules, which in turn serve as the templates for translation of some 10^{10} fibroin polypeptides—enough to spin an entire cocoon!

Developmentally Expressed Globin Genes Are Arranged in Sequence Along the Human Genome[68]

Although there is only one gene specifying the α or β polypeptide chain of hemoglobin molecules found in the adult mouse or human, this does not mean that a single α- or β-globin gene suffices during development of a complex mammalian organism. In fact, the requirement for slightly different hemoglobin proteins at various stages of development is fulfilled by having a whole family of related α-globin genes and another family of related β-globin genes. The expression of these genes is turned on and off coordinately and sequentially in response to changing oxygen tension, pH, and intracellular milieu as the embryo becomes a fetus and the fetus is born.

In humans, the α-like globin genes are clustered together on chromosome 16, while the β-like globin genes reside on chromosome 11 (Figure 20-24). Between the genes are long stretches of nontranscribed DNA devoid of other known genes. Also included are **pseudogene** sequences ($\psi\beta1$, $\psi\zeta$, and $\psi\alpha1$), regions that are highly homologous to the globin genes but do not give rise to detectable mRNA. Whether these sequences are former genes that have been mutated to nonfunctionality or whether they represent regions undergoing evolution to some new function is not known. Overall, both the α- and β-globin gene families encompass many kilobases of DNA, most of which cannot be considered to be genetic in the strictest sense of the word. This situation contrasts sharply with the much more economical usage of DNA in the genomes of procaryotes (like E. *coli*) and even lower eucaryotes (like yeast). There, the spaces between genes are generally much more modest in length.

In both the α-globin and β-globin gene clusters in humans, the genes themselves are arranged in order of their expression during development, with the embryonic genes followed by genes that function during fetal and adult life (see Figure 20-22). A similar order of genes is found in most other mammals, although the total size of the α- and β-globin gene clusters vary considerably. In contrast, chicken β-globin genes expressed in the adult are flanked by the embryonic genes, showing that there is nothing obligatory about this particular gene order.

Sequences of the several α-globin genes or β-globin genes contained in each cluster are very similar within their protein-coding regions. For instance, the two fetal β-like polypeptide chains (G_γ and A_γ) differ in only one amino acid out of 141. In contrast, their noncoding regions (both introns and leader and trailer sequences) hardly appear related, except that the second intervening sequence is always larger than the first. These findings suggest that the two globin gene clusters arose by repeated duplication and divergence of original α- and β-globin genes, which themselves were apparently derived from a common ancestral gene. We expect the protein-coding sequences to be less divergent than the noncoding sequences because protein function (including the ability of each kind of α-globin chain to form heterologous dimers with each kind of β-globin) is a major constraint during evolution.

Other Families of Related Genes Also Show Common Evolutionary Origin[65, 69, 70]

Intensive work on both protein and DNA sequences has revealed that many proteins in higher organisms are encoded by families of highly homologous but nonidentical genes. The different members of a gene family may be switched on and off at various stages of development, as in the case of globins. Or, the needs of different tissues within an organism may call for the synthesis of slightly different variants of a protein at the same time but in different cells.

Both these demands are met by the actin gene family in mammals. At least six closely related actin polypeptides are synthesized in various cells at various times. Actin is a major component of the contractile apparatus in muscle cells; but one type of α actin is specific for skeletal muscle, another α actin is found only in cardiac muscle, and both types are found in the smooth muscle of the gut. Still other actins (called β and γ) are constituents of the cytoskeleton of nonmuscle cells. Furthermore, slightly different α actins are present in fetal and adult skeletal muscle. As many as 20 to 30 different actin-related sequences can be detected in the human genome by hybridization. These probably lie on different chromosomes. This situation contrasts with that in yeast, where only a single actin gene suffices in fulfilling the needs of this single-celled organism.

Other well-studied examples of proteins encoded by gene families in mammals are albumin and α fetoprotein (major components of blood plasma), the serine proteases, the interferons (which are instrumental in our defense against viral infection; Chapter 24), and the immunoglobulins (Chapter 23). In insects, the chorion proteins—which constitute major components of the egg shell—are encoded by

		Repeat length	Approximate repetition
Sea urchin (*P. miliaris*)	H1 H4 H2B H3 H2A	6300 bp	300–600
Fruit fly (*D. melanogaster*)	H1 H3 H4 H2A H2B	4800 bp	100
Newt (*N. viridescens*)	H1 H3 H2B H2A H4	9000 bp	600–800

families of genes with many divergent members specialized for different roles (Chapter 22). Sometimes, the members of a gene family all reside in a single cluster on one chromosome. In other cases, they have been dispersed through evolution onto two or more chromosomes. Examination of their coding sequences, however, makes it clear that all genes in a family, regardless of how prodigious their numbers, have evolved from a single ancestral gene.

Multiple Copies of the Histone Genes[71, 72]

In some gene families, not only is there a need to encode related proteins, but many copies of identical genes are required. The genes for the histone proteins that package DNA into nucleosomes (Chapter 21) are a prime example. In most eucaryotes, there are five major histones called H1, H2A, H2B, H3, and H4. Since higher eucaryotes have relatively large amounts of DNA per cell, single-copy genes cannot produce sufficient numbers of histone proteins to package all the newly synthesized DNA that is made during the S phase of the cell cycle. Instead, higher eucaryotic genomes contain multiple copies of their histone genes—ranging from 10 to 20 in birds and mammals, and to 600 to 800 in sea urchins and newts. Yeast, in contrast, possesses only two copies of each of its four histone genes (for H2A, H2B, H3, and H4).

Organisms such as sea urchins and amphibians have the most highly repetitious histone genes because of their need for rapid cell division (and accompanying rapid DNA synthesis) during early development. Histone gene copy number must be high to supply the enormous amounts of histone mRNAs both stored in the egg and synthesized soon after fertilization. In these organisms, the five histone genes are grouped together into a basic unit that is tandemly repeated in a large head-to-tail array (Figure 20-25). However, the organization of the five individual histone genes within the basic repeat differs from one organism to another (and even between species). Sometimes, all five genes are transcribed off the same strand of DNA; in other cases, the genes have different orientations (see Figure 20-25). Because the sequence of the five histone proteins has been very highly conserved through evolution, the DNA sequences of the histone-coding regions in the basic repeat are quite conserved (despite the degeneracy of the genetic code). However, the spacer regions that separate the genes within the basic repeat unit are very

Figure 20-25
Organization of highly repeated histone gene clusters in several organisms. The genes (colored) are separated by nontranscribed spacer regions. The direction of transcription is indicated by arrows. The clusters are repeated tandemly in the sea urchin and fruit fly; in the newt, they are interspersed with 10 to 50 kilobase pairs of satellite DNA. [Data from C. C. Hentschel and M. L. Birnstiel, *Cell* 25 (1981):301–313.]

different in different organisms and are at least as long as the protein-coding regions themselves.

In mammals and most other higher animals, cell division during early embryogenesis is not so rapid, and fewer copies of the histone genes are needed. The organization of these genes is quite variable. Although vertebrate histone genes are often clustered, the gene order is not fixed, nor are there tandem arrays. In addition, some individual histone genes are located far away from any other histonelike sequences. Nonetheless, in all eucaryotic organisms, histone genes share two features not shared by most other genes: Their transcripts almost always lack both introns and poly A tails.

Tandem Repeats of rDNA[73–77]

Even in *E. coli,* seven copies of the rRNA genes are required to produce the 15,000 ribosomes needed by a rapidly growing bacterial cell (Chapter 14). Not surprisingly, higher eucaryotic cells have many more identical copies (50 to 5000) of the genes that specify the 18S, 5.8S, and 28S rRNA components in their 10 million ribosomes. These rRNA genes are tandemly arranged in that order into huge blocks situated on one or more chromosomes. The regions that are transcribed into large pre-rRNAs range in size from 8 kilobases (in yeast, *Drosophila,* and *Xenopus)* to 13 kilobases (in mammals) and are separated by **nontranscribed spacers** (Figure 20-26). In some organisms like the toad *Xenopus,* repetitious shorter sequences within these spacers appear to be essential for controlling or activating rDNA transcription (Chapter 21). Like the transcribed regions, the nontranscribed spacers are highly conserved from one tandem repeat unit to the next within an organism. The rDNA repeat units are usually looped off the main chromosomal fiber masses as highly extended threads. These loops then coalesce with specific proteins to form the **nucleoli,** where rRNA synthesis and processing take place and where assembly of the processed rRNA with ribosomal proteins begins. Some organisms have only one nucleolus organizer (rDNA repeat) region per haploid complement, while others have several.

The genes for 5S rRNA (120 nucleotides) also exist in tandemly repeated units separated by nontranscribed spacer regions. However, in higher eucaryotes, they are not closely linked to the large blocks of rDNA. In *Drosophila,* for instance, there is only one cluster of about 160 5S rRNA genes with a basic repeat unit of 375 base pairs that varies only slightly in length and sequence from one unit to another (see Figure 20-26). The 5S genes in *Xenopus* are distributed into many smaller clusters located near the ends (telomeres) of perhaps 18 different toad chromosomes. Here, there are two different sets of 5S genes. One encodes the 5S rRNA found in oocyte ribosomes, while the other specifies the 5S rRNA in somatic cell ribosomes, with only eight nucleotide differences distinguishing their two 120-nucleotide-long sequences. There are several hundred copies of the somatic-type 5S genes, but over 10,000 copies of the oocyte variety. The difference in gene copy number reflects the need to synthesize the large amounts of 5S rRNA that are stored in the egg during oogenesis (Chapters 21 and 22). Depending on which species of *Xenopus* and which 5S gene is being examined, the repeating units can contain internally repetitious subrepeats of a short sequence or 5S pseudogenes, sequences that are highly homologous to real genes but do not give rise to detectable RNA products (see Figure 20-26).

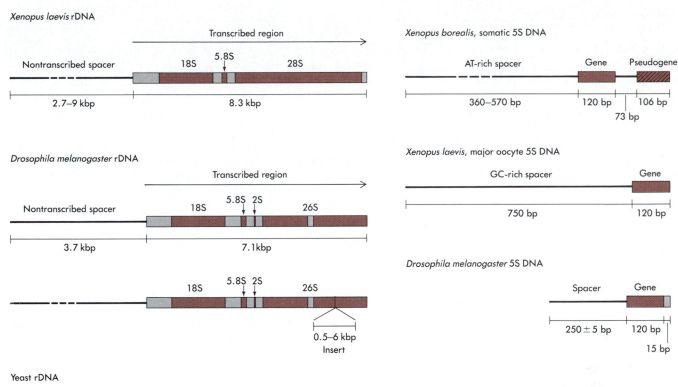

Xenopus laevis rDNA

Transcribed region

Nontranscribed spacer   18S   5.8S   28S

2.7–9 kbp   8.3 kbp

Drosophila melanogaster rDNA

Transcribed region

Nontranscribed spacer   18S   5.8S 2S   26S

3.7 kbp   7.1kbp

18S   5.8S 2S   26S

0.5–6 kbp
Insert

Yeast rDNA

Transcribed by Pol III   Transcribed by Pol I

5S   17S   5.8S   26S

1 kbp   8.1 kbp

Xenopus borealis, somatic 5S DNA

AT-rich spacer   Gene   Pseudogene

360–570 bp   120 bp   106 bp
73 bp

Xenopus laevis, major oocyte 5S DNA

GC-rich spacer   Gene

750 bp   120 bp

Drosophila melanogaster 5S DNA

Spacer   Gene

250 ± 5 bp   120 bp
15 bp

Figure 20-26

Arrangements of rRNA gene repeating units in various eucaryotic organisms. Note that in yeast, the repeating unit also contains the 5S gene, even though it is transcribed from the opposite strand by an RNA polymerase (III) different from the one (I) that synthesizes the large rRNA precursor. *Drosophila* rDNA is unusual in several ways. *Drosophila* 60S ribosomes contain an additional 2S rRNA species, and the 26S molecule is interrupted about in the middle. Moreover, insertions are present in a large fraction of *Drosophila* rDNA repeats, but are not introns, since these genes are not transcribed. The arrangement of the large rDNA repeat in mammals is similar to that in *Xenopus* except that both the transcribed and nontranscribed spacers are longer. [After N. V. Fedoroff, *Cell* 16 (1979):697–710.]

Clusters of Different tRNA Genes[78–82]

The eucaryotic cell's demand for tRNA molecules is so high that single genes do not suffice. Generally, ten to several hundred genes for each tRNA species are present in the haploid genome, depending on the organism examined. In organisms with lower numbers of tRNA gene copies, identical genes are usually widely dispersed and have flanking sequences that are totally unrelated—even in those immediately adjacent 5' and 3' regions that are transcribed as part of their respective tRNA precursor molecules. In organisms (like *Xenopus*) with more highly reiterated tRNA genes, tandem repeats of relatively long sequences containing the genes for several different tRNAs can be detected. For instance, two tRNA$_i^{Met}$ genes and one gene each for six other tRNAs lie on a 3.2-kilobase-pair repeated fragment (Figure 20-27). In other organisms, tRNA genes are often grouped into heteroclusters that contain the genes for several completely different tRNA species within regions of several thousand base pairs of DNA. Sometimes, tRNA pseudogene sequences are also included in these regions. Only in rare instances, however, does it appear that neighboring tRNA genes are cotranscribed into multimeric tRNA precursor molecules. (The well-documented examples are in yeast.) Hence, it is not at all clear why different tRNA genes are so clustered.

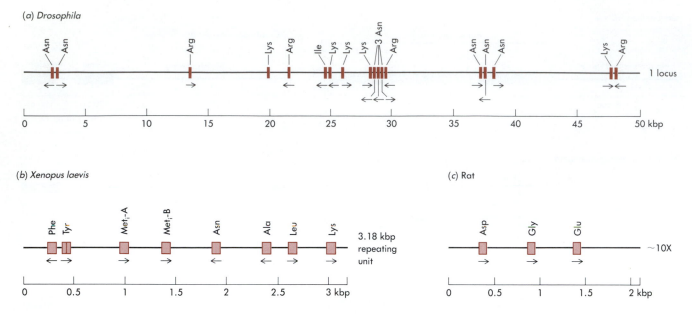

Figure 20-27

Arrangement of tRNA genes in higher eucaryotic organisms. (a) One of the 50 to 60 sites on *Drosophila* chromosomes where tRNA genes are clustered. [After P. H. Yen and N. Davidson, *Cell* 22 (1980):137–148.] (b) A 3.18-kilobase-pair unit from *Xenopus* DNA that is repeated about 150 times. The tRNA^Tyr gene contains an intron. One of the tRNA_i^Met genes differs from the other by only one base pair in the coding region, but it is believed to be a pseudogene. [After J. Fostel et al., *Chromosoma* 90 (1984):254–260.] (c) This region of rat DNA is present about ten times in the haploid genome. However, the exact sequences are nonidentical, and it is likely that one or more of the genes in some of the repeats are pseudogenes. [After T. Sekiya, Y. Kuchino, and S. Nishimura, *Nucleic Acid Res.* 9 (1981): 2239–2250.]

Gene Amplification in Response to Environmental Pressure[83, 84]

The existence in eucaryotic DNA of gene families containing many identical members can be explained by the repeated duplication of specific DNA segments over evolutionary time. Subsequent divergence (i.e., the accumulation of mutations in such duplicated regions) could similarly explain the creation of gene families in which the members are related to one another but not identical. Whether the members of a particular gene family are identical or simply related, we have seen that they tend to be clustered together in the genome and arranged in a head-to-tail fashion separated by stretches of spacer DNA (which can be anywhere from short to very long).

These observations raise the possibility that tandem gene duplication is a relatively frequent event in all cells, a consequence of occasional mistakes made during the normal processes of DNA replication and recombination. Usually, such DNA duplications will confer no selective advantage on the organism, and the extra gene copy will be rapidly lost. (A simple crossover event between adjacent homologous sequences would excise the extra gene copy as a small DNA circle; the circle would then be diluted out of the cell population because it cannot replicate by itself.)

Duplication (amplification) of a specific gene can, however, be selected and stabilized by applying appropriate external pressure. If, for instance, mammalian cells growing in culture are treated with the drug methotrexate (an inhibitor of the essential enzyme dihydrofolate reductase, DHFR), most of the cells will die. Among the rare survivors are cells that produce significantly larger amounts of DHFR and therefore have enough uninhibited enzyme to generate tetrahydrofolate for survival and growth. By stepwise selection, cell variants can eventually be obtained that have multiplied the normally single copy (per haploid genome) of the DHFR gene 100- to 1000-fold! The amplified genes (and concomitant drug-resistant state) can be either stable or unstable. In the unstably amplified state, the DHFR genes reside

on small, paired extrachromosomal elements known as **double minute chromosomes** (Figure 20-28). Because the double minutes lack
centromeres, they are rapidly lost when the selective drug (methotrexate) is removed. In contrast, stably amplified DHFR genes have
been incorporated directly into one of the cell's chromosomes, where
they often appear as elongated chromosomal bands called **homogeneously staining regions** (see Figure 20-28). These can occur either at
the natural site of the unamplified DHFR gene (resulting from amplification without excision) or on other chromosomes (resulting from
translocation, which could occur either before or after amplification).

Whether the amplified genes are located on normal chromosomes
or on double minutes, the DNA sequences surrounding the DHFR
gene are usually unchanged from the original unamplified gene, so
that regulation and transcription are normal. However, the size of the
amplified unit is highly variable, being sometimes as small as 25 kilobase pairs (just enough to encompass the DHFR gene itself) and
sometimes as large as 2000 kilobase pairs! We do not know whether
specific DNA sequences are involved in amplification, nor do we
understand the mechanism. However, we do know that even in the
absence of drug treatment, the rate of spontaneous duplication of the
DHFR gene is about 1×10^{-3}, so that one in a thousand cells carries at
least two DHFR gene copies. This observation could explain the tendency of eucaryotic organisms to meet the demand for increased gene
product during the course of evolution by gene amplification (forward mutation rate 10^{-3}) rather than by devising a more powerful
promoter or by stabilizing the gene product (forward mutation rate of
less than 10^{-5}).

Figure 20-28
Double minute chromosomes and a
homogeneously staining region (HSR)
in Chinese hamster cells that have
amplified the dihydrofolate reductase
(DHFR) gene. Metaphase chromosome
spreads from (a), a methotrexate-sensi-
tive cell, are compared to (b) and (c),
two resistant cells that contain extrachromosomal elements (indicated by
arrow). [Courtesy of R. T. Schimke *et
al.* (1981) *Cold Spring Harbor Symp.
Quant. Biol.* 45, 785–797. Used with
permission.] In (d) we see trypsin-
Giemsa banded chromosomes from a
(stably) methotrexate-resistant cell with
a large HSR located on one end of
chromosome 2 (diagrammed in e).
[Courtesy of J. L. Biedler and B. A.
Spengler, *Science* 191 (1976):185–187.
Used with permission.]

Selective amplification of many eucaroytic genes has been observed after treatment with drugs or other toxic substances. For example, exposure of tissue culture cells to poisonous doses of cadmium (Cd^{2+}) selects for cells that have amplified the gene for metallothionein, a protective metal-binding protein. Programmed gene amplification is also occasionally utilized by higher organisms as a mechanism for synthesizing large amounts of a specific gene product in response to special needs at a particular stage of development. Examples are the amplification of rDNA in amphibian oocytes, of chorion genes in *Drosophila*, and of many protozoan genes during the formation of the macronucleus that directs vegetative growth (Chapter 22). The polytenization of dipteran chromosomes in the larval salivary glands can also be considered a case of gene amplification, although in that case, almost the entire chromosomal DNA is uniformly amplified (Chapter 21).

Gene Sequences Drift Through Time

Despite the existence of sophisticated cellular DNA repair mechanisms, all DNA sequences in the genome, and thus all genes, are constantly subject to mutation. Mutations are caused by errors made during DNA replication and recombination, by chemical DNA-damaging agents, by the chemical instability of certain groups within DNA, and even by errors made by DNA repair mechanisms themselves. Many of these mutations represent single-base substitutions (transitions and transversions; Chapter 12). Other mutations arise from insertions or deletions of a few base pairs. However, some mutations represent gross changes in the DNA, resulting from large deletions and insertions, from inversion, or from translocations (Chapter 12). Given a relatively high rate of mutation, how are the sequences of genes conserved through evolutionary time?

We have no trouble understanding how natural selection can maintain a functional single-copy gene like globin or insulin. If the gene product is defective in any serious way, the organism producing it will be immediately subjected to a selective disadvantage; it will either die prematurely or produce fewer progeny than its unmutated siblings. Thus, natural selection does not preserve the DNA sequence of the gene per se but rather its functional product. In the case of protein-coding genes, even significant changes in the DNA sequence may not affect the functional product because of the degeneracy of the genetic code (Chapter 15). In the case of structural RNAs like ribosomal RNA or transfer RNA, the RNA transcript *is* the final gene product, and the rate of sequence change tolerated is often much lower than that for protein-coding genes. Natural selection even acts on the parts of genes that are noncoding because certain sequences are critical for function. For example, splice junction signals (lying partly within introns) must be maintained, and spurious initiation codons for protein synthesis must not be introduced into the 5' untranslated leaders of mRNAs.

But we must also remember that gene sequences do ultimately change through time. New genes may be assembled from fragments of preexisting genes, rather than created de novo. In some cases, poorly understood recombination events result in tandem duplication of DNA sequences. The vertebrate α- and β-globin clusters (see Figure 20-24) clearly arose by some such mechanism. Once a gene has been duplicated, one copy is sufficient for survival of the organism, while the other gene copy is free to change. These changes will generally be useless or even harmful, and the extra gene copy will probably be lost. However, occasionally one of the gene copies will mutate in an advantageous way, and natural selection will preserve both the new gene and the old. In other cases, several mutations may be paired or grouped in such a way that the gene product (whether protein or RNA) is significantly changed without affecting its function. For example, the folding (and function) of structural RNA species depends critically on the existence of specific internally base-paired regions, but the actual RNA sequence within the stems may be relatively unimportant (Chapter 14). In such cases, *compensatory* mutations occurring on both strands of a duplex region can preserve the base pairing. Similar compensatory changes can occur in the amino acid sequence of a protein as well (intragenic suppressor mutations), so that overall structure and function are preserved (Chapter 15). Thus, the molecular evolution of a gene like globin or insulin resembles the predicament of the Red Queen in *Alice in Wonderland,* who had to run fast in order to stay in place. Mutations are constantly changing gene sequences, while natural selection keeps pace by preserving functional gene products.

The Apparent Paradox of Multigene Families[85]

Even more baffling than how a gene duplication originates is how virtual identity can be maintained among the many gene copies that constitute a multigene family. We have seen how natural selection can maintain the sequence of an essential single-copy gene like α- or β-globin; any deleterious mutation in the gene immediately places the organism at a selective disadvantage relative to its normal (wild-type) competitors. However, in the case of a large multigene family like ribosomal RNA (500 gene copies in the human genome) or amphibian oocyte 5S ribosomal RNA (20,000 gene copies in the *Xenopus* genome), it is very difficult to imagine how a mutation in any single gene copy would endanger the survival of the organism. In fact, our biological intuition rejects the notion that a defect in one out of 500 gene copies (0.2 percent of the genes) could have any significant effect on the fitness of the organism. Transcription of the remaining 499 unmutated copies would still produce 99.8 percent of the normal level of rRNA required to assemble functional ribosomes. Moreover, if a mutation in one out of 500 genes is harmless, why should mutations in 2 or even 20 out of 500 gene copies (96 percent normal genes) be harmful? Thus, we might have expected that large multigene families would quickly accumulate mutations because the many wild-type genes would shelter the organism from deleterious changes in any particular gene copy. In addition, because many (and perhaps most) mutations are selectively "neutral" rather than harmful (i.e., they do not affect the function of the gene product), we might also have expected the many gene copies in a multigene family to accumulate many neutral as well as occasional deleterious mutations. But in sharp contrast to such predictions, most stable cellular RNA species encoded by multigene families (including 18S, 5.8S, 28S, and 5S rRNAs, and U1 and U2 small nuclear RNAs) have been found to be essentially homogeneous at the RNA sequence level. This discrepancy between our theoretical expectation (that genes within a multigene family will diverge from each other) and experimental observations (that genes within a multigene family are usually homogeneous) is the apparent paradox of multigene families.

Keeping Multigene Families Homogeneous[85-88]

We can resolve the paradox of multigene families by postulating that eucaryotic organisms have one or more mechanisms for maintaining homogeneity among the members of repeated DNA sequence families. In fact, there is both circumstantial and experimental evidence for the existence of such homogenization mechanisms.

One convincing reason for believing that eucaryotic cells have mechanisms for homogenizing tandemly repeated sequences is that the length of the basic repeat unit in most tandemly repeated multigene families is much larger than the protein- or RNA-coding sequences within it. For example, the 10 to 20 human genes for U2 small nuclear RNA are tandemly repeated in a single cluster on chromosome 17, but the 188-base-pair U2-coding region constitutes only 3 percent of the 6000-base-pair repeat unit! How could natural selection maintain the essentially perfect homogeneity of the remaining 97 percent of the repeat unit? One might have argued that this 5880-nucleotide DNA sequence is conserved because it contains important signals controlling the start and stop of transcription as well as the RNA processing of the primary U2 transcript to the size of mature U2 snRNA. However, we shall see in Chapter 21 that such regulatory signals in the DNA and RNA of eucaryotes occupy remarkably few nucleotides, just as do comparable regulatory signals in bacteria (Chapters 13, 14, and 16). Thus, essential control regions could conceivably account for conservation of another tiny fraction of the 6000-nucleotide U2 repeat unit, but there is no obvious way that selective pressure could bring about homogenization of the remaining 5700 nucleotides. The inescapable conclusion is that any DNA that is organized in a tandem array will automatically be homogenized, regardless of the genetic information contained within it. Thus, spacer sequences that cannot be subject to direct natural selection (because mutations within them would not harm the organism) are automatically cohomogenized along with the selectable sequences (coding and regulatory regions) to which they are attached. Although the mechanism of homogenization is a matter of great controversy among geneticists, it is clear that such a mechanism must exist to account for the homogeneity of tandemly repeated gene families, as well as of tandemly repeated "satellite" sequences having no obvious function.

More compelling evidence for a gene homogenization mechanism comes from recent experimental work on tandemly repeated genes encoding ribosomal rRNA in yeast. Located between each transcription unit encoding the 35S precursor of yeast 17S, 5.8S, and 25S ribosomal RNAs is a sequence that serves as a localized stimulator of recombination. Although the mechanism by which this sequence works is still unknown, it is not difficult to imagine how such recombinational hot spots could ensure homogenization of the tandemly repeated rDNA array by promoting gene conversion, unequal sister chromatid exchange (homologous but out-of-phase recombination between repeat units), or even gene amplification (Figure 20-29).

But even multigene families that are not tandemly repeated are kept homogeneous. For example, the eight identical genes encoding tyrosine tRNA in yeast are located on eight different chromosomes.

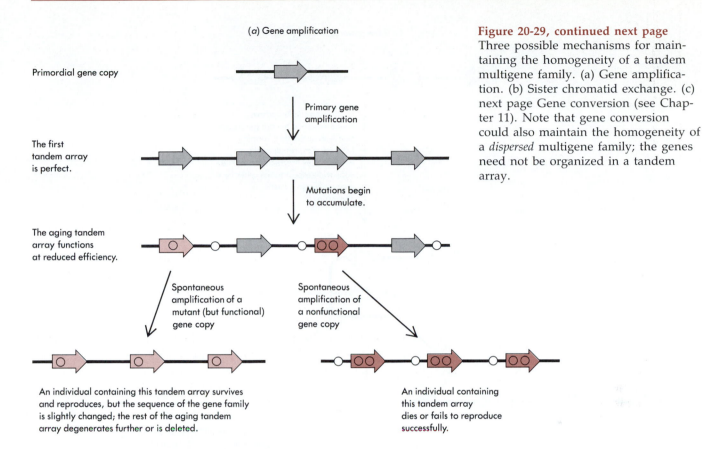

(a) Gene amplification

Primordial gene copy

Primary gene amplification

The first tandem array is perfect.

Mutations begin to accumulate.

The aging tandem array functions at reduced efficiency.

Spontaneous amplification of a mutant (but functional) gene copy

Spontaneous amplification of a nonfunctional gene copy

An individual containing this tandem array survives and reproduces, but the sequence of the gene family is slightly changed; the rest of the aging tandem array degenerates further or is deleted.

An individual containing this tandem array dies or fails to reproduce successfully.

Figure 20-29, continued next page
Three possible mechanisms for maintaining the homogeneity of a tandem multigene family. (a) Gene amplification. (b) Sister chromatid exchange. (c) next page Gene conversion (see Chapter 11). Note that gene conversion could also maintain the homogeneity of a *dispersed* multigene family; the genes need not be organized in a tandem array.

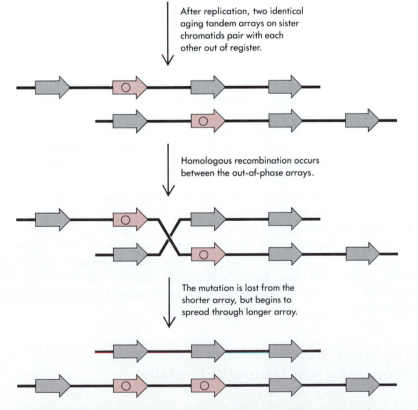

(b) Sister chromatid exchange

After replication, two identical aging tandem arrays on sister chromatids pair with each other out of register.

Homologous recombination occurs between the out-of-phase arrays.

The mutation is lost from the shorter array, but begins to spread through longer array.

Further cycles of sister chromatid exchange can either (a) expand the short (but perfect) tandem array to its original length or (b) spread the mutation through the longer array, eventually producing a new tandem array with a slightly different sequence.

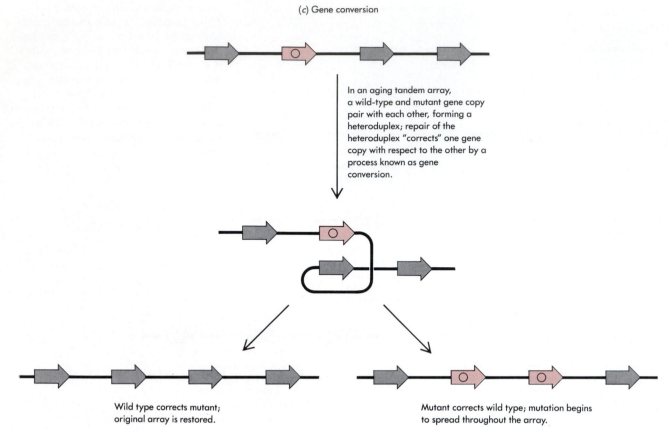

Figure 20-29
(Continued)

Although the tRNA-coding sequence and the 14-nucleotide interven-ing sequence are identical except at one position at each locus (see Figure 20-8), the 5' and 3' flanking sequences display no obvious similarities. How can we account for such localized homogeneity when our biological intuition would anticipate that the eight individ-ual tRNA genes in this small multigene family would have diverged slightly from each other through the accumulation of various minor or neutral mutations? The absence of such divergence indicates that these eight genes are **coevolving;** that is, their sequences all change together. But how might eight dispersed genes communicate with each other? The alert reader will have noticed that gene conversion, as detailed in Figure 20-29c for a tandemly repeated multigene family, could also homogenize a dispersed multigene family; tandem repeti-tion may simply accelerate the rate of gene conversion by increasing the local concentration of related genes. Repeated rounds of pairwise gene conversions may also help to explain why the 5' and 3' flanking sequences of the eight tyrosine tRNA genes are divergent. Although gene conversion can homogenize two closely related DNA se-quences, gene conversion is unlikely to proceed through unrelated sequences.

Tandemly Repeated DNA Sequences Are Naturally Unstable

We might have guessed that tandem repeats of an identical DNA sequence would be an ideal substrate for *intra*chromosomal recombi-nation (i.e., recombination within a single chromosome) and that tandemly repeated multigene families would therefore be rapidly lost from a eucaryotic genome (in much the same way that bacteriophage λ excises from the bacterial chromosome by homologous recombination between the two *att* sites; Chapter 17). In fact, certain mutants of the fruit fly *Drosophila* are known to have deleted a part of their tandemly repeated array of rDNA. Such flies grow slowly, because they cannot synthesize rRNA at the normal rate; however, over the course of many generations, the mutant flies gradually expand the number of rDNA repeat units and regain a normal growth rate. We can understand this type of expansion by noting that during unequal sister chromosome exchange (i.e., *inter*chromosomal exchange) one daughter chromosome acquires the repeat units lost by the other (Figure 20-29b). This process constitutes one mechanism for regaining the original number of repeated units within a tandem gene cluster that has undergone accidental contraction. Another such expansion mechanism is provided by gene amplification (Figure 20-29a). Thus, natural selection can maintain the overall gene copy number within a tandem gene cluster at an approximately constant level to meet the needs of the organism.

Abundance of Pseudogenes for Small Nuclear RNAs[89, 90]

The small RNAs like U1, U2, and U3, which are such numerous constituents of the nuclei of higher eucaryotic cells, are (not surprisingly) encoded by multiple copies of identical genes. Usually, the genes for these RNAs are organized into tandem arrays that contain only one type of small nuclear RNA (snRNA) gene surrounded by relatively conserved spacer DNA varying in length from 800 base pairs to greater than 45 kilobase pairs, depending on the organism.

In humans and other mammals, a most unexpected feature of the genomic regions complementary to the snRNAs is that pseudogenes outnumber real genes by at least ten to one. There are some 30 true U1 snRNA genes located on human chromosome 1; yet 500 to 1000 other regions hybridize to U1 RNA but cannot encode it. Some pseudogenes differ from real genes by as little as one nucleotide (out of 164 in the case of U1 snRNA) and have homologous flanking sequences. Other pseudogenes possess only a portion of the snRNA sequence and are flanked by DNA that is totally unrelated to the sequences adjacent to real genes. The 500 to 1000 U1 pseudogenes are dispersed at many locations in the human genome. The abundance of pseudogenes for U6 snRNA may be even greater.

The most intriguing snRNA pseudogenes are those that lack the region specifying the 3' end of their respective RNA and are surrounded by DNA sequences unrelated to those next to real genes. Such pseudogenes are usually flanked by short (6 to 20 base pairs), nearly perfect direct repeats. This suggests that these pseudogenes may have arisen by insertion of a partial copy of the snRNA sequence between staggered breaks in genomic DNA (as occurs during the movement of transposable elements; Chapter 11). For instance, a frequently occurring variety of U3 pseudogene exhibits homology to only the first 70 nucleotides at the 5' end of the RNA molecule. This is just the size predicted if the 3' terminus of U3 RNA had been utilized as a self-primer for a reverse transcriptase-like activity, producing a partial cDNA copy of U3 sequences that was then inserted into the human genome (Figure 20-30).

Figure 20-30
Schemes for the generation of a frequently occurring type of U3 snRNA pseudogene. The resulting pseudogene exhibits homology only to the first 70 nucleotides of the RNA and is flanked by short direct repeats. [After L. B. Bernstein, S. M. Mount, and A. M. Weiner, *Cell* 32 (1983):461–472.]

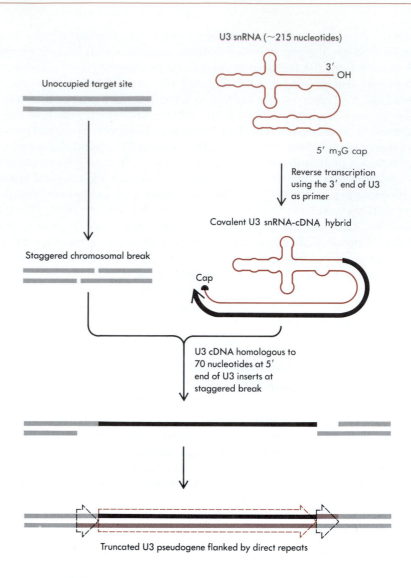

U3 snRNA (~215 nucleotides)

3' OH

5' m₃G cap

Unoccupied target site

Reverse transcription using the 3' end of U3 as primer

Covalent U3 snRNA-cDNA hybrid

Staggered chromosomal break

Cap

U3 cDNA homologous to 70 nucleotides at 5' end of U3 inserts at staggered break

Truncated U3 pseudogene flanked by direct repeats

Processed Genes: Intronless Replicas of Protein-Coding Genes[91–94]

As a mechanism for producing pseudogenes, reverse transcription does not appear limited to snRNAs. An increasing number of pseudogenes identified in the course of analyzing mammalian gene families bear distinctive hallmarks of RNA processing. These pseudogene sequences have lost their introns precisely and also possess 3'-terminal poly A tracts. The mRNA-like sequence is flanked by short direct repeats and lacks the upstream promoter sequences required for transcription of the corresponding real gene. In most cases, **processed pseudogenes** have also accumulated point mutations and small deletions or insertions. For example, the human β tubulin gene family consists of about 15 to 20 members. Seven of the 10 of these that have been studied are pseudogenes, of which 5 are intronless, contain poly A, and are flanked by short direct repeats. In contrast to the ability of U3 snRNA to prime its own reverse transcription (see Figure 20-30), it is not obvious how synthesis of a full-length cDNA copy of a polyadenylated mRNA could have been primed beginning on the poly A tract. For this reason, the proposal has been made that reverse transcription might be primed by the 3' end of the DNA at the chromosomal break into which the cDNA will be inserted. However, un-

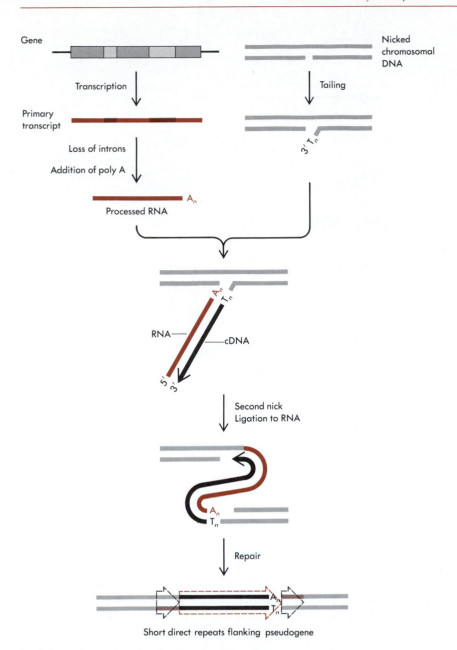

Gene

Primary

Loss of introns

Addition of poly A

Processed RNA
A_n

Nicked
chromosomal
DNA

Tailing

$3'$ T_n

A_n
T_n

RNA

cDNA

$5'$ $3'$

Second nick
Ligation to RNA

A_n
T_n

Repair

Short direct repeats flanking pseudogene

Figure 20-31
A proposed mechanism for generating "processed" pseudogenes. [After J. H. Rogers, *Int. Rev. Cytol.* 93 (1985):188–279.]

less we postulate further that such chromosomal breaks almost always occur within tracts of T (thymine) residues or that there is a mechanism for tailing the DNA with Ts (Figure 20-31), it is difficult to understand why priming occurs on the poly A tract of the mRNA rather than internally. Nonetheless, reinsertion of copies of processed RNAs provides an appealing explanation for the existence of these curious sequences in mammalian DNA. Interestingly, processed pseudogenes appear less frequently in other eucaryotic genomes, even those of lower vertebrates.

Several Classes of Transposable Elements Cause Mutations in *Drosophila*[95–98]

Bacterial and lower eucaryotic (yeast) genomes are not the only genomes inhabited by transposons. In *Drosophila*, for example, at least three distinct types of mobile elements have been found to be respon-

Figure 20-32

Structures of three classes of *Drosophila* transposable elements. The repeated sequences that make up the ends of each of these elements are shown in color. They all give rise to short direct repeats of target DNA upon insertion. Open reading frames (ORF) are shown for a representative copia element (other copias have different gene orders) and the complete P factor; gag (a nucleic acid binding protein), int (integrase), and RT (reverse transcriptase) refer to sequences homologous to retroviral genes; LTR stands for long terminal repeat. [After G. M. Rubin, in *Mobile Genetic Elements*, ed. J. A. Shapiro (New York: Academic Press, 1983), p. 329; D. Finnegan, *Int. Rev. Cytol.* 93 (1985):281–326; S. M. Mount and G. M. Rubin, *Mol. Cell Biol.* 5 (1985):1630–1638.]

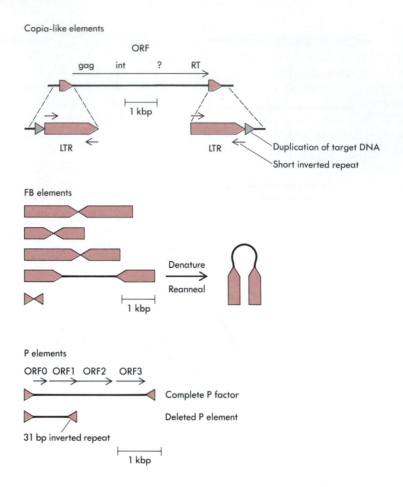

sible for mutations and other genetic rearrangements. These DNA sequences move repeatedly and can distribute themselves, apparently at random, around the genome. As the DNAs of more and more higher organisms are closely scrutinized, sequences similar to one or another of these categories continue to be recognized, arguing that all genomes are subject to remodeling by transposition.

Copia elements in *Drosophila melanogaster* possess long terminal direct repeats and are in many ways analogous to the Ty1 transposable elements of yeast (Chapter 18) and the integrated proviral forms of RNA tumor viruses (Chapter 24). Actually, there are several families (more than 11) of similar "copia-like" elements; the members of each family are well conserved and are located at 5 to 100 different sites in the *Drosophila* genome. Evidence that copia-like sequences are mobile comes not only from observations that the number and location of any particular element varies between *Drosophila* strains but also from the finding that stable mutations in a number of genes are due to their insertion. Copia-like elements are about 5000 base pairs long, with long terminal repeats (LTRs) of several hundred base pairs that vary in length and sequence between families (Figure 20-32). At the extreme ends of each element are short imperfect inverted repeats (about 10 base pairs). As with all other movable elements, insertion of copia-like elements into a new site is accompanied by duplication of a short stretch (3 to 6 base pairs) of target DNA; the length, but not the sequence, of the direct repeats that consequently appear immediately before and after the element is the same for all members of the same family. A distinctive property of copia-like elements is that their entire length is transcribed at a very high rate to produce polyadenylated RNAs; curiously, many of these transcripts are confined to the cell nucleus. Recent sequencing of a complete copia element has revealed that it possesses one long open reading frame that encodes proteins homologous to those of RNA tumor viruses. Homologies to reverse transcriptase, integrase, and nucleic acid-binding proteins are seen (Chapter 24), constituting strong evidence that copia moves via RNA, as does the yeast Ty1 element (Chapter 18). A transposable element with properties very similar to copia also occurs in the mouse genome; called the IAP element, it clearly encodes a protein that encapsulates its own RNA into noninfectious viruslike structures called **intracisternal A-type particles.**

A second distinct class of *Drosophila* transposable elements is known as the **fold-back (FB) elements.** These possess long inverted repeats either directly adjacent to one another or separated by up to several thousand base pairs of DNA (see Figure 20-32). When such sequences are denatured, they readily fold back to form hairpin structures. In contrast to copia, FB elements are very heterogeneous in both length and sequence and cause unstable mutations, resulting in a high frequency of deletion and chromosome rearrangement. The variably sized inverted repeats of FB elements are made up of internally repetitious sequences; yet, their extreme ends are highly conserved for about 30 base pairs and routinely produce 9-base-pair duplications of target-site DNA. There is evidence that any segment of DNA flanked by FB elements can transpose, underscoring the potential importance of this class of transposons to genome evolution. Although several percent of the genome of *Drosophila* and other higher eucaryotes is made up of inverted repeat sequences that rapidly fold back, probably only a small fraction of such sequences represents genuine FB transposable elements. The remaining sequences may be more similar to the *Alu* repeats of primate genomes.

P elements are a third class of *Drosophila* transposable elements. They are responsible for a genetic phenomenon known as **hybrid dysgenesis** and have recently been developed as potent vectors for introducing genes into new locations in the *Drosophila* genome (Chapter 22). P elements reside at multiple sites in most *Drosophila* DNA and are 0.5 to 1.4 kilobase pairs in size, bounded by perfect inverted repeats of 31 base pairs (see Figure 20-32). They represent internally deleted versions of a larger element of about 3 kilobase pairs called a P factor, which occurs in one or a few copies only in so-called P strains of *Drosophila*. Sequencing has revealed that the large P factors possess four major open reading frames, all of which are required for transposition. They presumably specify functions (e.g., transposase and repressor) similar to those required for the movement of bacterial transposons (Chapter 11). When P strain males (whose DNA contains complete P factors as well as P elements) are mated with M strain females (which contain only deleted P elements, if any), massive mobilization of the many copies of P elements results, presumably because of the sudden production of transposase encoded by the entering male DNA (Figure 20-33). In addition to frequent mutations caused by new P element insertions, the progeny of such dysgenic crosses exhibit lowered fertility and chromosome aberrations. Crosses of M strain males with P strain females (or P × P crosses) are not dysgenic, since transposition of the many short P elements is repressed when complete P factor(s) resides on the female genome: Presumably, the egg cytoplasm is already filled with repressor-type molecules. Upon insertion into a new site in *Drosophila* DNA, both P factors and P elements create 8-base-pair duplications of the target

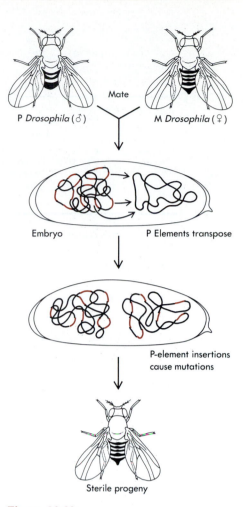

Figure 20-33
The phenomenon known as hybrid dysgenesis results from the mobilization of DNA sequences called P elements in *Drosophila* embryos. When a sperm from a P-carrying strain fertilizes an egg from a non-P (M) strain, the P elements transpose throughout the genome, usually disrupting vital genes. [After J. D. Watson, J. Tooze, D. T. Kurtz, *Recombinant DNA: A Short Course* (New York: W. H. Freeman, 1983).]

Figure 20-34

Comparison of a complete Ac element with several Ds elements. Note that one Ds element contains a completely unrelated sequence between its terminal 11-base-pair inverted repeats. All open reading frames (C1–C4 and D1–D5) encoding proteins greater than 10 kdal are indicated. [After R. F. Pohlmann, N. V. Fedoroff, and J. Messing, *Cell* 37 (184):635–643.]

sequence. P factors and other transposon systems that manifest hybrid dysgenesis appear to play an important role in the process of speciation because they effectively serve to isolate breeding populations and thus allow rapid incorporation of new mutations into the germ line.

Use of P factors as vectors is accomplished by inserting a gene into the middle of a deleted P factor and injecting this DNA into an M strain embryo along with a helper P element made defective for integration by deleting one terminal inverted repeat. A remarkably high fraction of cells in the germ line of the resulting fly will have acquired at some location(s) in their genomes the new gene, always flanked by P factor sequences (see Chapter 22 for details).

Controlling Elements in Maize Are Also Transposons[99, 100, 101]

As early as the first decades of this century, corn geneticists recognized the existence of unstable mutations that affect kernel color in maize. Today we realize that all the characteristics manifested by the "controlling elements" responsible for these mutations are precisely those exhibited by transposable elements in other systems. Insertion of a controlling element into or adjacent to a gene inhibits its expression and produces a colorless kernel. Excision of the element reverses the effect and results in colored spots on a colorless background. Even when their insertion does not affect gene expression, the controlling elements can be detected as sites of frequent chromosome breakage.

One of the best-studied maize controlling elements is the **Ac-Ds system** (Figure 20-34). Just as P elements in *Drosophila* are internally deleted verions of P factors, Ds elements come in several different sizes (0.4 to 4 kilobase pairs) and usually represent deleted forms of a larger complete element called Ac (4.5 kilobase pairs). Whereas Ac elements can move independently, Ds elements remain stationary unless an Ac element is also present. Sequencing a complete Ac element and several (shorter) Ds elements has revealed perfect inverted repeats of 11 base pairs at the termini flanked by 6- to 8-base-pair direct repeats of the target-site DNA. Interestingly, the open reading frames in the Ac element diverge from a central intergenic region, similar to the bacterial transposon Tn3 (Chapter 11). When an Ac or Ds element excises, it leaves behind imperfect but recognizable duplications of the 6- to 8-base-pair target sequence—a clear footprint of its former presence at a particular site in maize DNA.

Satellite DNA Sequences Near the Centromere[102, 103]

The chromosomal region adjacent to the centromere in higher eucaryotes is composed of very long blocks of **highly repetitive DNA,** in which simple sequences are repeated thousands of times or more. In many cases, these repeating simple sequences have compositions unlike that of most of the organism's DNA and so are easily separated by centrifuging slightly fragmented DNA in a cesium chloride (CsCl) density gradient (Figure 20-35). DNA separated by this process is referred to as **satellite DNA.** Often, several different satellites of

Figure 20-35
The satellite DNAs of *Drosophila virilus.* Fragmented DNA was centrifuged to equilibrium in a cesium chloride gradient. [After J. G. Gall and D. D. Atherton, *J. Mol. Biol.* 85 (1974):633–634.]

centromeric origin occur in a given species. The three satellites from *Drosophila virilis* have highly related sequences. One has a 5′-ACAAACT-3′ basic repeat; a second contains 5′-ATAAACT-3′; and the sequence repeated in the third is 5′-ACAAATT-3′. Why several sequences have so evolved is not clear, and none of these satellites is a completely pure repeat of the basic sequence. Often, satellites present in one species appear to be absent even in a closely related species, but this can be explained, since a change in only one base pair of the short repeating unit can significantly alter its base composition and therefore its density in cesium chloride.

Blocks of satellite sequences are readily localized to regions around the centromeres of metaphase chromosomes by a technique called **in situ hybridization** (Figure 20-36). Here, radioactively labeled RNA copies of the satellite sequence made in vitro are hybridized to squashed metaphase cells after they have been treated with alkali to

Figure 20-36
How satellite sequences can be localized to centromeres by in situ hybridization. (a) The experimental procedure. (b) An actual autoradiograph of a mouse chromosome preparation after hybridization. Note that mouse chromosomes have terminal centromeres and thus do not form conventional x-shaped metaphase chromosomes. [Courtesy of M. L. Pardue and J. G. Gall, *Chromosomes Today* 3 (1972):47–52. Used with permission.]

(a)

(b)

denature the chromosomal DNA. Subsequently, the position of the satellite sequences can be visualized by autoradiography.

In *Drosophila,* some 25 percent of the total DNA is of the simple sequence variety, with most higher eucaryotes having 10 to 20 percent of their genomes as such DNA. Satellite DNAs are usually not transcribed, for their restricted nucleotide sequences lack promoter sites on which RNA polymerase can initiate RNA chains. In contrast, there must be nucleotide sequences that are recognized by DNA polymerase, since centromeric DNA is replicated as rapidly as "unique sequence" DNA. Presumably, satellite sequences near the centromere are involved in binding particular proteins essential for centromere function. This includes holding the two daughter chromatids together during metaphase, assembling the kinetochores that allow subsequent attachment to the spindle, and guiding topoisomerases to the site(s) where they must act to separate the intertwined DNA duplexes of sister chromatids at the conclusion of replication. Perhaps the highly repetitive nature of satellite sequences at the centromeres of higher eucaryotes is related to the fact that many spindle fibers attach to each chromosome. In yeast, only one spindle fiber attaches to a centromere, each of which has a unique, nonrepetitive sequence (Chapter 18). Satellite sequences are thus clear examples of what might be called structural rather than genetic regions of the DNA contained in eucaryotic genomes. The nicked, not so highly repetitive sequences that appear at the ends of chromosomes (telomeres) are another example (Chapters 18 and 22).

Other Highly Repetitive Sequences Are Dispersed Throughout the Genome[94, 104–110]

The genomes of almost all higher eucaryotes also contain highly repetitive sequences that are not clustered together as are satellite sequences at centromeres. Rather, they are more evenly distributed throughout the genome, interspersed with longer stretches of unique (or moderately repetitive) DNA.

In the human genome, the majority of such sequences belong to a single family called the ***Alu* family.** Its members are about 300 base pairs long and are recognizably related but not precisely conserved in sequence. Their name derives from the fact that most contain a single site of cleavage for the restriction enzyme *Alu* I near their middle. Almost a million *Alu* sequences are present in the human genome, accounting for some 3 to 6 percent of the total DNA. Thus, any particular DNA segment of 5000 base pairs or longer has a high probability of containing an *Alu* sequence. Accordingly, many introns and sequences adjacent to genes (in both transcribed and nontranscribed regions) harbor *Alu* sequences.

The widespread dispersion of *Alu* family repeats throughout the human genome is believed to have occurred in an RNA-mediated fashion, similar to the mechanism envisioned for the creation of processed genes and of many small nuclear RNA pseudogenes. *Alu* sequences terminate at their 3' ends with repetitive A-rich tails and are often flanked by direct repeats of 8 to 19 base pairs. Although they are not transcribed in vivo in proportion to their abundance in the genome, isolated *Alu* sequences are usually excellent templates for in vitro transcription by RNA polymerase III. This suggests a scenario in which a rare *Alu* sequence transcript could be reverse transcribed by self-priming. Its U-rich 3' end (U-rich sequences are found at the ex-

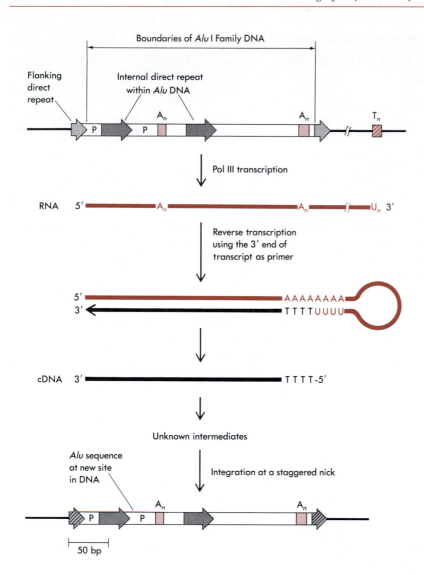

Figure 20-37
How *Alu* sequences may have multiplied and dispersed themselves throughout the human genome. A salient point is that sequences important for RNA synthesis (P) are carried within the sequence itself (mobile promoter). [After S. W. Van Arsdell, R. A. Denison, L. B. Bernstein, and A. W. Weiner, *Cell* 26 (1981):11–17; P. Jagdeeswaran, B. G. Forget, and S. M. Weissman, *Cell* (1981):141–142.]

treme 3′ ends of all RNA polymerase III products; Chapter 21) would fold back on its lengthier A-rich tail (Figure 20-37). The resulting cDNA copy would then insert between staggered breaks at a new site in the genome. Since the DNA sequences important for initiation of transcription by RNA polymerase III reside within the transcribed region rather than in upstream DNA (Chapter 21), an *Alu* cDNA would carry its transcription signals along to its new location. Thus, it would be capable of repeated rounds of transcription, cDNA synthesis, and insertion. The structure of many *Alu* family members can be explained if insertion occurs preferentially into the A-rich tails of pre-existing *Alu* sequences.

A most curious property of *Alu* sequences in the human genome (and of related sequences in other mammalian genomes) is that they are in part homologous to an abundant small RNA known as 7SL. This 300-nucleotide-long RNA associates with six proteins to form a small cytoplasmic RNA—protein complex (called SRP) that is essential for the translocation of newly synthesized proteins across membranes of the rough endoplasmic reticulum in higher eucaryotic cells (Chapter 21). 7SL RNA is transcribed by RNA polymerase III from three or four identical copies of its gene in the human genome. It exhibits homology to *Alu* sequences at both its 5′ and 3′ ends, but is interrupted by 155 nucleotides of unrelated sequence in its middle (Figure 20-38). This suggests that mammalian *Alu* sequences may

Figure 20-38

The relationship between human 7SL RNA and human *Alu* DNA. The different shadings indicate areas of sequence homology between the RNA and the DNA. The right half of human *Alu* DNA contains an insertion (uncolored) not present in the left half. The dark gray areas correspond to the most perfectly repeated regions, which are indicated by arrows in Figure 20-37. [After E. Ullu and C. Tschudi, *Nature* 312 (1984):171–172.]

have been derived from 7SL RNA (or DNA) by deletion of the central 7SL-specific sequence. Recently, the 7SL RNA of frogs and flies has been shown strongly to resemble mammalian 7SL RNA, although these lower eucaryotes completely lack *Alu* sequences. Hence, human *Alu* sequences have all the earmarks of a very large family of "processed 7SL pseudogenes" that arose subsequent to the divergence of mammals.

In addition to members of the *Alu* family, there are other highly repetitive sequences (present in more than 10^4 copies) widely dispersed in mammalian genomes and other higher eucaryotic genomes. Some are similar to *Alu* in that they are relatively short (less than 500 base pairs) and are frequently flanked by direct repeats. Included in this group are sequences recently discovered to be homologous to tRNA genes, suggesting that they may have arisen via tRNA intermediates. Other repetitive sequences are much longer (greater than 5 kilobase pairs), have A-rich 3' ends, and sometimes contain open reading frames. Thus, the distinction between bona fide transposons, pseudogenes, and repetitive DNA sequences in eucaryotic genomes has become progressively blurred as DNA sequence analysis has revealed the details of their structure. Determining whether and how the several types of repetitive sequences move is a current challenge for molecular biologists.

Variation in DNA Amounts Between Closely Related Species[111]

Before much was known about the actual sequences that make up higher eucaryotic genomes, it was expected that the amount of DNA per haploid genome would increase in proportion to the biological complexity of the organism. Higher plants and animals do have much more DNA than lower forms (see Figure 20-1), and we have already seen that this is in part due to the much more frequent occurrence of introns in the genes of higher as compared to lower eucaryotes. Yet, no one anticipated the finding that certain fish and amphibia would have 25 times more DNA than any mammalian species. And as more and more animals and plants are examined, closely related species are sometimes found to vary in their DNA content by a factor of five to ten (Figure 20-39).

These variations usually do not arise because of increases in chromosome number but rather because of increases in the amount of DNA per chromosome. The amounts of simple sequence satellite DNA, mildly repetitious DNA, and unique sequence DNA all increase, sometimes in roughly proportional amounts. Very large genomes need to be studied in more detail before we can understand

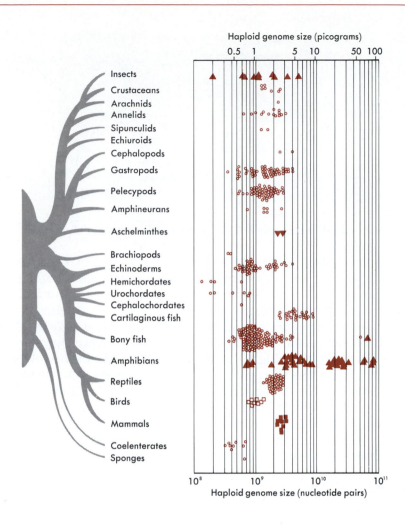

Haploid genome size (picograms)

Haploid genome size (nucleotide pairs)

Figure 20-39
Variation in the genome size among related members of evolutionarily related animal groups. [After R. J. Britten and E. H. Davidson, *Quant. Rev. Biol.* 46 (1971):111–138.]

the true meaning of such variations in genome size. However, we do know that the salamander *Necturus,* which has some 25 times more DNA than the toad, contains only single copies of its various hemoglobin genes. Hence, our best current guess is that the extra DNA simply represents longer spacer regions between true genes; these spacers are subject to little selective pressure and provide an ample home for all types of DNA sequences.

Selfish DNA?[112, 113, 114]

We have seen that the unexpected structures that contribute to the large size of higher eucaryotic genomes take many forms. The existence of introns multiplies the number of base pairs required to encode a protein by as much as tenfold. In addition, many mRNAs possess mysteriously long untranslated regions (particularly at their 3' ends) of as yet unknown function. Many protein-coding genes have given rise to gene families producing several closely related products, while other gene families consist of tens or thousands of identical gene copies. Within the lengthy sequences that separate genes are found recognizable repeated sequence elements: transposons of various types, other mobile repetitive sequences like the *Alu* family, pseudogenes, and complete or partial copies of certain viral genomes (Chapter 24). Furthermore, some DNA is devoted to specialized structural roles such as satellite sequences at centromeres

Figure 20-40

Increase in occurrence of P elements in *Drosophila melanogaster* collected world-wide. Plotted are the fractions of flies exhibiting P or M phenotypes upon mating (see Figure 20-33). The number of strains actually tested is given above each point on the graph. [After M. G. Kidwell, *Proc. Nat. Acad. Sci.* 80 (1983):1655–1659.]

or the shorter nicked repeated sequences at the ends of chromosomes (telomeres).

Still, we cannot account for all the sequences that make up a eucaryotic genome nor do we understand the enormous variations in DNA amount that occur between closely related species. This has led some scientists to speculate that DNA is inherently "selfish," constantly endeavoring to expand itself to the limit tolerated by each organism. Surely, transposons and *Alu* family sequences have properties most closely fitting what might be expected of **selfish DNA**: built-in mechanisms for reproduction and movement. Yet, even their proliferation seems to be limited in some way. For instance, the spread of P elements through wild-type strains of *Drosophila* has been a quite recent and remarkably rapid occurrence (Figure 20-40). Nevertheless, examination of the genomes of individual P strain flies reveals only one to several complete P factors and less than 50 shorter P elements, rather than the hundreds or thousands of copies that might be expected if their multiplication were unchecked. As with all parasites, advantages gained by replication must be balanced by ensuring survival of the host organism. Therefore, to what extent purely selfish DNA sequences have molded eucaryotic genomes remains to be determined.

Summary

The genomes of higher eucaryotes are much bigger than those of bacteria. Far more DNA is present than needed to encode even a hundredfold more proteins. We now realize that this excess DNA represents a number of different kinds of sequences. Most of them are unique and unanticipated features of higher eucaryotic genomes.

Cloning and sequencing higher eucaryotic genes led to the discovery of introns (intervening sequences), interruptions that expand the sizes of genes as much as 20-fold. Introns are highly variable in length and sequence. They can occur in any part of a protein-coding gene: 5′ leader, 3′ trailer, or the body. Although most protein-coding genes in vertebrates are split by one or more introns, the presence of introns is not required. Some eucaryotic tRNA and rRNA genes also contain introns.

In all cases, split genes are transcribed into precursor RNA molecules that contain intron as well as exon sequences. Exons are those parts of the gene that are represented in its final RNA product (mRNA, tRNA, or rRNA). Introns are removed by a process called RNA splicing. At least three different splicing systems can be distinguished

on the basis of the location of the signals recognized in the precursor and the biochemistry of splicing reaction. Ribosomal RNA splicing is an RNA-catalyzed reaction for which certain intron sequences (and their precise folding into a three-dimensional structure) are critical. Transfer RNA introns (which always reside in the anticodon loop) are removed by two separate (protein) enzymes, an endonuclease and a ligase; they recognize some aspect of the folded tRNA (exon) structure. Pre-mRNAs are spliced by a complex machinery that includes small nuclear RNA-protein complexes, which bind to conserved sequences at the junctions between introns and exons; an important splicing intermediate is a lariat structure in which the 5′ end of the intron forms a branch with a sequence near the 3′ end. Some introns in mitochondrial DNA encode proteins essential for splicing, but RNA catalysis (as with rRNA) appears to be fundamental to intron removal.

Introns in protein-coding genes often occur at boundaries between separately folded domains or functionally distinct regions of the polypeptide chain. Evolution of new proteins may therefore be accelerated by mixing strategies or by adding (or deleting) material at intron-exon junctions as a consequence of mutations that create (or abolish) splice sites.

Many genes in higher eucaryotic genomes belong to multigene families. Either different members encode related, but nonidentical gene products that fulfill slightly different functions in a multicellular or developing organism, or multiple copies of identical genes are necessary simply to synthesize massive amounts of a particular gene product. Multigene families can be either dispersed or clustered into a single or tandemly repeating array. The nontranscribed spacer regions that separate genes are lengthier than those in bacterial genomes.

Amplification of segments of higher eucaryotic DNA containing particular genes readily occurs when cells are subjected to selective growth conditions. Amplification was probably instrumental in the original creation of multigene families and greatly contributes to their maintenance. Tandem arrays of identical DNA sequences are particularly susceptible to contraction (resulting from intrachromosomal crossing over). Lost copies could be regained either by amplification or by unequal sister chromosome exchange. Amplification and unequal sister chromosome exchange, along with gene conversion, may play further roles in ensuring the homogeneity of tandemly repeated DNA sequence families.

Pseudogenes are another unexpected feature of higher eucaryotic genomes. These are sequences that exhibit high homology to real genes, but lack signals necessary for expression (e.g., promoters). Often their presence can be rationalized by assuming that a copy of a processed RNA transcript of the original gene became inserted into the genome.

Transposons also inhabit higher eucaryotic genomes. They are present in multiple copies (but not very large numbers). They fall into two main classes: those that move as DNA and encode transposase- and repressor-like proteins and those that move via RNA and encode reverse transcriptase and sometimes viral capsidlike proteins.

Some highly repetitive sequences in higher eucaryotic DNA reside at the centromeres, where they serve structural roles in chromosome replication and sorting. Other repetitive sequences are dispersed randomly throughout the genome.

Together, all these different kinds of nongenetic DNA account for the large and variable sizes of higher eucaryotic genomes. It may be that DNA seeks to expand itself to the maximum tolerated by each organism.

Bibliography

General References

Apirion, D., ed. 1984. *Processing of RNA*. Boca Raton, Fla.: C.R.C. Press. Comprehensive reviews on all types of RNA processing.

Busch, H. ed. *The Cell Nucleus*. New York: Academic Press. A series of volumes containing review articles on nuclear structure and function. The most valuable here is volume X (1982) on rDNA.

MacLean, N., S. P. Gregory, and R. A. Flavell, eds. 1983. *Eukaryotic Genes: Their Structure, Activity and Regulation*. London: Butterworths. A compendium of excellent reviews on the structure and function of eucaryotic genes.

Schimke, R. T., ed. 1982. *Gene Amplification*. Cold Spring Harbor, N.Y.: Cold Spring Harbor Laboratory. Gives details on all aspects of amplification phenomena.

Shapiro, J. A., ed. 1983. *Mobile Genetic Elements*. New York: Academic Press. Includes several valuable chapters on eucaryotic transposons, including those in plants.

Structures of DNA. 1982. *Cold Spring Harbor Symp. Quant. Biol.* 47. Includes valuable articles on gene families, repetitive DNA sequences, and pseudogenes.

Cited References

1. Britten, R. J., and E. H. Davidson. 1969. "Gene Regulation for Higher Cells: A Theory." *Science* 165:349–357.
2. Thomas, C. A., Jr. 1971. "The Genetic Organization of Chromosomes." *Ann. Rev. Genetics* 5:237–256.
3. Darnell, J. E. 1976. "mRNA Structure and Function." *Prog. Nucl. Acid Res. Mol. Biol.* 19:493–511.
4. Darnell, J. E., Jr. 1983. "The Processing of RNA." *Sci. Amer.* 249:89–99.
5. Berget, S. M., C. Moore, and P. A. Sharp. 1977. "Spliced Segments at the 5' Terminus of Adenovirus 2 Late mRNA." *Proc. Nat. Acad. Sci.* 74:3171–3175.
6. Chow, L. T., R. E. Gelinas, T. R. Broker, and R. J. Roberts. 1977. "An Amazing Sequence Arrangement at the 5' Ends of Adenovirus 2 Messenger RNA." *Cell* 12:1–8.
7. Berget, S. M., A. J. Berk, T. Harrison, and P. A. Sharp. 1977. "Spliced Segments at the 5' Termini of Adenovirus-2 Late mRNA: A Role for Heterogeneous Nuclear RNA in Mammalian Cells." *Cold Spring Harbor Symp. Quant. Biol.* 42:523–529.
8. Broker, T. R., L. T. Chow, A. R. Dunn, R. E. Gelinas, J. A. Hassell, D. F. Klessig, J. B. Lewis, R. J. Roberts, and B. S. Zain. 1977. "Adenovirus-2 Messengers—An Example of Baroque Molecular Architecture." *Cold Spring Harbor Symp. Quant. Biol.* 42:531–553.
9. Thomas, M., R. L. White, and R. W. Davis. 1976. "Hybridization of RNA to Double-Stranded DNA: Formation of R-loops." *Proc. Nat. Acad. Sci.* 73:2294–2298.
10. Breathnach, R., J. L. Mandel, and P. Chambon. 1977. "Ovalbumin Gene Is Split in Chicken DNA." *Nature* 270:314–319.
11. Doel, M. T., M. Houghton, E. A. Cook, and N. H. Carey. 1977. "The Presence of Ovalbumin mRNA Coding Sequences in Multiple Restriction Fragments of Chicken DNA." *Nucleic Acid Res.* 4:3701–3713.
12. Brack, C. and S. Tonegawa. 1977. "Variable and Constant Parts of the Immunoglobulin Light Chain Gene of a Mouse Myeloma Cell are 1250 Nontranslated Bases Apart." *Proc. Nat. Acad. Sci.* 74:5652–5656.
13. Lai, E. C., S. L. Woo, A. Dugaiczyk, J. F. Catterall, and B. W. O'Malley. 1978. "The Ovalbumin Gene: Structural Sequences in Native Chicken DNA Are Not Contiguous." *Proc. Nat. Acad. Sci.* 75:2205–2209.
14. Jeffreys, A. J., and R. A. Flavell. 1977. "The Rabbit Beta-Globin Gene Contains a Large Large Insert in the Coding Sequence." *Cell* 12:1097–1108.
15. Tilghman, S. M., D. C. Tiemeier, J. G. Seidman, B. M. Peterlin, M. Sullivan, J. V. Maizel, and P. Leder. 1978. "Intervening Sequence of DNA Identified in the Structural Portion of a Mouse Beta-Globin Gene." *Proc. Nat. Acad. Sci.* 75:725–729.
16. Breathnach, R., and P. Chambon. 1981. "Organization and Expression of Eucaryotic Split Genes Coding for Proteins." *Ann. Rev. Biochem.* 50:349–383.
17. Chambon, P. 1981. "Split Genes." *Sci. Amer.* 244:60–71.
18. Crouse, G. F., C. C. Simonsen, R. N. McEwan, and R. T. Schimke. 1982. "Structure of Amplified Normal and Variant Dihydrofolate Reductase Genes in Mouse Sarcoma S180 Cells." *J. Biol. Chem.* 257:7887–7897.
19. Yamada, Y., V. E. Avvedimento, M. Mudryj, H. Ohkubo, G. Vogeli, M. Irani, I. Pastan, and B. de Crombrugghe. 1980. "The Collagen Gene: Evidence for Its Evolutionary Assembly by Amplification of a DNA Segment Containing an Exon of 54 bp." *Cell* 22:887–892.

20. Wozney, J., D. Hanahan, V. Tate, H. Boedtker, and P. Doty. 1981. "Structure of the Pro Alpha (2) (I) Collagen Gene." *Nature* 294:129–135.

21. Goodman, H. M., M. V. Olson, and B. D. Hall. 1977. "Nucleotide Sequence of a Mutant Eucaryotic Gene: The Yeast Tyrosine-Inserting Ochre Suppressor SUP4-o." *Proc. Nat. Acad. Sci.* 74:5453–5475.

22. Guthrie, C., and J. Abelson. 1982. "Organization and Expression of tRNA Genes in *Saccharomyces cerevisiae*." In *The Molecular Biology of the Yeast Saccharomyces: Metabolism and Gene Expression*, ed. J. N. Strathern, E. W. Jones, and J. R. Broach. Cold Spring Harbor, N.Y.: Cold Spring Harbor Laboratory, p. 487–528.

23. Abelson, J. 1979. "RNA Processing and the Intervening Sequence Problem." *Ann. Rev. Biochem.* 48:1035–1069.

24. Tilghman, S. M., P. J. Curtis, D. C. Tiemeier, P. Leder, and C. Weissmann. 1978. "The Intervening Sequence of a Mouse Beta-Globin Gene Is Transcribed Within the 15S Beta-Globin mRNA Precursor." *Proc. Nat. Acad. Sci.* 75:1309–1313.

25. Roop, D. R., J. L. Nordstrom, S. Y. Tsai, M. J. Tsai, and B. W. O'Malley. 1978. "Transcription of Structural and Intervening Sequences in the Ovalbumin Gene and Identification of Potential Ovalbumin mRNA Precursors." *Cell* 15:671–685.

26. Alwine, J. C., D. J. Kemps, and G. R. Stark. 1977. "Method for Detection of Specific RNAs in Agarose Gels by Transfer to Diazobenzyloxymethyl-Paper and Hybridization with DNA Probes." *Proc. Nat. Acad. Sci.* 74:5350–5354.

27. Hopper, A. K., F. Banks, and V. Evangelidis. 1978. "A Yeast Mutant Which Accumulates Precursor tRNAs." *Cell* 14:211–219.

28. Cech, T. R. 1983. "RNA Splicing: Three Themes with Variations." *Cell* 34:713–716.

29. Greer, C. L., C. L. Peebles, P. Gegenheimer, and J. Abelson. 1983. "Mechanism of Action of a Yeast RNA Ligase in tRNA Splicing." *Cell* 32:537–546.

30. Filipowicz, W., and A. J. Shatkin. 1983. "Origin of Splice Junction Phosphate in tRNAs Processed by HeLa Cell Extract." *Cell* 32:547–557.

31. Laski, F. A., A. Z. Fire, U. L. RajBhandary, and P. A. Sharp. 1983. "Characterization of tRNA Precursor Splicing in Mammalian Extracts." *J. Biol. Chem.* 258:11974–11980.

32. Cech, T. R. 1985. "Self-Splicing RNA: Implications for Evolution." *Int. Rev. Cytol.* 93:3–22.

33. Kruger, K., P. J. Grabowski, A. J. Zaug, J. Sands, D. E. Gottschling, and T. R. Cech. 1982. "Self-Splicing RNA: Autoexcision and Autocyclization of the Ribosomal RNA Intervening Sequence of *Tetrahymena*." *Cell* 31:147–157.

34. Cech, T. R., N. K. Tanner, I. Tinoco, Jr., B. R. Weir, M. Zuker, and P. S. Perlman. 1983. "Secondary Structure of the *Tetrahymena* Ribosomal RNA Intervening Sequence: Structural Homology with Fungal Mitochondrial Intervening Sequences." *Proc. Nat. Acad. Sci.* 80:3903–3907.

35. Lazowska, J., C. Jacq, and P. P. Slonimski. 1980. "Sequence of Introns and Flanking Exons in Wild-Type and BOX-3 Mutants of Cytochrome *b* Reveals an Interlaced Splicing Protein Coded by an Intron." *Cell* 22:333–348.

36. Borst, P., and L. A. Grivell. 1981. "One Gene's Intron Is Another Gene's Exon." *Nature* 289:439–440.

37. Lamb, M. R., P. Q. Anziano, K. R. Glaus, D. K. Hanson, H. J. Klapper, P. S. Perlman, and H. R. Mahler. 1983. "RNA Processing Intermediates in *cis*- and *trans*-Acting Mutants in the Penultimate Intron of the Mitochondrial Gene for Cytochrome *b*." *J. Biol. Chem.* 258:1991–1999.

38. Davies, R. W., R. B. Waring, J. A. Ray, T. A. Brown, and C. Scazzocchio. 1982. "Making Ends Meet: A Model for RNA Splicing in Fungal Mitochondria." *Nature* 300:719–724.

39. Michel, F., and B. Dujon. 1983. "Conservation of RNA Secondary Structures in Two Intron Families Including Mitochondrial-, Chloroplast-, and Nuclear-Encoded Members." *EMBO J.* 2:33–38.

40. Garriga, G., and A. L. Lambowitz. 1984. "RNA Splicing in Neurospora Mitochondria: Self-Splicing of a Mitochondrial Intron *in Vitro*." *Cell* 39:631–641.

41. Mount, S. M. 1982. "A Catalogue of Splice Junction Sequences." *Nucleic Acid Res.* 10:459–472.

42. Treisman, R., S. H. Orkin, and T. Maniatus. 1983. "Specific Transcription and RNA Splicing Defects in Five Cloned β-Thalassaemia Genes." *Nature* 302:591–596.

43. Wieringa, B., F. Meyer, J. Reiser, and C. Weissmann. 1983. "Unusual Splice Sites Revealed by Mutagenic Inactivation of an Authentic Splice Site of the Rabbit Beta-Globin Gene." *Nature* 301:38–43.

44. Chu, G., and P. A. Sharp. 1981. "A Gene Chimaera of SV40 and Mouse Beta-Globin Is Transcribed and Properly Spliced." *Nature* 289:378–382.

45. Lerner, M. R., and J. A. Steitz. 1979. "Antibodies to Small Nuclear RNAs Complexed with Proteins Are Produced by Patients with Systemic Lupus Erythematosus." *Proc. Nat. Acad. Sci.* 76:5495–5499.

46. Busch, H., R. Reddy, and L. Rothblum. 1982. "SnRNAs, SnRNPs, and RNA Processing." *Ann. Rev. Biochem.* 51:617–654.

47. Mattaj, I. W. 1984. "snRNA: From Gene Architecture to RNA Processing." *Trends in Biochem. Sci.* 9:435–437.

48. Lerner, M. R., J. A. Boyle, S. M. Mount, S. L. Wolin, and J. A. Steitz. 1980. "Are snRNPs Involved in Splicing?" *Nature* 283:220–224.

49. Rogers, J., and R. Wall. 1980. "A Mechanism for RNA Splicing." *Proc. Nat. Acad. Sci.* 77:1877–1879.

50. Krämer, A., W. Keller, B. Appel, and R. Lührmann. 1984. "The 5' Terminus of the RNA Moiety of U1 Small Nuclear Ribonucleoprotein Particles Is Required for the Splicing of Messenger RNA Precursors." *Cell* 38:299–307.

51. Mount, S. M., I. Pettersson, M. Hinterberger, A. Karmas, and J. A. Steitz. 1983. "The U1 Small Nuclear RNA-Protein Complex Selectively Binds a 5' Splice Site *in Vitro*." *Cell* 33:509–518.

52. Keller, W. 1984. "The RNA Lariat: A New Ring to the Splicing of mRNA Precursors." *Cell* 39:423–425.

53. Ruskin, B., A. R. Krainer, T. Maniatis, and M. R. Green. 1984. "Excision of an Intact Intron as a Novel Lariat Structure During Pre-mRNA Splicing *in Vitro*." *Cell* 38:317–331.

54. Padgett, R. A., M. M. Konarska, P. J. Grabowski, S. F. Hardy, and P. A. Sharp. 1984. "Lariat RNAs as Intermediates and Products in the Splicing of Messenger RNA Precursors." *Science* 225:898–903.

55. Brody, E., and J. Abelson. 1985. "The 'Spliceosome'; Yeast Premessenger RNA Associates with a 40S Complex in a Splicing Dependent Reaction." *Science* 228:963–967.

56. Grabowski, P. J., S. R. Seiler, and P. A. Sharp. 1985. "A Multicomponent Complex Is Involved in the Splicing of Messenger RNA Precursors." *Cell* 42:345–353.

57. Frendewey, D., and W. Keller. 1985. "The Stepwise Assembly of a Pre-mRNA Splicing Complex Requires U-snRNPs and Specific Intron Sequences." *Cell* 42:355–367.

58. Black, D. L., B. Chabot, and J. A Steitz. 1985. "U2 as Well as U1 Small Nuclear Ribonucleoproteins Are Involved in Premessenger RNA Splicing." *Cell* 42:737–750.

59. Go, M. 1981. "Correlation of DNA Exonic Regions with Protein Structural Units in Haemoglobin." *Nature* 291:90–92.

60. Branden, C.-I., H. Eklund, C. Cambillau, and A. J. Pryor. 1984. "Correlation of Exons with Structural Domains in Alcohol Dehydrogenase." *EMBO J.* 3:1307–1310.

61. Gilbert, W. 1978. "Why Genes in Pieces?" *Nature* 271:501.

62. Doolittle, W. F. 1978. "Genes in Pieces: Were They Ever Together?" *Nature* 272:581–582.

63. Crick, F. 1979. "Split Genes and RNA Splicing." *Science* 204:264–271.

64. Craik, C. S., W. J. Rutter, and R. Fletterick. 1983. "Splice Junctions: Association with Variation in Protein Structure." *Science* 220:1125–1129.

65. Gilbert, W. 1985. "Genes-in-Pieces Revisited." *Science* 228:823–824.

66. Stein, J. P., J. F. Catterall, P. Kristo, A. R. Means, and B. W. O'Malley. 1980. "Ovomucoid Intervening Sequences Specify Functional Domains and Generate Protein Polymorphism." *Cell* 21:681–687.

67. Suzuki, Y., J. P. Gage, and D. D. Brown. 1972. "The Genes for Silk Fibroin in *Bombyx mori*." *J. Mol. Biol.* 70:637–649.

68. Maniatis, T., P. Melon, V. Parker, N. Proudfoot, and B. Seed.

1982. "Molecular Genetics of Human Globin Gene Expression." In *Molecular Genetics of Neuroscience*, ed. F. O. Schmitt, S. J. Bird, and F. E. Bloom. New York: Raven Press, pp. 87–101.

69. Vandekerckhove, J., and K. Weber. 1979. "The Complete Amino Acid Sequence of Actins from Bovine Aorta, Bovine Heart, Bovine Fast Skeletal Muscle, and Rabbit Slow Skeletal Muscle. A Protein-Chemical Analysis of Muscle Actin Differentiation." *Differentiation* 14:123–133.

70. Engel, J. N., P. W. Gunning, and L. Kedes. 1981. "Isolation and Characterization of Human Actin Genes." *Proc. Nat. Acad. Sci.* 78:4674–4678.

71. Maxson, R., R. Cohn, and L. Kedes. 1983. "Expression and Organization of Histone Genes." *Ann. Rev. Genetics* 17:239–277.

72. Hentschel, C. C., and M. L. Birnstiel. 1981. "The Organization and Expression of Histone Gene Families." *Cell* 25:301–313.

73. Birnstiel, M. L., M. Chipchase, and J. Speirs. 1971. "The Ribosomal RNA Cistrons." *Prog. Nucleic Acid Res. Mol. Biol.* 11:351–389.

74. Sinclair, J. H., and D. D. Brown. 1971. "Retention of Common Nucleotide Sequences in the Ribosomal Deoxyribonucleic Acid of Eukaryotes and Some of Their Physical Characteristics." *Biochemistry* 10:2761–2769.

75. Long, E. O., and I. B. Dawid. 1980. "Repeated Genes in Eukaryotes." *Ann. Rev. Biochem.* 49:727–764.

76. Fedoroff, N. V. 1979. "On Spacers." *Cell* 16:697–710.

77. Planta, R. J., and J. H Meyerink. 1980. "Organization of the Ribosomal RNA Genes in Eukaryotes." In *Ribosomes: Structure, Function and Genes*, ed. G. Chambliss, G. R. Craven, J. Davies, K. Davis, L. Kahan, and M. Nomura. Baltimore, Md.: University Park Press, pp. 871–887.

78. Fostel, J., S. Narayanswami, B. Hamkalo, S. G. Clarkson, and M. L. Pardue. 1984. "Chromosomal location of a major tRNA gene cluster of *Xenopu laevis*." *Chromosoma* 90:254–260.

79. Clarkson, S. G. 1983. "Transfer RNA genes." In *Eukaryotic Genes: Their Structure, Activity and Regulation*, ed. N. Maclean, S. P. Gregory, R. A. Flavell. London: Butterworths, pp. 239–261.

80. Yen, P. H., and N. Davidson. 1980. "The Gross Anatomy of a tRNA Gene Cluster at Region 42A of the *D. melanogaster* Chromosome." *Cell* 22:137–148.

81. Hovemann, B., S. Sharp, H. Yamada, and D. Soll. 1980. "Analysis of a *Drosophila* tRNA Gene Cluster." *Cell* 19:889–895.

82. Sekiya, T., Y. Kuchino, and S. Nishimura. 1981. "Mammalian tRNA Genes: Nucleotide Sequence of Rat Genes for tRNA^Asp, tRNA^Gly and tRNA^Glu." *Nucleic Acid Res.* 9:2239–2250.

83. Schimke, R. 1984. "Gene Amplification in Cultured Animal Cells." *Cell* 37:705–713.

84. Stark, G. R., and G. M. Wahl. 1984. "Gene Amplification." *Ann. Rev. Biochem.* 53:447–491.

85. Weiner, A. M., and R. A. Denison. 1982. "Either Gene Amplification or Gene Conversion May Maintain the Homogeneity of the Multigene Family Encoding Human U1 Small Nuclear RNA." *Cold Spring Harbor Symp. Quant. Biol.* 47:1141–1149.

86. Keil, R. L., and G. S. Roeder. 1984. "*Cis*-Acting, Recombination-Stimulating Activity in a Fragment of the Ribosomal DNA of *S. cerivisiae*." *Cell* 39:377–386.

87. Jeffreys, A. J., V. Wilson, and S. L. Thein. 1985. "Hypervariable Minisatellite Regions in Human DNA." *Nature* 314:67–73.

88. Van Arsdell, S. W., and A. M. Weiner. 1985. "Human Genes for U2 Small Nuclear RNA Are Tandemly Repeated." *Mol. Cell. Biol.* 4:492–499.

89. Bernstein, L. B., T. Manser, and A. M. Weiner. 1985. "Human U1 Small Nuclear RNA Genes: Extensive Conservation of Flanking Sequences Suggests Cycles of Gene Amplification and Transposition." *Mol. Cell. Biol.* 5:2159–2171.

90. Bernstein, L. B., S. M. Mount, and A. M. Weiner. 1983. "Pseudogenes for Human Small Nuclear RNA U3 Appear To Arise by Integration of Self-Primed Reverse Transcripts of the RNA into New Chromosomal Sites." *Cell* 32:461–472.

91. Hollis, G. F., P. A. Hieter, O. W. McBride, D. Swan, and P. Leder. 1982. "Processed Genes: A Dispersed Human Immunoglobulin Gene Bearing Evidence of RNA-Type Processing." *Nature* 296:321–325.

92. Lee, M. G.-S., S. A. Lewis, C. D. Wilde, and N. J. Cowan. 1983. "Evolutionary History of a Multigene Family: An Expressed Human Beta-Tubulin Gene and Three Processed Pseudogenes." *Cell* 33:477–487.

93. Soares, M. B., E. Schon, A. Henderson, S. K. Karathanasis, R. Cate, S. Zeitlin, J. Chirgwin, and A. Efstratiadis. 1985. "RNA-Mediated Gene Duplication: The Rat Preproinsulin I Gene Is a Functional Retroposon." *Mol. Cell Biol.* 5:2090–2103.

94. Rogers, J. H. 1985. "The Origin and Evolution of Retroposons." *Int. Rev. Cytol.* 93:188–279.

95. Rubin, G. M. 1983. "Dispersed Repetitive DNAs in *Drosophila*." In *Mobile Genetic Elements*, ed. J. A. Shapiro. New York: Academic Press, pp. 329–361.

96. Finnegan, D. J. 1985. "Transposable Elements in Eukaryotes." *Int. Rev. Cytol.* 93:281–326.

97. Mount, S. M., and G. M. Rubin. 1985. "Complete Nucleotide Sequence of the *Drosophila* Transposable Element Copia: Homology Between Copia and Retroviral Proteins." *Mol. Cell Biol.* 5:1630–1638.

98. Karess, R. E., and G. M. Rubin. 1984. "Analysis of P Transposable Element Functions in *Drosophila*." *Cell* 38:135–146.

99. Fedoroff, N. 1983. "Controlling Elements in Maize." In *Mobile Genetic Elements*, ed. J. A. Shapiro. New York: Academic Press, pp. 1–63.

100. Pohlmann, R. F., N. V. Fedoroff, and J. Messing. 1984. "The Nucleotide Sequence of the Maize Controlling Element *Activator*." *Cell* 37:635–643.

101. Sutton, W. D., W. L. Gerlach, D. Schwartz, and W. J. Peacock. 1984. "Molecular Analysis of Ds Controlling Mutations at the Adhl Locus of Maize." *Science* 223:1265–1268.

102. Gall, J. G., and D. D. Atherton. 1974. "Satellite DNA Sequences in *Drosophila virilis*." *J. Mol. Biol.* 85:633–634.

103. Pardue, M. L., and J. G. Gall. 1970. "Chromosomal Localization of Mouse Satellite DNA." *Science* 168:1356–1358.

104. Jelinek, W. R., and C. W. Schmid. 1982. "Repetitive Sequences in Eukaryotic DNA and Their Expression." *Ann. Rev. Biochem.* 51:813–844.

105. Van Arsdell, S. W., R. A. Denison, L. B. Bernstein, and A. W. Weiner. 1981. "Direct Repeats Flank Three Small Nuclear RNA Pseudogenes in the Human Genome." *Cell* 26:11–17.

106. Jagdeeswaran, P., B. G. Forget, and S. M. Weissman. 1981. "Short Interspersed Repetitive DNA Elements in Eucaryotes: Transposable DNA Elements Generated by Reverse Transcription of RNA Pol III Transcripts." *Cell* 26:141–142.

107. Ullu, E., and C. Tschudi. 1984. "Alu Sequences Are Processed 7SL RNA Pseudogenes." *Nature* 312:171–172.

108. Weiner, A. M., P. L. Deininger, and A. Efstratiadis. 1986. "Nonviral Retroposons: Genes, Pseudogenes, and Transposable Elements Generated by the Reverse Flow of Genetic Information." *Ann. Rev. Biochem.* 55: 631–661.

109. Singer, M. F., and J. Skowronski. 1985. "Making Sense Out of LINES: Long Interspersed Repeat Sequences in Mammalian Genomes." *Trends in Biochem. Sci.* 10:119–122.

110. Sakamoto, K., and N. Okada. 1985. "Rodent Type 2 *Alu* Family, Rat Identifier Sequence, Rabbit C Family, and Bovine or Goat 73-bp May Have Evolved From tRNA Genes." *J. Mol. Evolution* 22:134–140.

111. Britten, R. J., and E. H. Davidson. 1971. "Repetitive and Non-Repetitive DNA Sequences and a Speculation on the Origins of Evolutionary Novelty." *Quant. Rev. Biol.* 46:111–138.

112. Orgel, L. E., and F. H. C. Crick. 1980. "Selfish DNA: The Ultimate Parasite." *Nature* 284:604–607.

113. Doolittle, W. F., and C. Sapienza. 1980. "Selfish Genes, the Phenotype Paradigm and Genome Evolution." *Nature* 284:601–603.

114. Kidwell, M. G. 1983. "Evolution of Hybrid Dysgenesis Determinants in *Drosophila melanogaster*." *Proc. Nat. Acad. Sci.* 80:1655–1659.

CHAPTER 21

The Functioning of Higher Eucaryotic Genes

The careful reader will have noticed that in our discussion of the functioning of bacterial and bacteriophage genes, many statements were qualified; our understanding of how bacterial cells control the synthesis of proteins is less complete than our knowledge of the mechanisms of transcription and of protein synthesis. Yet, concentration on bacterial systems has hardened many previous speculations into facts. The more we learn, the more we appreciate the ingenious multiplicity of mechanisms that bacteria use to fine-tune the expression of each of their genes precisely to meet the cell's momentary needs.

To gain an understanding of the functioning of higher eucaryotic genes at the same level of detail is a much more formidable task. Studies with yeast (Chapter 18) have informed us of some of the fundamental aspects of gene expression, which are different in procaryotes and eucaryotes, since procaryotes lack internal membranes, whereas eucaryotes have a nuclear membrane that segregates the transcription of the genetic material from the translational machinery in the cytoplasm. But in another sense, yeast cells are not so different from bacteria. Both are unicellular organisms that grow by adjusting their metabolism (gene expression) to the availability of nutrients in a changing environment. Moreover, although there is a clear leap in size between bacterial cells and yeast cells (20- to 50-fold), as well as between the bacterial genome and the yeast genome, the differences can largely be accounted for by the increased complexity of the eucaryotic cell.

The cells of higher eucaryotic organisms differ from yeast cells in ways that place many further demands on the functioning of their genes. We have already seen that their amounts of DNA are enormous, varying between species in ways that are not yet completely understood. Their genomes are composed of different classes of sequences: true gene regions interrupted by numerous and lengthy introns, structural sequences necessary for chromosome sorting during cell division, and various kinds of repetitive sequences that seem to have a life of their own. Moreover, higher eucaryotic cells are significantly larger than yeast cells (up to about 50 times as large), requiring that the end products and intermediates in gene expression be transported much greater distances. Perhaps most important, different cells in different tissues of the same organism serve different functions, demanding that they make quite different protein products. Thus, cells in a vertebrate organism are programmed not so much to cope with a quickly changing environment as to stably express only a

676

subset of the cell's genes, for example, those needed to make a muscle cell, or a bone cell, or a skin cell. We calculate that only about 1 percent of the genome is being expressed in any given higher eucaryotic cell, at a given time. But that 1 percent is made up of different— though overlapping—sets of genes in different cells of the same organism.

Eucaryotic DNA Is Normally Packaged in Chromatin[1,2,3]

Each individual eucaryotic chromosome contains a single DNA molecule, and the huge size of these molecules demands that the DNA be folded in some orderly fashion. This is especially true during mitosis and meiosis, when the DNA must be compacted into separate chromosomes that can be tidily segregated pairwise into the newly forming daughter cells. But even during interphase (the G1, S, and G2 phases of the cell cycle), some folding is necessary simply to pack such long molecules into the cell's nucleus. Moreover, slight differences in the folding in different regions of the genome appear to play an important role in determining which genes are active in any particular cell. DNA folding is the task of specific nuclear proteins. The tight complexes formed between eucaryotic DNA and these proteins are referred to as **chromatin.**

The key proteins in chromatin that serve to fold eucaryotic DNA are the **histones.** These are relatively small proteins that contain a high proportion of positively charged amino acids (lysine and arginine), helping them to bind tightly to the negatively charged DNA double helix. Histones are among the most highly conserved proteins known. They are present in enormous amounts in virtually all eucaryotic cells, their combined mass being approximately equal to that of the cell's DNA. There are five basic types of histone molecules, called **H1, H2A, H2B, H3,** and **H4** (Table 21-1). They occur in equal numbers, except for H1, which is present at one-half the level of each of the others.

When chromatin is viewed in the electron microscope, it often has a beaded appearance built up of spherical, 100 Å particles connected by thin fibers (Figure 21-1a). The amount of DNA associated with each of the so-called **nucleosome** beads is approximately 200 base pairs. (The number varies from about 150 to 250 base pairs, depending on the organism and tissue being examined.) This periodicity can be easily demonstrated by lightly treating chromatin with DNase (micrococcal nuclease). Since the **linker DNA** that connects adjacent nucleosomes in the chromatin fiber is most available to the nuclease, a ladder of DNA fragments is generated, with each fragment differing in length

Table 21-1   Properties of Histones from Calf Thymus

Histone	Composition	MW	Relative Molar Abundance
H1	Lys rich	21,000	1
H2A	Slightly Lys rich	14,500	2
H2B	Slightly Lys rich	13,700	2
H3	Arg rich	15,300	2
H4	Arg rich	11,300	2

Figure 21-1

Demonstration that each nucleosome is associated with approximately 200 base pairs of DNA. (a) An electron micrograph of chromatin in the extended beads-on-a-string configuration. (Courtesy of Victoria Foe.) (b) The products of progressive digestion of chromatin with micrococcal nuclease. First, nucleosomes containing H1 (whose position is shown by a dotted line) and an average length of 200 base pairs of DNA are produced. Further digestion releases H1 and yields core particles containing 146 base pairs of DNA. (c) A ladder of DNA fragments differing by 200 base pairs in length is produced upon light digestion with nuclease. [Photograph courtesy of R. D. Kornberg.]

from the next by 200 base pairs (Figure 21-1b and c). More extensive digestion then chews away the remaining tails of linker DNA, yielding highly protected DNA segments of 146 base pairs, regardless of which organism or tissue is examined. This DNA fragment and its bound histones are called the **nucleosome core particle.** It contains two each of histones H2A, H2B, H3, and H4 (the **histone octamer**), as revealed by protein crosslinking analyses. The single molecule of histone H1 associated with each intact nucleosome is lost when the linker DNA is degraded (see Figure 21-1b).

In addition to the histones, chromatin contains an almost equal mass of other proteins (many acidic), more or less firmly attached to the repeating nucleosome array. These nonhistone chromosomal proteins are of many different types and must include various DNA and RNA polymerases as well as regulatory proteins. Since they do not participate in the basic chromatin structure, both the abundance and the identity of nonhistone chromosomal proteins can vary greatly from one cell type to another.

Irregular Bending of the DNA Double Helix Around the Nucleosome Core Particle[4, 5, 6]

In the nucleosome, the DNA winds around the outside of the octamer of histone molecules. Hence, it is quite accessible for interaction with other molecules, such as regulatory proteins. One side of the DNA helix, however, is protected by close association with the histones. We know this because treatment of nucleosomes with high concentrations of DNase I nicks each of the two DNA strands preferentially on the outer, unprotected side of the helix, producing a ladder of single-stranded fragments differing in length by 10 nucleotides (one turn of the helix) (Figure 21-2).

In wrapping about the outside of the disk-shaped histone octamer, the DNA makes nearly two full turns. This compacts it by a factor of about 5: 146 base pairs of extended DNA double helix would be about 500 Å long compared to the 100 Å diameter of the nucleosome particle. Wrapping also introduces one left-handed (negative) superhelical turn into the DNA associated with each nucleosome (Chapter 9). The histone molecules contact the 146 base pairs of DNA in the following order: H2A, H2B, H4, H3, H3, H4, H2B, and H2A. Thus, the nucleosome core particle clearly has a bipartite structure (Figure 21-3).

One of the most fascinating realizations to emerge from the recent three-dimensional analysis of nucleosome core particles is that the DNA does not take a perfectly regular path in winding about the histone octamer. Instead, the double helix is bent fairly sharply at several specific locations (Figure 21-4). The histones interact with the phosphodiester backbone at nearly every turn of the DNA on the inner face of the superhelix. The sharp bends occur adjacent to points of substantial contact between the DNA and histones H3 and H4. At these sites, the regular B helix conformation is distorted over several neighboring base pairs rather than being kinked abruptly.

Sharp bending of the DNA at certain points in its path around the nucleosome may explain the observation that nucleosomes are "phased" on certain DNA sequences. **Phasing** means that along any particular stretch of the genome, nucleosomes in all cells of a tissue or organism are positioned in exactly the same way relative to the nucleotide sequence. Although nucleosomes in many cases appear to be randomly placed, nucleosome phasing has been detected, not only in regions of repeating sequence, such as satellite DNA or tandemly repeated gene families, but also on nonrepetitive DNA. Perhaps the relative flexibility of some nucleotide sequences allows them to be preferentially accommodated at the points where the helix bends in its path around the nucleosome, while other sequences tend to be excluded. Particularly if the DNA sequence repeats every 200 base pairs (or a multiple thereof), such preferences could serve to locate nucleosomes at precise repeating positions. Phasing can also arise adjacent to sites where assembly of the DNA into a nucleosome is prevented by the tight binding of some nonhistone protein(s) to a specific sequence. We currently do not know whether phasing is a fortuitous consequence of the properties of nucleosomes, or whether phasing preferentially exposes certain DNA sequences for interaction with regulatory factors.

(a)

(b)

Figure 21-2
How nuclease digestion studies first suggested that DNA is wrapped around the outside of the nucleosome. (a) The sites most susceptible (exposed) to nuclease (DNase I) are spaced an average of ten nucleotides apart on each strand of the DNA. (b) Digested DNA shows a ladder of bands differing by about ten nucleotides in length after fractionation on a polyacrylamide gel under denaturing conditions. (Courtesy of R. D. Kornberg.)

(a)

Figure 21-3

The nucleosome core particle has a bipartite structure. (a) An electron micrograph of a nucleosome core particle preparation used for crystallographic studies. Nucleosome cores are flat, disk-like objects (about 57 × 110 Å) that stack in arcs because they are wedge shaped. (Courtesy of J. Finch.) (b) A model for the nucleosome core particle derived from low-resolution (22 Å) crystallographic studies and later confirmed by high-resolution (7 Å) studies. The DNA makes two turns about the histone octamer, giving the particle its bipartite structure. (The dyad axis, which intersects the 146 base pairs of DNA in its middle, is shown.) The histone positions were assigned on the basis of histone-histone and histone-DNA cross-linking results. [After R. D. Kornberg and A. Klug, *Sci. Amer.* 244 (1981): 52–64.]

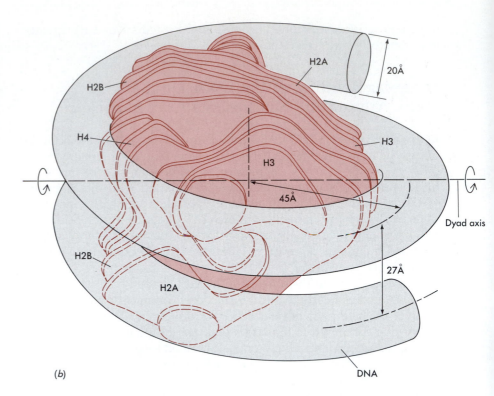

(b)

Higher-Order Chromatin Structure and the Role of Histone H1[1, 2, 7]

The beads-on-a-string structure, created when the DNA is wrapped around individual nucleosomes, is only the first order of packing experienced by eucaryotic DNA. When interphase nuclei are gently broken open, most chromatin is seen as a 300 Å filament in which the nucleosomes are further assembled into a structure of higher order (Figure 21-5a). Local decondensation of the 300 Å filament occurs during transcription of the DNA. Accordingly, the amount of chromatin existing as 300 Å filament provides a rough measure of the relative level of gene *in*activity in cells from various tissues.

Critical for the assembly of nucleosomes into the compact 300 Å filament is the presence of histone H1. This histone molecule is very elongated, with both N- and C-terminal arms extending from a more globular central core. In chromatin, these arms may extend along the linker DNA between nucleosomes. Relative to the nucleosome core particle, H1 is positioned such that it serves to seal the two turns of DNA about the histone octamer. Histone H1 is the most variable of all histone molecules and is found in several different subtypes. (One of these is a histone called H5, which condenses the inactive chromatin in nucleated avian erythrocytes.) The existence of histone H1 variants may underlie the differences in linker lengths seen in chromatin from different tissues or organisms.

Although the exact structure of the 300 Å filament has not yet been determined, most evidence strongly favors the model shown in Figure 21-5b. Here the chromatin is compacted by winding into a solenoid containing six nucleosomes per turn. In this model, the linker DNA is on the inside, as are the molecules of histone H1, which are known to contact one another closely in chromatin. By assembling

(a)

(b)

nucleosomes into the 300 Å filament, the DNA is compacted overall by a factor of about 40 (significantly greater than in the extended beads-on-a-string conformation).

Condensation of Inactive DNA into Heterochromatin[8]

Heterochromatin is the term frequently used to designate highly compacted chromatin that remains visible in the light microscope during interphase. Early genetic experiments suggested that such

Figure 21-4
Irregular bending of the DNA around the nucleosome core. (a) The structure of the right-handed DNA double helix deduced from high-resolution (7 Å) crystallographic studies of nucleosome core particles. Only one-half of the DNA is shown, but the remainder is related to the portion shown by dyad (twofold) symmetry. The DNA is essentially in the B-form helix but is distorted at several points. (b) The region at the center of the 146 base pairs of DNA where the two H3 histones contact the DNA. M stands for the major and m for the minor groove of the DNA helix. Long rods of density labelled B_1 and B_2 in the protein (presumably α helices) make extensive contacts with the minor groove of the DNA one-half turn of the helix away on either side of the dyad axis. [Courtesy of T. J. Richmond, J. T. Finch, B. Rushton, D. Rhodes, and A. Klug, *Nature* 311 (1984):532–537. Used with permission.]

Linker DNA and histone HI on inside

110 Å

300 Å

(a)

(b)

Figure 21-5
Structure of the 300 Å filament. (a) An electron micrograph of chromatin filaments. (Courtesy of B. A. Hamkalo and J. B. Rattner.) (b) The model most consistent with all available data for how nucleosomes pack into the 300 Å filament. [After Alberts et al., *Molecular Biology of the Cell* (New York: Garland, 1983), Fig. 8-5.]

regions contain very few genes, giving rise to the idea that heterochromatic DNA does not have a genetic role. Indeed, visibly compacted DNA is not transcribed into RNA, whereas the much more open form of chromatin called **euchromatin,** which is difficult to visualize during interphase, is vigorously copied into RNA chains. Heterochromatin and euchromatin contain approximately the same DNA/histone ratio, confirming that the DNA is packaged into the basic nucleosome structure in both cases. The biochemical basis for the difference between heterochromatin and euchromatin thus remains unknown; but during mitosis, euchromatin is temporarily converted into a form that is almost indistinguishable from heterochromatin.

Not only is heterochromatin relatively inactive, but genes can be inactivated by chromosomal rearrangements that insert them into a heterochromatic region. This suggests that there are control devices that can turn on or off the functioning of a very long section of a given chromosome. This point was first demonstrated for *Drosophila*, but holds equally well for mammalian systems.

The most striking findings concern the sex chromosomes. Initially, the unexpected discovery was made that in female mammals, the two homologous X chromosomes look quite different. One always appears highly condensed (hinting that it does not function), whereas the other is extended. This suggestion was confirmed by biochemical analysis, which showed that in each female cell, only the genes on one of the two X chromosomes are active. Surprisingly, the inactive chromosome varies from one cell to another; so female tissue is in reality a mosaic containing mixtures of two different cell types. Again, the molecular basis for this bizarre phenomenon remains unknown, although it may be related to methylation of the DNA (discussed later). Nevertheless, X chromosome inactivation is important for demonstrating that there are devices that can specifically block the expression of an entire chromosome.

While the other eucaryotic chromosomes do not show all-or-none effects, they also can have extended regions of heterochromatin, whose length depends on when in the lifetime of the organism they are examined. For example, many regions that are completely heterochromatic in later life appear as euchromatin early in development. Such regions probably contain genes programmed to function in one of the early embryonic stages. DNA adjacent to the centromere, which contains highly repetitive satellite sequences (Chapter 20), is an example of heterochromatin that is never transcribed but seems instead to play a role in chromosome structure and maintenance.

Organizing Chromatin onto the Metaphase Scaffold[9, 10]

As in yeast, each chromosome of a higher eucaryotic cell is believed to contain just one enormously long DNA molecule (approximately 5 cm in length for an average mammalian DNA). Yet, the metaphase chromosomes that become visible during mitosis are only about 5 μm long, the diameter of a typical mammalian cell nucleus. This means that there must be several additional higher orders of DNA folding beyond those provided by packaging first into nucleosomes and then into 300 Å filaments. Again, histone H1 may play a pivotal role, for it is found to be extensively phosphorylated (at up to five or six serine residues per molecule) in mitotic chromosomes. Following comple-

(a)   (b)

Figure 21-6
Electron micrographs of histone-depleted metaphase chromosomes from human HeLa cells showing (a) the familiar morphology of the residual scaffold and (b) that the DNA is attached to the scaffold in loops. [Courtesy of J. R. Paulson and U. K. Laemmli, *Cell* 12 (1977):817–828. Used with permission.]

tion of mitosis, H1 phosphorylation drops back to less than 20 percent of its peak value and again reaches a maximum level only late in interphase, just before the next cycle of chromosome condensation.

It has become clear that the incredible compaction observed in metaphase chromosomes is accomplished by the organization of the chromatin into large, looped domains attached at their bases to a protein scaffold called the **metaphase scaffold.** This has been demonstrated in experiments in which histones and most of the nonhistone proteins have been removed from a metaphase chromosome by treatment with polyanions (such as heparin or dextran sulfate), which compete with DNA for protein binding (Figure 21-6a). Such depleted chromosomes still retain much of their familiar metaphase morphology, but are surrounded by a halo of unfolded DNA. Upon closer examination (Figure 21-6b), the halo can be seen to consist of many discrete DNA loops, each anchored to the scaffold at its base. The lengths of most of the DNA loops are 10 to 30 μm (35 to 100 kilobase pairs). Examination of the protein content of the histone-depleted scaffold structure reveals a predominance of two proteins of high molecular weight (170 and 135 kdal). These proteins presumably form a network that constitutes the central axis linking together the bases of the DNA loops. How this elegant metaphase scaffold assembles is not known, but the recent realization that the 170 kdal protein is topoisomerase II (Chapter 9) argues that a change in the superhelicity of the DNA may be critical to the process.

Protamines Replace Histones to Compact DNA into Sperm Heads[11]

Although higher eucaryotic genomes remain packaged as nucleosomes throughout all stages of the cell cycle in most cells of an organism, there is one situation in which this is not true. The compaction required to condense an entire genome's worth of DNA down to the restricted volume of a sperm head is so great that even histones are

(a)

(b)

(c)

Figure 21-7
Model showing how protamine molecules can crosslink DNA double helices. (a) A representative protamine molecule (Salmine A1) folding into four α-helical segments connected by partially flexible joints. Each segment is about 8 amino acids long; the positions of glycine (G) and proline (P), which are known helix-breakers, are shown. Black dots represent arginines. (b) Two orientations showing how the side chains of four consecutive arginine residues would radiate out from an α-helical segment. (c) A model showing how protamine molecules could wrap around DNA helices in the major groove. Each α-helical segment could use its arginine side chains to make two contacts across the major groove (gray) of the DNA molecule it has wrapped around, while making two others across the minor groove of an adjacent DNA helix. Alternatively, different α-helical segments of a single protamine molecule could interact with two adjacent DNA helices using only major groove contacts.
[After R. W. Warrant and S.-H. Kim, *Nature* 271 (1978):130–135.]

too bulky and must be displaced. Instead, **protamines,** another special class of small, basic, DNA binding proteins, serve to bring the double helices into the closest possible proximity.

Nearly two-thirds of the amino acid residues in protamines are basic (mostly arginine). These are usually found clustered, four or five together along the polypeptide chain. After their synthesis in the cytoplasm, protamines are phosphorylated on their serine residues by a special protamine kinase. They then migrate to the cell nucleus, where they replace histones and become dephosphorylated, concomitantly condensing the DNA. In the final protamine-DNA complexes, the number of positive charges on the protamine amino acid side chains just balances the number of negative charges on the phosphate backbone of the DNA.

Most satisfying is the finding that protamine molecules undergo a dramatic conformational change as they complex with nucleic acid. (The crystal structure of a protamine-tRNA complex is what has been analyzed.) Whereas protamine polypeptide chains occur as random coils free in solution, they become largely α-helical upon binding. Specifically, they form three or four short α-helical segments (depending on the particular protamine) connected by two or three flexible joints (Figure 21-7a). Each helical domain contains four or more consecutive arginine residues and can lie along one of the grooves of double-helical DNA (the major groove is more likely). As a result, the two arginines facing the groove form hydrogen bonds to and neutralize one negatively charged phosphate group from each of the two strands of the double helix. The two remaining arginines in the same α-helical domain of the protamine molecule are then ideally situated to interact in the same way with the backbone phosphates of a neighboring DNA helix (Figure 21-7c). Hence, the protamine molecules serve to crosslink the DNA helices, effecting the high degree of condensation accomplished in sperm heads.

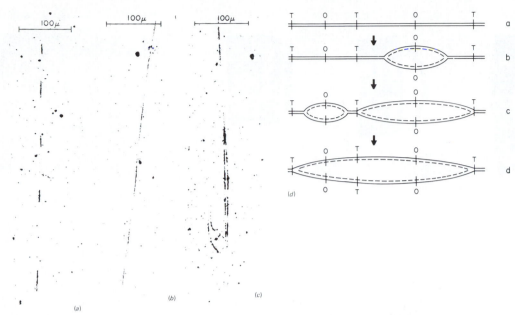

Different DNA Sequences Are Scheduled to Replicate at Different Times During S Phase[12, 13]

As in yeast, the DNA of higher eucaryotic genomes is (usually) replicated bidirectionally starting from multiple origins (Figure 21-8). This means that as many as 100 replication forks are operating on each mammalian chromosome (DNA molecule) at any time during S phase of the cell cycle. Adjacent origins are 50 to 300 kilobase pairs apart, approximately the same length as the DNA loops that extend from histone-depleted metaphase chromosomes. Thus, individual replication units or **replicons** may correspond directly to the independently supercoiled domains of DNA that lie between the attachment points to the metaphase scaffold.

As the DNA is duplicated, histones are quickly deposited and ensure that each new daughter strand becomes properly packaged into nucleosomes. Although little is known about the assembly of nucleosomes in vivo, the one fact that seems clear is that the old nucleosomes segregate conservatively at the fork. Despite the existence of an axis of twofold symmetry in the histone octamer (H2A, H2B, H3, H4)$_2$, nucleosomes do not split down the middle (semiconservatively like DNA), with half of each nucleosome going to each daughter duplex. Instead, one daughter duplex receives the old nucleosome intact, while the other daughter duplex acquires an entirely new nucleosome.

Not all replication origins are activated at the same time during S phase. Rather, clusters of adjacent replicons (between 25 and 100 replicons per cluster in mammalian cells) initiate and terminate replication at a specific time that is characteristic of that particular cluster. Again, a striking relationship between the organization of interphase and metaphase chromatin is evident, for each cluster of coordinately timed replicons appears to correspond to a distinct metaphase chromosome band. After staining with Giemsa (a DNA binding dye), some 2000 light and dark bands can be visualized alternating along mammalian metaphase chromosomes. Adjacent areas stain differently only in part because of differences in base composition; instead,

Figure 21-8
Bidirectional growth of a mammalian chromosome. (a) An autoradiograph of a Chinese hamster DNA molecule from a cell that has briefly been labeled with tritiated thymidine. The tandem arrays of exposed grains indicate the existence of several replication points in the portion of the DNA fiber under view. (b) Tandem arrays seen after a pulse exposure to label followed by a chase period in nonradioactive medium. Here the grain density declines from the center to the ends, suggesting that the growing points move in opposite directions. (c) An example of sister replicating molecules held together by a replicating fork. (d) A schematic model interpreting pictures (a) through (c). [Photographs (a), (b), and (c) are courtesy of J. Huberman; they originally appeared in J. A. Huberman and A. D. Riggs, *J. Mol. Biol.* 32 (1968):327.]

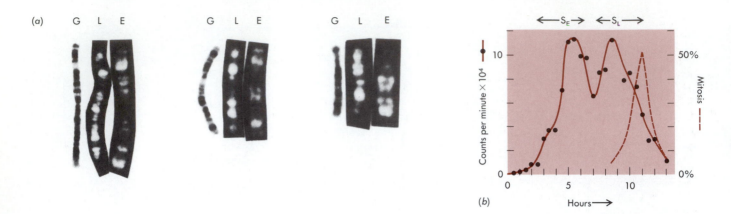

Figure 21-9
Different chromosome bands replicate at different times during S phase.
(a) Fluorescence micrographs of three representative hamster metaphase chromosomes, showing correspondence between Giemsa staining and late-replicating bands (L). The chromosomes labeled G were treated with trypsin and Giemsa. Late-replicating bands were visualized by staining with acridine orange after incorporating BUdR (bromodeoxyuridine) early in S phase; BUdR quenches the green acridine orange fluorescence of the labeled (early-replicating) DNA. Conversely, early-replicating bands (E) were visualized by incorporating BUdR late. (b) The bimodality of DNA synthesis during S phase. Synchronized cells were pulsed with ^{3}H-thymidine and the incorporation plotted against time. [Photographs courtesy of G. P. Holmquist et al., *Cell* 31 (1982):121–129. Used with permission.] To ascertain when a specific gene is replicated, BUdR-containing (heavy) DNA can be fractionated from nonisotopically labeled DNA in a density gradient and the fractions hybridized with the appropriate cloned probe to establish whether the gene incorporated BUdR early or late during S phase. [For more information, see M. A. Goldman, G. P. Holmquist, M. C. Gray, L. A. Caston, and A. Nag, *Science* 224 (1984):686–692.]

the bands clearly give an indication of slightly different modes of DNA compaction. When cells are exposed to isotopic DNA precursors (BUdR) during the first half of S phase, the light-staining chromosome bands become labeled; during the second half of S phase, the dark bands label (Figure 21-9a). Between the two halves of S phase, the rate of DNA synthesis dips significantly (Figure 21-9b). It has been known for some time that blocks of heterochromatin, including regions near the centromere (satellite sequences) and the inactive X chromosome in female mammals, correspond to dark-staining chromosome regions and replicate late during S phase. Conversely, genes that are transcriptionally active in all cells (often called "housekeeping genes") replicate early. Tissue-specific genes usually replicate late in cells in which they are not expressed, but early in cells in which they are potentially expressed. Hence, gene activity seems to be correlated with early replication and light-staining bands. However, how the timing is controlled and what ensures that every origin of replication is utilized only once during each S phase remain to be deciphered.

Specific Attachment Sites for Interphase DNA to the Nuclear Scaffold[14]

In contrast to the networks of fibrous structures (microtubules, microfilaments, etc.) that are readily visualized crisscrossing the eucaryotic cytoplasm, direct examination of nuclei reveals no such filamentous components (Figure 21-10). This, however, does not mean that the interphase nucleus is totally devoid of defined structures. There are one or more nucleoli (singular, nucleolus), discrete bodies that reform after each mitosis at chromosome regions containing repeated rDNA sequences. The nucleoli are responsible for the transcription, processing, and protein packaging of ribosomal RNAs. Also present is the nuclear lamina, a fibrous protein meshwork lying on the inner surface of the nuclear membrane. It appears to hold certain regions of the interphase chromatin (predominantly highly condensed heterochromatin) tightly against the membrane. Yet, chromatin is specifically excluded from the regions around the nuclear pores (Chapter 18), clearing a path for the passage of molecules to and from the cytoplasm. Beyond such examples, no clear organizing principles are evident in the remainder of the interphase nucleus (**nucleoplasm**). Nevertheless, there has long been a strong belief that some as-yet-undefined underlying structure must impose order on nuclear contents and processes.

Thus, it is very satisfying that solid evidence has recently emerged for stable attachment of specific DNA sequences to some sort of scaffolding structure throughout the various stages of the cell cycle. When nuclei from *Drosophila* tissue culture cells are exposed to lithium diiodosalicylate, the histones and many nonhistone proteins are extracted. Subsequent treatment of the remaining **nuclear scaffold** preparation with restriction endonucleases reveals that the DNA is attached at specific sites along the genome. For instance, if we examine the tandemly repeated histone gene family by asking which restriction fragments remain with the residual scaffold and which are released, tight binding can be localized to the AT-rich nontranscribed spacer region between the H1 and H3 genes (Figure 21-11a). Only one attachment point per histone gene repeat (made up of the H1, H3, H4, H2A, and H2B genes, in that order) is detected. This site must be the same in all 100 histone repeat units, since the relevant restriction fragments are either completely retained or completely released from the scaffold. Moreover, since the cells examined have not been synchronized, this attachment site must be stable throughout most stages of the cell cycle.

With several other *Drosophila* genes examined so far, attachment to the nuclear scaffold likewise occurs within five kilobase pairs upstream of (and sometimes also 3′ to) the transcribed region. One or more genes may reside between adjacent attachment sites. In general, active or potentially active genes possess an attachment site in close proximity, whereas certain unexpressed genes are not closely associated with the scaffold. Current topics of investigation are: defining more exactly the attachment sites (do they correspond to recognition sequences for topoisomerase II?); asking whether the interphase attachment points correspond to the bases of the DNA loops associated with the metaphase scaffold; determining if the attachment points around specific genes change during the development of an organism; and defining the protein composition of the preparations to which the DNA remains specifically attached.

Figure 21-10

Electron micrograph of a thin section of the nucleus from a human fibroblast cell. Within the nucleolus, the fibrillar centers contain the rDNA, the dense fibrillar component represents rRNA transcripts, whereas the granular component is composed of maturing ribosomal subunits. [Courtesy of E. G. Jordan and J. McGovern.]

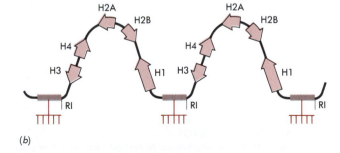

Figure 21-11

One specific attachment site to the nuclear scaffold exists in each repeat of the cluster of *Drosophila* histone genes. (a) A restriction endonuclease map of the major histone repeating unit is shown. Only the 657-base-pair *Hin* fI/*Eco* RI fragment remains attached to scaffold preparations, while all other DNA fragments generated by the enzymes indicated are released into the supernatant. (b) A schematic representation showing how the *Drosophila* histone repeat (present in about 100 tandem copies on chromosome 2) might be attached to the nuclear scaffold. [After J. Mirkovitch, M.-E. Mirault, and U. K. Laemmli, *Cell* 39 (1984):223–232.]

Right arm of
chromosome 3

X chromosome

Normal mitotic
chromosomes at
the same scale

Chromosome 4

Region where the 2
homologous chromosomes
are separated

Chromocenter

Left arm of
chromosome 3

Left arm of
chromosome 2

20 μm

(a)

(b)

10 μm

Figure 21-12

Drosophila polytene chromosomes.
(a) Sketch of the entire set of polytene chromosomes in one *Drosophila* salivary cell. These chromosomes have been spread out for viewing by squashing them against a microscope slide. Note that there are four different chromosome pairs present. Each chromosome is tightly paired with its homolog, and the four chromosome pairs are linked together by regions near their centromeres that have aggregated to create a single large "chromocenter." Here, this chromocenter has been split into two halves by the squashing procedure used. [After T. S.

Painter, *J. Hered.* 25 (1934):465–476.] (b) Light micrograph of polytene chromosome 3L showing the distinct patterns recognizable in different chromosome bands. These bands occur in interphase chromosomes and are a special property of the giant polytene chromosome. They should not be confused with the much coarser bands seen in Figure 21-9, which are revealed by special staining techniques used on normal mitotic chromosomes. Bands that are puffed are indicated by arrows. (Courtesy of Joseph G. Gall.)

hsp 28   hsp 23   1   hsp 26   hsp 22   4   5

1 kbp

Most attractive is the idea that this newly described nuclear scaffold functions to lure and retain proteins important for transcription. Since the genome is attached near promoters, but not within transcribed regions (Figure 21-11b), we can envision RNA polymerases proceeding along the free DNA loop regions, being recaptured by the scaffold after terminating RNA synthesis, and then—with the help of appropriate transcription factors—reinitiating at sites near the scaffold.

Polytene Chromosomes Allow Visualization of Gene Expression[15-19]

The salivary glands of *Dipteran* flies like *Drosophila* and *Chironomus* contain giant chromosomes that are visible during the interphase stage of the cell cycle. When these chromosomes are stained and examined in the light microscope, an alternating pattern of dark and light bands (bands and interbands) is revealed (Figure 21-12). Band sizes vary tremendously. Some are so thin that they are hardly visible, whereas others occupy several percent of the length of a given chromosome. Each such giant chromosome, called a **polytene chromosome**, consists of a large number of partially replicated chromosomes neatly stuck together in lateral array. Polytene chromosomes arise by repeated cycles of DNA duplication, uninterrupted by separation of the daughter chromosomes or cell division, so that most chromosomal regions are replicated some ten times to yield 2^{10} (or 1024) copies. Remaining unreplicated is the centromeric DNA, while the genes specifying ribosomal RNA are duplicated less often (probably because they are already present in multiple tandemly repeated copies).

In salivary gland cells, some 5000 bands (and interbands) can be detected in the four polytene chromosomes that make up the *Drosophila* genome. The different staining properties of the bands and interbands are highly suggestive of different states of compaction or organization of the chromatin in these regions. Thus, it was initially guessed that each band would encode a single protein. Indeed, early exhaustive genetic mapping of selected regions of several *Drosophila* chromosomes revealed that the number of complementation groups corresponds very closely to the number of visible bands. However, the average band contains some 30,000 base pairs of DNA, much more than enough to encode a single polypeptide chain (even if the gene's transcript contains several lengthy introns). Recent analyses using cloning and sequencing have in fact revealed that many bands do contain several (up to at least seven) genes, which give rise to independent transcripts and are separated by the usual nontranscribed regions (Figure 21-13). Such genes are frequently related, being members of the same gene family. But this is not always the case, nor are the genes in a single band always coordinately expressed, leaving the basis of their clustering a mystery. Whatever the reason, we can now confidently state that the number of genes in the

Figure 21-13
The clustering of seven genes within 15 kilobase pairs at cytological locus 67B of the *Drosophila* genome. Four of these genes are vigorously transcribed after heat shock and give rise to heat-shock proteins hsp 28, 23, 26, and 22. Other subsets of this gene cluster are known to be developmentally regulated: hsp 26 and hsp 28 are expressed during oogenesis, while hsp 22, hsp 26, and transcripts 1, 4, and 5 are expressed maximally in the early pupa and late third instar larva. The directions of transcription are indicated by arrows when known. Vertical arrows above the line indicate DNase hypersensitive sites mapped by Keene et al., *Proc. Nat. Acad. Sci.* 78 (1981):143–146. [Data from K. Sirotkin and N. Davidson, *Dev. Biol.* 89 (1982):196–210, and V. Corces, R. Holmgren, R. Freund, R. Morimoto, and M. Meselson, *Proc. Nat. Acad. Sci.* 77 (1980):5390–5393.]

Figure 21-14
An autoradiogram that illustrates the location of ^{3}H-uridine incorporation sites on polytene chromosomes. The highest grain densities are found over expanded regions on the chromosomes (chromosome puffs), indicating that these are the most active sites of RNA synthesis. [Courtesy of J. J. Bonner and M. L. Pardue, *Cell* 12 (1977):227–234. Used with permission.]

10 μm

Drosophila genome probably does not exceed the number of polytene chromosome bands by more than a factor of 5 or 10.

A particularly useful property of the giant polytene chromosomes is that obvious changes in morphology can be seen when the genes in a particular band become activated. The DNA decondenses into a much more open state, forming a distinctive **puff**. The larger and more diffuse a puff appears, usually the higher the rate of its transcription. This can be shown by briefly labeling cells with a radioactive precursor of RNA (^{3}H-uridine) and localizing the newly made transcripts by autoradiography. Whereas a compacted DNA band synthesizes very little RNA, intense labeling occurs over the largest puffs (Figure 21-14). As the organism progresses through development, puffs arise and regress (recondense) in a defined order as the synthesis of different mRNAs is turned on and off. Experimentally, puffing of different sets of bands can be induced by heat shock or by the addition of the insect hormone ecdysone, which stimulates the synthesis of proteins required for molting and pupation (also see Chapter 22). Electron microscopic examination of serial sections through large puffs can reveal the ribonucleoprotein structure of the nascent RNA transcripts (Figure 21-15). Use of antibodies directed against various nuclear proteins can show that molecules such as RNA polymerases, hnRNP proteins, snRNPs (Chapter 20), and topoisomerases (Figure 21-16) become specifically concentrated in puff regions.

DNase I Sensitivity Characterizes Active Chromatin Regions[20–27]

The changes seen in *Drosophila* polytene chromosomes make it obvious that active (or potentially active) genes undergo some dramatic change in their chromatin structure relative to quiescent regions of the genome. The decondensation of the DNA contained in a particular band to form a puff is often a necessary, but not a sufficient, condition for transcription. (For instance, the multigene band shown in Figure 21-13 puffs both developmentally and upon heat shock, but different subsets of its genes are expressed under the two conditions.) Likewise in vertebrate cells, the two distinguishable forms of chromatin do not correlate directly with gene activity. Heterochromatin does

Figure 21-15
Electron micrograph of a portion of a loop within the Balbiani ring 2 of chromosome 2 of a salivary gland cell of *Chironomus tentans*. It shows spherical ribonucleoprotein granules attached by "stalks" to the thin chromatin thread (arrow). Almost simultaneous with RNA synthesis, specific proteins attach to the nascent RNA chains, leading to the formation of ribonucleoprotein granules whose shapes reflect the secondary structure of their underlying RNA chains. [Reproduced with permission from B. Daneholt, *Cell* 4 (1975):1.]

remain both visibly condensed and inactive in RNA synthesis during interphase. Yet, the remainder of the DNA (the euchromatin) cannot all be transcriptionally active, since on average only about 7 percent of the sequences in the genome of a typical higher eucaryotic cell are ever copied into RNA molecules. Thus, we must ask whether other methods can detect changes in the structure of euchromatin in those regions expressing specific genes.

One important indication of altered chromatin structure within active gene regions is a heightened sensitivity to DNase I. Recall that when bulk chromatin is subjected to digestion using low concentrations of nuclease, a ladder of fragments differing by about 200 base pairs in length is generated (see Figure 21-1c). This pattern reflects the rather regular placement of successive nucleosomes along the chromatin fiber. One way of determining whether any particular gene region is organized into such a repeating nucleosome array is to hybridize the ladder of DNA fragments with radioactive DNA derived from the cloned gene. In tissues where the gene of interest is inactive, a defined ladder appears (Figure 21-17). In contrast, hybridizing fragments are much smaller on the average and irregularly sized in tissues where the gene is actively expressed. This greater sensitivity to

(a)

(b)

(c)

85F 87A 87C   93D

85F 87A 87C   93D 95D

Figure 21-16
Specific association of topoisomerase I with transcriptionally active regions of *Drosophila* polytene chromosomes. Antibodies prepared against *Drosophila* topoisomerase I were incubated with polytene chromosome preparations and their binding visualized by applying a fluorescently tagged second antibody directed against the anti-topoisomerase antibodies. A phase-contrast image (c) of part of chromosome 3 is compared to fluorescent images of the same chromosome region from (a) a larva grown at 25°C and (b) a larva heat shocked at 37°C for 20 minutes prior to dissection. 87A, 87C, 93D, and 95D are known locations of heat-shock puffs; their fluorescence is clearly enhanced in (b) relative to (a). [Courtesy of G. Fleishmann et al., *Proc. Nat. Acad. Sci.* 81 (1984):6958–6962. Used with permission.]

Figure 21-17

Demonstration of heightened DNase sensitivity in actively expressed versus unexpressed ribosomal DNA genes of *Dictyostelium*. Nuclei from vegetative cells (V) synthesizing ribosomal RNA or from slug cells (S), which synthesize little ribosomal RNA, were digested for various times with micrococcal nuclease. The stained gels reveal regularly spaced ladders of fragments, indicating that most DNA is organized into nucleosomes in both types of cells. Examining rRNA genes with a specific probe, however, shows that rDNA fragments are much smaller and more irregularly sized in vegetative cells than in slug cells. (Especially compare the 60-second patterns.) [Courtesy of J. P. Ness, P. Labhart, E. Banz, T. Koller, and R. W. Parish, *J. Mol. Biol.* 166 (1983):361–381. Used with permission.]

nuclease strongly suggests an unfolding or disorganization of the chromatin upon gene activation. The sensitive region always extends some distance upstream and downstream from the sequences being copied into RNA. If the nuclease used is DNase I, sensitivity is observed for all active genes, regardless of their extent of expression, and it is detected not only when a gene is being transcribed, but also in tissues where a gene is potentially activatable. When micrococcal nuclease is used, the results are more variable. Thus, as with puffing, DNase I sensitivity appears to be a necessary but not sufficient condition for gene expression.

Superimposed on the generally heightened DNase I sensitivity of active chromatin are so-called **hypersensitive sites**. These are localized regions (limited to less than 100 to 200 base pairs) where the nuclease has an extremely high probability of cutting either strand of

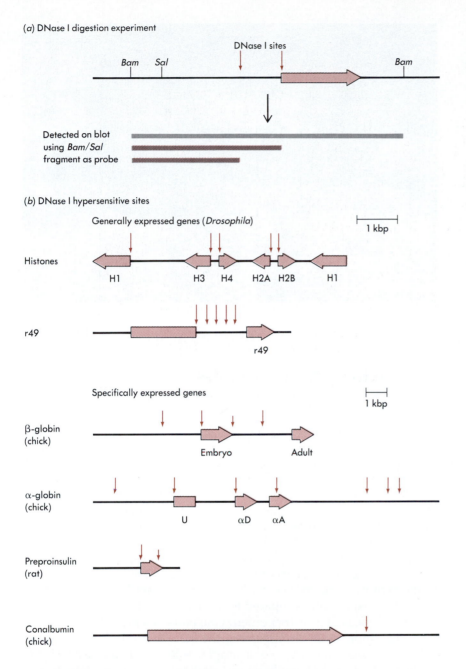

(a) DNase I digestion experiment

DNase I sites

Bam Sal

Bam

Detected on blot
using *Bam/Sal*
fragment as probe

(b) DNase I hypersensitive sites

Generally expressed genes (*Drosophila*)

1 kbp

Histones

H1   H3 H4 H2A H2B   H1

r49

r49

Specifically expressed genes

1 kbp

β-globin
(chick)

Embryo   Adult

α-globin
(chick)

U   αD   αA

Preproinsulin
(rat)

Conalbumin
(chick)

Figure 21-18
The location of DNase I–hypersensitive
sites in the vicinity of various genes.
(a) How the positions of sensitive sites
are mapped. Nuclei are first treated
with DNase I, and then the isolated
DNA is digested with a suitable restriction enzyme (here *Bam*). After fraction-
ating and transferring the digest to nitrocellulose, the Southern blot is probed
with a shorter subfragment (here *Bam/
Sal*) located near the gene of interest.
In addition to the full-length *Bam/Bam*
fragment, shorter bands with one end
generated by DNase I will appear in
the blot. (b) Some hypersensitive sites
are within or 3', as well as 5', to the
genes examined. See also Figure 21-3
for hypersensitive sites appearing near
some *Drosophila* heat-shock genes.
[After S. C. R. Elgin, *Cell* 27 (1981):413–
415.]

the DNA molecule (Figure 21-18). Hypersensitive sites are generally
found within 1000 base pairs flanking the 5' end of genes that are
active or can be activated. Usually, more than one such site is present
per gene. The first clues as to how DNase hypersensitivity might arise
came from examination of the minichromosome of SV40 (Figure 21-
19). Here, the hypersensitive sites are all clustered within a region of
the DNA that is free of nucleosomes. The absence of nucleosomes
could reflect either an altered DNA conformation, which cannot bind
nucleosomes (such as the formation of a Z helix), or the tight binding
of other proteins to the DNA. Thus, it is intriguing that hypersensitive sites correspond very closely to the locations of several types of
upstream control signals that are known to bind activator proteins
required for RNA polymerase II transcription.

Figure 21-19
A nucleosome-free gap in SV40 chromatin is localized at the origin of DNA replication (ORI). In the electron micrograph, several "minichromosomes" display a gap (indicated by arrows). The gap was localized by cleaving the circular minichromosomes with each of several restriction enzymes to produce the expected patterns of linearized complexes shown below. A variety of DNases preferentially cleave SV40 chromatin within several hundred base pairs of ORI, showing that the nucleosome-free and DNase-hypersensitive sites correspond. [Photograph courtesy of S. Saragosti, G. Moyne, and M. Yaniv, *Cell* 20 (1980):65–73. Used with permission.]

Gene Activity Also Correlates with Undermethylation of the DNA[20, 28, 29, 30]

In most higher eucaryotic genomes, including those of mammalian cells, approximately 5 percent of the C residues are modified by methylation at the 5 position of the cytosine ring. Such methylations occur almost exclusively at CpG sequences to yield mCpG (mC = 5-methylcytosine) and appear symmetrically on both strands of the DNA (since the CpG dinucleotide is always base-paired with another CpG sequence in the antiparallel double helix). Curiously, CpG sequences are themselves underrepresented in genomes that contain mC, being present at only about one-third the expected frequency.

What has become apparent in recent years is that certain CpG sites in the vicinity of many genes are undermethylated in tissues where the gene is expressed, compared to tissues where the gene is inactive. The use of particular restriction enzyme pairs has provided a convenient and simple assay for assessing the state of methylation of some subsets of CpG sequences. For example, both *Hpa* II and *Msp* I cut the sequence CCGG, but methylation renders the site resistant to the *Hpa* II nuclease only. Thus, by hybridizing *Hpa* II and *Msp* I digests of genomic DNA with probes corresponding to a particular gene region, the degree of methylation at nearby CCGG sequences can be easily measured. Demethylation of active genes is not complete. Up to 30 percent of the CpG sequences in transcribed regions are methylated, compared to about 70 percent methylation of all CpG sequences in animal cell DNA. As with DNase sensitivity, undermethylation is usually detected over a significantly larger region than the transcribed portion of a particular gene and is often found in tissues where the gene is not (yet) active but potentially activatable.

Although demethylation and expression are not coupled for every gene that has been examined, in most cases the correspondence is quite striking. This prompts the question of whether changes in DNA methylation are the cause or effect of gene activation. Methylation could regulate gene expression in two main ways. First, addition of a

Figure 21-20
The difference between C and 5-azaC. Although 5-azaC lacks a methylatable atom in the 5 position, the extent of demethylation it causes in DNA far exceeds the amount of 5-azaC incorporation. This may reflect the fact that DNA methylases are processive enzymes, so that the presence of a single 5-azaC interferes with methylation at nearby C residues.

5-methyl group to cytosine could increase (or decrease) the interaction between the DNA and proteins such as repressors and activators, for the 5-methyl group protrudes into the major groove of the double helix, where specific DNA-protein recognition often takes place (Chapters 9 and 16). Second, because the addition of a 5-methyl group to C leads to molecular crowding in the major groove, methylation tends to shift the conformational equilibrium of the DNA away from the standard B-form toward other forms (notably the Z form, in which a larger major groove can relieve some of the steric constraints; Chapter 9). Since DNA binding proteins are generally sensitive to the configuration of the sugar-phosphate backbone as well as to the base sequence at their recognition sites, such conformational changes could dramatically alter repressor (or activator) binding.

Some evidence that demethylation may indeed induce gene activation comes from studies using 5-azacytidine (5-azaC), an analog of C whose incorporation into DNA inhibits methylation in vivo (Figure 21-20). Strikingly, 5-azaC induces differentiation (turning on of genes) in cultured mouse embryo cells. Similarly in the case of a number of cloned genes that have been reintroduced into mammalian cells, the unmethylated form proves to be more active than the deliberately methylated gene. However, there are also examples where demethylation lags behind gene expression. Thus, the exact relationship between methylation and gene expression remains unclear, perhaps because only a few of the potential methylation sites in DNA are critical for gene control. Also unclear is how selective demethylation of a critical site might be accomplished when gene activation is desired. A final puzzle arises from the realization that some eucaryotic organisms (notably insects) regulate their genes perfectly well without any DNA methylation whatsoever.

Active Genes Have Altered Nucleosomes or None at All[20, 31–37]

Underlying the heightened DNase sensitivity of active gene regions must be some dramatic change in the wrapping of successive 200-base-pair stretches of DNA around nucleosome beads. This disorganization cannot be ascribed solely to the passage of RNA polymerase molecules, since neither the size of the affected regions nor the time of appearance of DNase sensitivity is strictly coupled to transcription. Instead, either an unfolding of the histone octamers or a disruption of the regular spacing between nucleosomes must occur. Experimentally, both biochemical fractionation and examination in the electron microscope have detected differences in the nucleosome structure of "active" and "inactive" chromatin.

Differentiating active from inactive chromatin are changes in the protein composition of nucleosomes, as well as certain covalent modifications of the histone proteins. The critical role played by histone H1 in condensing the basic beads-on-a-string configuration of chromatin into an inactive 300 Å filament has already been discussed. Consistent is the finding that most (about 80 percent) of the mC in mammalian cell DNA is localized in nucleosomes that contain histone H1. On the other hand, nucleosomes recovered from transcriptionally active regions are deficient in histone H1 and often contain instead the highly abundant nonhistone chromosomal proteins HMG 14 or 17. (HMG stands for **high mobility group**; these proteins are both small and unusually highly charged, therefore moving rapidly upon gel electrophoresis.) Some of the known covalent modifications of the core histone proteins (H2A, H2B, H3, H4) occur selectively in "active" nucleosomes. Included are acetylation, which may decrease the tendency of neighboring nucleosomes to aggregate, and the attachment of a polypeptide called ubiquitin (76 amino acids) to a fraction of H2A molecules, perhaps targeting them for specific protein degradation.

Examination by electron microscopy suggests that even more dramatic changes occur in the packaging of some active genes. For instance, the genes specifying ribosomal RNA—which are among the most transcriptionally active known—appear to be completely devoid of nucleosomes (Figure 21-21). Similarly, in the case of at least one intensely transcribed protein-coding gene region—the Balbiani ring 2 of *Chironomus* polytene chromosomes—the DNA is studded with RNA polymerase molecules, not nucleosomes. These observations raise the question of whether, in biochemical experiments, the "active" genes examined might have been truly active in only a relatively small fraction of the cells. Or, it may be that with anything less than the very highest rates of transcription, the DNA can occasionally be (partially) assembled into nucleosomes between successive passages of RNA polymerase molecules. Of course, just because nucleosomes cannot be visualized does not mean that the DNA is entirely free of bound protein(s). Much remains to be clarified concerning the molecular changes that accompany the transitions of chromatin from the inactive to the potentially active to the transcribing state.

Positive Versus Negative Control of Gene Expression in Higher Eucaryotes[38]

So far, we have presumed that most gene regulation in higher eucaryotes takes place at the level of transcription, as it does in bacteria (Chapter 16). Evidence that this is true has come from examining the functioning of many genes. Yet, also as in bacteria, control of gene expression at other levels cannot be overlooked. Later we will consider how regulatory mechanisms involving RNA processing, mRNA stability, translation, and even protein degradation serve to fine-tune the levels of gene products available for the functioning of higher eucaryotic cells.

Nevertheless, since RNA synthesis is the first step in gene expression, its control overrides regulation at all other levels. Most relevant,

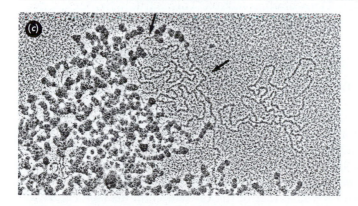

Figure 21-21
Active nucleolar chromatin from *Xenopus laevis* oocytes has a smooth, DNA-like appearance within the transcribed region (arrows in (a) and (c)).
(a) Oocyte nuclei were lysed and the samples centrifuged through a sucrose step gradient onto an electron microscope grid revealing Christmas-tree–like structures. (b) Comparison of presumptive RNA polymerases (P) to nucleosomes (N) in an RNase treated sample.
(c) Comparison of the nucleolar chromatin fiber to purified DNA. Note the similar appearance between the chromatin axis (arrow) and the coprepared circular viral DNA on the right side of the figure. [Courtesy of T. Koller. See P. Labhart and T. Koller, *Cell* 28 (1982):279–292, for details.]

therefore, is the question of whether positive or negative mechanisms determine which RNAs are transcribed from higher eucaryotic genomes. Bacteria utilize both activators (e.g., cAMP binding protein) and repressors (e.g., *lac* repressor) to control their RNA synthesis. These regulatory proteins recognize specific DNA sequences upstream from genes, thereby either preventing or facilitating the binding of RNA polymerases (Chapters 13 and 16). In addition, they have a general but much lower affinity for nonspecific (random) DNA sequences and spend part of their time binding to irrelevant stretches of the genome. To circumvent this problem, bacteria simply adjust the numbers of regulatory proteins they synthesize so that although some repressor (or activator) molecules are bound to random DNA, others are available to repress (or activate) transcription completely by binding to the correct specific DNA sequence.

The molecular problems posed by positive versus negative regulatory strategies differ for higher eucaryotic cells with their huge DNA contents. First, much greater numbers of repressor (activator) molecules are required to compensate for regulatory proteins being sopped up by more nonspecific DNA. Here, the correspondingly larger sizes of higher eucaryotic cells can probably accommodate the necessary increase in numbers of molecules. And second, the tremendously large size of the genome makes it much more probable that DNA sequences identical (or very similar) to a specific binding site will appear more than once (and perhaps in front of a gene that should not be controlled by that particular regulatory molecule).

There are two obvious ways out of this dilemma. One is to invoke the ability of histones and other chromosomal proteins to conceal DNA sequences. (Recall that only 7 percent of the DNA in the genome of a typical higher eucaryotic cell is ever copied into RNA.) But in this hall of mirrors, how would the chromosomal proteins know what sequences to cover up? The other way out is to invoke *positive* control of gene expression executed by the simultaneous binding of several regulatory factors to nearby specific sites on the DNA. Only if all sites are appropriately occupied would an RNA chain be initiated. And the probability of several regulatory molecules being bound in the correct juxtaposition by random DNA would be low. Negative control, on the other hand, would not be made more specific by multiple sites, because repressor binding to just one sequence would probably be enough to block RNA polymerase access. As we shall see, most (but not all) well-characterized transcriptional control mechanisms in higher eucaryotes do turn out to be positive and to involve multiple DNA binding factors. This is true whether the gene is transcribed by RNA polymerase I, II or III.

The Nucleolar Location of Ribosome Biogenesis[39, 40, 41]

Just as ribosomes are factories for making proteins in all cells, nucleoli are factories for making ribosomes in eucaryotic cells. These distinctive nuclear substructures form in somatic cells at the one or more chromosomal sites where multiple copies of rRNA genes are present in tandem array (Chapter 20). Although no membrane separates a nucleolus from the remainder of the nuclear contents (see Figure 21-10), nucleoli have distinctive protein compositions representing high concentrations of all the enzymes required for synthesizing and processing precursor ribosomal RNA into the 28S, 18S, and 5.8S molecules found in mature ribosomes. 5S rRNA molecules synthesized elsewhere in the nucleus are somehow specifically transported to the nucleolus for assembly into nascent ribosomes, as are almost all of the 70 to 80 different ribosomal proteins. They, like all other proteins, are synthesized on previous generations of ribosomes present in the cell cytoplasm. Assembly of these many components begins while the ribosomal RNA is still incompletely synthesized and processed, and apparently proceeds along a defined pathway similar to (but more

Mammalian 45S pre-rRNA (13,000n)

Mature rRNAs

complicated than) that deduced for *E. coli* ribosomes (Chapter 14). One important difference may be the participation of proteins that are not present in the final structures but remain in the nucleolus to catalyze further rounds of ribosome assembly. Nearly completed 40S and 60S subunits are then exported by unknown mechanisms from the nucleolus to the cytoplasm. There they acquire a few remaining proteins to become functional ribosomes.

Although we are still largely ignorant of the details of eucaryotic ribosome assembly, many insights into the mechanisms of rRNA synthesis and processing have been gained in recent years. Examination of spread nucleolar chromatin in the electron microscope (Figure 21-22) reveals tandem arrays of rRNA genes studded with closely packed RNA polymerase I molecules from which emanate growing rRNA precursor chains (about 100 per gene). At the point of each Christmas tree–like formation is the promoter region for that unit, while the base of the tree corresponds to the 3' ends of the rRNA transcripts. The active units are separated by the so-called **nontranscribed spacer regions** of the rDNA repeat (Chapter 20).

The 18S, 5.8S, and 28S sequences are transcribed into a single rRNA precursor molecule (in that order), ensuring that these three rRNAs will be available in equimolar amounts for assembly into ribosomes. The length of the pre-rRNA chain in mammalian cells is about 13,000 nucleotides (45S), of which about half (denoted **transcribed spacer**) is discarded during rRNA processing (see Figure 21-22). Recently, it has been discovered that the 5.8S RNA molecule (156 nucleotides) present in all eucaryotic ribosomes is homologous in both sequence and structure to the 5' end of bacterial 23S rRNA (Figure 21-23). Accordingly, the region specifying 5.8S RNA is located just upstream from 28S rRNA sequences in the rDNA, and 5.8S RNA very likely base-pairs with 28S RNA in the ribosome, using sequences at

Figure 21-22
A tandemly arranged series of nucleolar rRNA genes from *Triturus viridescens*, showing growing 45S rRNA precursor chains. The inset at the lower left shows the steps in the transformation of the mammalian 45S pre-rRNA molecule into mature rRNA molecules. ETS stands for external transcribed spacer and ITS for internal transcribed spacer. (Micrograph courtesy of O. L. Miller and B. Beatty.)

Figure 21-23

Proposed universal secondary structure model for 5.8S rRNA and its correspondence to the 5' end of *E. coli* 23S rRNA. Both secondary structures were based on extensive phylogenetic data. The numbering in the 23S sequence corresponds to residues from the 5' end. The sizes of the rRNAs in the inset are not drawn to scale. [After J. C. Vaughn, S. J. Sperbeck, W. J. Ramsey, and C. B. Lawrence, Nucleic Acid Res. 12 (1984):7479–7502.]

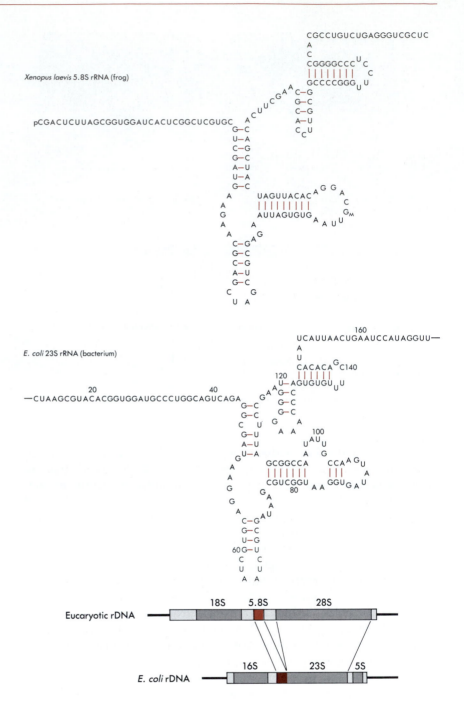

both its 5' and 3' ends. Why eucaryotes clip off this section of the molecule to form a separate RNA chain is unknown.

Whereas mature rRNA molecules show remarkable similarity between lower eucaryotes (yeast) and higher eucaryotes (mammals) and even between procaryotes and eucaryotes (Chapter 14), both the sizes and the sequences of the transcribed spacer segments are highly diverged. The number of nucleotides discarded during processing of the pre-rRNA increases from about 1400 in *E. coli* to 6800 in humans. Lacking in eucaryotic spacer regions are the unusual double-stranded stems that bring together distant sequences of *E. coli* pre-rRNA for recognition and cleavage by the enzyme RNase III (Chapter 14). Instead, even the first cleavages experienced by eucaryotic pre-rRNAs seem to occur in single-stranded regions and often directly generate

(a)

(b)

Figure 21-24
(a) Electron micrograph of a 45S pre-rRNA molecule spread in 80 percent formamide and 4 *M* urea to preserve the stronger hairpin loops. (b) A tracing of (a), which shows the regions destined to become 28S and 18S rRNA. [Reproduced with permission from P. Wellauer and I. Dawid, *Cold Spring Harbor Symp. Quant. Biol.* 38 (1974):325.]

the 5' and 3' termini of the mature rRNA species. The highly conserved, stable RNA secondary structures observed in vertebrate 45S pre-rRNAs are confined to interior portions of the 28S and spacer sequences (Figure 21-24). It has been proposed that a small RNA-protein complex (Chapter 20) containing U3 RNA, which is specifically confined to the nucleolus, participates in rRNA processing; but how it might act has not been firmly established.

Species-Specific Signals for rDNA Transcription by RNA Polymerase I[42–50]

In contrast to RNA polymerases II and III, each of which is responsible for the transcription of many different genes with many different sequences, RNA polymerase I synthesizes only a single product, the pre-rRNA transcript. Thus, it need recognize only one promoter and one terminator signal, but must do so very efficiently to achieve the high rate of transcription typical of active rRNA genes. For these reasons, it was initially assumed that it would be easy to decipher the sequences required for initiation and termination of synthesis by RNA polymerase I simply by comparing rDNA sequences from a number of eucaryotic species.

This strategy, however, quickly failed when completely different sequences were discovered around rRNA start sites in the DNA from different organisms (Figure 21-25). Strong homologies are detected

Figure 21-25
Limited sequence homologies around start sites for RNA polymerase I in rDNA of various eucaryotic organisms. No recognizable homologies appear where the sequence is not given. [After J. Sommerville, *Nature* 310 (1984):189–190.]

	−40	−30	−20	−10	+1	+10
Mouse		ATCTTT	TATTG		TACTGACACGC	
Rat		ATCTTT	TATTG		TACTGACACGC	
Human	ATCTTT		TTTTGG		TGCTGACACGC	
Xenopus laevis		CTACGCTTTT	GTGCGGGCAGGAAGGTAGGG			
Xenopus clivii		CTACGCGTTT	GTGCCGACAGGAAGGTAGGG			
Xenopus borealis		CTACGCTTTT	GTGCGGACAGGAAGGTAGGG			
Drosophila melanogaster		TTT	G		TAGGT	
Tetrahymena pyriformis			G		T	
Dictyostelium discoideum			T	ATACATATA		
Physarum polycephalum			A	ATACATATA		
Saccharomyces carlsbergensis			A	AGGTACTTCATGCGAAAGC		
Saccharomyces rosei			G	AGGAACTTCATGCGAAAGC		

only for very closely related species (e.g., two mammals or two kinds of toads). Yet, an essential role for sequences located in the immediate vicinity of the point where rRNA synthesis initiates (between about −45 and +10) has been established by examining the effects of deletions coming closer and closer to the start site. Single-base mutations have further pinpointed important residues within this region. For instance, in mammalian rDNA, a TTT sequence around position −30, a conserved G at position −16, and the first few transcribed nucleotides may all act together to achieve accurate and efficient initiation by RNA polymerase I. These sequences surrounding the rRNA start site are sufficient for the initiation of transcription by RNA polymerase I and can therefore be said to constitute a promoter for RNA polymerase I.

The rate of rRNA transcription is also dramatically influenced by more distant sequences. This aspect of rRNA gene expression has been most thoroughly analyzed in several species of the toad *Xenopus*, but similar situations probably exist in other eucaryotes, including *Drosophila* and even mammals. In *Xenopus laevis*, the nontranscribed spacers separating rRNA genes are of variable length and are composed mainly of several types of repeated DNA sequences. One type of repeated sequence (about 200 base pairs long) is a 90 percent perfect duplication of the rRNA promoter itself (here defined to lie between positions −142 and +6). This repeat appears as many as eight times within a single nontranscribed spacer. Between the reduplicated promoters and preceding the true rRNA promoter are even more copies of either a 60- or 81-base-pair repeated unit, both of which contain a 42-base-pair sequence that appears between positions −73 and −114 of the rRNA promoter (Figure 21-26). Not only have the 60/81-base-pair repeats been shown to enhance initiation at a downstream rRNA promoter, but the promoter duplications can (at least under certain conditions) initiate the synthesis of short transcripts. That the 60/81-base-pair repeats may bind some limiting factor essential for rRNA expression is suggested by the observation that rRNA promoters preceded by more of these elements outcompete templates with fewer repeats in in vitro transcription experiments. Such competition may in fact be the molecular basis for the phenomenon of **nucleolar dominance** observed when *Xenopus laevis* is mated with *Xenopus borealis*. The hybrid embryos express only *laevis* rRNA genes, whose spacer regions contain many more of the repeating 60/81-base-pair elements than *borealis* rDNA spacers. Whether the putative binding protein participates directly in transcription initiation or (more likely) prevents the binding of interfering molecules (e.g., histones) to DNA upstream from the promoter is not yet known.

On the other hand, differences in actual transcription initiation factors (and in the DNA sequences they recognize) are probably the basis of the species specificity observed in rRNA synthesis when components from two more distantly related species are mixed. Mouse cell extracts transcribe only mouse rDNA, and human cell extracts transcribe only human rDNA. But when the extracts are fractionated, it is not the RNA polymerase I itself that is species-specific, but rather some additional required factor(s). At least one required factor (perhaps in conjunction with RNA polymerase I) binds to the DNA in the vicinity of the start site, forming a complex so stable that a subsequently added rDNA promoter fragment cannot be utilized for transcription. It remains to be worked out exactly how many factors there are, which are species-specific, and what rDNA sequences they recognize. Even without this knowledge, it seems safe to ascribe to such transcription initiation factors a second type of nucleolar dominance that is observed when cells from different species (such as humans and mouse) are fused in the test tube. The resulting **somatic cell hybrids** (which have been widely used for gene-mapping experiments) specifically lose chromosomes of one parental species; in hybrids between human and rodent cells, human chromosomes are gradually lost. Although both types of rRNA genes are often retained, most hybrid cells express only the rRNA of the parent whose chromosomes are not lost. Loss of the chromosome(s) containing the gene(s) coding for the species-specific rRNA transcription factor(s) may be responsible.

How has the species specificity of RNA polymerase I transcription factors evolved? We have seen in Chapter 20 that tandemly repeated multigene families are generally very homogeneous because eucaryotic organisms possess mechanisms for homogenizing closely related DNA sequences. Thus, a random mutation in any individual rDNA repeat unit will have a finite chance of spreading to all the other repeat units. If such a mutation occurs in a region of the rDNA that binds an essential transcription factor, only those organisms that have sustained a compensatory mutation in the factor survive. In this fashion, mutations in the template select for mutations in the binding factors (and vice versa), so that the DNA and protein factors *coevolve*. Increasing (but coupled) divergence of both promoter sequences and interacting transcription factors ultimately leads to species specificity.

There is not enough information about rRNA termination signals to ascertain whether these, too, are species specific or can be generally recognized by RNA polymerase I. The terminators in vertebrate rDNA reside several hundred base pairs beyond the end of 28S sequences and are only now being characterized.

A most tantalizing question that has not yet begun to be tackled is how the production of all the components of eucaryotic ribosomes is coordinated. Ribosome synthesis responds to many conditions, such as the growth state of the cell, nutrient starvation, and viral infection. Surely there must be some ingenious mechanism to couple rDNA transcription by RNA polymerase I not only to 5S RNA production by RNA polymerase III but also to the synthesis of 70 to 80 ribosomal protein mRNAs by RNA polymerase II.

Promoter Sequences for RNA Polymerase II Reside Upstream from the Start Site[51-57]

In sharp contrast to RNA polymerase I, which synthesizes only a single type of RNA (pre-rRNA), RNA polymerase II is responsible for

Figure 21-26
A typical nontranscribed spacer region from *Xenopus laevis* rDNA, showing that the majority of its length is composed of sequences related to the true promoter. The small black boxes are imperfect copies of a 42-base-pair sequence that appears between positions −72 and −114 relative to the transcription start site. The white boxes (which include a black box) are imperfect reduplications of the entire promoter region consisting of about 200 base pairs. At the end of the spacer that abuts the site of transcription termination are other unrelated repetitive sequences. [After R. H. Reeder, *Cell* 38 (1984):349–351.]

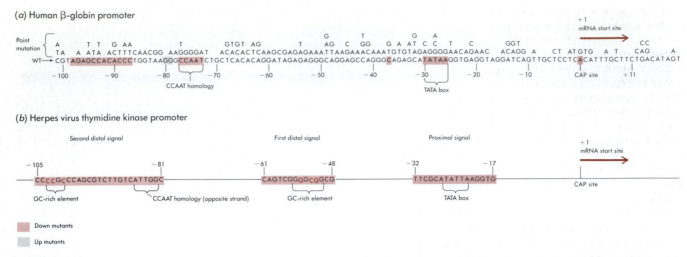

Figure 21-27

Essential sequences in two of the best-characterized mammalian RNA polymerase II promoters. (a) How saturation mutagenesis of the human β-globin promoter revealed that sequences homologous to those in other promoters are required for transcription. Point mutations were introduced into cloned DNA using nitrous acid, formic acid, or hydrazine, and the level of RNA synthesis measured after introduction of the mutant promoters into HeLa cells. Positions where mutations caused lower promoter activity are shown in color; positions where mutations enhanced RNA synthesis are shown in gray. All other substitutions gave wild-type transcription levels. Note that the RNA start site, the TATA box, and the CCAAT homology are all promoter regions sensitive to mutation. [Data courtesy of R. M. Myers, K. Tilly, and T. Maniatis; see also R. M. Myers, L. S. Lerman, and T. Maniatis, *Science* 229 (1985):242–247.] (b) How clustered and single-base substitution mutagenesis revealed the location of essential sequences in the herpes virus thymidine kinase promoter. Clustered base substitutions were made by the linker scanning technique [see S. L. Mc-Knight and R. Kingsbury, *Science* 217 (1982):316–324], and single-base substitutions (bold letters) by oligonucleotide mutagenesis [see S. L. McKnight and R. C. Kingsbury, *Cell* 37 (1984): 253–262]. Mutant DNAs were assayed for transcriptional level by injection into *Xenopus* oocytes (see Figure 21-36a). Mutations within the three regions whose sequences are shown resulted in lower promoter activity; mutation of regions indicated by straight lines caused no reduction in transcription efficiency. In this promoter, regions sensitive to mutation include two GC-rich regions, the TATA box and the CCAAT homology. Binding of the GC-rich sequences by transcription factor SP1 (see Figure 21-29) and the CCAAT homology by some other cellular protein has recently been demonstrated in vitro [See K. R. Jones, K. R. Yamanoto, and R. Tjian, *Cell* 42 (1985):559–572.]

the regulated synthesis of a multitude of different transcripts in all eucaryotic cells. RNA polymerase II acts on genes specifying certain small nuclear RNAs (U1, U2, U3, etc.) as well as on every known protein-coding gene, whether the gene is always expressed (a constitutive or **housekeeping gene**) or is expressed only at a particular stage of development in one tissue of a multicellular organism (a regulated gene). Perhaps not surprising, therefore, are two basic features of RNA polymerase II promoters. First, essential sequences lie almost entirely outside of (5' to) the region copied into RNA. Hence, the structure (and function) of RNA polymerase II transcripts can vary almost infinitely. Second, certain sequences within promoter regions are strikingly conserved across species boundaries. Since many thousands of different genes are transcribed by RNA polymerase II, many mutations would have to occur coordinately for species specificity to evolve; the probability of such complex coevolution is very low.

The sequences of hundreds of protein-coding genes from many organisms reveal certain homologies lying within several hundred nucleotides 5' to the initiation site for RNA polymerase II. A frequently occurring homology, the so-called **TATA box**, is centered around position −25 (Figure 21-27). It appears to be essential for accu-

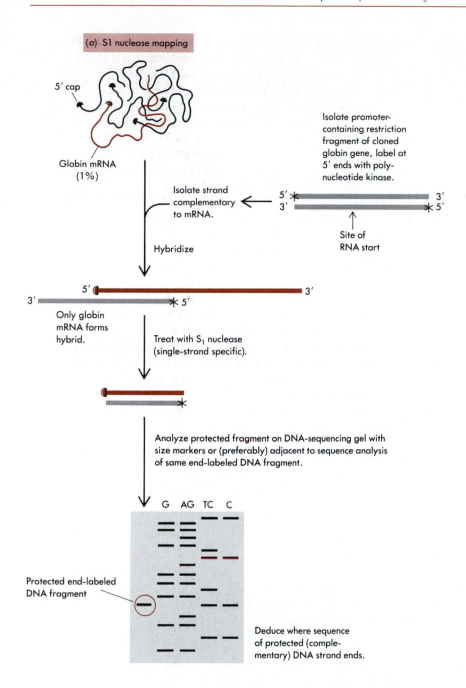

(a) S1 nuclease mapping

5' cap

Globin mRNA
(1%)

Isolate promoter-containing restriction fragment of cloned globin gene, label at 5' ends with poly-nucleotide kinase.

Isolate strand complementary to mRNA.

Site of RNA start

Hybridize

Only globin mRNA forms hybrid.

Treat with S₁ nuclease (single-strand specific).

Analyze protected fragment on DNA-sequencing gel with size markers or (preferably) adjacent to sequence analysis of same end-labeled DNA fragment.

G   AG   TC   C

Protected end-labeled DNA fragment

Deduce where sequence of protected (complementary) DNA strand ends.

Figure 21-28, continued next page
Two indirect methods for locating the 5' end of a particular transcript in a mixture of RNA molecules (a) **S₁ nuclease mapping,** (b) **primer extension.** Note that S1 nuclease mapping can also be used to locate the 3' end of an RNA by labeling the DNA probe at its 3' (instead of 5') end. This method is also widely used to locate the position of splice sites in an mRNA. (See Figure 9-47 for DNA sequencing by the Maxam-Gilbert method.)

rately positioning the start of transcription (Figure 21-28) and is in this sense analogous to the −10 region of bacterial promoters (Chapter 13) but different from the TATA sequences in yeast promoters (Chapter 18). A purine is usually (but not always) the first nucleotide incorporated into an RNA chain by RNA polymerase II (again similar to bacterial RNA polymerases; Chapter 13). The 5' end of the transcript then invariably becomes modified by the rapid addition of a cap moiety (Chapter 18).

Other sequence homologies found for RNA polymerase II promoters reside in more variable positions farther upstream from the start site. This is quite different from the stringent spacing requirement between the −10 and −35 regions displayed by bacterial promoters (Chapter 13). For instance, a CCAAT homology occurs in both promoters of Figure 21-27, but it resides on different strands of the DNA.

Figure 21-28
(*Continued*)

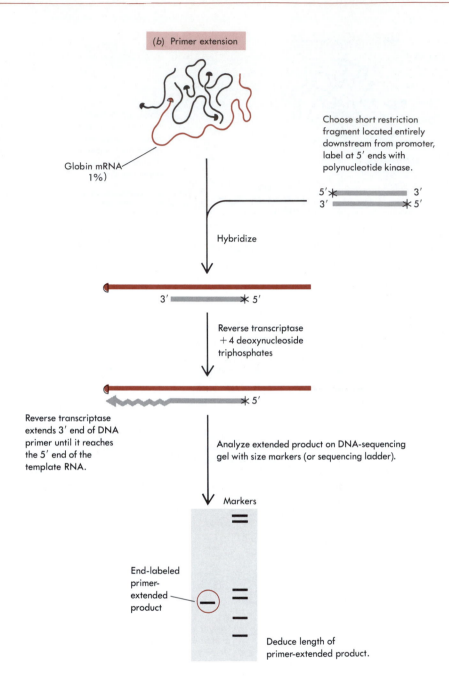

(*b*) Primer extension

Globin mRNA
1%)

Choose short restriction fragment located entirely downstream from promoter, label at 5' ends with polynucleotide kinase.

5' ✱ ▬▬▬▬▬ 3'
3' ▬▬▬▬▬ ✱ 5'

Hybridize

3' ▬▬▬▬▬ ✱ 5'

Reverse transcriptase + 4 deoxynucleoside triphosphates

Reverse transcriptase extends 3' end of DNA primer until it reaches the 5' end of the template RNA.

✱ 5'

Analyze extended product on DNA-sequencing gel with size markers (or sequencing ladder).

Markers

End-labeled primer-extended product

Deduce length of primer-extended product.

Also included among the upstream promoter elements are specialized signals that seem to characterize certain subsets of genes. Evidence is accumulating that all "housekeeping" genes are preceded by multiple copies of the GC-rich element shown in Figure 21-27b.

That these conserved sequences are essential for binding and initiation by RNA polymerase II when they do appear has been demonstrated by examination of the effects of deleting them or altering them by single or clustered point mutations (see Figure 21-27). An emerging rule is that at least two upstream sequence domains, in addition to the −25 (TATA) region, are required for efficient transcription of every gene. Sometimes, the position and orientation of these elements relative to one another (and to the start site) are critically important; sometimes not. Only now are we beginning to comprehend how these disparate promoter sequences and the proteins they bind contribute to the initiation of transcription by RNA polymerase II. That we still have much to learn is amply demonstrated by the all too frequent identification of yet another gene that seems to lack all conserved sequences previously identified.

Sorting Out Factors for Initiation by RNA Polymerase II[58–64]

The existence of multiple essential sequence elements whose positions are not stringently fixed within RNA polymerase II promoters suggests that the binding of more than one component is necessary for initiating transcription. Only through in vitro fractionation studies can we hope to arrive at a complete picture of the protein-DNA and protein-protein interactions that allow expression of the vast majority of eucaryotic genes. For example, we would like to know which (if any) of the conserved sequences shown in Figure 21-27a and b are recognized directly by RNA polymerase II (which itself contains multiple polypeptide chains; Chapter 18) and which bind auxiliary factors. How many factors are there? Which are generalized transcription factors required by all RNA polymerase II promoters, and which are specific for certain subsets of genes? Is there a defined order of addition of components required for the assembly of an active transcription complex? In addition to positively acting factors, are there negative regulatory proteins? Are cofactors, as well as nucleotide substrates, required for transcription?

The earliest fractionation studies showed that RNA polymerase II cannot by itself recognize a promoter and selectively initiate transcription. At least several separable factor fractions are also required. One fraction that has been partially purified from both mammalian and *Drosophila* cell extracts contains DNA binding protein(s) that interact with sequences from about −40 to +30 relative to the RNA start site. Whether this factor is needed by all RNA polymerase II promoters or is one that recognizes the TATA box (included in this region) is not yet clear. Another highly purified mammalian factor called SP1 is required for the transcription of genes whose promoters include multiple copies of the GC-rich regions shown in Figure 21-27b and binds tightly to these sequences (Figure 21-29). Thus, it is tempting to speculate that the observed undermethylation of C residues within such clustered GC-rich sequences in mammalian DNA might be due to this factor's binding to the promoters of constitutively active "housekeeping" genes. Finally, it has recently become clear that some factor binds the CCAAT sequence that resides between the CG-rich elements in the thymidine kinase promoter (see Figure 21-27b); isolation and characterization of this DNA binding component are under way.

In vitro experiments using fractionated systems have also shown that stable complexes including RNA polymerase II and factors form on the DNA prior to initiation of the RNA chain. This means that after a certain combination of factors have bound, they cannot be easily displaced by the addition of a second competing promoter. Interestingly, the hydrolysis of ATP to ADP is required for some very early step in transcription; this reaction is distinct from the incorporation of ATP into the nascent RNA chain. Though we are still far from a

0 ————————> 0

Increasing amounts of SP1

(b)

(c)

Figure 21-29

Specific binding of SP1 transcription factor to the early promoter of SV40. (a) The nucleotide sequence of the SP1-bound region, showing G residues either protected from (color) or experiencing enhanced (+) methylation by dimethyl sulfate in the presence of the factor. [After D. Gidoni, W. S. Dynan, and R. Tjian, *Nature* 312 (1984):409–413.] (b) DNase footprint (see Figure 16-4 for details of the method) illustrating protection of the six GC-rich regions that make up the three 21(22)-base-pair repeat sequences. (Courtesy of R. Tjian.) (c) How the contacts made by SP1 cluster on one strand of the DNA and are arranged similarly in the major groove of both the SV40 promoter (above) and an SP1-bound promoter from monkey DNA (below). (Courtesy of R. Tjian.)

complete description of the initiation of transcription by RNA polymerase II and its requirement for various factors, we can be confident that a clearer picture will soon emerge.

Transcriptional Enhancers Are Tissue-Specific and Can Act at a Distance[65–70]

In addition to the sequences that constitute a promoter itself, there are outside elements that drastically alter the efficiency of transcription by RNA polymerase II. The most intriguing of these are the activating regions known as **enhancers**. Enhancer sequences were first discovered in DNA tumor viruses such as SV40 and polyoma, where they are located near the origin of DNA replication, which is also the region that initiates RNA synthesis both early and late after infection (Figure 21–30). Deletion of the enhancer region (which is composed primarily of a tandem repeat of 72 base pairs) from SV40 DNA reduces early gene expression by a factor of over 100. Subsequently, it was found that the SV40 enhancer region augments transcription not only from its own promoters but also from almost any other promoter to which it is linked. Included are a variety of mammalian and avian genes as well as genes from other viruses. Thus, enhancer sequences have been exploited as experimental tools; by linking on an enhancer, genes with intrinsically weak promoters can be readily investigated.

Several remarkable features distinguish enhancer elements from the sequences essential for initiation by RNA polymerase II:

- They appear to be relatively large elements, often including several hundred base pairs and sometimes containing repeated sequences that are independently functional (see Figure 21-30).
- They can act over considerable distances, up to several thousand base pairs.
- They function in either orientation.
- They are position-independent, and they need not reside 5' to a gene, but can be located within or even downstream from the transcribed region. Yet, the enhancer element and the gene must always be present on the same DNA molecule. If several promoters are nearby, the enhancer acts preferentially on the closest.
- A particular enhancer element functions preferentially or exclusively in certain cell types. This tissue specificity (which also manifests itself as species specificity) may in part explain the host range of some animal viruses (Chapter 24). Likewise, tissue-specific enhancers could provide the basis for differential expression of genes during development or in the various cell types of a mature organism.

One stunning example of enhancer action is the turning on of immunoglobulin gene expression after DNA rearrangement brings a promoter into the proximity of a B-cell-specific enhancer element (Chapter 23). Another is the action of steroid hormones, which is discussed later.

Accounting for the properties of enhancers must be the presence of specific cellular proteins whose interaction with an enhancer element somehow renders the DNA in the vicinity of a nearby promoter region more accessible for the binding of RNA polymerase II or its associated factors. Indeed, the presence of an enhancer element is usually correlated with the location of a DNase I hypersensitive site (e.g., the origin region of SV40 DNA; see Figure 21-19). In one attractive model, the specific interaction of enhancer sequences with protein(s) creates a three-dimensional assemblage that acts as a high-affinity target site for polymerase, somewhat like a nest. Alternatively, a gyraselike enzyme might be bound and introduce supercoils, making a localized region of DNA easier for the polymerase to melt. Either model would

Figure 21-30
Schematic diagram of the elements essential for SV40 early gene expression, including the well-studied SV40 enhancer. The region of SV40 DNA shown corresponds to the nucleosome-free gap in SV40 chromatin (Figure 21-19) and includes the origin of DNA replication, the TATA box, three 21(22)-base-pair repeats, and the 72-base-pair repeat (which represents most of the enhancer region). The start points for early and late transcripts are shown. The GC-rich sequences duplicated within each 21-base-pair repeat are conserved essential promoter elements analogous to those present in the herpes virus thymidine kinase gene promoter (Figure 21-27). The sequence indicated at the 5' end of each 72-base-pair repeat is essential for enhancer function and is homologous to sequences present in other viral enhancer elements (e.g., polyoma, murine sarcoma virus). Also included are the binding sites for T antigen (Chapter 24), one of the rare transcriptional repressors known in a higher eucaryotic system. [After W. S. Dynan and R. Tjian, *Nature* 316 (1985):774–777; and W. Schaffner, E. Serfling, and M. Jasin, *Trends in Genetics* 1 (1985):224–230.]

Figure 21-31

A control region in the center of the 5S gene directs specific initiation of transcription. The *Xenopus* somatic 5S rRNA gene is diagrammed with the extent of deletions impinging on the gene from its 5' or 3' end indicated by horizontal bars. Those deletion mutations that support initiation of transcription in vitro are designated +. Note that when upstream (5') deletions extend into the gene (but not into the internal control region), RNA synthesis is initiated on vector sequences. When the T-rich sequence at the 3' end of the gene (see inset) is deleted, transcription proceeds beyond the point where termination normally occurs (around position 120; 5S RNAs ending with from one to four U residues have been detected). [After D. F. Bogenhagen, S. Sakonju, and D. D. Brown, *Cell* 19 (1980):27–35.]

Sequence at 3' end of gene

G T C G T A G G C T T T T G C A C T T T T

110   120   130

explain how enhancers can work at a distance, in either orientation, and even downstream of a promoter; protein assemblies could bring together components bound to disparate sites on the DNA, and the strain of supercoiling could be transmitted in either direction from the point of gyrase action along the DNA. Experimental tests of these models have so far been hampered by the fact that enhancer elements exert only marginal effects on in vitro transcription. Very recently, however, proteins that bind selectively to portions of enhancer regions have been detected. Thus, research aimed at deciphering how these important controlling elements activate genes should now proceed rapidly.

Internal Control Regions Direct Initiation by RNA Polymerase III[57, 71–79]

Like RNA polymerase II, RNA polymerase III acts on a number of different cellular genes. It synthesizes ribosomal 5S RNA, all tRNAs, and certain other small RNAs (like the 7SL RNA present in a small RNA-protein complex called SRP, described later). In addition, some genes carried by viral DNAs are transcribed by RNA polymerase III: Adenovirus specifies the two VA RNAs, and Epstein-Barr virus specifies two EBER RNAs (Chapter 24). Yet, RNA polymerase III products are not nearly so diverse as RNA polymerase II transcripts:

- They are almost always short, usually containing fewer than 300 nucleotides.

- In no known case do they encode proteins. Rather, they fulfill structural roles, in most cases as components of the cellular protein-synthesizing machinery.

- They are usually (but not always) transcribed from multiple copies of identical genes (Chapter 20). Thus, they are quite abundant in cells (10^4 to 10^7 copies).

Gene | Location of internal control regions and sequence homologies

Figure 21-32
Summary of the internal control regions required for genes transcribed by RNA polymerase III. Numbering is from the 5' end of the mature RNAs. In the consensus sequences for "A" block and the "B" block, the nucleotides in color are invariant in all non-5S genes, whereas those in gray are invariant in all tRNA genes. Regions defined as essential for transcription by deletion analyses are designated by rectangles. The broken line over the 5S gene shows the region protected from DNase by factor A (TFIIIA) binding. [After A. B. Lassar, P. L. Martin, and R. G. Roeder, *Science* 222 (1983):740–748.]

- They exhibit a common structural feature: It is nearly always possible to propose an RNA secondary structure that brings together the 5' and 3' ends of the molecule to form a stable base-paired stem [e.g., tRNAs (Figure 14-2) and 5S rRNA (Figure 14-15)]. Yet, some of these RNAs (like tRNAs) have undergone extensive processing, including trimming at both their 5' and 3' termini, splicing to remove intervening sequences, and nucleotide modification. Other RNAs function as primary transcripts (like VA and EBER RNAs) and retain the 5' triphosphate and 3' U-rich termini, hallmarks of initiation and termination by RNA polymerase III.

Since bacterial promoters, as well as promoters for eucaryotic polymerases I and II, all reside upstream from the start site, it was quite unexpected to learn that the synthesis of RNA polymerase III transcripts requires sequences located inside of the genes themselves. This was first discovered upon analysis of deletions that come closer and closer to the 5' end of the *Xenopus* 5S rRNA gene (Figure 21-31). No effect on the level of transcription was observed until the deletions extended more than 50 nucleotides into the 5S coding region itself! When the deletion boundary lay between positions +1 and +50 of the 5S gene, RNAs were simply initiated on the newly introduced DNA to yield a transcript of approximately normal length. Likewise, deletions impinging on the 5S gene from its 3' side lowered the transcription level only upon removal of more than 40 nucleotides of the 5S sequence itself. In this way, an **internal control region** lying between positions +50 and +84 of the 5S gene was defined as essential for initiating synthesis of the 5S RNA molecule.

Other genes transcribed by RNA polymerase III also possess internal control elements. Two stretches of conserved sequences (called the A and B blocks) have been demonstrated to be important by the introduction of deletions and point mutations into them. Their sequences and positions within various genes are shown in Figure 21-32. The distance between A and B blocks can vary considerably (for instance, when an intervening sequence is present in the anticodon loop region of a tRNA gene). It is most intriguing that the A and B

Figure 21-33
Scheme for interaction of the 5S rRNA gene with transcription factors based on in vitro analyses. [After J. J. Bieker, P. L. Martin, and R. G. Roeder, *Cell* 40 (1985):119–127.]

blocks correspond to sections of tRNA molecules that are highly conserved not only in eucaryotes but also in procaryotes, where tRNAs are transcribed from upstream promoters. Whichever arrangement evolved first, this situation in eucaryotic tRNA genes underscores the versatility of nucleic acid sequences. The very same region can function structurally (in the RNA product) and biosynthetically (in the DNA template).

The signals that direct termination by RNA polymerase III also reside within the transcribed region. A group of three or more T residues on the nontemplate strand, often preceded (or surrounded) by GC-rich sequences, specifies the 3' end of the RNA chain (see Figure 21-31). As for bacterial RNA polymerases, termination at such a sequence is believed to be facilitated by the intrinsic instability of a DNA-RNA hybrid made up of polyribo U and polydeoxy A (Chapter 13).

In addition to internal control elements, specific upstream and downstream flanking sequences are essential for the transcription of some genes by RNA polymerase III. Both these external regions and the better-studied internal signals appear to serve as recognition sites for proteins. Which of these DNA binding proteins are factors that cooperate directly with RNA polymerase III to achieve accurate transcription and which might act simply to render the gene more accessible (for instance, by preventing nucleosome formation) are now under intense scrutiny in in vitro systems.

Oocyte 5S rRNA Synthesis Is Autoregulated by a Transcription Factor That Binds to the Gene Product[78, 79]

A most important realization to emerge from in vitro studies with RNA polymerase III is that stable complexes formed between factors

Figure 21-34

Transcription factor TFIIIA binds both to the 5S gene (above) and to its product, 5S rRNA (below). On the DNA sequence, regions protected from DNase action by factor binding are boxed in color; guanine residues that interfere, when methylated, with binding of the transcription factor are indicated by underlined letters; and phosphate residues that make contact with the factor are marked by solid triangles. Thus, TFIIIA interacts specifically with the internal control region of the oocyte-specific 5S gene [see D. R. Engelke, S.-Y. Ng, B. S. Shastry, and R. G. Roeder, *Cell* 19 (1980):717–728; S. Sakonju and D. D. Brown, *Cell* 31 (1982):395–405] making closest contacts with the sense (RNA-like) strand. On the 5S RNA structure, the regions protected from the nuclease α-sarcin [see I. Wool, *Trends in Biochem. Sci.* 9 (1984):14–17] by TFIIIA binding are shown. Dark shading indicates the striking repetition of the sequence CCUGG in double-stranded regions of both the RNA and DNA binding sites; in fact, another such pentamer appears at positions 104 to 108 of the 5S gene. Note that although the sequence of the somatic 5S gene (on which the DNA studies were done) differs in a few positions from that of the oocyte 5S gene (the oocyte 5S RNA-protein complex was studied), the contacts made by the factor with the oocyte 5S gene are very likely to be the same as those shown. [After S. Sakonju and D. D. Brown, *Cell* 31 (1982):395–405, and courtesy of I. Wool.]

and the DNA allow many rounds of RNA synthesis to occur before dissociation. Such stable transcription complexes have been best characterized for the oocyte-specific 5S rRNA genes of the African clawed toad *Xenopus* (Figure 21-33). Recall that the *Xenopus* genome has two different sets of genes encoding 5S ribosomal RNA (Chapter 20). The oocyte-specific 5S RNA genes are expressed exclusively during the long period of oogenesis, when the maturing oocyte stores various macromolecular components for use after fertilization. The somatic 5S genes are expressed primarily after fertilization and throughout subsequent development of the frog. (Because of its aquatic existence, molecular biologists have fallen into the habit of calling this toad a frog.) How is the expression of oocyte-specific 5S RNA turned on during oogenesis and turned off in the mature egg?

Transcription of oocyte-specific 5S genes requires three factors in addition to RNA polymerase III. Two of these (called TFIIIB and TFIIIC) are general factors that also function with other genes such as tRNA or VA genes (see Figure 21-32). In contrast, the third transcription factor (called TFIIIA) is an oocyte-specific protein (40 kdal) that binds tightly to the internal control region, the middle third of the 5S gene (see Figures 21-31 and 21-32). Once the immature oocyte has synthesized a sufficient quantity of TFIIIA, its binding to oocyte-specific 5S genes activates them for transcription by RNA polymerase III (see Figure 21-33). But how are these same genes turned off in the mature oocyte?

For many years, it was known that oocyte-specific 5S RNA was stored in the cytoplasm of the egg as a ribonucleoprotein (RNP) particle consisting of 5S RNA complexed with a specific protein molecule. Remarkably, the 5S storage protein and TFIIIA turn out to be one and the same. This suggests an extremely elegant mechanism for turning the oocyte-specific genes off as soon as a sufficient quantity of 5S RNA has been synthesized (Figure 21-34). Early in oogenesis, when there is an excess of TFIIIA over 5S RNA, most of the factor will bind to 5S genes and stimulate transcription. Later in oogenesis, as 5S RNA accumulates, TFIIIA will be sequestered in the form of a 5S storage particle, and 5S RNA synthesis will be proportionately reduced. Eventually, almost all of the TFIIIA will be present in storage particles and 5S RNA synthesis will cease. Ultimately, after fertilization, the 5S RNA is used to build ribosomes, and the TFIIIA is somehow programmed for destruction.

Thus, we see that the transcription of oocyte-specific 5S rRNA genes is a direct consequence of the amount of TFIIIA available for DNA binding. But once a controlling factor such as TFIIIA has been identified, we are then obliged to ask how *its* synthesis is regulated. In other words, finding the "cause" for one "effect" leads us to wonder about the "cause" of the "cause," and this can only be found if we delve deeper into the mechanisms of eucaryotic gene expression.

Transfer RNA Populations Are Optimized for the Synthesis of Particular Protein Products in Specialized Tissues[80, 81, 82]

That transcription by RNA polymerase III can be controlled in a tissue-specific manner is also evident from examining tRNA populations in different cells of the same organism. In Chapter 15, we saw that the relative abundance of various tRNA species in microorganisms (bacteria and yeast) correlates very well with codon usage in highly expressed genes. Thus, it appears that the tRNA and mRNA promoters in these genomes have coevolved to yield levels of products that maximize the rates of synthesis of abundant proteins. But in a multicellular organism, different tissues are programmed to synthesize different abundant proteins, and a tRNA population tailored to the needs of one cell type might be quite unsuitable for another. Hence, a mechanism for differentially adjusting tRNA populations has evolved in higher eucaryotes. It exists in every tissue examined that is specialized for the production of protein(s) with unusual amino acid compositions.

The most dramatic example of a skewed tRNA population is found in the silk gland of *Bombyx mori* (the silkworm). This tissue is designed for the massive synthesis of two classes of silk polypeptides (called fibroin and sericin), both of which are extremely rich in certain amino acids (Gly, Ala, and Ser in fibroin; Ser, Gly, Asp, and Gly in sericin). At the time that fibroin and sericin mRNAs are being produced, the silk gland radically alters its relative transcription of tRNA genes. Over 30-fold increases in the amounts of tRNA^{Gly} and tRNA^{Ala} result, including the appearance of a new silk-gland-specific species of tRNA^{Ala}. Although the mechanism by which these tRNA genes are turned on has not been worked out to the same extent as the control of oocyte-specific 5S RNA synthesis, some sort of tissue-specific transcription factor must be involved. Again we must ask, What turns on the synthesis of this factor? Could it be that a single silk-gland-specific factor activates both the fibroin and sericin genes for RNA polymerase II transcription and the relevant tRNA genes for RNA polymerase III transcription?

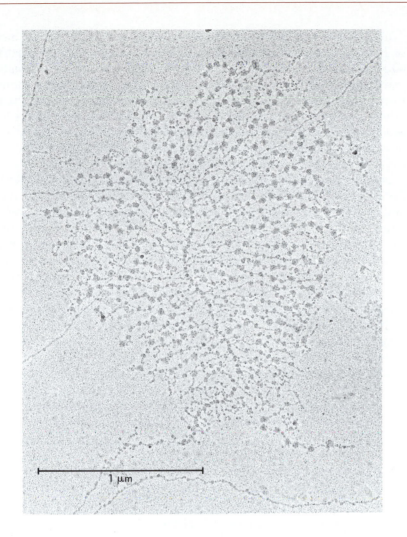

1 μm

Figure 21-35
The packaging of nascent RNA transcripts into hnRNP produces a beads-on-a-string appearance. An active gene with emanating transcripts from a *Drosophila melanogaster* follicle cell is shown. (Courtesy of Y. Osheim and A. Beyer.)

Messenger RNA Precursors Are Packaged by Specific Nuclear Proteins[83, 84, 85]

Even while pre-mRNAs are nascent and remain attached to their DNA templates via RNA polymerase II molecules, they quickly become (noncovalently) associated with certain RNA binding proteins. This ubiquitous protein-bound state of nuclear pre-mRNA is referred to as **heterogeneous nuclear ribonucleoprotein (hnRNP),** derived from hnRNA (*h*eterogeneous *n*uclear *R*NA). It is in this form that pre-mRNAs undergo splicing and the other processing reactions that convert them into mature mRNA molecules. When viewed in the electron microscope, hnRNP has the appearance of beads on a string, much like the nucleosome arrays observed in extended chromatin (Figure 21-35). But the hnRNP particles are larger than nucleosomes (about 200 Å across) and more heterogeneous in size. They are connected by RNase-sensitive links. Mild ribonuclease digestion converts hnRNP into individual particles that sediment as a surprisingly homogeneous peak at 30S to 40S, much like small ribosomal subunits. Yet, in contrast to ribosomes, hnRNP particles are very protein rich, containing about three-fourths protein and only one-fourth RNA (corresponding to an approximately 600-nucleotide-long fragment of pre-mRNA per 30S particle).

In higher eucaryotes, the major proteins present in hnRNP are a set of highly conserved polypeptides with molecular weights of 30 to 40 kdal. They contain unusual modified amino acids (e.g., dimethylarginine) and are very closely related to each other, since monoclonal antibodies directed against one protein often cross-react with others. Not surprisingly, the hnRNP polypeptides are among the most abundant nonhistone proteins of the cell nucleus. Somehow they know to associate only with pre-mRNA and not with rRNA or RNA polymerase III transcripts.

How and why are pre-mRNA molecules packaged into hnRNP? Because hnRNP complexes are relatively unstable during isolation and analysis (compared to nucleosomes, for example), much less is known about their structural organization. What seems certain is that the bound RNA fragment wraps several times about the exterior of an essentially proteinacious particle. In the few cases where a specific pre-mRNA has been examined, both intron and exon segments were found in hnRNP. There are recent indications that assembly into hnRNP is essential for splicing and the other processing events that fashion mature mRNA molecules. Alternatively, hnRNP proteins may act primarily to prevent tangling of the RNA; this could make the pre-mRNA more accessible to the processing machinery, or could enable the cell to more easily degrade introns once they are excised. Whatever the case, the job of hnRNP proteins must be strictly nuclear, for they are replaced upon mRNA export from the nucleus. In the cytoplasm, a completely different set of RNA binding proteins associates with mRNA molecules to form RNA-protein complexes called **mRNP** (*m*essenger *ribon*ucleo*p*rotein).

Cleavage and Polyadenylation, Rather Than Termination, Generate the 3′ Ends of Most mRNAs[52, 86–89]

The tracts of A residues found at the 3′ ends of the majority of eucaryotic mRNAs (and their nuclear precursors) are not encoded in the DNA. Rather, they are added by a template-independent enzyme called poly A polymerase after the gene has been copied into pre-mRNA by RNA polymerase II. Some 10 to 30 nucleotides upstream from the resulting poly A tail, within the genome-specified part of the RNA, the sequence AAUAAA (or a slight variant of it) always appears. However, this hexanucleotide cannot be the complete signal for polyadenylation, since mRNAs and pre-mRNAs contain many other AAUAAA sequences that are not followed by poly A tails.

Originally, it was assumed that RNA polymerase II responds to a transcription termination signal somewhere in the vicinity of the correct polyadenylation site, releasing the pre-mRNA transcript with a 3′ end that was then acted on by poly A polymerase. That this is not the case has recently become apparent. Instead, RNA polymerase II continues elongating the RNA chain beyond the polyadenylation site (sometimes for thousands of nucleotides); exactly what DNA sequences do signal transcription termination are not yet known. The nascent pre-mRNA is then cut by a specific endonuclease, which creates the correct 3′ terminus for polyadenylation and releases downstream sequences for rapid degradation. In this process, the 5′-AAUAAA hexanucleotide upstream from the cut site is recognized as part of the signal essential for cleavage; other relatively conserved sequences downstream from (3' to) the cut site are probably also involved. What recognizes these signals is only now being investigated with the aid of in vitro systems that accurately cleave and polyadenylate pre-mRNAs. Perhaps an essential role is played by one or more of the small nuclear RNA-protein complexes of the class to which U1 snRNPs belong (Chapter 20).

The 3' Ends of Histone mRNAs Are Produced by RNA Processing Involving a Small RNA-Protein Complex[87]

Unlike other eucaryotic mRNAs, histone gene transcripts in most organisms do not acquire poly A tails, nor do they have near their 3' termini the hexanucleotide AAUAAA that signals polyadenylation. Instead, histone mRNAs from a wide variety of species terminate with a highly conserved palindromic sequence that presumably forms a hairpin loop near the 3' end of RNA (Figure 21-36). In addition, an ACCA sequence often occurs just beyond the hairpin loop preceding the 3' end. Hints that these conserved features might be signals for RNA processing rather than for transcription termination by RNA polymerase II were first obtained by progressively deleting the 3' flanking sequence of the H2A histone gene. These studies showed that sequences as far as 80 nucleotides downstream beyond the mRNA terminus were required to generate the 3' end correctly and efficiently. Included in this portion of the spacer DNA between histone genes is another sequence that is highly conserved (CAAGAAAGA), particularly among the well-studied sea urchin histone genes.

Conclusive proof that the 3' ends of histone mRNA are generated by RNA processing came from experiments in which cloned H2A histone genes were used to synthesize long, radioactively labeled transcripts that possessed the correct 5' end of histone mRNA but proceeded far beyond the mature 3' end of the message. When these elongated histone mRNA precursors with extra 3' nucleotides were injected into *Xenopus* oocytes, the oocytes were able to convert them to mature mRNA (see Figure 21-36).

Correct 3'-end processing of histone mRNAs requires the participation of a small nuclear RNA-protein complex (snRNP) containing a 60-long RNA called U7. This has been established in experiments where sea urchin histone genes were injected into *Xenopus* oocytes (see Figure 21-36). Paradoxically, it was most fortunate that one of the sea urchin transcripts (the histone H3 precursor) *fails* to be processed correctly in this mixed system. This provided a convenient assay for isolating from sea urchin extracts the essential factor missing from frog oocytes. Activity has been traced to the U7 snRNP, which shares common antigenic proteins with the U1 and U2 snRNPs implicated in pre-mRNA splicing (Chapter 20). The sequence of U7 RNA suggests that it may interact with the pre-mRNA by forming base pairs with the conserved sequences just preceding and following the 3' end of the mature histone mRNA (see Figure 21-36). Since the mRNA precursors for all five histone species share almost identical 3'-terminal stem-and-loop structures and 3' flanking CAAGAAAGA sequences, it remains a mystery why *Xenopus* oocytes can successfully process four of the sea urchin histone mRNA precursors (H2A, H2B, H4, and

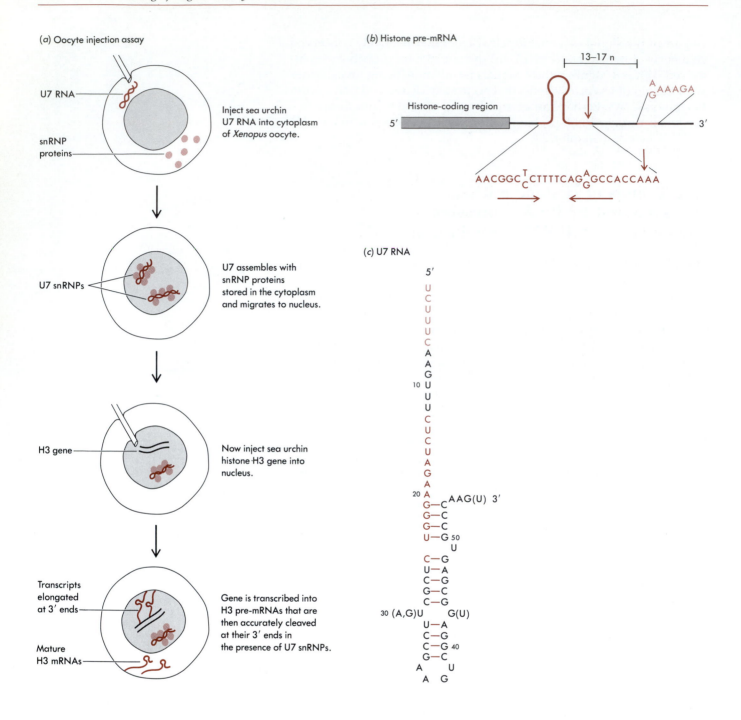

(a) Oocyte injection assay

U7 RNA

snRNP proteins

Inject sea urchin U7 RNA into cytoplasm of *Xenopus* oocyte.

U7 snRNPs

U7 assembles with snRNP proteins stored in the cytoplasm and migrates to nucleus.

H3 gene

Now inject sea urchin histone·H3 gene into nucleus.

Transcripts elongated at 3′ ends

Mature H3 mRNAs

Gene is transcribed into H3 pre-mRNAs that are then accurately cleaved at their 3′ ends in the presence of U7 snRNPs.

(b) Histone pre-mRNA

13–17 n

Histone-coding region

5′

3′

AACGGC$_C^T$CTTTTCAG$_G^A$GCCACCAAA

(c) U7 RNA

Figure 21-36

Processing the 3′ ends of histone pre-mRNAs requires U7 snRNA. (a) How it was established that U7 is involved in histone mRNA processing. (b) A schematic diagram of a histone pre-mRNA, showing the two consensus sequences that are conserved from sea urchins to mammals. The protein-coding region is not drawn to scale. The arrow indicates the site cleaved to generate the 3′ termini of histone mRNAs. The distance between the palindrome (base of the hairpin) and the downstream conserved sequence is also highly conserved. (c) The sequence of sea urchin U7 RNA with the regions complementary to the two histone pre-mRNA consensus sequences indicated by matching colors. [After M. L. Birnstiel, M. Busslinger, and K. Strub, *Cell* 41 (1985):349–359; C. Birchmeier, W. Folk, and M. L. Birnstiel, *Cell* 35 (1983):433–440; and K. Strub, G. Galli, M. Busslinger, and M. L. Birnstiel, *EMBO J.* 3 (1984):2801–2807.]

H1) but not the fifth (H3). Whether the RNA processing nuclease is contained in the snRNP itself and whether its action is endo- or exonucleolytic should soon be determined.

Different mRNAs from the Same Gene in Different Tissues: Exploiting Differences in Both Transcription and RNA Processing[90, 91, 92]

The existence of enhancer elements and putative tissue-specific factors that recognize them can easily explain how the various members of a gene family might be differentially expressed in different tissues or at different stages of development. Thus, the discovery that eucaryotic organisms have also evolved the capacity to use single genes to encode different protein products in different tissues was unanticipated. As with the related proteins specified by different members of a multigene family (e.g., the α- or β-globins), the two or more polypeptides encoded by a single gene are homologous but distinct— presumably tailoring them to the individual needs of the cells in which they are made. We now realize that this additional level of versatility exploits tissue-specific differences in both the transcriptional apparatus and the RNA processing machinery of eucaryotic cells.

If a single gene is preceded by two promoter regions, each with its own tissue-specific regulatory element(s), then two transcripts that differ in length at their 5' ends will be produced in a tissue-dependent fashion. If, further, the extra 5' nucleotides include splicing signals (5' and 3' junction sequences) and coding regions, two polypeptides differing in their N-terminal amino acid sequences may be produced in the relevant tissues. This is precisely the strategy adopted by the chicken genome to generate two different myosin light chains (important muscle proteins) from a single gene. Cardiac muscle contains one type of light chain (called LC_1) and gizzard muscle another (called LC_3), whereas skeletal muscle contains both LC_1 and LC_3. As shown in Figure 21-37, two promoters about 10,000 base pairs apart account for the synthesis of two pre-mRNAs; these undergo differential splicing, yielding the mRNAs for LC_1 and LC_3, which have identical 3' but distinct 5' coding regions. The finding that different myosin chains in rats are likewise encoded by single genes whose expression is identically regulated (by differential transcription and splicing) argues that this ingenious scenario evolved even before the divergence of birds and mammals.

In the case of the myosin LC_1/LC_3 gene, it is unclear whether structural differences between the two pre-mRNA transcripts dictate differential splice site utilization or whether tissue-specific changes in the splicing machinery are also involved. A growing list of other genes, however, strongly hints that the same transcript can be processed differently in different tissues to yield mRNAs with different coding potentials. One example is the rat calcitonin gene transcript, which is polyadenylated at one site in cells of the thyroid gland and at another site in the brain (see Figure 21-37). The two transcripts are spliced in distinct ways. We do not yet know whether calcitonin gene expression is controlled primarily by the choice of splice sites or by the choice of polyadenylation sites. If polyadenylation precedes splicing (as seems likely from the study of many other genes), then the cell might control polyadenylation in a tissue-specific manner, and the (*a*) Two promoters and different splicing patterns: the chicken myosin alkali light-chain gene.

(*b*) Two poly A sites and different splicing patterns: the rat calcitonin gene.

(*c*) Two different splicing patterns: the rat troponin T gene

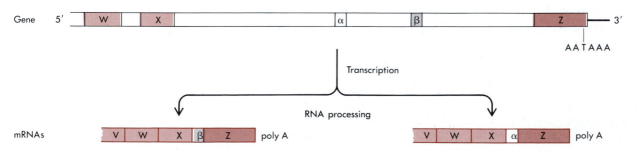

Figure 21-37
Three different strategies for producing more than one protein product from the same gene. In the case of the calcitonin gene, the protein product is further processed at the basic residues indicated to yield either calcitonin or a neuropeptide called calcitonin gene-related peptide (CGRP).

structure of the resulting mRNA precursor would dictate the subsequent choice of splice sites. Alternatively, if splicing can sometimes occur on the nascent mRNA transcript, then the cell might control the choice of splice sites, and the structure of the nascent transcript would influence the subsequent choice of polyadenylation site. In any case, the end result is two calcitonin mRNAs with three common 5' exons but different 3' exons, thus specifying different polypeptides (and eventually different hormones) in the two cell types.

Finally, there are situations where both the 5' end (promoter site) and 3' end (polyadenylation site) of a pre-mRNA appear to be identical in different tissues, but alternative splicing produces two different mRNAs. For example, mammalian troponin T, a regulatory protein of the contractile apparatus, occurs in several forms whose appearance depends on the developmental stage and the type of striated muscle examined. Part of this heterogeneity can be traced to a divergent troponin T gene family, including at least three separate genes. But one of these genes, in rats, has been shown to encode two different polypeptides whose amino acid sequences differ only in a small internal section (positions 229 to 242 in the 259-residue chain). This difference results from the mutually exclusive incorporation of one or another internal exon into the mRNA (see Figure 21-37). Tissue-specific mRNA processing could reflect fundamental (albeit subtle) changes in the splicing machinery of different cell types. However, a simpler model would be that the splicing machinery itself is universal, but different cell types contain different factors (protein or RNA or both) that can bind to specific sequences (or structures) within mRNA precursors, thereby facilitating or blocking the activity of the splicing machinery on particular splice junctions.

We now suspect that similar manipulation of RNA processing pathways may generate multiple RNAs from the transcripts of genes that play pivotal roles in the developmental program of *Drosophila* (the homeotic loci; Chapter 22). Likewise, immunoglobulin-producing cells utilize differential RNA processing to convert antibody molecules from their membrane-bound to secreted forms (Chapter 23). Introns and RNA splicing therefore confer on eucaryotic organisms a unique potential for the functional diversity of single genes, as well as an advantage in evolving new genes (Chapter 20).

Nearly All RNA Processing Is Nuclear

So far, all the processes we have considered in higher eucaryotic gene expression take place in the nucleus. The synthesis of RNA chains must occur there, since the cell's genome is confined to this compartment. RNA processing proceeds both during and directly after completion of chain growth. Thus, it is within the nucleolus that rRNA molecules are extensively trimmed and that they receive certain nucleotide modifications concomitant with ribosomal subunit assembly. Likewise, the nucleoplasm is where pre-mRNA molecules undergo

Figure 21-38
The nuclear, nonnucleolar location of both U1 snRNPs and hnRNP, as revealed by indirect immunofluorescence. Cells have been incubated with either (a) a patient serum containing autoantibodies directed against U1 snRNP protein components, (b) a monoclonal antibody directed against ribosomal RNA, or (c,d) a monoclonal antibody reactive with several major proteins (C proteins) of the hnRNP. [Note that (d) is a phase-contrast micrograph of the same field as in (c).] Then, a fluorescently tagged antibody directed against the first antibody is applied, and the cells are visualized in a fluorescent microscope. (a) and (b) show African green monkey kidney cells, whereas (c) and (d) show human *HeLa* cells. Note that the nucleolar and cytoplasmic location of rRNA is the exact inverse of the location of snRNP and hnRNP proteins. [Parts (a) and (b) courtesy of E. Lerner; (c) and (d) courtesy of G. Dreyfuss.]

splicing and acquire their 5′ caps and some internal m^6A residues (of unknown function); indirect immunofluorescence studies demonstrate that both hnRNP proteins and snRNPs (some of which act in splicing) are localized there (Figure 21-38). Polyadenylation could be linked to export (and therefore take place on the nuclear envelope), but this seems unlikely for several reasons. First, some mRNAs (notably histone messengers) lack poly A, so the two processes cannot be obligatorily coupled. Second, poly A tails are always complexed with specific proteins, but these are distinct in the nucleus and the cytoplasm; why should there be a special nuclear poly A binding protein if poly A is acquired only during export? What is clear is that the export apparatus somehow surveys mRNA molecules to ensure that every intron that is meant to be removed has been removed before passage to the cytoplasm. How this is accomplished remains a mystery. Finally, in the case of tRNA splicing, there are hints that introns may be removed only as tRNA molecules pass into the cytoplasm; both the splicing endonuclease and ligase appear to be concentrated on the nuclear membrane. Whether other processing enzymes that act on RNA polymerase III transcripts are also so located is not yet known. There are only a few exceptions to the general rule of nuclear RNA processing; for example, certain tRNA base modification reactions are known to occur in the cytoplasm.

Messenger RNA Lifetimes in Higher Cells[52, 86]

Compared to what we know about bacterial messengers, our knowledge of the lifetimes of functioning mRNA in higher cells is relatively sparse. It seems clear that many mRNA species have a limited lifetime, and average half-life estimates of 3 hours are frequently given for the mRNA in rapidly growing vertebrate cells, which divide about once a day. If these figures are correct, then their *relative* lifetimes are not unlike those of many bacterial mRNA molecules, which turn over some ten times during a cell cycle.

Although many cellular mRNAs encode "housekeeping" proteins that must be present at all times, the synthesis of a significant number of proteins (and RNAs?) must be turned on and off in a strict order to enable the cell to proceed with regularity through the cell cycle. This is particularly true for yeast, where over 50 genes are known to control the cell division cycle (CDC). Temperature-sensitive lesions in any of these CDC genes cause the cells to arrest in a defined stage of the cell cycle at elevated temperatures (Chapter 18). Comparable CDC genes must occur in higher eucaryotes, and it is likely that at least some of them encode unstable mRNAs whose brief existence advances the cell from one stage in the cell cycle to the next.

In contrast, much of the mRNA of the terminally differentiated cells of higher animals is relatively stable. Reticulocytes (immature red blood cells) provide a good example. These cells produce virtually no RNA while they are synthesizing their principal protein, hemoglobin. If their mRNA molecules were rapidly made and broken down, it would be possible to measure incorporation of RNA precursors into RNA; none is detected. This stability has an obvious advantage, particularly since the very constant environment of reticulocytes (as contrasted with the highly fluctuating growth conditions of bacteria) makes rapid changes in the mRNA population unnecessary. Reticulocytes synthesize mostly (more than 90 percent) hemoglobin. There is no reason to break down hemoglobin mRNA chains only to resynthesize them. Similarly, most of the mRNAs found in the cytoplasm of the adult liver encode plasma proteins that are released into the circulatory system. As expected, liver cell mRNA also appears to be relatively stable.

We understand almost nothing about what determines mRNA half-life. As a general rule, polyadenylated mRNA species live longer than nonpolyadenylated mRNAs (such as histone messengers). But we do not know whether the extraordinary stability of certain mRNAs (e.g., hemoglobin mRNA in a reticulocyte) is an intrinsic property of the mRNA itself, or whether it is acquired by complexing with specific stabilizing proteins (e.g., mRNP proteins). Most recent evidence favors the latter alternative and raises the intriguing possibility that mRNA (and even pre-mRNA) stability may be yet another level on which gene expression is actively controlled. For example, we will soon see that the mechanism of steroid hormone action may include the stabilization of specific mRNAs.

Myriad Initiation Factors for Eucaryotic Protein Synthesis[93, 94, 95]

In contrast to the three well-defined factors required to start translation in *E. coli* are a multitude of **eucaryotic initiation factors** (designated **eIFs**). How they contribute to the initiation of protein chains is diagrammed in Figure 21-39. The process is in some ways similar to that in procaryotes. The small ribosomal subunit (here 40S) is the first to form an initiation complex with the mRNA and initiator tRNA. Also, GTP is cleaved only upon joining of the large ribosomal subunit

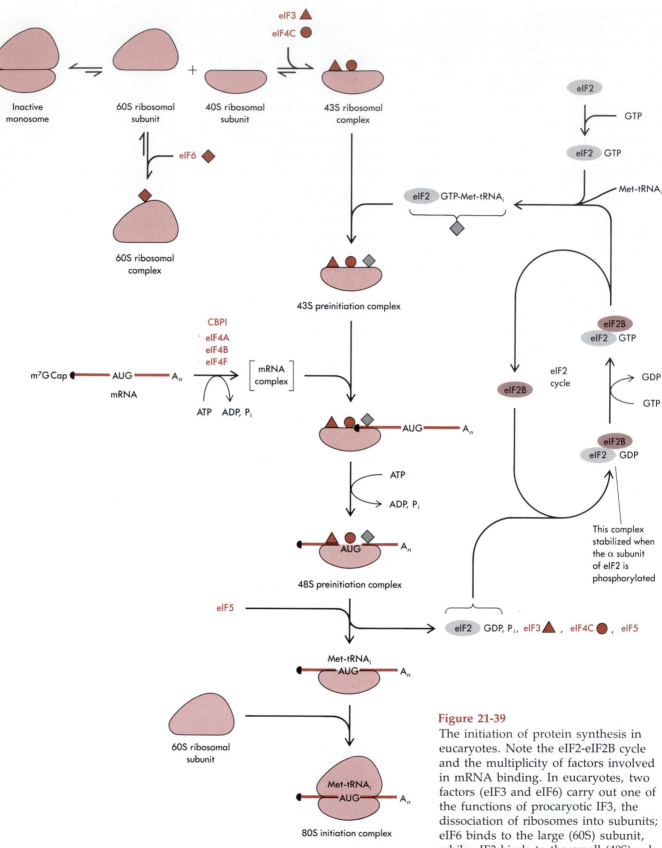

Figure 21-39

The initiation of protein synthesis in eucaryotes. Note the eIF2-eIF2B cycle and the multiplicity of factors involved in mRNA binding. In eucaryotes, two factors (eIF3 and eIF6) carry out one of the functions of procaryotic IF3, the dissociation of ribosomes into subunits; eIF6 binds to the large (60S) subunit, while eIF3 binds to the small (40S) subunit. CPB stands for cap binding protein; eIF4F is also called CPBII. (Scheme courtesy of W. Merrick.)

(60S) to form a complex (80S) that possesses an open tRNA binding site (A site) and can thus enter the elongation phase of protein synthesis.

However, there are many essential ways in which eucaryotic differs from procaryotic initiation (see Figure 14-30):

- Again, a special tRNA^{Met} (here called Met-tRNA$_i^{Met}$) is utilized, but it does not acquire an N-terminal formyl group.
- The initiator Met-tRNA$_i^{Met}$ and GTP form an isolatable complex with eIF2, independent of the small ribosomal subunit. (Recall that in procaryotes, IF2 binds initiator tRNA on the 30S ribosome.)
- The initiator tRNA is always bound to the 40S ribosomal subunit prior to the mRNA (rather than in either order).
- Hydrolysis of ATP to ADP is required for mRNA binding.
- Not only are there many more initiation factors, but many eIFs are themselves multisubunit proteins. Most striking is eIF3 (involved in mRNA binding, as is IF3 in bacteria); it has about ten subunits and weighs 10^6 kdal (one-third as much as the 40S subunit itself)!
- The presence of a capped 5' end on the mRNA is required for initiation, except for a very few viral mRNAs which are active without a cap (Chapter 24).

Pivotal to the regulation of eucaryotic protein synthesis is a final difference in the initiation process: The factor eIF2 undergoes GTP-GDP exchange via a cycle involving another protein called **eIF2B,** also called **GEF,** for **GTP-GDP exchange factor** (Figure 21-39). Enzymatically, the steps are virtually identical to the EF-Tu/Ts cycle in procaryotic chain elongation (see Figure 14-31). After eIF2 has deposited the initiator tRNA on the ribosome, it is released in an inactive form containing bound GDP. GDP is displaced only when eIF2 interacts with eIF2B to form an eIF2-eIF2B complex. In this complex, eIF2 has a heightened affinity for GTP and can thereby reenter the active pool and again bind Met-tRNA$_i^{Met}$. The progress of eIF2 through this cycle is sensitive to the phosphorylation of one of its three polypeptide subunits (called eIF2α): Phosphorylated eIF2 interacts more stably with eIF2B, and therefore both factors become sequestered into an eIF2-eIF2B complex. The consequent lowered availability of eIF2 for participation in initiation forms the common basis of several mechanisms in which seemingly disparate agents control mammalian protein synthesis.

40S Ribosomal Subunits Almost Always Select the First AUG[94]

In only one respect does the initiation of protein synthesis appear simpler in eucaryotes than in procaryotes, and that is in choosing where to start on the mRNA. First, only AUG is an efficient initiator codon (whereas GUG, UUG, and even AUU are also utilized in *E. coli*). Second, the AUG located nearest the 5' end of the mRNA (usually within 10 to 100 nucleotides) is almost always selected. The simplest model consistent with these facts proposes that 40S ribosomes attach to the capped 5' terminus and scan along the mRNA, searching for the first AUG and consuming ATP in the process. Stopping would be the consequence of correct codon-anticodon interaction between AUG and the Met-tRNA$_i^{Met}$, which is invariably prebound to the 40S subunit (see Figure 21-39). That AUG is exclusively recognized can be explained by the presence of a bulky modified A residue just 3′ to the anticodon of eucaryotic tRNA$_i^{Met}$. (This modification apparently disallows first-position wobble in codon-anticodon interactions; recall that it is lacking in *E. coli* tRNA$_F^{Met}$, allowing more flexibility in its interaction with initiator codons. See Figure 14-26.) Moreover, an additional upstream signal (such as the procaryotic mRNA region that base-pairs with the 3′ terminus of 16S rRNA; see Figure 14-28) would not be needed if ribosomes cannot bind directly to internal initiation sites.

Yet, closer scrutiny of the mechanisms underlying initiation site selection by eucaryotic ribosomes uncovers certain complexities. There are a few (< 10%) eucaryotic mRNAs where it is not the first AUG but a downstream AUG codon that begins the open reading frame. Such exceptions to the first-AUG rule can be explained by the effect of sequence context, which was revealed by comparing hundreds of mRNA sequences. Namely, ^A_GNNAUGG is optimal for initiation by 40S ribosomes; if pyrimidines appear instead at position −3 or +1 relative to the AUG, the site has a high probability of being bypassed. However, in an mRNA constructed to possess four tandemly reiterated copies of the same ribosome-binding site, only the first was used. This brings out the importance of proximity to the 5′ end of the mRNA for identifying a functional initiator codon. Another unresolved question is how 40S ribosomes load onto the mRNA. One possibility is that they literally thread on, for it has been shown that a free end on the mRNA (whether capped or not) is absolutely required; eucaryotic ribosomes are completely incapable of initiating on circular mRNAs, whereas procaryotic ribosomes do so quite nicely. Finally, do 40S ribosomes really move physically from the cap to the first AUG, propelled by the energy of ATP hydrolysis? Such a scanning mechanism is consistent with all available data. An alternative model suggests that several of the initiation factors (including a cap binding protein) assemble in an ATP-dependent process onto the mRNA to create a pathway for easy access of the 40S ribosome to the first AUG (see Figure 21-39).

Elongation and Termination by 80S Ribosomes[93, 96–99]

Compared to initiation factors, the factors required for elongating and terminating polypeptide chains in eucaryotes are disarmingly simple. Their contributions are summarized in Table 21-2.

EF1α (50 kdal) replaces EF-Tu of procaryotes. This factor forms a complex with GTP and aminoacyl-tRNA and then donates the latter to ribosomes. Less well characterized is an EF-Ts-like factor (called **EF1βγ**) that catalyzes GDP-GTP exchange and readies EF1α for another cycle of aminoacyl-tRNA binding. The eucaryotic factor **EF2** (about 100 kdal) corresponds to procaryotic EF-G. It hydrolyzes GTP and catalyzes translocation of the aminoacyl-tRNA from the A site to the P site on the ribosome with concomitant movement of the message.

The termination of polypeptide chains by 80S ribosomes involves only a single factor called **RF** (115 kdal). It replaces both RF1 and RF2 of procaryotes in that it can recognize all three stop codons (UAA, UGA, and UAG). Eucaryotic termination involves GTP hydrolysis,

Table 21-2 Factors Involved in the Elongation and Termination of Polypeptide Chains in Eucaryotes

Factor	Function	Comparable Procaryotic Factor*
EF1α	Binds animoacyl-tRNA to ribosomes	EF-Tu
EF1$\beta\gamma$	Assists the recycling of EF1α	EF-Ts
EF2	Translocation	EF-G
RF	Chain release	RF1
		RF2

*See Figures 14-31 and 14-32.

which is required for RF dissociation from the ribosome after the factor has activated peptidyl transferase to release the polypeptide chain.

An unusual property of the aminoacyl-tRNA synthetases from higher eucaryotic cells is that they are generally isolated as aggregates. These high-molecular-weight complexes contain multiple synthetase activities, as well as other components whose identity suggests that synthetases may be membrane-bound in vivo. An association with the cell membrane for at least some fraction of the synthetase population makes sense, since higher eucaryotic cells often depend on amino acids supplied externally (by the circulatory system). This location could also explain the remarkable observation that mammalian cells incorporate exogenously added amino acids directly into proteins without mixing with the large amino acid pools already present.

Specific Inhibitors of Eucaryotic Translation[100, 101, 102]

Not surprisingly, some antibiotics that inhibit translation in procaryotes are equally effective against eucaryotic protein synthesis. Puromycin, for example, also fools 80S ribosomes into attaching it to the growing end of polypeptide chains, thereby prematurely terminating translation (see Figure 14-33). Another inhibitor of both systems is fusidic acid, which prevents the release of procaryotic EF-G from the ribosome after it has functioned in translocation; it likewise inhibits EF2 release, suggesting a high degree of evolutionary conservation between these homologous GTP binding factors.

Other inhibitors are quite specific for eucaryotic protein synthesis. In their modes of action, they mainly exploit differences between the procaryotic and eucaryotic translational apparatus. For instance, m7Gp inhibits initiation only in eucaryotic in vitro systems, because it competes with the 5' ends of mRNAs for cap binding protein. A multitude of other agents bind specifically to some essential component of the eucaryotic machinery. They therefore block progression to the next step in the translation pathway, as summarized in Figure 21-40. Whereas most of these inhibitors are small molecules, some are polypeptides (e.g., ricin, a specific N-glycosidase that halts elongation by inducing cleavage of the 28S RNA in the 60S ribosomal subunit).

The deadly toxin produced by the diphtheria bacterium *Corynebacterium diphtheriae* is a well-studied polypeptide inhibitor of eucaryotic protein synthesis. Diphtheria toxin is a 65 kdal protein encoded not

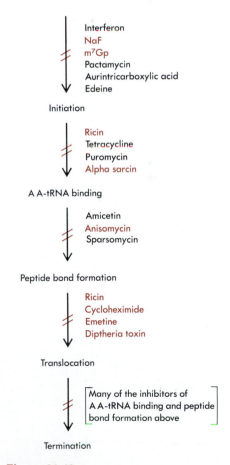

Figure 21-40
Sites of action of some inhibitors of eucaryotic protein synthesis. Inhibitors that target the eucaryotic translation system exclusively are shown in color.

Figure 21-41
The ADP-ribosylation reaction that inactivates EF2. Cells probably contain endogenous ADP-ribosyltransferases as well as inhibitors of this reaction that prevent catastrophic inactivation of EF2.

Dipthamide (modified histidine) in EF2

NAD$^+$

Diphtheria toxin (or other ADP-ribosyltransferase)

ADP-ribosyl modified dipthamide

Histidine

Nicotinamide

by the genome of the bacterium, but by a phage that lysogenizes it (called phage β). The toxin is excreted by the bacteria, becomes transported throughout the body of the infected victim, and gains entry into the cells of virtually all tissues. Once inside, it catalyzes the attachment of ADP-ribose to elongation factor EF2. The reaction is shown in Figure 21-41. ADP-ribose is donated by NAD$^+$ and becomes attached to an unusually modified histidine residue in EF2. In vitro, the reaction can be readily reversed by the addition of nicotinamide. Once ADP-ribosylated, EF2 is completely inactive in elongating polypeptide chains. Since the toxin acts catalytically, only a tiny amount (perhaps as little as one molecule) is effective in blocking a cell's entire protein synthesis capability and producing cell death. Presumably, the special modified histidine residue in EF2 is not present simply so that diphtheria toxin can kill mammalian cells; it may also be the target of cellular mechanisms designed to regulate protein synthesis in meaningful ways.

Finally, interferon must be mentioned as an important inhibitor of the initiation of translation. Because interferon operates on virus- infected cells, its synthesis and mode of action are described in some detail in Chapter 24. At least two components required for polypeptide chain initiation are targets in interferon-treated cells. Viral mRNAs are cleaved by a special endonuclease; and initiation factor eIF2 is inactivated by phosphorylation (see Figure 21-39).

Translational Controls: Fact, Not Fantasy[103]

The importance of translational controls to the lives of eucaryotic cells has been long debated. The obvious question is, Why waste energy making a messenger you may never use? Regulation of translation makes some sense for bacteria with their characteristically polycistronic mRNAs; if one protein product is needed and another not, then expression can only be controlled at the translational level (Chapters 16 and 17). By contrast, mRNAs are monocistronic in nucleated cells. Moreover, their production requires many more steps, several of which are already known to be regulated (promoter utilization, RNA splicing, polyadenylation) and others that may be (pre-mRNA stability, export from the nucleus). Thus, regulating eucaryotic translation might seem unnecessary. Yet, it provides a very rapid way to control gene expression and is used widely by higher eucaryotic organisms.

Sometimes, translational levels are manipulated by changing essential components of the cell's translational machinery. We would expect such alterations to affect the activity of every mRNA in the cell. For instance, it appears that phosphorylation of ribosomal proteins (particularly S6, in the 40S subunit) is correlated with higher polysome levels after mammalian cells are stimulated by a variety of different growth factors. Likewise, upon fertilization, the sudden appearance of a previously missing initiation factor activity mobilizes all the mRNAs stored in the brine shrimp egg for incorporation into polysomes (see also Chapter 22). Or, as we shall see, modifying initiation factor eIF2 is a device used by the reticulocyte to link its protein synthesis to the availability of heme (the prosthetic group of hemoglobin).

Other strategies are employed to alter the translation of only one mRNA (or some subset of cellular messengers). Many such mechanisms act at or before the initiation step. For instance, there is the inherent variability in eucaryotic mRNA lifetimes, as well as instances where mRNA stability is selectively altered in response to certain agents. Alternatively, mRNAs can exist stably in the cytoplasm but be somehow sequestered away from the translational apparatus (perhaps by mRNA binding proteins). Polio virus has been postulated to shut off host cell protein synthesis by destroying the cap binding protein essential for initiation on capped mRNAs, while its own mRNAs are not 5'-capped (Chapter 24). That the elongation phase of mammalian protein synthesis can also be controlled in a messenger-specific way is neatly demonstrated by the example of SRP action, which we will discuss shortly.

Thus, there can no longer be any doubt that translational controls do contribute importantly to the functioning of eucaryotic cells. Already it is apparent that such mechanisms operate on a variety of components of the translational apparatus. Presumably, it is only a matter of time until some regulatory strategy affecting every conceivable stage of eucaryotic protein synthesis is uncovered.

Translational Control of Hemoglobin Synthesis by Heme[104, 105]

One of the earliest documented examples of regulation at the level of translation occurs in mammalian reticulocytes (immature red blood cells). These cells have lost their nuclei, but retain high levels of stable mRNAs encoding mostly hemoglobin chains. Reticulocytes engage in exceedingly high rates of protein synthesis. Hence, reticulocyte lysates have long been the favorite system of biochemists interested in eucaryotic translation. Indeed, much of what we know about the mechanisms involved in protein synthesis is based on studies using reticulocyte lysates.

In reticulocyte lysates, protein synthesis starts at a high rate but declines sharply unless heme is added. (In cells, heme is synthesized by mitochondria, which are lacking in the in vitro system.) The decline occurs because heme deficiency activates an inhibitor of protein synthesis initiation called **heme-controlled inhibitor** (**HCI**). In the presence of heme, HCI remains inactive; in its absence, HCI becomes activated. Although its mechanism of action was mysterious for many years, we now know that HCI is a specific kinase that phosphorylates the α subunit of eIF2. Because of the eIF2 GTP-GDP exchange cycle depicted in Figure 21-39, phosphorylation of only a fraction (20 to 30 percent) of eIF2 is sufficient to stop initiation of protein chains. Apparently, all of the eIF2B (which is present in lower molar amounts than eIF2) becomes sequestered into eIF2-eIF2B complexes and is no longer available to recycle the remaining (unphosphorylated) eIF2. Moreover, as already explained, the phosphorylated eIF2 contained in the complex with eIF2B has a lowered affinity for GTP.

Thus, we see that hemoglobin translation is regulated at the level of initiation by phosphorylation of eIF2. The intermediary is a heme-sensitive kinase. There also exists in mammalian cells another kinase that acts similarly but is distinct. It is activated by double-stranded RNA and is present in elevated amounts in interferon-treated cells (Chapter 24). What role it plays in normal cells has not been established. Yet, the characterization of two such kinases strongly hints that there may be others, activated by different agents but all exerting their regulatory effect by phosphorylating the same factor (eIF2) crucial for the initiation of polypeptide chains.

Signal Recognition Particle (SRP): A Small RNA-Protein Complex That Initiates Export of Secreted Proteins[106, 107]

All cells must not only synthesize proteins in a controlled fashion but also deliver them to the appropriate site inside (or outside) the cell. In so doing, eucaryotic cells face many more difficulties than bacteria. First, the cells themselves are much larger. Hence, specific transport mechanisms may be required where bacteria can rely on simple diffusion. Second, there are numerous membrane-bound compartments, each with its own distinctive protein composition. Thus, for many newly synthesized eucaryotic proteins, a decision must be made about *which* membrane to cross, rather than just whether or not to cross a membrane.

Elongation arrest domain   Signal recognition domain

The eucaryotic proteins whose delivery to a particular cellular compartment is best understood are those that are synthesized by polysomes bound to the extensive intracellular membrane system called the **endoplasmic reticulum** (**ER**). This class includes secreted proteins, lysosomal proteins, and certain membrane proteins. Translocation across the membrane of the endoplasmic reticulum is initiated while the polypeptide is still incompletely synthesized (**cotranslational secretion**). A common feature of virtually all proteins that enter the endoplasmic reticulum is a special N-terminal extension called the **signal sequence** (or **signal peptide**). Eucaryotic signal sequences, like those in procaryotes (see Table 14-2), do not possess a particular sequence, but are all about 15 to 30 amino acids long and are rich in hydrophobic amino acids. They presumably fold into a helical hairpin conformation that readily inserts itself into the membrane bilayer to initiate transport (see Figure 14-35). Sometime during the subsequent passage of the growing polypeptide chain through the membrane, the signal sequence is clipped off by a special protease called **signal peptidase.** Once inside the lumen of the endoplasmic reticulum, the protein is further modified by the addition of carbohydrate (and other) groups that facilitate its ultimate delivery to the proper final destination.

That a small RNA-protein complex plays a pivotal role in this translocation process has been appreciated only recently. The **signal recognition particle** (**SRP**) is so named because it interacts specifically with the signal sequences of nascent secretory proteins. SRP is made up of 7SL RNA (about 300 nucleotides) (see Figure 20-38) and six different tightly bound protein components, ranging in size from 9 to 72 kdal (Figure 21-42). A fascinating aspect of its mode of action is SRP's ability to bind to ribosomes and halt protein synthesis (Figure 21-43). Translational arrest occurs not at initiation but when the nascent polypeptide chain is about 70 amino acids long. This is just the length at which the signal sequence will have emerged completely from the ribosome. (Recall that 30 to 40 amino acids adjacent to the growing carboxyl terminus are sequestered within the ribosome.) SRP does not arrest translation of all mRNAs, but specifically acts on those encoding to-be-secreted proteins. The translational block is reversed only when the SRP makes contact with the **docking protein** (72 kdal), an integral component of the ER membrane. Then, SRP diffuses away to initiate transport of another polypeptide chain. During translocation, the attachment of polysomes to the membrane of the endoplasmic reticulum is primarily mediated by the nascent polypeptide chains being translocated, aided by ribosome-receptor proteins

Figure 21-42
Schematic diagram of the mammalian signal recognition particle (SRP), showing which proteins associate with the two domains of the folded RNA. *Alu* refers to sequences in 7SL RNA that exhibit homology to the *Alu* repetitive sequence family in primate genomes, while S indicates the unique central region (see Figure 20-38). The inset shows a dark-field electron micrograph of unstained SRP; the long dimension of the rod-shaped particle is nearly equivalent to the diameter of a ribosome. [SRP model courtesy of P. Walter; micrograph courtesy of D. Andrews, P. Walter, and P. Ottensmeyer, *Proc. Nat. Acad. Sci.* 82 (1985):785–789. Used with permission.]

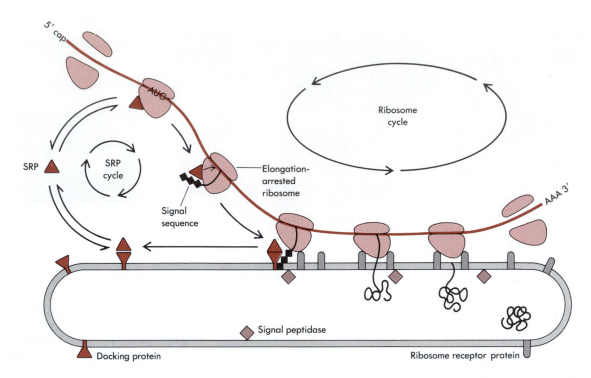

Figure 21-43

The functioning of SRP and the docking protein (also called the SRP receptor) in the translocation of proteins across the endoplasmic reticulum. [After P. Walter, R. Gilmore, and G. Blobel, *Cell* 38 (1984):5–8.]

(sometimes called ribophorins) in the endoplasmic reticulum that interact specifically with the large ribosomal subunit (see Figure 21-43). SRP is present in very high amounts in the cytoplasm, despite its only transitory role in the translocation process (about 10^6 copies, or one-tenth the number of ribosomes in mammalian cells).

SRP's activity as a negative regulator of translation has an important biological consequence. Many proteins exported by mammalian cells are degradative enzymes (e.g., nucleases and proteases). Even their occasional presence within the cytoplasm might be expected to wreak havoc within the cell that made them. By stopping the translation of such proteins midway, SRP ensures that they will not be completed until the correct membrane (across which they can be translocated) is at hand. Most recently, it has been found that the two activities of the SRP can be assigned to separate domains of the oblong-shaped particle (see Figure 21-42). Clipping the RNA with nuclease produces a large subcomplex that retains SRP's ability to bind to ribosomes (perhaps via RNA-RNA interactions) and to facilitate secretion; but this subcomplex cannot carry out translational arrest. It is fascinating that the two SRP domains correspond almost precisely to the two types of sequences present in 7SL RNA (see Figure 20-38).

Delivery of Proteins to Other Cellular Compartments[108–115]

Analyses of protein transport into other compartments of eucaryotic cells are only now beginning to catch up with our understanding of translocation into the endoplasmic reticulum. Best understood is the nature of import into mitochondria and plant cell chloroplasts; nuclei, because of their nuclear pores, present a somewhat different problem. Yet each of these compartments is surrounded by a double-membrane system (making protein translocation more complex than with the single membrane of the endoplasmic reticulum). Moreover, their outer membranes can be observed to be studded with ribosomes. Does this mean that protein translocation into these three compartments is also cotranslational? Is a special small RNA-protein complex likewise involved? (There do exist a number of small cytoplasmic RNA-protein complexes distinct from SRP; their structures have been defined to varying extents, but their functions remain unknown.) So far, it appears that a different mechanism of delivery is employed in each case.

Import of most proteins into mitochondria involves the following steps: synthesis as a precursor polypeptide (with an N-terminal extension that is usually positively charged); binding of this precursor to a receptor on the mitochondrial surface; and translocation into or across the mitochondrial membranes. There are differences, however, depending on whether the protein is destined for the outer membrane, the intermembrane space, or the inner membrane/matrix. Import of both intermembrane and inner membrane/matrix proteins requires an electrochemical potential across the inner membrane. The initial "targeting" interaction with a protein receptor in the outer mitochondrial membrane appears to be mediated by the precursor's N-terminal extension, which is cleaved off after import. The cleavages experienced by intermembrane proteins, however, are often more complex than those of proteins bound for the inner membrane/matrix. By contrast, delivery of proteins to the outer mitochondrial membrane neither requires an energized inner membrane nor is accompanied by proteolytic processing. All three mechanisms require much additional scrutiny. What is clear in each case is that import can occur after translation is complete. Moreover, mitochondrial proteins are synthesized principally on free (rather than membrane-bound) polysomes. Only the most indirect evidence hints at the involvement of a small RNA-protein complex in protein translocation into mammalian mitochondria.

Specific transport of plant proteins into chloroplasts is superficially similar to mitochondrial import. Chloroplast proteins are likewise synthesized as longer precursors, and insertion takes place after translation is complete. Once the additional sequences in the precursor have functioned—presumably in the recognition of specific receptors on the chloroplast surface—they are removed by chloroplast protease(s) to generate mature proteins. Interestingly, transport into chloroplasts is stimulated by light. This apparently reflects a requirement for energy in the form of ATP, which is produced in chloroplasts by photosynthetic phosphorylation. Much still remains to be learned about the specificity of precursor-receptor interactions and the passage of polypeptide chains into the chloroplast's interior.

The recurring motif that characterizes proteins delivered to the endoplasmic reticulum, to mitochondria, and to chloroplasts—a polypeptide extension that is critical for transport but is later discarded to form the mature protein—does not hold for proteins destined for the nucleus. Instead, nuclear proteins are synthesized in their mature form. Many nuclear proteins (e.g., histones, hnRNP proteins, and polymerases and their associated factors) bind strongly to either DNA or RNA. Hence, one theory is that they arrive in the nucleus by free diffusion through the pores and are selectively retained because of their high affinity for nuclear components. That this is probably not the case is now emerging from the results of deletion analyses of nuclear proteins. Instead, part of the amino acid sequence—which can in some cases be quite short—appears to target the protein for the nucleus. Whether there is a defined sequence common to all nuclear proteins is not yet clear. One intriguing possibility

Figure 21-44
The amino acid sequence of rat proproinsulin II, showing the sequences lost during its processing to the mature hormone, which contains only the A chain (30 amino acids) and the B chain (21 amino acids). Why the sequence of the released C peptide (31 amino acids) has been conserved over evolution is not known. Arrows show the sites of proteolytic cleavage during processing.

is that some proteins may be transported into the nucleus by specifically associating with another protein that does possess the nuclear targeting sequence. For instance, nucleoplasmin (29 kdal), a highly abundant protein of the *Xenopus* oocyte that forms specific complexes with histones, may serve to carry the histones into the nucleus as well as to assemble them onto the DNA. RNAs may also be transported into the nucleus by protein carriers. Small nuclear RNAs, which travel to the cytoplasm during their maturation, are an example. They do not become properly relocalized in the nucleus if their binding sites for certain associated proteins (snRNP proteins) have been mutated.

Polyprotein Precursors of Peptide Hormones[116–119]

Once synthesized, a eucaryotic protein is much more likely than its procaryotic counterpart to undergo extensive covalent modification. In both cases, the initiating methionine residue is normally removed by a special exopeptidase. Likewise, N-terminal signal sequences are cut off concomitant with translocation across membranes in all cell types. But additionally in eucaryotic cells, membrane and exported proteins acquire numerous carbohydrate moieties during their transit through the endoplasmic reticulum. Further, endopeptidases frequently cut specific internal peptide bonds to generate active polypeptides from larger precursors.

Cleavage of precursor proteins is a recurring theme in the synthesis of peptide hormones. A classic example is that of insulin (Figure 21-44), which has been known for several decades to be constructed of

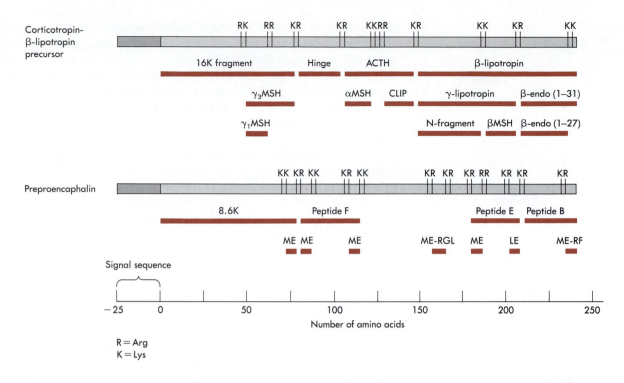

Signal sequence

R = Arg
K = Lys

two polypeptide chains. Only much later was it discovered that insulin is initially synthesized as an 86-amino-acid single chain (preproinsulin). After removal of its signal sequence to create proinsulin (62 amino acids) and folding into a specific three-dimensional shape stabilized by disulfide (S–S) bonds, internal cuts excise 31 amino acids from the middle of the chain to create the active insulin hormone.

Even more dramatic is the protein processing that occurs in pituitary and adrenal gland cells to generate the opiate peptides (natural hormones whose activity is partially mimicked by certain plant metabolites). Here, it has been shown that lengthy polypeptides (e.g., the corticotropin-β-lipotropin precursor and preproencephalin) are cleaved to yield a number of different hormones. As shown in Figure 21-45 (see also Figure 21-37), cleavage always occurs adjacent to pairs of basic amino acids (Lys-Arg or Lys-Lys). The majority of amino acid residues present in the original multihormone precursor are then discarded. Although the cotranslation of so many active molecules into one precursor would appear to be a device for producing equal amounts of hormones, this is not necessarily so. The pattern of peptide production in the fetus is different from the pattern in the adult, and there are even differences in different lobes of the pituitary gland. Disturbance of the normal cleavage pathways in the adult could also conceivably lead to altered neuronal activity and mental states. Hence, proteolytic processing is yet another control point for gene expression in higher eucaryotic cells.

Targeted Protein Degradation Is Essential for Cell Survival and Growth[120–123]

The amount of a gene product available for a cell's functioning depends not only on the many control mechanisms that operate during its synthesis. Protein lifetimes, like RNA lifetimes, are highly variable and subject to manipulation when a cell's environment changes. For instance, upon severe nutritional deprivation, significant proteolysis

Figure 21-45
The structures of two bovine multihormone precursors: The corticotropin-β-lipotropin precursor and preproencephalin. In both cases, the data are based on cloned cDNA sequences. The active peptide hormones are always bounded by paired basic amino acid residues, which are the sites of protein processing. The abbreviations for the various component peptides of the corticotropin-β-lipotropin precursor are: ACTH = corticotropin, endo = endorphin, MSH = melanotropin, and CLIP = corticotropin-like intermediate lobe peptide. Preproencephalin contains four copies of Met-encephalin (Tyr-Gly-Gly-Phe-Met) (ME in the figure) and one copy each of Leu-encephalin (which substitutes Leu for Met at the C terminus) (LE in the figure) and of two extended Met-encephalins (one a heptapeptide and one an octapeptide). [After S. Nakanishi et al., *Nature* 278 (1979):423–427; and R. E. Mains, B. A. Eipper, C. C. Glembotski, and R. M. Dores, *Trends in Neurosci.* 4 (1983):229–235.]

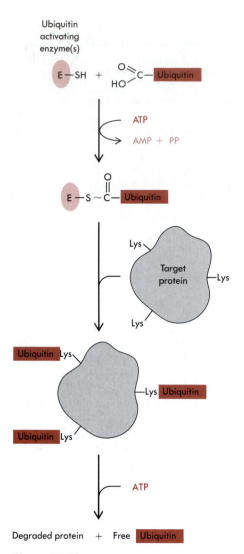

Figure 21-46
Sequence of events in ATP- and ubiquitin-dependent degradation of proteins. Recent work suggests that the first step of the process (the conjugation of ubiquitin to certain substrates) requires tRNA. [After A. Hershko and A. Ciechanover, *Ann. Rev. Biochem.* 51 (1982):335–364.]

of *intra*cellular proteins occurs in the lysosomes, cytoplasmic organelles normally devoted to the degradation of proteins and other molecules taken up from the exterior. But even during normal cell growth and maintenance, protein turnover is high. It is not unreasonable to expect that certain proteins needed at one stage of the cell cycle might have to be eliminated at another, or that a protein's rapid disappearance might be as crucial for survival as its rapid appearance when conditions change. The systems involved in such selective protein turnover are nonlysosomal. Generally, they require ATP, whose energy of hydrolysis is presumably channeled into making degradation specific (since peptide bond cleavage is not an energy-consuming reaction).

One important ATP-requiring proteolytic system in mammalian cells utilizes an unusual protein called **ubiquitin** to target other proteins for destruction. Ubiquitin is a 76-amino-acid polypeptide, whose sequence is nearly invariant from yeast to human. (It is perhaps the most highly conserved eucaryotic protein known.) It becomes activated for attachment to a targeted protein in a series of reactions similar to those catalyzed by the aminoacyl-tRNA synthetases (Figure 21-46). Once activated, usually several ubiquitin molecules are transferred to the protein substrate, forming isopeptide bonds between the ϵ-amino group of lysine residues and the carboxy-terminal glycine of ubiquitin. The ubiquitinated protein is subsequently degraded, releasing the ubiquitin molecules intact for use in further rounds of protein degradation. Both ubiquitin attachment and the less well characterized degradation steps require ATP (see Figure 21-46). The possible involvement of this proteolytic system in turning on gene expression is suggested by the finding that 10 to 20 percent of histone H2A molecules are ubiquitinated. This modification occurs preferentially in nucleosomes associated with transcribed regions, suggesting that selective histone degradation may contribute to the transition from inactive to active chromatin. Further, a mouse cell cycle mutant that arrests at the S/G2 boundary at high temperatures has been shown to be temperature-sensitive for ubiquitin activation (and hence for protein degradation). Most intriguing is the recent demonstration that ubiquitin-dependent proteolysis requires tRNA(s), conceivably linking this protein degradation system to amino acid levels in mammalian cells.

Histone Gene Expression Is Cell Cycle–Dependent[124,125]

Histones provide a conspicuous example of proteins that are needed in bulk during only one stage of the cell cycle. This is S phase, when an additional full complement of histone molecules must be supplied to package the newly replicated DNA into nucleosomes. Rather than storing histones for later use, eucaryotic cells synthesize histone molecules almost exclusively during S phase. The coupling with DNA replication appears to be direct, for inhibition of DNA synthesis abruptly terminates histone protein synthesis as well. Although the coupling mechanism is not understood in detail, growing evidence indicates that control of histone production is exerted at both the transcriptional and posttranscriptional levels.

There is little if any histone mRNA present in the cytoplasm just after mitosis. The dramatic increase observed when the cell enters S phase is due first to an increase (about tenfold) in the rate of synthesis of histone mRNAs. This has been traced (at least in part) to the ap-

C..GAA..TTC..G   Heat-shock consensus sequence

Figure 21-47
Binding of the heat-shock factor to sites upstream from the *Drosophila* hsp 70 gene. [After W. S. Dynan and R. Tjian, *Nature* 316 (1985):774–777; footprint courtesy of C. Parker.]

pearance of S phase–specific transcription factor(s) that selectively stimulate initiation by RNA polymerase II at histone gene promoters. (Whether there are different or common factors for the various histone genes is not yet known.) Second, once synthesized, histone mRNAs experience slightly different fates, depending on the stage of the cell cycle. Because they are not polyadenylated, histone mRNAs have unusually short half-lives, which contribute to their rapid disappearance when S phase is complete. But during S phase, their half-lives are lengthened (from 8 minutes to 40 minutes in cultured human HeLa cells) by as yet unknown mechanisms. Contrasting with the behavior of most histone mRNAs are mRNAs for minor, variant histones, whose synthesis and stability are not linked to DNA replication. They may provide a low level of new histone synthesis required during interphase because of DNA repair or nucleosome turnover accompanying gene expression.

Turning On of Heat-Shock Genes: Control at Both the Transcriptional and Translational Levels[126-131]

The cells of all organisms, from bacteria to human, respond to a sudden elevation of temperature (**heat shock**) by changing their pattern of gene expression. Usually, the synthesis of a few specific proteins (**hsp,** or **heat-shock proteins**) is dramatically turned up, while that of the majority of cellular polypeptides is sharply curtailed. In bacteria, a new σ factor accounts for an altered transcriptional pattern (Chapter 16). In eucaryotes, the heat-shock phenomenon has been most intensively scrutinized in *Drosophila*. Within one minute after temperature increase, puffs are induced at a few distinct sites along the polytene chromosomes (the heat-shock loci). Simultaneously, other puffs active at the time of heat shock regress. Indeed, we now know that the puffing reflects a conversion of the transcriptional apparatus to a state in which the seven heat-shock genes are almost exclusively transcribed. An accompanying change in the cell's translational machinery ensures that the heat-shock mRNAs are also preferentially translated.

The activation of heat-shock gene transcription is mediated by specific sequence elements present in the heat-shock promoters. Deletion analyses initially revealed that a 20-base-pair region located about 20 nucleotides upstream of (5' to) the TATA box was essential for induction of one *Drosophila* heat-shock gene *(hsp70)*. A consensus sequence for this regulatory element was derived by comparing sequences upstream of all seven heat-shock genes (Figure 21-47). It shows dyad symmetry, characteristic of sites recognized by multimeric DNA binding proteins. The one heat-shock gene that does not conform well to this consensus is only weakly inducible. That a positively acting transcription factor binds to the DNA region containing this consensus sequence has been suggested by recent DNase footprinting studies (see Figure 21-47). In fact, upstream of the *hsp70* gene there are four regions, each with its own consensus sequence, to which the factor(s) bind(s); binding to the two contiguous sites (sites 1 and 2) is cooperative and appears to be closely linked to transcriptional activation. How factor activity itself is regulated upon heat shock remains to be clarified.

The translational discrimination that occurs in heat-shocked cells is not yet as well understood as the transcriptional regulation. What is clear is that preexisting mRNAs are stable, but are less efficiently utilized probably because of lower rates of initiation. In fact, some signal contained near the 5′ end of the heat-shock mRNA seems to confer efficient translation on the induced heat-shock mRNAs. How the heat-shock proteins themselves help cells combat the deleterious effects of high temperature remains unknown. Most remarkable is the evolutionary conservation of the components of the heat-shock regulatory system: *Drosophila* genes introduced into mammalian cells are faithfully induced upon heat shock, but respond at a temperature characteristic of mammalian cells. Complete elucidation of the mechanisms underlying the heat-shock response therefore promises to provide general insights into gene regulation in all cells.

Positive Control of Gene Expression by Steroid Hormones: Enhancer Action and Messenger Stabilization[132, 133, 134]

Steroid hormones are important to the development and physiological regulation of animals ranging from water molds to humans. When a sensitive target cell is treated with a steroid, the specific synthesis of new RNAs is stimulated, usually within minutes. This initial response appears to be direct, for inhibitors of protein synthesis have no effect. Steroids, however, do not act by themselves. Instead, they bind tightly to highly specific receptor proteins (called **steroid hormone receptors**) that are present in several thousand copies per cell. The hormone-receptor complexes then interact with particular genomic loci to turn on transcription of the appropriate gene(s).

Best understood is the action of the glucocorticoid receptor, a large protein made up of four identical 94 kdal subunits. When complexed with dexamethasone, a synthetic glucocorticoid hormone, it binds to specific DNA sequences that can be considered hormone-dependent transcriptional enhancers. In the well-studied mouse mammary tumor virus genome, there are five such binding regions (one upstream and four downstream from the start point of transcription; Figure 21-48). In vitro, binding of the hormone-receptor complex occurs independently to these sites. In vivo, these receptor-binding sequences confer stimulation by glucocorticoids when fused in either orientation to another promoter whose expression is not normally hormone regulated. Several alterations in the chromatin structure of the hormone-activated gene(s) can be detected. First, general DNase I sensitivity increases upon dexamethasone addition and persists long after hormone withdrawal. Second, discrete DNase I–hypersensitive regions develop near the sites of hormone-receptor binding; these disappear when hormone is removed. Near genes whose expression is regulated by more than one steroid hormone (e.g., the chicken lysozyme gene responds to both glucocorticoids and progesterone), there are distinct (but sometimes overlapping) sites for the binding of

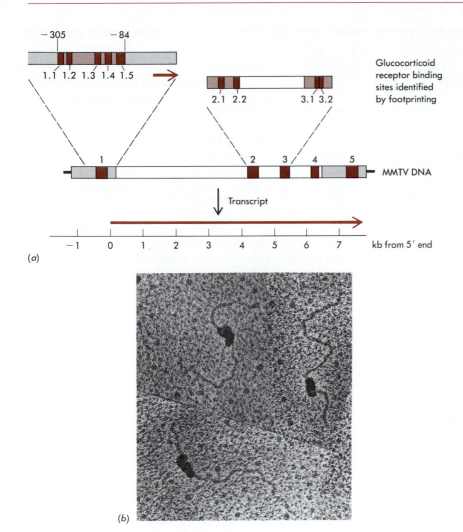

(a)

(b)

Glucocorticoid
receptor binding
sites identified
by footprinting

MMTV DNA

Transcript

kb from 5' end

Figure 21-48
Each of the five glucocorticoid receptor binding regions in mouse mammary tumor virus (MMTV) DNA contains multiple interaction sites. (a) Binding sites for the purified receptor protein were mapped by DNase footprinting. Each site (colored boxes) contains sequences closely related to a consensus octanucleotide $AGA_T^ACAG_T^A$. Certain mutations introduced into the protected sequences have been shown to alter both receptor binding and enhancer activity. Gray boxes at the ends of the DNA denote the right and left LTR (long terminal repeat) of the MMTV genome. [After K. R. Yamamoto, *Molecular Developmental Biology: Expressing Foreign Genes*, ed. L. Bogorad and G. Adelman (New York: Alan Liss, in press); and F. Payvar et al., *Cell* 35 (1983):381–392.] (b) Electron micrograph of receptor bound to region 1 DNA. Both the size and shape of the bound protein cluster suggest that many subunits are involved. (Courtesy of K. Yamamoto.)

the different hormone-receptor complexes. All these observations once again raise the question of how enhancer elements operate. Yet, they strongly argue that the glucocorticoid and other steroid hormone receptor proteins can be considered prototypes of the class of tissue-specific proteins that activate transcription through their interaction with enhancer-like elements.

Also contributing to steroid hormone regulation are mechanisms acting at other levels of gene expression. Most striking of these is the selective stabilization of hormone-stimulated transcripts against cytoplasmic degradation. For instance, in *Xenopus* liver, the half-life of vitellogenin mRNA (vitellogenin is an egg yolk precursor protein) is about 3 weeks in the presence of the inducing hormone estrogen; it drops to 16 hours after estrogen withdrawal. Again, it seems that the hormone acts through some sort of receptor protein, which appears to be different from the receptor that binds to the vitellogenin gene region to activate transcription. Selective mRNA stabilization has also been observed for a number of other genes that are stimulated by steroid hormones. In addition, after hormone treatment of sensitive cells, RNA processing has been reported to be altered, and increased rates of overall RNA synthesis are sometimes observed, leading to proliferative growth. Thus, we see that steroid hormones achieve massive accumulation of specific mRNAs by concerted manipulation of both transcriptional and posttranscriptional processes. Deciphering the precise underlying mechanisms is a current challenge for molecular biologists.

Summary

The huge genomes of higher eucaryotic cells are associated with specific proteins that serve to pack and organize the DNA within the nucleus. The first level of packing involves the four core histone proteins that wrap approximately 150 base pairs of DNA around the exterior of the nucleosome core particle. The next higher order of packing requires in addition histone H1, which associates preferentially with linker DNA between nucleosomes and yields a 300 Å filament. Even more highly condensed forms of eucaryotic DNA are found in interphase heterochromatin, in the metaphase scaffold, and in sperm heads, where the histones are replaced by another class of DNA binding proteins.

During interphase, different segments of the genome are replicated at different times. Replication early in S phase is correlated with transcriptional activity (or potential activity) and light-staining Giemsa bands on metaphase chromosomes. During all stages of the cell cycle, the DNA remains attached to an underlying structure called the nuclear scaffold. Points of attachment are located at specific sites in repeating gene clusters, close to but not within sequences that are copied into RNA.

Active chromatin is distinguished from inactive regions of higher eucaryotic genomes by a number of morphological and biochemical criteria. In the giant polytene chromosomes of *Drosophila* salivary glands, bands puff visibly as they become transcriptionally active. However, since a single band may contain several genes that are not cotranscribed, puffing and transcription are not strictly linked. Examination in the electron microscope has suggested that the regular packing of DNA into nucleosomes is altered or absent in very actively transcribed genes. Biochemically, active chromatin exhibits a heightened general sensitivity to DNase, hypersensitivity to DNase at a few more localized sites in or near the gene, and (usually) undermethylation of C residues relative to other parts of the genome.

The synthesis, processing, and assembly of ribosomal rRNA occur in nucleoli. Promoter regions and other activating sequences for RNA polymerase I lie upstream from the start site of transcription. Both in vivo and in vitro, the initiation of rRNA synthesis is highly species-specific because one or more required transcription factors have coevolved to recognize promoters only in the rDNA of the same species.

Several types of conserved sequence elements are present in promoter regions for RNA polymerase II, which synthesizes all cellular messengers. They are located upstream from the start site and are tightly bound by specific factors required for initiating transcription. More distant activating sequences known as enhancers can function in either orientation, can reside even within or downstream from the transcribed region, and are bound by tissue-specific proteins that somehow facilitate transcription from a nearby promoter.

Essential sequences within the genes themselves are a hallmark of promoters for RNA polymerase III. These internal control regions are bound by factors to form stable transcription complexes that synthesize multiple copies of product before dissociation from the DNA. The existence of gene-specific transcription factors that selectively activate certain 5S or tRNA genes in particular tissues is exemplified by a *Xenopus* factor that binds both to the oocyte-specific 5S gene and to its 5S rRNA product.

Transcripts synthesized by RNA polymerase II undergo a number of covalent modifications before export to the cytoplasm as mature messenger RNAs. Their 3' ends do not arise by transcription termination but from specific cleavage and (in most cases) the addition of poly A tracts. Small RNA-protein complexes participate in these processing events as well as in splicing. Meanwhile, the pre-mRNA is packaged by a special set of nuclear RNA binding proteins. Differences in promoter utilization, in splice site selection, and in 3'-end placement can enable a single gene to yield a number of different mRNAs (and protein products) in a tissue-specific or developmentally specific fashion.

Protein synthesis by 80S ribosomes proceeds along the same general pathway as with bacterial ribosomes. But the number of factors involved (particularly for initiation) is much greater. Usually, only one AUG per message serves as an initiator codon. Polypeptide chain initiation is regulated by a number of agents using a common mechanism: Phosphorylation of one of the initiation factors prevents its reutilization in bringing the initiator tRNA to the ribosome.

Newly synthesized proteins are specifically delivered to a variety of membrane-bound compartments within the higher eucaryotic cell. A small RNA-protein complex called SRP recognizes an N-terminal signal sequence to initiate the transfer of secreted proteins into the endoplasmic reticulum; SRP blocks completion of these polypeptide chains if the correct membrane is not present. For delivery to chloroplasts, to mitochondria or to the nucleus, other specific signals within the protein are essential, but their role and the targeting mechanisms involved are only now being worked out.

Even after their translation is complete, the proteins of higher eucaryotic cells are often modified further. Some are extensively cleaved to yield smaller functional products, such as peptide hormones. Also operating are energy-dependent systems that selectively degrade particular proteins; changes in cell or tissue metabolism requires not only the appearance of new proteins but also the disappearance of preexisting ones.

Gene function can be regulated by manipulating any of the many steps required for gene expression in higher eucaryotic cells. Clear evidence has emerged for control at both transcriptional and posttranscriptional levels for several well-studied gene systems. Histone mRNAs are both transcribed at higher rates and selectively stabilized during the S phase of the cell cycle. The *Drosophila* genes that are selectively turned on after heat shock require a specific transcription factor that binds to multiple sites on the DNA upstream from each RNA start point; the translational machinery also discriminates in favor of the mRNAs encoding the heat-shock proteins. Steroid hormones interact with special receptor proteins, which then bind DNA regions that function as enhancers to activate transcription from a nearby promoter. Striking stabilization of hormone-induced mRNAs as well as changes in RNA processing also contribute to the hormone response.

Bibliography

General References

Alberts, B., D. Bray, J. Lewis, M. Raff, K. Roberts, and J. D. Watson. *Molecular Biology of the Cell.* 2nd edition, in press. New York: Garland. This valuable text presents all aspects of the functioning of higher eucaryotic cells, as well as a more detailed consideration of the structures and functional states of chromosomes.

Busch, H., ed. *The Cell Nucleus.* New York Academic Press. A whole series of volumes (the first published in 1970) containing review articles on nuclear structure and function. Especially relevant here are Vol. IV–VIII on chromatin, Vol. VIII and IX on nuclear particles, and Vol. XI and XII on nucleolar and other rDNA-containing chromatin.

Cold Spring Harbor Symp. Quant. Biol. XLVII. 1982. *Structures of DNA.* Cold Spring Harbor, N.Y.: Cold Spring Harbor Laboratory. A two volume publication that includes many articles on chromatin structure as well as its organization into transcribing genes.

Gluzman, Y., ed. 1985. *Eukaryotic Transcription: The Role of cis- and trans-acting Elements in Initiation.* Cold Spring Harbor, N.Y.: Cold Spring Harbor Laboratory. An even more current version of the next reference (*Enhancers and Eucaryotic Gene Expression*) that focuses not only on enhancers, but also on other transcription factors and the signals that they recognize.

Gluzman, Y., and T. Shenk. 1983. *Enhancers and Eucaryotic Gene Expression.* Cold Spring Harbor, N.Y.: Cold Spring Harbor Laboratory. Current research articles on all aspects of enhancer structure and function.

Hadjiolov, A. A. 1985. *The Nucleolus and Ribosome Biogenesis.* New York, N.Y.: Springer-Verlag. A complete and informative treatise on the nucleolar structure and function, but lacking the very most recent references.

Perez-Bercoff, R., ed. 1982. *Protein Biosynthesis in Eucaryotes.* New York: Plenum. A compendium of valuable contributions on all aspects of eucaryotic translation.

Razin, A., H. Cedar, and A. D. Riggs, eds. (1984) *DNA Methylation: Biochemistry and Biological Significance.* New York, N.Y.: Springer-Verlag. An authoritative and balanced treatment of the molecular biology of DNA methylation.

Schlesinger, M. J., M. Ashburner, and A. Tissières, eds. 1982. *Heat Shock, from Bacteria to Man.* Cold Spring Harbor, N.Y.: Cold Spring Harbor Laboratory. Everything about the heat-shock response as of 1982.

Cited References

1. Kornberg, R. D., and A. Klug. 1981. "The Nucleosome." *Sci. Amer.* 244:52–64.
2. McGhee, J. D., and G. Felsenfeld. 1980. "Nucleosome Structure." *Ann. Rev. Biochem.* 49:1115–1156.
3. Igo-Kemenes, T., W. Horz, and H. G. Zachau. 1982. "Chromatin." *Ann. Rev. Biochem.* 51:89–121.
4. Richmond, T. J., J. T. Finch, B. Rushton, D. Rhodes, and A. Klug. 1984. "Structure of the Nucleosome Core Particle at 7 Å Resolution." *Nature* 311:532–537.
5. Wang, J. C. 1982. "The Path of DNA in the Nucleosome." *Cell* 724:724–726.
6. Simpson, R. T., and D. W. Stafford. 1983. "Structural Features of a Phased Nucleosome Core Particle." *Proc. Natl. Acad. Sci.* USA 80:51–55.
7. Thoma, F., T. Koller, and A. Klug. 1979. "Involvement of Histone H1 in the Organization of the Nucleosome and of the Salt-Dependent Superstructures of Chromatin." *J. Cell Biol.* 83:403–427.
8. Brown, S. W. 1966. "Heterochromatin." *Science* 151:417–425.
9. Paulson, J. R. and U. K. Laemmli. 1977. "The Structure of Histone-Depleted Metaphase Chromosomes." *Cell* 12:817–828.
10. Lewis, C. D., and U. K. Laemmli. 1982. "Higher-Order Metaphase Chromosome Structure: Evidence for Metalloprotein Interactions." *Cell* 29:171–181.
11. Warrant, R. W., and S.-H. Kim. 1978. "α-Helix-Double Helix Interaction Shown in the Structure of a Protamine-Transfer RNA Complex and a Nucleoprotamine Model." *Nature* 271:130–135.
12. Hand, R. 1978. "Eucaryotic DNA: Organization of the Genome for Replication." *Cell* 15:317–325.
13. Goldman, M. A., G. P. Holmquist, M. C. Gray, L. A. Caston, and A. Nag. 1984. "Replication Timing of Genes and Middle Repetitive Sequences." *Science* 224:686–692.
14. Mirkovitch, J., M.-E. Mirault, and U. K. Laemmli. 1984. "Organization of the Higher-Order Chromatin Loop: Specific DNA Attachment Sites on Nuclear Scaffold." *Cell* 39:223–232.
15. Ashburner, M., and G. Richards. 1976. "The Role of Ecdysone in the Control of Gene Activity in the Polytene Chromosomes of *Drosophila*." In *Insect Development*, ed. P. A. Lawrence. New York: Wiley, pp. 203–225.
16. Bonner, J. J. and M. L. Pardue. 1977. "Polytene Chromosome Puffing and *in Situ* Hybridization Measure Different Aspects of RNA Metabolism." *Cell* 12:227–234.
17. Lamb, M. M., and B. Daneholt. 1979. "Characterization of Active Transcription Units in Balbiani Rings of *Chironomus tentans*." *Cell* 17:835–848.
18. Sirotkin, K., and N. Davidson. 1982. "Developmentally Regulated Transcription from *Drosophila melanogaster* Chromosomal Site 67B." *Dev. Biol.* 89:196–210.
19. Corces, V., R. Holmgren, R. Freund, R. Morimoto, and M. Meselson. 1980. "Four Heat Shock Proteins of *Drosophila melanogaster* Coded Within a 12-Kilobase Region in Chromosome Subdivision 67B." *Proc. Nat. Acad. Sci.* 77:5390–5393.
20. Weisbrod, S. 1982. "Active Chromatin." *Nature* 297:289–294.
21. Elgin, S. C. R. 1981. "DNAase I-Hypersensitive Sites of Chromatin." *Cell* 27:413–415.
22. Weintraub, H., and M. Groudine. 1976. "Chromosomal Subunits in Active Genes Have an Altered Conformation." *Science* 193:848–856.
23. Garel, A., and R. Axel. 1976. "Selective Digestion of Transcriptionally Active Ovalbumin Genes from Oviduct Nuclei." *Proc. Nat. Acad. Sci.* 73:3966–3970.
24. Wu, C., 1980. "The 5' Ends of *Drosophila* Heat Shock Genes in Chromatin Are Hypersensitive to DNase I." *Nature* 286:854–860.
25. Saragosti, S., G. Moyne, and M. Yaniv. 1980. "Absence of Nucleosomes in a Fraction of SV40 Chromatin Between the Origin of Replication and the Region Coding for the Late Leader RNA." *Cell* 20:65–73.
26. Elgin, S. C. R., 1984. "Anatomy of Hypersensitive Sites." *Nature* 309:213–214.
27. Emerson, B. M., C. D. Lewis, and G. Felsenfeld. 1985. "Interaction of Specific Nuclear Factors with the Nuclease-Hypersensitive Region of the Chicken Adult β-Globin Gene: Nature of the Binding Domain." *Cell* 41:21–30.
28. Felsenfeld, G., and J. McGhee. 1982. "Methylation and Gene Control." *Nature* 296:602–603.
29. Doerfler, W. 1983. "DNA Methylation and Gene Activity." *Ann. Rev. Biochem.* 52:93–124.
30. Bird, A. P. 1984. "DNA Methylation—How Important in Gene Control?" *Nature* 307:503–504.
31. Allfrey, V. G. 1977. "Post-Synthetic Modifications of Histone Structure: A Mechanism for the Control of Chromosome Structure by the Modulation of Histone-DNA Interactions." In *Chromatin and Chromosome Structure*, ed. H. J. Li and R. Eckhardt. New York: Academic Press, pp. 167–191.
32. Ball, D. J., D. S. Gross, and W. T. Garrard. 1983. "5-Methylcytosine Is Localized in Nucleosomes That Contain Histone H1." *Proc. Nat. Acad. Sci.* 80:5490–5494.
33. Weintraub, H. 1985. "Assembly and Propagation of Re-

pressed and Derepressed Chromosomal States." *Cell* 42:705–711.

34. Varshavsky, A., L. Levinger, O. Sundin, J. Barsoum, E. Ozkaynak, P. Swerdlow, and D. Finley. 1983. "Cellular and SV40 Chromatin: Replication, Segregation, Ubiquitination, Nuclease-Hypersensitive Sites, HMG-Containing Nucleosomes, and Heterochromatin-Specific Protein." *Cold Spring Harbor Symp. Quant. Biol.* 47:511–520.

35. Labhart, P., and T. Koller. 1982. "Structure of the Active Nucleolar Chromatin of *Xenopus laevis* Oocytes." *Cell* 28:279–292.

36. Widmer, R. M., R. Lucchini, M. Lezzi, B. Meyer, J. M. Sogo, J.-E. Edstrom, and T. Koller. 1984. "Chromatin Structure of a Hyperactive Secretory Protein Gene (in Balbiani Ring 2) of *Chironomus*." *EMBO J.* 3:1635–1641.

37. Foe, V. E., L. E. Wilkinson, and C. D. Laird. 1976. "Comparative Organization of Active Transcription Units in *Oncopeltus fasciatus*." *Cell* 9:131–146.

38. Lin, S., and A. D. Riggs. 1975. "The General Affinity of lac Repressor for *E. coli* DNA: Implications for Gene Regulation in Prokaryotes and Eukaryotes." *Cell* 4:107–111.

39. Hadjiolav, A. A. and N. Nikolaev. 1976. "Maturation of Ribosomal Ribonucleic Acids and the Biogenesis of Ribosomes." *Prog. Biophys. Mol. Biol.* 31:95–144.

40. Vaughn, J. C., S. J. Sperbeck, W. J. Ramsey, and C. B. Lawrence. 1984. "A Universal Model for the Secondary Structure of 5.8S Ribosomal RNA Molecules, Their Contact Sites with 28S Ribosomal RNAs, and Their Prokaryotic Equivalent." *Nucleic Acid Res.* 12:7479–7502.

41. Bowman, L. H., W. E. Goldman, G. I. Goldberg, M. B. Hebert, and D. Schlessinger. 1983. "Location of the Initial Cleavage Sites in Mouse Pre-mRNA." *Mol. Cell. Biol.* 3:1501–1510.

42. Sommerville, J. 1984. "RNA Polymerase I Promoters and Transcription Factors." *Nature* 310:189–190.

43. Reeder, R. H. 1984. "Enhancers and Ribosomal Gene Spacers." *Cell* 38:349–351.

44. Moss, T. 1982. "Transcription of Cloned *Xenopus laevis* Ribosomal DNA Microinjected into *Xenopus* Oocytes, and the Identification of an RNA Polymerase I Promoter." *Cell* 30:835–842.

45. Grummt, I. 1982. "Nucleotide Sequence Requirements for Specific Initiation of Transcription by RNA Polymerase I." *Proc. Nat. Acad. Sci.* 79:6908–6911.

46. Yamamoto, O., N. Takakusa, Y. Mishima, R. Kominami, and M. Muramatsu. 1984. "Determination of the Promoter Region of Mouse Ribosomal RNA Gene by an *in Vitro* Transcription System." *Proc. Nat. Acad. Sci.* 81:299–303.

47. Boseley, P., T. Moss, M. Machler, R. Portmann, and M. Birnstiel. 1979. "Sequence Organization of the Spacer DNA in a Ribosomal Gene Unit of *Xenopus laevis*." *Cell* 17:19–31.

48. Reeder, R. H., and J. C. Roan. 1984. "The Mechanism of Nucleolar Dominance in *Xenopus* Hybrids." *Cell* 38:39–44.

49. Miesfeld, R., and N. Arnheim. 1984. "Species-Specific rDNA Transcription Is Due to Promoter-Specific Binding Factors." *Mol. Cell. Biol.* 4:221–227.

50. Learned, R. M., S. Cordes, and R. Tjian. 1985. "Purification and Characterization of a Transcription Factor That Confers Promoter Specificity to Human RNA Polymerase I." *Mol. Cell Biol.* 5:1358–1369.

51. Breathnach, R., and P. Chambon. 1981. "Organization and Expression of Eucaryotic Split Genes Coding for Proteins." *Ann. Rev. Biochem.* 50:349–383.

52. Nevins, J. R., 1983. "The Pathway of Eukaryotic mRNA Formation." *Ann. Rev. Biochem.* 52:441–466.

53. McKnight, S. L. and R. Kingsbury. 1982. "Transcriptional Control Signals of a Eukaryotic Protein-Coding Gene." *Science* 217:316–324.

54. Myers, R. M., L. S. Lerman, and T. Maniatis. 1985. "A General Method for Saturation Mutagenesis of Cloned DNA Fragments." *Science* 229:242–247.

55. Cochran, M. D. and C. Weissmann. 1984. "Modular Structure of the β-Globin and the TK Promoters." *EMBO J.* 3:2453–2459.

56. McKnight, S. L., R. C. Kingsbury, A. Spence, and M. Smith. 1984. "The Distal Transcription Signals of the Herpes Virus *tk* Gene Share a Common Hexanucleotide Control Sequence." *Cell* 37:253–262.

57. Shenk, T. 1981. "Transcriptional Control Regions: Nucleotide Sequence Requirements for Initiation by RNA Polymerase II and III." In *Current Topics in Microbiology and Immunology*, vol. 93. New York: Springer-Verlag, pp. 25–46.

58. Dynan, W. S., and R. Tjian. 1985. "Control of Eukaryotic Messenger RNA Synthesis by Sequence-Specific DNA Binding Proteins." *Nature* 316:774–777.

59. Parker, C. S., and J. Topol. 1984. "A *Drosophila* RNA Polymerase II Transcription Factor Contains a Promoter-Region-Specific DNA-Binding Activity." *Cell* 36:357–369.

60. Davison, B. L., J.-M. Egly, E. R. Mulvihill, and P. Chambon. 1983. "Formation of Stable Preinitiation Complexes Between Eukaryotic Class B Transcription Factors and Promoter Sequences." *Nature* 301:680–686.

61. Gidoni, D., W. S. Dynan, and R. Tjian. 1984. "Multiple Specific Contacts Between a Mammalian Transcription Factor and Its Cognate Promoters." *Nature* 312:409–413.

62. Bird, A., M. Taggart, M. Frommer, O. J. Miller, and D. Macleod. 1985. "A Fraction of the Mouse Genome That Is Derived from Islands of Nonmethylated, CpG-Rich DNA." *Cell* 40:91–99.

63. Fire, A., M. Samuels, and P. A. Sharp. 1984. "Interactions Between RNA Polymerase II, Factors, and Template Leading to Accurate Transcription." *J. Biol. Chem.* 259:2509–2516.

64. Sawadogo, M., and R. C. Roeder. 1984. "Energy Requirement for Specific Transcription Initiation by the Human RNA Polymerase II System." *J. Biol. Chem.* 259:5321–5326.

65. Khoury, G., and P. Gruss. 1983. "Enhancer Elements." *Cell* 33:313–314.

66. Weiher, H., M. Konig, and P. Gruss. 1983. "Multiple Point Mutations Affecting the Simian Virus 40 Enhancer." *Science* 219:626–631.

67. Schaffner, W., E. Serfling, and M. Jasin. 1985. "Enhancers and Eukaryotic Gene Transcription." *Trends in Genetics* 1:224–230.

68. Sergeant, A., D. Bohmann, H. Zentgraf, H. Weiher, and W. Keller. 1984. "A Transcription Enhancer Acts *in Vitro* Over Distances of Hundreds of Base-Pairs on Both Circular and Linear Templates but Not on Chromatin-Reconstituted DNA." *J. Mol. Biol.* 180:577–600.

69. Sassone-Corsi, P., A. Wideman, and P. Chambon. 1985. "A *trans*-Acting Factor Is Responsible for the Simian Virus 40 Enhancer Activity *in Vitro*." *Nature* 312:458–463.

70. Scholer, H. R., and P. Gruss. 1984. "Specific Interaction Between Enhancer-Containing Molecules and Cellular Components." *Cell* 36:403–411.

71. Ciliberto, G., L. Castagnoli, and R. Cortese. 1983. "Transcription by RNA Polymerase III." *Curr. Top. Dev. Biol.* 18:59–88.

72. Sakonju, S., D. F. Bogenhagen, and D. D. Brown. 1980. "A Control Region in the Center of the 5S RNA Gene Directs Specific Initiation of Transcription I: The 5' Border of the Region." *Cell* 19:13–25.

73. Bogenhagen, D. F., S. Sakonju, and D. D. Brown. 1980. "A Control Region in the Center of the 5S RNA Gene Directs Specific Initiation of Transcription II: The 3' Border of the Region." *Cell* 19:27–35.

74. Hall, B., S. G. Clarkson, and G. Tocchini-Valenti. 1982. "Transcription Initiation of Eucaryotic Transfer RNA Genes." *Cell* 29:3–5.

75. Stillman, D. J., and E. P. Geiduschek. 1984. "Differential Binding of a *S. cerevisiae* RNA Polymerase III Transcription Factor to the Promoter Segments of a tRNA Gene." *EMBO J.* 3:847–853.

76. Wilson, E. T., D. Larson, L. S. Young, and K. U. Sprague. 1985. "A Large Region Controls tRNA Gene Transcription." *J. Mol. Biol.* 183:153–163.

77. Schaack, J., S. Sharp, T. Dingermann, D. J. Burke, L. Cooley, and D. Soll. 1984. "The Extent of a Eukaryotic tRNA Gene." *J. Biol. Chem.* 259:1461–1467.

78. Brown, D. D. 1984. "The Role of Stable Complexes That Repress and Activate Eucaryotic Genes." *Cell* 37:359–365.

79. Lassar, A. B., P. L. Martin, and R. G. Roeder. 1983. "Transcription of Class III Genes: Formation of Preinitiation Complexes." *Science* 222:740–748.

80. Garel, J. P., R. L. Garber, and M. A. Siddiqui. 1977. "Transfer RNA in Posterior Silk Gland of *Bombyx mori*: Polyacrylamide Gel Mapping of Mature Transfer RNA, Identification and Partial Structural Characterization of Major Isoacceptor Species." *Biochemistry* 16:3618–3624.

81. Sprague, K. U., O. Hagenbuchle, and M. C. Zuniga. 1977. "The Nucleotide Sequence of Two Silk Gland Alanine tRNAs: Implications for Fibroin Synthesis and for Initiator tRNA Structure." *Cell* 11:561–570.

82. Goldsmith, M. R., and F. C. Kafatos. 1984. "Developmentally Regulated Genes in Silkmoths." *Ann. Rev. Genetics* 18:443–487.

83. Beyer, A. L., M. E. Christensen, B. W. Walker, and W. M. LeStourgeon. 1977. "Identification and Characterization of the Packaging Proteins of Core 40S hnRNP Particles." *Cell* 11:127–138.

84. Martin, T. E., J. M. Pullman, and M. D. McMullen. 1980. "Structure and Function of Nuclear and Cytoplasmic Ribonucleoprotein Complexes." In *Cell Biology, a Comprehensive Treatise*, vol. 4, ed. L. Goldstein and D. N. Prescott. New York: Academic Press, pp. 137–174.

85. Dreyfuss, G., Y. D. Choi, and S. A. Adam. 1984. "Characterization of Heterogeneous Nuclear RNA-Protein Complexes *in Vivo* with Monoclonal Antibodies." *Mol. Cell. Biol.* 4:1104–1114.

86. Darnell, J. E., Jr. 1982. "Variety in the Level of Gene Control in Eukaryotic Cells." *Nature* 297:365–370.

87. Birnstiel, M. L., M. Busslinger, and K. Strub. 1985. "Transcription Termination and 3′ Processing: The End Is in Site!" *Cell* 41:349–359.

88. Falck-Pedersen, E., J. Logan, T. Shenk, and J. E. Darnell, Jr. 1985. "Transcription Termination Within the E1A Gene of Adenovirus Induced by Insertion of the Mouse β-Major Globin Terminator Element." *Cell* 40:897–905.

89. Moore, C. L. and P. A. Sharp. 1985. "Accurate Cleavage and Polyadenylation of Exogenous RNA Substrate." *Cell* 41:845–855.

90. Amara, S. G., V. Jonas, M. G. Rosenfeld, E. S. Ong, and R. M. Evans. 1982. "Alternative RNA Processing in Calcitonin Gene Expression Generates mRNAs Encoding Different Polypeptide Products." *Nature* 298:240–244.

91. Nabeshima, Y., Y. Fujii-Kuriyama, M. Muramatsu, and K. Ogata. 1984. "Alternative Transcription and Two Modes of Splicing Result in Two Myosin Light Chains from One Gene." *Nature* 308:333–338.

92. Breitbart, R. E., H. T. Nguyen, R. M. Medford, A. T. Destree, V. Mahdavi, and B. Nadal-Ginard. 1985. "Intricate Combinational Patterns of Exon Splicing Generate Multiple Regulated Troponin T Isoforms from a Single Gene." *Cell* 41:67–82.

93. Moldave, K. 1985. "Eukaryotic Protein Synthesis." *Ann. Rev. Biochem.* 54:1109–1119.

94. Kozak, M. 1983. "Comparison of Initiation of Protein Synthesis in Procaryotes, Eucaryotes, and Organelles." *Microbiol. Rev.* 47:1–45.

95. Shatkin, A. J. 1985. "mRNA Cap Binding Proteins: Essential Factors for Initiating Translation." *Cell* 40:223–224.

96. Weissbach, H. 1980. "Soluble Factors in Protein Synthesis." In *Ribosomes: Structure, Function and Genetics*, ed. G. Chambliss, G. R. Craven, J. Davies, K. Davis, L. Kahan, and M. Nomura. Baltimore: University Park Press, pp. 377–411.

97. Hershey, J. W. B. 1980. "The Translational Machinery: Components and Mechanism." In *Cell Biology, a Comprehensive Treatise*, vol. 4, ed. L. Goldstein and D. N. Prescott. New York: Academic Press, pp. 1–68.

98. Caskey, C. T. H. 1980. "Peptide Chain Termination." *Trends in Biochem. Sci.* 5:234–237.

99. Deutscher, M. P. 1984. "The Eukaryotic Aminoacyl-tRNA Synthetase Complex: Suggestions for Its Structure and Function." *J. Cell. Biol.* 99:373–377.

100. Vazquez, D., E. Zaera, H. Dolz, and A. Jimenz. 1982. "Action of the Inhibitors of Protein Biosynthesis." In *Protein Biosynthesis in Eucaryotes*, ed. R. Perez-Bercoff. New York: Plenum, pp. 311–337.

101. Pappenheimer, A. M., Jr., and D. M. Gill. 1973. "Diphtheria." *Science* 182:353–358.

102. Lee, H., and W. J. Iglewski. 1984. "Cellular ADP-Ribosyltransferase with the Same Mechanism of Action as Diphtheria Toxin and *Pseudomonas* Toxin A." *Proc. Nat. Acad. Sci.* 81:2703–2707.

103. Hunt, T. 1985. "False Starts in Translational Control of Gene Expression." *Nature* 316:580–581.

104. Ochoa, S. 1983. "Regulation of Protein Synthesis Initiation in Eucaryotes." *Archives of Biochem. and Biophysics* 223:325–349.

105. Safer, B. 1983. "2B or Not 2B: Regulation of the Catalytic Utilization of eIF-2." *Cell* 33:7–8.

106. Walter, P., R. Gilmore, and G. Blobel. 1984. "Protein Translocation Across the Endoplasmic Reticulum." *Cell* 38:5–8.

107. Meyer, D. I., E. Krause, and B. Dobberstein. 1982. "Secretory Protein Translocation Across Membranes—the Role of the 'Docking Protein.'" *Nature* 297:647–650.

108. Schatz, G., and R. A. Butow. 1983. "How Are Proteins Imported into Mitochondria?" *Cell* 32:316–318.

109. Grossman, A., S. Bartlett, and N. H. Chua. 1980. "Energy-Dependent Uptake of Cytoplasmically Synthesized Polypeptides by Chloroplasts." *Nature* 285:625–628.

110. Dingwall, C., S. V. Sharnick, and R. A. Laskey. 1982. "A Polypeptide Domain That Specifies Migration of Nucleoplasmin into the Nucleus." *Cell* 30:449–458.

111. Kalderon, D., B. L. Roberts, W. D. Richardson, and A. E. Smith. 1984. "A Short Amino Acid Sequence Able To Specify Nuclear Location." *Cell* 39:499–509.

112. Hall, M. N., L. Hereford, and I. Herskowitz. 1984. "Targeting of *E. coli* β-Galactosidase to the Nucleus in Yeast." *Cell* 36:1057–1065.

113. Silver, P. A., L. P. Keegan, and M. Ptashne. 1984. "Amino Terminus of the Yeast GAL4 Gene Product Is Sufficient for Nuclear Localization." *Proc. Nat. Acad. Sci.* 81:5951–5955.

114. Mattaj, I. W., and E. M. DeRobertis. 1985. "Nuclear Segregation of U2 snRNA Requires Binding of Specific snRNP Proteins." *Cell* 40:111–118.

115. Wolin, S. 1985. "Small Cytoplasmic Ribonucleoproteins." *Trends in Genetics* 1:201–204.

116. Steiner, D. F. 1984. "The Biosynthesis of Insulin: Genetic, Evolutionary, and Pathophysiologic Aspects." Harvey Lecture, Series 78. New York: Academic Press, pp. 191–228.

117. Noda, M., Y. Furutani, H. Takahashi, M. Toyosato, T. Hirose, S. Inayama, S. Nakanishi, and S. Numa. 1982. "Cloning and Sequence Analysis of cDNA for Bovine Adrenal Preproenkephalin." *Nature* 295:202–206.

118. Mains, R. E., B. A. Eipper, C. C. Glembotski, and R. M. Dores. 1983. "Strategies for the Biosynthesis of Bioactive Peptides." *Trends in Neurosci.* 4:229–235.

119. Douglass, J., O. Civelli, and E. Herbert. 1984. "Polyprotein Gene Expression: Generation of Diversity of Neuroendocrine Peptides." *Ann. Rev. Biochem.* 53:665–715.

120. Hershko, A., and A. Ciechanover. 1982. "Mechanisms of Intracellular Protein Breakdown." *Ann. Rev. Biochem.* 51:335–364.

121. Busch, H. 1984. "Ubiquitination of Proteins." *Meth. Enzymol.* 106:238–262.

122. Ciechanover, A., D. Finley, and A. Varshavsky. 1984. "The Ubiquitin-Mediated Proteolytic Pathway and Mechanisms of Energy-Dependent Intracellular Protein Degradation." *J. Cell. Biochem.* 24:27–53.

123. Ciechanover, A., S. L. Wolin, J. A. Steitz, and H. F. Lodish. 1985. "Transfer RNA Is an Essential Component of the Ubiquitin- and ATP-Dependent Proteolytic System." *Proc. Nat. Acad. Sci.* 82:1341–1345.

124. Heintz, N., and R. G. Roeder. 1984. "Transcription of Human Histone Genes in Extracts from Synchronized HeLa Cells." *Proc. Nat. Acad. Sci.* 81:2713–2717.

125. Sittman, D. B., R. A. Graves, and W. F. Marzluff. 1983. "Histone mRNA Concentrations Are Regulated at the Level of Transcription and mRNA Degradation." *Proc. Nat. Acad. Sci.* 80:1849–1853.

126. Ashburner, M., and J. J. Bonner. 1979. "The Induction of Gene Activity in *Drosophila* by Heat Shock." *Cell* 17:241–254.

127. Pelham, H. R. B. 1982. "A Regulatory Upstream Promoter Element in the *Drosophila* Hsp 70 Heat-Shock Gene." *Cell* 30:517–528.

128. Parker, C. S., and J. Topol. 1984. "A *Drosophila* RNA Polymerase II Transcription Factor Binds to the Regulatory Site of an Hsp 70 Gene." *Cell* 37:273–283.

129. Ballinger, D. G., and M. L. Pardue. 1983. "The Control of Protein Synthesis During Heat Shock in *Drosophila* Cells Involves Altered Polypeptide Elongation Rates." *Cell* 33:103–114.

130. Klemenz, R., D. Hultmark, and W. Gehring. 1985. "Selective Translation of Heat Shock mRNA in *Drosophila Melanogaster* Depends on Sequence Information in the Leader." *EMBO J.* 4:2053–2060.

131. McGarry, T. J., and S. Lindquist. 1985. "The Preferential Translation of *Drosophila hsp70* mRNA Requires Sequences in the Untranslated Leader." *Cell* 42:903–911.

132. Yamamoto, K. R. 1985. "Hormone-Dependent Transcriptional Enhancement and Its Implications for Mechanisms of Multifactor Gene Regulation." In *Molecular Developmental Biology: Expressing Foreign Genes*, 43rd Symposium of the Society for Developmental Biology, ed. L. Bogorad and G. Adelman. New York: Alan Liss, pp. 131–148.

133. Zaret, K. S., and K. R. Yamamoto. 1984. "Reversible and Persistent Changes in Chromatin Structure Accompany Activation of a Glucocorticoid-Dependent Enhancer Element." *Cell* 38:29–38.

134. Brock, M. L., and D. J. Shapiro. 1983. "Estrogen Stabilizes Vitellogenin mRNA Against Cytoplasmic Degradation." *Cell* 34:207–214.

IX

GENE FUNCTION IN SPECIALIZED EUCARYOTIC SYSTEMS

CHAPTER 22

The Molecular Biology of Development

Embryology began as a descriptive science, growing out of the fascination that biologists experienced as they watched the remarkable events of early development in organisms as diverse as the fruit fly, the sea urchin, and the mouse. For example, the female frog deposits her eggs on the bottom of a pond, and the male frog fertilizes them. Shortly thereafter, tiny tadpoles can be seen freely swimming about, searching for food and shelter, completely on their own in a very large (and often dangerous) world. Viewed under the microscope, both the huge, nearly spherical frog egg and the much smaller, flagellated sperm have a characteristic morphology, but neither of these germ cells resembles the tadpole or the adult frog in any obvious way. How, then, do these two highly specialized germ cells fuse and develop so quickly into an independent organism with a functional digestive tract and a nervous system capable of coordinating rapid swimming motions? We know that the instructions for how the egg develops into an adult are written in the linear sequence of bases along the DNA of the germ cells. However, this genetic information would be useless if the fertilized egg could not express the information in an orderly fashion. During development, gene expression must therefore be regulated both in space (an adult fly must not have a leg in the middle of its forehead) and in time (the larval fly must not prematurely develop wings). In addition, it is not sufficient for the DNA within each cell to be properly expressed; the cells must also interact with one another so as to build complicated multicellular structures such as wings and legs. Embryologists search for the general principles of development by concentrating on the early developmental stages in the life of the organism when cellular differentiation and interaction are most apparent.

The Heart of Embryology Is the Problem of Cell Differentiation

All higher plants and animals are constructed from a large variety of cell types (e.g., nerve cells, muscle cells, thyroid cells, and blood cells) that must arise in an exquisitely coordinated way. In some organisms, specialization begins with the first few cell divisions after fertilization. In other organisms, a large number of divisions occur before any progeny cell is fixed in its fate. Regardless of the exact time that **differentiation** occurs, however, it always results in the transformation of the parental cell into a large number of morphologically different progeny cell types.

747

Classical developmental biologists examined differentiation from three viewpoints. First, what are the external (and internal) influences acting on the original undifferentiated cell that might initiate a chain of events resulting in two progeny cells of different constitution? Sometimes, asymmetrically acting external forces are easy to perceive. For example, gravity forces the yolk of an amphibian egg to the bottom. Thus, after the first few cell divisions subdivide the fertilized egg, some of the embryonic cells have more yolk than others.

Second, are the molecular differences between differentiated cells extreme, or does morphological differentiation arise from the presence of only a few unique proteins in especially large numbers? We now know that each type of differentiated cell contains many molecular species peculiar to that cell type. Thus, a complete description of differentiation at the molecular level would necessarily be a most formidable task, and would not necessarily help us to understand the essential mechanisms responsible for differentiation.

Third, are the various changes that bring about differentiation irreversible, and if so, how they are perpetuated in a heritable fashion from one cell to the next? As we shall see, these are not easy questions to answer. Whether differentiation is reversible or irreversible depends not only on which cell type we are talking about, and in which organism, but also to a surprising extent on the precise details of the experiment. Moreover, we are only just now beginning to have vague hints of the mechanisms by which cells can maintain or change their state of differentiation.

Until quite recently, embryology was largely studied as an isolated subject, apart from modern genetic or biochemical ideas. Now, however, it is clear that the morphological tools of the classical embryologist cannot give satisfying answers by themselves. Instead, as in genetics, fundamental answers require analysis at the molecular level. Thus, just as recent methodological advances have made certain aspects of biochemistry and genetics indistinguishable, so embryology is being transformed by progress in biochemistry and genetics brought about by the recombinant DNA revolution (Chapter 19).

A Hierarchy of Genes Controls Development[1, 2, 3]

How, then, can we understand development at the molecular level when differentiation involves so many changes in the protein composition of the cell (together with parallel changes in key metabolic pathways and intermediates, as well as in RNA, lipid, polysaccharide, and perhaps even ionic composition)? If each of the many hundreds of gene products that distinguish a liver cell from a muscle cell had to be studied in minute detail before we could begin to understand the molecular causes of development, the situation would be virtually hopeless. However, we shall see that for many, and perhaps all, developing organisms (both procaryotic and eucaryotic), the gene products responsible for development can be arranged in a hierarchy, with some genes controlling the expression of other genes. This conclusion should not really surprise us. For example, we have already seen that the λ CI protein (repressor) controls the developmental decision to grow lytically or to lysogenize (see Figure 17-16). Similarly, the *MAT* locus in yeast controls the expression of a large array of genes responsible for the differences between the two mating types and the diploid (see Figure 18-31). These are only a few obvious examples. Thus, the many hundreds of gene products that distinguish one mammalian cell from another are not all equal; some are more important than others, because they control the initial decisions to differentiate along one developmental pathway or another. The goal of molecular biologists studying development is therefore to discover which genes control the expression of other genes.

Traditionally, one of the problems that has confounded and confused the molecular study of development is that cells usually become "committed" or "determined" to differentiate long before any actual morphological differentiation is apparent. For example, stem cells in the bone marrow of mammals give rise to at least two different kinds of progenitor cells—those that proliferate and differentiate into oxygen-carrying red blood cells and those that proliferate and differentiate into antibody-producing lymphocytes. Yet, in the very early stages of proliferation, the two kinds of progenitor cells (although committed to different fates) are difficult to distinguish. Once we realize that genes are arranged in a hierarchy, it becomes clear that in most cases, **commitment** or **determination** corresponds to the expression of a controlling gene, while differentiation reflects the myriad molecular consequences of that initial developmental decision.

Necessity of Finding Good Model Systems for Studying Differentiation

We have already noted (Chapter 20) that the relative amount of genomic DNA increases by a factor of approximately 800 from *E. coli* to a mammalian cell, but not all of this additional DNA represents a real increase in genetic complexity. Satellite sequences, moderately repeated DNA sequence families, and introns account for much of the genome in higher cells. Nonetheless, any particular mammalian cell type (say, liver) does contain a much larger number of distinct protein species than does *E. coli*. In addition, although all of the several hundred different cell types in the body share a core of "housekeeping" proteins that are necessary for metabolism and replication, different mammalian cell types (say, liver and muscle) express very different subsets of the total protein-coding capacity of the genome.

Faced with such biological complexity, how are we to choose the best organism for studying selected aspects of differentiation? Historically, rapid progress in understanding *E. coli* was due primarily to the invention of powerful genetic techniques for dissecting complicated biochemical events; these experimental advantages, in turn, attracted the concentrated efforts of many scientists. Similarly, with only a few important exceptions, molecular biologists have tended to concentrate on organisms where relatively powerful genetic techniques can be brought to bear on otherwise intractable developmental problems. Thus, even if we are ultimately curious about human biology, common sense may direct us to work on a simpler or more convenient organism such as bakers' yeast (*Saccharomyces cerevisiae*), the fruit fly (*Drosophila melanogaster*), the soil nematode worm (*Caenorhabditis elegans*), or the laboratory mouse (*Mus musculus*). However, it is important to keep in mind that biologists have always been experimental opportunists, studying whatever systems promised to yield interesting results most readily, and today's molecular biologists are no different. Some organisms, such as the African clawed toad (*Xenopus laevis*) and the cellular slime mold (*Dictyostelium discoideum*), have

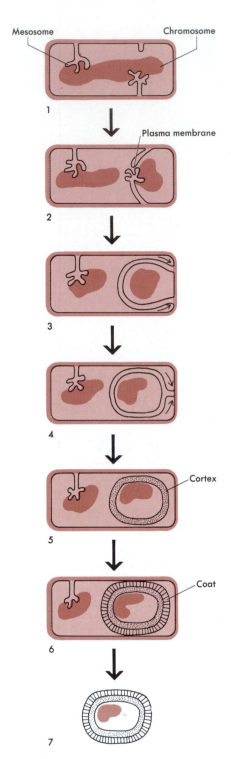

Figure 22-1
Diagrammatic representation of the stages in sporulation. (1) Preseptation: cell with axial chromosome. (2) Septation: plasma membrane folds in and cuts off an area containing intact chromosome. (3, 4) Stages of protoplast envelopment. (5) Cortex formation: material laid down between two membranes. (6) Coat formation and maturation. (7) Eventually the spore is released.

not been as extensively characterized genetically, but, as we shall see, offer unique experimental advantages for analyzing particular developmental phenomena.

To illustrate the advantage of using a model system, consider the problem of studying the mechanism of hormone action in a human (or even a mouse), where few (if any) mutants are available and where both the hormone and the hormone-responsive cells are often difficult to obtain in sufficient quantities. In contrast, work on the premier model system—induction of mating competence in yeast **a** cells by the mating hormone (pheromone) α factor (see Figure 18-32)—is moving swiftly because yeast is at least two orders of magnitude easier and cheaper to work with than higher cells.

THE MEASUREMENT OF TIME DURING DEVELOPMENT

Bacterial Sporulation Is the Simplest of All Model Systems[4]

The colossal magnitude of the task of attempting to understand the molecular basis of a complex differentiation process (e.g., the origin of a nerve cell) has led many people to look for cell systems much, much simpler than those studied by the classical embryologists. Some biochemists have gone so far as to focus their attention on the formation of bacterial **spores,** highly dehydrated cells that form inside certain bacteria (e.g., the genus *Bacillus*) as a response to a suboptimal environment. As spores possess virtually no metabolic activity, they resemble seeds of higher plants. Compared to vegetative bacteria, they are highly resistant to adverse conditions such as extreme heat or dryness. Thus, those bacteria that can produce spores have a much higher capacity to survive the extreme environmental conditions under which many forms of bacteria must live.

Some steps in the development of a spore are shown in Figures 22-1 and 22-2, which indicate that an early stage involves invagination of a portion of the bacterial membrane to enclose the chromosome and a small amount of cytoplasm. Afterward, a very thick and tough surface layer of protein is laid down on the outside. This coat has a completely different composition from the surface of the corresponding vegetative forms. Accompanying these morphological transformations are important changes in enzyme composition. The various cytochrome molecules responsible for aerobic metabolism completely disappear, and a new electron transport system appears. Ribosome content is also reduced, with an even greater decrease in the number of messenger RNA molecules. Small numbers of the enzymes necessary for protein synthesis remain; these are necessary for the resumption of protein synthesis when the spore germinates under the stimulus of more favorable nutritional conditions.

Sporulation thus requires cessation of the synthesis of almost all the proteins necessary for vegetative existence; conversely, germination converts spores to a condition in which they can make the vegetative proteins. So the making of a spore must entail an inhibition of most, if not all, vegetative mRNA synthesis, replacing it with the synthesis of mRNA specific for sporulation-specific proteins.

(a)

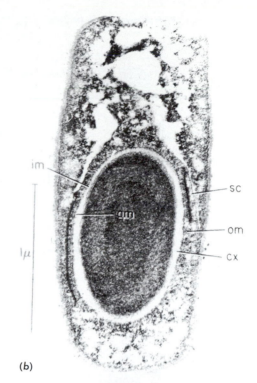

(b)

A Cascade of σ Factors for RNA Polymerase[5, 6]

Until recently, the molecular basis of the changeover from vegetative to sporulation-specific transcription was a total mystery. Now, a combination of genetic and biochemical experiments tells us that sporulation is accompanied by the appearance of new RNA polymerase specificity factors that read the genes necessary for sporulation. The overall structure of RNA polymerase in *B. subtilis* is essentially identical to that in *E. coli* ($\alpha_2\beta\beta'\sigma$), although the precise molecular weights of the various subunits differ between the Gram-positive and Gram-negative bacteria. However, vegetatively growing *B. subtilis* contains at least four different σ factors, called σ^{55}, σ^{37}, σ^{32}, and σ^{28} (with molecular weights of 55, 37, 32, and 28 kdal). Each of these different σ factors can bind interchangeably to the core RNA polymerase ($\alpha_2\beta\beta'$) and confer upon it the ability to recognize a different set of promoter sequences. Factors σ^{37} and σ^{32} seem to be involved in switching on genes that are expressed at the very onset of sporulation. When sporulation in *B. subtilis* is initiated by starvation for carbon, nitrogen, or phosphorus, at least two new sporulation-specific factors are synthesized: one is a σ factor known as σ^{29}, and the other is a factor that dissociates the vegetative σ^{55} (and probably σ^{37}) from the core RNA polymerase. In this way, σ^{29} replaces σ^{55} on core polymerase and allows the transcription of sporulation-specific promoters to begin. (Surprisingly, the factor that dissociates σ^{55} from core RNA polymerase does not inactivate the σ^{55} and σ^{37}, since these can be recovered in active form from extracts of sporulating bacteria.)

Interestingly, the existence of multiple σ factors that determine the expression of specialized sets of genes is emerging as a general feature of bacteria as diverse as *E. coli, Bacillus,* and *Streptomyces.* For example, the expression of heat-shock genes and nitrogen-regulated genes in *E. coli* is now known to be dictated by specific σ factors encoded by genes *htpR* and *ntrA* (Chapter 16). Also, multiple σ factors are a feature of most complex bacteriophages such as T4, λ, T7, μ, and *B. subtilis* phage SPO1 (where novel σ species were actually first discovered) and are responsible for the temporal cascade from early to late phage gene expression (Chapter 17).

(c)

Figure 22-2
Electron micrographs of thin sections of sporulating *Bacillus cereus* cells showing (a) septation, (b) cortex development, and (c) coat formation. Note that CHR designates the chromosome; m, the mesosome; sps, the transverse spore septum; cx, the cortex; sc, the spore coat; ex, the exosporium; im and om, the inner and outer membranes. (Courtesy of Dr. W. G. Morrell.)

(a) RNA polymerase core ($\alpha_2\beta\beta'$)

σ^{55}

Vegetative promoter

−35 region   −10 region

Altered conformation of RNA polymerase core

σ^{29}

Sporulation-specific promoter

−35   −10

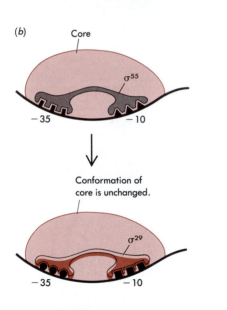

(b) Core

σ^{55}

−35   −10

Conformation of core is unchanged.

σ^{29}

−35   −10

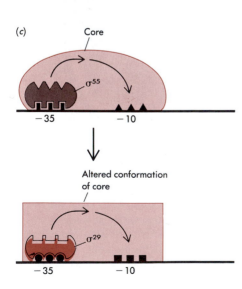

(c) Core

σ^{55}

−35   −10

Altered conformation of core

σ^{29}

−35   −10

Figure 22-3

How do σ factors change the specificity of RNA polymerase? (a) A "core-conformation" model. σ^{29} displaces σ^{55} from RNA polymerase core. Binding of σ^{29} could change the conformation of core polymerase in such a way that it recognizes a different −35 region on DNA. This model is difficult to envision because core enzyme can interact with at least six different σ factors. (b) A simultaneous "direct-contact" model. σ^{55} and σ^{29} could be elongated molecules that touch the −10 and −35 regions simultaneously. Experiments supporting this model show that *E. coli* σ can be chemically crosslinked to both the −35 and −10 regions. (c) A sequential "direct-contact" model. σ factors could bind first to the −35 region (closed promoter complex) and then to the −10 region (open complex). This model is also consistent with the chemical crosslinking experiments, but would require that RNA polymerase core undergo a significant conformational change. [After R. Losick and J. Pero, *Cell* 25 (1981):582–584.]

The σ factors are truly remarkable, because the binding of a single additional polypeptide chain can confer a new promoter specificity on the general RNA-synthesizing machinery of core RNA polymerase. How can a σ factor do this? Current evidence suggests that σ factors are bifunctional proteins that can simultaneously bind to core RNA polymerase and recognize specific promoter sequences in the DNA. When a comparison was made between many different promoters, all recognized by *E. coli* core RNA polymerase complexed with σ^{70} (a complex called a holoenzyme), it was possible to deduce an average (or consensus) sequence for the σ^{70}-specific promoter (see Figure 13-8). Similarly, when the consensus promoters for *B. subtilis* RNA polymerase bearing the σ^{55}, σ^{37}, and σ^{28} bacterial factors or phage SPO1 σ factors were aligned, it immediately became apparent that each of these different σ factors enables core polymerase to recognize a different combination of sequences centered approximately around positions −35 and −10 upstream from the start site of transcription. These are, of course, the same regions recognized by *E. coli* RNA polymerase holoenzyme. Since it is difficult to imagine that σ factors can alter the conformation of core RNA polymerase in such specific ways as to change the precise sequence of nucleotides recognized as a promoter, it seems more likely that σ factors are themselves very highly elongated molecules that touch DNA sequences as far apart as 16 to 19 base pairs (−35 and −10 regions) when the factor is bound to RNA polymerase (Figure 22-3). However, binding to the −35 region may be relatively weak or less essential, since the bacteriophage T4-encoded σ factor required for reading the late phage genes appears to bind only to the −10 region (Chapter 13).

Multicellular Organisms Must Have Devices to Control When Genes Act

In multicellular organisms, as in bacteria, not all genes in a cell function at the same time. Something must dictate that muscle cells, for example, selectively synthesize the various proteins used to construct muscle fibers. The understanding of embryology will thus, in one sense, be the understanding of how genes function selectively. Moreover, we must ask not only what causes two daughter cells to synthesize different proteins (differentiation), but also what enables their daughters to continue synthesizing the same particular group of proteins (stable inheritance of the differentiated state). With the problem phrased in this way, it is clear that no one will ever be able to work out all the chemical details that accompany the development of any higher plant or animal. For even a modest approach to a comprehensive understanding, we would have to look at the behavior of many hundreds of different proteins. Still, we should be able to discover certain general principles governing the selective expression of specific proteins. In fact, there is increasingly good evidence that in multicellular organisms, as in bacteria, differentiation is often controlled at the chromosomal level by devices that regulate the transcription of specific genes.

Embryologists are now asking whether the fundamental control mechanisms act negatively, like the lactose repressor and the *MATα2* gene product in yeast (see Figure 18-31), or positively, like the bacteriophage-specific SPO1 and T4 σ factors and the steroid receptors of higher cells (Figure 21-48). The dilemma, however, is that most of the systems that embryologists study are hopelessly complex from the biochemical viewpoint and cannot be correlated with the occurrence of a few well-defined enzymatic reactions. For example, although nerve cells are easy to identify morphologically, we are just beginning to understand their molecular structure, and only a few of the hundreds of protein molecules in nervous tissue have been well characterized. Although we are much more familiar with the muscle proteins, again large gaps in our knowledge are likely to make precise analysis of the function of each protein a most tricky endeavor.

The Measurement of Biological Time[7]

To control when genes act, cells must be able to measure time. In fact, the regularity of the cell cycle in both eucaryotic microorganisms (Chapter 18) and higher cells (Chapter 25) suggests that eucaryotes as well as procaryotes have accurate clock mechanisms at the cellular level. We are accustomed to thinking about time as a quantity that can be measured by a simple physical device, like the escapement of a mechanical clock, or a simple electronic device, like the crystal oscillator in a solid-state chip. But we have comparatively few clues as to how biological systems are able to measure and mark elapsed time.

In principle, biological systems could measure time in two very different ways. One possibility is that cells have an independent (autonomous) internal clock, together with devices to ensure that key reactions occur at certain times (much like an alarm clock). In this case, cellular events *a*, *b*, and *c* are *independent* of each other. Event *b* is scheduled for time *b*, and event *c* for time *c*; but if event *b* fails to occur for whatever reason, event *c* can still take place as scheduled. Alternatively, key reactions may be designed so that event *c* cannot take place unless event *b* has already occurred. For example, event *c* might be a chemical reaction that requires a product of event *b*. In this case, each cellular event is *dependent* on the previous events, and all events are linked together in a fixed order. The very sketchy evidence now available suggests that both independent and dependent clocks exist in eucaryotic development.

In Chapter 17, we saw that most complex bacteriophages use a cascade of positive and negative transcriptional regulators to control the order of gene expression during phage infection. The λ antitermination proteins N and Q function positively, while the T7 gene 2 protein (and possibly the gene *0.7* protein) functions negatively by

Figure 22-4
A mechanism by which sequential synthesis of different specificity factors might be used to count time.

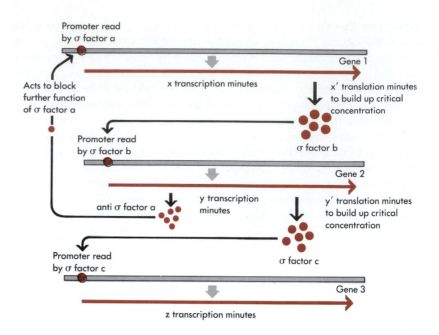

inactivating the bacterial RNA polymerase. In contrast, σ factors (encoded by the SPO1 genes *28*, *33*, and *34* and the T4 genes *33* and *55*) function both positively (to turn on expression of later genes) and negatively (to turn off expression of the earlier genes by displacing the old σ factor). Thus, even the very small genome of a bacteriophage can use a sequence of dependent events to count time. However, as a consequence of the relatively slow rate at which RNA polymerase molecules move along DNA templates (DNA replication occurs some 100 times faster!), bacteriophages can also use transcription as an independent measure of biological time. For example, at 37°C, the rate of movement of *E. coli* RNA polymerase is only 40 nucleotides per second, and so complete transcription of the λ genome by a single polymerase molecule would require over 25 minutes. Since several operons are transcribed simultaneously, this much time is not required for production of infectious phage. Nevertheless, transcription of the very long "late" operon encoding the capsid proteins counts out some 15 minutes, with the head proteins being made substantially before the tail proteins, as expected from the gene order (Figure 17-15). (Why the lysis genes *S* and *R precede* all of the capsid genes in the late transcription unit remains a mystery.)

Similar analysis applied to the *E. coli* chromosome gives a total possible transcription time of 33 hours. Thus, only 1 percent of its total genome would have to be used to code for periodic events occurring once every 20-minute division cycle. And even the 3 to 4 hours required for a *Bacillus* spore to germinate could easily be directly measured out on a DNA tape. There is, of course, no reason why such relatively long intervals must be coded along one continuous DNA segment. Several genes, each coding for a unique specificity factor necessary for reading the subsequently transcribed gene, would also do the job (Figure 22-4).

Although the division cycles of higher cells are much longer than those of bacteria or eucaryotic microorganisms, their DNA contents are correspondingly much greater; one RNA polymerase molecule would require a thousand days at 37°C to completely transcribe the haploid complement of human genes. Thus, much of the timing required for a 24-hour human cell cycle can easily be imagined to occur at the transcriptional level. Such speculation will probably receive its first experimental test from analysis of yeast *CDC* mutants (Chapter 18), which become blocked at specific stages of the cell cycle when they are subjected to slightly elevated temperature. Although yeast is a very simple eucaryote, it will often be the model system of choice for genetic and biochemical characterization of basic metabolic processes that it shares with higher eucaryotes.

The timing of certain crucial steps in the embryological development of higher organisms may also be controlled by transcription. We will see shortly that some of the homeotic genes responsible for morphogenesis in the fruit fly *Drosophila* have huge introns, and it is possible that the function of these very long transcription units (approximately 100 kilobases) is to delay expression of the gene product until a particular time in early embryogenesis.

Concentration on Organisms with Easily Observable Cleavage Divisions[8]

The mechanisms by which fertilized eggs develop into multicellular organisms have been a continuous source of mystery to biologists for almost a hundred years. During this time, experimental studies have been concentrated on echinoderms (especially the sea urchin) and amphibians such as frogs, salamanders, and toads. The reason for such emphasis lay in the ease with which echinoderm and amphibian eggs could be procured, fertilized, observed, and experimentally manipulated as they passed through the normal stages of development. In contrast, birds and mammals—where all the embryological stages take place within a thick-shelled egg or female parent—are much more tricky to work with. Only recently has real progress been made in opening up the early stages of mouse embryogenesis to experimental manipulation.

A very important feature of both urchin and frog embryology is that all the nutrients (mostly stored in the form of yolk) required for development through the blastula and gastrula stages must be present in the unfertilized egg. Only when the developing embryo becomes able to feed itself—as the frog does, for example, at the tadpole stage—does real growth in mass occur. Consequently, echinoderm and amphibian eggs are very, very large in comparison with their respective adult cells, and the cell divisions that occur after fertilization progressively produce smaller and smaller cells: hence the term **cleavage divisions** (Figure 22-5).

How the Frog Egg Counts Twelve Cell Divisions[9, 10, 11]

Although we have firm proof that transcription can measure biological time in bacteriophage, biologists have had only one major success in explaining the measurement of time during embryogenesis in a complex higher eucaryote. The crucial experiments took advantage of the extraordinarily large size of the frog egg (about 1 mm, or $\frac{1}{25}$ inch, in diameter): This egg is so large that foreign substances can easily be microinjected through a hollow glass needle without inflicting any damage on the developing embryo.

Immediately after fertilization, the frog egg undergoes 12 very rapid and almost synchronous cleavage divisions, characterized by

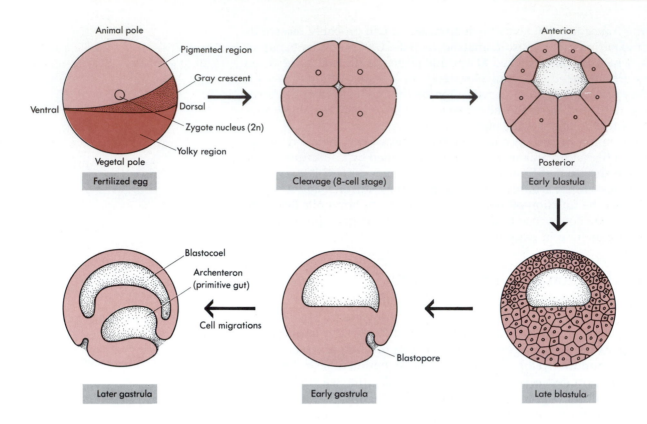

Figure 22-5
Early stages in the development of an amphibian egg.

short interphase periods when DNA synthesis occurs at a very accelerated rate. There is no measurable transcription of newly synthesized DNA during this time, perhaps because the presence of RNA polymerase molecules on the DNA would interfere with rapid DNA replication. After the twelfth division, when the fertilized egg has transformed itself into a blastula containing approximately 2^{12}, or 4096, cells, the rate of division slows down to a pace more typical of normal somatic cells, and transcription begins by RNA polymerases II and III (RNA polymerase I begins slightly later). This abrupt physiological change is called the **midblastula transition,** because it happens midway through **blastulation** (formation of a solid sphere of cells) but before **gastrulation** (when the embryonic cells become motile and the blastula invaginates to form various tissue layers such as endoderm, mesoderm, and ectoderm).

How do cells within the blastula know when they have undergone exactly 12 divisions? There are two obvious possibilities. The blastula could count each round of cell division directly (for example, by sequential modification of a site on the DNA once per division cycle), or it could count a product of the cell division (for example, the total amount of DNA in the entire blastula). We now believe the second possibility.

Experimentally, it has been found that the midblastula transition occurs earlier than usual in fertilized eggs that have been microinjected with foreign DNA. *Any* foreign DNA will work (plasmid, bacteriophage, or bacterial DNA), and the greater the quantity of DNA microinjected, the earlier the transition will occur. These observations strongly suggest that the mature frog egg contains a preformed factor (presumably a protein) that binds *stoichiometrically* to DNA: Twice as much DNA binds twice as much factor. The unfertilized egg contains exactly the right amount of this substance to bind the total mass of DNA in approximately 4096 cells, and there is no additional synthesis

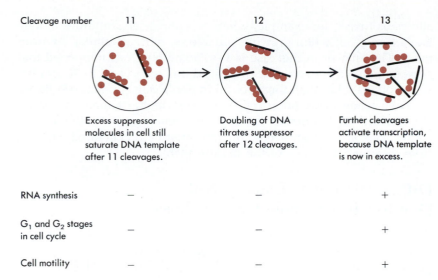

Cleavage number	11	12	13
	Excess suppressor molecules in cell still saturate DNA template after 11 cleavages.	Doubling of DNA titrates suppressor after 12 cleavages.	Further cleavages activate transcription, because DNA template is now in excess.
RNA synthesis	−	−	+
G_1 and G_2 stages in cell cycle	−	−	+
Cell motility	−	−	+

Figure 22-6
A titration model for the onset of transcription at the midblastula transition. [After M. Kirschner and J. Newport, *Cell* 30 (1982):687–696.]

of the factor between fertilization and the midblastula transition. Each cleavage division increases the amount of DNA in the developing blastula, and when the last trace of preformed factor is bound by newly synthesized DNA, the disappearance of free (unbound) factor triggers both transcription and the onset of cell motility required for gastrulation (Figure 22-6). We might speculate that the factor is a DNA binding protein that binds loosely and nonspecifically to any DNA, thereby preventing transcription. Initially, a high concentration of free factor in the egg would saturate all potential transcription units with factor. Only when free factor had been fully titrated by 4096 cells worth of new DNA could transcription resume. Mechanistically, regulation of the midblastula transition resembles regulation of the amount of oocyte-specific 5S RNA by titration of preformed transcription factor TFIIIA with the 5S gene product (Chapter 21).

Titration of preformed factors in the egg cytoplasm is an elegant, satisfying, and potentially general mechanism for timing many different developmental events. Nevertheless, it is extremely difficult to imagine how the maturing frog oocyte (egg) manages to synthesize exactly the right amount of the preformed factor. This dilemma brings us face to face with the central philosophical problem of all developmental biology. How can we claim to have explained the behavior of the developing frog blastula in terms of the properties of the unfertilized egg, when we do not know how the egg itself was made? Thus, the molecular study of development appears to be an endless hall of mirrors. Because all cells come from preexisting cells, and all organisms from preexisting organisms, almost every answer gives rise to a prior question. Our fundamental faith as molecular biologists is that by amassing a great many such partial answers, we will eventually be able to come full circle and escape from the paradox of the chicken and the egg.

THE NATURE OF DIFFERENTIATION

Cellular Differentiation Is Often Irreversible

At present, it is possible to isolate a variety of differentiated cells and grow them (like bacteria) outside living organisms under well-defined nutrient conditions. This technique of "cell culture" (Chapter 25) allows us to ask, for example, whether a nerve cell continues to look like a nerve cell when growing outside its normal cellular environment: The answer in this case is yes. Something has happened that appears to have permanently destroyed this cell's capacity to synthesize proteins other than those found in nerve cells. Similar results are obtained when many other cell types from higher animals are studied. But with higher plants, the opposite answer is more often found. As we shall see, a complete plant can often be regenerated, starting from differentiated cells of the mature plant.

Differentiation Is Usually Not Due to Chromosome Gain or Loss

Perhaps the most obvious hypothesis to explain irreversible differentiation is that only a fraction of the genes of the fertilized egg are passed on to a nerve cell, a muscle cell, and so on. This scheme, however, appears to be completely wrong. As far as we can tell, most cells of most organisms, with the obvious exception of the haploid sex cells, contain roughly the same chromosomal complement. Virtually all cell divisions are preceded by a regular mitosis, so that daughter cells receive an identical set of chromosomes. We cannot say, however, that no permanent changes have occurred at the level of individual genes. Even though the mass of evidence gathered by detailed characterization of many developmentally regulated genes in many organisms overwhelmingly supports the idea that differentiation rarely involves an irreversible change in genomic DNA, important exceptions are already known. For example, the mammalian immune response depends on both irreversible genomic rearrangements and selective mutation of lymphocyte genes responsible for antibody production and regulation (Chapter 23).

Differentiation Is Usually Not Irreversible at the Nuclear Level[12]

Nuclear transplantation experiments provide conclusive evidence against the notion that the selective synthesis of a restricted set of gene products in a differentiated cell requires irreversible changes in genomic DNA. In these experiments, diploid nuclei from differentiated cells were transplanted into unfertilized eggs whose haploid nuclei had been previously removed. The resulting genetically complete diploid eggs were then artificially induced to divide and grow, occasionally forming adult organisms whose chromosomal makeup was derived entirely from clonal reproduction of donor nuclei (Figure 22-7). Although nuclei derived from larval intestinal cells support clonal reproduction, nuclei from an adult frog do not work. This suggests that the pluripotency of some larval intestinal cells (the ability to *de*differentiate completely) is lost upon maturation. So far, the frog is the most complex organism for which nuclear transplantation has allowed clonal reproduction. Success in this case owes much to the very large size of the amphibian egg, which allows conventional microsurgical removal of the maternal nuclei.

Cellular Memory and the Replication Fork[13]

Just as bacteria divide to produce more bacteria and elephants give birth to elephants, so most differentiated cells in our own bodies breed true. Liver cells give rise to more hepatocytes, and fat cells

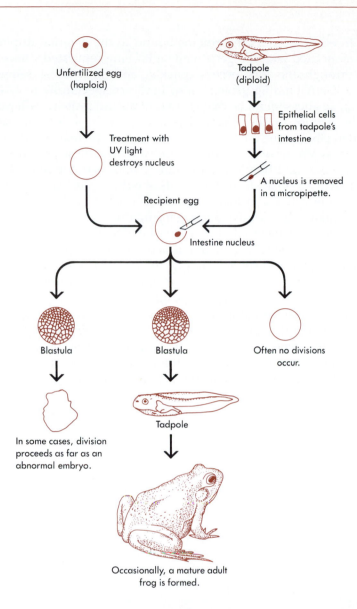

Unfertilized egg
(haploid)

Tadpole
(diploid)

Treatment with
UV light
destroys nucleus

Epithelial cells
from tadpole's
intestine

A nucleus is removed
in a micropipette.

Recipient egg

Intestine nucleus

Blastula

Blastula

Often no divisions
occur.

In some cases, division
proceeds as far as an
abnormal embryo.

Tadpole

Occasionally, a mature adult
frog is formed.

Figure 22-7
Steps in the creation of a clonal frog.
[Redrawn from J. R. Gurdon, *Sci. Amer.*
219 (1968):24–35.]

produce more adipocytes. Since different cell types express very different (but overlapping) sets of proteins, maintenance of the differentiated state must reflect mechanisms that allow both daughter cells to inherit the same pattern of active and inactive genes found in the parent.

Semiconservative DNA replication provides both daughter cells with identical copies of each gene, but the DNA sequences by themselves cannot specify whether a gene is to be turned on or off in the daughter cells. Instead, regulation of individual genes is accomplished by a variety of mechanisms (Chapter 21): diffusable regulatory factors, nondiffusable covalent modifications of the DNA (methylation of the dinucleotide CpG), and local changes in histone and nonhistone chromosomal proteins leading to active and inactive chromosomal domains. Each of these mechanisms could in principle contribute to the inheritance by each daughter cell of a characteristic pattern of active and inactive genes.

Unfortunately, a crucial gap in our knowledge of the enzymology of eucaryotic DNA replication prevents us from becoming too specific regarding the mechanisms that might allow the inheritance of active and inactive genes: We do not know what happens to chromatin (the complex of DNA with protein) as it passes through the replication fork. In other words, can proteins bound to the parental duplex remain associated with the gene while it is being replicated? Since passage through the fork occurs quickly, and since most sequence-specific DNA binding proteins also bind nonspecifically to DNA, a reasonable guess might be that some (but not all) proteins will remain loosely bound to the sugar-phosphate backbone of the single-stranded DNA while the bases are used as a template by DNA polymerase for synthesis of the complementary strand; these proteins could then resume tight binding to the newly reformed daughter duplex.

Our ability to make firm statements about the mechanisms for perpetuation of active and inactive genes is further qualified by lack of precise quantitative knowledge regarding the speed with which various proteins are able to find their binding site on DNA (Chapter 16). For example, consider a gene that is activated in the parental cell by a positive regulatory protein (e.g., the steroid receptor protein). Whether this gene remains active in the daughter duplexes after it passes through the replication fork may in some cases reflect a competition: Can the regulatory protein bind (or rebind) to the gene specifically before other proteins (perhaps histones) bind nonspecifically and shut the gene down? Thus, if the parental cell has a sufficient concentration of the regulatory protein, both gene copies could be activated after replication and the active state of the gene would be inherited by both daughter cells.

Similar arguments can be advanced regarding the perpetuation of active and inactive chromosomal domains during DNA replication; however, such speculation must be tempered by our lack of knowledge about the structure of active chromatin, the behavior of the nucleosomes at the replication fork, and the rate of nucleosome assembly onto naked DNA. The one fact that seems clear is that nucleosomes segregate conservatively at the fork, one daughter duplex receiving the old nucleosome intact while the other daughter duplex acquires an entirely new nucleosome. Since active nucleosomes appear to differ structurally from inactive nucleosomes (Chapter 21), a very important question is whether the deposition of old nucleosomes can occur preferentially on the leading or lagging strand.

DNA Methylation and Inheritance of the Differentiated State[14, 15, 16]

The DNA of most higher eucaryotes is modified by methylation at the 5 position of cytosine in the dinucleotide sequence ^mCpG (where ^mC denotes 5-methylcytosine) (Chapters 9 and 21). We have seen that many active or potentially active genes are undermethylated at certain CpG sequences, but examples are also known where decreases in methylation accompany rather than precede gene activity. Although the exact molecular basis for these effects is unknown, our current knowledge suggests that the presence of a methyl group in the major groove of the DNA can regulate gene expression by either increasing or decreasing the affinity of the DNA for specific regulatory proteins (repressors and activators). Because certain organisms, such as insects, lack detectable DNA methylation, methylation cannot be indispensable for regulation of eucaryotic genes. Nonetheless, the experimental evidence correlating DNA methylation with changes in gene expression in other organisms, and particularly in mammalian cells, is very impressive. Thus, we are justified in assuming that DNA methylation does play a role in regulation of gene expression, but alternative mechanisms must exist that can accomplish the same tasks.

During the S phase that precedes cell division, when a DNA replication fork passes through an ^{m}CpG doublet, each of the daughter helices is transiently hemimethylated (methylated only on the old DNA strand). If there were no way of restoring the symmetrical methyl group, methylation at this particular genomic locus would quickly be diluted out of the population through successive generations. However, we know that methylation is heritable, and thus it is not surprising that cells contain a maintenance methylase that is capable of methylating the cytosine opposite an ^{m}CpG doublet soon after it emerges from the replication fork.

When we ponder the existence of maintenance methylases in eucaryotes, together with the ability of methylation to regulate gene activity, it is very tempting to speculate that methylation at CpG doublets may be one of the mechanisms by which eucaryotic cells maintain the differentiated state. If genomic DNA in germ cells or in some very early stage of embryogenesis were automatically restored to a kind of methylation ground state corresponding to no differentiation, methylation and demethylation at particular chromosomal sites could play a role in programming expression of genes throughout the course of development (Figure 22-8). The main objection to such a model is that it is difficult to imagine how either methylation or demethylation would itself be regulated (the hall of mirrors effect, again). All the known eucaryotic maintenance methylases appear to be nonspecific in their activity: *Any* accessible hemimethylated CpG doublet will be methylated. As a result, we are forced to invoke other factors, such as the binding of specific regulatory proteins or changes in chromatin structure (Chapter 21), that could affect the accessibility of critical cytosine residues to methylation. However, once methylation or demethylation has taken place at a particular regulatory site, maintenance of the differentiated state could then be entrusted to a *non*specific methylase activity.

Irreversible Cytoplasmic Differentiation Concomitant with Loss of Ability to Divide

Many cells contain nuclei that, under normal conditions, will never divide again. Often, they represent cases of extreme differentiation, where a very large fraction of the total synthesis is devoted to making just a few different proteins. Immature red blood cells, or reticulocytes, are perhaps the most striking example. These cells never divide, having as their only function the production of hemoglobin molecules able to combine reversibly with oxygen. Moreover, in some species, including all mammals, the nuclei of these cells eventually disintegrate, so that the mature erythrocytes (red blood cells) are unable to make any more hemoglobin. Behaving somewhat similarly are the adult plasma cells, which produce specific antibodies. They likewise never divide, but secrete their antibodies for several days and then die. During this time, their nuclei remain fully functional, synthesizing the mRNA molecules coding for their antibody products. In contrast, many other specialized cells (e.g., nerve cells), though never dividing, may live many, many years. The functioning of such long-

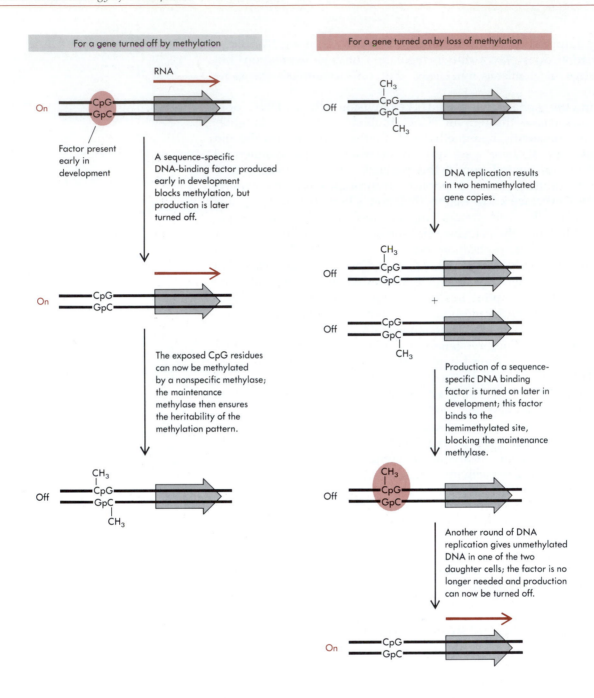

Figure 22-8
Methylation (or lack of it) is heritable and could maintain the differentiated state.

lived cells always depends on the presence of a nucleus able to make mRNA. Some protein synthesis, even at a very low level, seems to be necessary for the continued functioning of all cells. This synthesis may replace aging proteins targeted for destruction by the ATP- and ubiquitin-dependent proteolytic system (Chapter 21).

Stable Transcription Complexes Could Also Maintain the Differentiated State[17]

Experimental evidence suggests that genes transcribed by all three eucaryotic RNA polymerases are capable of forming stable transcription complexes with their respective transcription factors. In the case of RNA polymerase III, it is an essential transcription factor (*not* the

RNA polymerase III itself) that forms a stable complex with the DNA (Chapter 21). Once the transcription factor binds stably to the gene, that particular locus is activated and can be repeatedly transcribed by RNA polymerase III at least until DNA replication occurs. Recent experiments suggest that much the same situation holds true for both RNA polymerases I and II. Thus, stable transcription complexes provide an attractive possible mechanism for maintenance of the differentiated state during interphase (i.e., between cell divisions).

Stem Cells Divide Asymmetrically[18]

Stable transcription complexes can also help to explain the heritability of gene activity during DNA replication and subsequent cell division if the complexes can pass unharmed through the DNA replication fork. This possibility suggests an explanation for the behavior of stem cells. Not all cells give rise to two identical daughters. Stem cells divide to give one daughter identical to the parent, thus perpetuating the stem cell line itself, and another daughter committed to differentiate. For example, the highly differentiated erythrocytes circulating in the human body have a lifetime of only 120 days, after which they wear out and are destroyed in the spleen. Red blood cells are continuously replaced by repeated divisions of hematopoietic stem cells in the bone marrow, which give rise to an identical daughter stem cell and an immature erythrocyte (reticulocyte).

Recent experiments have shown that the RNA polymerase III transcription factor TFIIIA (Figure 21-34) binds preferentially to the noncoding strand of the 5S ribosomal RNA gene; other stably bound transcription factors for RNA polymerases I and II may also have a strong strand preference. We can speculate that when the DNA replication fork passes through such an activated transcription unit, the transcription factor(s) may remain bound to the parental noncoding strand. The daughter chromosome that inherits this parental DNA strand will therefore inherit the factor, and along with it the gene in a transcriptionally activated state. But the gene on the daughter chromosome that inherits the parental coding strand may simply be packaged into nucleosomes like any unremarkable stretch of DNA and thus be repressed (Figure 22-9). The conservative segregation of parental nucleosomes at the replication fork provides yet another possible mechanism for producing two daughter cells, only one of which retains the gene in an active (or inactive) state.

DNA REARRANGEMENTS AND AMPLIFICATION DURING DEVELOPMENT

Extensive DNA Rearrangements Occur During the Life Cycle of Ciliates[19–22]

In certain rare cases, differentiation does require rearrangement of DNA. For example, the bacterium *Salmonella* alternates between two antigenically distinct flagellar proteins (flagellins) by inversion of a specific promoter-containing DNA segment (Figure 11-26). Bakers' yeast switches between the **a** and α mating types by unidirectional gene conversion of the *MAT* locus, using one of the two silent

Active nucleosome
or transcription factor

Inactive nucleosome
or repressor

RNA

Parental stem cell

As DNA replication fork
passes through, DNA-binding
factors segregate conservatively
to one strand of DNA.

Replisome at
replication fork

New factors bind
to other DNA strand.

This daughter is
identical to parent,
thus renewing stem
cell lineage.

This daughter
differentiates,
because pattern of
gene expression is
changed.

For illustrative purposes
only, DNA-binding factors
are arbitrarily shown as
segregating to the
leading strand.

cassettes known as *HMR* and *HML* as a template (Figure 18-25). Both antigen alternation in *Salmonella* and mating-type switching in yeast are reversible DNA rearrangements that do not affect the total genetic information inherited by daughter cells. Similarly, as we shall see in Chapter 23, the production of immunoglobins and certain antigen-specific cell-surface proteins during the mammalian immune response involve complex rearrangements of the corresponding genes. However, these rearrangements occur exclusively in certain somatic cells called B and T lymphocytes, never in the germ line. Thus, these DNA rearrangements, although irreversible, cannot be passed on to the next generation.

Here we will consider the curious predicament of the ciliates, whose somatic and germ line nuclei share a common cytoplasm. These unicellular aquatic organisms provide by far the most dramatic exception to the general rule that differentiation occurs without rearrangement of DNA. The two nuclei within each ciliate cell serve very different functions. The larger **macronucleus** is the workaday nucleus responsible for all transcription during ordinary vegetative growth and asexual reproduction by binary fission. This somatic nucleus is polyploid and contains many copies of the genes required for vegetative growth and binary fission. In contrast, the smaller germ line **micronucleus** is diploid and transcriptionally silent during the asexual life cycle; this nucleus functions only when opposite mating types come together for sexual conjugation. Then, the diploid micronucleus undergoes meiosis, and the resulting haploid micronuclei serve as germ cells. After the haploid micronuclei of opposite mating type fuse to form a diploid zygotic nucleus, the old macronuclei degenerate. Finally, through a complex succession of programmed nuclear divisions and disintegrations, the zygotic nucleus gives rise to a new haploid micronucleus and an immature macronucleus (Figure 22-10).

In many, and perhaps most, species of ciliates, polyploidization of the macronucleus occurs in two stages. First, repeated rounds of DNA replication without subsequent nuclear division produce many daughter chromatids in perfect lateral register. In some species (especially the hypotrichs such as *Oxytricha*), the number of rounds of DNA replication is sufficient to produce visible polytene chromosomes that bear a striking resemblance to those found in the salivary gland and fat body of the fruit fly *Drosophila* (see Figure 21-12). However, unlike the *Drosophila* polytene chromosomes, which remain intact, the newly polytenized chromosomes in the immature ciliate macronucleus are immediately transected into fragments. Fragmentation is more extreme in some species than in others; in *Oxytricha nova*, the fragments range in size from 1000 to 20,000 base pairs. Transection is followed by massive degradation of as much as 95 percent of the polytenized DNA in certain ciliate species. Those pieces that survive degradation seem to contain the genes required for vegetative growth and asexual reproduction; DNA that is not necessary for the asexual phase of the life cycle may simply be discarded.

In the second stage of polyploidization, the surviving linear DNA fragments in the maturing macronucleus replicate to still higher copy number (about 45 copies per macronucleus in the case of *Tetrahymena*). Thus, each linear fragment is capable of autonomous replication and must be equipped with an origin of DNA replication and two telomeres (see Figures 18-14 and 18-15). As expected, the origins of DNA replication appear intact in germ line micronuclear DNA, but interestingly, the telomeres do not. Instead, macronuclear telomeres (composed of tandem repeats of the simple sequence C_4A_2 or C_4A_4, depending on the species of ciliate) appear to be polymerized de novo onto the ends of the fragments by a T_2G_4- or T_4G_4-adding enzyme, which is primer- but not template-dependent. In other words, depending on the species from which the enzyme is isolated, the telomere-adding enzyme will polymerize additional tandem repeats of T_2G_4 or T_4G_4 onto the free 3' hydroxyl group of a single-stranded primer with the sequence $(T_2G_4)_n$ or $(T_4G_4)_n$, where n is greater than or equal to 1. Presumably, the complementary strand containing C_4A_2 or C_4A_4 repeats is then synthesized by a template-dependent DNA polymerase, which copies the new T_2G_4 or T_4G_4 repeats. (The CCA-adding enzyme that polymerizes *ribo*nucleotides onto the 3' end of tRNA is another example of a primer-dependent but template-independent reaction; see Chapter 14.)

The linear fragments of DNA in the macronucleus are not true chromosomes, because they lack the centromere required for attachment of spindle fibers to the chromosome (Chapter 18). Without a spindle apparatus to ensure that each daughter cell receives an equivalent number of macronuclear DNA fragments, the macronucleus must divide by uneven budding rather than by conventional mitosis.

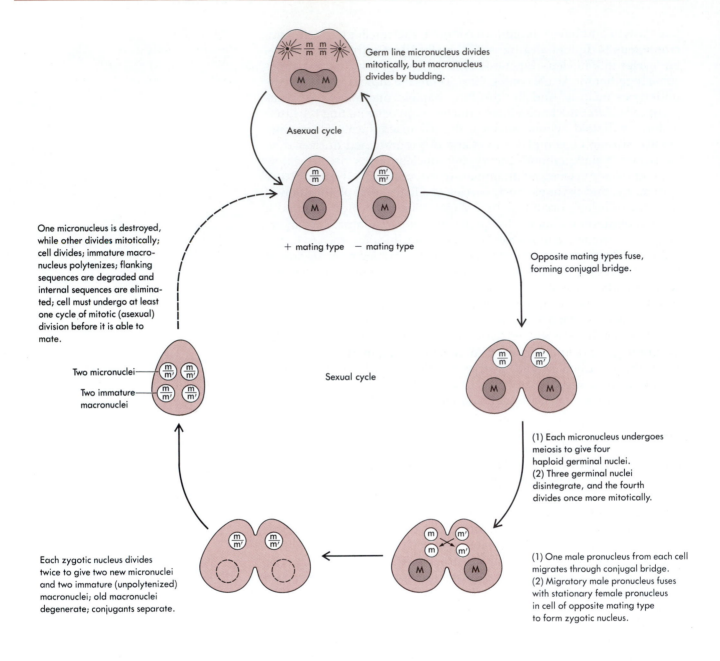

Germ line micronucleus divides mitotically, but macronucleus divides by budding.

Asexual cycle

+ mating type   − mating type

One micronucleus is destroyed, while other divides mitotically; cell divides; immature macronucleus polytenizes; flanking sequences are degraded and internal sequences are eliminated; cell must undergo at least one cycle of mitotic (asexual) division before it is able to mate.

Opposite mating types fuse, forming conjugal bridge.

Two micronuclei

Two immature macronuclei

Sexual cycle

(1) Each micronucleus undergoes meiosis to give four haploid germinal nuclei.
(2) Three germinal nuclei disintegrate, and the fourth divides once more mitotically.

Each zygotic nucleus divides twice to give two new micronuclei and two immature (unpolytenized) macronuclei; old macronuclei degenerate; conjugants separate.

(1) One male pronucleus from each cell migrates through conjugal bridge.
(2) Migratory male pronucleus fuses with stationary female pronucleus in cell of opposite mating type to form zygotic nucleus.

Figure 22-10

The complex nuclear life cycle of the ciliate *Tetrahymena*. The polyploid macronucleus is denoted by M; the diploid micronucleus, by m/m. As many as seven different mating types are known in this organism, but for simplicity, only two mating types are shown here (arbitrarily designated + and −). Because the genetic determination of mating type is complex, cells derived from a mating (exconjugants) may have a different mating type from either parent (the conjugants). Thus, the dotted arrow in the last stage of the sexual life cycle is meant to indicate that the exconjugants may differ from the parental strains in mating type as well as in genetic constitution. Cells of any mating type can divide mitotically, but for clarity, the asexual life cycle is illustrated for the + mating type alone.

Perhaps the initiation of DNA replication on the linear DNA fragments is subject to feedback inhibition, so that each macronuclear fragment can be maintained at constant copy number despite unequal distribution of macronuclear DNA fragments to the daughter macronuclei during budding.

Figure 22-11
Internal sequences are eliminated from genes during macronuclear development in the ciliated protozoan *Oxytricha nova*. Gray segments show micronuclear sequences that are preserved in the macronucleus; heavy hollow line, internal sequences eliminated during regeneration of the macronucleus; thin line, flanking sequences degraded during macronuclear regeneration; colored segments, telomeres presumably added after site-specific cleavage of micronuclear DNA. [After L. A. Klobutcher, C. L. Jahn, and D. M. Prescott, *Cell* 36 (1984):1045–1055.]

Why is the macronucleus polyploid? Unicellular ciliates are complex and gigantic cells by normal standards; such size and complexity have many advantages, perhaps the most obvious of which is that ciliates are able to engulf and consume smaller organisms. Many gene copies are required to synthesize enough structural RNA (rRNA, tRNA, etc.) and mRNA to stock the cytoplasm with sufficient quantities of each gene product. Thus, polyploidization enables the ciliate to achieve large size and complexity without solving the many new developmental problems associated with multicellularity.

Do Ciliates Splice DNA?[23]

Very recently, it has become clear that the macronuclear DNA fragments are not always colinear with the micronuclear DNA sequences from which they are derived. In addition to elimination of DNA sequences *between* genes during transection and selective degradation of the polytene chromosomes, internal DNA sequences *within* protein-coding regions are also precisely excised (Figure 22-11). From the limited data now available, such **internally eliminated sequences (IESs)** can be as short as 32 base pairs or as long as several kilobase pairs. Since some IESs interrupt protein-coding regions, they must be eliminated precisely lest nonsense proteins be synthesized. IESs are flanked by short, direct repeats and contain internal inverted repeats that could potentially function as signals for excision at either the RNA or DNA level; however, IESs do not resemble introns found in mRNA, tRNA, or rRNA, all of which are removed by RNA splicing. IESs might be removed by direct DNA splicing; by reverse transcription of a spliced RNA, as has been postulated for the generation of processed genes in mammalian genomes (Figures 20-30 and 20-31); or by gene conversion of the DNA using a spliced RNA (or its reverse transcript) as template. Only further experiments can decide between these unprecedented possibilities.

Trypanosomes Evade the Host Immune Response by Switching Surface Antigens

Trypanosomes are flagellated unicellular protozoan parasites that live in the bloodstream of infected mammals and (like the malarial parasite) are spread from one animal to another by an insect vector, in this case the tsetse fly. The most common human disease caused by trypanosomes is African sleeping sickness, a name that graphically describes

Figure 22-12
Activation and replacement of trypanosome VSG genes at the telomeric expression sites. (a) Early after infection, only one particular VSG antigen is expressed; three or more chromosomes have telomeric VSG genes, but only the expression site is actively transcribed. (b) Later during infection, directed gene conversions replace one or more of the telomeric VSG genes with new variant genes drawn from the transcriptionally silent basic copies. (c) During first relapse, a new variant VSG is expressed when the old telomeric expression site is turned off and a new site is activated.

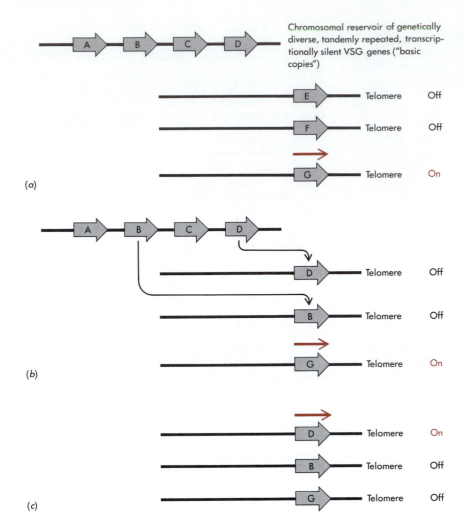

the complete fatigue accompanying persistent infection with an aggressive parasite. In Africa and South America, trypanosomes frequently infect cattle, severely lowering the yield of milk and meat and increasing susceptibility of the animals to other diseases. The lives of millions of people in underdeveloped countries are crippled by trypanosomiasis, and the development of practical treatments to combat this dread endemic disease is now the focus of a major worldwide effort.

The surface of each bloodstream trypanosome is concealed beneath a dense coat of a **variant surface glycoprotein** (**VSG**). The host immune system recognizes this VSG as a foreign antigen and makes antibodies against it in an effort to target the parasite for destruction. The host immune response would succeed in eliminating the trypanosome infection were it not for the amazing ability of the parasite to replace the first VSG with a second antigenically unrelated VSG. Such **antigen switching** apparently occurs quickly enough to keep the host on the defensive, and thereby effectively outwit the remarkable resources of the mammalian immune system (Chapter 23). Moreover, the trypanosome's genome contains a repertoire of perhaps a thousand different VSG genes. Thus, it is impossible to manufacture a conventional vaccine to guard against trypanosome infection, because any particular vaccine would react with only one of the many potential VSGs. The hope is that a better understanding of the basic molecular biology of trypanosome infection will suggest new approaches for eradicating the disease.

In Chapter 11, we saw how the bacterium *Salmonella* can switch from one surface antigen to another (a process called phase variation for historical reasons) by inverting a small promoter-containing DNA segment that can drive transcription of either of two divergently transcribed surface antigen genes (Figure 11-26). The trypanosome, however, can switch between at least 100 different VSGs. Clearly, a simple promoter inversion cannot explain this switching behavior.

In fact, trypanosome antigen switching is a much more difficult logistical operation than bacterial phase variation for two main reasons. First, the protozoan cell is far larger than the tiny bacterium, and the particular VSG being expressed at any one time constitutes at least 10 percent of the total trypanosome protein. To make so much VSG protein, either the VSG must be encoded by multiple genes, or transcription must take place at an extraordinary rate, or translation of the VSG messenger RNA must be unusually efficient, or the VSG mRNA must be unusually stable. Second, the trypanosome must be able to switch VSG disguises at a relatively high frequency (estimated to be about 10^{-4} to 10^{-5} per generation). In this way, in any given population of bloodstream parasites expressing one VSG antigen, at least one individual parasite will switch to a new VSG antigen before the host immune response is able to kill all members of the population. Moreover, trypanosomes expressing the new VSG antigen must not continue to express the old VSG, or they will be the target for attack by antibodies directed against the previous dominant VSG antigen. Such rapid antigenic switching requires that the VSG mRNA be relatively *un*stable. The necessary instability of the VSG mRNA only emphasizes the importance of producing huge amounts of translationally active VSG mRNA.

Multiple Mechanisms of Trypanosome Antigen Switching[24]

Antigen switching in the parasite *Trypanosome brucei* can be studied in the laboratory using infected rabbits. After initial innoculation, the rabbits become sick as *T. brucei* bearing a particular VSG multiply within the bloodstream. The rabbits recover as their immune system mounts a response against the first VSG, but soon relapse as trypanosomes bearing a new VSG multiply and eventually become the dominant VSG type in the bloodstream. By cloning mRNA from the bloodstream parasites after the first, second, and third cycles of parasitemia, it is possible to assemble a collection of VSG mRNAs corresponding to successive switching from one VSG gene to the next within a single cell lineage. When these cDNA clones are used as probes for the genomic organization of the VSG genes in the three successive parasite populations, a truly remarkable pattern emerges (Figure 22-12).

VSG genes are found in two places in the trypanosome genome: as tandemly repeated copies at internal positions within the chromosomes, and as individual copies at the chromosomal ends (telomeres). Telomeric VSG genes can be found on at least three different chromosomes, and quite possibly on more. (The situation is complicated because trypanosomes have eight to ten relatively large chromosomes of 200 to 2000 kilobase pairs and about a hundred minichromosomes ranging in size from about 50 to 150 kilobase pairs.) The expressed VSG gene is always located at one of the telomeric sites, so these are called **expression sites.** Only one telomeric VSG gene is activated at a time, but the mechanism responsible for such mutually exclusive activation remains a mystery. The several thousand VSG gene copies located internally within the chromosome are transcriptionally silent and can only be expressed after they are duplicated and moved from an internal tandem array to a telomeric expression site (presumably by a directed gene conversion). Quite possibly, basic copies of the VSG genes are continually moved from the internal tandem arrays to telomeric sites to replenish the variety of VSG genes awaiting activation at a telomeric expression site.

The similarity between switching of yeast mating types (Chapter 18) and trypanosome variant surface antigens should be obvious. The silent mating type cassettes at *HMR* and *HML* are analogous to the tandem repeats of transcriptionally silent basic VSG gene copies. The directed gene conversion that moves silent genetic information from *HMR* or *HML* to *MAT* (the expression site) is analogous to antigen switching. And the analogy may extend still further. In the yeast mating switch, not only must the genetic information at *MAT* be expressed, but the silent gene copies at *HMR* and *HML* must also be repressed. The selective repression of the *HMR* and *HML* loci (but not *MAT*) is the job of the *SIR* (silent information repression) gene products, which appear to alter the chromatin structure surrounding the silent cassettes. Perhaps a similar mechanism, based on the difference between active and inactive chromatin (Chapter 21), may allow trypanosomes to express only one telomeric VSG gene at a time.

Interestingly, the first 35 base pairs of all expressed VSG mRNAs are identical, regardless of antigenic type. However, the DNA encoding this 35-base-pair "mini-exon" sequence is not part of the basic VSG gene copy that is moved to the telomeric expression site, nor does the expression site contain a 35-base-pair mini-exon-coding sequence that can be used by whichever basic VSG gene copy happens to occupy the expression site. Instead, about 200 copies of the 35-base-pair sequence are present elsewhere in the trypanosome genome, largely in the form of a 1.35-kilobase-pair tandemly repeated DNA segment (Figure 22-13). In particular, at least one telomeric expression site is located on a 550-kilobase-pair chromosome completely lacking any 35-base-pair mini-exon sequences. Thus, transcription of a complete VSG mRNA must be discontinuous; the 35-base-pair mini-exon sequence is transcribed independently of the body of the VSG mRNA and subsequently added to it.

How is the 35-base-pair mini-exon sequence assembled onto the body of the expressed VSG mRNA? We now know that the 35-base-pair mini-exon-coding regions are transcribed to give 140-base RNAs known as "mini-exon-derived RNAs" or medRNAs. We also know that both the 140-base medRNAs and the mature VSG mRNA possess the 5' cap structure typical of all eucaryotic mRNAs. Thus, there are two obvious possible mechanisms for the joining reaction (see Figure 22-13). Either the medRNAs could serve as primers for transcription of the body of the VSG mRNA, or the medRNAs could be joined to the body of the VSG mRNA by an unusual *trans* (intermolecular) RNA splicing reaction. Experiments now in progress will soon distinguish between these possibilities.

Although first discovered in VSG mRNAs, the evidence now suggests that most (and perhaps all) trypanosome mRNAs share the same 5' mini-exon sequence. Why trypanosomes have resorted to such an exotic procedure for initiating mRNA transcription remains a mystery, but recent work suggests that *trans* splicing may not be confined to trypanosomes. In extracts of human cells, *trans* splicing can

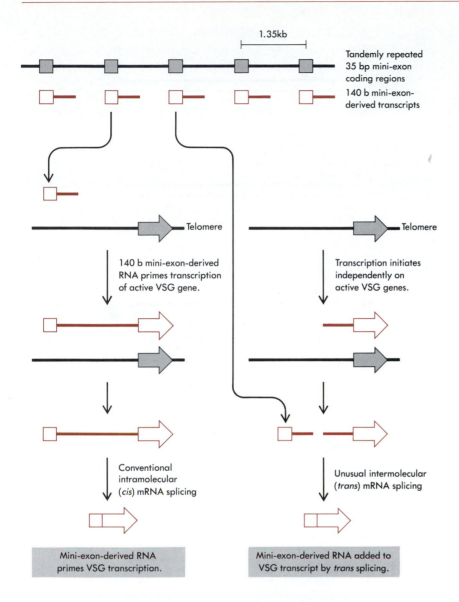

Figure 22-13
Two models for assembly of the common 35-nucleotide mini-exon sequence onto the body of the expressed trypanosome VSG gene (not drawn to scale).

proceed efficiently if the two separate RNA molecules carrying the 5′ and 3′ exons are constructed in such a way that base pairing between complementary sequences in the intron can hold the two molecules together. No natural instance of *trans* splicing has yet been documented in higher eucaryotes, but it seems unlikely that evolution has failed to take advantage of a reaction that can proceed so easily in vitro.

Unusual Excision and Amplification of the Gene Encoding *Tetrahymena* Ribosomal RNA[29, 30]

In Chapter 20, we discussed the ability of the ribosomal RNA precursor (pre-rRNA) in the ciliate *Tetrahymena* to function as an enzyme (ribozyme). However, autocatalytic splicing is not the only extraordinary aspect of the rRNA in this ciliate. In most eucaryotes ranging from yeast to mammals, the genes encoding ribosomal RNA constitute a large multigene family with several hundred germ line copies per haploid genome (Chapter 20). Remarkably, the germ line micronucleus of *Tetrahymena* contains only a single copy of the rDNA gene.

Figure 22-14

Multiple copies of the 21-kilobase linear extrachromosomal palindromic rDNA in the *Tetrahymena* macronucleus are derived by excision and amplification from a single 10.5-kilobase half palindrome in the germinal micronucleus. A plausible mechanism for this process is shown here. To distinguish between the two strands of the rDNA, one of them is drawn as a thick line. The flanking chromosomal sequences in the micronucleus are represented by wavy lines. M represents the 42-nucleotide inverted repeats on either side of the future center of palindromic symmetry. A′ represents the 20-nucleotide inverted repeats flanking the rDNA. Excision involves breakage of the chromosome at or near the two ends of the rDNA. The A′ sequence might serve as the recognition sequence for breakage. Some flanking sequences are lost from both breakage sites, probably by exonuclease activity. These events lead to the formation of four free ends, at least three of which are subsequently modified by addition of the telomeric T_2G_4 repeat sequences to one strand (black boxes) and C_4A_2 repeats to the other strand (white boxes). The 5′ end of the rDNA, which is not modified by the addition of telomeres, folds back to allow pairing between the two inverted M sequences. Folding back is possible because the micronuclear half palindrome is actually a little larger than half and includes the future center of palindromic symmetry. Intramolecular homologous recombination then joins the free end of the thick line to an internal site on the thin line, producing one short and one long hairpin molecule. The short hairpin molecule primes replication of the long one, and after one complete round of replication, the first mature extrachromosomal 21-kilobase palindrome is formed. The rRNA transcription units are shown only on this first mature rDNA molecule (large shaded arrows). The figure is not drawn to scale. [After M.-C. Yao, S.-G. Zhu, and C.-H. Yao, *Mol. Cell. Biol.* 5 (1985):1260.]

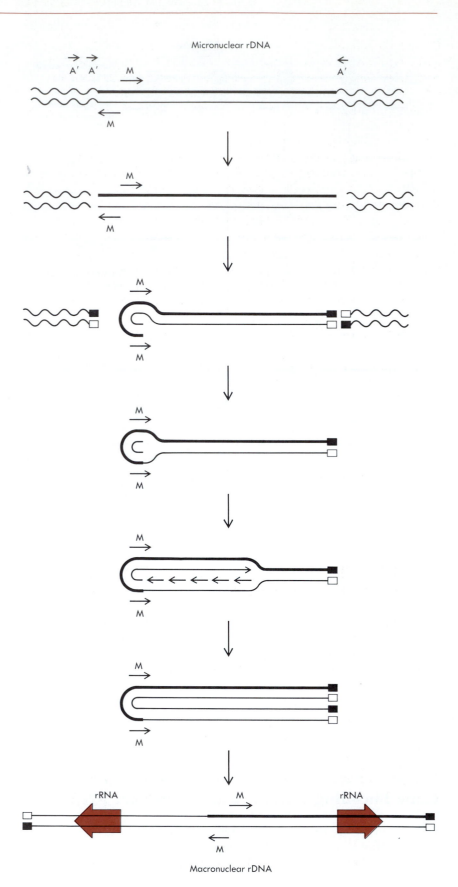

During regeneration of the new macronucleus after mating, this single rDNA copy is excised from micronuclear DNA as a 10.5-kilobase-pair linear molecule, transformed into a nearly perfect 21-kilobase-pair palindromic dimer, and subsequently amplified to higher copy number much like any other macronuclear DNA fragment (Figure 22-14).

Early studies had established that the palindromic rDNA dimer, like other linear DNA molecules in the macronucleus, has both an origin of DNA replication and telomeres. The origin is located very near (or even at) the center of symmetry, as judged by the distribution of replication bubbles visualized in the electron microscope (Chapter 21). Telomeres (which also function in yeast; see Figure 18-14) consisting of 40 to 70 tandem repeats of the simple sequence C_4A_2 were found at either end of the palindrome. In principle, then, we might have guessed that the single micronuclear (germ line) copy of the gene encoding rRNA would be excised as a simple half palindrome, where one endonuclease cutting site would be defined by the center of rDNA symmetry and the other site by tandem germ line repeats of C_4A_2. However, these expectations proved to be naive.

The first surprising result was that the DNA sequence surrounding the origin of replication near the center of the palindrome was not perfectly symmetrical. This was a clue that the extrachromosomal palindromic rDNA could not be generated by simple excision and dimerization of an exact germ line half palindrome (monomer). In fact, after conjugation, the earliest detectable rDNA excision product was not cut at the center of (imperfect) symmetry but rather 42 base pairs away (see Figure 22-14). Apparently, after excision of the germ line rDNA monomer, this 42-base-pair sequence can fold back to form an imperfect hairpin that primes DNA replication, thereby converting the newly excised linear monomers into mature palindromic dimers.

The second surprising result was that the germ line rDNA monomer contained only a single C_4A_2 where the telomere should be. Thus, the mature telomeres are probably synthesized de novo by the T_2G_4-adding enzyme together with DNA polymerase. (For historical reasons, telomere sequences are conventionally given as the C_nA_m-containing strand, although the properties of the T_mG_n-adding enzyme appear to determine the exact sequence of the telomeres in each species.) In principle, the solitary germ line C_4A_2 could serve as the recognition sequence for excision by a C_4A_2-specific endonuclease. Alternatively, excision could occur beyond the germ line C_4A_2 sequence, and an exonuclease would then nibble back the free end until the exposed germ line T_2G_4 could be recognized by the T_2G_4-adding enzyme. Recently, the germ line rDNA monomer was found to be flanked on either side by inverted repeats of a 20-base-pair sequence. This suggests that the 20-base-pair sequence (which is also found at many other places in the *Tetrahymena* genome) is the recognition site for an endonuclease that fragments micronuclear DNA at specific sites during regeneration of the macronucleus.

Excision and Selective Amplification of the rRNA Genes Within Amphibian Oocytes[31]

The number of rRNA genes within an amphibian oocyte is about a hundred- to a thousandfold greater than within somatic cells of the same organism. But this extra DNA does not reflect an increase in chromosome number from the 4N characteristic of oocytes prior to

Figure 22-15
Electron micrograph of an extrachromosomal nucleolar core from one oocyte of the African toad *Xenopus laevis*. About ten rRNA genes are tandemly arranged along the circular DNA molecule. The arrows point to individual genes. (Courtesy of O. L. Miller and B. R. Beatty, Oak Ridge National Laboratory.)

the first meiotic division. Instead, almost all the increase is due to large numbers of extrachromosomal nucleoli. Each nucleolus contains one or more circular DNA molecules with length varying from 12 to about 210 kilobase pairs. This corresponds to 1 to 20 tandemly arranged rRNA genes, each coding for the 45S rRNA precursor. Figure 22-15 shows a most elegant electron micrograph of one of these extrachromosomal nucleoli to which are attached over a thousand growing rRNA precursor molecules.

Very large numbers of extrachromosomal nucleolar circles are needed to produce sufficient RNA components (18S, 5.8S, and 28S rRNA) for the massive stock of ribosomes found in a mature oocyte, the largest of all cells. This is especially true because structural RNA species such as ribosomal RNA are the final product of synthesis, and unlike messenger RNA, the product cannot be further amplified by multiple rounds of translation. All the ribosomes present before gastrulation are synthesized before fertilization in the oocyte. During growth of the frog oocyte, for example, some 10^{12} ribosomes must be made in the course of only a few weeks. The number of rRNA genes within the oocyte chromosomes, however, reflects the requirements of the ordinary frog somatic cell, which is some 10^5 times smaller than the oocyte. Only by extensive rDNA amplification can oocyte growth proceed at a biologically acceptable rate.

Most of the nucleolar circles in mature oocytes arise by a rolling circle mechanism (Chapter 10) from previously made circles (Figure 22-16). Still very unclear, however, is the origin of the first extrachromosomal gene copies. Genetic evidence obtained from hybrid frogs shows that the oocyte's extrachromosomal rRNA genes must be derived from the tandemly repeated chromosomal rRNA genes. Conceivably, the first extrachromosomal gene copies could originate by a crossing-over mechanism similar to that which releases λ prophages from the *E. coli* chromosome. But this scheme requires the later reinsertion of rDNA so that the number of chromosomal rRNA genes will remain relatively constant. Another possibility is the enzymatic creation of specific single-strand cuts in chromosomal rDNA, which lead to the displacement of long, single-stranded sections. Such displaced chains could then be converted into double-stranded circles by a process of the rolling circle type. This matter will most likely be settled only when such circles can be made in vitro by oocyte extracts.

The Frog Oocyte 5S RNA Genes Are "Preamplified"

In all eucaryotes, the 18S, 5.8S, and 28S rRNA components of the ribosome are derived by RNA processing from a single primary transcript synthesized by RNA polymerase I; however, the 5S RNA component of ribosomes is transcribed from an entirely separate family of genes by RNA polymerase III (Chapter 20). Some provision must therefore be made during oogenesis for producing the same amount of 5S RNA as there is 18S plus 5.8S plus 28S rRNA synthesized by the thousandfold amplified extrachromosomal rDNA. As we saw in Chapter 20, the frog accomplishes this by having two different sets of 5S RNA genes, one coding for the 5S RNA found in oocyte ribosomes, the other for the 5S RNA of somatic cells. While there are only 500 somatic 5S genes, over 10,000 copies of the oocyte variety are present per haploid genome; both types of genes are organized into separate tandemly repeated arrays. Thus, no extrachromosomal amplification of 5S DNA need occur during oogenesis; the genes are in effect "preamplified." However, some control mechanism must selectively turn on the much larger gene set needed for oocyte ribosome synthesis.

The several hundred germ line copies of the 12,000-base-pair rDNA repeat unit normally account for only 0.2 percent of frog DNA, but after thousandfold amplification, the extrachromosomal rDNA accounts for 75 percent of the total oocyte DNA. This explains why developmentally regulated amplification of the rDNA is a better biological strategy than carrying a thousandfold excess of rDNA genes in every somatic cell. In contrast to the 12,000-base-pair rDNA repeat unit, the 5S gene repeat unit is only about 500 base pairs long. Consequently, even 20,000 germ line copies of the oocyte 5S genes still account for only 0.01 percent of the frog DNA, a tolerable amount.

In contrast to the tandemly repeated germ line rDNA genes, which are located on only one or a few frog chromosomes, the tandemly repeated 5S genes are located near (or at) one of the telomeres of almost every chromosome (Figure 22-17). We might attempt to rationalize the telomeric location of both oocyte and somatic 5S genes by noting that the major steps in ribosome assembly are completed *within* the nucleolus using newly synthesized ribosomal proteins imported from the cytoplasm (Chapter 21). Moreover, the many extrachromosomal nucleoli in an oocyte are physically located just inside the nuclear membrane, perhaps to minimize the distance that incoming ribosomal proteins and outgoing ribosomes must traverse during the massive accumulation of ribosomal subunits. Since telomeres are often associated with the nuclear membrane, the telomeric location of oocyte 5S genes might facilitate the efficient incorporation of newly synthesized 5S RNA into ribosomes. Unfortunately for this hypothesis, 5S genes in the fruit fly *Drosophila melanogaster* are located internally on the same chromosome that carries the rDNA genes. This discrepancy suggests that it may be a long time before we understand why these (and other) genes are located where they are on chromosomes.

Lampbrush Chromosomes Fulfill the Massive Transcriptional Demands of Large Oocytes[32]

To make 10^{12} ribosomes for use in early embryogenesis, the *Xenopus* oocyte must synthesize more than just rRNAs. A correspondingly high rate of transcription (and translation) of the mRNAs coding for ribosomal proteins is required. Also provisioned for the egg during the few months of oogenesis are huge stores of many other proteins (e.g., enough histones to package 4096 genomes worth of DNA), as well as most of the mRNAs that are translated only after fertilization during the early stages of development (maternal mRNAs).

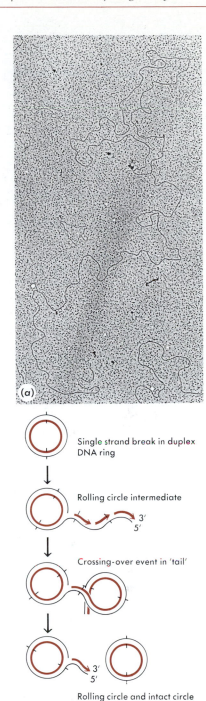

(a)

Single strand break in duplex DNA ring

Rolling circle intermediate

3′
5′

Crossing-over event in 'tail'

3′
5′

Rolling circle and intact circle

(b)

Figure 22-16
(a) An electron micrograph of a ribosomal DNA rolling circle from a *Xenopus laevis* oocyte. The circumference of the template circle corresponds to three ribosomal RNA genes and the attached tail to 3.4 ribosomal RNA genes. [Reproduced with permission from D. Hourcade, D. Dressler, and J. Wolfson, *Cold Spring Harbor Symp. Quant. Biol.* 38 (1973):537–550.] (b) Ribosomal DNA amplification by a rolling circle mechanism from an extrachromosomal nucleolar circle. Compare with Figure 10-21.

Purified 5S DNA

Tritium-labeled
RNA precursors + RNA polymerase

Multiple labeled
RNA copies of 5S
DNA are transcribed
in vitro.

NaOH-denatured
chromosome is
incubated with
labeled RNA.

Autoradiography
shows labeled
RNA annealed at
the ends of the long
chromosome arms.

Figure 22-17
The demonstration that the multiple copies of 5S RNA genes are located at or near the telomere region on the long arms of most chromosomes in the toad *Xenopus*. Labeled RNA copies of the 5S gene (shown), or in vivo labeled 5S RNA molecules, anneal with denatured DNA in the chromosomes and are revealed by autoradiography.

To meet such incredible demands for RNA synthesis, the chromosomes of large oocytes assume a bizarre extended configuration. As shown in Figure 22-18, much of the DNA is compacted into tight masses called **chromomeres** that contain the main chromosomal axes. But a significant fraction of the genome (from 5 to 30 percent, depending on the species) is extended into very long lateral loops that contain up to 100 kilobase pairs; hence the name **lampbrush chromosomes.** (Lampbrushes were used to clean soot out of the globe of an oil or gas lamp before the advent of the incandescent electric bulb.) While the condensed chromomeric regions are largely inactive in transcription, the majority of loops are transcribed at nearly maximal rates. That is, a nascent RNA chain emanates about every 100 base pairs (Figure 22-19) from the looped-out DNA helix. This can be compared to the average, highly transcribed gene region in somatic cells, where one RNA transcript is seen only every 10,000 base pairs. The presence of lampbrush chromosomes is nicely correlated with large oocyte size, which can occur in amphibians, reptiles, birds, and insects, but not in mammals. Among echinoderms, starfish have lampbrush chromosomes, whereas sea urchins do not; urchin eggs are several hundredfold smaller than those of starfish.

The most curious aspect of lampbrush chromosomes, which distinguishes them from somatic chromosomes, is that many of the lampbrush transcripts are gigantic. Even huge loops are often transcribed

Figure 22-18
A partial view of a pair of highly extended lampbrush chromosomes from an oocyte of the newt *Triturus viridescens*. Two chiasmata are seen in this micrograph, taken in the diplotene phase of meiosis. The arrow points to a section where the double-stranded character of each premeiotic chromosome is revealed. (Courtesy of J. Gall, Carnegie Institution of Washington.)

into a single RNA species. Such loops have a distinctively polar appearance caused by the association of proteins to form RNP particles on the nascent RNA chains (see Figure 22-19). Alternatively, two or more transcription units can be distinguished on some loops. Accounting for the generation of the unusually long transcripts is a failure in transcription termination and/or RNA processing. While the transcripts appear to initiate at bona fide promoters, they proceed far past the normal 3' ends and thus read into repetitive sequences of various sorts. Late in oogenesis, readthrough increases, giving rise to a higher fraction of such transcripts. Many of these RNAs are exported to the cytoplasm, and it is not known whether they find some eventual use. Meanwhile, however, vast quantities of functional mRNAs have also been produced both for translation during oogenesis and for storage in the egg.

Figure 22-19
A section along a loop of a *Triturus* lampbrush chromosome, showing a gradient of increasingly long RNA chains attached to their DNA template. The black dots at the sites where the RNA chains attach are RNA polymerase molecules. (Courtesy of O. L. Miller and B. Beatty, Oak Ridge National Laboratory.)

The *Drosophila* Chorion Genes Can Be Amplified Without Excision[33]

We have seen how ciliates achieve selective gene amplification in the somatic macronucleus by a curious and roundabout strategy: The entire chromosome complement is first polytenized, and then DNA segments that are unnecessary for vegetative growth are degraded. Amphibian oocytes achieve selective gene amplification more directly by excision of chromosomal rDNA and subsequent extrachromosomal amplification. However, as we shall see for the chorion genes of the fruit fly, developmentally regulated gene amplification need not involve chromosome fragmentation or extrachromosomal replication.

The *Drosophila* egg is such a large cell that a single haploid nucleus cannot possibly be transcribed at a sufficiently high rate to stock the entire egg cytoplasm in the brief 5- to 7-day time span available for oogenesis. Instead, thousands of follicle cells surround and actively nourish the maturing egg. In addition, 15 nurse cells located at the anterior end of the egg build cytoplasmic bridges to secrete various factors (including mRNA!) into the egg. Thus, we can think of the

Figure 22-20
Scanning electron micrograph of the surface of a mature eggshell (chorion) from another insect, the silk moth *Antheraea polyphemus* (1400×). The single micropyle for sperm entry is visible in the foreground; numerous breathing tubes (aeropyles) occupy the background. Note the extraordinary degree of structural organization in this tiny egg. (Courtesy of G. D. Mazur, Harvard University.)

mature egg as a group enterprise in which the egg itself plays a major role but the nurse and follicle cells do much of the hard work. As the egg matures, the surrounding follicle cells construct an eggshell, or **chorion,** to protect the egg from physical stress and drying out. During the 5 hours of choriogenesis, the follicle cells secrete a carefully orchestrated succession of some 20 cysteine-rich chorion proteins that assemble and crosslink to form a complex, multilayered eggshell. The only holes in the eggshell are the breathing tubes that allow respiration and the single anterior micropyle that permits sperm entry during fertilization (Figure 22-20). The mature eggshell (chorion) has a recognizable ventral and dorsal side, as well as an anterior and posterior end (shown schematically in Figures 22-24 and 22-27).

In *Drosophila*, the chorion proteins must be synthesized so quickly, and in such large quantities, that single gene copies are not enough; instead, the follicle cells are developmentally programmed to selectively amplify their chorion genes at the appropriate time in oogenesis. The genes encoding the chorionic proteins are located in two clusters, one at cytological band 7F1-2 on the *Drosophila* X chromosome and the other at band 66D12-15 on the third chromosome. Each gene is present as only a single copy. However, in follicle cells that are midway through oogenesis, an origin of DNA replication located within each chorion gene cluster is programmed to fire three to six times, eventually leading to a 32- to 64-fold amplification (2^6) of the adjacent chorion genes. (Note that these unusual developmentally regulated origins fire during interphase in follicle cells. We do not know whether the same origins also function during S phase in other cell types.) Replication proceeds bidirectionally, but each successive round of replication is less extensive than the one before: Perhaps topological constraints limit the number of replication forks that can lie close together on the DNA. The final structure resembles an onion-skin and remains an integral part of the chromosome. The gradient of

Figure 22-21

Developmentally regulated amplification of the *Drosophila* chorion genes in follicle cells. (a) Diagram showing first three rounds of interphase DNA replication at the 66D locus on chromosome 3. The three small arrows, shown only for the unamplified DNA, represent three well-characterized chorionic genes in this cluster. The polarity of a fourth chorionic gene, and the precise location of the origin relative to the genes, is not yet known. (b) A graphical representation of the final levels of gene amplification as a function of position along the chromosome. The genes are shown above the symmetrical amplification gradient to emphasize that the boundaries of the amplified DNA are much larger than the chorionic protein transcription units within it. (c) Chromatin strands from the "onionskin" of an early-stage 11-follicle cell, showing multiple replication forks (arrows). [(b) After Figure 5 and (c) reproduced from Figure 8 of Y. N. Osheim and O. L. Miller, Jr., *Cell* 33 (1983):543–553, with permission.]

amplification on the third chromosome extends for some 50 kilobase pairs in either direction and has been visualized directly in the electron microscope (Figure 22-21).

Developmental Problems
May Have Multiple Solutions[34]

Why does *Drosophila* selectively amplify the chorion genes just before they are expressed, rather than creating a germ line multigene family that is selectively turned on during choriogenesis? We might have guessed that selective amplification is part of the mechanism by which the genes are turned on, but the data tend to rule out this attractive possibility: In the normal course of development, chorion genes are amplified *before* they are turned on. Still worse for any simple rationalization of chorion gene amplification, in two silk moths (*Bombyx mori* and *Antheraea polyphemus*), the chorion genes are in fact present in multiple germ line copies; their expression, but not their copy number, is developmentally regulated.

Both the fruit fly and the silk moth regulate expression of the chorion genes, but the moth maintains amplified genes in the germ line, while the fly subjects the genes to somatic amplification just prior to use. This discrepancy may illustrate a general developmental truth: Organisms solve their developmental problems in any way that works, whether sloppy or elegant. And since evolution proceeds by sequential modification of preexisting developmental patterns, even a sloppy or awkward developmental mechanism may become so intricately entangled in the overall developmental program of an organism that it cannot be changed thereafter. Viewed in this light, developmental programs are partly logical, and partly mere historical expedients that have been frozen into place by subsequent evolution. A corollary of this truth is that there may be no universal developmental mechanism, but rather a multiplicity of mechanisms that can be combined and adapted in many surprising ways.

Stored mRNAs in the Egg May Be Regulated by Translational Control[35-42]

A mature oocyte (egg) may be huge, but it is still only a single cell. The fertilized egg must therefore undergo many cleavage divisions before there are enough cells for morphogenesis (blastulation and gastrulation; see Figure 22-5). In organisms where the life cycle puts a premium on rapid development (a toad egg may hatch in a mud puddle that is rapidly drying out in the hot sun), these cleavage divisions take place very fast. Such rapid cell divisions leave no time for an increase in embryo mass (growth) or for transcription, and protein synthesis during this time must use preformed mRNAs present in the oocyte at the moment of fertilization (stored "maternal" mRNAs). (Sperm cells are very small, have little cytoplasm, and carry essentially no mRNA into the egg.) Only after the midblastula transition, when RNA polymerase II and III are activated (see Figure 22-6) and mRNA synthesis commences, is the paternal contribution to the embryonic genome transcribed and expressed for the first time.

Not surprisingly, the transition from the mature oocyte to the newly fertilized egg is accompanied by large changes in protein synthesis. Many proteins specifically required for building the oocyte are no longer synthesized, while many stored maternal mRNAs that encode proteins required for early embryogenesis must be activated. How this switch from oocyte-specific to embryo-specific translation occurs is only just now beginning to be understood. In the surf clam *Spisula solidissima,* the switch is not accompanied by any detectable changes in the population of cellular mRNAs. Instead, utilization of different subsets of the total mRNA population at the two developmental stages is accomplished by translational control. Experimentally, it was found that in vitro translation of deproteinized mRNA from the egg and embryo yielded the same pattern of protein synthesis. However, when the mRNA was translated in vitro without prior deproteinization, egg mRNA ribonucleoprotein particles (mRNPs; Chapter 21) were translated into egg proteins, and embryo mRNPs were translated into embryonic proteins. This suggests that translation of specific mRNA species can be negatively regulated (masked) by some of the proteins that package mRNA. Other evidence suggests that positive translational control may be responsible for preferential expression of hsp70 protein (Chapter 21) during the heat-shock response in frog oocytes and in *Drosophila* tissue culture cells. Presumably, both positive and negative translational control reflect the sequence-specific recognition of mRNA by proteins, but recognition of mRNA by complementary ("antisense") RNA species remains a possibility (see the model for regulation of Tn10 expression in Figure 16-26). In addition, the level of posttranscriptional modifications such as methylation of 5' cap structures and polymerization of 3' poly A tails may selectively influence the translation of different subpopulations of mRNA.

What proteins might the early embryo need in large amounts? Rapid DNA replication requires both the deoxyribonucleotide triphosphate building blocks for DNA and histone proteins to package the newly made DNA into chromatin. Thus, one of the most abundant stored maternal mRNAs that is activated during early embryogenesis of the surf clam encodes the small subunit of ribonucleotide reductase, a key metabolic enzyme required for synthesizing the deoxyribonucleotide precursors of DNA. Similarly, some of the most abundant stored maternal mRNAs in frog and sea urchin eggs encode histones. Although the egg does contain a stock of histone proteins, this stock is quickly exhausted by the synthesis of new DNA. Translation of the stored mRNA population must then sustain the oocyte until mRNA synthesis resumes after the midblastula transition. In contrast to the situation in normal somatic cells (where histone mRNA is synthesized only during DNA replication and afterward is rapidly degraded in the cytoplasm), histone mRNA is stored in the nucleus of the oocyte before fertilization and is stable in the absence of DNA replication. The relative stability of histone mRNA in the cytoplasm after fertilization makes sense, since DNA replication in the early embryo is so fast that the cell is effectively always in S phase. However, the maternal stock of histone mRNA is ultimately degraded when DNA synthesis slows down after the midblastula transition, and the intervals between successive S phases increase.

Although the early embryo synthesizes many proteins just prior to use, other macromolecular components must be stored in massive amounts in the oocyte cytoplasm because there is simply no time to make them during embryogenesis. We have already seen that amplification of the genes encoding the large rRNAs and "preamplification" of the 5S rRNA genes allow the oocyte to accumulate a huge stock of ribosomes during oogenesis. In fact, from the study of mutant frogs lacking rDNA ("anucleolate," or *nu*⁻, frogs), we know that this ribosome stock is sufficient to carry the developing embryo all the way through to the tadpole stage! In contrast to ribosomes, which are stored as mature ribonucleoprotein complexes (ribosomal subunits), the oocyte cytoplasm stores snRNP proteins rather than mature snRNPs. When snRNAs (U1, U2, etc.) and mRNA precursors begin to be synthesized by RNA polymerase II after the midblastula transition, newly synthesized U1 and U2 snRNA assemble with the stored snRNP proteins to make mature U1 and U2 snRNPs that are capable of splicing the newly made mRNA precursors (Figures 20-18 and 20-19). Both ribosomes and snRNPs are ribonucleoprotein complexes, and ribosomes outnumber snRNPs by a factor of about 5 in a typical somatic cell. Why, then, does the oocyte store complete ribosomal subunits, but only snRNP proteins? Perhaps this illustrates once again that similar developmental problems often have very different solutions.

A Second Independent Clock in Fertilized Amphibian Eggs[43]

We have seen previously that titration of a preformed (protein?) factor tells the cleaving frog embryo when it has synthesized about 4000 cell equivalents of DNA; titration then triggers the many changes known collectively as the midblastula transition. This titration mechanism is not the only clock in the early embryo. Among the stored maternal mRNAs in the frog oocyte is the mRNA encoding fibronectin, a major protein of the extracellular matrix that plays a role in cell attachment and migration. Like many other changes in the embryo, translation of the stored fibronectin mRNA is normally scheduled to begin at the midblastula transition. Taking advantage of an experimental trick discovered by the pioneering embryologists of the last century, it was possible to demonstrate recently that translation of fibronectin is controlled by a second *independent* clock. The trick is that merely pricking an amphibian egg with a needle will activate it to begin the early developmental program. Apparently, a small hole in the egg can mimic the major physiological changes accompanying normal fertilization, such as the rapid increase in cytoplasmic pH and the release of bound calcium ions. These pricked eggs synthesize little or no DNA and consequently can never proceed beyond the midblastula transition; however, fibronectin synthesis begins on schedule, just as though the egg had been fertilized naturally. Thus, the egg contains a second independent clock that does not require DNA synthesis and counts time from the moment of activation, whether it occurs by natural means or by experimental trickery.

GENETIC CONTROL OF MORPHOGENESIS

Cyclic AMP Is a Morphogen in Slime Mold Development[44–47]

Clearly, the most interesting microorganisms to the embryologist are those in which cell differentiation is particularly prominent. Thus, much attention has been devoted to the cellular slime mold, *Dictyostelium discoideum*, whose life cycle is illustrated in Figure 22-22. The cycle begins with spores germinating in damp earth to yield amoeboid cells called myxamoebas, which feed by engulfing bacteria and reproduce by binary fission. When the multiplying amoebas exhaust their local food supply, they aggregate to form a multicellular slug, which migrates in a search of more food and better environmental conditions. Eventually, the slug undergoes a complex morphogenesis to construct a fruiting body that elevates the spores above soil level for most efficient dispersal to new locations.

Both aggregation and subsequent construction of the fruiting body

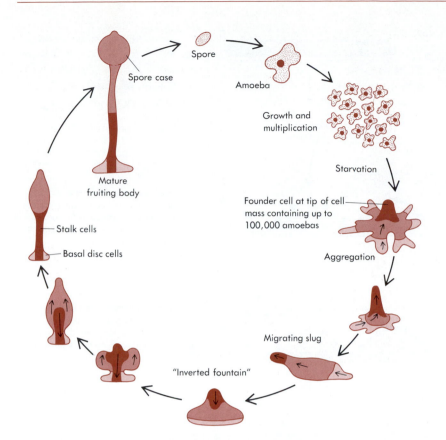

Spore

Spore case

Amoeba

Growth and
multiplication

Starvation

Founder cell at tip of cell
mass containing up to
100,000 amoebas

Aggregation

Mature
fruiting body

Stalk cells

Basal disc cells

Migrating slug

"Inverted fountain"

Figure 22-22
The life cycle of the cellular slime mold
Dictyostelium discoideum. Small arrows
show morphogenetic movement of
cells. The first amoebas to aggregate are
arbitrarily shown in the darkest color.

seem to be governed by an intracellular signaling mechanism that
uses cAMP as a chemical attractant. Within 6 to 8 hours after the
onset of starvation, each amoeba has become competent to sense a
transient increase in local cAMP concentration (reception) and to secrete a short pulse of cAMP in response (transmission). This is accomplished by a membrane-bound cAMP receptor protein that measures
cAMP concentration and then triggers a burst of activity by a membrane-bound adenylate cyclase. In addition, the cells possess a cAMP
phosphodiesterase to destroy excess extracellular cAMP in the medium and thereby prevent the signaling from being jammed by accumulation of chemical "noise."

Note that the center of each morphogenetic field is defined by the
first cell that spontaneously secretes a burst of cAMP: All the cells
become competent to signal at about the same time, but whichever
cell signals first (the **founder cell**) effectively summons all the others.
The evidence for this mechanism comes from a simple experiment in
which a glass micropipette filled with a solution of cAMP is used to
deliver small drops of cAMP to a random position within a field of
competent amoebas that have not yet begun to aggregate. The amoebas immediately begin to migrate toward the pipette, as though it
were a founder cell!

Mathematical models have been proposed to show that the ability
of each individual amoeba to secrete a pulse of cAMP shortly after
sensing one should lead to aggregation, but perhaps the clearest evidence for signaling comes from an elegant experiment that directly
visualizes concentric waves of cAMP surrounding the founder cell at
early stages of aggregation (Figure 22-23). Later, the formation of intercellular contacts between cell surface glycoproteins causes the
amoebas to adhere to one another and to stream toward the center.

Most important is the realization that both the size of an aggregation field and the speed with which it forms rule out the possibility

(a)

Figure 22-23
Visualizing a gradient of the slime mold morphogen, cyclic AMP, by an isotope dilution method. (a) An autoradiograph. (b) The experimental procedure. [Photograph reproduced with permission from K. J. Tomchik and P. N. Devreotes, *Science* 212 (1981):443–446.]

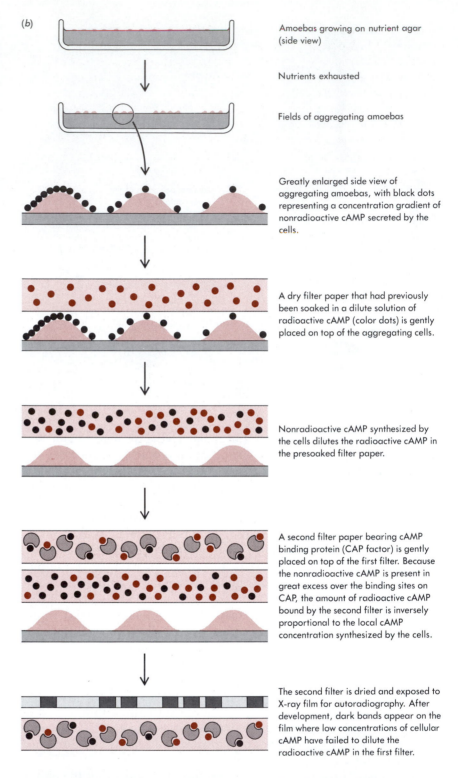

(b)

Amoebas growing on nutrient agar (side view)

Nutrients exhausted

Fields of aggregating amoebas

Greatly enlarged side view of aggregating amoebas, with black dots representing a concentration gradient of nonradioactive cAMP secreted by the cells.

A dry filter paper that had previously been soaked in a dilute solution of radioactive cAMP (color dots) is gently placed on top of the aggregating cells.

Nonradioactive cAMP synthesized by the cells dilutes the radioactive cAMP in the presoaked filter paper.

A second filter paper bearing cAMP binding protein (CAP factor) is gently placed on top of the first filter. Because the nonradioactive cAMP is present in great excess over the binding sites on CAP, the amount of radioactive cAMP bound by the second filter is inversely proportional to the local cAMP concentration synthesized by the cells.

The second filter is dried and exposed to X-ray film for autoradiography. After development, dark bands appear on the film where low concentrations of cellular cAMP have failed to dilute the radioactive cAMP in the first filter.

that cAMP secreted by the founder cell is merely diffusing outward and telling amoebas at the edge of the field to migrate inward toward the source of the signal. Such passive diffusion would be much too slow, and the cAMP would be destroyed much too quickly by the phosphodiesterase. Like a bucket brigade, the morphogenetic signal must be actively relayed from one amoeba to the next, until the entire field behaves as a single unit.

Cyclic AMP does more than simply bring together a mass of cells with radial symmetry; it also allows them to organize into a multicellular organism with a distinct front and back end (an anterior-posterior axis). The first cells to aggregate are at the center and form the front tip of the slug when the conical mass topples over on its side. Similarly, the last cells to join the aggregate form the tail of the slug. This cellular organization is preserved while the slug migrates, but then undergoes a curious inversion during morphogenesis to form a fruiting body. When the slug comes to rest, the front tip (organizer region) travels down through the rest of the slug to form an "inverted fountain." These migrating tip cells differentiate into a tough stalk, where the cells then suffer a programmed cell death (see the section Invariant Cell Lineages for more on this fascinating subject). Cells from the remainder of the slug are elevated by stalk formation and differentiate into spores that can begin a new life cycle. Except for the final steps of sporulation and stalk cell death, differentiation is completely reversible. For example, when stalk cells are isolated from the fruiting body before cell death begins, they may transform back into the ameoboid phase and proceed to divide by fission.

Strictly speaking, we do not know whether the behavior of slime mold amoebas is a good model for, say, cell movement during gastrulation of an amphibian or human embryo. However, it is an article of faith for experimental biologists that radically different organisms may solve similar problems in similar ways. The basis for this fundamental faith is that natural selection tends to (but need not) conserve basic developmental mechanisms in the course of evolution. Paradoxically, this faith does not conflict with the knowledge that similar developmental problems have often been solved in multiple (and surprisingly different) ways. Thus, we may hope that gradients of small chemical morphogens like cAMP or small polypeptides may direct cellular movements during the development and differentiation of tissues in higher organisms.

Embryogenesis Establishes the Basic Body Plan[48]

Perhaps the greatest challenge for modern embryologists is to explain in molecular terms how the simple, one-dimensional array of genetic information in DNA can be transformed into the complex, three-dimensional organization of the adult. Classical embryologists had already realized that the egg and the sperm could not be entirely "shapeless." Somehow, the germ cells had to contain either a miniaturized version of the adult body plan or a biochemical mechanism for regenerating that body plan afresh with each new generation. In the 1700s, this led to the theory of preformation, which held that the egg and the sperm are actually tiny adults (a homunculus in the case of *Homo sapiens*). The main function of development was then to make this hidden little creature bigger. We can look back on such speculation with a comfortable sense of superiority, but there was truth in the basic idea. We now know that certain developmental patterns *are* preformed within the egg, and other shape-determining mechanisms are in fact actuated by fertilization.

For insects, as for amphibians, early embryogenesis takes place so fast that there is no time for cell growth. Instead, the task of early embryogenesis is to subdivide the very large egg into many smaller cells so that **morphogenesis** can begin. Drosophila and other insects go about this task in a curious way (Figure 22-24). Immediately after fertilization in *Drosophila*, the zygotic nucleus undergoes roughly 9 very rapid divisions to yield approximately 512 nuclei without accom-

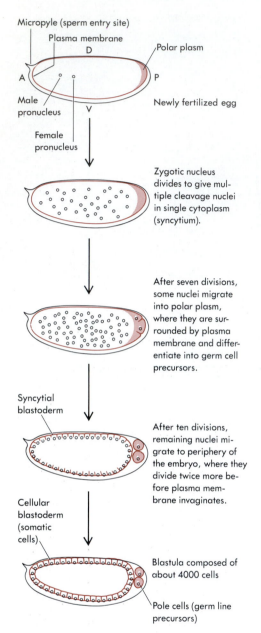

Figure 22-24
Early embryogenesis of the fruit fly *Drosophila*. A stands for anterior; P, posterior; D, dorsal; V, ventral. [Modified from W. J. Gehring, *Sym. Soc. Dev. Biol.* 31 (1973):103–128; and V. E. Foe and B. M. Alberts, *J. Cell. Sci.* 61 (1983):31–70.]

panying cellularization: The egg becomes a giant syncytium (literally, "one cell") filled with nuclei. After the ninth division, the nuclei migrate to the surface of the egg and divide four more times in synchrony. The plasma membrane of the egg invaginates to enclose each nucleus in a separate plasma membrane, thus forming the cellular **blastoderm.** Complex folding of this hollow ellipsoidal ball of cells then gives rise to a segmented larva known as the first **instar,** which has a head, three thoracic segments (T1–T3), and eight abdominal segments (A1–A8; Figure 22-25).

This instar is the first of three larval stages, separated by molts, during which the wormlike creature eats voraciously to accumulate sufficient biomass to construct an adult fly. After the third instar, the larva pupates to provide a protected environment for metamorphosis. During metamorphosis, the larva digests many of its own tissues and replaces them with adult structures; however, the body of the adult fly retains the segmental boundaries of the larval instar forms. Surprisingly, structures on the surface of the adult fly (such as legs, wings, and antennas) are constructed *not* from differentiated larval cells but rather from small groups of proliferating but relatively undifferentiated cells called **imaginal discs** (Figure 22-26). (An imago is an adult fly.) The 19 imaginal discs (9 paired and 1 fused genital disc) are set aside very early in embryogenesis, and each is committed to a particular developmental pathway (e.g., left leg, right leg, left eye, or right eye). The imaginal discs remain in a relatively undifferentiated state near the surface of the larva until changing hormonal levels during pupation cause them to differentiate and undergo morphogenesis. Thus, much of the basic body plan of the fly is established in embryogenesis, long before overt differentiation of the imaginal discs creates the adult cuticular structures that clothe the internal organs of the animal.

Genes That Encode Shape-Determining Factors in the Fruit Fly

To understand morphogenesis, modern biologists almost always start by identifying genes and gene products that affect the visible appearance of the organism. This is the strategy that was used to analyze morphogenesis of the bacteriophage λ and T4 capsids (Chapter 17). The dorsalizing genes of the fruit fly *Drosophila* have recently become a very promising focus for understanding the mechanisms that govern morphogenesis in higher eucaryotes.

We have seen that the unfertilized *Drosophila* egg is surrounded by a highly structured eggshell (chorion) whose head and tail (anterior and posterior) as well as front and back sides (ventral and dorsal) are readily distinguishable. Such external differentiation suggests that the egg within already possesses anterior-posterior and dorsal-ventral polarity, and this was confirmed by genetic and biochemical analysis.

The products of at least ten different genes are now known to be required for constructing an egg with proper dorsoventral polarity. If the mother lacks a wild-type copy of any of these genes, the dorsalized embryo cannot be restored to health by a wild-type gene copy contributed by the sperm at the time of fertilization. Evidently, it is too late to reconstruct the egg using the wild-type gene product. For this reason, such genes are said to exhibit a strict **maternal effect:** The genotype of the father is completely irrelevant.

For recessive mutations in the ten dorsalizing genes, maternal effects are only observed in the eggs of a homozygous female where both gene copies are defective. Eggs of a wild-type or heterozygous female, where a wild-type gene copy can supply the normal gene product, are completely normal. The reason for this is that essentially all of the mRNA and factors present in the mature egg are secreted into the egg cytoplasm by the 15 diploid nurse cells; *dorsal* mRNA, for example, can be identified in nurse cells before it appears in the oocyte. Transcription from the haploid egg nucleus itself makes little, if any, contribution toward stocking the huge egg cytoplasm.

A Morphogenetic Gradient May Establish Dorsoventral Polarity[49]

The appearance of a fully dorsalized embryo is stunning: It seems to be all back (dorsal) and no front (ventral)! The embryo cannot survive, because it lacks a variety of dorsolateral structures such as Filzkörper (breathing tubes) and ventral structures such as denticle bands (rows of small bumps that help the larva to crawl). Instead, dorsal structures extend all around the embryo. Furthermore, geneticists have been able to identify weaker mutations in some of these genes, which lead to loss of some, but not all, ventral structures. The existence of such intermediate (partially dorsalized) embryos suggests that dorsoventral polarity in the unfertilized egg is defined by a gradient of some sort. The collective job of the various dorsalizing genes may be to generate this gradient or to enable cells to interpret their position within the gradient.

Surprisingly, mutations in some of the ten dorsalizing genes can be partially or completely "rescued" by microinjection of cytoplasm taken from an early wild-type embryo before cellularization of the blastoderm (Figure 22-27). The rescuing activity for at least five of the mutants appears to be mRNA. Moreover, RNA from a mutant

Figure 22-25

A sequence of twelve scanning electron micrographs showing early embryogenesis in *Drosophila* from 3.5 to 14 hours after fertilization. The eggshell, or chorion, has been chemically removed to reveal the details of morphogenesis. The embryo is about 0.4 mm long, and individual cells are visible as small bumps. Note that all photographs are reproduced at the same scale: The first larval instar is no bigger than the egg, but morphologically it appears to be immensely more complex. (1) The syncytial blastoderm stage, with the plasma membrane invaginating to enclose individual cells. (2) Pole cells are clearly visible at the posterior of the embryo, and the remains of the micropyle can be seen at the anterior. The first furrows appear in the cellular blastoderm, indicating that morphogenesis has begun. (8) Intricate folding of the blastoderm is well under way, and segmentation has become apparent. (12) The late embryo, about to hatch as a first larval instar. Only at this late stage does it become difficult to discern individual cells in the embryo. (Courtesy of R. Turner and A. Mahowald, Case Western Reserve Medical Center.)

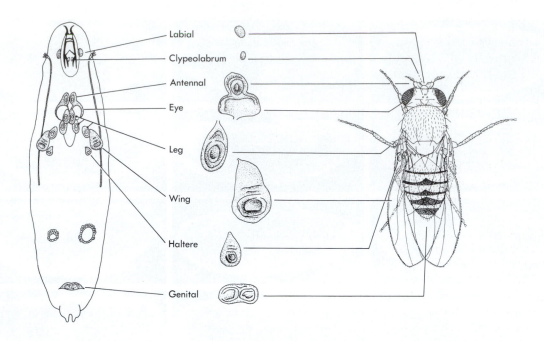

Labial

Clypeolabrum

Antennal

Eye

Leg

Wing

Haltere

Genital

Figure 22-26
The imaginal discs of *Drosophila*. The adult eye and antennal structures are both derived from a single pair of large, complex discs. The larva has three pairs of leg discs and also a pair of dorsal thoracic discs that are not shown here. Segmental boundaries have been omitted for clarity. [After J. W. Fristrom, R. Raikow, W. Petri, and D. Stewart, in *Problems in Biology: RNA in Development*, ed. E. W. Hanly. (Salt Lake City: University of Utah Press, 1969), p. 382.]

embryo with a defect in one dorsalizing gene will rescue a mutant embryo with a defect in a different dorsalizing gene, and vice versa. Thus, each of the ten or more dorsalizing genes appears to contribute a different gene product to the complex molecular machinery responsible for generating dorsoventral polarity.

Especially striking is the observation that rescue of ventral structures in embryos with a homozygous mutation in the *dorsal* gene occurs only when wild-type cytoplasm is injected into the ventral side of the mutant embryo (see Figure 22-27). Thus, the *dorsal* embryo has already established a gradient of positional information along the dorsoventral axis, but in the absence of the *dorsal* gene product, the cells are unable to interpret their position within this gradient and to differentiate accordingly. Equally interesting is the observation that injection of wild-type cytoplasm into the posterior ventral region of a *dorsal* embryo (where breathing tubes form in the wild type) rescues

Figure 22-27
Much of the information specifying the dorsoventral polarity of the *Drosophila* embryo is stored as maternal mRNA. Note that some mutations can be rescued by injections of wild-type cytoplasm but not wild-type mRNA, and that mutations in some of the ten dorsalizing genes are more completely rescued by this protocol than others. Note also that the table does not apply to homozygous *Toll* embryos. As described in the text, no matter where wild-type cytoplasm is injected into a *Toll* embryo, the site of injection defines the ventral side of the rescued embryo. This suggests that the *Toll* gene product is itself a morphogen or provides the source for a gradient of a morphogen.

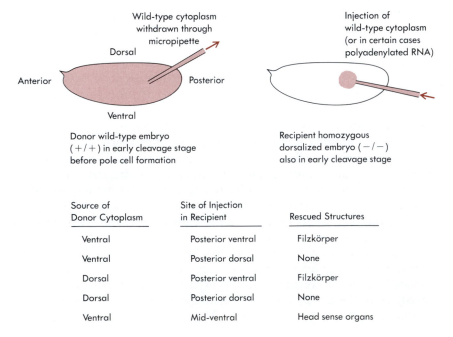

Wild-type cytoplasm withdrawn through micropipette

Dorsal

Anterior   Posterior

Ventral

Donor wild-type embryo (+/+) in early cleavage stage before pole cell formation

Injection of wild-type cytoplasm (or in certain cases polyadenylated RNA)

Recipient homozygous dorsalized embryo (−/−) also in early cleavage stage

Source of Donor Cytoplasm	Site of Injection in Recipient	Rescued Structures
Ventral	Posterior ventral	Filzkörper
Ventral	Posterior dorsal	None
Dorsal	Posterior ventral	Filzkörper
Dorsal	Posterior dorsal	None
Ventral	Mid-ventral	Head sense organs

Filzkörper, whereas injection into the mid-ventral region (where head structures form in the wild type) rescues head sense organs. Thus, the mature *Drosophila* egg has also established a gradient of positional information along the anterior-posterior axis. Moreover, since defects in dorsoventral polarity do not seem to affect the anterior-posterior gradient of information, these two gradients appear to be independent of each other.

The *Toll* Gene Product Determines the Polarity of a Morphogenetic Gradient[50]

The notion that the dorsoventral polarity of the egg is determined by a gradient of some sort is reinforced by the behavior of mutations in the dorsalizing gene known as *Toll*. Recessive mutations in *Toll*, like those in the other nine dorsalizing genes, produce embryos lacking ventral and lateral structures. However, the *Toll* gene product must behave differently from the products of the other dorsalizing genes, because it has been possible to isolate dominant mutations in *Toll*. Amazingly, these dominant *Toll* mutations cause ventral structures to extend all the way to the dorsal side of the embryo! This suggests that the *Toll* gene product may actually be the morphogen (or generate the morphogen) that causes ventral structures to form. Dominant (ventralizing) *Toll* mutations would result in overproduction of the *Toll* gene product at the normal ventral site or inappropriate expression of the *Toll* gene product throughout the egg, so that cells are told to make ventral structures in the wrong place.

Experiments using wild-type cytoplasm to rescue mutant *Toll* embryos support the interpretation that the *Toll* gene product is (or produces) a morphogen. As we have seen, the *Drosophila* embryo normally displays the same polarity as the eggshell (chorion) around it. When mutations in any of the other nine dorsalizing genes are rescued by microinjection of wild-type cytoplasm, the rescued (or partially rescued) embryo always retains the same polarity as the eggshell. But for *Toll* mutations, the polarity of the rescued embryo is determined by the site of injection of the wild-type cytoplasm: Dorsal injection reverses the normal polarity of the egg with respect to the shell. Thus, unlike the products of the other nine dorsalizing genes, which simply help cells of the blastoderm to interpret their position within a preexisting dorsoventral gradient, the *Toll* gene product appears to determine the polarity of the gradient itself.

We do not know whether this positional information might be encoded by a gradient of a morphogen (like cAMP in the slime mold); by an electrical potential akin to the proton motive force in bacteria (Chapter 6) or the propagation of a neural impulse in higher eucaryotes; by polar interactions between cell surface glycoproteins on adjacent cells; by large cytoskeletal structures that might span the entire embryo; or by mechanisms that are currently unimaginable.

Using recombinant DNA technology, the various dorsalizing genes can now be cloned and mutagenized in vitro. The ability of such mutagenized gene products to rescue dorsalized embryos can then be tested either by microinjecting the embryo with mutant mRNA synthesized in vitro, or by putting the mutagenized gene back into the *Drosophila* germ line using P element–mediated transformation. These studies may reveal the physical nature of positional information and the way cells interpret their position within a developmental gradient.

Figure 22-28

Homeotic mutations in the fruit fly *Drosophila* can convert one body part into another. (a) Head of a wild-type fly. (b) A relatively weak *Antennapedia* mutation called *Aristapedia* transforms the distal part of the antenna (the arista) into the corresponding distal part of a leg. (c) A strong *Antennapedia* mutation completely transforms the antenna into a leg. (d) Diagram showing the precise correspondence between parts of antenna and leg. [After J. H. Postlethwait and H. A. Schneiderman, *Dev. Biol.* 25 (1971):606–640; and B. Alberts, D. Bray, J. Lewis, M. Raff, K. Roberts, and J. D. Watson, *Molecular Biology of the Cell* (New York: Garland, 1983), p. 840.]

Homeotic Mutations in the Fruit Fly Change One Body Part into Another

The appearance of a dorsalized embryo may be curious, but strangest of all developmental derangements in the fly are **homeotic mutations,** which transform one part of the adult body into another. For example, the effect of the dominant homeotic mutation called *Antennapedia* (literally, "antenna-foot") is to transform the antenna on the head of the fly into a perfectly formed but absurdly misplaced leg (Figure 22-28). A less extreme mutation known as *Aristapedia* transforms only the outermost part of the antenna (the arista) into the corresponding parts of the leg (tarsus and claw). Because many *Antennapedia* mutations have visible chromosomal abnormalities (translocations, inversions, deletions) at a single locus, we know that the DNA responsible for *Antennapedia* is located at band 84B on the right arm of chromosome 3. Thus, the existence of the *Antennapedia* mutation immediately implies that the product of a single gene (or small group of genes) can work like a simple molecular switch to select one developmental fate or another for a related group of cells in the eye-antennal imaginal disc (see Figure 22-26). Weaker mutations, such as *Aristapedia,* further suggest that the switch can be thrown independently in different subgroups of these imaginal disc cells. Thus, the differentiated state of a cell, or group of cells, may simply reflect which combination of molecular switches has been thrown.

Molecular Clones of the Homeotic Genes Have Been Identified[51, 52]

Even the body plan of so small and inconsequential a creature as the fruit fly is immensely complicated, and it should come as no surprise that *Antennapedia* is not the only homeotic locus in *Drosophila.* There are, in fact, two large clusters of homeotic genes known as the **Antennapedia complex** (ANT-C) and the **bithorax complex** (BX-C). As shown in Figure 22-29, genes of the Antennapedia complex regulate the development of the head and the three thoracic segments (T1, T2, and T3) of the fly, while the bithorax complex controls development of two thoracic segments (T2 and T3) as well as all eight abdominal segments (A1–A8). BX-C contains three complementation groups (*Ultrabithorax, abdominal-A,* and *Abdominal-B;* abbreviated *Ubx, abd-A,* and *Abd-B*), each containing more than one gene. ANT-C contains at least four genes (*Deformed, Sex combs reduced, fushi tarazu,* and *Antennapedia;* abbreviated *Dfd, Scr, ftz,* and *Antp*). Mutations in these homeotic genes usually cause such severe defects that the fly dies in late embryogenesis. Occasionally, however, nonlethal mutations allow survival to adulthood, and these flies display visible abnormalities. In Figure 22-30, three separate nonlethal homeotic mutations in the *Ubx* complementation group (*abx, bx³,* and *pbx*) have been combined by genetic crosses to produce a four-winged fly where T3 is completely transformed into T2. (T3 normally bears two diminutive winglike balancing organs called halteres, and these are transformed into a nearly normal pair of wings in the triple mutant.)

Advances in recombinant DNA technology have made it possible to obtain molecular clones for several different (and perhaps all) homeotic genes in the Antennapedia and bithorax complexes. Although many subtle genetic tricks were required to assemble a complete array of recombinant clones spanning each locus, the basic experimental strategy for cloning the homeotic genes was simple (Chapter 19). Salivary gland polytene chromosomes were probed by in situ hybridization with randomly cloned fragments of *Drosophila* DNA, and in a few lucky cases, a fragment was found to be "near" the cytological location of a known homeotic gene (within a few hundred thousand base pairs!). It then became thinkable for scientists of sufficiently strong constitution to "walk along the chromosome" from the initial site of hybridization to the homeotic gene by generating a series of as many as 100 overlapping genomic clones. The first such chromosomal walk of 150 kilobase pairs took place within the bithorax complex of *Drosophila*, and it represented both a mental and a technical breakthrough. Steady progress in recombinant DNA technology has now made it an increasingly straightforward matter to obtain molecular clones for almost any chromosomal region of interest in almost any organism. Inevitably, the power of the recombinant DNA revolution has driven molecular geneticists to conceive of ever more ambitious projects. The 170-kilobase-pair genome of Epstein-Barr Virus (EBV) was recently sequenced in its entirety, and total DNA sequence analysis of the 3000-kilobase-pair circular chromosome of *E. coli* is no longer completely unthinkable. Although complete DNA sequence analysis of even the smallest human chromosome (about 50,000 kilobase pairs) must still be considered a dream, it should be possible to construct a complete restriction map for some human chromosomes. In fact, a complete restriction map for the 80,000-kilobase-pair genome of the small nematode worm *Caenorhabditis elegans* is now nearing completion.

The Genetic Control of Segmentation[53, 54]

Although the *ftz* gene is located within the Antennapedia complex between homeotic genes that affect development of the head and thorax (see Figure 22-29), mutations in the *ftz* gene are not strictly homeotic. Instead, *ftz* belongs to a fascinating group of at least 15 genes responsible for proper segmentation of the embryo. These segmentation genes fall into three classes, *ftz* itself belonging to the pair-rule class (Figure 22-31). In a homozygous *ftz* embryo, the posterior half of every "even" segment (T2, A1, A3, A5, and A7) and the anterior half of every "odd" segment (T1, T3, A2, A4, A6, and A8) is

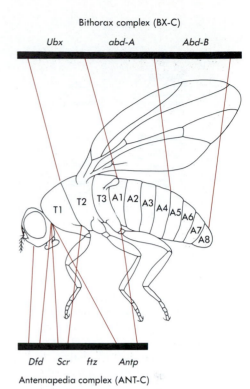

Figure 22-29
Primary domains of homeotic gene function. After metamorphosis, the body of the adult fly retains the segmental boundaries of the larval instar forms. The colored lines show the primary domain of function for each of the three complementation groups in the bithorax complex (BX-C) and each of the genes in the Antennapedia complex (ANT-C). The genetic maps are not drawn to scale. Note the wings on the second thoracic segment (T2) and the halteres, or balancing organs, on the third thoracic segment (T3). [After K. Harding, C. Wedeen, W. McGinnis, and M. Levine, *Science* 229 (1985):1236.]

Figure 22-30
A four-winged fruit fly resulting from mutations in three separate genetic loci (*abx*, *bx³*, and *pbx*) of the *Ultrabithorax* complementation group. This fly has two essentially normal second thoracic segments (T2) and no third thoracic segment (T3), because the combined effect of the three mutations is to transform T3 into T2 without affecting any other parts of the fly. (Courtesy of E. B. Lewis, California Institute of Technology.)

Figure 22-31

Three classes of genes govern embryonic segmentation in *Drosophila*. For each class, the schematic drawing shows a normal larva on the left and a mutant on the right. Dotted regions indicate denticle bands; dotted lines, the segmental boundaries; hatched boxes, the regions missing from the mutant larva; and transverse lines, corresponding parts of the normal and mutant larvas. [Reproduced from C. Nüsslein-Volhard and E. Wieschaus, *Nature* 287 (1980):795.]

deleted. The remaining half-segments appear to be fused so that the resulting embryo has half the normal number of segments (*fushi tarazu* is Japanese for "not enough segments"). Mutations in genes of the second class delete several adjacent segments (gap mutations), whereas mutations in genes of the third class delete half of each segment and replace it with a mirror image of the remaining half-segment (segment polarity mutations).

The behavior of many segmentation mutants was initially very puzzling, because defects frequently did not affect a single segment or group of segments, but rather half of one segment and the adjacent half of the next segment. Now, however, it has become apparent that the basic units of organization in the body of the larval and adult fly may not be the obvious morphological segments (visually defined by constrictions of the body) but rather parasegments (which extend from the middle of one morphological segment to the middle of the next, with the constrictions falling within each parasegment). This illustrates a common source of confusion in the molecular study of development: It is very easy to mistake a visible boundary (such as a segmental constriction, or the edge of a wing) for a domain of gene function.

The observation that only a few subtle mutations in controlling genes are sufficient to transform one body segment of the fly into another (see Figure 22-30) or to reduce the total number of body segments in the larva (see Figure 22-31) has tantalizing evolutionary implications. Higher insects have comparatively few body segments and either two or four wings, but are known from the fossil record to have evolved from annelids such as the earthworm, with many identical body segments. Thus, the evolution of the insects appears to reflect naturally occurring mutations no different from those identified experimentally by the geneticist in the laboratory. The ability of single mutations to dramatically change the body plan of an organism helps us to understand how the accumulation of many independent mutations has led, over the course of evolution, to the incredible variety of contemporary life forms.

Some Homeotic Genes Are Very Long[51, 52, 55]

Several exciting generalizations have already emerged from the preliminary analysis of the DNA sequence, transcription patterns, and developmental regulation of homeotic genes from the Antennapedia

Figure 22-32
The alternative splicing of *Antennapedia* transcripts is developmentally regulated. (Upper) Right arm of salivary gland polytene chromosome 3, showing Antennapedia complex at band 84B and the initial site of hybridization of the starting plasmids for the chromosomal walk at band 84F more than 2000 kilobase pairs away. (Lower) Some alternative splicing patterns. The most downstream exon contains the homeobox homology. [After R. L. Garber, A. Kuroiwa and W. J. Gehring, *EMBO J.* 2 (1983):2027–2036.]

and bithorax complexes. First, several of these homeotic genes appear to be huge in comparison with most other genes in the fruit fly. Of course, all eucaryotes have introns in some of their protein-coding genes, but in general, lower eucaryotes have fewer and smaller introns (Chapter 20). Surprisingly, at least one embryonic mRNA precursor encoded by the Antennapedia complex is nearly all intron, so that mRNAs of only a few kilobases are derived by splicing from a primary transcript of approximately 100 kilobases! Similarly, the primary transcript of the ultrabithorax region of the bithorax complex is about 75 kilobases, although the mature mRNAs derived from it are not unusual in size.

Why are some homeotic transcripts so long? One idea, discussed at the beginning of this chapter, is that both procaryotes and eucaryotes may use long transcription units to control when genes act. This possibility is especially attractive for eucaryotes, since the mature mRNA cannot be completely spliced out of the primary transcript until the most downstream exon has been transcribed. Another possibility, shown in Figure 22-32, is that multiple mRNA species can be spliced out of a single primary transcript in a developmentally regulated fashion. For example, one embryonic *Antennapedia* mRNA includes at least four exons, while an mRNA species synthesized later during the pupal stage includes two of these embryonic exons together with an additional pupal-specific exon. Since many exons appear to encode protein domains with distinct structural or enzymatic functions (Chapter 20), alternative splicing patterns like those seen in the Antennapedia and bithorax complexes may represent a mechanism for organisms to adapt one basic protein structure (encoded by core exons) to different but related developmental purposes. By adding or subtracting functional protein domains encoded by optional developmentally regulated exons, the structural and enzymatic properties of the homeotic gene product could be subtly modified, and the ability of the protein to interact with other cellular components could be changed as development progresses.

Extraordinary Conservation of a Protein-Coding Domain in Homeotic Genes[56-59]

The most startling insight to emerge from DNA sequence analysis of genes controlling pattern formation in *Drosophila* is the existence of a short, 180-base-pair sequence called the **homeobox,** which is found within the 3' exon of many homeotic and segmentation genes, including the *Ubx, abd-A,* and *Abd-B* regions of the bithorax complex and the *Dfd, Scr, ftz,* and *Antp* genes of the Antennapedia complex. The sequences of the individual homeoboxes are nearly identical and evidently code for part of a larger polypeptide: When the homeobox DNA sequences are translated in the correct reading phase, all the minor sequence divergence can be accounted for by conservative substitutions of one amino acid for another in a way that would presumably not affect the enzymatic activity or structural properties of the gene product. But even more amazing is the recent observation that when the *Antennapedia* homeobox is used as a radioactive probe to look for comparable DNA sequences in other eucaryotes, the African clawed toad *Xenopus,* the mouse, and humans were all found to have multiple related sequences. In fact, one of the *Xenopus* homeoboxes is so similar to the *Antennapedia* homeobox in *Drosophila* that only 1 amino acid out of 60 has changed despite millions of years of evolutionary separation between the two species. This raises the possibility that some of the same molecular mechanisms that play a role in morphogenesis of the fruit fly are at work in vertebrates as well, and that the gene products (or parts of them) have been subject to the most extreme selective pressure imaginable during evolution. There is even a preliminary indication that the *MAT*a1 and *MAT*α2 transcripts in yeast (Chapter 18) share slight homology with a part of the fly and vertebrate homeoboxes. Since the *MAT* locus is a well-characterized binary switch determining whether the yeast mating type is **a** or α, this finding fuels speculation that homeotic genes may be turned on and off in groups of cells in order to generate a code capable of specifying the developmental fate of major body parts.

Localization of Homeotic Transcripts in the Early Embryo[60]

The simplest interpretation of the available evidence is that a homeotic gene like *Antennapedia* affects the initial commitment of all or part of one imaginal disc to become all or part of a particular structure on the surface (cuticle) of the fly. An alternative but somewhat more complicated interpretation is that the homeotic gene product might govern not the initial commitment of the disc cells in embryogenesis, but rather the ability of the embryonic cells to retain that commitment through the three larval instars that precede metamorphosis. In either case, as long as the homeotic gene product is turned on and off at the level of transcription or RNA processing (Chapter 21), we might expect that particular homeotic gene transcripts will be found in certain regions of the embryo and not in others. That this is true can be shown dramatically by probing a cross section of a developing embryo with a radioactively labeled DNA fragment from a homeotic gene (Figure 22-33). This in situ hybridization technique is essentially the same as that used to map the location of genes on a chromosome, except here the radioactively labeled probe anneals

(a)   (b)

with single-stranded mRNA in the thin section of tissue, rather than with denatured DNA in the partially disrupted chromosome (see Figure 22-17). We can see that a DNA fragment from the *Dfd* gene of the Antennapedia complex detects a concentration of transcripts just after formation of the cellular blastoderm. The mRNA is distributed in a belt about one-third of the way back from the anterior tip of the embryo, in a region known to form the posterior head and first thoracic segments of the fly. This is the same region identified by geneticists as the target affected by mutations in the *Dfd* gene.

Transdetermination: Differentiation as a Hierarchy of Developmental Switches

An ingenious set of serial transplantation experiments provides strong support for the notion that commitment of an imaginal disc to become a specific cuticular structure depends on the particular combination of developmental switches thrown in that disc (Figure 22-34). A specific imaginal disc (say, an eye disc) is surgically excised from the larva and transplanted into the abdomen of an adult fly. There, bathed in hemolymph (insect blood), the disc cells survive and proliferate, outliving their host, so that for long-term culture the discs must be cut into smaller pieces and serially transplanted from one adult fly to another. The stable hormonal environment of the adult fly (lacking the molting hormone, ecdysterone) effectively mimics that of the larva, so the cultured disc cells, although committed to a particular fate, do not differentiate. Only when a cultured disc is transplanted back into a third instar larva on the verge of pupation can its developmental potential be tested. When this is done, the usual and not very surprising result is that after metamorphosis, the transplanted discs have retained their commitment to become a specific adult body part. Commitment and subsequent differentiation appear irreversible.

Occasionally, however, an imaginal disc will partially or completely transdetermine to a different body part as though the original commitment of the disc to its particular fate was not completely stable over the long term in culture. By analyzing many such **transdetermination** events, transdeterminations were found to obey strict rules (Figure 22-35). For example, an antenna disc can become a leg disc, and vice versa, but an antenna disc cannot become an eye disc except by two successive transdeterminations. Occasionally, a disc will transdetermine twice or more during serial transplantation—say, from genital to antennal disc, and then from antennal to wing disc. However, a genital disc is never observed to transdetermine *directly* to

Figure 22-33

Homeotic transcripts are localized in the *Drosophila* embryo. Serial sections of a wild-type embryo at the cellular blastoderm stage were hybridized with a radioactively labeled DNA probe from the *Dfd* gene in the Antennapedia complex. To do this, the tissue sections were spread out on a microscope slide and then overlaid with liquid photographic emulsion. After a sufficient amount of time had elapsed for radioactive disintegrations from the hybridized probe to expose the emulsion, the slide was developed. (a) Bright-field photomicrograph of the embryo after the in situ hybridization. A denotes anterior; P, posterior; pc, pole cells. (b) Dark-field autoradiograph of the same embryo, with silver grains showing the location of mRNA which hybridized with the labeled *Dfd* probe. Arrows in panel (a) show the boundaries of hybridization in panel (b). [Reproduced with permission from W. McGinnis, M. S. Levine, E. Hafen, A. Kuroiwa, and W. J. Gehring, *Nature* 308 (1984):428–433.]

Figure 22-34
When cultured in the abdomen of an adult fly, imaginal discs generally remain committed to their original fate, but occasionally transdetermine partially or completely to a disc of a different developmental fate. [After B. Alberts, D. Bray, J. Lewis, M. Raff, K. Roberts, and J. D. Watson, *Molecular Biology of the Cell* (New York: Garland, 1983, p. 839.]

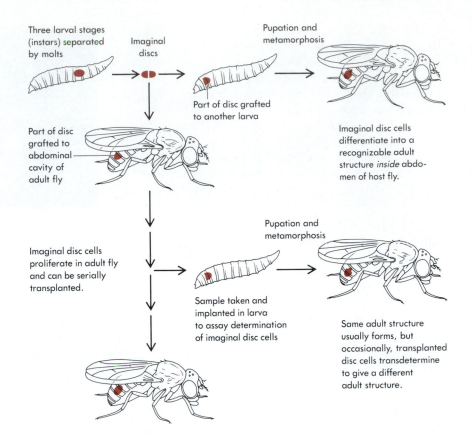

Three larval stages (instars) separated by molts

Imaginal discs

Pupation and metamorphosis

Part of disc grafted to another larva

Imaginal disc cells differentiate into a recognizable adult structure *inside* abdomen of host fly.

Part of disc grafted to abdominal cavity of adult fly

Imaginal disc cells proliferate in adult fly and can be serially transplanted.

Sample taken and implanted in larva to assay determination of imaginal disc cells

Pupation and metamorphosis

Same adult structure usually forms, but occasionally, transplanted disc cells transdetermine to give a different adult structure.

Figure 22-35
Only certain transdeterminations are allowed, and many of these resemble known homeotic mutations such as *Antennapedia*. [After E. Hadorn, *Sci. Amer.* 219 (1968):110–120. © 1968 by Scientific American, Inc. All rights reserved.]

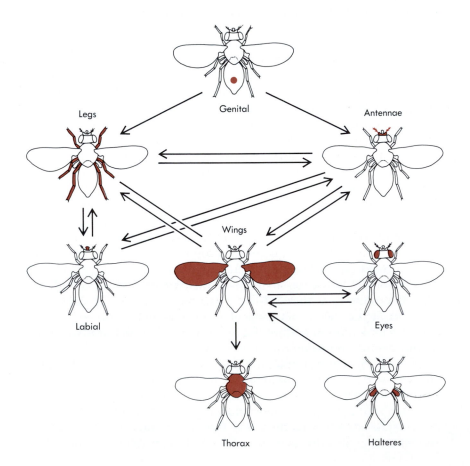

Genital

Legs

Antennae

Wings

Labial

Eyes

Thorax

Halteres

a wing disc. This suggests that each transdetermination corresponds to throwing a single molecular switch, and the ultimate developmental fate of each disc is determined by the combination of molecular switches thrown in it. Transdetermination bears a suspicious resemblance to the behavior of homeotic mutations such as *Antennapedia*, as though the consequence of certain homeotic mutations was to increase the frequency of experimentally induced transdetermination events from very low to essentially 100 percent. Taken together, these observations suggest that the products of homeotic genes are responsible for the initial commitment of imaginal discs, or the maintenance of that commitment, in the normal development of the fly.

Polyclones: Cells Can Communicate and Throw the Same Developmental Switch Together[61, 62]

We have seen that an imaginal disc consisting of many thousands of cells can partially or completely transdetermine from one body part to another (say, an antenna to a wing). In principle, this rather startling phenomenon could be explained by either of two contrasting mechanisms—clonal or polyclonal transdetermination—and these have radically different implications for development in general. In clonal transdetermination, a single cell in the antenna disc would first transdetermine in a heritable fashion into a wing cell and then rapidly outgrow the neighboring antenna cells. The transdetermined wing tissue would constitute a **clone**, the progeny of a single initial transdetermined cell. Alternatively, many cells within the antenna disc could transdetermine simultaneously and then proliferate so that the wing tissue would be a **polyclone**, the progeny of many different initial cells.

Clonal and polyclonal transdeterminations can be distinguished experimentally when a very small number of cells in the initial antenna discs are genetically marked with a special visible mutation that makes the tissue yellow (Figure 22-36). If the transdetermined wing tissue is always found to be exclusively wild-type or yellow, then a single transdetermined cell gives rise to the entire mass of transdetermined tissue. But if the transdetermined tissue is sometimes a mixture of wild-type and yellow cells, then a group of cells (both wild-type and mutant) must have transdetermined together. In fact, transdetermined wing tissue is often observed to be genetically mixed, so many different cells in the initial disc must have some way of communicating with one another in order to throw the same molecular switch (antenna to wing).

How do cells within a polyclone communicate with one another? We might speculate that ionic currents like a nerve impulse, or chemical morphogens like those in *Dictyostelium*, are able to pass through extracellular spaces or even through intracellular cytoplasmic bridges. Or perhaps cells recognize other cells within the same compartment by means of compartment-specific cell surface glycoproteins. (We shall see in Chapter 23 that cells of the mammalian immune system make extensive use of such cell surface recognition elements.) We also do not know the nature of the molecular switches that enable cells within a polyclone to remember their state of determination in a heritable fashion. The memory might correspond to a physical change in the DNA such as a chromosomal inversion (as in *Salmonella*

Antennal disc in larva

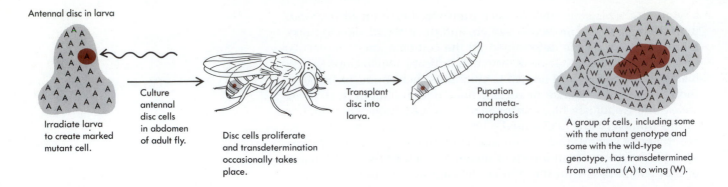

Irradiate larva to create marked mutant cell. → Culture antennal disc cells in abdomen of adult fly. → Disc cells proliferate and transdetermination occasionally takes place. → Transplant disc into larva. → Pupation and metamorphosis → A group of cells, including some with the mutant genotype and some with the wild-type genotype, has transdetermined from antenna (A) to wing (W).

Figure 22-36
Groups of unrelated cells (polyclones) transdetermine together. [After B. Alberts, D. Bray, J. Lewis, M. Raff, K. Roberts, and J. D. Watson, *Molecular Biology of the Cell* (New York: Garland, 1983, p. 839.]

phase variation), directed gene conversion (as in the yeast mating-type switch), or methylation/demethylation of specific DNA sequences. Another possibility is that the memory reflects a heritable change in the chromosomal proteins that package and regulate expression of DNA. A still more remote possibility is that the memory is retained within the nucleoplasm or cytoplasm by metabolic feedback circuits or self-replicating cytoskeletal structures, with no direct chromosomal record even being made.

Development of the Fruit Fly Is Modular[63, 64]

An immense wealth of genetic evidence has now been amassed to show that pattern formation in the developing *Drosophila* embryo is determined by a hierarchy of controlling genes. Maternal genes expressed in nurse cells during oogenesis are responsible for establishing gradients of positional information along the major developmental axes of the egg (e.g., the 10 or more dorsalizing genes). Genes expressed in the early embryo itself determine the initial folding of the cellular blastoderm (e.g., the 15 or more "segmentation" genes). And genes expressed still later in embryogenesis can partially or completely transform one adult body part into another by transdetermining the fate of a polyclone within an imaginal disc.

A hierarchy of developmental decisions may also help to explain the behavior of a homeotic mutation like *Aristapedia*, where only the most distal part of the antenna is transformed into the most distal part of the leg (see Figure 22-28). Why is the effect of the *Aristapedia* mutation confined to the distal part of the antenna, while the *Antennapedia* mutation transforms the entire antenna? One attractive (but hypothetical) explanation is that just as a set of molecular switches determines whether the cells within an imaginal disc are fated to become one adult cuticular structure or another, so another set of switches may determine whether particular subsets of cells within the imaginal disc (polyclones) are fated to be nearer the body of the adult fly (proximal) or farther away (distal). Seen in this way, "antenna" and "leg" would be alternative positions for one switch, "proximal" and "distal" alternative positions for a second switch, and the switching mechanism would cause these two switches to interact with each other. The *Aristapedia* mutation might prevent the first switch from being on "antenna" whenever the second switch was on "distal," or alternatively, *Aristapedia* might cause the first switch to reverse from "antenna" to "leg" in cells that threw the second switch to "distal." In contrast, *Antennapedia* might lock the first switch on "leg" regardless of the position of the second switch.

Thus, we see that the fruit fly *Drosophila* is likely to be put together from many modules, each differentiating in response to a hierarchy of developmental switches thrown in succession. Lest the word "switch" itself imply any particular molecular mechanism, we should remember that biological switches can be as different as the competition between RNA polymerase and CI repressor during the early stages of infection by bacteriophage λ (see Figure 17-16) and the displacement of one σ factor by another during sporulation in *B. subtilis*. We currently have no idea what kind of molecular switches will be found to govern the hierarchy of genes controlling modular morphogenesis in the fruit fly.

We shall see shortly that this kind of modular behavior is carried to even greater extremes in the development of a small nematode worm named *Caenorhabditis elegans*. Although comparable modules are not readily apparent in the development of higher organisms (except for segmentation of the vertebrate spinal column and rib cage), it would be most surprising if modular construction did not turn out to be a universal developmental mechanism, albeit in modified or somewhat disguised form.

Positive Control of Gene Function by Steroids[65, 66]

At the genetic level, the best-understood steroid hormone is ecdysterone, the growth hormone responsible for the successful passage of developing insects from their larval to their pupal stages. In the early larval period, the puffing pattern of the polytene *Drosophila* salivary chromosomes is stable. But as soon as ecdysterone is released into the insect hemolymph, the preexisting puffs begin to regress, and over 100 new puffs quickly develop during the transition from the larval to the pupal stage. Each band always puffs at a precise time and remains active for a well-defined interval (Figure 22-37). The first such puffs (the "early" puffs) begin appearing within 5 minutes of hormone release. They appear to be direct responses to the hormone, since their appearance is unaffected by inhibitors of protein synthesis. In contrast, those "late" puffs that only appear some hours after ecdysterone is added do not appear when protein synthesis is blocked. Presumably, a protein product(s) encoded by one or more of the early puffs induces the formation of the late puffs.

The ability of ecdysterone to induce very rapid expansion of the early puffs initially suggested that the hormone binds directly to specific sites on chromatin. This tends to be confirmed by experiments showing that ecdysterone can be photochemically crosslinked to the early puff sites. Thus, ecdysterone may function in insects just as other steroids do in higher cells. The hormone first binds to a specific receptor protein, and this hormone-receptor complex then binds directly to specific DNA sequences, where it modulates transcription. Although steroids usually function as positive activators of transcription (Chapter 21), this may not always be the case. In addition to inducing the expansion of early puffs, ecdysterone also causes a different group of preexisting puffs to regress (recondense). Moreover, ecdysterone appears to bind directly to these regressing puffs, just as it does to the expanding hormone-inducible puffs. These interesting observations suggest that at least in insects, and perhaps in higher eucaryotes, steroid hormones may also function as negative regulators of transcription. This possibility should not really surprise us. In bacteria, repressors work by physically blocking the sites on DNA

Figure 22-37
Schematic representation of the action of ecdysterone in the sequential production of early and late chromosome puffs.

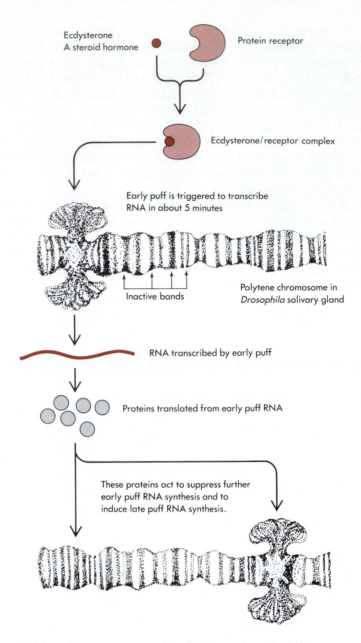

Ecdysterone
A steroid hormone

Protein receptor

Ecdysterone/receptor complex

Early puff is triggered to transcribe RNA in about 5 minutes

Inactive bands

Polytene chromosome in *Drosophila* salivary gland

RNA transcribed by early puff

Proteins translated from early puff RNA

These proteins act to suppress further early puff RNA synthesis and to induce late puff RNA synthesis.

where RNA polymerase must bind. Thus, in principle, any protein that functions as a positive activator of transcription by recognizing a specific DNA sequence in one gene can also function as a repressor of another gene. For the activator protein to function as a repressor, the protein binding site in the second gene need only be located in a region that is essential for transcription. Thus, the position of a protein binding site relative to the promoter may determine whether binding of a particular protein modulates transcription positively or negatively.

Bigger Is Not Always Better— Or, Why Study a Worm?

The nematode *Caenorhabditis elegans* is a very small worm that has recently come to occupy a very large place in the molecular biology of development. There are several reasons for the sudden scientific pop-

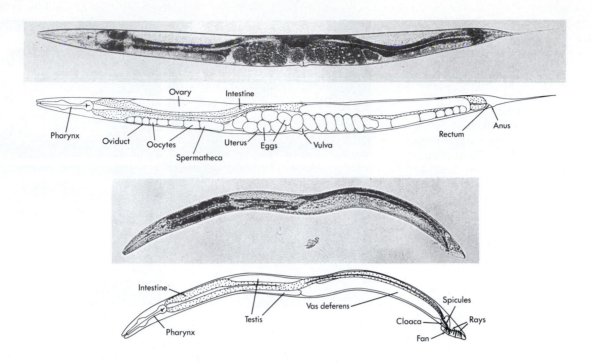

Figure labels (top): Ovary, Intestine, Pharynx, Oviduct, Oocytes, Spermatheca, Uterus, Eggs, Vulva, Rectum, Anus

Figure labels (bottom): Intestine, Pharynx, Testis, Vas deferens, Cloaca, Fan, Spicules, Rays

ularity of this organism. The worm is small (about 1 mm long) and has a very simple body plan (Figure 22-38). Nematodes can be grown in large numbers on Petri dishes seeded with bacteria (their favorite food), and they reproduce with a remarkable generation time of a little over 3 days from conception to maturity. There are two sexes (hermaphrodites and males) and a large number of progeny (300 per parent) so that the organism can be analyzed genetically with great ease. Perhaps most important of all, the worm is transparent at all stages of the life cycle from egg to adult. As a result, the cellular anatomy of the developing worm can be examined and recorded with absolute precision in the living animal by using a special light microscope equipped with Nomarksi interference optics. Interference optics display differences in refractive index between various cellular organelles (Figure 22-39). By following both the movements and divisions of the highly refractile cell nuclei starting immediately after fertilization, it has been possible to reconstruct the entire anatomical history of each cell as the fertilized egg develops into a multicellular adult.

Figure 22-38
The sexes of *Caenorhabditis elegans* are the self-fertilizing hermaphrodite (above) and the male (below). The reproductive system of the adult hermaphrodite produces both sperm and eggs. Eggs from the ovary pass through the oviduct and are fertilized in a specialized sperm-producing region of the oviduct called the spermatheca. The fertilized eggs then enter the uterus, where embryogenesis takes place. The juvenile worms leave the parent through the vulva. Eggs that escape fertilization within the hermaphrodite can be fertilized by sperm that the male deposits in the vulva of the hermaphrodite. [Reprinted with permission from J. E. Sulston and H. R. Horvitz, *Dev. Biol.* 56 (1977):110–156.]

Invariant Cell Lineages[67, 68]

Two amazing conclusions emerged when the information gathered from detailed anatomical studies of living worms was combined with more classical anatomical studies obtained by electron microscopy of serial thin sections of the worm at different developmental stages. First, each cell in the adult worm is derived from the zygote by a virtually invariant series of cell divisions, cell migrations, and occasional programmed cell deaths, collectively called a **cell lineage** (Figure 22-40). Second, as a direct consequence of the invariant cell lineages, individual nematodes are anatomically invariant carbon copies of each other. The mature adult hermaphrodite is always put together from exactly 945 cells (the actual number of nuclei is slightly larger, because some tissues such as the intestine are partially syncytial). If

Figure 22-39

Early development of the nematode *Caenorhabditis elegans* as viewed by Nomarski interference optics. (a) Fertilized egg just before the first cleavage. The male and female pronuclei have not yet fused completely. (b) The beginning of gastrulation, marked by cell movements and the formation of primitive tissues. For example, Ea and Ep are the anterior and posterior daughters of the first intestinal precursor cell called the E cell, D is a mesodermal precursor, and P4 gives rise to germ cells. (c) By 7 hours, the embryo starts to move, and individual cells are more difficult to discern. (d) The late embryo is clearly a miniature worm. The white arrow points to the mouth, and the two black arrows to germ cells. [Reprinted with permission from J. E. Sulston, E. Schierenberg, J. G. White, and J. N. Thomson, *Dev. Biol.* 100 (1983):64–119.]

we focus on the nervous system for the purpose of illustration, 302 of the 945 cells are always neurons. These 302 neural cells are always derived from 407 neural precursor cells, of which 105 always suffer a programmed cell death. Of the 302 surviving neurons, 8 are always dopaminergic (responding to the neurotransmitter dopamine) and 2 are always serotonergic (responding to serotonin). The remaining 643 nonneural cells in the adult hermaphrodite can be described with the same precision as those of the nervous system. No other eucaryotic organism can currently rival this degree of anatomical detail.

The 302-element neural circuitry of the adult worm may be as simple (and is certainly as reproducible) as the printed circuits in a mass-produced portable calculator. Such small size and simplicity will speed molecular analysis. Knowing how a worm coordinates the wiggling movements of its body musculature or how neural events trigger the mating of a male with a hermaphrodite (there is no female) should give us general insights into neural function. With luck, these insights will lay the groundwork for understanding the vastly more complex wiring pattern in our own brains, where literally billions of cells are hooked up in a variable and somewhat sloppy fashion.

Laser Microsurgery Complements Classical Genetic Analysis[69]

Because *C. elegans* grows readily on Petri dishes, tens of thousands of individual animals can be visually surveyed through a low-power microscope. This makes it exceptionally easy to identify and isolate mutants with obvious behavioral abnormalities in movement, thermotaxis, chemotaxis, avoidance of high osmotic strength, touch sensitivity, egg laying, and synthesis and degradation of neurotransmitters. For example, *unc* mutants wiggle in an *unc*oordinated fashion or fail to wiggle at all; these worms have a defect in either the nerves, muscles, or neuromuscular junctions required for smooth and efficient gliding motions. Characterization of *unc* mutants has already led to the identification of many genes, including those for actin and myosin.

Intensive genetic analysis can reveal the developmental consequences of a deficiency or alteration in a single gene product, or a mutation in a regulatory DNA sequence. But for studying the molecular mechanisms of development, it is sometimes experimentally advantageous simply to eliminate a particular cell from the embryo. There is no systematic way to isolate such mutants genetically, so biologists have traditionally resorted to physical techniques. Classical biologists surgically transplanted (or excised) relatively large multicellular chunks of embryo. In this way, for example, the dorsal lip of the amphibian blastopore (see Figure 22-5) was shown to be largely responsible for organizing the initial stages of blastulation: After the grafting of the dorsal lip from one embryo into the blastocoel of another, the extra lip region would instruct the surrounding cells to form a secondary embryo alongside the first. Modern developmental biologists can use a sharply focused laser beam to destroy (ablate) single cells in the embryonic nematode without any apparent damage to neighboring cells.

Mosaic Versus Regulative Development

The results of extensive laser ablation experiments, together with genetic analysis, are consistent with the notion that differentiation can proceed in at least two profoundly different styles—mosaic and regulatory—which presumably reflect profoundly different molecular mechanisms. In **mosaic development** (so called because the organism is assembled from independent parts), the differentiation of a cell does not depend on the behavior or even the existence of neighboring cells: Laser ablation eliminates its target cell but does not affect the behavior or differentiation of any other cells. Apparently, in mosaic development, internal events within each cell instruct that cell what to do; such events could be triggered by cell division itself or by the ticking of an internal biological clock that is set in motion by fertilization. In **regulatory development,** differentiation is partially or completely dependent on the interaction between neighboring cells. For example, during embryogenesis, the lone gonadal anchor cell induces a certain subset of ventral hypodermal cells to form a vulva (the opening through which eggs are laid in the hermaphrodite). If the anchor cell is destroyed by laser ablation, no vulva develops; but if almost all the gonadal cells other than the anchor cell are ablated, a normal vulva can still be formed.

Mosaic development governed by strict cell lineages is the overwhelming rule in *C. elegans,* and regulative development the exception (occurring only in vulval morphogenesis and differentiation of germ cells within the gonad). In most other organisms (ourselves included), the reverse is almost certainly true. Mosaic development may dominate in the nematode because survival of the species requires the life cycle to unfold with the utmost possible speed. Thus, there may not be sufficient time for complex intercellular signaling and interaction (many hours are required to establish a slime mold aggregation field; see Figure 22-23). Instead, each cell is on automatic pilot, with programmed instructions so precise that the independently developing cells can all fit together neatly into a functional multicellular organism. The price of such mosaicism may be very small size (perhaps only a limited number of cell divisions can be so rigidly programmed) and only modest complexity (cell-cell interactions may be required to construct a more elaborate anatomy).

Figure 22-40

The complete cell lineage tree of the adult hermaphrodite sex of the nematode *Caenorhabditis elegans*. Vertical axis represents time, increasing downward (about 50 hours at 20°C). Horizontal lines represent cell divisions; vertical lines ending in "X" represent cell deaths (131 total); other vertical lines represent surviving cells (945 total). As summarized diagrammatically in the inset, the fertilized egg (zygote) initially divides into the AB, MS, E, C, D, and P₄ founder cells from which the major cell lineages are

Hours after
fertilization

Zygote

MS

E

C

D

P4

Pharynx

Pharynx

Intestine

Body
muscle

Body
muscle

Germ line

0

10

20

30

40

50

derived. For illustrative purposes, only the lineages that generate the pharynx (mouth), intestine, germ line, and parts of the body muscle are labeled in the complete lineage tree. The germ line is derived from the P_4 cell, and the intestine from the E cell. Only the first embryonic division of the P_4 cell is shown, since its subsequent lineage is indeterminate (the number and pattern of cell divisions are not fixed). The pharynx is composed of cells from both the AB and MS lineages. The body musculature is derived from four different founder cells (AB, MS, C, and D), but only the D lineage gives rise exclusively to muscle cells. Most of the unlabeled cells in the complete lineage represent neurons or hypodermal cells. Hypodermal cells (i.e., cells lying just below the skin) secrete the protective cuticle that covers the surface of the worm and serves as an attachment site for the body muscles. Note that left and right in the figure do *not* correspond to anterior and posterior; for example, the pharynx at the front of the worm is composed of cells from both sides of the figure. [After a figure by Dr. J. E. Sulston, MRC Laboratory of Molecular Biology, Cambridge; and J. E. Sulston, E. Schierenberg, J. G. White, and J. N. Thomson, *Dev. Biol.* 100 (1983):64–119.]

What Can Cell Lineages Tell Us About Molecular Mechanisms of Development?

Most organisms with the anatomical complexity of the nematode have perhaps 10 to 1000 times as many cells. For the nematode to achieve the same level of biological complexity with as few as 945 cells means that a very high percentage of the cell divisions in the invariant cell lineages must produce daughter cells committed to different developmental fates. Consequently, cell lineages alone appear to determine cell fate (with the aforementioned exception of regulative development in certain vulval and germ line cells). Consider, for example, the posterior daughter of the anterior daughter of the posterior daughter of the zygotic nucleus. This particular cell (the E cell) will be the precursor of all the intestinal cells in the worm, and only cells descended from this cell can contribute to formation of the gut (see Figure 22-40).

How might cell fate be progressively determined by an unfolding cell lineage? Since each lineage is defined by a succession of cell divisions, we might speculate that it is DNA replication or cell division itself that results in commitment of the daughter cells. Current evidence suggests that there are at least four general mechanisms by which cell division might seal cell fate:

- Physical modification of the DNA (e.g., 5-methylation of cytosine) might be gained or lost during replication.
- Chromosomal proteins might preferentially associate with one strand of the parental duplex during semiconservative DNA replication.
- Passage of chromatin through the replication fork might strip old regulatory proteins from the DNA, and freshly replicated DNA emerging from the fork might be able to bind a new set of regulatory proteins.
- Cytoplasmic factors might segregate asymmetrically during cell division.

We have already discussed the first three possibilities as conceivable mechanisms for generating stem cell lineages; next we examine some of the evidence for cytoplasmic determination.

Insect Polar Plasm: Cytoplasmic Determinants of Nuclear Differentiation

Classical embryology abounds with instances of **cytoplasmic determinants**—cytoplasmic substances that tell a nucleus what to do. We have already seen that the female egg is not simply an undifferentiated, spherically symmetrical, haploid cell awaiting fertilization. Unfertilized eggs already possess complex internal architecture, reflecting extensive regional differentiation of the egg cytoplasm. Cytoplasmic determinants, which are distributed in a precise, three-dimensional pattern, are partitioned into different blastula cells by cleavage divisions following fertilization. These substances then control the subsequent differentiation of the cells (blastomeres) that receive them.

Among the best-studied cytoplasmic determinants are the **polar**

granules found in the posterior cytoplasm of insect eggs, which cause cleavage nuclei in their vicinity to differentiate into primordial germ cells (see Figure 22-24). Immediately after fertilization, the zygotic nucleus of a *Drosophila* egg undergoes eight nuclear divisions without cellularization. We already saw how most of the nuclei migrate to the outside of the egg after nine divisions, undergo four more synchronous divisions, and are then enclosed by invagination of the egg membrane to form a cellular blastoderm. These cells will give rise to the somatic (non–germ line) tissues of the fly. However, we did not discuss the remarkable observation that a few nuclei—destined to become the primordial germ cells—are set aside after only eight nuclear divisions to follow a distinct developmental pathway. At that time, nuclei that happen to be near small cytoplasmic structures called polar granules in the posterior of the egg are induced to differentiate into primordial germ cells. These **pole cells** are larger and rounder than the small cells of the blastoderm. The pole cells initially lie outside the sheet of cellular blastoderm, but later migrate to the gonads (a somatic tissue), where they give rise to thousands of progeny germ cells.

Taking advantage of the wealth of genetic mutations in *Drosophila*, an elegant experiment was devised to prove that the polar cytoplasm (containing polar granules) determines the differentiation of embryonic nuclei (Figure 22-41). The experimental design took advantage of a recognizable but genetically recessive mutation called yellow. The wild-type gene y^+ is dominant over y^- mutations, since one copy of the normal gene product is sufficient to confer normal body color on the fly. Thus, homozygous y^+/y^+ flies and heterozygous y^+/y^- flies have normal color, but homozygous y^-/y^- flies are a lighter yellow. Polar cytoplasm was withdrawn from the posterior end of a wild-type embryo with a tiny pipette and transplanted to the anterior end of another y^+/y^+ embryo (normal color). The transplanted polar cytoplasm induced formation of large, round cells resembling pole cells in the anterior of the second embryo, where no pole cells usually form. To test whether the presumptive pole cells were genetically functional, they were transplanted to yet a third embryo with the y^-/y^- constitution (yellow). These embryos developed into yellow flies because all their somatic tissues had the y^-/y^- genetic constitution, but some of their progeny were wild-type in appearance. Such wild-type flies could only have obtained the wild-type y^+ gene from y^+/y^+ nuclei in the second embryo. This proves that polar cytoplasm induced functional pole cells in the second embryo.

Cytoplasmic Determinants in the Amphibian Egg[70–73]

Polar plasm also exists in vertebrate eggs, although the experimental proof does not quite have the genetic elegance that is possible in *Drosophila*. Frog eggs contain a special **germinal cytoplasm** at the vegetal pole where primordial germ cells eventually form (see Figure 22-5). One of the active factors in this cytoplasm appears to be nucleic acid (presumably RNA, but conceivably DNA), because selective irradiation of the germinal cytoplasm with ultraviolet light prior to fertilization results in sterile but otherwise normal frogs. However, the irradiated eggs can be rescued by microinjecting unirradiated germinal cytoplasm from a second egg; tadpoles with fully formed gonads and germ cells are then obtained. Similar experiments have been

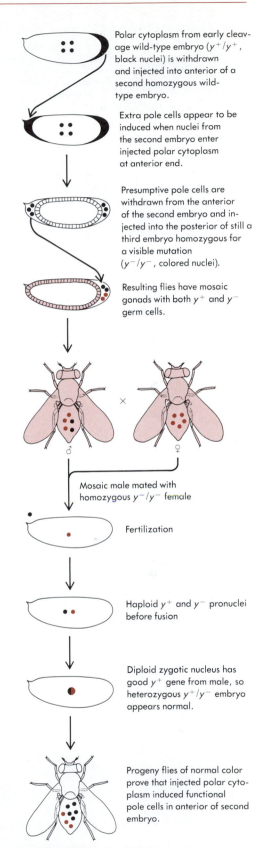

Polar cytoplasm from early cleavage wild-type embryo (y^+/y^+, black nuclei) is withdrawn and injected into anterior of a second homozygous wild-type embryo.

Extra pole cells appear to be induced when nuclei from the second embryo enter injected polar cytoplasm at anterior end.

Presumptive pole cells are withdrawn from the anterior of the second embryo and injected into the posterior of still a third embryo homozygous for a visible mutation (y^-/y^-, colored nuclei).

Resulting flies have mosaic gonads with both y^+ and y^- germ cells.

Mosaic male mated with homozygous y^-/y^- female

Fertilization

Haploid y^+ and y^- pronuclei before fusion

Diploid zygotic nucleus has good y^+ gene from male, so heterozygous y^+/y^- embryo appears normal.

Progeny flies of normal color prove that injected polar cytoplasm induced functional pole cells in anterior of second embryo.

Figure 22-41
Polar cytoplasm in the egg induces adjacent nuclei to differentiate into primordial germ cells. [After K. Illmensee and A. P. Mahowald, *Proc. Nat. Acad. Sci.* 71 (1974):1016–1020.]

(a)

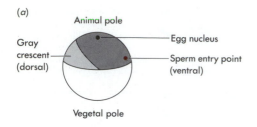

Animal pole

Gray crescent (dorsal)

Egg nucleus

Sperm entry point (ventral)

Vegetal pole

(b)

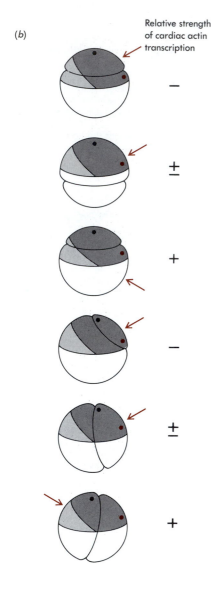

Relative strength of cardiac actin transcription

−

±

+

−

±

+

(c)

Figure 22-42

Cytoplasmic determinants are localized in the amphibian egg. (a) Axes of the fertilized frog egg. (b) Plane of ligation determines the level of cardiac actin transcription in the part of the embryo (denoted by arrow) that was grown and subsequently analyzed. (c) Schematic interpretation of results. Dots represent distribution of cytoplasmic substances that are responsible for activating cardiac actin genes in the developing embryo. [After J. B. Gurdon, T. J. Mohun, S. Fairman, and S. Brennan, *Proc. Nat. Acad. Sci.* 82 (1985):139–143.]

performed on eggs of the midge *Smittia* (an insect), with analogous results. The strong resemblance between germ cell determination in an insect and an amphibian suggests that there are highly conserved mechanisms to protect primordial germ cells from the events of somatic differentiation. We do not know why such protection is necessary, but we might speculate that somatic differentiation sometimes involves physical modification of the DNA, somatic rearrangement of the DNA, or changes in DNA packaging (chromatin structure) that cannot easily be reversed.

A related series of experiments has been performed to demonstrate that cytoplasmic determinants contained in the gray crescent of the amphibian egg are required for transcription of muscle-specific actin genes at the neurula stage (following gastrulation). Instead of using ultraviolet light to inactivate nucleic acid components in the egg, a tiny loop composed of a single human hair was carefully tied into a tight knot around the newly fertilized amphibian egg (Figure 22-42). Depending on the orientation of this ligature with respect to the natural axes of the egg (the animal and vegetal poles, and the gray crescent opposite the point of sperm entry), the embryo that develops from the nucleated half of the ligated egg will be deprived of different parts of the original egg cytoplasm. When the ligated eggs reached the equivalent of the neurula stage, the expression of muscle-specific actin genes was examined by measuring the ability of neurula RNA to form RNA-DNA hybrids with a cloned actin gene. The muscle actin genes were expressed only when ligated embryos retained a subequatorial region of the egg near the gray crescent area. Since the gray crescent normally becomes the dorsal side of the animal, where embryonic muscle first appears, we can see that the cytoplasm of the frog egg contains molecular determinants. These determinants (whose chemical nature is still unknown) instruct cleavage nuclei entering different regions of the cytoplasm to express different genes.

Cytoplasmic determinants and morphogenetic gradients are fundamentally different ways of determining the differentiation of embryonic cells. Cytoplasmic determinants are only able to instruct nearby nuclei to differentiate into a particular cell type. In contrast, because cells respond differently depending on where they lie within a gradient, gradients of positional information are able to influence a much larger field of cells. The advantage of cytoplasmic determinants may be that embryogenesis can proceed fastest when cells do not have to spend time creating and interpreting gradients of positional information. In the nematode, for example, where embryogenesis is extraordinarily rapid, cytoplasmic determinants direct one blastomere to differentiate into the E cell, which subsequently gives rise to the entire intestinal cell lineage (see Figure 22-40). The disadvantage of cytoplasmic determinants may be that only certain kinds of differentiation are simple enough to be determined solely by the influence of a single factor (or small group of factors) on an embryonic nucleus.

PUTTING CLONED GENES BACK INTO THE GERM LINE

The Triumph of Modern Embryology: Mutating Genes at Will

Recombinant DNA techniques enable us to mutate cloned genes in any way we wish, but we cannot learn everything about gene regulation just by studying the transcription of these mutant genes in vitro or by putting the mutant genes back into immortal somatic cell lines (Chapters 19 and 25). The real test of a mutant gene is to put it back into the organism and see how it functions in its natural environment. An additional (but more subtle) advantage of putting new genes into the germ line is that the genes will then experience the entire developmental history of the organism. This is important because cell lineage plays a role in determining cell fate (as we have seen in the nematode *C. elegans*), and this implies that sequential modification of gene or chromatin structure in the course of development may be necessary for proper developmental regulation. For each gene, then, development begins during spermatogenesis and oogenesis, when germ cell DNA first acquires its characteristic pattern of DNA modification and protein packaging. A newly introduced gene may have to begin at the beginning in order to be in its "natural" state at the moment when it should be turned on or off.

Classical Genetic Techniques Are Still Indispensable

We want to study the developmental consequences of replacing a normal gene in an organism with one that has been mutated in vitro in carefully engineered ways. This ambitious experimental program requires a subtle marriage of classical genetics and contemporary recombinant DNA technology. The power of classical genetics is that it asks a very general question: How many different kinds of mutations in how many different genes can produce the same or a related defect? In this way, we find out which genes are important for what, and we accumulate the knowledge necessary to clone each of these genes from the organism. For example, by looking for nematodes that are paralyzed or that move in an uncoordinated fashion *(unc)*, we assemble a catalog of all those genes involved in proper construction of the nervous system and body musculature. Similarly, by identifying fruit flies whose eggs fail to complete embryogenesis, we can gain access to a dazzling variety of genes, some of which determine the shape of the organism (e.g., *dorsal* embryos are all back and no front).

The major limitation of classical genetics is that the nature of the mutations (single base changes, insertions, deletions, inversions, and translocations) cannot be controlled. This limits our ability to design the most informative experiments. For example, one naturally arising single-base change in the homeobox of the *fushi tarazu* gene of the

Antennapedia complex is known to be lethal for development of the fly at slightly elevated temperature. However, there is no way to use classical genetic analysis to isolate additional mutations in this particular homeobox. Thus, it would be extremely helpful to be able to systematically vary the amino acid sequence of the homeoboxes in the *Drosophila* Antennapedia or bithorax gene complexes. Fortunately, the power of recombinant DNA technology is that it can make exactly such mutations in cloned DNA fragments. The trick, then, is not so much to clone the genes in the first place, nor to make interesting mutations in them in vitro, but to get the mutant genes back into the organism in a functional form.

Precise Gene Substitution Is Now Possible Only in Bacteria and Yeast[74a, b]

In yeast alone has it proved possible to use homologous recombination to precisely replace a resident wild-type gene with a mutant gene copy that has been genetically engineered in vitro (Chapters 18 and 19). We should expect no less from a eucaryotic microorganism that is as easy to work with and to analyze genetically as *E. coli.* So far, it has not proved possible to achieve gene replacement in any higher eucaryote than yeast, but this impasse may soon prove temporary. In any event, cloned genes have been successfully reintroduced into the germ line of both the fruit fly *Drosophila* and the laboratory mouse. Although integration occurs at apparently random chromosomal positions, it is particularly encouraging that these "misplaced" genes are in general subject to correct developmental regulation. Random integration is not as nice as precise gene replacement, because the resident wild-type (or mutant) gene is still present and may confuse the analysis; but at least a start has been made.

Putting Genes Back into *Drosophila* with P Element Vectors[75, 76]

We have already seen that the P elements in *Drosophila* are similar to many procaryotic and eucaryotic transposable elements (Figure 20-32). A complete P factor consists of 31-base-pair perfect terminal inverted repeats surrounding a 3-kilobase protein-coding region that encodes two functions: a transposase (which catalyzes movement of the element to new sites) and a repressor of transposition (which accumulates until further transpositions are blocked). During a dysgenic cross, sperm from a male fly bearing genomic P elements (P male) fertilize eggs from a female fly lacking P elements (M female). Genes in the tightly packed sperm head are transcriptionally silent, but P elements on the paternal chromosomes begin to be transcribed soon after entering the egg. Although P elements appear to be transcribed in all cells, transposition occurs exclusively in the germ line as a result of tissue-specific alternative mRNA splicing: The major P element transcript (spanning nearly the entire element) is spliced in somatic cells to produce an mRNA encoding the repressor function, but the same primary transcript can be alternatively spliced in germ line cells to produce an mRNA encoding the transposase. A variety of genetic evidence suggests that P element transposition occurs only in embryonic germ cells (pole cells); this could mean that the tissue-specific splice is restricted to pole cells, or that sufficient repressor accumulates in maturing cells of the germ line to block further transposition.

Tissue-specific alternative splicing of the major P element transcript raises the intriguing possibility that the transposase and repressor are related proteins that share one or more common exons. For example, a transposase-like protein that recognized the terminal inverted repeats of the P element, but lacked the active catalytic site of the transposase, could function as a repressor of transposition by competing with functional transposase for binding to the terminal repeats.

When insertion of a paternal P element happens to disrupt an essential gene (or chromosomal region), heritable new mutations are observed in the offspring of a P male and an M female. (The parental flies display no mutations, since transposition takes place only in the germ line.) The cumulative effect of such mutations (as well as accompanying chromosomal rearrangements) is called **hybrid dysgenesis,** because the progeny of the dysgenic cross are genetically less fit than the parents. The reciprocal cross of an M male with a P female does not lead to dysgenesis, since P elements in the haploid egg genome are repressed by high levels of accumulated repressor in the egg cytoplasm.

Interestingly, most P elements in the genome of a *Drosophila* P strain are internally deleted: Part or all of the transposase- and repressor-coding region is missing. However, as long as the 31-base-pair terminal inverted repeats are intact, these incomplete P elements can be mobilized when transposase is supplied by an intact P element. Apparently, the transposase need only recognize the inverted repeats of the P element. This observation suggested a strategy for putting genes back into *Drosophila* by using recombinant P elements as vectors to insert new genes into germ line DNA (Figure 22-43). Recombinant DNA technology was used to replace the internal protein-coding region of a complete P element with a foreign gene of interest. The resulting recombinant P element (carrying the foreign gene), together with an intact P element (to supply transposase activity), was then microinjected through a thin glass needle into a freshly fertilized nondysgenic egg (produced by crossing an M male with an M female). These injected P elements then behaved exactly like P elements that had been naturally introduced by sperm at the moment of fertilization. The transposase gene on the complete P element was transcribed in the embryo, the mRNA translated, and the transposase catalyzed transposition of both the complete P element and the recombinant P element into embryonic chromosomes.

To maximize the chance that the recombinant P element would integrate into germ line chromosomes, the DNA was microinjected near the posterior end of the fly, where cytoplasmic polar granules determine which blastula nuclei will become germ (pole) cells (see Figure 22-24). The microinjected flies (G_0 generation) did not generally express the new gene, since it was present in only a small percentage of their cells. However, some of the G_0 generation flies carried the new gene in germ line cells; thus, some of the progeny of these flies (G_1 generation) had the gene in every cell, so that the effects of the new gene could be properly studied. Flies that have successfully taken up the new gene are not difficult to recognize. For example, flies with a defect in the enzyme xanthine dehydrogenase (*rosy⁻*) are fatally poisoned by excess purines in their food. After microinjection of *rosy⁻* embryos with a wild-type *rosy⁺* gene, transformed flies expressing the new gene can be selected by growth on purine-enriched medium.

Figure 22-43
Use of recombinant P elements to introduce new germ line genes into *Drosophila*. (a) Structure of injected plasmids. (b) Protocol for introduction of a *rosy*⁺ gene into the germ line of a *rosy*⁻ strain.

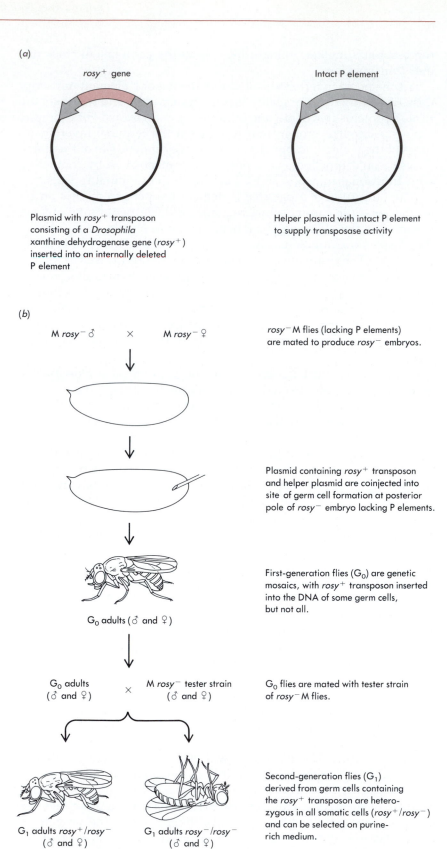

(a)

rosy⁺ gene

Intact P element

Plasmid with *rosy*⁺ transposon consisting of a *Drosophila* xanthine dehydrogenase gene (*rosy*⁺) inserted into an internally deleted P element

Helper plasmid with intact P element to supply transposase activity

(b)

M *rosy*⁻ ♂ × M *rosy*⁻ ♀

rosy⁻ M flies (lacking P elements) are mated to produce *rosy*⁻ embryos.

Plasmid containing *rosy*⁺ transposon and helper plasmid are coinjected into site of germ cell formation at posterior pole of *rosy*⁻ embryo lacking P elements.

G_0 adults (♂ and ♀)

First-generation flies (G_0) are genetic mosaics, with *rosy*⁺ transposon inserted into the DNA of some germ cells, but not all.

G_0 adults (♂ and ♀) × M *rosy*⁻ tester strain (♂ and ♀)

G_0 flies are mated with tester strain of *rosy*⁻ M flies.

G_1 adults *rosy*⁺/*rosy*⁻ (♂ and ♀) G_1 adults *rosy*⁻/*rosy*⁻ (♂ and ♀)

Second-generation flies (G_1) derived from germ cells containing the *rosy*⁺ transposon are heterozygous in all somatic cells (*rosy*⁺/*rosy*⁻) and can be selected on purine-rich medium.

Many Microinjected Genes Are Not Very Sensitive to Chromosomal Environment[33, 76–80]

Unlike targeted gene substitution in yeast (Chapter 18), microinjected P elements seem to insert randomly into new chromosomal sites. Initially, biologists had feared that correct expression of the newly introduced gene might be so strongly dependent on its immediate chromosomal environment that only genes that had integrated at or near the site of the resident wild-type gene would function properly. This worry was based on the existence of position effects in classical *Drosophila* genetics: Occasionally, genes were poorly expressed or even turned off completely after being moved by transposition or inversion to a new chromosomal location. Fortunately, the expression of new germ line genes introduced by recombinant P elements is generally not subject to strong position effects. For example, a recombinant P element carrying a cloned DNA fragment from the 66D12-15 chorion gene cluster will program correct developmentally regulated amplification of flanking DNA (see Figure 22-21) after insertion at a variety of different sites. Also almost independent of the site of chromosomal integration was the tissue-specific expression of metabolic enzymes such as alcohol dehydrogenase and dopa decarboxylase, as well as the inducibility of heat-shock genes by high temperature (Chapter 21).

Circadian Rhythms[79, 80, 81]

The accurate measurement of biological time is not restricted to the rapidly unfolding cell movements (morphogenesis) of early development. Many (and perhaps most) higher eucaryotes, from *Drosophila* to humans, exhibit daily **circadian rhythms** (from the Latin *circa die*, meaning "about one day long"). Psychologists have shown that when human subjects are isolated in a clockless environment lacking all visual and auditory clues about the passage of day and night, most individuals establish a "free-running clock" with a period of about 23 hours. This clock governs cyclic patterns of waking, sleeping, activity, and hunger. Evidently, our bodies "know" the length of the day, even without environmental clues; the alternation of day and night simply modifies or "entrains" our natural biological rhythm to a slightly longer period of 24 hours.

Circadian rhythms in the fruit fly *Drosophila* are not conspicuously different from our own. Behavioral mutants of the fly can be isolated from mutagenized populations, just as biochemical mutants are, so long as the investigator is patient enough to observe the behavior of individual flies or clever enough to devise a selection procedure. One particularly obvious selection for flightless mutants is to study those flies that do not fly away when the top is taken off the culture bottle. Similarly, flies can be isolated whose free-running clock is slower or faster than usual, or even completely irregular. These mutations map to a single genetic region called the *per* (for *period*) locus, and DNA from the locus has been cloned. Recently, a 7-kilobase-pair fragment of wild-type *per* DNA from *Drosophila* was introduced by P element transformation into an arhythmic fly, and the mutation was corrected to a normal 24-hour cycle (Figure 22-44). The 7-kilobase-pair fragment encodes a 4.5-kilobase mRNA, and parts of this mRNA are homologous to DNA in the chicken, mouse, and human genomes. Studies

Figure 22-44
Introduction of a wild-type *per* locus into the germ line of an aperiodic mutant fly using a recombinant P element restores normal circadian rhythms. Wild-type flies (*per$^+$*), an aperiodic mutant (*per^0*), and the *per^0* mutant transformed with a wild-type *per$^+$* locus, were grown in culture bottles on solid food. The flies reproduce under these conditions, and all stages of the life cycle—egg, larva, pupa, and adult—are represented. For 3 days before the start of the experiment, the flies were entrained to a 24-hour cycle with 12 hours of alternating light and dark. Beginning on day 1, the adults that had enclosed (emerged from the pupal cases after metamorphosis) were collected every 2 hours and counted; the graphs show the number of adults emerging in each 2-hour interval. Note that the alternating cycle of light and dark was discontinued after the fifth day, and the flies maintained thereafter in total darkness. Interestingly, the 24-hour circadian rhythm persists in darkness, showing that the fly's internal clock, once entrained, no longer requires daily stimulation. [After R. A. Bargiello, F. R. Jackson, and M. W. Young, *Nature* 312 (1984):752–754.]

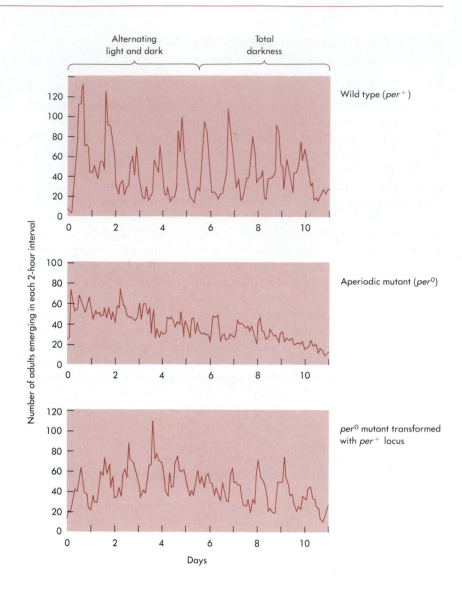

such as these form the basis for a new science: the molecular genetics of behavior. The inescapable conclusion is that we human beings, proud possessors of sophisticated intelligence, will find that our behavior is governed to some extent by elementary biochemical reactions.

Transgenic Mice: Mice with New Germ Line Genes[82–85]

The production (or construction, as molecular biologists sometimes like to say) of a transgenic mouse carrying a new germ line gene is superficially similar to transformation of *Drosophila* embryos with recombinant P elements; however, there is one important difference. No mammalian transposable element has yet been fully adapted to function as a recombinant vector analogous to P elements, although some very promising vectors derived from RNA tumor viruses (Chapter 24) are now being tested. Instead, transformation of mouse germ line DNA currently depends on the recombinational abilities of the fertilized mouse egg itself.

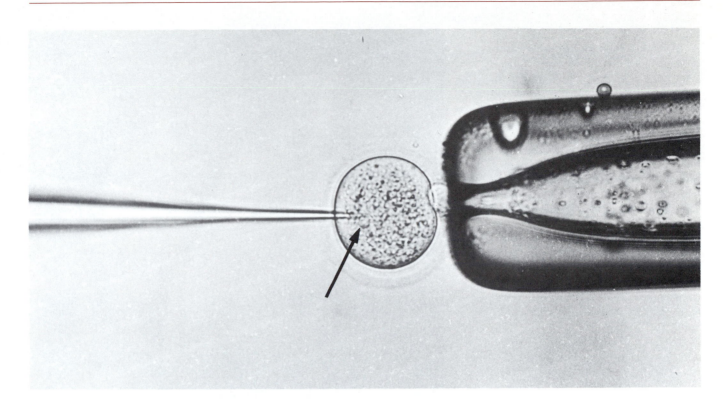

To make a transgenic mouse, eggs are first surgically removed from a mature female and then fertilized with sperm in the test tube. A recombinant plasmid carrying a recognizable new gene (one that can be distinguished from the mouse's own gene) is microinjected with a thin glass needle into the male pronucleus of the freshly fertilized egg before the haploid sperm and egg nuclei have fused to form a diploid zygotic nucleus (Figure 22-45). The circular plasmids generally recombine homologously with each other within the egg, forming a long, tandemly repeated concatemer; this concatemer then integrates randomly to give tandemly repeated genes at a single chromosomal site. The microinjected eggs are surgically implanted in a foster mother and allowed to come to term. The resulting mice are called **transgenic,** since part of their genome comes from another genetically unrelated organism (be it animal, vegetable, or bacterial). Because the foreign gene is introduced even before fusion of the egg and sperm nuclei, chromosomal integration takes place early, and the progeny transgenic mice often carry the new gene in some germ line and somatic cells. These mice (the G_0 generation) must then be bred to produce mouse lines in which all cells (both germ line and somatic) are genetically identical.

The initial experiments with transgenic mice have been extraordinarily rewarding. For example, elastase is a pancreatic protease that is normally secreted into the upper intestine to aid digestion. When the rat elastase gene was cloned into a plasmid and microinjected into fertilized mouse eggs, the transgenic mice expressed the rat elastase mRNA and protein only in pancreatic secretory cells, although the gene was present in the cells of all other tissues. Evidently, these two rodents are so closely related that control signals in rat DNA are correctly recognized by the corresponding mouse regulatory factors. Nonetheless, rats and mice are sufficiently different that the rat elastase mRNA and protein can be easily distinguished from the mouse products.

Figure 22-45
Injection of foreign DNA into a fertilized mouse egg before the male and female pronuclei have fused. The egg is first immobilized by applying mild suction to the large, blunt holding pipette (right). Several hundred copies of the gene of interest are then injected through the sharp end of the narrow microneedle (left). A large nucleolus can be seen near the tip of the microneedle (arrow). [Courtesy of J. Gordon and F. H. Ruddle, *Methods in Enzymology* 101 (1982):411–433.]

Figure 22-46
A "gigantic" male mouse grown from an egg injected with multiple copies of a rat growth hormone gene under the control of a mouse metallothionein promoter. A normal male sibling is shown for comparison. The transgenic mouse weighs 44 grams, the normal mouse 29 grams. The new gene is passed on to offspring, which grow two to three times faster than normal and reach a size up to twice normal. Note that both mice have short tails. This is not an inherited trait; rather, a small piece of tail has been chopped off to extract DNA for an analysis of the number and organization of the growth hormone gene ("transgene") in the normal and transgenic animals. (Courtesy of Dr. R. L. Brinster, University of Pennsylvania.)

The first pleasing surprise was that the foreign rat elastase gene could be correctly regulated despite the fact that its site of chromosomal integration differed in each of the transgenic mouse lines. Equally pleasing was the discovery that at least some regulatory sequences in mammalian DNA are quite small. When a series of transgenic mice was made where less and less upstream flanking sequence preceded the rat elastase gene, it was found that merely 204 base pairs were sufficient for correct tissue-specific expression. This suggests that even in as complicated a eucaryote as the mouse, control regions may generally lie upstream of the gene and need not be much (if at all) larger than those found in bacteria. Similarly encouraging results have been obtained showing tissue-specific expression of transgenic light and heavy immunoglobulin chains (Chapter 23). Perhaps the most dramatic experiment done so far with transgenic mice involved the construction of a fusion gene in which the coding region for rat growth hormone was attached to a powerful promoter derived from the mouse metallothionein gene (metallothioneins are small cysteine-rich proteins that transport essential heavy metals such as zinc and copper throughout the body). Transgenic mice carrying multiple copies of the highly expressed rat growth hormone gene grew very fast. The mouse with the most copies (approximately 35) grew fastest, to almost twice the size of a normal mouse (Figure 22-46); but sadly, this gigantic animal (by mouse standards) died a premature death. Although large mice are merely a scientific curiosity, the same kinds of experiments can also be performed on farm animals of obvious economic importance, such as sheep and goats.

Working with a Mammal Takes Time

Putting genes back into yeast is easy. One hundred million (10^8) cells can be transformed and plated on selective media in less than an hour's time, and the resulting transformants grow with a doubling time of 2 hours. The fruit fly *Drosophila* is not difficult to work with. Perhaps 100 successful microinjections can be done in a working day, and tens of thousands (10^4) of progeny flies can be grown in small bottles with a generation time of a few weeks and visually inspected for the desired mutants. Mice are the most tedious of all to work with. Only 5 to 10 successful microinjections can be done in a day, but not all of these necessarily express the foreign gene. Moreover, a standard mouse cage has room for at most a dozen mice, and their generation time is nearly 2 months. Fortunately, natural curiosity about our own species (humans are not so different genetically from mice) gives us the drive to carry out the laborious experiments of making many transgenic mice.

PLANT MOLECULAR GENETICS COMES OF AGE

The Large and Frequently Polyploid Genomes of Most Plants

Until recently, incisive work on the structure and expression of plant genomes lagged far behind similar work with animals. Perhaps in part this sorry state of affairs was caused by our natural preference (as animals) for studying other animals. But the main problem was that genetic analysis of higher plants is really quite difficult. Plants grow relatively slowly, and only a few generations of genetic crosses can be performed each year. In addition, most economically important plants are large and require expensive greenhouse facilities for growth in temperate climates. Even today, no plant has been as thoroughly analyzed at the genetic level as *Drosophila*. Genetic analysis of plants is further complicated because many important plants not only have large numbers of chromosomes but are highly **polyploid** and contain at least four copies of any particular chromosome. Multiple copies of the wild-type gene in a polyploid organism make recessive mutations (which are far more common than dominant mutations) exceedingly laborious to detect. Consequently, botanists and agronomists have usually depended on traditional plant-breeding techniques for crop improvement. These techniques involve the repeated crossing of large numbers of plants with desirable characteristics (disease resistance, large fruit, drought tolerance, etc.) in the hope that such characteristics can be stably combined in one germ plasm. Unfortunately, such breeding procedures are tedious; moreover, the results are often uncertain, because desirable traits may in fact be the product of several different genes, and genes may function differently in different genetic backgrounds.

The very large genomes of most plants represent yet another obstacle to understanding plant genetics at the molecular level. For example, the genome of corn is actually several-fold larger than the human genome, despite the obvious difference between the two organisms in biological complexity. Thus there is no doubt that most DNA in the corn genome does not encode gene products. Instead, as is also true for the genomes of higher animals (Chapter 20), plant genomes are full of functionally obscure DNA. Much of this "extra" DNA seems to arise from the insertion of repetitive DNA elements whose widespread presence often greatly complicates the cloning and sequencing of plant genes.

Chloroplasts Possess Their Own Unique DNA

The chloroplasts of green plants are cytoplasmic organelles that house the various pigments and enzymes of the light-harvesting photosynthetic apparatus. Even before the turn of the century, it was clear that green pigmentation was one of the easiest traits to observe in plant-breeding experiments. Although some pigmentation traits obeyed Mendel's laws, other color traits were only transmitted through the female parent that provided the cytoplasm of the zygote. These observations of **cytoplasmic** or **maternal inheritance** eventually led to the hypothesis that chloroplasts must carry genes.

We now know that chloroplasts contain a unique circular DNA genome that is completely different from the nuclear genome. Electron micrographs indicate that the chloroplast DNA is some ten to twenty times smaller than the *E. coli* chromosome. For example, the chloroplast genome of maize (corn) contains 140,000 base pairs of DNA. Such genomes are much too small to encode the approximately 1000 different proteins found in chloroplasts. Instead, biosynthesis of the chloroplast involves an intimate collaboration between the nuclear and chloroplast genomes. In fact, every known multimeric protein component of chloroplasts is a mixture of the products of both nuclear and chloroplast genes.

Most chloroplast proteins are encoded by nuclear DNA, translated in the cytoplasm, and imported into the chloroplast by a specific transport mechanism that enables polypeptides to cross the outer membrane of the organelle. However, some 100 chloroplast-specific proteins are synthesized within the chloroplast itself. These proteins are encoded by chloroplast DNA, transcribed by the chloroplast-specific RNA polymerase, and translated by the chloroplast-specific protein synthesizing machinery. Since RNA cannot cross the outer membrane of the chloroplast, chloroplast ribosomal RNAs and tRNAs must be encoded in chloroplast DNA.

Chloroplasts are not static organelles, but can adapt to different physiological conditions such as high or low levels of light. For example, when grown entirely in the dark, chloroplasts lack chlorophyll but retain carotenoid pigments. Thus many chloroplast genes are light-regulated, in certain cases by light-sensitive promoters!

Maize May Be the Best Understood Plant Genetically

Partly because of its economic importance, and partly because of many favorable biological characteristics, maize (corn) has become one of the best understood plants at the genetic level. Only an acre of land is needed to grow some 10,000 maize plants. Each plant bears several ears containing 500 or more kernels, and each kernel is the product of a separate genetic cross. A maize geneticist working with only 1 to 2 acres of corn can thus investigate the progeny of over a million genetic crosses in the course of a year. Most importantly, the male and female parts of maize are found on separate flowers. This simplifies controlled pollination, and it is the principle reason why maize genetics has advanced so far compared with the genetics of cereals such as wheat, rice, or barley.

Over 170 different genetic loci have been mapped onto the ten chromosomes of the maize genome. Gene mapping in maize is often facilitated by the existence of knobbed protuberances on meiotic chromosomes. These protuberances are frequently unique to each strain of maize, and make it possible to identify a chromosome (bearing a particular combination of wild-type and mutant genes) by rapid cytological observation rather than by slow genetic analysis. Another key to steady progress in the molecular genetics of maize is the behavior of the well-characterized transposable elements *Ac* and *Ds*. Many spontaneous mutations in maize involve insertion of these elements near or within genes; moreover, both *Ac* and *Ds* have been cloned and sequenced, so that they can be used as probes for the genes into which they have inserted (see Figure 20-34).

Using the Ti Plasmid to Introduce New Genes into Plants[86–93]

We have seen that the diploidy (and usually polyploidy) of most crop plants poses a substantial problem for classical genetic analysis, so that new plant varieties are often more easily made by classical plant-breeding procedures than by a molecular approach that first requires the identification and isolation of particular genes. A much simpler and more straightforward approach to crop improvement would be to introduce known genes into the chromosomes of a healthy plant in such a way that the genes could be strongly expressed at all times in all tissues, or only in certain tissues (say, leaves), or only under certain environmental conditions (say, strong light or high temperature). We shall see that this goal can now be achieved by using the Ti plasmid to introduce foreign or genetically engineered genes into healthy plants in a heritable fashion (Figure 22-47).

Plants can generally be divided into **monocots,** such as wheat and corn (where the sprouting seed has a single leaf), and **dicots,** such as tomatoes or apples (where the sprout has two leaves). Most dicotyledonous plants are naturally susceptible to crown gall disease, in which tumors (uncontrolled cell growths) occur at the site of a wound (often at the crown, separating the stem from the roots of the plant). These tumors are caused by infection with the Gram-negative soil bacterium *Agrobacterium tumefaciens*, which transforms plant cells at the wound site so that they have the capacity for autonomous growth in the absence of any further stimulation by the infecting bacteria. In addition, depending on the type of infecting *Agrobacterium*, the transformed plant cells acquire the ability to synthesize one of two unusual arginine derivatives known as octopine [*N*-alpha-

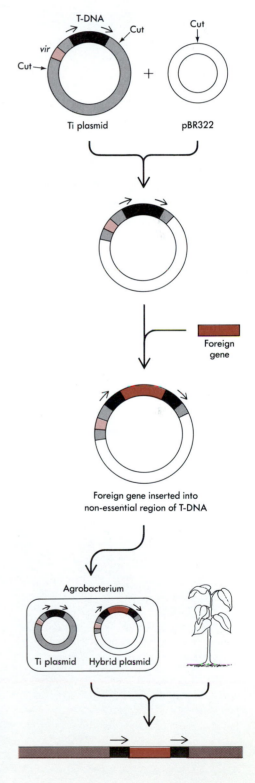

Figure 22-47
Recombinant Ti plasmids can introduce new genes into plants. Note that only the T-DNA is transferred to the plant genome, while a different region of the Ti plasmid known as *vir* encodes the enzymes required to mobilize the T-DNA. Small arrows above the T-DNA segment represent the 25-base-pair direct repeats that define the ends of the transposable element and are required for transposition. A large fragment of the Ti plasmid encompassing both the T-DNA and *vir* regions is first cloned into the common bacterial plasmid pBR322 (Figure 19-1). The new gene is then cloned into a nonessential region of the T-DNA and introduced into *A. tumefaciens* carrying an intact Ti plasmid. When plants are infected with these bacteria, the gene products of the *vir* region on the intact Ti plasmid mobilize the recombinant T-DNA, and the recombinant T-DNA integrates into the plant genome. [After J. D. Watson, J. Tooze, and D. T. Kurtz, *Recombinant DNA* (New York: Freeman, 1983), p. 171.]

(D-1-carboxyethyl)-L-arginine] or nopaline [*N*-alpha-(1,3-dicarboxy-propyl)-L-arginine]. These **opines** serve both as food and as an energy source for the infecting bacteria. Interestingly, the octopine-inducing bacteria can grow on octopine but not on nopaline, and the nopaline-inducing bacteria can grow on nopaline but not octopine. Thus, tumor formation can be seen as an evolutionary strategy devised by *Agrobacterium* to divert the metabolic resources of the plant into protecting, feeding, and perpetuating a bacterium from which it derives no obvious benefit.

In principle, the genes encoding the opine synthetases could have been present (but silent) in the uninfected plant cell genome and activated by infection with *Agrobacterium*. Alternatively, the opine synthetase genes could be carried by the bacterium and transferred to the plant cell genome during infection. We now know that the genes for both octopine and nopaline synthetase come from the infecting bacterium, where they are initially present on a giant plasmid known as the **Ti** (*tumor-inducing*) **plasmid.** Interaction between the infecting bacterium and the host plant cell stimulates the bacterium to excise a 30-kilobase-pair region of the Ti plasmid known as **T-DNA** (for transforming DNA) that is flanked by two directly repeated 25-base-pair sequences. Excision appears to occur by site-specific recombination between the two 25-base-pair terminal direct repeats and results in an extrachromosomal circle of T-DNA containing a single copy of the 25-base-pair sequence. Curiously, these T-DNA circles do not appear to be covalently closed and may be held together by proteins bound to the terminal repeats. These circular T-DNA molecules (or perhaps copies derived from them) are then transferred from the bacterium to the nucleus of the host plant cell, where the T-DNA integrates into the host nuclear genome, occasionally in the form of tandem repeats. In this way, the host acquires a variety of new genes carried on the T-DNA, including a number of genes responsible for host cell transformation, as well as the gene for an opine synthetase.

The gene products required for excision, transfer, and integration of the T-DNA into the host plant cell are not encoded by the T-DNA itself, but by a different part of the Ti plasmid known as the *vir* region (for virulence). *Vir* genes are induced by specific plant products, ensuring that the soil bacterium activates T-DNA transfer only in the presence of a potential plant host. Successful mobilization and integration of the T-DNA requires only the 25-base-pair terminal direct repeats; all internal T-DNA sequences are dispensable. Thus, the Ti plasmid (and the T-DNA within it) constitutes a natural vector for introducing new DNA sequences into the nuclear genome of somatic cells from susceptible plant species. The internal transforming genes of a cloned T-DNA can be removed by recombinant DNA techniques and replaced by any gene (whether of plant, animal, fungal, or bacterial origin) that can be expressed in plant tissues. Most commonly, the coding sequence of the foreign gene is substituted for the coding region of the opine synthetase gene: In this way, the natural promoter and polyadenylation signals of the opine synthetase gene confer high-level expression of the foreign protein. More recently, it has proved possible to replace the constitutive opine synthetase promoter with a light-inducible promoter derived from a gene encoding the small subunit of the chloroplast enzyme ribulose 1,5-bisphosphate carboxylase. The occurrence of the heat-shock response in plants as well as all other kingdoms (Chapter 21) promises to permit heat-inducible expression of foreign genes as well.

Fertile Plants Can Be Regenerated from Somatic Cells[94]

Introduction of new genes into plant somatic cells would be useful for basic studies of plant gene expression, but not for crop improvement, had not classical botanists discovered earlier that normal plants can often be regenerated from somatic tissues. Many plants (including such common house plants as the rex begonia) can be readily regenerated from a small disk of somatic tissue punched out of a leaf and placed on moist nutrient medium (Figure 22-48). To the naked eye, roots and shoots appear to sprout directly from the leaf disk; but on a microscopic level, differentiated cells within the leaf must first dedifferentiate, proliferate, and finally redifferentiate into the many tissues required to construct a complete plant. More technically challenging, but useful in many ways, is the regeneration of a complete plant from a **callus** (a disorganized, undifferentiated cell mass originally excised from a mature plant and then grown indefinitely in tissue culture). Treatment with the proper plant hormones induces differentiation of the callus into roots and shoots, and eventually a complete plant with fertile flowers can be regenerated (see Figure 22-48). No animal of apparently comparable biological complexity can yet be regenerated from mature somatic cells. (The clonal frog in Figure 22-7 was derived from intestinal nuclei of a tadpole.)

In principle, it should be possible to transform both leaf disks and callus cultures with T-DNA vectors, and then to regenerate complete transgenic plants (see Figure 22-48). (Although the top and bottom of a leaf disk are covered by a waxy cuticle, the open sides of the disk serve as wounds for T-DNA transformation.) However, several problems currently complicate the use of T-DNA vectors. Only some plant species can be persuaded to grow in callus culture, and not all of these can be regenerated from the callus; other plant species will not regenerate from a leaf disk; and lastly, the T-DNA vector itself does not infect leaf disks or callus cells of all plant species. Given the current pace of research in plant molecular genetics, such technical obstacles may soon be surmounted. Interestingly, transformation of somatic plant cells with wild-type T-DNA blocks regeneration, because the transforming genes subvert the normal mechanisms of growth control and development. Thus, it is essential that the transforming genes be altered or deleted in the T-DNA vector carrying the new genetic information into the plant genome.

The one major problem remaining in the use of genetically engineered Ti plasmids for crop improvement is that many important crop plants are monocots originally thought to be resistant to infection by *Agrobacteria*. Fortunately, recent evidence suggests that *Agrobacteria* can infect some monocots such as lily and asparagus (but not yet corn or wheat). Opine synthetase genes are expressed, but the transforming genes of the T-DNA are unable to work effectively in the monocot environment. Presumably, tricks will be devised in the near future to infect economically important monocots.

Crop Improvement Through Genetic Engineering [86, 95]

What new genes would we want to introduce to improve important crop plants? The seed proteins of such staple crops as wheat, beans,

(a)

(b)

(c)

Figure 22-48
Plants can be regenerated from mature somatic cells. (a) Regeneration of shoots from an undifferentiated mass of callus cells derived from the common garden flower *Petunia hybrida*. The callus on the left was maintained on normal plant tissue culture medium, whereas the callus on the right was grown on a special regeneration medium that allows organogenesis. (b) Shoots regenerating directly from the edge of a leaf disk on medium that supports both callus growth and organogenesis. (c) Transformed, kanamycin-resistant petunia shoots regenerating from a leaf disk grown on regeneration medium containing kanamycin. Only the disk on the left was infected with a genetically engineered strain of *Agrobacterium tumefaciens* that expresses the bacterial gene for kanamycin resistance in the transformed plant cells. (Courtesy R. B. Horsch, Monsanto Company.)

and corn are often deficient in particular amino acids; corn is especially poor in lysine, threonine and methionine. Genes encoding the major seed storage proteins have already been identified and cloned, and it should not prove difficult to use recombinant DNA technology to insert nucleotide sequences encoding the deficient amino acids into the correct reading frames. The structure of seed proteins (which are broken down to feed the germinating plant embryo) are presumably not so critical that insertion of a few new amino acids would interfere with function. Since abundant seed proteins are generally encoded by large multigene families, the genetically engineered seed protein genes will have to be introduced into the plant genome in relatively high copy number in order to have a significant effect on the nutritional characteristics of the seed. This presents no technical problem, because T-DNA can integrate in the form of long tandem repeats. However, because T-DNA integrates randomly into the plant genome (as do P elements in *Drosophila* and microinjected DNA in fertilized mouse eggs), one initial worry was that position effects might prevent genes on T-DNA vectors from being properly regulated in their new chromosomal environment. Fortunately, as is the case for transgenic flies and mice, the immediate chromosomal environment usually has little effect on regulation of the transgene. For example, soybean seed storage proteins are subject to proper developmental regulation when introduced into tobacco plants: The transgenes are expressed in seeds, but not in roots, shoots, or leaves.

One potential use for T-DNA vectors is in the development of insect-resistant plant lines. Certain bacteria such as *Bacillus thuringiensis* are known to produce peptides that are highly toxic to many insects. The genes encoding such natural insecticides could be incorporated directly into plant genomes, rendering repeated spraying with chemical insecticides unnecessary. However, in a pilot experiment where the gene encoding the *Bacillus thuringiensis* toxin was incorporated into a tobacco plant, the toxin produced by the newly introduced gene was inactivated by the ultraviolet component of sunlight! This unexpected result warns us that biological systems can be very subtle, and it will often be easier to dream up revolutionary applications of recombinant DNA technology than to put these ideas into practice.

Another potential use for recombinant T-DNA vectors stems from the observation that many soil bacteria are able to degrade various herbicides enzymatically. In principle, these bacterial genes could be cloned and expressed in new plant lines, so that herbicides could replace mechanical cultivation as a method for weed control. Still other bacteria may be found to produce antifungal agents, the biosynthetic gene(s) for which could be incorporated into crop plants. However, ecosystems are often delicately balanced, and the environmental impact of introducing transgenic plants will sometimes require serious scrutiny.

Agrobacterium tumefaciens Can Also Transform Chloroplasts[96]

Chloroplasts are thought to be derived from symbiotic photosynthetic procaryotes that colonized the cytoplasm of primitive eucaryotes. As predicted by this **endosymbiont hypothesis,** chloroplasts resemble bacteria in many ways. For example, ribosomes found in the cytoplasm of eucaryotes are resistant to the antibiotic chloramphenicol, but both procaryotic and chloroplast ribosomes are sensitive.

The differing antibiotic sensitivity of cytoplasmic and chloroplast ribosomes was used in an ingenious experiment designed to test whether T-DNA can insert into chloroplast as well as nuclear genomes. A gene fusion was constructed between the promoter of the nopaline synthetase gene and the coding region of the bacterial chloramphenicol resistance gene. Because the nopaline synthetase promoter was already known to function in bacteria, and the chloramphenicol resistance gene was itself derived from bacteria, it seemed possible that the gene fusion might function in chloroplasts. When the gene fusion was introduced into chloramphenicol-sensitive tobacco cells using a recombinant T-DNA vector, mature tobacco plants could be regenerated in the presence of chloramphenicol. The enzyme chloramphenicol transacetylase (which degrades chloramphenicol and is the product of the chloramphenicol resistance gene) was found exclusively in chloroplasts of the resistant plants; moreover, the chloramphenicol resistance trait was maternally inherited. These observations confirmed that the T-DNA had integrated directly into the chloroplast genome.

The ability of T-DNA to insert into the chloroplast genome has important agricultural implications. Many herbicides work by uncoupling light-driven electron transfer from production of high-energy chemical bonds during photosynthesis, and resistance to these herbicides often resides in chloroplast DNA. (Remember that the macromolecular complexes responsible for photosynthesis are generally put together from gene products encoded by both the nuclear and chloroplast genomes.) Thus it should prove technically possible to transfer resistance to various kinds of herbicides from one species of plant to another, using T-DNA as the vector to move the appropriate region of chloroplast DNA from the resistant to the sensitive chloroplasts. Thus herbicide-resistant crop plants may soon be a reality.

Plant Molecular Geneticists Need Better Model Organisms[97]

Recently, there has been renewed interest in a small flowering plant called *Arabidopsis thaliana*, which has been studied by classical genetic techniques for over 40 years. This harmless weed of the mustard family has no economic or nutritional value, but it has many other qualities that commend it to the plant molecular biologist. *Arabidopsis* has a generation time of only 5 weeks, produces 10^4 seeds per plant, and is so small that tens of thousands of plants can be grown in a small laboratory room. The plants are self-fertile (so that new mutations are naturally made homozygous) and can also be cross-fertilized by hand (so that genes can be mapped and multiple mutations can be introduced into a single plant stock). Over 75 mutations have already been mapped on the *Arabidopsis* genome. In addition, *Arabidopsis* is susceptible to infection with *Agrobacterium tumifaciens*, so it should eventually be possible to introduce foreign genes into the chromosomes of the organism.

Perhaps most important for the molecular biologist are the small genome size and comparative lack of repetitive DNA in *Arabidopsis*. The smaller the genome of an organism, the easier it is to clone genes, because there is less DNA to sift through. Also, DNA sequences that are repeated many times in the genome are an impediment to many recombinant DNA manipulations because these sequences tend to hybridize with each other. The genome of *Arabidopsis* is only 10 percent as large as that of cotton, 5 percent as large as that of tobacco, and 1 percent as large as that of wheat. Nonetheless, this "small" plant genome (7×10^7 base pairs) is still about 20 times larger than that of *E. coli*. Since distantly related organisms often share similar mechanisms of gene regulation, studies of *Arabidopsis* may prove useful or even directly applicable to work on economically important but genetically less tractable plants.

Symbiotic Development: Nitrogen-Fixing Bacteria in the Roots of Leguminous Plants[98]

Until now, we have discussed the development of individual organisms in isolation. The nitrogen-fixing bacteria of the genus *Rhizobium* that live in the roots of legumes such as beans, clover, and alfalfa provide a striking example of complex developmental interactions between two different organisms. Most plants can use only fixed nitrogen in the form of nitrates (NO_3^-). Although molecular nitrogen (N_2) constitutes about 70 percent of the atmosphere, this potential food stuff would be unavailable to either plants or animals if it were not for the nitrogen-fixing bacteria that can convert atmospheric molecular nitrogen into fixed nitrates. Some nitrogen-fixing bacteria are free-living, but most of the economically important species of *Rhizobium* live within specialized root nodules of legumes. The formation of such a nodule must require an intimate genetic collaboration between the bacteria and their plant host, since many mutations are known in both the bacterial and plant genomes that can block development of a functional nitrogen-fixing nodule.

Over the past fifteen years, there has been enormous progress in understanding the precise details of the genetic collaboration between rhizobia and legumes. In addition to the intrinsic biological fascination of the subject, these studies have also been motivated by the realization that many high-yielding cereal crops (such as corn) quickly exhaust the soil they grow in. The fixed nitrogen content of the soil can be rebuilt by time-consuming crop rotation with leguminous plants (alfalfa, clover, soybeans), or the soil can be artificially renewed by heavy applications of chemical fertilizers rich in phosphates and nitrates. But the cost of such fertilizer is almost prohibitive for many underdeveloped countries, and the ecological consequences of chemical runoff from croplands into local streams and lakes can be serious, even in a technologically advanced nation. These problems might be overcome if cereal and other crops could be naturally bred, or genetically engineered, to become functional hosts for rhizobia.

Symbiosis begins with attachment of rhizobia in the soil to the root hair of a germinating legume (Figure 22-49). Because each species of *Rhizobium* has a narrow **host range** (for example, *Rhizobium melliloti* will grow on alfalfa but not on clover), this initial attachment is thought to depend on a cell recognition event between surface molecules (possibly glycoproteins) on the bacteria and the plant. Following attachment, the bacterium is encapsulated within a pocket of the plant cell wall and carried deep inside the root by formation of an infection thread consisting primarily of cellulose. Cortical cells on the outside of the root dedifferentiate to form a growing meristem that bulges out into a nodule. Proliferating infection threads then release

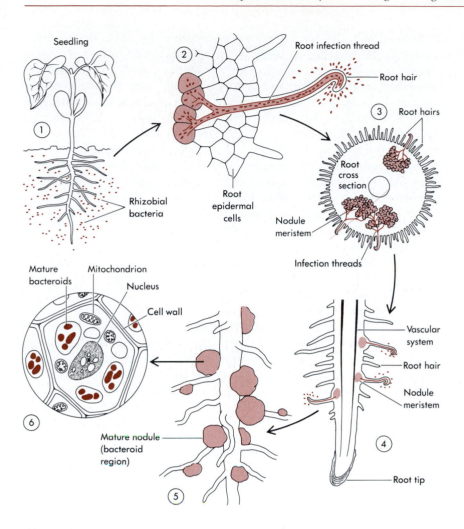

Figure 22-49
Development of a legume nodule.

the bacteria into the cytoplasm of the host plant cells, where the bacteria—after they are covered with a special membrane synthesized by the host—are known as **bacteroids.** Within this extremely protective environment, the bacteria finally differentiate into a nitrogen-fixing state by derepressing the genes for nitrogenase and bacteroid-specific cytochromes.

One of the most cogent examples of the degree of cooperation between rhizobia and the host plant is the biosynthesis of leghemoglobin, the plant's oxygen binding protein. The molybdenum-containing active site of the nitrogenase enzyme complex is specifically designed to reduce molecular nitrogen (N_2), but it can easily be poisoned by free molecular oxygen (O_2). Yet, the bacteroids also need a steady supply of oxygen for their general metabolism. The host plant meets these apparently contradictory demands by synthesizing large quantities of leghemoglobin (Chapter 20), which carries oxygen to the bacteroids while maintaining the overall oxygen tension at a sufficiently low level so as not to harm the nitrogenase. Remarkably, although the plant synthesizes the leghemoglobin polypeptide, the heme cofactor is supplied by the bacteria themselves. Coordinating the relative amounts of heme and leghemoglobin under these circumstances may not be a simple matter. In vertebrate erythrocytes, the synthesis of heme and hemoglobin is coordinated at the translational level; heme is required for formation of an initiation complex between the mRNA and the ribosome (Chapter 21). Perhaps a similar mechanism is at work within the legume cytoplasm.

Interestingly, bacteroids are approximately the same size as mitochondria, and in this sense they can be considered to be a nitrogen-fixing, plant cell organelle. In fact, the analogy with mitochondria is not only morphological but also metabolic. Mitochondria reduce atmospheric oxygen, and bacteroids reduce atmospheric nitrogen. Mitochondria donate the bulk of the ATP they generate to the general cellular metabolism, although a small fraction of the ATP is consumed by the mitochondrial DNA, RNA, and protein-synthesizing systems. Rhizobia, curiously, are completely unable to assimilate the ammonia they make by reducing molecular nitrogen. Instead, the ammonia is exported to the surrounding plant cytoplasm, where it is incorporated into larger metabolites such as amino acids, which can be imported and assimilated by the bacteria.

Both mitochondria and chloroplasts are thought to have descended from initially free-living photosynthetic bacteria that were engulfed by, or developed a symbiotic relationship with, a primitive nonphotosynthetic plant cell (endosymbiont hypothesis). Once the bacterial cell was lodged within the eucaryotic host, mutations within the bacterial genome could have deprived it of the ability to live outside its host. Subsequent mutations in both the nuclear and algal genomes could have slowly stripped the proto-mitochondrion and proto-chloroplast of all but the most indispensable functions. Conceivably, the symbiotic relationship between today's rhizobia and legumes may be giving us a glimpse of a similar kind of evolution in progress. Perhaps in the distant future, some species of rhizobia may lose the ability to live outside their plant hosts and thus become permanent nitrogen-fixing organelles. The same result might also be achieved deliberately (and hopefully sooner) by genetic engineering, if no unforeseen problems are encountered. Perhaps the most exciting challenge would be to use T-DNA transformation to extend the host range of *Rhizobium* species to include nonleguminous plants such as corn and wheat. The intimate and complex nature of the genetic collaboration between the Rhizobium bacteria and the host plant cell (see Figure 22-49) suggests that this task may be very difficult; but the potential benefits to humankind are so overwhelming that there is no shortage of effort directed toward accomplishing this goal.

The Genetics of Nitrogen Fixation

An investigation of the genes responsible for development is difficult, and genetic characterization of the cooperation between two different organisms will be doubly so. A second problem is that free-living rhizobia do not differentiate into bacteroids outside the plant host, nor does the host plant construct nodules in the absence of infecting bacteria. Thus, mutations in either the bacterial or the plant genes that are required for functional nodulation can only be identified by the laborious procedure of inoculating individual plants with individual bacterial strains. A third problem is that the genetic analysis of plants has not been easy. Plants grow slowly; techniques had to be developed for the rapid and convenient growth of large numbers of plants on a small laboratory scale; and plants are frequently polyploid. Ironically, the very difficulty of plant genetics has simplified the prevailing experimental approach to nitrogen fixation. Molecular biologists have chosen to take advantage of the power of bacterial genetics and to concentrate on the bacterial genes required for functional nodule formation. A deeper understanding of these genes should provide future leverage for investigating the genetic contribution of the host plant.

Cloning the Symbiosis Genes of Rhizobia[98, 99]

The bacterial genes required for formation of functional *symbiotic* nitrogen-fixing nodules are known as *sym* genes. The *sym* genes have been provisionally divided into three classes, depending on which step in the nodulation process appears to be blocked by mutation. Additional analysis will surely lead to further subdivision of these classes. The *nod* genes are required for initiation and the early steps of *nodule* formation. The *fix* genes are required for later steps in the construction of a nodule capable of nitrogen *fix*ation. And finally, the *nif* genes encode the three highly conserved subunits of the nitrogenase (which reduces molecular nitrogen to ammonia) as well as auxiliary proteins required for the nitrogenase to function. No good estimate can yet be made for the total number of *sym* genes, except for the *nif* genes, which number at least 17 in the enteric bacterium *Klebsiella pneumoniae*. The *sym* genes are thought to be located on a very large autonomously replicating plasmid, but a chromosomal location has not been strictly excluded.

The bacterial *sym* genes can be identified and cloned using the technique of **transposon tagging** (Chapter 11). The bacterial transposon Tn5 is a mobile genetic element carrying a gene that confers resistance to the antibiotic kanamycin. The kanamycin resistance acts as a molecular tag that allows geneticists to follow the movement of the transposon from one chromosomal site to another within a bacterium, or from one bacterium to another during conjugation. Tn5 inserts randomly during transposition. If the insertion site lies within the coding or controlling regions of a gene, the gene will be inactivated. To identify *sym* genes in *Rhizobium*, a strain of *E. coli* carrying Tn5 in its genome (and therefore resistant to kanamycin) is conjugated with a wild-type strain of *Rhizobium* lacking Tn5 (and therefore sensitive to kanamycin). After mating, rhizobia with a newly inserted Tn5 can be selected by growth on kanamycin; rhizobia lacking Tn5 will be killed by the antibiotic. The surviving bacterial colonies are then tested for mutations in the *sym* genes by inoculation onto wild-type seedlings of the appropriate plant species. Because the *sym* genes constitute only a small percentage of the total number of genes in the rhizobial chromosome, most of the Tn5 insertions will not affect the nitrogen-fixing genes. However, between 1 in 100 and 1 in 1000 seedlings tested is blocked at some stage preceding formation of a fully functional nodule. Although growth and examination of 1000 little plants is not a trivial exercise, the procedure has proved to be manageable.

Transposon mutagenesis yields a collection of mutant rhizobial strains, each carrying a Tn5 insertion in one of the *sym* genes. Molecular clones of these *sym* genes can easily be obtained by using the Tn5 DNA sequences as a molecular tag. First, radioactive Tn5 DNA is used as a probe to fish out from the mutant rhizobial DNA a large DNA fragment spanning the Tn5 insertion. Since Tn5 is an insertional mutagen, the rhizobial DNA sequences flanking the Tn5 insertion belong to the inactivated *sym* gene. Fragments of the inactivated *sym* genes are then used as radioactive probes to fish out an undamaged copy of the *sym* from the unmutated parental strain of *Rhizobium*.

Recently, expression of the *nod* genes was found to be induced by soluble plant metabolites that are excreted into the medium when plants are grown in culture. This important discovery helps to explain why the bacteria normally do not differentiate outside the plant host. Moreover, since the *nod* genes can now be turned on simply by treating bacteria with culture medium from growing host plants, this discovery also gives geneticists a way to study certain aspects of nodulation using the rapid techniques of classical microbial genetics, rather than the laborious inoculation of thousands of plants.

Summary

The expression of many genes must be exquisitely coordinated during differentiation and development. These developmentally regulated genes appear to be arranged in a hierarchy, with genes at the top of the hierarchy controlling the expression of those below them. Thus, the best hope for understanding complex developmental processes lies in identifying the controlling genes.

Genes are turned on and off during development by a variety of mechanisms. Displacement of one σ factor by another determines the order of gene expression during sporulation of Gram-positive bacteria like *B. subtilis*. Titration of a preformed factor in the frog egg by DNA synthesized during embryogenesis determines the timing of the midblastula transition. And the extraordinarily long transcription units of some homeotic genes in *Drosophila* may be a mechanism for delaying expression of the gene product until a specified time after transcription resumes in the newly fertilized egg.

We are just beginning to understand what causes cells to differentiate and how the differentiated state is maintained and stably inherited. Differentiation can be reversible or irreversible, depending on the cell type and the organism. Differentiation usually does not involve DNA rearrangement or loss, but rather may reflect changes in chromatin structure, the binding of specific regulatory proteins (repressors and activators), the formation of stable transcription complexes, and perhaps methylation and demethylation at the dinucleotide CpG. In addition, conservative segregation of nucleosomes at the replication fork provides an attractive (albeit hypothetical) explanation for how a stem cell can divide asymmetrically, giving rise to a differentiated daughter and another stem cell.

Sometimes, DNA rearrangements do accompany differentiation. In ciliates, extensive tracts of germ line DNA are eliminated from micronuclear DNA during regeneration of a new macronucleus, and many of the remaining DNA fragments are then precisely joined (perhaps by DNA splicing). Trypanosomes vary their surface antigens, probably by a directed gene conversion that moves a transcriptionally silent copy of the antigen-coding region on one chromosome to a transcriptionally active expression site on another chromosome.

Developmentally regulated gene amplifications are a way to provide extra gene copies in specific cell types where the demand for the gene product is high, but without imposing on every cell of the organism the genomic burden of carrying a large multigene family. During amphibian oogenesis, the ribosomal RNA genes are amplified extrachromosomally. In contrast, repeated rounds of localized DNA replication lead to chromosomal amplification of the chorion genes in follicle cells during *Drosophila* oogenesis.

Rapid progress is being made in understanding the genetic control of pattern formation during embryogenesis. The various dorsalizing genes appear to establish a gradient of dorsoventral polarity in the *Drosophila* egg, and the *Toll* gene product may actually be (or produce) a morphogen. The *Aristapedia* mutation in the fly causes only the distal part of the antenna to differentiate into a leg, as though the cells were differentiating in response to their position in some kind of gradient. Yet, except for the role of cyclic AMP as the morphogen that directs aggregation of starved *Dictyostelium* amoebas, we have no idea what the molecular nature of such a gradient might be, or how the cells might perceive their position within it.

Both internal and external forces play a role in determining whether a cell will differentiate along one developmental pathway or another. In a few simple cases, cytoplasmic determinants present in the unfertilized egg cause the embryonic cells containing them to differentiate into a specific cell type. Cytoplasmic determinants cause the formation of germ cells in insect and amphibian embryos, intestinal cells in the nematode, and muscle cells in the frog. In most cases, however, determination seems to reflect a much more complex interaction between many different gene products. Nonetheless, we suspect that the early stages of cellular differentiation correspond to expression of a few master genes high in the controlling hierarchy. Subsequent differentiation may simply reflect these initial decisions. Thus, the *Drosophila* segmentation genes determine whether the pattern of embryonic segments is correctly formed, but subsequent differentiation of these segments (e.g., into a thoracic segment bearing wings or an abdominal segment bearing reproductive organs) is the responsibility of other subordinate genes.

In mosaic development, interactions between cells play little or no role in determining cell fate. Instead, cytoplasmic determinants and cell lineage determine the nature and course of cellular differentiation. Although embryogenesis in the nematode worm *Caenorhabditis elegans* is almost exclusively mosaic, proper development of the gonad is clearly regulative and depends on interactions between the gonadal anchor cell and the surrounding ventral hypodermal cells. Mosaic development appears to be a special strategy for small, relatively simple organisms, whose survival depends on rapid embryogenesis. Mammalian development, which is much slower and more complex, is likely to be largely regulative.

Organisms often solve identical developmental problems in similar ways, because evolution tends to conserve useful molecular mechanisms. However, evolution can also preserve natural variations, so that related organisms may solve an identical problem in different ways. Thus, the follicle cells in *Drosophila* amplify the chorion genes shortly before secreting the eggshell around the developing oocyte, whereas the chorion genes in silk moths are present in the genome of all cells as a huge ("preamplified") multigene family.

Cloned DNA sequences can now be inserted into the germ line of essentially any organism. Precise replacement of a wild-type gene with a mutant gene by homologous recombination is currently possible only in bacteria and yeast. However, foreign DNA can be randomly inserted into the germ line of higher organisms by microinjecting the DNA into a fertilized egg. Transposable elements, such as the P element of *Drosophila* and certain mammalian retroviruses, have the potential to serve as universal vectors for introduction of foreign DNA by microinjection. Because the newly introduced gene experiences the entire developmental history of the transgenic animal, the effect of specific mutations on the behavior of the gene can be tested realistically. The production of transgenic sheep and goats demonstrates the potential of this technology for animal husbandry, but the ethical problems associated with human "gene therapy" should not be underestimated.

The potential social and economic benefits of genetically modifying important crop plants has provided much of the impetus for recent progress in plant molecular biology. Bacteria of the genus *Rhizobium* induce the formation of functional nitrogen-fixing nodules only on the roots of leguminous plants such as beans. The genetics of rhizobial nitrogen fixation and the biochemistry of the leguminous plant host are now under intensive investigation. The hope is that strains of nitrogen-fixing bacteria can be developed that can colonize the roots of nonleguminous plants. Other goals include development of corn strains with nutritionally adequate levels of lysine, threonine, and methionine, and the introduction of herbicide and insect resistance into crop plants. The Ti plasmid of *Agrobacterium tumefaciens* is being used in many of these studies as a vector to insert foreign DNA sequences into the plant germ line.

Bibliography

General References

Alberts, B., D. Bray, J. Lewis, M. Raff, K. Roberts, and J. D. Watson. 1983. *Molecular Biology of the Cell*. New York: Garland. Excellent discussions of communication between cells, morphogenesis in a variety of systems, germ cell formation, fertilization, differentiation, and the special features of plant cells.

Browder, L. W. 1980. *Developmental Biology*. Philadelphia: Saunders. A more classical textbook of early embryology, with an especially nice treatment of plants.

Davidson, E. H. 1986. *Gene Activity in Early Development*. 3rd ed. New York: Academic Press. A newly revised monograph covering all aspects of gene expression in the oocyte and embryo.

Loomis, W. F. 1986. *Developmental Biology*. New York: Macmillan. A new comprehensive text.

Losick, R., and L. Shapiro, eds. 1984. *Microbial Development*. Cold Spring Harbor, N.Y.: Cold Spring Harbor Laboratory. Excellent monographs on procaryotic and eucaryotic microbes as developmental systems.

Ptashne, M. 1986. *A Genetic Switch*. Palo Alto, Calif.: Blackwell. An approachable treatise on the molecular biology of control circuits in procaryotes and eucaryotes.

Raff, R. A., and T. C. Kaufman. 1983. *Embryos, Genes, and Evolution*. New York: Macmillan. An advanced but immensely readable and entertaining exploration of how mutations in DNA cause changes in embryogenesis that eventually lead to the evolution of new species.

Cited References

1. Donahue, T. F., R. S. Daves, G. Lucchini, and G. R. Fink. 1983. "A Short Nucleotide Sequence Required for Regulation of HIS4 by the General Control System of Yeast." *Cell* 32:89–98.
2. Johnson, A. D., and I. Herskowitz. 1985. "A Repressor (MAT α2 Product) and Its Operator Control Expression of a Set of Cell Type Specific Genes in Yeast." *Cell* 42:237–247.
3. Miller, A. M., V. L. MacKay, and K. A. Nasmyth. 1985. "Identification and Comparison of Two Sequence Elements That Confer Cell-Type Specific Transcription in Yeast." *Nature* 314:598–603.
4. Losick, R., and P. Youngman. 1984. "Endospore Formation in *Bacillus*." In *Microbial Development*, ed. R. Losick and L. Shapiro. Cold Spring Harbor, N.Y.: Cold Spring Harbor Laboratory, pp. 63–88.
5. Losick, R., and J. Pero. 1981. "Cascades of Sigma Factors." *Cell* 25:582–584.
6. Cowing, D. W., J. C. A. Bardwell, E. A. Craig, C. Woolford, R. W. Hendrix, and C. A. Gross. 1985. "Consensus Sequence for *Escherichia coli* Heat Shock Gene Promoters." *Proc. Nat. Acad. Sci.* 82:2679–2683.
7. Nurse, P. 1985. "Cell Cycle Control Genes in Yeast." *Trends in Genetics* 1:51–55.
8. Heasman, J., C. C. Wylie, P. Hausen, and J. C. Smith. 1984. "Fates and States of Determination of Single Vegetal Pole Blastomeres of *X. laevis*." *Cell* 37:185–194.
9. Newport, J., and M. Kirschner. 1982. "A Major Developmental Transition in Early *Xenopus* Embryos: I. Characterization and Timing of Cellular Changes at the Midblastula Transition." *Cell* 30:675–686.
10. Newport, J., and M. Kirschner. 1982. "A Major Developmental Transition in Early *Xenopus* Embryos: II. Control of the Onset of Transcription." *Cell* 30:687–696.
11. Kirschner, M., J. Newport, and J. Gerhart. 1985. "The Timing of Early Developmental Events in *Xenopus*." *Trends in Genetics* 1:41–47.
12. Gurdon, J. B. 1968. "Transplanted Nuclei and Cell Differentiation." *Sci. Amer.* 219:24–35.
13. Riley, D., and H. Weintraub. 1979. "Conservative Segregation of Parental Histones During Replication in the Presence of Cycloheximide." *Proc. Nat. Acad. Sci.* 76:328–332.
14. Razin, A., and A. D. Riggs. 1980. "DNA Methylation and Gene Function." *Science* 210:604–610.
15. Weintraub, H. 1985. "Assembly and Propagation of Repressed and Derepressed Chromosomal States." *Cell* 42:705–711.
16. Sanford, J. P., V. M. Chapman, and J. Rossant. 1985. "DNA Methylation in Extraembryonic Lineages of Mammals." *Trends in Genetics* 1:89–93.
17. Brown, D. D. 1984. "The Role of Stable Complexes That Repress and Activate Eucaryotic Genes." *Cell* 37:359–365.
18. Miller, J., A. D. McLachlan, and A. Klug. 1985. "Repetitive

Zinc-Binding Domains in the Protein Transcription Factor IIIA From *Xenopus* Oocytes." *EMBO J.* 4:1609–1614.

19. Callahan, R. C., G. Shalke, and M. A. Gorovsky. 1984. "Developmental Rearrangements Associated with a Single Type of Expressed α-Tubulin Gene in *Tetrahymena*." *Cell* 36:441–445.

20. Howard, E. A., and E. H. Blackburn. 1985. "Reproducible and Variable Genomic Rearrangements Occur in the Developing Somatic Nucleus of the Ciliate *Tetrahymena thermophila*." *Mol. Cell. Biol.* 5:2039–2050.

21. Yao, M.-C., J. Choi, S. Yokoyama, C. F. Austerberry, and C.-H. Yao. 1984. "DNA Elimination in *Tetrahymena*: A Developmental Process Involving Extensive Breakage and Rejoining of DNA at Defined Sites." *Cell* 36:433–440.

22. Blackburn, E. H., and J. W. Szostak. 1984. "The Molecular Structure of Centromeres and Telomeres." *Ann. Rev. Biochem.* 53:163–194.

23. Klobutcher, L. A., C. L. Jahn, and D. M. Prescott. 1984. "Internal Sequences Are Eliminated from Genes During Macronuclear Development in the Ciliated Protozoan *Oxytricha nova*." *Cell* 36:1045–1055.

24. Van der Ploeg, L. H. T., A. W. C. A. Cornelissen, P. A. M. Michels, and P. Borst. 1984. "Chromosome Rearrangements in *Trypanosoma brucei*." *Cell* 39:213–221.

25. Guyaux, M., A. W. C. A. Cornelissen, E. Pays, M. Steinert, and P. Borst. 1985. "*Trypanosoma brucei*: A Surface Antigen mRNA is Discontinuously Transcribed from Two Distinct Chromosomes." *EMBO J.* 4:995–998.

26. Milhausen, M., R. G. Nelson, S. Sather, M. Selkirk, and N. Agabian. 1984. "Identification of a Small RNA Containing the Trypanosome Spliced Leader: A Donor of Shared 5' Sequences of Trypanosomatid mRNAs?" *Cell* 38:721–729.

27. Solnick, D. 1985. "*Trans* Splicing of mRNA Precursors." *Cell* 42:157–164.

28. Konarska, M. M., R. A. Padgett, and P. A. Sharp. 1985. "*Trans* Splicing of mRNA Precursors in Vitro." *Cell* 42:165–171.

29. Stark, G. R., and G. M. Wahl. 1984. "Gene Amplification." *Ann. Rev. Biochem.* 53:447–491.

30. Yao, M.-C., S.-G. Zhu, and C.-H. Yao. 1985. "Gene Amplification in *Tetrahymena thermophila*: Formation of Extrachromosomal Palindromic Genes Coding for rRNA." *Mol. Cell. Biol.* 5:1260–1267.

31. Hourcade, D., D. Dressler, and J. Wolfson. 1973. "The Amplification of Ribosomal RNA Genes Involves a Rolling Circle Intermediate." *Proc. Nat. Acad. Sci.* 70:2926–2930.

32. Diaz, M. O., G. Barsacchi-Pilone, K. A. Mahon, and J. G. Gall. 1981. "Transcripts from Both Strands of a Satellite DNA Occur on Lampbrush Chromosome Loops of the Newt *Notophthalmus*." *Cell* 24:649–659.

33. De Cicco, D. V., and A. C. Spradling. 1984. "Localization of a Cis-Acting Element Responsible for the Developmentally Regulated Amplification of *Drosophila* Chorion Genes." *Cell* 38:45–54.

34. Goldsmith, M. R., and F. C. Kafatos. 1984. "Developmentally Regulated Genes in Silkmoths." *Ann. Rev. Genetics* 18:443–487.

35. Rosenthal, E. T., T. Hunt, and J. V. Ruderman. 1980. "Selective Translation of mRNA Controls the Pattern of Protein Synthesis During Early Development of the Surf Clam, *Spisula solidissima*." *Cell* 20:487–494.

36. Standart, N. M., S. J. Bray, E. L. George, T. Hunt, and J. V. Ruderman. 1985. "The Small Subunit of Ribonucleotide Reductase Is Encoded by One of the Most Abundant Translationally Regulated Maternal RNAs in Clam and Sea Urchin Eggs." *J. Cell Biol.* 100:1968–1976.

37. Caldwell, D. C., and C. P. Emerson, Jr. 1985. "The Role of Cap Methylation in the Translational Activation of Stored Maternal Histone mRNA in Sea Urchin Embryos." *Cell* 42:691–700.

38. Bienz, M., and J. B. Gurdon. 1982. "The Heat-Shock Response in *Xenopus* Oocytes Is Controlled at the Translational Level." *Cell* 29:811–819.

39. McGarry, T. J., and S. Lindquist. 1985. "The Preferential Translation of *Drosophila* hsp70 mRNA Requires Sequences in the Untranslated Leader." *Cell* 42:903–911.

40. Klemenz, R., D. Haultmark, and W. J. Gehring. 1985. "Selective Translation of Heat Shock mRNA in *Drosophila melanogaster* Depends on Sequence Information in the Leader." *EMBO J.* 4:2053–2060.

41. Munro, S., and H. Pelham. 1985. "What Turns on Heat Shock Genes?" *Nature* 317:447–478.

42. Woodland, H. R., J. M. Flynn, and A. J. Wyllie. 1979. "Utilization of Stored mRNA in *Xenopus* Embryos and Its Replacement by Newly Synthesized Transcripts: Histone H1 Synthesis Using Interspecies Hybrids." *Cell* 18:165–171.

43. Lee, G., R. Hynes, and M. Kirschner. 1984. "Temporal and Spatial Regulation of Fibronectin in Early *Xenopus* Development." *Cell* 36:729–740.

44. Loomis, W. F., ed. *The Development of Dictyostelium discoideum*. New York: Academic Press. 1982.

45. Chisholm, R. L., D. Fontana, A. Theibert, H. F. Lodish, and P. Devreotes. 1984. "Development of *Dictyostelium discoideum*: Chemotaxis, Cell-Cell Adhesion, and Gene Expression." In *Microbial Development*, ed. R. Losick and L. Shapiro. Cold Spring Harbor, N.Y.: Cold Spring Harbor Laboratory, pp. 219–254.

46. MacWilliams, H. K., and C. N. David. 1984. "Pattern Formation in *Dictyostelium*." In *Microbial Development*, ed. R. Losick and L. Shapiro. Cold Spring Harbor, N.Y.: Cold Spring Harbor Laboratory, pp. 255–274.

47. Tomchik, K. J., and P. N. Devreotes. 1981. "Adenosine 3',5'-Monophosphate Waves in *Dictyostelium discoideum*: A Demonstration by Isotope Dilution-Fluorography." *Science* 212:443–446.

48. Foe, V. E., and B. M. Alberts. 1983. "Studies of Nuclear and Cytoplasmic Behavior During the 5 Mitotic Cycles That Precede Gastrulation in *Drosophila* Embryogenesis." *J. Cell Sci.* 61:31–70.

49. Anderson, K. V., and C. Nüsslein-Volhard. 1984. "Information for the Dorsal-Ventral Pattern of the *Drosophila* Embryo Is Stored as Maternal mRNA." *Nature* 311:223–227.

50. Anderson, K. V., L. Bokla, and C. Nüsslein-Volhard. 1985. "Establishment of Dorsal-Ventral Polarity in the *Drosophila* Embryo: The Induction of Polarity by the *Toll* Gene Product." *Cell* 42:791–798.

51. Scott, M. P. 1985. "Molecules and Puzzles from the Antennapedia Homeotic Gene Complex of *Drosophila*." *Trends in Genetics* 1:74–80.

52. Ingham, P. 1985. "The Regulation of the Bithorax Complex." *Trends in Genetics* 1:112–116.

53. Nüsslein-Volhard, C., and E. Wieschaus. 1980. "Mutations Affecting Segment Number and Polarity in *Drosophila*." *Nature* 287:795–801.

54. Ingham, P., A Martinez-Arias, P. A. Lawrence, and K. Howard. 1985. "Expression of *Engrailed* in the Parasegment of *Drosophila*." *Nature* 317:634–636.

55. Garber, R. L., A. Kuroiwa, and W. J. Gehring. 1983. "Genomic and cDNA Clones of the Homeotic Locus *Antennapedia* in *Drosophila*." *EMBO J.* 2:2027–2036.

56. McGinnis, W., M. S. Levine, E. Hafen, A. Kuroiwa, and W. J. Gehring. 1984. "A Conserved DNA Sequence in Homeotic Genes of *Drosophila* Antennapedia and Bithorax Complexes." *Nature* 308:428–433.

57. Carrasco. A. E., W. McGinnis, W. J. Gehring, and E. M. DeRobertis. 1984. "Cloning of an *X. laevis* Gene Expressed During Early Embryogenesis That Codes for a Protein Domain Homologous to *Drosophila* Homeotic Genes." *Cell* 37:409–414.

58. McGinnis, W., C. P. Hart, W. J. Gehring, and F. H. Ruddle. 1984. "Molecular Cloning and Chromosome Mapping of a Mouse DNA Sequence Homologous to Homeotic Genes of *Drosophila*." *Cell* 38:675–680.

59. Levine, M., G. M. Rubin, and R. Tjian. 1984. "Human DNA Sequences Homologous to Protein Coding Region Conserved Between Homeotic Genes of *Drosophila*." *Cell* 38:667–673.

60. Harding, K., C. Wedeen, W. McGinnis, and M. Levine. 1985. "Spatially Regulated Expression of Homeotic Genes in *Drosophila*." *Science* 229:1236–1242.

61. Brower, D. L. 1985. "The Sequential Compartmentalization of *Drosophila* Segments Revisited." *Cell* 41:361–364.

62. Edelman, G. M. 1985. "Cell Adhesion and the Molecular Processes of Morphogenesis." *Ann. Rev. Biochem.* 54:135–169.

63. Gould, S. J. 1985. "Geoffroy and the Homeobox." *Natural History* 94:12–23.

64. Hogan, B., P. Holland, and P. Schofield. 1985. "How Is the Mouse Segmented?" *Trends in Genetics* 1:67–74.

65. Gronemeyer, H., and O. Pongs. 1980. "Localization of Ecdysterone on Polytene Chromosomes of *Drosophila melanogaster*." *Proc. Nat. Acad. Sci.* 77:2108–2112.

66. Gronemeyer, H. 1985. "Photoaffinity Labelling of Steroid Hormone Binding Sites." *Trends in Genetics* 1:264–267.

67. Sulston, J. E., E. Schierenberg, J. G. White, and J. N. Thomson. 1983. "The Embryonic Cell Lineage of the Nematode *Caenorhabditis elegans*." *Dev. Biol.* 100:64–119.

68. Horvitz, H. R., H. M. Ellis, and P. W. Sternberg. 1982. "Programmed Cell Death in Nematode Development." *Neurosci. Commentaries* 1:56–65.

69. Sternberg, P. W., and H. R. Horvitz. 1984. "Genetic Control of Cell Lineage During Nematode Development." *Ann. Rev. Genetics* 18:489–524.

70. Gurdon, J. B., T. J. Mohun, S. Fairman, and S. Brennan. 1985. "All Components Required for the Eventual Activation of Muscle-Specific Actin Genes Are Localized in the Subequatorial Region of an Uncleaved Amphibian Egg." *Proc. Nat. Acad. Sci.* 82:139–143.

71. Rebagliati, M. R., D. L. Weeks, R. P. Harvey, and D. A. Melton. 1985. "Identification and Cloning of Localized Maternal RNAs from *Xenopus* Eggs." *Cell* 42:769–777.

72. Laufer, J. S., P. Bassicalupo, and W. B. Wood. 1980. "Segregation of Developmental Potential in Early Embryos of *Caenorhabditis elegans*." *Cell* 19:569–577.

73. Strome, S., and W. B. Wood. 1983. "Generation of Asymmetry and Segregation of Germ-Line Granules in Early *C. elegans* Embryos." *Cell* 35:15–25.

74a. Gutterson, N. I., and D. E. Koshland, Jr. 1983. "Replacement and Amplification of Bacterial Genes with Sequences Altered *In Vitro*." Proc. Nat. Acad. Sci. 80:4894–4898.

74b. Smithies, O., R. G. Gregg, S. S. Boggs, M. A. Koralewski, and R. S. Kucherlapati. 1985. "Insertion of DNA Sequences into the Human Chromosomal β-Globin Locus by Homologous Recombination." *Nature* 317:230–234.

75. Laski, F. A., D. C. Rio, and G. M. Rubin. 1986. "Tissue Specificity of *Drosophila* P Element Transposition Is Regulated at the Level of mRNA Splicing." *Cell* 44:7–19.

76. Spradling, A. C., and G. M. Rubin. 1983. "The Effect of Chromosomal Position on the Expression of the *Drosophila* Xanthine Dehydrogenase Gene." *Cell* 34:47–57.

77. Goldberg, D. A., J. W. Posakony, and T. Maniatis. 1983. "Correct Developmental Expression of a Cloned Alcohol Dehydrogenase Gene Transduced into the *Drosophila* Germ Line." *Cell* 34:59–73.

78. Scholnick, S. B., B. A. Morgan, and J. Hirsh. 1983. "The Cloned Dopa Decarboxylase Gene Is Developmentally Regulated When Reintegrated into the *Drosophila* Genome." *Cell* 34:37–45.

79. Bargiello, R. A., F. R. Jackson, and M. W. Young. 1984. "Restoration of Circadian Behavioral Rhythms by Gene Transfer in *Drosophila*." *Nature* 312:752–754.

80. Zehring, W. A., D. A. Wheeler, P. Reddy, R. J. Konopka, C. P. Kyriacou, M. Rosbash, and J. C. Hall. 1984. "P Element Transformation with *Period* Locus DNA Restores Rhythmicity to Mutant, Arrhythmic *Drosophila melanogaster*." *Cell* 39:369–376.

81. Shin, H.-S., T. A. Bargiello, B. T. Clark, F. R. Jackson, and M. W. Young. 1985. "An Unusual Coding Sequence from a *Drosophila* Clock Gene Is Conserved in Vertebrates." *Nature* 317:445–448.

82. Palmiter, R. D., and R. L. Brinster. 1985. "Transgenic Mice." *Cell* 41:343–345.

83. Palmiter, R. D., R. L. Brinster, R. E. Hammer, M. E. Trumbauer, M. G. Rosenfeld, N. C. Birnberg, R. M. Evans. 1982. "Dramatic Growth of Mice That Develop from Eggs Microinjected with Metallothionein–Growth Hormone Fusion Genes." *Nature* 300:611–615.

84. Alt, F. W., T. K. Blackwell, and G. D. Yancopoulos. 1985. "Immunoglobulin Genes in Transgenic Mice." *Trends in Genetics* 1:231–236.

85. Hammer, R. E., V. G. Pursel, C. E. Rexroad, Jr., R. J. Wall, D. J. Bolt, K. M. Ebert, R. D. Palmiter, and R. L. Brinster. 1985. "Production of Transgenic Rabbits, Sheep and Pigs by Microinjection." *Nature* 315:680–683.

86. Schell, J., and M. Van Montagu. 1983. "The Ti Plasmids as Natural and as Practical Gene Vectors for Plants." *Bio/Technology* 1:175–180.

87. Koukolikova-Nicola, Z., R. D. Shillito, B. Hohn, K. Wang, M. Van Montagu, and P. Zembryski. 1985. "Involvement of Circular Intermediates in the Transfer of T-DNA from *Agrobacterium tumefaciens* to Plant Cells." *Nature* 313:191–196.

88. Peralta, E. G., and L. W. Ream. 1985. "T-DNA Border Sequences Required for Crown Gall Tumorigenesis." *Proc. Nat. Acad. Sci.* 82:5112–5116.

89. Barton, K. A., A. N. Binns, A. J. M. Matzke, and M.-D. Chilton. 1983. "Regeneration of Intact Tobacco Plants Containing Full Length Copies of Genetically Engineered T-DNA, and Transmission of T-DNA to R1 Progeny." *Cell* 32:1033–1043.

90. Herrera-Estrella, L., G. Van den Broeck, R. Maenhaut, M. Van Montagu, J. Schell, M. Timko, and A. Cashmore. 1984. "Light-Inducible and Chloroplast-Associated Expression of a Chimaeric Gene Introduced into *Nicotiana tabacum* Using a Ti Plasmid Vector." *Nature* 310:115–120.

91. Lamppa, G., F. Nagy, and N.-H. Chua. 1985. "Light-Regulated and Organ-Specific Expression of a Wheat Cab Gene in Transgenic Tobacco." *Nature* 316:750–752.

92. Morelli, G., F. Nagy, R. T. Fraley, S. G. Rogers, and N.-H. Chua. 1985. "A Short Conserved Sequence Is Involved in the Light-Inducibility of a Gene Encoding Ribulose 1,5-Bisphosphate Carboxylase Small Subunit of Pea." *Nature* 315:200–204.

93. Kimpel, J. A., and J. L. Key. 1985. "Heat Shock in Plants." *Trends in Biochem. Sci.* 10:353–357.

94. Horsch, R. B., J. E. Fry, N. L. Hoffmann, D. Eichholtz, S. G. Rogers, and R. T. Fraley. 1985. "A Simple and General Method for Transferring Genes into Plants." *Science* 227:1229–1231.

95. Jones, M. G. K. 1985. "Transformation of Cereal Crops by Direct Gene Transfer." *Nature* 317:579–580.

96. De Block, M., J. Schell, and M. Van Montagu. 1985. "Chloroplast Transformation by *Agrobacterium tumefaciens*." *EMBO J.* 4:1367–1372.

97. Meyerowitz, E., and R. E. Pruitt. 1985. "*Arabidopsis thaliana* and Plant Molecular Genetics." *Science* 229:1214–1218.

98. Ausubel, F. M. 1984. "Developmental Genetics of the *Rhizobium*/Legume Symbiosis." In *Microbial Development*, ed. R. Losick and L. Shapiro. Cold Spring Harbor, N.Y.: Cold Spring Harbor Laboratory, pp. 275–298.

99. Mulligan, J. T., and S. R. Long. 1985. "Induction of *Rhizobium meliloti nodC* Expression by Plant Exudate Requires *nodD*." *Proc. Nat. Acad. Sci.* 82:6609–6613.

CHAPTER 23

The Generation of Immunological Specificity

How vertebrates are able to generate specific immunological responses has intrigued biologists throughout this century. At first, their interest had purely practical considerations: The better we understand immune responses, the easier it might be to make them more effective against dangerous disease agents. In recent years, however, much attention has been focused upon immunology because of the realization that the induction of specific immune responses involves specific irreversible cell differentiation. Thus, understanding the immune response at the molecular level would be a profound breakthrough, with implications not only for medicine, but also for all aspects of the biology of multicellular organisms.

By the 1950s, it was clear that at the heart of specific immune responses is the body's ability to synthesize millions of structurally related but slightly different proteins, called antibodies. For the next three decades, the central issue for molecular immunology was to understand the genetic basis for this ability. The first insights came in the 1960s from determining the amino acid sequences of many different antibodies. Then, about ten years ago, the application of molecular cloning technology exploded the field, revealing quickly and precisely the nature of the genes that encode the vast array of different antibody molecules an organism can produce. Most revolutionary was the finding that antibody genes, in contrast to other genes we know of, are not inherited intact, but are assembled in somatic cells by DNA recombination. Thus, in the immune system, differentiation and specific gene expression are accomplished by irreversible changes in the DNA.

AN INTRODUCTION TO THE VERTEBRATE IMMUNE RESPONSE

The Immune Response Is Highly Specific, Distinguishes Self from Nonself, and Has Memory [1–5]

The immune response is a highly sophisticated defense system of vertebrates that evolved to combat invading pathogenic microorganisms and certain forms of cancer. Without it we would quickly die

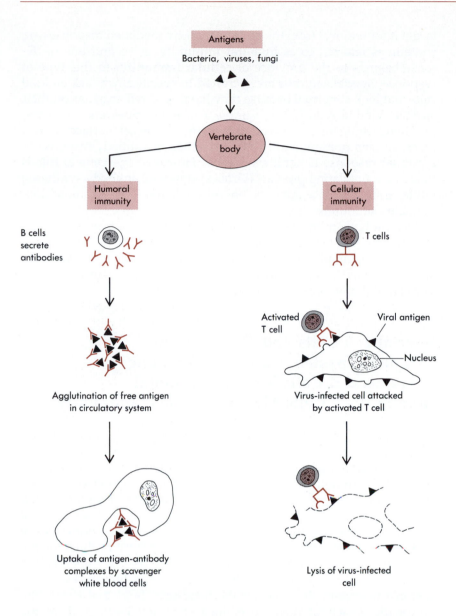

Figure 23-1
Diagrammatic view of an immune response to an antigen. Two lines of defense are mobilized. Humoral immunity culminates in the synthesis, by B cells, of circulating antibody molecules; cellular immunity culminates in activated T cells that attack the antigen, usually when it is present on the surface of the body's own cells.

from bacterial, viral, or fungal infections. Foreign substances, such as viruses, that enter the body's circulation and elicit an immune response are called **antigens.**

The ability to develop immunity to invading microorganisms was recognized long before the nature of the immune system was known and even before microorganisms themselves were understood. A person who recovered from a viral infection such as smallpox was known to become immune to the disease for life, although remaining susceptible to other virally induced diseases. This result tells us two things about the immune response. First, it is highly specific, since the smallpox virus is recognized as distinct from any other virus. Second, the immune response has memory, since it will provide specific immunity to an antigen (smallpox virus in our example) for the rest of an individual's life.

The immune system works through two types of defenses, either or both of which can be mounted against an invading antigen, such as a microorganism (Figure 23-1). First, it can produce circulating proteins, called **antibodies,** that bind to the antigen. When antibodies bind to a microorganism, they can either cause it to be inactivated or target it for removal from the body. Antibody-mediated responses are sometimes referred to as **humoral immunity.** A second line of defense belongs to the province of **cellular immunity.** In this type of response, specialized cells are recruited to specifically attack and kill cells that look abnormal because they have acquired antigens on their surface. Viral infections, for example, cause the appearance of novel, and hence antigenic, virus coded proteins on the cell surface. Why a given antigen predominantly leads to either an antibody response or a cellular response is not always clear. However, the general rule is that freely circulating proteins (viruses) stimulate antibody synthesis, while antigenic determinants bound to living cells promote cell-mediated immunity.

Because both humoral and cellular immune responses are so destructive to their targets, the immune system must be extraordinarily careful to distinguish foreign substances from the body's own cells and proteins. The ability to recognize "self" and *not* respond to it immunologically is called **tolerance.**

Specialized Cells and Organs of the Lymphoid System Carry Out Immune Responses: B Cells Mediate Humoral Immunity, T Cells Mediate Cellular Immunity[6, 7]

The need to constantly monitor the entire body for invading substances and to cope with the invaders at many different sites means that the cells and tissues of the immune system must be widely distributed throughout the body. Specialized white blood cells called **lymphocytes** carry out specific immune responses. They are small, round cells that circulate between blood vessels and the lymph system. The lymph system is a network of lymph nodes connected by lymphatic vessels (Figure 23-2). The vessels resemble those of the circulatory system except that they terminate as tiny, open-ended capillaries in the tissues. A colorless fluid, the lymph, circulates within the system. It is fed by the interstitial fluid that enters the lymphatic vessels at their extremities after first bathing all of the body's tissues and sweeping up foreign substances for removal to the lymph nodes. The lymph nodes are highly organized cavities where antigens are collected and where cells of the immune system congregate to interact with the antigens. Sooner or later, most people experience the swollen lymph nodes that signal an energetic immune response occurring during viral or bacterial infections.

There are two types of lymphocytes, B cells and T cells. The **B cells** synthesize antibodies, while the **T cells** carry out cellular immune responses. T cells interact with antigens through special membrane-anchored cell surface proteins called **T cell receptors,** which are analogous to antibodies (see Figure 23-1). B cells and T cells arise from common stem cell precursors located in the bone marrow, which is referred to as a primary lymphoid organ. In mammals, B cells mature within the bone marrow, while T cell precursors migrate out and mature in a specialized organ, the thymus. Once they have matured to a stage where they can cope with invading antigens, B and T cells leave the bone marrow and thymus, respectively, and travel to the secondary lymphoid organs. These include the peripheral lymph nodes, the tonsils, the spleen, and specialized tissue in the intestine

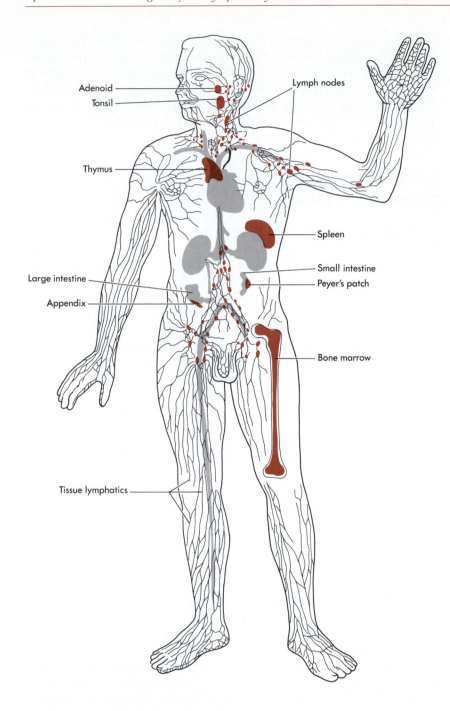

Adenoid

Tonsil

Thymus

Large intestine

Appendix

Tissue lymphatics

Lymph nodes

Spleen

Small intestine

Peyer's patch

Bone marrow

Figure 23-2
Diagram of the human lymphoid system. The system consists of the lymphoid organs shown here and of millions of lymphocytes that circulate between the lymphoid system and the bloodstream. Lymphocytes originate in the bone marrow, which is present in the long bones (one of which is indicated in the picture). After maturing in the bone marrow (B cells) or thymus (T cells), the lymphocytes migrate out and congregate in the other (secondary) lymphoid organs. From "The Immune System" by N. K. Jerne. 1973, *Sci. Amer.* All rights reserved.

called Peyer's patches (see Figure 23-2). In these sites, lymphocytes may be activated by exposure to the antigens that have been gathered up by the lymph and concentrated in the secondary lymphoid organs. From here the lymphocytes, some of them now primed for attack, begin their circulation to the blood vessels and back again to the lymph system, seeking out and responding to antigens as they go.

In addition to lymphocytes, there are other types of accessory cells involved in immune responses. These include macrophages, monocytes, and Langerhans cells. Many accessory cells are located in the lymph nodes, but others are well distributed throughout the body, including in the skin. Some of these cells perform the critically important function of accumulating antigens. Macrophages, for example,

Figure 23-3
Diagrammatic view of an antigen. The symbol R represents a single determinant of immunological specificity, the actual group that combines with an antibody molecule.

are scavengers that gobble up foreign substances and cell debris. They do not simply destroy the things they ingest. Rather, they reposition fragments of the ingested material at their surface. As we shall learn later, many lymphocytes can only recognize antigens and so respond to them when the antigens are displayed in this way. Accessory cells that gather up antigens and help lymphocytes to interact with them are called antigen-presenting cells.

Certain Properties Make Substances Antigenic[3, 8, 9, 10]

While the primary function of the immune system may be to save us from disease-causing microorganisms, the system itself cannot distinguish potentially pathogenic from nonpathogenic invaders. Rather, it is because they are foreign, or nonself, that invading microorganisms are antigenic. An object is potentially antigenic when it possesses an arrangement of atoms at its surface that differs from the surface configuration of any normal host component.

Being foreign is not the only property that makes an object antigenic, however. Another major requirement is that an antigen must either be a macromolecule or be built up from macromolecules (e.g., a virus particle). Most proteins and some polysaccharides and nucleic acids are antigens. Small molecules by themselves can seldom induce specific circulating antibodies. This is not due to a lack of specificity, since many small molecules, nonantigenic by themselves, change the antigenic properties of a larger molecule (e.g., a protein) when covalently coupled to it. The resulting antibodies then become partially directed against the small molecule. Small molecules that become antigenic when coupled to protein carriers are called **haptens.**

It is unlikely that the entire surface of a large molecule is necessary for its antigenicity. The immunological system probably responds to specific groups of atoms, called **antigenic determinants** or **epitopes,** located at a number of sites about a molecular surface (Figure 23-3). In proteins, an epitope is composed of five to eight amino acids. So a given protein molecule usually possesses many antigenic determinants and induces the formation of many types of antibodies. Of course, objects the size of bacteria possess a very large number of different antigenic determinants.

Although we are uncertain as to exactly how many unique antigenic determinants exist, the number is certainly large—larger than 10,000 and probably on the order of millions. The number 10,000 is based on experiments that test whether antibodies induced by a given protein ever accidentally combine with a completely unrelated protein. If there were only a limited number of antigenic determinants, we would expect the same determinant to be found on many different proteins, leading to unexpected antigenic homologies between unrelated proteins. But, in fact, such cross-reactions almost never occur.

Antibody Specificity (T Cell Receptor Specificity) Resides in Amino Acid Sequence[11]

The most striking property of the vertebrate immune system is its ability to specifically recognize thousands, perhaps even millions, of foreign antigens. This recognition is performed by the specialized proteins—antibody molecules and T cell receptors—that specifically bind to antigenic substances in much the same way that enzyme molecules bind specifically to their substrates. For a long time, various theories vied to explain how an organism could possibly be born with the ability to produce proteins capable of specifically recognizing the many different antigenic determinants that may be encountered in its lifetime. According to one long-dominant theory, only a small number of different antibody molecules were required, since the antigen itself would determine which antibody should be formed by combining with a newly synthesized antibody chain before it had folded into the final three-dimensional form. The interaction would allow the antibody molecule to fold around the antigen, automatically creating a complementary shape between the two molecules. Because, according to this scheme, the antigen directly determines the shape of the antibody, this model was called the *instructive theory* of antibody formation.

However, the instructive theory was ruled out many years ago. Convincing evidence against it came first from experiments in which the three-dimensional structure of an antibody with known specificity was temporarily destroyed (by denaturation) and then allowed to re-form in the absence of antigen: The antibody molecules resumed their original specificity! Now we know that the vast diversity and remarkable specificity of immune responses are due to the fact that vertebrates synthesize a vast number of different antibody molecules, each with a different amino acid sequence. Among this vast repertoire of molecules there are invariably some that can interact with any of the vast array of antigens the organism will meet with in its lifetime.

Far less is known about the structure of T cell receptor proteins than about antibody molecules, since the former were only detected biochemically a few years ago, whereas antibody molecules became available for study several decades ago. However, there is evidence that different T cell receptors have different amino acid sequences, and no doubt their antigen binding specificity is also determined by their amino acid sequence.

Clonal Selection Explains Specific Immune Responses and Immunological Memory[12, 13]

Fundamental to the way immune responses work is the fact that each B lymphocyte produces antibody molecules of a single specificity, and each T cell carries a single type of T cell receptor molecule. Rather than having each lymphocyte produce antibodies or T cell receptors that can recognize many different antigens, diversity is achieved by having a very large number of B and T cells, each synthesizing a unique antigen recognition molecule. Before B cells differentiate to mature antibody-secreting cells, called **plasma cells,** they carry their antibodies as membrane-bound surface molecules. In this form, the antibody serves as a receptor for antigen, much like the receptors on T cells. When an antigen collides with an appropriate antibody receptor on an immature B cell, the specific binding, along with other cellular interactions, triggers the proliferation and maturation of that particular B cell. The end result is to generate plasma cells that secrete antibodies of the same specificity as the original membrane-bound

Figure 23-4
The theory of clonal selection as it applies to B cells and T cells. Among the millions of B and T lymphocytes present in the vertebrate body, there are always some that can interact with any antigen that enters the body. These particular cells, whose antibody receptors or T cell receptors can bind the incoming antigen, are triggered by the binding to undergo proliferation and maturation. The result is expanded clones of activated T cells that can attack antigen-bearing cells, and expanded clones of B cells that secrete antibodies.

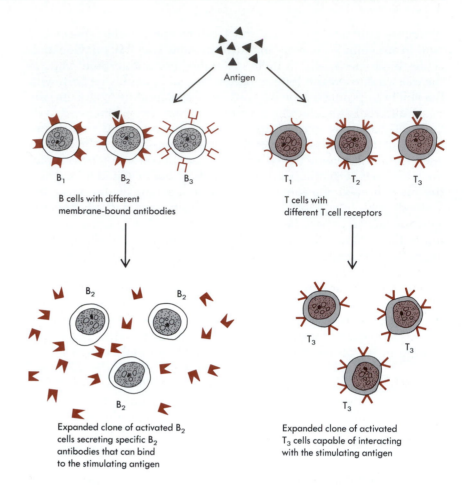

B cells with different
membrane-bound antibodies

T cells with
different T cell receptors

Expanded clone of activated B_2
cells secreting specific B_2
antibodies that can bind
to the stimulating antigen

Expanded clone of activated
T_3 cells capable of interacting
with the stimulating antigen

antibody carried by the immature B cell (Figure 23-4). The idea that antigens stimulate division and maturation of the B cells whose antibodies they can bind to was first predicted in the 1950s and is known as the **theory of clonal selection.** (A clone is a group of cells all descended from a common ancestor.) No longer regarded as just a theory, clonal selection is one of the cornerstones of immunology.

Antigen binding to a T cell receptor also triggers cellular proliferation (clonal selection) and maturation of the immature cell to cells capable of carrying out T cell functions, such as the destruction of virally infected cells. Thus, the theory of clonal selection applies equally well to both T cells and B cells. Mature, functional T cells and plasma cells are called **effector cells.** The maturation of a B cell from a resting small lymphocyte to an activated plasma cell involves a considerable increase in size, and the endoplasmic reticulum is amplified enormously to cope with the massive synthesis of antibody molecules (Figure 23-5). Activated T cells are also enlarged relative to resting T cells, but they lack the extensive endoplasmic reticulum found in plasma cells.

Clonal selection can explain immunological memory. When an animal first encounters an antigen, it produces a relatively slow **primary immune response,** because only a small number of its cells have membrane-bound antibody molecules or T cell receptors that can respond to the antigen. However, the next encounter with the same antigen leads to a far more rapid and intense **secondary response,** because cells that can bind that particular antigen were stimulated to divide by the previous encounter (Figure 23-6). The proliferation and

Figure 23-5
Successive stages in the development of mature plasma cells. Immature B cells are small, round cells that carry their antibodies as membrane-anchored receptors. Following antigen stimulation, at least eight cell generations and five days of growth are required before the appearance of cells producing a great deal of antibody. The most noticeable feature of the mature cell is the extensive endoplasmic reticulum, whose internal cavity is filled with antibody molecules. The electron micrographs show the cell as it would appear at the beginning and end of the process. Note that the two cells are shown at the same magnification. [Diagram after G. J. V. Nossal. 1964, *Sci. Amer.* All rights reserved. Electron micrographs courtesy of Dr. Dorthea Zucker-Franklin, NYU Medical Center.]

maturation of particular B or T lymphocytes induced by the first exposure to antigen lead not only to mature antibody-producing plasma cells or functional T cells, but also to expanded clones of long-lived, not yet mature B and T cells called **memory cells.** Memory cells can respond to the specific antigen the next time it is encountered, again differentiating to produce functional B and T effector cells (Figure 23-7). The human body has about 10^{12} lymphocytes. These cells are thought to arise from about 10^6 to 10^8 lymphocytes with different receptor specificities. Among the 10^{12} cells, lymphocytes carrying any particular receptor may be represented only once, or a particular cell may have been stimulated to divide by an encounter with an antigen and so have become a clone of as many as 10^6 cells.

The Immune System Actively Acquires and Maintains Tolerance to Self Antigens[14, 15, 16]

Just as remarkable as the immune system's ability to respond specifically to thousands of different antigens is its failure to respond to the host's own cells and proteins, the phenomenon known as tolerance. Tolerance is not due to any intrinsic lack of antigenicity in our own cells and proteins, since if they are transplanted to another

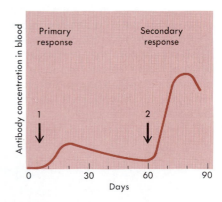

Figure 23-6
A graph showing the primary and secondary responses of antibody production in a rabbit. Arrows 1 and 2 represent the times of injection with the antigen. The secondary response is both more rapid and more intense than the primary response.

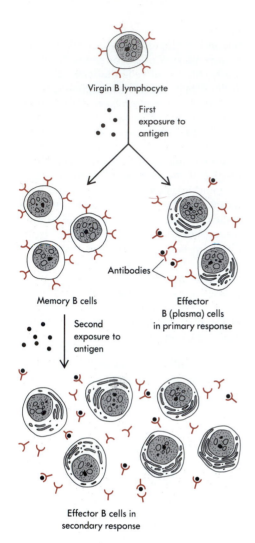

Figure 23-7
The cellular basis for the greater speed and intensity of a secondary antibody response. The first encounter of a virgin B lymphocyte with its cognate antigen causes the cell to proliferate and to differentiate into two types of cells: effector B plasma cells that synthesize antibodies for the primary response, and memory B cells that can differentiate into effector cells the next time the same antigen is encountered.

Labels within figure:
Virgin B lymphocyte
First exposure to antigen
Antibodies
Memory B cells
Effector B (plasma) cells in primary response
Second exposure to antigen
Effector B cells in secondary response

individual, that individual's immune system soon mounts a response to them and rejects them. Tolerance is not inherited, but rather is acquired and then actively maintained by the immune system throughout an individual's life. This maintenance requires the continued presence of the substances to which the individual is tolerant.

Our understanding of tolerance first came from experiments with cattle performed in the 1940s. Whereas tissue transplants between most individuals were generally unsuccessful owing to the immune response, transplants between dizygotic cattle twins succeeded. Dizygotic twins are genetically related as siblings, but unlike fraternal twins, they share a common blood supply and hence exchange cells in utero. This sharing apparently results in their being tolerant to one another's cells, proving that tolerance is acquired during embryonic development. Later experiments showed that almost any antigen can be made to induce tolerance experimentally. Whether a foreign substance acts as an antigen and induces an immune response or acts as a **toleragen** and induces tolerance depends on the amount of the substance and the route by which the substance enters an animal's body. Also, it is easier to induce tolerance in younger animals, rather than in adults.

One can imagine two plausible mechanisms by which tolerance may be achieved. Either those specific clones of T and B cells that can interact with self antigens are destroyed before they mature, or they mature but are then prevented from responding to self. Apparently, both mechanisms are used to achieve tolerance, but *how* these mechanisms work is still a mystery.

THE B CELL RESPONSE

Typical Antibody Molecules Are Tetramers Composed of Two Identical Light Chains and Two Identical Heavy Chains[17, 18, 19]

Typical antibody molecules are proteins of about 150,000 daltons that consist of four polypeptide chains. There are two identical **light chains,** containing about 220 amino acids each, and two identical **heavy chains,** containing about 440 amino acids each. Each light chain is linked to a heavy chain by one covalent S–S (disulfide) bond, and S–S bonds also run between the two heavy chains to give the schematic picture shown in Figure 23-8. Each chain also has regularly spaced intrachain disulfide bridges. Carbohydrate moieties are attached to the heavy chains, making antibody molecules glycoproteins.

The four polypeptide chains are arranged so that each antibody molecule contains two identical, widely separated antigen binding sites, each of which can combine with an antigen, using secondary bonds to hold their complementary surfaces together. Visualization of antibodies in the electron microscope reveals Y-shaped or T-shaped molecules with one such active site at the end of each arm. Each arm is effectively hinged to the central axis with the distance between active sites sometimes increasing when they are bound to antigen.

Figure 23-8
A model of the structure of a typical human antibody molecule, composed of two light and two heavy polypeptide chains. Interchain and intrachain disulfide bonds are indicated, and CHO marks the approximate position in the heavy chain of the carbohydrate moiety. Two identical active sites for antigen binding are located in the arms of the molecule.

The existence of two identical binding sites permits a single antibody molecule to link together two similar antigens. This feature is of great advantage in allowing antibodies to defeat an infection by a microorganism, since all microorganisms contain a large number of identical antigenic determinants. In the presence of specific antibodies, a microorganism thus becomes linked to a large number of similar microorganisms through antibody bridges (Figure 23-9), and these aggregates then tend to be taken up and destroyed by macrophages.

Some of the earliest information about antibody structure and its relationship to function came from breaking the antibody molecule apart with the proteolytic enzymes papain and pepsin. As shown in

Figure 23-9
A view of how antigens and antibodies combine to form large aggregates.

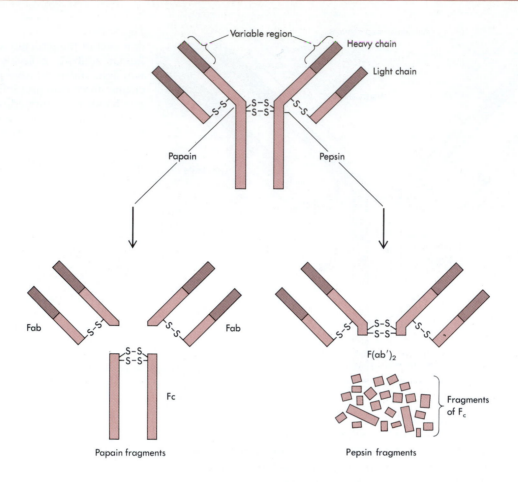

Figure 23-10
Creation of Fab, Fc, and F(ab')₂ fragments by specific proteolytic digestion of intact antibody molecules.

Figure 23-10, papain produces two identical antigen binding fragments, called **Fab fragments,** each with a single antigen binding site, and an **Fc fragment,** whose name reflects its ability to crystallize readily. Pepsin treatment yields an F(ab')₂ fragment that has two antigen-combining sites and is still capable of crosslinking antigen.

The stem portion of the Y-shaped antibody molecule, the Fc region, is designed to carry out the so-called effector functions of the molecule. **Effector functions** are the various activities that antibodies must perform in addition to antigen binding. For example, antigen-antibody complexes must interact with phagocytic cells that will ingest the complexes, and this interaction occurs between the Fc regions and the ingesting cells. Another effector function of these regions is interaction with a set of blood proteins collectively called the **complement system.** When complement proteins bind to antigen-antibody complexes, they can bring about destruction (lysis) of the antigen (Figure 23-11).

Myelomas and Hybridomas Provide Pure Antibodies for Amino Acid Sequencing[20, 21, 22]

Major insights into the molecular basis of immunological diversity and specificity first came from determining the amino acid sequences of the light and heavy polypeptide chains of many different antibody molecules. These studies require preparations of pure antibodies. Although antibodies account for about 20 percent of the protein in blood

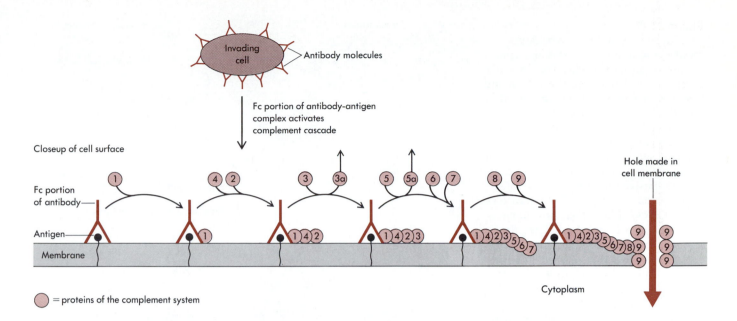

Figure 23-11
Antibody bound to antigen can activate the complement cascade and bring about lysis of the invading cell (micro-organism). Complement is a system of about 20 proteins that are present in the blood. In the presence of antibody molecules that are bound to antigen, particularly those on the surface of a cell, the first component of the cascade, C1, becomes activated. In this form, it both binds to the antigen-antibody complex on the cell surface and also activates the next component of the complement pathway. The cascade continues until a complex called C9 is generated. This complex can make a hole in a cell membrane. The hole allows the influx of ions and water into the cell, causing it to swell and then burst.

plasma, at any given time the antibodies present in an individual are a collection of many, many different species directed against a large variety of different antigens. Even after continued immunization with a single strong antigen, no single antibody species is present in pure enough form to allow its isolation. Traces remain of the many different antibodies previously induced in that individual's history. Even more important, a given single antigen always seems to promote synthesis of several different antibodies, each with a different amino acid sequence, yet able to specifically bind the same antigen.

Single antibody species can, however, be obtained in two ways. They were first obtained from the blood of certain humans (and mice) afflicted with the bone marrow disease multiple myeloma. (In mice, the disease is called myeloma or plasmacytoma.) A form of cancer, this disease involves the uncontrolled multiplication of antibody-producing cells. Most important, like all cancers, it arises from a single cell; thus, a given specific tumor produces only one antibody type characterized by a specific amino acid sequence. Because the tumor cells proliferate wildly and continue producing their unique antibody, a myeloma patient becomes a source of a highly pure, single type of antibody. As expected, the antibodies produced by different myeloma tumors each have different amino acid sequences, permitting the comparative study of many different antibody sequences. In the case of multiple myelomas, it is usually impossible to know what antigen could potentially bind to the antibody produced by any particular tumor. Such an antibody is usually called an **immunoglobulin** (**Ig**), a term that refers to any antibody-like molecule regardless of its antigen binding specificity.

The amount of a given myeloma protein made in a myeloma patient is very large, easily permitting protein sequence studies. Moreover, many myeloma patients excrete large amounts of specific proteins, called **Bence-Jones proteins** after their discoverer. A given Bence-Jones protein is identical to the light chain of the corresponding myeloma immunoglobulin. Its excretion into the urine is the result of overproduction of immunoglobulin light chains relative to heavy chains by the myeloma cells. The presence of very large amounts of Bence-Jones proteins makes their isolation extremely

Figure 23-12
Procedure for making a hybridoma. A mouse is immunized to stimulate proliferation of desired B cells. The animal's B cells are isolated from the spleen and fused to myeloma cells using polyethylene glycol (PEG) to promote fusion. The myeloma cell line used in the procedure is unusual in two ways: It has stopped synthesizing antibodies, and it is a mutant, called HGPRT$^-$, that cannot synthesize the enzyme hypoxanthine-guanine phosphoribosyl transferase (HGPRT). When these myeloma cells are fused to normal B cells, the resulting hybrid cells, but neither parental cell alone, can grow in selective hypoxanthine aminopterin thymidine (HAT) medium. HAT medium contains a drug, aminopterin, that blocks one pathway for nucleotide synthesis, making the cell dependent on a salvage pathway that uses the HGPRT enzyme. Since the myeloma cells lack this enzyme, all cells that fail to fuse to a normal B cell die in HAT medium. Normal B cells that fail to fuse to a myeloma cell die because they lack the tumorigenic property of immortal growth. Surviving hybridomas, which inherit the HGPRT gene from the normal B cell and the cancerous property of immortal growth from the myeloma cell, are screened to find those that inherited the antibody genes of the normal B cell and are secreting monoclonal antibodies that can bind to the immunizing antigen.

simple, and the first important results with antibody amino acid sequences arose from the study of these proteins.

Our general inability to know which antigen a myeloma protein recognizes is a major drawback for many types of studies. In contrast, a recent and enormously powerful technique has made it possible to produce pure antibodies, called **monoclonal antibodies,** with specificity to bind to any desired antigen. In this method, a mouse is repeatedly immunized with an antigen of choice. As a result, B cells making antibodies specific for that antigen proliferate. Then all the animal's B lymphocytes, now highly enriched for B cell clones that recognize the immunizing antigen, are removed and fused to a mutant myeloma cell whose own antibody synthesis has shut off. When cells fuse, their nuclei can also fuse to yield a hybrid cell that contains chromosomes from both parents. Hybrids between the myeloma and normal B cells are selected and screened for the synthesis of antibodies to the immunizing antigen (Figure 23-12). Usually, about one

(a)

(b)

Figure 23-13
Location of variable and hypervariable (CDR) regions in light and heavy chains. (a) Variable (V) regions of light and heavy chains are about 110 amino acids long and are located at the amino-terminal ends of the polypeptides. (b) Diagram of V regions of light and heavy chains, showing positions of hypervariable (complementarity-determining) regions. Numbers indicate typical amino acid position, and S–S represents a disulfide bridge. CDRs are flanked by the framework portions of the V regions.

in several hundred cell hybrids produces an antibody of the desired specificity, and this cell is purified as the desired **hybridoma.** The role of the myeloma cell in this process is to provide the normal B lymphocytes with the cancerous property of uncontrolled growth, thereby allowing hybridomas to grow easily both in tissue culture and after injection into mice. Normal noncancerous B cells would quickly die in either situation.

Light and Heavy Chains Have Variable Regions for Antigen Recognition and Constant Regions to Perform Effector Functions[23, 24, 25]

The first experiments to examine the differences in amino acid sequence between closely related light chains or heavy chains isolated from different antibody molecules were done by the method of tryptic peptide mapping. These experiments showed that similar types of chains from different molecules have many peptides in common and differ in others. As soon as amino acid sequence data were obtained for two very similar light chains, it was clear that the variability was all at the amino-terminal end of the chains, while the carboxyl halves were identical. Further sequencing soon showed the same situation in heavy chains: **variable (V) regions** at the amino terminus and **constant (C) regions** in the carboxyl portion (Figure 23-13).

As more and more amino acid sequences of antibody molecules were determined, it became clear that variability is not evenly distributed through the V regions. It is concentrated in three segments, called **complementarity determining regions (CDRs)** or **hypervariable regions,** and this is true in both the light- and heavy-chain variable segments (Figures 23-13 and 23-14). It is these hypervariable regions that line the walls of the antigen binding cavities in the native

Figure 23-14

The amino acid sequences of the variable regions of nine different human heavy chains. Unbroken lines mean identity with the first protein, called Tie. Dashes in the brackets mean that the corresponding protein has no amino acid at this position. These gaps were introduced to maximize the homology. Sequence variations (shaded area) are clustered in the three complementarity-determining regions, CDR1, CDR2, and CDR3. A fourth region of considerable variation, between amino acid residues 84 and 88, is not located in the antigen binding site, and its function is unknown. [After J. D. Capra and J. M. Kehoe, *Proc. Nat. Acad. Sci.* 71 (1974):4032.]

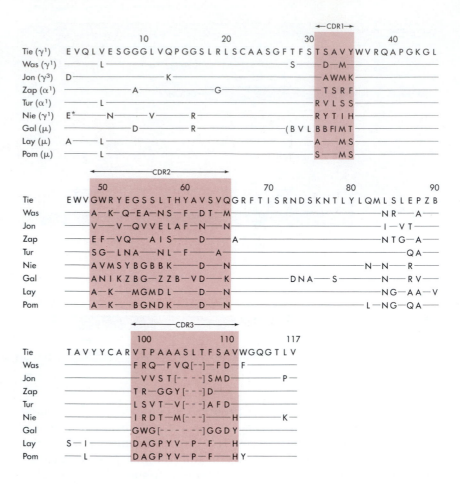

antibody molecule. The more highly conserved portions of variable regions are called the **framework.**

The first complete amino acid sequence of an antibody molecule revealed important sequence relationships between the variable regions of light and heavy chains and also between their constant regions. The sequences also revealed the presence of repeated domains within heavy-chain constant regions. Examination of the heavy-chain sequence strongly suggests that the gene encoding the constant portion of this polypeptide arose by successive duplication of a primitive antibody gene. This can be seen clearly by dividing the sequence of a heavy-chain constant region into four parts. The first, the variable portion, shows similarity to the variable part of the light chain. In turn, the three regions of the constant portion, C_H1, C_H2, and C_H3, show strong homologies both to each other and to the constant region of a light chain (Figure 23-15). Thus, the variable regions of light and heavy chains consist of one type of **homology unit,** while the constant regions of light chains and the three C_H segments consist of a second type of homology unit. Each homology unit is about 110 amino acids long, and each has one centrally located S–S bond, with the two sulfur atoms separated by some 65 or so amino acids. Even though V and C homology units have no apparent amino acid sequence homology to one another, their similar length and the comparable location of the S–S bonds suggest that they, too, arose from a common ancestor way back in evolutionary time.

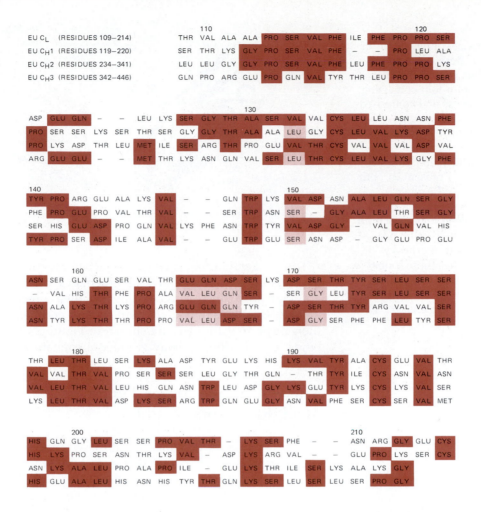

X-Ray Diffraction Reveals the Three-Dimensional Structure of Antibody Molecules[26-29]

To determine the three-dimensional arrangement of a protein, the protein must first be crystallized and then subjected to X-ray diffraction analysis. For antibody molecules, this was done first with Fab and Fc fragments (see Figure 23-10). Later, it was achieved with a complete immunoglobulin. From this analysis, it is clear that V and C homology units are folded in similar configurations, called the **immunoglobulin fold;** an important difference is that the V regions have extra polypeptide loops, composed of the hypervariable segments, for antigen binding (Figure 23-16). Each V or C homology unit contains two layers of extended β sheets that are roughly parallel to each other and that surround an interior portion in which are packed the side groups of hydrophobic amino acids. The V_L and V_H units then pack together to form the variable domains of the completed antibody molecule, while C_L and C_{H1} units lie together, as do the remaining pairs of C_H homology units (Figure 23-17). Each pair of homology units thus forms a discrete globular structural domain. This organizational tactic explains why V regions must have conserved framework portions and carefully localized hypervariable segments: If their structure were allowed to diverge too much, they would be unable to associate with another V homology unit and thereby take up the proper three-dimensional configuration within the molecule.

Figure 23-15
Sequence homology in the constant regions of the light and heavy chains of an antibody molecule, called EU, the first antibody molecule whose complete sequence was determined. Deletions indicated by dashes have been introduced to maximize the homology. Identical residues are shaded. Both dark and light shadings indicate identities that occur in pairs in the same positions. The figure emphasizes the presence of repeating domains within the heavy-chain constant region and their similarity to the light-chain constant region. [After G. M. Edelman et al., *Proc. Nat. Acad. Sci.* 63 (1969):78.]

Figure 23-16
The three-dimensional configuration of a single light chain. Both the variable and constant regions exhibit the characteristic immunoglobulin fold consisting of two β-pleated sheets, one composed of three strands and one of four strands. The strands within the two sheets are indicated in different colors. The characteristic intrachain disulfide (S–S) bonds, which are present in both the V and C regions of the chain, bridge the two β sheets in each domain. The hypervariable regions within the V region form loops that lie near similarly placed loops in the heavy-chain V region. Together these loops form the antigen binding site. [After M. Schiffer, R. L. Girling, K. R. Ely, and A. B Edmundson, *Biochemistry* 12 (1970):4620.]

The antigen binding cavities (active sites) formed by the hypervariable region loops of V_L and V_H segments vary in size and shape from antibody to antibody, depending on their precise amino acid sequence. Examples of two different cavities and their interactions with their respective antigens are shown in Figure 23-18. An antigen that binds to a specific antibody is called a **cognate antigen.**

There Are Two Major Types of Light Chains, λ and κ[19]

After the amino acid sequences of many different light chains were determined, it was apparent that in each vertebrate species, they can be assigned to one of two clearly distinct types, called **κ** (kappa) and **λ** (lambda), based on the amino acid sequences of their constant regions. κ and λ constant regions are about 30 to 40 percent homologous (Figure 23-19). There is only one type of κ light-chain constant region (C_κ) in an animal. However, there can be several types of λ

Figure 23-17
The three-dimensional structure of an antibody molecule. Each small sphere represents one amino acid. The carbohydrate chain attached to the C_H2 portion of the heavy chains is shown as large spheres. [After E. W. Silverton, M. A. Navia, and D. R. Davies, *Proc. Nat. Acad. Sci.* 74 (1977):5140.]

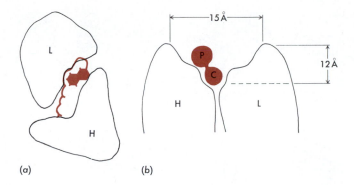

(a)   (b)

Figure 23-18
Schematic drawings of two antigen binding cavities interacting with their cognate antigens. (a) Top view of the shallow cleft (16 x 7 x 6 Å) between the heavy and light variable chains of the human myeloma protein NEW, which specifically binds the hapten vitamin K_1OH (shown in color). (b) Phosphoryl-choline (PC, in color) binding in the larger (15 x 20 x 12 Å) cavity formed by the hypervariable regions of the heavy and light variable chains of a mouse myeloma protein McPC 603. This side view reveals that most antibody-antigen interaction involves the heavy chain in this case. [(a) after F. F. Richards et al., in *The Immune System*, Third ICN-UCLA Symposium on Molecular Biology (New York: Academic Press, 1974). (b) after E. A. Padlan et al., in *The Immune System*, Third ICN-UCLA Symposium on Molecular Biology (New York: Academic Press, 1974).]

chains. For example, in the mouse, there are three types, λ_1, λ_2, and λ_3, whose constant regions are nearly identical in amino acid sequence. Any particular B cell synthesizes only one type of light chain, either κ or λ (and if λ, then only λ_1, λ_2, or λ_3). The variable regions of λ and κ chains can also usually be distinguished by characteristic amino acids in their framework portions (see Figure 23-19).

There Are Five Major Types of Heavy Chains (α, δ, ε, γ, μ) That Correspond to the Five Classes of Antibodies (IgA, IgD, IgE, IgG, and IgM)[19]

Antibody molecules can be assigned to different classes, depending on the amino acid sequence of the constant region of their heavy chains. As a result of these differences in amino acid sequence, antibody molecules belonging to some classes are composed of more than one light-heavy chain tetramer, and molecules belonging to different classes can also differ in the positions of the disulfide bridges that link their two heavy chains. The most important consequence of the differences in amino acid sequence is that distinct biological functions are conferred to different antibody classes. There are five major classes of antibodies, named **IgA, IgD, IgE, IgG**, and **IgM**, and several of these can be further divided into subclasses (e.g., IgG$_1$). The heavy-chain constant regions that correspond to the different classes of antibody molecules are called α, δ, ε, γ, and μ. Their amino acid sequences are about 40 percent homologous. The structures of the different classes of antibody molecules are shown in Figure 23-20, and their properties are summarized in Table 23-1. When more than one

Figure 23-19
Selected amino acid sequences from the V and C regions of several human λ (top) and κ (bottom) light chains. Not all chains have exactly the same number of amino acids, and the sequences are aligned for maximum homology. [After H. Eisen, *Immunology* (New York: Harper & Row, 1974).]

Variable	Hypervariable	Variable		Constant

A D Y Y C N S R D S S G K H V L F G G G T K L T V L G Q P K A A P S V T L F P P S S E E L Q A N K A
A D Y Y C S S Y V D N N N F X V F G G G T K L T V L R Q P K A A P S V T L F P P S S E E L Q A N K A
A H Y H C A A W D Y R L S A V V F G G G T Q L T V L R Q P K A A P S V T L F P P S S E E L Q A N K A
A D Y Y C Q A W D S S L N A V V F G G G T K V T V L G Q P K A A P S V T L F P P S S E E L Q A N K A
A D Y Y C Q A W D − − S M S V V F G G G T R L T V L S Q P K A A P S V T L F P P S S E E L Q A N K A
A T Y Y C Q Q Y D − − T L P R T F G Q G T K L E I K R T Y − A A P S V F I F P P S N E Q L K S G T A
A T Y Y C Q Q F D − − − N L P L T F G G G T K V D F K R T Y − A A P S V F I F P P S D E Q L K S G T A
G V Y Y C Q M R L − − E I P Y T F G Q G T K L E I R R T Y − A A P S V F I F P P S D E Q L K S G T A
G V Y Y C M Q A L − − Q T P L T F G G G T N V E I K R T Y − A A P S V F I F P P S B Z Z L K S G T A
A V Y Y C Q Q Y G − − S S P S T F G Q G T K V E L K R T Y − A A P S V F I F P P S D E Q L K S G T A

A = ala   E = glu   I = ile   N = asn   S = ser   Y = tyr
B = asx   F = phe   K = lys   P = pro   T = thr   Z = glx
C = cys   G = gly   L = leu   Q = gln   V = val
D = asp   H = his   M = met   R = arg   W = trp

Figure 23-20
The subunit structures of different classes of human immunoglobulins. Disulfide bridges are indicated by –S–S–. IgG1, IgG2, IgG3, and IgG4 are subclasses of IgG. IgA1 and IgA2 are monomeric subclasses of IgA. The J polypeptide chain joins the polymeric form of IgA and may initiate the joining of the five subunits that comprise extracellular IgM molecules. [Adapted from J. Gally in *The Antigens*, M. Sela (Ed.), Academic Press, New York, 1973, p. 209.]

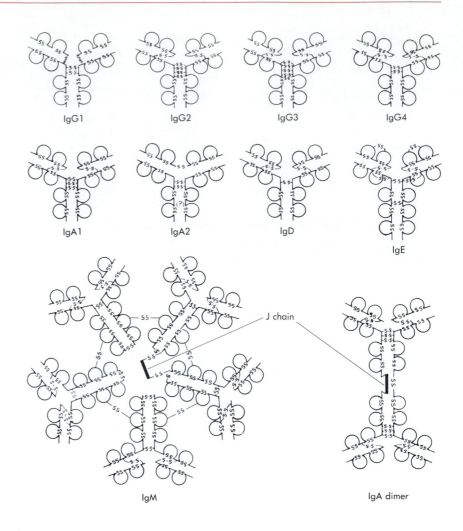

tetramer is present in a molecule (e.g., the IgM pentamer), the tetramers are held together by disulfide bridges, and some of these are joined by a special polypeptide called the **J** (**joining**) **chain.**

Different antibody classes have evolved to perform specialized functions. For example, IgM pentamers appear early during primary immune responses; then, during secondary responses, they are largely replaced by the more common IgG antibodies. IgG antibodies are the only ones that can cross the placenta to travel from mother to fetus during pregnancy. IgA molecules are most common in the secretions (tears, saliva, milk, etc.). The different behavior of the molecules is due to their heavy-chain constant regions. For example, specific receptors on placental cells recognize the constant region of IgG molecules, causing them to be ingested by endocytosis, shuttled across the cell, and released into the fetal blood on the other side of the cell. As we noted before, these diverse functions that antibodies perform are called effector functions.

Random Association of Heavy and Light Chains Can Help Explain Antibody Diversity

An antibody molecule of a particular class can have either a λ or a κ light chain. In other words, any type of light chain can combine with any type of heavy chain, a phenomenon known as **combinatorial**

Table 23-1   Properties of the Major Classes of Antibodies in Humans

Property	Antibody Class				
	IgM	*IgD*	*IgG*	*IgA*	*IgE*
Heavy chain	μ	δ	γ	α	ε
Light chain	κ or λ	κ or λ	κ or λ	κ or λ	κ or λ
Number of four-chain units	5	1	1	1 or 2	1
Structure	$(\mu_2\kappa_2)_5$ or $(\mu_2\lambda_2)_5$	$\delta_2\kappa_2$ or $\delta_2\lambda_2$	$\gamma_2\kappa_2$ or $\gamma_2\lambda_2$	$(\alpha_2\kappa_2)_{1-2}$ or $(\alpha_2\lambda_2)_{1-2}$	$\varepsilon_2\kappa_2$ or $\varepsilon_2\lambda_2$
Molecular weight (daltons)	900,000	185,000	150,000	160,000 320,000	200,000
% total serum immunoglobulin	6	1	80	13	0.002
% carbohydrate content	12	13	3	8	12
Biological function:					
Activates complement	+	–	+	–	–
Crosses placenta	–	–	+	–	–
First antibody to appear after immunization	+	–	–	–	–
High in secretions	–	–	–	+	–
Allergic reactions due to interaction with other cell types	–	–	–	–	+

+ = possesses the property. – = lacks the property.

association. This simple fact has extremely important implications. Because antigen binding cavities are formed by the variable regions of both the light and heavy chains, if any light chain can associate with any heavy chain, then the number of different antigen binding cavities that can be formed by 1000 different light chains and 1000 different heavy chains is $10^3 \times 10^3 = 10^6$. Of course, we have no way of knowing whether all light- and heavy-chain combinations can form functional antigen binding sites. Nonetheless, it seems possible that many combinations might result in molecules that can recognize some antigen or other. Thus, it is believed that the random association of light and heavy chains probably contributes substantially to antibody diversity.

The probable existence of combinatorial association provided the first major explanation of how an organism might produce a vast number of different antibody molecules while using only a relatively small number of genes.

Germ Line Versus Somatic Theories to Explain the Genetic Basis of Antibody Diversity[30, 31]

As soon as the instructive theory of antibody diversity was laid to rest and it was clear that antibody specificity and diversity are achieved by the synthesis of thousands of different antibody molecules, the central puzzle of immunology was to understand the genetic basis for this ability. The accumulation of amino acid sequence data only served to highlight the problem, since virtually every light- or heavy-chain variable region had a different sequence. Two theories vied to explain the genetic basis of antibody diversity. According to the *germ line theory*, a different gene for each antibody light or heavy chain would be inherited in the DNA. This theory was repugnant to those who realized the vast number of possible antibody molecules an animal can make and who believed that far too much of the genome would have to be devoted to the immune system. According to rival *somatic theories*, a smaller number of light and heavy chain genes would be inherited in the germ line. Then in B cells, these genes would undergo some sort of genetic alteration in their V-coding portions to generate novel variable regions and hence novel antigen binding specificities.

As we shall see in the following sections, the somatic theory turned out to be correct, although aspects of the germ line theory also help explain diversity. In all the vertebrates that have been studied so far, not a single intact antibody gene is inherited in the germ line. Rather, antibody genes are inherited as fragments that must be assembled in somatic (B) cells to generate functional genes. Because the numerous DNA fragments that encode variable regions can be assembled in many combinations, this mechanism of modular gene construction contributes greatly to V-region diversity. In addition, fully assembled antibody genes undergo further somatic mutation in their V-region-coding portions to increase diversity even more.

Antibody Structure Implies That Separate Genes Code for Variable and Constant Regions[32]

Before direct proof was possible, two considerations led some insightful immunologists to postulate that separate genes must encode the variable and constant regions of antibody polypeptide chains. The first problem was trying to imagine a mechanism that could generate so much diversity in one half of a protein while conserving the sequences of the remainder. Although elaborate schemes could be concocted, some researchers thought that the simplest hypothesis involved a few genes for the different constant regions of the heavy and light chains (e.g., one gene for C_μ, one gene for $C_{\gamma 1}$, and one gene for $C_{\lambda 2}$), many genes for the variable regions, and a mechanism that could join the two types of genes together.

A second observation that pointed to a two genes–one polypeptide model for antibodies came from the phenomenon known as **class switching,** first observed in a patient with multiple myeloma. This individual's tumor secreted antibodies with a single type of light chain but two types of heavy chains. The heavy chains had identical variable regions, but they differed in their constant regions. Later, normal B cells were found to undergo a transition from IgM production to the synthesis of other heavy-chain classes as part of their differentiation program. Once again, the same heavy-chain variable region was present before and after the class switch (Figure 23-21). At the time these observations were made, RNA splicing, which could have provided a plausible mechanism for the observations, had not yet been discovered; so the best hypothesis was that a gene coding for a particular heavy-chain variable region could associate first with one type of C_H (e.g., C_μ) and later with a different type. Remarkably, this explanation for class switching turned out to be essentially correct, although some class switches occur by changes in RNA processing rather than as a result of DNA rearrangements.

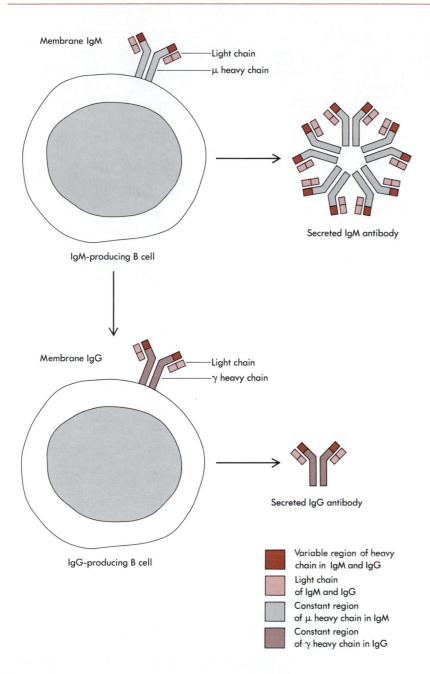

Figure 23-21
Attachment of the same V_H region to a different C_H region during an IgM to IgG class switch.

DNA Coding for Variable Regions Moves Closer to Constant-Region DNA to Create Functional Antibody Genes in B Cells[33]

In 1976, direct experimental evidence showed that antibody genes are assembled by recombination events that move DNA coding for variable regions closer to DNA encoding the constant regions. The idea behind the experiment was to compare the structure of a κ light-chain gene in two types of cells: embryonic mouse cells that do not synthesize antibodies, and a mouse myeloma that synthesizes a κ light chain. A change in the gene structure would imply that a DNA rearrangement had occurred to produce the functional gene. First, DNAs isolated from the two types of cells were digested with a restriction endonuclease and the resulting fragments separated by electrophoresis.

Figure 23-22
An experiment showing that antibody genes undergo DNA rearrangement in B cells that synthesize antibodies.

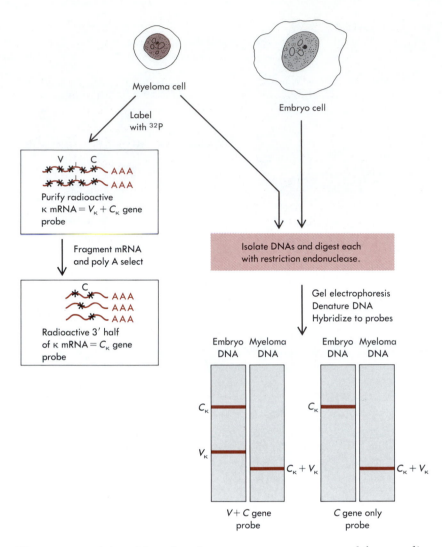

Then two nucleic acid probes for a κ gene were prepared from radioactively labeled κ mRNA isolated from the myeloma (Figure 23-22). One probe, the full-length mRNA, could hybridize to DNA encoding both the variable and constant regions, while the other, consisting of just the 3' half of the mRNA, could only hybridize to DNA coding for the constant region. When each probe was hybridized to the digested embryonic DNA, the full-length mRNA probe detected two different-size fragments (one tentatively the *C* gene, one the *V* gene), while the smaller probe detected only one of these (presumably the *C* gene). In contrast, in the myeloma DNA, the two probes hybridized to the same-size DNA fragment, indicating that in this DNA, the *V* and *C* genes were close together, in fact, close enough to lie on the same DNA fragment.

Similar experiments soon showed that DNA rearrangements are also involved in generating functional genes for λ light chains and for heavy chains. More recently, the same mechanism has been found for T cell receptor genes. So far, these are the only genes we know of in vertebrates that are assembled in somatic cells. All the rest are inherited intact, although, as we have seen, most genes of higher eucaryotes undergo editing at the RNA level to remove their intervening sequences. Assembling genes at the DNA level in specific cell types is a novel form of gene regulation, and, as we might expect, the process is closely tied in with B and T cell differentiation.

Three Unlinked Multigene Families Encode the λ and κ Light Chains and Heavy Chains[34]

If DNA coding for V regions moves near to C region genes in B cells, the question immediately arises as to the physical arrangement of the genes before they are moved. Are they near one another? Do they lie on separate chromosomes? If they lie together, what becomes of the DNA between them when they are rearranged? The first information about the arrangement of antibody genes came from genetic studies. The studies showed that genes coding for λ, κ, and heavy chains lie on separate chromosomes, but that in each case the V genes are near to their respective C genes.

The genetic studies that first mapped antibody genes were possible because antibody molecules, being proteins, can themselves be antigenic. As a result, it is possible to raise antisera to antibodies, and the antisera can then be used to follow the inheritance of particular antibody types. Antigenic determinants on variable regions are called **idiotypes.** Antigenic determinants that distinguish different types of constant regions (e.g., C_μ from C_γ, $C_{\lambda 1}$ from $C_{\lambda 2}$) are called **isotypes.** And antigenic determinants that distinguish allelic genes are called **allotypes.** By following the inheritance of idiotypic, isotypic, and allotypic determinants on antibody molecules, it was shown that the genes for λ and κ chains and heavy chains are genetically unlinked, but at each locus, V genes are linked to their respective C genes. Because there are several hundred variable-region genes and because there are vast distances between coding segments, the heavy-chain locus alone probably occupies at least 2000 kilobases of DNA. The λ-, κ-, and heavy-chain loci have been mapped to specific chromosomes in different vertebrate species (Table 23-2).

Light-Chain Genes Are Assembled by *V-J* Joining[35, 36, 37]

The discovery that antibody genes are assembled by DNA rearrangements still did not answer the question of whether antibody V-region diversity is explained by the germ line or somatic theories. This issue was resolved only when molecular cloning techniques made it possible to study the detailed structure of germ line and rearranged antibody genes. Then it was clear that the mechanism of antibody gene assembly generates a great deal of amino acid sequence diversity, while mutation of rearranged genes generates the rest: The somatic theory was the correct one.

Extensive data now exist for the structure of the antibody genes of humans and mice. As we shall see in this and following sections, the general principles for how λ-, κ-, and heavy-chain genes are arranged and how they are assembled by DNA recombination are quite similar. However, only a single DNA recombination event is needed to make a light-chain gene, while two, and sometimes three, DNA recombination events are involved in assembling functional heavy-chain genes. We can begin by describing the structure and assembly of the λ light-chain genes of the mouse, since they are the simplest.

Figure 23-23 shows the structure of a mouse λ light-chain gene as it appears in most cells of the body, as well as the rearrangement it undergoes as it becomes a functional gene in a B cell. The gene is inherited in four segments. Two of these segments are exons, and

Table 23–2 Chromosome Locations of the Heavy-Chain Genes and the λ and κ Light-Chain Genes in Mouse and Human

Polypeptide chain	Chromosome	
	Human	*Mouse*
H	14	12
κ	2	6
λ	22	16

Figure 23-23
Diagram of a λ light-chain gene as it appears in germ line DNA, and the DNA rearrangement that could occur to produce a functional gene in a B cell. Except for the gap between *V* and *J*, the gene is drawn to scale. Boxes in DNA and RNA molecules represent exons or parts of exons. White portions of exons encode 5' or 3' untranslated regions of the final mRNA. These are designated 5' UT and 3' UT, respectively. L_λ encodes amino acids -19 to -5 of the leader peptide; V_λ encodes amino acids -4 to 0 of the leader peptide and 1 to about 98 of the variable region; and J_λ encodes amino acids 99 to 110 of the variable region.

two are incomplete fragments that must be brought together in the DNA to complete a third exon. The C_λ segment is an exon that encodes the complete constant region of the chain and also includes sequences 3' to the coding region that are transcribed to yield a 3' untranslated region (of unknown function) on the λ mRNA. The other three segments, L_λ, V_λ, and J_λ, are involved in specifying the variable region. Most important are the V_λ and J_λ segments. These two fragments must be brought together at the DNA level, a process known as **V-J joining,** to create a functional exon and thus a functional antibody gene. During this rearrangement, the J_λ DNA segment and the C_λ exon retain their relative configurations. The additional exon, L_λ, which lies just 93 bases upstream from the V_λ segment, also maintains its relationship to V_λ during the rearrangement. This is important, since a transcriptional promoter for the gene lies just in front of the L_λ segment. L_λ specifies a 5' untranslated region of the λ mRNA followed by sequences that encode most of a 15–20 amino acid–long hydrophobic leader peptide that guides the nascent light chain through the membrane of the rough endoplasmic reticulum but is then promptly cleaved from the chain. V_λ encodes the last four amino acids of the leader peptide and the N-terminal 98 amino acids of the variable region. J_λ encodes the last 12 or so amino acids of the approximately 110–amino acid–long V_λ region.

It is apparent from Figure 23-23 that the newly assembled *V-J* exon is still separated from the single C_λ exon by about 1.2 kilobase pairs. This intervening sequence is like any other conventional intron and is removed from the primary λ gene transcript by RNA splicing. The 93-base gap between L_λ and the *V-J* exon is also removed at the RNA level by splicing.

In the description just given (and in Figure 23-23), we have implied that the L_λ and V_λ segments lie upstream from the J_λ and C_λ segments. While this is almost certainly the case, the distance between the two pairs of segments is so large that it has not been possible to clone a fragment of DNA from mouse (or human) embryonic cells carrying all four segments. (This has been achieved for a λ gene of the chicken, however, where the distance from V_λ to J_λ turns out to be much shorter than it is in the mouse or human.)

So far, we have described only a single mouse λ light-chain gene. The complete mouse λ light-chain locus has the structure shown in Figure 23-24. There are two clusters of *J* and *C* segments, although the distances between them are not known. In agreement with amino acid sequences of mouse λ light chains, there are three functional C_λ exons. A fourth C_λ exon is present in the DNA, but it has suffered mutations that make it nonfunctional.

What happens to the DNA between a *V* and *J* segment when the two are brought together by recombination? In many cases, it can be shown that the DNA is lost as if it had been excised during the recombination process. How this might occur is indicated in Figure 23-23.

We have already noted that an antibody-producing plasma cell synthesizes only one light chain. Once a DNA rearrangement generates a functional light-chain gene (e.g., a λ_1 gene), further rearrangements of light chain genes are apparently inhibited.

The strategy for constructing functional κ light-chain genes is just the same as for λ genes, but the overall structure of the mouse κ multigene family is quite different from that of the mouse λ locus. Figure 23-25 shows the κ gene family. Most important, there are at

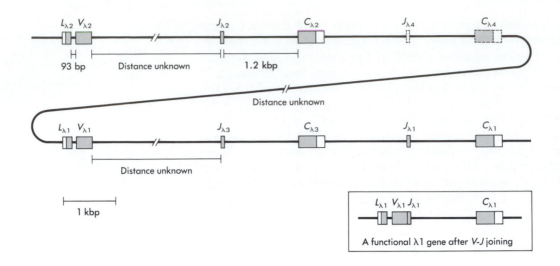

Figure 23-24

The probable structure of the mouse λ light-chain locus in its germ line configuration. The relative positions of the two *V-J-C* clusters are not known with certainty. Except for the unknown distances, the figure is drawn to scale. The structure of one possible light-chain gene after DNA rearrangement is shown. Note that $V_{\lambda 1}$ can also join with $J_{\lambda 3}$ to make a functional λ_3 light-chain gene. Boxes represent exons or parts of exons. White portions of some exons encode 5′ or 3′ untranslated regions of mRNA.

least several hundred V_κ segments, each with its adjacent L_κ exon. Furthermore, there are five J_κ segments that lie between the V_κ segments and the single C_κ exon (although one of the five J_κ segments is defective and cannot be used to make a functional gene). Clearly, the possibility for making different κ light chains is very great. This is because (as far as we know) within a particular cell, any one of the V_κ DNA segments may be joined to any J_κ segment. If we estimate that there are 300 V_κ segments, then 300 $V \times 4\,J = 1200$ possible V_κ exons that can be generated by *V-J* joining. The random joining of any *V* to any *J* is called **combinatorial joining.** It is an important source of diversity in the immune system, both in the κ genes and, in a more limited way, in the mouse λ light-chain genes. In humans, both the λ and κ loci resemble the mouse κ locus with its multiple *V* segments. It seems that inbred mice may have suffered a genetic catastrophe in which they simply lost many of their λ V-coding fragments.

The hundreds of V_κ segments can be divided into subfamilies according to their nucleotide sequence homology. These relationships are shown by preparing radioactively labeled copies (probes) of different V_κ segments and hybridizing them one at a time to mouse DNA that has been digested with a restriction endonuclease and separated by gel electrophoresis (the technique known as Southern blotting). Each V_κ probe will usually hybridize to five to ten different-size fragments of cellular DNA.

Heavy-Chain Variable-Region Exons Are Assembled by Two DNA Recombination Events: *D-J* Joining and *V-(D-J)* Joining[38, 39, 40]

Not surprisingly perhaps, the basic strategy for constructing functional heavy-chain genes is similar to that used at the light-chain loci. But there are additional complexities that give heavy-chain polypeptides their special characteristics. The structure of the mouse heavy-chain multigene family is shown in Figure 23-26. Three types of segments, V_H, D, and J_H, are joined at the DNA level to generate a *V-D-J* exon that encodes the mature heavy-chain variable region. *D* stands for diversity, since this segment encodes the third hypervariable re-

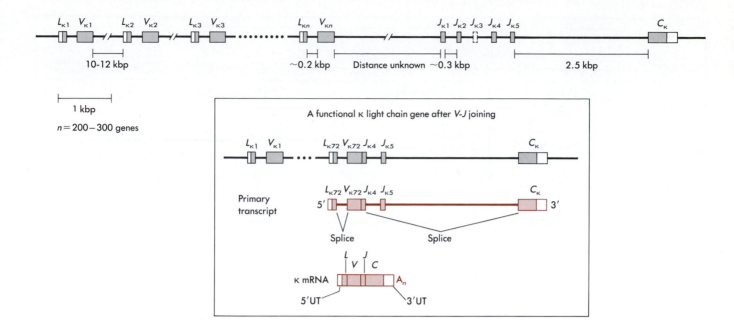

gion, long known to be particularly variable in amino acid sequence and in length from one heavy chain to another (see Figure 23-14). There are several hundred V_H segments. As with the light-chain genes, each has an attendant L_H exon that encodes the hydrophobic leader peptide. These L_H segments do not contribute to diversity, of course, since the leader is removed during maturation of the polypeptide. There are 10 to 12 D segments that encode between 3 and 14 amino acids, and there are 4 to 6 J segments that each encode about 15 amino acids. To create a functional heavy-chain gene, one D segment, presumably selected at random, is first joined to a J_H segment. Then a second DNA recombination event joins a V_H segment to the D-J_H piece. If we take the current, and perhaps conservative, estimate of 250 V_H segments, then assuming that combinatorial joining will allow all combinations of V_H-D-J_H, we can calculate that random joining of these segments can generate approximately $250\,V \times 10\,D \times 4\,J =$ 10,000 different V-D-J exons for heavy-chain variable regions!

The heavy-chain V_H-coding segments probably lie upstream of the constant region exons, which in turn are quite close to the J_H segments (see Figure 23-26). As you will recall, there are five major types of heavy-chain constant regions—α, δ, ε, γ, and μ—and some of these types have additional subtypes (γ_1, γ_2, etc.). Immunologists had long suspected a separate germ line "gene" for each type, although as we now know, these "genes" are really exons. In contrast to the much smaller light-chain C regions encoded by single exons, the heavy-chain C regions are each coded for by 5 to 7 exons. How an assembled V-D-J exon chooses to associate with a particular heavy-chain constant region will be discussed later in the chapter. In the example shown in Figure 23-26, a V-D-J exon is using C_μ, whose exons lie closest to the V_H genes. In this case, a primary transcript containing the V-D-J and C_μ exons is spliced to produce a functional heavy-chain mRNA. As we shall see, association of a V-D-J exon with some C_H "genes" requires that DNA recombination first bring the two closer together. The heavy-chain constant-region "genes" occupy more than 150 kilobase pairs of DNA, and the distance between adjacent "genes" can be as great as 55 kilobase pairs.

Figure 23-25
The κ light-chain locus and a functional κ light-chain gene. There are 200 to 300 V_κ segments, each with an attendant L_κ exon, five J_κ segments (of which one is inactive in the mouse), and one C_κ exon. Boxes represent exons or parts of exons. White portions of boxes encode 5' or 3' untranslated regions of mRNA. Except for the gaps, the figure is drawn approximately to scale. Note that the $L + V$ segments may be spread over 2000 to 3000 kilobase pairs of DNA, since the distance between at least some of these segments is 10 to 12 kilobase pairs.

Figure 23-26
The heavy-chain locus as it appears in germ line DNA and as it appears after *D-J* and *V-D-J* DNA rearrangements that produce a functional heavy-chain gene in a B cell. Each different heavy-chain class has a constant region that is encoded by multiple exons, but for the sake of simplicity and space, only the exons of C_μ are shown in detail. In the figure, the DNA between *D* and *J* or between *V* and *D-J* segments that join is shown as being circularized and discarded as part of the recombination process, but this is only hypothetical. As will be discussed later, the carboxyl terminus of the membrane-bound and secreted forms of μ heavy chains are encoded by different exons, and different RNA-processing patterns determine which type of protein will be made. In the figure, a secreted μ chain is being made. Because of the vast size and complexity of the heavy-chain locus, the figure is not drawn to scale; but some distances are indicated.

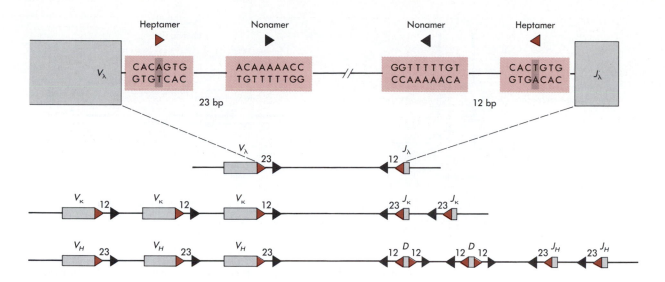

Figure 23-27
Nucleotide sequences that specify joining of V_L-J_L and V_H-D-J_H DNA segments in antibody genes. The top line of the figure is a close-up view of nucleotide sequences at the borders of a *V* and *J* segment that could be joined to make a functional λ light-chain gene. Note that the segments are bordered by a palindromic heptamer that is oriented in opposite directions at the two edges that will be joined. The heptamers are followed by either 23 or 12 nucleotides, whose sequence is not important, and then by a nonamer sequence. Like the heptamers, the nonamers are oriented in opposite directions. Color and black triangles represent the heptamer and nonamer, respectively, with the direction of the triangle giving the orientation of the sequence. Below the enlargement showing the sequence are abbreviated diagrams of the λ, κ, and heavy-chain V-region-coding DNA. Color and black triangles represent the positions of heptamers and nonamers that flank these sequences. The 12- or 23-base-pair spacing between them is indicated. These heptamer and nonamer sequences are consensus sequences, and the actual sequences can vary slightly from those given.

The heavy-chain V_H DNA segments, like the V_κ segments, can be divided into subfamilies based on their ability to hybridize to one another. As we discussed earlier in the chapter, antibody genes must arise by successive gene duplications. Presumably, V_H segments that belong to homologous families represent more recent duplications in evolutionary time, while families whose nucleotide sequences are less closely related diverged long ago.

Heptamer-Nonamers and the 12-23 Spacer Rule Govern *V-J* and *V-D-J* Joining[41, 42]

A specialized recombination system is probably needed for the unusual task of joining *V-J* and *V-D-J* DNA segments to make light- and heavy-chain genes. Based on other types of specialized recombination, we might expect to find unique nucleotide sequences serving as recognition signals in the system. Special sequences are indeed found at the borders of germ line *V*, *D*, and *J* segments, just near the sites where recombination events can take place. Figure 23-27 shows typical sequences that flank V_L and J_L segments and V_H, *D*, and J_H segments of light- and heavy-chain genes, respectively. What can be seen in this picture, and confirmed by analyzing many additional such sequences, is that each segment border that can participate in recombination has two highly conserved sequences, one seven bases long (a **heptamer** with the consensus sequence CACAGTG) and the other nine bases long (a **nonamer** with the consensus sequence ACAAAAACC), that are separated by a short nonconserved sequence. (Note that the heptamer sequence is a palindrome around the central AT base pair.) At two edges that can be joined (e.g., V_λ and J_λ in Figure 23-27), the conserved sequences are related to one another by a 180° rotation. Also very striking is the length of the nonconserved sequence that lies between the haptamer and the nonamer. When two segments can join, one always has a 12 (±1)-nucleotide–long spacer between its conserved heptamer and nonamer, while the other has a 23 (±2)-nucleotide–long spacer. Thus, three specific features characterize the sequences that specify *V-J* and *V-D-J* joining: the presence of the conserved heptamer and nonamer sequences, the orientation of the heptamer-nonamer to the segments

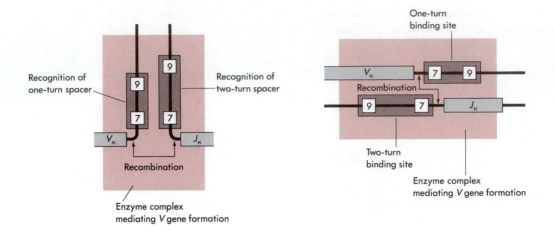

being joined, and the length of the spacer between the heptamer and nonamer.

According to the rules just described, we can see from the configuration of sequences that lie on the 3′ sides of V_H segments, on both sides of D_H segments, and on the 5′ sides of J_H segments why V_H segments never mistakenly skip over D_H sequences to join directly with J_H segments (see Figure 23-27). V_H-J_H joining would break the 12-23 spacer rule, while joining the 5′ side of *D* to J_H would require recombination between incorrectly oriented heptamer-nonamers. However, it is also apparent that not all undesirable recombinations are prohibited by these rules alone. For example, recombinations could occur between segments belonging to different gene families (λ and κ, heavy and light). Perhaps this is why the three gene families lie on different chromosomes: to decrease the possibility of such a mistake.

The lengths of the spacers between the heptamers and nonamers, 12 and 23 base pairs, correspond approximately to one and two turns of a DNA double helix. This has led to the idea that perhaps it is a combination of specific sequences and three-dimensional arrangement that is recognized by the putative antibody gene recombinase. Figure 23-28 shows some models that have been drawn to help clarify the nature of the substrate for this still-hypothetical enzyme.

Variation in the Precise Positions of *V-J* and *V-D-J* Recombinations Contributes Further to Antibody Diversity (Junctional Diversity)[43, 44]

When amino acid sequences of the variable regions of, for example, different κ light chains are compared with those predicted from nucleotide sequences of V_κ and J_κ DNA, it is apparent that there is much more amino acid sequence variability at the position encoded by the *V-J* joint than we would expect. Further scrutiny of the nucleotide sequences reveals that the extra variability can be accounted for very easily if the recombination events that join *V* and *J* segments do not occur at a precise nucleotide relative to the heptamer-nonamer recognition sequence. An example of the flexibility of *V-J* joining is shown in Figure 23-29 for a V_κ-J_κ pair. This one joining can occur in four different ways to generate four different *V-J* nucleotide sequences that would encode three different amino acid sequences. When we

(a)

(b)

(c)

Figure 23-29
An example of how variation in the precise position of *V-J* joining generates amino acid sequence variation (junctional diversity) at the position of the joint. (a) Edges of a V_κ and a J_κ segment that are about to be joined together. (b) One DNA strand of a hypothetical substrate for the *V-J* joining enzyme. The heptamer sequence is being used to generate a stem structure. Nucleotides that can participate in the recombination are numbered. (c) The four possible nucleotide sequences that can be generated by recombination events occurring at slightly different positions result in three different amino acids at position 96 in the mature K light chain. [After Gottlieb, P. D. *Molec. Immunol.* 17 (1980): 1423; based on H. Sakano et al., *Nature* 280 (1970):288 and E. E. Max, et al., *Proc. Nat. Acad. Sci.* 76 (1979):3450.]

recall that there are five J_κ segments and that each of these can probably undergo joining with any of several hundred V_κ segments, it is immediately apparent that variability in *V-J* joining, called **junctional diversity,** can make an important contribution to antibody diversity. Junctional diversity is found at both the κ and λ light-chain loci. The DNA at *V-J* joints encodes a portion of the third hypervariable, or CDR, region of light chains, a portion long known from amino acid sequence data to be the most variable region of the chain and also known to form a part of the antigen binding cavity.

Junctional diversity also occurs in heavy-chain genes, both at V_H-D joints and at D-J_H joints. However, in the case of heavy chains, even

Figure 23-30

Model for *D-J* joining, showing how N nucleotides may be added at the joint in a template-free fashion. (1) Two segments that are to be joined are lined up with the help of the hypothetical DNA-joining enzymes. Nicks are introduced into all four DNA strands at the ends of the heptamers adjacent to the coding sequences (positions are indicated by arrows). (2) The two heptamers are joined end-to-end. The coding sequences do not join yet, but are held near to one another by the joining proteins. (3) Some nucleotides may be removed from the *D* and *J* ends by an exonuclease. (4) One or more nucleotides are added at the *D* and *J* ends in a template-free fashion. This reaction may be carried out by terminal deoxynucleotidyl transferase, an enzyme that can polymerize random deoxynucleotides at 3′ ends. (5) Joining is completed by a DNA polymerase that copies the added nucleotides and by a ligase that seals the structure. [After F. Alt and D. Baltimore, *Proc. Nat. Acad. Sci.* 79 (1982):4118.]

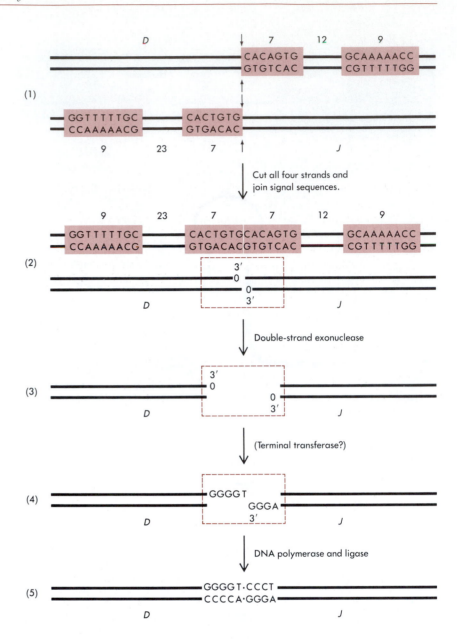

this mechanism of increasing variability can not account for all the diversity seen in V_H regions at the positions corresponding to the joints. There are cases where extra nucleotides seem to have been inserted between the V_H-D or D-J_H segments. These extra nucleotides are called **N regions**. Presently, it is thought that N nucleotides may be added in a template-free fashion as the gene is being assembled (Figure 23-30). An enzyme called terminal transferase, with the capacity to add nucleotides to the 3′ end of a DNA strand, is known to exist, so perhaps it or some other enzyme with similar ability is involved. In heavy chains, the most variable part of the V region is the third hypervariable segment. Appropriately, this turns out to be encoded by the *D* segment and by the V_H-D and D-J_H joints.

The variability in the precise position of *V-J* and *V-D-J* joining leads to some recombination events that generate stop codons in the rearranged genes. Presumably, these genes are not functional. Such inactive rearrangements are seen frequently in antibody-producing cells. Apparently, the cell keeps on rearranging its DNA until a functional gene is formed. Presumably, it is worth this extra effort in order to have the extra variability that can be generated by junctional diversity.

Mutation of Fully Assembled Antibody Genes Further Increases Diversity[45–48]

From the preceding discussion, we might conclude that molecular biologists can now predict the complete amino acid sequence of every possible mouse λ light chain and, if they had the energy to clone and sequence enough DNA, every κ light chain as well. (Of course, even if they cloned and sequenced all the heavy-chain genes, they still could not predict all the heavy chains an animal can synthesize if indeed there is template-free addition of nucleotides into the genes.) However, when we compare the predicted amino acid sequences, for example, of mouse λ_1 light chains with the actual amino acid sequences of mouse λ_1 light chains, there are still more amino acid differences in the variable regions than can be accounted for by *V-J* joining and junctional diversity at the *V-J* joints. The additional diversity consists of single amino acid differences scattered here and there, and they can usually be explained by the introduction of single-base changes in the *V-J* exon. Additional experiments have shown that point mutations are also introduced into fully assembled V_H genes. This source of diversity is called **somatic hypermutation.** The mechanism is unknown, but remarkably, antibody genes that have been subjected to it only carry mutations in a limited region encompassing the variable-region exon, with their constant-region exons apparently not subjected to this mutational process.

There are biologically compelling reasons why somatic hypermutation may be advantageous. First, antibodies that appear during the later stages of an immune response frequently have a higher affinity for the immunizing antigen than do the antibodies found during the primary response. If the first B cells triggered by antigen binding undergo somatic hypermutation of their rearranged antibody genes as they divide, then starting from a set of antibodies that are *already* able to recognize the incoming antigen, one may more readily generate antibodies with an "improved fit" (higher affinity) for the antigen. Experimental evidence supports the idea that mutation of fully assembled antibody genes does indeed generate antibodies with higher affinity for an immunizing antigen.

A second reason why somatic hypermutation may be highly advantageous is that despite the vast number of different antibodies an animal can make by all the possible arrangements of its germ line information, this is still a fixed number, predetermined at birth. In contrast, the world of antigens, which consists of living microorganisms, is continually changing as the organisms grow and mutate. To keep up with the changing world of antigens, perhaps we need an immune system that can also change.

Mechanisms of Antibody Diversity Reviewed[49, 50]

For two decades, immunologists puzzled over how an organism could be genetically programmed to produce perhaps as many as a million different antibody molecules. Now that we know the answer,

Table 23-3 The Sources of Diversity in Mouse Antibody Molecules

	Heavy Chains	Light Chains	
		κ	λ
1. Multiple germ line gene segments for variable regions			
$\quad V$	200–300 V_H	100–300 V_κ	2 V_λ
$\quad D$	10 D	0	0
$\quad J$	4 J_H	4 J_κ	3 J_λ
2. Combinatorial (random) joining of V-J or V-D-J segments	$250 \times 10 \times 4 = 10,000$	$250 \times 4 = 1000$	$2 \times 3 = 6$
3. Junctional diversity (variability in position) in V-J or V-D-J joining	$10,000 \times 3 \times 3 = 90,000$	$1000 \times 3 = 3000$	$6 \times 3 = 18$
4. N nucleotide addition at V-D-J joints	>90,000	—	—
5. Somatic hypermutation of rearranged genes	>>90,000	>3000	>18
6. Combinatorial joining of any light chain with any heavy chain	(>>90,000) $\times$	[>3000 $+$	>18] $=$
		$>>>>2.7 \times 10^8$ possible antibodies	

Numbers of germ line V_H and V_κ segments are approximate. The calculations in the table are based on the assumption that any V, D (if present), or J at a particular locus can be joined, and also that any light chain can associate with any heavy chain to make a functional antibody molecule.

it is apparent that researchers had seriously underestimated the magnitude of the problem! Using the strategies described in the preceding sections, we can calculate that the average mouse or human can synthesize more than 200 million different antibody molecules, even before somatic hypermutation goes to work on the genes. Given somatic hypermutation as well, the number of possibilities becomes almost unlimited.

Combinatorial diversity provided the first simplification of the diversity problem. Since two polypeptide chains form the antigen binding cavity, if we assume that any light chain can associate with any heavy chain, the number of antibody molecules with different active sites is equal to the number of different light chains multiplied by the number of different heavy chains. So the question of antibody diversity becomes the number of possible light- and heavy-chain variable regions an animal can make. (Since constant regions do not contribute to the diversity of antibody specificities, they are excluded from this discussion.) For the mouse, the calculations are as follows and as shown in Table 23-3.

First let us consider the light chains. At the κ locus, there are estimated to be 100 to 300 V_κ segments and 4 functional J_κ segments. Using the number 250 for V_κ segments, the random association of V_κ segments and J_κ segments (combinatorial joining) would yield 1000 different V_κ regions. If we estimate that variability in the precise point of V-J joining (junctional diversity) can lead to three different amino acids at this position, then the number of V_κ genes rises to 3000.

There are only 2 V_λ segments and 3 functional J_λ segments in the mouse; so considering combinatorial joining and junctional diversity, there are 18 possible V_λ-coding sequences, too small a number to contribute significantly in our calculations. (It should be noted that most antibodies in a mouse have κ rather than λ light chains.)

For heavy chains, there are about 250 V_H segments, 10 D segments, and 4 functional J_H segments. Combinatorial joining of these segments would produce 10,000 V_H-region genes. If we add in the three amino acids for junctional diversity at both the V-D and D-J joints, we can calculate the possible number of V_H genes as $10,000 \times 3 \times 3 =$

90,000. This is probably an underestimate of the number of possible V_H sequences, since it does not account for the greater junction variability that can arise from template-free addition of N nucleotides at the *V-D* and *D-J* borders.

Multiplying the 90,000 heavy chains by the 3000 light chains, we get 2.7×10^8 possible antibody molecules. Since somatic hypermutation can introduce any of 20 different amino acids into many different sites in the V regions of these molecules, it is apparent that the number 2.7×10^8 is only a small fraction of the total number of antibodies an animal is likely to synthesize in its lifetime.

For many years, immunologists argued over whether antibody diversity is encoded in the germ line or generated by genetic events in somatic (B) cells. Now it is clear that aspects of both theories are correct, and the question has more or less been laid to rest. Still, it is significant that even though there are many germ line V-region segments, there are still only a few hundred for light chains and a few hundred for heavy chains. It is what happens to these segments in B cells that turns approximately 300 light-chain V and J segments into approximately 3000 V_L segments, and about 300 heavy-chain V, D, and J segments into about 90,000 V_H segments. Also, the fact that antibodies are made of two chains gives these 93,000 chains the potential to form more than 200 million different antibody molecules.

A Tissue-Specific Transcription Enhancer Element Is Located in the Intron Between *J* and *C* DNA[51, 52, 53]

In a plasma cell synthesizing antibodies, about 10 percent of the cell's mRNA is heavy-chain message. Unrearranged V_H segments are not transcribed; only the V_H segment that has been joined to *D-J* to make a functional gene is active. The DNA rearrangements that bring V_H segments and their adjacent L_H exons next to $D\text{-}J_H$ segments do not alter the promoter elements that lie in front of the L_H exons, so the question arises as to why juxtaposition of *V-D-J* segments leads to such a high level of gene expression. An important part of the answer came from experiments that showed that this rearrangement brings the promoter of the chosen *V* segment near to a powerful enhancer element that is located between the *J* and *C* segments.

When a molecularly cloned rearranged heavy-chain gene was introduced by DNA transfection into a mouse myeloma cell, the introduced gene was transcribed very efficiently. This made it possible to experimentally delete portions of the gene to see which part was important for the high level of transcription. It turns out that an important set of sequences, about 200 nucleotides long, lies between *J* and *C* (Figure 23-31). This sequence has properties of the transcriptional enhancer elements that had already been discovered in the promoter regions of DNA and RNA tumor viruses: It can increase transcription when it is placed in either orientation relative to a gene, and it can influence transcription from a promoter located several kilobase pairs away. This would be a long enough range to enhance transcription from the promoter of a rearranged *V* gene, but not from the promoters lying next to *L* and *V* segments that are still in their germ line configuration. Moreover, the heavy-chain enhancer element is tissue specific: Although it enhances transcription dramatically in lymphoid cells of the B lineage, it does not seem to work in fibroblasts at all.

Figure 23-31
DNA rearrangements that join *V-D-J* to make a heavy-chain *V* gene bring the transcriptional promoter for the rearranged *V* gene under the influence of a powerful transcriptional enhancer element located in the intron between *J* and *C*. Note that before the DNA rearrangements, the nearest promoter to the enhancer element is at least 100 kilobase pairs away; but after *V-D-J* joining, the needed promoter is within about 2 kilobase pairs of the enhancer.

In addition to the enhancer, a promoter element lying 5′ of each *L* segment is also important for heavy-chain gene transcription. Recently, this element has also been shown to have tissue specificity, functioning more efficiently in B cells. A critical sequence in the element is a conserved octamer, with the consensus sequence ATGCAAAT, that lies just upstream of the TATA boxes in front of many L_H exons (see Figure 23-31). Interestingly, an octamer with the complementary sequence (ATTTGCAT) is present in the promoter regions of light-chain *V* genes. In addition, the κ light-chain locus, like the heavy-chain locus, has been found to contain an enhancer element in the intron separating the J_κ cluster and the κ constant-region exon.

Antibody Gene Rearrangements Occur in a Developmentally Regulated Fashion[54, 55]

B cell differentiation culminates in a plasma cell that synthesizes a single type of light chain and a single type of heavy chain. This means that although vertebrates are diploid, only one of the two heavy-chain loci and one of the four light-chain loci are used in a given cell, a phenomenon known as **allelic exclusion,** which is unique to antibody genes (Figure 23-32). Antibody gene rearrangements follow a carefully ordered sequence that is intimately tied to, and in fact, helps to define, stages of B cell differentiation. V_H segments are rearranged first, then κ light-chain genes. At least in the mouse, it appears that λ light-chain genes are assembled only if rearrangements at the κ locus fail to produce a functional κ light chain. The first antibodies a B cell produces are membrane-bound IgM molecules. Later, the process of class switching will join the same heavy-chain *V-D-J* exon to other C_H-region exons. Researchers can study mouse cells at different stages of differentiation by using cancer viruses to transform the cells into permanently growing cell lines that are often frozen at a particular stage along the developmental pathway.

The developmentally regulated rearrangements and expression of antibody genes are diagrammed in Figure 23-33. In the **null cells,** the earliest cells to show evidence of antibody gene rearrangements, heavy-chain genes have begun to rearrange, but only *D-J* recombinations have occurred (see Figure 23-33). Although only one chromosome will ultimately carry a functional heavy-chain gene, both chro-

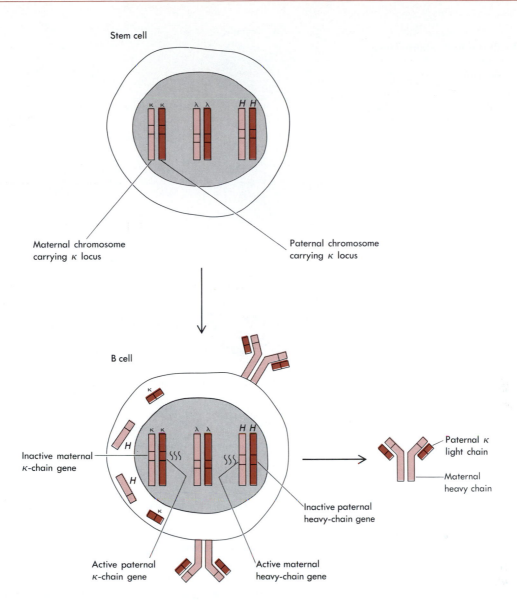

Figure 23-32
Allelic exclusion at the antibody loci. A particular B cell synthesizes just one type of heavy chain and one type of light chain. Once a functional gene for a heavy chain (and then for a light chain) is assembled, further DNA rearrangements are inhibited, perhaps by the end product of the rearrangement. For example, synthesis of a μ heavy-chain polypeptide may inhibit further DNA rearrangement at the other heavy-chain locus.

mosomes undergo D-J_H rearrangements. The next cell in the pathway is called a **pre–B cell.** Immature pre–B cells undergo V_H to D-J_H rearrangements. As soon as a functional V-D-J exon is generated, further heavy-chain gene rearrangements stop and the cell synthesizes heavy chains with a μ constant region. These chains remain in the cytoplasm until light chains are made. Once heavy-chain gene expression has begun, the μ-positive pre–B cell goes on to rearrange its κ light-chain genes. If a functional κ gene is made, apparently all V-region DNA rearrangements cease. Then the κ and μ chains assemble and move to the membrane, where they are anchored and serve as receptors for antigens. If rearrangements at the two κ loci fail to produce a functional gene, then λ-chain genes undergo rearrangement, and if successful, the B cell will have receptor molecules that are λ-μ tetramers. (Note that although secreted IgM antibodies are made up of five heavy-light-chain tetramers, membrane-bound IgM antibodies are simple tetramers.)

All the events described in the preceding paragraph occur independently of antigen stimulation. The end result is to generate a set of **virgin B_μ cells,** whose membrane-bound IgM receptors probably

Figure 23-33
A model for B cell differentiation. There is experimental evidence for the different steps in the picture, except for the progression from B_μ to $B_{\mu+\delta}$ and back to B_μ and for the possible role of membrane-bound IgD in signaling the elimination of self-reacting cells. [After *Immunology*, 2d edition by L. E. Hood, I. L. Weissman, W. B. Wood, and J. H. Wilson. 1984 by The Benjamin/Cummings Publishing Co.]

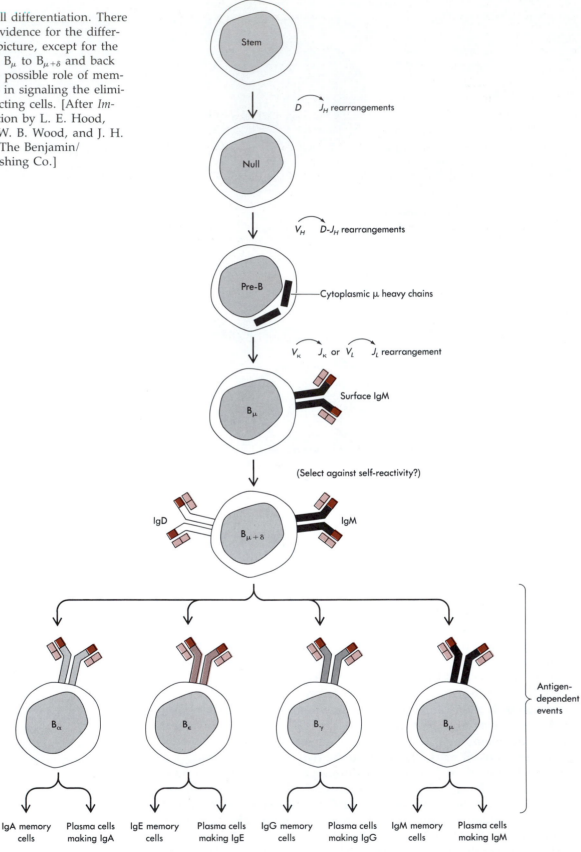

include the complete repertoire of antibody V regions an animal can generate before somatic hypermutation intervenes. The next step in B cell maturation is a $B_{\mu+\delta}$ **cell,** which produces both IgM and IgD antibody receptors. The two types of molecules have identical V regions, differing only in their heavy-chain constant regions. As we shall see, they arise by changes in RNA processing of the heavy-chain gene. Finally, in response to antigen binding, the $B_{\mu+\delta}$ cells differentiate into **blast cells,** which divide and secrete IgM and IgD antibodies. These cells also retain IgM and IgD as membrane-bound receptors. The activated blast cells then differentiate to yield short-lived plasma cells and long-lived memory cells.

Plasma cells secrete antibodies at the rate of about 10,000 molecules per cell per second. Because of class switching, which can join any heavy-chain V region exon to any heavy-chain constant region, a single B cell can give rise to plasma cells that synthesize antibodies of any class. Memory cells do not secrete antibodies, but rather keep them as membrane-bound receptors. Memory cells can also have antibodies belonging to any heavy-chain class. The memory cells will provide a prompt response during future encounters with their cognate antigens.

As discussed earlier, the differentiation of a resting B cell to a miniature antibody factory, the plasma cell, involves a massive expansion of the cell's cytoplasm and of the rough endoplasmic reticulum, where antibody synthesis occurs (see Figure 23-5).

Partially Rearranged Heavy-Chain Genes Are Transcribed and Translated to Yield D-J-C Products[56, 57, 58]

An exciting model to explain the ordered rearrangement of antibody genes and allelic exclusion holds that the protein product of the first rearrangement is involved in regulating the second rearrangement, that the product of the second rearrangement controls the third, and so on. For example, the synthesis of a complete μ heavy chain may signal the cell to stop further rearrangements of heavy-chain genes, while synthesis of a light chain that can combine with the cytoplasmic μ heavy chains to form a complete antibody molecule may put an end to light-chain gene rearrangements. This model could explain allelic exclusion, since gene rearrangements would be blocked as soon as one functional gene of each type was made. Support for this model comes from transgenic mice in which a functional κ light-chain gene was introduced into mouse embryo cells. In the B cells of adult animals that express the introduced gene, endogenous κ-chain genes do not undergo DNA rearrangement. Apparently, the cell is fooled into thinking it has already rearranged its light-chain genes, since a functional light chain is present.

How could the model explain the fact that D-J rearrangements at the heavy-chain locus must occur before V_H segments can join to D-J segments? In other words, could a gene product with a regulatory role be generated as a result of D-J joining? Recently, RNA species that contain D, J_H, and C_μ sequences but lack V_H have been found in immature B cells. Thus, there must be transcriptional promoters between V_H and D segments. Furthermore, examination of the nucleotide sequences of these transcripts shows that there are AUG (methionine initiation) codons near their 5′ ends, and, at least in some cases, these are the start of open reading frames that extend through D, J_H, and C_μ. In fact, D-J-C polypeptides have been detected in some types of mouse B lymphoid tumors that have undergone D-J but not yet V-D-J recombination. Conceivably, once these D-J-C_μ products are synthesized, they signal to the cell that now it is time to begin joining V_H to D-J segments.

At first it was believed that V-region segments are transcriptionally dormant until they are joined to D-J segments and thereby brought near to enhancer elements. Now, however, transcription of unrearranged V_H segments has been detected, although the level of transcription is lower than that of complete heavy-chain genes. V_H transcription occurs in pre–B cells that have made D-J rearrangements and are at the stage of undergoing V_H to D-J_H rearrangements. Once a successful V-D-J rearrangement has occurred, however, the small germ line V_H transcripts are shut off. It is not known why transcription of the V_H segments is temporally related to V to D-J rearrangements. Perhaps the transcription simply reflects the presence of open (and hence active) chromatin, which could be an essential substrate for the putative recombinase involved in V-D-J DNA rearrangements. Alternatively, the small V_H transcripts could be translated to yield V_H polypeptides that play some part in the V to D-J DNA rearrangement.

Expression of D-J-C_μ transcripts does not show allelic exclusion; rather, transcripts from both chromosomes can be present in the same cell. Apparently, allelic exclusion occurs only after a V_H segment has successfully joined to the DNA of a rearranged D-J segment to generate a complete antibody gene.

Differential RNA Processing Explains the Progression from Membrane-Bound to Secreted Antibody[59, 60]

When B cells are first activated by interaction with their cognate antigens, they begin to secrete IgM antibodies, which, until then, were present only as membrane-bound receptors. This progression results from a change in RNA processing. Surface IgM is anchored to the cell membrane by the carboxyl end of its μ heavy chain. The 41 hydrophobic amino acids that make up the transmembrane anchor are encoded by two small exons that lie beyond four exons that encode the bulk of the C_μ chain. The switch to synthesis of secreted IgM results from a change in RNA processing that appears to activate an RNA termination site lying between the fourth C_μ exon and the two membrane-anchor-coding exons (Figure 23-34). As a result, RNA transcripts terminate before these two exons and a new carboxyl terminus, one lacking a transmembrane anchor, is specified by the fourth C_μ exon. This segment of the fourth C_μ exon is skipped over during RNA splicing to join the two membrane-anchor-coding mini exons.

How the binding of antigen to surface IgM brings about an alteration in RNA processing is a complete mystery at the moment. Furthermore, in activated B cells, both membrane-bound and secreted antibody are synthesized, while in the more mature plasma cells, only secreted antibody is made. Thus, this method of gene expression can be fine-tuned to match the needs of the cell.

All antibody classes can serve as membrane-bound receptors as

Figure 23-34

Synthesis of membrane-anchored or secreted μ heavy chains is a consequence of differential RNA processing. Membrane-anchored and secreted μ chains have identical amino acid sequences up to their carboxyl termini, where they diverge: The last 41 amino acids of the membrane form are replaced by 20 different amino acids in the secreted form. The membrane-anchored form is made from an RNA that splices out the end of the fourth C_μ exon. This portion of the fourth exon encodes the carboxyl terminus of the secreted form and the 3' untranslated region of the exon. To make secreted μ chains, either this splice must be suppressed or RNA synthesis must terminate at the first poly A addition site so that transcripts never reach the membrane-anchor-coding mini-exons. The amino acid sequences of the carboxyl termini of the two forms of μ heavy chain are shown at the bottom of the figure. The hydrophobic sequence in the membrane-bound form, which could serve to anchor the chain, is underlined. The secreted form has a cysteine residue (boxed) that can disulfide-bond to the J chain in secreted μ (see Figure 23-20), as well as an asparagine-linked carbohydrate chain (N–CHO). Boxes in DNA or RNA molecules represent exons. [After *Immunology*, 2d edition by L. E. Hood, I. L. Weissman, W. B. Wood, and J. H. Wilson. 1984 by The Benjamin/Cummings Publishing Co.]

well as being secreted; and as we might expect, all the heavy-chain constant-region genes have membrane-anchoring exons and alternative sequences that are used to encode the carboxyl-terminal ends of secreted chains.

IgM (μ)→IgD (δ) class switch by RNA processing

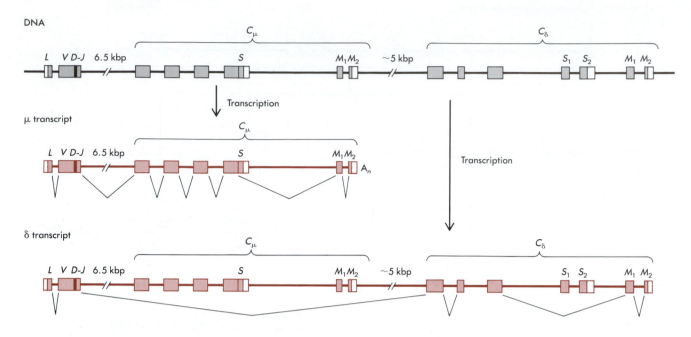

Figure 23-35

IgM to IgD class switching is accomplished by a change in RNA processing. Differential RNA processing is used to join the transcript of an assembled heavy-chain variable-region "gene" to either a μ or a δ constant-region transcript. Note that DNA that encodes both the μ and δ constant-regions consists of exons that encode both a membrane and a secreted form of the chain. In this figure, the splicing patterns indicated would result in synthesis of the membrane-anchored form of each chain. Boxes represent exons in DNA or RNA with white portions of boxes representing 5' and 3' untranslated regions.

Differential RNA Processing Allows the Synthesis of IgM and IgD Antibodies by a Single Plasma Cell[61, 62]

As we said earlier, heavy-chain constant regions perform the biologically important effector functions, such as localizing antibodies to certain sites in the body, interacting with complement components, and binding to macrophages. It is highly desirable for an organism to be able to combine its complete V-region repertoire with each of these different functions. Thus, antibody gene assembly includes mechanisms for joining fully assembled *V-D-J* exons to different C_H exons, the phenomenon of class switching. There are two mechanisms for class switching, and their use is developmentally regulated. One mechanism involves differential RNA splicing (for IgM to IgD), and the other involves DNA switch recombination (for IgM to any other class but IgD).

The first class switch that takes place in B cell differentiation allows a given *V-D-J* exon to acquire a δ constant region in addition to μ. Some unknown signal triggers virgin B_μ cells that express just surface IgM to progress to mature $B_{\mu+\delta}$ cells that have both membrane-bound IgM and IgD. The two types of molecules have heavy chains with identical variable regions, and of course, both are associated with the same light chain, since only one light chain is made in a given cell. Exons that encode C_δ lie downstream from C_μ coding exons (Figure 23-35), and differential RNA splicing of a long transcript allows the use of the downstream C_δ exons. This change in RNA splicing may result from a change in RNA termination. In a virgin B_μ cell, RNA transcripts may terminate before the C_δ sequences, while in $B_{\mu+\delta}$ cells, some primary transcripts may extend past C_μ and through C_δ. These transcripts can then be spliced to produce mRNAs using the C_δ exons. At present, it is not clear whether poly A addition or RNA

Figure 23-36

Class switching from IgM to IgG, IgA, or IgE involves DNA recombination at switch signals. In this figure, an assembled *V-D-J* exon is being moved from a position adjacent to the μ constant-region exons to a position next to the $C_{\gamma1}$ exons to create a functional $\gamma1$ heavy-chain gene. For simplicity, the multiple exons of each type of heavy-chain constant region are abbreviated in the figure except for the final picture of $C_{\gamma1}$ and not all the constant regions are indicated. The special sequences involved in switch recombination are indicated as horizontal black boxes. Note that there are no such sequences in front of the C_δ exons, which makes sense, since the switch from μ to δ is accomplished by differential RNA processing rather than by DNA recombination. Note also that the transcriptional enhancer element (black circle) retains its position relative to the gene's promoter after the switch occurs.

splicing is the means of regulating the μ to δ progression, or whether both processes are regulated to achieve the desired control.

The progression from a B_μ cell to $B_{\mu+\delta}$ usually must take place before B cells are competent to interact with foreign antigens. At some point after B_μ cells appear, they may need to be screened to remove or inactivate those cells that can react with self antigens, so that the animal will be tolerant to self antigens. Conceivably, the appearance of membrane-bound IgD antibodies in addition to IgM is a signal that this screening process has taken place and it is now safe for the cell to participate in immune responses.

Antibody Class Switching Beyond IgD Involves DNA Recombination in the Switch Regions of the Heavy-Chain Constant-Region Genes[63, 64, 65]

Once triggered by antigen recognition, $B_{\mu+\delta}$ cells have the potential to differentiate to produce memory cells and plasma cells that synthesize antibodies belonging to all possible heavy-chain classes. DNA rearrangements are involved in class switches that transfer *V-D-J* exons from C_μ to other heavy-chain types except, as we have just discussed, to C_δ. The DNA rearrangements involve recombination within special segments, called **switch regions,** that lie about 2 to 3 kilobase pairs upstream of all the heavy-chain constant-region genes, except C_δ (Figure 23-36). For example, a switch region in front of the C_μ exons is used to bring a *V-D-J* segment near to the $C_{\gamma1}$ exons. Notice in Figure 23-36 that this DNA rearrangement leaves the heavy-chain enhancer element in the same location relative to the gene's transcriptional promoter as it was before the switch took place.

Switch regions are quite long, on the order of 1 to 2 kilobase pairs, but they are composed of many, many copies of very short sequences. The repeated sequences are mostly different for different C_H genes, and they vary in length from 5 to 80 nucleotides. It is not clear whether the recombination events that lead to class switching occur

Figure 23-37

Diagram of how a helper T cell may be activated to help a B cell produce its antibodies. Antigen is taken up by accessory cells (called antigen-presenting cells) and processed. Fragments of the antigen appear on the surface of the accessory cell. A helper T cell whose antigen-specific T cell receptor can recognize this antigen binds to the accessory cell and becomes activated. The activated T cell then interacts with a B cell whose antibody can recognize the same antigen that stimulated the helper T cell. The B cell also has processed antigen on its surface. The interaction of T and B cells, along with antigen stimulation, triggers the B cell to divide and differentiate to an antibody-producing plasma cell. [After *Immunology*, 2d edition by L. E. Hood, I. L. Weissman, W. B. Wood, and J. H. Wilson. 1984 by The Benjamin/Cummings Publishing Co.]

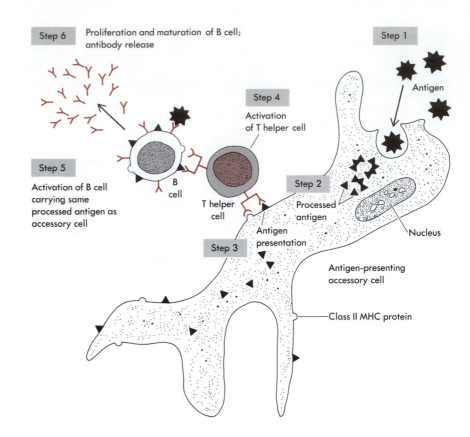

Step 6 — Proliferation and maturation of B cell; antibody release

Step 1 — Antigen

Step 4 — Activation of T helper cell

Step 5 — Activation of B cell carrying same processed antigen as accessory cell

B cell

T helper cell

Antigen presentation

Step 2 — Processed antigen

Step 3

Nucleus

Antigen-presenting accessory cell

Class II MHC protein

within a single DNA molecule or between different chromosomes. Perhaps both types of events are allowed.

Recently, cells that synthesize both IgM and IgG, or both IgM and IgE, or both IgM and IgA have been observed. Because allelic exclusion forbids the expression of heavy-chain genes from two chromosomes within one cell, these cells are presumably using two heavy-chain C-region "genes" located on the same chromosome. The DNA rearrangements we just described would preclude this situation, and now it is thought that in some cases, instead of DNA rearrangement, extremely long transcripts may be made from a single heavy-chain locus and spliced to allow simultaneous synthesis of two heavy-chain classes (other than IgM with IgD).

Complex Interactions Involving B Cells, T Cells, and Macrophages Help B Cells Produce Their Antibodies[66, 67, 68]

Antigen stimulation of B cells to divide and secrete antibodies is usually a far more complex process than a simple collision between freely circulating antigen and a membrane-bound immunoglobulin receptor. An animal's response to an invading antigen mobilizes a whole network of cells. Much remains to be learned about this process, but a rough sequence of events is clear. First, as described earlier in this chapter, accessory cells, such as macrophages, trap the incoming antigen and *process* it for *presentation* to a subset of T cells called **helper T cells** (**T$_H$ cells**) (Figure 23-37). (As we shall see, helper T cells are distinct from the T cells described earlier, which directly attack and destroy antigens or antigen-bearing cells.) **Antigen processing** seems to involve partially digesting the antigen and returning fragments of it to the cell surface. **Antigen presentation** probably involves positioning these fragments at the cell surface until they are recognized by a helper cell carrying an appropriate T cell receptor. Interaction of the T_H cell with the antigen-charged accessory cell activates the helper. Then the activated helper stimulates B cells to divide and become plasma cells. This interaction is antigen-specific. Apparently, B cells can also process antigens and present them on their surface; so a T helper cell whose receptor first recognized a particular antigen on a macrophage will recognize the same antigen on a B cell.

How the helper T cell stimulates the B cell is not yet clear, but it probably results, at least in part, from the release of growth factors. As will be discussed in Chapter 25, growth factors are polypeptides that bind to specific receptors on the cell surface and by so doing somehow trigger cell division. Different cell types require different growth factors. The factors released by helper T cells that promote the differentiation of B cells to plasma cells are not yet clearly defined. One may be the growth factor called *interleukin 2* (Il-2), which is also a growth factor for T cells. Others, called helper factors, are produced in small amounts and are likely to remain elusive until their genes are cloned. Growth factors are only part of the trigger that sends the B cell on its way. Of course, binding to its cognate antigen is also a trigger. The two stimuli may be related, however, since an important consequence of antigen binding may be an increase in the expression of growth factor receptors on the cell surface.

One clue as to how antigen binding may trigger B and T cells to divide comes from studying the nonspecific stimulation of B and T cell division by mitogens such as lectins and lipopolysaccharides (Table 23-4). These substances bind to the surface of lymphocytes and trigger several rounds of cell division. When they bind, they can induce a phenomenon known as **capping,** in which cell surface receptors become crosslinked and move laterally in the membrane to coalesce as a cap (Figure 23-38). One hypothesis is that capping can send a signal to the cell nucleus that triggers events that lead to cell division. Since many antigens are multimeric and hence able to engage more than one antibody molecule, they, too, can induce capping, and this might be one way they participate in triggering B cell division and maturation.

Why shouldn't interaction with circulating antigen be a sufficient trigger for B cells to differentiate and secrete antibodies? Why do they need helper T cells, too? Speculation is that the need for several cells to interact provides a way to regulate the process of antibody synthesis and also B cell division. If antigen alone were the trigger, then activated B cell clones might divide and differentiate out of control or might not divide enough to cope with the antigen load. As we shall see in the next section, another group of T cells, called suppressor T cells, can block the proliferation and terminal differentiation of B cells. According to another model, the need for T cells to stimulate B cells to produce antibodies may be involved in achieving tolerance.

Although helper T cells are required for all IgG, IgE, and IgA immune responses, some IgM responses are T_H cell–independent. These responses are usually directed against antigens composed of identical repeating determinants; but the biological significance of these T_H-independent responses is obscure.

Table 23-4   Mitogens Specific for B Cells or T Cells

B Cells	T Cells
Lipopolysaccharide (LPS)*, from bacterial cell walls	Concanavalin A (Con A), a plant lectin Phytohemagglutinin (PHA), a plant lectin

*Mitogenic for mouse but not human B cells.

Figure 23-38
Capping. (a) Diagram of how a multimeric antigen can act to cluster mobile immunoglobulin in the cell membrane of a small lymphocyte. (b) Autoradiographs of a B cell binding to its radioactively labeled cognate antigen: (1) Uniform binding of the antigen following incubation at 0°C for 30 minutes. (2) Aggregation (capping) of antigen at one cell pole after incubation at 37°C for 15 minutes. Such capping results from the cross-aggregation of the surface immunoglobulin receptors by the polymeric antigen. Note that only one of the four cells in this picture has receptors for this particular antigen. [Reproduced with permission from E. Diener and V. H. Paetkau, *Proc. Nat. Acad. Sci.* 69 (1972):2364.]

(a)

(b)

THE T CELL RESPONSE

There Are Different Types of T Cells Called Cytotoxic, Helper, and Suppressor Cells[69, 70, 71]

Two very different types of cellular immune responses are mediated by T lymphocytes. One is to destroy cells that display foreign antigens on their surfaces, and it is performed by **cytotoxic T lymphocytes (CTLs)**. Most often, the targets are the animal's own cells when, for example, they become infected with a virus and so express viral-coded proteins on their surfaces. Cytotoxic T cells can also destroy cells in skin grafts or organ transplants between individuals. Cytotoxic T cells are the cellular immune system's equivalent of antibodies. They recognize antigens via their membrane-bound T cell receptors, which have striking similarities to antibody molecules. Each cytotoxic T cell carries a receptor with a different antigen binding specificity. Interaction with an antigen that specifically binds to the receptor triggers the cell's maturation to a killer and also causes the cell to proliferate and generate memory cells that will respond to future encounters with the same antigen. The primary difference between humoral immunity and cytotoxic T cell responses is that antibodies are secreted, circulate freely, and cope primarily with freely circulating antigens. In contrast, T cell antigen receptors remain

Table 23-5 Some Usefel Cell Surface Antigens on Mature T Cells
of Mice and Human

Mouse	Thy-1	L3T4	Lyt-2,3
Helper T cells (T$_H$)	+	+	−
Cytotoxic T cells (CTL)	+	−	+
Suppressor T cells (T$_S$)	+	−	+
Human		**T4**	**T8**
Helper T cells		+	−
Cytotoxic T cells		−	+
Suppressor T cells		−	+

L3T4 and T4 are equivalent, as are T8 and Lyt-2,3.

+ = possesses the antigen. − = lacks the antigen.

membrane-anchored and seem to direct their attention primarily to antigens on cell surfaces.

The second function of cellular immunity is to regulate the responses of B cells and cytotoxic T cells. As we have already seen, specialized T lymphocytes called helper T cells interact with B cells to trigger their maturation to antibody-producing plasma cells. Helper T cells are also needed to help cytotoxic T cells mature into effector cells. A second type of regulatory T cell, the **suppressor T cell** (T$_S$), has the opposite effect: It blocks B and T cell responses. All these responses are antigen-specific. For example, each helper T cell recognizes a unique antigenic determinant and only stimulates a B cell that recognizes the same determinant. (How helper T cells may interact with B cells to stimulate antibody synthesis was discussed earlier and diagrammed in Figure 23-37). Helper T cells have antigen receptors similar to those of cytotoxic T cells, and they are probably encoded by the same genes. The receptors for suppressor T cells have not yet been clearly defined.

The various types of T lymphocytes (and also B cells) are small, round cells that are difficult to distinguish in the microscope. They can be identified, however, by the presence of specific cell surface proteins. These proteins were discovered serologically, and in most cases their functions are still unknown. They are enormously useful, however, both for identification of lymphocyte subsets and for separating different cell types for experimentation. All mouse T cells carry the Thy-1 antigen on their surface. Helper T cells of humans have the T4 antigen, originally defined by its ability to react with a particular monoclonal antibody prepared against human T cells. The gene encoding T4 has recently been molecularly cloned. Table 23-5 lists antigens that are currently used to identify different types of mouse and human T lymphocytes. The best marker for B cells, of course, is surface immunoglobulin.

Pluripotent stem cells in the bone marrow give rise to all the different types of blood cells: the B and T lymphocytes, as well as the red blood cells and other cells of the myeloid lineage. In humans and mice, immature T cell precursors leave the bone marrow and migrate to the thymus. There they undergo massive proliferation and, after a sequence of events largely shrouded in mystery, emerge as helper, cytotoxic, and suppressor T cells.

Figure 23-39
The experiment that showed that T cells recognize antigens only in combination with normal cell surface proteins. An inbred mouse of strain A was immunized with a virus. Its T cells were then removed and tested for their ability to kill target cells in vitro. The target cells were fibroblasts infected with the same virus used in the immunization, but they were derived either from a mouse of strain A or from a different inbred strain with a different *MHC* gene, designated strain B in the figure. Despite the presence of the same viral antigens on both target cells, the T cells from mouse strain A could only kill target cells derived from a mouse of strain A.

T Cells Recognize Foreign Antigens in Combination with MHC Gene Products[72–76]

Cytotoxic T cells respond to antigens present on cell surfaces, so it may be important to prevent them from interacting with freely circulating antigens. About ten years ago, an important experiment unexpectedly revealed the mechanism that probably restricts T cells to interactions with cell-bound antigens. The key is that T cells can only "see" antigens when they can also "see" certain normal cell surface proteins simultaneously. In the experiment that led to this discovery (Figure 23-39), T cells were removed from a mouse that had first been immunized with a virus so as to activate and stimulate the proliferation of cytotoxic T cells directed against cells infected with this particular virus. These T cells could kill fibroblasts infected with the virus in tissue culture. Surprisingly, however, they could not kill fibroblasts infected with the virus if the fibroblasts came from a genetically different mouse. Thus, cytotoxic T cells do not just recognize foreign antigens; they recognize foreign antigens in combination with determinants present on their host's own cells.

By using mouse genetics, it was possible to show that the normal determinants that T cells see are cell surface proteins coded for by **major histocompatibility complex** (**MHC**) genes, in particular, the so-called class I *K* and *D* genes. (In humans the corresponding genes are called *HLA-A*, *HLA-B*, and *HLA-C*.) These proteins are present on the surface of all cell types in the body. Thus, cytotoxic T cells are probably able to eliminate any cell type infected with a virus. MHC class I proteins are extremely polymorphic. This means that there are many different *K* and *D* alleles in the mouse population. Polymorphism explains why any two mice are likely to have different MHC proteins and why, in the aforementioned experiment, the cytotoxic T cells could not kill the virally infected cells from a different mouse. T cell receptors concentrate on the polymorphism of class I products, using it to distinguish self from nonself. Retrospectively, we see that **MHC restriction** of T cell function makes a lot of sense. It ensures that cytotoxic T cells will restrict their activities to cell surfaces and not clog up their receptors with freely circulating antigens.

The ability of T cells to recognize foreign antigens only in combination with the host's own MHC gene products is acquired as the T cells mature in the thymus, a learning process called **thymic education.** In many ways, it resembles tolerance, the process by which T cells learn not to destroy the host's own normal cells. Tolerance is also learned, rather than being preprogrammed into the T cells. Both processes involve a selection that results in a pool of T cells whose receptors have just the right recognition properties, namely, they see altered self but not self alone and not antigen alone. T cells failing the selection are presumably destroyed before they can leave the thymus or are held in check after they mature.

The fact that T cells recognize antigen + MHC immediately raises the question of whether a single T cell receptor looks at both molecules or whether there are two receptors, one specific for antigen, the other for self MHC (Figure 23-40). Although we do not yet know the answer, at present there is more evidence in favor of a single receptor. One experiment that argues for a single receptor involves fusing two T cells with different specificities to form a hybrid cell. One of the initial cells can recognize antigen A + MHC 1, while the second cell can recognize antigen B + MHC 2. If there were a distinct T cell receptor for each antigen and another T cell receptor for each MHC type, then we would expect the four types of receptors to mix freely, and the hybrid cell would have new recognition capabilities, namely, antigen A + MHC 2 and antigen B + MHC 1, as well as the original capabilities of its parent cells. This is not the case. The hybrid cell has only two specificities, the same ones as its parent cells. Since T cell receptors and the genes that code for them have recently been isolated, the question of whether one or two receptors are involved in T cell recognition of antigens and MHC should soon be answered. If it is true that a single T cell receptor recognizes antigen and MHC simultaneously, then antigens and MHC proteins must lie close together on the cell surface; in fact, they probably have to touch each other. It is hard to see how a vast array of different antigens could all interact with the same MHC proteins, but somehow this would have to be the case, and recently, evidence of such interaction has been obtained. Perhaps only weak, fleeting interactions are required between the antigen and MHC protein in order for them to be seen by the T cell receptors.

Helper T cells are also MHC-restricted, but unlike cytotoxic T cells, they recognize products of class II MHC genes. These genes specify cell surface proteins that are present only on certain cells of the immune system, including antigen-presenting cells. Before helper T cells can regulate the responses of other cells, they must first be activated by their cognate antigens. Accessory cells such as macrophages present the processed antigens to helper cells (see Figure 23-37), and this process requires the presence of class II antigens on the presenting cell. In most cases, helper T cells must also recognize class

Two-receptor model

One-receptor model
(dual specificity version)

One-receptor model
(altered self version)

Figure 23-40
Are there separate receptors on T cells for antigen and MHC recognition, or does a single receptor look at both molecules? Two models for the single-receptor hypothesis are shown. In one (dual specificity version), the single receptor is represented as having its two specificities distinctly separate; in the other, the single receptor recognizes antigen because the antigen binds to MHC and so makes the MHC protein look abnormal.

II antigens on the B cells they stimulate. Forcing cells to look at particular cell surface markers seems to be how the immune system controls cellular interactions. MHC gene products are like sign posts that tell cells when they have reached the correct destination.

The T Cell Receptor Is Composed of Two Polypeptide Chains, α and β[77-84]

In striking contrast to antibody molecules, which were among the first proteins whose structure was determined, the T cell receptor proved to be difficult even to find biochemically. This is because there was no T cell equivalent of the myeloma tumor with its abundant supply of single-specificity antibody molecules. For a long time, researchers supposed that T cell receptors were encoded, at least in part, by the same genes that code for antibodies. Since the two types of proteins perform the same function of recognizing innumerable foreign substances, it seemed reasonable to assume that the same genes were involved. The favored hypothesis was that T cell receptors are specified by the same variable-region genes as antibodies, but that they probably have their own constant-region genes, since they require completely different effector functions than antibodies. Like many good ideas, this one turned out to be completely wrong. The genes for T cell receptors are entirely distinct from those that code for antibodies. However, the two families of genes are clearly related evolutionarily.

T cell receptor proteins were finally identified several years ago, when it became possible to grow many types of cloned T cells in tissue culture. When certain cloned T cells are injected into rabbits or mice, they can elicit antibodies that react specifically with the immunizing T cell but not with any other T cell of a similar type. This extreme specificity for a single clone of T cells argues that the molecule(s) recognized by the antisera are the T cell receptor, in particular, its specific antigen binding portion. Furthermore, these antisera can frequently block T cell function, such as helper activity. Some of these antisera can precipitate a protein from the T cells. These proteins, now firmly identified as T cell receptors, are glycoproteins with molecular weights of 80 to 90 kdal. They are dimers that consist of one α polypeptide chain and one β polypeptide chain, each about 40 to 50 kdal, linked by a disulfide bond as well as noncovalent bonds.

The α and β Chains of the T Cell Receptor and a Related Polypeptide Called γ Resemble Light and Heavy Chains of Antibodies

At the same time that some immunologists were trying to purify T cell receptor proteins, others were trying to clone the genes that encode them. This was difficult, since no nucleic acid probes were available to hybridize with the desired genes and since virtually nothing was known about the T cell receptor protein at the time the studies began. The breakthrough came in 1984, when cDNA clones of first β-chain and then α-chain mRNAs were obtained. Figure 23-41 shows how this was done for a mouse β chain using a method called **subtraction hybridization.** The idea behind the experiment was to isolate cDNA clones of mRNAs that are expressed in T cells but not in B cells. These

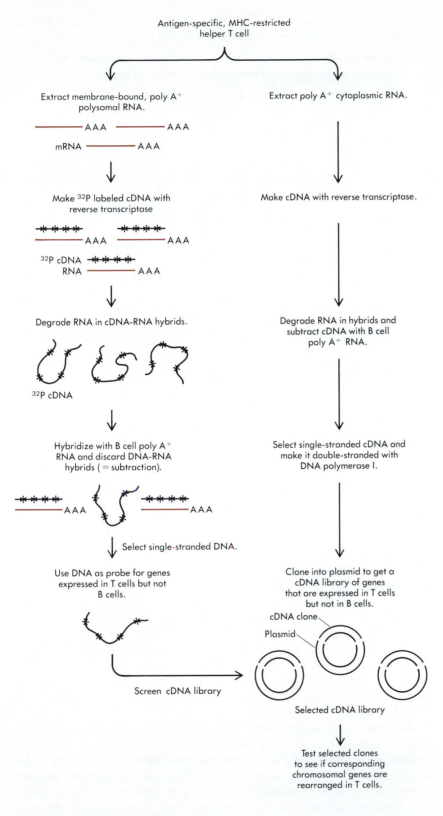

Figure 23-41
The experiment that first led to the isolation of a cDNA clone of the β chain of the mouse T cell receptor. The strategy used makes three assumptions about genes that encode the polypeptides that make up the T cell receptor: Since they encode membrane-anchored polypeptides, their mRNAs will be present on membrane-bound polysomes; they are expressed in T cells but not in B cells; and they undergo DNA rearrangements in T cells relative to embryonic cells or other cell types in the body.

two cell types express a very similar set of genes, sharing 98 percent or more of their mRNAs. Thus, if cDNA prepared from T cell mRNA is hybridized to B cell mRNA and all the cDNA strands that can hybridize are discarded (subtracted), the remaining cDNA should be greatly enriched for copies of T cell–specific mRNAs, including those for the T cell receptor polypeptides.

Figure 23-42

Proposed primary structure of the β chain of the T cell receptor and conformation of the α-β dimer believed to make up the receptor. The figure is based on data derived from nucleotide sequencing of DNA clones of the α- and β-chain genes. The leader peptide is removed from the mature chain. D and J indicate regions of the variable region of the β chain that are encoded by the D and J segments of DNA. So far, α-chain variable regions have been found to contain J but not D segments. The transmembrane and cytoplasmic domains of the chains are indicated. [After M. K. Davis, Y. Chien, N. R. J. Gascoigne, and S. M. Hedrick, *Immunol. Revs.* 8 (1984):235–258.]

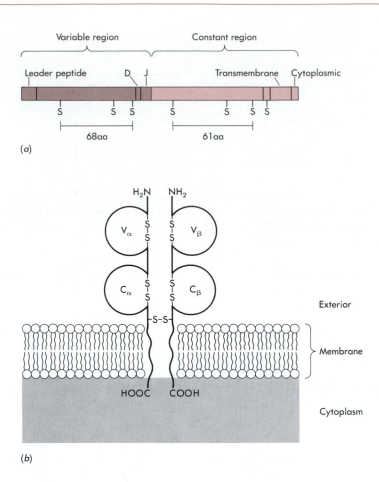

In addition to being expressed exclusively in T cells, genes coding for T cell receptors might be expected to be rearranged in T cell DNA much the way that antibody genes are rearranged in B cells. Some of the cDNA clones obtained in the preceding experiment turned out to correspond to genes with this property. Furthermore, when these clones were sequenced, they were found to encode proteins that had structural similarity and even amino acid sequence homology to antibody light and heavy chains. They could not be antibody genes, however, because they were not expressed in B cells and because their nucleotide sequences were clearly distinct from those of antibody genes. The definitive proof that they corresponded to T cell receptor genes came when amino acid sequences of the recently purified T cell receptor chains were finally obtained. These were very similar to those predicted from nucleotide sequences of the T cell–specific cDNA clones.

As soon as several different cDNA clones of α- and β-chain mRNAs from different T cells were sequenced, it was apparent that both chains have variable regions in their amino-terminal halves and constant regions in their carboxyl halves. These are followed on the carboxyl end by transmembrane domains that anchor the chains to the cell surface and then a short cytoplasmic tail (Figure 23-42). As with antibody molecules, the variable and constant regions are each about 110 amino acids long, as predicted by nucleotide sequences, and each of these homology units has cysteine residues that can form intra-chain disulfide bonds at positions similar to those in antibody variable and constant homology units. While the T cell receptor chains are more similar to antibody light chains in size, they are also similar to the membrane-bound form of heavy chains in that they, too, have membrane anchors. Moreover, both the α and β chains of the T cell receptor are glycosylated. From all this evidence, there is virtually no doubt that T cell receptors and antibodies are evolutionarily related proteins.

A completely unexpected finding to emerge from subtraction hybridization experiments was the isolation of another T cell–specific cDNA clone that also encodes an antibody-like protein and whose corresponding genes are also rearranged in T cells. It does not seem to correspond to the α or β chains of the T cell receptor proteins that have been purified, since it lacks canonical Asn-X-Thr/Ser sequences that signal addition of the N-linked sugar residues that are known to be present on these chains. The function of this third T cell receptor-like gene, named the **γ gene,** is a mystery at the moment. An intriguing clue to its function is that γ-gene mRNA is present at high levels in immature fetal thymocytes, while in contrast, α mRNA is low. In populations of mature T cells, the reverse is true. The β-gene mRNA is present in comparable amounts in both cell populations. These results suggest that the γ gene may be important during T cell differentiation. Conceivably, it plays a role in thymic education.

T Cell Receptor Genes Look Like Antibody Genes and Are Assembled by *V-D-J* Joining[85-90]

As soon as cDNA clones were available, it was easy to isolate the corresponding chromosomal genes for the α, β, and γ polypeptides, both in their germ line configurations and in their rearranged, functional forms in T cells. Information about the structure of these multigene families has been accumulating rapidly. The most striking thing about them is how similar their organization is to that of antibody gene families: There are multiple V, D, and J segments that must be assembled by DNA recombination to make functional V-region exons, and these lie upstream from just one or a few constant-region genes. The α, β, and γ multigene families lie on different chromosomes (Table 23-6).

The β-chain gene family resembles heavy-chain genes in having V, D, and J segments, but it is also reminiscent of the λ light-chain locus in having J segments in front of each of its constant-region genes (Figure 23-43). It is easy to see how extensive V-region diversity can

Table 23-6 Chromosome Locations of the T Cell Receptor Genes

Polypeptide Chain	Chromosome	
	Human	*Mouse*
α	14	14
β	7	6
γ	7	13

Figure 23-43
Current picture of the mouse T cell receptor β-chain locus. The *D, J,* and *C* segments are drawn approximately to scale, but there is little evidence available concerning the distance or even the orientation of most of the *V* segments relative to the two *D, J, C* clusters, although recently one *V* segment was found on the 3' side of C_{β_2} and in the opposite transcriptioned orientation (not shown). This particular exon is joined to a *D* segment by an inversion at the DNA level. To draw the top part of the figure approximately to scale, the *D* and *J* elements are represented as black lines, as are the second exons of the two constant region genes. A close-up (inset box) shows the orientation of the heptamer-nonamers that flank *V, D,* and *J* segments and the nucleotide distance between them (± 1 nucleotide). The structure of a functional β-chain gene obtained after *D-J* and *V-D-J* joining is indicated.

be generated in the β chains. Although they have fewer *V* segments than heavy-chain genes (there may be as few as 20 V_β segments, compared to 200 to 300 V_H segments), β genes appear to have multiple D_β segments, and they have more J_β segments than do heavy-chain genes. Assuming random association of these segments, great variability is already possible. Furthermore, as with antibody genes, there is variability in the positions of the *V-D* and *D-J* joints, which can lead to amino acid sequence variability at these positions. Another mechanism for generating diversity, not seen with antibody genes, is that the D_β segments can be read in all three reading frames. In addition, as with heavy-chain genes, there is evidence of N nucleotide addition at the *V-D* and *D-J* joints in rearranged β genes. Whether fully rearranged β-chain genes undergo further mutation (somatic hypermutation) in their V-coding portions is not yet known.

Very similar recombination systems are probably used to join the *V*, *D*, and *J* segments of T cell receptor genes and antibody genes. Similar sequences flank the segments that undergo joining: namely, heptamer-nonamers separated by 12 or 23 base pairs, as we have already described for antibody genes (see Figure 23-27). However, if the same joining rules that have been defined for antibody gene segments are used, then V_β segments would be able to skip over D_β segments and join directly to J_β segments (see Figure 23-43). Whether this is used to increase diversity remains to be seen.

There are two β-chain constant-region genes that encode almost identical constant-region polypeptides. Each is composed of four exons that correspond closely to the structural domains of the constant region: One exon encodes the amino-terminal 125 amino acids that are outside the cell; the second, a mini-exon, encodes the 6-amino-acid hinge region containing the cysteine residue probably involved in disulfide bonding to the α chain; the third specifies 36 hydrophobic amino acids of the transmembrane domain; and the fourth encodes the 5 cytoplasmic amino acids as well as the 3' untranslated region of the mRNA. The β chain of the T cell receptor is expressed in both helper and cytotoxic T cells; however, it does not seem to be used in suppressor T cells.

The α-chain genes include just one constant region "gene," which is composed of four exons. There are multiple V_α segments and an unusually large number of J_α segments (at least 18). So far, there is no evidence that D_α segments are present. Most striking is the enormous distance between J_α segments and the constant-region exons: Some fully assembled α-chain genes would generate primary RNA transcripts of 60 kilobases or more. The γ-chain genes seem to have a smaller number of *V* segments than α genes and have several *J-C* clusters. Whether they have D_γ segments is still uncertain. Simplified diagrams of the mouse α and γ genes are shown in Figure 23-44. The β-chain genes and the three antibody gene loci are included for comparison.

Antigen Binding Primes T Cells to Respond to the Growth Factor Interleukin 2[91, 92, 93]

When T cells leave the thymus, they are ready to respond to antigens; but as with B cells, this response usually cannot occur by a simple collision between, for example, a cytotoxic T cell and a virus-infected fibroblast. Rather, antigen-stimulated **blast transformation** of a T cell must take place at the surface of an accessory cell such as a macrophage. These accessory cells, which are located throughout the body but particularly concentrated in secondary lymphoid organs such as lymph nodes, present antigens in some special way that primes the T cell for further maturation. As we discussed for B cells, this presentation may resemble the action of lymphocyte mitogens. In any case, seeing antigen in this form induces the T cell to express cell surface receptors for the T cell growth factor interleukin 2. Antigen also stimulates T cells to release Il-2. Helpers are particularly potent producers, and this in part may explain how they stimulate cytotoxic T cells to divide.

Tissue Rejection Antigens and Immune Response Genes Map to the MHC Locus[94–97]

The major histocompatibility complex (MHC) is a cluster of genes that occupies some 2000 to 4000 kilobase pairs of DNA. Four types of genes, discovered independently, map to the complex. The first, discovered more than 45 years ago, encodes the *transplantation antigens* that cause rapid rejection of skin and organ grafts between individuals and thus gives the *MHC* locus its name (*histo* means "tissue"). As we have seen, these are the class I antigens that cytotoxic T cells use to distinguish self from nonself. The second set of genes, originally called *immune response genes* (Ir), determines whether animals can mount a strong or a weak response to particular antigens. Immune response genes encode the class II MHC proteins that helper T cells look at when they see antigens on accessory cells and on B lymphocytes (and, very rarely, class I gene products). Presumably, some antigens cannot interact well with certain MHC proteins. As a result, these antigens are unable to stimulate helper T cells, and the animal is unable to mount a good immune response. The third set of genes, called *Tla* and *Qa* genes, encodes proteins that were discovered as cell surface antigens on lymphocytes. They are classified as class I *MHC* genes because their products have structural similarity to the class I transplantation antigens, but the function of the *Tla* and *Qa* genes is not yet known. The fourth type of gene to map to the *MHC* locus encodes components of the *complement system*. Maps of the *MHC* locus in mouse and human are shown in Figure 23-45. In mice, the

Figure 23-44
Simplified diagram of the genes encoding antibodies and T cell receptors. Boxes indicate V-region-coding segments or C "genes." (The latter have been simplified so that a single box represents the several exons that encode most C regions.) Vertical lines indicate D- or J-gene segments. It is not yet known whether the α and γ genes have D segments. Distances are approximately to scale unless they are too large or are unknown. In these cases, they are indicated by breaks in the lines. [After Winoto et al., *Nature* 316 (1985):832–836.]

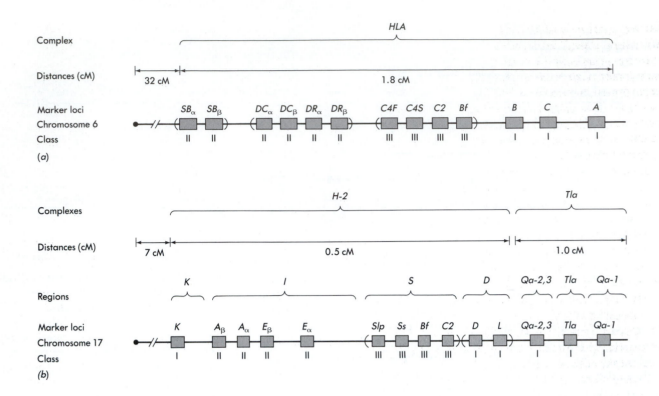

Figure 23-45

(a) Genetic map of the human *HLA* complex located on chromosome 6. Distances are in centimorgans. Black circle at left end of chromosome represents the centromere. Marker loci were defined genetically, and the order of those within parentheses is not known. Note that the boxes do not represent single genes. Class I transplantation antigens are designated A, B, and C. Class II antigens are designated SB, DC, and DR. Class III genes encode the complement components designated C2, C4, and *Bf*. (b) Genetic map of the mouse *H-2* complex located on chromosome 17. Distances in centimorgans. Organization of the locus is strikingly similar to that of the human *HLA*, except that class I genes, *K, D,* and *L*, are separated. There are also *Tla* and *Qa* genes, whose possible equivalent is now being discovered in humans. Class I genes encode transplantation antigens. Class II genes encoded by the *I* region include immune response genes. Class III genes of the *S* region, *Ss, Slp, C2,* and *Bf*, encode some serum proteins of the complement system, important in immune responses. [After *Immunology*, by L. E. Wood, I. L. Weissman, W. B. Wood, and J. H. Wilson. 1984. The Benjamin/Cummings Publishing Co.]

MHC is referred to as *H-2*, while in humans, it is called **HLA (human leukocyte antigen complex)**.

Except for the complement components, *MHC* genes all specify cell surface proteins. Furthermore, many of these proteins serve as signals by which the animal distinguishes self from nonself, and they are intimately involved in responding to nonself. One of the most striking features of this recognition system is that it is incredibly polymorphic. The *K* gene of the mouse, for example, specifies a class I transplantation antigen, and more than 55 alleles of *K* have been identified. Thus, even though each inbred mouse has a single type of *K*-gene product, any two mice are likely to have different *K* genes. This extreme polymorphism is also seen at the mouse class I *D* and *L* genes, the human class I *HLA-A, B,* and *C* genes, and certain class II genes. Since cells express several of these genes simultaneously (e.g., *K, D,* and *L* in the mouse), and since animals express the *MHC* genes from both chromosomes (one from the mother and one from the father), it is apparent why the cells from each individual are likely to be different.

It seems unlikely that *MHC* genes are polymorphic just to ensure graft rejection between individuals. However, we can imagine one very good reason why it is advantageous for *MHC* genes to be so polymorphic. Antigens are seen in the context of *MHC* gene products, and different antigens would be expected to interact effectively with some but not all MHC proteins. By having extensive polymorphism in MHC genes, there is a better chance that each individual will carry at least one protein that can interact with any particular antigen and so result in a good immune response to this substance. Even more important, polymorphism greatly increases the probability that at least some individuals in the population will survive should a new antigen, such as a killer virus, suddenly appear on the scene.

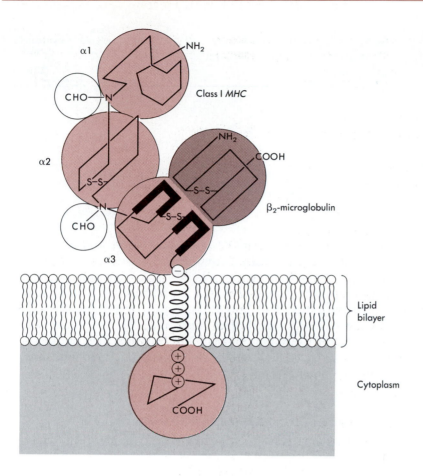

α1

NH₂

CHO—N

Class I *MHC*

α2

NH₂

COOH

S–S

S–S

β₂-microglobulin

N

CHO

S–S

α3

⊖

Lipid
bilayer

⊕
⊕
⊕

Cytoplasm

COOH

Figure 23-46
Model of a class I molecule in association with β_2 microglobulin. Three external domains and the cytoplasmic domain of the class I polypeptide are indicated by large circles. Small circles represent carbohydrate side chains. Charged amino acids, important in locking the transmembrane domain into the lipid bilayer, are indicated. Intrachain disulfide bridges in the third external domain and in β_2 microglobulin are similar in position to those in antibody polypeptides, and there is amino acid sequence homology to antibody molecules in these two domains. [From Strominger et al., *Scand. J. Immunol.* 11 (1980):573–592.]

Genes and Proteins of the MHC[98–103]

An enormous surprise came when the structure of HLA and H-2 class I proteins were determined. They turned out to have striking structural similarities to antibody molecules! Class I gene products are integral membrane glycoproteins with molecular weights of 40,000–45,000. They are associated noncovalently with a 96-amino-acid-long peptide, **β_2 microglobulin.** β_2 microglobulin has amino acid sequence homology to heavy-chain constant-region domains, although it is encoded by a gene that is entirely separate from antibody genes. The transplantation antigens have five distinct regions: Three are external to the cell, one spans the membrane, and one is the cytoplasmic carboxyl terminus of the protein. Each of the three external domains is 90 amino acids long, about the length of an antibody homology unit, and two of these have the antibody-like disulfide bridges spanning about 65 amino acids (Figure 23-46). β_2 microglobulin associates with the domain closest to the cell surface, and this domain, like β_2, has amino acid sequence homology to a heavy-chain constant-region homology unit. The structure of this paired unit probably resembles a comparable unit in an antibody molecule.

The genes that encode class I proteins have multiple exons, usually eight. As with antibody genes and T cell receptor genes, the coding capacity of these exons corresponds closely to distinct protein domains (Figure 23-47). Class II *MHC* gene products have structural similarities to class I molecules and to antibodies, too. They are heterodimers composed of α and β chains. Each chain has two domains outside the cell that are about 90 amino acids long, and three of these four domains have centrally placed disulfide bridges.

Recently, over 800 kilobase pairs of DNA from the mouse *MHC* locus have been molecularly cloned on cosmids. A surprising result from analyzing this DNA is that there are many more class I *MHC* genes in the *Tla-Qa* region than there are in the *H-2* region (Figure 23-48). It seems likely that these genes, too, will turn out to encode proteins important in cell-cell interactions, perhaps interactions that occur during development.

Evolution of the Immunoglobulin Supergene Family[103]

It has been known for more than two decades that light and heavy chains of antibody molecules are built up by the repetition of structurally related homology units. As we learned, the basic unit is about 110 amino acids long, has a disulfide bond that bridges about 65 amino acids, and can adopt a three-dimensional configuration known as the immunoglobulin fold. Furthermore, pairs of homology units can fit together to make domains with distinct functions, for example, antigen binding or effector functions. In antibody genes, single exons encode approximately one V or one C homology unit, and it is virtually impossible to look at the light- and heavy-chain gene clusters without imagining their evolution by repeated duplications of V and C exons.

Not surprisingly, T cell receptors for antigen also are built up from homology units of the immunoglobulin type. Far more surprising, however, the MHC class I and II molecules also employ this structure, and so does the class I bound polypeptide β_2 microglobulin (Figure 23-49). Thy-1, the surface antigen used for T cell identification in mice, is yet another member of the family, being composed of a single membrane-anchored immunoglobulin-like homology unit. In addition, three other surface proteins—the poly-Ig receptor that me-

Figure 23-48
Location of individual genes that have been cloned within the genetically defined region of the mouse *H-2/Tla* complex. Genes are indicated as color or gray boxes. Color boxes represent functional genes, while gray boxes represent genes that may be nonfunctional. Genes that are linked by a line have been physically linked by cloning. The order of gene clusters in separate regions is not known. Note that there are many more genes in the *TlA/Qa* region than in the *K* and *D* regions. [Adapted from A. Winoto, M. Steinmetz, and L. Hood, *Proc. Natl. Acad. Sci.* 80 (1983):3425.]

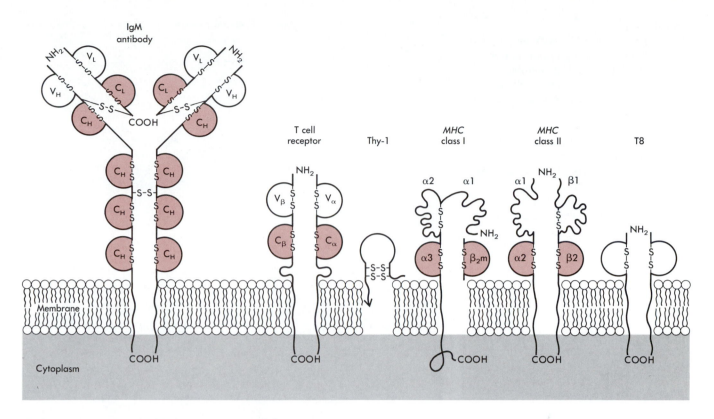

Figure 23-49

A highly schematic representation of the related domains and membrane orientations of six proteins encoded by members of the immunoglobulin supergene family. Each circle represents a domain of about 110 amino acids, usually with an intrachain disulfide bridge (S–S), as indicated. Domains with homology to constant-region domains of antibody polypeptides are shaded. Note that the antibody molecule shown is an IgM type. IgG, a more common immunoglobulin molecule in serum, would have only three C_H domains, the second domain having perhaps diverged to become the hinge region.

diates the transport of polymeric immunoglobulin molecules across endothelial cells and L3T4 and T8, cell surface proteins on certain T cells—are also constructed from this basic unit. These striking structural similarities imply that some primitive gene gave rise to the genes that encode this whole array of proteins.

A multigene family is a group of homologous genes with similar functions. The name **supergene family** has been given to a set of multigene families and single-copy genes that are related by sequence but not necessarily related in function. The proteins described in the preceding paragraphs belong to such a family. Figure 23-50 shows a scheme that has been proposed to explain the evolution of the immunoglobulin supergene family. The scheme is based on similarities in sequence and on the exon-intron structure of the present-day members of the family. The primordial gene is imagined to have encoded a cell surface protein, perhaps with a leader peptide and a transmembrane domain as well as an immunoglobulin homology unit. An early event may have been the partial duplication of this gene to yield separate genes for V and C homology units, since these have little amino acid sequence homology remaining today. This early duplication would allow separate regions to evolve that are specialized for membrane anchoring or for recognition.

Even though members of the immunoglobulin supergene family have evolved to have different functions, many encode proteins that interact with one another. For example, poly Ig receptors bind to antibodies, and T cell receptors may bind to MHC proteins. Did the fact that immunoglobulin homology units have a three-dimensional structure that gives them the ability to bind to one another contribute to the evolution of this very complex multicomponent system? More structural analysis of the members of this family will be needed to answer this intriguing question. The immunoglobulin supergene

Figure 23-50
Hypothetical scheme to explain the evolution of the immunoglobulin supergene family. The earliest member of the family may have encoded a cell surface protein. An early step in the evolution was the duplication of this gene and the divergence of the two resulting genes to encode V and C homology units. The protein structure of the present-day products of the genes are indicated at the bottom of the figure. The T8 molecule of humans has a mouse equivalent called Lyt 2. [From L. Hood, M. Kronenberg, and T. Hunkapiller, *Cell* 40 (1985):225–229.]

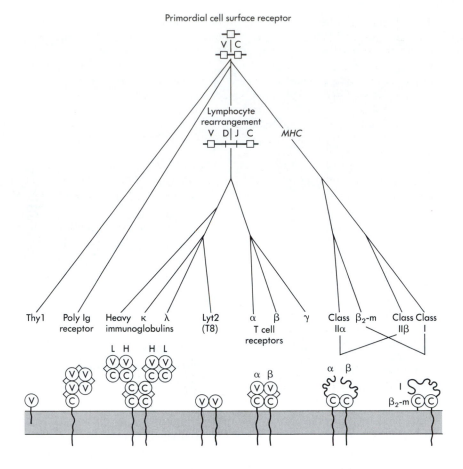

family occupies a significant fraction of the vertebrate genome, perhaps as much as 0.1 to 0.5 percent. If other complex biological systems evolved in a similar way, then we may ultimately be able to trace the evolution of vertebrate protein-coding genes to just a thousand or so ancestral genes.

Summary

Multicellular organisms have many defenses against invasions by microorganisms. Many of these defenses (e.g., an impermeable outer skin or the antiviral agent interferon) are relatively nonspecific. But in addition, vertebrates have evolved an immune system that provides specific defenses against almost any microorganism or other foreign substance that enters the body. Foreign substances that elicit an immune response are called antigens. The immune system has two types of defenses against antigens (microorganisms): Circulating proteins, called antibodies or immunoglobulins (Ig), bind to antigens and inactivate them or cause them to be destroyed; and specialized cells with surface receptors capable of recognizing specific antigens attack and destroy the antigens. The first type of response is referred to as humoral immunity; the second is called cellular immunity.

Immune responses are carried out by lymphocytes. These are small, round cells that originate in the bone marrow of adult animals and, after they mature, circulate within the lymph and circulatory systems. There are two major classes of lymphocytes: B lymphocytes synthesize antibodies, while T lymphocytes carry out cellular immune responses using membrane-anchored, antigen-specific T cell receptors. The ability of antibody molecules and T cell receptors to interact specifically with antigens is a consequence of their particular amino acid sequences. The immune system can recognize tens of thousands of different antigens. Even before it meets up with an antigen, the body synthesizes millions of antibody molecules and T cell receptors with slightly different amino acid sequences. Among this vast repertoire of molecules are invariably some that can bind to any incoming antigen.

According to the now-accepted theory of clonal selection, each B lymphocyte synthesizes an antibody of just one specificity, and each T cell has just one type of antigen receptor. Before encountering the antigen capable of binding its particular antibody, each B cell carries its antibodies as membrane-bound receptors. An encounter with antigen stimulates the cell to divide and also to differentiate to a mature plasma cell that secretes antibody capable of binding the stimulating antigen. This sequence of events applies to T cells as well, except that their antigen receptors remain bound to the cell surface after antigen stimulation.

Gene cloning finally provided the explanation for the genetic basis of immunological diversity and specificity, but the first insights came from comparing the amino acid sequences of many different antibody molecules. This was possible because cancers of B cells (called myelomas) yield large populations of cells producing a single type of antibody molecule.

Typical antibody molecules are tetramers composed of two identical light chains, each about 220 amino acids long, and two identical heavy chains, each about 440 amino acids long. The molecule is Y or T shaped and has two identical antigen binding sites in the arms of the Y. These sites are formed by the combination of part of the light chain and part of the heavy chain. When light and heavy chains of different antibody molecules are sequenced, it is apparent that there is extensive amino acid sequence variability in their amino-terminal regions (approximately the first 110 amino acids), while the remainder of the molecule belongs to one of a few types. The amino-terminal portion is called the variable, or V, region; the remainder is the constant, or C, region. Appropriately, it is the V regions of the light and heavy chains that together form the antigen binding sites. The most variable portions of the V regions, called the hypervariable regions or complementarity determining regions (CDRs), line the antigen binding pocket.

In the case of light chains, there are two major types of constant regions that define two types of light chains, κ (kappa) and λ (lambda). In the case of heavy chains, there are five major types of constant regions that define five classes of antibodies (IgM has a μ heavy chain defined by the sequence of its constant region, IgD has a δ heavy chain, IgG has a γ chain, IgA has an α chain, and IgE has an ε chain). The classes defined by the heavy-chain constant regions are important because these regions impart distinct biological functions to the different types of molecules.

Amino acid sequencing made it clear that antibody specificity and diversity are achieved by the joining of thousands and thousands of different variable regions to a few constant regions of either the light- or heavy-chain type. This led some immunologists to postulate that there might be separate genes for the variable and constant regions of the proteins. Genetic studies showed that the genes for heavy chains, κ light chains, and λ light chains lie on different chromosomes and that the genes for V regions are near to the genes for C regions. The central question was, Are there separate genes for all the different V regions, or is the variability in V-region sequence generated by some sort of mutational process within B cells? These two contrasting explanations for antibody diversity were called the germ line theory and the somatic theory.

When the genes encoding λ and κ light chains and heavy chains were cloned, it was clear that the somatic theory was correct. No antibody genes are inherited intact: They are assembled in somatic cells from multiple fragments of DNA and the assembly process generates enormous diversity in the sequences encoding variable regions. Consistent in spirit with the germ line theory, there are many inherited gene fragments that can take part in the assembly, and this, too, contributes to diversity. DNA encoding V regions lies near to, and probably upstream of, DNA encoding constant regions. However, the multigene families that encode the light and heavy chains are very large, with the heavy-chain locus alone probably occupying several thousand kilobase pairs of DNA.

Light-chain V-region "genes" are assembled by a single DNA recombination event that joins a V_L fragment encoding about 90 percent of the V region to a J_L fragment encoding about 10 percent. After this joining, the assembled V-region gene (which is really an exon) is still separated (by about 2 kilobase pairs) from a single exon encoding the constant region. The two are joined at the level of RNA by the standard RNA-splicing mechanism. Heavy-chain V-region genes are assembled by two DNA recombination events. The first joins a small D segment to a J_H segment, while the second joins a V_H segment to the rearranged D-J fragment. The resulting V-D-J exon is separated from the closest constant region exons by several kilobase pairs. There is a separate constant-region gene for each type of heavy chain (8 in the mouse), although these genes really are clusters of exons. In general, each exon encodes a structurally or functionally discrete portion of the constant region.

There are two ways in which information encoding a heavy-chain V region can be joined to a constant-region gene. An assembled V-D-J exon can be joined to either of the two closest C-region genes, μ or δ, by RNA splicing of long precursor transcripts. Alternatively, class switching, the joining of a V-D-J exon to other constant-region exons (γ, α, or ε) can occur by a third DNA rearrangement, the switch recombination, which moves the V region exon near to exons encoding these constant regions.

V-J or V-D-J joining involves recombination at specific sites, and specific sequences govern the event. Each DNA segment that can be joined is bordered by a conserved palindromic sequence of 7 bases, the heptamer, followed by either 12 (± 1) or 23 (± 2) nucleotides whose sequence is unimportant, followed by a conserved 9-base nonamer. Two fragments can only be joined if their heptamer-nonamers are oriented in opposite directions and are separated by 12 or 23 base pairs. These rules prevent incorrect fragments from joining. The recombination event does not always occur at the same nucleotide relative to the heptamer sequence. As a result, joining of the same two DNA segments can lead to different amino acid sequences, a phenomenon called junctional diversity that contributes greatly to V-region diversity. In addition, at D-J and V-D-J joints, N nucleotides can be added in a template-free fashion to further increase diversity.

In mice, at the heavy-chain locus, there are 200 to 300 V_H segments, about 10 D segments, and 4 J_H segments. At the κ light-chain locus, there are 100 to 300 V_κ segments and 4 J_κ segments. At the λ locus, there are 3 V_λ fragments and 3 J_λ fragments. As far as we know, any V segment can join to any D segment, for example. This random association, called combinatorial joining, generates great diversity. Since the D segment and the D-J and V-D-J joints or the V-J joint in light chains encode the third hypervariable regions, a part of the chain that makes the antigen binding cavity, it is clear why the above strategies generate so much diversity in antigen recognition.

A last genetic source of diversity in amino acid sequences is the subjection of fully assembled antibody genes to mutational processes that lead to single amino acid changes scattered randomly about the V region. This process is called somatic hypermutation.

So far, we have listed five sources of diversity in amino acid sequence of light and heavy chains: multiple germ line V, D, and J segments, random association of these fragments (called combinatorial joining) variability in the position of joining (called junctional diversity), N nucleotide addition in heavy chains, and somatic hypermutation. Finally, since antigen binding sites are constructed from the V regions of both light and heavy chains, if any light chain can associate with any heavy chain, then the number of antibody molecules with different antigen binding sites would be the number of different light chains multiplied by the number of heavy chains. In the mouse, this theoretical number greatly exceeds a staggering 300 million!

The rearrangement and expression of antibody genes are carefully regulated in conjunction with B cell differentiation. First, a stem cell undergoes *D-J* rearrangements at both of its heavy-chain loci. Then V_H segments are joined to the *D-J* fragment. As soon as a functional gene is made, further rearrangements stop. The first chains that are made use the μ constant region, so the cell is now a μ-positive pre–B cell. Light chains are then assembled by *V-J* joining. As soon as a functional chain that can associate with the cytoplasmic μ heavy chain is made, further rearrangements cease, and the assembled antibodies move to the cell surface. The cell with its surface immunoglobulin is now called a B_μ cell. At this point, the μ chains terminate with membrane-anchoring amino acids. All of the preceding steps occur in the absence of antigen. The next step seems to be to a $B_{\mu+\delta}$ cell with two types of antibody molecules on its surface. These two types have the same light-chain and heavy-chain V regions, but alternative RNA processing allows the heavy-chain V region to be joined to either the μ or δ constant region. It may be these $B_{\mu+\delta}$ cells, rather than B_μ cells, that are now competent to respond to antigens. In any case, in response to antigen, B_μ or $B_{\mu+\delta}$ cells differentiate to plasma cells and secrete μ antibodies. They also differentiate to produce memory cells. In addition, they will undergo class switching to transfer their heavy-chain V regions to γ, α, and ε constant regions so that antibodies of the same specificity but different classes can be made.

The progression of a competent virgin B cell to an antibody-secreting plasma cell requires more than a collision with its cognate antigen. In addition, T lymphocytes called helper T (T_H) cells must be present in most cases. These cells use their T cell receptors to interact with antigens that have been taken up and presented to them by a class of accessory cells called antigen-presenting cells. Thus, three cells interact to stimulate the final differentiation of the B cell.

Cytotoxic T cells are the cellular immune system's equivalent of antibody molecules, and they carry antigen-specific surface T cell receptors. In contrast to antibodies, however, T cell receptors can only see antigens when the T cell can also see normal cell surface proteins simultaneously. This mechanism ensures that T cells will be restricted to dealing with antigens located on cell surfaces and will not be confused by freely circulating antigens. Whether the same receptor looks at both molecules—antigen and normal cell surface protein—is not yet certain. The normal surface proteins that T cells see are coded for by the major histocompatibility complex (MHC). Cytotoxic T cells look at proteins coded for by the class I *MHC* genes. Helper T cells, which also can see antigens only if they look at normal cell surface proteins simultaneously, look at products of class II *MHC* genes.

Recently, T cell receptors for antigen and the genes that code for them have been isolated. The receptors are dimers composed of one α chain and one β chain. They have variable regions in their amino-terminal halves and constant regions in the carboxyl termini. The genes that encode them look very much like antibody genes and are clearly related evolutionarily. They undergo DNA rearrangement to produce functional genes. In addition to genes for the α and β chains, a third set of genes called γ has been discovered, and these genes also undergo rearrangement in T cells. The function of these genes is unknown. Possibly they are involved in T cell differentiation and thymic education.

The *MHC* is a huge multigene locus that should probably be considered as much a part of the immune system as the three antibody gene loci and the T cell receptor gene loci. It occupies several thousand kilobase pairs of DNA. It encodes the cell surface proteins that are essential to T cell function. Remarkably, the class I and II *MHC* gene products have structural similarity to antibodies. This has led to the notion that there is an extended gene family, called a supergene family, that may have evolved from a single primordial gene.

Bibliography

General References

Benacerraf, B., and E. R. Unanue. 1979. *Textbook of Immunology*. Baltimore: Williams & Wilkins.

Eisen, H. 1980. *Immunology*. 2nd ed. New York: Harper & Row.

Hood, L. E., I. L. Weissman, W. B. Wood, and J. H. Wilson. 1984. *Immunology*. 2nd ed. Menlo Park, Calif.: Benjamin/Cummings.

Kimball, J. W. 1983. *Introduction to Immunology*. New York: Macmillan.

Nisonoff, A. 1985. *Introduction to Molecular Immunology*. 2nd ed. Sunderland, Mass.: Sinauer.

Cited References

1. Hood, L. E., I. L. Weissman, W. B. Wood, and J. H. Wilson. 1984. *Immunology*. 2nd ed. Menlo Park, Calif.: Benjamin/Cummings.

2. Nisonoff, A. 1985. *Introduction to Molecular Immunology*. 2nd ed. Sunderland, Mass.: Sinauer.

3. Eisen, H. 1980. *Immunology*. 2nd ed. New York: Harper and Row.

4. Kimball, J. W. 1983. *Introduction to Immunology*. New York: Macmillan.

5. Benacerraf, B., and E. R. Unanue. 1979. *Textbook of Immunology.* Baltimore: Williams & Wilkins.

6. Gowans, J. L., and E. J. Knight. 1964. "The Route of Re-Circulation of Lymphocytes in the Rat." *Proc. Roy. Soc. London (B)* 159:257–282.

7. Miller, J. F. A. P. 1961. "Immunological Functions of the Thymus." *Lancet.*

8. Landsteiner, K. 1945. *The Specificity of Serological Reactions.* Rev. ed. Cambridge, Mass.: Harvard University Press. Reprinted by Dover, New York, 1962.

9. Westhof, E., D. Altschuh, D. Moras, A. C. Bloomer, A. Mondragon, A. Klug, and M. H. V. Van Regenmortel. 1984. "Correlation Between Segmental Mobility and the Location of Antigenic Determinants in Proteins." *Nature* 311:123–126.

10. Tainer, J. A., E. D. Getzoff, H. Alexander, R. A. Houghten, A. J. Olson, R. A. Lerner, and W. A. Hendrickson. 1984. "The Reactivity of Anti-Peptide Antibodies Is a Function of the Atomic Mobility of Sites in a Protein." *Nature* 312:127–134.

11. Haber, E. 1964. "Recovery of Antigen Specificity After Denaturation and Complete Reduction of Disulfides in a Papain Fragment of Antibody." *Proc. Nat. Acad. Sci.* 52:1099–1106.

12. Burnet, F. M. 1959. *The Clonal Selection Theory of Acquired Immunity.* Cambridge, Eng.: University Press.

13. Ada, G. L., and P. Byrt. 1969. "Specific Inactivation of Antigen-Reactive Cells with I^{125}-Labeled Antigen." *Nature* 222:1291–1292.

14. Owen, R. D. 1945. "Immunogenic Consequence of Vascular Anastomoses Between Bovine Twins." *Science* 102:400–401.

15. Billingham, R. E., L. Brent, and P. B. Medawar. 1953. "Actively Acquired Tolerance of Foreign Cells." *Nature* 172:603–606.

16. Felton, L. D., G. Kauffmann, B. Prescott, and B. Ottinger. 1955. "Studies on the Mechanism of Immunological Paralysis Induced in Mice by Pneumococcal Polysaccharides." *J. Immunol.* 74:17–26.

17. Edelman, G. M. 1970. "The Structure and Function of Antibodies." *Sci. Amer.* 223:34–42.

18. Porter, R. R. 1973. "Structural Studies of Immunoglobulins." *Science* 180:713–716.

19. Nisonoff, A., J. E. Hopper, and S. B. Spring. 1975. *The Antibody Molecule.* New York: Academic Press.

20. Kohler, G., and C. Milstein. 1975. "Continuous Cultures of Fused Cells Secreting Antibody of Predefined Specificity." *Nature* 256:495–497.

21. Milstein, C. 1980. "Monoclonal Antibodies." *Sci. Amer.* 243:66–74.

22. Potter, M. 1972. "Immunoglobulin-Producing Tumors and Myeloma Proteins of Mice." *Physiol. Rev.* 52:631–719.

23. Hilschmann, N., and L. C. Craig. 1965. "Amino Acid Sequence Studies with Bence-Jones Proteins." *Proc. Nat. Acad. Sci.* 53:1403–1409.

24. Wu, T. T., and E. A. Kabat. 1970. "An Analysis of the Sequences of the Variable Regions of Bence Jones Proteins and Myeloma Light Chains and Their Implications for Antibody Complementarity." *J. Exp. Med.* 132:211–250.

25. Edelman, G. M. 1969. "The Covalent Structure of an Entire γG-Immunoglobulin Molecule." *Proc. Nat. Acad. Sci.* 63:78–85.

26. Amzel, L. M., and R. J. Poljak. 1979. "Three Dimensional Structure of Immunoglobulins." *Ann. Rev. Biochem.* 48:961–967.

27. Silverton, E. W., M. A. Navia, and D. R. Davies. 1977. "Three-Dimensional Structure of an Intact Human Immunoglobulin." *Proc. Nat. Acad. Sci.* 74:5140–5144.

28. Edmonson, A. B., K. R. Ely, E. E. Abola, M. Schiffer, and N. Panagiotopoulos. 1975. "Rotational Allomerism and Divergent Evolution of Domains in Immunoglobulin Light Chains." *Biochemistry* 14:3953–3961.

29. Colman, P. M., J. Diesenhofer, R. Huber, and W. Palm. 1976. "Structure of the Human Antibody Molecule 'Kol', an Electron Density Map at 5 A Resolution." *J. Mol. Biol.* 100:257–278.

30. Hood, L. E., and D. W. Talmadge. 1970. "Mechanism of Antibody Diversity: Germ Line Basis for Variability." *Science* 168:325–334.

31. Lennox, E. S., and M. Cohn. 1967. "Immunoglobulins." *Ann. Rev. Biochem.* 36:365–406.

32. Dreyer, W. J., and J. C. Bennett. 1965. "The Molecular Basis of Antibody Formation: A Paradox." *Proc. Nat. Acad. Sci.* 54:864–869.

33. Hozumi, N., and S. Tonegawa. 1976. "Evidence for Somatic Rearrangement of Immunoglobulin Genes Coding for Variable and Constant Regions." *Proc. Nat. Acad. Sci.* 73:3628–3632.

34. Fudenberg, H. H., A. C. Wang, J. R. L. Pink, and A. S. Levin. 1971. *Ann. N.Y. Acad. Sci.* 190:501–506.

35. Brack, C., M. Hirama, R. Lenhard-Schuller, and S. Tonegawa. 1978. "A Complete Immunoglobulin Gene Is Created by Somatic Recombination." *Cell* 15:1–14.

36. Seidman, J. G., E. E. Max, and P. Leder. 1979. "A k-Immunoglobulin Gene Is Formed by Site-Specific Recombination Without Further Somatic Mutation." *Nature* 280:370–375.

37. Eisen, H. N., and E. B. Reilly. 1985. "Lambda Chains and Genes in Inbred Mice." *Ann. Rev. Immunol.* 3:337–365.

38. Early, P., H. Huang, M. Davis, K. Calame, and L. Hood. 1980. "An Immunoglobulin Heavy-Chain Variable Region Gene Is Generated from Three Segments of DNA: V_H, D and J_H." *Cell* 19:981–992.

39. Sakano, H., R. Maki, Y. Kurosawa, W. Roeder, and S. Tonegawa. 1980. "Two Types of Somatic Recombination Are Necessary for the Generation of Complete Immunoglobulin Heavy-Chain Genes." *Nature* 286:676–683.

40. Sakano, H., Y. Kurosawa, M. Weigert, and S. Tonegawa. 1981. "Identification and Nucleotide Sequence of a Diversity DNA Segment (D) of Immunoglobulin Heavy-Chain Genes." *Nature* 290:562–565.

41. Sakano, H., K. Huppi, G. Heinrich, and S. Tonegawa. 1979. "Sequences at the Somatic Recombination Sites of Immunoglobulin Light-Chain Genes." *Nature* 280:288–294.

42. Max, E. E., J. G. Seidman, and P. Leder. 1979. "Sequences of Five Potential Recombination Sites Encoded Close to an Immunoglobulin k Constant Region Gene." *Proc. Nat. Acad. Sci.* 76:3450–3454.

43. Kurosawa, Y., and S. Tonegawa. 1982. "Organization, Structure, and Assembly of Immunoglobulin Heavy Chain Diversity DNA Segments." *J. Exp. Med.* 155:201–218.

44. Alt, F., and D. Baltimore. 1982. "Joining of Immunoglobulin Heavy Chain Gene Segments: Implications from a Chromosome with Evidence of Three D-J_H Fusions." *Proc. Nat. Acad. Sci.* 79:4118–4122.

45. Cohn, M., B. Blomberg, W. Geckeler, W. Raschke, R. Riblet, and M. Weigert. 1974. "First Order Considerations in Analyzing the Generator of Diversity." In *The Immune System Genes, Receptors, Signals,* ed. E. Sercarz, A. R. Williamson, and C. F. Fox. New York: Academic Press, pp. 89–117.

46. Bernard, O., N. Hozumi, and S. Tonegawa. 1978. "Sequences of Mouse Immunoglobulin Light Chain Genes Before and After Somatic Changes." *Cell* 15:1133–1144.

47. Kim, S., M. Davis, E. Sinn, P. Patten, and L. Hood. 1981. "Antibody Diversity: Somatic Hypermutation of Rearranged V_H Genes." *Cell* 27:573–581.

48. Siekevitz, M., S. Y. Huang, and M. L. Gefter. 1983. "The Genetic Basis of Antibody Production: A Single Heavy Chain Variable Region Gene Encodes All Molecules Bearing the Dominant Anti-Arsonate Idiotype in the Strain A Mouse." *Eur. J. Immunol.* 13:123–132.

49. Tonegawa, S. 1983. "Somatic Generation of Antibody Diversity." *Nature* 302:575–581.

50. Kindt, T. J., and J. D. Capra. 1984. *The Antibody Enigma.* N.Y.: Plenum Press.

51. Gillies, S. D., S. L. Morrison, V. T. Oi, and S. Tonegawa. 1983. "A Tissue-Specific Transcription Enhancer Element Is Located in the Major Intron of a Rearranged Immunoglobulin Heavy Chain Gene." *Cell* 33:717–728.

52. Banerji, J., L. Olson, and W. Schaffner. 1983. "A Lymphocyte-Specific Cellular Enhancer Is Located Downstream of the Joining Region in Immunoglobulin Heavy Chain Genes." *Cell* 33:729–740.

53. Falkner, F. G., and H. G. Zachau. 1984. "Correct Transcription of an Immunoglobulin-k Gene Requires an Upstream Fragment Containing Conserved Sequence Elements." *Nature* 310:71–74.

54. Alt, F. W., G. D. Yancopoulos, T. K. Blackwell, C. Wood, E. Thomas, M. Boss, R. Coffman, N. Rosenberg, S. Tonegawa, and D. Baltimore. 1984. "Ordered Rearrangement of Immunoglobulin Heavy Chain Variable Region Segments." *EMBO J.* 3:1209–1219.

55. Alt, F., N. Rosenberg, S. Lewis, E. Thomas, and D. Baltimore. 1981. "Organization and Reorganization of Immunoglobulin Genes in A-MULV-Transformed Cells: Rearrangement of Heavy But Not Light Chain Genes." *Cell* 27:381–390.

56. Reth, M. G., and F. W. Alt. 1984. "Novel Immunoglobulin Heavy Chains Are Produced from DJ$_H$ Gene Segment Rearrangements in Lymphoid Cells." *Nature* 312:418–423.

57. Yancopoulos, G. D., and F. W. Alt. 1985. "Developmentally Controlled and Tissue-Specific Expression of Unrearranged V$_H$ Gene Segments." *Cell* 40:271–281.

58. Ritchie, K. A., R. L. Brinster, and U. Storb. 1984. "Allelic Exclusion and Control of Endogenous Immunoglobulin Gene Rearrangement in κ Transgenic Mice." *Nature* 312:517–520.

59. Early, P., J. Rogers, M. Davis, K. Calame, M. Bond, R. Wall, and L. Hood. 1980. "Two mRNAs Can Be Produced from a Single Immunoglobulin mu Gene by Alternative RNA Processing Pathways." *Cell* 20:313–319.

60. Rogers, J., P. Early, C. Carter, K. Calame, M. Bond, and R. Wall. 1980. "Two mRNAs with Different 3' Ends Encode Membrane-Bound and Secreted Forms of Immunoglobulin mu Chain." *Cell* 20:303–312.

61. Maki, R., W. Roeder, A. Traunecker, C. Sidman, M. Wabl, W. Raschke, and S. Tonegawa. 1981. "The Role of DNA Rearrangement and Alternative RNA Processing in the Expression of Immunoglobulin Delta Genes." *Cell* 24:353–365.

62. Blattner, F. R., and P. W. Tucker. 1984. "The Molecular Biology of Immunoglobulin D." *Nature* 307:417–422.

63. Davis, M. M., K. Calame, P. W. Early, D. L. Livant, R. Joho, I. L. Weissman, and L. Hood. 1980. "An Immunoglobulin Heavy-Chain Gene Is Formed by at Least Two Recombinational Events." *Nature* 283:733–739.

64. Maki, R., A. Traunecker, H. Sakano, W. Roeder, and S. Tonegawa. 1980. "Exon Shuffling Generates an Immunoglobulin Heavy Chain Gene." *Proc. Nat. Acad. Sci.* 77:2138–2145.

65. Kataoka, T., T. Miyata, and T. Honjo. 1981. "Repetitive Sequences in Class-Switch Recombination Regions of Immunoglobulin Heavy Chain Genes." *Cell* 23:357–368.

66. Mitchison, N. A. 1971. "The Carrier Effect in the Secondary Response to Hapten-Protein Conjugates. II. Cellular Cooperation." *Eur. J. Immunol.* 1:18.

67. Sprent, J. 1978. "Role of H-2 Gene Products in the Function of T Helper Cells from Normal and Chimeric Mice in Vivo." *Immunol. Rev.* 42:108–137.

68. Unanue, E. R. 1984. "Antigen-Presenting Function of the Macrophage." *Ann. Rev. Immunol.* 2:395–428.

69. Cantor, H., and E. A. Boyse. 1975. "Functional Subclasses of T Lymphocytes Bearing Different Ly Antigens. I. The Generation of Functionally Distinct T-cell Subclasses Is a Differentiation Process Independent of Antigen." *J. Exp. Med.* 141:1377–1389.

70. Gershon, R. K., and K. Kondo. 1972. "Infectious Immunological Tolerance." *Immunology* 21:903–914.

71. Reinherz, E., and S. F. Schlossman. 1980. "Differentiation and Function of Human T Lymphocytes." *Cell* 19:821–827.

72. Zinkernagel, R. M., and P. C. Doherty. 1974. "Restriction of in Vitro T Cell-Mediated Cytotoxicity in Lymphocytic Choriomeningitis Within a Syngeneic or Semiallogeneic System." *Nature* 248:701–702.

73. Zinkernagel, R. M. 1976. "H-2 Compatibility Requirement for Virus Specific, T-Cell-Mediated Cytolysis. The H-2K Structure Involved Is Coded by a Single Cistron Defined by H-2K^b Mutant Mice." *J. Exp. Med.* 143:437–443.

74. Bevan, M. J., and P. Fink. 1978. "The Influence of Thymus H-2 Antigens on the Specificity of Maturing Killer and Helper T Cells." *Immunol. Rev.* 42:3–19.

75. Kappler, J. W., B. Skidmore, J. White, and P. Marrack. 1981. "Antigen-Inducible, H-2-Restricted, Interleukin-2-Producing T Cell Hybridomas. Lack of Independent Antigen and H-2 Recognition." *J. Exp. Med.* 153:1198–1214.

76. Babbitt, B. P., P. M. Allen, G. Matsueda, E. Haber, and E. R. Unanue. 1985. "Binding of Immunogenic Peptides to Ia Histocompatibility Molecules." *Nature* 317:359–361.

77. Allison, J. P., B. W. McIntyre, and D. Bloch. 1982. "Tumor-Specific Antigen of Murine T-Lymphoma Defined with Monoclonal Antibody." *J. Immunol.* 129:2293–2300.

78. Meuer, S. C., O. Acuto, T. Hercend, S. F. Schlossman, and E. L. Reinherz. 1984. "The Human T-Cell Receptor." *Ann. Rev. Immunol.* 2:23–50.

79. Haskins, K., J. Kappler, and P. Marrack. 1984. "The Major Histocompatibility Complex-Restricted Antigen Receptor on T Cells." *Ann. Rev. Immunol.* 2:51–66.

80. Yanagi, Y., Y. Yoshikai, K. Leggett, S. P. Clark, I. Aleksander, and T. W. Mak. 1984. "A Human T Cell-Specific cDNA Clone Encodes a Protein Having Extensive Homology to Immunoglobulin Chains." *Nature* 308:145–149.

81. Hedrick, S. M., E. A. Nielsen, J. Kavaler, D. I. Cohen, and M. M. Davis. 1984. "Sequence Relationships Between Putative T-Cell Receptor Polypeptides and Immunoglobulins." *Nature* 308:153–158.

82. Chien, Y-h., D. M. Becker, T. Lindsten, M. Okamura, D. I. Cohen, and M. M. Davis. 1984. "Third Type of Murine T Cell Receptor Gene." *Nature* 312:31–35.

83. Saito, H., D. M. Kranz, Y. Takagaki, A. C. Hayday, H. N. Eisen, and S. Tonegawa. 1984. "A Third Rearranged and Expressed Gene in a Clone of Cytotoxic T Lymphocytes." *Nature* 312:36–40.

84. Saito, H., D. M. Kranz, Y. Takagaki, A. C. Hayday, H. N. Eisen, and S. Tonegawa. 1984. "Complete Primary Structure of a Heterodimeric T-Cell Receptor Deduced from cDNA Sequences." *Nature* 309:757–762.

85. Davis, M. M. 1985. "Molecular Genetics of the T Cell-Receptor Beta Chain." *Ann. Rev. Immunol.* 3:537–560.

86. Malissen, M., K. Minard, S. Mjolsness, M. Kronenberg, J. Goverman, T. Hunkapiller, M. B. Prystowsky, Y. Yoshikai, F. Fitch, T. W. Mak, and L. Hood. 1984. "Mouse T Cell Antigen Receptor: Structure and Organization of Constant and Joining Gene Segments Encoding the Beta Polypeptide." *Cell* 37:1101–1110.

87. Hayday, A. C., D. J. Diamond, G. Tanigawa, J. S. Heilig, V. Folsom, H. Saito, and S. Tonegawa. 1985. "Unusual Organization and Diversity of T-Cell Receptor α-Chain Genes." *Nature* 316:828–832.

88. Winoto, A., S. Mjolsness, and L. Hood. 1985. "Genomic Organization of the Genes Encoding Mouse T-Cell Receptor α-Chain." *Nature* 316:832–836.

89. Yoshikai, Y., S. P. Clark, S. Taylor, U. Sohn, B. I. Wilson, M. D. Minton, and T. W. Mak. 1985. "Organization and Sequences of the Variable, Joining and Constant Region Genes of the Human T-Cell Receptor α-Chain." *Nature* 316:837–840.

90. Raulet, D. H., R. D. Garman, H. Saito, and S. Tonegawa. 1985. "Developmental Regulation of T Cell Receptor Gene Expression." *Nature* 314:103–107.

91. Morgan, D. A., F. W. Ruscetti, and R. Gallo. 1976. "Selective in Vitro Growth of T-Lymphocytes from Normal Human Bone Marrows." *Science* 193:1007–1009.

92. Robb, R. J., A. Munck, and K. A. Smith, 1981. "T-Cell Growth Factor Receptors: Quantitation, Specificity and Biological Relevance." *J. Exp. Med.* 154:1455–1474.

93. Smith, K. A. 1984. "Interleukin-2." *Ann. Rev. Immunol.* 2:319–333.

94. Gorer, P. A. 1936. "The Detection of Antigenic Differences in Mouse Erythrocytes by the Employment of Immune Sera." *Br. J. Exp. Pathol.* 17:42–50.

95. McDevitt, H. O., and B. Benacerraf. 1969. "Genetic Control of Specific Immune Responses." *Adv. Immunol.* 11:31–74.

96. Klein, J. 1975. "Biology of the Mouse Histocompatibility-2 Complex." New York: Springer.

97. Klein, J., A. Juretic, C. N. Baxevanis, and Z. A. Nagy. 1981. "The Traditional and New Version of the Mouse H-2 Complex." *Nature* 291:455–460.

98. Orr, H. T., J. A. Lopez de Castro, P. Parham, H. L. Ploegh, and J. L. Strominger. 1979. "Comparison of Amino Acid Se-

quences of Two Human Histocompatibility Antigens, HLA-A2 and HLA-B7: Location of Putative Alloantigenic Sites." *Proc. Nat. Acad. Sci.* 76:4395–4399.

99. Nathenson, S. G., H. Uehara, B. M. Ewenstein, T. J. Kindt, and J. E. Coligan. 1981. "Primary Structural Analysis of the Transplantation Antigens of the Murine H-2 Major Histocompatibility Complex." *Ann. Rev. Biochem.* 50:1025–1052.

100. Sood, A. K., D. Pereira, and S. M. Weissman. 1981. "Isolation and Partial Nucleotide Sequence of a cDNA Clone for Human Histocompatibility Antigen HLA-B by Use of an Oligode-oxynucleotide Primer." *Proc. Nat. Acad. Sci.* 78:616–620.

101. Ploegh, H. L., H. T. Orr, and J. L. Strominger. 1980. "Molecular Cloning of a Human Histocompatibility Antigen cDNA Fragment." *Proc. Nat. Acad. Sci.* 77:6081–6085.

102. Hood, L, M. Steinmetz, and B. Malissen. 1983. "Genes of the Major Histocompatibility Complex of the Mouse." *Ann. Rev. Immunol.* 1:529–568.

103. Hood, L., M. Kronenberg, and T. Hunkapillar. 1985. "T Cell Antigen Receptors and the Immunoglobulin Supergene Family." *Cell* 40:225–229.

CHAPTER 24

The Extraordinary Diversity of Eucaryotic Viruses

Viruses are well known agents of disease in humans and other animals, as well as in plants. They can cause relatively benign infections, such as the common cold and warts, or they can induce severe conditions, such as cancer, polio, and AIDS. Quite apart from their importance as pathogens, however, animal viruses have served a critical role in biological research as probes to the inner workings of eucaryotic cells. Long before gene-cloning techniques were available, viruses provided an easy source of pure genes whose functioning obeys the rules of the cell in which they replicate. Since viral replication often occurs at the expense of normal cellular processes, viral gene expression is frequently more efficient, and thus easier to study, than cellular gene expression. Many major discoveries about the molecular biology of higher cells came first from studying their viruses: RNA splicing, 5' caps and 3' poly A tracts on mRNAs, the pathway for synthesis of cell surface proteins, the role of enhancer elements in transcription, and reverse transcription of RNA into DNA are just a few examples. Even today, the remarkably numerous strategies that viruses use to parasitize host cells continue to provide important insights into the biochemical functions of eucaryotic cells. But quite apart from the commendable goals of conquering disease and probing the cell, the study of animal viruses has become an end in itself, illuminating the extraordinary diversity and adaptability of even the simplest organisms.

AN INTRODUCTION TO ANIMAL VIRUSES

Viruses Come in Many Sizes and a Few Shapes[1–5]

A virus is an intracellular parasite that can exist in two fundamentally different forms, one inside the cell and one outside the cell. The extracellular form is called a **virion,** or virus particle. Virions consist of the viral genome, either RNA or DNA, associated proteins, and sometimes other chemical constituents, such as lipids. The proteins (and lipids) protect the genes from the ravages of the extracellular environment, ensure their entry into the host cell, and often carry out some initial steps of the replication cycle. During infection, virion structure is largely or completely destroyed, so that the genome is free to be transcribed and replicated.

At first sight, virions of animal viruses give the impression of great variety (Figure 24-1). Their diameter ranges from about 10 to 200 nm, which corresponds to a volume difference of almost 10,000-fold. They can also have exotic projections at their surface. This first impression is misleading, however, because all but the most complex viruses are constructed according to one of two basic plans. This simplicity is necessitated by the small size of most viruses. To construct a sphere the size of a small virus (like polio) from average-size proteins requires about 5.5 million daltons of protein. However, a genome that can fit within such a sphere is only big enough to code for 300 kdal of protein, and not all of this is available for building the virion. Clearly, the portion of a virus particle that is made of virus-coded proteins, the portion known as the **capsid,** must be made from just one or a few different molecules repeated over and over. Given this restriction, virion capsids are limited to one of two basic organizations: a hollow helix of undefined length with the genome in the center, resembling a rod, or a quasi-spherical *icosahedral* structure (Chapter 17).

Further contributing to the varied appearance of animal virus virions is the presence or absence of an outer membranous coat. Virions that contain only a genome and capsid are considered *naked.* In contrast, some viruses have a lipid bilayer, the **envelope,** surrounding their capsid. In such viruses, the capsid plus genome is called the **nucleocapsid.** The envelope is derived from the cell membrane in a two-step process called **budding** (Figure 24-2). First, virus-coded glycoproteins are inserted into the membrane. Then virion capsids may interact with the cytoplasmic ends of the glycoproteins, causing the membrane to wrap around the capsid and eventually to pinch off from the cell. The cell membrane can seal itself behind the budding virus. Some enveloped viruses can be released continuously by infected cells for many generations without harmful effects.

Unlike phages, animal viruses never possess the elaborate tails needed for injecting genes through the bacterial cell wall. Rather, capsid proteins of naked virions or viral glycoproteins protruding from the surface of enveloped virions interact with receptors on animal cell membranes, causing the virions to be engulfed and subsequently uncoated inside the cell.

Virus Genomes Can Be Single- or Double-Stranded, Linear or Circular, RNA or DNA[6]

In contrast to cells, whose genomes are always double-stranded DNA, viral genomes can have an array of configurations and can be RNA or DNA. Both RNA and DNA genomes can be single- or double-stranded. All known double-stranded RNA genomes are *segmented;* that is, they consist of several RNA molecules, each carrying a different gene, rather than a continuous, covalent polynucleotide. Some single-stranded RNA genomes are also segmented. Double-stranded DNA viral genomes can be either linear or circular. The circular DNA can be covalently closed on both or only one of its strands. One linear double-stranded DNA virus genome, that of the poxvirus vaccinia, has its two strands covalently joined at their ends. As a result, poxvirus DNA forms a single-stranded circle upon denaturation. There are no known segmented DNA viruses, and except for the plant viroids discussed at the end of this chapter, there are no known eucaryotic viruses with circular RNA genomes.

Figure 24-1

Electron micrographs and sketches of representative members of the important families of animal viruses. Sketches show relative sizes of different viruses; ss = single-strand, ds = double-strand. [Sketches after R. E. F. Matthews, *Intervirology* 12 (1979):3–5. Electron micrographs courtesy of Frederick A. Murphy.]

Rhabdorvirus (Vesicular Stomatitis) 216,000 ×

Pox virus (Vaccinia) 200,000 ×

Orthomyxovirus (Influenza A) 135,000 ×

Reovirus 135,000 ×

Herpes (Herpes Simplex Type 1) enveloped (left) and naked capsids (right) 189,000 ×

Adenovirus 234,000 ×

Togavirus (Semlikiforest) 120,000 ×

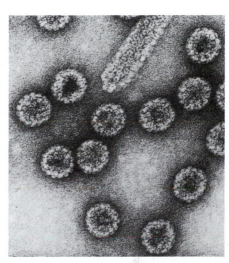

Papovavirus (Papillomavirus; human wart virus) 135,000 ×

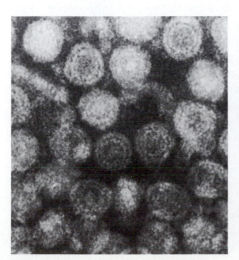

Hepadnavirus (Hepatitis B) 225,000 ×

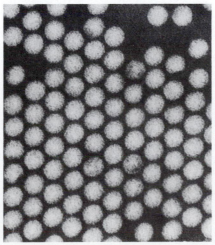

Picornavirus (Polio virus) 180,000 ×

Parvovirus 171,000 ×

Figure 24-2
Release of virions by budding. Specific association between the cytoplasmic ends of the viral spike glycoproteins and the capsid protein may cause the membrane to enclose around the nucleocapsid.

RNA virus genomes vary in size within a narrow range, about 7 to 20 kilobases. DNA virus genomes, on the other hand, can be as small as 3 kilobases or as large as 200 kilobase pairs. There may be a good reason for the upper limit on RNA genome size. Cells have mechanisms to repair DNA but not RNA. Unless the virus were to evolve sophisticated proofreading systems to correct errors made during replication, it would be almost impossible to copy a very large RNA molecule without introducing deleterious mutations. In fact, replication of RNA virus genomes is accompanied by very high mutation rates (10^{-3} to 10^{-4} is common). At this rate, a genome of 100 kilobases would suffer 10 to 100 mutations every time it was copied, and almost no progeny would be viable.

Cells in Culture Revolutionized the Study of Animal Viruses[7, 8, 9]

A number of animal viruses, such as rabies and smallpox, were well known in the last century, long before the discovery of bacterial viruses. A good understanding of animal viruses, however, lagged well behind that of bacteriophages. Essential to the development of modern animal virology was the discovery in the early 1950s that animal viruses can grow on tissue culture cells in vitro. Until then, virtually all work with viruses had to be performed in living animals. Besides being slow and expensive, the use of whole animals made it essentially impossible to carry out quantitative molecular biological studies.

Virologists use three types of cell culture systems. **Primary cultures** consist of cells freshly removed from an animal and placed in a dish with cell culture medium. While closest to the conditions of the living host, these cultures are usually complex mixtures of cell types, many of which are incapable of further division. After growing for a few generations in vitro, primary cultures give rise to a second type of culture, **cell strains,** which usually consist mostly of **fibroblasts,** relatively nondifferentiated precursors to connective tissue. They will grow for about 50 generations in the laboratory, but then they usually die. The third type of culture, **cell lines,** grow indefinitely in vitro. Many frequently used cell lines, such as the human HeLa cell line or mouse L cell line, were derived from naturally occurring tumors, while others, such as the mouse 3T3 cell line, grew out of cell strains, probably as the result of rare mutations. Although the cells of cell lines are not really normal cells, they are by far the easiest to work with of the three types of cultures and they provide the most reproducible results; thus, they are strongly preferred by virologists. In fact, viruses that grow well on established cell lines were the first to be studied extensively at the molecular level.

As with bacteriophages, animal virus replication in cell culture is often detected because the virus kills, or **lyses,** the host cell. A single virus particle can give rise to a **plaque,** just as a phage can. Some animal viruses that do not lyse the cells they grow on can be detected in cell culture by biochemical or immunological methods or by observation of altered cellular morphology.

A Large Number of Animal Viruses Have Been Isolated[10]

By now, there are many thousands of animal virus isolates. At first they seemed to embrace a diverse and confusing array of characteristics, at least in terms of the nature and severity of the diseases they cause, the host animal species they infect, and their mode of transmission. But with the advent of modern biochemical approaches, it became apparent that the thousands of isolates can be classified into a reasonably small number of virus *families* based on fundamental characteristics of the virion: namely, particle morphology, genome structure, and the presence of common antigenic determinants on virion proteins. The important families of animal viruses are presented in Table 24-1.

Although there is little or no relationship between virus families, individual isolates within a given family are strikingly similar to one another. All adenoviruses, for example, have naked icosahedral capsids of 252 subunits with prominent spikes at the vertices and a double-stranded DNA genome about 36,000 base pairs long. This precise description applies to adenoviruses isolated from species as diverse as frogs, turkeys, and humans, and from countries all over the world, although the precise nucleotide sequence of the genome varies considerably. These observations tell us that the evolution of a successful virus type is probably a rare event. Once achieved, however, rapid replication and high mutation rates allow the virus to adapt itself readily to grow in the maximum number of hosts.

Mechanisms of Genome Expression Provide a Molecular Biologist's Classification of Animal Viruses[11–17]

Although the thousands of known animal viruses can be grouped by evolutionary relationships, the resulting number of families, with their diversity of replication and expression strategies, can still overwhelm the beginning virologist. Around 1970, it became clear that although the nature of the genome and its mode of replication vary widely among families, all viruses have a common need for cellular machinery for the synthesis of virus-coded proteins. Thus, viral mRNA is the starting point for a common pathway that all viruses

Table 24-1 Important Families of Animal Viruses

Family	Examples	Genome Size (kb)	Virion
1. dsDNA Viruses			
Papovavirus	Polyoma, SV40, wart viruses	5–8	Naked, icosahedral
Adenovirus	Many human and animal adenoviruses	35–40	Naked, icosahedral
Herpesvirus	Herpes simplex I, II, varicella-zoster (chickenpox), Epstein-Barr virus	120–200	Enveloped, icosahedral
Poxvirus	Variola (smallpox), vaccinia	120–300	Enveloped, complex
Iridovirus	Mostly insect viruses	150–300	Enveloped, icosahedral
2. ssDNA Viruses			
Parvovirus	Adeno-associated virus, minute virus of mice, canine, feline, and human parvovirus	4–5	Naked, icosahedral
3. (+) RNA Viruses			
Picornavirus	Polio, common cold viruses, foot and mouth disease, many enteric viruses	7	Naked, icosahedral
Togavirus	Sindbis virus, Semliki forest virus, yellow fever virus, rubella virus, encephalitis viruses	12	Enveloped, icosahedral
Coronavirus	Human common-cold–like diseases, mouse hepatitis virus	16–21	Enveloped, helical
4. (−) RNA Viruses			
Rhabdovirus	Rabies, vesicular stomatitis virus	12–15	Enveloped, helical
Paramyxovirus	Newcastle disease, measles, mumps, respiratory syncytial virus, Sendai virus	15	Enveloped, helical
Orthomyxovirus	Influenza	14 (8 segments)	Enveloped, helical
Bunyavirus	Bunyamwera, Uukuniemi, LaCrosse, Rift Valley fever viruses	15–20 (3 segments)	Enveloped, helical

Table 24-1 (*Continued*)

Family	Examples	Genome Size (kb)	Virion
Arenavirus	Lassa virus, lymphocytic choriomeningitis virus	12–15 (2 segments)	Enveloped, helical
5. dsRNA Viruses			
Reovirus	Reovirus, human and animal diarrhea viruses	18–30 (10 segments)	Naked, icosahedral
6. RNA-DNA Viruses			
Retrovirus	Rous sarcoma, avian, feline, and murine leukemia, mouse mammary tumor virus, HTLV	7–10 (diploid)	Enveloped, icosahedral
7. DNA-RNA Viruses			
Hepadnavirus	Hepatitis B virus of humans, birds, and rodents	3	Enveloped, icosahedral

ultimately use. Recognition of the central role of mRNA leads to a molecular biological classification of animal viruses based on the relationship between the viral genome and mRNA and also on the genome's mode of replication. The system is shown in Figure 24-3.

It is easy to understand the relationship of double-stranded DNA virus genomes to the synthesis of mRNA, since this process is identical to that used by the host. In fact, most DNA viruses replicate in the nucleus, where they can have access to the cell's RNA polymerase, as well as capping, splicing, and poly A–adding enzymes, to process their transcripts. Single-stranded DNA genomes also have a fairly conventional relationship to mRNA, since they quickly become double-stranded DNA after infection and can follow the same fate as cellular genes. But what about viruses with RNA genomes? Like the RNA phages discussed in Chapter 17, some animal viruses have RNA genomes that correspond to mRNA and that can function as messages even in vitro. These viral genomes are called *positive* or *plus (+) strand* genomes. Many other RNA viruses have *negative* or *minus (−) strand* genomes, meaning they are complementary to the sense or mRNA strand. Since animal cells lack enzymes to copy RNA, and since the negative strands cannot be translated, − strand RNA is functionally dead. Viruses with − strand genomes must encode a **transcriptase** that can make mRNA from a − strand RNA template. Furthermore, the enzyme must be packaged in the virion in association with the viral genome. Then after infection, the genome-associated transcriptase synthesizes viral mRNA, allowing the replication cycle to begin. At the end of the cycle, newly synthesized molecules of transcriptase are again packaged along with the genome, making the next cycle of infection possible.

Viruses with double-stranded RNA genomes must also carry a transcriptase within their particles, since, like − strand RNA, double-

Figure 24-3
Classification of viruses by replication mechanism and relationship of genome to mRNA. Arrows inside the boxes show the information flow during replication. Arrows to mRNA originate at template for mRNA synthesis.

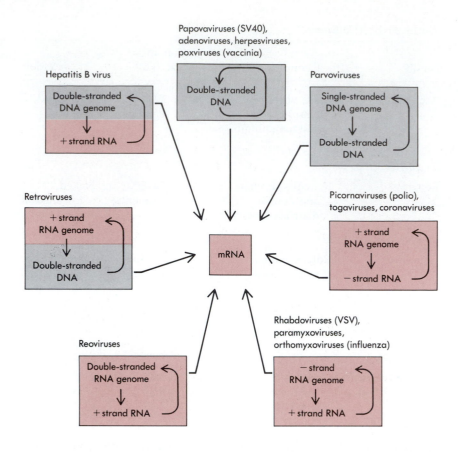

stranded RNA is not a normal template for replication or transcription in animal cells and cannot be translated. That virus-coded enzymes are packaged in virions was one of the most important discoveries in molecular animal virology. Curiously, the first enzyme so discovered was found in a poxvirus, the only known class of DNA viruses that replicates in the cytoplasm instead of in the nucleus, necessitating the production of many enzymes that other DNA viruses borrow from their host.

Using the principles discussed so far, animal viruses can be assigned to seven different groups (see Figure 24-3):

1. *Double-strand DNA viruses* (except the poxviruses) use cellular RNA polymerase for mRNA synthesis. All these viruses encode both capsid proteins and nonvirion proteins required for replication of their genomes.

2. *Single-strand DNA viruses* are either − stranded or made of both + and − strands, each one packaged in a separate virion. Unlike double-strand DNA viruses, single-strand DNA viruses encode capsid proteins only. They rely on cellular systems for both their transcription and their replication.

3. *+ strand RNA virus* genomes, which serve as mRNA soon after infection, have 3' poly A tails and usually some sort of 5'-end modification like cellular mRNAs. They always specify enzymes that can replicate RNA via a − strand intermediate, since animal cells lack such an enzymatic activity.

4. *− strand RNA virus* genomes serve as a template for the synthesis of mRNA by virion transcriptase soon after infection. Not surprisingly, − strand RNA virus genomes lack the terminal modifica-

tions that characterize eucaryotic mRNAs.

5. *Double-strand RNA viruses* also carry virus-coded enzymes within the virion that can synthesize mRNA from this type of template.

6. *Retroviruses* have single, + strand RNA genomes that can serve as mRNAs. The genome's first act after infection, however, is not to serve as mRNA, but rather to serve as a template for the synthesis of double-stranded DNA. This DNA integrates into host chromosomal DNA, from which it serves as a template for transcription of viral mRNA and progeny genomes. **Reverse transcriptase,** a virus-coded enzyme carried inside the virion, makes viral DNA from the genomic RNA template.

7. *Hepatitis B viruses,* also known as *Hepadnaviruses,* have genomes with a full-length − DNA strand and a short + strand. Virions contain a DNA polymerase that completes the + strand soon after infection, resulting in a double-stranded DNA molecule that can be transcribed by the cell to yield viral mRNAs. As we shall see later, however, the replication of these viruses is quite different from most double-stranded DNA viruses and resembles the replication of retroviruses.

It is often quite easy to categorize a new virus isolate using the preceding information. For example, the presence of a transcriptase, easily detected in virion preparations, indicates a − strand or double-stranded RNA virus. Positive strand viruses or double-stranded DNA viruses have genomes that are infectious in the absence of virion proteins. Thus, just one experiment can sometimes reveal much of the life cycle of a virus.

Animal Viruses Solve Common Problems of Replication and Gene Expression in Diverse Ways[18–21]

In animal cells, in contrast to bacterial cells, most mRNAs are monocistronic. In other words, ribosomes only initiate translation at an AUG codon near the 5′ end of the RNA. Consider, for example, a + strand RNA virus where all the genes are encoded on the same strand of RNA. How can such an animal virus synthesize the numerous proteins needed for a successful infection? RNA viruses have several novel solutions to this problem:

- Protein cleavage: Many RNA viruses direct the synthesis of multiple proteins by first synthesizing giant polyproteins, which are then clipped by a protease to release individual proteins. In the most extreme case, that of poliovirus, the entire genome is translated into a single polypeptide, which is cleaved to ten final proteins.

- Segmented genomes: Some RNA viral genomes consist of multiple pieces of RNA, each of which encodes one or a few distinct proteins.

- Nested mRNAs: Some viruses express internal genes by generating a nested set of mRNAs with identical 3′ ends but different 5′ portions. Each of these yields a different protein, which is translated from the reading frame nearest the 5′ end of the mRNA.

- Sequential synthesis: In some cases plus RNA is synthesized as one long unit, and separate mRNAs for each protein are generated by cleavage or termination and reinitiation.

Of course, DNA viruses also face the problem of monocistronic mRNAs. While they can solve this problem by having transcriptional promoters in front of every gene, sometimes they use complex RNA-splicing patterns to join the identical 5' end to multiple transcripts.

DNA viruses face a special problem with replication, which arises from the fact that DNA polymerases cannot initiate synthesis of new DNA strands, but can only extend synthesis from primers. This causes an acute problem at the ends of linear genomes, since removal of the primer, which would normally be a short RNA chain, would always leave a 3'-end single-stranded tail (Chapter 10). Solutions to this problem involve circular genomes, terminal self-complementary structures that can self-prime the synthesis of new DNA chains, and terminal proteins that, when covalently attached to a nucleoside, serve as primers and remain covalently attached to the 5' end of each DNA strand.

Cell Surface Glycoproteins Are Receptors for Animal Viruses[22–24]

The first event in a successful infection is interaction of a virion with a specific *receptor* molecule on the cell surface. This interaction is mediated by virus-coded proteins on the surface of the virion and is highly specific. The presence or absence of appropriate receptors often determines whether a cell is *permissive* for virus infection, so the host range of a virus, including the type of tissue it infects, can be determined by the composition of receptors on the cell surface. Receptors for animal viruses are normal cell proteins and are not present for the convenience of the virus. For example, the ability of AIDS virus to infect a certain type of lymphocyte (T helper cell) most likely reflects its use of a cell surface protein characteristic of this cell type as its receptor.

A cell usually has 10^4 to 10^5 receptors for a particular virus. Most receptors are glycoproteins. Several virus groups (e.g., influenza viruses) recognize the terminal residue of the polysaccharide chain, usually sialic acid, so treatment of a cell with sialidase (**neuraminidase**) destroys its ability to be infected with such viruses (Figure 24-4). The specificity of virion-receptor interactions is demonstrated by the example of influenza virus. Influenza does not interact with receptors that have an $\alpha2\rightarrow3$ instead of an $\alpha2\rightarrow6$ linkage between the sialic acid and the adjacent sugar (see Figure 24-4). Influenza virus mutants that have acquired the ability to use receptors with the $\alpha2\rightarrow3$ linkage can readily be obtained, however, and these have a single amino acid change in their receptor binding protein. Thus, virus-receptor interactions display a stereospecificity similar to that of enzyme-substrate interactions.

Receptors for certain viruses are present on red blood cells. This permits an easy virus assay, since interaction of a virus with more than one cell at a time causes them to visibly **agglutinate,** or clump. In such a case, the viral protein that binds to the receptor is called **hemagglutinin.**

Figure 24-4
The terminal sugars of the oligosaccharide side chains used as part of the receptor for many animal viruses. If the $\alpha2\rightarrow6$ linkage (dark color) between the terminal N-acetylneuramic (or sialic) acid residue and the adjacent sugar (usually N-acetylgalactosamine, as shown) is hydrolyzed by viral neuraminidase, the receptor is no longer recognized. A single point mutation in the influenza hemagglutinin can alter the recognition specificity from $\alpha2\rightarrow6$ to $\alpha2\rightarrow3$.

Viruses Enter Cells by Fusion or Endocytosis and Often Uncoat in Acidic Endosomes[25, 26]

Two well-documented mechanisms by which viruses get through the cell membrane and into the cytoplasm are fusion and endocytosis. The fusion mechanism is illustrated by paramyxoviruses, enveloped viruses that include mumps and Sendai viruses. These viruses can catalyze direct fusion between viral and cellular membranes. This ability resides in a viral envelope glycoprotein, called the F or fusion protein. F protein is generated by cleavage of a precursor polypeptide. The cleavage exposes a hydrophobic region in F protein that is necessary for interaction with the cell membrane. Once a virus is bound to its receptor, the fusion activity makes holes simultaneously in the viral envelope and the cell membrane. The two membranes then seal together, and as a result, the capsid is introduced into the cell. Cells that lack a protease capable of cleaving the F protein precursor release virions that are neither infectious nor capable of inducing cell fusion. Treatment of these virions with trypsin in vitro restores both their infectivity and their cell-fusing ability, so fusion activity appears to be essential for infectivity. The simultaneous interaction of a virion that has fusion activity with two or more cells can cause the cells to fuse into a large, multinucleated cell. Such cells are often seen in virus-infected cultures.

The alternative mechanism for virus penetration, endocytosis, uses host cell machinery that is normally used for the uptake of macromolecules. Infection by Semliki Forest virus, a togavirus, has provided a model system for this type of penetration. Semliki Forest virus is an enveloped virus with surface glycoproteins. After attaching to its specific cell surface receptor, the virus, along with its receptor, becomes engulfed by a **coated pit** into a vesicle (Figure 24-5). Coated pits are areas of the cell membrane, lined on the cytoplasmic side with the protein **clathrin,** where endocytosis of receptors and their ligands occurs. Even after the coated vesicle with its adsorbed virion is inside the cytoplasm, the virus itself is still functionally outside the cytoplasm, surrounded by the membranous walls of the pit. Once inside the cell, the pit with its attached virus loses its clathrin coat and fuses with large cellular vesicles called endosomes. The low pH provided

Figure 24-5
Uptake of viruses by endocytosis and fusion. Following receptor-mediated uptake of virus via coated pits, the acidic pH due to fusion of the vesicle with an endosome stimulates fusion of the viral envelope with the vesicle membrane resulting in release of the nucleocapsid into the cytoplasm. [After "How an Animal Virus Gets Into and Out of Its Host Cell," by K. Simons, H. Garoff, and A. Helenius. *Sci. Amer.* 246 (1982). Used with permission.]

by the contents of these endosomes activates membrane-fusing activity latent in the viral glycoprotein spikes, causing the viral and endosomal membranes to fuse together and release the contents of the virion into the cytoplasm. This fusion activity can be induced in vitro by incubating Semliki Forest virus with cells and lowering the pH. Then direct fusion of the virus envelope with the plasma membrane occurs, in much the same way that paramyxoviruses normally fuse to cells at physiological pH. Infection of cells by viruses that enter by endocytosis can be blocked by chemicals such as chloroquine or by ammonium ions, which can form weak bases that raise the pH of endosomes and so prevent activation of viral fusion proteins.

RNA VIRUSES

Poliovirus: A Model + Strand RNA Virus[27-30]

Poliovirus is the best-studied *picornavirus*. It has a single + strand RNA genome, about 7400 nucleotides long, whose complete nucleotide sequence is known. The naked icosahedral virion is composed of

Figure 24-6
The 5′ end of the poliovirus genome. The 22-amino-acid VPg protein is shown linked via a phosphodiester bond between the tyrosine residue and the 5′ uridine of the poliovirus genome. A cellular enzyme removes VPg from polio mRNA by hydrolysis at the position shown. [After N. Kitamura et al., *Nature* 291 (1981):547.]

four virus-coded proteins—VP1, 2, 3, and 4—each present in 60 copies per virion. VP1 is probably responsible for attachment to cellular receptors that are found only on primate epithelial cells, fibroblasts, and nerve cells, accounting for the narrow host range of poliovirus. The entire life cycle of poliovirus occurs in the cytoplasm. This can be shown dramatically by the ability of the virus to replicate in cells whose nucleus has been removed by treatment with the drug cytochalasin B. The poliovirus life cycle is one of the most rapid known for an animal virus. Within 6 to 8 hours, an infected cell releases about 10^5 progeny virions.

Because it is a + strand genome, polio RNA that enters the cell first serves as mRNA for protein synthesis. This is essential, since replication of the viral RNA requires virus-coded enzymes, and none are present in polio virions. As we might expect, RNA isolated from polio virions can serve as mRNA in vitro. Like cellular mRNAs, the polio genome has a poly (A) sequence at its 3′ end. Unlike cellular mRNA and unlike most other viral mRNAs, it is not capped at the 5′ end. Rather, it has a tiny, 22-amino-acid-long protein, called VPg, covalently attached to its 5′ end via a phosphodiester bond between the 5′ uridine nucleotide and a tyrosine residue (Figure 24-6).

The Polio Genome Encodes Only One Polyprotein[31, 32, 33]

When newly made poliovirus proteins inside an infected cell were first examined by gel electrophoresis, it was immediately apparent that the sum of their molecular weights exceeded by two or three times the coding capacity of the genome. This paradox was explained by the finding that most of the proteins represent proteolytic cleavage products of one another. When serving as mRNA, polio RNA is translated into a single polyprotein from which all the viral proteins are derived by cleavage (Figure 24-7). Thus, there is only one ribosome entry point on the entire viral genome. These conclusions are supported by the nucleotide sequence of polio, which reveals a 6620-nucleotide open reading frame occupying the vast majority of the genome (see Figure 24-7). The poliovirus genome was the first large mRNA active in eucaryotic cells that was found to be monocistronic. At the time, the discovery represented a surprising departure from phage and bacterial systems, where polycistronic mRNAs are com-

Figure 24-7

The poliovirus genome and its translation products. The cleavage of the full-length precursor, NCVP, to P1, P2 and P3 is nascent, occurring before translation is complete. All of the cleavage events are catalyzed by polio-encoded proteins.

mon; but as we have seen, it turned out to be typical of eucaryotic mRNAs. The protein-coding region in the polio genome is preceded to the 5′ side by an untranslated region 740 nucleotides long. While eucaryotic ribosomes frequently initiate translation at the first AUG codon they come to on an mRNA, eight AUG codons at the 5′ end of polio RNA are bypassed in favor of the ninth at nucleotide 741. To the 3′ side of the protein-coding region is a short untranslated region and finally the poly A tail.

The 240 kdal poliovirus polyprotein precursor does not normally exist in infected cells, because it is proteolytically cleaved at two sites before its synthesis is complete. Only if these **nascent cleavages** are blocked can the complete protein, called NCVP, be observed. The two initial cleavages divide the molecule into three regions. Unexpectedly, the amino-terminal region, which is thus the first part of the protein to be made, contains the virion proteins, while the carboxyl-terminal region is further cleaved to yield proteins essential for replication, including the RNA-dependent RNA polymerase, called **replicase.** In addition to replicase, this region also yields VPg, as well as a protease responsible for most of the polyprotein cleavages, specifically those that occur between glutamine-glycine (Gln-Gly) pairs. This protease is active even while it is still in precursor form, so several cleavages are autocatalytic. Not all cleavages of poliovirus polyproteins are at Gln-Gly pairs. Adjacent tyrosine-glycine (Tyr-Gly) and asparagine-serine (Asn-Ser) pairs are also cleaved, and now evidence suggests that the Tyr-Gly cleavages are made by a second viral protease encoded by the central portion of the genome. Not all Gln-Gly or Tyr-Gly pairs found in the poliovirus polyprotein are cleaved proteolytically. The enzymes involved in cleavage probably recognize specific features of protein conformation as well as particular pairs of amino acids.

The fact that all the poliovirus gene products are obtained by processing a single precursor implies that all the products will be synthesized in equimolar amounts. At most, one replicase or VPg molecule is required to make a new genome, whereas 60 molecules of each capsid protein are required to encapsidate it. This means that each infected cell synthesizes over 5 million unnecessary molecules of replicase in order to generate enough capsid protein for the 10^5 progeny virions. As we shall see, other more complex + strand RNA viruses have devised more efficient strategies of gene expression.

Poliovirus Replication May Be Initiated by a Protein Covalently Attached to the 5' Ends of RNA[34–37]

After serving as mRNA to direct the synthesis of proteins needed in its replication, the incoming polio genome must disengage from the ribosomes in order to serve as a template for RNA replication. Replication begins with the synthesis of − strands, which are generated by copying the + strand from an initiation site near its 3' end. Poliovirus replication has two unusual features. First, since the 3' end of the poliovirus genome consists of poly A, the 5' end of the − strand has the complementary sequence, poly U. Unlike virtually all other eucaryotic mRNAs, the poly A of polio RNA is not added after transcription; instead, it is replicated along with the rest of the genome. Second, unlike most enzymes responsible for RNA synthesis, but like all known DNA polymerases, polio replicase can synthesize RNA by adding nucleotides to an existing primer bound to a template, but it cannot initiate synthesis on a template with a ribonucleoside triphosphate. Thus, replicase is inactive when presented only with poliovirus RNA as template: It must have a primer as well. The primer for polio replicase does not seem to be a short oligonucleotide, the common primer for cellular DNA polymerases. Rather, it is thought to be U residues that have been covalently attached to the tiny protein VPg (see Figures 24-8 and 24-6). This would explain why VPg is covalently attached to the 5' ends of all newly synthesized polio + and − strands.

As infection proceeds, there is an accumulation of up to about 600 **replicative intermediates** per cell. These structures have a full-length − strand template with five or so + strands being made on them simultaneously (see Figure 24-8). The newly made + strands have one of three fates: They can serve as mRNA for further protein synthesis; they can be templates for further − strand synthesis; or they can become genomes in new virions. Probably the amounts of replicase and capsid protein that have built up inside the cell determine which of the pathways will be chosen. Thus, early in infection, + strands predominantly become mRNA after losing their terminal VPg by proteolytic cleavage. As replicase becomes available, + strands are increasingly used to direct − strand synthesis. And finally, as newly synthesized capsids appear, the + strands are incorporated into virions at the expense of the other uses.

Poliovirus Infection Stops the Synthesis of Cell Proteins[38, 39]

Very soon after poliovirus infection, cellular protein synthesis slows down dramatically, decreasing to 10 to 20 percent of normal levels 2 hours after infection. Inhibition of cellular protein synthesis is a common feature of many animal virus infections, but the mechanism has been studied most extensively with poliovirus.

Figure 24-8
Poliovirus RNA replication. Newly synthesized positive strands may serve as mRNA or as templates for further rounds of replication, or they may be packaged into virions. Note that − strands are used only as templates for synthesis of more + strands.

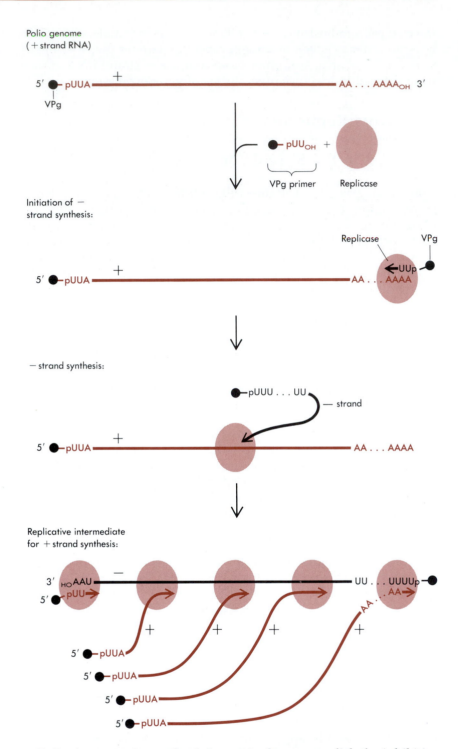

Poliovirus protein synthesis is required to accomplish the inhibition of host translation. Cellular mRNAs are not degraded or modified in polio-infected cells, but their translation is blocked at the initiation step. Polio-infected cell extracts proteolytically cleave and thereby inactivate a translation initiation factor, called eIF-4B, whose normal function is to recognize the capped, methylated 5′ termini of eucaryotic mRNAs and participate in binding mRNA to ribosomes. Since eucaryotic mRNAs are capped but polio mRNAs are not, it is apparent why inactivation of eIF-4B selectively inhibits cellular protein synthesis. Among known animal viruses, only the picornaviruses have capless mRNAs, so other viruses must use different mechanisms to selectively inhibit cellular but not viral protein synthesis.

Poliovirus Assembly Involves Proteolytic Cleavage of Capsid Protein Precursors[40, 41]

One of the earliest nascent cleavages of the giant polyprotein NCVP releases an amino-terminal cleavage product, P1, that gives rise to the four capsid proteins. This occurs in a series of steps that involve proteolytic cleavage and association of the cleavage products with one another (Figure 24-9). Initially, five copies of the P1 precursor associate into pentamers. Then the P1 molecules are cleaved into VP0, VP1, and VP3, after which groups of 12 pentamers aggregate into empty capsidlike structures called **procapsids.** Insertion of a genome into the procapsid leads to an immature structure called a **provirion,** which becomes a mature virion by proteolytic cleavage of VP0 to VP2 and VP4. The early cleavages are all at Gln-Gly bonds, but the final cleavage of VP0 is at an Asn-Ser bond and may be made by VP0 itself. This mode of assembly in which aggregation of precursor molecules is followed by cleavage into distinct proteins is found in other virus groups as well. It provides a simple mechanism for generating sequential changes in conformation as they are needed. For example, the final cleavage of VP0 might change the virion structure from one that is sufficiently open to allow the virus genome to get in, to one that is sufficiently tight to prevent the genome from falling out and to protect it against even extremely low pH.

Other + Strand RNA Viruses Have More Elaborate Strategies of Gene Expression[42–45]

Togaviruses have + strand RNA genomes of about 12,000 bases and icosahedral capsids within an envelope. In contrast to picornaviruses like polio, togavirus genomes are dicistronic (Figure 24-10). The cistron nearest the 5' end encodes the viral replicase, and the cistron nearest the 3' end encodes a precursor to the virion proteins. Since the internal initiation site is not seen by ribosomes, the genome itself can serve as mRNA only for the replicase protein. Capsid proteins do not appear until later in infection, and they are translated from a subgenomic mRNA consisting of the 3' end of the genome. This mRNA is made by initiating replication at an internal site on the − strand. Since this mRNA lacks the entire replicase cistron, the initiation codon for translating capsid protein is now exposed and available for translation. This simple scheme allows the virus to regulate its gene expression both *temporally* (coat protein is not made until it is needed) and *quantitatively* (by the relative amounts of the two mRNAs and their relative efficiency as messengers).

A still more extreme case of nested mRNAs is shown by coronaviruses (Figure 24-11). The 16-kilobase genome of these viruses is expressed via no less than seven distinct mRNAs, each containing the 3' end of the genome and each serving as messenger for a different gene product. Interestingly the different mRNAs have a common 72-base long, 5' leader sequence. It is doubtful that this sequence is joined to the body of each mRNA by conventional splicing of a precursor RNA. Instead, newly synthesized leader segments may be used to prime internal initiations of transcription. Thus coronavirus mRNAs may result from discontinuous RNA synthesis.

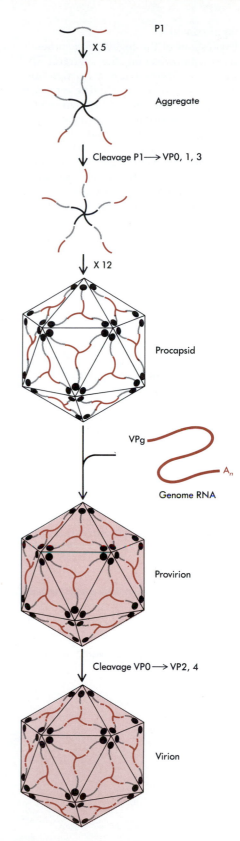

Figure 24-9
Model for polio virion assembly.

Figure 24-10
Expression of the togavirus genome. Early after infection, the + strand RNA genome is translated to yield two non-structural proteins, including replicase (top portion of figure). Later in infection, a subgenomic mRNA is synthesized by use of an internal initiation site on the − RNA strand. This exposes a previously inaccessible AUG codon, permitting translation of a precursor to all virion proteins (the capsid protein and three membrane glycoproteins).

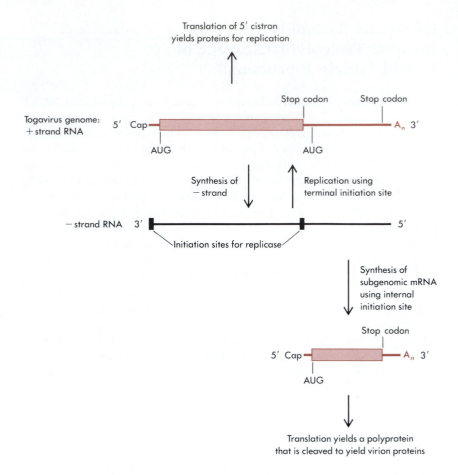

Figure 24-11
Nested mRNAs of coronaviruses. The seven distinct cistrons (not drawn to scale) are each translated from AUG codons near the 5′ end of a distinct subgenomic mRNA, generated by internal initiation from a common − strand template.

Lipid bilayer

G, transmembrane glycoprotein

N, nucleocapsid protein

L and NS, enzyme complex for RNA synthesis

M, matrix protein

Figure 24-12
The VSV virion. The bullet shape is distinctive for rhabdoviruses.

VSV: A Model − Strand RNA Virus[46]

Rhabdoviruses are small viruses with − strand RNA genomes and enveloped virions with a distinctive bullet shape (Figure 24-12). The most well known and deadly member of the family is rabies virus, but the best-studied is vesicular stomatitis virus (VSV), which is a mild animal pathogen that grows well in cell culture. VSV codes for only five proteins, and all of them are found in the virion. The *nucleocapsid (N) protein* encases the genome in a flexible helical capsid found neatly coiled within the virion. The *matrix (M) protein* lies between the capsid and envelope and makes specific contacts with both the N protein and the internal end of the virion *glycoprotein (G)*, which is also the protein responsible for interaction with cellular receptors. Since the genome is − strand RNA, it lacks the cap and poly A characteristic of + strand genomes. It cannot be translated, of course, so enzymes necessary for mRNA synthesis are present in the virus particle. The last two virion proteins, *L (large)* and *NS (nonstructural)*, fill this role. The L-NS complex is called transcriptase, since its role is to synthesize + strand transcripts as soon as the viral genome gets into a cell.

VSV Transcriptase Is a Multifunctional System That Catalyzes Both mRNA Synthesis and Replication[47–51]

With VSV virions, as with most − strand viruses, synthesis of mRNA can be readily duplicated in vitro by using a mild detergent to remove the envelope and then adding an appropriate mix of nucleoside triphosphates, buffers, and salts. Remarkably, the ensuing reaction not only makes + strands, but makes them in exactly the way they are found in the infected cell: as five distinct mRNAs, each encoding one of the virion proteins and each capped, polyadenylated, and in the same proportions relative to one another as found in the infected cell. Thus, the relatively simple L-NS enzyme complex can catalyze six distinct reactions: template directed RNA synthesis, polyadenylation, and the four enzymatic steps involved in capping reactions (Chapter 21).

The template for the L-NS complex is highly specific. Free RNA molecules are inactive, and only VSV RNA complexed with N protein works. Synthesis of the VSV mRNAs proceeds in an ordered fashion

Figure 24-13
Transcription and replication of the VSV genome. The choice between the two processes seems to be determined by the availability of N protein.

from the 3′ to the 5′ end of the genomic template (Figure 24-13). One transcriptase complex binds to the 3′ end of the genome and first synthesizes a 48-base leader fragment of RNA. This fragment is released from the template, and the complex then moves on to synthesize the five mRNAs sequentially, polyadenylating each at its 3′ end before initiating synthesis of the next molecule. Between each transcript, the transcriptase has about a 30 percent chance of falling off the template. Since the enzyme can only get on the template at its 3′ end, each successive mRNA molecule is about 30 percent less abundant than its predecessor. Examination of the order of the genes along the VSV genome reveals that there will be the most mRNA for N protein and the least for L and NS. This makes sense, since N protein coats the whole genome, but only a few L and NS molecules are needed to copy it.

A novel mechanism is used to add poly A to the five VSV mRNAs. The − strand genome contains a specific sequence at the region corresponding to the junction between each mRNA. This sequence includes a stretch of seven U residues corresponding to the position of

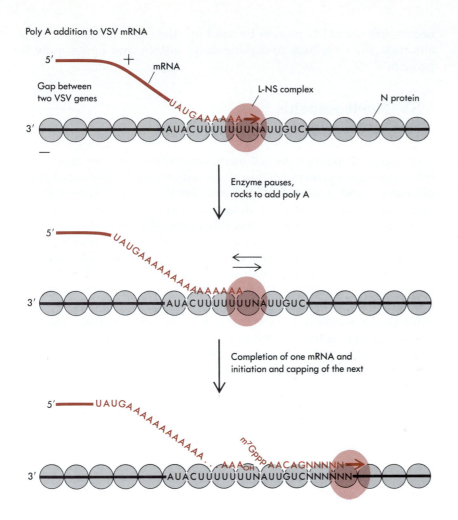

Poly A addition to VSV mRNA

Figure 24-14
Polyadenylation of VSV mRNAs. The sequence shown is found at the border of all VSV genes. It is still uncertain whether the transcriptase terminates following poly A addition and reinitiates synthesis, or whether it synthesizes a continuous precursor, which includes the poly A, that is rapidly cleaved. The result would be the same. Note that two bases (NA) are not copied into mRNA. (N represents any nucleotide.)

the mRNA poly A. The sequence apparently allows the poly A to be added by *slippage* (Figure 24-14), involving a back-and-forth motion of transcriptase over the region with subsequent generation of a long sequence of nothing but A residues.

As VSV replication proceeds, there must be a shift from the synthesis of mRNA to the synthesis of full-length + strands that can serve as templates for the synthesis of progeny genomes. This shift is apparently regulated by the concentration of N protein. The 48-base leader RNA molecule synthesized as the first + strand product contains the specific binding initiation site for N. If sufficient N protein is available, it will bind initially to this site and then rapidly to itself, building a + strand nucleocapsid immediately behind the advancing transcriptase. Coating nascent + strands with N protein prevents the polyadenylation, termination, and reinitiation reactions from occurring and so causes the synthesis of a complete + strand molecule.

The + strand–N protein complexes are not incorporated into virions. Rather, they serve as templates from which L-NS enzyme synthesizes progeny − strand genomes. Since N binding can initiate only on leader RNA, once this fragment is made and released from the template, N can no longer bind, and the transcriptase is committed to synthesizing mRNAs instead of full-length + strands. This regulatory scheme permits the replication to remain nicely in balance: Once N has reached an adequate concentration, genome replication occurs; but should N protein be used up, then the transcriptase will automatically shift back to making more mRNA and hence more N protein.

VSV Nucleocapsids Acquire Envelopes by Budding[52, 53]

VSV G protein reaches the cell membrane by the same route as most cell surface glycoproteins. It is synthesized on membrane-bound polyribosomes, and a short hydrophobic signal sequence causes membrane attachment and then directs the growing protein into the lumen of the rough endoplasmic reticulum. The protein is glycosylated at two sites soon after it passes through the membrane of the endoplasmic reticulum. Twenty hydrophobic amino acids near the carboxyl terminus of the protein serve as a membrane anchor, leaving a short carboxyl-terminal tail protruding on the cytoplasmic face of the rough endoplasmic reticulum. Once G protein is transported through the membrane, the signal peptide is proteolytically removed and the protein is transported to the plasma membrane via the Golgi apparatus, where the oligosaccharide chains are matured.

Patches of cell membrane that will become envelopes are modified by the insertion of G protein, whose C-terminal cytoplasmic end associates with M protein. Most cellular membrane proteins are excluded from the region. Association between nucleocapsids and M proteins is followed by budding of mature virions. The final size of the virion is determined entirely by the size of the nucleocapsid, which itself is a function of genome length. Mutants of VSV with much shorter genomes yield virions of the same diameter but proportionally shorter length than wild-type virions.

Influenza Is a Segmented − Strand RNA Virus[54, 55]

Like VSV, influenza virus is a − strand enveloped virus; it therefore contains a virion transcriptase associated with its genome. In other respects, however, influenza is quite different from VSV and other − strand viruses. Most notably, the genome is divided into eight segments of distinct sizes, each encoding a separate gene product. (There are actually ten gene products, as we will soon see.) Each segment is encased within its own helical nucleocapsid and is associated with its own transcriptase enzyme. The other major difference between influenza and other − strand (and also + strand) RNA viruses is its dependence on the host nucleus for replication. Influenza will not replicate in cells from which the nucleus has been removed by experimental manipulation or which have been treated with drugs (like actinomycin D) that prevent DNA-dependent RNA synthesis. As will become apparent later, this dependence reflects the bizarre method that influenza transcriptase uses to initiate transcription.

Influenza virions are roughly spherical with two distinct types of surface glycoproteins that can be seen as spikes of different appearance in electron micrographs (Figure 24-15). One type of spike, composed of the hemagglutinin (HA) protein, recognizes cell surface receptors and also contains the fusion potential that allows the virus to penetrate the cell membrane after endocytosis. The other, composed

Figure 24-15
The influenza virion.

of the neuraminidase (NA) protein, removes terminal sialic acid residues from cell surface proteins. The virions also contain a matrix (M) and nucleocapsid (N) protein like VSV. Four additional proteins together probably constitute the viral transcriptase. Each virus-coded protein has been associated with a specific RNA segment, and there is a rough correspondence in size, so that the larger segments encode the larger proteins (Figure 24-16).

Influenza Initiates Transcription with Stolen Caps[56, 57, 58]

For a long time, the dependence of influenza virus replication on the nucleus was a puzzle. Also unexplained was the fact that the virion transcriptase was essentially inactive in vitro, in sharp contrast to that of VSV, for example. These observations made sense when it was discovered that the influenza transcriptase, like the polio replicase, requires a primer, and the primer is found in the cell's nucleus. The primer for influenza transcriptase can be supplied in vitro by adding full-length cellular mRNAs (such as globin mRNA) to a mixture of disrupted virions and nucleoside triphosphates. The product of this reaction is a faithful copy of each − strand segment, except that the initial few nucleotides have been replaced with a short fragment derived from the 5′ end of the cellular (globin) mRNA, including the capping group. The RNA was initiated with a 5′ capping group stolen from another mRNA!

In infected cells, influenza mRNAs have heterogeneous 5′ ends, with variable numbers of nucleotides preceding the start of the influenza sequence. Rather than having machinery for synthesizing capping groups, the influenza transcriptase has evolved a mechanism for removing them from cellular mRNAs. In the nuclei of infected cells, the transcriptase complex cleaves newly made mRNAs, usually just after an A or a G residue a dozen or so nucleotides from the cap. These fragments then immediately serve as primers by base-pairing between the A or G and the terminal U of each genome segment and thus allow subsequent polymerization using the genome RNA as template. This mechanism explains the dependence of influenza virus replication on cellular RNA synthesis. Only newly made mRNAs can serve as cap donors, so inhibition of cellular mRNA synthesis is followed rapidly by a decline in influenza mRNA synthesis.

Figure 24-16
Electrophoretic separation on a polyacrylamide gel of the eight genomic RNA segments purified from two strains of influenza virus (A/PR/8/34 (PR8) and A/Hong Kong/8/68 (HK)). Corresponding genome segments have slightly different mobilities in the two viruses. Proteins specified by each segment are indicated. Note the double coding of the two smallest RNAs. [Photo courtesy of Dr. Peter Palese.]

NS₁ mRNA

NS₂ mRNA

Region removed
by RNA splicing

Figure 24-17
Structure and coding of NS1 and NS2 mRNAs. Dotted lines at the 5′ ends represent caps plus a short additional sequence stolen from the 5′ end of cell mRNAs. The translated portions are shown by boxes, with different shading indicating different reading frames.

Since influenza mRNA is made in the nucleus, it is available to cellular enzymes that process normal cellular mRNAs. The virus takes advantage of this unique situation in several ways. Most important, two of its mRNAs are spliced to generate additional messages. The virus does not, however, parasitize its host's poly A–adding equipment. Rather, poly A addition is accomplished by a slippage mechanism similar to that of VSV.

Some Influenza Virus Genome Segments Encode More Than One Protein[59]

RNA splicing explains the observation that although the influenza genome consists of only eight pieces of RNA, more than eight viral proteins can be detected in infected cells, and none is a cleavage product of another. The two smallest genome segments each give rise to two partially overlapping mRNAs and two proteins. In both cases, one of the mRNAs is generated from the other by RNA splicing; and in both cases, splicing occurs so that the carboxyl-terminal portion of the protein encoded by one mRNA is read in a different reading frame than that encoded by the parent mRNA.

Influenza genome segment 8 codes for two nonstructural proteins called NS_1 and NS_2. Excluding the cannibalized cellular 5′ cap fragment, NS_1 mRNA is a colinear, although incomplete, copy of genome segment 8 (Figure 24-17). About 55 virus-specific nucleotides at the 5′ end of NS_2 mRNA are shared with NS_1 mRNA. Then, relative to NS_1 mRNA, there is a 473-nucleotide deletion generated by splicing using host cell machinery. The NS_1 and NS_2 mRNAs are then colinear, although translationally out of phase in their 3′ regions. As a result of these mRNA structures, NS_1 and NS_2 proteins share nine amino-terminal amino acids but differ completely in the remainder of their amino acid sequences (see Figure 24-17). The two mRNAs derived from genome segment 7 have similar relationships and give rise to two proteins that also share nine amino-terminal amino acids but then diverge in sequence.

High-Frequency Recombination Results from Reassortment of Genome Segments[60]

A consequence of having segmented genomes is that in a cell infected by two different influenza viruses, novel viruses can arise by reassortment of segments. As shown in Figure 24-16, equivalent genome segments present in different strains of influenza often have slightly different sizes and encode proteins with slightly different molecular weights or immunological properties. When a cell is infected with two such genetically distinct influenza viruses, a high frequency of the progeny are recombinant for genes located on separate RNA segments. This high-frequency recombination results from the intracellular reassortment of genome segments and provides a rapid means for the evolution of novel strains of influenza.

The segmentation of viral genomes raises the question of how a virus particle selects one copy of each segment for packaging. Clearly, a virion that packaged, say, two copies of segment 1 instead of one copy each of segments 1 and 2, would not be capable of initiating a complete infection cycle. There must be some mechanism for ensur-

Figure 24-18
Primary structure of the influenza hemagglutinin precursor. The initial product (HA_0) is cleaved by a cellular enzyme to HA_1 and HA_2, which remain linked by a disulfide (S–S) bond. Trimers of this protein make up the HA spikes on the surface of the virion. [After Don C. Wiley, *Virology*. Ed. B. N. Fields et al. New York: Raven Press, 1985.]

ing that each virion gets a complete set of segments. Further, the random assortment of segments following infection with two different viruses shows that the segments do not remain physically associated during replication. The problem is a general one for all viruses with segmented genomes, and at present, virtually nothing is known about how a complete set of genome segments is selected for incorporation into virions.

The Structure of Influenza Hemagglutinin Glycoprotein Is Helping to Reveal the Molecular Basis of Flu Epidemics[61, 62, 63]

The influenza hemagglutinin (HA) glycoprotein is similar to the receptor-recognizing proteins of many viruses. It has two functions essential for infectivity. First, it recognizes and binds sialic acid–containing host cell receptors. Second, after virus uptake into endosomes, it mediates the fusion of virus and vesicle membranes to allow penetration of the influenza nucleocapsids into the cytoplasm. Because of its surface location and role in infection, hemagglutinin is a primary target for the host's immune system. HA elicits antibodies that, by binding to specific regions on the protein, neutralize virus infectivity. Viral mutants that alter the amino acid sequence of these regions arise frequently, and they can infect previously immune individuals. This variability in amino acid sequence of the hemagglutinin (as well as similar changes in the other viral glycoprotein, the neuraminidase) is probably the major cause of recurring flu epidemics.

HA spikes are trimers of identical molecules. Each monomer is synthesized as a typical membrane glycoprotein. After insertion into the membrane, the protein is cleaved to produce two peptides, HA_1 and HA_2, that are held together by disulfide bonds and anchored into the cell membrane via a C-terminal hydrophobic region (Figure 24-18). The amino terminus of HA_2 contains nonpolar amino acids that are not exposed at neutral pH. Acidic pH alters the conformation of HA and must allow the nonpolar sequence of HA_2 to interact with viral and endosomal membranes to promote their fusion.

X-ray crystallography shows that the HA trimer is an elongated cylinder to which each (HA_1 + HA_2) monomer contributes a long fibrous stem and a globular head (Figure 24-19). The receptor binding site is visible as a cleft in the head. Sites for antibody binding are also on the heads, and amino acid substitutions leading to new antigenic types are found in four different regions. Apparently, changes in all four regions are required to produce a new epidemic strain of flu. About every ten years, worldwide outbreaks (called **pandemics**) of influenza occur. These result from flu strains with so many more

(a)

135 Å

Membrane

■ = Antibody binding site

▽ = Carbohydrate group

(b)

Figure 24-19

Structure of the influenza hemagglutinin spike based on data from X-ray crystallography. (a) A schematic representation of the peptide backbone of one of the three identical ($HA_1 + HA_2$) dimers. HA_2 is shaded. Cylinders show regions of α-helical structure. The colored portion indicates the hydrophobic NH_2-terminal region of HA_2. Regions that undergo mutation to create new epidemic strains of flu are boxed. They lie on the head of the molecule, far from the membrane anchor. (b) A highly schematic drawing of the HA trimer. Each of the three units represents a ($HA_1 + HA_2$) dimer. R indicates receptor binding regions. Other boxes, shown only for one unit of the trimer, as in (a). [After I. A. Wilson, J. J. Skehel, and D. C. Wiley, *Nature* 289 (1981): 366–373.]

amino acid substitutions in their HA that almost no one is immune to infection. At least one such strain arose by reassortment when the HA gene of a duck influenza virus was acquired by a human flu strain.

While we can now begin to understand how changes in influenza HA protein allow the virus to circumvent our immune system, it is not yet clear why a single infection with many other viruses (e.g., mumps or measles) provides life-long immunity. It seems that the receptor binding proteins of these viruses should be able to mutate as rapidly as influenza HA. Conceivably, one explanation is that the most exposed, and hence most antigenic, sites on influenza glycoproteins are relatively unimportant to the function of the proteins, whereas in other viruses, the antigenic parts are close to critical functional domains, thus preventing them from varying rapidly. Also, its segmented genome allowing rapid reassortment is another difference important to the emergence of dramatically new strains of influenza and hence for recurring flu infections.

Reovirus Has a Double-Stranded, Segmented RNA Genome[64, 65, 66]

Reoviruses are characterized by double-stranded RNA genomes in ten segments. The virions are icosahedral and naked, and they contain an unusual two-layered capsid. Since double-stranded RNA cannot serve as mRNA or as a template for mRNA synthesis in uninfected cells, there is an associated virus-coded transcriptase in the reovirus virion. As with influenza virus, each reovirus genome segment encodes a different protein.

Reovirus virions enter cells by endocytosis and are carried to sites near the nucleus. During their passage through endocytic vesicles, about half their capsid proteins are removed by lysosomal proteases. This is the total extent of virion uncoating. The loss and cleavage of the outer capsid proteins activates the transcriptase in the virion and creates pores that permit the entry of nucleotides and the release of newly made mRNAs. This process can readily be duplicated in the test tube by treating reovirus virions with a protease and then adding ribonucleoside triphosphates. A remarkably efficient and prolonged synthesis of mRNAs ensues, with the virions continuing to make mRNAs for many days, extruding the new molecules through pores, at the vertices of the capsid (Figure 24-20). The reaction is asymmetric and conservative: Only + strands are made, and both incoming double strands remain intact and stay within the infecting, partially uncoated, subviral particle. The mRNAs are also unusual in that they are capped but do not have poly A. The virion transcriptase is intimately associated with the capsid, and so far, it has not been possible to purify the enzyme in a soluble form and still retain transcriptase activity.

The + strand RNAs that are spewed out of the reovirus virions not only are highly efficient mRNAs, but also serve as intermediates in replication. Later on in infection, newly made capsid proteins assemble with + strand RNAs into capsidlike structures called **precores.** The + strands are converted to double-stranded form by the virion polymerase within the precores. Precores serve to synthesize new mRNAs, which continue the replication process until large areas of replicating virus, called *factories,* are built up inside the cell. Very late in infection, the precores are converted into new virions by the addition of outer shell proteins.

As with influenza virus, random reassortment of reovirus genome segments is seen when two genetically distinct viruses infect the same cell. Again, the mechanism that directs the encapsidation of a correct set of ten genome segments is completely unknown.

DNA VIRUSES

The Challenge of Being a DNA Versus an RNA Virus[67]

There are advantages in an animal virus having a DNA genome instead of an RNA genome. The virus can use its host's transcriptional apparatus and so evolve complex patterns of gene expression. It can

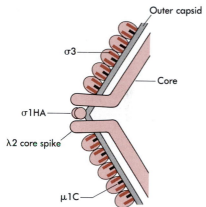

Figure 24-20
Electron micrograph of a stained and shadowed preparation of a reovirus core caught in the process of RNA synthesis. Twelve chains of RNA are thought to be extruded from the vertices of the partially uncoated virion. RNA synthesis took place for 1 minute at 37°C. [From N. M. Bartlett, S. C. Gillies, S. Bullivant, and A. R. Bellamy. 1974. *J. Virology* 14:315–326.] Diagram shows one vertex of a reovirus virion with possible locations of some viral proteins. [From B. N. Fields and M. I. Greene. 1982. *Nature* 300:19.]

also use the DNA repair systems of the host and thereby evolve larger genomes than the RNA viruses. It would seem that another advantage might be the ability to use host enzymes involved in DNA replication. While host enzymes are used in many cases, the relationship of viral and host DNA synthesis is complex and actually presents special problems for the DNA viruses.

DNA synthesis in animal cells takes place in a restricted period of the cell cycle, the S phase. Furthermore, cells in many tissues are in a nondividing state most of the time and thus have concentrations of deoxynucleotides and DNA metabolic enzymes (polymerases, ligases, etc.) too low to support many rounds of DNA virus replication. RNA viruses do not face a comparable problem, since almost all cells, even those that are not dividing, have a constant need for RNA synthesis and thus always have high concentrations of the appropriate precursors (ribonucleoside triphosphates). To overcome the shortage of enzymes and precursors needed for DNA synthesis, DNA viruses must either replicate exclusively in dividing cells or devise means to induce conditions that will permit DNA synthesis. The simplest of DNA viruses, the parvoviruses, are completely incapable of replicating except in cells that are already actively dividing or in which replication of another DNA virus (such as adenovirus) is already occurring. Other more complex DNA viruses that can infect nondividing cells (e.g., papovaviruses and adenoviruses) stimulate cells to initiate DNA synthesis and thus to provide for adequate supplies of precursors and enzymes needed for viral replication. In addition, these viruses encode some proteins used in their replication. The most complex DNA viruses, herpesviruses and poxviruses, also encode a number of enzymes involved in nucleotide metabolism, which allows them to directly stimulate the production of deoxynucleotides by the infected cell. Interestingly, the most complex DNA viruses, which have the most control over their replication needs, also have the fastest life cycles.

SV40: A Small Double-Strand DNA Virus with a Circular Genome[68, 69, 70]

SV40 (simian virus number 40) belongs to the *papovavirus* family, which also includes the wart *(papilloma)* viruses. Another member of the family is the mouse virus polyoma, which is very similar to SV40. SV40 was isolated as a contaminant in poliovirus vaccines. The vaccines were being prepared using monkey cells instead of human cells because the latter were considered a potential source of dangerous cancer-causing viruses. It was a surprise when SV40 turned out to cause tumors in newborn rodents, but it was of great interest to molecular cancer biologists, who made SV40 (and polyoma, which causes tumors in hamsters) among their favorite objects of study (Chapter 26).

SV40 virions are naked icosahedral structures composed of three virus-coded proteins: VP1; the major protein, and lesser amounts of VP2 and VP3. The DNA of SV40 is a closed circular structure about 5200 base pairs in length, and its complete sequence is known. In virions and also inside the cell, the SV40 genome resembles cellular DNA in that it is associated with cellular histones and packaged into nucleosomes (see Figure 9-27 and Figure 21-19). When the SV40 genome is deproteinized, it adopts a negative superhelical configuration

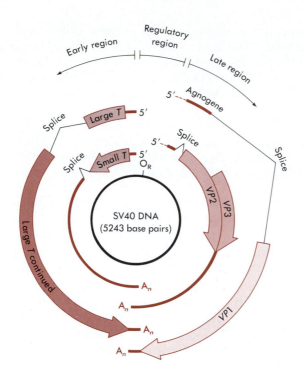

Figure 24-21
SV40 genome and transcription map. The translated portion of each mRNA is shown by large arrows, with different shades of color for each reading frame. The role of the protein encoded by the agnogene at the 5′ end of late mRNA is uncertain. O_R is the origin of DNA replication.

(form I DNA). A single nick in either DNA strand can relax the super-coil and result in a circular (form II) molecule, while a double-strand break produces a linear molecule (form III DNA). Supercoils can readily be separated from other viral or cellular DNAs because their condensed shape gives them unusual physical properties (Chapter 9).

Different SV40 Genes Are Expressed at Different Times During Lytic Infection[71, 72, 73]

As with most DNA viruses, expression and replication of the SV40 genome takes place in the nucleus and makes extensive use of cellular polymerases. The genome can be divided into three regions (Figure 24-21). The **early region** and the **late region,** each almost half the genome in length, encode protein. The third region, which lies between them, is a complex regulatory region containing promoters and regulators of expression as well as the site at which DNA synthesis begins, the origin of replication. Soon after SV40 DNA (still associated with histones) moves to the nucleus, cellular RNA polymerase recognizes the efficient early promoter region and transcribes the early region. The late region, which encodes capsid proteins, is silent at this time and is not expressed until after viral DNA replication begins.

The early region provides functions necessary for initiation of viral DNA replication, although it does not encode a DNA polymerase. The region consists of a single transcription unit, but encodes two proteins called **T proteins,** or *T antigens* (T stands for "tumor" and reflects their initial discovery in rodent tumor cells induced by SV40 infection; Chapter 26). As with many other DNA viruses, this multiple use of a single transcript is accomplished by differential RNA splicing of the primary transcript using cellular machinery. The two resulting early mRNAs have identical 5′ and 3′ ends, but differ in the size of an intron that is removed (see Figure 24-21). Since the larger RNA with the smaller spliced-out region retains a translational terminator, it encodes a relatively small protein, called *small t* (approximately 20 kdal). The other spliced mRNA has a larger region removed, including the terminator, and thus is translated into a much larger protein of about 90 kdal, called *large T*. The two proteins thus share about 80 amino-terminal amino acids, but each has a unique carboxyl-terminal region. The role of the small t protein in the virus life cycle is not well understood. The large T protein is found mostly in the nucleus and is a complex, much-studied protein with multiple functions: It regulates its own synthesis, stimulates cellular DNA synthesis, is involved in initiating viral DNA synthesis, and is needed for late gene expression.

The late region of the SV40 genome, which encodes the capsid proteins, is expressed only after DNA synthesis begins, about 12 to 14 hours after infection. It is transcribed in the opposite direction, hence from the opposite strand, as the early region. Like the T proteins, the late proteins (VP1, VP2, and VP3) are translated from differentially spliced, overlapping RNAs. VP2 and VP3 are translated in the same reading frame, starting at different points, whereas VP1, the major virion protein, is translated from sequences that overlap those encoding VP2 and VP3, but in a different reading frame. A fourth late gene is present in the leader of the late mRNAs. It encodes a protein 61 amino acids long called the *agno* protein, whose function is unknown.

Regulatory Sequences for SV40 Gene Expression[74-82]

A stretch of DNA lying between the 5′ ends of the SV40 early and late mRNAs contains sequences that regulate both early and late gene transcription as well as DNA replication. In part because it is transcribed so efficiently, the SV40 early region has been intensively studied as a model for eucaryotic gene expression. Several sets of sequences upstream of the sites of initiation of early SV40 mRNA synthesis are important for early gene transcription (see Figure 21-30). Approximately 20 to 30 bases upstream from the cap sites is an AT-rich region, the Goldberg-Hogness or TATA box, with the sequence TATTTAT in SV40. This sequence is not essential for transcription to occur, but is important for positioning RNA polymerase II. If it is removed, the same number of transcripts is made, but their 5′ ends are heterogeneous. Just 5′ of the TATA sequence is a nearly perfect triple tandem repeat of 21 base pairs containing six copies of the sequence 5′-GGGCGG-3′. This repeated sequence is essential for efficient transcription and binds a specific cellular transcription factor, Sp1 (see Chapter 21). Between about 100 and 250 bases 5′ of the cap site is a perfect tandem repeat of a 72-base-pair sequence that contains a transcriptional enhancer element (see Chapter 21). As with other enhancers, the number of bases between the SV40 enhancer and the early region whose transcription it influences can be varied to a considerable extent without deleterious effect on the level of transcription, and its orientation relative to the early genes can be reversed. Enhancers from other viral or cellular genes can be inserted in place of the SV40 enhancer and the resulting virus still functions.

Even though the transcription of SV40 genes is performed by cellular enzymes and factors, expression of both the early and late regions is regulated by the large T protein. There are three specific binding regions for T in the control segment. Interaction of T with these sites has three effects. First, it shuts down synthesis of early mRNA (and therefore itself); thus, T protein synthesis is autoregulated, just as the synthesis of the phage λ repressor is. Second, it directs the initiation of viral DNA synthesis. Finally, it turns on the synthesis of late mRNAs, either directly or indirectly as a result of stimulating DNA synthesis.

The late SV40 promoter is unusual in that it contains no identifiable TATA sequence. The late SV40 mRNAs are extremely heterogeneous at their 5' ends, presumably reflecting the alternative use of a large number of different initiation sites, consistent with the idea that the role of the TATA sequence is to align the initiation site of transcription, not to provide an initial binding site for RNA polymerase (see Figure 21-30). How sequences in the control region directly regulate late gene expression is not yet well understood.

Large T Protein Is Involved in Initiating Bidirectional DNA Replication from a Unique Origin on the SV40 Genome[83–87]

SV40 replicates its DNA bidirectionally with forks growing in opposite directions from a unique origin of replication. Replicating molecules purified from infected cells contain two circles of equal length composed of replicating DNA, as well as a superhelical loop of still unreplicated DNA (see Figure 10-6). The replication mechanism is similar in many ways to that of bacterial and plasmid genomes. The two replication forks advance from a common origin, and one strand is synthesized as a continuous piece in a 5'→3' direction, whereas the other is synthesized as a series of small RNA-primed Okazaki fragments. Replication is complete when the two moving forks reach each other opposite the origin. The precise sequence at the termination point is unimportant for completion. The initiation signal, on the other hand, has been defined by mutants as a precise sequence about 30 nucleotides long lying between the early and late genes (see Figure 21-30).

SV40 relies almost entirely on cellular enzymes to replicate its DNA, except that large T protein is essential for initiation of replication. However, its role in this process is not yet understood. It specifically and sequentially binds to three tandem sites: One is contained entirely within the initiation sequence for replication, while the other two overlap additional portions of this sequence (see Figure 21-30). Now that a system of in vitro replication that is dependent on T antigen is available, the mechanism of T antigen function should soon be worked out (see Chapter 10).

Adenoviruses Have Linear, Double-Stranded DNA Genomes[88, 89]

Adenovirus genomes are linear, double-stranded DNA molecules about 36 kilobase pairs long. Each strand has a short sequence (about 100 base pairs long) at one end that is complementary to the sequence at the other end, an inverted terminal repeat structure. As a result, when the DNA is denatured, the single strands self-anneal to form panhandle structures. The DNA found in virions has a 55 kdal

Figure 24-22

The adenovirus genome and its RNA transcripts. (Data are for human adenovirus.) The 40 or so adenovirus transcripts are divided into many families (both early and late) sharing the same promoter, indicated by the bracket at the 5′ end of each family. The proteins encoded by each are listed above and below, named by function or molecular weight. The roman numerals indicate virion proteins. Large arrow heads are present on late transcripts, small arrowheads on early transcripts. [After J. Tooze, ed., *Molecular Biology of Tumor Viruses: DNA Tumor Viruses*, 2nd ed. (Cold Spring Harbor, N.Y.: Cold Spring Harbor Laboratory), with permission.]

protein, the *terminal protein*, covalently attached to the 5′ end of each strand.

Like SV40, adenoviruses replicate their DNA in the nucleus and assemble their progeny virions there; and like SV40, they display both temporal control of gene expression and the use of multiple overlapping spliced mRNAs to pack a large amount of genetic information into their genomes.

Adenoviruses Use Alternative Splicing Patterns, Overlapping Genes, and Both DNA Strands to Maximize Use of Their Genomes[90-99]

Early in infection, during the 6 to 8 hours before viral DNA replication begins, adenovirus mRNAs are made that represent about 25 to 30 percent of the genome. These are derived from scattered regions of the genome called E1 to E4 (E for early) (Figure 24-22). All these regions give rise to multiple transcripts. In each case, the transcripts are derived from common precursor RNAs by differential splicing, and they retain the 5′ and 3′ ends of the precursor mRNAs but have different internal portions. Sometimes, this results in overlapping mRNAs that encode proteins that are a subset of one another, since they use the same reading frame. Sometimes, it results in the use of different overlapping reading frames in such a way that the resulting proteins share amino acids at their amino termini but diverge in their carboxyl portions. Early transcripts are derived from both strands of the DNA. Some, but not all, of the numerous polypeptides predicted to arise from the many early mRNAs have been identified in infected cells or by in vitro translation.

The expression of early genes itself is regulated, with some genes being expressed before others. Transcription of the E1A region is followed by transcription of regions EIB, E3, E4, and, last of all, region E2. Most important in this sequential regulation is the EIA region at the left of the genome. EIA encodes a protein that is required for expression of almost all other adenovirus genes. In the absence of this protein, no RNAs from EIB, E2, E3, or E4 are made, and infection aborts. The EIA gene product stimulates the rate of initiation of transcription, but the mechanism is unknown. Stimulation does not appear to involve an interaction with specific enhancers or promoters and is not even specific for adenovirus genes, since the level of transcription of some other viral and even cellular genes can be affected by EIA.

The E2 region encodes three proteins directly involved in DNA replication: a DNA binding protein, an 80 kdal protein that is cleaved to yield the terminal protein on the ends of the genome, and the 140 kdal adenovirus DNA polymerase. Thus, unlike the simpler papovaviruses, which must rely heavily on cellular enzymes, adenovirus-coded proteins fill many of the major roles in DNA synthesis.

Once viral DNA synthesis begins, late gene transcription becomes dominant in the cell. The late RNAs principally encode the 12 or so virion capsid proteins. Most of the late mRNA originates from a single promoter, the major late promoter (see Figure 24-22). This extremely active promoter directs the synthesis of a primary transcript that can extend almost to the extreme right end of the adenovirus genome, a length of over 28 kilobases. This long primary transcript is processed to yield at least 20 different mRNAs. All share the identical 5' terminus, as well as a 5'-end sequence about 200 nucleotides long known as the tripartite leader. The leader is assembled by RNA splicing that joins three small untranslated regions. The messages differ at their 3' ends owing to differential cleavage and polyadenylation at one of five possible sites that lie downstream of typical poly A addition sequences (AAUAAA) (Figure 24-23). Each of these five poly A–tailed transcripts then serves as the progenitor of a set of spliced mRNAs with common 5' and 3' termini. The different mRNAs in each set are made by joining the leader to one of several different sites on each transcript. This mechanism permits a wide variety of transcripts to be regulated together at the level of transcription and to receive the identical 5' sequence. The sequence encoded by the tripartite leader is a very efficient one for translation, and thus the mRNAs that contain it will be translated at a high level.

The early and late mRNAs described here are all transcribed by RNA polymerase II, and all have 5' capping groups and 3' poly A tails. However, not all adenovirus transcripts are made by RNA polymerase II or serve as mRNAs. Small adenovirus transcripts about 150 nucleotides long called virus-associated, or VA, RNAs are produced by RNA polymerase III, and like other genes transcribed by this enzyme, have promoter sequences within the genes themselves (see Figure 24-22). VA RNAs are made in very large amounts late in infection and seem to be necessary for translation of late mRNAs, possibly by blocking the effects of interferon (discussed later in this chapter).

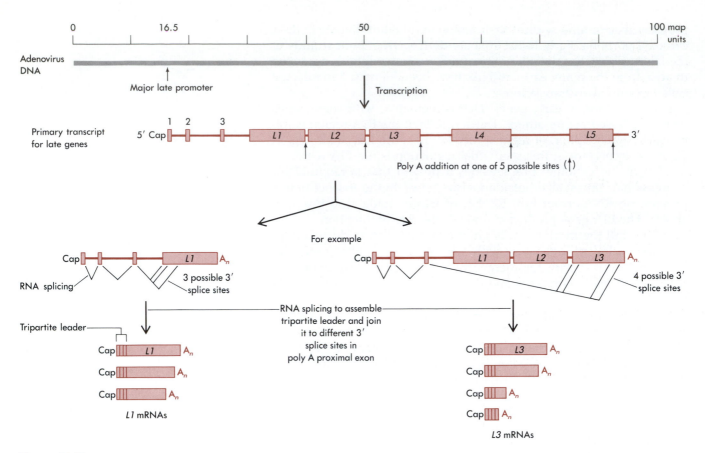

Figure 24-23
A model for adenovirus late mRNA synthesis. Variety is generated both by choice of polyadenylation site and by choice of 3′ splicing site to which the tripartite leader is joined. Note that only the late mRNAs derived from the L1 and L3 bodies are shown.

The entire adenovirus genome has been sequenced, and analysis of the sequence reveals numerous additional open reading frames that could encode peptides of up to 10 kdal for which neither mRNAs nor the corresponding proteins have yet been found. Given the dense packing of genetic information in adenoviruses and the fact that RNA splicing makes it difficult to identify all the possible genes from DNA nucleotide sequences alone, it will probably be a long time before all the mRNAs and proteins specified by any adenovirus are known.

An Adenovirus Protein Covalently Binds dCMP to Initiate DNA Replication at the Inverted Terminal Repeats of the Adenovirus Genome[100–104]

Adenovirus DNA replication is unusual in that only one strand is synthesized at a time and all synthesis is continuous. Synthesis begins at either end of the linear molecule (or at both ends of the same molecule) and proceeds 5′→3′ until the daughter strand is completed and one parental strand displaced (or until both parental strands are copied) (see Figure 10-31). There is no discontinuous replication involving Okazaki fragments. Rapid progress is being made in identifying proteins involved in adenovirus DNA replication. The first breakthrough came with the discovery of the virus-coded terminal protein of 55 kdal covalently attached to each 5′ end of adenovirus DNA and the recognition of the role of this protein as a primer for DNA synthesis. A second breakthrough in the study of adenovirus replication was the finding that if extracts of adenovirus-infected cells were given deoxynucleotides, ATP, and Mg^{2+}, they could initiate and complete semiconservative replication in a test tube, using exogenously added adenovirus DNA as a template. This has permitted the biochemical identification of essential components and a more detailed analysis of the events involved in replication (see Chapter 10 and Figure 10-31).

Parvoviruses Are Small, Single-Strand DNA Viruses That Replicate Using Alternative Hairpin Structures[105, 106]

Parvoviruses are very small animal viruses with single-stranded DNA genomes of about 4500 to 5000 bases. Their virions are composed of three capsid proteins that have overlapping amino acid sequences and that are derived by translation of differentially spliced mRNAs. Unlike other DNA viruses, which either encode enzymes necessary for DNA replication or stimulate the cell to provide them, parvoviruses are dependent on other means. One group of parvoviruses can grow only on cells that are also infected with a more complex virus, such as adenovirus. The remaining parvoviruses are autonomous in that they do not depend on infection by another virus; however, they can only replicate in rapidly dividing cells. In nature, such viruses infect only certain cell types, such as blood-forming cells, which proliferate rapidly. They also infect tumors, since these also contain rapidly dividing cells.

Parvovirus genomes have unusual terminal nucleotide sequences that can fold into special T-shaped terminal structures. These structures provide yet another solution to the problem of replicating the ends of DNA molecules. As shown in Figure 24-24, the 3' end of the genome can serve as a primer. Simple addition of deoxynucleotides generates a copy of the genome, and refolding at the ends creates new primers as necessary. With the introduction of specific single-strand breaks at key points, new genomes can be synthesized.

The Large DNA Viruses Have More Complex Genomes and Replication Cycles[107-115]

The two families of very large DNA viruses, the herpesviruses and the poxviruses, have genomes 80 to 200 kilobase pairs long that encode 50 to 200 different proteins, including many enzymes involved in nucleotide metabolism. Except for size and complexity, the two groups of viruses are quite different. Herpesviruses have enveloped virions with icosahedral capsids and replicate in the nucleus, while poxviruses have complex, seemingly asymmetric capsids and are the only DNA animal viruses that replicate in the cytoplasm. A remarkable technical achievement was the recent determination of the complete nucleotide sequence of the genome of the herpesvirus known as Epstein-Barr virus. Its genome is 172,282 base pairs long!

Herpesviruses include some well-known infectious agents such as herpes simplex types 1 and 2; varicella-zoster virus, which causes chickenpox and shingles; and the Epstein-Barr virus, which causes infectious mononucleosis. Although the virions of all these viruses are indistinguishable from one another, the genomes vary considerably in size and organization (Figure 24-25). All herpesviruses contain

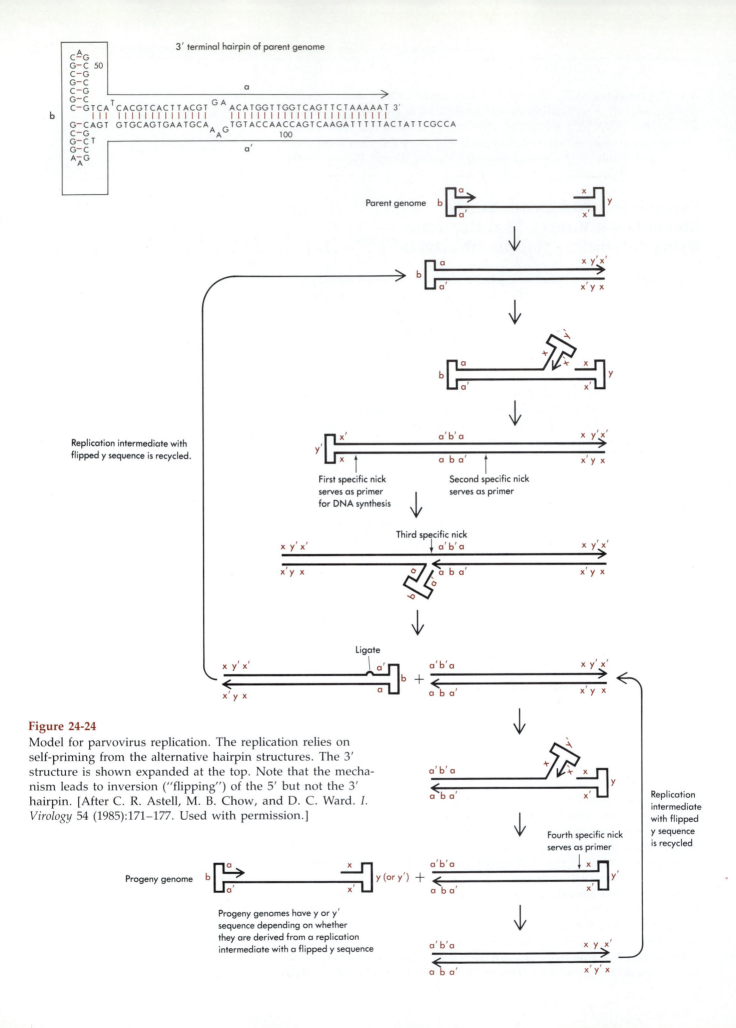

Figure 24-24

Model for parvovirus replication. The replication relies on self-priming from the alternative hairpin structures. The 3' structure is shown expanded at the top. Note that the mechanism leads to inversion ("flipping") of the 5' but not the 3' hairpin. [After C. R. Astell, M. B. Chow, and D. C. Ward. *I. Virology* 54 (1985):171–177. Used with permission.]

3' terminal hairpin of parent genome

Replication intermediate with flipped y sequence is recycled.

First specific nick serves as primer for DNA synthesis

Second specific nick serves as primer

Third specific nick

Ligate

Fourth specific nick serves as primer

Replication intermediate with flipped y sequence is recycled

Parent genome

Progeny genome

Progeny genomes have y or y' sequence depending on whether they are derived from a replication intermediate with a flipped y sequence

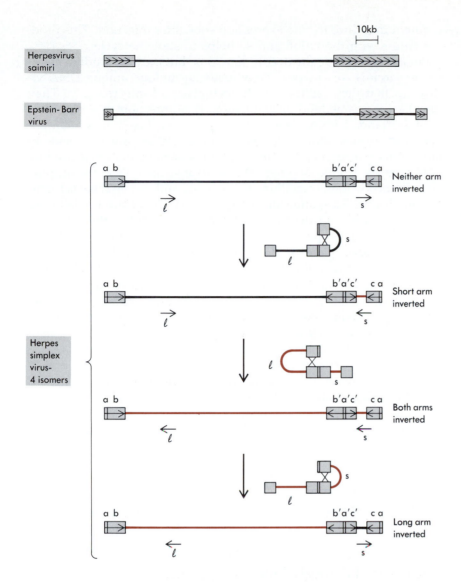

Figure 24-25
Some herpesvirus genomes. The boxes enclose repeated sequences, shown as arrows. Note the extensive short tandem repeats at the ends of the herpes Saimiri and Epstein-Barr virus genomes. The herpes simplex genome can be thought of as consisting of two arms, long (l) and short (s), each flanked by a long inverted repeat. Recombination between the inverted repeat sequences leads to inversion of one arm or the other.

repeated sequences of various kinds. In Epstein-Barr virus, the ends of the genome have a large number of short (500-base-pair) repetitions of the identical sequence as well as an internal sequence consisting of half a dozen repeats of a 3-kilobase-pair sequence. The total number of these repeated sequences does not vary greatly from molecule to molecule, but the distribution between the ends differs from one genome to another. The organization of herpes simplex genomes is quite different. These genomes can be viewed as having two covalently joined parts (arms), each flanked by long inverted repeats (see Figure 24-25). Recombination between the inverted repeat sequences leads to reorganization, or *isomerization*, of the genome. Since each arm can be in one of two orientations, there are four possible isomers, and all four can be found soon after infection with only one isomer.

During infection, the expression of the 50 or so herpes simplex genes is tightly regulated. Three classes of genes are known, called α, β, and γ (or immediate early, early, and late). Expression of the α genes is required to induce β gene expression; expression of β genes both induces γ genes and shuts off α genes; and expression of γ genes turns off β genes. Thus, there are three distinct waves of herpesvirus gene expression during replication. Interestingly, the process seems to be circular in that at least one virion protein (a γ gene product) is required to induce α gene expression soon after infection. This product enters with the virion and so helps to start the cycle.

Poxviruses include the deadly smallpox, but the best-studied member is vaccinia, or cowpox. Poxviruses are unique among DNA viruses in that they replicate in the cytoplasm of infected cells. They thus face the same kind of problem early in infection as − strand or double-stranded RNA viruses: The cytoplasm does not contain enzymes for nucleic acid synthesis, yet viral mRNAs must be made to initiate the infection cycle. The solution is analogous to that used by RNA viruses. Unlike all other DNA animal viruses we know of, poxviruses encode their own RNA polymerase, and this enzyme is found within virions. Soon after infection, vaccinia virions are partially uncoated to cores, and the RNA polymerase transcribes the 100 or so early genes, corresponding to about half the genome. Included among the early gene products are enzymes of nucleotide metabolism, such as thymidine kinase; enzymes for DNA replication, like DNA polymerase; and enzymes involved in synthesis and processing of transcripts, such as RNA polymerase, and capping and poly A−adding systems. Also included are enzymes that complete the uncoating process, releasing free DNA for replication. The function of most of the early genes, however, remains unknown.

The poxvirus genome and its replication are unusual. DNA of vaccinia virus is about 186 kilobase pairs in length. It has an inverted terminal repeat of about 10 kilobase pairs. Its most unusual feature is that the ends are covalently joined by phosphodiester bonds. Thus, if the genome is denatured, it forms a continuous, single-stranded circle. How this genome structure is replicated is still speculative with definitive proof awaiting further experiments (see Figure 10-23).

RETROVIRUSES AND HEPATITIS B VIRUS

Retroviruses: RNA Genomes That Replicate Through DNA Intermediates[116–120]

Retroviruses have a uniquely intimate association with the host cell. Although the viral genome is + strand RNA, the intermediate in replication is a double-stranded DNA copy of the genome, called the **provirus.** The provirus resembles a cellular gene and must even integrate into host chromosomes in order to serve as a template for transcription of new viral genomes. New genomes are therefore synthesized and processed in the nucleus by unmodified cellular machinery. Although viral genome RNA looks like mRNA, it does not serve as such early after infection. Rather, it is used as a template for DNA synthesis. Since an enzyme capable of carrying out this reaction is not usually present in the cell, it must enter with the virion. This virus-coded enzyme is called *reverse transcriptase*, and its presence in virions is diagnostic for retroviruses. Progeny genomes transcribed from integrated proviral DNA move to the cell membrane to assemble into virions. These bud from the cell membrane without killing the cell. A schematic view of the retrovirus life cycle is shown in Figure 24-26.

Retroviruses were first isolated early in this century as agents that could cause cancer in chickens. One of the first viruses found, Rous sarcoma virus (RSV), was isolated by Peyton Rous in 1911 and is still intensively studied as a model cancer-causing agent. Many naturally

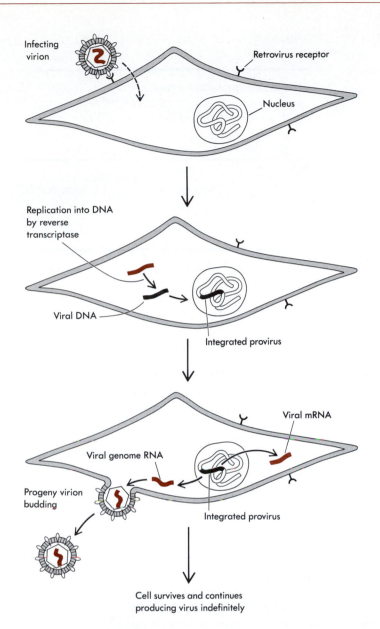

Infecting virion

Retrovirus receptor

Nucleus

Replication into DNA by reverse transcriptase

Viral DNA

Integrated provirus

Viral mRNA

Viral genome RNA

Progeny virion budding

Integrated provirus

Cell survives and continues producing virus indefinitely

Figure 24-26
An overview of retrovirus replication. Only one DNA provirus is made per infecting virion; this is replicated by the normal cell division process, following integration into the cell genome. An infected cell is resistant to superinfection by another virus once it starts synthesizing viral glycoprotein, which binds to the receptors and so blocks infection. Usually the cell is not killed.

occurring cancers of animals, as well as some other diseases, have been traced to retroviruses. Most recently, several human isolates have been found: one (HTLV-I) causes a relatively rare, but invariably fatal, cancer of T lymphocytes, while another (HIV, also known as HTLV-III or LAV) has been identified as the cause of acquired immunodeficiency syndrome (AIDS). As we shall see in Chapter 26, the study of retroviruses has led to major insights into the molecular basis of cancer and growth control.

Retrovirus virions have icosahedral nucleocapsids surrounded by an envelope. The viral genome is unique in being diploid: Each virion contains two identical copies of the seven- to ten-kilobase genome. The reason for a diploid genome is not clear. Perhaps it permits the repair of damage and high levels of recombination during replication.

The genetic organization of retrovirus genomes is known in great detail, and the complete nucleotide sequence of well over a dozen genomes of different retrovirus isolates have been determined (Figure 24-27). Typical retroviruses have three protein-coding genes: *gag* (so named because it encodes a group-specific antigen) encodes a precur-

Figure 24-27

Some retrovirus genomes. The boxes indicate open reading frames. Boxes on different lines are in different frames. Note that the open reading frames of *tat* are separated by an intron in the AIDS virus. Most naturally occurring retroviruses resemble murine leukemia virus in structure, encoding only *gag*, *pol*, and *env* genes. mRNAs for *env*, *src*, *tat*, and probably *sor* and *orf* are generated by RNA splicing.

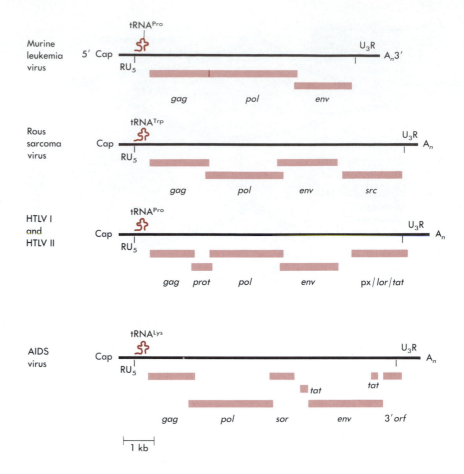

sor polyprotein that is cleaved to yield the capsid proteins; *pol* is cleaved to yield reverse transcriptase and an enzyme involved in proviral integration; and *env* encodes the precursor to the envelope glycoprotein (analogous to the G protein of VSV or the influenza HA protein). Recently a fourth type of gene (called *pX*, *lor*, or *tat*) has been found at the 3′ end of HTLV -I and -II genomes. This gene seems to encode a transcriptional activator (similar at least superficially to the E1A protein of adenovirus), which greatly increases expression of the provirus. The AIDS virus contains still other genes not present in most retroviruses. In addition, a few retrovirus isolates have an additional gene, called an *onc* gene, that gives them the ability to rapidly induce certain types of cancer (Chapter 26).

Retrovirus genomes contain long untranslated regions at both their 5′ and 3′ ends. These sequences include signals needed for replication and also for transcription. Particularly noticeable is a short, directly repeated sequence (R) at each end of the genome, as well as a tRNA molecule that is held in place by base-pairing to a sequence near the 5′ end (see Figure 24-27). The genomic sequence complementary to a 3′ portion of the tRNA is referred to as the primer binding (PB) site. The tRNA is stolen from the host cell during replication and serves as a primer for DNA synthesis soon after infection. This is essential because, like all other DNA polymerases, reverse transcriptase cannot initiate DNA chains, but can only extend chains from a 3′-OH primer.

Reverse Transcriptase Generates
Long Terminal Repeats in Proviral DNA[121–128]

Reverse transcriptase that enters the cell in a retrovirus virion has the problem of copying a single-stranded RNA molecule into a double-stranded DNA molecule suitable for integration into host cell DNA and capable of being transcribed by RNA polymerase II. The solution is remarkably complex. The integrated provirus is not a precise copy of the RNA genome. Rather, it is longer at each end owing to the addition of sequences, called U_5 and U_3 (for unique to the 5' or 3' end), derived from the other end of the genome (see Figures 24-27 and 24-28). This results in a perfect direct repeat, called the **long terminal repeat** (LTR), at each end of the provirus. The LTR, with the structure U_3-R-U_5, contains signals for both initiation of transcription by RNA polymerase II and for 3' cleavage and polyadenylation of transcripts.

Viral DNA synthesis takes place in the cytoplasm shortly after the virion enters the cell. Reverse transcriptase begins by adding deoxynucleotides to the tRNA primer. After adding just 100 to 200 nucleotides, corresponding to the R region and to U_5, which lies between R and the tRNA primer, the enzyme comes to the 5' end of the genome. To continue synthesis of this − strand DNA, the RNA that has just been copied is removed by a specific nuclease, RNase H, which is part of the reverse transcriptase molecule. (RNase H degrades the RNA in a DNA-RNA hybrid.) Removal of the RNA permits the newly made DNA copy of R to base-pair with the R sequence at the 3' end of the genomic RNA. Thus, a new template-primer pair is formed, which allows reverse transcriptase to proceed and copy the remainder of the genome (see Figure 24-28). The newly made single-stranded DNA, with its covalently attached primer tRNA, must now be converted to double-stranded DNA. To do this, a specific nick is made in the RNA immediately to the left of U_3. The 3'-OH end of the RNA generated in this way now serves as primer for + strand DNA synthesis. This time, the − strand DNA just made is copied, as is part of the tRNA primer. A second jump then occurs, similar to the first, but this time the + strand DNA copy of the primer can base-pair with the − strand DNA copy of the primer binding site, again following removal of genome RNA by RNase H. Completion of both DNA strands produces the linear double-stranded structure U_3-R-U_5···U_3-R-U_5, where U_3-R-U_5 is the long terminal repeat.

The complicated process of proviral DNA synthesis is carried out entirely by reverse transcriptase. The enzyme needs at least four enzymatic activities for the task: RNA-directed DNA synthesis, degradation of RNA in RNA-DNA hybrids (the RNase H activity), DNA-directed DNA synthesis, and specific cleavage of RNA at the 5' end of U_3. The product of synthesis is linear DNA. This linear DNA is transported to the nucleus, where the ends are joined, possibly by a cellular enzyme, to make covalently closed circles. These serve as precursors for integration. The mechanism of integration of viral DNA is not fully understood. The circle must be opened at a specific site, since the ends of the integrated provirus are always the same. By contrast, the site of integration into the cellular DNA is apparently not sequence specific. At the joints between viral and cellular DNA, two bases are always removed from each end of the provirus, and in addition, four to six bases of cell DNA at the integration site are duplicated. For example, integration of a provirus into the sequence

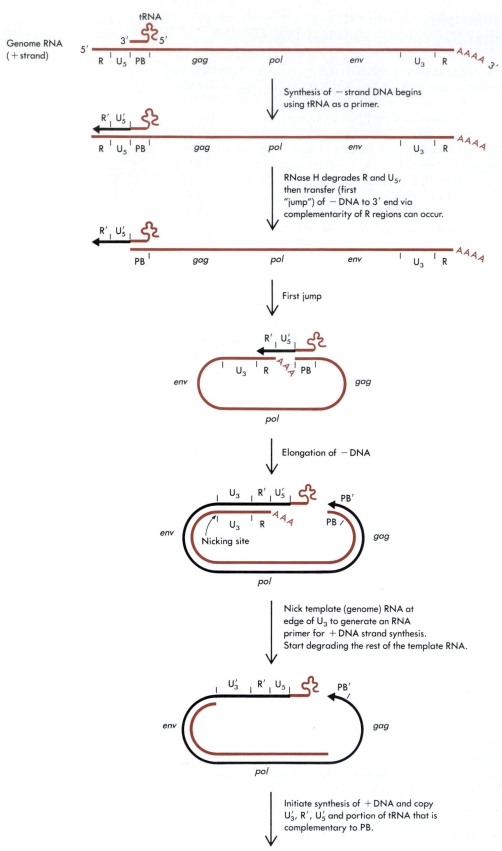

Figure 24-28
Synthesis of retrovirus DNA. All the steps are carried out by reverse transcriptase.

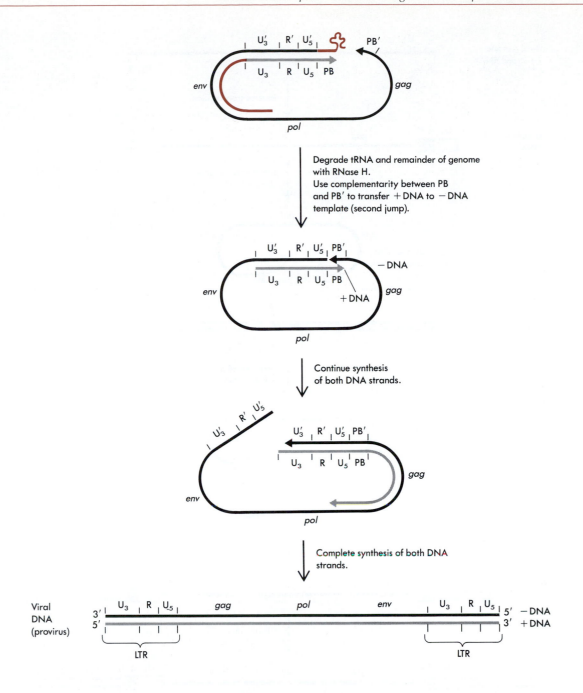

. . . TAGTCG . . . can give the structure TAGTC-U_3-R-U_5-AGTCG . . . as a result of duplicating the sequence AGTC. This process is poorly understood, but it might occur as shown in Figure 24-29. At least one viral enzyme, an endonuclease sometimes called the viral *integrase*, is essential for the integration process.

Once the provirus is integrated into the cell DNA, it is stable and is replicated along with the host's DNA. Unlike bacteriophage λ, for example, proviruses are never excised from the site of integration, although they can be lost as a result of deletions. As we already noted, retrovirus infections usually do not harm the cell, and infected cells continue to divide, with the integrated provirus serving as a template to direct viral RNA synthesis. A notable exception to this outcome is seen with the AIDS retrovirus. For some reason, the specific population of helper T lymphocytes that can be infected by this

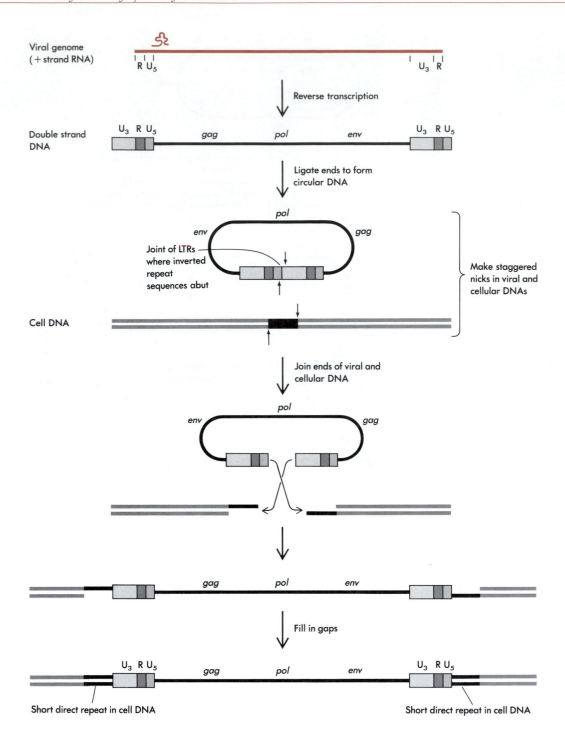

Figure 24-29
A model for integration of viral DNA. These events are inferred from the structures observed. The process is probably irreversible; the virus-coded endonuclease specifically recognizes the sequence formed by joining the inverted repeats, and the last two base pairs from each end are lost during integration, thus this sequence cannot be reassembled.

virus frequently are killed by the infection. It is due to the depletion of helper T cells that infected individuals are unable to mount vigorous immune responses.

LTRs Direct Transcription of the Provirus[129–133]

The LTRs not only provide ends for integration, but also supply transcriptional signals. These can be very efficient. With some retroviruses, as much as 10 percent of the mass of the mRNA in an infected cell can be viral RNA! The U_3 portion of the LTR contains sequences that control initiation of transcription: Typical TATA and CCAAT sequences are about 25 and 80 bases upstream, respectively, of the cap site at the U_3-R junction (Figure 24-30). Still farther upstream is a transcriptional enhancer, often present as a directly repeated sequence 70 to 100 bases long. Retrovirus enhancers are analogous to the enhancer of SV40, and some can even substitute for it.

As with most DNA viruses, the RNA transcripts synthesized by RNA polymerase II from the integrated retrovirus provirus are processed entirely by cell machinery. They are cleaved and polyadenylated at the 3′ end of R and have several different fates. A fraction of the full-length RNAs is transported from the nucleus to the cytoplasm and assembled in budding virions to become new viral genomes. Another fraction of full-length RNAs is transported to the cytoplasm to serve as mRNA for the *gag* and *pol* genes (Figure 24-31). Finally, a portion of the RNA is spliced to yield mRNAs for *env* (and for the *pX/lor* gene if present).

Figure 24-30
Features of LTRs. The example shown is from murine leukemia virus. Virtually all retroviruses (and many "transposable elements" of flies and yeast) share these features. Note that the relationship between the cap site and the poly A site (determined by the relevant signals) sets the position of the terminally redundant (R) sequence in the transcript.

Figure 24-31
Synthesis of retrovirus RNAs and proteins. Only a fraction of the primary transcripts are spliced to yield mRNA for *env*. The remainder serve either as new genomes or as mRNA for *gag*. A small fraction serve as mRNA for the *gag-pol* product by translational suppression of a termination codon at the end of *gag*.

Proviruses Can Become Established in the Germ Line as Normal Genes[134–140]

According to the replication cycle we have described, retroviruses can be propagated not only as infectious agents, but also as cellular genes. If, for example, a virus infection occurs in a germ cell (a cell destined to become a sperm or egg cell), then the resulting provirus can be passed on to the progeny and inherited as if it were a normal cellular gene. Such genetic elements, called **endogenous proviruses,** already exist and in fact are readily detectable in vertebrate DNA, including that of humans. In some animals, such as the mouse, there are literally thousands of endogenous proviruses. In fact, as much as 0.5 percent of mouse DNA is estimated to consist of endogenous retrovirus proviruses.

Endogenous proviruses are identical in structure to proviruses of exogenously infecting retroviruses, although usually they are not transcribed. Some can be induced to yield infectious virus following treatment of the cells with various chemicals (bromodeoxyuridine, or BUdR, and iododeoxyuridine, or IUdR, are particularly effective). The failure to be transcribed is due in part to extensive methylation of endogenous proviruses at CG sequences, and inducing agents frequently reverse this methylation.

As with exogenous retroviruses, endogenous proviruses are found integrated into an apparently random variety of locations in chromosomal DNA. Since different individuals often have the provirus in different locations, endogenous proviruses must have been introduced into the germ line relatively recently. We might expect that insertions of proviruses would sometimes inactivate genes and so give rise to a detectable phenotype. At least one such mutation is known, at the *d* locus in mice. It results in a distinctive coat color known as dilute brown, which is characteristic of a number of strains of inbred mice. Revertants of the *d* mutation are easily detected (by their coat color). They occur when the provirus is deleted, and their frequency indicates that the rate of this event is about 3×10^{-6} per generation.

Now it is possible to insert new proviruses deliberately into the germ line of laboratory mice by infecting early mouse embryos in vitro and reimplanting them into foster mothers, where they mature normally. The resulting animals have proviruses at a variety of locations in many different cell types, sometimes including the germ cells. Further breeding of such mice can lead to strains of mice in which a provirus derived from the original infection is inherited as a Mendelian gene. Once present, such elements behave much like endogenous proviruses. Like endogenous proviruses, newly introduced proviruses sometimes inactivate genes when they integrate, thereby causing mutations.

It should be apparent from the preceding discussion that endogenous proviruses have many characteristics also associated with transposable elements: They form a related group of DNA sequences with a characteristic structure, including terminal repeats, and are found at different locations in the DNA of different individuals. Had they first been discovered as DNA sequences rather than as virus-related sequences, they would have been labeled transposons. Recently, it has been recognized that many elements originally identified as eucaryotic transposable elements have striking similarity to proviruses, including the major hallmarks of LTR structure and function. These

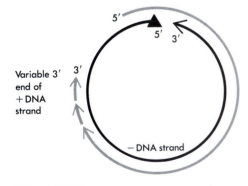

Figure 24-32
The DNA genome of hepatitis B virus (3200 base pairs). The genome is packaged into virions with a DNA polymerase that can further elongate the 3' end of the incomplete + strand. The triangle is a protein molecule covalently bound to the 5' end of the − strand. [After J. Summers and W. S. Mason, *Cell* 29 (1982):403–415.]

elements include the copia family of *Drosophila* and the Ty elements of yeast (Chapter 18) and have been referred to as retrotransposons. In both cases, there is direct evidence that these elements are duplicated and inserted by mechanisms virtually identical to those of retroviruses, except, probably, for the absence of an extracellular virion form.

Hepatitis B Virus Is a DNA Virus That Replicates Like a Retrovirus[17, 141, 142]

Two distinct groups of viruses cause most serious infectious liver disease in humans and some other mammals and birds. Hepatitis A virus, also called infectious hepatitis virus, is an RNA virus belonging to the picornavirus group. Hepatitis B (once known as serum hepatitis) virus is an enveloped virus with a small DNA genome and a unique mode of replication that resembles a retrovirus replication cycle turned inside out. Hepatitis B–like viruses have been isolated from humans, woodchucks, ground squirrels, and ducks, and together these isolates form the Hepadnavirus (for *hepatitis DNA virus*) family.

The hepatitis B genome is circular, but both of its DNA strands are linear (Figure 24-32). A gap in the complete − (or L for long) strand is bridged by an incomplete complementary + (or S for short) strand. The − strand has a protein covalently bound at its 5′ end. The + strands found in virions are in various degrees of completion, resulting in a single-strand gap extending from a fixed point at the 5′ end of the + strand to a variable point at its 3′ end. A DNA polymerase present in the nucleocapsid core can elongate the 3′ ends of incomplete strands.

Cells infected with hepatitis B virus contain large amounts of − DNA strands that are not attached to + DNA, and many are shorter than full genome length. A search for the template for their synthesis resulted in the unexpected finding that − DNA is synthesized in immature corelike particles from an RNA template that is degraded as synthesis proceeds. Thus, hepatitis B viruses replicate by reverse transcription, just like the retroviruses. The RNA template for − DNA strand synthesis presumably is full length and is designated the **pregenome**. Details of this replication cycle remain to be worked out, but a possible scheme for hepatitis B replication is shown in Figure 24-33. Unlike retroviruses, initiation of − strands seems to involve the protein covalently attached to their 5′ ends, a situation similar to adenovirus DNA replication. The + DNA strands would be synthesized using the free single-stranded − DNA as template and probably a + strand RNA fragment as primer.

INTERFERON

Viral Infections Induce Interferons[143–147]

Interferon was discovered accidentally about 30 years ago, when it was observed that culture medium from cells that had been exposed to partially inactivated viruses could protect other cells from infection with live viruses. The active component in the medium was called

Figure 24-33
A model for hepatitis B virus DNA replication. Details of process not yet clarified. [After J. Summers and W. S. Mason, *Cell* 29 (1982):403–415.]

interferon and was subsequently identified as a family of glycoproteins. Early experiments revealed that interferon synthesis is frequently induced in cells by virus infection. Although the infected cells usually die, they release interferons, which act on adjacent cells and protect them against viral infections. Interferon is an important defense mechanism in vertebrates. It is induced within hours of virus infection, a far more rapid response than the highly specific immune responses involving antibodies and activated cytotoxic T cells (Chapter 23). Without interferon, we would probably not survive many virus infections.

Infection of cells with most DNA and RNA viruses induces the synthesis and release of similar interferons. This similarity of response to a very diverse set of viral agents raises the question of what common aspect of viral infection signals the cell to produce interferon. Although the answer is not fully understood, we know that at least in some cases, the key component is double-stranded RNA. Almost all virus infections involve the synthesis of at least some RNA of both + and − strands. Although double strands are rarely directly involved in virus replication, they can form by chance if complementary single strands are present. Supporting the hypothesis that double-stranded RNA is the key inducer of interferon is the fact that synthetic double-stranded RNAs are very efficient inducers of interferon.

Interferons are extraordinarily active proteins. They are usually detected by their ability to reduce the number of virus plaques obtained following infection with some standard virus (often VSV). Concentrations of interferon as low as 3×10^{-14} *M* are detectable by this assay, and fewer than 50 molecules are apparently sufficient to protect a single cell.

There are three known types of interferons, and they are produced by different cell types under different conditions. In response to viral infection, lymphocytes synthesize primarily, but not exclusively, α interferon, sometimes called leukocyte interferon, whereas infection of fibroblasts usually induces β interferon. The two species of interferon are related proteins that share approximately 20 percent of their amino acid sequences. By contrast, γ interferon, sometimes called immune interferon, is scarcely related to the other two in amino acid sequence and is not induced by virus infection. Rather, it is synthesized by lymphocytes in response to *mitogens* (agents that stimulate cells to divide; Chapter 23).

Because they are released in such small amounts and because of their heterogeneity, for a long time interferons were difficult to study. Fortunately, the problem was overcome, first by advances in large-scale cell culture and purification techniques and, most importantly, by recombinant DNA techniques, which made it possible to synthesize very large amounts of active interferon in *E. coli*. These advances have raised hopes that interferon may one day provide effective treatment for viral infections.

Interferon Induces the Antiviral State[148, 149]

Interferon does not directly protect cells from virus infection. Rather, it alters the metabolism of the cell in a number of ways that make it less suitable as a host for viral replication—a complex condition called the **antiviral state.** Interferon that is chemically coupled to large beads

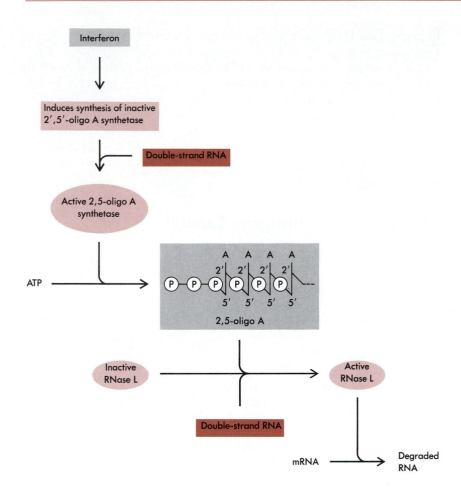

Figure 24-34
The 2,5-oligo A pathway for interferon
action. Note that the final effect depends both on the induction of 2,5-
oligo A synthetase by interferon and on
the presence of double-stranded RNA.

so it cannot get into cells can still induce the antiviral state. Apparently, interferons act by binding to specific receptors on the cell surface. The activated receptors then transfer a signal to the inside of the
cell, a situation analogous to the way hormones or growth factors
trigger cellular responses. How this occurs in the case of interferon to
bring about the antiviral state is unknown.

Viral infection in interferon-treated cells is blocked at various stages
of infection following attachment and penetration. The level of viral
genome and mRNA synthesis is reduced, and translation is also de-
pressed. Whether replication and virus assembly are also directly affected by the antiviral state is not clear. Two molecular mechanisms
have been deduced that help to explain the antiviral state. Both mechanisms involve the induction of a cellular enzyme, and in both cases,
the enzyme is activated by double-stranded RNA.

Interferon induces a protein kinase, which, in the presence of
double-stranded RNA, specifically phosphorylates the α subunit of
initiation factor eIF-2. This causes eIF-2B to be sequestered in eIF-2-
eIF-2B complexes (see Chapter 21) and hence to become inactive in
protein synthesis.

The second enzyme induced by interferon is called 2', 5'-oligo A
synthetase. It is part of a complex pathway that ultimately leads to the
degradation of single-stranded RNA, and hence of mRNA (Figure
24-34). The 2', 5'-oligo A synthetase polymerizes ATP residues into
short oligomers consisting of 2 to 15 adenosine monophosphate residues linked by phosphodiester bonds from the 2' position of one
ribose to the 5' position of the next (in contrast to the common 3'–5'

linkage). Like the protein kinase induced by interferon, 2′, 5′-oligo A synthetase is only active in the presence of double-stranded RNA. The 2′, 5′-oligo A does not affect the cell directly. Rather, it binds to and activates a specific ribonuclease, called RNase L (L for latent, since the enzyme is always present in the cell but normally inactive). RNase L then cleaves single-stranded RNA. It seems to be equally active on both cellular and viral mRNAs. This lack of specificity is probably not a disadvantage, however, since the host will benefit if a virus-infected cell dies rather than releasing even a small burst of progeny virus particles.

The Interferon Multigene Family[150–154]

The genes encoding α and β interferons are very similar in organization. Each encodes a glycoprotein 166 to 169 amino acids long after removal of signal peptides. In humans, neither α nor β interferon genes contain introns; the absence of introns is a feature shared with only a few other genes. Regions of 200 or so bases lying immediately 5′ of the coding regions of the α and β genes are also related and contain sequences that regulate gene expression. Remarkably, when a fragment as short as 46 bases from this region is placed in front of an unrelated gene (such as globin) and then introduced back into a cell, it renders the gene inducible by virus infection, just like the interferon gene from which the fragment was derived.

Humans and most other mammals contain multiple α interferon genes that encode closely related but distinct proteins with slightly different biological properties. In humans, there are 15 functional α genes belonging to two major types, α_I and α_{II}, as well as 10 or more pseudogenes. (Pseudogenes are closely related to functional genes, but have suffered mutations so that they cannot be expressed.) All of these genes, as well as the single β interferon gene, are located on chromosome 9. The nucleotide sequences of many of these genes are known, and from their relationships, we can infer the evolutionary relationships between interferon genes. The α and β genes probably diverged from a common ancestor 250 to 500 million years ago, about the time of the divergence of mammals and birds. About 120 million years ago, around the time of the origin of mammals, the α_I and α_{II} subtypes separated. Most recently, since the divergence of mammalian species, the α_I gene family has been generated, presumably by tandem duplication. Similar sorts of diversity have arisen independently among interferon genes of other species as well.

The diverse biological properties of the different interferons include the specificity of interaction with different types of cells and the potency of induction of the antiviral state. Presumably, this diversity is important for ensuring an effective response to viral infections.

Some Viruses Fight Interferon by Synthesizing Small RNAs[155]

Whatever mechanisms host organisms devise to combat viral infections, viruses seem to come up with ways to circumvent these defenses. So it is with certain viruses and interferon. Recent evidence indicates that both adenoviruses and Epstein-Barr virus produce small

RNAs, such as the adenovirus VA RNAs discussed earlier (see Figure 24-22), whose role may be to reverse the effects of interferon. The situation has been studied most with adenovirus VA RNAs.

VA RNAs are present in exceedingly high amounts late in infection with adenoviruses (probably 10^8 VA$_I$ and over 10^6 VA$_{II}$ molecules per cell). Adenovirus mutants from which the VA$_I$ genes have been deleted have a defect in the initiation of translation of their late mRNAs. In vitro, this initiation defect can be overcome by the addition of the translation factor eIF-2B to infected cell extracts. These observations suggest that the recycling of initiation factor eIF-2 is somehow affected by the presence of the VA$_I$ RNAs.

The following scenario is consistent with all the known facts and could explain why adenovirus infections are relatively refractile to interferon. Adenovirus infection induces interferon, which binds to its receptors and induces the antiviral state, activating the double-stranded RNA dependent kinase. Although the kinase requires double-stranded RNA to function, only *low* levels of such RNA and only RNAs that have perfect double-stranded regions of a certain minimum length are effective. In fact, high levels of RNA or imperfect duplex RNA can inhibit the activation of the enzyme. VA RNAs possess much double-stranded character but are interrupted by many bulges and loops. Furthermore, they are made at very high levels. So the present hypothesis is that VA RNA prevents the activation of the interferon-induced protein kinase that would otherwise inhibit translation by phosphorylating the α subunit of eIF-2.

PLANT VIRUSES

The Viruses of Plants Also Present a Variety of Genomic Organizations[156]

Like animal viruses, plant viruses exist in a variety of forms; they contain either DNA or RNA as their genetic material, have either rod or polyhedral shapes, and can be transmitted either by insect bites or by contact upon wounded regions. Most known plant viruses contain single + stranded RNA as their genetic material. Some like tobacco mosaic virus (TMV) have rod shapes while other have polyhedral shapes: for example, tomato bushy stunt virus is structurally remarkably similar to poliovirus, suggesting a common origin long ago in the evolutionary past. + stranded plant viruses can further be divided into those which possess a single RNA chain and those which have several RNA chains, each necessary for viral infectivity and which are separately encapsulated into separate virions. Cowpea mosaic virus, for example, contains two RNAs, one encoding several proteins including a terminal protein and a protease, with the other chain coding for capsid products. There also exist segmented double-strand RNA plant viruses whose genome structure somewhat resembles that of the reoviruses. The best known of these is wound tumor virus (WTV) which contains 12 different segments and which can replicate in either insect or plant cells. While infection of insect cells by WTV is nonsymptomatic, its replication within its plant vectors gives rise to a variety of symptoms including tumor formation. Many passages of WTV in plant cells often cause it to lose its ability to multiply in insect

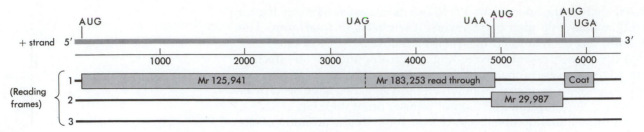

Figure 24-35
TMV coding sequences. The three reading frames of the positive strand are presented, with their respective coding regions, drawn to scale as open boxes with indicated polypeptide molecular weights. Initiation and termination codons of these coding sequences are displayed above the positive strand. (After Goelet et. al., 1982.)

cells, a state that correlates with genetic deletion in two RNA segments which encode outer capsid polypeptides.

Less variety exists in known DNA plant viruses. Only two known classes exist, one of which contains double strand DNA and which is polyhedral in shape. The best understood of this class is cauliflower mosaic virus (CaMV). The second class of DNA plant viruses are the geminiviruses that consist of paired capsids held together like twins (gemini) with each capsid containing a circular single-stranded DNA molecule of some 2500 nucleotides. In some cases, the two paired genomes are genetically identical while in other cases, the two DNA genomes bear almost no sequence resemblance.

TMV Remains the Best-Known Plant Virus[157, 158]

The single + stranded TMV RNA of 6395 nucleotides encodes four polypeptides. Sixty-nine residues from the 5' capped end is an AUG initiation codon that is followed by an open reading frame (terminated by an UAG stop codon) that specifies a 125,941 MW polypeptide. Occasionally (2–3%) readthrough occurs past this stop codon to produce a larger 183,253 MW polypeptide (Figure 24-35). Both proteins are thought to be part of the enzymatic complex responsible for RNA replication. Slightly overlapping the terminal codons of the readthrough polypeptide is the reading frame coding for the third polypeptide (MW 29,987) whose function remains unknown. The third gene stops two nucleotides before the initiation codon of the fourth protein, the coat protein (17,604 MW). Synthesis of these two latter polypeptides initiates at internal AUG codons using two nested sub-genomic mRNAs with identical 3' ends. These sub-genomic mRNAs may be made like those of the animal toga viruses, by initiating replication at internal sites on the − strand template. In this way the AUG initiation codons for translating the 29,987 and 17,604 (coat protein) genes are exposed.

Assembly of the coat protein subunits around the newly made + strands to produce progeny TMV particles occurs in a non-obvious way (Figure 24-36). A unique internal site within the gene coding for the 29,987 polypeptide serves as a nucleation center which binds to disk-like aggregates of the coat protein. Subsequent disk addition leads to growth in the 5' direction while assembly in the 3' direction involves addition of single coat protein monomers. There are sequence homologies between the nucleation assembly site and the section of the coat protein gene which codes for those amino acids which bind to the RNA. So, conceivably the 29,987 polypeptide also can bind to RNA.

Replication of the DNA Containing Cauliflower Mosaic Virus Involves Reverse Transcription[156]

The chromosome of the double-stranded DNA virus, cauliflower mosaic virus (CaMV), contains some 8000 bp, the sequence of which has been totally worked out. Its two chains (α and β) both contain single-stranded breaks with the 3' ends slightly overlapping the 5' ends to produce triple-stranded regions. Two major transcripts, one of which is slightly more than full length, are made off the α strand. Six major open reading frames exist in the full-length transcript. At first, it seemed likely that CaMV might replicate like SV40, but now it appears that DNA replication starts by reverse transcription of the larger mRNA transcript to yield a DNA-RNA hybrid helix that in turn is converted into double helical DNA identical to that found in the infectious parental CaMV virion. The replication of CaMV thus resembles that of the hepatitis B viruses.

Frontiers of Virology: Viruses Without Genes?[159-164]

By now, we understand the nature of most viruses and their role in disease. There remain, however, a few infectious diseases whose cause has not been resolved. Most prominent of these are a set of related conditions of different species that lead to progressive degeneration of brain tissue and eventually to death. The group includes two human diseases known as *kuru* and *Creutzfeldt-Jakob disease*, as well as a disease of sheep, called *scrapie*. These are clearly infectious diseases: Kuru was transmitted by cannibalism until the custom ceased, and Creutzfeldt-Jakob disease is usually passed on by improperly sterilized surgical instruments or by organ transplants from an affected donor. Scrapie is extremely difficult to work with in the laboratory. It has been adapted to mice and hamsters, decreasing the assay time for disease from years to months, but no cell culture system yet exists. It is readily transferred from animal to animal, however, by injections of infected brain.

The most remarkable feature of scrapie is that no one has been able to obtain biochemical evidence that it contains nucleic acid. The infectious agent in scrapie preparations is highly resistant to treatments such as ultraviolet light or chemical crosslinking reagents, potent inactivators of RNA or DNA genomes. It is, however, sensitive to chemicals such as phenol, which denature protein. Furthermore, the scrapie agent seems to have a molecular weight far below the usual virus when determined chromatographically. The agent that causes scrapie has been highly purified (7000-fold) from crude brain extracts. This material contains no detectable nucleic acid and consists largely of a single protein of about 27 kdal.

These observations have led to the suggestion that scrapie is a novel kind of infectious agent, called a **prion** (for proteinaceous infectious particle). The 27 kdal protein is called PrP (prion-related protein). Although PrP seems to be inseparable from the infectious scrapie agent, it has recently been shown that PrP is probably a normal cell protein. DNA complementary to mRNA for a protein closely

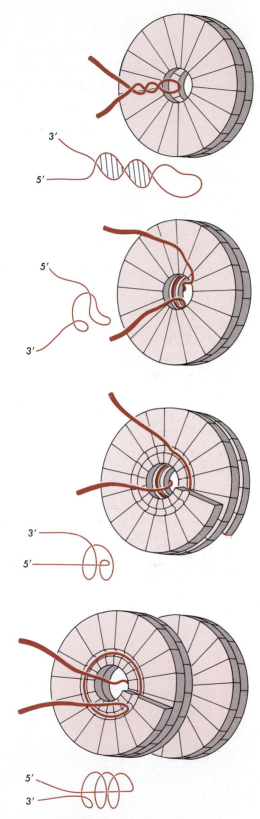

Figure 24-36
A possible mechanism for nucleation of assembly of TMV. The protein and RNA are drawn diagrammatically, as is the predicted configuration of the RNA backbone at each stage.

related or identical to PrP has been cloned. Both infected and uninfected brains express the RNA. Given this, what can scrapie be?

If, indeed, scrapie contains no nucleic acid, then conceivably it is a cell protein that stimulates its own synthesis, even to a level high enough to kill the cell. Since PrP is expressed in uninfected cells, however, this possibility seems rather unlikely. Further complicating the hypothesis that scrapie lacks nucleic acid is the apparent existence of distinct strains of the agent with different biological properties that breed true upon repeated passage. This fact suggests some sort of genetic system able to generate, propagate, and express such alterations. Why, then, has the nucleic acid in scrapie not been detected? To both escape detection and be resistant to treatments that inactivate nucleic acid, such a "genome" must be exceptionally small, and probably otherwise unusual in structure as well.

There is a precedent for extremely small viruses with unusual genomes. They are the well-characterized **viroids** of plants. Viroids violate many of the principles established for animal viruses. They consist only of RNA, about 240 to 350 nucleotides long. The RNA is a single-stranded circle that can form extensive base pairs within itself so that it assumes a stiff, nearly double-helical structure (Figure 24-37). The small size means that viroids present only very small targets for inactivation by ultraviolet light, and are thus highly UV resistant. The double-stranded structure makes them resistant to ribonuclease, which usually cleaves only at unpaired bases. The nucleotide sequence displays little or no potential for coding proteins: In potato spindle tuber viroid (PSTV), for example, there are no AUG initiation codons, and frequent termination codons occur in all reading frames.

Clearly, viroid RNAs cannot code for proteins to replicate their genomes. Unlike animal cells, plant cells must therefore contain enzymes capable of replicating RNA. Indeed, such enzymes have been detected. Synthesis of RNA strands both identical and complementary to viroid genomes can readily be observed in infected cells in culture. These strands include full-length genomes as well as multiple-genome-length molecules. Although details are scanty, replication seems to occur, at least in part, by a rolling circle type of mechanism in which synthesis continues around and around the template. The enzymes involved in viroid RNA replication and their role in the life of the plant cell remain to be explored.

The extent to which viroids may be models for diseases like scrapie remains unknown. At the very least, such small and seemingly improbable entities as viroids serve to reemphasize the great diversity of eucaryotic viruses.

Figure 24–37
The structure of a viroid. The nucleotide sequence shown is that of the potato spindle tuber virus (PSTV). Extensive base pairing allows this sequence to form an almost correct double helix. [After H. L. Sänger, in *The Microbe 1984. I. Viruses,* ed. B. W. J. Mahy and J. R. Pattison (New York: Cambridge University Press, 1984), pp. 282–334.]

Summary

Eucaryotic viruses have been studied intensively by molecular biologists both because of their ability to cause disease and because they serve as probes to the study of eucaryotic gene expression. The development of cell cultures on which viruses could grow made the molecular biology of many eucaryotic viruses feasible.

The extracellular form of a eucaryotic virus is called a virion. It consists of a protein coat, the capsid, surrounding the nucleic acid genome. In some viruses, the capsid is surrounded by an envelope that is derived from cellular membranes by budding. Capsids can have one of two basic shapes: rodlike or roughly spherical (icosahedral). Rod-shaped capsids can be folded up within an outer envelope to yield virions of varied appearance, and surface spikes can add further diversity.

The virus life cycle consists of attachment to specific receptors on the cell surface, entry of the virion into the cell by fusion or endocytosis, uncoating, transcription and replication of the viral genome, and virion assembly.

Thousands of animal viruses have been isolated. They can be classified into families based on their particle morphology and similarity of genome structure. Viral genomes can be RNA or DNA, single-stranded or double-stranded, linear or, in the case of some double-stranded DNA genomes, circular. Single-stranded RNA virus genomes can correspond to the mRNA or sense strand, in which case they are called positive or plus (+) strand RNA, or they can correspond to the complement of mRNA, in which case they are called negative or minus (−) strand RNA. So far, the known single-stranded DNA viruses are either − strand, or they consist of complementary strands, but each strand is packaged in a separate virion.

A useful system for classifying animal viruses is based on the chemical nature of the genome, the way in which it replicates, and the relationship of the genome to the synthesis of viral mRNA. This classification scheme emphasizes molecular biological principles that underlie animal virus life cycles. According to the system, there are seven major types of animal viruses: (1) Double-strand DNA viruses use cellular enzymes to transcribe their mRNAs; (2) single-strand DNA viruses convert their DNA to double-stranded DNA and then use cellular enzymes to synthesize their mRNAs; (3) + strand RNA viruses can act as mRNA and so direct the synthesis of enzymes that can replicate RNA and also transcribe RNA to yield viral mRNAs; (4) − strand RNA viruses must carry an enzyme inside their virions that can synthesize mRNA from the − strand genome; (5) double-strand RNA viruses must carry an enzyme in the particle that can synthesize mRNA from a double-strand RNA template; (6) retroviruses have a + strand RNA genome, but carry a reverse transcriptase in the virion that converts the RNA to double-stranded DNA; the DNA integrates into a host chromosome and is transcribed by cellular RNA polymerase to make viral mRNA and progeny genomes; (7) hepatitis B virus has a double-stranded DNA genome, but the DNA replicates via an RNA intermediate, thus resembling a retrovirus.

Most RNA viruses (except influenza) replicate in the cytoplasm. They can grow to high titers in a short period of time on many cell types, since they encode their own enzymes for replication and since all cells, even those that are not dividing, are always synthesizing RNA and thus have high concentrations of the needed triphosphates.

Most DNA viruses (except poxviruses) replicate in the nucleus where they can use the cell's transcriptional machinery, and most use cellular enzymes to replicate their DNA as well. However, they face a problem with DNA replication: Most cells in an animal are not dividing at any one time and thus lack the high concentrations of enzymes and precursors needed for extensive DNA replication. To overcome this problem, most DNA viruses have devised ways to turn on the replicative capacity of the host cell.

A problem common to RNA and DNA viruses is how to express numerous genes when most eucaryotic mRNAs are monocistronic.

Poliovirus (a picornavirus) has a + strand RNA genome about 7.4 kilobases long in a naked icosahedral capsid. The genome serves as mRNA (usually after removal of the tiny 5' terminal VPg protein). The genome encodes a single protein, the polyprotein, from which about a dozen proteins are generated by sequential proteolytic cleavages. Critical to the early events of the life cycle are the virus-coded proteins involved in synthesizing RNA. These include VPg and replicase. VPg is thought to become covalently attached to U residues, which then can prime synthesis of polio RNA from either a + or − strand template. This is important, since replicase, like DNA polymerases, can extend but not initiate polynucleotide chains on a template. In contrast to the poly A on cellular mRNAs, the polio poly A is copied during RNA replication. Replicative intermediates consist of a single − strand with several + strands in various states of completion dangling off them. Virion assembly occurs when aggregates of a precursor undergo successive proteolytic cleavages to yield four final virion polypeptides: VP1, 2, 3, and 4.

Vesicular stomatitis virus (VSV) (a rhabdovirus) is a − strand RNA virus. The genome is encased in viral coded N protein, and the rodlike capsid is wound up inside an envelope to form a bullet-shaped particle. Four other virus-coded proteins are present in the virion: G, an envelope glycoprotein that binds the virus to specific receptors on cells; M, a matrix protein that lies between the envelope and the capsid; and L and NS, which together form an enzyme complex that synthesizes mRNAs and replicates RNA. VSV virions that are treated with a mild detergent (to allow triphosphates to get in) can synthesize mRNAs in vitro. Five mRNAs are produced, each with a 5' cap and 3' poly A tail. Thus, the L-NS complex can carry out the multiple enzymatic reactions involved in capping, RNA synthesis, and poly A addition. Poly A addition to the viral RNA is performed by a method entirely different from cellular poly A addition. Between each VSV gene is a conserved nucleotide sequence that includes a string of seven U residues. When the L-NS complex comes to this sequence, it pauses and stutters to add a long string of A residues. Then it either falls off the template (30 percent of the time) or moves on to synthesize the next mRNA.

In vivo, the choice of whether to synthesize VSV mRNAs or to synthesize a full-length + strand that can serve as a template for the synthesis of more progeny genomes is determined by the concentration of N protein that has built up inside the cell. N protein initiates binding to + strands at a site near the 5' end on a short leader RNA. In the absence of N, the leader RNA is freed, and mRNAs are synthesized. In the presence of N protein, binding to the leader occurs and continues immediately behind the advancing L-NS enzyme, preventing the synthesis of mRNAs and allowing the synthesis of full-length + strand RNA. VSV virions are assembled by budding.

Influenza virus (an orthomyxovirus) is a segmented, − strand RNA virus. Influenza virions are enveloped and have two surface, virus-coded glycoproteins, a hemagglutinin and a neuraminidase. The genome consists of eight fragments of RNA. In contrast to other RNA viruses, influenza replicates in the nucleus. There it initiates viral RNA transcription by clipping 5' caps from newly made cellular RNAs and using them to initiate transcription of its eight genome fragments. It also uses cellular RNA-splicing machinery to generate spliced mRNAs from two of its genome segments, so the eight segments of flu give rise to ten mRNAs and ten virus-coded proteins. The three-dimensional structure of the HA (hemagglutinin) protein (and also of the neuraminidase) has been determined. The primary HA polypeptide is cleaved to yield two polypeptides (HA_1 and HA_2) that are held together by a disulfide bond. Three ($HA_1 + HA_2$) monomers associate to form the mature HA protein, which has a cylindrical shape. Grooves on the top of the cylinder recognize receptors on cells, and four sites have been identified on the surface of the protein that are altered to produce antigenically new strains of flu. The ability to change its coat, sometimes by genome reassortment, allows flu to induce recurring epidemics.

Reovirus has a double-stranded, segmented RNA genome in a naked complex double capsid. When reovirus virions enter cells, they are only partly uncoated. They lose proteins from the outer shell, but the inner capsid remains intact, and RNA synthesis commences inside of it. Synthesis is asymmetric and conservative. Only + strands are made at this point, and the incoming double strands remain intact. The + strands are extruded through pores at the 12 vertices of the capsid. + strands serve as mRNA and as intermediates in replication. They assemble with new capsid proteins into precores, where they are converted to double-stranded RNA. The precores can synthesize additional + strand RNA until late in infection, when they acquire their outer shell.

SV40 (a papovavirus) has a small, double-stranded, circular DNA genome. The genome is about 5.2 kilobases long and is divided into three regions: About half is the early region; the other half is the late region; and a small segment between the two is a control region containing the origin of DNA replication and transcriptional signals to regulate early and late gene expression. Transcription of the early and late genes occurs in opposite directions from different DNA strands. The shift from early to late gene expression occurs after DNA replication. Two early gene proteins,

small t and large T proteins, are made from overlapping spliced mRNAs. The function of small t is unknown, while large T has multiple functions. It turns on cellular DNA synthesis, is needed for initiation of viral DNA replication, shuts down synthesis of early genes, and is needed to help turn on late gene expression. It binds to three adjacent sites in the regulatory region of the genome. Transcriptional signals that control early gene expression include a TATA sequence, a GC-rich 21-base-pair repeat, and an enhancer element.

Adenoviruses have linear, double-stranded DNA genomes, about 36 kilobase pairs long, in a naked icosahedral virion with spikes. The virus replicates in the nucleus. At least seven regions of the genome are expressed early in infection, and both strands of DNA are used. The transcription map for early and late genes is extraordinarily complex. Multiple mRNAs are generated from single segments of DNA by complex splicing patterns. Almost all late genes are expressed from a single very strong promoter, the major late promoter. A primary transcript, about 28 kilobases long, gives rise to more than 20 late mRNAs by differential poly A addition and RNA splicing. The 5' end of each strand of adenovirus DNA is covalently attached to a 55 kdal terminal protein. Its precursor, preterminal protein, becomes covalently attached to a C residue, which then serves as a primer for DNA replication. During replication, only one strand at a time is synthesized, and synthesis is continuous. Adenovirus DNA replication can now be achieved in vitro, so the various proteins involved are rapidly being identified and purified.

Parvoviruses have single-stranded, linear DNA genomes. These simple viruses do not encode proteins that can stimulate host DNA synthesis and so can only replicate in dividing cells. They have unusual sequences at their termini that allow the ends to fold and form primers for DNA synthesis.

Poxviruses have huge double-stranded DNA genomes. The poxvirus genome is linear, but its two strands are covalently joined at each end. These are the only known DNA viruses that replicate in the cytoplasm. Their genomes are so large that they can encode all the enzymes needed to make DNA and RNA transcripts without the help of cellular enzymes in the nucleus.

Retroviruses have a single + strand RNA genome, but for unknown reasons, they are diploid: There are two identical copies of the RNA in each virion. Virions are enveloped. After entering the cell, reverse transcriptase within the virion makes a double-stranded DNA copy of the genome. This linear DNA moves to the nucleus, where it is circularized and integrated into a nonspecific site in the cellular DNA. Integration occurs at a specific site on the viral genome. The viral DNA, called the provirus, is longer than the RNA genome because of the presence of long direct repeats (several hundred nucleotides long) at each end of the DNA. These are called the long terminal repeats or LTRs, and they contain the signals for initiation and termination of transcription of viral RNA. RNA polymerase II transcribes the provirus from its integrated position in the host's DNA. Full-length transcripts become progeny genomes or serve as mRNA for the synthesis of two polyproteins. Proteolytic cleavage of the *gag* gene product gives rise to the capsid proteins, while proteolytic cleavage of the *pol* gene product releases reverse transcriptase, a protease involved in the cleavages, and a nuclease believed to serve as an integrase for proviral DNA. A spliced mRNA is translated to yield a surface glycoprotein responsible for attachment of the virus to specific receptors on cells. Progeny virions are budded from the cell membrane without killing the cell, and most virus-infected cells grow indefinitely without any apparent harm from the virus.

All vertebrates inherit multiple copies of retrovirus proviruses in their genomes. Most of these are silent, but some are expressed to yield viral proteins and in some cases complete infectious viruses. Had proviruses been discovered in genomic DNA before they were known to be retroviruses, they would have been called transposable elements, since their structure is similar to a class of transposable elements found in yeast and *Drosophila*.

Hepadnaviruses (hepatitis B virus) have a double-stranded, circular DNA genome, but both strands in the circle are linear. The virus replicates its DNA through an RNA intermediate. Thus, it resembles a retrovirus, except that the DNA rather than the RNA phase of the life cycle is packaged in the virions.

Viral infections often induce the synthesis of a set of cellular glycoproteins called interferons. Double-stranded RNA is also a powerful inducer of interferon, and the double-stranded RNA formed during many types of viral replication is probably the actual inducer of interferon in viral infections. Interferons are released from the infected cell, bind to receptors on this cell and on adjacent cells, and induce a complex series of changes that render cells less infectable by many types of viruses. This set of changes is called the antiviral state. Two aspects of the antiviral state are understood in some detail. First, interferon induces a protein kinase, which, when activated by double-stranded RNA, specifically phosphorylates the α subunit of the initiation factor eIF2, thereby inactivating it. Second, interferon induces the synthesis of an enzyme called 2',5'-oligo A synthetase. The enzyme is activated by double-stranded RNA and then synthesizes 2',5'-oligo A from ATP. 2',5'-oligo A binds to a specific ribonuclease, RNase L, and thus activates it to cleave single-stranded RNA and hence mRNA. The enzyme does not seem to discriminate between cellular and viral mRNA; however, it may not matter that the infected cell dies, as long as it does not release a large burst of progeny viruses. Interferons are like hormones in many ways. They are effective at very low concentrations, and they exert complex effects on cells by binding to specific cell surface receptors. There are three types of interferons, called α, β, and γ. There are multiple genes for α interferons.

There are certain neurological diseases (scrapie in sheep, kuru and Creutzfeldt-Jakob disease in humans) that seem to be caused by viruses, but the viruses have not yet been identified. These diseases may be caused by agents resembling plant viroids, tiny RNA molecules without capsids and apparently without protein-coding genes as well, or they may be caused by viruses that consist of protein but no nucleic acid (prions).

Bibliography

General References

Armstrong, J., E. J. Atencio, B. E. Bergman, H. S. Bilofsky, L. B. Brown, C. Burks, G. N. Cameron, M. J. Cinkosky, U. Elbe, C. E. England, J. W. Fickett, B. T. Foley, W. B. Goad, G. H. Hamm, J. A. Hayter, D. Hazledine, M. Kanehisa, L. Kay, G. G. Lennon, F. I. Lewitter, C. R. Linder, A. Leutzenkirchen, P. McCaldon, M. J. McLeod, D. L. Melone, G. Myers, D. Nelson, J. L. Nial, H. M. Perry, W. P. Rindone, L. D. Sher, M. T. Smith, G. Stoesser, C. D. Swindell, and C.-S. Tung, eds. 1985. *Nucleotide Sequences 1985. IV. Viral and Synthetic Sequences*. Washington, D.C.: IRL Press.

Fields, B. N., D. M. Knipe, R. M. Chanock, J. L. Melnick, B. Roizman, and R. E. Shope, eds. 1985. *Virology*. New York: Raven Press.

Mahy, B. W. J., and J. R. Pattison, eds. 1984. *The Microbe 1984. I. Viruses*. New York: Cambridge University Press.

Tooze, J., ed. 1980. *Molecular Biology of Tumor Viruses: DNA Tumor Viruses*. 2nd ed. Cold Spring Harbor, N.Y.: Cold Spring Harbor Laboratory.

Weiss, R. A., N. Teich, H. E. Varmus, and J. M. Coffin, eds. 1985. *Molecular Biology of Tumor Viruses: RNA Tumor Viruses*. 2nd ed. 2 vols. Cold Spring Harbor Laboratory, N.Y.

Cited References

1. Wiley, D. C. 1985. "Viral Membranes." In *Virology*, ed. B. N. Fields, D. M. Knipe, R. M. Chanock, J. L. Melnick, B. Roizman, and R. E. Shope. New York: Raven Press, Chap. 4, pp. 45–68.

2. Harrison, S. 1985. "Virus Structure." In *Virology*, ed. B. N. Fields, D. M. Knipe, R. M. Chanock, J. L. Melnick, B. Roizman, and R. E. Shope. New York: Raven Press, Chap. 3, pp. 27–44.

3. Hogle, J. M., M. Chow, and D. J. Filman. 1985. "Three-Dimensional Structure of Poliovirus at 2.9 Å Resolution." *Science* 229:1358–1365.

4. Rossmann, M. G., E. Arnold, J. W. Erickson, E. A. Frankenberger, J. P. Griffith, H. J. Hecht, J. E. Johnson, G. Kramer, M. Luo, A. G. Mosser, R. R. Rueckert, B. Sherry, and G. Vriend. 1985. "Structure of a Human Common Cold Virus and Functional Relationship to Other Picornaviruses." *Nature* 317:145–153.

5. Simons, K., H. Garoff, and A. Helenius. 1982. "How an Animal Virus Gets Into and Out of Its Host Cell." *Sci. Amer.* 246:58–66.

6. Reanney, D. 1984. "The Molecular Evolution of Viruses." In *The Microbe 1984. I. Viruses*, ed. B. W. J. Mahy and J. R. Patterson. New York: Cambridge University Press, pp. 175–196.

7. Enders, J. F., T. H. Weller, and F. C. Robbins. 1949. "Cultivation of the Lansing Strain of Poliomyelitis Virus in Cultures of Various Human Embryonic Tissues." *Science* 109:85–87.

8. Todaro, G., and H. Green. 1963. "Quantitative Studies on the Growth of Mouse Embryo Cells in Culture and Their Development into Established Lines." *J. Cell. Biol.* 17:299–313.

9. Dulbecco, R. 1952. "Production of Plaques in Monolayer Tissue Cultures by Single Particles of an Animal Virus." *Proc. Nat. Acad. Sci.* 38:747–752.

10. Matthews, R. E. F. 1982. "Classification and Nomenclature of Viruses." 4th report of the ICTV. *Intervirology* 17:1–199.

11. Baltimore, D. 1971. "Expression of Animal Virus Genomes." *Bacteriol. Revs.* 35:235–241.

12. Kates, J. R., and B. R. McAuslan. 1967. "Poxvirus DNA-Dependent RNA Polymerase." *Proc. Nat. Acad. Sci.* 58:134–141.

13. Mayor, H. D., K. Torikai, J. Melnick, and M. Mandel. 1969. "Plus and Minus Single-Stranded DNA Separately Encapsidated in Adeno-Associated Satellite Virions." *Science* 166:1280.

14. Shatkin, A. J., and J. D. Sipe. 1968. "RNA Polymerase Activity in Purified Reovirus." *Proc. Nat. Acad. Sci.* 61:1462–1469.

15. Temin, H. M., and S. Mizutani. 1970. "RNA-Dependent DNA Polymerase in Virions of Rous Sarcoma Virus." *Nature* 226:1211–1213.

16. Baltimore, D. 1970. "RNA-Dependent DNA Polymerase in Virions of RNA Tumor Viruses." *Nature* 226:1209–1211.

17. Kaplan, P. N., R. L. Greenman, J. L. Gerin, R. H. Purcell, and W. S. Robinson. 1973. "DNA Polymerase Associated with Human Hepatitis B Antigen." *J. Virology* 12:995–1005.

18. Summers, D. F., and J. V. Maizel. 1968. "Evidence for Large Precursor Proteins in Poliovirus Synthesis." *Proc. Nat. Acad. Sci.* 59:966–971.

19. Duesberg, P. H. 1968. "The RNA's of Influenza Virus." *Proc. Nat. Acad. Sci.* 59:930–934.

20. Ball, L. A., and C. N. White. 1976. "Order of Transcription of Genes of Vesicular Stomatitis Virus." *Proc. Nat. Acad. Sci.* 73:442–446.

21. Simmons, D. T., and J. H. Strauss, Jr. 1972. "Replication of Sindbis Virus. I. Relative Size and Genetic Content of 26S and 49S RNA." *J. Mol. Biol.* 71:599–614.

22. Dalgleish, A. G., P. C. L. Beverley, P. R. Clapham, D. H. Crawford, M. F. Greaves, and R. A. Weiss. 1984. "The CD4(T4) Antigen Is an Essential Component of the Receptor for the AIDS Retrovirus." *Nature* 312:763–767.

23. Klatzmann, D., F. Barre-Sinoussi, M. T. Nugeyre, C. Dauguet, E. Vilmer, C. Griscelli, F. Brun-Vezinet, C. Rouzioux, J. C. Gluckman, J.-C. Chermann, and L. Montagnier. 1984. "Selective Tropism of Lymph-Adenopathy Associated Virus (LAV) for Helper-Inducer T Lymphocytes." *Science* 225:59–63.

24. Rogers, G. N., J. C. Paulson, R. S. Daniels, J. J. Skehel, I. A. Wilson, and D. C. Wiley. 1983. "Single Amino Acid Substitutions in Influenza Hemagglutin Change Receptor Binding Specificity." *Nature* 304:76–78.

25. Scheid, A., and P. W. Choppin. 1974. "Identification of Biological Activities of Paramyxovirus Glycoproteins. Activation of Cell Fusion, Hemolysis and Infectivity by Proteolytic Cleavage of Inactive Precursor Protein of Sendai Virus." *Virology* 57:475–490.

26. White, J., and A. Helenius. 1980. "pH Dependent Fusion Between the Semliki Forest Virus Membrane and Liposomes." *Proc. Nat. Acad. Sci.* 77:3273–3277.

27. Kitamura, N., B. Semler, B. G. Rothberg, G. R. Larsen, C. J. Adler, A. J. Dorner, R. A. Emini, R. Hanecak, J. J. Lee, S. van der Werf, C. W. Anderson, and E. Wimmer. 1981. "Primary Structure, Gene Organization and Polypeptide Expression of Poliovirus RNA." *Nature* 291:547–553.

28. Racaniello, V. R., and D. Baltimore. 1981. "Molecular Cloning of Poliovirus cDNA and Determination of the Complete Nucleotide Sequence of the Viral Genome." *Proc. Nat. Acad. Sci.* 78:4887–4891.

29. Crocker, T. T., E. Pfendt, and R. Spendlove. 1964. "Poliovirus Growth in Non-Nucleate Cytoplasm." *Science* 145:401–403.

30. Nomoto, A., Y. F. Lee, and E. Wimmer. 1976. "The 5' End of Poliovirus mRNA Is Not Capped with m⁷G(5')ppp(5')-Np." *Proc. Nat. Acad. Sci.* 73:375–380.

31. Jacobson, M., and D. Baltimore. 1968. "Polypeptide Cleavages in the Formation of Poliovirus Proteins." *Proc. Nat. Acad. Sci.* 61:77–84.

32. Palmenberg, A. C., M. A. Pallansch, and R. R. Rueckert. 1979. "Protease Required for Processing Picornaviral Coat Protein Resides in the Viral Replicase Gene." *J. Virology* 32:770–778.

33. Palmenberg, A. C., and R. R. Rueckert. 1982. "Evidence for Intramolecular Self-Cleavage of Picornaviral Replicase Precursors." *J. Virology* 41:244–249.

34. Baltimore, D., and M. Girard. 1966. "An Intermediate in the Synthesis of Poliovirus RNA." *Proc. Nat. Acad. Sci.* 56:741–748.

35. Flanegan, J. B., and D. Baltimore. 1977. "Poliovirus-Specific Primer Dependent RNA Polymerase Able To Copy Poly(A)." *Proc. Nat. Acad. Sci.* 94:2677–2680.

36. Lee, F. Y., A. Nomoto, B. M. Detjen, and E. Wimmer. 1977. "A Protein Covalently Linked to Poliovirus Genome RNA." *Proc. Nat. Acad. Sci.* 74:59–63.

37. Yogo, F., and E. Wimmer. 1973. "Poly(A) and Poly(U) in Poliovirus Double-Stranded RNA." *Nature New Biol.* 242:171–174.

38. Penman, S., and D. Summers. 1965. "Effects on Host Cell Metabolism Following Synchronous Infection with Poliovirus." *Virology* 27:614–620.

39. Etchison, D., J. Hanse, E. Ehrenfeld, I. Edery, N. Sonenberg, S. Milburn, and J. W. B. Hershey. 1984. "Demonstration *in Vitro* That Eukaryotic Initiation Factor 3 Is Active but That a Cap-Binding Protein Complex Is Inactive in Poliovirus-Infected HeLa Cells." *J. Virology* 51:832–837.

40. Fernandez-Tomes, C. B., and D. Baltimore. 1973. "Morphogenesis of Poliovirus. II. Demonstration of a New Intermediate, the Provirion." *J. Virology* 12:1122–1130.

41. Rueckert, R. R., A. K. Dunker, and C. M. Stoltzfus. 1969. "The Structure of Maus-Elberfeld Virus: A Model." *Proc. Nat. Acad. Sci.* 62:912–919.

42. Strauss, E. G., C. M. Rice, and J. H. Strauss. 1984. "Complete Nucleotide Sequence of the Genomic RNA of Sindbis Virus." *Virology* 133:92–110.

43. Wirth, D. F., F. Katz, B. Small, and H. F. Lodish. 1977. "How a Single Sindbis Virus mRNA Directed the Synthesis of One Soluble Protein and Two Integral Membrane Proteins." *Cell* 10:253–263.

44. Stern, D. F., and S. I. T. Kennedy. 1980. "Coronavirus Multiplication Strategy. II. Mapping of the Avian Infectious Bronchitis Virus Intracellular RNA Species to the Genome." *J. Virology* 36:440–449.

45. Lai, M. M. C., C. D. Patton, R. S. Baric, and S. A. Stohlman. 1983. "Presence of Leader Sequences in the mRNA of Mouse Hepatitis Virus." *J. Virology* 46:1027–1033.

46. Schubert, M., G. G. Harmison, and E. Meier. 1984. "Primary Structure of the Vesicular Stomatitis Virus Polymerase (L) Gene: Evidence for a High Frequency of Mutations." *J. Virology* 51:505–514.

47. Ball, L. A., and G. W. Wertz. 1981. "VSV RNA Synthesis: How Can You Be Positive?" *Cell* 26:143–144.

48. Iverson, L. E., and J. K. Rose. 1981. "Localized Attenuation and Discontinuous Synthesis During Vesicular Stomatitis Virus Transcription." *Cell* 23:447–484.

49. Emerson, S. U., and Y.-H. Yu. 1975. "Both NS and L Proteins Are Required for *in vitro* RNA Synthesis by Vesicular Stomatitis Virus." *J. Virology* 15:1348–1356.

50. Blumberg, B. M., M. Leppert, and D. Kolakofsky. 1981. "Interaction of VSV Leader RNA and Nucleocapsid Protein May Control VSV Genome Replication." *Cell* 23:837–845.

51. Herman, R. C., M. Schubert, J. D. Keene, J. D. and R. Lazzarini. 1980. "Polycistronic Vesicular Stomatitis Virus RNA Transcripts." *Proc. Nat. Acad. Sci.* 77:4662–4665.

52. Katz, F. N., J. E. Rothman, V. R. Lingappa, G. Blobel, and H. F. Lodish. 1977. "Membrane Assembly *in Vitro*: Synthesis, Glycosylation, And Asymmetric Insertion of a Transmembrane Protein." *Proc. Nat. Acad. Sci.* 74:3278–3282.

53. Huang, A. S., J. W. Greenwalt, and R. R. Wagner. 1966. "Defective T Particles of Vesicular Stomatitis Virus. I. Preparation, Morphology and Some Biologic Properties." *Virology* 30:161–172.

54. Palese, P., and J. L. Schulman. 1976. "Mapping of the Influenza Virus Genome: Identification of the Hemagglutinin and Neuraminidase Genes." *Proc. Nat. Acad. Sci.* 73:2142–2146.

55. Barry, R. D., D. R. Ives, and J. G. Cruickshank. 1962. "Participation of Deoxyribonucleic Acid in the Multiplication of Influenza Virus." *Nature* 194:1139–1140.

56. Bouloy, M., S. J. Plotch, and R. M. Krug. 1978. "Globin mRNA's Are Primers for the Transcription of Influenza Viral RNA *in Vitro." Proc. Nat. Acad. Sci.* 75:4886–4890.

57. Braam, J., I. Ulmanen, and R. M. Krug. 1983. "Molecular Model of a Eukaryotic Transcription Complex: Functions and Movements of Influenza P Proteins During Capped RNA Primed Transcription." *Cell* 34:609–618.

58. Robertson, J. S., M. Schubert, and R. A. Lazzarini. 1981. "Polyadenylation Sites for Influenza mRNA." *J. Virology* 38:157–163.

59. Lamb, R. A., and P. W. Choppin. 1979. "Segment 8 of the Influenza Virus Genome Is Unique in Coding for Two Polypeptides." *Proc. Nat. Acad. Sci.* 76:4908–4912.

60. Burnet, F. M., and M. Edney. 1951. "Recombinant Viruses Obtained from Double Infections with the Influenza A Viruses MEL and Neuro-WS." *J. Exp. Biol. Med. Sci.* 39:353–362.

61. Wilson, I. A., J. J. Skehel, and D. C. Wiley. 1981. "Structure of the Hemagglutinin Membrane Glycoprotein of Influenza Virus at 3 Å Resolution." *Nature* 289:366–373.

62. Wiley, D. C., I. A. Wilson, and J. J. Skehel. 1981. "Structural Identification of the Antibody-Binding Sites of Hong Kong Influenza Hemagglutinin and Their Involvement in Antigenic Variation." *Nature* 289:373–378.

63. Skehel, J. J., P. M. Bayley, E. B. Brown, S. R. Martin, M. D. Waterfield, J. M. White, I. A. Wilson, and D. C. Wiley. 1982. "Changes in the Conformation of Influenza Virus Hemagglutinin at the pH Optimum of Virus Mediated Membrane Fusion." *Proc. Nat. Acad. Sci.* 79:968–972.

64. Gomatos, P. J., and I. Tamm. 1963. "Animal and Plant Viruses with Double Helical RNA." *Proc. Nat. Acad. Sci.* 50:878–885.

65. Skehel, J. J., and W. K. Joklik. 1969. "Studies on the *in Vitro* Transcription of Reovirus RNA Catalyzed by Reovirus Cores." *Virology* 39:822–831.

66. Acs, G., H. Klett, M. Schonberg, J. Christman, D. H. Levin, and S. C. Silverstein. 1971. "Mechanism of Reovirus Double-Stranded Ribonucleic Acid Synthesis *in Vitro* and *in Vivo." J. Virology* 8:684–689.

67. Roizman, B. 1985. "Multiplication of Viruses: An Overview." In *Virology*, ed. B. N. Fields, D. M. Knipe, R. M. Chanock, J. L. Melnick, B. Roizman, and R. E. Shope. New York: Raven Press, Chap. 5, pp. 69–76.

68. Sweet, B. H., and M. R. Hillerman. 1960. "The Vacuolating Virus SV40." *Proc. Soc. Exp. Biol. Med.* 105:420.

69. Fiers, W., R. Contreras, G. Haegeman, R. Rogiers, A. van de Voorde, H. Van Heuverswyn, J. Van Herreweghe, G. Volkaert, and M. Ysebaert. 1978. "Complete Nucleotide Sequence of SV40 DNA." *Nature* 273:113–120.

70. Reddy, V. B., B. Thimmappaya, R. Dhar, K. N. Subramanian, B. S. Zain, J. Pan, P. K. Ghosh, M. L. Celma, and S. M. Weissman. 1978. "The Genome of Simian Virus 40." *Science* 200:494–502.

71. Crawford, L. V., C. N. Cole, A. E. Smith, E. Paucha, P. Tegtmeyer, K. Rundell, and P. Berg. 1978. "Organization and Expression of Early Genes of SV40." *Proc. Nat. Acad. Sci.* 75:117–121.

72. Sambrook, J., B. Sugden, W. Keller, and P. A. Sharp. 1973. "Transcription of Simian Virus 40. III. Orientation of RNA Synthesis and Mapping of 'Early' and 'Late' Species of Viral RNA Extracted from Lytically Infected Cells." *Proc. Nat. Acad. Sci.* 70:3711–3715.

73. Jay, G., S. Nomura, C. W. Anderson, and G. Khoury. 1981. "Identification of the SV4-Agno Gene Product: A DNA Binding Protein." *Nature* 291:346–349.

74. Benoist, C., and P. Chambon. 1981. "The SV40 Early Promoter Region: Sequence Requirements *in Vivo." Nature* 290:304–310.

75. Ghosh, P., P. Lebowitz, F. Frisque, and Y. Gluzman. 1981. "Identification of a Promoter Component Involved in Positioning the 5' Termini of SV40 Early mRNAs." *Proc. Nat. Acad. Sci.* 78:100–104.

76. Fromm, M., and P. Berg. 1982. "Deletion Mapping of DNA Regions Required for SV40 Early Region Promoter Function *in Vivo." J. Mol. Appl. Genetics* 1:457–481.

77. Dynan, W. S., and R. Tjian. 1985. "Control of Eukaryotic Messenger RNA Synthesis by Sequence-Specific DNA-Binding Proteins." *Nature* 316:774–778.

78. Banerji, J., S. Rusconi, and W. Schaffner. 1981. "Expression of a Beta-Globin Gene Is Enhanced by Remote SV40 Sequences." *Cell* 27:299–308.

79. Gruss, P., R. Dhar, and G. Khoury. 1981. "Simian Virus 40 Tandem Repeated Sequences as an Element of the Early Promoter." *Proc. Nat. Acad. Sci.* 78:943–947.

80. Tjian, R. 1978. "The Binding Site on SV40 DNA for a T-Antigen Related Protein." *Cell* 13:165–179.

81. Ghosh, P., V. Reddy, J. Swinescoe, P. Lebowitz, and S. Weissman. 1978. "Heterogeneity and 5' Terminal Structure of the Late RNA's of SV40." *J. Mol. Biol.* 126:813–846.

82. Ryder, K., E. Vakalopoulou, R. Mertz, I. Masriangelo, P. Hough, P. Tegtmeyer, and E. Fanning. 1985. "Seventeen Base Pairs of Region I Encode a Novel Tripartite Binding Signal for SV40 T Antigen." *Cell* 42:539–548.

83. Sebring, E. D., T. J. Kelly, Jr., M. M. Thorsen, and N. P. Salzman. 1971. "Structure of Replicating Simian Virus 40 Deoxyribonucleic Acid Molecules." *J. Virology* 8:478–490.

84. Danna, K., and D. Nathans. 1972. "Bidirectional Replication of Simian Virus 40 DNA." *Proc. Nat. Acad. Sci.* 69:3097–3100.

85. Tegtmeyer, P. 1972. "Simian Virus 40 Deoxyribonucleic Acid Synthesis: The Viral Replicon." *J. Virology* 10:591–598.

86. Fareed, G. C., and N. P. Salzman. 1972. "Intermediate in SV40 DNA Chain Growth." *Nature New Biol.* 238:274.

87. Shortle, D. R., R. F. Margolskee, and D. Nathans. 1979. "Mutational Analysis of the SV40 Replicon: Pseudorevertants of Mutants with Defective Replication Origin." *Proc. Nat. Acad. Sci.* 76:6128–6131.

88. Garon, C. F., K. W. Berry, and J. A. Rose. 1972. "A Unique Form of Terminal Redundancy in Adenovirus DNA Molecules." *Proc. Nat. Acad. Sci.* 69:2391–2395.

89. Rekosh, D. M. K., W. C. Russell, A. J. D. Bellet, and A. J. Robinson. 1977. "Identification of a Protein Linked to the Ends of Adenovirus DNA." *Cell* 11:283–295.

90. Sharp, P. A., P. H. Gallimore, and S. J. Flint. 1974. "Mapping of Adenovirus 2 RNA Sequences in Lytically Infected Cells and Transformed Cell Lines." *Cold Spring Harbor Symp. Quant. Biol.* 39:457–474.

91. Berget, S. M., C. Moore, and P. A. Sharp. 1977. "Spliced Segments at the 5' Terminus of Adenovirus 2 Late mRNA." *Proc. Nat. Acad. Sci.* 74:3171–3175.

92. Chow, L. T., R. E. Gelinas, T. R. Broker, and R. J. Roberts. 1977. "An Amazing Sequence Arrangement at the 5' Ends of Adenovirus 2 Messenger RNA." *Cell* 12:1–8.

93. Chow, L. T., T. R. Broker, and J. B. Lewis. 1979. "Complex Splicing Patterns of RNAs from the Early Regions of Adenovirus 2." *J. Mol. Biol.* 134:265–303.

94. Ricciardi, R. P., R. L. Jones, C. L. Cepko, P. A. Sharp, and B. E. Roberts. 1981. "Expression of Early Adenovirus Genes Requires a Virus-Coded Acidic Polypeptide." *Proc. Nat. Acad. Sci.* 78:6121–6125.

95. Stillman, B. W., J. B. Lewis, L. T. Chow, M. B. Mathews, and E. Smart. 1981. "Identification of the Gene and mRNA for the Adenovirus Terminal Protein Precursor." *Cell* 23:497–508.

96. Montell, C., E. F. Fisher, M. H. Caruthers, and A. J. Berk. 1982. "Resolving the Functions of Overlapping Genes by Site-Specific Mutagenesis at a mRNA Splice Site." *Nature* 295:380–384.

97. Ziff, E. B., and R. M. Evans. 1978. "Coincidence of the Promoter and Capped 5' Terminus of RNA from the Adenovirus 2 Late Transcription Unit." *Cell* 15:1463–1476.

98. Thomas, G. P., and M. B. Mathews. 1980. "DNA Replication and the Early to Late Transition in Adenovirus Infection." *Cell* 22:523–533.

99. Price, R., and S. Penman. 1972. "A Distinct RNA Polymerase Activity Synthesizing 5.5S and 4S RNA in Nuclei Isolated from Adenovirus 2 Infected HeLa Cells." *J. Mol. Biol.* 70:435–450.

100. Pearson, G. D., and P. C. Hanawalt. 1971. "Isolation of DNA Replication Complexes from Uninfected and Adenovirus-Infected HeLa Cells." *J. Mol. Biol.* 62:65–80.

101. Sussenbach, J. S., P. C. Van der Vliet, D. J. Ellens, and H. S. Jansz. 1972. "Linear Intermediates in the Replication of Adenovirus DNA." *Nature* 239:47–49.

102. Challberg, M. D., and T. J. Kelly. 1979. "Adenovirus DNA Replication *in Vitro*." *Proc. Nat. Acad. Sci.* 76:655–659.

103. Challberg, M. D., S. V. Desiderio, and T. J. Kelly. 1980. "Adenovirus DNA Replication in *Vitro*: Characterization of a Protein Covalently Bonded to Nascent Chains." *Proc. Nat. Acad. Sci.* 77:5105–5109.

104. Richy, J. H., M. S. Horowitz, and J. Hurwitz. 1981. "Formation of a Covalent Complex Between the 80,000 Dalton Adenovirus Terminal Protein and 5' dCMP *in Vitro*." *Proc. Nat. Acad. Sci.* 78:2678–2682.

105. Astell, C. R., M. Smith, M. B. Chow, and D. C. Ward. 1982. "Structure and Replication of Minute Virus of Mice DNA." *Cold Spring Harbor Symp. Quant. Biol.* 47:751.

106. Tattersall, P., and D. C. Ward. 1976. "Rolling Hairpin Model for Replication of Parvovirus and Linear Chromosomal DNA." *Nature* 263:106–109.

107. Hayward, G. S., R. J. Jacob, S. C. Wadsworth, and B. Roizman. 1975. "Anatomy of Herpes Simplex Virus DNA: Evidence for Four Populations of Molecules That Differ in the Relative Orientations of Their Long and Short Components." *Proc. Nat. Acad. Sci.* 72:4243–4247.

108. Baer, R., A. T. Bankier, M. D. Biggin, P. L. Deininger, P. J. Farrel, J. J. Gibson, G. Hatfull, G. S. Hudson, S. C. Satchwell, C. Seguin, P. S. Tofnell, and B. G. Barell. 1984. "DNA Sequence and Expression of the B95-8 Epstein-Barr Virus Genome." *Nature* 310:207–211.

109. Honess, R. W., and B. Roizman. 1974. "Regulation of Herpesvirus Macromolecular Synthesis. I. Cascade Regulation of the Synthesis of Three Groups of Viral Proteins." *J. Virology* 14:8–19.

110. Batterson, W., and B. Roizman. 1983. "Characterization of the Herpes Simplex Virion-Associated Factor Responsible for the Induction of α Genes." *J. Virology* 46:371–377.

111. Kates, J., and B. R. McAuslan. 1967. "Messenger RNA Synthesis by a 'Coated' Viral Genome." *Proc. Nat. Acad. Sci.* 57:314–320.

112. Oda, K., and W. K. Joklik. 1967. "Hybridization and Sedimentation Studies on 'Early' and 'Late' Vaccinia Messenger RNA." *J. Mol. Biol.* 27:395–419.

113. Belle Isle, H., S. Venkatesan, and B. Moss. 1981. "Cell-Free Translation of Early and Late mRNAs Selected by Hybridization to Cloned DNA Fragments Derived from the Left 14 Million to 72 Million Daltons of the Vaccinia Virus Genome." *Virology* 112:306–317.

114. Geshelin, P., and K. Berns. 1974. "Characterization and Localization of the Naturally Occurring Cross-Links in Vaccinia Virus DNA." *J. Mol. Biol.* 88:785–796.

115. Baroudy, B. M., S. Bankatesan, and B. Moss. 1982. "Incompletely Base-Paired Flip-Flop Terminal Loops Link the Two DNA Strands of the Vaccinia Virus Genome into One Uninterrupted Polynucleotide Chain." *Cell* 28:315–324.

116. Rous, P. 1911. "A Sarcoma of the Fowl Transmissible by an Agent Separable from the Tumor Cells." *J. Exp. Med.* 13:397–411.

117. Poeisz, B. J., F. V. Ruscetti, A. F. Gazdar, P. A. Bunn, J. D. Minna, and R. C. Gallo. 1980. "Detection and Isolation of Type C Retrovirus Particles from Fresh and Cultured Lymphocytes of a Patient with Cutaneous T-Cell Lymphoma." *Proc. Nat. Acad. Sci.* 77:7415–7419.

118. Barré-Sinoussi, F., J. C. Chermann, F. Rey, M. T. Nugeyre, S. Chamaret, J. Gruest, C. Danguet, C. Axler-Blin, F. Vézinet-Brun, C. Rouzioux, W. Rozenbaum, and L. Montagnier. 1983. "Isolation of T-Lymphotropic Retrovirus from a Patient at Risk for Acquired Immune Deficiency Syndrome (AIDS)." *Science* 220:868–870.

119. Popovic, M., M. G. Sarnagadharan, E. Read, and R. C. Gallo. 1984. "Detection, Isolation, and Continuous Production of Cytopathic Retroviruses (HTLV-III) from Patients with AIDS and Pre-AIDS." *Science* 224:497–500.

120. Billeter, M. A., J. T. Parsons, and J. M. Coffin. 1974. "The Nucleotide Sequence Complexity of Avian Tumor Virus RNA." *Proc. Nat. Acad. Sci.* 71:3560–3564.

121. Varmus, H. E. 1982. "Form and Function of Retroviral Proviruses." *Science* 216:812–821.

122. Dahlberg, J. E., R. C. Sawyer, J. M. Taylor, A. J. Faras, W. E. Levinson, H. M. Goodman, and J. M. Bishop. 1974. "Transcription of DNA from the 70S RNA of Rous Sarcoma Virus. I. Identification of a Specific 4S RNA Which Serves as Primer." *J. Virology* 13:1126–1133.

123. Coffin, J. M., and W. A. Haseltine. 1977. "Terminal Redundancy and the Origin of Replication of Rous Sarcoma Virus RNA." *Proc. Nat. Acad. Sci.* 74:1908–1912.

124. Sodroski, J. G., C. A. Rosen, and W. A. Haseltine. 1984. "Trans-Acting Transcriptional Activation of the Long Terminal Repeat of Human T Lymphotropic Viruses in Infected Cells." *Science* 225:381–385.

125. Shank, P. R., S. Hughes, H. J. Kung, J. E. Majors, N. Quintrell, R. V. Guntaka, J. M. Bishop, and H. E. Varmus. 1978. "Mapping Unintegrated Forms of Avian Sarcoma Virus DNA: Both Termini of Linear DNA Bear a 300 Nucleotide Sequence Present Once or Twice in Two Species of Circular DNA." *Cell* 15:1383–1395.

126. Gilboa, E., S. W. Mitra, S. Goff, and D. Baltimore. 1979. "A Detailed Model of Reverse Transcription and Tests of Crucial Aspects." *Cell* 18:93–100.

127. Panganiban, A. T., and H. M. Temin. 1984. "Circles with Two Tandem LTRs Are Precursors to Integrated Retrovirus DNAs." *Cell* 36:673–679.

128. Moelling, K., D. Bolognesi, H. Bauer, W. Büsen, H. W. Plassmann, and P. Hausen. 1971. "Association of Viral Reverse Transcriptase with an Enzyme Degrading the RNA Moiety of RNA-DNA Hybrids." *Nature New Biol.* 234:240–243.

129. Yamamoto, T., B. deCrombrugghe, and I. Pastan. 1980. "Identification of a Functional Promoter in the Long Terminal Repeat of Rous Sarcoma Virus." *Cell* 22:787–798.

130. Laimins, L. A., G. Khoury, C. Gorman, B. Howard, and P. Gruss. 1982. "Host-Specific Activation of Transcription by Tandem Repeats from Simian Virus 40 and Moloney Murine Sarcoma Virus." *Proc. Nat. Acad. Sci.* 79:6453–6457.

131. Gorman, C. M., G. T. Merlino, M. C. Willingham, I. Pastan, and B. H. Howard. 1982. "The Rous Sarcoma Virus Long Terminal Repeat Is a Strong Promoter When Introduced into a Variety of Eukaryotic Cells by DNA-Mediated Transfection." *Proc. Nat. Acad. Sci.* 79:6777–6781.

132. Hayward, W. S. 1977. "Size and Genetic Content of Viral RNAs in Avian Oncovirus Infected Cells." *J. Virology* 24:47–63.

133. Vogt, V. M., and R. Eisenman. 1973. "Identification of a Large Polypeptide Precursor of Avian Oncornavirus Proteins." *Proc. Nat. Acad. Sci.* 70:1734–1738.

134. Vogt, P. K., and R. R. Friis. 1971. "An Avian Leukosis Virus Related to RSV(0): Properties and Evidence for Helper Activity." *Virology* 43:223–234.

135. Lowy, D. R., W. P. Rowe, N. Teich, and J. W. Hartley. 1971. "Murine Leukemia Virus: High-Frequency Activation in Vitro by 5-Iododeoxyuridine and 5-Bromodeoxyuridine." *Science* 174:155.

136. Groudine, M., R. Eisenman, and H. Weintraub. 1981. "Chromatin Structure of Endogenous Retroviral Genes and Activation by an Inhibitor of DNA Methylation." *Nature* 292:311–317.

137. Astrin, S. 1978. "Endogenous Viral Genes of White Leghorn Chickens: A Common Site of Residence as Well as Sites Associated with Specific Phenotypes of Viral Gene Expression." *Proc. Nat. Acad. Sci.* 75:5941–5945.

138. Jenkins, N. A., N. G. Copeland, B. A. Taylor, and B. K. Lee. 1981. "Dilute (d) Coat Colour Mutation of DBA/2J Mice Is Associated with the Site of Integration of an Ecotropic MuLV Genome." *Nature* 293:370–374.

139. Breindl, M., K. Harbers, and R. Jaenisch. 1984. "Retrovirus-Induced Lethal Mutations in Collagen I Gene of Mice Is Associated with an Altered Chromatin Structure." *Cell* 38:9–16.

140. Boeke, J. D., D. J. Garfinkel, C. A. Styles, and G. R. Fink. 1985. "Ty Elements Transpose Through an RNA Intermediate." *Cell* 40:491–500.

141. Summers, J. A., A. O'Connell, and I. Millman. 1975. "Genome of Hepatitis B Virus: Restriction Enzyme Cleavage and Structure of DNA Extracted from Dane Particles." *Proc. Nat. Acad. Sci.* 72:4597–4601.

142. Summers, J., and W. S. Mason. 1982. "Replication of the Genome of a Hepatitis B-Like Virus by Reverse Transcription of an RNA Intermediate." *Cell* 29:403–415.

143. Isaacs, A., and J. Lindenmann. 1957. "Virus Interference. I. The Interferon." *Proc. Roy. Soc. London (B)* 147:258–263.

144. Lengyel, P. 1982. "Biochemistry of Interferons and Their Actions." *Ann. Rev. Biochem.* 51:251–282.

145. Lampson, G. P., A. A. Tytell, A. K. Field, M. M. Nemes, and M. R. Hilleman. 1967. "Inducers of Interferon and Host Resistance. 1. Double-Stranded RNA from Extracts of *Penicillium Funiculosum*." *Proc. Nat. Acad. Sci.* 58:782–789.

146. Taniguchi, T., M. Sakai, Y. Fujii-Kurigama, M. Muramatsu, S. Kobayashi, and T. Sudo. 1979. "Construction and Identification of a Bacterial Plasmid Containing the Human Fibroblast Interferon Gene Sequence." *Proc. Jpn. Acad.* 55:464–469.

147. Nagata, S., T. H. Taira, A. Hall, L. Johnsrud, M. Streuli, J. Escodi, W. Ball, K. Cantell, and C. Weissmann. 1980. "Synthesis in *E. coli* of a Polypeptide with Human Leukocyte Interferon Activity." *Nature* 284:316–320.

148. Kerr, I. M., and R. E. Brown. 1978. "pppA2' p5'A2'p5'A: An Inhibitor of Protein Synthesis Synthesized with an Enzyme Fraction from Interferon Treated Cells." *Proc. Nat. Acad. Sci.* 73:520–523.

149. Lebler, B., G. C. Sen, S. Shaila, D. Cabrer, and P. Lengyel. 1976. "Interferon, Double-Stranded RNA and Protein Phosphorylation." *Proc. Nat. Acad. Sci.* 73:3107–3111.

150. Owerbach, D., W. J. Rutter, T. B. Strauss, P. Gray, D. V. Goeddel, and R. M. Lawn. 1981. "Leucocyte and Fibroblast Interferon Genes Are Located on Human Chromosome 9." *Proc. Nat. Acad. Sci.* 55:1133–1140.

151. Fujita, T., S. Ohno, H. Yasumitsu, and T. Taniguchi. 1985. "Delimitation and Properties of DNA Sequences Required for the Regulated Expression of Human Interferon-β Gene." *Cell* 41:489–496.

152. Ryals, J., P. Dierks, H. Ragg, and C. Weissmann. 1985. "A 46-Nucleotide Promoter Segment from an IFN-α Gene Renders an Unrelated Promoter Inducible by Virus." *Cell* 41:497–507.

153. Goodbourn, S., K. Zinn, and T. Maniatis. 1985. "Human β-Interferon Gene Expression Is Regulated by an Inducible Enhancer Element." *Cell* 41:509–520.

154. Miyata, T., and H. Hayashida. 1982. "Recent Divergence from a Common Ancestor of Human IFN-α Genes." *Nature* 295:165–168.

155. Kitajewski, J., R. J. Schneider, B. Safer, S. M. Munemitsu, C. E. Samuel, B. Thimmappaya, and T. Shenk. 1986. "Adenovirus VAI RNA Antagonizes the Antiviral Action of Interferon by Preventing Activation of the Interferon-Induced eIF-2α Kinase." *Cell* 45:195–200.

156. Robertson, H. D., S. H. Howell, M. Zaitlin, and R. L. Malmberg, editors. 1983. "Plant Infectious Agents," *Viruses, Viroids, Virusoids, and Satellites.* Cold Spring Harbor Laboratory Press.

157. Goelet, P., G. P. Lomonossoff, P. J. G. Butler, M. E. Akam, M. J. Gait, and J. Karn. 1982. "Nucleotide Sequence of Tobacco Mosaic Virus RNA." *Proc. Nat. Acad. Sci.* 79:5818–5822.

158. Klug, A. 1979. The Assembly of Tobacco Mosaic Virus: Structure and Specificity." The Harvey Lectures 74:141.

159. Gajdsek, C. 1985. "Unconventional Viruses Causing Subacute Spongiform Encephalopathies." In *Virology*, ed. B. N. Fields, D. M. Knipe, R. M. Chanock, J. L. Melnick, B. Roizman, and R. E. Shope. New York: Raven Press, Chap. 63.

160. Priusner, S. B. 1982. "Novel Proteinacious Infectious Particles Cause Scrapie." *Science* 216:136–144.

161. Oesch, B., D. Westaway, M. Wälchli, M. P. McKinley, S. B. H. Kent, R. Abersold, R. A. Barry, P. Tempst, D. B. Teplow, L. E. Hood, S. B. Prusiner, and C. Weissmann. 1985. "A Cellular Gene Encodes Scrapie, Pr-P 27-30 Protein." *Cell* 40:735–746.

162. Diener, T. 1971. "Potato Spindle Tuber Virus." IV. A Replicating, Low Molecular Weight RNA." *Virology* 45:411–428.

163. Gross, H. J., H. Domdey, C. Lossow, P. Jank, M. Raba, H. Albern, and H. L. Sänger. 1978. "Nucleotide Sequence and Secondary Structure of Potato Spindle Tuber Viroid." *Nature* 273:203–208.

164. Branch, A. D., H. D. Robertson, and E. Dickson. 1981. "Longer-Than-Unit-Length Viroid Minus Strands Are Present in RNA from Infected Plants." *Proc. Nat. Acad. Sci.* 78:6381–6385.

CANCER AT THE
GENETIC LEVEL

CHAPTER 25

The Control of Cell Proliferation

Biologists have long known that the control of cell division is one of the most basic aspects of multicellular existence. Throughout embryological development, as well as through all of adult life, many differentiated cells have the choice to divide or not, and only if a programmed series of correct decisions is made can the respective organisms continue to function normally. But mere desire to come to grips with the important questions does not necessarily generate clean answers; we would be less than candid if we did not admit that understanding the control of cell proliferation has been one of the more slowly moving fields in biology. Today, however, key facts are at last emerging, and it is likely that the next few years of research will provide the intellectual framework to understand proliferation control in a large variety of seemingly unrelated animal cell types. An early key to progress in higher eucaryotes was the realization that experiments need not be restricted to intact organisms, but that real answers can come from growing animal cells in culture. Now much hope lies in our ability to use molecular cloning techniques to identify and isolate genes that control cell growth.

Always intertwined with any study of cell proliferation is the nature of cancer. This collection of horrific diseases by definition involves cells that divide when they should not, usually to produce the contiguous cellular masses called **tumors.** The different classes of cancer arise by changes in the various forms of differentiated cells, with the resulting cancer cells exhibiting many of the morphological and functional characteristics of their normal precursors. Most often, they arise in cells undergoing frequent division (e.g., epithelial cells of the skin), and so their essence is not the fact that they can divide rapidly, but that they lack normal control systems to shut off unwanted cell division. So a prime scientific goal for most of this century has been to find out the essential biochemical differences between normal and cancer cells.

USE OF TISSUE CULTURE CELLS TO STUDY CELL PROLIFERATION

Establishing Cells in Culture[1, 2]

Until the past decade, there was much mysticism about growing cells in culture. Success apparently demanded very precise experimental conditions; moreover, not everyone seemed to possess the necessary magic touch. For a long time, only masses of excised tissue would consistently grow, and then only when implanted upon very undefined serum clots or embryonic extracts. Such "tissue cultures" occasionally multiplied indefinitely if microbial contamination could be prevented and if regular changes were made in the surrounding nutrient medium. Although there were repeated efforts to grow cultures from dissociated cells, successes were very infrequent until small amounts of the proteolytic enzyme trypsin were used to dissociate tissue masses into their component individual cells. Then the dissociated cells (**primary cells**) usually grew well when seeded onto culture plates at high cell densities. In contrast, initial attempts to induce single cells to grow into clones did not succeed. At best, the individual cells multiplied several times and then died. It is thought that the failure of these early cloning attempts is due to the inherent leakiness of most vertebrate primary cells in culture. Because vital nutrients diffuse out, growth occurs best when cells are seeded at high density, making cross-feeding possible. Hence, the practice arose of placing single primary cells in very small drops of culture fluid. Here, extensive dilution is not possible, and tiny clones result. Another way to increase cloning efficiency is to place dissociated single cells upon a "feeder" layer of cells sterilized by large X-ray doses. None of the irradiated feeder cells can multiply, but they remain metabolically active and release sufficient nutrients to allow most of the seeded primary cells to form visible colonies.

Cell cultures arising from the transfer and subsequent multiplication of primary cells are called **secondary cells.** Such cells generally do not have infinite lives, but divide only a finite number of times before dying out. For example, most secondary chicken cells multiply 20 to 40 times and then invariably die. Likewise, cultures of human skin cells usually divide no more than 50 to 100 times before they die. Interestingly, this number coincides with the number of divisions that such cells make during the life span of their respective vertebrate source. Furthermore, skin cells or fibroblasts derived from older individuals have shorter lifetimes, as measured in cell generations, than do those from younger donors. The temptation thus exists to wonder whether this seemingly programmed cell death is related to the normal vertebrate aging process.

Fortunately, not all animals yield primary cultures all of whose progeny die. Although most mouse and hamster cells also die after dividing some 30 to 50 times, a few progeny pass through the "crisis period" and acquire the ability to multiply indefinitely. Progeny derived from such exceptional cells form the continuous **cell lines** (Table 25-1). Cells that can grow indefinitely in culture are said to be **established** and frequently are referred to as **immortal.** Although no "normal" human cell line has yet been developed, a number of cell lines have been derived from human tumors. Similarly, a few chicken tumors have been established into cell lines. The possession of the cancerous phenotype often allows easier adaptation to the conditions of cell culture. In part, this property may relate to the fact that both cancer cells and cells adapted to permanently grow in culture often have an increased chromosome number, frequently nearly twice the normal diploid number. For example, cells from the human cancer cell line *HeLa* have some 70 to 80 chromosomes, compared to the normal 46 chromosome complement (Figure 25-1). Likewise, mouse 3T3 cell lines, although derived from normal mouse embryos, have about 65 chromosomes rather than the normal diploid number of 40. Apparently, the possession of extra chromosomes (**aneuploidy**) frequently helps cells to grow in culture.

Table 25-1 Some Properties of Several of the More Important Cell Lines

| | Origin | | Characteristics | | |
	Species	Tissue	Morphology	Growth in Suspension	Chromosomal Makeup
3T3	Mouse 2N = 40	Endothelial	Fibroblast	No	Heteroploid (differs from cell to cell) (mode = 65)
L	Mouse 2N = 40	Connective	Fibroblast	Yes	Heteroploid (mode = 65)
CHO	Chinese hamster 2N = 22	Ovary	Epithelial	Yes	Pseudodiploid
BHK 21	Syrian hamster 2N = 44	Kidney	Fibroblast	Yes	Diploid
BSC	Monkey 2N = 42	Kidney	Epithelial	No	Diploid
MPC	Mouse 2N = 40	Bone marrow myeloma cell	Lymphoid	Yes	Heteroploid
RPH	Frog 2N = 26	Egg	Epithelial	No	Haploid (N = 13)
HeLa	Human 2N = 46	Cervical tumor	Epithelial	Yes	Heteroploid (mode = 75)
KB	Human 2N = 46	Nasopharyngeal tumor	Epithelial	Yes	Heteroploid (mode = 75)

The Obscure Origin of Many Cell Lines[1, 3]

Usually primary cultures are prepared from large tissue masses, if not from organs or whole embryos. Initially, they consist of a large variety of differentiated cells (*fibroblasts, macrophages, epithelial* and *endothelial* cells, etc.), and so the cellular origin of a given cell line is frequently obscure. Under the conditions that have been used most often for growth in culture, the cells that multiply best have the spindle shape of connective tissue cells and so are called fibroblasts (Figure 25-2a). In the past decade, however, much effort has gone toward the development of culture conditions that selectively promote the growth of other cell types and inhibit the growth of fibroblasts. For example, an early success with epithelial cells (Figure 25-2b) took advantage of their ability to use D-valine as their sole valine source (epithelial cells possess the enzyme D-amino-acid oxidase, which allows the transformation of D-valine into L-valine). Fibroblasts, however, do not contain this enzyme and so cannot grow when D-valine is the only source of valine in the growth medium.

The shape that cells of a given line assume in culture is not invariant. For example, some strains of HeLa cells growing in rich (high-serum-content) media have the stretched, elongated fibroblast shape, whereas in poorer media, they pack close together and have a rounded, cobblestone-like appearance (Figure 25-3). Thus, while we may refer to one cell line as epithelial cells and another as fibroblasts, we must not always assume that their origin was from epithelial (connective) tissue.

Sometimes, however, the presence of specific cell products makes the origin of certain cell lines unambiguous. In particular, the origin of cells secreting specific hormones is easy to ascribe. Cells from tumors of the testes secrete androgens, while ACTH and growth hormones are made by cell lines derived from pituitary tumors. Like-

(a)

(b)

Figure 25-1
(a) The 73 chromosomes of a HeLa cell contrasted with (b) the normal 46 complement of a cell from a normal human male. (Courtesy of Dr. T. T. Puck, University of Colorado, Denver.)

wise, cells derived from many liver tumors preferentially secrete the major liver product albumin. Equally clear is the cellular origin of cultured **neuroblastomas** (tumors of nervous tissue), since their cells have excitable membranes and under certain nutritional conditions send out numerous axonal processes (Figure 25-4). Lymphoid cells growing in culture can be distinguished by the presence of antibodies (B cells) or characteristic surface antigens (T cell types or B cells), and the epithelial origin of many cell lines now can be routinely established through their possession of one to several keratins, the fibrous proteins used to construct their intermediate (100 Å) filaments.

Not all the properties of differentiated cells are retained in their cultured progeny. Many established cell lines seem less differentiated than their in vivo precursors and often show similarities to the undifferentiated embryonic cells present in early development. Many so-called "normal" cell lines thus closely resemble cancer cells, many types of which also represent partially dedifferentiated or incom-

(a)   (b)

Figure 25-2
Outgrowth of secondary cells starting from primary human lung tissue. Selective proliferation of (a) fibroblasts in L-valine and (b) epithelial cells in D-valine. [Reproduced with permission from S. Gilbert and B. Migeon, *Cell* 5 (1975):11.]

(a)

(b)

Figure 25-3
Dependence of the shape of HeLa cells on growth conditions. (a) In low serum, they have a rounded cobblestone shape, while (b) in enriched serum, they have the spindle shape of fibroblasts. (Reproduced with permission from T. Puck, *The Mammalian Cell as a Microorganism,* Holden-Day, 1972.)

pletely differentiated states. For example, most liver-derived hepatomas fail to make many normal liver enzymes. This becomes more plausible when we realize that continued cell growth in culture may be achieved by a loss of some of the devices that slow down, if not inhibit, cell division in the multicellular organism, and whose loss may result in the cancerous state. The isolation of many cell lines may thus have been dependent on the selection of cell mutants less able to respond to signals halting cell division.

Attachment to Solid Surfaces Versus Growth in Suspension

Most "fibroblasts" and "epithelial" cells grow best, if not exclusively, when attached to solid surfaces, forming thin cell lawns one to several layers thick. Under optimal conditions, primary epithelial cells form layers one cell thick (monolayers), while the lawns formed by fibroblasts are frequently two to three cells thick. Only rarely can cell lines be derived that multiply well in suspension. Such lines usually originate from tumors (transformed cells). In general, cultures containing cells that grow well in suspension produce many more cells than those whose growth is restricted to surfaces. So the cell lines that have been most frequently used to probe the molecular biology of vertebrates are those like HeLa and KB, which have suspended growth patterns. Recently, it has been possible to increase the yield of cells that can only grow when they are attached to surfaces by using microcarrier beads (Figure 25-5). Microcarrier beads provide a large surface area for cell attachment within a small volume of medium, thereby producing high concentrations of dividing cells, similar to a suspension culture.

Most lymphoid cells grow best, or exclusively, in suspension, either as single cells or in tiny clumps (Figure 25-6). Generally, resting lymphocytes can be coaxed to divide only in the presence of certain mitogens and growth factors or following infection with certain viruses. Many resulting lymphoid lines are immortal, and gram amounts of such cells are easily obtainable. Growth of other blood cell

Figure 25-4
Phase-contrast photomicrograph of mouse neuroblastoma cells growing attached to a solid surface. Many elongated processes (neurites) extend from each cell. When such cells grow in suspension, they have the rounded cancerous appearance characteristic of the cells that form tumors; but when they attach to a solid surface, they become flattened and begin to send out large numbers of neurites. (Courtesy of Dr. Gunter Albrecht-Bühler, Cold Spring Harbor Laboratory.)

types in culture has become possible with the discovery of factors that promote the differentiation of erythrocytes, granulocytes, and macrophages.

"Normal" Cell Lines[4, 5]

In the early days of cell culture, the suspicion existed that cells became capable of growing in culture only when they acquired mutations that gave them the cancer cell's capability for unrestricted growth. This belief arose not only because it was much easier to make cell lines out of cancer cells, but also because many so-called "normal" cell lines had the roundish cancerous morphology, had the property of disorganized growth (Figure 25-7), and could cause tumors when injected into animals. Ideas flip-flopped 20 years ago, however, when the growth properties of mouse cell lines were found to depend on the cell density at which secondary cells are grown. If secondary cultures are always grown to high cell density, then cell lines that emerge after crisis form cultures of disorganized cells several layers thick. Such cells, though of normal origin, generally look like cancer cells and form tumors when injected into immunologically identical mice. If, however, the secondary mouse cultures are always kept at the low densities in which cell-cell contact is infrequent, the resulting cell lines have normal fibroblast morphology and form the thin organized monolayers characteristic of "normal" cells (Figure 25-8). Correspondingly, these cells only infrequently form tumors when injected into appropriate mice.

Realization of the powerful selective forces that exist during the "crisis period" now allows isolation of cell lines with desired growth properties. By keeping the cell number low, we remove a selection in favor of cancerlike cells, which can grow well even under crowded conditions. Thus, there are now a number of cell lines, the prototype being mouse 3T3 lines, that have many of the growth properties of the normal cell and so can be used for in vitro studies of transformation to the cancerous state. However, it is clear that most established cell lines, even the 3T3 type, are not completely normal. Besides their enhanced growth potential relative to primary cells, they frequently have highly abnormal chromosome numbers, as mentioned earlier.

Figure 25-5
Scanning electron micrographs of pig kidney cells growing on microcarrier beads. The cells are used for the production of foot-and-mouth disease vaccine. (Courtesy of B. Meignier and J. Tektoft, Institut Merieux, Lyon, France.)

Figure 25-7
The morphologies of normal (cell line 9) versus transformed (cell line 14b) hamster embryo cells. (a, b) The phase-contrast photomicrographs (250×) show the spread-out, flat character of the normal hamster cell growing on a flat surface. The cell in the center is moving toward the right with its fan-shaped leading edge containing many ruffles (arrow). Five hours elapsed between the two photos. (c, d) Corresponding views (300×) of cancerous hamster cells resulting from transformation by adenovirus 2. The cells have a much rounder appearance and exhibit very little motility (arrow points to cells that have not moved in the 19 hours separating the two photos). Daughter cells tend to stay close together, resulting in tiny islands of cells. [Reproduced with permission from R. D. Goldman, C.-M. Chang, and J. Williams, *Cold Spring Harbor Symp. Quant. Biol.* 39 (1975):601.]

Cell Transformation[6,7,8]

The sudden change of a cell possessing normal growth properties into one with many of the growth properties of the cancer cell is called **cell transformation.** Spontaneously, this phenomenon occurs only rarely, although the frequency can be significant (10^{-5} to 10^{-6}) in certain established 3T3-like cell lines. As we shall see in Chapter 26, transformation occurs at much higher frequencies after infection with a tumor virus or after exposure to a carcinogenic (cancer-causing) chemical or radiation source. Irrespective of its origin, however, the transformed cell, besides often having the potential to grow into tumors, usually possesses additional characteristics: It has a much rounder shape, reflecting less organized arrangement of cytoplasmic microfilaments and microtubules; it is much less subject to contact inhibition of movement; it grows to higher cell densities; it is often less restricted to growth on solid surfaces; and it appears nutritionally less fastidious, particularly in its serum requirements. Not all transformed cells, however—not even those derived from the same cell line—are morphologically identical. As we shall detail later, the cancerous phenotype probably arises by a number of quite distinct biochemical changes.

(a)   (b)   (c)   (d)

Normal cells   Cancer cells

Figure 25-8
Schematic comparison of the multiplication of a normal cell and of a cancer cell upon a solid surface. The normal cells divide until they form a solid monolayer. Cancer cells, however, can form irregular masses several layers deep.

Working Out Nutritional Requirements[9, 10]

Well-characterized cell lines allow controlled experiments on the nutritional requirements of vertebrate cells. A semisynthetic medium consisting of 13 different amino acids, 8 vitamins, and a mixture of inorganic salts, together with glucose as an energy source and dialyzed horse serum (Table 25-2), permits HeLa and L cells to grow optimally to high densities. Additional requirements (e.g., serine, inositol, and pyruvate) sometimes emerge when cells grown at low density leak too many essential metabolites. Continuously passed cell lines tend to grow better at lower densities than newly derived lines,

Table 25-2 Basal Media for Growth of the HeLa Cell Line in Culture

L-Amino Acid (mM)		Vitamins (mM)		Salts (mM)	
Arginine	0.1	Biotin	10^{-3}	NaCl	100
Cysteine	0.05	Choline	10^{-3}	KCl	5
Glutamine	2.0	Folic acid	10^{-3}	$NaH_2PO_4 \cdot H_2O$	1
Histidine	0.05	Nicotinamide	10^{-3}	$NaHCO_3$	20
Isoleucine	0.2	Pantothenic acid	10^{-3}	$CaCl_2$	1
Leucine	0.2	Pyridoxal	10^{-3}	$MgCl_2$	0.5
Lysine	0.2	Thiamine	10^{-3}		
Methionine	0.05	Riboflavin	10^{-4}		
Phenylalanine	0.1				
Threonine	0.2				
Tryptophan	0.02				
Tyrosine	0.1				
Valine	0.2				

Miscellaneous

Glucose	5 mM
Penicillin	0.005%*
Streptomycin	0.005%*
Dialyzed horse serum	5%

*To prevent microbial contamination.
SOURCE: After H. Eagle, *Science* 122 (1955):501.

Figure 25-9

Phases in the life cycle of a mouse hepatoma cell growing in tissue culture and dividing once every 24 hours.

suggesting that cloning probably exerts selection pressure for diminished leakiness.

The requirement for serum is almost universal for all vertebrate cell lines, hinting that one or more serum factors may play key roles in regulating in vivo cell multiplication. In vitro, the final cell densities reached by specific cell lines are often a function of the serum concentration. Certain cell lines that form a one-cell-thick monolayer in 1 percent serum grow layers three to four cells thick in 10 percent serum. Also very important is the exact pH at which cells grow. The optimal pH can vary from one cell line to another, though most normal cells grow best at a pH of around 7.6.

Growth of Cells in Completely Defined Media[11, 12]

Although amino acid, vitamin, salt, and energy requirements for growing most vertebrate cells in culture were worked out over 30 years ago, the identification of factors in serum that allow cells to grow has taken far longer. Even today, most cell culture is performed with serum-containing medium, since despite some successes, completely defined media remain difficult to work with. Once again, the existence of cell lines with their great hardiness and often reduced serum requirement plays an important part in the development of completely defined media. Serum can sometimes be replaced by a combination of specific polypeptide growth factors and hormones, carrier proteins, and, for certain cell types, lipids and proteins that facilitate cell attachment. For example, mouse 3T3 cells, which are fibroblastic cell lines, will divide in basal medium (Table 25-2) supplemented with transferrin, platelet-derived growth factor (PDGF), insulin or insulin-like growth factors (IGF) called somatomedins, and epidermal growth factor (EGF). Transferrin, a protein that transports ferric iron (Fe^{3+}) into cells, is required for growth of almost all cultured cells and is present at a high concentration in serum. Growth factors are polypeptides that stimulate the growth of specific cell types. Thus, different factors are often needed for different cells.

Different Requirements for Growth of Different Cell Types[12]

The reason that fibroblastic cells frequently take over primary cultures is that the 10 percent fetal calf or calf serum employed in standard culture conditions contains PDGF and other factors for fibroblast growth, but lacks specific growth factors required by many other cell types and even contains substances that inhibit the growth of some cells. Little is known about the inhibiting substances, but many specific growth factors have been identified now. For example, growth of certain T lymphoid cells requires a T cell growth factor called interleukin-2. Thus, the correct combination of protein supplements, in combination with an appropriate basal medium, will allow the growth of different cell types in vitro. Whether the cells have the same requirements in vivo is not known in most cases. Although defined media will support the growth of specific cell lines or cell types, most researchers still prefer to grow T cells, for example, in a medium containing serum to which interleukin-2 was added.

Despite considerable progress in tissue culture systems, we are probably still a long way from identifying all the stimulatory and inhibitory macromolecules that regulate the growth of all the different vertebrate cell types in vivo.

The Cell Cycle[13–16]

The length of the cell cycle under optimal nutritional conditions varies from one cell line to another. At 37°C, a few lines can divide once every 10 to 12 hours, but the majority require some 18 to 24 hours. Now, most biologists divide cell cycles into the M phase (mitosis), G_1 phase (period prior to DNA synthesis), S phase (period of DNA synthesis), and G_2 phase (period between DNA synthesis and mitosis) (Figure 25-9). Collectively, G_1, S, and G_2 are called **interphase,** the period of the cell cycle distinct from division of the nucleus (*mitosis*) and cytoplasm (**cytokinesis**). M is the shortest stage, usually lasting only an hour or so. All M phase cells, irrespective of their previous shapes, become completely spherical with accompanying depolymerization of the microtubule skeleton. The length of the S phase is remarkably similar in many different cells, as is G_2, while the greatest variation is seen in the length of G_1. Often, G_1 and S are about equal, with an 8- to 10-hour duration; but in some cells, G_1 lasts only 1 to 2 hours. At some point late in G_1, called the **restriction** or **R point,** a cell becomes committed to traverse the remainder of the cell cycle. Thus, variations in cell cycle time are mostly due to variations in the length of G_1 up to the R point.

A most important tool in the analysis of the cell cycle is the synchronization of growth to place virtually all cells in the same phase. One of the best methods of synchronization uses the fact that spherical M phase cells (Figure 25-10) attach much less firmly to glass surfaces than do interphase cells. Thus, by shaking the cultures, we can isolate large numbers of uncontaminated M phase cells and subsequently measure the rates at which different important compounds are made throughout the cell cycle. In HeLa cells, RNA and protein synthesis occur throughout interphase (G_1, S, G_2), but both show abrupt decrease during the mitotic phase, when the chromosomes contract and line up along the spindle fibers. While the cessation of RNA synthesis may be a consequence of shape changes in chromatin concomitant with chromosome contraction, why protein synthesis declines is not clear. The effect is primarily due to translational control, and synthesis resumes promptly at the start of G_1.

Fusion of Cells in Different Phases of the Cell Cycle[17, 18, 19]

Fusion of a cell in interphase (G_1, S, G_2) with an M phase cell quickly results in the appearance of condensed chromosomes in the nucleus of the interphase cell (Figure 25-11). Mitotic cells thus possess a chemical factor that can promote the condensation of interphase chromosomes. When G_1 cells are used in such experiments, their chromosomes are thinner upon condensation than those of their M phase partners, reflecting their single chromatid construction. In contrast, when G_2 cells are fused, their chromosomes, upon condensation, more closely resemble the normal double-chromatid mitotic chromosome. And when S phase cells are fused, the resulting condensed chromosomes are usually highly fragmented, perhaps because they were caught in the process of chromosome duplication and so were highly susceptible to nuclease attack.

(a)

(b)

(c)

Figure 25-10
Scanning electron micrographs of normal hamster cells (cell line 9). (a) A cell rounded up for mitosis. Its surface is covered with microvilli, a characteristic feature of all M phase cells. (b) Two daughter cells in the process of separation. Note that both are covered with blebs, very characteristic protrusions of the late M, early G_1 period of normal cells. (c) A cell in late G_1. No blebs are visible; instead, a number of microvilli are present, especially in the more central, less spread out region. When the spreading is complete and S phase commences, the outer surface becomes very smooth and only the occasional microvillus can be seen. [Reproduced with permission from R. D. Goldman, C.-M. Chang, and J. Williams, *Cold Spring Harbor Symp. Quant. Biol.* 39 (1975):601.]

(a)

(b)

(c)

Figure 25-11
Appearance of the chromosomes in hybrid cells formed by the fusion of two HeLa cells phased in different parts of the cell cycle. (a) M and G_1. (b) M and S. (c) M and G_2. (Reproduced with permission from T. Puck, *The Mammalian Cell as a Microorganism,* Holden-Day, 1972.)

Initiation of DNA Synthesis[14, 20]

Once the decision is made to reenter the proliferation cycle, a sequence of preprogrammed operations is initiated long before actual DNA synthesis begins. For example, the levels of the deoxyribonucleoside triphosphate precursors (dATP, dGTP, dCTP, and dTTP) are very low at the beginning of G_1 and remain that way until some 30 minutes prior to initiation of DNA synthesis. Then they rise rapidly to peak amounts during S phase, remaining at relatively high levels until M ends and G_1 commences. How these levels are enzymatically controlled remains to be worked out. Likewise, the enzymology of eucaryotic DNA synthesis itself is still poorly understood. We do know that multiple replication forks, moving bidirectionally, are needed to complete the replication of all the DNA within the 8- to 10-hour S phase (see Figure 21-8). In reproducible succession, distinct areas of chromosomes, called **replication units,** become populated with multiple replication forks.

How does the cell keep track of all the DNA and thus ensure that each chromosome is replicated just once during each S phase? Fusion of cells in the various phases of the cell cycle has provided a preliminary answer to this question. When an S phase nucleus and a G_1 nucleus are brought together, the G_1 nucleus also begins to make DNA, indicating the presence of a positive initiating factor(s). However, fusion between an S phase cell and a G_2 cell does not lead to DNA synthesis in the G_2 nucleus or to its inhibition in the S nucleus. The refractoriness of the G_2 chromosomes might conceivably to be due to the rapid coating of newly synthesized DNA by concomitantly synthesized nondiffusible chromosomal proteins. These proteins would then be the mechanism for keeping track of replicated DNA and preventing it from being duplicated again during the S phase. Cell cycle–related changes would have to occur in one or more of these chromosomal proteins before a new round of DNA synthesis could start. Since G_1 nuclei, fresh out of mitosis, will undergo DNA synthesis when fused with S phase nuclei, the change must occur during M phase.

Mutations in Cells Growing in Culture[21, 22, 23]

One approach to studying the control of cell growth is to isolate mutants conditionally blocked in their growth or possessing altered growth properties. Because cell lines are diploid (heteroploid), we would anticipate that most mutations that are not sex-linked will be only infrequently expressed. Thus, much effort was initially spent on finding subdiploid cell lines, where one or more of the chromosomes of the normal diploid chromosomal complement are missing. Much work has been done with the Chinese hamster, which not only has a normal $2N$ complement of only 22 chromosomes, but also frequently gives rise to cell lines in which one or more of the autosomes are missing. Using these lines, researchers have found mutations that block the biosynthesis of several key metabolites, such as glycine, adenine, and inositol. Moreover, the routine employment of specific mutagens greatly increases the frequency of such mutations.

Vital to such experimentation is the employment of techniques that selectively isolate desired mutants from a large background of normal cells. The best technique utilizes the fact that exposure to visible light rapidly kills cells that have incorporated the base analog 5-BUdR into their DNA (Figure 25-12). Since only actively multiplying cells incorporate 5-BUdR (or thymidine) into their DNA, those mutagenized cells that are unable to grow are not killed by an exposure to visible light.

The BU visible light technique has been particularly successful in the isolation of **temperature-sensitive (ts) mutants,** which are unable to multiply at high (or low) temperature, but which multiply normally at low (or high) temperature. This class of mutants should include virtually all cellular functions, since their creation depends only on induction of amino acid changes that lead to loss of protein function at either the high or low temperatures. Already there are ts mutants defective in DNA replication, RNA synthesis, protein synthesis, and ion transport, as well as others unable to traverse specific stages in the cell cycle. There are, in addition, mutant cells that can express the transformed state only at low temperature (30°C), as well as other mutants that can express the transformed phenotype only at high temperature (37°C) (Figure 25-13).

Stopping in Early G_1[14, 15]

Examination of fibroblasts that have stopped dividing upon formation of a confluent monolayer reveals that all the cells are blocked in early G_1. (This resting point for quiescent cells is sometimes referred to as G_0.) Similarly, when the growth of such cells is prematurely blocked by removal of essential growth factors like isoleucine, glutamine, phosphate, or serum, the starved cells do not stop randomly in the cell cycle, but always proceed through mitosis, to yield cells with the $2N$ amount of DNA. In each case, growth is stopped very early in G_1, for DNA synthesis is commenced about 8 hours after reintroduction of the limiting growth factors (about the length of the G_1 stage in these cells). When so blocked in early G_1, quiescent cells remain healthy, even if the needed nutrients are withheld for many generation times. This picture parallels the situation of most in vivo animal cells, which can remain viable and metabolically active even after many months in a nonproliferating state. In contrast, if poisons like colchicine or hydroxyurea are added to the cells, they do not block the cells at G_1: Colchicine stops cells in M phase, while hydroxyurea interrupts S phase DNA synthesis. Cells so inhibited do not remain viable for long and usually die within 24 hours after exposure to these poisons.

Many cancer cells have lost the control mechanism that sends nutritionally limited normal cells into G_1 quiescence. So when, for one reason or another, such cancer cells do not divide, a spectrum of cancer cells blocked in all phases of the cell cycle is produced. As such, they are much less healthy than their normal equivalents and die off more rapidly. This concept provides a rationale for the frequent observation that the presence of antimetabolites differentially kills off cancer cells.

Activation of G_1 Quiescent Cells by Mitogenic Stimuli[24]

Exposure to any of a large number of different stimuli, that collectively are called **mitogens,** triggers many quiescent G_1-blocked cells into division. Some mitogens, like certain hormones, or growth

Figure 25-12
Use of the 5-BUdR visible light technique for the isolation of mutant cells unable to multiply as the result of either a nutritional deficiency or the possession of a temperature-sensitive vital enzyme or structural protein. A mutagenized mixed population is exposed to 5-BUdR under conditions where only wild-type cells grow and incorporate 5-BUdR into DNA. Subsequent exposure to visible light kills only the wild-type cells, leaving the mutant cells untouched. Subsequently, an enriched medium is added (or the temperature lowered or raised) to allow the mutant cells to grow into colonies.

(a)

(b)

Figure 25-13

Photomicrograph of temperature-sensitive mutants of a transformed Chinese hamster lung cell line that prevent the expression of the transformed state at high temperature. (a) The transformed growth pattern of cells growing at 34°C. (b) Normal growth pattern of cells growing at 39°C. [Reproduced with permission from K. Miyashita and T. Kakynaga, *Cell* 5 (1975):131.]

factors present in normal serum, are probably normal physiological effectors. But other strong mitogens, like some bacterial lipopolysaccharides and certain cell agglutinating proteins (*lectins*) isolated from plant seeds, normally may never provoke cell division. Yet their addition invariably sets certain quiescent cells into division. Likewise, proliferation often results when certain cell types are treated with small amounts of the proteolytic enzyme trypsin or with neuraminidase, the enzyme that removes the complex amino sugar sialic acid from surface proteins. Much effort now goes toward understanding how mitogens stimulate cell proliferation. Once again, the use of tissue-culture-grown cells makes the studies possible.

FACTORS THAT REGULATE CELL PROLIFERATION, FUNCTION, AND DIFFERENTIATION

Platelet-Derived Growth Factor (PDGF), Insulin-Like Growth Factor (IGF), and Epidermal Growth Factor (EGF) Together Stimulate the Growth of 3T3 Cells[25-34]

Much of our current understanding of how growth factors work comes from the study of tissue-culture-grown 3T3 fibroblasts and peripheral blood lymphocytes. A key concept to come from these studies is that multiple growth factors are required for optimal cell growth and division. In the case of 3T3 cells, it was found that animal sera prepared from unclotted blood did not support the growth of cells nearly as well as did animal sera prepared from clotted blood. A mitogen found in platelets, the **platelet-derived growth factor** (**PDGF**), can restore most of the growth-promoting activity to the serum prepared from unclotted blood. This factor is released from platelets during the clotting reaction. However, PDGF by itself is a rather poor mitogen for 3T3 cells. Only in combination with the growth factors found in platelet-poor plasma does it give an optimal mitogenic response.

Treatment of 3T3 cells with PDGF makes the cells *competent* to respond to the growth factors found in platelet-poor plasma. Cells remain competent for several hours, even after PDGF is removed, but they do not progress any closer in time to the onset of DNA synthesis (S phase). Exposure to platelet-poor plasma allows the competent cells to progress through the 10- to 12-hour G_1 phase of the cell cycle and on to DNA synthesis. For 3T3 cells, the progression factors present in platelet-poor plasma can be replaced with the defined growth factors, **epidermal growth factor** (**EGF**) and **insulin-like growth factor** (**IGF**).

Most polypeptide growth factors are small proteins with molecular weights that vary from 1 to 40 kdal. PDGF is a glycoprotein dimer of about 30 kdal, and genes that encode it have been cloned (Table 25-3). EGF is a 53-amino-acid-long polypeptide. A cDNA that encodes it has been molecularly cloned and sequenced. Interestingly, this sequence reveals that EGF is encoded as part of a large polypeptide, about 1200 amino acids in length, containing several EGF-like sequences (Figure 25-14). The sequence also reveals that the precursor may be able to

Table 25-3 Properties of Some Polypeptide Growth Factors Whose Corresponding Genes/mRNAS Have Been Cloned (See also Table 25-5)

Name	Structure	Biological Properties
PDGF (platelet-derived growth factor)	Two disulfide-bonded chains; glycoprotein	Mitogen for cultured cells of mesenchymal origin; repair of vascular system in vivo
EGF (epidermal growth factor)	53 amino acids (preproEGF 1200 amino acids)	Mitogen for cells of ectodermal and mesodermal origin; role in early development in vivo?
TGF-α (transforming growth factor)	50 amino acids; (preproTGF-α 160 amino acids); homology with EGF	Mitogen like EGF but can make normal cells grow in agar; binds to EGF receptor; high levels in some cancer cells
TGF-β (transforming growth factor)	Homodimer of MW 25,000; no homology with TGF-α or EGF	Mitogen for some cells, growth inhibitor for others; stimulates growth in agar; synergy with TGF-α
NGF (nerve growth factor)	Two identical β chains of 118 amino acids; low homology to insulin	Development and maintenance of sympathetic and embryonic neurones
IGF-I (insulin-like growth factor-I; somatomedin C)	70 amino acids (preproIGF-I 130 amino acids); 45% homology with insulin	Intermediate in the action of growth hormone; skeletal elongation; mitogen for cells in vitro
IGF-II (insulin-like growth factor-II)	67 amino acids (preproIGF-II 180 amino acids); homology with insulin	In vivo role uncertain
GH (growth hormone; somatotropin)	191 amino acids	Synthesized in anterior pituitary; stimulates liver to produce somatomedins
GHRF (growth hormone releasing factor; somatoliberin)	44 amino acids (precursor about 100 amino acids	Hypothalamic peptide; regulates synthesis of growth hormone in the pituitary
Insulin	Two disulfide-bonded chains; A is 21 amino acids, B is 30 amino acids; A, B cleaved from preproinsulin of 110 amino acids	Stimulates glucose uptake, protein synthesis, and lipid synthesis; can be mistaken for growth factor in vitro

travel to the plasma membrane and be anchored there like a receptor molecule, but the significance of this is unknown. The IGF that promotes 3T3 growth, called IGF-1 or somatomedin C, is a protein of about 7 kdal. It is closely related to insulin and shares some of its biological effects (Figure 25-15 and Table 25-3). Insulin and IGF-1 are about 45 percent homologous at the amino acid level and can bind to each other's receptors on the cell surface. The affinity of the two proteins for each other's receptors is about a hundred- to a thousandfold

Figure 25-14

Diagram of mRNA for epidermal growth factor (EGF) as deduced from the nucleotide sequence of a cDNA clone of the mouse EGF mRNA. The boxed region represents the open reading frame that encodes the 1217-amino-acid-long preproEGF from which the 53-amino-acid-long EGF is proteolytically cleaved. Colored areas to the 5' side of EGF encode EGF-like sequences. [After A. Gray, T. J. Dull, and A. Ullrich, *Nature* 303 (1983):722–725; and J. Scott et al., *Science* 221 (1983):236–240.]

Figure 25-15
Sequence and schematic representation of the structure of IGF-1 and comparison to insulin. (a) Amino acids in white circles are the same as those in identical positions in the A and B chains of human insulin. (b) Representation of insulin is based on X-ray analysis of insulin crystals; representation of proinsulin and IGF-1 is based on model building. Note that insulin consists of an A and a B chain that are generated by proteolytic cleavage. [After T. L. Blundell, S. Bedarkar, S. Rinderknecht, and R. E. Hubel, *Proc. Nat. Acad. Sci.* 75 (1978)183.]

(a)

(b) Insulin   Proinsulin   IGF

less than for their own, but at high concentrations, insulin mimics IGF-1 and so can easily be mistaken for a growth factor.

In Vivo PDGF May Promote Wound Healing, While IGF Mediates the Action of the Pituitary Growth Hormone[35, 36, 37]

The ability of a polypeptide to stimulate growth of a particular cell type in vitro does not prove that it serves the same function in vivo. The role that many growth factors play in the whole organism is only beginning to be understood.

In vitro, platelet-derived growth factor (PDGF) is the major polypeptide mitogen in serum for cells of mesenchymal origin, such as fibroblasts, smooth muscle cells, and glial cells. In vivo, PDGF does not circulate freely in blood, but is stored in the α granules of blood platelets. During blood clotting and platelet adhesion, the contents of these granules are released, often at sites of injured blood vessels. PDGF probably has an important role in the repair of blood vessels. PDGF can stimulate the migration of arterial smooth muscle cells from the medial to the intimal layer of the artery, where they then proliferate as an early response to injury.

Insulin-like growth factor IGF-1 (somatomedin C) is a polypeptide growth factor that controls animal size and height. Growth hormone from the pituitary had long been known to control animal size, but its mechanism remained elusive. Then, when hypophysectomized animals (those from whom the pituitary gland had been removed) were treated with growth hormone, their serum was found to contain a factor that could stimulate the incorporation of SO_4^{-2} into chondroitin (cartilage) slices, as well as the oxidation of fat and the synthesis of DNA. Since growth hormone itself generates no such activity when added to serum from pituitary-free animals, searches were made for the intermediate. It turned out to be a factor of about 8 kdal and was first named somatomedin because of its ability to regulate the growth of most forms of somatic tissue (Figure 25-16). Somatomedin turned out to be a mixture of polypeptides. The polypeptides have significant homology to the pancreatic hormone insulin and so were renamed insulin-like growth factors, or IGFs. Besides stimulating somatomedin production, growth hormone may contribute to tissue growth by promoting the differentiation of certain types of precursor cells, which then become targets for somatomedin.

The in vivo role of EGF is unclear. It was first identified and purified on the basis of its effects on prenatal and neonatal mouse tissues, including ability to promote precocious eye opening and incisor eruption in mouse embryos. It also inhibits release of hydrochloric acid (HCl) from the intestinal mucosa. EGF is present at particularly high concentration in the submaxillary gland and in human urine.

Polypeptide Growth Factors Bind to Specific Cell Surface Receptors[38–45]

How polypeptide growth factors stimulate cell division has always been a major question in growth factor research. The process begins when the factors bind to specific cell surface receptors located at the plasma membrane. Most factors can exert mitogenic functions at very low concentrations, in the nanogram-per-milliliter range. To act at such low concentrations, growth factors have receptors of very high affinity. For a given growth factor, the number of receptors on the cell surface ranges from about 5000 to 500,000. Not every cell expresses the same receptors on its surface, which accounts for the specificity of growth factors for some cell types but not others. Growth factor receptors are frequently large glycoproteins. Recently, genes encoding a number of growth factor receptors and the related insulin receptor have been molecularly cloned (Figure 25-17).

Growth factor receptors usually have three identifiable domains (see Figure 25-17). There is an extracellular domain, usually highly glycosylated, to which the polypeptide hormone (the ligand) binds; a short hydrophobic domain, which is the transmembrane portion of the receptor; and an intracellular domain, which may be responsible for transmitting signals to the cell that a ligand has bound to the receptor. One model of receptor function holds that a receptor is similar to an allosteric enzyme: When the growth factor (allosteric effector) binds to the extracellular domain of the receptor, it causes a change in the three-dimensional structure such that the enzymatic activity of the receptor is activated in the cytoplasmic domain (Chapter 5).

In many cases, the enzymatic activity stimulated by growth factor binding is a tyrosine kinase activity. Upon ligand binding, the receptors for PDGF, EGF, IGF, and the hematopoietic factor CSF-1 all become phosphorylated on tyrosine residues located in their cytoplasmic domains (Figure 25-18). The kinase activity responsible for this

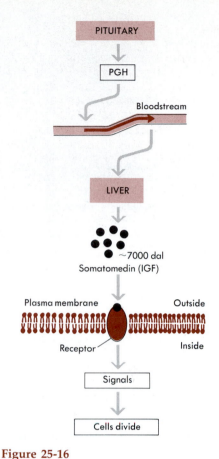

Figure 25-16
Schematic view of how the pituitary growth hormone (PGH) acts through somatomedins (IGFs). PGH also acts directly to increase the number of target cells that respond to somatomedins.

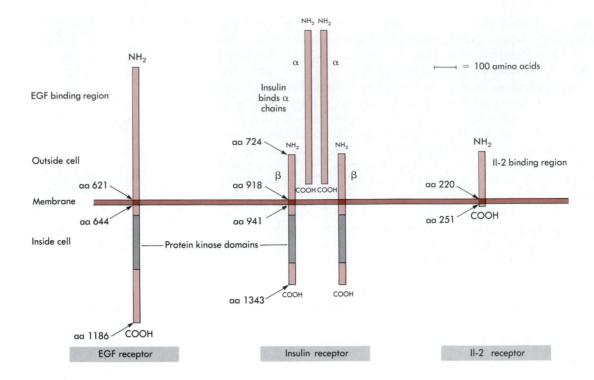

Figure 25-17
Diagrams of the EGF, insulin, and Il-2 cell surface receptors based on nucleotide sequences of cDNA clones of their mRNAs. All three are glycoproteins. The EGF and insulin receptors have protein (tyrosine) kinase activity in their cytoplasmic domains.

Figure 25-18
Phosphotyrosine.

phosphorylation is in the receptor itself. Thus, the reaction is autocatalytic. This result is in striking contrast to most cellular protein phosphorylations, which occur on serine or threonine rather than tyrosine residues.

Growth factor receptors that possess tyrosine kinase activity may be capable of phosphorylating other cellular proteins as well as themselves. Although this has not yet been directly demonstrated, the number of cellular proteins phosphorylated on tyrosine dramatically increases after the addition of growth factors whose receptors have tyrosine kinase activity.

After binding, the receptor–growth factor complex is internalized, transported to lysosomes, and degraded (Figure 25-19). By this mechanism, occupied receptors are cleared from the cell surface. With time, new receptors are synthesized and expressed on the cell surface. It is thought that the important signals resulting from growth factor–receptor binding are not the result of internalization of the receptor-ligand complex but rather are transmitted from the cell surface. This is because in some cases, antibody to a receptor can mimic the effect of the receptor's normal ligand. However, there is no definitive evidence that would rule out the possibility that some of the effects of growth factors result from the internalization and degradation of the factor-receptor complexes.

Growth Factor–Receptor Interactions Have Pleiotropic Effects on Cells, Including Changes in Gene Expression[34, 46–49]

The responses of cells to growth factors are numerous. Table 25-4 lists some responses of fibroblasts to the platelet-derived growth factor. Some effects are "immediate," being detectable within seconds or minutes of growth factor addition. These changes include rapid ionic changes, cytoskeletal redistributions, protein phosphorylations, membrane lipid metabolism changes, and alterations in the protein synthetic machinery. Other effects of growth factors are not quite so rapid. These are the so-called early events, which can be measured within the first few hours of growth factor addition. They include induction of specific new gene transcription and expression of new proteins.

Although different growth factors elicit different cellular responses in a given cell, there are many overlapping responses as well. For example, in 3T3 cells, insulin, IGF-1, EGF, and PDGF all stimulate the phosphorylation of the ribosomal protein S6. However, only PDGF leads to a reorganization of the actin filaments.

A single growth factor can elicit different responses from different cell types. Furthermore, in addition to stimulating cell growth, some growth factors can trigger cellular differentiation, enhance cell survival, initiate cell migration, and stimulate the secretion of tissue-specific hormones from differentiated cell types.

It is estimated that the early PDGF induced genes are about 0.1 to 0.3 percent of the total genes transcribed in 3T3 cells. This would correspond to roughly 10 to 100 genes. An important question is, Which of these genes are critical to the mitogenic response? A strong candidate for such a gene has come from the finding that a variety of mitogens stimulate expression of a cellular gene, called c-*myc*, in several different cell types. An increase in c-*myc* mRNA is seen after PDGF treatment of quiescent 3T3 cells, after con A treatment of T lymphocytes, after lipopolysaccharide treatment of B cells, after EGF treatment of an epidermal carcinoma cell line, and as an early response in liver regeneration. And as we shall learn in Chapters 26 and

Figure 25-19
Binding and internalization of EGF monitored with fluorescein-labeled EGF on A-431 cells. Cells were incubated with fluorescein-conjugated EGF for 45 minutes at 6°C, washed to remove unbound EGF, then either fixed (a and b) or incubated at 37°C for 2 minutes (c and d), 10 minutes (e and f), or 20 minutes (g and h) and then fixed. Photos (a), (c), (e), and (g) are fluorescence micrographs taken under epi-illumination. Photos (b), (d), (f), and (h) are photographs of the same cells by Nomarski optics. [Courtesy of H. Haigler, J. F. Ash, S. J. Singer, and S. Cohen, *Proc. Nat. Acad. Sci.* 75 (1978):3317.]

Table 25-4 The Mitogenic Response to PDGF

Immediate Events: 1–10 min and/or Transcription-Independent	Early Events: 30–180 min and/or Transcription-Dependent
1. Tyrosine-specific phosphorylations	1. PDGF can be removed from culture medium and cells still divide
2. Inhibition of EGF binding	2. Acquisition of intracellular signal as shown by cell fusion
3. Inhibition of guinea pig insulin binding	3. Induction of rare gene sequences encoding low-abundance mRNAs
4. Stimulation of phospholipase A_2 and prostaglandin release	4. Appearance of rare cytoplasmic proteins
5. Stimulation of polysome formation	5. Stimulation of somatomedin binding
6. Stimulation of phosphatydylinositol turnover	6. Increase in low-density lipoprotein receptor content
7. Reorganization of actin filaments	7. Stimulation of amino acid A transport system

SOURCE: C. D. Stiles, "The Molecular Biology of Platelet-Derived Growth Factor," *Cell* 33 (1983):653–655.

27, c-*myc* expression is altered in many types of cancer cells, strongly implicating c-*myc* as an important gene in cell growth control.

Further experimental confirmation that the c-*myc* gene is at least partly responsible for the mitogenic response of certain growth factors comes from the observation that microinjection of *myc*-encoded protein into quiescent 3T3 cells causes them to enter the cell cycle even in the absence of PDGF. However, just like cells treated with PDGF, the presence of the progression-type growth factors present in platelet-poor plasma is also needed for entry of the microinjected cells into S phase. The *myc* gene product is located in the nucleus, but how it may function to make cells competent to respond to progression factors is not known.

Another gene whose expression increases rapidly in response to PDGF is c-*fos*. Like c-*myc*, c-*fos* encodes a nuclear protein and is also known to be involved in the cancerous transformation of cells (Chapter 26).

What Are the Second Messengers for Growth Factor Receptors?[50–57]

At least three types of mechanisms are now being considered to explain how growth factor–receptor interactions exert such pleiotropic effects. These include the generation of ion fluxes, the activation of protein kinases, and the enzymatic generation of small hydrophilic molecules that can freely diffuse away from the plasma membrane. That these mechanisms are operating as growth factor second signals or messengers is suggested by the observations that experimental perturbation of these metabolic pathways can have dramatic effects on cell growth.

T lymphocytes can be activated independently of antigen or concanavalin A treatment by simultaneous treatment of the cells with calcium ionophores and phorbol esters. Calcium ionophores allow calcium to cross the cell membrane, increasing the intracellular concentration of calcium. The increased calcium then activates the calcium binding regulatory protein calmodulin. Calmodulin, with its bound calcium, can then bind to and regulate the activity of many different cell proteins, including kinases. That such calcium fluxes are physiologically relevant to the growth of T cells is further indicated by

Figure 25-20
Phosphatidylinositol turnover leading to the appearance of diacylglycerol and inositol triphosphate, two second messengers for hormone stimulation. Extracellular signals (e.g., hormones or factors) may transmit signals that alter cell behavior by stimulating the turnover. Inositol triphosphate causes release of Ca^{2+} from intracellular stores and also activates a kinase. Diacylglycerol activates protein kinase C.

recent findings that activation of T cells by antigen opens up calcium channels in the cell membrane.

As already mentioned, calcium ionophores are not sufficient for the mitogenic activation of T lymphocytes. They must be used in conjunction with phorbol esters to achieve full mitogenic effect. Phorbol esters are compounds extracted from croton oil that have long been known to act as tumor promoters (agents that in combination with other so-called initiating carcinogens can cause cancer). Phorbol esters have a wide variety of effects on a great many cell types. In some fibroblasts, phorbol esters have a mitogenic effect very similar to that of PDGF. They even activate the cellular *myc* gene. At least some of their effects on cells are mediated by binding to and activating the calcium-dependent, lipid-dependent **protein kinase C** at the cell membrane. This protein kinase has a specificity for serine and threonine residues on protein substrates. Protein kinase C is found in most mammalian tissues, but its physiological substrates are largely uncharacterized.

The normal endogenous cellular activator of protein kinase C is diacylglycerol (Figure 25-20). Diacylglycerol is produced transiently in cell membranes as a result of the breakdown of phosphatidylinositol. Phosphatidylinositol itself is a minor membrane phospholipid whose breakdown is catalyzed by a membrane-bound phospholipase C activity (Figure 25-20). Phospholipase C is in turn activated by a variety of extracellular stimuli, many of which are growth factors. Thus, PDGF treatment of fibroblasts, antigen or lectin treatment of T cells, and fertilization of eggs all lead to an activation of phospholipase C and a concomitant rapid turnover of membrane phosphatidylinositol.

Figure 25-21
Synthesis and degradation of cAMP.

The diacylglycerol released from the breakdown of phosphatidylinositol is then able to transiently activate protein kinase C. By this mechanism, protein kinase C is thought to act as one of the signal transduction pathways for polypeptide growth factors.

The other product formed by the breakdown of phosphatidylinositol is the hydrophilic group inositol triphosphate (Figure 25-20). This compound diffuses away from the plasma membrane and also acts as a second messenger. Inositol triphosphate can provoke the release of intracellular calcium from the endoplasmic reticulum. The result is a transient increase of the calcium concentration within the cell and indirect effects on other ion fluxes as well. Inositol triphosphate also seems to be able to activate cellular protein kinases.

Another important second messenger for polypeptide hormones is *cyclic AMP (cAMP)*. Some polypeptide hormones can exert their effects by either activating or inactivating the inner plasma membrane–bound enzyme *adenylate cyclase* (Figure 25-21). This alters the level of cAMP inside the cell, which then determines the activity of yet another set of kinases known as the *cAMP-dependent protein kinases*. These enzymes transfer phosphate from ATP to the serine or threonine residues of a small group of specific target proteins. These proteins, as a result of being phosphorylated, can be either activated or inactivated and so bring about changes in cellular metabolism, including changes in gene expression. The level of cAMP also regulates the enzyme *phosphoprotein phosphatase*, which dephosphorylates the substrates of the cAMP-dependent protein kinases. Apparently, different cells can express different sets of kinases or different sets of target proteins, so changes in cAMP levels can have very different effects on different cell types.

The effect of hormone receptor complexes on adenylate cyclase is indirect. It is mediated by a third membrane protein called the GTP binding protein, or *G protein*. G protein binds GDP on a cytoplasmic face until it collides with a specific hormone-receptor complex. Then it is activated, exchanges its GDP for GTP, and as a result binds to and alters the activity of adenylate cyclase. The reaction terminates when a GTPase activity intrinsic to the G protein hydrolyzes the GTP back to GDP. The activation of phospholipase C by ligands is also thought to be mediated by a G-type protein.

Despite much effort and precedent from many biological systems, efforts to relate cAMP levels to growth factor–induced responses have not yet yielded clear-cut answers.

A Different Mechanism of Action for Steroid Hormones[58, 59]

The mechanism of action of steroid hormones is different from that of the polypeptide growth factors, even though the types of changes they induce in cells can appear similar. The steroid hormones, such as estrogen, do not have any direct effect on cyclic nucleotide levels, nor are their receptors latent kinases. Their hydrocarbon nature allows them to pass through the plasma membrane to interact with their specific protein receptors, which are located inside the cells. The resulting steroid-receptor complexes then combine with specific chromosomal sites. The activation of specific RNA sythesis that follows is the direct result of the binding of the receptor-steroid complex to chromatin.

Molecular Cloning of the Polypeptide Factors That Stimulate Growth, Maturation, and Function of Blood Cells[60-66]

Most of the red and white cells that circulate in the blood have a limited life span. They are constantly replaced through the division of cells located in the bone marrow. The process of blood cell formation is called **hematopoiesis,** and its scale is enormous. For example, red cells are formed at a rate of more than 100 million cells a minute in an adult human. There are many different types of blood cells belonging to distinct cell lineages (Figure 25-22). Along each lineage are cells at different stages of maturation, with many of the different types existing in close proximity within the bone marrow.

All the different types of blood cells arise from a small population of stem cells formed early in the embryo. The existence of these **pluripotent stem cells** can be demonstrated by heavily irradiating a mouse so that it can no longer form its own blood cells. The animal will die if left untreated; but it can be saved by injection of bone marrow cells from a second mouse. A tiny fraction of the donor cells settle in the bone marrow of the recipient animal and differentiate to produce cells belonging to all the blood cell lineages. If the donor cells are marked by lightly irradiating them to introduce random chromosome breaks into different cells, a specific chromosome abnormality can sometimes be found in cells of all lineages within a recipient animal. This result indicates that all the cell types can arise from a single donor cell.

Pluripotent stem cells injected into an irradiated recipient can migrate to the spleen and give rise to colonies containing cells of the different blood cell lineages. Hence, they are called **colony-forming units–spleen** (**CFU-S**). As can be seen in Figure 25-22, pluripotent stem cells give rise to **progenitor cells** that will yield a more limited array of cell types. A major breakthrough in the study of hematopoiesis was the finding, beginning about 20 years ago, that many types of progenitor cells will grow and differentiate in vitro. Progenitor cells are named according to the type of progeny they yield. For example, a colony-forming unit–erythroid (CFU-E) is a cell capable of forming a colony containing only cells of the erythroid lineage.

Maintaining the correct ratios of blood cell types in an animal would seem to be an awesome task. Only rarely does the process become deranged and result in illnesses like anemia or leukemia. Some of the mechanisms that control hematopoiesis involve cell-cell interactions within the complex cellular architecture of the bone marrow. In addition, a group of glycoprotein factors have been shown to control cell growth, commitment to differentiation, and the functioning of hematopoietic cells.

Factors that stimulate hematopoietic cells have been identified by removing the supernatant fluids from different cultured cell types and showing that they can promote the growth or function of a particular type of blood cell in vitro. Often, the factors were named first according to the biological effect they elicit: for example, mast cell growth factor, macrophage-granulocyte inducer, macrophage-inhibiting factor, and thymocyte mitogenic factor. In many cases, the cells producing the factors and the target cells the factors affect are not pure populations. It is not always possible to be certain how many

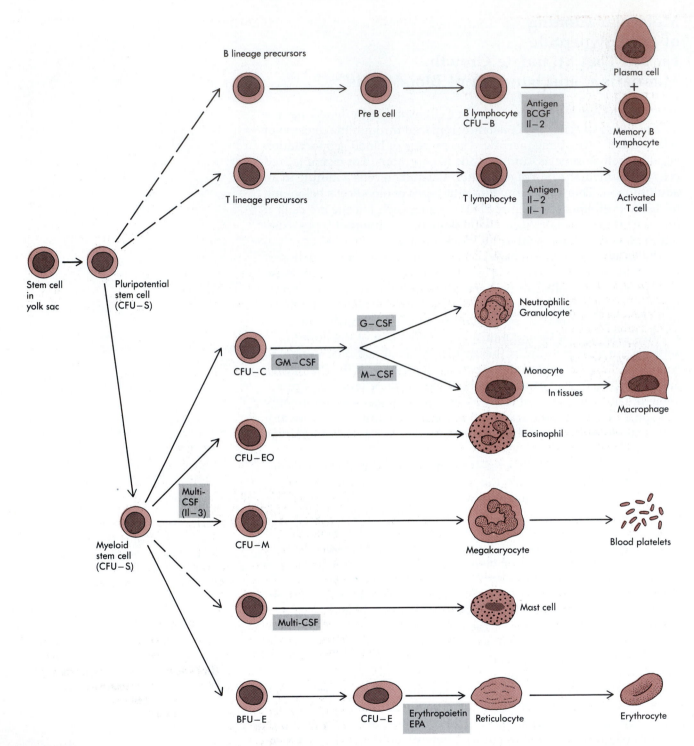

Figure 25-22
Schematic representation of hematopoiesis. Stages thought to be regulated by particular growth factors are indicated. The scheme is still undergoing revisions and additions.

different factors are present in the culture fluids or which cell type produces them. Gradually, researchers have managed to purify hematopoietic growth and differentiation factors and more recently to molecularly clone the genes that encode them. Since some factors are active at low concentrations and since it is difficult to purify the proteins biochemically, the ability to clone the corresponding genes and

Table 25-5 Protein Factors That Stimulate and/or Are Made by Blood Cells

Name	Structure	Biological Properties
Il-2 (interleukin 2; also known as T cell growth factor, TCGF)	133 amino acids; MW 15,000	Proliferation of activated T cells; also stimulates B cells and thymocytes; made by activated T cells
Multi-CSF or Il-3 (multi-lineage colony stimulating activity; also known as BPA, HCGF, MCGF, and PSF)	133 amino acids (mouse) (from 166-amino-acid precursor); glycoprotein	Stimulates growth of stem cells, myeloid progenitor cells of all types, and mast cells
GM-CSF (granulocyte-macrophage colony-stimulating factor)	118 amino acids (mouse); glycoprotein; MW 23,000	Needed continuously for proliferation and differentiation of progenitors of granulocyte-macrophages
G-CSF (granulocyte colony-stimulating activity)	glycoprotein; MW 25,000	Stimulates formation of granulocytes
M-CSF (macrophage colony-stimulating factor; also known as CSF-1)	glycoprotein; MW 70,000	Stimulates production of macrophages from bipotential cells
Erythropoietin (Epo)	166 amino acids; glycoprotein	Growth and differentiation of erythrocyte precursors
Il-1 (interleukin 1); defined in mouse, same as Il-1α of human	271 amino acids; may be processed to active 159-amino-acid form	Diverse effects; stimulates Il-2 release, B cell growth, prostaglandin release; may be endogenous pyrogen
Lymphotoxin	171 amino acids; forms multimers; glycoprotein	Released by activated lymphocytes; antitumor activity
TNF (tumor necrosis factor; cachectin)	157 amino acids; forms multimers; glycoprotein; 30% homology with lymphotoxin	Produced by activated monocytes; antitumor activity; cachectin inhibits enzymes of fat utilization; causes emaciation
γ-interferon	146 amino acids; glycoprotein	A lymphokine with antitumor activity; member of a family of proteins with antiviral activity

then to produce purified proteins by recombinant DNA technology is a major advance in the field. Often, factors that were previously assigned different names turn out to be identical. Combined with systems for in vitro cell growth, it is at last becoming possible to develop clean systems to study the function and mode of action of many of the factors that influence the behavior of blood cells. A list of some of the cleanly identified hematopoietic growth factors is presented in Table 25-5.

Erythropoietin Induction of Red Blood Cell Production[67, 68]

The level of red blood cells (erythrocytes) and hence the level of oxygen in the blood is under the specific control of **erythropoietin,** a glycoprotein with a molecular weight of about 37 kdal. Somehow, cells in the kidney and liver that synthesize erythropoietin can detect a drop in the level of erythrocytes in the blood and respond by releasing the factor. Erythropoietin raises the red cell number by inducing the selective proliferation and differentiation of erythrocyte precursors. An early precursor that will grow in vitro in response to high concentrations of erythropoietin plus a second factor, erythroid-potentiating activity (EPA), is called an **erythrocyte burst-forming unit,** or **BFU-E** (Figure 25-22). The BFU-E undergoes about 12 divisions before it becomes a mature erythrocyte. **Erythrocyte colony-forming units (CFU-Es)** can also grow in vitro and respond to lower concentrations of erythropoietin. These cells are only about 5 cell divisions from a mature erythrocyte. The erythropoietin-stimulated cascade of divisions produces an ever-increasing number of red blood cells from a small number of progenitors. It is estimated that in vivo one BFU-E can give rise to 2^6 CFU-Es and 2^{11} mature erythrocytes!

An early effect of erythropoietin on cells is the selective stimulation of RNA synthesis. It occurs within 1 hour of erythropoietin addition and involves synthesis of tRNA, rRNA, and mRNA. The mRNA molecules that code for globin itself appear later, indicating that hemoglobin formation is the end result of an involved differentiation process that leads not only to hemoglobin formation but also to the mitotic events that increase the number of erythropoietin-sensitive cells.

The gene for erythropoietin has been cloned, and a cDNA clone of its mRNA has been isolated and sequenced. The first step in the cloning was to purify erythropoietin from human urine. Then the protein was digested with trypsin, and the amino acid sequence of tryptic fragments was determined. These sequences made it possible to synthesize corresponding oligonucleotide probes that could hybridize to the erythropoietin gene, and these probes were used to select the gene from a library of human DNA.

Glycoprotein Factors That Control the Production, Differentiation, and Function of Granulocytes and Macrophages[69, 70]

The granulocyte and macrophage components of blood form by a differentiation-mitotic process starting from a common bipotential precursor cell in the bone marrow (Figure 25-22). These events can be studied in vitro, where the progenitor cells will grow into clones in semisolid media if they are given specific protein inducers, called **colony-stimulating factors** (**CSFs**). Progenitor cell division and survival is completely dependent on the continued presence of one or another of the CSFs.

So far, four CSFs from mice have been purified—G-CSF, M-CSF, GM-CSF, and multi-CSF—and DNA encoding all of these have been cloned (see Table 25-5). All four factors are glycoproteins, and so far they show no amino acid sequence homology to one another. Each factor binds to a specific receptor. A given granulocyte-macrophage progenitor cell can respond to more than one CSF, with different cells requiring different concentrations of the factors for proliferation. Although the factors were discovered because they promote cell proliferation, they also have three other important effects on responsive cells: maintenance of cell survival, promotion of differentiation commitment, and stimulation of differentiated cell functions. How the factors influence differentiation may be concentration-dependent. For example, a high concentration of GM-CSF causes a bipotential cell to enter the granulocytic pathway, while a low concentration favors formation of macrophages. CSFs can stimulate the functions of mature cells, for example, phagocytosis of bacteria by granulocytes and macrophages. How survival, proliferation, differentiation, and function are separately controlled by these factors is not yet understood.

At least some CSFs have additional targets besides granulocyte-macrophage precursors. In particular, multi-CSF, also known as interleukin 3 or Il-3, can stimulate proliferation and differentiation of multipotential stem cells leading to the production of all the major blood cell types.

Many different cell types seem to be capable of making one or another CSF, but the biologically important sources and the mechanisms that ensure correct CSF levels in vivo are not yet known.

T Cell Growth Factor (Interleukin 2) Stimulates Growth of Activated T Lymphoid Cells[71-75]

For a long time, it was thought that mitogen (antigen) alone could stimulate mature T or B cells to divide in the presence of standard media containing serum; but now it is believed that additional specific polypeptide factors are also needed. One example is the **T cell growth factor (TCGF),** called *interleukin 2* (Il-2), which is required for the proliferation of mature T cells. The term *interleukin* denotes factors made by leukocytes (white blood cells) that also act on leukocytes. Interleukin-2 can also be classified as a *lymphokine,* a term used to describe factors made by lymphocytes, one type of white blood cell.

Mature T cells taken from peripheral blood will grow indefinitely in culture if they are stimulated with antigen (mitogen) and supplied with Il-2. Antigen stimulates the appearance of Il-2 receptors, and it also stimulates the synthesis of some Il-2. This explains why antigen alone can appear to be a complete mitogen for T cells (Figure 25-23). How antigen binding to its receptor (the T cell antigen receptor discussed in Chapter 23) triggers expression of Il-2 receptor genes and Il-2 is not known, but, as noted, an early event following antigen binding may be the change in the level of intracellular calcium.

Both Il-2 and its receptor have been molecularly cloned (see Figure 25-17). How Il-2 binding triggers T cell growth is unknown. The receptor has such a small cytoplasmic domain (13 amino acids) that it is doubtful that there is an intracellular enzymatic activity intrinsic in the cytoplasmic portion of the molecule, such as the kinase activity associated with the cytoplasmic portion of the EGF, PDGF, or insulin receptors.

Figure 25-23
Stimulation of T cell growth by antigen and Il-2. Antigen (often a preparation of heterologous spleen cells) induces expression of Il-2 receptors.

T lymphocyte
T cell antigen receptor
Il–2 receptor
Antigen +MHC
Il–2
Antigen-presenting cell
Antigen stimulation of T cell
Activated T cell
Proliferating T cells

Transforming Growth Factors Permit Cells to Grow Without Attachment[76, 77]

Normal fibroblasts or 3T3 cells must be attached to a surface to grow in vitro. Some transformed cells produce factors that can reversibly allow fibroblasts to grow in suspension. Since the ability to grow in suspension is often associated with the tumorigenic phenotype, these factors are called **transforming growth factors.** Two types have been separated: TGF-α and TGF-β. TGF-α is structurally related to the epidermal growth factor (EGF) and binds to the same receptor. EGF can substitute for TGF-α in culture. TGF-β is a distinct growth factor that, like PDGF, is found in platelets; yet, it binds to a receptor distinct from PDGF, EGF, or IGFs. TGF-α and TGF-β are not by themselves sufficient to support the growth of fibroblasts in suspension: PDGF is required along with the TGFs.

Do Negative Growth Factors Exist?[78–81]

Although much cell proliferation in vivo may be regulated by controlling the availability of growth factors to target cells, in some cases it seems that it would be useful for an organism to be able to shut off cell proliferation directly. This idea has prompted the search for physiological factors that may prevent cells from dividing. The identification of such factors has been arduous and often controversial because of the difficulties in distinguishing between a negative growth regulator and a compound that simply has a toxic effect on the cell. However, there is now widespread agreement that negative growth regulators do exist.

Astonishingly, TGF-β is a growth inhibitory factor. Although TGF-β can enhance the growth of fibroblasts in culture, it can be inhibitory to other cell types. TGF-β inhibits the growth of some epithelial cell lines in culture. Interestingly, it is also able to inhibit the growth of various cancer cells in culture, raising hopes that it might have use in the treatment of cancer.

Another negative growth factor may be γ-interferon, a member of the interferon family of proteins whose synthesis is induced in response to viral infections (Chapter 24). One effect of interferon is to cause target cells to stop cycling and make them refractory to the actions of mitogenic stimuli. This growth-arrested state is reversible, and the cells return to their normal growth factor responsiveness a short time after removal of interferon.

What Are Lymphotoxin and Tumor Necrosis Factor (Cachectin)?[82–85]

Sometimes, cancer patients that get severe bacterial infections experience tumor regressions, and this led to attempts, almost a hundred years ago, to cure cancer by injections of bacteria or bacterial toxins. A paradox was that these substances did not destroy tumor cells in vitro. The reason became clear when it was found that they probably work indirectly by inducing the synthesis of naturally occurring proteins with antitumor activity. Mice or rabbits injected with an immunostimulatory bacterium (*Mycobacterium bovis* strain bacillus Calmette-Guerin, or **BCG**) and subsequently with endotoxin produce a protein in their serum called **tumor necrosis factor (TNF)**, which preferentially kills tumor cells both in vitro and in vivo. Tumor necrosis factor is produced by macrophages. A second factor, a lymphokine with similar properties, is produced by mitogen-stimulated lymphocytes and is called **lymphotoxin.** Complementary DNA clones for both substances have been obtained and sequenced (see Table 25-5). Human TNF and lymphotoxin share about 30 percent of their amino acid sequences.

An intriguing finding is that tumor necrosis factor appears to be identical to a hormone called **cachectin,** also produced by stimulated macrophages. A molecule believed to be mouse cachectin has recently been purified and partially sequenced, and its amino acid sequence is almost identical to the corresponding region of human TNF. Furthermore, the substance has potent TNF activity. Cachectin suppresses enzymes involved in fat storage and breakdown in cultured adipocytes and may be responsible for the hypertriglyceridemic state sometimes seen in animals during infections. Cachectin has also been implicated as the cause of the starvation-like appearance of cancer patients. Clearly, cachectin must affect normal as well as tumor cells, and in fact, both types of cells have receptors for cachectin. Why cachectin preferentially kills cancer cells is unclear. Whether TNF (cachectin) or lymphotoxin will be effective in cancer treatment should soon be determined, now that large amounts of pure material can finally be made by recombinant DNA technology.

Specificity of the Nerve Growth Factor for Sympathetic Neurons[86]

Growth of the neurons of the sympathetic nervous system requires a specific factor called the **nerve growth factor** (**NGF**), a protein with a molecular weight of 26,500 daltons that is composed of two identical polypeptide chains. NGF can be isolated both from a variety of tissue sources and from serum, but has so far been found in largest amounts in certain snake venom and the salivary glands of mice. Antibodies against NGF, when injected into mice, cause the specific degeneration of the sympathetic nervous system, showing that it plays a vital role in the maintenance of the nervous system. In vitro studies, using sympathetic ganglion, show that addition of NGF quickly stimulates a variety of anabolic processes that lead to outgrowth of neurites from the ganglionic masses (Figure 25-24). NGF binds only to surface receptors found on sympathetic nerve cells and shows no affinity for receptors isolated from a variety of other cell types (e.g., heart or kidney). Its extreme tissue specificity is the result of the localization of its receptor to one very specific type of nerve cell.

Examination of the amino acid sequence of NGF shows unmistakable homologies to the sequence of the insulin precursor proinsulin, suggesting a common evolutionary origin (Figure 25-25). NGF, however, shows no affinity for the insulin receptor, nor does insulin bind to NGF receptors.

Growth Signals from Extracellular Matrix and Cell Shape[87–90]

To form tissues in vivo, cells must form contacts with other cells or with an extracellular matrix, or both. Extracellular matrices such as the basal lamina, bone, and cartilage are secreted cellular products.

(a)

(b)

Figure 25-24
Outgrowth of neurites from sympathetic ganglion cells upon stimulation by the nerve growth factor. (a) Chick embryo ganglion in a control medium. (b) Ganglion in a medium supplemented with salivary NGF at a concentration of 0.01 μg/ml. [Reproduced with permission from R. Levi-Montalcini, *Science* 143 (1964):105.]

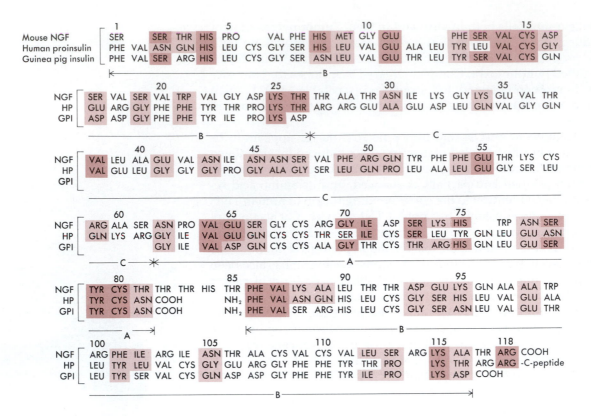

Figure 25-25

The sequence homology between nerve growth factor (NGF) and insulin. The amino acid sequence of the primary subunit of mouse nerve growth factor is aligned with that of human proinsulin and guinea pig insulin. Solid color blocks indicate sets of identical residues, and shaded color blocks indicate favored amino acid replacements (one base change in codon). The letters beneath the rows indicate the limits of each A, B, and C chain of insulin. Mature insulin is composed of an A and B chain joined by disulfide bonds. The similarities in primary sequence also lead to a similar three-dimensional conformation for the two molecules. [After W. A. Frazier, R. H. Angeletti, and R. A. Bradshaw, *Science* 176 (1972):483.]

They consist of complex and variable arrays of collagens, glycosaminoglycans, proteoglycans, and glycoproteins. The extracellular matrix provides the bulk, shape, and strength of many tissues, and it feeds back upon the behavior of the cells it surrounds, influencing their shape, positioning, metabolism, state of differentiation, and division.

In vitro, many animal cells can only metabolize and divide if they are spread out; and to spread out, they must first attach to a surface. Some cells can attach and spread well on glass or certain treated plastics. Others will only spread out well on protein-coated surfaces or even on a basal lamina. A much-studied attachment protein secreted by animal cells is **fibronectin,** a large, ubiquitous glycoprotein that is a prominent constituent of extracellular matrices in vivo as well as a major serum protein. Fibronectin-coated surfaces are ideal for many cells to attach to, and addition of fibronectin to rounded cells in culture can cause them to flatten out. The flattening of the cells is also related to the return of an organized cytoskeleton, and so the extracellular matrix and the intracellular cytoskeleton are somehow connected through the plasma membrane.

Both attachment and shape seem to be important for the growth of anchorage-dependent cells. Controlled shape changes can be induced in normally flat cells by altering the adhesiveness of the surface to which they attach. When this is done, untransformed endothelial cells and fibroblasts stop growing when only slightly rounded; in contrast, transformed lines can be more rounded before growth stops. Forcing normally flat cells to be round has specific effects on cell metabolism. Fibroblasts placed in suspension undergo a shutdown of DNA, RNA, and protein synthesis. Understanding how attachment and shape can regulate a cell's ability to receive growth signals may be of great importance in understanding cancer, where such control is frequently lost. In some cases, the loss may be related to the fact that some cancer cells secrete TGFs that somehow allow cells to bypass the requirement for anchorage.

Conversion of Fibroblasts into Adipose Cells Is Enhanced by Azacytidine[91, 92, 93]

The conversion of fibroblasts into adipose (fat) cells depends on two stimuli: polypeptide factors, called adipogenic factors, and conditions where cell division is not possible (e.g., excessive crowding). While multiplying, 3T3 cells always maintain a fibroblast shape. But upon reaching confluence, they have an increasing probability of converting into very large, round adipose cells (Figure 25-26). Not all arrested fibroblasts become so differentiated, but certain 3T3 strains (preadipose strains) have high probabilities of undergoing the adipose conversion. It is lack of growth, not contact between arrested cells, that is important for fat accumulation. This is shown by suspending single 3T3 cells in methylcellulose gels. Under such conditions, they remain small and cannot make new DNA or divide. If the cells differentiate into adipose cells, they enlarge and synthesize neutral fat (triglyceride) until the center of the cell is filled with a huge fat droplet, which pushes the nucleus into an eccentric position near the cell membrane (Figure 25-27). As long as the adipose cell maintains its huge fat droplet, it is incapable of cell division. But under the right conditions, young adipose cells can lose their triglyceride content and regain their fibroblast appearance. Then they are capable of further cell division.

While cessation of cell division is a requirement for the fibroblast-to-fat cell conversion, it is not a sufficient signal. Conversion also requires a polypeptide with the properties of an adipogenic factor. One such factor is growth hormone.

An intriguing nonphysiological stimulus of the fibroblast-to-fat cell conversion is azacytidine (aza C), which can substitute for cytosine residues in cellular DNA. Treatment of embryonic fibroblasts with aza C leads to the appearance of fat cells, as well as muscle cells and chondrocytes. The mechanism for this effect is thought to be a change in gene expression brought about by a change in methylation patterns of the DNA. DNA that is substituted with aza C cannot be properly methylated, and the extent of methylation is known to be correlated with gene expression (Chapter 21). Which particular genes are important in bringing about the conversion of fibroblasts to fat cells remain to be determined.

Maintenance of Myoblasts in Continuous Cell Culture[94]

It is possible to routinely culture **myoblasts,** the cellular precursors of the multinucleated striated muscle cells. Although myoblasts appear essentially undifferentiated, in culture they usually divide only several times before they begin to aggregate and fuse into postmitotic, multinucleated muscle cells that soon acquire a cross-striated appearance and develop a contractile potentiality (Figure 25-28). By this stage, the differentiation process is irreversible, and no further DNA synthesis is possible. The resulting multinucleate cells, however, retain the capacity to make the mRNA for the many proteins whose continued synthesis is essential for normal muscle functioning.

Figure 25-26
Formation of fat cells within a confluent layer of mouse fibroblasts (3T3 L1). Three weeks after the colonies became confluent, the culture was stained with Oil Red O (shows black in this photograph), which selectively stains fat deposits red. The marked variation in size and staining intensity reflects differences in the proportion of cells undergoing adipose conversion in the different colonies. [Reproduced with permission from H. Green and M. Meuth, *Cell* 3 (1974):127.]

Figure 25-27
A thin section of a mature adipose cell, showing a thin cytoplasmic rim surrounding the large central fat droplet. The nucleus has the characteristic eccentric location of adipose cells. [Electron micrograph by Elaine Lenk of MIT; reproduced with permission from H. Green and M. Meuth, *Cell* 3 (1974):127.]

Figure 25-28

Fusion of myoblasts and the development of striated muscle fibers set within colonies of multiplying myoblasts. (a) View of a Petri dish upon which 150 myoblasts had been plated 13 days previously. (b) Higher magnification of a portion of a single colony, showing large numbers of nuclei surrounding each developing fiber. (c) A still higher magnification, showing the striated nature of the individual fibers. [Reproduced with permission from D. Yaffee, *Proc. Nat. Acad. Sci.* 61 (1968):477.]

Such irreversible differentiation, however, is not a necessary fate for all dividing myoblasts. Circumstances have been found that can maintain myogenic cells in continuous cultures. The key elements are conditions that inhibit the cell-fusing process (e.g., low Ca^{2+} or periodic additions of trypsin), together with employment of highly enriched serum of fetal origin. Thus, the transmission of a specific differentiation potential to progeny cells is not necessarily dependent on a visible morphological expression. So all the descendants of a single cloned myoblast can give rise to muscle-forming colonies.

Skin in Culture[95, 96]

It is now possible to grow epidermal and other types of epithelial cells in culture. Keratinocytes derived from any type of stratified squamous epithelium, those of the epidermis, cornea, mouth, tongue, etc., will divide in vitro to form colonies and eventually a continuous epithelium (Figure 25-29). One of the tricks to getting the cells to grow was the addition of irradiated fibroblast feeder cells that presumably supply needed growth factors. Also important was to add agents that increase the intracellular content of cAMP.

In real skin, keratinocytes in the basal layer divide, and one daughter cell moves up and out of the layer to begin terminal differentiation. The cells enlarge as they differentiate, and they alter the pattern of keratins they synthesize. Finally, as they reach the outer layers of the skin, the cells destroy most of their proteins and organelles, including the nucleus, and become flattened **squames** packed with keratin fibrils. Squames are ultimately sloughed off from the outer skin layers. A very similar process occurs in vitro, except that the final stages involving loss of the nucleus do not occur unless special steps are taken. One way to ensure that complete differentiation occurs is to detach the cells and place them in a suspended state. Another method is to reduce the level of vitamin A in the medium. Yet another method is to enzymatically detach the entire sheet of cells from the plastic surface of the tissue culture dish and graft it onto an animal.

The ability to lift up sheets of cultured skin cells and graft them has led to the remarkable use of tissue-culture-grown skin to treat people with severe burns. Because skin cells grow so well in culture, it is possible to take a small piece of an individual's skin and within two or three weeks, grow enough skin in the laboratory to cover the entire body. Since the skin is derived from the patient's own cells, there is no danger of immune rejection. However, the skin is not completely normal, since it lacks hair follicles and sweat glands.

CANCER CELLS

Searching for Chemical Differences Between Normal and Cancer Cells[1]

Almost as soon as biologists began to describe the molecules within normal cells, they looked to see whether those same molecules could be found in cancer cells. Likewise, when a new chemical reaction (or enzyme) was described in a cancer cell, a search was often made to see if the same reaction occurred in the normal cell. Often, these searches initially seemed to lead somewhere, with the tumor cells containing much more (or less) of a particular compound than their normal equivalents. Further analysis, however, usually failed to justify the initial hope that the respective change was at the heart of the cancerous phenotype. A fundamental difficulty in this approach has always been the inability to decide whether a given change is the primary metabolic disturbance that creates the cancer, or whether it is a minor secondary response to the altered metabolism caused by the primary change. Moreover, in many such experiments, it is impossible to select a good control with which to compare cells from a growing tumor: Often, we cannot be sure in what type of normal cell the cancerous transformation occurred, and often the normal cell equivalent is difficult to obtain in sufficient quantity or purity to study. Also, comparisons have frequently been made between cells from actively growing animal tumors and their so-called normal counterparts growing in culture. But many of the so-called "normal" cell lines may have undergone a number of genetic changes during their adaptation to growth in the artificial environment of cell culture.

Therefore, many of the differences we associate with the cancerous transformation were discovered by comparing a "normal" cell line (e.g., 3T3) with the transformed cell line created by exposure of cells from the "normal" cell line to a cancer-causing virus. Even here, it has been necessary to guard against being misled by new sets of spontaneous mutations that occur during the further growth in cultures of the respective normal and transformed cell lines. One of the most reliable systems for probing cancer biochemistry has been cell lines transformed by mutated cancer viruses (Chapter 26). The resulting cells have a cancerous behavior when grown at a low temperature (say, 32°C) but a normal behavior when grown at a higher (e.g., 38°C) temperature.

Warburg and the Meaning of Increased Glycolysis

The first convincing biochemical difference between normal cells and cancer cells was discovered over 60 years ago by the German biochemist Otto Warburg. He observed that virtually every type of cancer cell that forms a solid tumor (as opposed to cancers of the blood, like leukemia) excretes much larger quantities of lactic acid than does its normal counterpart. The cancer cells do this both when growing as solid tumors in animals and when multiplying as single cells in culture. Since the original observation in the early 1920s, the meaning of lactic acid overproduction (often called the **Warburg effect**) has been studied again and again, but its real significance remains elusive. The excessive lactic acid, however, arises from glucose via the glycolytic pathway (Chapter 2). But this increase in fermentation (anaerobic metabolism) does not result from an insufficiency of any enzymes involved in the various oxidative pathways (e.g., oxidative phosphorylation proceeds normally). Nonetheless, much more glucose is consumed by these tumor cells than they need to grow and multiply. This suggests the loss of a normal control device that regulates the rate at which glucose is taken into a cell, and in fact, many transformed cells can be shown to have an increased rate of glucose transport.

Contact Inhibition of Movement[97]

Strong evidence that the plasma membranes of cancer cells are very different from those of normal cells comes from comparing their behaviors after their ruffled edges touch each other. Such contacts

Figure 25-29
Six-day colony consisting of about 300 human epidermal cells. Supporting peripheral 3T3 cells are lethally irradiated. (Courtesy of Y. Barrandon and H. Green.)

between normal cells, but not between cancer cells, frequently result in cell adhesion, accompanied by a slowing down of random amoeboid movements by which the cells move along surfaces (contact inhibition of movement). Early in this process (10 to 20 seconds after contact), actin-containing cables begin to form in the seemingly disorganized cytoplasmic region beneath the ruffled edges (Figure 25-30). At the same time, the ruffled edges start to lose their morphological identity, soon becoming indistinguishable from less active regions of the plasma membrane. Generation of new intracellular actin-containing cables in response to cellular contacts may thus be an essential aspect of contact inhibition of movement.

Malignancy as a Loss of Normal Cellular Affinities[98]

The "sticky" quality of normal cells that leads to adhesion displays considerable specificity. A given type of cell (e.g., a liver cell) prefers to stick to others of its own kind (e.g., other liver cells) and shows very little, if any, affinity for other types (e.g., kidney cells). This type of specificity has been elegantly demonstrated in experiments in which small amounts of the proteolytic enzyme trypsin are used to break apart organs such as the liver or the kidney into their single-cell components. If these isolated cells are then incubated in the absence of trypsin, they reaggregate to form tissue fragments similar to those in the intact organ, that is, small fragments of liver tissue and small fragments of kidney tissue. When kidney and liver cells are mixed together, small fragments of liver and kidney are again detected. Thus, a kidney cell prefers to stick to a kidney cell, and a liver cell to a liver cell.

If this experiment is repeated with cancer cells, however, the normal cellular affinities no longer hold. For example, the mixing of cells from a skin cancer with normal kidney cells results in aggregates in which the skin cancer cells are interspersed among the kidney cells. Conceivably, the inability of cancer cells to form tight adhesive junctions allows them to lie adjacent to almost any cell type. This may be one reason why malignant cells invade a variety of normal organs.

The nature of the sticky substance(s) that normal cells use to attach to each other is still unclear. As already discussed, however, fibronectin appears to be a glue at least for cell-matrix interactions, and it provides a good substrate for cell attachment and spreading in vitro. Fibronectin is frequently decreased or absent in transformed cells, and this may contribute to the cells' loss of adhesiveness. Electron micrographs of cells that have stuck to glass (or to each other) reveal dense adhesive regions (plaques) 60 to 80 Å thick that are located just inside the plasma membrane (see Figure 25-30). One component of adhesion plaques is a protein called vinculin, which may serve as an anchor for cytoskeletal cables.

Alterations of the Cytoskeleton of Transformed Cells[99-101]

Examination of many cancer cells reveals that they usually display a much less organized arrangement of the various cytoskeletal proteins than is found in their normal equivalents. Normal fibroblasts, for example, contain large numbers of fibers, cables in which the various

(a)   (b)

(c)

Figure 25-30
(a, b) Electron microscopic views of the "adhesive plaques" (arrows) that form within 10 to 20 seconds of contact between ruffled edges. Organized groups of actin-containing filaments form beneath the plaques within less than a minute. Bar represents 0.5 μm. [Reproduced with permission from J. Haeysman, *CIBA Foundation Symp.* 14 (new series) (1973):190.] (c) An electron micrograph in which the plane of thin section cuts just above the region of hamster embryo cell surface contact. This section reveals the crystalline-like lattices of the adhesive plaques (dark patches) (42,000×). (Courtesy of R. D. Goldman, Carnegie-Mellon University.)

cytoskeletal proteins (e.g., actin, myosin, and tropomyosin) are regularly arranged. In contrast, transformed cells contain many fewer such fibers, and those that are present are generally much thinner (Figure 25-31). Correspondingly, the touching of the cancer cells does not lead to cessation of cell movement, but instead frequently induces

(a)   (b)

(c)   (d)

Figure 25-31
A comparison of the organization of muscle proteins within normal and transformed mouse cells. Immunofluorescent techniques are used to visualize (a) actin within a "normal" 3T3 cell, (b) myosin within a 3T3 cell, (c) actin in a 3T3 cell transformed by SV40 virus (SV3T3), and (d) the myosin of an SV3T3 cell. [Reproduced with permission from R. Pollack, M. Osborn, and K. Weber, *Proc. Nat. Acad. Sci.* 72 (1975):994.]

Figure 25-32

Blebs on the surface of cancer cells. (a, b) In these cancerous hamster cells that originated through transformation by adenovirus 5, blebs are a characteristic feature throughout the cell cycle except during M phase. As shown in these photos taken 1 minute apart, blebs are very unstable protrusions, generally having lifetimes of less than a minute. In other types of cancer cells, blebs are less frequent. In their place are found large numbers of microvilli and ruffles, which, like blebs, are constantly thrown out and then retracted. [Reproduced with permission from R. D. Goldman, C.-M. Chang, and J. - Williams, *Cold Spring Harbor Symp. Quant. Biol.* 39 (1975):601.] (c) An electron micrograph of a thin section of a portion of a chicken cell transformed by Rous sarcoma virus, showing a large bleb. It is largely filled with free polyribosomes. (Courtesy of Lan Bo Chen and Elaine Lenk, MIT.)

(a) (b)

(c)

the uncoordinated generation of many more ruffles and blebs than are found on their normal cellular equivalents (Figure 25-32). Why cancer cells are less able to form large numbers of highly organized fibers is not yet clear. The reason, however, may be due to alterations in the expression of some of the cytoskeletal proteins. For example, alterations in the expression of tropomyosin are a feature of some transformed cells. These alterations in tropomyosin expression appear to correlate with the rearrangement of stress fibers and this may in part be responsible for the morphological alterations observed in transformed cells. Furthermore, expression of a mutant actin gene has been found in some transformed cells. In addition, certain cancer cells may have an alteration in one or more components at the plasma membrane (e.g., vinculin), which may hinder the attachment of individual actin filaments to the inner membrane surface.

Molecular Changes at the Cell Surface That Accompany Cell Transformation[87, 102, 103]

Only recently have the chemical techniques been available for us to critically compare the plasma membranes of normal cells and transformed cells. So, despite intense interest in this problem, current knowledge is still very incomplete. Making this problem inherently difficult is the complexity of most plasma membranes, most of which at a minimum must easily contain hundreds of different components. Of necessity, attention has been focused either on the major structural components or on those relatively few membrane-bound enzymes that we can conveniently assay. However, an essential change that defines the cancer cell membrane might occur in a minor structural component, so it is not clear whether we could reasonably expect to find critical cancer changes by membrane biochemistry alone.

Moreover, it is becoming increasingly clear that normal plasma membranes are not invariant in structure. The exact groupings that protrude on their outer surfaces not only change with the different phases of the cell cycle, but are very much a function of the contacts that they make with other cells. Great caution must be shown in comparing normal cells with their transformed equivalents unless we are sure that both types have been harvested in the same point in the cell cycle, at the same density, and under identical conditions of growth.

Now it seems fairly certain that the cancerous transformation does not change the relative amounts of the four main phospholipids that form the basic lipid bilayer. Differences, however, and sometimes seemingly quite specific ones, have been seen in both the nature and kind of the glycolipids and glycoproteins that are inserted into the bilayer. In particular, much study has gone toward the gangliosides and other glycolipids. Not only are the more complex of them specifically diminished in amounts in many cancers, but the enzymes involved in their biosynthesis are also specifically reduced. It is not at all clear, however, that such changes are primary events in the formation of a cancer cell. Not only do such changes often lag behind the assumption of the distinctive cancer cell morphology when viral transformation occurs, but, perhaps even more important, other transformed cell lines have a perfectly normal glycolipid content. So we must postulate either that changes in a variety of quite distinct molecular components each can produce the distinctive cancer cell membrane, or that the alterations in ganglioside content are irrelevant features that are somehow selected for by a prior primary change.

Most recently, use has been made of monoclonal antibodies to probe the difference between normal and transformed cell surfaces. About 10 percent of the monoclonal antibodies raised against cell surfaces are directed against glycolipids. Interestingly, it is some of these antibodies that are found to be specific for the transformed cell surface. The antigens detected are not strictly limited to tumor cells, however, since they are frequently found on embryonic cells as well.

Much attention has gone toward examination of the major proteins that are inserted in the outer plasma membrane, especially the glycoproteins whose exterior protruding surfaces are covered with sugar groups. Here at least one change common to many types of cancer cells has already been found: the reduction or loss of the large external glycoprotein, fibronectin. This protein, composed of disulfide-bonded subunits of about 230 kdal, is found as a fibrillar network above, beneath, and around cells in culture. It disappears upon transformation of many cell types by RNA or DNA tumor viruses. Readdition of purified fibronectin to transformed cells in culture causes them to flatten and elongate, to adhere more tenaciously, and to acquire more organized microfilament bundles (Figure 25-33). However, the cells can still grow in agar and to high cell densities. So whether fibronectin plays a critical role in tumorgenicity, possibly adhesion and metastasis, remains to be seen.

(a)

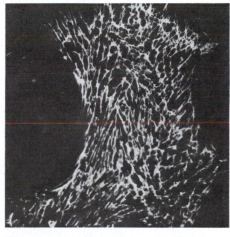

(b)

Figure 25-33

(a) Diagram showing how the synthesis or external addition of fibronectin is associated with the flattened shape of the fibroblast. (b) Immunofluorescent micrograph of fibronectin on the surface of a Ni1 8 hamster fibroblast growing in vitro. Antibodies to fibronectin were conjugated to fluorescein to make this picture. [Courtesy of Richard O. Hynes.]

Decreased Serum Requirements of Cancer Cells and Secretion of Growth Factors[76, 104–107]

Many cancer cells can grow in culture medium supplemented with much less serum than that required by the corresponding normal cells. This difference reflects a lower need for one or more of the growth-controlling factors that serum supplies. The mouse cell line 3T3, for example, multiplies faster and to higher density in growth medium containing 10 percent serum than in a 1 percent serum-supplemented medium. But transformed 3T3 cells grow equally well in 1 percent serum and 10 percent serum.

Many years ago, it was found that some cells that require little or no serum to grow produce their own growth factors. Culture fluids from these cells can stimulate the growth of serum-requiring cells. This finding and the fact that cancer cells require lower concentrations of serum led to the notion that cancer may arise when cells begin to uncontrollably produce growth factors for which they also express the corresponding receptors, the autocrine hypothesis of tumor formation. Now there is evidence that at least some tumor-virus-induced cancers arise by this mechanism (Chapter 26). Other cancers may involve mechanisms that allow cells to bypass the need for specific growth factors.

TGFs were first found in the medium of certain cell lines transformed by tumor viruses. As discussed earlier in this chapter, not only do they stimulate growth, but TGF-βs confer the cancerous property of anchorage-independent growth when added to the culture medium of normal cells (Figure 25-34). The effects of TGF-βs are completely reversible: Twenty-four hours after they are removed from the medium, normal cells revert to a normal phenotype.

What the normal function of TGF-β might be is anyone's guess, but it has been postulated that certain embryological processes might transiently require cells to have cancerlike phenotypes. Whether secretion of TGF-β is a common feature of spontaneously arising human cancers remains to be seen.

Many human small-cell lung cancer cell lines secrete polypeptides that belong to the family of bombesin-like peptides (BLPs), and these cells also possess receptors for BLPs. Bombesin was originally isolated from the skin of the frog *Bombina bombina*. It is a peptide that can stimulate pancreatic enzyme secretion. The related peptides of mammals, which include gastrin-releasing peptide, cause a variety of physiological responses: They can induce gastrin cell hyperplasia, increase pancreatic DNA content in vivo in rats, and stimulate proliferation of mouse 3T3 cells in vitro. Interestingly, monoclonal antibodies that bind to bombesin and block its binding to cellular receptors can inhibit the growth of human small-cell lung cancer cells growing in vitro or growing in vivo following their subcutaneous injection into mice. These results have been interpreted as further support for the autocrine hypothesis of tumor formation.

Differential Gene Expression in Normal Versus Transformed Cells[47, 108, 109]

Powerful new tools of molecular biology are providing brute force approaches to defining biochemical differences between normal cells and cancer cells. Popular and promising is the preparation of cDNA

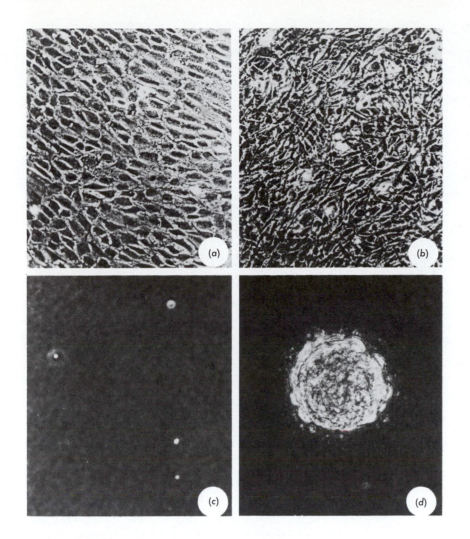

Figure 25-34
Effect of TGFs on the morphology and growth in soft agar of normal rat kidney cells. (a) Cells growing in standard medium versus (b) similar cells with partially purified TGF-α added to the medium for 6 days. (c) Cells fail to grow in agar unless (d) treated with partially purified TGF-α, in which case they grow into colonies of well over 500 cells within 2 weeks. [Courtesy of G. J. Todaro, J. E. DeLarco, H. Marquardt, M. L. Bryant, S. A. Sherwin, and A. H. Sliski, *Hormones and Cell Culture*, ed. G. H. Soto and R. Ross (Cold Spring Harbor, N.Y.: Cold Spring Harbor Laboratory, 1979).]

clones that represent mRNAs present exclusively or more abundantly in transformed cells than in their normal cell counterparts. Approximately 3 to 5 percent of mRNAs, corresponding to roughly a few hundred genes, seem to fall in this category. How long it will be before this method of defining cancer proves fruitful is hard to guess.

Another modern approach to defining cancer cells is the use of high-resolution two-dimensional gel electrophoresis to analyze the proteins of normal cells and cancer cells. This remarkable technique, which separates proteins on the basis of their isoelectric points and on their sizes, can resolve more than a thousand proteins in mammalian cells. When appropriate pairs of normal and transformed cells are compared, specific differences often can be seen (Figure 25-35); but determining the importance of such differences to the cancerous phenotype is usually difficult.

The Daunting Incompleteness of Eucaryotic Cell Biochemistry

The fact that only a small percent of the biochemical reactions of a cell are known might make us wonder if we will soon be able to understand biological processes as complex as growth control and cancer.

Figure 25-35

Proteins in (a) normal and (b) SV40-transformed cells of the rat fibroblast cell line REF52 analyzed by two-dimensional gel electrophoresis. Cells were grown in the presence of 35S-methionine for 2 hours prior to the analysis. Direction of migration in the first and second dimensions of the gels is indicated. In (b) some of the proteins whose amounts increase in SV40-transformed cells are indicated by upward pointing arrows, while those whose amount decreases are indicated by downward pointing arrows. About 12% of the proteins are altered two-fold or more by viral transformation, but no entirely new proteins can be detected and no proteins disappear completely. Viral T antigens are present in too low an amount to be seen on gel. PCNA, proliferating cell nuclear antigen: the rate of synthesis of this nuclear protein is tightly coupled to the rate of proliferation in normal cells but not in transformed cells. [Courtesy of James I. Garrels and B. Robert Franza, Jr., Cold Spring Harbor Laboratory.]

As we have seen, however, reproducible biochemical differences between normal cells and cancer cells have been found; and as we shall learn in Chapters 26 and 27, genes that can cause cancer and so induce at least some of these changes have been isolated. Most incredibly, some cancer-causing genes have been identified as genes that were already known to be involved in growth control, namely, genes encoding growth factors and their receptors. Thus, the incompleteness of eucaryotic biochemistry—while daunting—is not insurmountable when the right experiments are performed. However, we must admit that the many black boxes of cellular biochemistry still pose a substantial challenge to those who wish to fully understand growth control and cancer.

Summary

To learn how the cells of multicellular animals control their proliferation rates has long been a key objective of experimental biology. With such understanding, we will be in a much stronger position to probe both the embryological development and the correct adult functioning of higher animals. And we may at last be able to come to grips with the essence of the various forms of cancer, hereditary changes in cells that lead to uncontrolled cell proliferation.

Now almost all productive approaches to the control mechanisms governing cell proliferation utilize the in vitro growth of cells in culture. The cells excised from an animal to start a culture are called primary cells, while their descendants present several division cycles later are known as secondary cells. Most secondary cell cultures fail to multiply more than 20 to 50 times, and it is only the rare cell that acquires the ability to multiply indefinitely to form a cell line. Why only the exceptional cell has unlimited growth potential is unclear. Up to now, cell lines have been obtained only from selected animal species (e.g., mouse, rat, hamster, human, and rarely the chicken). And in general, it proves to be much easier to obtain cell lines from tumor cells than from their normal equivalents.

The greater potential of tumor cells to yield cell lines led for many years to the belief that perhaps all cells that grow as cell lines do so because they have acquired by mutation(s) many of the essential properties of the cancer cell. However, we now realize that only those fibroblasts (epithelial cells) continuously passaged at high cell density later give rise to cell lines that grow to the high cell density characteristic of cancer cells. If cells are selected at low density, the respective cell lines grow to much lower densities and generally stop dividing after they have formed organized cell layers. Such cell lines, the prototype of which is mouse 3T3 lines, are not completely normal, however, since they have enhanced growth potential relative to normal cells and they often have highly abnormal numbers of chromosomes.

Possession of well-defined cell lines, like the human cancer cell line HeLa and the mouse L cell line, made it possible to work out the detailed nutritional requirements of cells growing in culture. In general, growth not only requires a well-defined collection of amino acids and vitamins, but also depends on protein factors present in blood serum. The nature of the "serum factors" long remained a mystery; but now it is apparent that most serum factors are polypeptide hormones that act by binding specific receptors on the surface of their target cells. Each cell type requires two to five specific factors that may have to function in a specific sequence. Some, like transferrin, which transports ferric iron into cells, are almost universally required; in contrast, PDGF affects cells of mesodermal origin, and interleukin 2 appears to be highly specific for certain lymphoid cells. Polypeptide growth factors stimulate cells by binding to specific cell surface receptors. The receptors for some growth factors (PDGF and EGF) and hormones (insulin) have tyrosine kinase activity that is activated when their corresponding ligands bind. This may result in phosphorylation and thus activation of cellular proteins, which in some unknown way triggers altered gene transcription and ultimately cell division.

Some polypeptide hormones, by binding to their receptors, modify the activity of the membrane-bound enzyme adenylate cyclase, using the GTP binding protein (G protein) as an intermediate, so that the level of cAMP in the cell is altered. This, in turn, affects the activity of cAMP-dependent protein kinases that phosphorylate specific target proteins, thereby regulating their activity. How cAMP levels are related to the cell division cycle is still unclear. Other possible second messengers to carry signals from growth factor receptors to the cell include calcium and diacylglycerol.

In the absence of growth-promoting factors (mitogens), most cells enter a resting state in the G_1 phase of the life cycle and so contain a diploid amount of DNA. Once a cell has begun to replicate its DNA, it continues through the G_2 phase and obligatorily passes into mitosis. Once into G_1, a new mitogenic stimulus must be applied before the cell will start another cell cycle. Otherwise, the cell enters a quiescent state, which often leads into a differentiation process that may create a highly specialized cell (e.g., nerve) unable to initiate a new cycle of growth and cell division. Differentiation per se, however, does not lead to an inability to divide. For example, the commonly used fibroblasts, epithelial cells, and lymphoblasts all represent clearly defined states of differentiation. So it is only when the physiological role of a given cell makes further division unwanted (as with nerve cells and red blood cells) that the capacity to divide is lost.

Transformation is the process by which normal cells acquire many of the morphological as well as growth properties of cancer cells. In general, transformed cells have a more spherical shape than their normal counterparts and show many similarities to cells just emerging from mitosis. For example, many tumor cells show excessive blebbing throughout the cell cycle, a characteristic of normal cells only in early G_1. Likewise, many of the membrane proteins of cancer cells appear more mobile than those found in normal interphase cells, conceivably reflecting a much less organized arrangement of their cytoskeletons. Transformed cells also have lost many of the normal adhesive properties, leading to an inability to be sited next to appropriate cellular neighbors. In addition, many tumor cells have lost the normal control of glucose utilization. They consume much more glucose than can be metabolized efficiently, thereby leading to massive lactic acid secretion.

Many tumor cells have lower serum requirements than their normal cell equivalents, and some tumor cells secrete factors that alter the growth, and in some cases the morphology, of normal cells. To what extent the turning on of these factors may mediate the transformed phenotype in many cancers is unknown.

Molecular biologists are now applying powerful new technologies to analyze and clone genes whose expression is altered in transformed cells relative to normal cells, or in cells following mitogen stimulation. One very interesting finding so far is that PDGF added to quiescent 3T3 cells leads to increased levels of mRNA for genes called c-*myc* and c-*fos*, which, as we shall learn in the next chapter, are aberrantly expressed in certain cancers.

Bibliography

General References

Alberts, B., D. Bray, J. Lewis, M. Raff, K. Roberts, and J. D. Watson. 1983. *Molecular Biology of the Cell.* New York: Garland.

Baserga, R. 1985. *The Biology of Cell Reproduction.* Cambridge, Mass., and London: Harvard University Press.

Feramisco, J., B. Ozanne, and C. Stiles. 1985. *Growth Factors and Transformation.* Cold Spring Harbor, N.Y.: Cold Spring Harbor Laboratory.

Littlefield, J. W. 1976. *Variation, Senescence, and Neoplasia in Cultured Somatic Cells.* A Commonwealth Fund Book. Cambridge, Mass., and London: Harvard University Press.

Metcalf, D. 1984. *The Hematopoietic Colony Stimulating Factors.* Amsterdam, London, and New York: American Elsevier.

Mitchison, J. M. 1971. *The Biology of the Cell Cycle.* New York: Cambridge University Press.

Paul, J. 1975. *Cell and Tissue Culture.* 5th ed. New York: Churchill Livingstone.

Pollack, R., ed. 1975. *Readings in Mammalian Cell Culture.* 2nd ed. Cold Spring Harbor, N.Y.: Cold Spring Harbor Laboratory.

Cited References

1. Pollack, R., ed. 1975. *Readings in Mammalian Cell Culture.* 2nd ed. Cold Spring Harbor, N.Y.: Cold Spring Harbor Laboratory.
2. Hayflick, L. 1965. "The Limited in Vitro Lifetime of Human Diploid Cell Strains." *Exp. Cell Res.* 37:614–636.
3. Buonassisi, V., G. Sato, and A. I. Cohen. 1962. "Hormone-Producing Cultures of Adrenal and Pituitary Tumor Origin." *Proc. Nat. Acad. Sci.* 48:1184–1190.
4. Todaro, G. J., and H. Green. 1963. "Quantitative Studies of the Growth of Mouse Embryo Cells in Culture and Their Development into Established Lines." *J. Cell Biol.* 17:299–313.
5. Aaronson, S. A., and G. J. Todaro. 1968. "Basis for Acquisition of Malignant Potential by Mouse Cells Cultivated in Vitro." *Science* 162:1024–1026.
6. Temin, H. M., and H. Rubin. 1958. "Characteristics of an Assay for Rous Sarcoma Virus and Rous Sarcoma Cells in Tissue Culture." *Virology* 6:669–688.
7. Todaro, G. J., and H. Green. 1964. "An Assay for Cellular Transformation by SV40." *Virology* 23:117–119.
8. Macpherson, I., and L. Montaignier. 1964. "Agar Suspension Culture for the Selective Assay of Cells Transformed by Polyoma Virus." *Virology* 23:291–294.
9. Eagle, H. 1955. "Nutrition Needs of Mammalian Cells in Tissue Culture." *Science* 122:43–46.
10. Ham, R. G. 1965. "Clonal Growth of Mammalian Cells in a Chemically Defined, Synthetic Medium." *Proc. Nat. Acad. Sci.* 53:288–293.
11. Hayashi, I., and G. Sato. 1976. "Replacement of Serum by Hormones Permits Growth of Cells in Defined Medium." *Nature* 259:132–134.
12. Sato, G., A. Pardee, and D. Sirbasku, eds. 1982. "Growth of Cells in Hormonally Defined Media." *Cold Spring Harbor Conference on Cell Proliferation* 9. Cold Spring Harbor Laboratory, Cold Spring Harbor, New York.
13. Alberts, B., D. Bray, J. Lewis, M. Raff, K. Roberts, and J. D. Watson. 1983. *Molecular Biology of the Cell.* New York: Garland.
14. Baserga, R. 1985. *The Biology of Cell Reproduction.* Cambridge, Mass., and London: Harvard University Press.
15. Mitchison, J. M. 1971. *The Biology of the Cell Cycle.* New York: Cambridge University Press.
16. Pardee, A. B., R. Dubrow, J. L. Hamlin, and R. F. Kletzien. 1978. "Animal Cell Cycle." *Ann. Rev. Biochem.* 47:715–750.
17. Harris, H., and J. F. Watkins. 1965. "Hybrid Cells Derived from Mouse and Man: Artificial Heterokaryons of Mammalian Cells from Different Species." *Nature* 205:640–646.
18. Rao, P. N., and R. T. Johnson. 1970. "Mammalian Cell Fusion: Studies on the Regulation of DNA Synthesis and Mitosis." *Nature* 225:159–164.

19. Rao, P. N., and R. T. Johnson. 1974. "Induction of Chromosome Condensation in Interphase Cells." *Adv. Cell & Mol. Biol.* 3:136.
20. Huberman, J. A., and A. D. Riggs. 1968. "On the Mechanism of DNA Replication in Mammalian Chromosomes." *J. Mol. Biol.* 32:327–341.
21. Kao, F. T., and T. T. Puck. 1968. "Genetics of Somatic Mammalian Cells. VII. Induction and Isolation of Nutritional Mutants in Chinese Hamster Cells." *Proc. Nat. Acad. Sci.* 60:1275–1281.
22. Patterson, D., F. T. Kao, and T. T. Puck. 1974. "Genetics of Somatic Mammalian Cells: Biochemical Genetics of Chinese Hamster Cell Mutants with Deviant Purine Metabolism." *Proc. Nat. Acad. Sci.* 71:2057–2061.
23. Basilico, C. 1978. "Selective Production of Cell Cycle Specific ts Mutants." *J. Cell Physiol.* 95:367–376.
24. Nowell, P. C. 1960. "Phytohemagglutinin: An Initiator of Mitosis in Cultures of Normal Human Leukocytes." *Cancer Res.* 20:462–466.
25. Cohen, S. 1962. "Isolation of a Mouse Submaxillary Gland Protein Accelerating Incisor Eruption and Eyelid Opening in the New-Born Animal." *J. Biol. Chem.* 237:555–562.
26. Balk, S., J. F. Whitfield, T. Youdale, and A. C. Braun. 1973. "Roles of Calcium, Serum, Plasma, and Folic Acid in the Control of Proliferation of Normal and Rous Sarcoma Virus-Infected Chicken Fibroblasts." *Proc. Nat. Acad. Sci.* 70:675–679.
27. Ross, R., J. A. Glomset, B. Kariya, and L. Harker. 1974. "A Platelet-Dependent Serum Factor That Stimulates the Proliferation of Arterial Smooth Muscle Cells in Vitro." *Proc. Nat. Acad. Sci.* 71:1207–1210.
28. Kohler, N., and A. Lipton. 1974. "Platelets as a Source of Fibroblast Growth-Promoting Activity." *Exp. Cell Res.* 87:297–301.
29. Pledger, W. J., C. D. Stiles, H. N. Antoniades, and C. D. Scher. 1977. "Induction of DNA Synthesis in BALB/c 3T3 Cells by Serum Components: Reevaluation of the Commitment Process." *Proc. Nat. Acad. Sci.* 74:4481–4485.
30. Pledger, W. J., C. D. Stiles, H. N. Antoniades, and C. D. Scher. 1978. "An Ordered Sequence of Events Is Required Before BALB/c-3T3 Cells Become Committed to DNA Synthesis." *Proc. Nat. Acad. Sci.* 75:2839–2843.
31. Antoniades, H. N., C. D. Scher, and C. D. Stiles. 1979. "Purification of Human Platelet-Derived Growth Factor." *Proc. Nat. Acad. Sci.* 76:1809–1813.
32. Gray, A., T. J. Dull, and A. Ullrich. 1983. "Nucleotide Sequence of Epidermal Growth Factor cDNA Predicts a 128,000-Molecular Weight Protein Precursor." *Nature* 303:722–725.
33. Scott, J., M. Urdea, M. Quiroga, R. Sanchez-Pescador, N. Fong, M. Selby, W. Rutter, and G. I. Bell. 1983. "Structure of a Mouse Submaxillary Messenger RNA Encoding Epidermal Growth Factor and Seven Related Proteins." *Science* 221:236–240.
34. Cochran, B. 1985. "The Molecular Action of Platelet-Derived Growth Factor." *Adv. Cancer Res.* 45:183–216.
35. Evans, H. M., and J. A. Long. 1921. "The Effect of the Anterior Lobe Administered Intraperitoneally Upon Growth, Maturity and Oestrous Cycles of the Rat." *Anat. Rec.* 21:62.
36. Clemmons, D. R., and J. J. Van Wyk. 1981. "Somatomedin: Physiological Control and Effects on Cell Proliferation." In *Handbook of Experimental Pharmacology,* vol. 57, ed. R. Baserga. New York: Springer-Verlag, pp. 161–208.
37. Zapf, J., E. R. Froesch, and R. E. Humbel. 1981. "The Insulin-Like Growth Factors (IGF) of Human Serum: Chemical and Biological Characterization and Aspects of Their Possible Physiological Role." *Curr. Top. Cell Regul.* 19:257–307.
38. Hunter, T., and B. M. Sefton. 1980. "Transforming Gene Product of Rous Sarcoma Virus Phosphorylates Tyrosine." *Proc. Nat. Acad. Sci.* 77:1311–1315.
39. Ushiro, H., and S. Cohen. 1980. "Identification of Phosphotyrosine as a Product of Epidermal Growth Factor-Acti-

vated Protein Kinase in A-431 Cell Membranes." *J. Biol. Chem.* 255:8363–8365.

40. Kasuga, M., Y. Zick, D. L. Blithe, M. Crettaz, and C. R. Kahn. 1982. "Insulin Stimulates Tyrosine Phosphorylation of the Insulin Receptor in a Cell-Free System." *Nature* 298:667–669.

41. Ek, B., B. Westermark, A. Wasteson, and C. H. Heldin. 1982. "Stimulation of Tyrosine-Specific Phosphorylation by Platelet-Derived Growth Factor." *Nature* 295:419–420.

42. Nishimura, J., J. S. Huang, and T. F. Deuel. 1982. "Platelet-Derived Growth Factor Stimulates Tyrosine-Specific Protein Kinase Activity in Swiss Mouse 3T3 Cell Membranes." *Proc. Nat. Acad. Sci.* 79:4303–4307.

43. Jacobs, S., F. C. Kull, J. H. S. Earp, M. E. Svoboda, J. J. Van Wyk, and P. Cuatrecasas. 1983. "Somatomedin-C Stimulates the Phosphorylation of the Beta-Subunit of Its Own Receptor." *J. Biol. Chem.* 258:9581–9584.

44. Stoscheck, C. M., and G. Carpenter. 1984. "Down Regulation of Epidermal Growth Factor Receptors: Direct Demonstration of Receptor Degradation in Human Fibroblasts." *J. Cell Biol.* 98:1048–1053.

45. Sherr, C. J., C. W. Rettenmeier, R. Sacca, M. F. Roussel, A. T. Look, and E. R. Stanley. 1985. "The c-*fms* Proto-Oncogene Product Is Related to the Receptor for the Mononuclear Phagocyte Growth Factor CSF-1." *Cell* 41:665–676.

46. Armelin, H. A., M. C. S. Armelin, K. Kelly, T. Stewart, P. Leder, B. H. Cochran, and C. D. Stiles. 1984. "Functional Role for c-*myc* in Mitogenic Response to Platelet-Derived Growth Factor." *Nature* 310:655–660.

47. Cochran, B. H., A. C. Reffel, and C. D. Stiles. 1983. "Molecular Cloning of Gene Sequences Regulated by Platelet-Derived Growth Factor." *Cell* 33:939–947.

48. Kelly, K., B. H. Cochran, C. D. Stiles, and P. Leder. 1983. "Specific Regulation of the c-*myc* Gene by Lymphocyte Mitogens and Platelet-Derived Growth Factor." *Cell* 35:603–610.

49. Chambard, J. C., A. Franchi, A. LeCam, and J. Pouyssegur. 1983. "Growth Factor-Stimulated Protein Phosphorylation in G_0-G_1-Arrested Fibroblasts." *J. Biol. Chem.* 258:1706–1713.

50. Castagna, M., Y. Takai, K. Kaibuchi, K. Sano, U. Kikkawa, and Y. Nishizuka. 1982. "Direct Activation of Calcium-Activated, Phospholipid Dependent Protein Kinase by Tumor-Promoting Phorbol Esters." *J. Biol. Chem.* 257:7847–7851.

51. Cooper, J. A., D. F. Bowen-Pope, E. Raines, R. Ross, and T. Hunter. 1982. "Similar Effects of Platelet-Derived Growth Factor and Epidermal Growth Factor on the Phosphorylation of Tyrosine in Cellular Proteins." *Cell* 31:263–273.

52. Schuldiner, S., and E. Rozengurt. 1982. "Na^+/H^+ Antiport in Swiss 3T3 Cells: Mitogenic Stimulation Leads to Cytoplasmic Alkalinization." *Proc. Nat. Acad. Sci.* 79:7778–7782.

53. Rozengurt, E., P. Stroobant, M. D. Waterfield, T. F. Deuel, and M. Keehan. 1983. "Platelet-Derived Growth Factor Elicits Cyclic AMP Accumulation in Swiss 3T3 Cells: Role of Prostaglandin Production." *Cell* 34:265–272.

54. Berridge, M. J., J. P. Heslop, R. F. Irvine, and K. D. Brown. 1984. "Inositol Triphosphate Formation and Calcium Mobilization in Swiss 3T3 Cells in Response to Platelet-Derived Growth Factor." *Biochem. J.* 222:195–201.

55. Gilman, A. G. 1984. "G Proteins and Dual Control of Adenylate Cyclase." *Cell* 36:577–579.

56. Nishizuka, Y. 1984. "The Role of Protein Kinase C in Cell Surface Signal Transduction and Tumor Promotion." *Nature* 308:693–698.

57. Truneh, A., F. Albert, P. Goldstein, and A.-M. Schmitt-Verhulst. 1985. "Early Steps of Lymphocyte Activation Bypassed by Synergy Between Calcium Ionophores and Phorbol Ester." *Nature* 313:318–320.

58. Yamamoto, K. R., and B. M. Alberts. 1976. "Steroid Receptors: Elements for Modulation of Eukaryotic Transcription." *Ann. Rev. Biochem.* 45:721–746.

59. Payvar, F., O. Wrange, J. Carlstedt-Duke, S. Okret, J.-A. Gustafsson, and K. R. Yamamoto. 1981. "Purified Glucocorticoid Receptors Bind Selectively in Vitro to a Cloned DNA Fragment Whose Transcription Is Regulated by Glucocorticoids in Vivo." *Proc. Nat. Acad. Sci.* 78:6628–6632.

60. Metcalf, D. 1984. *The Hematopoietic Colony Stimulating Factors.* Amsterdam, London, and New York: Elsevier.

61. Till, J. E., and E. A. McCulloch. 1961. "A Direct Measurement of the Radiation Sensitivity of Normal Mouse Bone Marrow Cells." *Radiat. Res.* 14:213–222.

62. Bradley, T. R., and D. Metcalf. 1966. "The Growth of Mouse Bone Marrow Cells in Vitro." *Aust. J. Exp. Biol. Med. Sci.* 44:287–300.

63. Wu, A. M., J. E. Till, L. Siminovitch, and E. A. McCulloch. 1967. "A Cytological Study of the Capacity for Differentiation of Normal Hemopoietic Colony-Forming Cells." *J. Cell Physiol.* 69:177–184.

64. Abramson, S., R. G. Miller, and R. A. Phillips. 1977. "The Identification in Adult Bone Marrow of Pluripotent and Restricted Stem Cells of the Myeloid and Lymphoid Systems." *J. Exp. Med.* 145:1567–1579.

65. Johnson, G. R., and D. Metcalf. 1977. "Pure and Mixed Erythroid Colony Formation in Vitro Stimulated by Spleen Conditioned Medium with No Detectable Erythropoietin." *Proc. Nat. Acad. Sci.* 74:3879–3882.

66. Keller, G. M., and R. A. Phillips. 1982. "Detection in Vitro of a Unique Multipotent Hematopoietic Progenitor." *J. Cell Physiol. Suppl.* 1:31–36.

67. Goldwasser, E. 1975. "Erythropoietin and the Differentiation of Red Blood Cells." *Fed. Proc.* 34:2285–2292.

68. Heath, D. S., A. A. Axelrad, D. L. McLeod, and M. M. Shreeve. 1976. "Separation of the Erythropoietin-Responsive Progenitors BFU-E and CFU-E in Mouse Bone Marrow by Unit Gravity Sedimentation." *Blood* 47:777–792.

69. Metcalf, D. 1985. "The Granulocyte-Macrophage Colony-Stimulating Factors." *Science* 229:16–22.

70. Ihle, J. N., J. Keller, L. Henderson, F. Klein, and E. Palaszynski. 1982. "Procedures for the Purification of Interleukin-3 to Homogeneity." *J. Immunol.* 129:2431–2436.

71. Morgan, D. A., F. W. Ruscetti, and R. C. Gallo. 1976. "Selective in Vitro Growth of T-Lymphocytes from Normal Human Bone Marrows." *Science* 193:1007–1008.

72. Taniguchi, T., H. Matsui, T. Fujita, C. Takaoka, N. Kashima, R. Yoshimoto, and J. Hamuro. 1983. "Structure and Expression of a Cloned cDNA for Human Interleukin-2." *Nature* 302:305–309.

73. Cantrell, D. A., and K. A. Smith. 1984. "The Interleukin-2 T-Cell System: A New Cell Growth Model." *Science* 224:1312–1316.

74. Leonard, W. J., J. M. Depper, G. R. Crabtree, S. Rudikoff, J. Pumphrey, R. J. Robb, M. Kronke, P. B. Svetlik, N. J. Peffer, T. A. Waldman, and W. C. Greene. 1984. "Molecular Cloning and Expression of cDNAs for the Human Interleukin-2 Receptor." *Nature* 311:626–631.

75. Nikaido, T., A. Shimizu, N. Ishida, H. Sabe, K. Teshigawara, M. Maeda, T. Uchiyama, J. Yodoi, and T. Honjo. 1984. "Molecular Cloning of cDNA Encoding Human Interleukin-2 Receptor." *Nature* 311:631–635.

76. DeLarco, J. E., and G. J. Todaro. 1978. "Growth Factors from Murine Sarcoma Virus Transformed Cells." *Proc. Nat. Acad. Sci.* 75:4001–4005.

77. Roberts, A. B., C. Frolik, M. A. Anzano, and M. B. Sporn. 1983. "Transforming Growth Factors from Neoplastic and Non-Neoplastic Tissues." *Fed. Proc.* 42:2621–2626.

78. Holley, R. W., R. Armour, and J. H. Baldwin. 1978. "Density-Dependent Regulation of Growth of BSC-1 Cells in Cell Culture: Growth Inhibitors Formed by the Cells." *Proc. Nat. Acad. Sci.* 75:339–341.

79. Tucker, R. F., G. D. Shipley, H. L. Moses, and R. W. Holley. 1984. "Growth Inhibitor from BSC-1 Cells Closely Related to Platelet Type β Transforming Growth Factor." *Science* 226:705–707.

80. Iversen, O. H. 1981. "The Chalones." In *Tissue Growth Factors,* ed. R. Baserga. New York: Springer-Verlag, pp. 491–450.

81. Sager, R. 1985. "Genetic Suppression of Tumor Formation." *Adv. Cancer Res.* 44:43–68.

82. Granger, G. A., and W. P. Kolb. 1968. "Lymphocyte in Vitro Cytotoxicity: Mechanisms of Immune and Nonimmune Small Lymphocyte Mediated Target L Cell Destruction." *J. Immunol.* 101:111–120.

83. Carswell, E. A., L. J. Old, R. L. Kassel, S. Green, N. Fiore, and B. Williamson. 1975. "An Endotoxin-Induced Serum Fac-

tor That Causes Necrosis of Tumors." *Proc. Nat. Acad. Sci.* 72:3666–3670.

84. Pennica, D., G. E. Nedwin, J. S. Hayflick, P. H. Seeburg, R. Derynck, M. A. Palladine, W. J. Kohr, B. B. Aggarwal, and D. V. Goeddel. 1984. "Human Tumor Necrosis Factor: Precursor Structure, Expression and Homology to Lymphotoxin." *Nature* 312:724–729.

85. Beutler, B., D. Greenwald, J. D. Hulmes, M. Chang, Y.-C. E. Pan, J. Mathison, R. Ulevitch, and A. Cerami. 1985. "Identity of Tumor Necrosis Factor and the Macrophage-Secreted Factor Cachectin." *Nature* 316:552–554.

86. Levi-Montalcini, R. 1976. "The Nerve Growth Factor: Its Role in Growth, Differentiation and Function of the Sympathetic Adrenergic Neuron." *Prog. Brain Res.* 45:235–258.

87. Hynes, R. O. 1982. "Fibronectin and Its Relation to Cellular Structure and Behavior." In *Cell Biology of Extracellular Matrix,* ed. E. D. Hay. New York: Plenum.

88. Hynes, R. O., and K. M. Yamada. 1982. "Fibronectins: Multifunctional Modular Glycoproteins." *J. Cell Biol.* 95:369–377.

89. Folkman, J., and A. Moscona. 1978. "Role of Cell Shape in Growth Control." *Nature* 273:345–349.

90. Ben-Ze'ev, A., S. Farmer, and S. Penman. 1980. "Protein Synthesis Requires Cell-Surface Contact While Nuclear Events Respond to Cell Shape in Anchorage-Dependent Fibroblasts." *Cell* 21:365–372.

91. Green, H., and M. Meuth. 1974. "An Established Pre-Adipose Cell Line and Its Differentiation in Culture." *Cell* 3:127–133.

92. Taylor, S. M., and P. A. Jones. 1979. "Multiple New Phenotypes Induced in 10T1/2 and 3T3 Cells Treated with 5-Azacytidine." *Cell* 17:771–779.

93. Nixon, B. T., and H. Green. 1984. "Growth Hormone Promotes the Differentiation of Myoblasts and Preadipocytes Generated by Azacytidine Treatment of 10T1/2 Cells." *Proc. Nat. Acad. Sci.* 81:3429–3432.

94. Yaffe, D. 1968. "Retention of Differentiation Potentialities During Prolonged Cultivation of Myogenic Cells." *Proc. Nat. Acad. Sci.* 61:477–483.

95. Rheinwald, J. G., and H. Green. 1975. "Serial Cultivation of Strains of Human Epidermal Keratinocytes: The Formation of Keratinizing Colonies from Single Cells." *Cell* 6:331–344.

96. Gallico, G. G. III, N. E. O'Connor, C. C. Compton, O. Kehinde, and H. Green. 1984. "Permanent Coverage of Large Burn Wounds with Autologous Cultured Human Epithelium." *N. Eng. J. Med.* 311:448–451.

97. Abercrombie, M., and J. E. M. Heaysman. 1954. "Observations on the Social Behavior of Cells in Tissue Culture." *Exp. Cell Res.* 6:293–306.

98. Hynes, R. O., ed. 1979. *Surfaces of Normal and Malignant Cells.* New York: Wiley.

99. Pollack, R. E., M. Osborn, and K. Weber. 1975. "Patterns of Organization of Actin and Myosin in Normal and Transformed Cultured Cells." *Proc. Nat. Acad. Sci.* 72:994–998.

100. Matsumura, F., J. J. C. Lin, S. Yamashiro-Matsumura, G. P. Thomas, and W. C. Topp. 1983. Differential expression of tropomyosin forms in the microfilaments isolated from normal and transformed rat cultured cells. *J. Biol. Chem.* 258:13954–13964.

101. Leavitt, J., G. Bushar, T. Kakunaga, H. Hamada, T. Hirakawa, D. Goldman, and C. Merril. 1982. Variations in mutant α-actin expression accompanying incremental increases in human fibroblast tumorigenicity. *Cell* 28:259–268.

102. Hynes, R. O. 1973. "Alteration of Cell-Surface Proteins by Viral Transformation and by Proteolysis." *Proc. Nat. Acad. Sci.* 70:3170–3174.

103. Hakomori, S., 1986. Glycosphingolipid. *Sci. Am.* 254:44–53.

104. Temin, H. M. 1967. "Control by Factors in Serum of Multiplication of Uninfected Cells and Cells Infected and Converted by Avian Sarcoma Virus." *Wistar Inst. Symp.* 7:103.

105. DeLarco, J. E., and G. J. Todaro. 1978. "Growth factors from Murine sarcoma virus transformed cells." *Proc. Nat. Acad. Sci.* 75:4001–4005.

106. Sporn, M. B., and A. B. Roberts. 1985. "Autocrine Growth Factors and Cancer." *Nature* 313:745–747.

107. Cuttitta, F., D. N. Carney, J. Mulshine, T. W. Moody, J. Fedorko, A. Fischler, and J. D. Minna. 1985. "Bombesin-Like Peptides Can Function as Autocrine Growth Factors in Human Small-Cell Lung Cancer." *Nature* 316:823–826.

108. Linzer, D. I. H., and D. Nathans. 1983. "Growth-Related Changes in Specific mRNAs of Cultured Mouse Cells." *Proc. Nat. Acad. Sci.* 80:4271–4275.

109. O'Farrell, P. H. 1975. "High Resolution, Two-Dimensional Electrophoresis of Proteins." *J. Biol. Chem.* 250:4007–4021.

CHAPTER 26

The Genetic Basis of Cancer

For the first half of this century, much of cancer research was focused on identifying agents that can induce cancers in experimental animals. The first agents that were found were viruses. Eventually, certain chemicals and also radiation were shown to induce tumors in animals and transform cells in tissue culture. With a deeper understanding of the mechanism of action of viruses and of chemicals and radiation as mutagens, it became apparent that the majority of agents that can induce cancer act at the level of DNA. Today, it is widely accepted that cancer is a disease of malfunctioning cellular genes or unwanted viral gene expression. Until very recently, when cloning technologies allowed the isolation of cellular genes from spontaneously arising tumors, the only way to get at cancer-causing genes was to study those present in the genomes of certain DNA and RNA tumor viruses. The study of these genes and the mechanisms by which they transform cells provided a key to understanding the genetic basis of cancer.

AGENTS THAT CAUSE CANCER IN ANIMALS

Cancer as a Hereditary Change[1-4]

When a cancer cell divides, the two progeny cells are usually morphologically identical to the parental cell. The factors that give cancer cells their essential quality of unrestrained growth are thus regularly passed on from parent to progeny cells. These changes persist not only in tumors growing in intact animals but also in tumor cells growing in tissue culture. Hundreds of generations of growth can occur in tissue culture without appreciable reversion to a normal state. The permanence of such changes is shown not only by perpetuation of a typical morphology, but also by the ability of progeny cells to cause new tumors when injected into a tumor-free animal of genetic composition similar to the one from which the original tissue culture was obtained.

The heritability of the changes allowing unrestrained growth suggests that genetic changes within chromosomal DNA may underlie the cancerous phenotype. An early line of support for the notion of cancer as a genetic change came from experiments in which cancer cells were fused with normal cells. The 4*N* tetraploid cells that resulted from nuclear fusion often, though not always, had a normal phenotype and did not form tumors when injected into genetically identical hosts. Most 4*N* cells, however, generally are not stable and frequently divide to produce cells with lower chromosome numbers. Many of the subtetraploid cells that arose from noncancerous hybrids of normal and cancer cells were cancerous, indicating that the loss of one or more cellular chromosomes leads to the reexpression of the cancerous phenotype. Besides indicating that the cancerous phenotype is a stable property encoded in the genes of cancer cells, this result also implies that the phenotype is a recessive genetic trait, perhaps resulting from a loss of normal genetic material. Now, however, we know that some genetic changes leading to cancer are dominant while others are recessive. In some cases dominant genes have been identified that can make cells cancerous, and normal chromosomes that can suppress the cancerous phenotype have also been identified.

Somatic Mutations Are Probable Causes of Cancer[5]

The essential changes that make cells cancerous are now thought to be primarily somatic mutations, that is, mutations occurring in cells not destined to become sex cells. One line of evidence for this hypothesis comes from the fact that many cancers have one or more distinctive, abnormal chromosomes, usually resulting from the breakage and rejoining of two other chromosomes. In these cases, every cell in the tumor possesses the aberrant chromosome, but usually none of the individual's normal cells do. Presumably, the abnormal chromosome arises in a somatic cell and contributes to the cell's development into cancer. Since somatic mutations, including chromosome abnormalities, do occur, it would be surprising if some of them did not upset the normal control devices regulating cell division and thus cause cancer.

There is now substantial evidence that several somatic mutations are necessary to cause cancer. This idea arose in part from the fact that the incidence of cancer greatly increases with age. This phenomenon could be explained if a particular cell had to accumulate several mutations, each occurring randomly in time, before becoming a full-fledged cancer cell (Chapter 27).

Cancer Can Be Induced by Radiation[6]

Indirect evidence that somatic mutations are important causal factors in the induction of cancer also came from the realization that a very large number of the agents that increase the frequency of cancer are strong mutagens. Collectively, all cancer-causing agents are called **carcinogens**. Among the most potent carcinogens are the ultraviolet and ionizing forms of radiation, agents long known to be highly mutagenic. For example, exposure of skin to ultraviolet light causes skin cancer, while X-rays applied to the thyroid induce thyroid cancer. None of these various forms of radiation, however, can be used to cause all the cells in an exposed population to become cancerous. This is because they also cause many other deleterious changes, many of which lead to cell death. A radiation dose large enough to make virtu-

Figure 26-1
Benzpyrene (benzo[a]pyrene), benzanthracene, and aflatoxin B₁ (found in moldy peanuts).

Figure 26-2
Conversion of benzpyrene by enzymes of the liver into a carcinogenic (mutagenic) epoxide derivative.

ally all cells cancerous would also kill practically every cell. Not only can radiation cause simple mutations, but the ionizing form also breaks chromosomes, frequently resulting in large deletions of genetic material. Thus, radiation causes a variety of genetic changes, many of which may contribute to the cancerous state.

The deleterious effect of the ultraviolet component of sunlight on skin is exaggerated in certain individuals with a rare inherited disorder called *xeroderma pigmentosum*. These people are homozygous for a recessive mutation that inactivates a gene involved in repairing thymine dimers, the major damage caused by ultraviolet radiation, and so they develop multiple skin cancers. Presumably sufferers of xeroderma pigmentosum are unable to repair somatic mutations, some of which cause a cell to become cancerous. The increased incidence of skin cancers in people with xeroderma pigmentosum provides further support for the mutational origin of cancer.

Chemical Carcinogens Can Be Converted into Strong Mutagens in Vivo[7,8,9]

An extremely diverse set of seemingly unrelated organic and inorganic compounds have long been known to be potential carcinogens, but a firm connection between many chemical carcinogens and mutagens did not emerge until the 1960s. While certain compounds (e.g., the nitrogen mustards) that mimic ionizing radiation in their physiological consequences had long been known to be both mutagens and carcinogens, many of the most powerful carcinogens long appeared devoid of any mutagenic potential. Particularly striking was the absence of any apparent mutagenicity within a variety of highly carcinogenic fused-ring hydrocarbons, such as benzpyrene, benzanthracene, and aflatoxin (Figure 26-1). These have no special affinity for DNA, and it was therefore postulated that their carcinogenicity somehow resulted from their binding to and inactivation of key proteins. More recent evidence, however, shows that when these carcinogens enter certain cells, they become metabolized into derivatives that are not only highly mutagenic but also much more powerfully carcinogenic than their precursors. For example, benzpyrene itself causes no genetic changes, but in cells is converted into the very potent mutagen benzpyrene-7,8-diol-9,10-epoxide (Figure 26-2).

Much of the mutagenicity of such activated carcinogens arises from their ability to chemically modify DNA. These modifications lead to base substitutions during DNA replication and also to breaks in the polynucleotide backbones, which may in turn generate more extensive genetic rearrangements. In general, compounds that are transformed into strong carcinogens likewise become strong mutagens, and vice versa. There thus seems to be little doubt that much carcinogenesis is the result of changes in DNA.

Long unexplained were strikingly different responses to the same carcinogen often shown by different individuals within a species. Recent evidence indicates that such differences are due in part to genetically determined levels of the enzymes that metabolize these carcinogens. Among the most important of these carcinogen-metabolizing enzymes are the cytochromes P-450, a family of hydroxylases. While these enzymes act on most hydrocarbons and other foreign compounds to produce harmless metabolites, occasionally their metabolites are chemically reactive and, as mentioned above, subsequently react with DNA. Now there is intense interest in the regulation of expression of the cytochromes P-450. Understanding this regulation may help to explain the wide variation of metabolizing activity within the human population. Study of this variation is commonly referred to as pharmacogenetics.

The discovery that carcinogens are often activated to mutagens by enzymes present in the liver led to the development of several important assays for the potential carcinogenicity of various chemicals. The first and still most widely used of these is the Ames test (Figure 26-3; see also Chapter 12). In this test, the ability of a compound to be activated to a mutagen is estimated by incubating an indicator strain of bacteria with the compound in the presence of an extract of liver cells. The most frequently used bacteria are a strain of *Salmonella* with a mutation at the *his* locus. Bacteria that can grow and form colonies on histidine-free medium arise from revertants at this locus induced by the carcinogen, which in turn was activated by the enzymes present in the liver extract. Since almost all known carcinogens are revealed as mutagens in this test, it is now used extensively for preliminary screening of compounds that will come in contact with humans, replacing much more expensive and time-consuming animal testing.

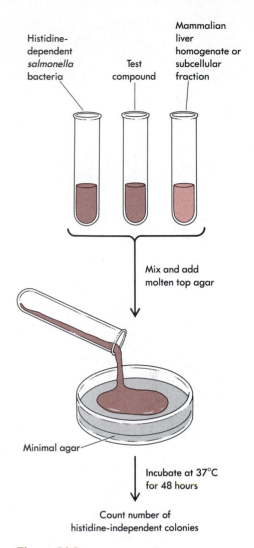

Figure 26-3
Schematic representation of the bacterial mutagenesis assay used in the Ames test.

Tumor Promoters Act Synergistically with Carcinogens[10]

While most of the chemicals described above are probably carcinogenic because they induce changes in DNA (either directly or after chemical modification), other chemicals contribute to cancer induction by a different mechanism. One group of such agents was first identified in experiments involving carcinogenesis of mouse skin. If mouse skin is painted with a carcinogen like benzpyrene, then few tumors will develop, even with multiple treatments. If, however, a single exposure to benzpyrene is followed by multiple applications of one of a variety of agents, such as the phorbol ester *12-O-tetradecanoyl phorbol-13-acetate (TPA)* (Figure 26-4), which itself is not a carcinogen, tumors will eventually develop. In this situation, the benzpyrene is

Figure 26-4
The phorbol ester tumor promoter 12-O-tetradecanoyl phorbol-13-acetate (TPA).

referred to as an *initiator* and the TPA as a *promoter*. Initiators are carcinogens that act by inducing mutations in the cell's DNA. Promoters, on the other hand, are not mutagens and alone they never induce tumors. Rather they mimic the effects of growth factors in that they transiently stimulate both cell division and altered gene expression (Chapter 25). In the case of phorbol esters this is thought to occur by activation of the cellular enzyme protein kinase C, which acts as a receptor for phorbol esters. Protein kinase C, which can phosphorylate appropriate substrate proteins, is normally activated by phospholipids, Ca^{2+}, and diacylglycerol (Chapter 25). However, phorbol esters can substitute for diacylglycerol (with which they have some structural homology), apparently by increasing the affinity of protein kinase C for Ca^{2+}. Since the role of protein kinase C in cell growth is still poorly understood, how its activation may cause an initiated cell to become a malignant one is not known. Perhaps it allows the initial mutation to become fixed in a cell and its descendants and provides a setting appropriate for the occurrence of additional mutations.

Viruses as a Cause of Cancer[11,12]

Athough the first known tumor virus, Rous sarcoma virus, was discovered in 1911, over 70 years ago, examples of additional viruses that could induce tumors in animals at first appeared so infrequently that it was generally believed that viruses could not be a major cause of cancer. But as soon as newborn animals began to be used as test systems during the early 1950s, it became much easier to show that a virus has oncogenic potential. Also initially preventing many people from accepting a viral causation of cancer was the fact that electron micrographs do not reveal the presence of viruses in most tumor cells. Now, however, we realize that under many circumstances, the genetic material of a virus may be inserted into a host chromosome in a "proviral" form without any production of virus particles. Thus, the absence of detectable viruses does not provide evidence either for or against a viral origin for a given cancer cell. We also realize now that not all viruses spread horizontally as highly contagious agents and also that viruses that are widespread in the population may contribute to cancer in only a small percent of infected individuals, so that epidemiological evidence alone cannot easily rule out viral involvement in many cancers. Furthermore, the latent period between infection and appearance of the cancer is often far greater with virus-induced cancer than with many other infectious diseases, so that the connection may be very obscure.

Viruses and Chemicals Can Transform Cells in Culture[13,14,15]

It is highly desirable to use cell culture systems to study the transformation of normal cells into malignant cells. Fortunately, some tumor viruses and chemical carcinogens can induce the appearance of cells with altered phenotypic characteristics consistent with properties of cancer cells (Chapter 25). Of particular importance from an experimental standpoint are the alterations in morphology and the loss of various kinds of growth control, since these allow transformed cells to be recognized and separated from their normal neighbors. Most carcinogenic agents (in contrast to some very efficient tumor viruses) can transform only a very small fraction of the cells exposed to them (sometimes one in a million, or less), so analysis of the changes they induce would be impossible if the transformed cells could not be readily isolated.

Transformed cells are usually isolated in one of two ways: either as **foci** of cells arising from a single transformed cell that overgrows its neighbors, or as a colony of cells grown in suspension culture (Figure 26-5). These two methods rely on two different specific alterations: loss of **contact inhibition** of growth in the case of foci, and loss of **anchorage independence** of growth in the case of suspension cells. A third property that can be used to isolate transformed cells in culture is their immortality. Primary cultures of animal cells usually have a limited life in culture and die after a maximum of about 50 generations. Many tumor cells and transformed cells, by contrast, will grow indefinitely in culture. A fourth property of transformed cells that permits their selective growth relative to normal cells is their ability to grow in low concentrations of serum. These four properties of altered growth do not always occur together during transformation, and some are due to the action of distinct genes.

Figure 26-5
The two most frequently used assays for transformation of animal cells by tumor viruses. (a) Focus assay. Photograph of a group (focus) of chicken cells transformed by RSV infection. The spherical transformed cells are easily distinguished from the background of normal cells. (Courtesy of H. Rubin, Department of Molecular Biology, University of California, Berkeley.)
(b) Colony formation in agar or methocel. A colony of rat embryo cells (REF52 line) transformed by the *ras* and adenovirus *E1A* genes. Individual cells can be seen at the borders of the colonies. (Photograph by Kazuo Maruyama.)

Use of Newborn Animals and Nude Mice to Demonstrate the Oncogenic Potential of Cancer Cells and Viruses[16,17]

A breakthrough in studying the tumorgenicity of viruses and virus-transformed cells was the use of very young, even newborn, animals. The reason is that virus-coded proteins are usually expressed on the surface of virus-transformed cells, making the cells targets for destruction by the immune system. In many species, the immunological system does not become very effective until just after birth, so in newborn animals, the oncogenicity of the virus can be detected in the absence of immunological interference.

Certain tricks can be used to temporarily suppress the immune system and thus study the oncogenicity of antigenic tumor cells in older animals as well. Large doses of ionizing radiation or immunosuppressive drugs like the cortical steroids can be applied before cells of uncertain oncogenic potential are to be tested. In recent years, increasing use has been made of *nude mice,* a mutant strain of bald mice that congenitally contain very little thymus tissue and so are incapable of mounting the strong T cell responses involved in rejecting tumor cells. In these mice, the tumorgenicity of cells (including human cells) that have different histocompatibility antigens on their surface can be tested. Interestingly, even when tumor cells are injected into genetically identical animals or into nude mice, it is usually necessary to inject over a million tumor cells to provoke a tumor. Whether this is due to residual immunological responses or to other factors that involve the survival of tumor cells in a new environment is unclear.

Antigens that appear on the cell surface and that mediate tumor rejection are called **tumor antigens** or, if they are unique to tumor cells and absent from normal cells, **tumor-specific transplantation antigens** (**TSTAs**). Tumors induced by the same virus often possess the same TSTA, while different viruses induce distinct TSTAs. It is possible to determine whether two different tumors have the same TSTA by determining whether killed cells of one tumor can elicit immune protection against live cells of a second tumor. First, the killed cells of a tumor are used to immunize an animal; then, several weeks later, live tumor cells of a second tumor are injected into the animal. If the prior immunization leads to decreased growth (transplantation rejection) of the live cells, then the two tumors are said to share TSTA. If there is no effect on the growth of the live tumor cells, then the two tumors are considered to possess distinct TSTAs.

An interesting finding is that tumors induced in rodents by chemicals or radiation possess strong tumor antigens and that each antigen is unique. The molecular basis of these carcinogen-induced TSTAs and thus the basis for their diversity is unknown.

A question of considerable importance is whether spontaneously arising tumors, particularly in humans, possess tumor-specific surface antigens and thus whether the immune system plays an important role in protecting us from cancer. At present, the best evidence for **immune surveillance** against cancer comes from studying the incidence of cancer in individuals whose immune system is defective, either as a result of a genetic defect, as a result of deliberate drug-induced suppression to prevent rejection of transplanted organs, or as a result of diseases such as AIDS that destroy immune function. The finding is that the most common forms of cancer are not more frequent in these individuals but that they have significantly higher incidences of cancers that are not widely seen in the immunologically competent population. These include Kaposi's sarcoma and B cell lymphoma. Now it is suspected that these tumors may have a viral origin and so possess tumor antigens that normally cause their rejection. In contrast, more common cancers may not possess strong TSTAs, and this may be why we are more susceptible to them. In any case, it is clear that the evolution of vertebrates with their relatively long lives must have been dependent on the development of an immunological response against certain forms of cancer.

Viral Transformation Is Mediated by Oncogenes[18]

Although studies of carcinogen- and radiation-induced tumors have yielded much information on the chemical mechanisms involved in the alteration of DNA, until recently they revealed nothing about the specific genes responsible for transforming normal cells into cancer cells. Far more insight into these genes came first from the study of viruses that cause cancer in experimental animals and transform cells in culture. This is because the viruses usually carry the cancer-inducing genes, or *oncogenes*, as part of their genome.

So far, viruses that can cause cancer and also transform cells in culture include several groups with DNA genomes—papovaviruses (SV40, polyoma, and papilloma viruses), adenoviruses, and herpesviruses—as well as the RNA-containing retroviruses. In general, the oncogenes of DNA viruses are virus-coded gene products that are also required for the normal replication of the virus. In contrast, most of the retrovirus oncogenes that have been studied are not required for replication. Rather, they are cellular genes that were transduced by chance onto the viral genome. Normal cellular genes with the potential to become oncogenes are called **proto-oncogenes** or **cellular oncogenes** (**c-oncs**).

Besides carrying oncogenes, tumor viruses that induce stable transformation usually must have the capacity to form stable associations with the infected cell, often with the cell's genome. In addition, they must not impair the ability of the cell to divide. As discussed in Chapter 24, these characteristics are typical of retrovirus infections. With most oncogenic DNA viruses, however, integration of viral DNA into the cell genome is not part of the normal replication cycle, and replication of the virus usually leads to cell death. In this case, stable transformation occurs only when the cell is nonpermissive for virus DNA replication or when the viral genome is damaged so that replication cannot occur. Also, in this case, integration of viral DNA occurs only rarely and by mechanisms that are not well understood.

DNA TUMOR VIRUSES

The Small DNA Viruses SV40 and Polyoma Induce Tumors in Newborn Animals[11,19,20]

Polyoma and SV40 were discovered by accident about 30 years ago. Polyoma turned up as a contaminant in an extract from a leukemic mouse. Rather than inducing leukemias when injected into newborn mice, the extract induced parotid (salivary gland) adenocarcinomas. The active agent was identified as a virus and named polyoma ("many tumors") because of its ability to induce tumors occasionally in the thymus, ovaries, mammary glands, adrenals, and liver, as well as inducing the more common parotid tumors. Polyoma is common in wild mice, where it apparently behaves as a harmless passenger in adults. It causes tumors only when injected into newborn mice and also newborn hamsters, rats, and rabbits.

SV40 was a contaminant in the rhesus monkey cell cultures used to prepare the Salk and Sabin polio vaccines in the late 1950s. Although it has no visible effects on rhesus monkey cells in culture, it kills cultured cells derived from African green monkeys. Ironically, monkey cells were used to prepare polio vaccines because human cells were considered more likely to harbor potentially dangerous viruses! Although SV40 was soon found to induce tumors when injected into newborn hamsters, no one has ever found evidence that it induced cancer in the early recipients of polio vaccine.

Lytic Versus Transforming Response[21,22]

To be transformed by a virus, a cell must survive the infection in good health; however, papovaviruses are frequently lytic. Cells in which a specific virus multiplies are called *permissive* cells, whereas cells in which viral growth does not occur are known as *nonpermissive* cells. Permissive cells are usually derived from an animal that is the natural host for a virus. In contrast, nonpermissive cell lines usually originate from animals incapable of multiplying a given virus. Thus, depending on the specific virus, a cell can be either permissive or nonpermissive. For example, the very well known 3T3 mouse cell lines are permissive for polyoma, a mouse virus, but nonpermissive for SV40, a monkey virus. When SV40 infects a 3T3 cell, it never multiplies but occasionally induces cell transformation. When polyoma infects the nonpermissive cell line BHK (baby hamster kidney), it can transform but not multiply, and so BHK cells are frequently used in polyoma transformation experiments.

Both the lytic and transforming responses require that only a single virus particle infect a cell. This can be deduced from the fact that when SV40 particles are added to cells, the number of cells transformed (or lysed) is directly proportional to the amount of input virus. Provided enough particles are added, virtually all the exposed cells can be transformed, indicating that all the cells growing in the culture are capable of being transformed. The probability of a single SV40 particle causing transformation, however, is very, very low. Some tens of thousands of infecting particles ordinarily must be present to give the average cell a high probability of conversion to the cancer state.

Only the Early Region of the SV40 or Polyoma Genome Is Needed for Cell Transformation[23–28]

Lytic growth of SV40 or polyoma on permissive cells can be divided into an early stage, which lasts some 8 to 10 hours and culminates in the onset of viral DNA replication, and a late stage, which commences with viral DNA replication, involves synthesis of the viral capsid proteins that package the viral DNA, and terminates with the destruction of the cell. In the early stage of the cycle, only about half the viral genome, the so-called early region, is transcribed. SV40 has two early gene products, large and small T proteins, while polyoma has three, the large, middle, and small T proteins.

The SV40 and polyoma early gene products are involved in viral DNA replication, both directly by acting on viral DNA and indirectly by inducing cellular genes needed for DNA replication. Soon after viral infection of either permissive or nonpermissive cells, there is a dramatic increase in the amount of many of the host enzymes involved in DNA synthesis (Figure 26-6). This increase occurs even when SV40 infects G_1-blocked cells whose DNA synthesis has stopped. Growth in such inhibited cells can occur despite the fact that before infection, these cells lack many of the enzymes (such as CDP reductase, dTMP synthetase, and DNA polymerase) used to make various DNA nucleotide precursors and to link them together. Thus, successful SV40 infection must somehow unlock the cellular control device that normally shuts off the synthesis of the DNA enzymes when their presence is no longer wanted. The ability to induce these

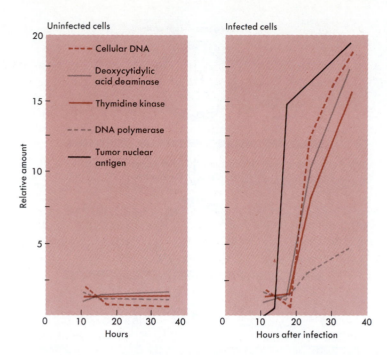

Figure 26-6
Induction of enzymes involved with DNA synthesis following infection of permissive monkey cells by SV40 virus. Before these cell-coded enzymes appear, synthesis of the virus-specific large T protein has begun. (After Dulbecco, *Sci. Amer.*, April 1967.)

enzymes means the virus can replicate in quiescent cells as well as cells that are already dividing. In the case of SV40, it is the large T antigen that is responsible for the induction of host enzymes, and purified T protein microinjected into cells can transiently produce the effect. Induction of the DNA enzymes initiates not only SV40 DNA synthesis, but also synthesis of host DNA.

Only the early region of the SV40 (or polyoma) viral genome is needed to transform cells. Numerous temperature-sensitive mutants have been isolated that prevent viral growth at high temperature. Among these mutants, those located in late genes do not block cellular transformation, while many of those in early-region genes do. Even more convincing are experiments in which fragments of the SV40 (or polyoma) genome prepared by restriction endonuclease digestion of viral DNA are introduced into cells. Fragments that encompass the early region of the genome are sufficient to induce transformation.

Cells that are nonpermissive for growth of SV40 or polyoma can only express viral early genes. Furthermore, viral DNA replication can not occur in these cells. It is the block to a successful lytic cycle and thus the failure to be lysed that allows nonpermissive cells to become transformed into cancer cells. Very rarely, SV40 (or polyoma) succeeds in transforming permissive cells. In such cases, the virus involved turns out to carry mutations that block its ability to undergo DNA replication. Without these mutations, the cell would be killed and transformation would not occur.

SV40 or Polyoma DNA Is Integrated into Chromosomal DNA in the Transformed Cell[29]

One can imagine a number of different mechanisms by which viruses might be carcinogenic. For example, the transient expression of viral gene products in a cell might initiate a cascade of events, including altered cellular gene expression, that end in transformation. Alternatively, viruses whose genomes integrate into cellular DNA might

Figure 26-7
Schematic illustration of how SV40 may integrate into a host chromosome to produce a transformed cell. Integration within the late region has the important consequence that intact late SV40 mRNA cannot be synthesized. In contrast, intact early mRNA molecules can be made, and so the cell is transformed.

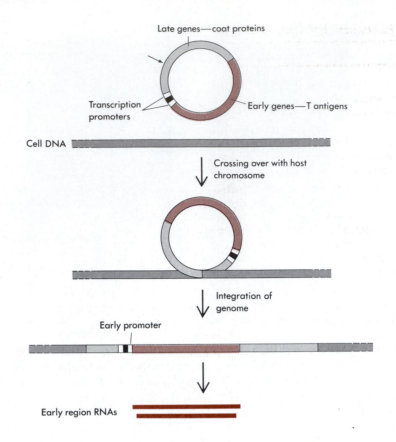

cause cancerous mutations in cellular genes in the process of integration. Yet another possibility is that viral genes become permanently associated with the cell either as independently replicating plasmids or as genes integrated into the host chromosomes and that their gene products are needed continuously to maintain the cancerous phenotype of the cell. SV40 and polyoma transform cells by the last mechanism.

DNA-RNA hybridization studies first revealed that cells transformed by SV40 or polyoma always contain viral DNA integrated into chromosomal DNA. The integrated SV40 DNA is present within the nucleus covalently bound to chromosomal DNA. The number and organization of integrated viral genomes vary from cell line to cell line, with different transformed cells containing from 1 to 20 SV40 genomes. The site of SV40 (or polyoma) DNA integration within the cell's DNA is random and does not involve sequence homology with the virus. Furthermore, when whole viral genomes integrate, the integration point on the circular viral DNA is random. However, only integration events that interrupt viral late genes can produce transformed cells, since the early genes must be intact and functioning to achieve transformation (Figure 26-7).

Stable transformation is a fairly rare event requiring large numbers of infecting virus particles. Far more common, however, is abortive transformation. In this case, viral infection of nonpermissive cells is followed by early gene expression, and the cell acquires a transformed appearance, but only transiently. Then the cell reverts to a normal phenotype and viral DNA is lost. Thus, integration of viral DNA is not necessary for transformation per se, but it is necessary for stable transformation.

Tandem copies of SV40 DNA

Figure 26-8
Tandem copies of SV40 DNA that could release complete viral genomes by recombination, and replication of the integrated SV40 DNA by the onionskin model following fusion of a transformed nonpermissive cell with a permissive cell. Recombination at identical sequences within tandem copies of the SV40 genome, for example at the two sequences within the circles, would release a complete viral genome that could replicate normally. (After P. Bullock and M. Botchan, *Gene Amplification.* Edited by R. T. Schimke, Cold Spring Harbor Laboratory. 1982. 215–224.)

Infectious Particles Are Liberated Following Fusion of Some Transformed Nonpermissive Cells with Nontransformed Permissive Cells[30,31,32]

Although SV40 (or polyoma)-transformed cells may contain a complete viral genome, replication of free viral DNA does not occur. That the viral genome is competent for replication is shown by rescue of infectious virus particles following fusion with normal permissive cells. For example, when mouse 3T3 cells transformed by complete SV40 viral genomes are fused with monkey cells, many of the fused cells yield SV40 progeny within 24 hours. Thus, something present in the permissive cell allows SV40 to carry out a complete multiplication cycle. This result is somewhat surprising, since most integrations of the SV40 genome interrupt late genes essential to the viral life cycle (see Figure 26-7). Furthermore, to yield infectious virus, a complete SV40 genome must be able to excise from the host chromosome and recircularize. Probably this can occur when more than one SV40 genome is present in tandem (Figure 26-8), a situation that could result from integration of a circular dimer. Recombination within such an integrated structure could release a complete circular viral genome, capable of replicating in the environment provided by the permissive cell. In addition, fusion with a permissive cell is thought to result in the "onionskin replication" of SV40 genomes, in which initiation occurs at the usual SV40 origin but on viral DNA still trapped within the host chromosome (Figure 26-8). The multiple copies of SV40 DNA that would result from these extra initiations would provide additional substrates for recombination events that could release complete viral genomes.

Two or Three Polyoma-Coded Early Proteins (One or Two SV40 Proteins) Maintain the Transformed State[22,33–37]

Not only is polyoma (or SV40) DNA invariably present in virus-transformed cells, but viral mRNA and early proteins can always be detected as well. These observations originally suggested that the continuing presence of viral gene products may be necessary to maintain transformation. Proof came when it was shown that when SV40 (or polyoma) early gene products are inactivated, the cell loses its trans-

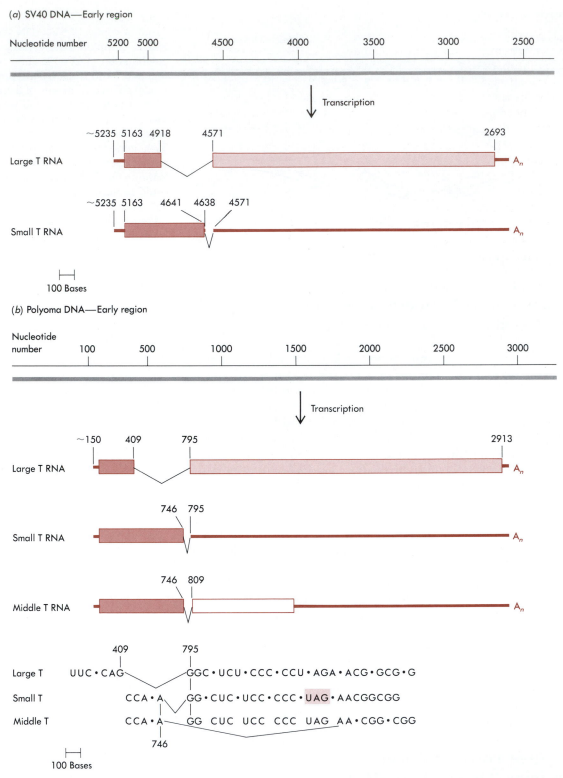

Figure 26-9

Early regions and early-region transcripts of SV40 and polyoma. (a) SV40 early region. Nucleotide numbers shown on the DNA are those assigned from the nucleotide sequence of the genome. The transcripts for large and small T proteins are shown just prior to splicing. Boxes represent coding portions of RNAs; lines represent untranslated portions. The splice in large T RNA joins two segments in different reading frames; the splice in small T RNA occurs after the stop codon that terminates translation of small T. [Based on data from V. B. Reddy et al., *Science* 200 (1978):494.] (b) Polyoma early region. All three reading frames are used to encode small, middle, and large T proteins. The nucleotide sequences at the splice points of the three transcripts are shown below the figure. The last few amino acids of small T are encoded by sequences lying just beyond the splice acceptor. [After Rassoulzadegan et al., *Nature* 300 (1982):713.]

formed phenotype. Inactivation of the viral early gene products can be accomplished if the virus used to transform the cell carries a temperature-sensitive mutation. The first successful experiments involved temperature-sensitive (ts) mutants in SV40 large T protein. Cells transformed at low temperature by some ts mutants of SV40 large T lose their transformed phenotype when placed in high temperatures and become transformed again when the temperature is lowered. This result, also accomplished for a number of other tumor virus transforming genes, is most revealing. Given the innumerable biochemical changes that distinguish normal cells from cancer cells, we might have thought that once begun, the tangle of events that result in a cancer cell could not be reversed. Instead, a few critical proteins may be sufficient to maintain these many changes.

More than 20 years have already gone into identifying the early gene products of SV40 and polyoma and determining which are needed for transformation and why. Despite this effort, we still do not have all the answers. Much of this work took place before the discovery of RNA splicing and overlapping genes. Now, sequencing of cDNAs from the early polyoma and SV40 mRNAs, in vitro translation of early mRNAs, and direct analysis of the proteins specified by the early mRNAs have led to the maps of the viral early regions shown in Figure 26-9. The three polyoma virus early proteins—small, middle, and large T—share amino-terminal amino acids, but differ in their carboxyl termini because of different splicing patterns employing different overlapping reading frames. The two SV40 early proteins—large and small T—also share amino-terminal amino acids and differ at their carboxyl termini. In this case, however, small T protein-coding sequences are contiguous, with a splice in the mRNA occurring after the stop codon that terminates small T protein.

The extensive overlapping of the polyoma and SV40 early genes has made it difficult to isolate viral mutants that affect just one gene product. Consider, for example, the much-studied polyoma mutants called **hr-t mutants.** They alter or abolish both small and middle T proteins, and although these mutants can still replicate in certain cell types, they are unable to transform cells. This result indicates that either small T, middle T, or both are involved in transformation. Also complicating the analysis is the finding that different cell types respond differently to the viruses. For example, deletion mutants that affect the SV40 small T protein can transform rat cells to produce dense-growing, morphologically altered cells, but the cells are unable to grow in the semisolid medium methocel. However, when hamster cells are used, these same mutants transform just as well as wild-type virus. While analysis of viral mutants has implicated all three polyoma early proteins and the two SV40 T proteins in the transformation of at least some cell types, a deeper understanding of the role of these proteins in transformation has come only recently from studying the transforming ability of the individual proteins separately.

Polyoma Large T Protein Immortalizes Cells, and Polyoma Middle T Protein Transforms Immortalized Cells[38-41]

The solution to separating the functions of the SV40 (and polyoma) transforming proteins from one another was accomplished by the technique of cDNA cloning, that is, making a DNA copy of an mRNA using reverse transcriptase and then cloning this DNA in a plasmid vector. Since the different T proteins are translated from differently spliced RNAs, any cDNA derived from an early-region mRNA can encode only one of the T proteins.

When a cDNA clone of the polyoma virus middle T mRNA was joined to a transcriptional promoter within an appropriate vector, the resulting construct was able to transform an established rat cell line in culture, and the transformed cells were tumorigenic when injected into animals. At first, this result was thought to mean that polyoma middle T is the only viral gene required for transformation and tumorigenicity. Then, when the same experiment was repeated using primary cells instead of established cell lines, middle T was unable to induce transformation. However, when a cDNA clone of polyoma large T mRNA was introduced into the cells along with the middle T cDNA clone, the primary cells were transformed. Infection with SV40 or polyoma virus has long been known to establish primary rodent cells into permanent cell lines as well as to transform them. In fact, permanently growing transformants would not be obtained if this were not the case, since primary cells in culture are destined to die off within 10 to 20 doublings. Alone, the cDNA clone of polyoma large T antigen can establish primary cells into perpetually growing cell lines that require much lower concentrations of growth factors than normal cells to divide. These cells, however, are not transformed in appearance and are not tumorigenic: Middle T protein is needed to bring about the transformation of the established cells. These experiments demonstrated that in polyoma, different genes bring about different aspects of the transformed phenotype, and they provided one of the first insights into why multiple mutations may be needed to produce a cancer cell.

The cDNA clones of polyoma small T also affect cell behavior in culture, causing cells to adhere less firmly to plastic substrates without altering their morphology significantly. The role of small T in creating a fully cancerous phenotype is still unclear, however, since middle and large T proteins together seem capable of inducing full transformation in at least some cell types.

Polyoma large T protein is a phosphoprotein of about 100 kdal. It is located almost exclusively in the nucleus of both infected and transformed cells. Only the amino-terminal 40 percent or so of large T is required for immortalization. Like SV40 large T protein, it is a DNA binding protein that regulates several key steps in the virus replication cycle, involving early and late transcription and viral DNA replication (Chapter 24).

Polyoma middle T protein has a molecular weight of about 56 kdal. It is phosphorylated, and at least some of it is located at the inner surface of the plasma membrane. In fact, middle T may have to be anchored in the membrane to be active, since mutants defective in membrane attachment are defective in their ability to transform. Polyoma middle T coprecipitates and copurifies from cells with a cellular protein called pp60$^{c\text{-}src}$. This protein is a tyrosine-specific kinase, and recent studies indicate that interaction with middle T may alter its enzymatic activity. As discussed later in this chapter, *src* (pronounced "sarc") is a proto-oncogene that can be mutated to become a transforming gene, and now it seems possible that the transforming activity of middle T is the result of its interaction with the c-*src* gene product.

The biochemical properties of the polyoma small T protein are only poorly understood, and as already noted, the protein may not always be required for transformation. Small T seems to be found in the nucleus.

SV40 Large T Protein Both Immortalizes and Transforms Cells[42, 43]

SV40 and polyoma DNAs do not cross-hybridize significantly, so until they were sequenced and found to encode related proteins, it was not realized how similar the two viruses are. Their early T proteins share extensive amounts of identical amino acids. SV40, however, has no middle T protein, and comparison of the DNA sequences of SV40 and polyoma shows that SV40 lacks the sequence that encodes the overlapping parts of polyoma small and middle T. Since polyoma middle T protein plays such a critical role in cell transformation, inducing both morphological alterations and anchorage-independent growth, in SV40 these functions must be taken on by SV40 large T (and possibly small T).

SV40 large T is a protein of about 90 kdal and most of it is located in the nucleus. The number of functions so far attributed to this one protein is remarkable: SV40 large T protein binds specifically to the SV40 genome (Chapter 24), playing a role in initiation of viral DNA replication and in control of early and late gene transcription; a C-terminal portion of the protein can serve a helper function for the translation of adenovirus late mRNAs in monkey cells that otherwise are nonpermissive for adenovirus growth; the protein has intrinsic ATPase activity; as already noted, SV40 large T induces the expression of host enzymes involved in DNA replication and causes a round of cellular DNA replication; and finally, a small amount of SV40 large T probably protrudes through the plasma membrane, where it serves as a tumor-specific transplantation antigen.

As with polyoma middle T, there is evidence that the transforming activity of SV40 large T protein might be exerted through its association with a cellular protein, in this case p53. A portion of large T purifies with the nuclear p53 protein. This association leads to increased stability of p53, thereby increasing the amount of p53 present in cells. At present, the normal role of p53 is unknown, and the protein has no known enzymatic function. There are, however, several lines of evidence that together suggest that the association with large T is important for transformation. First, the same cellular protein is also associated with another, completely different viral oncogene product—that of the adenovirus E1B gene, in adenovirus infected cells. Second, p53 is expressed at abnormally high levels in certain tumors. Third, under some conditions, the p53 gene can behave as an oncogene: When it is manipulated so that it is expressed at a high level, it can transform certain cell types.

Because there are so many functions in the SV40 large T polypeptide, as well as overlapping sequences coding for parts of little and large T, it is very difficult to use genetics to determine which of the activities of large T are responsible for cell transformation. Analysis of deletion mutants confirms that the ability of large T to induce S phase DNA synthesis in resting cells is a necessary but not sufficient activity for inducing transformation. It certainly makes sense that unregulated expression of a viral protein with the ability to drive a cell into S phase might be a transformation function, perhaps the basis for immortalization.

At present, there is no obvious function of SV40 that resembles the function of polyoma middle T. As with polyoma, SV40 small T is not required for transformation of many cell types.

SV40- and Polyoma-Transformed Cells Have Strong Tumor-Specific Transplantation Antigens[44,45,46]

The inability of SV40 and polyoma to induce tumors in adult animals is almost certainly because both viruses induce strong antigens on the surface of cells they transform. Some SV40 (or polyoma)-transformed cells can be serially passed in adult animals, but large numbers of cells must be used to induce progressive, fatal tumors. As expected, the SV40 transplantation antigens are distinct from those of polyoma but common to all SV40-transformed or tumor cells (likewise for polyoma). Large T protein is the SV40 TSTA, since injection of purified protein into an animal can induce immunity to SV40-transformed cells or to tumor induction by SV40 virus. By constructing deletions of the SV40 large T coding region, it has been shown that determinants of the SV40 TSTA are scattered along the large T protein. Why some of large T is stuck into the cell membrane and whether this is critical to cell transformation remain a mystery. In any case, its presence on the cell surface probably protects certain species of adult animals from what might otherwise be common and fatal tumors.

Only About 8 Percent of the Adenovirus DNA Genome Is Needed to Transform Cells[47,48]

Also intensively studied for their transforming ability are the larger DNA-containing adenoviruses. There are many isolates of human adenoviruses, and they normally multiply to produce common coldlike symptoms. But they can also transform nonpermissive or semipermissive rat and hamster fibroblasts and epithelial cells in culture, and some of them can induce tumors when injected into newborn hamsters. The adenovirus genome is a linear, double-stranded DNA molecule of about 35 kilobase pairs packaged in an icosahedral protein capsid. The molecular biology of adenovirus replication and gene expression was described in Chapter 24, and here we will focus only on the region of the genome involved in transformation and tumorigenicity.

Like SV40 and polyoma, adenovirus transformation involves stable integration of viral DNA into random locations in chromosomal DNA. Rat and hamster cells transformed by adenoviruses usually contain incomplete viral genomes. Since adenovirus lytic infection results in a complete takeover of host cell functions, including ultimately the shutoff of host protein synthesis and DNA replication, it may be that even in nonpermissive cells, a number of viral genes must be deleted to prevent killing the cells and thus allow stable cell transformation.

Adenovirus transformants always contain a segment of adenovirus DNA derived from the extreme left-hand end of the viral genome, often in multiple tandem arrays. Frequently, but not consistently, other parts of the genome are also present. The smallest restriction endonuclease fragment that can transform includes the left-hand 8 percent of the genome, about 3.2 kilobase pairs. This fragment contains two early genes called E1A and E1B (Figure 26-10).

There Are Multiple mRNAs from the Two Adenovirus Transforming Genes E1A and E1B[49-53]

E1A can be considered a pre–early gene because its expression is required for the full expression of all the other adenovirus early genes and so, in turn, the late genes. The effect of E1A on adenovirus early genes, including E1B, is at the level of transcription, so mutants defective in E1A transcribe only the E1A region efficiently. E1A is also responsible for stimulating host DNA synthesis, an effect seen transiently during lytic infection before cellular functions are shut down and an effect that may play an important role in cell transformation.

In a lytic infection, three polyadenylated mRNAs are made from the E1A region by differential splicing from a common precursor RNA (see Figure 26-10). The three mRNAs begin and end at the same nucleotide, but different splicing patterns give them different sizes and protein-coding potentials. The smallest is not an early mRNA, since it only appears after DNA replication, and it is not involved in transformation, since it is not expressed in transformed cells. The two larger E1A mRNAs are made early and are always present in transformed cells. They code for proteins that are identical in sequence except for an extra 43 amino acids in the middle of the larger one. Since cDNAs of both mRNAs have now been cloned, it is possible to test their biological activities independently. Remarkably, only the RNA encoding the larger protein can turn on early adenovirus genes. The role of the other in lytic infection is not yet known, but both mRNAs encode proteins that can play a role in cell transformation. The E1B region, like E1A, yields multiple mRNAs (see Figure 26-10). Two are present in transformed cells and encode 19 kdal and 55 kdal proteins.

Finding multiple mRNAs from the E1A and E1B regions made an understanding of adenovirus transformation seem more remote. As with polyoma and SV40, however, cDNA cloning now makes it possible to analyze the role of each gene product separately.

Figure 26-10
Region of the adenovirus genome (E1A and E1B) involved in transformation and transcripts derived from this region. The 9S RNA from E1A is made primarily late in a lytic infection and is often absent from transformed cells.

The E1A Gene Establishes Cells and Can Complement Polyoma Middle T for Transformation[54–59]

While DNA fragments making up the left-hand 8 percent of the adenovirus genome are needed to induce a fully transformed and oncogenic phenotype, DNA fragments that include just E1A, the left-hand 4.5 percent of the genome, can induce "atypical" transformation in some cells (in particular, primary rat cells). The atypical transformants are only partially transformed morphologically and are not tumorigenic, but like fully transformed cells, they are immortal and they are aneuploid (they have an abnormal number of chromosomes). Thus, E1A, like polyoma's large T gene, can rescue primary cells from senescence and cause them to grow indefinitely in culture, but E1A is not sufficient to induce full oncogenic transformation. Rather surprisingly, cDNAs for either of the large E1A mRNAs can induce prolonged growth in primary cells following their introduction by DNA transfection. Analysis of deletion mutants has localized this establishment function of E1A to within the amino-terminal 140 amino acids, a region common to the two proteins.

While the mechanism of action of the E1A gene products is unknown, it is now suspected that these nuclear proteins in some way alter the transcriptional machinery, so that expression of some genes is augmented while expression of other genes is repressed.

A DNA fragment carrying the E1B region cannot transform cells or immortalize them, even if it is first ligated to an SV40 promoter so that E1A is not needed for its transcription. However, E1B together with E1A induces complete transformation. Strikingly, a cDNA clone of polyoma middle T can also complement E1A to produce full transformation.

As yet, the role of the E1B proteins in transformation is poorly understood. Either the E1B 19K or 55K proteins seem sufficient, since mutations affecting just one or the other still allow complete morphological transformation in combination with E1A. The E1B 55K protein appears to be primarily nuclear and is associated with the same p53 protein as SV40 large T protein, while the 19K protein is located in the endoplasmic reticulum and also around, and probably within, the nucleus and may function to stimulate cell division following adenovirus infection. Studies on the roles of the adenovirus proteins in transformation are complicated by the dense packing of adenovirus genes and by the multiple uses to which single viral proteins are put to carefully regulate the complex lytic cycle. The ability of adenoviruses to transform cells is probably a small, almost inadvertent part of the viral life cycle, a consequence of their ability to turn quiescent cells into fertile environments for viral DNA replication.

Cells Transformed by a Highly Oncogenic Adenovirus May Escape Immune Surveillance by Eliminating Surface Histocompatibility Antigens[60]

Different human adenoviruses differ in their ability to induce tumors in animals. For example, adenovirus type 12 (Ad12) is strongly oncogenic, inducing tumors in rodents in 1 to 3 months, while Ad5 is nononcogenic. In vitro, however, both Ad12 and Ad5 viruses or their purified DNAs can transform rodent cells. But the tumorigenicity of

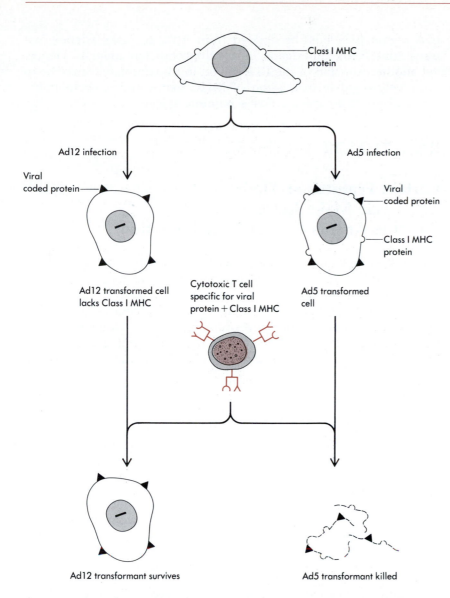

Figure 26-11
Highly oncogenic Ad12, but not weakly oncogenic Ad5, may suppress expression of class I MHC gene products on the cell surface, making transformed cells invisible to cytotoxic T cells and hence increasing their oncogenicity.

the resulting transformed cells reflects the oncogenicity of the transforming viruses. Thus, cells transformed by Ad12 will form tumors when injected into syngeneic rats, but cells transformed by Ad5 will not. This difference is also true of cells transformed by only the E1 regions of Ad12 or Ad5.

If the tumorigenicity of the Ad12 and Ad5 transformed cells is tested in immunodeficient animals, such as nude mice, then the Ad5 transformed cells fare better. They are still less tumorigenic than Ad12 transformants, but they can induce tumors in some animals. This last result suggests that the inability of Ad5 to induce tumors may in part be because the cells it transforms are immunologically rejected, while those transformed by Ad12 escape immune surveillance. Recently, a number of Ad12-transformed rat cells, but not Ad5 transformants, were shown to lack the rat class I major histocompatibility complex (MHC) gene product on their surface. Since cytotoxic T cells that kill virus-infected cells can only recognize viral antigens on the cell surface together with MHC class I antigens (Chapter 23), the absence of the MHC antigen would make the viral antigens, and hence the transformed cell, invisible to the killer T cells (Figure 26-11). The Ad12-transformed cells, but not the Ad5 transformants, lacked mRNA for the MHC class I heavy-chain gene. Furthermore, it is the

E1A region that must be derived from Ad12 to make adenovirus-transformed cells tumorigenic in immunocompetent animals. Hence, yet another function of the E1A region in transformation may be to negatively regulate expression of MHC genes and so help transformed cells to escape the host's immune system.

RNA TUMOR VIRUSES

Certain Features of Their Life Cycle Make Retroviruses Suitable Agents of Cell Transformation[61,62,63]

While the DNA tumor viruses encompass a variety of structurally diverse entities, those RNA viruses that have the capacity to cause cancer are all morphologically similar and probably are descended from a single ancestral virus. All belong to the retrovirus group. Unlike all other viruses, retroviruses replicate their RNA genome via a DNA intermediate that becomes permanently integrated into host cell DNA. Thus, like many other agents that cause cancer, oncogenic RNA viruses alter DNA.

A number of features of the retrovirus life cycle are central to the process of transformation (Figure 26-12; Chapter 24). Shortly after infection, the single-stranded viral RNA genome is copied into a linear molecule of double-stranded DNA. This DNA is longer than the viral RNA because of the presence of a specialized sequence, the long terminal repeat (LTR) of about 300 to 1000 base pairs found at each end. The LTR is derived from a combination of sequences present at the 3' end (U_3), the 5' end (U_5), or both ends (R) of the RNA genome and has the structure U_3–R–U_5. Following synthesis, the viral DNA is integrated into cellular DNA so that the ends of the LTR are directly joined to cellular sequences to form a stable structure, the provirus. Unlike DNA tumor virus integration, this is a highly regular and specific process essential for efficient replication of the virus. It is, however, random with respect to the site in cell DNA where the integration occurs.

Subsequent steps in viral replication are carried out by normal cell systems. Viral RNA is synthesized using cell RNA polymerase II, the enzyme responsible for cellular mRNA synthesis. Transcripts can serve either as mRNAs for virion proteins (in some cases after splicing) or as new genomes. Signals directing efficient transcription of the provirus are present in the U_3 portion of the LTR.

Several aspects of this cycle are of special importance to transformation by retroviruses. First, integration of viral DNA always occurs in an infected cell. Once integrated, the provirus is stable indefinitely. Second, infection of permissive cells does not kill them or noticeably impair their ability to divide. Thus, it is usual for a permissive cell to be transformed by a retrovirus and constantly release infectious virus. Third, defective viruses can be packaged with viral proteins supplied by a coinfecting virus and then can complete a new round of infection including provirus formation and expression in the absence of the helper virus. This means that all the viral genes (e.g., *gag*, *pol*, and *env*) can be missing from the genome, and yet, as long as its terminal sequences are intact, the genome can still be copied into viral DNA and the DNA can be integrated and expressed.

Figure 26-12
Replication of a retrovirus genome. The single + stranded RNA genome is copied to yield double-stranded DNA, which circularizes and integrates into cellular DNA. Integration is random with respect to the cellular sequence at the site of integration, but specific with respect to the viral genome. Transcription of the integrated provirus by cellular RNA polymerase II yields full-length RNA that serves either as mRNA for the synthesis of *gag* and *pol* gene products or as progeny genomes, as well as yielding a spliced *env* mRNA.

Only Oncogene-Carrying Retroviruses Transform Cells in Vitro[64,65]

Retroviruses are widespread in nature, and the many hundreds of isolates can be classified into groups according to biological properties, morphology, and genome structure (Table 26-1). Four types of particles, A, B, C, and D, have been identified by electron microscopy. Most of the cancer-causing viruses that have been studied extensively are C-type oncoviruses (Figure 26-13), although the important mouse mammary tumor viruses have the B-type morphology. The C-type virus group itself consists of a vast array of viruses. Most C-type viruses can neither transform cells in culture nor induce tumors in animals. Many viruses of this type are **endogenous viruses** (Chapter 24) whose proviruses inhabit the germ line of various species and which are frequently expressed as infectious but benign virions. It remains a matter of debate whether these viruses are ancestral to the oncogenic **exogenous viruses** or vice versa. Nevertheless,

Table 26-1 Classification of Members of the
Retrovirus Family: Subfamilies

Oncoviruses	Oncogenic and closely related nononcogenic retroviruses of reptiles, fish, birds, rodents, cats, primates, etc. Examples include Rous sarcoma virus and all other known retroviruses that carry cell-derived oncogenes; endogenous and exogenous ALVs (avian leukosis viruses) and MuLVs (murine leukemia viruses); MMTVs (mouse mammary tumor viruses); and HTLV-I (human T cell lymphotropic virus).
Lentiviruses	"Slow" viruses that induce neurological impairment and chronic pneumonias. Examples include visna virus of sheep and caprine encephalitis-arthritis virus (CAEV) of goats; HIV/HTLV-III/LAV (AIDS virus) may also belong to this group.
Spumaviruses	Viruses that induce persistent infection without evidence of pathogenesis in a number of mammalian species; also known as foamy viruses.

their continued existence as chromosomally inherited viruses requires that they be adapted to do little or no harm whatsoever to their host.

Figure 26-13
Electron micrographs of budding or extracellular retroviruses. (a) Budding B-type particle. (b) Extracellular B-type particle. (c) Budding C-type particle. (d) Extracellular C-type particle. (Courtesy of D. L. Fine and M. Gonda.)

Most oncogenic exogenous retroviruses belong to one of two types. By far the most common type has a genome structure indistinguishable from its benign endogenous relatives. Viruses of this type, called **nondefective viruses** or **replication-competent viruses**, usually induce leukemias after a fairly long latent period and have been studied extensively in chickens, mice, and cats. While clearly related to the nonpathogenic retroviruses, these oncogenic retroviruses differ from them by multiple mutations and recombinations, and as a result, some are quite highly infectious, horizontally transmitted agents. Feline leukemia virus, for example, is considered the second leading cause of death among pet cats (after automobile accidents).

The second type of oncogenic retrovirus differs fundamentally from the first in that members of this group carry oncogenes, which have nothing to do with virus replication. The oncogenes give the virus the ability to induce tumors in vivo after a short latent period, sometimes as little as a few days. They also usually give the virus the ability to transform cells in culture, a property that is almost essential to the molecular analysis of a transforming gene. This class of retroviruses is called **acute transforming viruses**. Rous sarcoma virus belongs to this group, and since its isolation over 70 years ago, about 50 more viruses of this type have been found in chickens, turkeys, mice, rats, cats, and monkeys (Table 26-2). About 25 different oncogenes have been identified in these viruses. Many of the acute transforming viruses, like the Rous sarcoma virus, were identified in extracts of spontaneously arising tumors as agents that could reproducibly induce specific forms of cancer. Others were obtained from rare solid tumors that arose after injecting an exogenous oncogenic C-type virus that normally would induce only leukemias.

The Oncogenes of Acute Transforming Retroviruses Code for Proteins That Maintain the Transformed State[66–69]

The integration of only a single provirus of Rous sarcoma virus (RSV) is sufficient to transform a susceptible normal cell into its cancerous equivalent. Transformation is rapid, and within 24 hours of an RSV infection, a culture of normal primary chicken fibroblasts can become synchronously converted into rounded transformed cells (see Figure 26-5). Compelling evidence that RSV codes for a protein that induces transformation came in 1970 from the isolation of RSV mutants that are temperature-sensitive for transformation. Unlike ts mutants of SV40 large T that affect both transformation and virus replication, the RSV mutants replicate normally at both 35°C and 41°C. If cells infected (but not transformed) at 41°C are placed at 35°C, they assume a transformed phenotype, and if they are shifted back to 41°C, they rapidly revert to a normal morphology. As with SV40, this profoundly important experiment showed that an RSV gene product not only initiates but also maintains the transformed phenotype.

Early biochemical definition of the transforming gene of RSV came from analyzing the genomes of RSV variants that cannot transform at any temperature. These viruses arise spontaneously at high frequency in all RSV stocks. They are called **transformation-defective (td) mutants**, and they have RNA genomes that are about 20 percent smaller than transformation-competent RSV. Thus, td mutants arise by deletions that occur during RSV replication. The gene defined by

Table 26-2 Some Acute Transforming Retroviruses and Their Oncogenes

onc	Retrovirus isolates	v-onc origin	v-onc protein	Virus disease
src	Rous sarcoma virus (RSV)	chicken	pp60^{src}	sarcoma
fps*	Fujinami sarcoma virus (FuSV)	chicken	P130$^{gag\text{-}fps}$	sarcoma
	PRCII-avian sarcoma virus (ASV)	chicken	P105$^{gag\text{-}fps}$	sarcoma
fes*	Snyder-Theilen feline sarcoma virus (FeSV)	cat	P85$^{gag\text{-}fes}$	sarcoma
	Gardner-Arnstein FeSV	cat	P110$^{gag\text{-}fes}$	sarcoma
yes	Y73-ASV	chicken	P90$^{gag\text{-}yes}$	sarcoma
fgr	Gardner-Rasheed FeSV	cat	P70$^{gag\text{-}actin\text{-}fgr}$	sarcoma
ros	UR2-ASV	chicken	P68$^{gag\text{-}ros}$	sarcoma
abl	Abelson murine leukemia virus (MLV)	mouse	P90-P160$^{gag\text{-}abl}$	pre-B cell leukemia
	Hardy-Zuckerman (HZ2)-FeSV	cat	P98$^{gag\text{-}abl}$	sarcoma
ski	SKV	chicken	P110$^{gag\text{-}ski\text{-}pol}$	squamous carcinoma
erbA	Avian erythroblastosis virus (AEV)	chicken	P75$^{gag\text{-}erb\text{-}A}$	Erythroblastosis and sarcoma
erbB			gp65^{erbB}	
fms	McDonough (SM)-FeSV	cat	gp180$^{gag\text{-}fms}$	sarcoma
fos	FBJ (Finkel-Biskis-Jinkins)-MSV	mouse	pp55^{fos}	osteosarcoma
mos	Moloney MSV	mouse	P37$^{env\text{-}mos}$	sarcoma
sis	Simian sarcoma virus (SSV)	monkey	P28$^{env\text{-}sis}$	sarcoma
	Parodi-Irgens FeSV	cat	P76$^{gag\text{-}sis}$	sarcoma
myc	MC29	chicken	P100$^{gag\text{-}myc}$	sarcoma, carcinoma, and myelocytoma
myb	Avian myeloblastosis virus (AMV)	chicken	p45^{myb}	myeloblastosis
	AMV-E26	chicken	P135$^{gag\text{-}myb\text{-}ets}$	myeloblastosis and erythroblastosis
rel	Reticuloendotheliosis virus (REV)	turkey	p64^{rel}	Reticuloendotheliosis
kit	HZ4-FeSV	cat	p80$^{gag\text{-}kit}$	sarcoma
raf*	3611-MSV	mouse	P75$^{gag\text{-}raf}$	sarcoma
H-ras	Harvey MSV (Ha-MSV)	rat	pp21^{ras}	sarcoma and erythroleukemia
	RaSV	rat	P29$^{gag\text{-}ras}$	sarcoma?
K-ras	Kirsten MSV (Ki-MSV)	rat	pp21^{ras}	sarcoma and erythroleukemia
ets	AMV-E26	chicken	P135$^{gag\text{-}myb\text{-}ets}$	See above for *myb*

*fps, the oncogene of several chicken sarcoma viruses, is derived from a chicken gene, c-fps, that is the chicken equivalent of c-fes, a gene of cats first discovered as the oncogene fes in several feline sarcoma viruses. It was not realized that fes and fps were homologous when they were assigned different names. Note that AGV and AMV-E26 each have two oncogenes.

SOURCE: R. Weiss, N. Teich, H. Varmus, and J. Coffin, eds., *RNA Tumor Viruses,* 2nd ed. (Cold Spring Harbor, N.Y.: Cold Spring Harbor Laboratory, 1985).

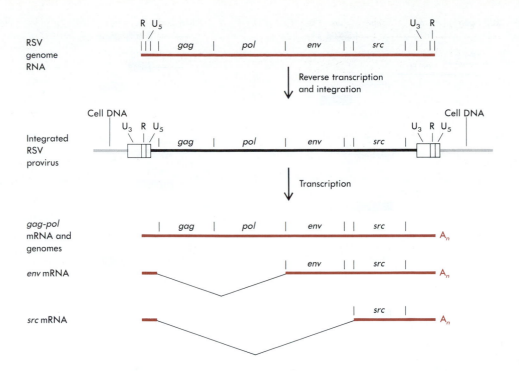

Figure 26-14
The Rous sarcoma virus genome
(Prague strain) and its mRNAs.

this deletion, which gives RSV its ability to transform cells and induce tumors, is called *src* or sometimes the v-*src* (for viral) oncogene. The *src* oncogene maps near the 3' end of the RSV genome, but different acute transforming retroviruses have different oncogenes in different locations within their genomes, and different oncogenes code for distinctly different proteins. Like *src*, these viral oncogenes are never involved in virus replication, only in cellular transformation.

The protein product of the *src* gene is translated from a spliced mRNA initiated, like viral RNAs, within the 5' LTR (Figure 26-14). Thus, cells infected by RSV contain three virus-containing RNA species: genome-length RNA, a spliced *env* RNA, and a spliced *src* RNA.

Except for Rous Sarcoma Virus, Transforming Retroviruses Are Defective and Require a Helper Virus to Grow[70–73]

Since RSV contains all replicative genes in addition to *src*, a cell infected by a single particle of RSV is transformed and also becomes a producer of RSV progeny. However, in all other retroviruses that carry oncogenes, the acquisition of the oncogene has been accompanied by the loss of viral genes needed for replication (Figure 26-15). As a result, all the acute transforming retroviruses (except RSV) are defective and can only replicate in the presence of a **helper virus** that supplies the missing gene products (Figure 26-16). A defective viral genome can be packaged perfectly well as long as it contains a special set of nucleotides near its 5' end required for recognition by virion proteins. Thus, stocks of acute transforming viruses are always mixtures, with some particles containing defective genomes and some containing helper virus genomes. When these mixtures are diluted so that a cell is infected by a single particle containing a defective viral genome, the resulting cell can be transformed. The incoming defec-

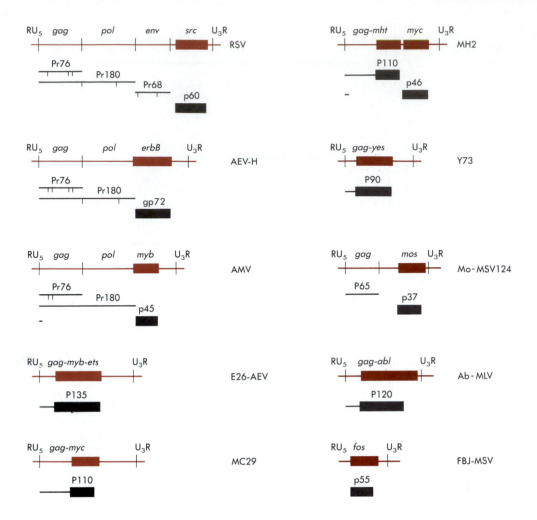

Figure 26-15
Genomes and encoded proteins of some transforming retroviruses of chickens and mice. Note variable locations of oncogenes within the viral genomes and varying deletions of *gag, pol,* and *env* genes in all but the nondefective Prague strain of Rous sarcoma virus (RSV). AEV-H (avian erythroblastosis virus): AMV (avian myeloblastosis virus): E26-AEV (E26 isolate of AEV): MC29 (avian myelocytomatosis virus): MH2 (another isolate of myelocytomatosis virus with altered pathogenicity relative to MC29): Y73 (an avian sarcoma virus): Mo-MSV124 (Moloney murine sarcoma virus): Ab-MLV (Abelson murine leukemia virus): FBJ-MSV (Finkel-Biskis-Jinkins mouse osteosarcoma virus). [After R. Weiss, N. Teich, H. Varmus, and J. Coffin, eds., *RNA Tumor Viruses,* 2nd ed. (Cold Spring Harbor, N.Y.: Cold Spring Harbor Laboratory, 1985), pp. 579 and 765.]

tive genome can be copied by reverse transcriptase as long as it possesses the 5′ end primer binding site for initiation of − strand DNA synthesis and the 3′ end origin of + strand DNA synthesis (Chapter 24). Furthermore, as long as its LTRs are intact, its proviral DNA can be integrated normally and will be transcribed into RNA. Thus, the cell will be transformed because the viral oncogene is expressed, but the transformed cell does not produce infectious progeny. When such a **nonproducer** transformed cell is infected with nondefective helper virus, it begins to produce a mixture of defective transforming viruses and nondefective viruses and so becomes a permanent source of infectious transforming virus.

Retrovirus Oncogenes Are Derived from Normal Cell Genes[74,75]

Before molecular cloning made it easy to get large quantities of pure retrovirus oncogenes, the best way to zoom in directly on an oncogene was to use reverse transcriptase to make DNA complementary to the genome of a virus like RSV that contains an oncogene. This DNA could then be hybridized with an excess of viral RNA from an RSV td mutant lacking an oncogene, and the unhybridized DNA could be recovered and used as a probe. When this oncogene-specific probe for the RSV *src* gene was hybridized to chicken DNA, it became

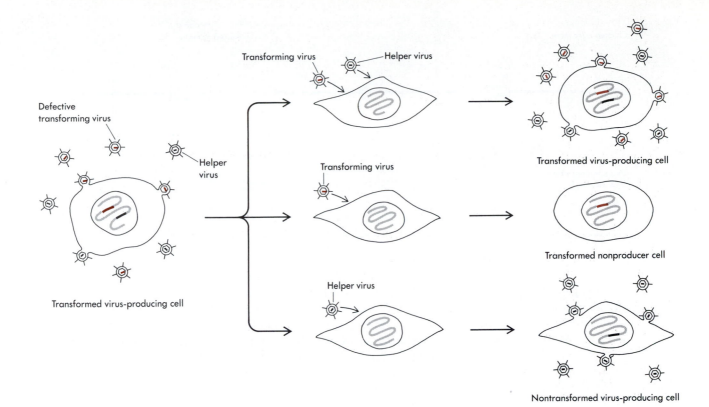

clear that a *src* gene was present in all chickens, in both normal cells and cancer cells. The cellular *src*-related gene, called c-*src*, is not part of an endogenous retrovirus but is a normal cellular gene that has been transduced into the RSV genome in some manner analogous to bacteriophage transduction of a bacterial gene. The detection of oncogenes in normal cell DNA, first seen for the cellular counterpart of v-*src* in 1976, made it clear that long before gene-cloning techniques were invented, acute transforming retroviruses had isolated and handed us cellular genes with the potential to cause cancer.

Most Cellular Oncogenes Contain Intervening Sequences That Are Absent in Their Viral Counterparts[76,77,78]

The v-*src* oncogene of RSV is about 1700 nucleotides long and is transcribed from the viral promoter in the 5' LTR to yield a spliced mRNA containing viral sequences joined to *src* sequences. Not surprisingly, c-*src* mRNA, about 4 kilobases in size, has no retroviral sequences on it. Furthermore, when the chicken c-*src* gene was cloned, it was found to be a large gene consisting of 13 exons and spanning some 8 kilobases (Figure 26-17). Like most normal cell genes, almost all the proto-oncogenes that have been examined so far have intervening sequences. Not surprisingly, viral oncogenes (v-oncs) never have whole introns. Since they are transmitted as part of a viral RNA, any intron initially derived from its c-onc parent would be removed during normal processing, with the final intron-free mRNA becoming the viral genome and template for subsequent DNA synthesis and proviral integration.

Figure 26-16
Transformation of cells by viruses obtained from a cell producing a defective transforming virus and a nondefective helper virus. Infection by a defective transforming virus alone generates a transformed nonproducer cell; infection by a helper virus alone generates a nontransformed, virus-producing cell; and infection by both viruses generates a cell that again produces both types of progeny. Transforming viruses can be rescued from nonproducer cells by infection with a helper virus.

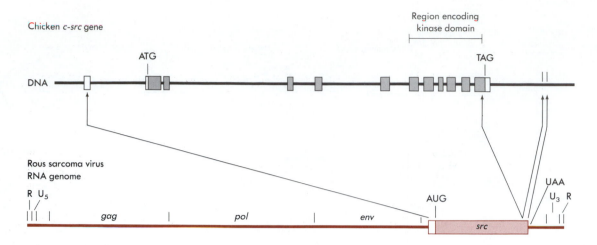

Figure 26-17
Structure of the chicken c-*src* gene and structure of the Rous sarcoma virus genome, showing the relationship of nucleotide sequences present in c-*src* and v-*src*. Exons of c-*src* are shown as boxes, with white portions representing untranslated portions of exons; arrows indicate regions of the c-*src* gene and flanking sequences from which v-*src* was generated, primarily by removal of introns. A short sequence that does not encode part of c-*src* and that lies 3′ of the gene is found at the 3′ end of the v-*src* gene, and part of the most 3′ exon of c-*src* is absent from v-*src*. A stop codon (UAA) for the translation of v-*src* is present in the viral sequences that lie just 3′ of the v-*src* insert. [Based on data from T. Takeya and H. Hana-fusa, *Cell* 32 (1983):881–890.]

Many of the oncogenes found in retroviral genomes are substantially altered relative to their cellular counterparts. Often, transduction of c-oncs involves loss of part of the cellular gene and fusion of the remainder to a viral gene, frequently to *gag* (see Figure 26-15). The mechanisms by which retroviruses transduce c-oncs is not known with certainty, but a highly plausible model is shown in Figure 26-18. According to this model, integration of a provirus in the vicinity of a cellular oncogene is followed by a deletion event that brings the 5′ half of the provirus near to, or even fuses it to, c-onc sequences. An RNA transcript of this DNA would carry the retrovirus packaging sequence at its 5′ end and so would be efficiently packaged. In a particle containing this RNA and a helper virus genome, reverse transcriptase can copy both genomes by a copy choice type mechanism so as to transfer viral 3′ end sequences needed for replication to the novel oncogene-carrying genome.

Cellular Oncogenes Are Highly Conserved in Evolution[79–82]

RSV was isolated from a chicken, and its v-*src* gene was derived from chicken c-*src*. Nonetheless, soon after it was first made, the RSV v-*src*–specific probe was found to hybridize to the DNA of all vertebrates, and later, even to the DNA of flies. Thus, all organisms from insects to humans have a c-*src* gene. By evaluating the thermal stability of the hybrids formed between the cDNA *src* probe and various vertebrate cellular DNAs, it could be estimated that the c-*src* genes of chickens and humans have diverged no more than 10 to 15 percent in nucleotide sequence. The astounding conservation of the *src* gene across such vast evolutionary distance implies that proto-oncogenes must play essential fundamental roles in cellular metabolism.

Essentially similar results have been found with the other viral oncogenes, although the extent of nucleotide sequence homology of c-oncs across species is different for different genes. As expected, however, amino acid sequences are frequently far more conserved than nucleotide sequences. As a result, homologies between oncogenes that could not be detected by nucleic acid hybridization can become apparent when DNA sequences are used to predict amino acid sequences of oncogene proteins.

Recently the v-*ras* oncogene of Harvey sarcoma virus, a gene derived from a rat c-*ras* gene, was found to hybridize to yeast DNA under conditions of very low stringency. Using v-*ras* as a probe, two homologous yeast *RAS* genes were cloned. Although the yeast *RAS* proteins are larger than the vertebrate ones, the amino acid sequence homology in the first two-thirds of the proteins is an astounding 75 percent (Chapter 27).

The fact that oncogenes are conserved explains why acute transforming viruses isolated from different species sometimes rescue homologous oncogenes (see Table 26-2). Thus, v-*fps* present in a chicken sarcoma virus is the chicken homolog of v-*fes*, present in a cat sarcoma virus. Similarly, the oncogene *abl* was rescued from the mouse genome by the Abelson mouse virus that rapidly induces B cell leukemias, and many years later from the cat genome by a feline sarcoma virus.

Figure 26-18
Model for how a transforming retrovirus may be generated. The first step is integration of a provirus in the vicinity of a cellular proto-oncogene. This might be followed by a deletion of DNA. Alternatively a read through transcript made from the structure at the top of the figure might be aberrantly spliced to generate a novel mRNA encoding a *gag-onc* fusing protein.

Some Cellular Oncogenes Can Be Activated by Ligation to Efficient Transcriptional Promoters[83–88]

Not only are proto-oncogenes present in normal cells, but many of them are expressed in many different cell types in a tissue-specific and developmentally specific fashion. Clearly, proto-oncogenes have an important function, and this normal function is not to cause cancer. Why, then, are these genes able to do so when they reside in retrovirus genomes? Two fundamentally different but not necessarily exclusive explanations have been found. First, when they reside in retroviral genomes, v-oncs fall under the control of viral transcriptional signals located in the LTR. Thus, instead of carefully regulated gene expression, they may be expressed constitutively and often at extremely high levels. This might convert a normal gene to an oncogene by leading either to increased concentration of the gene product or to its expression in an incorrect cell type. Second, alterations in the structure of v-onc proteins relative to their c-onc relatives are frequently found. These convert the harmless cellular genes to oncogenes—for example, by altering the regulation of an enzyme or its intracellular location. Examples of both types of oncogene activation are known.

Evidence that high levels of expression of normal cellular oncogenes can transform cells first came from experiments with c-*mos*, which is the cellular homolog of the oncogene (v-*mos*) of a mouse sarcoma virus. When the cloned c-*mos* gene is ligated to a retrovirus LTR and the DNA transfected into 3T3 cells, transformed foci appear. The transformed cells contain the transfected *mos* genes and elevated levels of the corresponding mRNAs, indicating that the abnormal expression of a normal gene can transform a cell.

The type of experiment just described does not work well, however, when c-*src* is used in place of c-*mos*. High levels of c-*src* expression do not transform cells efficiently. Analysis of the transforming ability of recombinants between c-*src* and v-*src* shows that v-*src* has undergone alterations in amino acid sequence relative to c-*src* and that these changes, as well as a high level of expression, are what make v-*src* a transforming gene.

How Many Mechanisms Are There of Oncogene Action?[89,90]

To summarize, using RSV as a model, we have seen that retroviruses that transform cells in culture do so because they have acquired oncogenes. Retroviral oncogenes are not required for virus replication. Rather, they are modified normal cellular genes that have come under the control of viral transcription signals and have often suffered mutations that in some way alter their activity to permit cell transformation. Thus, the availability of these oncogenes provided researchers with the first useful tools to learn the biochemical reactions involved when a cell is transformed into a cancer cell. Initially, it was thought that once the proteins encoded by oncogenes were known, it would be relatively simple to learn what their normal and altered functions were. In practice, this has proved to be a very complex task. We now know of about 30 oncogenes, but we know the normal function of only a few. Nevertheless, the study of oncogene products is a very active area of investigation that has already yielded considerable insight into mechanisms regulating the growth of normal cells.

Although there are now about 30 known cellular genes that can be converted to oncogenes, the number of mechanisms by which they act is probably considerably smaller than this. This was first suggested by comparing nucleotide and predicted amino acid sequences, which, surprisingly, revealed that a number of oncogene products thought to be unrelated, in fact had similar amino acid sequences.

Table 26-3 Oncogene Families

Oncogene	Subcellular location of protein	Properties or normal function of protein
Class I: Protein kinases		
src	Plasma membrane	Tyrosine-specific protein kinase
yes	Plasma membrane	Tyrosine-specific protein kinase
fgr	?	Tyrosine-specific protein kinase
abl	Plasma membrane	Tyrosine-specific protein kinase
fps (fes)	Cytoplasm	Tyrosine-specific protein kinase
erbB	Plasma membrane (transmembrane)	EGF receptor/tyrosine-specific protein kinase
fms	Plasma membrane (transmembrane)	CSF-1 receptor/tyrosine-specific protein kinase
ros	Plasma membrane (transmembrane)	Tyrosine-specific protein kinase
kit	Plasma membrane	
mos	Cytoplasm	Serine/threonine protein kinase
raf (mil)	?	Serine/threonine protein kinase
Class II: GTP binding proteins		
H-ras	Plasma membrane	Guanine nucleotide binding protein with GTPase activity
K-ras	Plasma membrane	Guanine nucleotide binding protein with GTPase activity
Class III: Growth factors		
sis	Secreted	Derived from a gene that encodes PDGF
Class IV: Nuclear proteins		
myc	Nucleus	
myb	Nucleus	
fos	Nucleus	
ski	Nucleus	
Class V: Hormone receptor		
erbA	Cytoplasm	Thyroid hormone receptor
Unclassified:		
rel	?	
ets	?	

From this analysis, we can divide oncogenes into several distinct families of related protein sequences (Table 26-3). Oncogenes in the same family encode proteins with similar enzymatic activity or intracellular location, suggesting that their role in transformation may be similar.

src **Encodes a 60 kdal Phosphoprotein with Protein Kinase Activity**[86,91–95]

Although their existence had been known for many years, it was only in 1977 that the first oncogene protein, the v-src gene product, was detected. The crucial step was the development of antisera to the protein. When RSV induces tumors in rabbits, the rabbits produce antibodies to the src gene product as well as to viral structural gene products. The tumor sera could precipitate a 60 kdal protein from

Figure 26-19

Detection of the viral *src* gene product. Normal cells (chicken embryo fibroblasts), cells infected with Rous sarcoma virus, or cells infected with a transformation defective (td) mutant of Rous sarcoma virus that has a deletion of its *src* gene were grown in medium containing ^{35}S methionine so that their proteins would be radioactively labeled. Lysates of the cells were then mixed either with serum from normal rabbits, serum from rabbits that had been injected with Rous sarcoma virus and had developed tumors (TBR serum), or TBR serum that was mixed with disrupted virions to adsorb all the antibodies that can bind to viral proteins (the *src* gene product is not present in virions). Immune complexes were then precipitated, collected, and the proteins separated by electrophoresis on SDS polyacrylamide gels. The photo shown is of an autoradiogram of such a gel. When rabbits develop tumors induced by Rous sarcoma virus, they produce antibodies to all the viral coded proteins, even those that are only present inside the transformed cell (like the *src* gene product). Presumably this occurs because some tumor cells lyse and their contents are then seen by the immune system. T7: Molecular weight marker proteins. Lanes 1, 2, 3: normal cell proteins precipitated with (1) normal serum, (2) TBR serum, (3) TBR serum-blocked with viral proteins. Lanes 4, 5, 6: td sarcoma virus infected cell proteins precipitated with (4) normal serum, (5) TBR serum, and (6) TBR serum-blocked. Lanes 7, 8, 9: nondefective sarcoma virus transformed cell proteins precipitated with (7) normal serum, (8) TBR serum, and (9) TBR serum-blocked. In the autoradiogram shown, only the 60K protein band in lanes 8 and 9 fits the criteria for a product of the viral *src* gene: it is present in RSV transformed cells but not in cells infected by the td virus and its precipitation is not inhibited when viral proteins are used to block the TBR serum. Most of the other bands that are seen in viral infected cells are *gag, pol,* and *env* gene products. [From A. R. Purchio, E. Erikson, J. S. Brugge, R. L. Erikson, *Proc. Nat. Acad. Sci.* 75 (1978):1567–1571.]

RSV-transformed cells (but in early experiments, not from uninfected or td RSV-infected cells), while antisera to viral structural proteins failed to precipitate this protein (Figure 26-19). Furthermore, the same protein could be identified as a product of in vitro translation from RNA fragments of the 3′ third of the RSV genome.

The *src* gene product is a phosphoprotein and is designated pp60^{src}. Its amino terminus is modified by myristylation (covalent attachment of fatty acid), and in cells, the bulk of the protein is found attached to the inner surface of the plasma membrane. This attachment is essential for the oncogenic function of pp60^{src}, since mutants in the N terminus that cannot be myristylated do not attach well to the membrane and are defective in transformation. If cells transformed by RSV with a ts *src* mutation have their nucleus removed by exposure to the drug cytochalasin B, the remaining intact **cytoplasts** still have a temperature-sensitive transformed morphology. This remarkable result shows that at least some of the effects of *src* are not achieved via an effect on cellular gene expression or DNA replication, either directly or indirectly.

Early experiments failed to detect a c-*src* product in uninfected cells, but later, some tumor antisera detected low levels of pp60^{c-src} in normal, uninfected cells. This is consistent with the presence of a few copies of c-*src* mRNA in virtually all uninfected cells. The pp60^{c-src} differs from pp60^{v-src} in just a few amino acids, mostly at its C terminus, and, like v-*src*, is localized at the inner surface of the plasma membrane.

Given a protein that can transform normal cells to cancer cells, it is, to say the least, not easy to guess what its function is. Soon after it

Figure 26-20
Demonstration of protein kinase activity in immunoprecipitates of pp60^{src}. Methodology is diagrammed in part (a), experimental results are shown in part (b). Cell extracts were prepared from chick embryo fibroblast cultures that were either uninfected [tracks 1a and 1b of part (b)], infected with avian leukosis viruses that lack a transforming gene [tracks 2], or infected with Rous sarcoma virus [tracks 3]. Each extract was mixed with either normal rabbit serum [a lanes in part (b)] or serum obtained from a rabbit bearing a Rous sarcoma virus induced tumor (TBR serum) (b lanes). Immune complexes were collected by binding to protein A–bearing *S. aureus* bacteria, washed, and the *S. aureus*–immune complexes suspended in a reaction buffer containing ^{32}P-ATP. After incubation to allow phosphorylation to occur, antibody and other proteins were separated from the bacteria, the bacteria were removed, and the remaining proteins were separated by gel electrophoresis. The heavy band in lane 3b is antibody to pp60^{src}, which has become radioactive because it was phosphorylated by pp60^{src}. Phage T7 virion proteins serve as molecular weight markers. [Part (b) is from M. S. Collett and R. L. Erikson, *Proc. Nat. Acad. Sci.* 75 (1978):2021–2024.]

was first detected, however, and long before it was purified, pp60^{v-src} was found to have protein kinase activity. By chance, a suitable in vitro substrate for pp60 is the immunoglobulin used to precipitate it. Thus, incubating immunoprecipitates with radioactive ATP causes the anti-pp60^{src} immunoglobulin to become radioactively labeled (Figure 26-20). Since then, many additional proteins, including pp60^{src} itself, have been found to be appropriate substrates for *src* kinase activity in vitro, and a plethora of normal cell proteins have been seen as targets in vivo.

All protein kinases that had been described before pp60^{src} phosphorylated either threonine or serine residues. Thus, researchers were quite surprised when analysis of the specificity of pp60^{src} showed that the protein catalyzes the transfer of the terminal phosphoryl group of ATP to the hydroxyl group of a tyrosine residue in its substrates. Then a careful search soon revealed a very low level of phosphotyrosine (P-tyrosine) in normal cells; RSV-transformed cells have about a tenfold higher level. Most important, in cells transformed by temperature-sensitive *src* mutants, both the transformed phenotype and the level of P-tyrosine in the cell shift rapidly in response to temperature shifts. Similarly, temperature-sensitive *src* mutants have reduced kinase activity in vitro. These results indicate that the kinase activity reflects, and may indeed be synonymous with, transforming ability.

Because v-*src* is under the transcriptional control of the viral LTR, it is expressed at much higher levels than c-*src*. As we noted, however, it is not just the higher level of v-*src* protein in cells that leads to transformation. When c-*src* is joined to an LTR (or other strong promoter) and introduced into cells, it is expressed at high levels but the cells are not transformed. Furthermore, the level of P-tyrosine in the cell is not elevated significantly. Thus, even though pp60$^{c\text{-}src}$ has kinase activity in vitro, apparently it has very little activity in vivo, in sharp contrast to pp60$^{v\text{-}src}$. Somehow, the mutations in v-*src* must enhance its kinase activity in vivo. These findings again point to the kinase activity of v-*src* as the means (or at least a correlate of the means) by which it transforms normal cells to cancerous ones.

Substrates for pp60^{src} [96–101]

The finding that pp60^{src} has a highly unusual protein kinase activity produced euphoria in the field. Protein phosphorylation is a major mechanism for inducing pleiotropic effects in cells, and so it seemed a most appropriate activity for pp60^{src}, a protein with the ability to induce the myriad changes associated with cancer. It seemed that by identifying cellular proteins phosphorylated on tyrosine residues after RSV infection, one would quickly begin to unravel the sequence of events that makes cells cancerous. However, while experimental results have come rapidly, insights have come more slowly. A major problem is that a very large number of different cellular proteins (Table 26-4) seem to be phosphorylated under the influence of pp60, and it is unlikely that more than a few of these are important to transformation. In most cases, either we do not know what the proteins do or we do not know why altering them would make a cell cancerous. A possible exception is vinculin. In normal cells, vinculin is located in adhesion plaques, the sites of cellular attachment to surfaces and sites where pp60^{src} is also concentrated, and is thought to serve as a connecting anchor between actin cables and the membrane. In transformed cells, vinculin is found diffusely throughout the cell. Conceivably, phosphorylation causes vinculin redistribution, and this might help to bring about the transformed cell morphology.

A second problem in identifying important protein substrates for

Table 26-4 Some In Vivo Substrates for pp60^{src}

p36 (an abundant cell protein found on the inner surface of the plasma membrane; highest concentration in epithelial and endothelial cells, particularly the terminal web of the brush border of columnar epithelial cells of the intestine)
Vinculin
Enolase
Phosphogylcerate mutase
Lactate dehydrogenase

pp60 has been that many of the simplest technologies that can be applied to the problem (e.g., the use of two-dimensional protein gels) would only be able to detect relatively abundant substrates.

Recently, evidence has indicated that pp60^{src} may indirectly stimulate the inositol lipid pathway (Chapter 25). The phosphorylation of inositol lipids is significantly increased in RSV-transformed cells, although probably not directly by pp60^{src}. The synthesis and breakdown of inositol lipids is correlated with rapid cell proliferation. Furthermore, certain factors (e.g., vasopressin, prostaglandin F2, bombesin) that stimulate cell proliferation when they bind to specific cell surface receptors stimulate breakdown of the final product of the inositol biosynthetic pathway, phosphatidylinositol-4,5-biphosphate (see Figure 25-21). This breakdown releases diacylglycerol and inositol triphosphate. Both molecules are thought to be intracellular messengers. Diacylglycerol activates protein kinase C, thought to be a cellular target for tumor promoters. Inositol triphosphate may mobilize calcium ions (Ca^{2+}) from an intracellular store. Ca^{2+} is a well-known second messenger in cells (Chapter 25). Thus, according to this hypothesis, v-*src* may transform cells because it helps to drive the inositol lipid pathway, which in turn may generate natural signals for cell proliferation.

The study of *src* has run up against the poorly understood issue of how signals to proliferate are transmitted from the outer reaches of the cell (either the inner or outer surfaces of the plasma membrane) to the nucleus. While it was once thought that understanding transformation would require a prior understanding of the regulation of cell growth, the opposite now seems to be true. Analysis of oncogenes has in fact provided considerable new information on mechanisms regulating cell growth.

The *sis* Oncogene of Simian Sarcoma Virus Is Derived from the Cellular Gene for Platelet-Derived Growth Factor[102–106]

The fastest way to figure out what an oncogene does would be to identify the gene as one whose function is already known. Given that there are some 50 to 100,000 protein-coding genes in vertebrates and that we know the function of only a couple of thousand, this at first seemed like a remote possibility. Incredibly, though, it has already happened for several oncogenes.

The identity of the v-*sis* oncogene of simian sarcoma virus (SSV), with a cellular gene coding for platelet-derived growth factor (PDGF), was found quite by chance. PDGF is a protein secreted by platelets during clot formation and also by activated monocytes. It may promote wound healing by stimulating growth of nearby cells (Chapter 25). When the amino acid sequences of several fragments of human PDGF were published, one astute reader entered the sequences into a computer to look for related proteins. Out came the oncogene v-*sis*, whose nucleotide and predicted amino acid sequence had already been published, and with it a most important discovery in cancer research. Finding that a cellular gene known to induce normal cell proliferation was the progenitor of a viral oncogene with the capacity to induce abnormal cell proliferation was overwhelmingly reassuring to retrovirologists and growth factor researchers. They must indeed be on the right track (as they had known all along).

The v-*sis* oncogene is the oncogene of simian sarcoma virus, which was isolated from a fibrosarcoma of a pet woolly monkey. (The gene is also present as the oncogene of a feline sarcoma virus; see Table 26-2.) SSV induces fibrosarcomas when injected into monkeys, and it transforms fibroblasts in vitro. The v-*sis* oncogene consists of about 1 kilobase of woolly monkey cellular sequences recombined into the *env* gene of a helper virus to form SSV. The v-*sis* oncogene, which encodes a 28 kdal protein, p28^{sis}, replaces most of *env* and is fused to sequences encoding its amino terminus. Since these sequences include a signal involved in directing the *env* gene product to the cell surface, the resulting *env-sis* fusion protein is also directed to the surface and is secreted. However, fusion to *env* is not essential for the activation of the *sis* gene to a transforming oncogene. If a molecular clone of c-*sis* (which probably possesses its own amino-terminal signal sequence) is ligated to potent transcriptional signals (an LTR), it, too, can now transform fibroblasts in vitro.

PDGF is composed of dimers of two polypeptide chains, sometimes called A (or 1) and B (or 2), although it is not yet clear if PDGF is a heterodimer (AB) or a mixture of homodimers (AA + BB). Both A and B homodimers appear to possess biological activity. A and B are derived from different genes, but have related amino acid sequences, and v-*sis* is derived from the gene encoding the B polypeptide chain. Until the relationship of the A and B chains is better understood, the relationship of v-*sis* and c-*sis* will also remain somewhat confused.

Interestingly, v-*sis* cannot transform epithelial cells. It can only transform cells that possess PDGF receptors. Transformation by SSV seems to be an example of transformation by autocrine stimulation (Chapter 25): The cell that expresses both PDGF and PDGF receptors constitutively, continuously stimulates its own division.

Some Oncogenes (*erbB* and *fms*) Are Derived from Cellular Genes That Encode Receptors for Polypeptide Growth Factors[107,108]

Although there are numerous known growth factors, so far PDGF is the only one observed to have been converted to an oncogene. However, there appears to be a related mechanism for generating an oncogene, namely, mutation of a gene encoding a receptor for a polypeptide growth factor. Presumably, such mutations cause the encoded receptor to behave as though stimulated, even in the absence of its ligand.

The epidermal growth factor (EGF) receptor is a large glycoprotein with intrinsic kinase activity (Chapter 25). When EGF binds to the receptor, the kinase becomes active, and the receptor, as well as other cellular proteins, is rapidly phosphorylated. The discovery that this phosphorylation occurs on tyrosine residues immediately caused researchers to wonder whether growth factors and oncogenes might stimulate cell growth by similar mechanisms. Pursuing this line of thought, and spurred on by the finding that *sis* encodes PDGF, researchers generated antibodies to the EGF receptor, allowing enough of the protein to be purified from a human tumor cell line that makes unusually large amounts, and the receptor was partially sequenced. The remarkable discovery was immediately made (again by comparison with sequences in computer data banks) that this protein must be the product of the human analog of the *erbB* oncogene. The *erbB* oncogene, first found in the genome of avian erythroblastosis virus (AEV), is responsible for the ability of AEV to induce erythroleukemia in chickens and to transform both erythroblasts and fibroblasts in vitro.

The EGF receptor has at least three functional domains (see Figure 25-18): a large glycosylated portion outside the cell that binds EGF, a short transmembrane domain, and a cytoplasmic domain that has both kinase activity and the tyrosine sites of autophosphorylation. The v-*erbB* oncogene is missing most of the amino-terminal portion of the EGF receptor, a consequence of the recombination event that placed the gene in the retroviral genome and has also sustained mutation in its C-terminal coding portion. Somehow, the truncated, mutant EGF receptor of AEV must have lost not only its EGF binding capacity but also its normal controls, so that it constantly signals the cell to divide even in the absence of EGF.

Interestingly, the tyrosine kinase domain of *erbB* has amino acid sequence homology to the *src* gene product. However, as we have seen, in contrast to the EGF receptor, pp60^{src} is located entirely within the cell, at the inner face of the plasma membrane, and lacks any transmembrane or extracellular domains.

Analogous to the case of *erbB* is that of the oncogene *fms*, found in two cat sarcoma viruses. The *fms* oncogene is a transmembrane protein that turns out to be derived from the cellular gene for the receptor for CSF-1 (also known as M-CSF), a polypeptide growth factor that specifically stimulates the growth and differentiation of macrophages (Chapter 25). In contrast to *erbB*, v-*fms* does not differ from the authentic receptor by the loss of part of the extracellular domain. However, v-*fms* does possess an altered carboxyl terminus, including alteration of a conserved tyrosine residue (the tyrosine corresponding to Tyr-527 in pp60^{src}) that is an important site of autophosphorylation in this family of proteins. It is now thought that the loss of this site, and the resulting inability of the residue to be phosphorylated, is important in altering the regulation of activity of the protein, possibly leaving it turned on in the absence of CSF-1.

Two other oncogenes, *ros* (Table 26-3) and *neu* (Chapter 27) are also transmembrane proteins with tyrosine kinase domains and *kit* probably also shares these properties. All three may encode growth factor receptors, although the factors they bind have not yet been identified.

Members of the *ras* Gene Family Encode 21 kdal Proteins That Bind Guanine Nucleotides[109,110,111]

The *ras* oncogenes have received particularly intense attention because of their probable involvement in many human tumors (Chapter 27). The *ras* genes constitute a small multigene family with three functional members: H-*ras* (or c-H-*ras*, the cellular homolog of v-H-*ras*, the oncogene present in Harvey sarcoma virus); K-*ras* (or c-K-*ras*, present in Kirsten sarcoma virus as v-K-*ras*); and N-*ras* (which is not present on any known retrovirus genome and whose significance will be discussed in Chapter 27; see Figure 27-4). All three functional *ras* genes encode highly related proteins of 21 kdal called p21^{ras} or p21.

The *ras* proteins, both viral and cellular, specifically and tightly bind GTP as well as other guanosine nucleotides. In addition, it has recently been found that normal c-*ras*-coded proteins have GTPase activity. The GTPase activity is greatly reduced in transforming p21 variants. Other cellular proteins that bind guanine nucleotides and have GTPase activity include the G proteins that modulate activity of the adenylcyclase system. These proteins are active when they are bound to GTP, but become inactive when their GTPase converts the GTP to GDP. Although *ras* proteins do not correspond to any known G proteins, it is now thought that it is the loss of GTPase activity in activated *ras*-coded proteins that may be critical to their transforming activity (Chapter 27).

The virus-coded p21s, but not the normal cell proteins, have an autokinase activity in that they can transfer a phosphate group from GTP to a threonine residue at position 59. As it turns out, the c-*ras* proteins have an alanine residue at the corresponding position, and so they cannot have the autokinase activity. The presence of Thr at position 59, however, is not essential for converting a c-*ras* proto-oncogene to a transforming protein.

The *myc* Gene Product Is a Nuclear Protein[112–119]

Unlike the *src, sis, erbB,* and *ras* products found inside, outside, or within the plasma membrane, several oncogene products are found in the nucleus. The *myc* oncogene is the oncogene of several defective avian retroviruses that can induce myelocytomas (tumors of myeloid cells), sarcomas, and carcinomas and in vitro can transform chicken (but not readily rodent) fibroblasts, myeloid cells, and epithelial cells. In some viruses, v-*myc* is fused to *gag* and translated as a *gag-myc* fusion protein. In one virus, it is produced from a subgenomic mRNA and not fused to a viral protein (see Figure 26-15). Either way, the *myc*-coded protein is found in the nucleus. So far no evidence exists that it functions by binding to DNA (even though it can bind DNA in vitro). Its function remains unknown.

Like E1A, *myc* alone cannot transform rodent fibroblasts efficiently, but it can transform primary cells efficiently if introduced along with an activated *ras* oncogene (Chapter 27). In contrast to E1A and polyoma large T, *myc* is not very efficient at promoting the establishment of primary cells, except, as noted, when another oncogene is also present.

The c-*myc* oncogene almost certainly plays a role in cell division, and it is expressed in normal growing cells as well as in tumor cells. In cells whose growth has been arrested, *myc* mRNA is greatly reduced. But if such cells are stimulated by mitogens (e.g., PDGF in the case of fibroblasts), *myc* mRNA appears within 1 to 2 hours. Mitogens, like PDGF, set off a chain of events by inducing a cell's "competence" to respond to additional growth factors (Chapter 25). Since v-*sis* is related to PDGF, the finding that PDGF turns on *myc* expression implies that the two oncogenes, *sis* and *myc*, may lie on the same growth control pathway. Consistent with a regulatory role, both the viral and cellular *myc* proteins are unusually labile and disappear from the cell with a half-life of 20 minutes or so.

The cellular *myc* gene has three exons, the first of which is not translated into protein (see Figure 26-22). It has been suggested that the first exon may have a regulatory role in c-*myc* expression, perhaps in the stability of the RNA or perhaps in translational control. In any case, in retroviruses, v-*myc* has both lost its first exon and fallen under LTR control. These two disruptions of its normal regulation presumably make an oncogene out of c-*myc*. Unlike many other oncogenes discussed so far, mutational change in the *myc*-encoded protein does not seem to be important for its activities as an oncogene.

The c-*myc* gene is a member of a gene family. So far, it is the only member of the family that has been found in a retrovirus genome, but other members, L-*myc* and N-*myc,* are amplified in certain naturally occurring tumors (Chapter 27).

Another oncogene whose product is localized in the nucleus is *fos,* the oncogene of the mouse osteosarcoma virus. The function of the 55 kdal phosphoprotein pp55$^{c\text{-}fos}$ is unknown. c-*fos* expression is quickly stimulated when quiescent cells are exposed to mitogens such as PDGF, and within minutes of exposure, unstable c-*fos* mRNA accumulates. An enhancer-like element lying 5′ of the *fos* gene is essential for this transcriptional activation in response to mitogens.

Oncogene Families: How Many Genes, How Many Functions?[89,120,121,122]

The fact that retroviruses isolated from different species have sometimes transduced homologous oncogenes was the first indication that the number of proto-oncogenes in cells might not be enormous. This possibility has received further support as cellular cancer genes detected in new ways (Chapter 27) have frequently turned out to be the homologs of known retrovirus oncogenes. Altogether, about 30 oncogenes are now known (see Table 26-3), and many people expect that the number will reach about 100. Fortunately, the known genes can be grouped by structure and function so that the number of oncogene classes is considerably less than 30. Moreover, now that several oncogenes have been tied to growth factor pathways, confidence that we can figure out their functions has soared.

About half the oncogenes in Table 26-3 turn out to resemble pp60^{src} in that they have protein kinase activity and share amino acid sequence homology with one another in the region that contains the active site for the kinase (Figure 26-21). In many cases, the homology at the amino acid level is so close that there is no doubt that the different oncogenes must have diverged from a common ancestor; and yet, the nucleotide sequences are very different, showing that the divergence of this family of oncogenes must have occurred early in eukaryotic evolution. For example, the products of *src* and *yes* have 82 percent amino acid homology, but the genes themselves have only 65 percent homology at the nucleotide level. Furthermore, *src* and *abl* are similarly related oncogenes, yet human c-*src* is more closely related to Drosophila c-*src* than either is to the corresponding c-*abl*. Thus, *src* and *abl* must have evolved from a common ancestor before the separation of vertebrates and insects. The full significance of this evolution into a highly conserved family will probably not become apparent until the normal functions of these proteins are better understood.

While most members of the protein kinase family of oncogenes are tyrosine-specific kinases, two, *mos* and *raf* (or *mil*) phosphorylate serine or threonine residues.

As already noted, in contrast to *erbB* and *fms* and also the more recently discovered *ros* and *neu,* and probably *kit* genes, most members of the tyrosine kinase family of oncogenes do not seem to encode proteins that protrude outside the cell and serve as receptors.

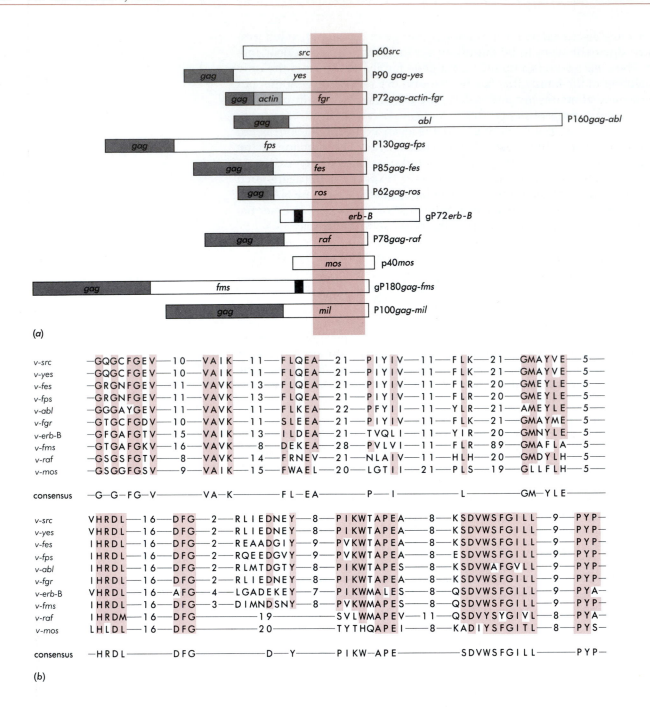

Figure 26-21

(a) Schematic representation of many of the proteins encoded by v-oncs belonging to the *src* family of oncogenes. Lengths are proportional to the number of amino acids in each protein. The NH₂ end of each chain is on the left. The colored strip represents a 250-amino-acid region of each protein that is similar in amino acid sequence [see part (b) of figure] and, by homology with pp60^{src}, is predicted to contain the protein kinase domain. Black bars represent the transmembrane domains of the *erb-B* and *fms* oncogenes, whose cellular homologs encode growth factor receptors. Note that most of the oncogenes are expressed as fusion proteins with proteins encoded by viral *gag* (and in one case *gag* plus a fragment of cellular actin). It is possible that in some cases, truncating the NH₂ end of the normal protein and/or joining it to a protein that is directed to the plasma membrane contributes to converting the proto-oncogene product into a transforming protein. [After T. Hunter, *Sci. Amer.* 251 (1984):70–79.] (b) Amino acid sequences from the regions of homology between members of the *src* oncogene family.

Nonetheless, the fact that *erbB* and *fms* encode growth factor receptors raises the possibility that the tyrosine kinase oncogene products that are entirely inside the cell may also be linked to extracellular signals, perhaps via interaction with other proteins that do have extracellular domains.

Another family of oncogenes, consisting of genes whose products are localized at the nucleus, is exemplified by *myc*. The *myc* gene product is remotely related in amino acid sequence to the *myb* and *ski* oncogene products, both of which are also nuclear proteins. As already noted, *fos* is also in the nucleus: Its function is totally unknown, and it is difficult to say whether it will continue to remain in the same gene family with *myc*, *myb*, and *ski*.

The *ras* genes constitute a third oncogene group, being localized at the inner surface of the plasma membrane, like many members of the *src* family, but binding guanine nucleotides and having GTPase activity. These properties are shared with the G proteins that interact with the adenylcyclase system to propagate signals initiated by polypeptide hormone-receptor interactions. However, *ras* proteins are different from any of the known G proteins, and their real function is still unknown.

The *erbA* gene may define a new group of oncogenes. Evidence indicates that this gene, which is present as a *gag–erbA* fusion in AEV, is derived from the cellular gene that encodes the receptor for thyroid hormone.

One of the v-oncs that did not originally fall into any of the classes just described was *sis*, but now its identification as a derivative of the gene encoding one chain of PDGF, as well as the connection of growth factors, their receptors, and their ability to turn on nuclear (*myc* and *fos*) oncogenes, provides the strongest link between the different classes of oncogenes. What acute transforming retroviruses have conveniently provided is a set of growth genes, or at least that subset of growth-controlling genes that can be altered so as to become oncogenes of retroviruses.

Some Retroviruses Have Two Oncogenes[123,124]

A striking difference between the DNA tumor viruses discussed earlier and many transforming retroviruses is that the former need two or three genes to cause cancer, whereas viruses like RSV seem to transform cells and cause tumors with just a single oncogene. While several exceptional retroviruses do have two different cell-derived genes (see Table 26-2), in these cases, related viruses that contain just one of the oncogenes are also capable of inducing tumors, so the significance of the extra gene is uncertain.

It is possible that retroviruses with a single transforming gene may only initiate tumors and that subsequent events may be needed to cause the infected cells to progress to a full-blown tumor. On the other hand, the rapid induction of tumors by acute transforming viruses (sometimes within a week or two), as well as the fact that the tumors often are not clonal but arise from many transformed cells, argues that a single transforming virus may indeed be able to transform cells in vivo. However, it must be noted that it is usually difficult to know whether additional mutations may have occurred in the tumor cells that contribute to their cancerous growth, and in most cases, such mutations have not been seriously looked for. Furthermore, many acute transforming retroviruses cause progressive fatal tumors only in certain highly susceptible, inbred strains of animals, so additional oncogenes might be necessary to allow the viruses to cause tumors in most animals. Clearly, the question of whether one oncogene is ever sufficient to fully transform a cell remains unanswered.

Slow Lymphoma-Inducing Retroviruses Activate Proto-Oncogenes in Vivo by Promoter or Enhancer Insertion[125–128]

Many retroviruses that can cause cancer lack oncogenes, are not defective, and encode only *gag, pol,* and *env*. These viruses, which are frequently found in chickens, mice, and cats, usually cause different forms of leukemia that typically take much longer to appear than tumors induced by transforming retroviruses. Because they never transform cells in culture, for a long time it was a mystery as to how these viruses induce tumors. The answer came from studying B cell leukemias induced by avian leukosis virus (ALV) in chickens.

First, every tumor induced by ALV has at least one provirus, suggesting the importance of viral DNA in maintaining the tumor. Second, mapping experiments using the Southern blot technique show that tumors contain proviruses integrated in preferred sites. Third, in some tumors, the ALV provirus is almost entirely deleted, although a viral LTR is always present, indicating that no viral gene product is necessary to maintain transformation, but the LTR is. And fourth, most ALV-induced tumors contain mRNA that can hybridize to cDNA complementary to the U_5 region of the LTR but not to other viral sequences. These facts came together to make spectacular sense when virtually all ALV-induced tumors were found to have ALV proviruses integrated near the cellular c-*myc* gene. Usually, integration occurs within the first c-*myc* exon or intron, and the virus is pointed in the same transcriptional orientation as c-*myc* (Figure 26-22). Furthermore, the virus usually has suffered a deletion, often near to or including its 5′ LTR. As a result, the viral 3′ LTR is often used to promote transcription, and an mRNA with U_5 sequences fused to c-*myc* sequences is made. Thus, c-*myc* is activated by **promoter insertion**. The level of the aberrant c-*myc* RNA in tumor cells can be 10 to 50 times higher than the normal level of c-*myc* RNA. This aberrant expression might reasonably be expected to disrupt growth control.

Sometimes, proviruses integrated near c-*myc* in retrovirus-induced tumors are oriented in the opposite transcriptional direction of c-*myc*. In these cases, *myc* mRNAs do not have viral sequences at their 5′ ends, and activation of the gene is thought to result from the proximity of an enhancer element in the provirus. Enhancer elements can function in either orientation relative to the genes whose expression they affect.

It is important to remember that proviral integration is random in each infected cell. Thus, only a rare integration that by chance occurs near c-*myc* (in the case of ALV-induced B cell tumors of chickens) will

Figure 26-22
Integration of an ALV provirus into the c-*myc* gene and activation of the gene by promoter insertion. The c-*myc* gene has three exons. The first exon is not translated but encodes the 5' untranslated portion of the *myc* mRNA. Arrows above the gene indicate actual sites of provirus integration that have been observed in different ALV-induced B cell tumors. In the example shown, the 5' LTR has been deleted. However, this is not usually the case, and it remains unclear why transcript that activate c-*myc* usually initiate in the 3' LTR rather than the 5' LTR. [Based on data from H. L. Robinson and G. C. Gagnon, *J. Virology* 57 (1986):28–36.]

ultimately result in a tumor. The probability that any single infected cell will be transformed is very low, but since there are at least as many cells at risk as possible integration sites, it is highly probable that a tumor will arise in the infected animal if viral replication is extensive.

Both Viral and Cellular Genes Determine What Type of Leukemia Is Induced by Nondefective Retroviruses[129,130,131]

Nondefective C-type viruses that induce leukemias usually cause predominantly one type of disease, such as tumors of B cells, tumors of T cells, or tumors of cells of the erythroid lineage. Given the mechanism of proto-oncogene activation by provirus insertion, one wonders how this specificity is achieved. Furthermore, why is a particular oncogene, for example, c-*myc*, preferentially activated in a particular type of virus-induced tumor? While the answers are not completely known, it is clear that the age of the recipient animal, the availability of replicating target cells, and the genetic constitution of both the host and the virus determine the outcome of an infection. For example, although ALV causes primarily B cell leukemias in some lines of chickens, in one particular strain it causes primarily erythro-leukemias. Remarkably, in the latter, ALV proviruses are integrated near the c-*erbB* gene rather than near c-*myc*.

One clear-cut determinant of disease specificity in some retroviruses is the LTR. DNA copies of different retroviruses can be molecularly cloned, and fragments of each can be exchanged to construct recombinant viruses. When the U_3 portions of LTRs are exchanged between two mouse leukemia viruses, one of which causes T cell leukemias and the other erythroleukemias, the viruses switch their disease specificity. The exchange of LTRs causes a change in the preferred site of virus replication in injected animals, and this change in organ tropism probably underlies the change in the type of leukemia induced. It is probably different tissue specificities of the transcriptional enhancers in the LTRs of different viruses that account for the results.

Mouse Mammary Tumor Virus (MMTV) Activates Proto-Oncogenes During Tumor Induction[132,133,134]

Oncogenic mouse mammary tumor viruses (MMTVs), like the leukemia viruses already discussed, are replication-competent, lack cell-derived oncogenes, and induce tumors after a long latent period (4 to 9 months). Like the nondefective leukemia viruses, they, too, transform cells by integrating adjacent to cellular proto-oncogenes. Two cellular genes that lie on different mouse chromosomes have been identified as targets of MMTV activation. The two genes, called *int-1* and *int-2*, do not correspond to any of the oncogenes that have been discovered in retroviral genomes. Proviral integrations that activate *int-1* or *int-2* can occur over a region of about 30 kilobases, and the proviruses are positioned so that their transcriptional orientation is outward from the direction of transcription of the *int* genes (Figure 26-23). This positioning is consistent with activation of expression of the cellular gene by enhancer insertion rather than by promoter insertion, and indeed, *int* mRNAs in tumor cells do not carry MMTV sequences. Some tumors carry integrated MMTV proviruses near both *int-1* and *int-2* genes, indicating that the virus can contribute to multistep carcinogenesis.

The MMTV LTR probably plays a role in determining the target tissue for transformation. Supporting this idea are the results of experiments involving **transgenic mice.** Transgenic mice are generated by introducing cloned genes into an embryo at a very early stage in development. If the gene is taken up and integrated into cellular DNA, it becomes part of the animal's genetic makeup. If the gene is integrated into cells that differentiate to become germ cells, it can be passed on to future generations of mice. When an LTR from a molecular clone of MMTV was ligated to the c-*myc* gene and the resulting DNA was introduced into embryos, the resulting (female) transgenic mice developed breast tumors after pregnancy. The specificity of disease induction was presumably conferred by the MMTV LTR. Interestingly, only a few breast cells became cancerous, suggesting that aberrant expression of c-*myc* alone may not be sufficient to cause a tumor. Some additional genetic event may have to occur in a particular cell to cause it to lose growth control.

Figure 26-23
Integration sites of MMTV proviruses that have been observed in the vicinity of the *int-1* and *int-2* genes in breast tumors of mice. Each black arrowhead represents an integration in a separate tumor. Arrows indicate transcriptional orientations of proviruses (black arrows) or of the *int* genes (open arrows). The *int-1* transcript is about 2.6 kilobases; the *int-2* transcript, about 3.2 kilobases. [The *int-1* data are from R. Nusse and H. E. Varmus, *Cell* 31 (1982):99–109; and R. Nusse, A. van Ooyen, D. Westaway, Y. K. Fung, H. E. Varmus, and C. Moscovici, *Cancer Cells* 2 (1984):205–210. The *int-2* data are from G. Peters, S. Brookes, R. Smith, and C. Dickson, *Cell* 33 (1983):369–377; and C. Dickson, G. Peters, R. Smith, and S. Brookes, *Cancer Cells* 2 (1984):195–203. Figure modified from R. Weiss, N. Teich, H. Varmus, and J. Coffin, eds., *RNA Tumor Viruses*, 2nd ed. (Cold Spring Harbor, N.Y.: Cold Spring Harbor Laboratory, 1984), p. 212.]

Inherited Retroviruses Cause Cancer in High-Leukemic Inbred Mice[135–143]

The discovery in the early 1930s that certain inbred strains of mice develop leukemias and lymphomas between the age of 6 and 12 months argued strongly that cancer is a genetic disease in mice. But in the 1950s, it was found that cell-free extracts of spontaneous tumors of high leukemic inbred mice could induce leukemia when they were injected into newborn mice that normally would not develop the disease. The active agent was soon shown to be an RNA-containing virus. Since reverse transcriptase had not yet been discovered or even guessed at, and since chromosomal genes were known to be DNA, not RNA, the paradox of how leukemia could be both inherited like a gene and caused by RNA viruses seemed incomprehensible. The discovery of reverse transcriptase provided a resolution to the paradox: Perhaps the DNA proviruses of oncogenic RNA viruses could be inherited as genes. While initially surprising, it was soon shown that genes that cause certain strains of mice to develop lymphoma are indeed DNA copies of RNA tumor viruses, and these genes are inherited in a Mendelian fashion.

Since all mice inherit retrovirus proviruses, what distinguishes inherited viruses that cause lymphoma? The answer seems to be that these particular inherited viruses are able to replicate very efficiently in their hosts, so high-leukemic mouse strains exhibit high titers of replicating viruses in their tissues soon after birth. Even so, 6 to 12 months elapse before disease develops. Now we know that the inherited viruses so critical to lymphoma development in high-leukemic inbred mice do not usually cause disease directly. They undergo recombination with other inherited retrovirus sequences to generate novel viruses, called MCFs, whose *env* genes and LTRs are acquired during the recombination process. The MCF viruses then initiate tumors, usually of T cells. As already described for avian leukosis virus, they do this by integrating adjacent to one, or even more, of several different proto-oncogenes, including c-*myc* or a newly discovered gene called *pim-1*.

Viruses Also Cause Cancer in Animals Outside the Laboratory[144,145,146]

Many of the tumor viruses we have described cause cancer under highly unnatural circumstances. For example, SV40 is found in monkeys but causes tumors when injected into newborn hamsters; and almost all AKR mice are doomed to die from retrovirus-induced thymomas, but the strain of mice was selected and inbred for high cancer incidence. What about animals outside the lab? Are they, too, succumbing to virus-induced cancers? Next to laboratory animals, the most readily available animals for study are house pets and farm animals. It is clear that in certain species, at certain times and places, viruses are a significant cause of cancer. With a few exceptions, such as Marek's disease virus, a herpesvirus responsible for lymphoma induction in chickens, retroviruses seem to be involved in most of the natural virus-induced cancers.

A striking example of viruses causing cancer occurs in household cats. Feline leukemia viruses (FeLV) are spread horizontally from cat to cat, being secreted in large amounts in saliva and passed between animals by scratching and biting. FeLV causes many diseases besides lymphomas, and in fact, infected cats are more likely to die from anemia, immunosuppression, thymic atrophy and other degenerative diseases, probably caused by decimation of particular lymphoid or erythroid cell populations by FeLV.

Chicken flocks are frequently afflicted by retrovirus (avian leukemia virus) infections. These cause leukemias and can spread, through infection of the reproductive tract and the eggs, to newly hatched chicks, which show wasting and runting syndromes as well as later lymphomas. Having been infected so young, the chick's immune system sees the replicating retrovirus as self and does not reject it.

Mice trapped in the wild have shown high incidences of retrovirus-induced lymphomas and also of a retrovirus-induced neurological disease that resembles the devastating Lou Gehrig's disease. Mice live in small, closely packed groups called *demes*, and in some of these demes, retroviruses have become established as milk-borne agents and are transmitted to suckling mice.

Summary

Much cancer research in this century has been devoted to identifying agents that induce cancer in experimental animals and then to determining the mechanisms by which such agents act. Most such agents, called carcinogens, act either directly or indirectly on DNA. Agents that cause cancer are usually chemicals or radiation that can induce mutations in DNA, or they are viruses that alter the expression of cellular genes. These findings led to the realization that cancer is primarily a genetic disease that arises when specific mutations (or, in some cases, a viral genome) accumulate within a single cell and cause it to lose growth control.

Many compounds that can induce tumors in animals at first did not appear to be mutagens. But in the 1960s, it became apparent that these carcinogens are converted to powerful mutagens by enzymes in the body. This discovery led to the development of tests to assay the potential carcinogenicity of compounds in the environment. Most widely used is the Ames test, in which the compound to be tested is incubated with an extract of liver cells, and its mutagenicity then tested on bacteria that carry a mutation at the *his* locus. If the compound has mutagenic activity or has been converted to a strong mutagen by exposure to the liver enzymes, bacterial revertants arise, and these can grow on media that lack histidine.

Very important to the study of carcinogens, and particularly to the study of viruses that induce cancers, was the development of cell culture systems in which transformation of normal cells to malignant cells can be observed. Transformed cells are detected as foci of cells with altered morphology or as cells with altered growth requirements that can grow under conditions where normal cells cannot.

Several types of DNA viruses can cause tumors in animals. These include papovaviruses (SV40, polyoma, and papilloma viruses), adenoviruses, certain herpesviruses, poxviruses, and hepatitis B viruses. All tumorigenic RNA viruses belong to the retrovirus group. The best-studied tumor viruses are SV40, polyoma, adenoviruses, and retroviruses, although papilloma viruses and hepatitis B viruses are now receiving more attention as well. In all known cases, when viruses transform cells, at least a portion of their genome becomes stably associated with the cell. This usually involves integration of viral genes into cellular DNA. There are two known mechanisms of transformation. Either the virus encodes genes whose products alter cellular growth, or integration of the viral genome occurs in such a way as to alter the expression of a cellular gene. Viral or cellular genes whose protein products can induce cancerous changes in cell behavior are called oncogenes. Normal cellular genes with the capacity to become oncogenes when they suffer mutations or alterations in their expression are called proto-oncogenes.

SV40 and polyoma viruses usually transform only those cells that are not permissive for the complete lytic cycle of the virus. Only the early regions of the small circular viral genomes (about half the genome) are needed for transformation. In polyoma, this region encodes three proteins whose genes overlap: small, middle, and large T proteins. In SV40, the early region encodes two proteins: large and small T. The role of the different SV40/polyoma early genes in transformation has been studied by synthesizing cDNA copies of their mRNAs, joining these to transcriptional promoters, introducing the constructs into cells, and observing the effect of the gene on the growth and morphology of the cell. The studies reveal that polyoma large T can induce indefinite growth in primary cells and reduce their serum requirements for growth, while middle T is needed to induce the remainder of the transformed phenotype, including morphological transformation. If established cell lines are used instead of primary cells, middle T alone can transform them. Middle T may act by altering the activity of the cellular *src* gene product, a protein with tyrosine kinase activity. SV40 large T protein seems to induce many aspects of transformation single-handedly, although perhaps less efficiently than polyoma large T plus middle T. SV40 large T protein may act via its interaction with a cellular protein, p53, located in the nucleus, whose stability and thus amount increase in SV40-transformed cells. The role (or even need) of the SV40 and polyoma small T proteins in transformation remains unclear.

Only the left-hand 8 percent of the linear adenovirus genome is needed to transform cells. This region includes the E1A and E1B genes, each of which encodes overlapping mRNAs. Two E1A mRNAs and one or two E1B mRNAs encode proteins that contribute to the transformed phenotype. The E1A gene induces indefinite growth of cells and also induces certain other aspects of the transformed phenotype. Although either of its two transforming proteins can induce indefinite growth, both products are needed for cellular transformation. In lytic growth, E1A is needed for full expression of other adenovirus genes, and it is thought that E1A acts by altering gene expression. The role of the E1B gene products in transformation remains obscure. Different adenoviruses differ in their ability to induce tumors in rodents. One reason seems to be that at least some highly oncogenic adenoviruses suppress the expression of surface MHC proteins, so that the virus-infected cell is not seen and killed by cytotoxic T cells.

There are two types of cancer-causing retroviruses. One type is nondefective, lacks oncogenes, is unable to transform cells in culture, and induces primarily leukemias with a long latent period (3 to 12 months). With the exception of the Rous sarcoma virus, the second type is always defective, carries an oncogene, induces tumors in susceptible animals with a short latent period (days or a few weeks), and can usually transform cells in culture. These viruses are often called acute transforming retroviruses.

The oncogenes of acute transforming retroviruses are acquired from normal cells by recombination, and usually the acquisition involves both a loss of viral genetic material and an alteration to the protein encoded by the cellular gene. As a result, the virus is defective and requires a nondefective helper to replicate. However, a single defective viral genome can transform a cell.

The first retroviral oncogene to be studied in detail was *src*. It encodes a protein of 60 kdal with tyrosine protein kinase activity. This activity is probably responsible for the ability of the viral *src* gene to transform cells, but the putative substrates that are affected by phosphorylation and that thus bring about the transformed phenotype remain undiscovered. The cellular c-*src* gene encodes a protein that possesses protein kinase activity; however, it is now thought that the activity is regulated in some way that has been lost in the v-*src* gene product as a result of mutations.

About 50 different transforming retroviruses have been isolated from different animal species, and about 20 to 25 different oncogenes have been discovered in the genomes of these viruses. The oncogenes can be divided into families of related genes. (1) One family consists of genes that encode protein kinases and that have amino acid sequence homology to the *src* gene product. Like pp60^{src}, most of these proteins are located at the inner surface of the plasma membrane. However, at least two of them, *erbB* and *fms*, encode transmembrane proteins that are derived from cellular genes that encode receptors for polypeptide growth factors. The c-*erbB* gene encodes the receptor for epidermal growth factor (EGF), and c-*fms* encodes the receptor for the macrophage colony-stimulating factor (M-CSF or CSF-1). (2) Members of the *ras* gene family encode 21 kdal proteins with very similar amino acid sequence that are located at the inner surface of the plasma membrane, bind guanine nucleotides, and have GTPase activity. These properties make *ras* proteins analogous to the G proteins that mediate interactions between receptors and the cAMP system. (3) One oncogene, *sis*, is derived from a gene that encodes the B chain of the platelet-derived growth factor (PDGF). (4) Several oncogenes (*myc*, *fos*, and *myb*) encode proteins that are located in the nucleus and which are thought to alter the expression of other cellular genes, although there is little strong evidence to support this hypothesis. (5) One oncogene, *erbA*, encodes a protein with weak amino acid sequence homology to steroid (glucocorticoid and estrogen) receptors. Several more oncogenes remain unclassified.

Retroviruses that lack oncogenes and that induce leukemias and lymphomas do so by integrating adjacent to cellular proto-oncogenes. In the case of B cell lymphomas induced by avian leukosis viruses, proviruses almost invariably integrate within the first (noncoding) exon or the first intron of c-*myc*, undergo a deletion of a portion of the viral genome, and use their 3' LTR to initiate novel *myc* transcripts that have viral R and U$_5$ sequences covalently joined to their 5' ends. Sometimes, these viruses induce erythroleukemias. In these tumors, proviruses are found in the vicinity of the *erbB* gene. Mouse mammary tumor viruses induce breast tumors in mice by integrating adjacent to either (or both) the *int-1* or *int-2* gene, cellular genes not previously identified as oncogenes of retroviruses.

DNA copies of retroviruses are inherited in many animals. While most of these are probably harmless to their hosts, in some inbred strains of mice the inherited viruses cause cancer. This is usually because these particular proviruses are expressed and yield viruses that are able to replicate extensively in their hosts. Besides the inherited retroviruses of mice, retroviruses can cause cancer in animals outside the laboratory. In cats, retroviruses spread horizontally through biting and scratching and induce lymphomas, sarcomas, and immunosuppressive disorders. In chickens,

the viruses can cause leukemias and be passed through the reproductive tract to the eggs and hence to newly hatched chicks.

Bibliography

General References

Bishop, J. M. 1985. "Viral Oncogenes." *Cell* 42:23–38.

Feramisco, J., B. Ozanne, and C. Stiles, eds. 1985. *Cancer Cells 3: Growth Factors and Transformation*. Cold Spring Harbor, N.Y.: Cold Spring Harbor Laboratory.

Gross, L. 1970. *Oncogenic Viruses*. 2nd ed. Elmsford, N.Y.: Pergamon Press.

Levine, A. J., G. F. Vande Woude, W. C. Topp, and J. D. Watson, eds. 1984. *Cancer Cells 1: The Transformed Phenotype*. Cold Spring Harbor, N.Y.: Cold Spring Harbor Laboratory.

Tooze, J., ed. 1980, 1981. *DNA Tumor Viruses. Molecular Biology of Tumor Viruses*. Rev. 2nd ed. Cold Spring Harbor, N.Y.: Cold Spring Harbor Laboratory.

Vande Woude, G. F., A. J. Levine, W. C. Topp, and J. D. Watson, eds. 1984. *Cancer Cells 2: Oncogenes and Viral Genes*. Cold Spring Harbor, N.Y.: Cold Spring Harbor Laboratory.

Varmus, H., and A. J. Levine, eds. 1983. *Readings in Tumor Virology*. Cold Spring Harbor, N.Y.: Cold Spring Harbor Laboratory.

Weiss, R., N. Teich, H. Varmus, and J. Coffin, eds. 1984. *RNA Tumor Viruses. Molecular Biology of Tumor Viruses*. 2nd ed. Cold Spring Harbor, N.Y.: Cold Spring Harbor Laboratory.

Weiss, R., N. Teich, H. Varmus, and J. Coffin, eds. 1985. *RNA Tumor Viruses. Molecular Biology of Tumor Viruses*. 2nd ed. Supplements and appendices. Cold Spring Harbor, N.Y.: Cold Spring Harbor Laboratory.

Cited References

1. Klein, G., U. Bregula, F. Wiener, and H. Harris. 1971. "The Analysis of Malignancy by Cell Fusion." *J. Cell Sci.* 8:659–672.
2. Stanbridge, E. J., C. J. Der, C-J. Doersen, R. Y. Nishimi, D. M. Peehl, B. E. Weissman, and J. E. Wilkinson. 1982. "Human Cell Hybrids: Analysis of Transformation and Tumorgenicity." *Science* 215:252–259.
3. Pereira-Smith, O. M., and J. R. Smith. 1983. "Evidence for the Recessive Nature of Cellular Immortality." *Science* 221:964–966.
4. Craig, R., and R. Sager. 1985. "Suppression of Tumorigenicity in Hybrids of Normal and Oncogene-Transformed CHEF Cells." *Proc. Nat. Acad. Sci.* 82:2062–2066.
5. Cairns, J. 1978. *Cancer, Science, and Society*. San Francisco: Freeman.
6. Upton, A. C. 1982. "Radiation Carcinogenesis." In *Cancer Medicine*, 2nd ed., ed. J. F. Holland and E. Frei. Philadelphia: Lea and Febiger, pp. 96–108.
7. Miller, E. C., and J. A. Miller. 1966. "Mechanisms of Chemical Carcinogenesis: Nature of Proximate Carcinogens and Interactions with Macromolecules." *Pharmacol. Rev.* 18:805.
8. Ames, B. N., W. E. Durston, E. Yamasaki, and F. D. Lee. 1973. "Carcinogens Are Mutagens: A Simple Test System Combining Liver Homogenates for Activation and Bacteria for Detection." *Proc. Nat. Acad. Sci.* 70:2281–2285.
9. Miller, E. C., and J. A. Miller. 1981. "Searches for Ultimate Chemical Carcinogens and Their Reactions with Cellular Macromolecules." *Cancer* 47:2327–2345.
10. Hecker, E., N. E. Fusenigne, W. Kunz, F. Marks, and H. W. Thielmann, eds. 1981. *Carcinogenesis: A Comprehensive Survey*. Vol. 7. *Cocarcinogenesis and Biological Effects of Tumor Promoters*. New York: Raven Press.
11. Tooze, J., ed. 1980, 1981. *DNA Tumor Viruses. Molecular Biology of Tumor Viruses*. Rev. 2nd ed. Cold Spring Harbor, N.Y.: Cold Spring Harbor Laboratory.
12. Weiss, R., N. Teich, H. Varmus, and J. Coffin, eds. 1984. *RNA Tumor Viruses. Molecular Biology of Tumor Viruses*. 2nd ed. Cold Spring Harbor, N.Y.: Cold Spring Harbor Laboratory.
13. Varmus, H., and A. J. Levine, eds. 1983. *Readings in Tumor Virology*. Cold Spring Harbor, N.Y.: Cold Spring Harbor Laboratory.
14. Temin, H. M., and H. Rubin. 1958. "Characteristics of an Assay for Rous Sarcoma Virus and Rous Sarcoma Cells in Tissue Culture." *Virology* 6:669–688.
15. Macpherson, I., and L. Montagnier. 1964. "Agar Suspension Culture for the Selective Assay of Cells Transformed by Polyoma Virus." *Virology* 23:291–294.
16. Gross, L. 1970. *Oncogenic Viruses*. 2nd ed. Elmsford, N.Y.: Pergamon Press.
17. Shin, S-I., V. H. Freedman, R. Risser, and R. Pollack. 1975. "Tumorigenicity of Virus-Transformed Cells in *Nude* Mice Is Correlated Specifically with Anchorage Independent Growth in Vitro." *Proc. Nat. Acad. Sci.* 72:4435–4439.
18. Bishop, J. M. 1985. "Viral Oncogenes." *Cell* 42:23–38.
19. Gross, L. 1953. "A Filterable Agent Recovered from Ak Leukemic Extracts Causing Salivary Gland Carcinomas in C3H Mice." *Proc. Soc. Exp. Biol. Med.* 83:414–421.
20. Sweet, B. H., and M. R. Hilleman. 1960. "The Vacuolating Virus, SV40." *Proc. Soc. Exp. Biol. Med.* 105:420–427.
21. Acheson, N. H. 1980, 1981. "Lytic Cycle of SV40 and Polyoma Virus in DNA Tumor Viruses." In *DNA Tumor Viruses. Molecular Biology of Tumor Viruses*. Rev. 2nd ed., ed. J. Tooze. Cold Spring Harbor, N.Y.: Cold Spring Harbor Laboratory, Chap. 3.
22. Topp, W. C., D. Lane, and R. Pollack. 1980. "Transformation by SV40 and Polyoma Virus." In *DNA Tumor Viruses. Molecular Biology of Tumor Viruses*. Rev. 2nd ed., ed. J. Tooze. Cold Spring Harbor, N.Y.: Cold Spring Harbor Laboratory.
23. Tegtmeyer, P., 1980, 1981. "Genetics of SV40 and Polyoma Virus." In *DNA Tumor Viruses. Molecular Biology of Tumor Viruses*. Rev. 2nd ed., ed. J. Tooze. Cold Spring Harbor, N.Y.: Cold Spring Harbor Laboratory, pp. 297–338d.
24. Graham, F. L. and A. J. Van der Eb. 1973. "A New Technique for the Assay of Infectivity of Human Adenovirus 5 DNA." *Virology* 52:456–467.
25. Kimura, G., and A. Itagaki. 1975. "Initiation and Maintenance of Cell Transformation by Simian Virus 40. A Viral Genetic Property." *Proc. Nat. Acad. Sci.* 72:673–677.
26. Fried, M. 1965. "Cell-Transforming Ability of a Temperature-Sensitive Mutant of Polyoma Virus." *Proc. Nat. Acad. Sci.* 53:486–491.
27. Benjamin, T. L. 1970. "Host Range Mutant of Polyoma Virus." *Proc. Nat. Acad. Sci.* 67:394–399.
28. Graham, F. G., P. J. Abrahams, C. Mudder, H. L. Heijneker, S. O. Warnoar, F. A. J. de Bries, W. Fiers, and A. J. Van der Eb. 1975. "Studies on in Vitro Transformation by DNA and DNA Fragments of Human Adenovirus and SV40." *Cold Spring Harbor Symp. Quant. Biol.* 39:637–650.
29. Sambrook, J., H. Westphal, P. R. Srinivasan, and R. Dulbecco. 1968. "The Integrated State of Viral DNA in SV40-Transformed Cells." *Proc. Nat. Acad. Sci.* 60:1288–1295.
30. Watkins, J. F., and R. Dulbecco. 1967. "Production of SV40 Virus in Heterokaryons of Transformed and Susceptible Cells." *Proc. Nat. Acad. Sci.* 58:1396–1403.
31. Botchan, M., J. Stringer, J. Mitchison, and J. Sambrook. 1980. "Integration and Excision of SV40 DNA from the Chromosome of a Transformed Cell." *Cell* 20:143–152.
32. Botchan, M. 1979. "Studies on SV40 Excision from Cellular Chromosomes." *Cold Spring Harbor Symp. Quant. Biol.* 43:709.

33. Hutchinson, M. A., T. Hunter, and W. Eckhart. 1978. "Characterization of T Antigens in Polyoma-Infected and Transformed Cells." *Cell* 15:65–80.

34. Paucha, E., A. Mellor, R. Harvey, A. E. Smith, R. M. Henwick, and M. Waterfield. 1978. "SV40 Large and Small Tumor Antigens from Simian Virus 40 Have Identical Amino Termini Mapping at 0.65 Map Units." *Proc. Nat. Acad. Sci.* 75:2165–2169.

35. Prives, C., Y. Gluzman, and E. Winocour. 1978. "Cellular and Cell-Free Synthesis of Simian Virus 40 T-Antigens in Permissive and Transformed Cells." *J. Virology* 25:587–595.

36. Staneloni, R. J., M. M. Fluck, and T. L. Benjamin. 1977. "Host-Range Selection of Transformation Defective 'HR-T' Mutants of Polyoma Virus." *Virology* 77:598–609.

37. Bouck, N., N. Beales, T. Shenk, P. Berg, and G. Di Mayorca. 1978. "New Region of the Simian Virus 40 Genome Required for Efficient Viral Transformation." *Proc. Nat. Acad. Sci.* 75:2473–2477.

38. Treisman, R., U. Novak, J. Favaloro, and R. Kamen. 1981. "Transformation of Rat Cells by an Altered Polyoma Virus Genome Expressing Only the Middle-T Protein." *Nature* 292:595–600.

39. Rassoulzadegan, M., A. Cowie, A. Carr, N. Glaichenhaus, R. Kamen, and F. Cuzin. 1982. "The Roles of Individual Polyoma Virus Early Proteins in Oncogenic Transformation." *Nature* 300:713–718.

40. Jat, P., and P. A. Sharp. 1986. "Large T Antigens of sV40 and Polyoma Virus Efficiently Establish Primary Fibroblasts." *J. Virology.* 59:745–750.

41. Courtneidge, S., and A. E. Smith. 1983. "Polyoma Virus Transforming Protein Associates with the Product of the c-*src* Cellular Gene." *Nature* 303:435–539.

42. Lane, D. P., and L. V. Crawford. 1979. "T Antigen Is Bound to a Host Protein in SV40-Transformed Cells." *Nature* 278:261–262.

43. Linzer, D. I. H., and A. J. Levine. 1979. "Characterization of a 54K Dalton Cellular SV40 Tumor Antigen Present in SV40-Transformed Cells and Uninfected Embryonal Carcinoma Cells." *Cell* 17:43–52.

44. Defendi, V. 1963. "Effects of SV40 Virus Immunization on Growth of Transplantable SV40 and Polyoma Virus Tumors in Hamsters." *Proc. Soc. Exp. Biol. Med.* 113:12–16.

45. Khera, K. S., A. Ashkenazi, F. Rapp, and J. L. Melnick. 1963. "Immunity in Hamsters to Cells Transformed in Vitro Among the Papovaviruses." *J. Immunol.* 91:604–613.

46. Tevethia, S. S., R. S. Greenfield, D. C. Flyer, and M. J. Tevethia. 1980. "SV40 Transplantation Antigen Relationship to SV40-Specific Proteins." *Cold Spring Harbor Symp. Quant. Biol.* 44:235–242.

47. Flint, S. J. 1980, 1981. "Transformation by Adenovirus." In *DNA Tumor Viruses. Molecular Biology of Tumor Viruses.* Rev. 2nd ed., ed. J. Tooze. Cold Spring Harbor, N.Y.: Cold Spring Harbor Laboratory, pp. 547–576e.

48. Trenton, J. J., Y. Yabe, and G. Taylor. 1962. "The Quest for Human Cancer Viruses." *Science* 137:835–841.

49. Berk, A. J., and P. A. Sharp. 1978. "Structure of the Adenovirus 2 Early mRNAs." *Cell* 14:695–711.

50. Chow, L. W., T. B. Broker, and J. B. Lewis. 1979. "Complex Splicing Patterns of RNAs from the Early Regions of Adenovirus-2." *J. Mol. Biol.* 134:265–303.

51. Perricaudet, M., G. Akusjarvi, A. Virtanen, and U. Pettersson. 1979. "Structure of Two Spliced mRNAs from the Transforming Region of Human Subgroup C Adenoviruses." *Nature* 281:694–696.

52. Virtanen, A., and U. Pettersson. 1983. "The Molecular Structure of the 9S mRNA from Early Region 1A of Adenovirus Serotype 2." *J. Mol. Biol.* 165:496–499.

53. Virtanen, A., and U. Pettersson. 1985. "Organization of Early Region 1B of Human Adenovirus Type 2: Identification of Four Differentially Spliced mRNAs." *J. Virology* 54:383–391.

54. Houweling, A., J. P. van den Elsen, and A. J. Van der Eb. 1980. "Partial transformation of primary rat cells by the leftmost 4.5% fragment of adenovirus DNA." *Virology* 105:537–550.

55. Van den Elsen, P., A. Houweling, and A. J. Van der Eb. 1983. "Expression of Region E1b of Human Adenovirus in the Absence of Region E1a Is Not Sufficient for Complete Transformation." *Virology* 128:377–390.

56. Land, H., L. P. Parada, and R. A. Weinberg. 1983. "Tumorigenic Conversion of Primary Embryo Fibroblasts Requires at Least Two Cooperating Oncogenes." *Nature* 304:596–602.

57. Ruley, H. E. 1983. "Adenovirus Early Region 1A Enables Viral and Cellular Transforming Genes to Transform Primary Cells in Culture." *Nature* 304:304–306.

58. Sarnow, P., Y. S. Ho, J. Williams, and A. J. Levine. 1982. "Adenovirus E1b-58kd Tumor Antigen and SV40 Large Tumor Antigen are Physically Associated with the Same 54kd Cellular Protein in Transformed Cells." *Cell* 28:387–394.

59. Kovesdi, I., R. Reichel, and J. R. Nevins. 1986. "Identification of a Cellular Transcription Factor Involved in E1A *Trans*-Activation." *Cell* 45:219–228.

60. Schrier, P. I., R. Bernards, R. T. M. J. Vaessen, A. Houweling, and A. J. Van der Eb. 1983. "Expression of Class I Major Histocompatibility Antigens Switched Off by Highly Oncogenic Adenovirus 12 in Transformed Rat Cells." *Nature* 305:771–775.

61. Coffin, J. M. 1985. "Structure of the Retroviral Genome." In *RNA Tumor Viruses. Molecular Biology of Tumor Viruses,* 2nd ed., ed. R. Weiss, N. Teich, H. Varmus, and J. Coffin. Cold Spring Harbor, N.Y.: Cold Spring Harbor Laboratory, pp. 261–368.

62. Coffin, J. M. 1985. "Genome Structure." In *RNA Tumor Viruses. Molecular Biology of Tumor Viruses,* 2nd ed., ed. R. Weiss, N. Teich, H. Varmus, and J. Coffin. Cold Spring Harbor, N.Y.: Cold Spring Harbor Laboratory, pp. 17–73.

63. Varmus, H. 1982. "Form and Function of Retroviral Proviruses." *Science* 216:812–821.

64. Bishop, J. M. 1983. "Cellular Oncogenes and Retroviruses." *Ann. Rev. Biochem.* 52:301–354.

65. Bishop, J. M., and H. E. Varmus. 1982. "Functions and Origins of Retroviral Transforming Genes." In *RNA Tumor Viruses. Molecular Biology of Tumor Viruses,* 2nd ed., ed. R. Weiss, N. Teich, H. Varmus, and J. Coffin. Cold Spring Harbor, N.Y.: Cold Spring Harbor Laboratory, pp. 999–1108.

66. Martin, G. S. 1970. "Rous Sarcoma Virus: A Function Required for the Maintenance of the Transformed State." *Nature* 227:1021–1023.

67. Duesberg, P., and P. Vogt. 1973. "RNA Species Obtained from Clonal Lines of Avian Sarcoma and from Avian Leukosis Virus." *Virology* 54:207–219.

68. Coffin, J. M., and M. A. Billeter. 1976. "A Physical Map of the Rous Sarcoma Virus Genome." *J. Mol. Biol.* 100:293–318.

69. Hayward, W. S. 1977. "Size and Genetic Content of Viral RNAs in Avian Oncovirus Infected Cells." *J. Virology* 24: 47–63.

70. Hanafusa, H., T. Hanafusa, and H. Rubin. 1963. "The Defectiveness of Rous Sarcoma Virus." *Proc. Nat. Acad. Sci.* 49:572–580.

71. Temin, H. 1963. "Separation of Morphological Conversion and Virus Production in Rous Sarcoma Virus Infection." *Cold Spring Harbor Symp. Quant. Biol.* 27:407–414.

72. Hartley, J. W., and W. P. Rowe. 1966. "Production of Altered Cell Foci in Tissue Culture by Defective Moloney Sarcoma Virus Particles." *Proc. Nat. Acad. Sci.* 55:780–786.

73. Vogt, P. K. 1971. "Spontaneous Segregation of Nontransforming Viruses from Cloned Sarcoma Viruses." *Virology* 46:939–946.

74. Stehelin, D., H. E. Varmus, J. M. Bishop, and P. K. Vogt. 1976. "DNA Related to the Transforming Gene(s) of Avian Sarcoma Viruses Is Present in Normal Avian DNA." *Nature* 260:170–173.

75. Scolnick, E. M., E. Rands, D. Williams, and W. P. Parks. 1973. "Studies on the Nucleic Acid Sequences of Kirsten Sarcoma Virus: A Model for Formation of a Mammalian RNA-Containing Sarcoma Virus." *J. Virology* 12:458–463.

76. Goldfarb, M., and R. A. Weinberg. 1981. "Generation of Novel, Biologically Active Harvey Sarcoma Viruses Via Apparent Illegitimate Recombination." *J. Virology* 38:136–150.

77. Swanstrom, R., R. C. Parker, H. E. Varmus, and J. M. Bishop. 1983. "Transduction of a Cellular Oncogene—The Genesis of Rous Sarcoma Virus." *Proc. Nat. Acad. Sci.* 80:2519–2523.

78. Takeya, T., and H. Hanafusa. 1983. "Structure and Sequence

of the Cellular Gene Homologous to the RSV *src* Gene and the Mechanism for Generating the Transforming Virus." *Cell* 32:881–890.

79. Spector, D. H., H. E. Varmus, and J. M. Bishop. 1978. "Nucleotide Sequences Related to the Transforming Gene of Avian Sarcoma Virus Are Present in DNA of Uninfected Vertebrates." *Proc. Nat. Acad. Sci.* 75:4102–4106.

80. Shilo, B., and R. A. Weinberg. 1981. "DNA Sequences Homologous to Vertebrate Oncogenes Are Conserved in *Drosophila melanogaster*." *Proc. Nat. Acad. Sci.* 78:6789–6792.

81. DeFeo-Jones, D., E. M. Scolnick, R. Koller, and R. Dhar. 1983. "*ras*-Related Gene Sequences Identified and Isolated from *Saccharomyces cerevisiae*." *Nature* 306:707–709.

82. Shibuya, M., T. Hanafusa, H. Hanafusa, and J. R. Stephenson. 1980. "Homology Exists Among the Transforming Sequences of Avian and Feline Sarcoma Viruses." *Proc. Nat. Acad. Sci.* 81:4697–4701.

83. Blair, D. G., M. Oskarsson, T. G. Wood, W. L. McClements, P. J. Fischinger, and G. F. Vande Woude. 1981. "Activation of the Transforming Potential of a Normal Cell Sequence: A Model for Oncogenesis." *Science* 212:941–943.

84. Chang, E. M., M. E. Furth, E. M. Scolnick, and D. R. Lowy. 1982. "Tumorigenic Transformation of Mammalian Cells Induced by a Normal Human Gene Homologous to the Oncogene of Harvey Murine Sarcoma Virus." *Nature* 297:479–483.

85. Iba, H., T. Takeya, F. R. Cross, T. Hanafusa, and H. Hanafusa. 1984. "Rous Sarcoma Virus Variants That Carry the Cellular *src* Gene Instead of the Viral *src* Gene Cannot Transform Chicken Embryo Fibroblasts." *Proc. Nat. Acad. Sci.* 81:4424–4429.

86. Parker, R. C., H. E. Varmus, and J. M. Bishop. 1984. "Expression of v-*src* and Chicken c-*src* in Rat Cells Demonstrates Qualitative Differences Between pp60^{v-src} and pp60^{c-src}." *Cell* 37:131–139.

87. Shalloway, D., P. M. Coussens, and P. Yaciuk. 1984. "Overexpression of the c-*src* Protein Does Not Induce Transformation of NIH 3T3 Cells." *Proc. Nat. Acad. Sci.* 81:7071–7075.

88. Miller, A. D., T. Curran, and I. M. Verma. 1984. "c-*fos* Protein Can Induce Cellular Transformation: A Novel Mechanism of Activation of a Cellular Oncogene." *Cell* 36:51–60.

89. Hunter, T. 1984. "The Proteins of Oncogenes." *Sci. Amer.* 251:70–79.

90. Kitamura, N., A. Kitamura, K. Toyoshima, Y. Hirayama, and M. Yoshida. 1982. "Avian Sarcoma Virus Y73 Genome Sequence and Structural Similarity of Its Transforming Gene Product to That of Rous Sarcoma Virus." *Nature* 297:205–207.

91. Brugge, J. S., and R. L. Erikson. 1977. "Identification of a Transformation-Specific Antigen Induced by an Avian Sarcoma Virus." *Nature* 269:346–348.

92. Collett, M. S., and R. L. Erikson. 1978. "Protein Kinase Activity Associated with the Avian Sarcoma Virus *src* Gene Product." *Proc. Nat. Acad. Sci.* 75:2021–2024.

93. Levinson, A. D., H. Oppermann, L. Levintow, H. E. Varmus, and J. M. Bishop. 1978. "Evidence That the Transforming Gene of Avian Sarcoma Virus Encodes a Protein Kinase Associated with a Phosphoprotein." *Cell* 15:561–572.

94. Hunter, T., and B. M. Sefton. 1980. "Transforming Gene Product of Rous Sarcoma Virus Phosphorylates Tyrosine." *Proc. Nat. Acad. Sci.* 77:1311–1315.

95. Cooper, J. A., K. L. Gould, C. A. Cartwright, and T. Hunter. 1986. "Tyr^{527} Is Phosphorylated in pp60^{c-src}: Implications for Regulation." *Science* 231:1431–1434.

96. Radke, K., and G. S. Martin. 1979. "Transformation by Rous Sarcoma Virus: Effects of *src* Gene Expression on the Synthesis and Phosphorylation of Cellular Polypeptides." *Proc. Nat. Acad. Sci.* 76:5212–5216.

97. Cooper, J. A., and T. Hunter. 1981. "Four Different Classes of Retroviruses Induce Phosphorylation of Tyrosine Present in Similar Cellular Proteins." *Mol. Cell Biol.* 1:394–407.

98. Sefton, B. M., T. Hunter, E. H. Ball, and S. J. Singer. 1981. "Vinculin: A Cytoskeletal Target of the Transforming Protein of Rous Sarcoma Virus." *Cell* 24:165–174.

99. Rohrschneider, L., M. Rosok, and K. Shriver. 1982. "Mechanism of Transformation by Rous Sarcoma Virus: Events Within Adhesion Plaques." *Cold Spring Harbor Symp. Quant. Biol.* 46:953–965.

100. Cross, F. R., E. A. Garber, and H. Hanafusa. 1985. "N-Terminal Deletions in Rous Sarcoma Virus p60^{src}: Effects on Tyrosine Kinase and Biological Activities and on Recombination in Tissue Culture with the Cellular *src* Gene." *Mol. Cell Biol.* 5:2789–2795.

101. Sugimoto, Y., M. Whitman, L. C. Cantley, and R. L. Erikson. 1984. "Evidence That the Rous Sarcoma Virus Transforming Gene Product Phosphorylates Phosphatidylinositol and Diacylglycerol." *Proc. Nat. Acad. Sci.* 81:2117–2122.

102. Doolittle, R. F., M. W. Hunkapiller, L. E. Hood, S. G. Devare, K. C. Robbins, S. A. Aaronson, and H. M. Antoniades. 1983. "Simian Sarcoma Virus Onc Gene, v-*sis*, Is Derived from the Gene (or Genes) Encoding a Platelet-Derived Growth Factor." *Science* 221:275–276.

103. Waterfield, M. D., G. J. Scrace, N. Whittle, P. Stroobant, A. Johnson, A. Wasteson, B. Westemark, C-H. Heldin, J. S. Huang, and T. F. Deuel. 1983. "Platelet-Derived Growth Factor is Structurally Related to the Putative Transforming Protein p28^{sis} of Simian Sarcoma Virus." *Nature* 304:35–39.

104. Gazit, A., H. Igarashi, I-M. Chiu, A. Srinivasan, A. Yaniv, S. R. Tronick, K. C. Robbins, and S. A. Aaronson. 1984. "Expression of the Normal Human *sis*/PDGF-2 Coding Sequence Induces Cellular Transformation." *Cell* 39:89–97.

105. Leal, F., L. T. Lewis, K. C. Robbins, and S. A. Aaronson. 1985. "Evidence That the v-*sis* Gene Product Transforms by Interaction with the Receptor for Platelet-Derived Growth Factor." *Science* 230:327–330.

106. C. Betsholz, A. Johnsson, C-H. Heldin, B. Westermark, P. Lind, M. S. Urdea, R. Eddy, T. B. Shows, K. Philpott, A. L. Mellor, T. J. Knott, and J. Scott. 1986. "cDNA Sequence and Chromosomal Localization of Human Platelet-Derived Growth Factor A-Chain and Its Expression in Tumor Cell Lines." *Nature* 320:695–699.

107. Downward, J., Y. Yarden, E. Mayes, G. Scrace, N. Totty, P. Stockwell, A. Ullrich, J. Schlessinger, and M. D. Waterfield. 1984. "Close Similarity of Epidermal Growth Factor Receptor and v-*erb*-B Oncogene Protein Sequences." *Nature* 307:521–527.

108. Sherr, C. J., C. W. Rettenmier, R. Sacca, M. F. Roussel, A. T. Look, and E. R. Stanley. 1985. "The c-*fms* Proto-Oncogene Product Is Related to the Receptor for the Mononuclear Phagocyte Growth Factor, CSF-1." *Cell* 41:665–676.

109. Ellis, R. W., D. DeFeo, T. Y. Shih, M. A. Gonda, H. A. Young, N. Tsvchida, D. R. Lowy, and E. M. Scolnick. 1981. "The p21 *src* Genes of Harvey and Kirsten Sarcoma Viruses Originate from Divergent Members of a Family of Normal Vertebrate Genes." *Nature* 292:506–511.

110. Shih, T. Y., A. G. Papageorge, P. E. Stokes, M. O. Weeks, and E. M. Scolnick. 1980. "Guanine Nucleotide-Binding and Autophosphorylating Activities Associated with the p21 Protein of Harvey Murine Sarcoma Virus." *Nature* 287:686–691.

111. Gibbs, J. M., I. S. Sigal, M. Poe, and E. M. Scolnick. 1984. "Intrinsic GPTase Activity Distinguishes Normal and Oncogenic *ras* p21 Molecules." *Proc. Nat. Acad. Sci.* 81:5704–5708.

112. Donner, P., I. Greiser-Wilke, and K. Moelling. 1982. "Nuclear Localization and DNA Binding of the Transforming Gene Product of Avian Myelocytomatosis Virus." *Nature* 296:262–265.

113. Hann, S. R., H. D. Abrams, L. R. Rohrschneider, and R. N. Eisenman. 1983. "Proteins Encoded by the v-*myc* and c-*myc* Oncogenes: Identification and Localization in Acute Leukemia Virus Transformants and Bursal Lymphoma Cell Line." *Cell* 34:789–798.

114. Eisenman, R. N., C. Y. Tachibana, H. D. Abrams, and S. R. Hann. 1985. "v-*myc* and c-*myc* Encoded Proteins Are Associated with the Nuclear Matrix." *Mol. Cell Biol.* 5:114–126.

115. Ralston, R., and J. M. Bishop. 1983. "The Protein Products of the Oncogene *myc*, *myb* and Adenovirus E1a Are Structurally Related." *Nature* 306:803–806.

116. Kelly, K., B. H. Cochran, C. D. Stiles, and P. Leder. 1983. "Cell-Specific Regulation of the c-*myc* Gene by Lymphocyte Mitogens and Platelet-Derived Growth Factor." *Cell* 35:603–610.

117. Van Beveren, C., F. van Straaten, T. Curran, R. Muller, and I. M. Verma. 1983. "Analysis of FBJ-MuSV Provirus and c-*fos* (Mouse) Gene Reveals That Viral and Cellular *fos* Gene Products Have Different Carboxy Termini." *Cell* 32:1241–1255.

118. Cochran, B. H., J. Zullo, I. M. Verma, and C. D. Stiles. 1984. "Expression of the c-*fos* Gene and of a *fos*-Related Gene Is Stimulated by Platelet-Derived Growth Factor." *Science* 226:1080–1082.

119. Treisman, R. 1985. "Transient Accumulation of c-*fos* RNA Following Serum Stimulation Requires a Conserved 5' Element and c-*fos* 3' Sequences." *Cell* 42:889–902.

120. Weinberger, C., S. M. Hollenberg, M. G. Rosenfeld, and R. M. Evans. 1985. "Domain Structure of Human Glucocorticoid Receptor and Its Relationship to the v-*erb-A* Oncogene Product." *Nature* 318:670–672.

121. Greene, G. L., P. Gilna, M. Waterfield, A. Baker, Y. Hort, and J. Shine. 1986. "Sequence and Expression of Human Estrogen Receptor Complementary DNA." *Science* 231:1150–1154.

122. Green, S., P. Walter, V. Kumar, A. Krust, J-M. Bornet, P. Argos, and P. Chambon. 1986. "Human Oestrogen Receptor cDNA: Sequence, Expression and Homology to v-*erb-A*." *Nature* 320:134–139.

123. Anderson, S. M., W. S. Hayward, B. G. Neel, and H. Hanafusa. 1980. "Avian Erythroblastosis Virus Produces Two mRNAs." *J. Virology* 36:676–683.

124. Mushinski, J. F., M. Potter, S. R. Bauer, and E. P. Reddy. 1983. "DNA Rearrangement and Altered RNA Expression of the c-*myb* Oncogene in Mouse Plasmacytoid Lymphosarcomas." *Science* 220:795–798.

125. Hayward, W. S., B. G. Neel, and S. M. Astrin. 1981. "Activation of a Cellular Onc Gene by Promoter Insertion in ALV-Induced Lymphoid Leukosis." *Nature* 290:475–480.

126. Payne, G. S., J. M. Bishop, and H. E. Varmus. 1982. "Multiple Arrangements of Viral DNA and an Activated Host Oncogene in Bursal Lymphomas." *Nature* 295:209–214.

127. Tsichlis, P. N., and J. M. Coffin. 1980. "Recombinants Between Endogenous and Exogenous Avian Tumor Viruses: Role of the C Region and Other Portions of the Genome in the Control of Replication and Transformation." *J. Virology* 33:238–249.

128. Robinson, H. L., and G. C. Gagnon. 1986. "Patterns of Proviral Insertion in Avian Leukosis Virus-Induced Lymphomas." *J. Virology* 57:28–36.

129. Nilsen, T. W., P. A. Maroney, R. G. Goodwin, F. M. Rottman, L. B. Crittenden, M. A. Raines, and H. Kung. 1985. "c-*erb-B* Activation in ALV-Induced Erythroblastosis: Novel RNA Processing and Promoter Insertion Result in Expression of an Amino-Truncated EGF Receptor." *Cell* 41:719–726.

130. Chatis, P. A., C. A. Holland, J. W. Hartley, W. D. Rowe, and N. Hopkins. 1983. "Role for the 3' End of the Genome in Determining Disease Specificity of Friend and Moloney Murine Leukemia Viruses." *Proc. Nat. Acad. Sci.* 80:4408–4411.

131. DesGroseillers, L., E. Rassart, and P. Jolicoeur. 1983. "Thymotropism of Murine Leukemia Virus Is Conferred by Its Long Terminal Repeat." *Proc. Nat. Acad. Sci.* 80:4203–4207.

132. Nusse, R., A. Van Ooyen, D. Cox, Y. K. T. Fung, and H. Varmus. 1984. "Mode of Proviral Activation of a Putative Mammary Oncogene (*int-1*) on Mouse Chromosome 15." *Nature* 307:131–136.

133. Peters, G., S. Brookes, R. Smith, and C. Dickson. 1983. "Tumorigenesis by Mouse Mammary Tumor Virus: Evidence for a Common Region for Provirus Integration in Mammary Tumors." *Cell* 33:369–377.

134. Stewart, T. A., P. K. Pattengale, and P. Leder. 1984. "Spontaneous Mammary Adenocarcinomas in Transgenic Mice That Carry and Express MMTV/*myc* Fusion Genes." *Cell* 38:627–637.

135. Furth, J., H. R. Seibold, and R. R. Rathbone. 1933. "Experimental Studies on Lymphomatosis of Mice." *Amer. J. Cancer* 19:521–604.

136. Gross, L., 1951. "Spontaneous Leukemia Developing in C3H Mice Following Inoculation, in Infancy, with AK-Leukemic Extracts, or AK-Embryos." *Proc. Soc. Exp. Biol. Med.* 76:27–32.

137. Rowe, W. P. 1973. "Genetic Factors in the Natural History of Murine Leukemia Virus Infection: G. H. Clowes Memorial Lecture." *Cancer Res.* 33:3061–3068.

138. Chattopadhyay, S. K., W. P. Rowe, N. M. Teich, and D. R. Lowy. 1975. "Definitive Evidence That the Murine C-Type Virus Inducing Locus *Akv-1* Is Viral Genetic Material." *Proc. Nat. Acad. Sci.* 72:906–910.

139. Hartley, J. W., N. K. Wolford, L. J. Old, and W. P. Rowe. 1977. "A New Class of Murine Leukemia Virus Associated with Development of Spontaneous Lymphomas." *Proc. Nat. Acad. Sci.* 74:789–792.

140. Cuypers, H. T., G. Selten, W. Quint, M. Zijlstra, E. R. Maandag, W. Boelens, P. Van Wezenbeek, C. Melief, and A. Berns. 1984. "Murine Leukemia Virus-Induced T-Cell Lymphomagenesis: Integration of Proviruses in a Distinct Chromosomal Region." *Cell* 37:141–150.

141. Tsichlis, P. N., P. G. Strauss, and L. F. Hu. 1983. "A Common Region for Proviral DNA Integration in Mo MuLV-Induced Rat Thymic Lymphomas." *Nature* 302:445–448.

142. Li, Y., C. A. Holland, J. W. Hartley, and N. Hopkins. 1984. "Viral Integration Near c-*myc* in 10–20% of MCF 247 Induced AKR Lymphomas." *Proc. Nat. Acad. Sci.* 81:6808–6811.

143. Steffen, D. 1984. "Proviruses Are Adjacent to c-*myc* in Some Murine Leukemia Virus-Induced Lymphomas." *Proc. Nat. Acad. Sci.* 81:2097–2101.

144. Hardy, W. D., Jr., P. W. Hess, E. G. MacEwen, A. J. McClelland, E. E. Zuckerman, M. Essex, S. M. Cotter, and O. Jarrett. 1976. "Biology of Feline Leukemia Virus in the Natural Environment." *Cancer Res.* 36:582–588.

145. Rubin, H., L. Fanshier, A. Cornelius, and W. F. Hughes. 1962. "Tolerance and Immunity in Chickens After Congenital and Contact Infection with an Avian Leukosis Virus." *Virology* 17:143–156.

146. Gardner, M. B., V. Klement, R. R. Rongey, P. McConahey, J. D. Estes, and R. J. Huebner. 1976. "Type C Virus Expression in Lymphoma-Paralysis-Prone Wild Mice." *J. Nat. Cancer Inst.* 57:585–590.

CHAPTER 27

The Origins of Human Cancer

During the past 30 years it has become increasingly clear that cancer is a genetic disease that results when multiple mutations (genetic events) accumulate within the DNA of a single somatic cell, causing the cell to lose growth control. But the question was, How could we identify the cancer-causing genes in human tumors? Furthermore, even if we could identify the genes, could we understand their function or discover how they had gone astray in the cancer cell? In the past few years, we have begun to obtain such convincing answers for several cancers that we can begin to use this knowledge to search for means both to prevent and cure these diseases. This chapter describes the types of research that suddenly came together in the 1970s and '80s to begin revealing the genetic basis of human cancer.

THE GENETIC BASIS OF HUMAN CANCER

Cancer Is Clonal and Arises More Frequently in Cell Types That Undergo Frequent Division[1]

The human body consists of about 10^{13} cells, and these are differentiated into the many different types that form the different organs: skin, liver, kidney, blood cells, and so on. In the mature body, some cell types divide rarely or even never (e.g., nerve cells), while others, like skin and the progenitors of the circulating blood cells, must divide throughout life in order to replace the billions of cells that die every day. Clearly, a very, very carefully controlled program must determine the growth of every type of cell in the body, since a few extra doublings or insufficient divisions would soon produce chaos. Tumors arise when a single cell loses its sense of these controls.

Evidence that most tumors arise from a single cell came first from analyzing tumors in females who inherit different isozymes of the X-chromosome-coded enzyme *glucose-6-phosphate dehydrogenase (G6PD)*. The human female has two X chromosomes, but only one of the two is active in each cell. Early in embryonic life, one X chromosome in each cell is modified so that it is transcriptionally silent. However, the particular chromosome (paternal or maternal) chosen for inactivation is different in different cells. Thus, females are **mosaics**: Most of their tissues consist of a mixture of cells, some using the paternal and some the maternal X chromosome. If a tumor arose in many cells, then different cells of the same tumor should express different isozymes of G6PD. However, tumors turn out to express only one isozyme, indicating their origin from a single cell.

There are two types of tumors, only one of which is cancerous. If a cell divides when it should not, but nonetheless stays within its normal location, then its growth produces a mass of cells called a *benign tumor,* which is not a cancer. Benign tumors can become very large so that they require surgical removal, but they are not usually life threatening. In contrast, some cells that begin to divide when they should not also acquire the ability to invade surrounding tissues and to move to alien sites in the body (i.e., **metastasize**). Thus, a tumor may begin, for example, in a single lung cell; but at some point after a mass of cells has formed, tumorous lung cells from the mass break off and migrate to the brain or other organs where they seed new tumor growths, called *metastases.* Metastatic cells, still identifiable as lung cells in the above example, multiply and destroy the architecture and ultimately the function of the invaded organ. Tumors that are highly invasive of surrounding tissues and that can metastasize are called *malignant tumors* or *cancers.* It is their ability to spread to vital organs and so escape the surgeon's knife that makes them deadly.

Cancers can arise in many types of differentiated cells and since cancers are classified according to the cell type in which they arise, there are over 200 different forms of human cancer, each named for the cell type of origin. For example, gliomas are cancers of glial cells; hepatomas, cancers of liver cells (hepatocytes); melanomas, cancers of the pigment-producing melanocytes that are particularly densely located in moles. The many different forms of human cancer are classified into one of three groups. **Carcinomas** include all the cancers that arise in cells of the sheets or epithelia that cover our surfaces (skin, gut, etc.). **Sarcomas** include cancers of the supporting tissues (bone, muscle, blood vessels, fibroblasts). **Leukemias** and **lymphomas** include cancers of the cells that produce the circulating cells of the blood and the immune system. More than 90 percent of human cancers are carcinomas. Sarcomas and leukemias together account for only about 8 percent of human tumors. Furthermore, among the carcinomas, cancers of the lung, breast, and colon are the most common fatal tumors in Western countries and account for more than half of cancer deaths, while liver cancer is most common in Asia and Africa. Although not perfectly correlated, cancers tend to arise in cells that continue to proliferate in adult life. In contrast, nerve cells, for example, never become cancerous in adults, although they can do so in infants, where they are still undergoing division. In addition, major sites of cancer are the external epithelia that are in contact with the environment: the air we breathe and the food and water we consume.

Human Cancers Are the Result of Multiple Mutations[2-5]

Although a few forms of cancer are associated with childhood, for most cancers incidence increases sharply with age. For example, the death rate from cancer of the large intestine increases more than a thousandfold between the ages of 30 and 80 (Figure 27-1), and this is quite typical of many cancers. The model that has been used to explain curves like the one in Figure 27-1 is that cancer arises when a cell

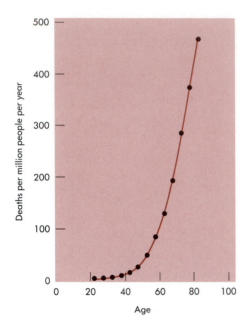

Figure 27-1

Annual death rate in relation to age due to cancer of the large intestine. Statistics are for the United States. The data are plotted on a linear scale. The death rate from this disease increases about a thousandfold between the ages of 30 and 80.

accumulates several specific mutations (genetic events). Since mutations can occur at any time, the probability that any particular cell will acquire a particular mutation increases in direct proportion to age. The probability that the cell has acquired mutations in all N of the genes needed to produce a cancer cell rises as the Nth power of age. Although it is very difficult to arrive at reliable values for N, extensive analysis of cancer incidence data has led to estimates of 2 to 7. Most leukemias are believed to result from the accumulation of two to four specific mutations (genetic events) within a single cell, while carcinomas may require anywhere from two to six or seven.

Other evidence that cancer results from multiple mutations comes from studying certain individuals who inherit a very high probability of developing a specific form of cancer, for example, a tumor called retinoblastoma. These people appear to have inherited one of the mutations that can lead to the cancer. However, only when a particular cell acquires one or more additional changes does the disease develop.

Epidemiology Reveals That Environmental Factors Cause Most Cancers[6]

Until about 1940, it was widely believed that cancer was an inevitable consequence of aging, that somehow the very process of living included the inevitable creation of cancer cells at some significant frequency. Since cell division requires DNA replication, and DNA replication produces mutations at a low rate, even today this remains an entirely reasonable hypothesis. Fortunately, however, it does not seem to be true for the majority of cancers that kill us. We know this because of epidemiological studies that have analyzed statistics of cancer deaths in relation to country, social group, occupation, living habits, and different time periods. The major finding from these studies is that there is great variation in the incidence of most cancers from place to place, from time to time, and depending on people's activities (Table 27-1). Furthermore, when individuals migrate to a new region, usually, within one or two generations, they acquire the cancer incidence profile of their new home, even when they continue to intermarry with genetically similar individuals. Thus, environmental factors rather than intrinsic, inevitable aging processes must play a large role in determining the probability of getting cancer.

Table 27-1 Geographic Variation in Incidence of Some Common Cancers

Type of Cancer	Region of Highest Incidence	Range of Variation*	Region of Lowest Incidence
Lung	Great Britain	35	Nigeria
Stomach	Japan	25	Uganda
Liver	Mozambique	70	Norway
Skin	Queensland	>200	Bombay
Colon	United States	10	Nigeria
Breast	United States	15	Uganda

*Range of variation is the highest incidence observed in any country divided by the lowest incidence.
SOURCE: Data from R. Doll, "Strategy for Detection of Cancer Hazards to Man," *Nature* 265 (1977): 589–596. Except for breast cancer, the data are for men.

It is not always easy to identify the factors in the environment, diet, or personal habits that cause cancer. However, clear-cut examples now include cigarette smoking, which probably causes 85 to 90 percent of all lung cancers; hepatitis B virus infection, which in some unknown way is thought to lead to 80 to 90 percent of liver cancers; asbestos, which causes mesotheliomas and, in combination with cigarette smoking, greatly increases the incidence of lung cancer: and sunlight, which induces skin cancers.

Mutagens and Certain Viruses Are Carcinogens for Humans[7, 8, 9]

The fact that the cancerous phenotype of a cell is stably inherited from generation to generation had suggested many years ago that genetic changes in DNA might underlie the loss of cellular growth control seen in cancers. However, this evidence alone was not at all compelling, since specific patterns of gene expression, as in differentiated cell types, can be stably inherited in the absence of changes to DNA base sequences. It was the realization that the majority of carcinogens are mutagens for DNA that first led to general acceptance of the idea that many cancers arise from changes at the DNA level. In addition, two lines of evidence had previously indicated that at least some human cancers involve very specific alterations in DNA. First, as already mentioned, a high probability of contracting certain very rare forms of cancer is inherited; and in some of these cases, specific chromosomal defects are found to be the basis for the inherited tendency. Second, malignant tumors frequently have specific chromosome abnormalities that are lacking in the normal cells of the same individual. As we shall see, advances in molecular cloning technologies have made it possible to isolate at least some of the specific genes that are altered to produce cancer cells.

Frequently in conflict with the notion of a mutational origin of human cancer has been the belief that viruses are a significant cause of human cancer. Now there is little doubt that both viruses (e.g., hepatitis B) and other types of carcinogens cause human cancers. Furthermore, some viruses may cause cancer precisely because they can cause mutations in cellular DNA. The only question is the extent to which viruses versus carcinogens or other agents contribute to the disease.

HUMAN CANCER GENES DETECTED BY DNA TRANSFECTION

Chromosomal DNA from Human Tumors Can Induce Transformed Foci in 3T3 Cells[10–15]

One type of experiment that led to the isolation of cellular genes involved in human cancer relies on the technique of DNA transfection. Many types of cells, from bacterial to mammalian cells, can take up DNA and, at a low frequency, integrate some of the ingested DNA into their own chromosomes. Furthermore, if the donor DNA encodes a genetic trait that the recipient cell lacks, the recipient cell can acquire that trait. In bacteria, the process is called **transformation**

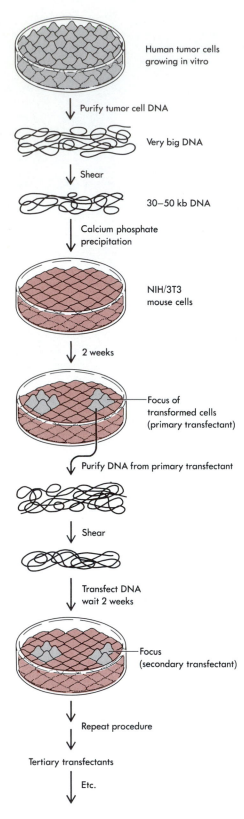

Figure 27-2
Assay for cellular oncogenes using transfection transformation of mouse 3T3 cells.

(confusingly, the same word used to describe the cancerous alteration of animal cells growing in tissue culture); in mammalian cells, it is usually called **transfection**. Although the frequency of DNA uptake is extremely low, it can be greatly increased through special techniques. In mammalian cells, transfection frequency increases dramatically if the DNA is fed to the recipient cells as a microscopic-size DNA–calcium phosphate coprecipitate.

If human cancer cells have altered genes that cause the cells to become cancerous, then, it was reasoned, DNA from cancer cells might be able to confer the cancerous phenotype on noncancerous recipient cells. To look for the hypothetical human cancer genes, DNA from human tumor cell lines was purified, sheared mechanically to break it into manageable pieces of about 30 to 50 kilobase pairs, dissolved in phosphate buffer, and then precipitated by the addition of calcium chloride (Figure 27-2). The solution was poured onto a monolayer of mouse 3T3 cells and the fine precipitate allowed to settle down onto the cells. 3T3 cells were selected as recipients because they are highly efficient at DNA uptake and also because they had been used for decades to assay cancer-causing genes of tumor viruses. Remarkably, DNA from human tumor cell lines was found to induce foci of transformed cells in the 3T3 monolayer, and these transformed cells could induce tumors in animals. Importantly, DNAs isolated from normal cells and used side by side as a negative control did not induce foci. This experiment indicated that human cancer cells have indeed sustained genetic alterations that can be assayed by DNA transfection. The number of foci obtained in this type of experiment increases linearly with the amount of DNA added, in-

Table 27-2 Human *ras* Oncogenes Detected by DNA Transfection

	Cell Lines (Positive/Tested)	Tumors	*ras* Oncogenes H	K	N
Carcinomas					
Bladder	2/7	2/25	3	1	—
Breast	1/19	0/18	1	—	—
Colon	1/2	2/2	—	2	1
Liver	1/4	1/8	—	—	2
Lung	6/12	1/5	1	5	1
Ovary	0/1	1/5	—	1	—
Pancreas	1/1	1/2	—	2	—
Sarcomas					
Fibrosarcoma	1/4	0/4	—	—	1
Rhabdomyosarcoma	1/3	1/2	—	1	1
Leukemias					
ALL	7/12	—	—	1	6
AML	1/2	—	—	—	1
APL	1/1	0/1	—	—	1
CML	1/2	0/3	—	—	1
Total:	24/70*	9/75	5	13	15

*The data are compiled from many published studies. In some reports, only cell lines giving positive results are described. Thus, in this table, the ratio of positive cell lines to those tested is an overestimate. Usually, about 10 to 20 percent of cell lines are positive for transfectable *ras* genes; the number varies considerably, depending on tumor type. SOURCE: M. Barbacid, "Human Oncogenes," in *Important Advances in Oncology*, ed. V. DeVita, S. Hellman, and S. Rosenberg (Philadelphia: Lippincott, 1986).

dicating that foci are induced by a single gene that must act in a dominant fashion.

Foci of 3T3 cells transformed by human tumor DNA are called primary transfectants. When DNA from these cells is purified, it, too, can pass the transformed phenotype on to fresh 3T3 cells, inducing foci called secondary transfectants. In contrast, DNA from normal 3T3 cells cannot induce foci. The procedure can be repeated serially over and over again, generating tertiary transfectants, and so on.

The first human tumor cells whose DNA was shown to induce foci in the 3T3 transfection transformation assay were certain established cell lines derived from tumors of the bladder, lung, colon, lymphoid cells, and a neuroblastoma. Of the many human tumor cell lines that have now been tested by this method, approximately 15 to 20 percent possess detectable transforming genes (Table 27-2). As soon as these results were obtained, it was immediately of great interest to know whether DNA isolated from primary human tumor material fresh from a patient rather than from established cell lines would also possess transforming activity in the 3T3 focus assay. Once again, about 15 percent of primary human tumor DNAs have been found to possess focus-inducing activity (see Table 27-2). Together these results show that at least some human tumors of unidentified cause possess genetic alterations that result in dominant transforming genes. Furthermore, since DNA from normal human cells lacks such transforming genes, the alterations must be related to the cancerous phenotype of the tumor cells.

Molecular Cloning of Transforming Genes from Human Tumors[16, 17]

To begin characterizing the human transforming genes biochemically, tumor DNAs were digested with various restriction endonucleases that cleave at specific 6-base sequences; the cleaved DNAs then were tested to see if they retained focus-inducing ability. We would expect a gene to possess cleavage sites for some enzymes but not others, and thus we would expect that some enzymes but not others would destroy its biological activity. This is precisely what was found for the human transforming genes. Each tumor DNA possesses a characteristic pattern of inactivation by restriction endonucleases, and furthermore, this pattern is retained through serial transfer from the tumor DNA to the DNA of all primary 3T3 foci, to secondary foci, and so on.

A most surprising result that soon emerged was that the same enzymes would destroy the transforming activity of DNAs from certain different human tumor cell lines, and in these cases, enzymes that failed to destroy the transforming gene of one line would also fail to destroy the transforming gene of the other lines. This result implied that the same genes were frequently involved in causing different cancers, a result that was soon verified by additional studies. In some cases tumors of the same cell type are somewhat more likely to harbor the same transforming gene, however, human tumors of different cell types can involve the same activated oncogene (see Table 27-2).

The next step in analyzing the human transforming genes was to clone them. The simplest molecular cloning procedures rely on having a hybridization probe for the desired gene. At first, since the identity of the transforming genes was unknown, it seemed that no such probes existed. Then an extremely useful observation supplied a hybridization probe that made it relatively easy to clone the human

Figure 27-3

Detection of human *Alu* containing DNA in mouse NIH/3T3 transfectants. Primary foci were induced in NIH/3T3 cells with DNA isolated from the human EJ bladder tumor cell line. DNA from a primary focus was used to induce secondary foci. DNAs from a primary focus and from five secondary foci were digested with a restriction endonuclease (*Eco* RI) and subjected to Southern transfer. Blots were probed with a human repetitive sequence DNA. Lane a: DNA from a primary focus. Lanes b–f: DNAs from five secondary foci. The differences in size of some *Alu*-containing bands in different foci is due to differences in degradation and integration of the EJ transforming gene during the transfection procedure. [Courtesy of C. Shih and R. A. Weinberg, *Cell* 29 (1982):161–169.]

transforming genes. Vertebrates possess short, highly reiterated sequences scattered throughout their genomes. Although these sequences are shared by all members of a species, they have diverged from species to species. Thus, these sequences can be used as hybridization probes to distinguish the DNA of, for example, humans and mice. In humans, the highly reiterated set of sequences is called the *Alu* family. Because approximately 300,000 copies of the *Alu* family sequence are scattered throughout the genome, there is likely to be such a sequence near any particular gene of interest, including, for example, the human cancer genes.

When a mouse cell transformed by human tumor cell DNA is digested with a restriction enzyme and then subjected to gel electrophoresis and Southern transfer of the DNA to nitrocellulose paper, the location of human DNA in the Southern blot can be seen with a radioactive probe for human *Alu* sequences, while the mouse DNA remains invisible (Figure 27-3). It is apparent from Figure 27-3 that primary 3T3 transfectants take up and integrate a great number of human genes. Usually, as much as 0.1 percent of the donor cell genome, the equivalent of about 3×10^3 kilobase pairs or, in this experiment, about 100 fragments of 30- to 50-kilobase-pair DNA, is incorporated. However, when DNA from the primary transfectant is used to induce foci a second time, even though once again about 100 pieces of donor DNA are taken up, since there is about 10^4 to 10^5 times as much mouse as human DNA in a primary 3T3 transfectant, most of the DNA taken up in the secondary transfectant is mouse DNA. The only human DNA that will tend to be found in the secondary transformant and in all subsequent transfectants, and thus the only DNA seen by the *Alu* probe, is that surrounding the human gene being selected because of its ability to induce transformed foci. This is what is found in Southern blots of DNA from secondary and tertiary 3T3 transfectants when probed with the human *Alu* probe. DNA from a secondary or tertiary 3T3 transfectant transformed by a human tumor gene contains only a few (or, depending on the gene and the restriction endonuclease used in the analysis, sometimes only one) *Alu* hybridizable bands (see Figure 27-3).

It is possible to clone many human transforming genes by constructing a λ library of chromosomal DNA isolated from a secondary or tertiary 3T3 transformant. The library is screened with the human *Alu* probe, and clones that contain this sequence are then tested for their ability to transform 3T3 cells. This approach was first used successfully to isolate the transforming gene of a human bladder carcinoma cell line, the so-called EJ cell line.

So Far, Most 3T3-Transforming Genes from Human Tumors Are H-*ras*, K-*ras*, or N-*ras* Genes[18–28]

As soon as the human EJ bladder oncogene was obtained, researchers wondered where this gene came from: Was it viral or, as expected, cellular in origin? A radioactive probe was prepared from a fragment of the gene (a fragment lacking repetitive *Alu* sequences) and used to analyze Southern blots of normal human DNA and of DNA from the EJ bladder tumor cell line. It was apparent that this gene was a normal cellular gene present in both DNAs. Furthermore, using a variety of restriction enzymes, no difference could be seen between the arrangement of the gene found in normal DNA and that found in bladder tumor cell DNA. Since normal human cell DNA is unable to transform 3T3 cells, this result indicated that whatever genetic change had activated the EJ bladder oncogene, it did not involve a major structural alteration of the gene.

Shortly after the human bladder oncogene was cloned, the startling finding was made that the gene was none other than the human H-*ras-1* gene, the human cellular homolog of the Harvey *ras* gene, already well known as the transforming gene of the highly oncogenic rodent retrovirus Harvey sarcoma virus (Chapter 26). At the same time, the transforming genes of a human lung tumor and of a colon cancer were found to be the human K-*ras-2* gene, the homolog of the transforming gene of Kirsten sarcoma virus (see Table 27-2). Although only the H- and K-*ras* genes were known from the study of transforming retroviruses, a new member of the family, N-*ras*, was soon found when a transforming gene from a human neuroblastoma was cloned. It turned out to have only very weak nucleotide homology to H- or K-*ras* genes, but when portions of the gene were sequenced, it was clear that it encoded a very similar protein.

As discussed in Chapter 26, vertebrate *ras* genes constitute a small multigene family. In humans, five *ras* genes have been found. H-*ras-1* and -2 are highly homologous in nucleotide sequence, but H-*ras-2* is a nonfunctional pseudogene and lacks introns. K-*ras-1* and -2 are highly homologous to each other and are only weakly homologous to the H-*ras* genes in nucleotide sequence; and K-*ras-1* is a nonfunctional pseudogene. Finally, as already noted, N-*ras* is only weakly homologous to the other *ras* genes. Functional *ras* genes (H-*ras-1*, K-*ras-2*, and N-*ras*) vary greatly in size (from about 5 to 45 kilobase pairs) owing to differences in intron size, but they have similar exon structures (Figure 27-4a). The K-*ras-2* gene differs from the other two, however, in having two fourth protein-coding exons (4a and 4b) that permit the formation of two proteins with different amino acid sequences near their carboxyl termini. Their extreme C-termini, however, share the structure found in all *ras* proteins (Cys-A-A-X, where A is a aliphatib amino acid; Figure 27-4b).

Despite differences in nucleotide sequence, the three functional *ras* genes encode highly related proteins, called p21s, that consist of 189 amino acids (188 when the more usual 4b exon of K-*ras-2* is used). The sequence of the first 86 amino acids is virtually identical among the three *ras* gene products; the next 79 amino acids are about 85 percent homologous between any two; and the carboxyl-terminal 24 amino acids are strikingly different in sequence between the three except for the final four residues (see Figure 27-4b). The *ras* proteins are post-translationally modified by covalent attachment of palmitic acid to the Cys residue present near their carboxyl termini. After modification, and probably as a result of it, they become associated with the inner surface of the plasma membrane. Although they are known to bind guanine nucleotides (GDP and GTP) and to possess GTPase activity, their biological function is unknown.

Since *ras* genes were already known to function as oncogenes in certain acute transforming retroviruses, the stunning finding that they could also become oncogenes in spontaneous human tumors provided strong circumstantial evidence that the human oncogenes detected by DNA transfection had indeed played a role in the development of their respective tumors. In addition, it provided unity between viral and spontaneous carcinogenesis.

It is not known why activated *ras* genes are the most common type of transfectable human cancer gene. In part, it may be that these

(a) 1 kbp

N-ras	Met	Thr	Glu	Tyr	Lys	Leu	Val	Val	Val	Gly	Ala	Gly	Gly	Val	Gly	Lys	Ser	Ala	Leu	Thr	Ile	Gln	Leu	Ile	Gln	Asn	His	Phe	Val	Asp	30
H-ras	*	*	*	*	*	*	*	*	*	*	*	*	*	*	*	*	*	*	*	*	*	*	*	*	*	*	*	*	*	*	
K-ras	*	*	*	*	*	*	*	*	*	*	*	*	*	*	*	*	*	*	*	*	*	*	*	*	*	*	*	*	*	*	

N-ras	Glu	Tyr	Asp	Pro	Thr	Ile	Glu	Asp	Ser	Tyr	Arg	Lys	Gln	Val	Val	Ile	Asp	Gly	Glu	Thr	Cys	Leu	Leu	Asp	Ile	Leu	Asp	Thr	Ala	Gly	60
H-ras	*	*	*	*	*	*	*	*	*	*	*	*	*	*	*	*	*	*	*	*	*	*	*	*	*	*	*	*	*	*	
K-ras	*	*	*	*	*	*	*	*	*	*	*	*	*	*	*	*	*	*	*	*	*	*	*	*	*	*	*	*	*	*	

N-ras	Gln	Glu	Glu	Tyr	Ser	Ala	Met	Arg	Asp	Gln	Tyr	Met	Arg	Thr	Gly	Glu	Gly	Phe	Leu	Cys	Val	Phe	Ala	Ile	Asn	Asn	Ser	Lys	Ser	Phe	90
H-ras	*	*	*	*	*	*	*	*	*	*	*	*	*	*	*	*	*	*	*	*	*	*	*	*	*	Thr	*	*	*	*	
K-ras	*	*	*	*	*	*	*	*	*	*	*	*	*	*	*	*	*	*	*	*	*	*	*	*	*	Thr	*	*	*	*	

N-ras	Ala	Asp	Ile	Asn	Leu	Tyr	Arg	Glu	Gln	Ile	Lys	Arg	Val	Lys	Asp	Ser	Asp	Asp	Val	Pro	Met	Val	Leu	Val	Gly	Asn	Lys	Cys	Asp	Leu	120
H-ras	Glu	*	*	His	Gln	*	*	*	*	*	*	*	*	*	*	Asp	*	*	*	*	*	*	*	*	*	*	*	*	*	*	
K-ras	Glu	*	*	His	His	*	*	*	*	*	*	*	*	*	*	Glu	*	*	*	*	*	*	*	*	*	*	*	*	*	*	

N-ras	Pro	Thr	Arg	Thr	Val	Asp	Thr	Lys	Gln	Ala	His	Glu	Leu	Ala	Lys	Ser	Tyr	Gly	Ile	Pro	Phe	Ile	Glu	Thr	Ser	Ala	Lys	Thr	Arg	Gln	150
H-ras	Ala	Ala	*	*	*	Glu	Ser	Arg	*	*	Gln	Asp	*	*	Arg	*	*	*	*	*	Tyr	*	*	*	*	*	*	*	*	*	
K-ras	Pro	Ser	*	*	*	Asp	Thr	Lys	*	*	Gln	Asp	*	*	Arg	*	*	*	*	*	Phe	*	*	*	*	*	*	*	*	*	

N-ras	Gly	Val	Glu	Asp	Ala	Phe	Tyr	Thr	Leu	Val	Arg	Glu	Ile	Arg	Gln	Tyr	Arg	Met	Lys	Lys	Leu	Asn	Ser	Ser	Asp	Asp	Gly	Thr	Gln	Gly	180
H-ras	Gly	*	Glu	*	*	*	*	*	*	*	*	*	*	*	Gln	His	Lys	Leu	Arg	*	Leu	Asn	Pro	Pro	Asp	Glu	Ser	Gly	Pro	Gly	
K-ras-4a	Arg	*	Glu	*	*	*	*	*	*	*	*	*	*	*	Gln	Tyr	Arg	Leu	Lys	*	Ile	Ser	Lys	Glu	Glu	Lys	Thr	Pro	Gly	Cys	
K-ras-4b	Gly	*	Asp	*	*	*	*	*	*	*	*	*	*	*	Lys	His	Lys	Glu		*	Met	Ser	Lys	Asp	Gly	Lys	Lys	Lys	Lys	Lys	

N-ras	Cys	Met	Gly	Leu	Pro	Cys	Val	Val	Met
H-ras	Cys	Met	Ser	Cys	Lys	*	Val	Leu	Ser
K-ras-4a	Val	Lys	Ile	Lys	Lys	*	Ile	Ile	Met
K-ras-4b	Lys	Ser	Lys	Thr	Lys	*	Val	Ile	Met

(b)

Figure 27-4

Structure of the human N-*ras*, H-*ras-1*, and K-*ras-2* genes and amino acid sequences of the polypeptides (p21s) they encode. (a) Exons that encode protein are shown as solid boxes; untranslated exons are white boxes. Whether there is a fifth untranslated exon (exon 0) for the N-*ras* gene is not yet known. Primary transcript of the K-*ras-2* gene and the two ways it can be spliced are indicated below the gene. The 4a exon can be skipped over and the 4b exon joined to exon 3. (b) Amino acids shared by all three proteins are indicated by asterisks. Arrows indicate boundaries of coding exons: these are identical in all three genes. Amino acids (12 and 61) that have been found to be altered and to lead to activated *ras* genes in human tumors are shaded. All *ras* proteins that have been studied terminate in the sequence cysAAX, where A stands for an aliphatic amino acid and X is any carboxyl terminal residue.

genes are more frequently involved in human cancers, possibly because of the simple mechanism by which they can become activated oncogenes. It is also possible that the 3T3 transformation assay is particularly sensitive to detecting this class of oncogene, while 3T3 cells are known to be relatively insensitive to morphological transformation by certain other oncogenes, for example, *myc* and *myb*.

Although the vast majority of transfectable human transforming genes have turned out to be one of the *ras* genes, a few other genes have been detected as well. A gene called *met* has been found in a chemically transformed human tumor cell line; and an activated *raf* gene (Chapter 26) has been detected in several human carcinomas. How many different human oncogenes will ultimately be detected by the 3T3 transfection transformation assay is hard to say, but now most researchers expect no more than about a dozen.

Activated *ras* Genes Result from a Single-Base Change That Leads to a Single-Amino-Acid Change[29–34]

The molecularly cloned normal human H-*ras-1* gene cannot transform 3T3 cells, although it can become a transforming gene if it is joined to a very strong transcriptional promoter so that abnormally high levels of p21 are synthesized in transfected cells. In contrast to these findings, the molecularly cloned EJ bladder oncogene can, of course, transform 3T3 cells efficiently without being joined to a new viral promoter. Thus, the EJ gene must carry a mutation relative to its normal counterpart.

With molecular clones of the normal and EJ H-*ras-1* genes available, it was easy to find the difference between them. The two genes were digested with restriction endonucleases; then fragments from each were rejoined in order to identify the region of the EJ gene that gave it transforming activity (Figure 27-5). A small fragment responsible for the property was soon found. When this fragment was sequenced, it turned out to differ from the normal gene by just one nucleotide; a G in the twelfth codon of the normal gene had become a T. The nucleotide difference leads to a change in the twelfth amino acid of the *ras* protein from glycine to valine. Analysis of other human transforming genes revealed that a single nucleotide change in the twelfth codon is a common mechanism by which a K-, H-, or N-*ras* gene is activated to become an oncogene. Single-base changes leading to an altered amino acid at position 61 or 13 have also been found (Table 27-3). *ras* proteins activated by a single-amino-acid change at position 12 often have altered electrophoretic mobility on SDS gels.

The technique of in vitro mutagenesis has been used to introduce mutations into many different sites of a molecularly cloned *ras* gene. The experiments show that in addition to amino acids 12 and 61, mutations that alter residues 13, 59, or 63 can also turn *ras* genes into transforming genes. In addition, in vitro mutagenesis has been used to introduce every possible amino acid change into position 12 of the H-*ras-1* gene. Replacement of glycine with any amino acid except proline results in an activated gene. Interestingly, *ras* proteins have homology to the guanine nucleotide binding, protein synthesis elongation factor, EF-Tu. The structure of EF-Tu is known, and it turns out that regions of good homology between *ras* and EF-Tu are in areas of

Figure 27-5

3T3 transforming ability of DNA molecules constructed from fragments of the normal human H-*ras-1* gene and the activated H-*ras-1* gene from the EJ bladder carcinoma. The H-*ras-1* gene with its four protein-coding exons is shown at the top of the figure. Each line below this represents a recombinant molecule. Portions of each molecule derived from the EJ H-*ras-1* gene are shown as black lines; portions derived from the normal H-*ras-1* gene are shown as gray lines. The restriction enzyme sites used to generate the recombinants are shown. The ability of each to induce transformed foci in 3T3 cells is shown at the right. For each pair of reciprocal recombinants (designated 1 and 2), only one member of the pair is active. The results identify the region important to oncogene activation as lying between the Xma I and the KpnI restriction endonuclease sites and encompassing the first protein-coding exon of the gene. [After C. J. Tabin et al., *Nature* 300 (1982):143–149.]

EF-Tu that bind guanine nucleotides. The amino acid of EF-Tu that corresponds to residue 12 in *ras* proteins lies on a loop, located between an α helix and a β strand, that includes the region that interacts with the phosphate group of the guanine nucleotides.

As soon as it was learned that single-amino-acid substitutions could activate *ras* proteins, viral *ras* genes were reexamined to see whether their transforming activity might be a result of mutations in the protein-coding region as well as a result of high-level expression of p21 due to the influence of potent viral transcriptional signals. Indeed, viral *ras* genes turn out to carry several mutations relative to cellular *ras* genes, and these include mutations at codons 12 and 59 (see Table 27-3). Thus, viral *ras* genes are altered in three ways, any one of which could give them transforming activity. Interestingly, the alteration at codon 59 introduces a threonine residue in both the *v*-H- and *v*-K-*ras* p21s, and this residue is phosphorylated in both proteins. A kinase activity latent in p21 itself is responsible for the phosphorylation. Normal *ras* gene products lack this residue and are not phosphorylated.

Table 27-3 Comparison of Amino Acids That Distinguish Normal and Transforming *ras* Proteins

| *ras* Gene | Amino Acid Positions | | | |
	12	13	59	61
H-ras-1-Derived Genes				
Human *c-H-ras-1*	Gly	Gly	Ala	Gln
Human EJ bladder tumor	Val			
Human HS242 breast tumor				Leu
Harvey virus *v-H-ras*	Arg		Thr	
K-ras-2-Derived Genes				
Human *c-K-ras-2*	Gly	Gly	Ala	Gln
Human Calu lung tumor	Lys			
Human SW480 colon tumor	Val			
Kirsten virus *v-K-ras*	Ser		Thr	
N-ras-Derived Genes				
Human N-*ras*	Gly	Gly	Ala	Gln
Human SK-N-SH neuroblastoma				Lys
Human HL60 promyelocytic leukemia				Lys
Human HT-1080 fibrosarcoma				Lys
Human AML 33		Asp		

Activated *ras* Genes
Have Decreased GTPase Activity[35–42]

How does a single-amino-acid change in p21 have such a dramatic effect on the biological function of the molecule? Until we know the normal cellular function of the *ras*-coded proteins, we cannot have a satisfactory answer to this question. However, one clue is that alterations at amino acid 12 reduce or abolish the GTPase activity inherent in p21. The mutations do not alter the ability of the protein to bind guanine nucleotides, however. How might a change in GTPase activity of p21 lead to alterations in growth control?

At the moment, our best clue to what the normal function of *ras* gene products may be is the knowledge that *ras* proteins bind guanine nucleotides and have GTPase activity. There are other classes of proteins with these properties, for example, the G proteins that interact with adenylate cyclase to alter cAMP levels in cells (Chapter 25). Normally, these proteins, which are located at the plasma membrane, are inactive; but when cells are stimulated by certain hormone-receptor interactions, the G proteins can become activated and bind GTP. In this form, they can stimulate or depress adenylate cyclase, thus changing the intracellular level of cAMP and altering cellular behavior. Then a GTPase activity present in the G protein hydrolyzes GTP to GDP, thereby returning the protein to its inactive form. Although *ras*-coded p21s do not correspond to any of the three subunits of known G proteins, this model provides a useful way of understanding how a mutation that destroyed the GTPase activity of a protein might cause the protein to remain in an active form and so to send signals relentlessly that affect cell behavior (Figure 27-6).

New clues to the function of *ras*-coded proteins in mammalian cells may come from studying *ras* gene products in yeast. In yeast, there

Figure 27-6
Highly schematized view of the inactive form of a *ras* protein binding GDP while the activated form binds GTP. If a mutation destroys the GTPase present in the molecule, then the molecule may remain in the active state.

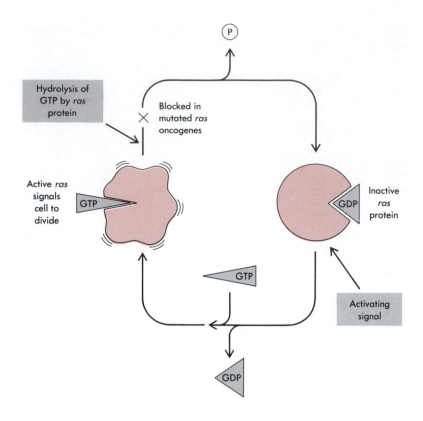

are two *ras* genes, *RAS1* and *RAS2*. The proteins they encode are considerably larger than mammalian p21s (309 and 322 amino acids), but they have a surprising degree of amino acid sequence homology to p21s (90 percent in the first 80 amino acids), including the same amino acids at the critical positions 12, 13, 59, 61, and 63, of mammalian *ras* gene products (Figure 27-7). While yeast cells can survive with just one of their two *RAS* genes, if both are destroyed by mutations, cells die. Remarkably, when the human H-*ras-1* gene is introduced into a yeast *ras1⁻ras2⁻* cell, it complements the defect, rescuing the cell and permitting vegetative growth. In addition, a modified yeast *RAS*-encoded protein with a mutation at amino acid 12 and a portion of its carboxyl terminus removed can induce transformed foci in 3T3 cells. These experiments provide evidence that yeast and mammalian *ras* proteins must serve at least some analogous functions in their respective cells, and so they encourage us to believe that powerful yeast genetics can be used to get at the normal cellular roles of mammalian as well as yeast *ras* genes.

Preliminary evidence that *ras* proteins play an essential role in controlling the mammalian cell cycle comes from studies using the technique of microinjection. Using very fine glass needles, purified p21 can be microinjected into mouse fibroblasts growing in vitro. The activated form of the protein is particularly effective in triggering cellular DNA synthesis. Furthermore, microinjection of monoclonal antibodies against p21 leads to G_1-induced arrest of cell growth. Resting 3T3 cells injected with anti-*ras* antibodies are unable to enter S phase when they are subsequently stimulated to divide, while dividing 3T3 cells injected with the antibody complete ongoing rounds of DNA synthesis but fail to initiate new rounds.

```
Ras 2   Met Pro Leu Asn Lys Ser Asn Ile Arg Glu Tyr Lys Leu Val Val Val Gly Gly [Gly][Gly] Val Gly Lys Ser Ala Leu Thr Ile Gln Leu   30
Ras 1    *  Gln Gly  *   *   *  Thr Ile Arg  *   *   *  Ile  *   *   *  Gly [ * ][ * ] *   *   *   *   *   *   *   *   *  Phe
H-ras   Met Thr  *   *  Leu  *   *   *  Ala [ * ][ * ] *   *   *   *   *   *   *   *   *   *  Leu

Ras 2   Thr Gln Ser His Phe Val Asp Glu Tyr Asp Pro Thr Ile Glu Asp Ser Tyr Arg Lys Gln Val Val Ile Asp Asp Glu Val Ser Ile Leu   60
Ras 1   Ile  *  Ser Tyr  *   *   *  Gly  *   *   *   *   *   *   *   *   *   *   *   *   *   *  Asp Lys Val Ser Ile  *
H-ras   Ile  *  Asn His  *   *   *  Glu  *   *   *   *   *   *   *   *   *   *   *   *   *   *  Gly Glu Thr Cys Leu  *

Ras 2   Asp Ile Leu Asp Thr [Ala] Gly [Gln] Glu [Glu] Tyr Ser Ala Met Arg Glu Gln Tyr Met Arg Asn Gly Glu Gly Phe Leu Leu Val Tyr Ser   90
Ras 1    *   *   *   *   *  [ * ]  *  [ * ]  *  [ * ]  *   *   *   *   *  Glu  *   *   *   *  Thr  *   *   *   *   *  Leu  *  Tyr Ser
H-ras    *   *   *   *   *  [ * ]  *  [ * ]  *  [ * ]  *   *   *   *   *  Asp  *   *   *   *  Thr  *   *   *   *   *  Cys  *  Phe Ala

Ras 2   Ile Thr Ser Lys Ser Ser Leu Asp Glu Leu Met Thr Tyr Tyr Gln Gln Ile Leu Arg Val Lys Asp Thr Asp Tyr Val Pro Ile Val Val   120
Ras 1   Val Thr Ser Arg Asn  *  Phe Asp Glu Leu Leu Ser  *  Tyr Gln  *   *  Gln  *   *   *   *  Ser  *  Tyr Ile  *  Val  *  Val
H-ras   Ile Asn Asn Thr Lys  *  Phe Glu Asp Ile His Gln  *  Arg Glu  *   *  Lys  *   *   *   *  Ser  *  Asp Val  *  Met  *  Leu

Ras 2   Val Gly Asn Lys Ser Asp Leu Glu Asn Glu Lys Gln Val Ser Tyr Gln Asp Gly Leu Asn Met Ala Lys Gln Met Asn Ala Pro Phe Leu   150
Ras 1    *   *   *   *  Leu  *   *  Glu Asn Glu Arg Gln Val  *  Tyr Glu Asp Gly Leu Arg Leu  *  Lys Gln Leu Asn Ala  *  Phe Leu
H-ras    *   *   *   *  Cys  *   *  Ala Ala Arg Thr Val Glu  *  Arg Gln     Ala Gln Asp Leu  *  Arg Ser Tyr Gly Ile  *  Tyr Ile

Ras 2   Glu Thr Ser Ala Lys Gln Ala Ile Asn Val Glu Glu Ala Phe Tyr Thr Leu Ala Arg Leu Val Arg Asp Glu Gly Gly Lys Tyr Asn Lys   180
Ras 1    *   *   *   *   *  Gln Ala Ile Asn  *  Asp Glu  *   *   *  Ser  *  Ile  *  Leu Val  *  Asp Asp Gly Gly Lys Tyr Asn Ser
H-ras    *   *   *   *   *  Thr Arg Gln Gly  *  Glu Asp  *   *   *  Thr  *  Val  *  Glu Ile  *  Gln His Lys Leu Arg Lys Leu Asn

Ras 2   Thr Leu Thr Glu Asn Asp Asn Ser Lys Gln Thr Ser Gln Asp Thr Lys Gly Ser Gly Ala Asn Ser Val Pro Arg Asn Ser Gly Gly His   210
Ras 1   Met Asn Arg Gln Leu Asp Asn Thr Asn Glu Ile Arg Asp Ser Glu Leu Thr  *  Ser  *  Thr Ala Asp Ile Glu Lys Lys Asn Asn Gly
H-ras   Pro Pro Asp Glu Ser Gly Pro Gly Cys Met Ser Cys Lys

Ras 2   Arg Lys Met Ser Asn Ala Ala Asn Gly Lys Asn Val Asn Ser Ser Thr Thr Val Val Asn Ala Arg Asn Ala Ser Ile Glu Ser Lys Thr   240
Ras 1   Ser Tyr Val Leu Asp Asn Ser Leu Thr Asn Ala Gly Thr Gly  *  Ser Ser Lys Ser Ala Val Asn His Asn Gly Glu Thr Thr  *  Arg
H-ras

Ras 2   Gly Leu Ala Gly Asn Gln Ala Thr Asn Gly Lys Thr Gln Thr Val Arg Thr Asn Ile Asp Asn Ser Thr Gly Gln Ala Gly Gln Ala Asn   270
Ras 1   Thr Asp Glu Lys  *  Tyr Val Asn Gln Asn Asn Asn Asn Glu Gly Asn  *  Lys Tyr Ser Ser Asn Gly Asn Gly Asn Arg Ser Asp Ile
H-ras

Ras 2   Ala Gln Ser Ala Asn Thr Val Asn Asn Arg Val Asn Asn Asn Ser Lys Ala Gly Gln Val Ser Asn Ala Lys Gln Ala Arg Lys Gln Gln   300
Ras 1   Ser Arg Gly Asn Gln Asn Asn Ala Leu Asn Ser Arg Ser Lys Gln Ser  *  Glu Pro Gln Lys  *  Ser Ser Ala Asn Ala Arg Lys Glu
H-ras

Ras 2   Ala Ala Pro Gly Gly Asn Thr Ser Glu Ala Ser Lys Ser Gly Ser Gly Gly Cys Cys Ile Ile Ser
Ras 1   Tyr                                         *   *   *   *   *  Ile Ile Cys
H-ras                                                                   *  Val Leu Ser
```

Activated *ras* Genes in Carcinogen-Induced Rodent Tumors[43–46]

Even before human tumor DNAs were shown to harbor genes capable of transforming 3T3 cells, DNAs from certain mouse and rat cells transformed by chemical carcinogens had been shown to possess transfectable transforming genes. These results provided evidence that chemical carcinogens, most of which are potent mutagens for DNA, may work by activating oncogenes directly.

One striking example of oncogene activation by chemical carcinogens involves mammary tumors induced in rats by the carcinogen N-*nitroso-N-methylurea (NMU)*. This carcinogen is known to induce G to A transitions in DNA by methylation at the oxygen 6 position. If 50-day-old rats, young enough to still be undergoing sexual development, are injected with a single dose of NMU, 6 to 12 months later they develop mammary tumors. More than 80 percent of these tumors contain an activated H-*ras-1* gene. Furthermore, 100 percent of these result from a G to A mutation in the second nucleotide of the twelfth codon, a change that causes the normal glycine to be replaced

Figure 27-7
Comparison of the amino acid sequences of proteins encoded by the yeast *RAS1* and *RAS2* genes and the human H-*ras-1* gene. The sequences are aligned for maximum homology. Amino acids that are common to all three proteins are indicated by an asterisk. Note that quite a few of the amino acid differences between the proteins are conservative amino acid changes. Amino acids that can be altered to produce activated mammalian *ras* oncogenes are shaded (positions corresponding to residues 12, 13, 59, 61, and 63 of the H-*ras-1* protein. [After S. Powers et al., *Cell* 36 (1984):607–612.]

Table 27-4 Transforming Genes Detected in Chemically Induced Rodent Tumors or Chemically Transformed Cell Lines

Tumor or Cell Line	Gene
NMU-induced mammary carcinomas (rat) DMBA-induced mammary carcinomas (rat) Benign and malignant DMBA/TPA-induced skin papillomas (mouse) MC-, BP-, and MNNG-transformed primary cells (guinea pig)	*c*-H-*ras-1*
MC-induced fibrosarcomas and MC-transformed fibroblasts (mouse) MC-induced thymic lymphoma (mouse) BP-induced fibrosarcoma (mouse)	*c*-K-*ras-2*
ENU-induced neuroblastoma cell lines (rat)	*neu*
MNNG-"retransformed" human osteosarcoma cell line	*met*

by a glutamic acid residue. NMU is unstable in vivo, and the total exposure time to the carcinogen may be no more than a few hours. Thus, within this time, random G to A mutations must occur. Presumably, some of these lie in H-*ras-1* genes of breast cells and so put these cells at risk to become tumors. Mammary tumors induced in rats by the carcinogen *DMBA (7,12-dimethylbenz(a)anthracene)* (which forms adducts with deoxyguanosine and deoxyadenosine that lead to subsequent excision repair) have a lower incidence of activated H-*ras-1* genes than NMU-induced tumors, and those that have been examined have mutations in codon 61.

Activated *ras* genes have been detected in DMBA-induced mouse skin carcinomas, in methylcholanthrene-transformed mouse fibroblasts, and in mouse thymomas induced by NMU and X-rays (Table 27-4). Interestingly, a different transforming gene is activated in rat neuroblastomas induced by exposing rat embryos in utero to *nitrosoethylurea.* This oncogene, called *neu,* is distinct from but related to the gene encoding the EGF receptor and may encode a receptor for a different polypeptide growth factor.

Activated *ras* Genes Alone May Not Cause Cancer[47]

If cancer results from the accumulation of several genetic alterations within a single cell, then we might expect that an activated *ras* gene alone would not be sufficient to induce cancer. Now numerous lines of evidence support this view.

Skin tumors can be induced experimentally in mice by sequential treatments with an initiator such as DMBA followed by a tumor promoter such as *12-O-tetradecanoyl-phorbol-13-acetate (TPA).* In this system, it is possible to obtain both premalignant, benign papillomas as well as malignant carcinomas. When DNA is isolated from either the precancerous lesions or a transplantable carcinoma and analyzed for the presence of activated *ras* genes, often they are found to be present. The fact that these genes are present in the benign papillomas argues that the presence of an activated *ras* gene alone does not make a cell cancerous. Thus, additional events must be required for the *ras* oncogene–containing cell to become a full cancer. The result also suggests that *ras* activation may be an early step in the development of these mouse skin tumors.

Direct Evidence That Two or More Oncogenes Cooperate to Produce Cancer[48–52]

If multiple mutations are required to cause cancer, then how can we explain the ability of a single activated *ras* gene to transform 3T3 cells to cancer cells? Insight into this paradox came from experiments in which attempts were made to use activated *ras* genes to transform primary rodent cells rather than established 3T3 cell lines. The *ras* genes were essentially unable to transform the primary cells.

As discussed in Chapter 25, although 3T3 cell lines are not tumorigenic, they nonetheless have enhanced growth potential relative to normal cells: They are established (immortal) and grow readily at low cell density. Previous studies with DNA tumor viruses, polyoma and adenovirus, had shown that although a single viral gene can induce transformation of established 3T3-like cell lines, two, and sometimes three, viral genes are required to transform primary cells. For example, polyoma large T protein alone induces primary cells to grow like established cells, but polyoma middle T protein is needed in addition to induce the rounded morphology and in vivo growth properties typical of a fully transformed cell. In at least some established cell lines, the need for the former gene is bypassed, and polyoma middle T alone can transform the cells. A similar situation may apply to cellular transformation by activated *ras* genes.

Activated *ras* genes alone are highly inefficient at transforming at least some types of primary rodent cells. However, if a second oncogene is present, then they readily induce transformed foci (Figure 27-8). Oncogenes that can complement *ras* include *myc*, the E1A gene of adenovirus, *myb*, and SV40 and polyoma large T (Chapter 26). Interestingly, these genes all share the ability to induce establishment

Figure 27-8
The activated H-*ras-1* bladder oncogene alone cannot induce visible foci of transformed cells in primary baby rat kidney cells. Cells were transfected with appropriate DNAs and fixed and stained about 3 weeks later. (a) Negative control (no oncogenes). (b) *E1a* gene of adenovirus. (c) Activated H-*ras-1* gene alone. (d) *Ela* plus activated H-*ras-1* genes. [Courtesy of H. E. Ruley. Reprinted with permission from *Nature* 304 (1983):602-606.]

or immortalization of primary cells, and they all encode proteins that are localized primarily in the cell nucleus.

Now some evidence indicates that even two oncogenes may not be sufficient to induce the fully cancerous phenotype. Some primary cells transformed by both *myc* and *ras* oncogenes fail to grow as progressive tumors, suggesting that additional oncogenes may also be needed.

If multiple oncogenes are needed to induce cancers, then how can the Harvey and Kirsten sarcoma viruses, which carry just a single activated *ras* oncogene, induce tumors so efficiently in mice? The answer to this question is not clear. One finding is that extremely high levels of activated *ras* genes may be sufficient to transform some types of primary cells in culture. As already noted, the *ras* genes of Harvey and Kirsten sarcoma viruses are activated in three ways: by point mutations in codons 12 and 59 and by the proximity to the *ras* gene of the strong retrovirus transcriptional signals. Conceivably, this makes the viral *ras* genes potent enough to transform cells single-handedly. Yet a second possibility is that cells infected by these viruses are at high risk to acquire the additional changes necessary to become cancer cells, since they already have one cancerous change. Or maybe only rare cells that already have aberrant gene expression are transformed by the virus.

If activated *ras* genes represent only one type of mutant gene that may contribute to the formation of human tumors, then clearly other genes remain to be found. Some of these genes are discussed in the following sections.

CHROMOSOME ABNORMALITIES ASSOCIATED WITH HUMAN TUMORS

Chromosome Abnormalities Are Associated with Many Forms of Human Cancer[53, 54]

Many human tumors have chromosome abnormalities, and frequently specific abnormalities are associated with specific forms of cancer. These findings have emerged gradually over the past 30 years as better and better techniques have become available for identifying and analyzing human chromosomes and for growing primary human tumor cells in vitro.

To visualize chromosomes, cells are usually captured in the metaphase stage, where the chromosomes are most condensed. A larger number of such cells can be obtained if the cells can be grown in vitro for a short time so that colchicine can be added to accumulate cells at mitosis. Next the cells are immersed in a hypotonic solution. This causes them to swell so that the chromosomes are farther apart and thus less likely to fall on top of one another when the cell is opened (see Figure 25-1). The cells are allowed to settle onto a slide, air-dried, fixed, and stained. Often, they are photographed and the 46 chromosomes cut out and aligned as shown in Figure 27-9, with the 22 pairs of autosomes arranged by size and followed by the two sex chromosomes.

Early chromosome preparations did not permit all the human chromosomes to be identified, since the only distinguishing characteris-

(a)

(b)

Figure 27–9
Typical preparation of human metaphase chromosomes before the discovery of banding techniques. (a) Metaphase spread from a lymphocyte of a normal male. (b) Arrangement of chromosomes in (a) into a standard karyotype. [Photo by J. J. Biesele, courtesy of H. E. Sutton, *An Introduction to Human Genetics,* 3rd ed. (Philadelphia: Saunders, 1980), p. 65.]

Figure 27-10
Examples of G banding and the nomenclature for chromosomes that has arisen from this methodology (400-band stage). The short arm of a chromosome is designated p and the long arm q. The arms are divided into one to three regions by landmarks. Landmarks are consistent features useful for identification such as the ends of the arm, the centromere, and certain very prominent bands. A region is the space between adjacent landmarks. Regions are numbered consecutively from the centromere, and bands are numbered consecutively within each region. Thus, for example, band 2q23 refers to chromosome 2, the long arm, region 2, band 3. [Courtesy of Dr. Jorge J. Yunis.]

tics available were length and centromere location. Then, starting in 1970, **banding** techniques were developed that allowed every chromosome and even regions within chromosomes to be distinguished. The first method used fluorescent derivatives of quinacrine, which, for unknown reasons, bind preferentially to some regions of chromosomes. Later, the *modified Giemsa stain* was developed, which results in banding patterns (G bands) similar to those produced by the quinacrine method (Q bands) but is simpler to perform. These methods have led to diagrammatic representations of human chromosomes and to the nomenclature for specific chromosomal regions described in Figure 27-10. Now it is also possible to apply the banding techniques to chromosomes at the less condensed prophase stage, where even more bands (as many as 3000) can be identified.

Chromosome abnormalities in human tumors can involve both the number and, more frequently, the structure of chromosomes. During mitosis, one or more chromosomes may fail to migrate properly at anaphase. As a result of this **nondisjunction**, both daughter cells have an abnormal chromosome number: One daughter cell receives both chromosomes of a pair and hence has a **trisomy** of that particular chromosome, while the other receives neither and so is **monosomic**. Structural abnormalities in chromosomes arise through errors in breakage and reunion and can involve deletions (del), inversions (inv), and reciprocal translocations (t). In addition, some cancer cells have enormous amplifications of specific DNA sequences. These amplifications may be located within a chromosome, where they appear as **homogeneously staining regions (HSRs)**, or they may be present on tiny chromosomes called **double minutes (DMs)**.

Since a specific chromosome abnormality may be a primary cause of a certain type of tumor, it may be present in the tumor from its inception. But in addition, a tumor that at first displays just one chromosome abnormality may acquire additional changes with time. In at least some cases, these changes go hand in hand with an increasingly poor prognosis for the patient's survival. Once under way, tumors seem to give rise to variants with less and less controlled growth, and at least some of these changes in behavior may result from changes in chromosomes.

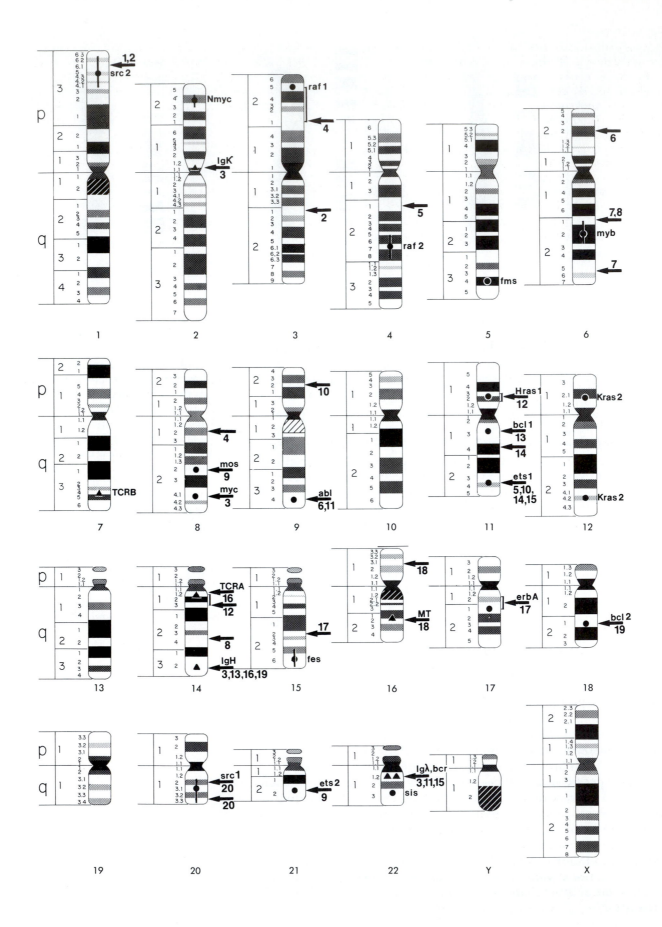

Figure 27-11

Human chromosome map showing 400 bands defined by Giemsa staining. The locations of 20 oncogenes (dot), 6 other cellular genes (triangle), and 2 breakpoints involved in specific chromosome rearrangements seen in cancers are marked. For oncogene abbreviations see Table 26-3. Other cellular genes are IgH, IgK, and Igλ: antibody genes for heavy chains, and for K and λ light chains, respectively. TCRA and TCRB: genes for the α and β chains of the T cell receptor. MT: metallothinein gene cluster. bcr: breakpoint cluster region. Rearrangements are numbered as follows: 1 = del 1p in neuroblastoma; 2 = t(1;3) in myelodysplasia (MDS); 3 = t(2;8), t(8;14), or t(8;22) in Burkitt's lymphoma; 4 = t(3;8) in mixed parotid gland tumor; 5 = t(4;11) in MDS; 6 = t(6;9) in ANLL; 7 = del 6q in nonHodgkin's lymphoma (NHL); 8 = t(6;14) in cystadenocarcinoma of ovary; 9 = t(8;21) in ANLL; 10 = t(9;11) in ANLL; 11 = t(9;22) in chronic myelogenous leukemia; 12 = t(11;14)(p13-14.1;q11.2-13) in T-cell acute lymphoblastic leukemia; 13 = t(11;14) × (q13.3;q32.3) in NHL; 14 = del 11q in NHL; 15 = t(11;22) in Ewing's sarcoma; 16 = inv(14) in T-cell chronic lymphocytic leukemia; 17 = t(15;17) in ANLL; 18 = inv(16) in ANLL; 19 = t(14;18) in NHL; 20 = del 20q in MDS. [Courtesy of J. J. Yunis. Reproduced with permission from *Important Advances in Oncology*, eds. V. deVita, S. Hellman, and S. Rosenberg (Philadelphia: Lippincott, 1986.)]

Some years ago, it was suggested that chromosome abnormalities might cause cancer by altering the expression of specific genes. In the past few years, specific translocations associated with several human tumors have been found to involve activation of the proto-oncogenes *c-myc* and *c-abl* (Chapter 26). In addition, amplified genes in HSRs and DMs of certain tumors are members of the *myc* gene family. Using molecular cloning technologies, the diverse mechanisms by which chromosome abnormalities lead to oncogene activation are rapidly being elucidated.

Some specific chromosome abnormalities associated with specific forms of cancer are summarized in Figure 27-11 and Table 27-5. It is easiest to analyze the chromosomes of blood cells and hence of leukemias. It is more difficult to separate the cells of solid tumors and to grow them in culture in order to accumulate mitotic cells. Hence, chromosome abnormalities have been recognized most often in various forms of leukemia, and only recently have specific abnormalities begun to be seen as common features of many solid tumors as well.

The Philadelphia Chromosome and *abl* Proto-Oncogene Activation in Chronic Myelogenous Leukemia[55–62]

The first specific chromosome abnormality to be associated with cancer was the *Philadelphia chromosome* (Ph[1]), named for the city in which it was discovered. This small chromosome is present in the leukemic cells of at least 90 percent of patients with *chronic myelogenous leukemia* (*CML*), an invariably fatal cancer involving uncontrolled multiplication of myeloid stem cells (Chapter 25). The abnormality is also seen in some patients with ANLL (acute nonlymphocytic leukemia) and ALL (acute lymphocytic leukemia). Ph[1] is derived from a chromosome 22 by a *reciprocal translocation* involving chromosome 9. (A reciprocal translocation is one in which no genetic material is lost when fragments of the chromosomes are exchanged.) A portion of the long arm of 22 is translocated to 9 while a small fragment from the tip of the long arm of 9 is translocated to 22 (Figure 27-12). Thus, there are actually two abnormal chromosomes present in CML cells: the short-

Table 27-5 Some of the Neoplasms with a Known Consistent
Chromosome Defect*

Disease	Chromosome Defect	Breakpoints or Deletion
Leukemias		
Chronic myelogenous leukemia	t(9;22)	9q34.1 and 22q11.21
Acute nonlymphocytic leukemia		
M1	t(9;22)	9q34.1 and 22q11.21
M2	t(8;21)	8q22.1 and 21q22.3
M3	t(15;17)	15q22 and 17q11.2
M4[†]	inv 16[‡]	p13.2 and q22
M4[†], M5[†]	t(9[§];11)	9p22 and 11q23
M1, M2, M4, M5, M6	del 5q	5q22q23
	del 7q	7q33q36
	+8	
Chronic lymphocytic leukemia	+12	
	t(11;14)[‡]	11q13 and 14q32
Acute lymphocytic leukemia		
L1-L2	t(9;22)	9q34.1 and 22q11.21
L2 [‖]	t(4;11)	4q21 and 11q23
L3	t(8;14)	8q24.13 and 14q32.33
Lymphomas		
Burkitt's, small noncleaved cell (non-Burkitt),[†] large-cell immunoblastic[†‡]	t(8;14)	8q24.13 and 14q32.33
Follicular small cleaved,[†] follicular mixed,[†] and follicular large cell[†]	t(14;18)	14q32.3 and 18q21.3
Small-cell lymphocytic[†]	+12	
Small-cell lymphocytic, transformed to diffuse large cell[†]	t(11;14)[‡]	11q13 and 14q32
Carcinomas		
Neuroblastoma, disseminated	del 1p	1p31p36
Small-cell lung carcinoma	del 3p	3p14p23
Papillary cystadenocarcinoma of ovary	t(6;14)	6q21 and 14q24
Constitutional retinoblastoma[†]	del 13q	13q14.13
Retinoblastoma[‡]	del 13q	13q14
Aniridia-Wilms' tumor[†]	del 11p	11p13
Wilms' tumor[‡]	del 11p	11p13
Benign Solid Tumors		
Mixed parotid gland tumor	t(3;8)[‡]	3p25 and 8q21
Meningioma	−22	22

*Subbands, denoted by a decimal-digit system, were defined by using high-resolution banding techniques. M1, M2, etc., and L1, L2, etc., represent distinct forms of nonlymphocytic leukemia and acute lymphocytic leukemia, respectively.
[†]Consistent chromosome defects revealed by high-resolution banding techniques.
[‡]Few cases described.
[§]Chromosomes 6, 10, 17, and 19 may serve as alternative receptor chromosomes.
[‖]Recently suspected to represent an undifferentiated form of ANLL-M4 leukemia, with the same breakpoint 11q23 as the 9;11 translocation.
SOURCE: Table compiled by J. J. Yunis and reprinted with modification from *Science* 221 (1983):227–236.

ened 22 (Ph^1) and an abnormal chromosome 9 called 9q$^+$. The translocation breakpoints involved in generating Ph^1 and 9q$^+$ always occur within the same bands: q34 on chromosome 9 and q11 on chromosome 22.

The first hint that the Philadelphia chromosome might lead to activation of the *abl* proto-oncogene came when somatic cell hybrids were used to map *c-abl* and it was found to lie on chromosome 9 (see Figure 27-13). Next, *c-abl* was shown to be present on the Ph^1 chromosome, indicating that it lay in the portion that is translocated to 22 to generate the aberrant Ph^1. Then an abnormally large *abl* mRNA was found in CML cells, suggesting that a structural alteration had indeed occurred in the immediate vicinity of the gene. Now, molecular cloning techniques have made it possible to analyze the 9-22 breakpoints in detail and have shown how *c-abl* expression is altered as a result of the translocation that generates the Philadelphia chromosome.

Two chromosome breaks are required to generate Ph^1 and 9q$^+$. One occurs within a 6-kilobase-pair region on chromosome 22 called the *breakpoint cluster region* (*bcr*). The second break occurs on chromosome 9 at the 5′ side of the *abl* gene. Apparently, this break can lie anywhere within a region of about 50 kilobase pairs. The bcr itself lies within a large gene, now called the *bcr* gene. When chromosomes 9 and 22 fuse, the 5′ half of the *bcr* gene ends up on the 5′ side of *abl*, with the two genes lying in the same transcriptional orientation (Figure 27-13). A large precursor RNA encompassing both genes is spliced so that the 5′ exons of the *bcr* gene are joined to a specific exon in the middle of *c-abl* (see Figure 27-13). This explains why most Philadelphia chromosomes, regardless of their precise breakpoints, generate novel *abl* transcripts of about the same size. These 8-kilobase-pair transcripts are translated to yield polypeptides of about 210 kdal (P210) instead of the normal 145 kdal. In P210, about 25 amino acids normally present at the amino terminus of *c-abl* are replaced with about 600 to 700 residues of the *bcr* protein.

The *abl* gene was first discovered as the transforming gene of the highly oncogenic Abelson murine leukemia virus. In the virus, *v-abl* is expressed as a fusion protein with the viral *gag* gene product (Chapter 26). The *abl* gene has amino acid sequence homology to the tyrosine kinase family of oncogenes, and the viral *gag-abl* fusion protein has detectable tyrosine protein kinase activity. In contrast, the tyrosine kinase activity presumed to be present in the product of the normal proto-oncogene, *c-abl*, is not detectable. P210, the *bcr-abl* fusion protein found in CML cells, resembles the viral *gag-abl* in having detectable tyrosine specific kinase activity. Although *c-abl* probably acts as a kinase in some appropriate circumstances in vivo, it seems likely that alterations of its amino terminus remove a regulatory mechanism that normally controls its activity and allow the kinase activity to be monitored experimentally. Presumably this is related to why the *bcr-abl* and *gag-abl* fusions are oncogenic.

A question of interest is the specific role of the *bcr* gene in activating *abl*. Perhaps the *bcr* portion of the *bcr-abl* fusion protein modulates the activity of the *abl* portion in such a way that the hybrid protein specifically acquires the ability to transform myeloid stem cells (recall that Abelson virus with its *gag-abl* protein primarily induces tumors of B cells). Alternatively, it is possible that the *bcr* gene is expressed specifically in stem cells. Thus, fusion to this gene might ensure that the aberrant oncogenic *abl* fusion protein is expressed efficiently in this

(*a*)　9　(9q$^+$)　　(*b*)　22　Ph^1

Figure 27-12
The Philadelphia chromosome in CML. (b) Appearance of normal and aberrant chromosomes 9 and 22 from a CML patient.

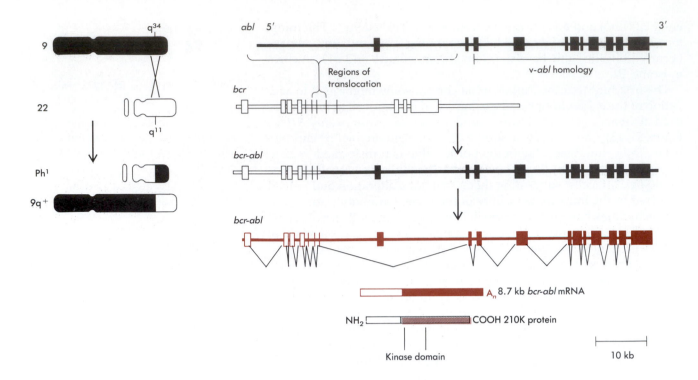

Figure 27-13

Diagram showing how the Philadelphia chromosome arises by a reciprocal translocation involving chromosomes 9 and 22 and how this event joins the *bcr* and *abl* genes. The *abl* gene is located on band q34 of chromosome 9; *bcr* is at q11 of chromosome 22. In the diagrams of the genes, boxes denote exons (these appear as vertical lines for small exons), and horizontal lines represent DNA. Not all the exons of *bcr* are defined yet, but breakpoints cluster within the bracketed region. The 5' exons of the very large *abl* gene are not yet completely defined. Thus, it is not clear whether the breakpoints on chromosome 9 fall within the gene or can occur 5' of the gene. The former possibility is more likely. Following translocation, a transcript of the 5' exons of *bcr* is apparently spliced to a specific internal exon of the *abl* gene. The resulting transcript is translated to yield the aberrant *bcr-abl* fusion protein. [After J. Adams, *Nature* 315 (1985):542.]

cell type and hence can participate in the cancerous transformation of such cells.

Translocation of the *myc* Proto-Oncogene to Immunoglobulin Gene Loci in B Cell Tumors of Mice and Humans[63–70]

Burkitt's lymphoma is a tumor of B cells that is common in Africa. It probably arises as a result of viral infection. Tumor cells generally secrete antibody, and they possess highly characteristic chromosome translocations. The translocations invariably involve chromosome 8 and one of the three chromosomes that carry the antibody light- or heavy-chain genes: chromosomes 14 (heavy-chain genes), 2 (λ light-chain genes), or 22 (κ light-chain genes). Furthermore, the translocation breakpoints are very specific, occurring at the same bands in different tumors (Figure 27-14). Translocations involving chromosomes 8 and 14 [designated t(8:14)] are by far the most common type in Burkitt's lymphoma and are present in 90 percent of the tumors.

The fact that antibody genes, which are so active in B cells, map to the same chromosomal regions involved in the specific translocations in Burkitt's tumors led to the suggestion that a specific oncogene on chromosome 8 might be activated if it were placed near to these genes as a result of a translocation. Considerable evidence now indicates that this hypothesis was fundamentally correct and that the proto-oncogene involved is *c-myc*.

c-myc is the cellular homolog of the transforming gene of the chicken retrovirus MC29 (Chapter 26). In humans, *c-myc* lies on the long arm of chromosome 8, at band q24, and spans about 5 kilobase pairs. The gene has three exons: the first specifies a long (550 base pairs) untranslated region of the mRNA, while the second two en-

Normal 8 Defective   Normal 14 Defective

Figure 27-14
The common t(8:14) translocation seen in Burkitt's lymphoma. Diagrams of normal and aberrant chromosomes are represented at the 1200 Giemsa band stage. The approximate locations of the c-*myc* and heavy chain genes are shown. [Courtesy of J. J. Yunis, *Science* 221 (1983):227–236.]

code the *myc* protein (439 amino acids). It is a phosphoprotein that is present in the nucleus. Two forms are detected in cells, and they migrate in SDS gels with molecular weights of 64 and 67 kdal. The normal function of c-*myc* is unknown. However, quiescent cells that are stimulated to divide by addition of growth factors or nonphysiological mitogens show a rapid, sharp increase in *myc* mRNA, suggesting a role for this gene in cell proliferation. In dividing cells, levels of *myc* mRNA seem to be fairly constant, suggesting that the gene does not function in a cell-cycle-dependent manner.

In translocations that involve chromosome 14, c-*myc*, along with the distal portion of chromosome 8, moves adjacent to DNA encoding the heavy-chain constant regions. The two loci are joined head to head in opposite transcriptional orientations. As discussed in Chapter 23, antibody genes show allelic exclusion, meaning that heavy-chain genes on only one of the two chromosomes are used. The *myc* gene is translocated to the nonfunctional heavy-chain locus, which explains the ability of the cells to synthesize antibodies. In the less frequent translocations involving antibody light-chain loci, *myc* remains on chromosome 8, and fragments of either 2 or 22 are brought nearby. In these cases, the antibody genes and *myc* are joined tail to head, in the same transcriptional orientation, with the antibody gene lying to the 3' side of *myc*.

It is now widely believed that translocations involving c-*myc* lead to altered transcription of the gene and that this is how they contribute to malignant transformation. One piece of evidence that indirectly supports this view is that the translocations involving c-*myc* do not alter the protein-coding portion of the gene. Translocated genes that have been completely sequenced usually have not sustained even a single-base change that might alter the activity of the encoded *myc* protein. It has proved difficult to demonstrate that *myc* mRNA is consistently elevated in different Burkitt's tumors. Part of the problem is that it is difficult to decide what a normal level of *myc* mRNA should be. If the B cell is supposed to be quiescent, awaiting antigen stimulation, then the presence of any *myc* mRNA might be sufficient to stimulate inappropriate cell division, so any disturbance of carefully controlled levels of *myc* expression might contribute to transformation.

How does the proximity of antibody genes change the transcription of a translocated *myc* gene? The answer to this question appears to be complex and may involve different mechanisms, depending on the precise breakpoints involved in the translocation. The relative orientation of the antibody genes and *c-myc* in the translocations tells us that activation cannot involve use of the antibody gene promoters to drive transcription of the *myc* gene. In some cases, the *myc* gene may fall under the influence of a powerful tissue-specific enhancer present near the antibody constant-region genes. In other cases, sequences 5' of *c-myc* may be lost that normally allow the gene to be negatively regulated. In still other cases, enhancer elements that have not yet been identified may turn out to lie near the rearranged *myc* gene.

In the case of translocations to the heavy-chain locus, most breakpoints on chromosome 14 occur within the switch regions lying in front of sequences that encode the heavy-chain μ, α, or γ constant regions, while the breakpoints near *c-myc* are widely scattered. The latter can occur from somewhere within the first *myc* intron to more than 25 kilobase pairs 5' of *c-myc*. In the case of the less frequent translocations involving antibody light-chain genes, *myc* and the antibody gene may be as much as 100 kilobase pairs apart.

Mouse plasmacytomas (tumors of mature B cells) also possess chromosome translocations that involve *c-myc* and one or another of the immunoglobulin loci. Except for some differences in the position of preferred breakpoints, many of the observations already described apply equally well to these translocations and in many cases were discovered first in the mouse system. Some other types of B cell tumors in humans also involve translocations of *c-myc* and antibody gene loci, revealing that these specific translocations are not limited to a single tumor type. B cell lymphomas induced in chickens by avian retroviruses that lack oncogenes usually involve activation of the *c-myc* gene by proviral insertion (Chapter 26).

Why is *c-myc* so frequently involved as an oncogene in these particular tumors, and why does the activation mechanism so frequently involve translocation to the antibody gene loci? So far, there is no adequate explanation for why a particular oncogene is frequently activated in a particular type of tumor. Furthermore, it should be noted that retroviruses that carry *myc* as an oncogene can induce a variety of tumors (Chapter 26), and there is evidence that *c-myc* can serve as an oncogene in human tumors of cell types other than B cells. As for the use of the antibody gene loci, there are two plausible reasons for their involvement. As originally proposed, the fact that they are transcriptionally active in B cells may be an essential aspect of their role in activating *c-myc*. Second, it seems likely that at least some, and possibly most, of the translocations in B cell tumors result when the recombination mechanisms involved in constructing functional antibody genes make a mistake and, instead of joining two segments of an antibody gene, join a proto-oncogene to an antibody gene. This explanation could account for the frequent cases where *c-myc* is introduced into a switch region of the heavy-chain locus.

As in the case of tumors with activated *ras* genes, more than one oncogene is probably involved in generating Burkitt's lymphomas. As we will discuss in detail shortly, viral infection may be an important first step in launching these tumors. Furthermore, some Burkitt's tumor cell lines have activated *ras* genes that can be detected in the 3T3 transfection assay, in addition to specific chromosome translocations.

Some B cell tumors have translocations that involve the antibody loci but oncogenes other than *c-myc*. Recently, two such genes have been molecularly cloned based on the assumption that sequences lying near the translocation breakpoints in the antibody loci might encode oncogenes. These new candidate oncogenes are designated *bcl-1* and *bcl-2*.

Retinoblastoma Involves the Loss of Both Copies of a Gene (*Rb-1*) on Band q14 of Human Chromosome 13[71,72]

Retinoblastoma is an eye tumor of children and is one of several heritable forms of childhood cancer. About half the cases of retinoblastoma are hereditary, while the remainder, called sporadic, occur in genetically normal individuals. Retinoblastoma is inherited as a highly penetrant autosomal dominant trait, and a child who inherits the single gene predisposing to retinoblastoma will almost certainly develop the tumor. These individuals often develop tumors in both eyes, while sporadic cases of retinoblastoma occur in only one eye. These findings led to the proposal that retinoblastoma results from the accumulation of at least two mutations within a single cell and that people with a high probability of contracting the disease inherit one of the necessary mutations.

Probably all forms of retinoblastoma involve abnormalities in band q14 on chromosome 13, and in some cases these are visible microscopically as a deletion of the q14 band. Tumor cells lack a normal chromosome 13 and are homozygous for the deleted version of chromosome 13. In addition, normal cells from individuals with a high tendency to the disease can sometimes be seen to carry one normal chromosome 13 and one with a deletion of band q14, while their tumor cells are homozygous for the abnormal chromosome. The deletions of 13q14 seen in different cases of retinoblastoma can vary at their endpoints, indicating that the deletion does not activate an oncogene by introducing breaks into a specific region of DNA, but rather that its effect may arise through the removal of a gene. Thus, retinoblastoma seems to be the result of a recessive mutation. Mutations (deletions) at the *Rb-1* locus are recessive to a normal gene and only contribute to tumorigenesis when the normal allele is inactivated. As shown in Figure 27-15, this can occur through errors of mitosis or through recombination events during mitosis that lead to loss of the normal chromosome, as well as through mutations of the normal allele.

Many years ago, experiments were performed to determine whether cancer is dominant or recessive. Normal cells and cancer cells were fused and the resulting cell hybrids tested for their tumorigenicity. Years of such experimentation have shown that hybrids can be either cancerous or normal, depending on the particular pair of cells involved in the fusion. These results suggest that at least some cancers are the result of recessive mutations. The study of retinoblastoma provides additional support for this idea. Clearly, it is important now to clone the *Rb-1* gene, since it may represent a new class of oncogene distinct from the dominant transforming genes such as acti-

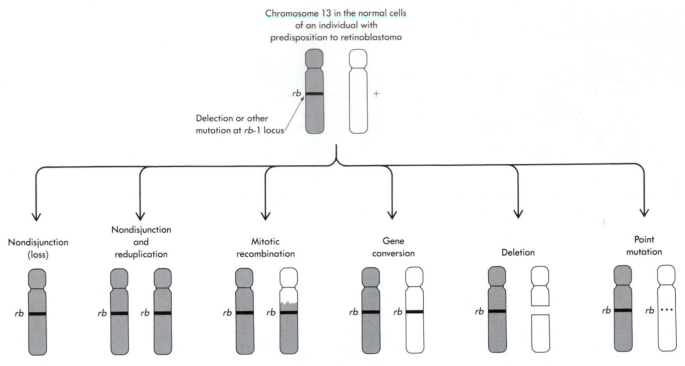

Chromosome 13 in the normal cells
of an individual with
predisposition to retinoblastoma

Deletection or other
mutation at *rb*-1 locus

Nondisjunction
(loss)

Nondisjunction
and
reduplication

Mitotic
recombination

Gene
conversion

Deletion

Point
mutation

Six possible chromosome 13 pairs that might be seen in
a retinoblastoma tumor arising in the individual shown above

Figure 27-15
Schematic representation of the chromosome 13s present in an individual with inherited predisposition to retinoblastoma, and the mechanisms that could lead to homozygosity of the recessive defect and hence to a tumor. [After W. K. Cavanee et al., *Nature* 305 (1983):779–784.]

vated *ras*. Conceivably, recessive cancer genes could turn out to be negative regulators of oncogene expression, but many other possibilities can also be imagined.

Mutations at *Rb-1* predispose to other forms of cancer as well as retinoblastoma. A fairly high percentage of patients that recover from heritable retinoblastoma (but not sporadic retinoblastoma patients) develop osteosarcoma. As in retinoblastoma, homozygosity of an abnormal chromosome 13 with a deletion at q14 appears to be involved. This result indicates that recessive mutations that predispose to cancer may be common to many different forms of disease, and a relatively small number of such genes may turn out to underlie many types of tumors.

Amplified Oncogenes in Double Minutes and Homogeneously Staining Regions in Certain Human Tumors[73–80]

Double minute chromosomes (DMs) and homogeneously staining regions (HSRs) within chromosomes are related karyotypic abnormalities seen in tumor cells. As discussed in Chapter 20, these abnormalities are also seen in cells following selection for drug resistance (see Figure 20-28). They result from the amplification of cellular genes. The precise structure of the amplified fragment within a particular DM or HSR is difficult to establish, since the fragments involved are enormous, ranging in size from 100 to more than 1000 kilobase pairs. DMs and HSRs apparently represent interchangeable forms of an amplified sequence, and a single tumor can give rise to cell lines that contain one or the other structure. In addition, HSRs can vary in size and number and can move to different chromosomal locations.

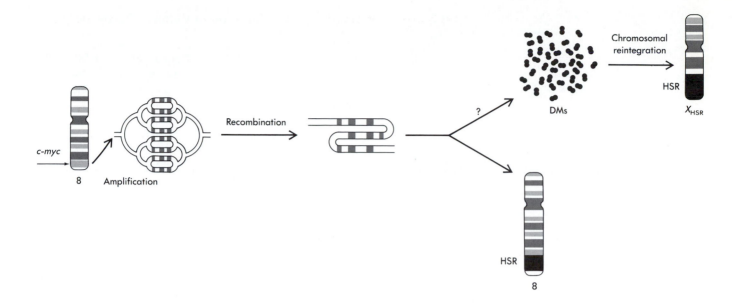

Although the sequence of events that underlies these alterations is not known, plausible scenarios are illustrated in Figure 27-16.

It is highly likely that DMs and HSRs contribute to the transformed phenotype of tumors that carry them, probably conferring a growth advantage. Otherwise, it is difficult to see why these structures would be retained. It was thought that the sequences amplified in DMs and HSRs might encompass cellular proto-oncogenes, since alteration in the level of expression, particularly increased transcription, was known to be one mechanism for activating some of these genes (e.g., *ras*, *mos*, and *myc*). This hypothesis has turned out to be correct. Several different oncogenes have been identified in the DMs and HSRs of different human tumors (Table 27-6). It is most unlikely that this is a coincidence. Rather, the increased expression of the oncogene resulting from its amplification almost certainly contributes to tumorigenicity and explains the persistence of DMs and HSRs in tumors.

Human neuroblastomas were one of the first tumors shown to harbor DMs and HSRs. When DNA from neuroblastoma cells was hybridized with radioactive viral oncogene probes, dark bands were observed in Southern blots when *myc* probes were used. When the homologous human sequences were cloned, they turned out to be a new member of the *myc* gene family, a gene now called N-*myc*. N-*myc* sequences can be amplified as much as a thousandfold in some neuroblastomas (Figure 27-17). Neuroblastomas, like many tumors, can undergo progression with time from less malignant to more malignant forms. Interestingly, N-*myc* amplifications are usually seen in the later stage III and IV tumors. It is these that also can be most readily established into cell lines in tissue culture.

Some small-cell lung carcinomas also have amplified *myc* genes in DMs and HSRs. The amplifications are found in late stages of tumor evolution, in the so-called variant cells that display a more malignant behavior. Amplifications can involve either c-*myc*, N-*myc*, or a new member of the *myc* gene family now designated L-*myc*.

Additional evidence that elevated levels of *myc* gene expression resulting from gene amplification are related to growth properties of tumor cells comes from studies with the HL60 cell line. HL60 is derived from a human promyelocytic leukemia. The cells have an ampli-

Figure 27-16
Model for how HSRs and DMs may arise and move from one chromosome to another. DNA encompassing an oncogene may be replicated excessively owing to reinitiations at a particular origin of replication (the onionskin model). Recombination within the onionskin structure generates tandem arrays of the sequence. These might leave the chromosome to become double minutes, which in turn might be able to reintegrate into new chromosomal locations, for example, the X chromosome in the case shown, giving rise to HSRs. [After K. Alitalo, et al., *Genes and Cancer* (Glen R. Lis, Inc., 1984) pp. 383–397.]

Table 27-6 Examples of Amplification of Cellular Oncogenes in Human Tumors

Amplified Gene	Tumor	Degree of Amplification	DM or HSR Present
c-myc	Promyelocytic leukemia cell line, HL60	20×	+
	Small-cell lung carcinoma cell lines	5–30×	?
N-myc	Primary neuroblastomas (stages III and IV) and neuroblastoma cell lines	5–1000×	+
	Retinoblastoma cell line and primary tumors	10–200×	+
	Small-cell lung carcinoma cell lines and tumors	50×	+
L-myc	Small-cell lung carcinoma cell lines and tumors	10–20×	?
c-myb	Acute myeloid leukemia	5–10×	?
	Colon carcinoma cell lines	10×	?
c-erbB	Epidermoid carcinoma cell line	30×	?
	Primary gliomas		?
c-K-ras-2	Primary carcinomas of lung, colon, bladder, and rectum	4–20×	?
N-ras	Mammary carcinoma cell line	5–10×	?

SOURCE: Modified from H. E. Varmus, *Ann. Rev. Genetics* 18 (1984):553–612.

fied *c-myc* gene and an activated N-*ras* gene that can be detected by transfection transformation of 3T3 cells. Addition of retinoic acid to HL60 cells induces them to differentiate into granulocyte-like cells that have only a very limited ability to proliferate. When this occurs, *c-myc* expression drops dramatically.

VIRUSES AND HUMAN CANCER

The Long Search for the Human Tumor Viruses

In the past 25 years, two discoveries were particularly important in fueling support for the viral origins of human cancer. One was the ability of certain DNA viruses, particularly the papovaviruses and adenoviruses, to transform cells in vitro and to induce tumors in rodents. The second was the discovery of reverse transcriptase and the subsequent explosion of the retrovirus field. Adenoviruses are extremely common in humans, where they cause colds and other types of infections, so it seemed possible that they might also be human tumor viruses. The realization that many forms of cancer in animals are caused by retroviruses and that these can be both inherited and horizontally spread made it seem very inevitable that retroviruses would be found to be a cause of one or more forms of human cancer.

Many human tumors were thus screened for the presence of integrated papovavirus and adenovirus DNA with almost entirely negakbp

2.0

Neuroblastoma cell lines Neuroblastoma tumors Controls Rb

Figure 27-17

A Southern blot showing amplification of N-*myc* in neuroblastoma cells and in a retinoblastoma. DNAs from cell lines or primary tumors were digested with a restriction endonuclease, and subjected to electrophoresis and transfer to nitrocellulose paper; the paper was hybridized to a radioactive probe prepared from a molecular clone of the N-*myc* gene. DNAs in lanes 1–8 were isolated from cell lines derived from neuroblastomas; lanes 9–12, from primary neuroblastomas; lanes 13–16, from other types of human tumors that also possess HSRs but which do not involve N-*myc*; lane 17, from normal human skin fibroblasts; and lane 18, from a human retinoblastoma cell line. Amplifications of N-*myc* range from about 5- to 8-fold (lane 8) to 20- to 25-fold (lanes 3 and 4) to 80- to 140-fold (lanes 1, 2, 5, 7, and 9). Lanes 13–17 show the appearance of hybridization to N-*myc* that is not amplified. [Courtesy of M. Schwab. From M. Schwab et al., *Nature* 305 (1983):245–248.]

tive results. Then more than a decade was spent trying to isolate the putative human retroviruses from tumors. Only one such virus, one that causes a rather rare form of T cell leukemia, has been found so far.

The several virus-cancer relationships that have emerged in the past decade have revealed just how difficult it can be to prove a causal relationship between virus infection and tumor development. No known human cancer appears to be caused by a virus that rapidly and single-handedly induces tumors. Rather, viruses that cause human cancers, like the carcinogens suspected of causing human cancers, act over long periods of time and probably only in combination with other events. It is for this reason that it is still difficult to know how frequently viruses may be involved in causing many types of human tumors.

Chronic Hepatitis B Virus Infection Leads to Liver Cancer[81–84]

Hepatitis B is a tiny DNA virus that infects humans and certain nonhuman primates and has a strong tropism for liver cells. When a nonimmune person becomes infected with hepatitis B, viral replication occurs in the liver. In about 95 percent of the cases, replication terminates when neutralizing antibodies develop and the individual becomes immune to reinfection. However, in about 5 percent of the cases, persistent infection of the liver is established and continues for life. Often, persistent infections are not associated with liver damage; but in some cases, symptomatic and progressive forms of liver damage occur. The latter condition is known as *chronic hepatitis B*. Persistent hepatitis B virus infections, particularly those associated with chronic hepatitis and cirrhosis, are believed to help cause 80 to 90 percent of primary hepatocellular carcinomas (also called hepatoma or liver cancer). Hepatitis B virus infection is estimated to cause about 500,000 cancer deaths a year worldwide.

The association of hepatitis B and liver cancer was discovered when the worldwide distribution of chronic hepatitis was determined and compared with the worldwide distribution of hepatoma. As shown in Figure 27-18, the two diseases show a striking overlap in geographic distribution. Further studies have shown that patients with hepatoma

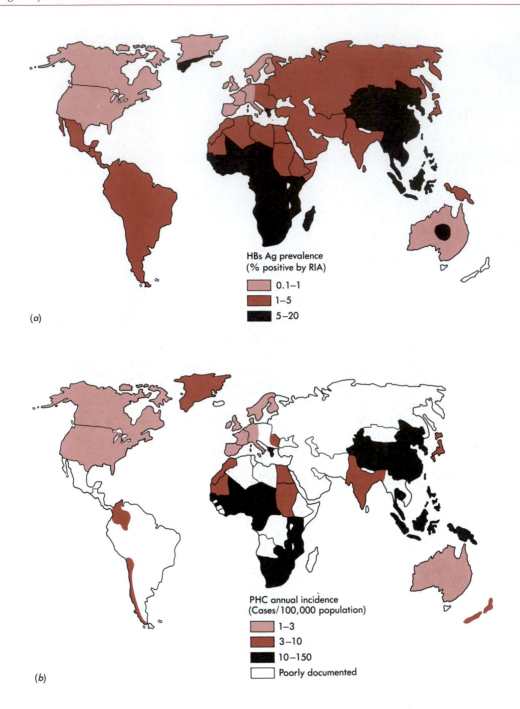

HBs Ag prevalence
(% positive by RIA)

▢ 0.1–1
▢ 1–5
▢ 5–20

(a)

PHC annual incidence
(Cases/100,000 population)

▢ 1–3
▢ 3–10
▢ 10–150
▢ Poorly documented

(b)

Figure 27-18

Maps showing the similarity in world-wide distribution of (a) hepatitis B virus infection and (b) annual incidence of primary hepatocellular carcinoma (PHC). Evidence for virus infection is presence of the hepatitis B surface antigen (HBsAg) in serum as determined by radioimmunoassay (RIA). [After P. Maupos and J. L. Melnick, *Prog. Med. Virol.* 27 (1981):1–5.]

are 5 to 70 times more likely to show evidence of hepatitis B infection than similar individuals without hepatoma. Although most hepatitis B virus carriers never develop hepatoma, the overwhelming majority of hepatoma patients show evidence of infection. Chronic hepatitis and cirrhosis combined with infection early in life create the greatest risk of developing hepatoma, a risk that can be 250-fold that of noncarriers. Primary hepatitis B infection usually precedes the onset of liver cancer by about 25 to 30 years.

Until recently, it was difficult to study the life cycle of hepatitis B virus because it could not be grown in tissue culture. However, molecular cloning of the genome and the discovery of hepatitis B–like viruses that induce liver cancer in woodchucks have allowed some important advances (Chapter 24).

What is the mechanism by which hepatitis B virus induces cancer?

Based on our knowledge of tumor viruses, we might suppose that hepatitis B carries a transforming gene, or that the virus integrates into cellular DNA and in the process activates a cellular proto-oncogene. So far, there is no evidence that hepatitis B virus possesses any type of transforming gene. Whether the virus activates cellular oncogenes is still an open question. Consistent with this possibility are the facts that hepatomas are clonal and viral DNA is found to be integrated into cellular DNA in tumor cells. However, no evidence for integrations near known proto-oncogenes has yet been obtained. Nonetheless, the fact that viral integrations occur leaves the possibility of oncogene activation viable. In addition, deletions at the sites of viral integrations have been observed in some tumor lines, so conceivably hepatitis B acts as a mutagen. In contrast to these models, some researchers now believe that hepatitis B may contribute to liver cancer by a different mechanism. Chronic infection is accompanied by chronic hepatic injury and regeneration. Perhaps this long-term trauma, like chronic alcohol abuse, which is also associated with liver cancer, in some way promotes the induction and growth of malignant cells.

Epstein-Barr Virus (EBV) and Its Relationship to Burkitt's Lymphoma, Nasopharyngeal Carcinoma, and Mononucleosis[85,86,87]

Burkitt's lymphoma, a tumor of mature B cells, is found throughout the world, but is most frequent among children in certain parts of East Africa. In the early 1960s, the physician Burkitt suggested that the tumor might be caused by a virus. Examination of tumor specimens soon led to the discovery of a herpes virus called **Epstein-Barr virus** (**EBV**). EBV infects B cells in vitro and immortalizes them. It establishes a latent state in which just some of its genes are expressed in most cells, although a rare cell expresses all the viral genes and produces infectious virus.

It has been difficult to obtain conclusive evidence that EBV causes Burkitt's lymphoma, because infection by EBV is not limited to the people or areas where the tumors are found but is present throughout the world. However, areas with a high incidence of the tumor have a higher incidence of infection, and infection occurs at an earlier age. For example, in some tropical areas where the tumor is frequent, more than 90 percent of the children are infected by six years of age, while in the United States, only 30 to 40 percent show evidence of infection by this age. In addition, more than 90 percent of the Burkitt's tumors that arise in Africa are associated with EBV, whereas other types of leukemias do not show such an association. But if the virus is so widespread, why is the tumor so much more frequent in certain tropical areas of the world? There is a good correlation between the affected areas and a high incidence of malaria, and many researchers think that it is the combination of EBV and malaria that produces a high incidence of Burkitt's lymphoma.

In contrast to the difficulty in proving an association of EBV with Burkitt's lymphoma, it was soon shown conclusively that EBV causes mononucleosis. This disease, probably more common in the United States than in Africa, results when individuals are infected by EBV at a later age. Mononucleosis is communicated through kissing or through use of saliva-contaminated utensils, since EBV replicates readily in the oropharyngeal region. Mononucleosis is a disease of excessive lymphoid cell proliferation. Many of the symptoms are probably caused by the immune system's response to EBV-infected cells.

As discussed earlier in this chapter, Burkitt's lymphomas possess specific chromosome translocations involving the *c-myc* proto-oncogene, and these are almost certainly involved in tumor induction. Thus, EBV may serve as just one factor in tumor formation. The role of the virus in the cancerous transformation is not known, but presumably, the viral genes that can immortalize B cells are involved. Whether the virus helps to maintain the tumor or simply helps it to get started is also unknown. That EBV can be a potent tumor virus is indicated by the fact that individuals who are immunosuppressed can develop lymphomas (as well as fatal mononucleosis) following EBV infection. These tumors lack the characteristic chromosome abnormalities seen in Burkitt's lymphoma and are polyclonal instead of monoclonal like the Burkitt's tumors. Thus, the immune system probably saves us from a high probability of getting EBV-induced B cell cancers.

Why do children without obvious immune deficiency get Burkitt's lymphoma? This is unclear, since their tumor cells often appear to be highly antigenic owing to the expression of viral antigens on their surface. A hypothesis is that simultaneous infection with malaria somehow interferes with immune rejection of the tumor cells.

EBV is also implicated as a causal agent in nasopharyngeal carcinoma (NPC), a tumor of adults that is quite rare except among the Chinese. There seems to be a genetic susceptibility factor involved, since Chinese people retain an elevated incidence of NCP irrespective of where they live. In addition, environmental factors such as diet (particularly cured fish), smoke, and chemicals may be involved in contributing to the onset of disease. EBV genomes and their transcripts are found in almost all undifferentiated NPC biopsies, regardless of the country or the genetic constitution of the host.

A Retrovirus Called HTLV I Causes Adult T Cell Leukemia[88–91]

So far, the only retrovirus that has been identified as the cause of a human cancer is the human *T cell leukemia/lymphoma virus type I (HTLV I)*, which causes a form of T cell cancer called *adult T cell leukemia-lymphoma (ATLL)*. This is a fatal tumor that is quite common in parts of Japan and the Caribbean. HTLV I was first detected in a continuously proliferating T cell line derived from a patient in the United States. Subsequently, the virus was found in T cell tumors from patients in Japan. An important advance was the ability to transmit the virus to fresh human cord-blood T cells in vitro by cocultivation with HTLV I–producing cells.

Several lines of evidence strongly implicate HTLV I as the cause of ATLL. First, the geographic distribution of the disease corresponds closely to areas where virus infection is frequent (the latter is determined by the presence of antibodies to the virus in serum). Second, all ATLL patients are infected with HTLV I. Third, the proviral DNA is always present in tumor cells, but not usually in other cells from the same patient. Finally, HTLV I can transform human T cells in vitro.

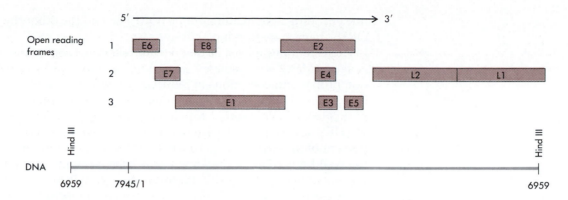

Figure 27-19
Genomic organization of bovine papillomavirus DNA. The circular 7945-base-pair genome is represented as a linear molecule. Open boxes denote regions of open reading frames (ORFs) and hence potential protein-coding segments. The segment of the genome that is transcribed in transformed cells and the direction of transcription are indicated by the arrow at the top of the figure. ORFs within this region are designated E1-E8. [After N. Sarver, M. S. Rabson, Y.-C. Yang, J. C. Byrne, and P. M. Howley, *J. Virology* 52 (1984):377–388.]

As in the case of the specific cancers associated with hepatitis and EBV infections, the majority of individuals who carry HTLV I virus do not develop T cell leukemia; and, as in the case of these viruses, incubation times between infection and disease are long.

Although integrated HTLV I proviral DNA is always present in tumor cells, the virus does not integrate into specific sites in chromosomal DNA or near proto-oncogenes. The HTLV I provirus has been completely sequenced, and the virus does not have a cell-derived oncogene. However, the virus can transform (immortalize) T cells in vitro, suggesting that it carries a virus-coded transforming gene. Thus, in terms of its transforming ability, HTLV I more closely resembles oncogenic DNA viruses than replication-competent retroviruses. The candidate transforming gene is encoded at the 3' end of the virus by a gene called *pX* or *lor*. This gene may encode a protein that functions as a *trans*-acting factor to increase viral transcription (Chapter 24). Conceivably, the *pX* gene product also activates transcription of specific cellular genes. For example, overexpression of genes that could stimulate excessive T cell division might put the cells at risk to accumulate additional cancer-causing mutations.

Papillomaviruses and Cervical Carcinoma[92–97]

Certain human genital cancers show epidemiological features characteristic of infectious disease; for example, cervical carcinoma is most common among women with multiple sex partners. Now evidence is accumulating that papillomaviruses may play a role in causing cervical carcinoma and also penile cancer, as well as benign genital warts.

Papillomaviruses are small DNA viruses belonging to the papovavirus group. They are found in a variety of animals, and isolates from rabbits (Shope papillomavirus) and cows (bovine papillomavirus, or BPV) have been particularly well studied. Papillomaviruses cause benign tumors (warts or papillomas) of the cutaneous and mucosal epithelia, including many types of warts common in humans. In some cases, certain papillomas can undergo malignant transformation. Furthermore, some papillomaviruses can transform cells in vitro. Papillomaviruses replicate as extrachromosomal circular DNA, maintaining a fairly constant number of genomes per infected cell. In warts, DNA is stably maintained as a plasmid in basal cells, and a complete viral replication cycle resulting in the production of progeny genomes occurs only in the outer layers of skin as cells undergo terminal differentiation.

Different papillomaviruses have similar genome organization (Figure 27-19). At least in some viruses, two regions of the genome, one encoding open reading frame (ORF) E6 and the other the E2 to E5 ORFS, have transforming activity.

There are now at least 30 distinct but related human papillomavirus isolates. Many were isolated using known viral DNAs as hybridization probes and conditions of reduced stringency to detect new members of the family. Most interesting has been the finding that two particular isolates, human papillomavirus type 16 (HPV-16) and type 18 (HPV-18) are particularly common in cervical carcinomas: Over 80 percent of invasive cervical cancers harbor one of these viruses. Even the well-known human cell line HeLa, which originated from a cervical carcinoma, harbors HPV-18. Interestingly, in tumors, viral DNA is integrated into cellular DNA and is usually transcriptionally active. At least some tumors have been found to express the E6 open reading frame.

It remains unproved that HPV causes cervical carcinoma, and as usual, it is difficult to rule out the possibility that the association of the virus with the tumor is just a result of an ability of HPV to replicate in tumor cells plus a high probability that sexually active individuals may harbor these viruses. Nonetheless, the circumstantial evidence implicating HPV as a cause of cervical cancer is compelling.

Summary

Most cancers probably arise when several mutations accumulate within the DNA of a single cell, causing it to lose growth control. Many lines of evidence support a genetic basis for cancer: The cancerous phenotype of a cell is stably inherited; the majority of agents that induce cancer in animals or transform cells in vitro are mutagens that act on DNA; and specific chromosomal abnormalities predispose to cancer or are present in tumors. Evidence that more than one genetic event is required to induce a tumor has come from analyzing cancer incidence as a function of age and from the study of inherited forms of cancer. These studies suggest that at least two and possibly six or seven genetic events are required to produce a cancer cell.

Carcinogens in the environment probably cause the majority of human cancers. While the particular substances remain to be identified in most cases, some clear-cut examples include tobacco (lung cancer), asbestos (mesothelioma), and sunlight (skin cancer).

Viruses are another important cause of some human cancers. They probably act in combination with other genetic events to bring about the cancerous transformation of cells. A still-unanswered question is the percent of human cancers that have a viral etiology, but now the best guess is probably about 10 to 20 percent.

For some time, a major goal of molecular biologists has been to identify the cellular genes involved in human tumors, the so-called human oncogenes. Several lines of research converged in the 1980s to make this goal achievable. Absolutely critical was the development of molecular cloning technologies. In addition, the analysis of oncogenes carried by tumor viruses (particularly the cell-derived oncogenes carried by retroviruses), the analysis of specific chromosome abnormalities in tumors, and the ability to assay for single-copy cellular genes using the technique of DNA transfection all played essential roles.

DNA isolated from about 15 to 20 percent of human tumors can induce transformed foci in mouse 3T3 cells, while DNA from normal cells does not induce foci under the same experimental conditions. The transformed phenotype can be serially passed via DNA from the primary transfectants to fresh 3T3 cells, then from the resulting secondary transfectants to additional cells, and so on. A single human gene that acts in a dominant fashion must be responsible for the transforming ability of the tumor DNA. These human oncogenes can be molecularly cloned from the DNA of secondary or tertiary 3T3 transfectants by using a human species-specific repeated DNA sequence, the *Alu* repeat, as a hybridization probe.

Remarkably, the majority of the human oncogenes that have been detected by DNA transfection of 3T3 cells turn out to be activated human *ras* genes, homologs of the genes first identified as the oncogenes of the Harvey and Kirsten mouse sarcoma retroviruses. There are three functional *ras* genes in human cells, H-*ras-1*, K-*ras-2*, and N-*ras*. Any one of these can become an oncogene as a result of mutation. Frequently a single-base change that alters the twelfth codon of the gene and consequently changes the twelfth amino acid from the normal glycine converts the normal cellular gene to an oncogene. Alterations of the twelfth codon can also activate N- and K-*ras* genes. Some tumors have *ras* genes that are activated as a result of mutations of the thirteenth or sixty-first codon.

Since the normal cellular function of *ras* genes is not known, it is difficult to say why a single-base change converts the gene to an oncogene. However, *ras* genes encode 21 kdal proteins, located at the inner surface of the plasma membrane, that bind guanine nucleotides and possess GTPase activity. Although activated *ras* genes have guanine nucleotide binding activity, they have sharply decreased GTPase activity. By analogy to other guanine nucleotide binding proteins, such as the G proteins that regulate cAMP levels, it is postulated that loss of GTPase activity leads to loss of a regulatory mechanism and results in the *ras* protein remaining in an active form much of the time.

A few other cellular oncogenes have been detected by DNA transfection. One, the *neu* gene (in carcinogen-induced neuroblastomas and as glioblastomas) is distinct from but related to the gene encoding the EGF receptor. It is not known why *ras* genes are the predominant type detected by the assay. It may reflect either their preferential activation in human tumors or the greater ability of *ras* oncogenes to be seen in the 3T3 transformation assay.

It is difficult to prove that activated *ras* genes detected by transfection were involved in causing the human tumors that harbor them. However, it seems likely that this is the case. Using carcinogen-induced breast tumors of rats as a model system, it has been shown that a single brief exposure to a carcinogen can induce tumors almost all of which have an alteration at codon 12 of the rat H-*ras*-1 gene.

Although activated *ras* genes can transform 3T3 cells, they are unable to transform some types of primary rodent cells efficiently. However, if they are transfected along with a second oncogene such as *myc* or the E1A gene of adenovirus, then transformants are obtained. This result provides direct evidence that more than one oncogene is needed to cause cancerous transformation.

It has been known for some time that human tumors possess chromosome abnormalities, but the field has exploded as better and better techniques have become available for preparing and analyzing chromosomes. Most of the work has been done with leukemias, but solid tumors, frequently more difficult to study, are now beginning to be analyzed in detail as well. Abnormalities can involve both chromosome number and structure. So far, most effort has gone to identifying specific abnormalities that are present in particular forms of cancer. However, superimposed on these almost invariable abnormalities are often numerous other changes, some common but not ubiquitous for a specific form of cancer. There is evidence that as tumors progress, often their chromosomes become increasingly aberrant, both in number and structure.

It was suggested that specific chromosome abnormalities might involve specific oncogene activation. Remarkably, several examples of this have now been found. The Philadelphia chromosome in chronic myelogenous leukemia, the first consistent abnormality associated with a human cancer, results from a translocation between chromosomes 9 and 22 that joins the 5' half of a gene called *bcr* to the 5' side of the *abl* oncogene. As a result, a long *bcr-abl* primary transcript is made. After splicing, this *bcr-abl* RNA is translated to yield a *bcr-abl* fusion protein that is probably involved in causing CML.

Burkitt's lymphomas, as well as certain other B cell lymphomas, have translocations that involve the immunoglobulin gene loci and the *c-myc* oncogene. Most common is a t(8;14) translocation in which the immunoglobulin heavy-chain constant region is fused head to head with *c-myc*. While the way in which this event activates *c-myc* is not yet completely clear, it is suspected that the fusions in some way alter c-*myc* gene expression and that this, in turn, contributes to the development of the tumor.

Certain tumors possess tiny chromosomes, called double minutes (DMs), that lack centromeres and can vary in number from cell to cell. These tumors frequently also have homogeneously staining regions (HSRs) within certain chromosomes. The two structures appear to be interchangeable forms of an amplified DNA sequence. Now it has been shown that the DMs and HSRs of neuroblastomas involve the amplification of a fragment of DNA carrying a gene called N-*myc*, while variants of small-cell lung tumors have amplifications of either *c-myc*, N-*myc*, or L-*myc*. In both cases, amplification appears to occur with tumor progression, and is present only in later stages of tumor development. The normal cellular functions of members of the *myc* gene family are not known, but the genes encode proteins that are localized to the nucleus and may be involved in regulating the expression of other cellular genes.

Deletion of a specific chromosomal band (band 14 on chromosome 13) is associated with a tumor called retinoblastoma. Tumors are thought to arise when both copies of a gene at this location, a gene called *Rb-1*, are deleted or mutated. Thus, *Rb-1* may represent a new form of oncogene that is recessive rather than dominant. The homozygous deletion that usually gives rise to retinoblastoma can probably cause other types of tumors as well. Other specific band deletions are associated with yet other forms of cancer.

Most searches for human tumor viruses have ended in frustration. However, there are now several cases where a viral cause of cancer seems either clear, probable, or at least possible. These include hepatitis B virus as a cause of liver cancer, a retrovirus called HTLV I as the cause of adult T cell lymphoma, papilloma viruses as a cause of cervical cancer, and possibly the herpes virus Epstein-Barr as a factor in causing nasopharyngeal carcinoma and Burkitt's lymphoma. In none of these cases is the mechanism by which the virus causes the disease known with certainty. In all cases, the majority of people infected with the virus do not contract cancer. There are many examples where immunosuppression is associated with rare forms of cancer, and these cancers may have a viral origin. Our immune system probably protects us from most viruses that are capable of causing cancer.

The next decade will surely see a revolution in our understanding of the molecular basis of cancer. When or how this information will allow us to control the disease is difficult to say. For now, the best approach is still prevention. Elimination of cigarette smoking and vaccination against hepatitis B virus are probably the major new "cancer cures" presently available.

Bibliography

General References

Bishop, J. M. 1982. "Oncogenes." *Sci. Amer.* 246:80–92.

Cairns, J. 1978. Cancer, Science and Society. San Francisco: Freeman.

DeVita, V., S. Hellman, and S. Rosenberg, eds. 1986. *Important Advances in Oncology.* Philadelphia: Lippincott.

Sendberg, A. 1980. *The Chromosomes in Human Cancer and Leukemia.* North Holland and Amsterdam: Elsevier.

Weinberg, R. A. 1983. "A Molecular Basis of Cancer." *Sci. Amer.* 249:126–142.

Cited References

1. Fialkow, P. J. 1974. "The Origin and Development of Human Tumors Studied with Cell Markers." *New Eng. J. Med.* 291:26–35.
2. Farber, E., and R. Cameron. 1980. "The Sequential Analysis of Cancer Development." *Adv. Cancer Res.* 31:125–226.
3. Knudson, A. G., H. W. Hethcote, and B. W. Brown. 1975. "Mutation and Childhood Cancer: A Probabilistic Model for the Incidence of Retinoblastoma." *Proc. Nat. Acad. Sci.* 72:5116–5120.
4. Foulds, L. 1975. *Neoplastic Development.* Vol. 2. New York: Academic Press.
5. Klein, G., and E. Klein. 1985. "Evolution of Tumors and the Impact of Molecular Oncology." *Nature* 315:190–195.
6. Hiatt, H. H., J. D. Watson, J. A. Winsten, eds. 1977. *Origins of Human Cancer. Cold Spring Harbor Conferences on Cell Proliferation.* Vol. 4. Cold Spring Harbor, N.Y.: Cold Spring Harbor Laboratory.
7. Ames, B. N., W. E. Durston, E. Yamasaki, and F. D. Lee. 1973. "Carcinogens Are Mutagens: A Simple Test System Combining Liver Homogenates for Activation and Bacteria for Detection." *Proc. Nat. Acad. Sci.* 70:2281–2285.
8. Yunis, J. J. 1983. "The Chromosomal Basis of Human Neoplasia." *Science* 221:227–236.
9. Essex, M., G. Todaro, H. zur Hausen, eds. 1980. *Viruses in Naturally Occurring Cancer. Cold Spring Harbor Conferences on Cell Proliferation.* Vol. 7. Cold Spring Harbor, N.Y.: Cold Spring Harbor Laboratory.
10. Hill, M., and J. Hillova. 1971. "Production Virale dans les Fibroblasts de Poule Traites per l'Acide Desoxyribonucleique de Cellules XC de Rat Transformees par le Virus de Rous." *Compt. Rend. Acad. Sci.* 272:3094–3097.
11. Graham, F. L., and A. J. van der Eb. 1973. "A New Technique for the Assay of Infectivity of Human Adenovirus 5 DNA." *Virology* 52:456–467.
12. Pellicer, A., D. Robins, B. Wold, R. Sweet, J. Jackson, I. Lowy, J. M. Roberts, G. K. Sim, S. Silverstein, and R. Axel. 1980. "Altering Genotype and Phenotype by DNA-Mediated Transfer." *Science* 209:1414–1422.
13. Cooper, G. M. 1982. "Cellular Transforming Genes." *Science* 217:801–806.
14. Weinberg, R. A. 1982. "Oncogenes of Spontaneous and Chemically Induced Tumors." *Adv. Cancer Res.* 36:149–163.
15. Barbacid, M. 1986. "Human Oncogenes." In *Important Advances in Oncology,* ed. V. DeVita, S. Hellman, and S. Rosenberg. eds. Philadelphia: Lippincott.
16. Gusella, J. F., C. Keys, A. Varsanyi-Breiner, F. Kao, C. Jones, T. T. Puck, and D. Housman. 1980. "Isolation and Localization of DNA Segments from Specific Human Chromosomes." *Proc. Nat. Acad. Sci.* 77:2829–2833.
17. Shih, C., and R. A. Weinberg. 1982. "Isolation of a Transforming Sequence from a Human Bladder Carcinoma Cell Line." *Cell* 29:161–169.
18. Ellis, R. W., D. De Feo, T. Y. Shih, M. A. Gonda, H. A. Young, N. Tsuchida, D. R. Lowy, and E. M. Scolnick. 1981. "The p21 *src* Genes of Harvey and Kirsten Sarcoma Viruses Originate from Divergent Members of a Family of Normal Vertebrate Genes." *Nature* 292:506–511.
19. Der, C., T. Krontiris, and G. M. Cooper. 1982. "Transforming Genes of Human Bladder and Lung Carcinoma Cells Are Homologous to the *ras* Genes of Harvey and Kirsten Sarcoma Viruses." *Proc. Nat. Acad. Sci.* 79:3637–3640.
20. Parada, L. F., C. J. Tabin, C. Shih, and R. A. Weinberg. 1982. "Human EJ Bladder Carcinoma Oncogene Is Homologue of Harvey Sarcoma Virus *ras* Gene." *Nature* 297:474–478.
21. Santos, E., S. R. Tronick, S. A. Aaronson, S. Pulciani, and M. Barbacid. 1982. "T24 Human Bladder Carcinoma Oncogene Is an Activated Form of the Normal Human Homologue of BALB- and Harvey-MSV Transforming Genes." *Nature* 298:343–347.
22. Capon, D., Y. Ellson, A. Levinson, P. Seeburg, and D. Goeddel. 1983. "Complete Nucleotide Sequences of the T24 Human Bladder Carcinoma Oncogene and Its Normal Homologue." *Nature* 302:33–37.
23. Shimizu, K., D. Birnbaum, M. Ruley, O. Fasano, Y. Suard, L. Edlund, E. Taparowsky, M. Goldfarb, and M. Wigler. 1983. "The Structure of the K-*ras* Gene of the Human Lung Carcinoma Cell Line Calu-1." *Nature* 304:497–500.
24. McGrath, J. P., D. J. Capon, D. H. Smith, E. Y. Chen, P. H. Seeburg, D. V. Goeddel, and A. D. Levinson. 1983. "Structure and Organization of the Human Ki-*ras* Proto-Oncogene and a Related Processed Pseudogene." *Nature* 304:501–506.
25. Taparowsky, E., K. Shimizu, M. Goldfarb, and M. Wigler. 1983. "Structure and Activation of the Human N-*ras* Gene." *Cell* 34:581–586.
26. Shih, T., A. Papageorge, P. Stokes, M. Weeks, and E. Scolnick. 1980. "Guanine-Nucleotide Binding and Autophosphorylating Activities Associated with the p21^src Protein of Harvey Murine Sarcoma Virus." *Nature* 287:686–691.
27. Aaronson, S. A., and S. R. Tronick. 1986. "The Role of Oncogenes in Human Neoplasia." In *Important Advances in Oncology,* ed. V. DeVita, S. Hellman, and S. Rosenberg. Philadelphia: Lippincott.
28. Shimizu, K., Y. Nakatsu, M. Sekiguchi, K. Hokamura, K. Tanaka, M. Terada, and T. Sugimura. 1985. "Molecular Cloning of an Activated Human Oncogene, Homologous to v-*raf*, from Primary Stomach Cancer." *Proc. Nat. Acad. Sci.* 82:5641–5645.
29. Tabin, C. J., S. M. Bradley, C. I. Bargmann, R. A. Weinberg, A. G. Papageorge, E. M. Scolnick, R. Dhar, D. R. Lowy, and E. H. Chang. 1982. "Mechanism of Activation of a Human Oncogene." *Nature* 300:143–149.
30. Reddy, E. P., R. K. Reynold, E. Santos, and M. Barbacid. 1982. "A Point Mutation Is Responsible for the Acquisition of Transforming Properties of the T24 Human Bladder Carcinoma Oncogene." *Nature* 300:149–152.
31. Fasano, O., T. Aldrich, F. Tamanoi, E. Taparowsky, M. Furth, and M. Wigler. 1984. "Analysis of the Transforming Potential of the Human H-*ras* Gene by Random Mutagenesis." *Proc. Nat. Acad. Sci.* 81:4008–4012.
32. Seeburg, P. H., W. W. Colby, D. J. Capon, D. V. Goedell, and A. D. Levinson. 1984. "Biological Properties of Human c-Ha-*ras* Genes Mutated at Codon 12." *Nature* 312:71–75.
33. Jurnak, F., 1985. "Structure of the GDP Domain of EF-Tu and Location of Amino Acids Homologous to *ras* Oncogene Proteins." *Science* 230:32–37.
34. McCormick, F., B. F. C. Clark, T. F. M. la Cour, M. Kjeldgaard, L. Norskov-Lauritsen, and J. Nyborg. 1985. "A Model for the Tertiary Structure of p21, the Product of the *ras* Oncogene." *Science* 230:78–82.
35. Gibbs, J. B., I. S. Sigal, M. Poe, and E. M. Scolnick. 1984. "Intrinsic GTPase Activity Distinguishes Normal and Oncogenic *ras* p21 Molecules." *Proc. Nat. Acad. Sci.* 81:5704–5708.
36. McGrath, J., D. Capon, D. Goeddel, and A. Levinson. 1984. "Comparative Biochemical Properties of Normal and Activated Human *ras* p21 Protein." *Nature* 310:644–655.
37. Dhar, R., A. Nieto, R. Koller, D. DeFeo-Jones, P. Robinson, G. Temeles, and E. M. Scolnick. 1984. "Nucleotide Sequence of Two H-*ras*-Related Genes Isolated from the Yeast *Saccharomyces cerevisiae*." *Nucleic Acid Res.* 12:3611–3618.

38. Powers, S., T. Kataoka, O. Fasano, M. Goldfarb, J. Strathern, J. Broach, and M. Wigler. 1984. "Genes in *S. cerevisiae* Encoding Proteins with Domains Homologous to the Mammalian *ras* Proteins." *Cell* 36:607–612.

39. Kataoka, T., S. Powers, S. Cameron, O. Fasano, M. Goldfarb, J. Broach, and M. Wigler. 1985. "Functional Homology of Mammalian and Yeast *RAS* Genes." *Cell* 40:19–26.

40. Hurley, J., M. Simon, D. Teplow, J. Robishaw, and A. Gilman. 1984. "Homologies Between Signal Transducing G Proteins and *ras* Gene Products." *Science* 226:860–862.

41. Feramisco, J. R., M. Gross, T. Kamata, M. Rosenberg, and R. W. Sweet. 1984. "Microinjection of the Oncogene Form of the Human H-*ras* (T24) Protein Results in Rapid Proliferation of Quiescent Cells." *Cell* 38:109–117.

42. Mulcahy, L. S., M. R. Smith, and D. W. Stacey. 1985. "Requirement for *ras* Proto-Oncogene Function During Serum-Stimulated Growth of NIH/3T3 Cells." *Nature* 313:241–243.

43. Shih, C., B. Z. Shilo, M. P. Goldfarb, A. Dannenberg, and R. A. Weinberg. 1979. "Passage of Phenotypes of Chemically Transformed Cells Via Transfection of DNA and Chromatin." *Proc. Nat. Acad. Sci.* 76:5714–5718.

44. Zarbl, H., S. Sukamar, A. V. Arthur, D. Martin-Zanca, and M. Barbacid. 1985. "Direct Mutagenesis of H-*ras*-1 Oncogenes by Nitroso-Methyl-Urea During Initiation of Mammary Carcinogenesis in Rats." *Nature* 315:382–385.

45. Guerrero, I., A. Villasante, A. Mayer, and A. Pellicer. 1984. "Carcinogen and Radiation-Induced Mouse Lymphomas Contain an Activated *c-ras* Oncogene." In *Cancer Cells: Oncogenes and Viral Genes*, ed. G. F. Vande Woude, A. J. Levine, W. C. Topp, and J. D. Watson. Cold Spring Harbor, N.Y.: Cold Spring Harbor Laboratory.

46. Schechter, A. L., D. F. Stern, L. Vaidyanathan, S. J. Decker, J. A. Drebin, M. J. Greene, and R. A. Weinberg. 1984. "The *neu* Oncogene: An *erbB*-Related Gene Encoding a 185,000-Mr Tumor Antigen." *Nature* 312:513–516.

47. Balmain, A., M. Ramsden, G. T. Bowden, and J. Smith. 1984. "Activation of the Mouse Cellular Harvey-*ras* Gene in Chemically Induced Benign Skin Papillomas." *Nature* 307:658–660.

48. Van der Eb, A. J., H. van Ormondt, P. I. Schrier, J. H. Lupker, H. Jochemsen, P. J. van der Elsen, R. J. De Leys, J. Maat, C. P. van Beveren, R. Dijkeme, and A. de Waard. 1980. "Structure and Function of the Transforming Genes of Human Adenoviruses and SV40." *Cold Spring Harbor Symp. Quant. Biol.* 44:383–399.

49. Rassoulzadegan, M., Z. Naghashfar, A. Cowie, A. Carr, M. Grisoni, R. Kamen, and F. Cuzin. 1983. "Expression of the Large T Protein of Polyoma Virus Promotes the Establishment in Culture of 'Normal' Rodent Fibroblast Cell Lines." *Proc. Nat. Acad. Sci.* 80:4354–4358.

50. Land, H., L. F. Parada, and R. A. Weinberg. 1983. "Tumorigenic Conversion of Primary Embryo Fibroblasts Requires at Least Two Cooperating Oncogenes." *Nature* 304:596–602.

51. Ruley, H. E. 1983. "Adenovirus Early Region 1A Enables Viral and Cellular Transforming Genes To Transform Primary Cells in Culture." *Nature* 304:602–606.

52. Spandidos, D. A., and N. M. Wilkie. 1984. "Malignant Transformation of Early Passage Rodent Cells by a Single Mutated Human Oncogene." *Nature* 310:469–475.

53. Rowley, J. D. 1982. "Identification of the Constant Chromosome Regions Involved in Human Hematologic Malignant Disease." *Science* 216:749–755.

54. Yunis, J. J. 1986. "Chromosomal Rearrangements, Genes, and Fragile Sites in Cancer: Clinical and Biological Implications." In *Important Advances in Oncology*, ed. V. DeVita, S. Hellman, and S. Rosenberg. Philadelphia: Lippincott.

55. Nowell, P. C., and D. A. Hungerford. 1960. "A Minute Chromosome in Human Granulocytic Leukemia." *Science* 132:125–132.

56. Rowley, J. D. 1973. "A New Consistent Chromosomal Abnormality in Chronic Myelogenous Leukemia Identified by Quinacrine Fluoresence and Giemsa Staining." *Nature* 243:290–293.

57. De Klein, A., A. G. van Kessel, G. Grosveld, C. R. Bartram, A. Hagemeijer, D. Bootsma, N. K. Spurr, N. Heisterkamp, J. Groffen, and J. R. Stephenson. 1982. "A Cellular Oncogene Is Translocated to the Philadelphia Chromosome in Chronic Myelocytic Leukemia." *Nature* 300:765–767.

58. Gale, R. P., and E. Canaani. 1984. "An 8-Kilobase *abl* RNA Transcript in Chronic Myelogenous Leukemia." *Proc. Nat. Acad. Sci.* 81:5648–5652.

59. Groffen, J., J. R. Stephenson, N. Heisterkamp, A. de Klein, C. R. Bartram, and G. Grosveld. 1984. "Philadelphia Chromosomal Breakpoints Are Clustered Within a Limited Region, *bcr*, on Chromosome 22." *Cell* 36:93–99.

60. Konopka, J. B., S. M. Watanabe, and O. N. Witte. 1984. "An Alteration of the Human *c-abl* Protein in K562 Leukemia Cells Unmasks Associated Tyrosine Kinase Activity." *Cell* 37:1035–1042.

61. Shtivelman, E., B. Lifchitz, R. Gale, and E. Canaani. 1985. "Fused Transcript of *abl* and *bcr* Genes in Chronic Myelogenous Leukemia." *Nature* 315:550–554.

62. Adams, J. 1985. "Oncogene Activation by Fusion of Chromosomes in Leukemia." *Nature* 315:542–543.

63. Ohno, S., M. Babonits, F. Weiner, J. Spira, G. Klein, and M. Potter. 1979. "Non-Random Chromosome Changes Involving Ig-Gene Chromosomes (Nos. 12 and 6) in Pristane Induced Mouse Plasmacytomas." *Cell* 18:1001–1008.

64. Shen-Ong, G. L. C., E. J. Keath, S. P. Piccoli, and M. D. Cole. 1982. "Novel *myc* Oncogene RNA from Abortive Immunoglobulin Gene Recombination in Mouse Plasmacytomas." *Cell* 31:443–452.

65. Dalla-Favera, R., S. Martinotti, R. Gallo, J. Erikson, and C. Croce. 1983. "Translocation and Rearrangements of the *c-myc* Oncogene Locus in Human Undifferentiated B-Cell Lymphomas." *Science* 219:963–997.

66. Leder, P., J. Battey, G. Lenoir, C. Moulding, W. Murphy, H. Potter, T. Stewart, and R. Taub. 1983. "Translocations Among Antibody Genes in Human Cancer." *Science* 222:765–771.

67. Corcoran, L. M., S. Cory, and J. M. Adams. 1985. "Transposition of the Immunoglobulin Heavy Chain Enhancer to the *myc* Oncogene in a Murine Plasmacytoma." *Cell* 40:71–79.

68. Hamlyn, P. H., and T. H. Rabbitts. 1983. "Translocation Joins *c-myc* and Immunoglobulin Gamma-1 Genes in a Burkitt Lymphoma Revealing a Third Exon in the *c-myc* Gene." *Nature* 304:135–139.

69. Seibenlist, U., L. Hennighausen, J. Battey, and P. Leder. 1984. "Chromatin Structure and Protein Binding in the Putative Regulatory Region of the *c-myc* Gene in a Burkitt Lymphoma." *Cell* 37:381–391.

70. Kelly, K., B. A. Cochran, C. D. Stiles, and P. Leder. 1983. "Cell-Specific Regulation of the *c-myc* Gene by Lymphocyte Mitogens and Platelet-Derived Growth Factor." *Cell* 35:603–610.

71. Knudson, A. G. 1985. "Hereditary Cancer, Oncogenes, and Antioncogenes." *Cancer Res.* 45:1437–1443.

72. Cavanee, W. K., T. P. Dryja, R. A. Phillips, W. F. Benedict, R. Godbout, B. L. Gallie, A. L. Murphree, L. C. Strong, and R. L. White. 1983. "Expression of Recessive Alleles by Chromosomal Mechanisms in Retinoblastoma." *Nature* 305:779–784.

73. Biedler, J. L. and B. A. Spengler. 1976. A Novel Chromosome Abnormality in Human Neuroblastoma and Antifolate-Resistant Chinese Hamster Cell Lines in Culture. *J. Nat. Cancer Inst.* 57:683–695.

74. Schimke, R. T., ed. 1982. *Gene Amplification*. Cold Spring Harbor, N.Y.: Cold Spring Harbor Laboratory.

75. Varmus, H. E. 1984. "The Molecular Genetics of Cellular Oncogenes." *Ann. Rev. Genetics* 18:553–612.

76. Collins, S., and M. Groudine. 1982. "Amplification of Endogenous *myc*-Related DNA Sequences in a Human Myeloid Leukemia Cell Line." *Nature* 298:679–681.

77. Schwab, M., K. Alitalo, K. Klemphauer, H. E. Varmus, J. M. Bishop, F. Gilbert, G. Brodeur, M. Goldstein, and J. Trent. 1983. "Amplified DNA with Limited Homology to *myc* Cellular Oncogene Is Shared by Human Neuroblastoma Tumor." *Nature* 305:245–248.

78. Brodeur, G. M., R. C. Seeger, M. Schwab, H. E. Varmus, and J. M. Bishop. 1984. "Amplification of N-*myc* in Untreated Human Neuroblastomas Correlates with Advanced Disease Stage." *Science* 224:1121–1124.

79. Nau, M. M., B. J. Brooks, J. Battey, E. Sausville, A. F. Gazdar, I. R. Kirsch, O. W. McBride, V. Bertness, G. F. Hollis, and J. D. Minna. 1985. "L-*myc*, a New *myc*-Related Gene Amplified and Expressed in Human Small Cell Lung Cancer." *Nature* 318:69–73.

80. Ullrich, A. L., L. Coussens, J. S. Hayflick, T. J. Dull, A. Gray, A. W. Tam, J. Lee, Y. Yarden, T. A. Libermann, J. Schlessinger, J. Downward, E. L. V. Mayes, N. Whittle, M. D. Waterfield, and P. H. Seeburg. 1984. "Human Epidermal Growth Factor Receptor cDNA Sequence and Aberrant Expression of the Amplified Gene in A431 Epidermoid Carcinoma Cells." *Nature* 309:418–423.

81. Szmuness, W. 1978. "Hepatocellular Carcinoma and the Hepatitis B Virus: Evidence for a Causal Association." *Prog. Med. Virol.* 24:40–69.

82. Summers, J., J. M. Smolec, and R. Snyder. 1978. "A Virus Similar to Human Hepatitis B Virus Associated with Hepatitis and Hepatoma in Woodchucks." *Proc. Nat. Acad. Sci.* 75:4533–4537.

83. Tiollais, P., C. Pourcel, and A. Dejean. 1985. "The Hepatitis B Virus." *Nature* 317:489–495.

84. Melnick, J. L. 1983. "Hepatitis B Virus and Liver Cancer." In *Viruses Associated with Human Cancer*, ed. L. A. Phillips. New York: Dekker, pp. 337–367.

85. Epstein, M. A., and B. G. Anchong. 1979. *The Epstein-Barr Virus*. New York: Springer-Verlag.

86. De The, G., A. Geser, N. E. Day, P. M. Tukei, E. H. Williams, D. P. Beri, P. G. Smith, A. G. Dean, G. W. Born Kamm, P. Feorino, and W. Henle. 1978. "Epidemiological Evidence for Causal Relationship Between Epstein-Barr Virus and Burkitt's Lymphoma: Results of the Ugandan Prospective Study." *Nature* 272:756–761.

87. Miller G. 1985. "Epstein-Barr Virus." In *Virology*, ed. B. N. Fields, D. M. Knipe, R. M. Chanock, J. L. Melnick, B. Roizman, and R. E. Shope. New York: Raven Press, pp. 563–589.

88. Poiesz, B. J., F. W. Ruscetti, A. F. Gazdar, P. A. Bunn, J. D. Minna, and R. C. Gallo. 1980. "Detection and Isolation of Type C Retrovirus Particles from Fresh and Cultured Lymphocytes of a Patient with Cutaneous T-Cell Lymphoma." *Proc. Nat. Acad. Sci.* 77:7415–7419.

89. Poiesz, B. J., F. W. Ruscetti, M. S. Reitz, V. S. Kalyanaraman, and R. C. Gallo. 1981. "Isolation of a New Type C Retrovirus (HTLV) in Primary Uncultured Cells of a Patient with Sezary T-Cell Leukemia." *Nature* 294:268–271.

90. Miyoshi, I., I. Kubonishi, S. Yoshimoto, T. Akagi, Y. Ohtsuki, Y. Shiraishi, K. Nagata, and Y. Hinuma. 1981. "Type C Virus Particles in a Cord Blood T-Cell Line Derived from Cocultivating Normal Human Cord Leukocytes and Human Leukaemic T Cells." *Nature* 294:770–771.

91. Wong-Staal, F., L. Ratner, G. Shaw, B. Hahn, M. Harper, G. Francini, and R. Gallo. 1985. "Molecular Biology of Human T-Lymphotropic Retroviruses." *Cancer Res.* 45:4539–4544.

92. Orth, G., F. Breitburd, M. Faure, and O. Croissant. 1977. "Papillomaviruses: A Possible Role in Human Cancer." In *Origins of Human Cancer*, ed. H. H. Hiatt, J. D. Watson, and J. A. Winsten. *Cold Spring Harbor Conferences on Cell Proliferation*. Vol. 4. Cold Spring Harbor, N.Y.: Cold Spring Harbor Laboratory.

93. Rapp, F., and F. J. Jenkins. 1981. "Genital Cancer and Viruses." *Gynecol. Oncol.* 12:25–41.

94. Howley, P. M., and T. R. Broker. 1985. *Papillomaviruses: Molecular and Clinical Aspects*. UCLA Symposia on Molecular and Cellular Biology, New Series. Vol. 32. New York: Liss.

95. Durst, M., L. Gissman, H. Ickenberg, and H. zur Hausen. 1983. "A Papillomavirus DNA from a Cervical Carcinoma and Its Prevalence in Cancer Biopsy Samples from Different Geographic Regions." *Proc. Nat. Acad. Sci.* 80:3812–3816.

96. Yang, Y-C., H. Okayama, and P. M. Howley. 1985. "Bovine Papillomavirus Contains Multiple Transforming Genes." *Proc. Nat. Acad. Sci.* 82:1030–1034.

97. Schwarz, E., U. K. Freese, L. Gissman, W. Mayer, B. Roggenbuck, A. Stremlau, and H. zur Hausen. 1985. "Structure and Transcription of Human Papillomavirus Sequences in Cervical Carcinoma Cells." *Nature* 314:111–115.

XI

EVOLUTION OF THE GENE

CHAPTER 28

The Origins of Life

Throughout the preceding chapters, we have seen that the more deeply we look at the molecular level, the more similar all known life forms appear to be. The evidence is therefore overwhelming that life on this planet must have descended from a single ancestral life form or **progenote**. But what this progenote was like and how exactly it first arose are terribly difficult questions.

The theory has occasionally been advanced that the first life forms did not originate on Earth, but arrived here from somewhere else in the universe. Even if this were true, we would still want to know how life originated elsewhere. In this chapter, we will assume, as do the vast majority of practicing biologists, that life originated on Earth.

The past 30 years have witnessed tremendous progress in our ability to conjecture about the origins of life. During this time, geologists have used advanced radioisotope dating techniques to show that the Earth first condensed from an interstellar cloud about 4.6 billion years ago. Chemists have shown that many of the essential building blocks of life (sugars, bases, and amino acids) could have assembled spontaneously from chemicals thought to have been present on the primitive Earth before the first living organisms arose (a process called **prebiotic synthesis**). **Paleobiologists**, who study ancient rocks and fossil sediments, have shown that the first fossil cells (**microfossils**) evolved from the prebiotically synthesized building blocks of life about 3 to 4 billion years ago, when the primitive Earth was only 1 to 1.5 billion years old. And molecular biologists have established that the many forms of life on Earth must share a common ancestor, since all cells are so similar at the molecular level.

In this chapter, we will see how recent advances in our knowledge of gene structure, protein structure, and RNA processing enable us to speculate with renewed confidence about the origin of life on Earth. In particular, we will emphasize the role that RNA enzymes might have played in the evolution of the first replicating biological systems, as well as in the early evolution of the translation apparatus and the genetic code.

Molecular Fossils Can Tell Us How Life Began[1, 2, 3]

A fossil is what remains after the slow process of mineralization has turned living matter into stone. Only the most durable parts of the organism (shells, bones, feathers, wood, or leaves) tend to survive this process, and thus most familiar fossils are large compared to cells. But under favorable circumstances, the process of mineraliza-

(a)

(b)

(c)

(d)

Figure 28-1
Fossil cells can be identified by using a light microscope to examine thin sections of ancient rocks. Procaryotic cells in rocks known to be (a) 3.5 billion years old from the Swartkoppie Formation in South Africa, (b) 2 billion years old from the Gunflint Iron Formation in Canada, and (c) 0.80 billion years old from the Spitsbergen Archipelago in Norway. The cells are much better preserved in the younger rocks. The cells in (b) represent a mixed colony consisting of both spheroidal and filamentous forms. Doublets in (a) and quadruplets in (c) are thought to be dividing cells. For comparison, (d) shows a dividing culture of the modern blue-green procaryotic alga *Aphanocapsa spheroides* at the same magnification as the specimen in (a). [See A. H. Knoll and E. S. Barghoorn, *Science* 198 (1977):396. Photographs courtesy of A. H. Knoll, Harvard University.]

tion can potentially preserve such small details as the wall of a single cell only a few microns in diameter. In the late 1950s, paleobiologists discovered that structures resembling modern algae and bacteria (microfossils) could be discerned by microscopic examination of extremely thin sections of certain ancient rocks (**stromatolites**) from Australia and South Africa (Figure 28-1). Stromatolites build up slowly in shallow seas when floating mats of algae or bacteria multiply and give rise to one layer of progeny after another.

There is strong evidence that the microfossils in stromatolites are really fossil cells, and not just bubbles or mineral deposits that happen to resemble a living organism (pseudofossils). First, deposits of blue-green algae resembling ancient stromatolites are still being formed in shallow bays today. Second, microfossils are comparable in size and shape to a variety of modern algal and bacterial cells. Third, different kinds of microfossils are often found together, as expected if different living species flourished side by side in a favorable environment (see Figure 28-1). And fourth, microfossils sometimes appear as doublets, suggesting that the cells perished in the act of division. Unfortunately, the process of mineralization preserves only the gross morphology of the cell wall, and most of the organic matter is destroyed as the rock undergoes cycles of heating and cooling. Thus, microfossils tell us little about the cell at the molecular level and provide no clues about the life forms that preceded the first microorganisms with tough cell walls.

We can reconstruct the earliest stages of evolution in either of two ways. One way is to ask what chemical or enzymatic reactions can occur when the conditions that may have prevailed on the primitive Earth are simulated in a modern laboratory. While these experiments cannot rigorously prove how life began, they can establish the plausibility of various scenarios.

The other way is to scrutinize modern cells for evidence of molecular fossils. **Molecular fossils** are structures (such as the placement of introns within a gene) or functions (such as the utilization of a particular metabolic pathway) that are shared by such diverse organisms that they must have been present in the first cells, as well as in earlier precellular forms of life. The abundance of such molecular fossils strongly suggests that living cells preserve within them a remarkably complete record of ancient evolution. Careful reading of this record may allow us to develop a coherent picture of primordial events which, until recently, were only fit for speculation. In this chapter, we will consider some current ideas about the origins of life, starting with the primitive "soup" and continuing to the present, with a special emphasis on what we can learn by reading the molecular fossil record.

The Formation of the Earth

Cosmologists agree that the solar system coalesced rather quickly from a whirling cloud of gas and interstellar dust about 4.6 billion years ago. As this nebulous cloud underwent gravitational collapse over a period of a few hundred million years, most of the mass of the cloud was concentrated in the sun. But for reasons that are still not well understood, some of the cloud did not fall into the sun and instead condensed into tiny **planetesimals** resembling meteorites. These planetesimals in turn aggregated to form the embryonic planets. The core of the primitive Earth melted from the heat of condensation, but the outer crust cooled quickly except where convection or volcanic eruptions brought heat and mass from the molten core to the surface. The primitive atmosphere was formed as various gases escaped from the interior of the earth (**outgassing**). However, it is a matter of contention whether outgassing produced an atmosphere that was highly reducing (H_2, NH_3, CH_4, and H_2O), mildly reducing (H_2O, H_2, N_2, CO, and CO_2), or nonreducing (H_2O, N_2, and CO_2). The crust and atmosphere of the primitive Earth cooled slowly, and the first shallow seas were formed as the less volatile gases, including water, rained back down to Earth. Lightning from these rainstorms, ultraviolet light from the intense solar radiation penetrating through the thin atmosphere, and alternating cycles of cold nights and hot days: These provided the chemical cauldron in which prebiotic synthesis created the essential building blocks of life.

Prebiotic Synthesis of the Building Blocks of Life

We normally think of evolution as the process by which new species of organisms arise from old, but we can extend the concept of evolution to include the process by which the first living cell evolved from

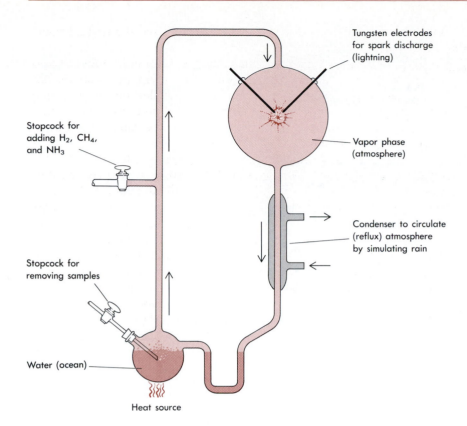

Stopcock for
adding H$_2$, CH$_4$,
and NH$_3$

Tungsten electrodes
for spark discharge
(lightning)

Vapor phase
(atmosphere)

Condenser to circulate
(reflux) atmosphere
by simulating rain

Stopcock for
removing samples

Water (ocean)

Heat source

Figure 28-2
A reflux apparatus used to simulate
prebiotic synthesis of organic molecules
under conditions thought to prevail on
the primitive Earth. [After S. L. Miller
and L. E. Orgel, *The Origins of Life on
Earth* (Englewood Cliffs, N.J.: Prentice-
Hall, 1974), Figure 7-1.]

the chemicals that were present on the primitive Earth. In this sense,
the evolution of life began long before the first living organisms existed.

It now seems obvious that the first living systems could not have
evolved unless the most essential building blocks of life (purines,
pyrimidines, sugars, amino acids, etc.) had already been synthesized
by prebiotic pathways, but this concept was revolutionary when proposed by Alexander Oparin in 1938. Prebiotic synthesis was first subjected to an experimental test in 1953. In an attempt to simulate possible conditions on the primitive Earth, high voltage sparks
("lightning") were passed through the gaseous phase ("atmosphere") of a refluxing mixture of CH$_4$, NH$_3$, H$_2$O, and H$_2$ (Figure
28-2). These four molecules had all been identified spectroscopically
in clouds of interstellar gas and were probably among the major constituents of the Earth's early atmosphere. Samples of the aqueous
phase ("ocean") were then withdrawn from the refluxing mixture
over a period of days and were found to contain at least ten different
amino acids as well as hydrogen cyanide (HCN) and various alde-
hydes. One possible pathway for amino acid synthesis under these
conditions is shown in Figure 28-3.

In other laboratory experiments, the nucleic acid base adenine was
synthesized by refluxing a concentrated solution of hydrogen cyanide
in ammonia (see Figure 28-3). Polymerization of HCN also gives rise
to orotic acid, which can be photochemically decarboxylated by sunlight to yield uracil. The synthesis of various sugars from formaldehyde (H$_2$CO) had been described as early as 1861, but was thoroughly
reinvestigated in the 1960s as interest in prebiotic synthesis grew.
Interestingly, the polymerization of formaldehyde results in ribose
but not deoxyribose. As we shall see, this observation is but one of
many suggesting that RNA preceded DNA in the evolution of life.

Amino acids

Nucleic acid bases

Sugars

Figure 28-3
A few of the many proposed pathways for prebiotic synthesis of amino acids, nucleic acid bases, and sugars by natural condensation of simple molecules present in the oceans and atmosphere of the primitive Earth. Each of these reactions yields other products besides the biologically relevant molecules shown here. For illustrative purposes, the atoms in the backbone of the adenine ring have been grouped arbitrarily to emphasize that adenine can be thought of as a polymer of hydrogen cyanide monomers. Similarly, ribose is shown as a polymer of formaldehyde.

The kind of prebiotic reactions shown in Figure 28-3 could have taken place in either a **reducing** or a **nonreducing atmosphere**, but not in the modern, strongly **oxidizing atmosphere** containing 21 percent O_2. However, calculations suggest that the abundance of atmospheric oxygen today is due solely to the accumulated synthesis of O_2 by living systems. Thus, the Earth's early atmosphere would have been reducing until the emergence of the first photosynthetic procaryotes.

The pathways of prebiotic synthesis remain speculative because we have so little definite knowledge about the exact conditions prevailing on the primitive Earth. In fact, many of the proposed prebiotic reaction pathways are mutually incompatible. For example, at the concentrations required to generate sugars in alkaline solution, formaldehyde would react rapidly with the amino groups of both amino acids and nucleic acid bases. Nonetheless, local variations in temperature, chemical composition, pH, and other factors might have permitted incompatible reactions to occur in different places or at different times on the primitive Earth.

The proposed prebiotic pathways pose other problems as well. The yield of biologically interesting molecules is often poor, side reactions abound, and synthesis often requires unusually high concentrations of the reactants. However, it is possible that repeated cycles of freezing and thawing, or cycles of evaporation and rehydration, might have concentrated either the reactants or the products.

The conclusion we can draw from more than 30 years of laboratory experiments is that a multitude of alternative pathways may have contributed to prebiotic synthesis. Our confidence in the notion of prebiotic synthesis is also greatly strengthened by recent chemical analysis of meteorites that have fallen to Earth from elsewhere in the solar system. Amino acids and other building blocks of life are found deep within these rocks, where they cannot be regarded as earthly contaminants.

The smallest building blocks of life (nucleic acid bases, sugars, and amino acids) were probably assembled into still larger molecules by sequential **condensation** (removal of water molecules). Water molecules could have been removed physically by heating or chemically by reaction with a water-hungry condensing agent like a polyphosphate. Such condensations would generate the nucleic acid precursors by forming a glycosidic bond between a nucleic acid base and a ribose ring. Condensations would also create **polymers** by forming phosphodiester bonds between the nucleic acid precursors and amide bonds between amino acids. Thus, as the prebiotic cauldron cooled, the resulting soup of organic chemicals probably contained oligonucleotides, peptides, and oligosaccharides.

THE RNA WORLD

Living Systems Must Replicate and Evolve

The first cells were clearly alive, but what about the precellular collections of molecules that preceded them? Can we draw a line between the inanimate world created by prebiotic synthesis and the very earliest stages in the evolution of life? To answer this question, we must first ask ourselves, What is the simplest collection of molecules that would qualify as a "living system"? Opinions will vary, of course, but most biologists would agree that every living system must at the very least be capable of both replication and evolution. Replication enables the parental copy of the system to give rise to two identical (or nearly identical) daughter copies of itself. Evolution implies that the process of replication is always slightly imperfect, so that variant daughter copies may have the potential to replicate faster than the parent or to adapt better to changing environmental conditions.

The First "Living Molecule" Was Almost Certainly a Nucleic Acid

We have seen in Chapter 3 that the self-complementary structure of a nucleic acid double helix is perfectly suited for the task of replication. One strand serves as the template surface for polymerization of a new daughter strand. Polypeptide chains, on the other hand, are not self-complementary and cannot serve as templates for their own replication. This profound distinction between nucleic acids and proteins constitutes a powerful argument that the first replicating structures (genomes) were made of nucleic acids. Moreover, the universality of nucleic acid genomes in contemporary cells and viruses makes it difficult to believe that the ancestral living systems did *not* have either an RNA or a DNA genome.

Until very recently, reconstructing the origin of the earliest life forms appeared to be a hopeless case of "the chicken and the egg." Only nucleic acids seemed capable of storing and replicating genetic information, and only proteins were known to function as catalysts. Thus, it was exceedingly difficult to imagine how the first nucleic acid genome (whether DNA or RNA) could be replicated without a protein catalyst, and equally difficult to imagine how an efficient protein catalyst could arise without a nucleic acid genome to encode it. In 1981, a startling discovery presented a solution to this paradox. The ribosomal RNA intron of the ciliate *Tetrahymena* was shown to excise itself from the rRNA precursor without the benefit of a protein catalyst (see Figures 20-13 and 20-14). The ability of the rRNA intron to function as an enzyme implied that the very first living molecule might have been an **RNA replicase** that catalyzed its own replication without help from a protein (Figure 28-4). This hypothetical RNA would therefore have functioned as both the genetic material and the replication enzyme. The discovery in 1983 of a second catalytic RNA species (the RNA component of RNase P; see Figure 14-19) lent strong support to this notion. Since 1983, many new instances of catalytic RNA have been discovered. As we shall see, the common features of these catalytic RNAs suggest that they are molecular fossils whose history dates back to the earliest life forms. Studies of catalytic RNA therefore play a major role in shaping our current views about the origins of life.

How did the first RNA replicase arise? Presumably, random prebiotic condensation of mononucleotides formed longer and longer polymers. Most of these random RNA polymers would have been functionless, but an occasional RNA molecule may have been able to work as a very primitive, error-prone RNA replicase. This first RNA replicase would have been able to use either itself or another nucleic acid as a template for polymerization of nucleic acid precursors found in the prebiotic soup. Of course, no replicase can copy its own active

Figure 28-4
Simplified scheme for an RNA replicase. The + and − strands of the RNA enzyme have similar secondary structures because the sequences are complementary. Only synthesis of a new + strand is shown, but − strands would be replicated in the same way. For clarity, several assumptions have been made in drawing the figure. The RNA polymerase is shown as a *processive* enzyme that remains bound to the template strand until a complete copy has been synthesized. However, the polymerase could also be *dissociative*, separating from the template after addition of each new nucleotide. The precursors are shown as mononucleotides, although we discuss the possibility that random oligonucleotides could have served the same purpose (see Figures 28-11 and 28-12). The nascent RNA strand is thought to refold as synthesis proceeds; otherwise, formation of a stable RNA duplex might block further replication.

site, so we must imagine that *two* RNA replicases arose simultaneously. This is not implausible because, as we shall see, very short RNA sequences can have distinct enzymatic activities (see Figure 28-7).

The hypothetical RNA replicase would not only replicate but also evolve, because some of the variant copies generated by error-prone replication would be able to replicate faster or more accurately than the parental molecule. We therefore regard this RNA replicase as the first "living system." In the section entitled The RNA World, we will see that recent progress in the field of RNA processing suggests that just such an RNA replicase was the not-so-distant ancestor of certain molecules found in living cells today.

Why RNA and Not DNA Was the First Living Molecule

As already discussed, new examples of RNA molecules with enzymatic activity are constantly being discovered (see Figure 28-19), but no enzymatic activity has ever been attributed to DNA. This is perhaps the most powerful argument that the first replicating nucleic acid was RNA rather than DNA. However, there are other substantial reasons for believing in the primacy of RNA:

- The difference between RNA and DNA is that the deoxyribose rings of DNA lack 2'-hydroxyl groups. Both the 2' and 3'-hydroxyl groups are known to play a structural role in tRNA, where they are involved in a number of unusual hydrogen bonds that fold the "cloverleaf" tRNA secondary structure into the final L-shaped tertiary structure (see Figures 14-5 through 14-8 and Figure 28-13). RNA enzymes, like protein enzymes, create a catalytic surface by folding into complex secondary and tertiary structures. Lacking 2'-hydroxyl groups, DNA may be less adept than RNA at folding into the complex three-dimensional shapes required for catalytic activity.

- The 2' hydroxyl of ribose also plays a direct catalytic role in some RNA enzymes. Evidence for this comes from studies of the self-splicing mitochondrial Group II introns (see Figures 28-19 and Table 28-1), as well as from theoretical consideration of the available functional groups in RNA (see Figure 28-6). Thus, DNA enzymes could never be as versatile as RNA enzymes.

- Although all cells today have DNA genomes, DNA precursors are always synthesized by reduction of RNA nucleoside diphosphate precursors using the highly conserved enzyme **ribonucleoside diphosphate reductase** (see Figure 28-20). This suggests that the biochemical pathway for ribonucleotide synthesis evolved first, when cells (or precellular living systems) had RNA genomes. Later, when DNA genomes replaced the more labile RNA genomes, the preexisting biochemical pathways were adapted for the synthesis of DNA precursors.

- Finally, ribose is much more readily synthesized than deoxyribose under simulated prebiotic conditions.

RNA Enzymes Work Like Protein Enzymes[4]

Regardless of whether it is made of protein, RNA, or even some other exotic material, every enzyme must have an active site. This site is a three-dimensional surface that binds the substrate and positions the functional groups of the enzyme around the bound substrate so that catalysis can proceed. When the crystal structure of tRNA was first

Table 28-1 Intron Boundary Sequences

Intron Type	5' Splice Junction	3' Splice Junction
Group I, mitochondrial + nuclear	U↓	G↓
Group II, mitochondrial	↓GUGCG	Y_nAU↓
Chloroplast mRNA (*Euglena*)	↓GUG^C_UG	Y_nA^U_C↓
Nuclear mRNA precursor (general)	^C_AAG↓GU^A_GAGU	Y_nAG↓
Nuclear mRNA precursor (yeast)	↓GUAUGU	Y_nAG↓

Underscored nucleotides are essentially invariant; other nucleotides represent a consensus. Arrows designate the actual 5' and 3' splice sites. [After T. R. Cech, *Cell* 44 (1986):207.]

determined in 1974, it became clear that RNA molecules could fold up into three-dimensional structures resembling those formed by proteins (see Figures 14-5 through 14-8). However, no one was sufficiently bold (or farsighted?) to suggest at the time that RNA might function as a catalyst. The reason for such universal hesitation was that only 4 relatively similar nucleotides (instead of 20 diverse amino acids) could be used to construct the active site of an RNA enzyme. Although we now know that RNA can function catalytically, we might still expect RNA enzymes (also called **ribozymes**) to be less versatile than protein enzymes, because RNA does not have as large a variety of functional groups.

Like protein enzymes, RNA enzymes almost certainly use **proton exchange reactions** to catalyze hydrolysis of RNA substrates (Figure 28-5). In fact, very similar mechanisms may be responsible for the hydrolysis of RNA by the protein RNase A and the catalytic RNA component of RNase P. Despite some superficial differences between the mechanisms (RNase A leaves a 3' phosphate and RNase P a 5' phosphate), in each case one functional group (a general base) removes a proton (H^+) while a nearby functional group (a general acid) adds a proton. A wealth of structural and enzymatic data suggest that the proton exchange reactions in RNase A are catalyzed by histidines 12 and 119. (Histidine is a particularly useful amino acid for catalysis because both nitrogens of the imidazole ring can bind protons.) Like proteins, RNA molecules also possess a variety of potential functional groups (Figure 28-6), but we are still far from understanding which groups in RNA are catalytically important.

Very Short RNAs Can Be Catalytic[5-9]

How realistic is it to suppose that random prebiotic condensation of ribonucleotides could actually fashion a crude RNA replicase? The answer to this question depends on the size of the smallest RNA that can function as a catalyst. If all catalytically active RNA molecules had to be as large and complex as the RNA component of RNase P (377 nucleotides in *E. coli;* see Figure 14-19) or the self-splicing intron of the *Tetrahymena* rRNA (413 nucleotides; see Figures 20-14 and 20-8), it would be very difficult to argue that even a very crude RNA replicase could arise by random condensation of ribonucleotides. However, recent progress in the study of plant viruses has demonstrated that as few as 52 nucleotides can act as an accurate RNA catalyst, and a crude catalyst might be simpler.

Hydrolysis of RNA by pancreatic RNase

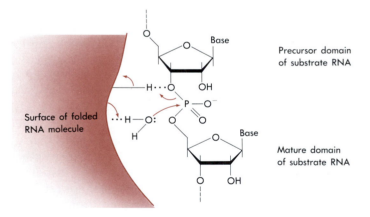

Figure 28-5
Both protein enzymes and RNA enzymes (ribozymes) catalyze hydrolysis of RNA substrates by donating and capturing protons in proton exchange reactions. Pancreatic RNase cuts after the pyrimidines U and C, producing a 3′ phosphate and a 5′ hydroxyl. The proton donor and acceptor groups (labeled A and B) are thought to be histidines 119 and 12 at nearly opposite ends of the 124-residue enzyme. Hydrolysis proceeds through a cyclic 2′,3′-monophosphate intermediate that is enzymatically hydrolyzed in a second step. [After F. M. Richards and H. Wyckoff, in *The Enzymes*, Third Ed. (P. D. Boyer, editor), Vol. 4:647.] Hydrolysis of RNA by RNase P produces a 3′ hydroxyl and a 5′ monophosphate. The identity of the proton donor and acceptor groups in the catalytic RNA molecule is not yet known. [After N. R. Pace and T. L. Marsh, *Origins of Life* 16 (1985):97.]

Hydrolysis of RNA by RNase P

Many plant viral pathogens have single-stranded RNA genomes (see Figure 24-37). The genome of Lucerne Transient Streak Virusoid (LTSV), for example, is a covalently closed single-stranded RNA molecule of 324 nucleotides. The most stable secondary structure for the infecting + strand of the virusoid appears to be a double-stranded rod with many "bulged" bases and unpaired loops (Figure 28-7a). Surprisingly, the viral + strand is capable of accurate self-cleavage if the molecule is first heated to melt the most stable rod-shaped structure and then quickly cooled to "freeze" the molecule in a less stable (but enzymatically active) alternative secondary structure. Equally surprising was the observation that the viral − strand can cleave itself at a site only six nucleotides away from the + strand self-cleavage site. The role of self-cleavage in the viral life cycle remained a mystery until the demonstration that viral replication proceeds by a rolling circle mechanism (Figure 28-7b). Then it became clear that self-cleavage is required to excise monomeric viral RNAs from the multimeric RNA tails produced by the rolling circle mechanism. The reason why the self-cleavage sites in the + and − strand multimeric tails assume

Histidine proton transfer

Base tautomerization

Lactam   lactim

3'-terminal ribose as a proton reservoir

C3' endo   C1' exo   C3' endo

Base tautomer activation of an internal ribose 2' OH

C2' endo   C3' endo

the enzymatically active secondary structure (Figure 28-7c) rather than the more stable structure (Figure 28-7a) is not yet understood. Perhaps the binding of host cell proteins stabilizes the catalytic conformation.

Self-cleavage is a common and perhaps universal feature of the many different small, single-stranded RNAs that infect plants. Comparison of the cleavage sites in different viral RNAs initially suggested that self-cleavage does not require a large or complex RNA structure. This was verified experimentally by testing the ability of very short fragments of LTSV RNA to cleave themselves in the test tube. The minimal structural requirements for self-cleavage are three short stems meeting in a loop; within this structure, 17 highly conserved nucleotides are thought to play a role in catalysis, while the remaining nucleotides presumably establish the proper secondary and tertiary structures for self-cleavage (see Figure 28-7c). Unfortunately, we know so little about the interactions responsible for RNA tertiary structure that there is no way to predict the three-dimensional structure of the self-cleaving RNA domain from the two-dimensional base-pairing scheme (see Figure 28-7c).

Figure 28-6
Possible mechanisms by which RNA might catalyze proton exchange. The enzymology of RNA catalysis is not yet sufficiently advanced to distinguish between these (and other) possibilities. For comparison, the figure also shows proton transfer by the imidazole side chain of histidine, a reaction that is thought to occur at the active site of pancreatic RNase (see Figure 28-5). [After N. R. Pace and T. L. Marsh, *Origins of Life* 16 (1985):97.]

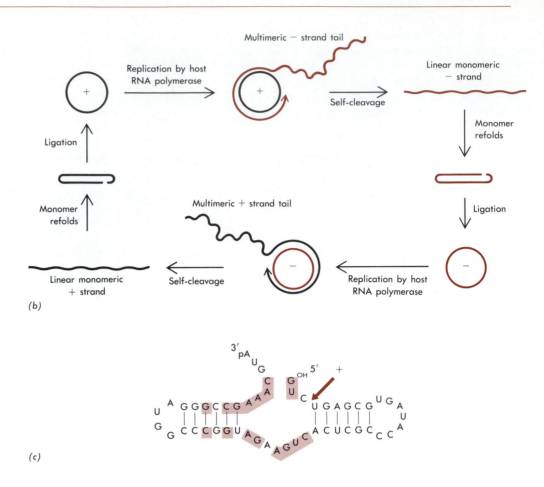

(b)

(c)

Figure 28-7

Only 52 bases are sufficient for self-cleavage of a plant virus RNA genome. (a) Secondary structure for the circular + strand of Lucerne Transient Streak Virusoid (LTSV) RNA. Sites of self-cleavage for the + and − RNA strands are denoted by color and black arrows, respectively, and the nucleotide sequence domains required for these reactions are indicated by the corresponding bars. (b) Rolling circle model for LTSV RNA replication. Both the + and − RNA strands must be capable of self-cleavage in order to excise monomeric units from the multimeric RNA tails of the rolling circle intermediates. Note how the secondary structure shown in (a) aligns the 5′ and 3′ ends of the linear monomer for ligation. (c) When the nucleotide domain indicated by the color bar in (a) assumes the alternative secondary structure shown here, it is enzymatically active. These 52 nucleotides alone are capable of cleaving the phosphodiester bond indicated by the color arrow. The boxed nucleotides are conserved in a large variety of other single-stranded RNA plant viruses and appear to be required for catalysis. [Courtesy of A. C. Forster and R. H. Symons, The University of Adelaide, Adelaide, Australia.]

The Self-Cleaving Transcripts of the Newt[10]

Virtually the same self-cleaving RNA sequence first discovered in plant viruses (see Figure 28-7) has recently been found in the newt *Notophthalmus viridescens*. The genome of this vertebrate has many blocks containing head-to-tail (tandem) repetitions of a 330-base-pair sequence. Some of these blocks of DNA are transcribed into very long nuclear RNA molecules that are subsequently exported to the cytoplasm. Unexpectedly, the cytoplasmic transcripts are all small RNAs whose length is an integral multiple of 330 nucleotides. Inspection of the DNA sequence revealed that the 330-base-pair DNA repeating unit contains the same conserved sequence elements as the plant viral self-cleavage site (see Figure 28-7c), implying that the long RNA transcripts were cleaving themselves at 330-base intervals.

How could a vertebrate have acquired tandem DNA copies of a 330-nucleotide sequence that is capable of self-cleavage as RNA? The most plausible explanation is that a circular, single-stranded RNA virus (like LTSV) infected an ancestor of modern newts. Virus replication then led to the production of a multimeric RNA tail (see Figure 28-7b) that was reverse-transcribed into DNA and inserted into the newt genome. The generation of pseudogenes (see Figure 20-30), processed genes (see Figure 20-31), and *Alu* sequences (see Figure 20-37) in mammalian genomes provides an attractive precedent for such a **reverse flow of genetic information** from cellular RNA into genomic DNA.

Insertion of the self-cleaving RNA sequence must have occurred long ago in the genome of an ancestral newt, since all species of the salamander family from Europe, North America, and Asia have homologous genomic sequences. A perplexing question then is why, after so many millions of years of evolution, a sequence that is now perpetuated as DNA has retained the capability for self-cleavage as RNA. In any event, the discovery of the 330-base-pair newt DNA sequence suggests that many more molecular fossils of catalytic RNA are lurking in unlikely places. Such fossils will continue to expand the library of important clues regarding the nature of the earliest living molecules on Earth.

Sequential Transesterification Reactions Promote RNA Self-Splicing[11, 12, 13]

We have seen that proton exchange reactions are responsible for cleavage of RNA by both protein and RNA enzymes (see Figure 28-5). Proton exchange reactions also account for the series of **phosphoester bond transfers**, or **transesterifications**, that lead to autocatalytic excision and subsequent reactions of the *Tetrahymena* rRNA intron (see Figures 20-13, 20-14, 28-8, and 28-19). Superficially, transesterification appears to be a far more sophisticated reaction than either the site-specific hydrolysis of tRNA precursors catalyzed by RNase P or the autocatalytic cleavage of plant viral RNAs. But mechanistically, both transesterification and hydrolysis involve nucleophilic attack of a hydroxyl group on a specific phosphoester bond. When the attacking group is a water molecule, the result is hydrolysis; when the attacking group is the sugar hydroxyl group of a ribonucleotide, the result is a transesterification (Figures 28-8 and 28-9). Since hydrolysis and transesterification are so closely related, small changes in the structure of an RNA enzyme could in principle convert a simple RNase (such as the plant viral self-cleavage domain; see Figure 28-7c) into an enzyme capable of catalyzing transesterifications, and vice versa.

The series of transesterifications that catalyze excision of the *Tetrahymena* rRNA intron all involve attack by the 3' hydroxyl of a ribonucleotide on a phosphoester bond. The 3' hydroxyl group of the rRNA

(a)

Figure 28-8
(a) Structure of the *Tetrahymena* riboso-mal RNA intron. For simplicity, the intron is shown here in two dimensions, but long-range interactions such as base pairing between the sequences marked *9R* and *9R'* (black arrow) are known to fold the intron into a complex three-dimensional structure. [After T. R. Cech, *Sci. Amer.* 255 (1986):64.] (b) Schematic representation of the reactions of the self-splicing intron. The initiating guanosine nucleotide is indicated by a shaded circle. Square brackets denote postulated intermediates that have not been isolated. Numbers represent the length in nucleotides of the intron and its fragments. In both (a) and (b), exon sequences are shown in color.

Figure 28-9
Proton exchange reactions also catalyze RNA transesterifications. The elaborate secondary and tertiary structure of the *Tetrahymena* intron (see Figures 20-14 and 28-8) are represented here and in the next figure by a simple curved line. Once the free guanine nucleotide binds to the intron, its 3'-hydroxyl group is activated to initiate a nucleophilic attack on the phosphoester bond at the 5' splice site. Nucleophilic attack by guanosine initiates the series of transesterifications (phosphoester transfer reactions) that excise the intron as a linear molecule and ligate the two exons (see Figures 20-13 and 28-8). Experimentally verified hydrogen bonds between the guanosine nucleotide and its binding site on the RNA are indicated by dotted lines; the hydrogen bond donors and acceptors of the guanosine binding site could include ribose rings and phosphates, as well as the bases themselves. Note that catalysis by the protein enzyme pancreatic RNase proceeds by a similar mechanism, with the 2'-hydroxyl group of a ribose ring being activated to attack the adjacent phosphate (see Figure 28-5). [After B. L. Bass and T. R. Cech, *Nature* 308 (1984):820.]

precursor itself cannot initiate the transesterifications, or the intron would become attached to the 3' end of the rRNA. Instead, the RNA enzyme binds a free guanosine nucleotide (GMP, GDP, or GTP), and the 3'-hydroxyl group of this mononucleotide initiates the sequence of transesterifications. As a result, all the products of the initial reaction are linear RNA molecules. (However, the 3' hydroxyl of the excised intron can attack an internal phosphoester bond to generate circular RNAs; see Figures 20-13, 28-8, and 28-19).

The transesterification reactions responsible for excising the ribosomal intron do not consume any high-energy bonds in the form of nucleotide diphosphates, triphosphates, or other high-energy dinucleotides such as NAD or FAD (see Figure 28-14). Instead, the same number of phosphoester bonds are made as are broken, and the energy of the existing phosphoester bonds is conserved. Thus, the intron autocatalytically rearranges existing phosphodiester bonds, rather than using high-energy intermediates to make new ones. In this sense, the intron can be thought of as an *RNA "rearrangease"* or *RNA isomerase*. As we shall see, the first RNA replicase may have been a similar RNA enzyme that was capable of rearranging prebiotically synthesized random oligonucleotides into a complementary copy of an RNA template strand.

Exploring the Fossil Evidence in the *Tetrahymena* rRNA Intron[14]

Studies of the *Tetrahymena* rRNA intron began in 1981 with the relatively modest goal of understanding how the RNA carries out catalytic reactions, but the work soon had important implications for the origin of life. In the next few sections, we will describe experiments showing that the excised *Tetrahymena* rRNA intron can function as a poly C polymerase (see Figure 28-11). We will then extrapolate from the evidence provided by this molecular fossil to the reactions and mechanisms of ancient RNA enzymes. Specifically, we will present plausible models for how a poly C polymerase might be further modified to function as an RNA replicase (see Figure 28-12) and perhaps even as a primitive tRNA synthetase (see Figure 28-17).

Figure 28-10

The internal guide sequence of the *Tetrahymena* ribosomal RNA intron selects the 5' splice site. A six-nucleotide internal guide sequence near the 5' end of the intron base-pairs with the 5' exon sequence just upstream from the 5' splice site. This aligns the active site of the intron with the phosphoester bond at the 5' splice site. The activated guanosine nucleotide with a free 3'-hydroxyl group (see Figure 28-9) is shown bound to the active site. Presumably, the 3' splice site is selected by a similar but more complicated mechanism that positions the 3' hydroxyl of the 5' exon next to the 3' splice site. Exon sequences are shown in color, and the internal guide sequence is boxed. The intron shown here is completely schematic; the actual structure is complex and brings the 5' and 3' exons close together (see Figures 20-14 and 28-8). [After G. Garriga, A. M. Lambowitz, T. Inoue, and T. R. Cech, *Nature* 322 (1986):86.]

We hope in this extended discussion of the autocatalytic *Tetrahymena* rRNA intron to demonstrate convincingly that RNA molecules could have functioned as the first replicating (and therefore living) systems. We will also describe a scenario in which variants of this RNA replicase gave rise first to the **RNA world**, a rudimentary living world where RNA functioned as the sole catalytic molecule, and then later, with the advent of protein synthesis, to a more complicated world where RNA and protein shared the catalytic functions (the **ribonucleoprotein**, or **RNP world**). This RNP world was the immediate precursor of our own **DNA world**.

The Core of the Self-Splicing Intron Can Function as an RNA Polymerase[15–19]

An enzyme is defined as a molecule that can catalyze a chemical reaction without itself being permanently altered in any way. The intact *Tetrahymena* rRNA intron is therefore not a true enzyme, because it is chemically changed by the very reaction it catalyzes. However, as shown in Figure 28-8, the sequential transesterification reactions that excise the intron as a linear molecule do not destroy the active site of the RNA: The guanosine residue at the 3' end of the excised linear intron can still attack an internal site near the 5' end of the intron in an autocyclization reaction. The cyclization joint is subsequently hydrolyzed by the RNA enzyme to liberate a linear form of the intron, which can in turn undergo yet another cycle of autocyclization and hydrolysis, thereby fragmenting the original intron. The role of autocyclization and fragmentation may be to ensure that excision of the intron is irreversible under normal circumstances.

Even the second round of autocyclization and hydrolysis does not destroy the active site of the RNA. The final product is a linear molecule in which the active site of the intron remains intact. Once it is deprived of an internal substrate, this intron core can use other RNA molecules as substrates.

A remarkable series of experiments revealed that the intron core can function as a true enzyme that is capable of polymerizing poly C from oligo C. In the normal self-splicing reaction, the G-rich **internal guide sequence** of the intron selects the 5' splice site by forming complementary base pairs with the 5' exon sequence (Figure 28-10). This base-pairing scheme positions the 5' splice site at the active site. The guide sequence plays a similar role when the intron core functions as a poly C polymerase (Figure 28-11): The G-rich internal guide sequence selects oligo C precursors from a mixture of oligonucleotides

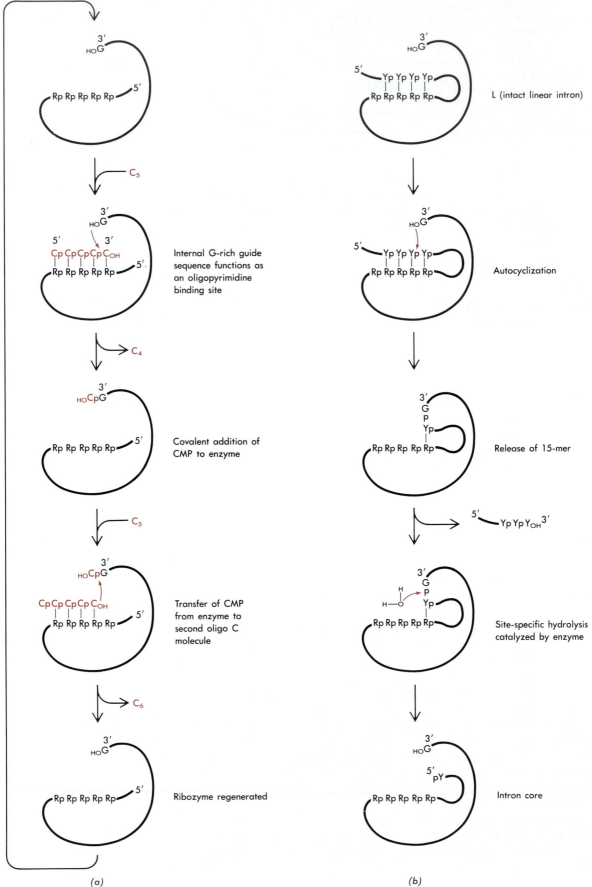

Enzymatic polymerization of oligo C by intron core

Internal G-rich guide sequence functions as an oligopyrimidine binding site

Covalent addition of CMP to enzyme

Transfer of CMP from enzyme to second oligo C molecule

Ribozyme regenerated

Analogous self-reactions of linear intron

L (intact linear intron)

Autocyclization

Release of 15-mer

Site-specific hydrolysis catalyzed by enzyme

Intron core

(a)

(b)

Figure 28-11
The *Tetrahymena* rRNA intron core can function as a poly C polymerase. Enzymatic polymerization of poly C from oligo C is shown in (a). The analogous self-reactions of the intact 413-nucleotide intron are shown in (b) to illustrate the strict correspondence between reactions that the intron can catalyze using itself or oligo C as substrate. The first round of autocyclization is shown as releasing a fragment 15 nucleotides long (15-mer); subsequent hydrolysis followed by a second round of autocyclization would release an additional 4-mer. Note that the G at the active site of the intron core is the 3' end of the original intron, not the guanosine cofactor that initiated excision. R represents a purine. [After A. J. Zaug and T. R. Cech, *Science* 231 (1986):470.]

in solution and properly positions them at the active site. The intron core then functions as a poly C polymerase by catalyzing the same transesterification reactions as in self-splicing; the only difference is that oligo C molecules take the place of the 5' splice site. The result is molecular rearrangement of the oligo C substrate: C residues are moved from one oligo C to another, so that some molecules become longer at the expense of others.

At first glance, it may strike some readers as false advertising to claim that the intron core functions as a poly C "polymerase." After all, the poly C "polymerase" merely rearranges existing oligonucleotides without creating any new phosphoester bonds, while modern RNA polymerases synthesize new RNA chains de novo by polymerizing nucleotide triphosphates one at a time. But modern polymerases require activated mononucleotide triphosphate precursors. These precursors could not have been plentiful in the primordial soup, because they are unstable and easily hydrolyzed to nucleotide diphosphates and monophosphates. Oligonucleotides, in contrast, are relatively stable and must have been able to accumulate under prebiotic conditions (how else could the first RNA replicase have arisen?). Thus, if the primordial soup was rich in random oligonucleotides but poor in activated mononucleotides, the ability to use oligonucleotides as substrates for RNA polymerization would have been a major advantage for the first RNA replicase.

A Detailed Model for the First RNA Replicase[20, 21]

Although the ability of the excised linear form of the *Tetrahymena* rRNA intron to function as a poly C polymerase could be fortuitous, a far more exciting possibility is that the intron core retains the ability to polymerize oligonucleotides because it is in fact a molecular fossil of the primordial RNA replicase. Viewed in this light, the essential task of the intron core originally was, and still remains, the polymerization of RNA; self-splicing is simply a fortuitous side reaction of the RNA polymerization activity.

Recent experiments have shown that the intron core can be changed from a poly C polymerase into a poly U polymerase simply by making the internal guide sequence A rich rather than G rich. This result suggests the tantalizing possibility that the internal guide sequence corresponds to the template of the original RNA replicase. If so, it may be possible to transform the intron core back into a template-dependent RNA polymerase by sophisticated genetic engineering:

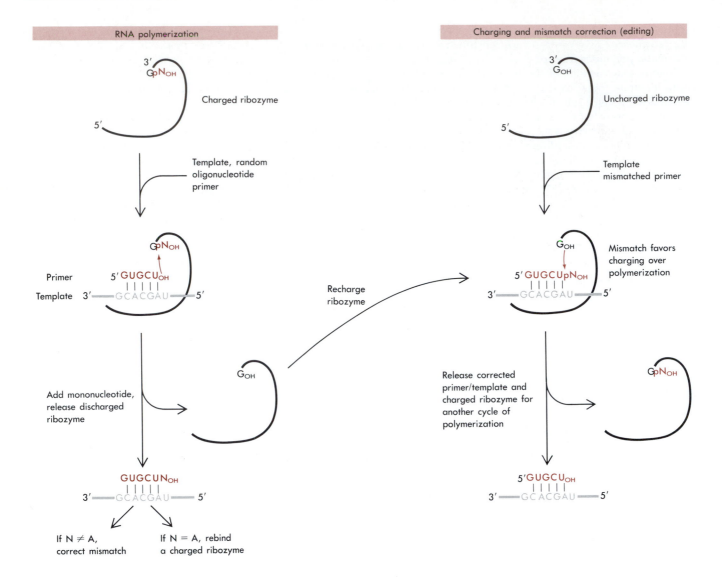

Figure 28-12

A scheme for RNA self-replication. Deletion of the internal guide sequence from the intron core might generate a template-dependent RNA polymerase able to copy an external template. Random oligonucleotides would serve not only as primers for RNA replication but also as the source of activated mononucleotides. Note that the polymerase must dissociate from the template after addition of each mononucleotide in order to be recharged. If the same or similar mechanism could add oligonucleotides to the growing RNA primer, the RNA enzyme might also function as an RNA ligase to join the randomly initiated RNA fragments into a full-length copy of the template RNA. The particular template sequence used here to illustrate the mechanism is arbitrary. [After T. R. Cech, *Proc. Nat. Acad. Sci.* 83 (1986):4360.]

The *internal* guide sequence of the intron core would have to be deleted and replaced by a separate single-stranded RNA template capable of functioning as an *external* guide sequence. The resulting template-dependent RNA polymerase might then be able to copy other RNA molecules (Figure 28-12).

Since the intron core already has the catalytic abilities of an RNA replicase, the major challenge in this genetic engineering project would be to facilitate an association between the catalytic domain of the intron core and the external guide sequence (template). The three-dimensional crystal structure of various tRNAs shows that folding of the negatively charged RNA backbone can create binding pockets for divalent cations such as Mg^{2+}, Ca^{2+}, and Zn^{2+} (Figure 28-13). This implies that the intron core could be redesigned as a **metalloenzyme** with several metal binding pockets on its surface. These bound metals would then hold the template RNA in place by forming ionic bridges to the negatively charged phosphates of the template backbone.

TψC–stem
TψC–loop
Acceptor stem
3' acceptor end
D–loop
Variable loop
D–stem
Anticodon stem
Anticodon
Anticodon loop

Coordinated Mg^{+2}
Spermine
Possible Mg^{+2}

Figure 28-13
The three-dimensional structure of yeast tRNA^{Phe} showing how the phosphate backbone of the folded RNA chain creates specific binding sites for the cations Mg^{2+} and spermine. (Spermine, which contains four positively charged nitrogen atoms, is a polyamine with the structure NH$_3^+$–(CH$_2$)$_3$–NH$_2^+$–(CH$_2$)$_4$–NH$_2^+$–(CH$_2$)$_3$–NH$_3^+$.) The position of the cations was determined by refining the tRNA^{Phe} crystal structure shown in Figure 14-6 to a resolution of 2.5 Å. As in Figure 14-6, the sugar-phosphate backbone is drawn as a continuous ribbon, base pairs are shown as complete crossbars, and unpaired bases are indicated by short bars extending from the backbone. [After G. J. Quigley, M. M. Teeter, and A. Rich, *Proc. Nat. Acad. Sci.* 75 (1978):64.]

All Reactions of the *Tetrahymena* rRNA Intron Are Fundamentally Similar[12, 18]

Viewed superficially, the excision, circularization, hydrolysis, and poly C polymerization reactions of the *Tetrahymena* rRNA intron seem very different (see Figures 28-8 through 28-11). But on closer examination, these transesterification reactions can all be seen to proceed by the same molecular mechanism: Invariably, the 3' hydroxyl of the intron core is activated to participate in a proton exchange reaction (see Figures 28-5 and 28-9). Thus, this same 3'-hydroxyl group always serves as the active residue of the RNA enzyme. The only apparent exception is the initial cut at the 5' splice site during intron excision, but here, a free guanosine nucleotide simply substitutes for the 3'-terminal guanosine hydroxyl of the intron itself.

Since the intron activates its own 3'-hydroxyl group, the secondary and tertiary structures of the intron core must be designed to surround the active residue with an appropriate array of proton donor and acceptor groups. The activated 3'-hydroxyl group can then attack any phosphoester bond that is held in the proper position by the internal guide sequence. No significant energy is gained or lost in the resulting *isoenergetic* transesterification reaction, so the reaction can proceed forward and backward with equal ease. A direct consequence of the *reversibility* of the reaction is that the activated hydroxyl group not only can attack a preexisting phosphoester bond ("charging"), but also can *be* attacked when it forms part of a phosphoester bond ("discharging"):

Uncharged 5' intron-OH 3' + substrate 5' pNpNpNpN-OH 3' $\longrightarrow$
charged 5' intron-pN 3' + substrate 5' pNpNpN-OH 3'

In most reactions of the intron, binding of the substrate RNA to the internal guide sequence positions a particular phosphoester bond at the active site for transesterification. Only when no RNA substrate is bound does hydrolysis of the charged enzyme occur. In this case, the active site inefficiently catalyzes the attack of water (H–OH) instead of a bound oligonucleotide (5′ pNpNpNpN–OH 3′) on the activated phosphoester bond (see Figures 28-8 and 28-11).

The ability of the *Tetrahymena* intron to catalyze so many different reactions may indicate that the active sites of RNA enzymes are generally more flexible than those of protein enzymes (also see section entitled Why Does RNase P Retain Its RNA?). This may provide a critical clue regarding the origin of protein synthesis, because it suggests that the activated phosphoester bond of a charged RNA replicase could react with chemical groups other than the 3′ hydroxyl of RNA (see section entitled A Model for How a Charged Ribozyme Might Have Specifically Activated an Amino Acid).

Metabolic Pathways Evolved To Synthesize Scarce Molecules[22]

We have seen that many molecules of biological importance were synthesized prebiotically under the harsh conditions prevailing on the youthful Earth. However, despite the best efforts of astronomers, geologists, chemists, and paleobiologists, our knowledge of these primordial conditions remains uncertain and fraught with speculation. Indeed, it would not be an exaggeration to say that every expert in the field of molecular evolution has a different notion of what exactly was in the prebiotic soup. This uncertainty compounds the difficulty of evaluating various scenarios for the origin of life.

Any replicating system requires a source of precursor molecules. In the particular scenario we have presented for the origin of life, the prebiotic soup would have to be rich in random oligonucleotides to serve as precursors for the expanding population of RNA replicases. But even if the first replicating system were composed not of RNA but of some other substance (protein, carbohydrate, or even inorganic crystals), the replicating molecules would eventually begin to exhaust the local precursor supply. This would be true whether the molecules were replicating and evolving on a lake bottom, in a tidal pool, or at the edge of a geothermal vent on the ocean floor.

Initially, nutrient limitation would simply decrease the replication rate, and the "starving" population of replicating molecules would encounter scarce precursor molecules less and less often. But unless nutrients could be supplied to the expanding population faster than the replicating molecules were naturally destroyed (e.g., by spontaneous hydrolysis), the population would eventually decline. Some replicating molecules might escape from the starving population, perhaps by diffusion or by a physical catastrophe such as a mud slide, flood, or volcanic eruption. But even if such emigrant molecules colonized a new niche where molecular precursors were still abundant, the problem of nutrient limitation would inevitably recur.

The only real solution to the problem of nutrient starvation would be the evolution of a subpopulation of RNA molecules capable of accelerating (catalyzing) synthesis of those precursors in shortest supply. For example, it is possible that both D-ribose and the free nucleic acid bases were in relatively abundant supply, but that formation of the glycosidic bonds to generate nucleosides required catalysis to keep pace with the expanding population of replicating RNA molecules. Thus, the most ancient enzymes of **intermediary metabolism** may have evolved to accelerate the *slowest* steps in the prebiotic synthetic pathways. (Intermediary metabolism consists of the synthesis of biological molecules, called **anabolism**, and the breakdown of these molecules, called **catabolism**.)

Later, as living systems became more complex and made still greater demands on the nutrient supply, additional steps in the hybrid prebiotic/biotic synthetic pathways would become rate limiting. Carried to its logical extreme, this point of view predicts that modern biosynthetic pathways are the direct descendants of prebiotic pathways. In fact, many experts in molecular evolution have attempted to deduce the nature of prebiotic synthetic pathways from the enzymatic pathways responsible for intermediary metabolism today.

Nucleotide Cofactors May Be Metabolic Fossils of the RNA World[23]

The only proteins in the primitive RNA world were random polypeptides assembled by prebiotic condensation of amino acids (see Figure 28-3). Admittedly, some of these random polymers might have possessed rudimentary catalytic abilities. For example, a positively charged polymer like polylysine could interact with negatively charged oligonucleotides to facilitate their polymerization by an RNA enzyme. But in the absence of a working translation system (see Figures 28-17 and 28-18), there would be no way to inherit the information encoding such a potentially useful polypeptide. As a result, the polypeptide could not be synthesized repeatedly and could not be improved upon by selecting for more efficient catalysis. Eventually, the expanding population of replicating RNA molecules would outstrip the availability of those few random polypeptides that were catalytically useful, and the population of RNA molecules would starve and decline.

There was only one way that the enzymes of intermediary metabolism could have replicated and evolved, and thereby kept pace with the demands of a growing population of RNA replicases: The first enzymes of intermediary metabolism had to be made of RNA and had to be replicated by the very RNA polymerase population whose replication they would in turn accelerate. In the resulting *mixed* population of RNA molecules, the RNA enzymes of intermediary metabolism would be selected not only for catalytic efficiency, but also for efficient replication.

The RNA enzymes we have discussed so far (RNase P and the *Tetrahymena* ribosomal intron) use RNA as the substrate. What reasons do we have for supposing that an RNA enzyme could catalyze steps in intermediary metabolism using a substrate other than RNA? First, RNAs contain an abundance of potential catalytic groups and are capable of folding into complex tertiary structures (see Figures 28-6, 28-8, and 28-13). Thus, there is no intrinsic reason why an RNA enzyme could not be nearly as versatile as a protein enzyme. Second, the original enzymes of intermediary metabolism might have evolved most readily from the RNA replicase itself. For example, if the nucleotide binding site of the RNA replicase was able to recognize the cyclic form of the ribose ring, a variant RNA replicase might have catalyzed conversion of the linear form of ribose to the cyclic form (see Figure 28-3) simply by holding the linear molecule in proper conformation (see Figure 2-8). Later, as living systems evolved and made more complex demands on intermediary metabolism, the number of distinct RNA enzymes would inevitably increase.

Is there any evidence that the first enzymes of intermediary metabolism were actually made of RNA? Meticulous characterization of metabolic pathways by the pioneers of experimental biochemistry during the late nineteenth and early twentieth century led to the realization that the catalytic abilities of many proteins are augmented by **cofactors** bound either covalently or noncovalently to the active site. While protein enzymes may have been the first to take advantage of the catalytic versatility of cofactors, the intriguing possibility exists that RNA enzymes had already adopted this strategy in the RNA world. A remarkable number of such enzyme cofactors are closely related to the nucleotides (Figure 28-14). Although other interpretations are possible, one attractive hypothesis is that **nucleotide cofactors** are metabolic fossils of the enzymes of intermediary metabolism in the RNA world. This would explain why the enzymatically *active* end of the nucleotide cofactor generally lies opposite an enzymatically *inactive* adenylate or adenosyl group: Perhaps the adenylate or adenosyl group served as a convenient "handle" for an RNA enzyme to hold onto a nucleotide cofactor, much as the structure of the *Tetrahymena* ribosomal intron creates a specific binding site for a guanine nucleoside (see Figure 28-9). Later in evolution, the development of a working translation system would have allowed more efficient protein catalysts to assume the functions of most RNA enzymes. An evolving protein enzyme would then have retained the cofactor as a key part of its active site, but would have replaced the rest of the ancestral RNA (or RNP) enzyme with a polypeptide chain (Figure 28-15).

RNA Genes May Have Been Assembled from Smaller Pieces

Consider a living system composed of an RNA replicase and auxiliary RNA species that contribute to the efficient replication of the replicase and are in turn replicated by it. Initially, these auxiliary RNAs (which arose by random polymerization of nucleotides) would be extremely simple, each having only a single rudimentary function. But later, as the living system increased in complexity, the pressure for more efficient catalysis would select for complex auxiliary RNAs, capable of performing two or more functions in a coordinated fashion.

Figure 28-14

The nucleic acid bases serve as structural components of some coenzymes. NAD$^+$/NADH and FAD/FADH$_2$ mediate oxidation-reduction reactions, CoA carries acyl groups, and coenzyme B$_{12}$ transfers hydride ions. Although the AMP residues in NAD$^+$, FAD, and CoA are not directly involved in catalysis, the 5'-deoxyadenosyl residue of coenzyme B$_{12}$ is covalently bound to the central cobalt atom and participates directly in hydride ion transfer. Note also the unusual base 5,6-dimethylbenzimidazole in coenzyme B$_{12}$, as well as the structural resemblance between the nucleic acid bases and the catalytically active nicotinamide (NAD$^+$) and isoalloxazine (FAD) rings. The nucleic acid components of the coenzymes are shown in color.

Nicotinamide adenine dinucleotide (oxidized form, NAD$^+$)

Flavin adenine dinucleotide (oxidized form, FAD)

Acyl-coenzyme A (acyl-CoA)

Coenzyme B$_{12}$ (5′-deoxyadenosylcobalamin)

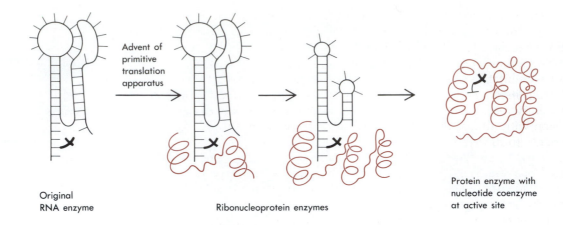

Advent of primitive translation apparatus

Original RNA enzyme

Ribonucleoprotein enzymes

Protein enzyme with nucleotide coenzyme at active site

Figure 28-15

A model for the evolution of coenzymes from RNA enzymes. At first, the cofactor was base-paired to the RNA enzyme through its adenosyl group. Later, after a working translation system had evolved, catalysis by the original RNA enzyme was enhanced by interaction with small polypeptides. As the translation system became more reliable, larger polypeptides gradually took over all the functions of the RNA except for the critical role of the cofactor in catalysis. Ultimately, only the coenzyme remained at the active site of a protein enzyme. [Adapted from H. B. White, III, in *The Pyridine Nucleotide Coenzymes* (1982):1–17. Academic Press, New York.]

The most obvious way to create a multifunctional RNA would be to fuse two preexisting auxiliary RNAs (say, a pyrimidine binding RNA and a ribose binding RNA) into a single RNA chain with a new activity (perhaps the ability to catalyze formation of the glycosyl bond between ribose and the base). The only alternative to such a simple **gene fusion** would be to reinvent both preexisting activities at the same time as part of a single RNA chain, a truly remote possibility.

How could two RNA chains be fused into a single bifunctional molecule in the primitive RNA world? One possibility is that this reaction occurred by the same prebiotic pathways that synthesized the first oligonucleotides. Another possibility is that such fusions might have been catalyzed by the RNA replicase itself. In the model for the first RNA replicase (see Figure 28-12), as well as in actual experiments showing that the intron core functions as a poly C polymerase (see Figure 28-11), polymerization proceeds by removing mononucleotides from the 3' end of one molecule and adding them to the 3' end of another. The replicase in the model therefore functions as an *RNA rearrangease* or *recombinase*, and thus it might have joined and rearranged *poly*nucleotides as well as mononucleotides. This possibility could actually be tested by experiments with the *Tetrahymena* intron core.

The idea that complex RNAs could have been assembled from separate preexisting structural or functional domains of RNA is completely analogous to the idea that split genes encoding proteins can accelerate evolution in the DNA world: Many examples are known in which exons encoding distinct structural or functional protein domains have been joined into a composite gene by recombination at the DNA level between the adjacent introns (see Figures 20-20 through 20-23, Figure 28-22, and Table 28-2).

Compartmentation in Biological Systems[25]

No living system could continue to replicate and evolve unless all the enzymes and precursors were maintained at a high concentration and protected from destructive forces. Otherwise, the replicating molecules would diffuse away from each other or be chemically degraded faster than they could accumulate. For example, a mixed population of interdependent RNA molecules in a lakeside puddle might be hopelessly diluted by a torrential rainstorm or chemically destroyed by repeated evaporation to total dryness in the heat of a blazing sun.

Arguments such as these suggest that life must have evolved in a protected environment, but the nature of that environment is a matter for pure speculation. All we can say is that at some very early point in the evolution of life (perhaps even before the first RNA replicase), **compartmentation** was necessary to separate the living system ("inside") from its environment ("outside"). This compartment had to be sufficiently resilient to withstand abuse, sufficiently flexible to allow division without loss of contents, and sufficiently permeable to allow nutrients to diffuse in while wastes diffused out.

Today, all living systems are compartmentalized by membranes built of lipid bilayers (see Figures 2-26 through 2-31). Lipid bilayers separate the inside of every cell from the external environment and also define the internal compartments of eucaryotic cells (nucleus, endoplasmic reticulum, mitochondrion, chloroplast, etc.). Unlike most other biologically important macromolecules (nucleic acids, proteins, and carbohydrates), lipid bilayers assemble *spontaneously* from mixtures of lipids in aqueous solution when the lipid concentration exceeds a certain critical value. Below this concentration, the individual lipid molecules are soluble in the aqueous phase. At higher concentrations, the lipid molecules aggregate with one another to form huge sheets. More importantly, fragments of the sheets spontaneously break off and reseal themselves as closed balloon-like **vesicles** (Figure 28-16; see also Figure 2-31).

Lipid vesicles can fuse with each other to form larger vesicles, or they can divide irregularly by pinching off daughter vesicles. These spontaneous transformations of a lipid bilayer bear a striking resemblance to the budding of a yeast cell (see Figure 18-2), the movement of vesicles through the Golgi apparatus (see Figures 18-9 and 18-16), the budding of enveloped viruses from eucaryotic cells (see Figure 24-2), and the release of secretory granules after fusion of the granule membrane with the plasma membrane (see Figure 18-16). The implication is that living systems have devised mechanisms for stabilizing, accelerating, and directing the natural transformations of lipid sheets.

The spontaneous formation of closed lipid vesicles suggests that lipidlike molecules may have played a crucial role in providing a protected environment for primitive living systems. The budding of lipid vesicles also suggests a crude mechanism for "cell" division: Primitive living systems, each surrounded by a membrane bilayer, would simply divide unevenly whenever the vesicle became too large to withstand the hydrodynamic forces acting upon it. Similarly, vesicle fusion could have served as a primitive form of mating. If we suppose that useful genes arose in two different (and possibly competing) living systems, each enclosed in a separate lipid vesicle, vesicle fusion could bring the two genes together into a single living system. This would accelerate evolution by ensuring that any useful RNA sequence was made available to the entire "breeding" population of vesicles.

One problem in thinking of the first "cell" as RNA encapsulated in a lipid vesicle is that both RNA and *phospho*lipids are negatively charged (see Figure 28-16). Repulsion between these negative charges might then be expected to make encapsulation a very inefficient process. Recent experiments, however, show that the addition of positively charged (basic) proteins increases the efficiency of encapsulation by reducing the repulsion between negative charges. This suggests that prebiotically synthesized basic peptides may have facilitated the encapsulation of the first living systems in lipid

(a) The simplest phospholipids:

Ester-linked fatty acyl chain — L-glycerol — Phosphate

Ether-linked branched aliphatic chain — D-glycerol — Phosphate

(b)

Lipids are soluble below critical concentration

Increased concentration / Decreased concentration

Lipids aggregate into bilayers above critical concentration

Flexible lipid bilayer can pucker

Spontaneous budding / Spontaneous fusion

Phospholipid vesicle

+

Bilayer

Figure 28-16
Many lipids spontaneously form bilayers that are relatively impermeable to water and are capable of fusing and budding. (a) All phospholipids have a hydrophobic aliphatic chain and a hydrophilic glycerophosphate group. (b) Bilayers form spontaneously because the hydrophobic fatty acyl chains tend to associate with each other, while the hydrophilic phosphate groups remain on the outside of the bilayer because they prefer to interact with water. In archaebacteria, the phospholipids are ether-linked to D-glycerophosphate rather than ester-linked to L-glycerophosphate as in eubacteria and eucaryotes (see Table 28-5).

vesicles. Alternatively, the most primitive lipids may have lacked phosphate head groups.

Although lipid membranes are a particularly attractive means for effecting compartmentation, there are other possibilities for the most primitive (and therefore least demanding) living systems. For example, many polymers and even some minerals can form porous gels with the consistency of pudding or gelatin. Such **coacervates** might have retained the larger functional RNA molecules while allowing free diffusion of smaller molecules such as precursors and wastes.

Unfortunately, the earliest forms of compartmentation could not have been preserved as microfossils (see Figure 28-1). However, the synthesis of ester-linked lipids has been achieved under simulated prebiotic conditions, and today all living organisms use functionally similar ester-linked or ether-linked lipids (Table 28-5). Taken together, the evidence is strong that lipid bilayers are molecular fossils of the original form of compartmentation in biological systems. Perhaps all living systems, from the first RNA replicase to the most sophisticated neuron in the human brain, have been surrounded by lipid bilayers.

Table 28-2 Many Animal Proteins Have Sequence Segments in Common*

EGF-type segment
  Epidermal growth factor (EGF)
  Tumor growth factors
  Low density lipoprotein receptor
  Blood clotting factor IX
  Blood clotting factor X
  Protein C
  Tissue plasminogen activator
  Urokinase
  Complement component C9
  Notch gene product (fruit fly)
  lin-12 gene product (nematode)
C9-type segment
  Complement component C9
  Low density lipoprotein receptor
  (Fibronectin?)
  Notch gene product (fruit fly)
  lin-12 gene product (nematode)
Fibronectin "finger"
  Fibronectin
  Tissue plasminogen activator
Proprotease "kringle"
  Plasminogen
  Tissue plasminogen activator
  Urokinase
  Prothrombin

*The terms "finger" and "kringle" refer to specific disulfide-linked modular units of protein construction. Note that some proteins are a mosaic of at least three different modular units found in other proteins.
SOURCE: R. F. Doolittle, D. F. Feng, M. S. Johnson, and M. A. McClure, *Cold Spring Harbor Symp. Quant. Biol.* 51 (1986):447–455.

THE RNP (RIBONUCLEOPROTEIN) WORLD

Translation Evolved in an RNA World

Because proteins are more versatile catalysts than RNA, intense selective pressure forced early living systems to devise an efficient mechanism for the mass production of useful proteins. And since a living system would be at a serious disadvantage if it could not pass on the "accomplishments" of one generation to the next, selective pressure likewise ensured that the production of such proteins would be *heritable*. However, only nucleic acids can store and replicate the genetic information specifying proteins; the proteins themselves cannot (see the section entitled The First "Living" Molecule Was Almost Certainly a Nucleic Acid). Thus, very early in evolution, the appropriate molecular machinery arose to replicate RNA molecules encoding useful proteins and to translate the RNA code into the correct polypeptide.

The Modern Ribosome Is Very Complicated[26-31]

Ever since the realization in 1956 that DNA makes RNA makes protein (the central dogma; Chapter 3), two great questions have dominated all discussions of molecular evolution: What was the nature of the first replicating system, and how did the first primitive ribosome translate RNA sequences into protein? The recent discovery and characterization of enzymatically active RNAs quickly persuaded most molecular biologists that the first replicating systems were made of RNA. Can we use our picture of the RNA world and our knowledge of contemporary protein synthesis to make reasonable (and perhaps even testable) proposals about the origins of translation and the genetic code?

Today, translation is carried out by the ribosome, a supremely sophisticated molecular machine that contains about 4500 bases of RNA and 50 different proteins (see Figures 14-13 through 14-16). Such sophisticated machinery did not arise all at once, but developed over perhaps a billion years as evolution tinkered one step at a time with the basic design of the original ribosome. Thus, the crude translation machinery that gave birth to the RNP world, a world made of both RNA and protein, must have been radically simpler than the ribosome of today. How, then, can we know whether *any* parts of the modern ribosome are remnants (molecular fossils) of the original translation system, or whether the original molecular machine has been rendered unrecognizable by subsequent tinkering?

In our discussions of the primitive ribosome, we will make three simplifying assumptions.

First, we assume that the modern translation apparatus is a vastly improved version of the original apparatus, but does not differ from it fundamentally. We make this assumption because molecular evolution is generally *conservative*, modifying and adapting old pieces of molecular machinery whenever possible, instead of devising totally new ones.

Second, we assume that the original translation system was very crude and could not discriminate reliably between chemically similar amino acids such as lysine and arginine, or leucine, isoleucine, and valine (see Figure 2-16). Thus, translation of any particular template RNA (mRNA) would have produced a mixture of chemically similar but nonidentical polypeptide chains of varying length. The ability to start and stop translation precisely and to distinguish as many as 20 different amino acids would then arise step by step as the ribosome improved.

Third, and most importantly, we assume that the first tRNA synthetase and possibly the first ribosome were derived from the RNA replicase or from one of the RNA enzymes that depended on it for replication. This assumption follows logically from our first assumption that molecular evolution is a conservative process in which existing molecules are adapted to ever more sophisticated functions. The first RNA replicase must necessarily have been crude (see Figures 28-4 and 28-7), but the descendants of this RNA enzyme would have increased in complexity and diversity as populations of RNA molecules emerged that could replicate more rapidly. Thus, it seems likely that the original apparatus for polymerization of amino acids would have been a *variant* of the most sophisticated molecular machinery then available, rather than an altogether new invention.

Assuming that the mechanism of modern protein synthesis is a molecular fossil that still resembles the original translation apparatus, we should be able to divide our speculations about the origin of protein synthesis into two separate questions: How were the primitive tRNAs specifically charged with the correct (cognate) amino acid, and how were these charged tRNAs aligned prior to peptide bond formation?

The Adaptor Hypothesis Revisited

In 1955, Francis Crick had a crucial insight into how the translation apparatus aligns specific amino acids on the template RNA molecule (mRNA) before polymerization. The simplest model was that the mRNA folded up to create cavities on its outer surface, each cavity specific for a particular amino acid; the amino acids would then polymerize spontaneously, if they were already activated, or be polymerized by an additional enzymatic activity. Crick pointed out that it is relatively difficult to imagine how the hydrophilic bases in RNA could form hydrophobic pockets capable of interacting specifically with the aliphatic and aromatic side chains of amino acids such as leucine, isoleucine, valine, phenylalanine, and tyrosine (see Figure 2-16). Crick therefore argued that a class of *bifunctional* "adaptor" molecules must exist that would recognize *both* the mRNA sequence *and* the corresponding amino acid side chains (see the section in Chapter 3 entitled The Adaptor Hypothesis of Crick).

The discovery of the role that tRNAs play in protein synthesis dramatically validated Crick's adaptor hypothesis, but the nature of these RNA adaptor molecules was more complex than anticipated. Transfer RNAs are indeed bifunctional: The 3'-hydroxyl group of the acceptor stem is specifically coupled to an amino acid (see Figure 14-6), while the anticodon recognizes mRNA sequences. However, a modern tRNA is not a simple adaptor that can function by itself: The tRNA is unable to recognize the appropriate amino acid side chain directly and therefore requires a second adaptor molecule (a specific aminoacyl-tRNA synthetase) that can recognize both the tRNA and the appropriate amino acid (see Figure 14-9).

A Model for How a Charged Ribozyme Might Have Specifically Activated an Amino Acid[31]

Modern aminoacyl-tRNA synthetases are protein enyzmes, but at the dawn of the RNP world, a tRNA synthetase could only have been made of RNA. Since the RNA replicase may well have been the most sophisticated enzyme at that time, let us examine the possibility that a variant RNA replicase might have served as the first tRNA synthetase (Figure 28-17).

We imagine the same series of reactions as in RNA polymerization by the RNA replicase (see Figure 28-12), but hypothesize that a particular amino acid has a fortuitous affinity for the active site of the charged enzyme. The 3' mononucleotide (NMP) of the charged enzyme could then be discharged by attack of the amino acid carboxyl group instead of the 3'-hydroxyl group of a polynucleotide (see Figure 28-19). This would generate an *activated* amino acid~NMP intermediate (AA~NMP), where the high energy of the broken phospho-

Figure 28-17

A variant RNA replicase may have given rise to the first tRNA synthetase. Since positively charged amino acids would be able to interact most easily with a variant RNA polymerase (see text), the hypothetical origin of lysine tRNA synthetase is shown. Later, as proteins gradually took over the function of most RNAs in the RNP world (see Figures 28-14 and 28-15), only the activated AA~AMP charging intermediate would remain as a molecular fossil of the original tRNA synthetase. [Courtesy of A. M. Weiner and N. Maizels.]

ester bond is conserved in the aminoacyl bond to the nucleotide phosphate. If the activated AA~NMP intermediate then remains bound to the enzyme, the resulting high-energy aminoacyl bond could be attacked by the 3'-hydroxyl group of yet another ribozyme with tRNA-like features. In order to promote stable binding of the tRNA-like ribozyme to the replicase, an internal guide sequence (3' UGG 5') would base pair with the 3' terminal CCA of the hypothetical tRNA (Figure 28-17). This series of reactions would charge the 3' end of the primitive tRNA with an amino acid (AA~RNA), using essentially the same sequence of reactions as are found in protein synthesis today (see Figure 14-9).

Prebiotically Synthesized Oligonucleotides as an Energy Source for Peptide Bond Formation[20, 21]

We assumed in this model for a primitive tRNA synthetase (see Figure 28-17), as we did in the model for the first RNA replicase (see Figure 28-12), that random prebiotically synthesized oligonucleotides served as the energy source for early biological polymerizations. But is this assumption energetically reasonable for protein synthesis? Does the phosphoester bond of the charged synthetase actually have sufficient energy to form the postulated high-energy AA~NMP charging intermediate? A simple calculation suggests that it does. Today, formation of the AA~AMP intermediate is driven by hydrolysis of a high-energy phosphoester bond in ATP (Equation 6-6):

$$\text{ATP} \longrightarrow \text{AMP} + \text{P}\sim\text{P} \qquad (\Delta G = -8 \text{ kcal/mol})$$

We have also seen that polymerization of nucleic acids from the nucleotide triphosphate precursors does not involve a large change in free energy (Equation 6-18):

$$\text{RNA } (n \text{ bases}) + \text{NTP} \longrightarrow$$
$$\text{RNA } (n + 1 \text{ bases}) + \text{P}\sim\text{P} \qquad (\Delta G = +0.5 \text{ kcal/mol})$$

Subtracting Equation 6-18 from Equation 6-6, we find that

$$\text{RNA } (n + 1 \text{ bases}) \longrightarrow$$
$$\text{RNA } (n \text{ bases}) + \text{NMP} \qquad (\Delta G = -8.5 \text{ kcal/mol})$$

This shows that the phosphoester bond of the charged synthetase contains at least as much energy as the high-energy bond in ATP that is hydrolyzed to form the AA~AMP intermediate today.

Protein synthesis is thus driven forward by *coupling* the synthesis of the low-energy peptide bond ($\Delta G = +0.5$ kcal/mol) to hydrolysis of a high-energy phosphoester bond ($\Delta G =$ about -8.5 kcal/mol). Although the ΔG for formation of the enzyme-bound AA~NMP intermediate may be positive, the large overall ΔG of about -8.0 kcal/mol favors protein synthesis (see Figure 6-3).

The First tRNA Synthetases May Have Activated Basic Amino Acids[31]

The very complexity of modern protein synthesis suggests that the first crude protein synthesizing apparatus would only have been able to synthesize short, irregular peptides. Since positively charged amino acids might preferentially bind to the negatively charged sugar-phosphate backbone of the tRNA synthetase, the first charging enzymes may have acted on basic amino acids like lysine, arginine, and histidine to produce basic AA ~ tRNAs. And since aminoacyl~RNA bonds are highly reactive, the aminoacyl~RNA bond of one such molecule could have been attacked by the amino terminal group of another aminoacyl~RNA. (As discussed later, even modern peptide bond formation may occur by such an "uncatalyzed" or "spontaneous" reaction.) The dipeptidyl~RNA resulting from spontaneous reaction of two aminoacyl~RNAs retains an activated aminoacyl bond, and so it could in turn be attacked by a third aminoacyl~RNA to generate a tripeptidyl~RNA, and so forth. The result would be random polymerization of basic amino acids into short positively charged peptides.

At the dawn of the RNP world, short basic peptides might have been useful in at least two ways. First, positively charged peptides could have neutralized the repulsion between negatively charged RNA chains, thereby stabilizing RNA enzymes or allowing them to bind other substrate RNA molecules more tightly. In this role, the basic peptides would resemble modern protamines, which help to pack DNA tightly within sperm heads by neutralizing the negative charges of the phosphate backbone and by forming crossbridges between adjacent DNA strands (Figure 21-7). Second, supposing the first lipids were *phospho*lipids, short basic peptides might also have stabilized lipid vesicles by interacting with the negatively charged phosphate head groups (Figure 28-16).

If indeed the ability to synthesize random basic peptides in an mRNA-independent and ribosome-independent reaction was sufficient to confer a selective advantage on the primitive living system, tRNA synthetases might have been the first components of the translation apparatus to arise, followed soon by the appearance of more sophisticated tRNAs specifically designed for protein synthesis. This would provide the driving force for the next great step in the evolution of protein synthesis: the invention of a ribosome with the ability to align the charged tRNAs on an mRNA, and to translocate the mRNA after each round of peptide bond formation.

The Elusive Peptidyl Transferase[26, 32]

What clues can the modern ribosome provide regarding the first template-dependent mechanism for polymerization of activated aminoacyl~tRNAs? In the modern bacterial ribosome, the 30S subunit binds the mRNA, while the 30S and 50S subunits contribute jointly to forming two tRNA binding sites on the surface of the 70S ribosome (Chapter 14). Evidently, the "fit" between the mRNA and the tRNAs on the ribosomal surface is so tight that only a tRNA with the correct anticodon can bind efficiently to the mRNA on the ribosome. This suggests that the task of the original ribosome might have been as simple as providing correctly oriented binding sites for the mRNA and the charged tRNA. Such binding sites would form a rigid "workbench" to accomplish the two essential functions of a primitive ribosome: to guarantee that the mRNA would base-pair with the tRNA anticodon rather than with some other part of the tRNA molecule, and to stabilize the transient base-pairing interaction between the charged tRNAs and the mRNA so that peptide bond formation could occur.

What is the mechanism of peptide bond formation? There have been numerous attempts to determine whether purified ribosomal proteins can catalyze polymerization of activated amino acids (AA~tRNAs), yet no single molecule has been identified as having peptidyl transferase activity. Of course, the explanation could be that the activity resides in several ribosomal proteins collectively, rather than in a single protein. But there are two other very real possibilities:

- The peptidyl transferase might be an RNA enzyme instead of a protein. This would imply that the large ribosomal RNA does not serve simply as a passive scaffold for ribosomal proteins, but actually catalyzes peptide bond formation. This hypothesis would help to rationalize the extraordinary conservation of ribosomal RNA structure and sequence (see Figure 14-15).

- The other possibility is that peptide bond formation is in fact "spontaneous." This would imply that the entire ribosome serves as a peptidyl transferase by aligning the activated AA~tRNA and the peptidyl~tRNA on the template mRNA, and not by chemically catalyzing the making and breaking of covalent bonds as is the case for most biological catalysts (see Figure 2-25).

The First Ribosomes Synthesized Short Peptides

The first primitive ribosomes (protoribosomes) probably arose to accelerate spontaneous peptide bond formation, perhaps by simply aligning two charged tRNAs and perhaps also providing peptidyl transferase activity. The protoribosome might have been an entirely new RNA enzyme, or it might have been derived from the tRNA synthetase itself by duplication of the tRNA binding site (Figure 28-18). (Duplication could be caused by a replication error, or by recombination between RNA molecules as discussed in the section entitled RNA Genes May Have Been Assembled from Smaller Pieces.) Originally, the pair of tRNA binding sites on the protoribosome might have interacted solely with the conserved CCA ends of the charged tRNAs; however, selective pressure for more stable and accurate alignment of the charged tRNAs would have led to additional interactions between the tRNAs and their protoribosome binding sites. The region on the tRNA that participated in these additional interactions would in effect become the anticodon, and the corresponding region of the protoribosome would function as a built-in mRNA (see Figure 28-18).

Coevolution of tRNA Synthetases and Peptide-Specific Ribosomes May Have Defined the Rudiments of the Genetic Code[31,33]

As long as there was only one kind of tRNA synthetase specific for a particular basic amino acid (say, lysine) the protoribosome would simply have made short lysine peptides. But eventually, the selective advantage of being able to make other useful peptides would have forced the first tRNA synthetases, the first primitive tRNAs, and the protoribosomes to diversify. For example, if a variant tRNA

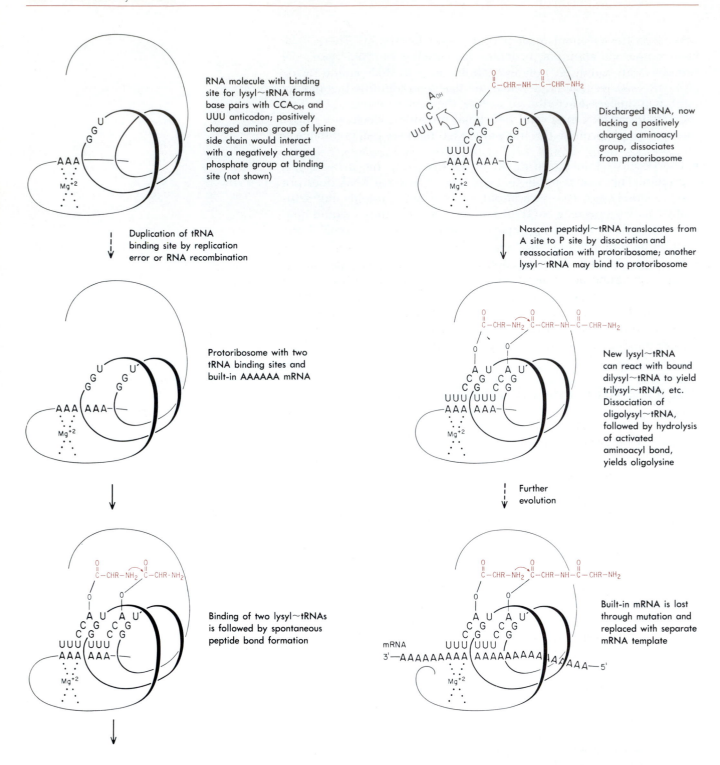

RNA molecule with binding site for lysyl~tRNA forms base pairs with CCA$_{OH}$ and UUU anticodon; positively charged amino group of lysine side chain would interact with a negatively charged phosphate group at binding site (not shown)

Duplication of tRNA binding site by replication error or RNA recombination

Protoribosome with two tRNA binding sites and built-in AAAAAA mRNA

Binding of two lysyl~tRNAs is followed by spontaneous peptide bond formation

Discharged tRNA, now lacking a positively charged aminoacyl group, dissociates from protoribosome

Nascent peptidyl~tRNA translocates from A site to P site by dissociation and reassociation with protoribosome; another lysyl~tRNA may bind to protoribosome

New lysyl~tRNA can react with bound dilysyl~tRNA to yield trilysyl~tRNA, etc. Dissociation of oligolysyl~tRNA, followed by hydrolysis of activated aminoacyl bond, yields oligolysine

Further evolution

Built-in mRNA is lost through mutation and replaced with separate mRNA template

Figure 28-18

The first ribosomes (protoribosomes) may have been peptide-specific. For simplicity, the primitive tRNAs are shown as hexanucleotides with the anticodon immediately adjacent to the 3' terminal CCA$_{OH}$; the actual tRNAs would have been more complicated. Since the first tRNA synthetases are likely to have been specific for basic amino acids like lysine (Figure 28-17), a lysine-specific protoribosome is illustrated here. [Courtesy of A. M. Weiner and N. Maizels.]

synthetase arose that could activate arginine better than lysine, selective pressure would ensure that the primordial tRNA structure evolved into two slightly different tRNA species. The lysine- and arginine-specific protoribosomes would distinguish between these two tRNAs based on the ability of the anticodons to form base pairs with their built-in mRNAs (Figure 28-18); the two tRNA synthetases might recognize some more general feature of the primitive tRNA structures. The result would be two different peptide-specific protoribosomes, each with a different internal guide sequence serving as a "built-in" mRNA, and each with a corresponding tRNA synthetase. In this way, the rudiments of the genetic code would be defined as the population of tRNA synthetases and peptide-specific protoribosomes coevolved.

Although the rudimentary genetic code may have begun as a "frozen accident," reflecting the fortuitous affinity of RNA enzymes for particular amino acids and oligonucleotides (primitive tRNAs), the evolving code must soon have become subject to many different kinds of selection. Examples of such fine tuning today are the use of related codons for chemically related amino acids, and degeneracy in the third position of the code, as ways to minimize the potentially harmful effects of transcriptional and translational errors (see Chapter 15).

The Unexpected Diversity of Modern tRNA Synthetases[34, 35]

All modern tRNA synthetases are made exclusively of protein, and they all catalyze the same reaction (aminoacylation) using structurally very similar substrates (tRNAs). Before any of these tRNA synthetases had been characterized at the molecular level, the common expectation was that they would turn out to be structurally similar variants of a common ancestral protein—the first protein that was able to aminoacylate a tRNA-like species. Thus it was extremely puzzling to find that some tRNA synthetases are large, and some are small; some are monomers, and some are dimers; and no extensive amino acid homology among them is detectable.

We have seen that all tRNA synthetases may indeed share a common ancestor, but that this ancestor was probably an RNA enzyme (the RNA replicase) rather than a protein. As proteins gradually replaced most RNA enzymes during evolution of the RNP world (Figure 28-15), the stepwise conversion of each tRNA synthetase (made of RNA) into a tRNA synthetase (made of protein) would have had to take place *independently*. Assuming that there are many different ways of converting an RNA enzyme to a protein enzyme, this would account for the puzzling diversity of modern tRNA synthetases.

Loss of the Internal Guide Sequence May Have Made mRNA-Dependent Translation Possible

The dependence of peptide-specific protoribosomes on built-in mRNAs would have severely limited the variety of peptides that could be made. This limitation would then have provided the driving force for the evolution of a true ribosome capable of using a separate mRNA molecule as the template for protein synthesis. In principle, replacement of the built-in mRNA of a peptide-specific protoribosome with a separate mRNA template is completely analogous to transformation of the *Tetrahymena* intron into a template-dependent RNA replicase by replacement of the internal guide sequence with an external guide sequence (Figure 28-13). Assuming that the negatively charged backbone of the built-in mRNA was held in place by interaction with bound metals ions on the surface of the peptide-specific protoribosome, simple deletion of the built-in mRNA region may have sufficed to allow binding of a separate mRNA species.

The Ancestral Ribosome Remains a Puzzle

The simple model presented here for the origin of protein synthesis (see Figures 28-17 and 28-18) has two main strengths: It demonstrates that a primitive RNA enzyme could in principle accomplish activation and attachment of an amino acid to a primitive tRNA-like molecule. It furthermore shows how selection for the synthesis of small peptides on a very primitive ribosome (protoribosome) may have played a major role in defining the rudiments of the genetic code and in initiating the stepwise evolution of the modern protein synthetic apparatus. On the other hand, the model does not even attempt to address the nature of the first true ribosome that was capable of translocating the mRNA as well as initiating and terminating protein synthesis.

The best current hope for reconstructing the ancestral ribosome is to try to understand the three-dimensional structures of both the small and large subunits of the modern procaryotic ribosome (see the section entitled The Modern Ribosome Is Very Complicated). This ambitious goal, unthinkable only a few years ago, may now be within reach. The approximate positions of all 21 proteins in the small 30S subunit of the *E. coli* ribosome are already known (see Figure 14-39), and these data are currently being refined. In addition, techniques developed in studying the small subunit are now being applied to the more difficult task of locating the 34 proteins in the 50S subunit. Even more promising is the very recent crystallization of the 50S ribosomes from a salt-tolerant bacterium. The resulting crystals yield precise X-ray diffraction data at a resolution of 6 Å leading to the expectation that perhaps within the next two decades the 50S subunit of the ribosome (and conceivably the 30S subunit) will be understood at the atomic level. By learning how the modern ribosome performs the task of protein synthesis, we may eventually be able to identify molecular fossils of the ancestral translation machine.

Nuclear mRNA Splicing May Have Evolved from Self-Splicing Introns[36, 37, 38]

Unlike the self-splicing of the *Tetrahymena* rRNA intron, the splicing of nuclear mRNA introns requires the U1, U2, U4, U5, and U6 small nuclear ribonucleoprotein particles, or snRNPs (see Figure 20-18) and proceeds by a lariat mechanism (see Figures 20-19 and 28-19). The presence of an snRNA component in each of the five different snRNPs required for nuclear mRNA splicing suggests that introns of this type first appeared in the RNP world. But how could such an

Intron type:	Group I	Group II	Nuclear mRNA
Catalysis:	Self-Splicing	Self-Splicing	pre-mRNA + snRNP

Mechanism:

Figure 28-19
Nuclear mRNA splicing may have evolved from a self-splicing Group I or Group II intron, since the three mechanisms have underlying similarities. Straight lines represent exon sequences, wavy lines intron sequences, and a hollow circle the site of ligation between the 5' and 3' exons. When the sequence of transesterification reactions is initiated by the 3'-hydroxyl group of a free mononucleotide, the excised intron will be linear (Group I introns). When the reaction is initiated by the 2'-hydroxyl group of an internal nucleotide within the intron, the excised intron will be a lariat (Group II and nuclear mRNA introns). In the course of evolution, nuclear mRNA introns appear to have lost some of the original *cis*-acting functional elements of the autocatalytic Group I and Group II introns; the lost functions may now be performed by the snRNA components of snRNPs acting in *trans*. [After T. R. Cech, *Cell* 44 (1986):207.]

intron arise? The recent discovery of a new class of self-splicing introns in fungal mitochondria suggests an answer (Figure 28-20). These Group II mitochondrial introns resemble nuclear mRNA introns in several ways. Both are excised by a lariat mechanism, and the intron boundary sequences are similar (see Table 28-1). The implication is that nuclear mRNA introns may have descended from Group II self-splicing introns.

In fact, the splicing mechanisms of the *Tetrahymena* rRNA intron (Group I), the fungal mitochondrial introns (Group II), and nuclear mRNA introns are all very similar (see Figure 28-19). The self-splicing of Group I introns is initiated by the 3' hydroxyl of a free nucleotide (see Figures 28-8 and 28-9), but there is no theoretical reason why the splicing reaction could not be initiated by nucleophilic attack of an internal 2' hydroxyl group from the intron itself. The sequence of transesterifications would then excise the intron as a lariat molecule with the branched A nucleotide engaging in three phosphodiester bonds, precisely what is seen for the splicing of Group II and nuclear mRNA introns (see Figures 20-19 and 28-19). Thus, it appears that Group I, Group II, and nuclear mRNA introns may descend from a common ancestor. Only the origin of tRNA introns is still mysterious.

Small Nuclear Ribonucleoprotein Particles (snRNPs) as Fossils of the RNP World[39-43]

Group II mitochondrial introns are capable of self-splicing in the test tube, but nuclear mRNA splicing requires snRNPs. Thus, if nuclear mRNA introns really did evolve from self-splicing Group II introns, certain essential structural or enzymatic functions originally carried out by internal intron sequences must have been gradually lost. These functions would then have been taken over by the snRNA components of the snRNPs, thereby freeing the internal intron sequences from any functional constraints. Ultimately, only the intron boundary sequences (and in some yeasts, the intron branch point) would be left to define the exact position of the nuclear mRNA introns (see Table 28-1).

Although the role of snRNP proteins in snRNP function is currently unknown, the close relationship between Group II introns and nuclear mRNA introns implies that the snRNAs rather than the proteins are likely to be the catalytic components of the snRNPs. In fact, one interesting possibility is that the snRNAs might correspond to *pieces* of a Group II self-splicing RNA; assembly of the snRNAs with conserved mRNA intron sequences would then reconstruct the original self-splicing RNA. In any event, snRNP proteins are likely to have evolved later to "fine-tune" or accelerate catalysis by the snRNAs.

The ability of the 5' end of U1 snRNA to recognize the 5' splice site of mRNA introns through base pairing (see Figure 20-18) suggests that the 3' splice site may also have been recognized originally by a complementary snRNA. Thus, the discovery that the 3' splice site in vertebrates is now recognized by a protein alone suggests that in certain cases, protein can completely take over an RNA function. Perhaps snRNPs are still actively evolving and not frozen forever as fossils of the RNP world.

Viruses Contain Clues About the Evolution of the Cell

The origin of bacteriophages and viruses is nearly as puzzling as the origin of the cell itself, and for much the same reason: Modern viruses, like modern cells, are such complicated molecular machines that it is exceedingly difficult to imagine how they could have arisen stepwise from simpler ancestral forms. Even as small a virus as MS2 bacteriophage, with a 3569 nucleotide single-stranded RNA genome encoding only 4 gene products, regulates expression of those 4 gene products in remarkably complex ways (Figures 17-31 to 17-41). Larger double-stranded DNA viruses like T7 bacteriophage (with a 39,936 nucleotide genome encoding 55 gene products; Figures 17-9 to 17-13) or the eucaryotic Epstein-Barr virus (with a 172,282 nucleotide genome encoding over 200 proteins; Figure 24-25) almost begin to rival the host cell itself in complexity.

Complicating any attempt to explain the evolution of viruses is the variety of viral life styles (Chapter 24). Some viruses have DNA genomes, and some RNA. Viral DNA and RNA genomes can be either single-stranded or double-stranded. Viral DNA genomes exist in both linear and circular forms, but only linear forms of viral RNA genomes are known. Retroviruses have linear single-stranded RNA genomes, but replicate through a double-stranded DNA intermediate; hepadnaviruses have circular single-stranded DNA genomes, but replicate through an RNA intermediate.

The extraordinary diversity of viral life styles suggests that most families of viruses evolved independently, so the origins of T7 may not be revealing about the origins of MS2. There is, however, one especially useful question that can be asked about *all* viruses, and that is whether the virus is old or new. A virus that evolved quite recently in a sophisticated host cell (perhaps resembling a modern procaryote or eucaryote) might be expected to have a life style resembling that of its host. In fact, many double-stranded DNA viruses may be of relatively recent vintage. Alternatively, a virus that evolved long ago in a simpler (perhaps even precellular) living system might be expected to retain certain hallmarks of the more primitive living world in which it arose. This suggests the fascinating possibility that some RNA viruses may be molecular fossils of the RNA or RNP world, and that retroviruses (whose RNA genomes replicate through a DNA intermediate) and hepadnaviruses (whose DNA genomes replicate through an RNA intermediate) may be fossils of a world in transition from RNA to DNA genomes. The implication is that viruses may be a rich source of information about early evolution, if only we can learn to decipher the molecular fossil record.

Segmented Viral RNA Genomes May Be Molecular Fossils of the RNP World

The genomes of many RNA viruses, including influenza and reovirus, are segmented (Table 24-1, Figure 24-16). Since there is good reason to think that genomes in the RNA and RNP world were also segmented, these viruses may be very ancient.

The DNA genomes of modern organisms contain thousands of genes lined up one after another. By recognizing transcriptional start and stop signals in the DNA, RNA polymerase subdivides these vast tracts of genetic information into more manageable units: mRNAs encoding only one or a few proteins, and noncoding RNA species like rRNA and the RNA component of RNase P. The translational apparatus then further subdivides the genetic information in mRNAs by recognizing the translational start and stop signals that identify protein-coding regions. In the relatively primitive RNP world, where neither the transcription nor the translation apparatus may have been sufficiently advanced to recognize *internal* sequences within an RNA molecule, it might have been advantageous to encode separate proteins (and functional RNAs) on separate RNA segments. End-to-end transcription followed by end-to-end translation would then yield the desired protein products.

The disadvantage of storing genetic information in segmented form is that progeny can only survive by inheriting at least one of each genomic segment. (This is as true for a modern eucaryote with multiple DNA chromosomes as for an RNA virus with a segmented genome.) Perhaps, by studying the (currently unknown) mechanism by which a modern RNA virus packages one copy of each genomic segment in the viral capsid, we can learn how the segments of RNA genomes might have been segregated to daughter cells in the RNP world.

THE DNA WORLD

DNA Fulfills the Need for a Stable Genetic Material[44]

Eventually, living systems became so complex that RNA was no longer a suitable genetic material. The reason is that RNA is far more vulnerable to "spontaneous" hydrolysis than is DNA: The 2'-hydroxyl group on RNA has the ability to catalyze cleavage of the adjacent phosphoester bond (see Figure 28-5), and this reaction is accelerated by the extreme conditions that may have been unavoidable on the primitive Earth (e.g., high pH, high temperatures, and the presence of certain divalent cations such as Zn^{2+}). In fact, nineteenth-century biochemists prepared DNA by heating cells in alkali under conditions where DNA survives but RNA is quickly reduced to oligonucleotides. Spontaneous hydrolysis therefore severely limited the complexity of RNA-based living systems.

The relative instability of RNA suggests that no organism as complex or at least as fast growing as a bacterium could evolve with an RNA genome, even if a major fraction of cellular metabolism were devoted to rapid repair of the spontaneous lesions. RNA is, however, a sufficiently stable material for structural components of the cell such as ribosomal RNA, tRNA, and snRNAs, which need not have a useful lifetime of more than a few cell generations.

When Did Genomes Become Double-Stranded?

Strictly speaking, we cannot know whether genomes first became double-stranded in the RNA, RNP, or DNA world. But there are two persuasive reasons for believing that double-stranded RNA genomes evolved before the birth of the DNA world. One reason is that RNA is subject to a high rate of spontaneous hydrolysis, but in a double-stranded RNA genome, such lesions could be healed by having the intact complementary strand serve as a splint for ligation of the two RNA fragments. An even better reason is that the existence of duplex genomes in the RNP world would have expedited the transition from RNA to DNA genomes.

Ribonucleoside Diphosphate Reductase Made the DNA World Possible

All organisms synthesize DNA precursors by reducing the 2'-hydroxyl group of ribonucleotide diphosphates (rNDPs) to make deoxyribonucleotide diphosphates (dNDPs), a reaction catalyzed by the enzyme ribonucleoside diphosphate reductase. Thus, the evolution of this enzyme was probably the key event that ushered in the DNA world (Figure 28-20).

But the actual conversion from RNA to DNA genomes would have been a complicated process, requiring at least four steps: the *synthesis of DNA precursors,* the *reverse transcription* of preexisting RNA genomes into DNA, the *replication* of the new DNA genomes by a DNA polymerase, and finally the *transcription* of the resulting genomic

Figure 28-20

The enzyme ribonucleoside diphosphate reductase performs the key step in the conversion of RNA precursors into DNA precursors. The base can be either A, C, G, or U. The structures of the coenzyme NAD$^+$ and NADH are given in Figures 2-3 and 28-14; in NADP$^+$ and NADPH, the 2' hydroxyl of the adenosyl residue has an additional phosphate. By reversible reduction of a cystine disulfide bridge to two cysteine sulfhydryl groups, the protein thioredoxin carries two reducing equivalents (hydrogens, in color) from NADPH to ribonucleoside diphosphate reductase (see Figure 2-19).

DNA segments back into functional RNA molecules. We cannot discount the possibility that four new enzyme activities—ribonucleoside diphosphate reductase, reverse transcriptase, DNA polymerase, and a DNA-dependent RNA polymerase—evolved simultaneously. But there is a simpler (and more attractive) scenario for conversion from RNA to DNA genomes. Once the RNP world reached a certain level of sophistication, the production of dNDPs by ribonucleoside diphosphate reductase might have enabled preexisting replication enzymes to copy RNA genomes into DNA and to transcribe the resulting DNA genomes back into functional RNA molecules.

How could preexisting enzymes copy RNA into DNA? If RNA genomes were already double-stranded in the RNP world, enzymes capable of replicating and transcribing double-stranded nucleic acids would already have existed. These enzymes would at first have used RNA precursors exclusively, but some of them may have been able to use DNA precursors, too, as these new molecules became available.

DNA Precursors Would Have Been Dangerous in an RNP World

The introduction of ribonucleoside diphosphate reductase into an RNP world was dangerous, because the incorporation of DNA precursors (lacking the 2'-hydroxyl group) into growing RNA chains would have seriously altered the structure and catalytic activity of the resulting RNA molecules. This implies that the polymerases responsible for producing functional RNAs had to be able to discriminate between RNA and DNA precursors. How could this be done?

In the RNA and RNP worlds, a clear distinction may already have existed between (double-stranded) genomic RNAs and (single-stranded) functional RNAs. In this case, the RNA replicase that copied genomic RNA may have been different from the RNA transcriptase that copied genomic RNA into functional RNA. If only the replicase, and not the polymerase, could accept DNA precursors, then the danger of having DNA precursors in an RNP world would have been diminished.

Why Does DNA Contain Thymine Instead of Uracil?[45]

Supposing that the first functional ribonucleotide reductase simply reduced each of the four RNA precursors to make DNA precursors, we would expect the original DNA genomes to have contained uracil. Why, then, does DNA today have thymine in place of uracil? Spontaneous deamidation of cytosine to uracil is known to occur at a low but significant rate (see Figures 12-9 and 12-10). The potentially harmful effects of these C to U mutations are prevented by the enzyme uracil-DNA glycosylase, which excises uracil (but not thymine) from DNA (see Figures 12-9 and 12-10). This suggests that uracil was replaced by thymine (5-methyluracil) after the initial conversion from RNA to DNA genomes, so that the increasingly efficient repair enzymes could recognize the hydrolysis products of cytosine in DNA.

We have seen that ribonucleoside diphosphate reductase converts the four rNDPs to dNDPs (Figure 28-20). Three of these dNDPs can be phosphorylated directly to yield DNA precursors (dATP, dCTP, and dGTP) but the fourth product (dUDP) must be broken down to dUMP before it can be converted to dTMP by the enzyme thymidylate synthase. (dTMP is subsequently phosphorylated to yield dTTP.) Thymidylate synthase is a dimer of identical subunits of molecular weight 35,000 daltons in organisms as varied as bacteria, bacteriophage, yeast, viruses, and vertebrates. The extraordinary conservation of this enzyme suggests that the transformation from uracil-containing to thymine-containing genomes occurred only once, very early in the DNA world.

Curiously, no RNA repair systems are known. Apparently the occasional deamidation of C to U in structural RNAs does little damage, perhaps because even the most stable structural RNA species are gradually diluted out by newly synthesized RNA as the cell multiplies.

Why Is the DNA Double Helix Right-Handed?

Two molecules are said to be **stereoisomers** if they are identical except for the geometric configuration of the four bonds surrounding a tetrahedral carbon atom. For a simple molecule like an amino acid, which has only one such asymmetric carbon atom, one stereoisomer will simply be the mirror image of the other (see Figure 5-11). Molecules with an asymmetric arrangement of bonds around a central carbon atom are able to rotate the plane of polarized light, a property known as optical activity. Thus, the two stereoisomers of a simple asymmetric molecule like glyceraldehyde are often referred to as either D or L, depending on whether the stereoisomer rotates polarized light in a right-handed (*dextrorotatory*) or left-handed (*levorotatory*) direction (Figure 28-21). For convenience, complex molecules with more than one asymmetric carbon atom are sometimes designated as D or L to show how they are related to a simple D or L molecule (see Figure 28-21), but then the D or L no longer indicates whether the molecule is dextrorotatory or levorotatory when tested experimentally.

Figure 28-21
D-ribose, D-glucose, and D-glyceraldehyde are geometrically related. Asterisks indicate asymmetric (tetrahedral) carbon atoms with four different substituents. Horizontal lines denote bonds that emerge from the plane of the page; vertical lines denote bonds that recede into the page. The asymmetric carbon atom shared by D-ribose, D-glucose, and D-glyceraldehyde is shown in color.

Today, all living organisms have D-ribose rather than L-ribose in their nucleic acids, and as a result, the typical nucleic acid duplex is right-handed rather than left-handed. (Neither the phosphates nor the bases are asymmetric.) Similarly, all contemporary living organisms have L- rather than D-amino acids in their proteins. Yet, every scheme for prebiotic synthesis of sugars or amino acids produces nearly equal mixtures of the D and L forms. Why, then, are biological systems not built from L-ribose and D-amino acids (a mirror image of our world), or from all D forms or all L forms? The reason surely lies in the detailed enzymology of the first RNA replicase and the first working translation system.

Whether we are considering a sophisticated modern RNA polymerase or the first crude RNA enzyme, all RNA polymerases depend on base pairing between the template and the pool of potential nucleotide precursors to select the correct nucleotide for addition to the growing daughter strand. Since active sites necessarily have a handedness, the first RNA enzyme would have preferred either the D precursors or the L precursors. As a result, the handedness of the active site of this crude RNA replicase would have been heritable, although the enzyme itself might have been composed of both D- and L-nucleotides. In fact, not only would all the immediate descendants of this first enzyme continue to discriminate between D-ribose and L-ribose, but natural selection for the most rapid and accurate RNA replicase would tend to increase the initial level of discrimination between stereoisomers. Still later, when the first DNA genomes were copied from preexisting RNA genomes, DNA inherited the right-handedness of its RNA predecessors. Thus, the handedness of our nucleic acids appears to be a "frozen" accident of biological history.

Why Are Proteins Made of L-Amino Acids Rather Than D-Amino Acids?

If the handedness of DNA reflects the handedness of the active site of the first RNA replicase, a similar argument could be used to explain why proteins are made of L-amino acids and not D-amino acids. The first successful polypeptide polymerase (and therefore all of its descendants) might simply have preferred L-amino acids. But there is another, far more interesting possibility. Perhaps once the fateful decision had been made to use nucleic acids containing D-ribose, then the first proteins *had* to be constructed from L-amino acids.

Since translation arose after RNA replication, the first primitive tRNA-like molecules and synthetases would have been built from ᴅ-ribose. If such RNAs were able to interact more easily with ʟ-amino acids than with ᴅ-amino acids, we could rationalize the use of ʟ-amino acids in living organisms today. This line of reasoning predicts that a mirror image of our own biological world (with ʟ-nucleotides and ᴅ-amino acids) could exist, perhaps elsewhere in the universe, but that an all ᴅ or ʟ biological world could not. Such intriguing but highly speculative ideas cannot be evaluated until we know much more about the origin of the translational apparatus.

DNA Metabolism Evolved in a World Rich in Protein Catalysts[46]

Early in the evolution of the RNP world, every sophisticated catalytic reaction had to be carried out by RNA; rudimentary proteins might have assisted the RNA catalysts, but were probably not catalysts themselves. As the RNP world became able to translate longer mRNAs accurately, proteins became more important (see Figure 28-15). Some RNA enzymes, such as those responsible for mRNA splicing and protein synthesis, eventually evolved into RNP machines that survive today (snRNPs and the ribosome); other RNA enzymes (the tRNA synthetases?) ultimately lost their RNA component. The complexity of RNP machines such as the ribosome and snRNPs suggests that the mature RNP world was very sophisticated.

Since proteins are more versatile catalysts than RNA (see Figure 28-6), accurate synthesis of long polypeptides enabled living systems to become far more complex. In addition, as new enzymes arose to meet new needs in the evolving RNP world, the superior catalytic abilities of protein guaranteed that most of these enzymes would be made of protein instead of RNA or RNP. Today, if an enzymatic function is carried out by an RNP complex, we can safely assume that the enzyme is a molecular fossil of the RNP world. But if a modern enzyme (e.g., the DNA replication machinery) is composed solely of protein, we cannot be absolutely certain that it began as a pure protein rather than as an RNP enzyme whose RNA was gradually replaced by protein (see Figure 28-15).

RNP Machines Survive in the DNA World[47]

Today, DNA replication is carried out by an efficient protein machine that evolved in a world rich in protein catalysts. In fact, DNA replication proceeds nearly as fast as deoxyribonucleotide triphosphate precursors can diffuse into the active site of the enzyme, about 750 nucleotides per second. The more ancient process of polypeptide chain elongation is carried out by the ribosome, a ribonucleoprotein fossil of the RNP world, and proceeds much more slowly, at a rate of about 15 amino acids per second.

The fact that nucleotides are polymerized 50 times faster than amino acids could be interpreted to mean that translation is inefficient compared to DNA replication and that RNP complexes are poor catalysts compared to proteins. But another interpretation is that translation is an intrinsically slower process than DNA replication because it must *decode* a nucleic acid rather than merely *copying* it. The immediate precursors of DNA are simply the four deoxynucleotide triphosphates; these small molecules diffuse rapidly and can be present in very high concentration without becoming a metabolic burden on the cell. The immediate precursors of protein synthesis are 20 different charged tRNAs; these larger molecules diffuse slowly, must be charged with the correct amino acid by the appropriate tRNA synthetase, and cannot be present in significantly higher concentrations than they already are (tRNA constitutes about 15 percent of total cellular RNA).

Recent calculations suggest that the peptide bond occurs very efficiently once the peptidyl-tRNA and the charged tRNA are correctly positioned on the ribosome. Thus, we should not mistake ribosomes and snRNPs for clumsy, inefficient fossils of the RNP world. Like the DNA replication machinery, these RNP complexes continued to evolve in a world rich in proteins and have undoubtedly been improved wherever possible by the addition of protein. The persistence of RNP machines in a DNA world where protein is the major catalyst therefore suggests that the RNA component of RNP complexes cannot easily be replaced by a polypeptide chain.

Why Does RNase P Retain Its RNA?[4]

Since the main task of ribosomes and snRNPs is to perform sequence-specific reactions using RNA templates and substrates, it is conceivable that RNP machines (which can recognize other RNA molecules by base pairing) might even be *more* efficient than equivalent machines made entirely of protein. Evidence for this comes from RNase P, the RNP enzyme responsible for recognizing and processing 30 to 40 different pre-tRNAs (see Figure 14-19). Although the catalytic function of the RNA component of RNase P is always the same, the sequence of the RNA is surprisingly different in different bacterial species. This implies that the RNA structure is subject to few constraints. But then why hasn't the function of the RNA been taken over by the protein component of this RNP enzyme? An obvious possibility is that RNA can do something that proteins cannot. Perhaps an RNA structure is more flexible than a protein structure and can more easily accommodate many slightly different pre-tRNA substrates.

The Great Antiquity of mRNA Introns[48, 49, 50]

The abundance of mRNA introns in vertebrate genes poses a straightforward question: Were preexisting protein genes broken up by the insertion of mRNA introns, or were mRNA introns gradually lost from an ancestral gene? Since mRNA introns are less abundant in lower eucaryotes than in vertebrates and are altogether absent in familiar bacteria like *E. coli*, we might be tempted to guess that these introns have been inserted into preexisting genes in higher eucaryotes. But given the strong evidence that bacteria and lower eucaryotes must streamline their genomes in order to reproduce as rapidly as possible (Chapters 18 and 20), it is impossible to tell whether these simpler organisms have discarded ancestral introns or have been under selective pressure to resist the insertion of new introns.

The only sure way to determine when and why mRNA introns arose is to trace the evolutionary history of an ancient gene encoding a ubiquitous highly conserved protein. This has been done for the enzyme triosephosphate isomerase, which carries out a basic step in both glycolysis (see Figure 4-16) and gluconeogenesis (the synthesis of glucose from smaller metabolites). Since the amino acid sequence of triosephosphate isomerase is very similar in both procaryotes and eucaryotes, this enzyme must have evolved into its final form in the progenote.

The gene encoding triosephosphate isomerase has no introns in *E. coli* or yeast, five introns in the fungus *Aspergillus*, six introns in chickens and humans, and eight introns in maize (Figure 28-22). Since the positions of five introns are strikingly conserved between maize (corn) and vertebrates, these introns must have been present before the divergence of plants and animals over a billion years ago. The position of the sixth conserved intron appears to have shifted by three amino acids, a process known as **intron sliding** (see Figure 20-23). However, the conservation of six intron positions between maize and vertebrates does not tell us whether the introns were inserted into the triosephosphate isomerase gene of the last common ancestor of plants and animals or whether they were present still earlier in evolution. Nor can this question be resolved by examining the triosephosphate isomerase gene from *Aspergillus*; the fungal gene shares only one intron with maize and chicken, and the remaining four introns in *Aspergillus* are so different that intron sliding cannot be distinguished from independent insertion.

Protein-Coding Genes Were Assembled by Exon Shuffling[50, 51]

The best evidence that introns may have been present in the very first triosephosphate isomerase gene comes from a comparison of the highly conserved three-dimensional structure of the enzyme with the intron positions in the various genes. Many of the introns are located between adjacent blocks of DNA encoding a basic unit of protein structure known as an α/β domain (see Figure 28-22). This strongly suggests that the first triosephosphate isomerase gene had introns because it was assembled from at least ten relatively small primordial exons, each encoding a tiny domain of protein structure. A corollary is that modern triosephosphate isomerase genes have been derived from the ancestral gene by differential intron loss. The alternative interpretation of the data—that introns were inserted into a preexisting triosephosphate isomerase gene as mutagenic agents—tends to be excluded because introns sometimes interrupt the codon for a highly conserved (and therefore functionally essential) amino acid like Trp_{169}.

Exons Correspond to Functional Domains of Proteins[50, 52]

Not long after the discovery of split protein genes, it became clear that modern exons sometimes encode relatively large domains of protein structure (Chapter 20). This observation gave rise to the hypothesis that the rate of evolution may be enhanced in the DNA world by **exon shuffling**: Recombination at the DNA level between different intron sequences could generate novel combinations of exons, and such

(a)

(b)

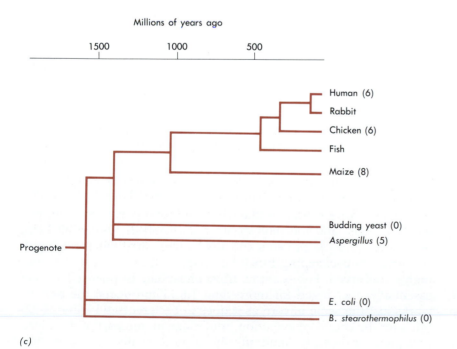

Millions of years ago

(c)

Figure 28-22

The ancestral triosephosphate isomerase gene may have had at least ten introns. (a) The three-dimensional structure of triosephosphate isomerase with coils representing α-helices (see Figure 2-22) and arrows representing β strands (see Figure 2-23). The eight β strands form a symmetrical β barrel (see Figure 5-21) surrounded by a sheath of α helices; the protein thus consists of eight similar α/β structural domains. Protein domains encoded by separate exons in maize are indicated by different colors. (b) The protein sequence with α-helical domains (arrows) and β-strand domains (rectangles) are colored as in (a). The positions of introns are shown in the protein sequence of maize (top), chicken (middle), and the fungus *Aspergillus* (bottom). (c) An evolutionary tree constructed as in Figure 28-24 from the amino acid sequence of triosephosphate isomerase; divergence times were calculated by assuming that amino acid substitutions accumulate in triosephosphate isomerase at a constant rate. Where known, the number of introns in the gene is indicated in parentheses. [After W. Gilbert, M. Marchionni, and G. McKnight, *Cell* 46 (1986):151.]

novel exon combinations will almost always produce translatable mRNAs because mRNA splice junction sequences are interchangeable (see Figures 20-20 through 20-23).

A single exon encoding a domain of epidermal growth factor provides an especially dramatic example of exon shuffling. This exon appears in four unrelated proteins including epidermal growth factor itself, the low-density lipoprotein receptor, and blood clotting factors IX and X (Table 28-2; see p. 1127). Dispersal of this exon must have taken place in the DNA world, since vertebrate proteins evolved comparatively recently. In contrast, triosephosphate isomerase is such an ancient protein that we cannot tell from the available data whether its gene was assembled from separate exons by RNA recombination in the RNP world or by DNA recombination in the DNA world.

Introns Can Be Precisely Deleted in the DNA World[53]

The gradual loss of ancestral introns over evolutionary time could occur by several different mechanisms. Intron loss caused by random deletion at the DNA level would generally be *imprecise*. As a result, a few amino codons would be gained, lost, or changed at the boundaries of the deletion. In fact, this might be one cause of intron sliding. *Precise* intron loss would most likely reflect the reverse flow of genetic information—that is, any process in which a spliced or partially spliced mRNA (or cDNA copy of the mRNA) served as the source of genetic information for generating the new gene copy. For example, an intron would be deleted if the mRNA (or a cDNA copy of the mRNA) served as the template for conversion of the parental gene (see Figure 20-29). Alternatively, introns would be lost if a cDNA copy of a spliced or partially spliced mRNA were integrated into the genome as a *functional* processed gene (see Figure 20-31). One of the two insulin genes of the rat appears to have lost an intron by just this mechanism.

The Red Queen Argument: Running Fast To Stay in Place[54, 55, 56]

In trying to understand evolution, we must avoid the simplistic notion that evolution always improves genes. Many genes need no improvement. Basic processes like DNA replication, transcription, protein synthesis, and metabolism are remarkably similar in all living organisms, and the proteins and RNAs that carry out these highly conserved **housekeeping** functions (page 704) are themselves often highly conserved. For example, triosephosphate isomerase has been essentially unchanged for more than 1.6 billion years; the enzyme works just the same in corn as it does in bacteria (see Figure 28-22). But over the course of evolution, mutations in one part of the triosephosphate isomerase molecule have necessitated compensatory changes elsewhere in the molecule to preserve protein function. RNA and protein sequences will therefore change with time (Chapter 20), even when the changes preserve gene function rather than modifying or improving it. The only exceptions will be gene products whose structure is so critical that the sequence cannot tolerate changes. The amino acid sequences of histones H3 and H4, for example, are identical in cows and garden peas!

Table 28-3   Some Human Proteins Are Much Older Than Others

Ancient Proteins

First editions. Essentially unchanged since the divergence of procaryotes and eucaryotes. Mostly mainstream metabolic enzymes. Example: triosephosphate isomerase (human and bacterial sequences are 46% identical).

Second editions. Human and procaryotic proteins have homologous sequences but apparently different functions. Example: glutathione reductase in human red blood cells and mercury reductase in the bacterium *Pseudomonas* (27% identical).

Middle-age Proteins

Proteins found in most eucaryotes but procaryotic counterparts are as yet unknown. Example: actin.

Modern Proteins

Recent vintage. Proteins found in animals or plants but not both. Not found in procaryotes. Example: collagen.

Very recent inventions. Proteins found in vertebrates but not elsewhere. Example: plasma albumin.

Recent mosaics. Modern proteins that are clearly the result of exon shuffling. Example: low density lipoprotein receptor.

SOURCE: R. F. Doolittle, D. F. Feng, M. S. Johnson, and M. A. McClure, *Cold Spring Harbor Symp. Quant. Biol.* 51 (1986):447–455.

Although many proteins have remained essentially unchanged over the entire course of evolution (Table 28-3, "first editions"), other proteins have slowly accumulated mutations that made it possible to put them to new uses ("second editions"). In addition, some novel proteins were invented relatively late in the course of evolution ("middle-age proteins" and certain "modern proteins") while still others were assembled from segments of preexisting proteins as we discussed previously ("recent mosaics").

The History of the DNA World
Is Written in Gene Sequences[57–60]

The slow accumulation of mutations over time makes it possible to measure the evolutionary distance between organisms by counting the number of accumulated changes (mutations) in a gene encoding a highly conserved protein or RNA. The evolution of the organisms from a common ancestor is then represented by a branched pathway called a **genealogical tree** (also known as a **phylogenetic** or **evolutionary tree**). The branching pattern of the tree is calculated using the **principle of parsimony** to determine the minimum number of genetic changes required to derive the sequence of the gene in each organism from a common ancestor (Figure 28-23).

The evolutionary distance separating organisms on a phylogenetic tree is usually expressed in units of accumulated amino acid or nucleotide changes rather than in units of millions of years. The reason for this is that gene sequences change at different rates in different organisms, and some genes will change faster than others, depending on how sensitive the gene product is to mutation (Table 28-4). (We saw previously that the RNA component of RNase P evolves very fast because it is quite *insensitive* to mutation.) Strictly speaking, then, the principle of parsimony can only establish the branching pattern of the phylogenetic tree. To convert such evolutionary distances into chron-

(a) The correct dendrogram minimizes total changes.

Ancestral sequence

(b) The total number of changes is greater for an incorrect dendrogram.

Ancestral sequence

Figure 28-23
Evolutionary (phylogenetic) trees can be established using the principle of parsimony. First, the sequence of a particular highly conserved gene or protein is obtained for all the organisms under consideration (A, B, C, D, E, and F). Then the organisms are arranged in a branching diagram (dendrogram) that minimizes the number of changes required to derive all the sequences from a single ancestral sequence. Numbers along the branches indicate additional changes occurring during that time interval. For example, organisms D and E have two and five changes, respectively, relative to organism C, which in turn differs from the ancestral A sequence by three changes. Thus, D and E differ from A by five and eight changes, respectively.

ological time, it is usually necessary to examine the fossil record to see when organisms of a certain type first appeared. However, it is sometimes reasonable to assume that mutation rates are comparable for closely related organisms. In such cases, a highly conserved protein like globin or cytochrome *c* can be used as a **molecular clock** to measure how long the species have been diverging from each other.

All Bacteria Look Very Much Alike[61]

No one today would seriously suggest that bats and birds are closely related simply because they both can fly, but until 1977, most biologists and molecular biologists were unaware that the generally accepted division of all life forms into only two kingdoms (procaryotes and eucaryotes) was based on equally superficial characteristics.

Bacteria lack nuclei and have rigid cell walls about 1 μm across (only the irregularly shaped mycoplasmas lack cell walls; see Figure 4-25). Their length varies slightly, depending on whether they are bacilli (rod shaped), cocci (spherical), or spirochetes (spiral). The uniform size, simple design, and lack of internal architecture in the bacterial cell present a striking contrast to the eucaryotic cell. Eucaryotic cells are larger (about 10 μm) and are characterized by a sophisticated nucleus, extensive membrane systems, internal organelles, and often the ability to participate in complex developmental pathways.

For many years, the apparent simplicity of bacteria persuaded biologists that modern procaryotes are more similar to the first living systems than are modern eucaryotes. Moreover, it seemed reasonable to think that bacteria evolved into eucaryotes by stepwise changes, the most critical of which would have been the invention of a nuclear envelope (separating transcription from translation), loss of the cell wall, an increase in overall size, and the acquisition of internal organelles such as mitochondria and chloroplasts.

The Bewildering Biochemical Diversity of Bacteria

Although bacteria are remarkably similar in appearance, they are remarkably diverse in all other respects, and serious problems arise

Table 28-4 Different Proteins Evolve at Different Rates

Protein	Point Mutations per 100 amino acids per 100 million years
Pseudogenes	400
Fibrinopeptides	90
Lactalbumins	27
Lysozymes	24
Ribonucleases	21
Hemoglobins	12
Acid Proteases	8
Triosephosphate isomerase	3
Phosphoglyceraldehyde dehydrogenase	2
Glutamate dehydrogenase	1

SOURCE: R. F. Doolittle, D. F. Feng, M. S. Johnson, and M. A. McClure, *Cold Spring Harbor Symp. Quant. Biol.* 51 (1986):447–455.

when an attempt is made to lump them all into a single procaryotic kingdom. Bacterial species inhabit an enormous range of environments. **Thermoacidophiles** (heat- and acid-loving bacteria) such as *Sulfolobus* live in hot sulfur springs, where the temperatures (90°C) are near the boiling point of water and the surroundings are extraordinarily acidic (pH 2). Another thermoacidophile known as *Thermoplasma* was discovered under smoldering piles of coal tailings; like mycoplasmas, this organism has no cell wall and is protected from its inhospitable habitat by only a lipid bilayer. In contrast, **halophiles** (salt-loving bacteria) such as *Halobacterium* not only tolerate but actually require nearly saturating concentrations of salt, such as those found in the Great Salt Lake and the Dead Sea.

Bacterial species also differ dramatically in metabolism. Most molecular biologists are only familiar with bacteria such as *E. coli* and *Salmonella*, which are medically important and experimentally tractable, and have been intensively studied at both the genetic and biochemical level. These bacterial species generate energy by fermentation in the absence of oxygen and by oxidative phosphorylation in the presence of oxygen (see Figure 2-12). Oxidative phosphorylation produces energy by establishing a proton gradient that is coupled to the generation of ATP (the **chemiosmotic theory**; see Figure 2-14). In this process, hydrogen atoms (H) from food molecules are used to reduce molecular oxygen (O_2) to water (H_2O).

Other bacteria known as **methanogens** generate energy by an analogous reaction in which hydrogen gas (H_2) reduces carbon dioxide (CO_2) to methane (CH_4). Because molecular oxygen poisons the enzymes that reduce carbon dioxide, methanogens live in **anaerobic** (oxygen-free) environments such as bogs and lake bottoms, where they are responsible for the production of flammable "marsh gas." In fact, methanogens flourishing in modern sewage treatment plants can produce commercially marketable amounts of industrial gas. There are even bacteria that produce energy by using hydrogen gas to reduce elemental sulfur to hydrogen sulfide (H_2S)!

Obligate anaerobes cannot tolerate exposure to air. These bacteria are hard to work with in the laboratory and have only recently begun to receive the attention they deserve. **Facultative anaerobes** can survive in both the presence and absence of oxygen and have been better studied. For example, bacteria of the nitrogen-fixing genus Rhizobia will grow on the laboratory benchtop and can be analyzed genetically. However, Rhizobia are induced to differentiate into nitrogen-fixing bacteroids only by exposure to plant metabolites within the protected anaerobic environment of root nodules (see Figure 22-49).

A Universal Phylogeny Based on 16S Ribosomal RNA[62]

The enzyme cytochrome *c* is a component of the respiratory chain of all eucaryotes and many procaryotes and thus has been particularly useful for building phylogenetic trees. Some of the most interesting and peculiar bacteria lack cytochrome *c*, however, and even in those bacterial genera with the enzyme, the proteins are often too different to establish reliable relationships between them. To remedy this serious gap in our understanding of the evolution of the simplest living organisms, Carl Woese and his colleagues decided in 1969 to examine the sequence of 16S ribosomal RNA from as many organisms as possible, with special emphasis on unusual bacteria that had previously eluded reliable phylogenetic placement.

Woese chose 16S rRNA for construction of a phylogenetic tree because it is truly universal and is so highly conserved in structure and function that phylogenetic trees are relatively easy to construct. In addition, 16S rRNA is an abundant RNA that can be quickly purified and analyzed even from small samples of cells. The early stages of 16S rRNA sequence analysis culminated in 1977 with a radical hypothesis. Woese proposed that procaryotes should be divided into two groups, called the archaebacteria and the eubacteria, which are as different from each other as either is from the eucaryotes. The clear implication of this proposal was that archaebacteria, eubacteria, and eucaryotes had all descended from an earlier common ancestor that did not survive. This hypothesis met with substantial resistance within the biological community because it contradicted two common but unfounded assumptions—that all bacteria are closely related and that bacteria more closely resemble the first living cells than do any eucaryotes.

Archaebacteria Assume Their Rightful Place[61-65]

Despite widespread scepticism about the value of dividing procaryotes into eubacteria and archaebacteria, proponents of the hypothesis continued to refine the universal phylogeny based on 16S rRNA (Figure 28-24) and to amass supporting biochemical evidence (see Table 28-5). Today, there is no longer any doubt that all living organisms belong to three coequal kingdoms, or **lines of descent**, and that none of these three kingdoms can be thought of as having given rise to the others (see Figure 28-24 and Table 28-5). Instead, all three have descended from an earlier living organism, or progenote, whose nature we can only infer by asking what archaebacteria, eubacteria, and the eucaryotic nucleus have in common. (The eucaryotic nucleus is directly descended from the progenote, but as we shall see, eucaryotic organelles such as the mitochondrion and chloroplast were derived by endosymbiosis of oxygen-fixing and photosynthetic eubacteria.)

The universal phylogeny based on 16S-like rRNA reveals other startling conclusions. Human beings (*Homo sapiens*) are in fact more closely related to corn (*Zea mays*) than a Gram-negative bacterium (*E. coli*) is to a Gram-positive bacterium (*Bacillus subtilis*) (see Figure 4-8 for the significance of Gram staining). Thus, the evolutionary distance separating two different bacteria can be greater than the distance between a sophisticated plant and the most sophisticated animal. The 16S-like phylogeny also provides definitive evidence for the endosymbiont hypothesis that mitochondria and chloroplasts are descended from eubacteria (see the section entitled The Endosymbiotic Origin of Mitochondria and Chloroplasts).

The Progenote (First Cell)
Differed from All Modern Cells

The universal phylogeny based on 16S-like rRNA tells us that the three great kingdoms of living organisms are all descended from a progenote. But what was this progenote like? The abundance of

ARCHAEBACTERIA

Halobacterium volcanii
Halococcus morrhuae } Halophiles

Sulfobolus solfataricus
Thermoproteus tenax } Thermoacidophiles

Methanospirillum hungatei
Methanobacterium formicicum
Methanococcus vanielii } Methanogens

Progenote

Homo sapiens (human)
Xenopus laevis (frog)
Zea mays (corn)
Saccharomyces cerevisiae
  (budding yeast)
Oxytricha nova (protozoan)
Dictyostelium discoideum
  (cellular slime mold)
Trypansoma brucei (trypanosome)

Flavobacterium heparinum
Pseudomonas testosteroni
Escherichia coli
Agrobacterium tumafaciens
Zea mays mitochondrion
Bacillus subtilis
Anacystis nidulans
Zea mays chloroplast

EUCARYOTES

EUBACTERIA

0.1 mutations
per sequence position

introns in archaebacteria and eucaryotes suggests that the progenote had introns but that these were lost during eubacterial evolution as the genome was streamlined for very rapid growth (see Table 28-5). Similarly, since eubacteria and eucaryotes have ester-linked un-branched lipids containing L-glycerophosphate, it is likely that the progenote did, too.

Table 28-5   A Few of the Known Differences Between Archaebacteria and Eubacteria

Archaebacteria	Eubacteria
Genomic rearrangements common	Genomes quite stable
Transposable elements often abundant	Few transposable elements
Some introns in rRNA and tRNA	No introns known
No peptidoglycan in cell wall	Peptidoglycan cell wall
Branched-chain fatty acids	Straight-chain fatty acids
Ether-linked lipids	Ester-linked lipids
Lipids contain D-glycero-phosphate	Lipids contain L-glycero-phosphate
rRNA, tRNA, and ribosomes share both eubacterial and eucaryotic features	
Larger multisubunit RNA polymerases resembling the eucaryotic enzymes	Simpler RNA polymerases
"Reverse" gyrases in thermophiles introduce + supercoils	Gyrases introduce only − supercoils
EF2 sensitive to diphtheria toxin as in eucaryotes	EF2 insensitive to diphtheria toxin

Figure 28-24
An evolutionary tree can be constructed by comparing the complete sequences of 21 different 16S and 16S-like ribosomal RNAs (rRNAs). The scale bar represents the number of accumulated nucleotide differences per sequence position in the rRNAs of the various organisms. Note that the scale bar cannot be recalibrated in billions of years without making the unjustified assumption that mutations accumulate in the DNA of all organisms at the same rate per unit time. [After N. R. Pace, G. J. Olsen, and C. R. Woese, *Cell* 45 (1986):325.]

But we cannot automatically assume that any trait shared by two of the three great kingdoms must reflect the nature of the progenote. Such a shared trait could also have arisen more than once as different organisms independently discovered its value. Independent evolution of the same characteristic in separate branches of a phylogenetic tree is called **convergent evolution**.

Bacteria Are More Highly Evolved than Higher Organisms

Efforts to deduce the nature of the progenote are also confounded by the fact that different organisms evolve at different rates. Although mutations arise in DNA throughout the life cycle of an organism, the effect of these mutations on fitness can only be tested in each *new* generation. As a result, rapidly multiplying organisms like bacteria and many lower eucaryotes have had a far greater opportunity to lose or modify the characteristics of the progenote than have more slowly growing higher organisms. This implies that many bacteria, although they are no more ancient than eucaryotes (see Figure 28-24), are actually more highly evolved.

The Endosymbiotic Origin of Mitochondria and Chloroplasts[66–70]

Eucaryotic cells contain a variety of internal organelles, each surrounded by a lipid bilayer. Many of these organelles (e.g., lysosomes, peroxisomes, and the endoplasmic reticulum) are relatively simple (see Figure 18-8). But two of them, mitochondria and chloroplasts, are about the same size as bacteria and, like bacteria, have circular DNA genomes (see Figure 15-17). Mitochondrial and chloroplast genomes encode the rRNA and tRNA components of the organellar translation apparatus, as well as mRNAs for organellar proteins that are synthesized within the organelle. The mitochondrial and chloroplast ribosomes are sensitive to antibiotics such as chloramphenicol, which kill many bacteria but do not affect the cytoplasmic ribosomes of eucaryotes.

The resemblance of mitochondria and chloroplasts to bacteria naturally led to the idea that these organelles began as free-living bacteria that had been engulfed by a primitive eucaryote (the **urcaryote**; see Figure 28-25). Once internalized, these symbiotic bacteria flourished within the host eucaryote as **endosymbionts**, while supplying the host with the ability to generate energy by oxidative phosphorylation and (in the case of plants) by photosynthesis. As the protomitochondrion and the protochloroplast slowly degenerated into specialized organelles, genes were transferred from organellar DNA to the nuclear genome of the host, leaving only a handful of essential genes behind in the organelle. As a result, most mitochondrial and chloroplast proteins are now encoded in the nuclear DNA, translated in the cytoplasm, and transported across the outer

Figure 28-25
A possible scheme for early evolution. [After J. E. Darnell and W. F. Doolittle, *Proc. Nat. Acad. Sci.* 83 (1986):1271.]

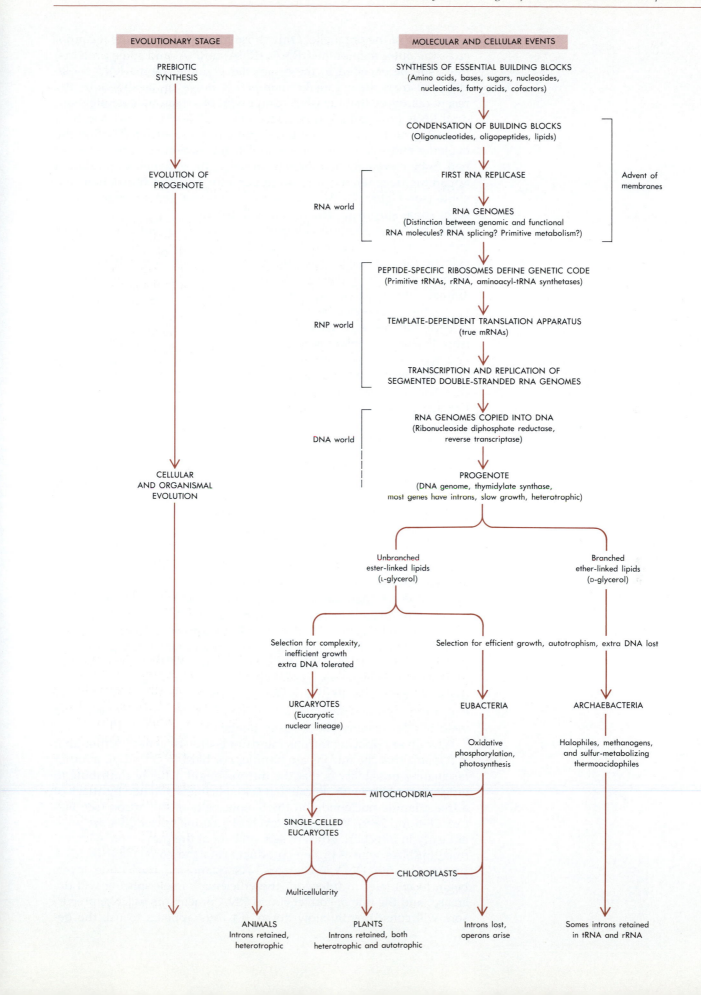

membrane of the organelle. Only those molecular species that cannot cross the outer membrane (rRNA, tRNA, mRNA, and some proteins) must still be encoded in the organellar genome. (Some RNA molecules *can* cross the organellar membrane, however, as shown by the recent discovery that the RNA component of a mammalian mitochondrial RNA processing enzyme resembling RNase P is encoded by a nuclear gene.) As fewer and fewer proteins were encoded within the evolving endosymbiont, the organellar translation system no longer had to be extremely accurate. Eventually, the organellar translation apparatus degenerated into an apparently minimal translation machine (see Figure 14-15) and the mitochondrial genetic code underwent some surprising changes (see Table 15-9).

Did mitochondria and chloroplasts evolve from eubacterial or archaebacterial progenitors? Comparison of bacterial and eucaryotic cytochrome *c* initially suggested that mitochondria might have descended from the purple photosynthetic eubacteria. Comparison of animal mitochondrial and eubacterial 16S rRNA sequences failed to prove this, however, because the animal mitochondrial rRNA sequences had diverged too extensively to permit a meaningful comparison. Fortunately, plant mitochondrial 16S rRNAs are less divergent, and in this case, the comparison led to a surprising result. Plant mitochondria descended from a group of purple eubacteria that includes rhizobacteria (see Figure 22-49), agrobacteria (see Figure 22-48), and rickettsias (see page 544). Even today, each of these procaryotes is able to live within or in very close association with eucaryotic cells. This makes the endosymbiont hypothesis all the more plausible and allows us to complete a tentative scheme for early evolution (Figure 28-25).

Several Bacteriophage T4 Genes Contain Self-Splicing Introns[71, 72]

No introns have ever been found in *E. coli*, the most intensively studied of all procaryotes. The discovery in 1984 of an intron in a bacteriophage of *E. coli* therefore came as quite a shock in 1984. Three different bacteriophage T4 genes are now known to have self-splicing introns resembling the *Tetrahymena* rRNA intron. Two of the genes encode enzymes that convert RNA precursors into DNA precursors (thymidylate synthase and the small subunit of ribonucleoside diphosphate reductase; see Figure 28-20). By expressing high levels of these two enzymes, T4 diverts the metabolic resources of the infected bacterium from making RNA to making DNA, thereby increasing the yield of DNA-containing progeny phage.

Why do self-splicing introns interrupt useful T4 genes? Although it is possible that the introns are harmless or hard to get rid of, another fascinating possibility is that the introns might actually contribute to efficient phage growth. Recall that self-splicing is initiated by attack of a free guanine nucleotide on the 5' splice site of the intron (see Figures 28-8 and 28-9). When high levels of guanine nucleotides are present early in infection, efficient self-splicing of the transcripts will produce mRNAs whose protein products catalyze conversion of RNA precursors into DNA precursors. As guanosine nucleotide levels begin to fall later in infection, the efficiency of self-splicing will decrease, and the rate of conversion of RNA precursors to DNA precursors will consequently slow down. T4 thus appears to use the dependence of self-splicing on the concentration of guanosine nucleotides as a regulatory device; the introns prevent the cellular ribonucleotide pool from being reduced to such a low level that the yield of progeny phage would suffer.

How Did Bacteriophage T4 Get Its Introns?[73, 74]

Similar self-splicing introns are found at three different sites in the bacteriophage T4 genome. Since complicated self-splicing introns are unlikely to have arisen independently at three different sites, the introns must be capable of transposing from one site to another. This suggests that T4 may have acquired its first intron by infecting a host cell whose genome already contained the intron. As previously described, mitochondria are descended from endosymbiotic bacteria (see Figure 28-24), and fungal mitochondria are known to contain self-splicing introns (see Figure 28-19). Thus, T4 bacteriophage may have acquired its introns by infecting the mitochondrion (or protomitochondrion) of a eucaryotic cell. Such **horizontal transmission** of a genetic element from one species of organism to another (as opposed to **vertical transmission** of an element from parent to daughter) may be responsible for the widespread occurrence of self-splicing introns in apparently unrelated organisms.

Fungal mitochondria have self-splicing introns (Table 28-1 and Figure 28-19) but the mitochondria of higher organisms do not. This is especially puzzling, because the eubacteria that gave rise to the mitochondria of higher plants (and possibly animals) lack introns today. One explanation would be that fungal mitochondria acquired self-splicing introns by horizontal transmission after endosymbiosis began. Alternatively, if eubacteria originally had introns, these introns may have been retained by the endosymbiont but lost by free-living eubacteria. Still a third possibility is that endosymbiosis occurred more than once among the eucaryotes. In this case, mitochondria would be **polyphyletic,** and fungal mitochondria might have descended from a different endosymbiont than the mitochondria of other eucaryotes.

Molecular Biology Is a Subdiscipline of Biology

The evolution of genes is necessarily complex because genes exist only within organisms, and organisms can interact with one another and with the environment in exceedingly subtle ways. Although molecular biology is concerned primarily with molecules, it must always be seen as a subdiscipline of biology. As the next two sections will illustrate, we simply cannot expect to understand the structure and function of genes unless we are prepared to understand the biology of the organisms in which those genes reside.

Wormy Mice in a Hybrid Zone[75]

Europe is inhabited by two distinct but interbreeding species of house mouse (Figure 28-26). *Mus musculus* resides in eastern Europe, and *Mus domesticus* in western Europe. Hybrid mice having traits derived from both species are found only in an extraordinarily narrow zone

Figure 28-26

A map of Europe showing the distribution of the two interbreeding species of house mouse, *Mus domesticus* and *Mus musculus*. The two species meet and hybridize in a narrow zone (heavy black line) extending from Denmark through central Europe to the Black Sea. Restriction maps were established for the mitochondrial DNA (mtDNA) of mice collected at each of the circled locations. Hollow circles indicate mice with *M. domesticus* mtDNA, color circles indicate mice with *M. musculus* mtDNA, and black circles indicate mice with *M. musculus* nuclear genes but mtDNA from *M. domesticus*. The data show that in parts of Scandinavia, mtDNA from the western European house mouse has invaded the eastern European house mouse and displaced the resident mtDNA. [After S. D. Ferris, R. D. Sage, C.-M. Huang, J. T. Nielsen, U. Ritte, and A. C. Wilson, *Proc. Nat. Acad. Sci.* 80 (1983):2290.]

about 20 km (12 miles) wide running from the Adriatic Sea to the North Sea. Why is the hybrid zone so narrow? The zone does not follow any major geographic feature, such as a mountain range, that might affect the climate, so we cannot argue that mice having *M. musculus* traits are unable to survive west of the zone, while mice with *M. domesticus* traits are unable to survive to the east. Nor can we argue that the zone is narrow because mice seldom migrate more than 20 km; each parent species has in fact spread over a range exceeding 1000 km. Instead, it would appear that the zone is narrow because the hybrid progeny of these two interbreeding species are somehow *unfit* and therefore reproduce more slowly than either parental species.

Inefficient reproduction of hybrid mice explains why a stable boundary exists between the two species and also why the hybrid zone is so narrow. *M. musculus* cannot move westward without mating with *M. domesticus*, thus producing hybrid mice at the boundary zone. The hybrid mice are somewhat unfit and cannot continue westward out of the zone because they are unable to compete with the resident *M. domesticus*. A similar argument explains why *M. domesticus* cannot move eastward. The resulting hybrid zone is self-perpetuating and serves as a "genetic sink" to block gene flow between the two populations. The width of the zone depends on how unfit the hybrid mice are.

Why do the hybrid progeny of *M. musculus* and *M. domesticus* reproduce poorly? To answer this question, mice were caught in barns and houses from 30 different towns within and around the hybrid zone and then examined for abnormalities. Curiously, mice from the hybrid zone were frequently infected with intestinal parasites such as pinworms (nematodes) and tapeworms. Since both parental species are relatively resistant to these parasites, the implication is that reassortment of parental chromosomes in the hybrid mice lowers resistance to infection. This suggests that resistance is a **polygenic trait** conferred by several different genes working together. Reassortment would therefore break up groups of cooperating genes or produce incompatible combinations of parental genes. In either case, the fitness of the hybrid mice would decrease, and they would reproduce more slowly.

This example emphasizes that genes are part of genomes, genomes produce organisms, and organisms have complex life cycles. Thus, natural selection on the gene occurs at the level of the organism, not at the level of the gene itself. As a result, a lowly pinworm parasite can block genetic exchange between two potentially interbreeding species.

Mitochondrial and Nuclear DNA Are Evolving Separately[76, 77]

Further molecular examination of the European house mouse population led to another interesting conclusion. As shown in Figure 28-26, the two interbreeding species of house mouse, *M. domesticus* and *M. musculus,* meet and hybridize in a narrow zone extending from Denmark through central Europe to the Adriatic Sea. Since mammalian mitochondrial DNA (mtDNA) is only 16.5 kilobases in size (see Figure 15-17) and is easy to purify, it was possible to compare the restriction maps of mtDNA from mice collected at many different locations. This comparison revealed that in parts of Scandinavia, mtDNA from the western European house mouse (*M. domesticus*) has "invaded" the eastern European house mouse (*M. musculus*) and displaced the resident mtDNA. Why this has happened is a mystery.

The *M. domesticus* mtDNA might conceivably confer a selective advantage on mice living in Scandinavia, perhaps because of minor differences in the local climate or because mtDNA from *M. domesticus* but not from *M. musculus* confers resistance to some disease or parasite found exclusively in Scandinavia. Alternatively, since mtDNA encodes only a small fraction of the mitochondrial proteins (Chapter 18), it is conceivable that mutation in a nuclear gene of Scandinavian mice is incompatible with the presence of *M. musculus* mitochondria.

The inheritance of mtDNA in European mice demonstrates that a group of genes may be selected simply because the genes are physically linked on the same replicon or because they work together to confer a selective advantage on the organism. Even though natural selection ultimately takes place at the level of the organism, the organization of genes on chromosomes evidently plays a crucial role in determining the variety of genetic variants that are available for natural selection to act on.

One Eve or Many?[78–82]

The earliest examples of anatomically modern human fossils have been found in Africa, suggesting that a band of the earliest humans left Africa and that their descendants migrated throughout Europe, Asia, and finally the Americas. Although this theory that modern humans (*Homo sapiens*) first emerged in Africa is widely believed, the fossil evidence is unlikely to be complete, and there is nothing to exclude the possibility that modern humans evolved on another continent (where the fossil evidence has yet to be discovered) and then migrated to Africa.

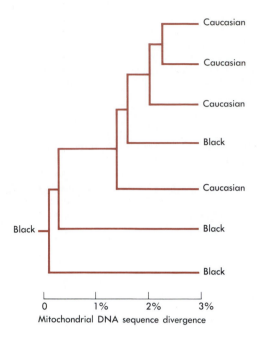

Caucasian

Caucasian

Caucasian

Black

Caucasian

Black — Black

Black

Black

|—————|—————|—————|—————|
0   1%   2%   3%

Mitochondrial DNA sequence divergence

Figure 28-27
A human genealogical tree based on the nucleotide sequences of human mitochondrial DNA. The most likely tree was deduced using the method of maximum parsimony as described in Figure 28-24. The root of the tree lies in Africa, and there are two primary lines of descent; one leads to Africans only, the other to some Africans and to all other populations. [After R. L. Cann, M. Stoneking, and A. C. Wilson, *Nature* 325 (1987):31.]

Can we address the question of where *Homo sapiens* originated by examining the *molecular* fossil record that is written in the DNA sequences of present-day humans? We have seen that highly conserved molecules like cytochrome *c*, triosephosphate isomerase, and 16S ribosomal RNA evolve slowly and can be used to establish evolutionary trees that span hundreds of millions of years (see Figures 28-22c to 28-24). But radioisotope dating of fossil remains suggests that our own species diverged from primitive hominoids only 100,000 to 140,000 years ago. Molecules like cytochrome *c* or 16S rRNA change too slowly to be of much help in constructing an evolutionary tree describing the emergence of the human races over such a brief time. Instead, we need a "fast clock" molecule that evolves quickly, and for this purpose, mitochondrial DNA seems to be ideal.

The mutation rate of mitochondrial DNA is about ten times higher than that of nuclear DNA, primarily because mitochondria have no DNA repair systems. As a consequence of such rapid evolution, mitochondrial DNA exhibits much greater variation from one person to another than does nuclear DNA. This has made it possible to construct a human genealogical tree by comparing mitochondrial DNA sequences from different individuals (see Figure 28-27), just as the phylogenetic tree leading back to the progenote was constructed by applying the principle of parsimony to many different 16S rRNA sequences (see Figures 28-23 and 28-24). Whether the mitochondrial DNAs from 7 or 147 individuals of different racial extraction were compared, the conclusion was the same. Most modern Africans are closer to the ancestral mitochondrial DNA sequence than are any modern Caucasians, implying that humans first emerged in Africa and then emigrated to other continents.

The human genealogical tree also suggests that all present-day humans may have shared a common ancestral mother (Figure 28-27). Mitochondrial DNA is known to be **maternally inherited**: Although both the oocyte and the sperm contain mitochondria, progeny inherit mitochondrial DNA from the mother only. The fact that 147 different mitochondrial DNA sequences can all be arranged in a single tree implies that there was indeed a single original mitochondrial DNA sequence. Supposing that the mitochondrial DNA of our most ancient ancestors exhibited the same degree of individual variation as mitochondrial DNA does today, this would mean that we all share a common mother, our mitochondrial Eve. The molecular fossil record therefore has the potential to tell us about events as ancient as the origin of the first replicating system or as recent as the emergence of human beings from the higher primates.

Summary

Molecular biology is an experimental rather than a theoretical science. Whether we want to know why globin genes are expressed only in red blood cells or why the fruit fly has a segmented body, we expect every theory to be tested experimentally and ultimately proved or disproved. We can afford to have such high expectations of molecular biology because the living systems that we study can be manipulated experimentally to test each theory.

Questions of molecular evolution are far trickier, because

it is usually harder, and sometimes even impossible, to do the right experiments. If we want to know, for example, whether humans are more closely related to chimpanzees or gorillas, we would really like to examine the missing links between the species. But these transitional organisms have not survived, and we must instead compare DNA sequences in the surviving species.

Questions of molecular evolution become even more daunting when we try to speculate about the very earliest life forms. We have seen that even the best fossils preserve only the durable parts of an organism; the macromolecules that most interest us are lost. As a result, we must attempt to reconstruct ancient organisms by examining living organisms (molecular fossils) and by designing model experiments that mimic the conditions thought to prevail on the primitive Earth. Unfortunately, it is impossible to obtain direct proof for any particular theory of the origin of life. The sobering truth is that even if every expert in the field of molecular evolution were to agree on how life originated, the theory would still be a best guess rather than a fact.

In this chapter, we have presented one possible scenario for the origin of the first "living molecule" that could replicate and evolve. Our purpose has been to show how the recent discovery of enzymatically active RNA species has made it possible to speculate usefully about the molecular

origins of life. Aside from the intellectual challenge of trying to imagine how life began, such speculation inevitably inspires specific experiments that sharpen and modify the speculations themselves. For example, the experiment showing that the *Tetrahymena* ribosomal RNA intron core can function as a poly C polymerase was the first in a series of experiments designed to test whether an RNA molecule could actually copy itself in the test tube today. This experiment also led directly to the model for how an RNA enzyme could function as an RNA replicase and even perhaps as a tRNA synthetase. Although the scenario we presented is most unlikely to be true in detail, it surely contains important elements of the truth. Showing how life could have begun is a first step toward learning how life actually began.

Bibliography

General References

Day, W. 1984. *Genesis on Planet Earth: The Search for Life's Beginning.* 2nd ed. New Haven, Conn.: Yale University Press. A lively but idiosyncratic treatment of the origins of life, drawing on data from such diverse fields as geology, prebiotic chemistry, paleobiology, and molecular biology.

Ferris, J. P., and D. A. Usher. 1983. "Origins of Life." In *Biochemistry*, ed. G. Zubay (coordinating author). Reading, Mass.: Addison-Wesley. A superb, eminently readable, and often witty introduction to the chemistry of prebiotic synthesis and the origins of metabolism.

Miller, S. L., and L. E. Orgel. 1974. *The Origins of Life on Earth.* Englewood Cliffs, N.J.: Prentice-Hall. A concise and immensely readable classic monograph by two leaders in the field, emphasizing the role of prebiotic synthesis in the origin of life.

Raff, R. A., and T. C. Kaufman. 1983. *Embryos, Genes, and Evolution.* New York: Macmillan. A delightfully written essay showing how changes in DNA that affect embryogenesis can ultimately lead to the evolution of new species.

Schopf, J. W., ed. 1983. *Earth's Earliest Biosphere: Its Origin and Evolution.* Princeton, N.J.: Princeton University Press. Although aimed almost exclusively at specialists, each of the advanced technical essays in this authoritative collection begins with a preface suitable for beginning students of the field.

Cited References

1. Knoll, A. H. 1985. "The Distribution and Evolution of Microbial Life in the Late Proterozoic Era." *Ann. Rev. Microbiol.* 39:391–417.
2. Knoll, A. H., and E. S. Barghoorn. 1985. "Archean Microfossils Showing Cell Division from the Swaziland System of South Africa." *Science* 198:396–398.
3. Dill, R. F., E. A. Shinn, A. T. Jones, K. Kelly, and R. P. Steinen. 1986. "Giant Subtidal Stromatolites Forming in Normal Salinity Waters." *Nature* 324:55–58.
4. Pace, N. R., and T. L. Marsh. 1985. "RNA Catalysis and the Origin of Life." *Origins of Life* 16:97–116.
5. Keese, P., and R. H. Symons. 1985. "Domains in Viroids: Evidence of Intermolecular RNA Rearrangements and Their Contribution to Viroid Evolution." *Proc. Nat. Acad. Sci.* 82:4582–4586.
6. Hutchins, C. J., P. D. Rathjen, A. C. Forster, and R. H. Symons. 1986. "Self-Cleavage of Plus and Minus RNA Transcripts of Avocado Sunblotch Viroid." *Nucl. Acid Res.* 14:3627–3640.
7. Keese, P., and R. H. Symons. 1986. "The Structure of Viroids

and Virusoids." In *Viroids and Viroid-Like Pathogens*, ed. J. S. Semanicik. Boca Raton: CRC Press, pp. 1–48.

8. Buzayan, J. M., W. L. Gerlach, and G. Bruening. 1986. "Non-enzymatic Cleavage and Ligation of RNAs Complementary to a Plant Virus Satellite RNA." *Nature* 323:349–353.
9. Dinter-Gottlieb, G. 1986. "Viroids and Virusoids Are Related to Group I Introns." *Proc. Nat. Acad. Sci.* 83:6250–6254.
10. Epstein, L. M., and J. G. Gall. 1987. "Self-Cleaving Transcripts of Satellite DNA from the Newt." *Cell* 48:535–543.
11. Cech, T. R. 1986. "RNA as an Enzyme." *Sci. Amer.* 255:64–75.
12. Cech, T. R., and B. L. Bass. 1986. "Biological Catalysis by RNA." *Ann. Rev. Biochem.* 55:599–629.
13. Bass, B. L., and T. R. Cech. 1984. "Specific Interaction Between the Self-Splicing RNA of *Tetrahymena* and Its Guanosine Substrate: Implications for Biological Catalysis by RNA." *Nature* 308:820–826.
14. Cech, T. R. 1985. "Self-Splicing RNA: Implications for Evolution." *Internat. Rev. of Cytol.* 93:3–22.
15. Zaug, A. J., J. R. Kent, and T. R. Cech. 1984. "A Labile Phosphodiester Bond at the Ligation Junction in a Circular Intervening Sequence RNA." *Science* 224:574–578.
16. Garriga, G., A. M. Lamobwitz, T. Inoue, and T. R. Cech. 1986. "Mechanism of Recognition of the 5' Splice Site in Self-Splicing Group I Introns." *Nature* 322:86–89.
17. Been, M. D., and T. R. Cech. 1986. "One Binding Site Determines Sequence Specificity of *Tetrahymena* Pre-rRNA Self-Splicing, *Trans*-Splicing, and RNA Enzyme Activity." *Cell* 47:207–216.
18. Szostak, J. W. 1986. "Enzymatic Activity of the Conserved Core of a Group I Self-Splicing Intron." *Nature* 322:83–86.
19. Zaug, A. J., and T. R. Cech. 1986. "The Intervening Sequence RNA of *Tetrahymena* Is an Enzyme." *Science* 231:470–475.
20. Cech, T. R. 1986. "A Model for the RNA-Catalyzed Replication of RNA." *Proc. Nat. Acad. Sci.* 83:4360–4363.
21. Haertle, T., and L. E. Orgel. 1986. "Template-directed Synthesis on the Oligonucleotide d(C_7–G–C_7)." *J. Mol. Biol.* 188:77–80.
22. Horowitz, N. H. 1945. "On the Evolution of Biochemical Syntheses." *Proc. Nat. Acad. Sci.* 31:153–157.
23. White, H. B., III. In *The Pyridine Nucleotide Coenzymes* (1982):1–17. New York: Academic Press.
24. Gilbert, W. 1986. "The RNA World." *Nature* 319:618.
25. Jay, D. G., and W. Gilbert. 1987. "Basic Protein Enhances the Incorporation of DNA into Lipid Vesicles: A Model for the Formation of Primordial Cells." *Proc. Nat. Acad. Sci.* 84. In press.
26. Moore, P. B. 1985. "Polypeptide Polymerase: The Structure and Function of the Ribosome in 1985." In *Proc. Robert A. Welch Found. Conf. Chem. Res.* 29:185–214.

27. Crick, F. H. C., S. Brenner, A. Klug, and G. Pieczenik. 1976. "A Speculation on the Origin of Protein Synthesis." *Origins of Life* 7:389–397.

28. Woese, C. R. 1970. "The Problem of Evolving a Genetic Code." *BioScience* 20:471–485.

29. Eigen, M., and R. Winkler-Oswatitsch. 1981. "Transfer-RNA, an Early Gene?" *Naturwissenschaften* 68:282–292.

30. Crothers, D. M. 1982. "Nucleic Acid Aggregation Geometry and the Possible Evolutionary Origin of Ribosomes and the Genetic Code." *J. Mol. Biol.* 162:379–391.

31. Weiner, A. M., and N. Maizels. 1987. "3' terminal tRNA-like structures tag genomic RNA molecules for replication: Implications for the origin of protein synthesis." *Proc. Nat. Acad. Sci.* 84:7383–7387

32. Burma, D. P., D. S. Tewari, and A. K. Srivastava. 1985. "Ribosomal Activity of the 16S·23S RNA Complex." *Arch. Biochem. Biophys.* 239:427–435.

33. Wong, J. T.-F. 1981. "Coevolution of Genetic Code and Amino Acid Biosynthesis." *Trends in Biochem. Sci.* 6:33–36.

34. Webster, T., H. Tsai, M. Kula, G. A. Mackie, and P. Schimmel. 1984. "Specific Sequence Homology and Three-Dimensional Structure of an Aminoacyl Transfer RNA Synthetase." *Science* 226:1315–1317.

35. Hoben, P., N. Royal, A. Cheung, F. Yamao, K. Biemann, and D. Söll. 1982. "*Escherichia coli* Glutaminyl-tRNA Synthetase. II. Characterization of the *glnS* Gene Product." *J. Biol. Chem.* 257:11644–11650.

36. Sharp, P. 1985. "On the Origin of RNA Splicing and Introns." *Cell* 42:397–400.

37. Cech, T. R. 1986. "The Generality of Self-Splicing RNA: Relationship to Nuclear mRNA Splicing." *Cell* 44:207–210.

38. Schmelzer, C., and R. J. Schweyen. 1986. "Self-Splicing of Group II Introns in Vitro: Mapping of the Branch Point and Mutational Inhibition of Lariat Formation." *Cell* 46:557–565.

39. Tazi, J., C. Alibert, J. Temsamani, I. Reveillaud, G. Cathala, C. Brunel, and P. Jeanteur. 1986. "A Protein That Specifically Recognizes the 3' Splice Site of Mammalian Pre-mRNA Introns Is Associated with a Small Nuclear Ribonucleoprotein." *Cell* 47:755–766.

40. Gerke, V., and J. A. Steitz. 1986. "A Protein Associated with Small Nuclear Ribonucleoprotein Particles Recognizes the 3' Splice Site of Premessenger RNA." *Cell* 47:973–984.

41. Zhuang, Y., and A. M. Weiner. 1986. "A Compensatory Base Change in U1 snRNA Suppresses a 5' Splice Site Mutation." *Cell* 46:827–835.

42. Sharp, P. A. 1987. "Splicing of Messenger RNA Precursors." *Science* 235:766–771.

43. Maniatis, T. and R. Reed. 1987. "The Role of Small Nuclear Ribonucleoprotein Particles in Pre-mRNA Splicing." *Nature* 325:673–678.

44. Reanney, D. C. 1986. "Genetic Error and Genome Design." *Trends in Genetics* 2:41–46.

45. Hardy, L. W., J. S. Finer-Moore, W. R. Montfort, M. O. Jones, D. V. Santi, and R. M. Stroud. 1987. "Atomic Structure of Thymidylate Synthase: Target for Rational Drug Design." *Science* 235:448–455.

46. Alberts, B. M. 1986. "The Function of the Hereditary Materials: Biological Catalyses Reflect the Cell's Evolutionary History." *Amer. Zoologist* 26:781–796.

47. Jacob, F. 1977. "Evolution and Tinkering." *Science* 196:1161–1166.

48. Marchionni, M., and W. Gilbert. 1986. "The Triosephosphate Isomerase Gene from Maize: Introns Antedate the Plant-Animal Divergence." *Cell* 46:133–141.

49. McKnight, G. L., P. J. O'Hara, and M. L. Parker. 1986. "Nucleotide Sequence of the Triosephosphate Isomerase Gene from *Aspergillus nidulans*: Implications for a Differential Loss of Introns." *Cell* 46:143–147.

50. Gilbert, W., M. Marchionni, and G. McKnight. 1986. "On the Antiquity of Introns." *Cell* 46:151–154.

51. Südhof, T. C., D. W. Russell, J. L. Goldstein, M. S. Brown, R. Sanchez-Pescador, and G. I. Bell. 1985. "Cassette of Eight Exons Shared by Genes for LDL Receptor and EGF Precursor." *Science* 228:893–895.

52. Brändén, C.-I. 1986. "Anatomy of α/β Proteins." In *Current Communications in Molecular Biology: Computer Graphics and Molecular Modeling*, eds. R. Fletterick and M. Zoller. Cold Spring Harbor, N.Y.: Cold Spring Harbor Laboratory, pp. 45–51.

53. Soares, M. B., E. Schon, A. Henderson, S. K. Karathanasis, R. Cate, S. Zeitlin, J. Chirgwin, and A. Efstratiadis. 1985. "RNA-Mediated Gene Duplication: The Rat Preproinsulin I Gene Is a Functional Retroposon." *Mol. Cell. Biol.* 5:2090–2103.

54. Van Valen, L. 1974. "Molecular Evolution as Predicted by Natural Selection." *J. Mol. Evol.* 3:89–101.

55. Zuckerkandel, E. 1976. "Evolutionary Processes and Evolutionary Noise at the Molecular Level." *J. Mol. Evol.* 7:269–311.

56. Smith, J. M. 1983. "The Genetics of Stasis and Punctuation." *Ann. Rev. Genetics* 17:11–25.

57. Britten, R. J. 1986. "Rates of DNA Sequence Evolution Differ Between Taxonomic Groups." *Science* 231:1393–1398.

58. Wilson, A. C. 1985. "The Molecular Basis of Evolution." *Sci. Amer.* 253:164–173.

59. Helm-Bychowski, K., and A. C. Wilson. 1986. "Rates of Nuclear DNA Evolution in Pheasant-Like Birds: Evidence from Restriction Maps." *Proc. Nat. Acad. Sci.* 83:688–692.

60. Vawter, L., and W. M. Brown. 1986. "Nuclear and Mitochondrial DNA Comparisons Reveal Extreme Rate Variation in the Molecular Clock." *Science* 234:194–196.

61. Woese, C. R. 1981. "Archaebacteria." *Sci. Amer.* 244:98–125.

62. Pace, N. R., G. J. Olsen, and C. R. Woese. 1986. "Ribosomal RNA Phylogeny and the Primary Lines of Evolutionary Descent." *Cell* 45:325–326.

63. Doolittle, W. F. 1985. "Archaebacteria Coming of Age." *Trends in Genetics* 1:268–269.

64. Hofman, J. D., L. C. Schalkwyk, and W. F. Doolittle. 1986. "ISH51: A Large, Degenerate Family of Insertion Sequence-Like Elements in the Genome of the Archaebacterium, *Halobacterium volcanii*." *Nucleic Acid Res.* 14:6983–7000.

65. Kjems, J., and R. A. Garrett. 1985. "An Intron in the 23S Ribosomal RNA Gene of the Archaebacterium *Desulfurococcus mobilis*." *Nature* 318:675–677.

66. Yang, D., Y. Oyaizu, H. Oyaizu, G. J. Olsen, and C. R. Woese. 1985. "Mitochondrial Origins." *Proc. Nat. Acad. Sci.* 82:4443–4447.

67. Chomyn, A., M. W. J. Cleeter, C. I. Ragan, M. Riley, R. F. Doolittle, and G. Attardi. 1986. "URF6, Last Unidentified Reading Frame of Human mtDNA, Codes for an NADH Dehydrogenase Subunit." *Science* 234:614–618.

68. De Block, M., J. Schell, and M. Van Montagu. 1985. "Chloroplast Transformation by *Agrobacterium tumefaciens*." *EMBO J.* 4:1367–1372.

69. Shinozaki, K., M. Ohme, M. Tanaka, T. Wakasugi, N. Hayashida, T. Matsubayasi, N. Zaita, J. Chunwongse, J. Obokata, K. Yamaguchi-Shinozaki, C. Ohto, K. Torazawa, B. Y. Meng, N. Sugita, H. Deno, T. Kamogashira, K. Yamada, J. Kusuda, F. Takaiwa, A. Kato, N. Tohdoh, H. Shimada, and M. Sugiura. 1986. "The Complete Nucleotide Sequence of the Tobacco Chloroplast Genome: Its Gene Organization and Expression." *EMBO J.* 5:2043–2049.

70. Chang, D. D. and D. A. Clayton. 1987. "A Mammalian Mitochondrial RNA Processing Activity Contains Nucleus-Encoded RNA." *Science* 235:1178–1184.

71. Gott, J. M., D. A. Shub, and M. Belfort. 1986. "Multiple Self-Splicing Introns in Bacteriophage T4: Evidence from Autocatalytic GTP Labeling of RNA in Vitro." *Cell* 47:81–87.

72. Sjöberg, B.-M., S. Hahne, C. Z. Mathews, C. K. Mathews, K. N. Rand, and M. J. Gait. 1986. "The Bacteriophage T4 Gene for the Small Subunit of Ribonucleotide Reductase Contains an Intron." *EMBO J.* 5:2031–2036.

73. Michel, F., and B. Dujon. 1986. "Genetic Exchanges Between Bacteriophage T4 and Filamentous Fungi?" *Cell* 46:323.

74. Colleaux, L. d'Auriol, M. Betermier, G. Cottarel, A. Jacquier, R. Galibert, and B. Dujon. 1986. "Universal Code Equivalent of a Yeast Mitochondrial Intron Reading Frame Is Expressed into *E. coli* as a Specific Double Strand Endonuclease." *Cell* 44:521–533.

75. Sage, R. D., D. Heyneman, K.-C. Lim, and A. C. Wilson. 1986. "Wormy Mice in a Hybrid Zone." *Nature* 324:60–63.

76. Ferris, S. D., R. D. Sage, C.-M. Huang, J. T. Nielsen, U. Ritte, and A. C. Wilson. 1983. "Flow of Mitochondrial DNA Across a Species Boundary." *Proc. Nat. Acad. Sci.* 80:2290–2294.

77. Wilson, A. C., R. L. Cann, S. M. Carr, M. George, U. B. Gyllensten, K. M. Helm-Bychowski, R. G. Higuchi, S. R. Palumbi, E. M. Prager, R. D. Sage, and M. Stoneking. 1985. "Mitochondrial DNA and Two Perspectives on Evolutionary Genetics." *Biol. J. Linnean Soc.* 26:375–400.

78. Cann, R. L., M. Stoneking, and A. C. Wilson. 1986. "Mitochondrial DNA and Human Evolution." *Nature* 325:31–36.

79. Latorre, A., A. Moya, and F. J. Ayala. 1986. "Evolution of mitochondrial DNA in *Drosophila subobscura*." *Proc. Nat. Acad. Sci.* 83:8649–8653.

80. Wainscot, J. 1987. "Out of the Garden of Eden." *Nature* 325:13.

81. Wainscoat, J. S., A. V. S. Hill, A. L. Boyce, J. Flint, M. Hernandez, S. L. Thein, J. M. Old, J. R. Lynch, A. G. Falusi, D. J. Weatherall, and J. B. Clegg. 1986. "Evolutionary Relationships of Human Populations from an Analysis of Nuclear DNA Polymorphisms." *Nature* 319:491–493.

82. Jones, J. S., and S. Rouhani. 1986. "How Small Was the Bottleneck?" *Nature* 319:449–450.

Index

This index covers both volume I and volume II.
Page numbers in bold face refer to a definition and major discussion of the entry. F after a page number indicates a figure; T after a page number indicates a table.